JOURNAL OF CHROMATOGRAPHY LIBRARY – volume 65

sample preparation in chromatography

JOURNAL OF CHROMATOGRAPHY LIBRARY – volume 65

sample preparation in chromatography

Serban C. Moldoveanu
Brown & Williamson Tobacco Corporation,
Macon, GA, USA

Victor David
Department of Analytical Chemisrtry,
University of Bucharest, Romania

2002

ELSEVIER
Amsterdam - Boston - London - New York - Oxford - Paris
San Diego - San Francisco - Singapore - Sydney - Tokyo

ELSEVIER SCIENCE B.V.
Sara Burgerhartstraat 25
P.O. Box 211, 1000 AE Amsterdam, The Netherlands

CHEM
o 12318073

First edition 2002

Library of Congress Cataloging in Publication Data
A catalog record from the Library of Congress has been applied for.

British Library Cataloguing in Publication Data

Sample preparation in chromatography. - (Journal of
    Chromatography ; v. 65)
    1.Chromatographic analysis 2.Sample introduction
    (Chemistry)
    I.Moldoveanu, Serban II.David, Victor
    543'.089

    ISBN 0444503943

ISBN: 0-444-50394-3
ISSN: 0301-4770 (Series)

PREFACE

Sample preparation is an essential step in many analyses. This step consists of a variety of preliminary operations performed on the sample to modify it and make it more suitable for analysis. Chromatographic methods use sample preparation extensively, and descriptions of sample preparation procedures are available with virtually any reported chromatographic method. This makes the subject of sample preparation for chromatography important and, at the same time, considerably broad. Because of the subject complexity, and not due to scientific obstacles, a presentation of sample preparation for chromatography is a rather difficult task. As practice shows, sample preparation for chromatography involves a wide variety of procedures that cover numerous and sometimes unrelated fields. This explains the discrepancy between the amount of information on sample preparation in original research papers and the limited number of books presenting this subject in a systematic manner.

The intention of the authors was to approach the topic of sample preparation in a methodical way, viewing it as a logical connection between sample collection and analytical chromatography. The book describes various ways to process the sample, explaining the principle, discussing the advantages and disadvantages, describing the applicability to different types of samples, and showing the fitness to specific chromatographic determinations. This provides a guide for choosing the appropriate sample preparation for a given analysis. The book also contains numerous literature references necessary for a more detailed understanding of each specific procedure. However, the diversity of sample preparation procedures is very wide, and new procedures are continuously developed and reported. For this reason, only the original literature offers full information regarding the analysis of a specific type of sample.

The first part of the book contains an overview of sample preparation, showing its relation to sample collection and to the core chromatographic analysis. The second part covers procedures that do not use chemical modifications of the analyte and includes methods for sample dissolution, concentration, and cleanup designed mainly for modifying the initial matrix of the sample. This part starts with conventional separations such as filtration and distillation and finishes with more advanced techniques such as solid phase extraction and electroseparations. The third part gives a description of the chemical modifications that can be performed on a sample either for fractionation purposes or to improve a specific property of the analyte. This part includes derivatizations, polymer chemical degradations, and pyrolysis.

The book is intended to be useful for a wide range of readers, including specialists in analytical chemistry and beginners in this field. Impossible to be exhaustive, the book covers the main aspects of sample preparation and provides guidance regarding the decision process for selecting an analytical method for a specific practical problem. The book also describes some applications of general interest. It was not the intention of the authors to write a "cook book" for sample preparation, but a material describing the reasons for which specific sample preparations are applied.

Help for the presentation of the material of this book was provided by the editor, Ms. Reina Bolt. Invaluable contribution came from Ms. Carol Benton, who made numerous corrections to the manuscript and prepared the index.

Table of Contents

Table of Contents

Table of Contents

Table of Contents

Part 3. Sample Preparation Techniques Involving Chemical Modifications

Table of Contents

PART 1

Introduction and General Considerations

CHAPTER 1

Preparatory Information

1.1   OVERVIEW OF A CHEMICAL ANALYSIS

Sample preparation is a process required for the transformation of a sample to make it amenable for chemical analysis or to improve the analysis. This is necessary when a given sample cannot be directly analyzed or when direct analysis generates poor results. Typical problems with analyses are interferences and low sensitivity. Sample preparation is usually needed to eliminate interferences and to increase sensitivity. Samples are made from two distinct parts, the analytes and the matrix. The *analytes* are the compounds of interest that must be analyzed, while the *matrix* is the remainder of the sample, which does not require analysis. Sample preparation may be done on the analytes, on the matrix, or on both to perform dissolution, cleanup, concentration, or chemical modifications of the sample for obtaining better analytical results.

Sample preparation is only one of the steps performed in a chemical analysis. Viewed as a combination of information and operations, a chemical analysis follows the typical scheme: input ⟶ process ⟶ output. The input consists of initial information about the sample, such as origin, nature, and purpose of analysis. The output is formed by the results, when the purpose of the analysis is achieved. The process consists of various steps, one of them being sample preparation. In chromatographic methods of analysis, these steps usually follow the sequence: sample collection ⟶ sample preparation ⟶ analytical chromatography ⟶ data processing. The analytical chromatography step can be considered the core of the process, and it includes the identification and measurement of the analytes. Instead of chromatography, other techniques can be applied as the core analytical step. Among these are various kinds of spectroscopy, thermal techniques, electrochemical techniques, etc. The classification of analytical methods is usually based on this core step. Chromatographic methods use analytical chromatography as the core step in the analytical process. A simplified flow diagram of a chemical analysis using analytical chromatography as a core step is given in Figure 1.1.1.

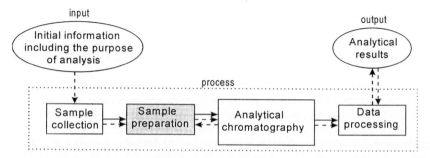

FIGURE 1.1.1. *Simplified diagram of a chromatographic method of analysis, with specific feedback for sample preparation and for the data processing steps (operation flow is indicated by* ⟶ *, and information flow is indicated by* ----► *).*

The diagram shows the position of sample preparation in relation to other operations of the analysis. Also, a flow of information occurs along the operations, the knowledge from one step influencing the others. Partial or final resulting data may be used as feedback for decisions in the process. The diagram in Figure 1.1.1 shows, for example, feedback to the sample preparation step from the analytical chromatography step and to the data processing step from the analytical results.

In practice, both simple and also complex analytical problems are encountered. The solution of complex problems cannot be obtained using a single analysis. Then, different chemical analyses in a chain, a battery, or a combination of the two are required to solve the problem. In these cases, more complex informational flow occurs along the analytical process.

The purpose of Part 1 of this book is to present the interrelation between various steps performed in a chemical analysis, to show the role played by sample preparation, and to provide some details about the other steps in the analysis. Also, some theoretical background applicable to various separation techniques is given in Part 1.

- *Initial information*

In the initial information about a chemical analysis is included the purpose of analysis and the expected utility of the results. This information determines the choice between the application of an existing analytical procedure and the development of a new technique. A new technique is typically needed to solve an analytical problem that has not been solved previously. At the other extreme, a laboratory is obliged to follow for the analysis all the steps included in an existing protocol of a standardized method; otherwise the analytical results are susceptible to being rejected [1]. Some analyses require the implementation of a known method of analysis but for a new type of sample. For some samples, the matrix may comprise most of the sample material, and the analytes may be present only in traces. The choice of a specific analytical method is determined not only by the nature of the analytes, but also by the nature of the matrix.

The choice of an analytical procedure requires the collection of as much information as possible about the available methods and techniques reported in literature for solving the given problem. It is the goal of this book to provide some of this initial information for the specific case of sample preparation when a chromatographic separation and measurement are used. (The terms *method* and *technique* are used with various meanings, but in general *method* designates a complete set of operations used for solving an analytical problem, and *technique* designates a known technical operation.)

Another component of the initial information is related to the knowledge about the sample itself, such as origin, nature, or matrix composition, expected level of the analytes, and number of samples that must be analyzed. Some of these data may not be known, and various preliminary tests may be needed for finding them. It is not uncommon that a qualitative analysis is needed before applying a more elaborate quantitative measurement. Also, a semiquantitative estimation of the analytes of interest may precede the quantitative analysis. In these cases, a chain of chemical analyses is used for the completion of the analytical process.

Information about the required accuracy and precision of the analysis is another important element for the choice of a given analytical method [2]. Also, external information not related to the technical aspects of the problem is always necessary, such as available instrumentation, manpower, expertise available in the laboratory, the cost of analysis, and funding. Even when comprehensive initial information is available, various choices for different analytical methods are possible [3], [4]. For this reason, when a method is new or modifications are needed for an existing procedure, the feedback for each step during the development stages of a chemical analysis is significantly more intensive compared to that of using a well-established method [5].

A summary of the desired preliminary information about a new chemical analysis is shown in Table 1.1.1.

TABLE 1.1.1. *Summary of necessary information about a new chemical analysis.*

Type of intended analysis and external information	Immediate purpose of analysis	Nature of sample (analytes and matrix)	Required quality of results
Is the analysis needed for a known type of sample or for a new type?	Is the analysis done only to determine elemental composition?	Is the sample solid, liquid (solution, colloid), gas or mixed phases? Is the sample homogeneous?	What is the utility of the results?
Is a new analytical method needed or can a reported one be used? (Full knowledge of reported methods required.)	Is the analysis a complete or a selective qualitative or semiquantitative measurement?	Is the sample inorganic, organic, mix, of biological origin, environmental?	Is the evaluation of qualitative results needed?
Is the analysis a "one time" need or a routine analysis?	Is the analysis qualitative, quantitative or both?	What is the number of samples and the number of analytes to measure?	Is high sensitivity required (known detection limit)?
What is the required turnaround time (rapid or long term analysis)?	Is quantitation needed for all or for specific components?	Is the matrix complex or simple? Are interferences expected?	What precision and accuracy are required?
A specific protocol must be followed, or no regulations are imposed?	Is any structural analysis, isomer analysis, MW of macromolecules needed?	Has the sample small molecules, ionic compounds, polymers, or is it a mix?	Is statistical analysis on various sets of samples required?
Is the analysis done for a process evaluation?	Is the sequence of a polymer or a weak interaction compound determined?	What is the estimated level of the analyte (major, trace, etc.)?	Is a comparison with info. in a data base or other laboratories necessary?
What instrumentation is available?	Is the analysis destructive or must be nondestructive?	What is the solubility of the sample? Is there a need for digestion?	What is the desired robustness of the analysis?
What are the cost, manpower, and funding?	Is the whole sample analyzed or only a specific part (surface, etc.)?	What is the thermal stability, perishability of the sample?	What is the desired ruggedness of the analysis?
What expertise is available?	Do other analyses follow?	What are the availability and the value of the sample?	
How certain is the info. about the sample?	Are there preparative purposes?	Are there hazard problems?	

Following the columns in Table 1.1.1 and having preliminary answers to as many questions as possible helps in optimizing the path to follow in a chemical analysis.

In addition to the initial information about the sample, the information available in literature regarding the work already done on the same or similar type of analysis is extremely valuable. This information can be found in numerous publications. A number of computer-assisted information retrieval systems also are available. Among these, Analytical Abstract (Royal Society of Chemistry) with the program WinSPIRS from SilverPlatter Software, NV, 1999, and Dialog from The Dialog Corp., Cary, 2001 (customer@dialog.com) accessing various data bases are very powerful tools for searching the available information on a specific problem. The literature search for solving a specific analytical problem is very important considering the large number of available publications and the fact that new procedures are continuously developed and reported.

- Sample collection

Sampling, or sample collection, is the first operation in the chemical analysis (after the gathering of initial information). The importance of this operation is crucial because improper sampling leads to incorrect information from the chemical analysis. Also, some sampling techniques are closely related to sample preparation. For these reasons, some aspects of sampling are discussed in Chapter 2. However, a full discussion on sampling is beyond the purpose of this book, and dedicated monographs are available (see e.g. [6], [7]).

An intermediate step between sample collection and sample preparation is the sample storage and preservation before the analysis. Some samples are perfectly stable, and their storage is not an issue. However, many samples such as gas samples containing vapors or particles, emulsions, and mainly biological samples may pose very difficult problems regarding storage and preservation. Specific operations may be needed for specific samples to be stored or preserved, such as lyophilization. The subject of sample storage and preservation may be split between sampling and sample preparation. As an example, the collection of blood on heparin or EDTA can be considered a sampling procedure, although chemical interactions occur during collection.

- Sample preparation

Sample preparation is common in many analyses and is developed to allow or to improve a specific analysis. This step may be the most time-consuming in an analysis and affects significantly the analytical information. Numerous procedures are described in literature for sample preparation. Sample preparation may target the matrix of the sample, the analytes, or both. One common operation in sample preparation is the dissolution of the sample if the sample is solid. Then, the matrix is usually modified during a cleanup, fractionation, and concentration of the sample. The analytes also can be modified by chemical reactions (derivatization, etc.) in order to obtain better properties for the chromatographic analysis.

Ideally, in an analytical method, as few as possible operations of sample preparation are utilized. More operations require longer processing time and may lead to more error sources and possibly less accurate analytical results. However, in many cases extensive sample preparation is applied for better results. It is not always easy to choose from a number of analytical methods developed for the same purpose the one with the correct balance between simplicity and quality of the results. Various techniques for sample preparation have been developed specifically for chromatography. General procedures for sample cleanup and concentration not specific for chromatographic analysis also are utilized as part of some analytical chromatographic methods.

Chapter 3 of this book presents an overview of sample preparation for chromatography, describing the path for selecting a specific sample preparation, the role of different sample preparation techniques, and the strategy to assemble a new analytical method. The technical details related to various sample preparation procedures are given in Part 2 and Part 3 of the book. The material is grouped in two general categories: a) procedures using physical processes to perform cleanup, concentration, and preliminary fractionation of the sample, which change the initial matrix without chemical modification of the analyte or of the matrix, these being discussed in Part 2, and b) chemical modifications of the sample performed either for the purpose of cleanup and fractionation or with the purpose of better chromatographic properties or detectability in the core chromatographic step, these being presented in Part 3. Although these operations are discussed separately, sample preparation combines physical and chemical procedures to achieve a specific purpose. Some applications of general interest are described for various sample preparation techniques. These applications cover fields of interest such as analyses of drugs, pollutants, food and beverages, etc. The examples are not particularly oriented toward life science. Subjects such as protein modifications/labeling, DNA sequencing, etc. were considered beyond the purpose of the book. (The terms *step, procedure,* and *operation* have various meanings, but usually the term *step* is used in this book to indicate sample collection, sample preparation, core chromatography, etc., the term *procedure* is used for a group of operations with a specific purpose such as cleanup, derivatization, etc., and *operation* indicates the performance of a practical work such as filtration, distillation, etc.)

Many sample preparation techniques such as cleanup, concentration, as well as techniques used for fractionation and selective collection of the analytes from the sample are based on separations. For this reason, a number of separation procedures are discussed. More information regarding separations may be obtained from dedicated monographs such as B. L. Karger et al., *An Introduction to Separation Science*, J. Wiley, New York, 1973. Also, the derivatization is a subject requiring a wide range of information, such as synthesis of the reagent, detailed description of the derivatization procedure, etc. More information regarding derivatization may be obtained from original publications or from dedicated monographs such as G. Lunn, L. C. Hellwig, *Handbook of Derivatization Reactions for HPLC*, J. Wiley, New York, 1998.

The end result of the sample preparation is usually a *processed sample* that is submitted to (or directly introduced into) the core analytical step. The success of a sample preparation can be evaluated based on the differences between the initial sample and the processed sample. These differences cannot be measured by one single parameter. The purpose of sample preparation is to have a processed sample that leads to better analytical results compared to the initial sample. This can be achieved by having a processed sample easier to introduce in the core analytical system, with fewer matrix components (cleaner sample), more concentrated in analytes, chemically modified to enhance the separation and the detection in the core analytical system, etc.

- *Analytical chromatography*

The core of any analytical process includes the identification and measurement of the analytes. This step follows the sample preparation. Peculiar for the use of analytical chromatography as a core step in the analysis is that it also includes a detailed separation process. In spite of this combined role of separation and detection, the analytical chromatography is viewed as a single technique. The most common chromatographic separations are gas chromatography (GC) and high performance liquid chromatography (HPLC). For GC and HPLC, the detectors are almost always incorporated in the instrumentation. These instruments perform a detailed separation of the sample, allow the direct introduction of the analytes into the detector, and accomplish the measurement. The capability to perform detailed separations associated with the high sensitivity, the need for only a small amount of sample for analysis, and the capability to perform compound identifications make chromatographic techniques to be preferred for a large number of analyses.

A chromatographic technique is usually selected for analysis of a particular sample when the pure compound of interest in a sample can be analyzed by that technique. For example, volatile compounds can be easily analyzed using gas chromatography. Therefore, a GC technique is usually selected for the analysis of volatile compounds even when these are present in a plant material. Molecules with high molecular mass are typically analyzed by HPLC.

A strong interrelation exists between sample preparation and the analytical chromatographic process. The choice of a sample preparation procedure depends on the type of chromatography selected for sample analysis. On the other hand, the choice of the type of chromatography depends not only on the nature of the sample, but frequently on the availability of a good sample preparation technique. The detection and measurement associated with the chromatographic process is a component with major impact on sample preparation. Based on the intended detection, a specific sample preparation procedure can be implemented. Also, the choice of a specific detection system may be determined by the availability of a specific sample preparation procedure. These interrelations between sample preparation ←→ chromatographic separation ←→ detection system play an important role in choosing the operations in a chemical analysis [8]. More discussions about the chromatography can be found in Chapter 4. Extensive literature is available on this subject, including books and dedicated journals (see e.g. [9], [10]).

In certain sample preparations a preliminary chromatographic separation is sometimes utilized. However, this is not followed by a measurement but by the collection of the cleaner sample. For this reason, preliminary chromatography is viewed as a sample preparation procedure and not as a core analytical step. It is common that sample preparation chromatography is significantly simpler and less advanced than analytical chromatography. It may be limited to the separation of only a few components or may have poor resolution (see Section 1.6). However, preliminary chromatography is used successfully in sample preparation. Chromatographic separations for sample preparation purposes are discussed in Chapter 4 of this book.

Certain analytical separations such as capillary electrophoresis (CE) or other electrophoretic techniques, which perform the separation based on the difference in the rate of migration in an electric field and not on a chromatographic process, have many similarities with HPLC and with electrochromatography. The same sample preparation techniques that were developed for chromatographic analysis can sometimes be used for CE, gel electrophoresis, etc.

- Data processing

Data processing is another important step in any chemical analysis. This step can be as simple as the calculation of a concentration from a calibration curve or may be much more complex involving electronic data processing, elaborate data searches in existent data bases, extensive statistical evaluations of multiple sets of data, etc. The data validation effort, where the results from known standards or samples previously analyzed by other techniques are compared with the results of the method to be validated, is also part of the data processing. Specific protocols are available for data validation, and there are various levels of certification [7]. The display of the information generated in an analytical procedure is also part of data processing (see e.g. [11]). In the multitude of analytical techniques and methods described in literature, the display of information varies considerably. The units expressing concentrations, precision, limit of detection, limit of quantitation, etc. may be different from method to method. In this book, the IUPAC (International Union of Pure and Applied Chemistry) recommended notations and units were followed [12]. The names of organic compounds also can vary depending on the nomenclature rules adopted by the authors. In this book, the names of the organic compounds as used by the original authors were, in general, followed. In other cases the IUPAC recommendations were applied [13].

Many criteria and parameters are applied in evaluating an analytical process based on the quality of the final results as well as on a variety of factors regarding the operations required to perform the analysis. Seen as a channel of processing information, the relation between input and output can be classified as linear or nonlinear, random or deterministic, continuous or discrete, time dependent or steady in time, with memory or without memory. Statistical procedures can be used for describing precision, accuracy, and informational entropy gained in the analysis [2]. Other quantitative parameters as well as qualitative descriptions are used to account for other analysis characteristics such as ease of operations or average time required to do the analysis. Some aspects regarding some statistical parameters common in analytical data processing are discussed in Section 1.11.

- Analytical results

The output of the analytical process consists of the analytical results. Once the analytical results are available, they may provide the necessary information and the analysis is complete. It is also possible that the analytical results indicate the need for other analyses or for the repetition of the same analysis on the same sample or on a new sample from the same material, similar materials, or associated materials. The decision about repeating the analysis or some steps of the analysis is indicated as a feedback process suggested in Figure 1.1.1 by the arrow between the analysis results and the data processing. Specific steps can be modified in the analytical process when an analysis is repeated, and based on the final information, the analytical process can be improved if necessary.

1.2. TYPES OF OPERATIONS USED IN SAMPLE PREPARATION

Almost every analytical method includes a sample preparation step. Sample preparation is usually a chain of several operations applied to the sample. Simple operations, such as filtration or dilution, and also more complex operations, such as SPME (solid phase microextraction) or column chromatography, can be included in this chain. The frequency of usage of different procedures for sample preparation has been estimated in a survey regarding trends in sample preparation conducted by *LC-GC* magazine. The most commonly used sample preparation operations are shown in Figure 1.2.1 [14].

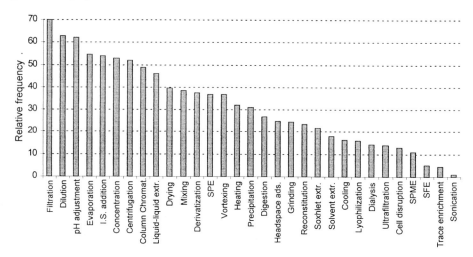

FIGURE 1.2.1. *Relative frequency of sample preparation operations obtained as the percentage of respondents that currently use the operation [14].* I.S. = internal standard, SPE = solid phase extraction, SPME = solid phase microextraction, SFE = supercritical fluid extraction.

Sample preparation operations can be classified based on a number of criteria, which include: a) the purpose of the operation, b) the physical nature of sample to be processed, c) the type of sample, d) the operational type of process that relates to the efficiency and the amount of sample utilized, e) the utilization of solvents, f) the type of chromatographic process for which the sample is processed, and g) the property to be enhanced by the operation. Some of these criteria are indicated in Table 1.2.1.

TABLE 1.2.1. *Criteria used for classification of sample preparation procedures.*

Purpose of the operation	Physical nature of the sample to be processed	Type of sample	Operational type of the process	Use or absence of solvents	Type of chromato- graphic process	Property to be enhanced
dissolution	gas	pharmaceutics	single contact	with solvents	GC	solubility
cleanup	liquid	pollutant	differential	solventless	HPLC	volatility
fractionation	solid	petroleum/oil	counter- current	using membranes	TLC	UV-Vis absorbance
concentration	heterogeneous	food/flavors	migration		other	fluorescence
chemical modification		other	etc.			etc.

It is difficult to classify sample preparation procedures based on any single criterion. For example, the use of the purpose of sample preparation as a classification criterion appears to be very useful for a classification. However, a clear differentiation between cleanup and fractionation is difficult to make, and some concentration processes where a solvent is eliminated from a sample can be considered fractionations.

Another example is the use of the classification based on the physical nature of the sample. This classification may lead to problems when applied to a gas sample containing solid particles as aerosols.

The type of sample also leads to a meaningful criterion for classification. The survey conducted by *LC-GC* magazine obtained the results shown in Figure 1.2.2 regarding the most commonly analyzed types of samples [14].

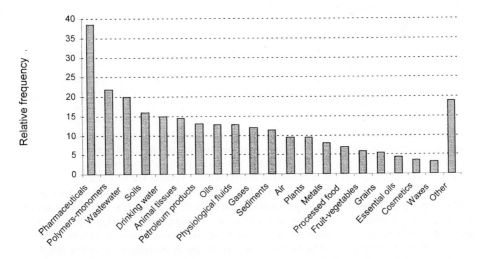

FIGURE 1.2.2. *Relative frequency of sample types submitted for analysis obtained as the percentage of respondents that analyze these types of samples [14].*

However, the same type of sample preparation can be used for various materials, and the specification of sample type may not be always relevant for the application of a specific technique.

A common deficiency encountered when using a unique criterion for classification of the operations in sample preparation is that some groups are very large while others are very limited. Using the criteria of purpose of sample preparation, sample dissolution, for example, is a much simpler process compared to cleanup. Consequently, a very large number of procedures for sample cleanup are reported, while only a few dedicated reports are limited to the description of sample dissolution.

The most general criterion applicable for organizing the presentation of sample preparation procedures can be considered the one based on the physical or chemical nature of the process. Even here, the nature of the process can be ambiguous, such as in ion exchange where chemical interactions take place during the separation. In the case of an unclear nature of the process, additional criteria can be used, such as the purpose of the operation. If this purpose is a separation, the operation can be classified as physical, and when the purpose is to obtain a new compound, the operation can be classified as chemical. The material in this book is organized in operations that use physical changes of the sample and those that use chemical changes. However, because this classification is not detailed enough, further differentiation is necessary.

- *Physical processes used in sample preparation*

Most physical operations are applied to a sample to achieve the cleanup and the concentration of the analytes. This is typically done using separations of the analytes from other sample constituents (matrix) or separations of certain matrix components from the rest of the sample. A large number of physical separation procedures are known, and they can be further classified based on various criteria such as the operational type, the utilization of solvents, or the mechanism of the process [15], [9].

A classification based on the operational type is usually done for separations [15] and distinguishes a) single-contact, b) differential, c) crosscurrent, d) countercurrent, e) countercurrent with reflux, f) differential migration, and g) zone refining. Single-contact processes take place in a single stage when two fractions remain in contact until equilibrium is (approximately) attained. The two fractions (phases) are then separated physically. In this type of separation are included crystallization, single step liquid-liquid extraction, etc. (A *phase* is defined as a physically distinct and homogeneous portion of a system that is separated from the rest of the system by a separation surface. The phase may be gaseous, liquid, or solid.) Single-step processes are restricted to binary separations. In differential processes a fraction is also separated in a single stage, but the produced fraction can be removed in increments. Typical for this separation is the simple distillation. Crosscurrent processes are multistep versions of the single-contact mode. Repeated distillation, recrystallization, and Soxhlet extraction are examples of this type of separation. Countercurrent processes are separations where two streams flow in opposite directions, such as in dialysis systems where the pure solvent flows in one direction and the solution to be extracted in the other direction along the dialysis membrane. Typical for countercurrent processes with reflux is fractional distillation, where a part of the distillate is returned to the fractionation column. Differential migration methods are typical for separations such as chromatography where the components of a mixture have different migration rates, as a fluid is constantly flowing through the separation medium. Zone refining is another separation that operates by moving a zone where separation takes place along a solid material subjected to separation. Zone melting (and crystallization) is the typical example of this type of separation. This classification is used when factors such as the amount of material, separation efficiency, speed of the process, etc. are important to consider.

The classification based on the mechanism of the process gives the following groups: a) mechanical operations, b) procedures involving phase transitions, c) procedures involving distribution equilibria, d) procedures based on differences in the migration rate through a barrier such as a membrane, and e) procedures based on the rate of transport in a field.

Mechanical operations include simple processes such as grinding, particle separations such as sieving, filtration, and sedimentation, but also more complicated operations such as ultrafiltration, etc. Filtration is a very common operation in many sample preparation methods. Centrifugation also can be included in this group and not viewed as based on the rate of transport in a field (gravific). Phase transitions involve the transfer of the analyte or of a component of the matrix from one of its phases into another phase. Examples include distillation, dissolution, crystallization/precipitation, etc. Phase equilibria are very common in many simple operations used in sample preparation. Some theoretical aspects of phase transitions are discussed in Section 1.3. Distribution equilibria are based on the differences in the distribution of the compounds to be separated between two phases. Distribution equilibria are involved in chromatography, solvent extraction, solid phase extraction, etc. These processes are common in both sample preparation as well as in the core chromatographic process. Some theoretical aspects of distribution equilibria are discussed in Section 1.4. The processes based on the differences in the transport rate through a barrier such as a membrane include ultrafiltration, dialysis, osmosis, etc. These separations may also involve a concentration gradient as a driving force for the transport process. The processes based on differences in the transport rate in a field such as electric or magnetic field include electrophoresis, thermal diffusion, etc.

For some physical sample preparation processes the classification based on the mechanism is straightforward. Others can be classified in more than one category. An example is ultracentrifugation, which can be considered a field separation and at the same time a mechanical separation. Size exclusion chromatography can be seen as a mechanical process and also as a separation based on the transport rate through a barrier. There are separations that involve more than one process in the same step. One typical example is precipitation, which combines a chemical reaction and the formation of solids from a solution due to the lack of solubility of one of the reaction products. (A *solution* can be defined as a phase containing more than one component.) Chromatography of ion pairs involves both a chemical reaction and a distribution equilibrium. Some chromatographic procedures also involve more than one mechanism for the separation process. For these reasons, this classification is useful only to organize the description of various separation procedures and not for indicating an unequivocal separation mechanism. Also, some separations are applied in combination with others. For example, precipitation is frequently used together with filtration or centrifugation. Distillation can be applied in a number of repeated operations, which is known as fractional distillation.

The description of many physicochemical properties of a system can be obtained from basic thermodynamic relations. A short introduction to some basic thermodynamic relations is therefore necessary when discussing physicochemical properties. One of these basic relations is the definition of Gibbs free enthalpy, which is expressed as follows[16]:

$$G = H - TS \tag{1.2.1}$$

where G is the free enthalpy, H is the enthalpy, T is the absolute temperature, and S is the entropy. The enthalpy H is further expressed by the formula:

$$H = E + pV \tag{1.2.2}$$

where E is internal energy, p is the pressure and V is the volume. In differential form these expressions are written as follows:

$$dG = dH - TdS - S\,dT \tag{1.2.3}$$

$$dH = dE + pdV + V\,dP \tag{1.2.4}$$

The first law of thermodynamics for a closed system (which does not exchange energy or mass with its surroundings) is expressed in the form:

$$dE = dQ + dW \tag{1.2.5}$$

where Q is the heat and W the work; and for a reversible system, the second law of thermodynamics is expressed by

$$dQ - TdS = 0 \tag{1.2.6}$$

For systems with the work resulting from volume modifications, the value of dW is expressed by the formula:

$$dW = -pdV \tag{1.2.7}$$

Using rel. (1.2.4) through (1.2.7) in the expression giving the variation of free enthalpy, rel. (1.2.3) becomes

$$dG = Vdp - SdT \tag{1.2.8}$$

For open systems with r components suffering changes, rel. (1.2.8) is modified in the form:

$$dG = Vdp - SdT + \sum_{j=1}^{r} dn_j \mu_j \tag{1.2.9}$$

where n_j is the number of moles and μ_j is the chemical potential of component "j". The chemical potential can be defined as the molar free enthalpy of a pure substance, or

$$\mu_j = \left. \frac{\partial G}{\partial n_j} \right|_{T,p,n_1,n_2,\ldots} \tag{1.2.10}$$

where the index $T,p,n_1,n_2,\ldots$ indicates constant temperature, pressure, and number of moles of the other components. Chemical potential is particularly important for equilibria at constant temperature and pressure when G is related only to μ_i. Rel. (1.2.9) is frequently applied in discussions on various physicochemical properties.

In a multicomponent system, the partial molar quantities for a component "i" in a phase can be defined for any extensive thermodynamic function Z (enthalpy, energy, entropy, etc.). The partial molar quantity Z is the change in Z for a change in n_i or

$$Z = \partial Z / \partial n_i \tag{1.2.11}$$

Partial molar quantities are intensive thermodynamic functions. For example, the partial molar volume is the increase of the volume per mole of component "i" when the number of moles of "i" are modified with an infinitesimal amount. Thus, the partial molar volume of component "i" is given by the expression:

$$\frac{\partial V}{\partial n_i}\bigg|_{T,p,nj} = V_i \tag{1.2.12}$$

On the other hand, from rel. (1.2.9) $V = \dfrac{\partial G}{\partial p}\bigg|_{T,ni}$, and therefore between V_i and μ_i there is the following relation:

$$V_i = \frac{\partial \mu_i}{\partial p}\bigg|_{T,n_i} \tag{1.2.13}$$

In equilibrium conditions, and using the ideal gas law, rel. (1.2.13) is equivalent with the expression:

$$d\mu_i = RT \frac{dp_i}{p_i} \tag{1.2.14}$$

where p_i is the partial pressure of the component "i" and R is the gas constant ($R = 8.31451$ J deg^{-1} mol^{-1} = 1.987 cal deg^{-1} mol^{-1}). The integration of rel. (1.2.14) leads to the expression:

$$\mu_i = \mu_i^{\circ} + RT \ln (p_i / p_i^{\circ}) \tag{1.2.15}$$

where p_i° is the (vapor) pressure of a pure liquid "i" (at the equilibrium temperature T), and μ_i° is the standard chemical potential for the compound "i". Rel. (1.2.15) can be written in the form:

$$\mu_i = \mu_i^{\circ} + RT \ln x_i \tag{1.2.16}$$

where it is assumed that Raoult's law is valid ($p_i = p_i^{\circ} x_i$). For liquid solutions, rel. (1.2.16) is still valid, and also can be written in the form:

$$\mu_i = \mu_i^{\circ} + RT \ln a_i \tag{1.2.17}$$

where a_i is the activity of solute "i" in solution. Rel. (1.2.17) is frequently used for the understanding of phase equilibria.

- Chemical processes used in sample preparation

Many chemical modifications are applied in sample preparation. Their general purpose is to obtain a sample with a different chemical composition that is easier to analyze. Chemical modifications may serve a number of purposes including: a) sample solubilization, b) cleanup or fractionation, and c) modification of the chemical structure to obtain specific desired properties of the new compound generated from the analyte.

A different classification of chemical reactions differentiates those involving small molecules and those involving polymers. The reactions performed on small molecules with the purpose of obtaining new compounds with properties used in the core chromatographic step are known as derivatizations. The chemical modification of polymers can further be differentiated as chemical degradations or pyrolysis.

Other classifications of the chemical process may be based on the nature of the chemical reaction, such as alkylation/arylation, silylation, acylation, addition, oxidation-reduction, etc. Also, the type of functional groups to be derivatized may offer a classification criterion. Another classification can be based on the property of the analyte that is intended to be enhanced, such as volatility/thermal stability for GC analysis, UV-Vis absorption, fluorescence, or chemiluminescence for HPLC analysis.

All these classifications have advantages and disadvantages. A typical problem is that the same operation sometimes can be included in different classes. For example, when the nature of the derivatized functional group in the analyte is chosen as a criterion for classification, the place of any multifunctional analyte becomes a problem. A unique criterion for the classification of sample preparation procedures using chemical reactions is not adequate. For this reason, Part 3 of the book discusses separately chemical modifications used for cleanup or fractionation, chemical modification applied to small molecules to enhance a specific property, and polymer chemical changes. Further, chemical modifications of small molecules are discussed from three different points of view: a) purpose, b) reaction type, and c) nature of derivatized functional group.

1.3.   SOME THEORETICAL ASPECTS OF PHASE TRANSFERS

The separations based on phase transfers are frequently applied for sample preparation. Phase transfers are common in simple operations such as distillation, crystallization/precipitation, or dissolution, where sample components undergo a change in physical state. These changes are used in sample preparation either for separating the analyte from the matrix or to eliminate or reduce some components of the matrix. More complex separation techniques such as zone melting are also based on phase transfers, but they are not common in sample preparation for chromatography.

In a phase transfer, the concentration difference for a component in two different phases or the departure from interphase equilibrium is the driving force for the mass transfer between phases. When the equilibrium is attained, the net transfer of material stops, and for practical purposes the separation does not occur anymore. However, at equilibrium a transfer of material still takes place in and out of each phase, but it occurs at the same rate in each direction, and the concentration for the specific component remains constant in each phase. The study of phase equilibria is based on Gibbs phase rule:

$$F = C - P + 2 \tag{1.3.1}$$

where F is the number of *degrees of freedom*, C is the *number of components*, and P is the *number of phases* in the system. The number of degrees of freedom F represents the number of variables that can be modified for the system. The number of components C represents the minimum number of chemically distinct constituents necessary to describe the composition of each phase.

The equilibria between phases are independent of the actual amounts (masses) of the phases that are present. For example, the vapor pressure of water above the liquid water is not dependent on the amount of water that generates the vapors. For this reason, only *intensive variables* such as concentration, pressure (p), and temperature (T) are chosen as variables F, and not those that represent the amount and are *extensive variables*.

Phase equilibria must comply with certain conditions. Thermal equilibrium requires that the temperatures of all phases are the same. For mechanical equilibrium it is necessary that the pressures of all phases are the same. Also, chemical equilibrium is assumed for the system. From these conditions, Gibbs phase rule results, as shown below. Considering a system with P phases (P1, P2, P3...) and C components {i}, the state of the system at equilibrium is specified if the temperature, the pressure, and the amount of each component is known in each phase. The total number of variables required to specify the state of the system is therefore $P\,C + 2$. Since the size of the system does not matter, instead of the amount of each component expressed as the number of moles n_i^P (index "i" for the component and "P" for the phase), it is sufficient to have the mole fractions x_i^P where

$$x_i^P = n_i^P \Big/ \sum_{i=1}^{C} n_i^P \tag{1.3.2}$$

The mole fractions for all components in a phase "P" satisfy the relation:

$$\sum_{i=1}^{C} x_i^P = 1, \text{ for every phase "P"} \tag{1.3.3}$$

and therefore, one variable for each phase can be obtained with knowledge of the values for all the others. The total number of variables to be specified is therefore $P\,C + 2 - P$. In addition, at the equilibrium, for each component the chemical potential μ_i^P must be the same in each phase (e.g. [16]). This leads to $C\,(P - 1)$ conditions:

$$\{\mu_i^{P1} = \mu_i^{P2} = \mu_i^{P3} = ...\}_{i = 1,2, ...C.} \tag{1.3.4}$$

The number of degrees of freedom that equal the total required variables to specify the state minus the constraints will be $F = P\,C + 2 - P - C\,(P - 1) = C - P + 2$, as expressed by rel. (1.3.1). In a monocomponent system $C = 1$, and three different cases are possible: $\{P = 1, F = 2\}$, $\{P = 2, F = 1\}$, and $\{P = 3, F = 0\}$. The water system is shown as an example in Figure 1.3.1. The diagram indicates the three areas for ice, liquid

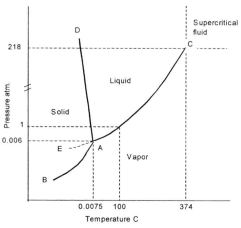

FIGURE 1.3.1. *Diagram showing the water system (not to scale).*

water, and water vapors. Point C is the critical point beyond which the liquid and gas are no longer distinguishable. At higher temperatures and pressures the fluid is present in supercritical state. The temperature of the critical point T_c and the pressure at the critical point p_c are characteristic for every compound and are important parameters in sample preparation techniques such as supercritical fluid extraction (SFE). The diagram also shows that at 100° C the corresponding vapor pressure of water is 1 atm., which is the boiling point of water. The curve AB is the sublimation curve for ice, and line AD separates the ice and liquid water regions (melting-point curve). The decrease of the melting point with the increase in pressure takes place only for water and very few other substances, the typical behavior being the increase in the melting point. All phases are in equilibrium at point A. Liquid water may be cooled below its freezing point without solidifying, the line AE being the curve for supercooled water.

The slopes of the lines in the diagram shown in Figure 1.3.1 are given by Clausius-Clapeyron equation:

$$\frac{d(\ln p)}{dT} = \frac{\Delta H}{RT^2}$$

(1.3.5)

where ΔH is the change in molar enthalpy (enthalpy per mole) for the phase change, and R is the gas constant. This equation can be obtained using the assumption that along the coexistence curve of two phases in equilibrium, for a change in T, a differential change in the chemical potential of one phase must be matched by an equal change in the chemical potential of the other phase. This leads to the expression:

$$d\mu^L(T,p) = d\mu^V(T,p)$$

(1.3.6)

where the index "L" and "V" denote two phases (such as liquid and vapor). For a system in equilibrium, dG =0, and from rel. (1.2.9) the following relation can be written:

$$d\mu = \frac{\partial \mu}{\partial T}\Big|_p dT + \frac{\partial \mu}{\partial p}\Big|_T dp = -SdT + Vdp$$

(1.3.7)

where $\frac{\partial \mu}{\partial T}\Big|_p = -S$ and $\frac{\partial \mu}{\partial p}\Big|_T = V$ with the index p denoting constant pressure and index T a constant temperature, and where S is the entropy per mole (molar entropy) and V is the volume per mole (molar volume). Based on rel. (1.3.6) and rel. (1.3.7), the following expression is valid:

$$-S^L dT + V^L dp = -S^V dT + V^V dp$$

(1.3.8)

Using the relations:

$$S^V - S^L = \Delta S_{vap} = \Delta H_{vap} / T_{vap} \tag{1.3.9}$$

where ΔH_{vap} is the molar enthalpy of phase change (vaporization), expression (1.3.8) leads to

$$\frac{dp}{dT} = \frac{\Delta H_{vap}}{T \Delta V_{vap}} \tag{1.3.10a}$$

Neglecting the volume of the liquid, $\Delta V_{vap} = RT / p$, and rel. (1.3.10a) becomes

$$\frac{dp}{dT} = \frac{\Delta H_{vap} \, p}{RT^2} \tag{1.3.10b}$$

which is equivalent with rel. (1.3.5). For a pure compound with p°_1 and p°_2 vapor pressures of pure liquid at temperatures T_1 and T_2, respectively, the integration of rel. (1.3.5) with ΔH_{vap} constant gives $\ln (p^\circ_2 / p^\circ_1) = - \Delta H_{vap} (1 / T_2 - 1 / T_1) / R$. This expression shows that the variation of vapor pressure as a function of temperature is exponential.

- *Gas-liquid equilibrium for two component systems*

For systems with two components, the phase rule becomes $F = 4 - P$, and the possible cases are $\{P = 1, F = 3\}$, $\{P = 2, F = 2\}$, $\{P = 3, F = 1\}$, and $\{P = 4, F = 0\}$. The maximum number of degrees of freedom is three, and therefore a complete graphical representation of a two-component system requires a three-dimensional diagram with pressure, temperature, and composition as coordinates. Keeping one coordinate constant, for example temperature, a bidimensional diagram is obtained, with pressure and mole fraction as variables. This type of diagram can be used, for example, to follow the pressure-composition diagram during the distillation at a constant temperature.

For a pressure/composition diagram for a solution with two components, the partial pressure of each component above the solution must be calculated. The partial vapor pressure of a component "i" above a solution in which its mole fraction is x_i^L is given by Henry's law (see also rel. 1.4.20), which states that the partial pressure is proportional with the mole fraction x_i^L in the liquid. For ideal solutions (see e.g. [16]), Henry's law takes the form of Raoult's law:

$$p_i = p^\circ_i \, x_i^L \tag{1.3.11}$$

where p°_i is the vapor pressure of pure liquid "i".

For a mixture of gases, the total pressure p_{Total} is given by Dalton's law, which states that the total pressure of a mixture of ideal gases is the sum of partial pressures. The partial pressure in the vapor phase is given by the formula:

$$p_i = x_i^V \, p_{Total} \tag{1.3.12}$$

where x_i^V is the molar fraction of component "i" in gas phase.

For a solution with two components "i" and "A" based on Dalton's law, using rel. (1.3.11) and $x_A^L = (1 - x_i^L)$, the total vapor pressure p_{Total} is equal to the sum of partial pressures:

$$p_{Total}^L = x_i^L\, p_i^o + (1 - x_i^L)\, p_A^o \tag{1.3.13}$$

In rel. (1.3.13), the total pressure is expressed as a function of the mole fraction in liquid phase and is noted p_{Total}^L ($p_{Total}^L \equiv p_{Total}$). Rel. (1.3.13) shows that p_{Total}^L varies linearly with the mole fraction x_i^L, as also shown in Figure 1.3.2. At the same time, the mole fraction x_i^V for "i" in vapor phase obtained from rel. (1.3.12) is given by the expression:

$$x_i^V = p_i / p_{Total}^L = p_i^o\, x_i^L / [x_i^L\, p_i^o + (1 - x_i^L)\, p_A^o] \tag{1.3.14a}$$

From rel. (1.3.14a) it can be seen that for the same total vapor pressure, x_i^V is not equal with x_i^L. The expression of x_i^L as a function of x_i^V can be obtained easily by rearranging rel (1.3.14a):

$$x_i^L = p_A^o\, x_i^V / [x_i^V\, p_A^o + (1 - x_i^V)\, p_i^o] \tag{1.3.14b}$$

The total pressure expressed as a function of x_i^V and noted p_{Total}^V ($p_{Total}^V \equiv p_{Total}$) has the expression:

$$p_{Total}^V = p_i^o\, p_A^o / [p_i^o - x_i^V\, (p_i^o - p_A^o)] \tag{1.3.15}$$

Rel. (1.3.15) is obtained by substitution in rel. (1.3.13) of the expression of x_i^L obtained as a function of x_i^V from rel. (1.3.14b).

With either x_i^L or x_i^V used on the independent variable axis, the variation of p_{Total} is indicated by two curves noted p_{Total}^L and p_{Total}^V, respectively, as shown in Figure 1.3.2a. The variation of x_i^L as a function of x_i^V for the same system is given in Figure 1.3.2b.

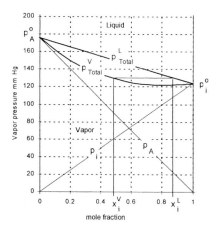

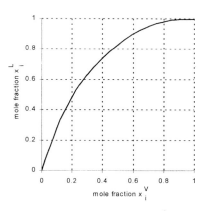

FIGURE 1.3.2a. *Composition of vapors x_i^V above an ideal solution and composition of the liquid x_i^L at a constant temperature.* The data are for a solution of $C_3H_6Br_2$ (i) in $C_2H_4Br_2$ (A) at $85°$ C [17].

FIGURE 1.3.2b. *Variation of x_i^L as a function of x_i^V for the system described in Figure 1.3.2a.*

As shown by rel. (1.3.14a) or (1.3.14b) and by the graph given in Figure 1.3.2b, the liquid and the vapors in equilibrium have different compositions, and the vapor is enriched in the more volatile compound ($x_i^V \leq x_i^L$). This difference allows the separation by distillation. During distillation, a liquid solution is boiled and the vapors are condensed. The separation is based on the condensation of the vapors for a concentration interval where the composition of the liquid and that of the vapors is different. Fractional distillation is also based on the differences in vapor pressure of the components of a solution (see Section 8.1).

A separation factor α can be used to determine the potential separation capability of distillation. For two liquids, the separation factor is defined as

$$\alpha = p_i^o / p_A^o \tag{1.3.16}$$

Eliminating p_i between rel. (1.3.11) and (1.3.12), it results $p_i^o\, x_i^L = x_i^V\, p_{Total}$, which leads to $p_i^o = x_i^V\, p_{Total} / x_i^L$ and similarly to $p_A^o = x_A^V\, p_{Total} / x_A^L$. Substituting p_i^o and p_A^o in rel. (1.3.16), the expression for α becomes:

$$\alpha = (x_i^V / x_i^L) / (x_A^V / x_A^L) \tag{1.3.17}$$

Distillation is typically described by a temperature/composition diagram where the pressure is set at a constant value. For the vapor pressure set at atmospheric pressure (p = 760 mm Hg = 760 Torr = 1 atm. = 101.325 kPa), the liquid boils, and the diagram shows the variation in the boiling point as a function of solution composition (see Section 8.1).

Many solutions do not follow the linear variation of p_{Total}^L shown in Figure 1.3.2 for an ideal solution. Raoult's law is replaced in these cases by the following dependence:

$$p_i = \gamma'_{i,A}\, p_i^o\, x_{i,A} \tag{1.3.18}$$

where $x_{i,A}$ is the mole fraction of solute "i" in solvent "A" and $\gamma'_{i,A}$ is the activity coefficient with the expression:

$$\gamma'_{i,A} = a_{i,A} / x_{i,A} \tag{1.3.19}$$

In rel. (1.3.19) $a_{i,A}$ is the activity of solute "i" in the solvent "A". Activity can be viewed as an "effective mole fraction" determining the extent to which a compound of mole fraction $x_{i,A}$ determines the equilibrium properties in solutions (for nonideal gases, the activity is replaced by fugacity f and an activity coefficient $\gamma = f / p$ is defined [16]). In this case, the variation of p_{Total}^L is given by an expression similar to rel. (1.3.13) based on Dalton's law, which can be written as follows:

$$p_{Total}^L = \gamma'_{i,A}\, x_{i,A}\, p_i^o + \gamma'_{A,i}\, (1 - x_{i,A})\, p_A^o \tag{1.3.20}$$

where $\gamma'_{i,A}$ and $\gamma'_{A,i}$ are dependent on the molar fraction $x_{i,A} \equiv x_i^L$.

For a diluted solution the mole fractions can be expressed using molar concentrations as shown below for $x_{i,A}$:

$$x_{i,A} = n_{i,A} / (n_{i,A} + n_A) \approx n_{i,A} / n_A \approx c_{i,A} M_A / 1000 \, \rho_A \qquad (1.3.21)$$

where $n_{i,A}$ and n_A are the number of moles of component "i" and solvent "A" respectively, M_A is the molecular weight of solvent "A", ρ_A is the density of solvent "A", and $c_{i,A}$ is the molar concentration of component "i" in phase "A". (Two different notations are further used to indicate molar concentrations; one is the notation $c_{i,A}$ for a species "i" in phase "A", and the other is [i] for species "i" in an unspecified solution.)

Using rel. (1.3.21), the expressions for the variation in total pressure as a function of mole fraction can be obtained as a function of concentration. The activity coefficient γ' relating the activity with the mole fraction also can be replaced with a different activity coefficient γ. This activity coefficient γ is dependent on the molar concentration $c_{i,A}$ and its expression is the following [16]:

$$\gamma_{i,A} = a_{i,A} / c_{i,A} \qquad (1.3.22)$$

The relation between the two activity coefficients $\gamma'_{i,A}$ and $\gamma_{i,A}$ can be obtained for diluted solutions by using $x_{i,A}$ from rel. (1.3.21) in rel. (1.3.19). The expression for $\gamma'_{i,A}$ can be written in this case in the form:

$$\gamma'_{i,A} = a_{i,A} \, 1000 \, \rho_A / c_{i,A} M_A \qquad (1.3.23)$$

which shows that the relation between $\gamma'_{i,A}$ and $\gamma_{i,A}$ is given by the expression:

$$\gamma'_{i,A} = \gamma_{i,A} \, 1000 \, \rho_A / M_A \qquad (1.3.24)$$

In phase transfer processes it is still common to use mole fractions and the activity coefficient defined by rel. (1.3.19) to express the composition of solutions. However, this is less common for distribution equilibria (see Section 1.4), where the molar concentrations and activity coefficient γ are typically used.

A nonlinear variation of p^L_{Total} as a function of x_i^L is shown in Figures 1.3.3a and 1.3.3b for nonideal solutions. The deviation from Raoult's law can be positive ($\gamma'_{i,A} > 1$) or negative ($\gamma'_{i,A} < 1$). Figure 1.3.3a shows the variation of p^L_{Total} for a solution with compounds showing positive deviation from Raoult's law. For some liquid solutions, there is a point where a specific liquid composition x_i^L gives a total pressure that corresponds to a vapor composition x_i^V equal to that of the liquid ($x_i^L = x_i^V$). This point is shown in Figure 1.3.3b and indicates the formation of an azeotropic mixture (liquid and vapors have the same composition). The components of a solution with azeotropic composition cannot be separated by distillation.

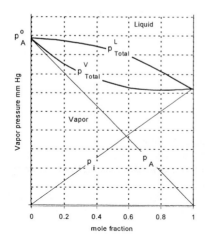

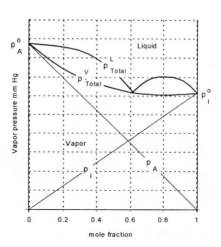

FIGURE 1.3.3a. *Diagram showing vapor-liquid equilibrium at a constant temperature for a nonideal solution.*

FIGURE 1.3.3b. *Diagram showing vapor-liquid equilibrium at a constant temperature for a nonideal solution that forms an azeotropic mixture.*

- *Liquid-solid equilibrium for two component systems*

Solid-liquid equilibrium for a two-component system in which the liquids are miscible and the solid-solid solubility is negligible is typically described by a diagram as shown in

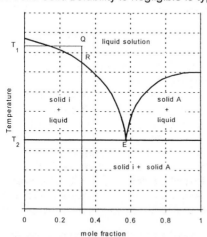

Figure 1.3.4. The pressure is kept constant in the process described by this type of diagram (the maximum number of degrees of freedom is three). For a point Q at temperature T_1, the material is melted. Decreasing the temperature, the composition remains unmodified up to the point R, where the solid "i" starts crystallizing and the composition is modified along the line RE up to the point E (eutectic point). When the temperature reaches T_2 corresponding to the eutectic point, the whole material becomes solid. The composition of the solution corresponding to point E is known as eutectic composition. The material below T_2 is a solid formed from crystals of pure "i" and microcrystalline mixture of "i" and "A" with the eutectic composition.

FIGURE 1.3.4. *Diagram for solid-liquid equilibrium of a two-component system in which the liquids are miscible and the solids-solid solubility is negligible.*

1.4. INTRODUCTORY APPROACH TO THEORY OF DISTRIBUTION EQUILIBRIA

A model for the distribution equilibrium is that of an analyte "i" in unspecified physical state (gas, liquid, or solid), which is distributed between two phases "A" and "B". The phases can be two (nonmiscible) liquids, a gas and a liquid, a gas and a solid, or a liquid and a solid. When the analyte forms solutions with both phases between which it is distributed, the process is known as *partition*. The distribution of an analyte between a gas and a solid is based on an *adsorption* process of the analyte on the surface of the solid, from the solution in a gas or a liquid. Distribution equilibria play a key role in chromatographic separations and other separation techniques. Some of the more common distribution equilibria including liquid-liquid partition, gas-liquid partition, gas-solid adsorption, and liquid-liquid partition of ion pairs are discussed in more detail in this section.

- *Liquid-liquid partition*

Liquid-liquid partition represents the mechanism of many separations. These separations include solvent extractions and various HPLC separations. In liquid-liquid partition the component "i" is distributed between two nonmiscible liquid phases "A" and "B" in an equilibrium of the type:

$$i_B \rightleftharpoons i_A$$

The values of the chemical potentials $\mu_{i,A}$ and $\mu_{i,B}$ of the component "i" in each of the two phases "A" and "B" must be equal at equilibrium (e.g. [16]). Using the common expression for the chemical potentials given by rel. (1.2.16), the equality of the chemical potentials for compound "i" in the two phases is written as follows:

$$\mu^{\circ}_{i,B} + RT \ln a_{i,B} = \mu^{\circ}_{i,A} + RT \ln a_{i,A} \tag{1.4.1}$$

In rel. (1.4.1) $\mu^{\circ}_{i,A}$ and $\mu^{\circ}_{i,B}$ are the standard chemical potentials that are constants, $a_{i,A}$ and $a_{i,B}$ are the activities of analyte "i" in the two phases "A" and respectively "B", T is absolute temperature, and R is the gas constant. The rearrangement of rel. (1.4.1) leads to the expression:

$$\ln (a_{i,A} / a_{i,B}) = - (\mu^{\circ}_{i,A} - \mu^{\circ}_{i,B}) / (RT) = - \Delta\mu_i^{\circ} / (RT) \tag{1.4.2}$$

Re. (1.4.2) shows that for a constant temperature T the ratio of the activities of the analyte "i" in the two phases is a constant expressed as follows:

$$K_i = a_{i,A} / a_{i,B} = \exp [- (\mu^{\circ}_{i,A} - \mu^{\circ}_{i,B}) / (RT)] \tag{1.4.3}$$

The constant K_i is the *thermodynamic distribution constant* for the partition process of component "i" between liquid phases "A" and "B".

The activities can be expressed as shown in Section 1.3 by the product between activity coefficients γ defined by rel. (1.3.22) and the molar concentrations c (a = γ c), such that rel. (1.4.3) can be written as follows:

$$K_i = (\gamma_{i,A} \, c_{i,A}) / (\gamma_{i,B} \, c_{i,B}) = \exp[-\Delta\mu_i^\circ / (RT)] \qquad (1.4.4)$$

where $\Delta\mu_i^\circ$ is the difference between the standard chemical potentials in solvents "A" and "B" for the solute "i". Rel. (1.4.4) is usually written in the form:

$$\log K_i = -\Delta\mu_i^\circ / (2.302 \, RT) \qquad (1.4.5)$$

Using the formula:

$$K_i = c_{i,A} / c_{i,B} \qquad (1.4.6)$$

the relation between K_i and K_i is the following:

$$K_i = [(\gamma_{i,A}) / (\gamma_{i,B})] \, K_i \qquad (1.4.7)$$

which shows that K_i is also a constant (depending on temperature and the nature of phases "A" and "B" and of compound "i"). The expression for K_i can be written as follows:

$$K_i = [(\gamma_{i,B}) / (\gamma_{i,A})] \, K_i \qquad (1.4.8)$$

and also:

$$K_i = [(\gamma_{i,B}) / (\gamma_{i,A})] \exp[-\Delta\mu_i^\circ / (RT)] \qquad (1.4.9)$$

The constant K_i is called *partition coefficient (or distribution constant)*, and it is commonly used to characterize analytical partition processes. It gives the ratio of the concentrations of "i" in phases "A" and "B" at equilibrium, as compared to K_i that gives the ratio of activities. The value of K_i is dependent on temperature and also on the chemical nature of component "i" and of the two solvents "A" and "B", although no chemical interaction is assumed in the system. Tables with K_i values for different systems and temperatures are available, and also extrathermodynamic techniques for practical estimations are reported (see Section 1.6). For complex problems, an *effective equilibrium constant* K^{eff} is sometimes used instead of individual values for each equilibrium [18].

A theoretical evaluation for K_i is also possible. For this purpose, the parameters in rel. (1.4.9) must be estimated. There are two common conventions used to choose the standard state of a component in a solution. Since $\mu_{i,A}^\circ$ and $\mu_{i,B}^\circ$ are independent of composition, one choice for the standard state is that of the pure liquid or solid "i" at the working temperature and pressure. In this case, $\mu_{i,A}^\circ = \mu_{i,B}^\circ = \mu^{Liquid}_i$, or $\mu_{i,A}^\circ = \mu_{i,B}^\circ = \mu^{Solid}_i$. Therefore for this choice, $K_i = 1$. Another convention is to choose the standard state to be that of the pure liquid only for a solvent, with all the other components having a "fictitious" state with the properties that pure "i" would have if its limiting low concentration properties in solution were to be retained in a pure substance [16].

The estimation of the activity coefficients, $\gamma_{i,A}$, can be done starting with two expressions of the chemical potential for a regular solution:

$$\mu_{i,A} = \mu_{i,A}^\circ + RT \ln \gamma_{i,A} \, c_{i,A} \qquad (1.4.10)$$

and

$$\mu_{i,A} = \mu^{\circ}_{i,A} + \Delta H_{i,A} + RT \ln c_{i,A} \qquad (1.4.11)$$

where $\Delta H_{i,A}$ is the enthalpy term (partial molar excess enthalpy) that accounts for deviation from an ideal solution. From rel. (1.4.10) and (1.4.11) the expression for $\gamma_{i,A}$ can be written as follows:

$$RT \ln \gamma_{i,A} = \Delta H_{i,A} \qquad (1.4.12)$$

For processes at constant volume, the enthalpies become energies, and $\Delta H_{i,A}$ is the heat of mixing a mole of pure compound "i" with a large quantity of "A" to form a dilute solution. The total energy of mixing can be expressed as the sum of net energy of vaporization of the two components ΔE_i^V and ΔE_A^V minus the energy of dissolution of vaporized components $\Delta E_{i,A}^S$:

$$\Delta H_{i,A} = \Delta E_i^V + \Delta E_A^V - \Delta E_{i,A}^S \qquad (1.4.13)$$

This relation can be expressed as a function of solubility parameters [16]. A solubility parameter δ_i for a compound "i" is defined [19] as the square root of the ratio of net energy of vaporization ΔE_i^V and V_i the molar volume of the component "i":

$$\delta_i = (\Delta E_i^V / V_i)^{1/2} \qquad (1.4.14)$$

From rel. (1.4.13) and (1.4.14) and with an estimation for $\Delta E_{i,A}^S$, it can be shown [15] that the expression for $\gamma_{i,A}$ from rel. (1.4.12) becomes

$$\gamma_{i,A} = \exp [\Delta H_{i,A} / (RT)] = \exp [V_i (\delta_i - \delta_A)^2 / (RT)] \qquad (1.4.15)$$

Including rel. (1.4.15) in the expression (1.4.9) for K_i and using the convention $\boldsymbol{K_i} = 1$, the result can be written as follows:

$$K_i = \exp \{V_i [(\delta_i - \delta_B)^2 - (\delta_i - \delta_A)^2] / (RT)\} \qquad (1.4.16)$$

Rel. (1.4.16) can be used for predicting the distribution constant K_i using the values for the molar volume and parameter δ that are tabulated (see Table 8.2.1) for a variety of compounds [15].

Although the distribution constant K_i for a given system appears to depend only on temperature, in practice it was found that K_i also may vary with the concentrations $c_{i,A}$ and $c_{i,B}$. The graph representing $c_{i,A}$ as a function of $c_{i,B}$ at a given temperature is called an isotherm and has the equation:

$$c_{i,A} = K_i c_{i,B} \qquad (1.4.17)$$

The isotherm is a straight line when K_i is a true constant, but it is a (nonlinear) curve when the distribution constant K_i varies with $c_{i,A}$ and $c_{i,B}$ (see also Section 1.8).

The application of liquid-liquid partition coefficient can be directly applied not only to solvent extraction processes, but also to chromatographic processes where the stationary phase is a liquid or a liquid like material.

- *Gas-liquid partition*

Gas-liquid partition is another type of distribution equilibrium with many practical applications. The theory developed for liquid-liquid partition can be easily extended to gas-liquid partition. If the component "i" is distributed between a gas (diluting gas) and a liquid "A" in an equilibrium of the type:

$$i_g \rightleftharpoons i_A$$

then rel. (1.4.3) can be written as follows:

$$K_i = a_{i,A} / a_{i,g} = \exp\left[-\left(\mu^o_{i,A} - \mu^o_{i,g}\right) / (RT)\right] \qquad (1.4.18)$$

The activity $a_{i,g}$ for the gas phase can be approximated with the expression $a_{i,g} = p_i / p^o_i$, where p_i is the partial vapor pressure of the component "i" in the gas phase and p^o_i is the vapor pressure above the pure component. Therefore, rel. (1.4.18) can be written as follows:

$$K_i = (\gamma'_{i,A}\, x_{i,A}) / (p_i / p^o_i) = \exp\left[-\left(\mu^o_{i,A} - \mu^o_{i,g}\right) / (RT)\right] \qquad (1.4.19)$$

Rel. (1.4.19) shows that K_i is a constant for a constant temperature, and therefore the ratio $x_{i,A} / p_i$ is also constant. This can be written in the form:

$$k = x_{i,A} / p_i \qquad (1.4.20)$$

Rel. (1.4.20) expresses Henry's law (see e.g. [17]). (Henry's law states in fact that $p_i = k\, x_{i,A}$ where $k = 1/k$.) Rel. (1.4.20) shows that increased pressure of the component "i" in gas phase increases the analyte molar fraction in the solvent. The chemical nature of the analyte "i" and of the solvent "A" also influence the value of k. From rel. (1.4.20) and using rel. (1.3.21) for $x_{i,A}$, the following relation can be written:

$$k' = c_{i,A} / p_i \qquad (1.4.21)$$

Rel. (1.4.21) includes in the constant k' the molecular weight and the density of the solvent "A" and shows that for the gas-liquid partition of an analyte, when the analyte does not interact chemically with the liquid, the concentration in the liquid is proportional with the partial pressure in gas phase. (Henry's law can be written in the form $p_i = k'\, c_{i,A}$.)

For a volume V_g of the gas and $n_{i,g}$ number of moles of "i" in the gas phase, the partial pressure p_i has the expression:

$$p_i = RT\,(n_{i,g} / V_g) \qquad (1.4.22)$$

Using rel. (1.3.21) for $x_{i,A}$, rel. (1.4.22) for p_i, and taking $p^o_i = 1$, the thermodynamic distribution constant for gas-liquid partition can be written as follows:

$$K_i = [c_{i,A} / (n_{i,g} / V_g)]\,(\gamma'_{i,A}\, M_A) / (1000\, \rho_A\, RT) \qquad (1.4.23a)$$

This expression can be written in a simplified form as follows:

$$K_i = K_i \, (\gamma'_{i,A} \, M_A) \, / \, (1000 \, \rho_A \, RT) = K_i \, \gamma_{i,A} \, / \, (RT) \qquad (1.4.23b)$$

where

$$K_i = c_{i,A} \, / \, (n_{i,g} \, / \, V_g) \qquad (1.4.24)$$

and the expression for $\gamma'_{i,A}$, is given by rel. (1.3.24).

This relation shows that K_i is also a constant (depending on temperature and the nature of compounds "A" and "i"). This constant K_i is called *partition coefficient* for the gas-liquid partition, and it is commonly utilized to characterize gas-liquid partition processes. Expressing the number of moles of analyte "i" per volume of gas phase as $c_{i,g} = n_{i,g} \, / \, V_g$ (or using $p_i = RT \, c_{i,g}$), the expression for K_i becomes:

$$K_i = c_{i,A} \, / \, c_{i,g} = K_i \, RT \, / \, \gamma_{i,A} \qquad (1.4.25)$$

Rel. (1.4.25) is analogous to rel. (1.4.9) for liquid-liquid partition. The values for K_i, for k and for Henry's law constant $\mathbf{k}$ are available for specific systems [9], [20] or can be calculated using extrathermodynamic estimation procedures (see Section 1.10).

The isotherm for the gas-liquid partition process will have the equation:

$$c_{i,A} = K_i \, c_{i,g} \qquad (1.4.26)$$

and rel. (1.4.26) represents a straight line when K_i is a true constant, but it is a (nonlinear) curve when the distribution constant K_i varies with $c_{i,A}$ and $c_{i,g}$. As seen in rel. (1.4.25), the constant value of K_i implies that $\gamma_{i,A}$ is constant at different concentrations, and this assumption is not always true.

Values for gas-liquid partition coefficients were determined experimentally for specific compounds such as n-octane, toluene, ethanol, methyl ethyl ketone, dioxane, and nitromethane as solutes "i" in various solvents "A" [21], [22].

Instead of concentration ratios described by rel. (1.4.18), the distribution equilibrium also can be described by the molar distribution of component "i" between the two phases "A" and "B". The total molar amount (mass) of solute "i" in phase "A" is $c_{i,A} \, V_A$, and the total molar amount of solute "i" in phase "B" is $c_{i,B} \, V_B$, where V_A is the volume of phase "A" and V_B is the volume of phase "B". The ratio of the number of moles in the two phases is given by

$$k_i = (c_{i,A} \, V_A) \, / \, (c_{i,B} \, V_B) \qquad (1.4.27a)$$

and the constant k_i is known as the *capacity factor*. For a gas-liquid partition, "B" is replaced by "g" and the capacity factor k_i is defined as:

$$k_i = (c_{i,A} \, V_A) \, / \, (c_{i,g} \, V_g) \qquad (1.4.27b)$$

Using rel. (1.4.27b) and (1.4.25), the relation between K_i and k_i can be expressed for a gas-liquid partition as follows:

$$k_i = K_i \, (V_A \, / \, V_g) \qquad (1.4.28)$$

Distribution equilibria are typically used for separations such as in various types of chromatography. The separation depends on the distribution of two different compounds "i" and "j" that must be separated between the phases "A" and "B". The comparison is done using a parameter known as *separation factor* α. The distribution of a component "i" between the two phases being described by $K_i = c_{i,A} / c_{i,B}$, and that of a compound "j" being described by $K_j = c_{j,A} / c_{j,B}$, the separation factor is given by

$$\alpha = K_i / K_j = (c_{i,A} / c_{i,B}) / (c_{j,A} / c_{j,B}) = (c_{i,A} \, c_{j,B}) / (c_{i,B} \, c_{j,A}) \qquad (1.4.29)$$

(By convention $\alpha \geq 1$, and if this is not the case, its definition becomes $\alpha = K_j / K_i$.) Both the capacity factor and separation factor are parameters used frequently in chromatography (see also Section 1.6). Rel. (1.4.29) for α and rel. (1.4.19) for K_i lead for diluted solutions to the expression:

$$\log \alpha = - (\Delta\mu_i^\circ - \Delta\mu_j^\circ) / (2.302 \, RT) \qquad (1.4.30)$$

Rel. (1.4.30) shows that the separation factor depends on the differences of the standard chemical potentials between phases "A" and "B" for the solutes "i" and "j".

Similarly to the liquid-liquid partition coefficient, the gas-liquid partition coefficient can be directly applied to chromatographic processes where the stationary phase is a liquid or a liquid like material.

- Gas-solid adsorption

Gas-solid adsorption is typically encountered in some gas chromatographic separations that take place on solid stationary phases. The term adsorption is frequently used with a more general meaning, indicating the process of concentrating a solute into a stationary phase that can be solid, semi-liquid, or even liquid. However, the accurate meaning is the adsorption on a solid surface not involving dissolution. The adsorption of a gaseous component from a diluting gas on a solid surface can occur either by physical interactions due to van der Waals forces between the gas molecules and the solid or due to chemical interactions leading to chemical bonds. Assuming that the adsorption of the molecules of a component "i" takes place on specific sites S by physical interactions, an equilibrium between the gas and the adsorbed form of "i" can be written as follows:

$$i_g + S \rightleftharpoons i_s$$

This equilibrium is governed by the equilibrium constant:

$$K = \frac{[i_s]}{[i_g][S]} \qquad (1.4.31)$$

where $[i_g]$ is the concentration of molecules "i" in gas phase (diluted in another gas), $[S]$ the concentration of unoccupied sites on the solid, and $[i_s]$ the concentration of occupied sites on the solid (molar concentration is indicated by $[\ ]$). The expression of the constant K also can be written in the form:

$$K = \frac{\theta \, k_B T}{(1 - \theta) \, p_i}$$

(1.4.32)

where θ is the fraction of surface covered and given by $\theta = q_i / q_m$ with q_i the amount of analyte "i" adsorbed on the solid phase and q_m the amount of "i" necessary to cover the whole adsorbent surface, k_B is Boltzmann constant ($k_B = 13.805 \; 10^{-24}$ J deg^{-1}), p_i is the partial gas pressure, and T the absolute temperature. This expression for the constant K is obtained from rel. (1.4.31) assuming that the adsorption occurs only in a monomolecular layer and taking $[i_g] = p_i / (k_B T)$, $[i_s] = \theta \, N$, and $[S] = (1 - \theta) \, N$ where N is the number of sites per unit area.

Rel. (1.4.32) can also be written in the form:

$$b = \frac{\theta}{(1 - \theta) \, p_i}$$

(1.4.33)

where $b = K / (k_B T)$. Rearranging rel. (1.4.33) it results

$$\theta = q_i / q_m = \frac{b \, p_i}{(1 + b \, p_i)}$$

(1.4.34)

Rel. (1.4.34) is known as a Langmuir adsorption isotherm. The graphs for θ as a function of p_i for two arbitrary values of b (b = 2 and b = 10) are shown in Figure 1.4.1a. From rel. (1.4.34) can be generated the formula:

$$q_i / p_i = \frac{b \, q_m}{(1 + b \, p_i)}$$

(1.4.35)

Although the ratio q_i / p_i is not in general a constant (even for constant temperature), when the value $b \, p_i$ is much smaller than 1 (very small p_i), the ratio can be considered constant (and Langmuir isotherm is approximated with a line). This leads to the relation:

$$K^A_i = q_i / p_i$$

(1.4.36)

The value K^A_i is the *adsorption constant* for the adsorption process at a constant temperature on a gas-solid interface. Rel. (1.4.36) is the equivalent for the adsorption process of rel. (1.4.7) for the partition.

The adsorption in a monomolecular layer on the surface of a solid is not the general case in practice. For the evaluation of adsorption in multimolecular layers, it is common to replace the quantities q_i and q_m with the volumes V_i and V_m representing the volume of the component "i" adsorbed on the surface and, respectively, the volume when "i" occupies all sites on the solid with a monomolecular layer. In this case, the equivalent of Langmuir adsorption isotherm is the adsorption isotherm described by the equation:

$$\frac{V_i}{V_m} = \frac{p_i^o \, p_i}{p_i^o - p_i} \frac{E}{p_i^o + (E-1)p_i}$$

(1.4.37)

In rel. (1.4.37) E has the expression:

$$E = \exp\left[(\Delta H^m - \Delta H^l) / (RT)\right]$$

(1.4.38)

where ΔH^m is the heat of adsorption for the component "i" in a monomolecular layer, and ΔH^l is the heat of liquefaction of the gas "i". The equation (1.4.37) is known as B.E.T. isotherm (Brunauer, Emmett, Teller) [9]. The graphs of B.E.T. isotherms for three arbitrary values of E are given in Figure 1.4.1b.

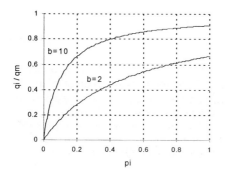

FIGURE 1.4.1a. *Langmuir adsorption isotherms for two arbitrary values b = 2 and b = 10.*

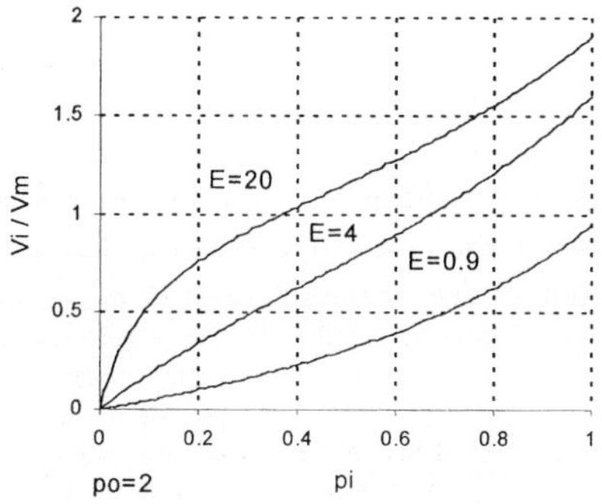

FIGURE 1.4.1b. *B.E.T. adsorption isotherms for p^o_i = 2 and three arbitrary values for E.*

The amount q_i of analyte "i" adsorbed on the solid phase is proportional with V_i, and the variation of the adsorption constant K^A as a function of p_i can be easily obtained from rel. (1.4.37). In this case, K^A can be approximated with a constant only when p_i is much smaller than p_i^o. In this case B.E.T. isotherms can be approximated with a line. As seen from the isotherms given in Figures 1.4.1a and 1.4.1b, the increase in the partial pressure of the component "i" leads to an increase in the adsorbed quantity of gas on the solid surface. The increase in temperature leads to a decrease in the value for b for a Langmuir isotherm and therefore to less material adsorbed on the solid. The same temperature dependence is valid for B.E.T. isotherm when E > 1. The chemical nature of the solid and of the gas "i" are also determining factors in the adsorption process, although no chemical reaction takes place. Values for adsorption constants K^A for different systems are available, and also these constants can be estimated using extrathermodynamic techniques (see Section 1.6).

- *Liquid-liquid partition of ion pairs*

A special case of liquid-liquid partition is that of the transfer between phases associated with a chemical reaction. This process is known as liquid-liquid partition of ion pairs and occurs for example when an ionic species X⁻ in an aqueous solution reacts with another ionic species Y^+ generating a compound XY easily extractable in an organic phase. This can be expressed by the equilibrium:

$$X^-_w + Y^+_w \; \rightleftharpoons \; XY_o$$

where the index "w" indicates aqueous solution and "o" organic solution, and where the equilibrium constant K is given by the expression:

$$K = \frac{[XY]_o}{[X^-]_w[Y^+]_w} \tag{1.4.39}$$

([X⁻] indicates molar concentration of species "X".) The distribution of a species X^- will be in this case given by the partition coefficient K_X, which has the expression:

$$K_X = \frac{[XY]_o}{[X^-]_w} = K \, [Y^+]_w \tag{1.4.40}$$

The ions X^-, Y^+ form an *ion pair*, and the pair extraction in the organic phase depends, as seen in rel. (1.4.40), on the equilibrium constant K and the concentration of Y^+ in aqueous phase. An ionic analyte X^-, therefore, can be transferred with good yield in an organic phase by adding in the aqueous phase a reacting species that forms a compound XY extractable in the organic phase and by increasing the concentration $[Y^+]$. This process has a wide range of applicability for analytes X^- when the analysis can be performed on the species XY, or when XY can easily regenerate X^-.

- *Liquid-liquid partition of compounds simultaneously in ionic and nonionic form*

A large number of compounds have either acidic or basic character, and they exist in both nonionic and ionic form. This is for example the case of many weak acids and bases. Because in distribution processes the two forms, ionic and nonionic, behave very differently, it is useful to understand the distribution process for these types of compounds. As an example, for a monoprotic acid that exists in both X^- and XH forms and for an aqueous solution, the equilibrium is described by the acidity constant K_a given by the expression:

$$K_a = \frac{[X^-]_w[H^+]_w}{[XH]_w} \tag{1.4.41}$$

the index w showing water solution. Two equilibrium constants can be defined using the expressions:

$$K_{X^-} = \frac{[X^-]_o}{[X^-]_w} \quad \text{and} \quad K_{XH} = \frac{[XH]_o}{[XH]_w} \tag{1.4.42}$$

The total equilibrium constant for the species XH and X- is given by the formula:

$$K = \frac{[X^-]_o + [XH]_o}{[X^-]_w + [XH]_w} \qquad (1.4.43)$$

Substituting in rel. (1.4.43) the value of $[X]_w$ from rel. (1.4.41) it is obtained

$$K = \frac{([X^-]_o / [XH]_w) + K_{XH}}{K_a / [H^+]_w + 1} \qquad (1.4.44)$$

and with further substitution of $[XH]_w$ from rel. (1.4.41) the formula for K becomes

$$K = \frac{K_{X^-}(K_a / [H^+]_w) + K_{XH}}{K_a / [H^+]_w + 1} \qquad (1.4.45)$$

Rel. (1.4.45) shows the dependence of the distribution constant of pH for a weak acid, and for very low values for K_{X^-} shows that the increase in acidity favors the distribution of the acid in the organic phase, while a decrease in acidity (basic solutions) favors the transfer of the acid in aqueous phase.

Similar calculations can be done for bases, showing the opposite process.

1.5. SOME ASPECTS OF MASS TRANSFER PROCESSES

During any separation a mass transport process occurs. The mass transport can take place by convection, diffusion, migration, or a combination of these processes. Convection consists of a physical movement at macroscopic scale, while diffusion and migration take place at molecular or atomic scale (except for turbulent diffusion). The diffusion process occurs as the result of a concentration gradient, a pressure gradient, or a temperature gradient, which determines a difference in the chemical potential in the system. The concentration gradient, for example, tends to be diminished by the random spreading at molecular level determining the mass transfer and equalizing the chemical potential. Migration takes place when the molecules (or atoms) are forced to move by a magnetic, electric, or gravific field.

- *Mass transfer by diffusion*

Mass transfer due to a concentration gradient is described by the two Fick's laws for diffusion. The first law indicates that the flux J of mass passing in a unit of time through a unit of surface area, due to the concentration gradient, in a given direction is expressed by the formula:

$$J = - D \nabla c \qquad (1.5.1a)$$

where D is the diffusion coefficient of the medium (given in area/time units), and c is the concentration of the diffusing species. For diffusion in the direction (x), $\nabla c = \dfrac{dc}{dx}$ and rel. (1.5.1a) becomes

$$J_x = -D \frac{dc}{dx}$$

(1.5.1b)

The minus sign in rel. (1.5.1a) and (1.5.1b) indicates that the flow occurs from the region with higher concentration to the region with lower concentration.

Fick's second law (that results from the first law) [16] has the expression:

$$\frac{dc}{dt} = D \frac{\partial^2 c}{\partial x^2}$$

(1.5.2)

where t is time, the concentration c is expressed in units of mass per units of length, and it was assumed that the diffusion process takes place only in one direction (x). This law can be directly applicable, for example, to the diffusion in a tube or a rectangular channel where the concentration c varies only along the channel and is the same across the channel. On the basis of the assumption that for t = 0 (initial condition) the concentration is described by a given function c (x,0), the solution of equation (1.5.2) can be written as follows (see e.g. [23]):

$$c(x,t) = \frac{1}{\sqrt{\pi Dt}} \int_{-\infty}^{+\infty} c(\eta,0) \exp[-(\eta - x)^2 / (4Dt)] d\eta$$

(1.5.3a)

With the assumption that D is constant and that the whole amount m of material was initially (at t = 0) contained in one point at x = 0, equation (1.5.3a) leads to the relation (see e.g. [15]):

$$c = \frac{m}{2\sqrt{\pi Dt}} \exp[-x^2 / (4Dt)]$$

(1.5.3b)

Rel. (1.5.3b) gives the concentration c at the point x and at time t of the diffusing substance. For any given t = constant, the mass m can be obtained from $m = \int_{-\infty}^{+\infty} c\, dx$, and the maximum concentration is expressed by the relation:

$$c_{max} = \frac{m}{2\sqrt{\pi D t}}$$

(1.5.4)

A graph showing the variation of c as a function of t and x for m =1 and D = 1 is given in Figure 1.5.1. (The units for c, m, x, t and D must be consistent. One choice is c in g/cm, m in g, x in cm, t in s, and D in cm^2/s. SI units can also be used [12].)

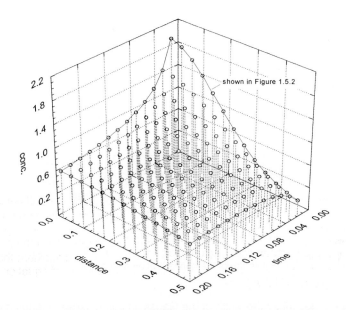

FIGURE 1.5.1. *The variation of the concentration c of a diffusing substance as a function of time and distance from the origin where the mass is m =1 and D =1 (all arbitrary units).*

By introduction in rel (1.5.3b) of the notation:

$$\sigma^2 = 2 D t \qquad (1.5.5)$$

the expression for c/m can be written as follows:

$$c / m = \frac{1}{\sqrt{2\pi\sigma^2}} \exp(-x^2 / 2\sigma^2) \qquad (1.5.6a)$$

The expression (1.5.6a) characterizes a typical Gaussian bell curve, describing a random process. Graphs of c/m as a function of the distance x are shown in Figure 1.5.2 for $\sigma = 0.2$ (corresponding for example to D = 1 and t = 0.02) and for $\sigma = 0.4$ (corresponding for example to D = 1 and t = 0.08).

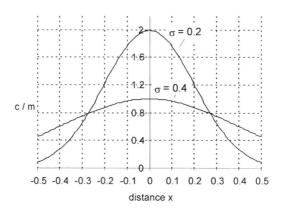

FIGURE 1.5.2. *The variation of c/m as a function of distance x for two values of σ, namely σ = 0.2 (corresponding for example to D = 1 and t = 0.02) and σ = 0.4.* The graphs show the Gaussian bell shape of the concentration distribution. The whole amount m of material was initially contained at x = 0.

The parameter σ describes the width of the Gaussian curve, larger σ leading to wider bell shapes as seen in Figure 1.5.2. Indeed, from the analytical expression of the Gaussian curve, it results that for any chosen height $0 < c < c_{max}$, the value for x is given by the expression:

$$x = \pm\sqrt{-2\sigma^2 \ln[(c/m)\sigma\sqrt{2\pi}]} \qquad (1.5.7)$$

From rel. (1.5.7) results the width $W = 2|x|$ as an increasing function of σ. Of particular importance in chromatography are the width at half height W_h and the width at the inflection point W_i of the Gaussian curve. Using $c = 1/2\, c_{max}$ in rel. (1.5.7), the result is

$$W_h = 2\,(2 \ln 2)^{1/2}\,\sigma \qquad (1.5.8)$$

From the second derivative of the expression (1.5.6a), it is easily obtained

$$W_i = 2\sigma \qquad (1.5.9)$$

In a chromatographic process, the diffusion is associated with the movement of the mobile phase and of the diffusing analyte along the chromatographic column. If ξ is the distance from the origin to the center of the moving diffusion zone, the expression for c/m becomes

$$c/m = \frac{1}{\sqrt{2\pi\sigma^2}}\exp[-(x-\xi)^2/(2\sigma^2)] \qquad (1.5.6b)$$

This expression can be used for the understanding of peak broadening in a chromatographic process (see Section 1.4).

The diffusion equation (1.5.2) for the diffusion in a tube or a rectangular channel also can be solved when at the initial moment $t = 0$ the concentration of the diffusing species is $c = c_0$ for $x \leq 0$, and $c = 0$ for $x > 0$. These conditions (boundary conditions) are written as follows:

$$c\big|_{t=0,x\leq0} = c_0 \qquad c\big|_{t=0,x>0} = 0 \qquad (1.5.10)$$

These conditions describe the case of a tube separated in half by a thin wall and filled with an ideal medium that has in the first half a compound with the concentration $c = c_0$ and in the other half the concentration $c = 0$. After removal of the separating wall, the compound starts diffusing in the whole tube, and the variation in its concentration must be calculated. The solution of this unidimensional diffusion problem described by rel. (1.5.2) and with the conditions (1.5.10) can be written as follows (see e.g. [15]):

$$c = \frac{c_0}{2}[1 - \frac{2}{\sqrt{\pi}}\int_0^{\frac{x}{2\sqrt{Dt}}}\exp(-\eta^2)d\eta] = \frac{c_0}{2}\operatorname{erfc}(\frac{x}{2\sqrt{Dt}}) \qquad (1.5.11)$$

where erfc $(z) = 1 - \operatorname{erf}(z)$ is the error-function complement, and erf (z) is the error-function defined by the expression:

$$\operatorname{erf}(z) = \frac{2}{\sqrt{\pi}}\int_0^z\exp(-\eta^2)d\eta \qquad (1.5.12)$$

Values for the error-function and for error-function complement are given in literature (see e.g. [24]) and can be obtained based on the expression of erfc (z) as a gamma function, erfc $(z) = [1/(\pi^{1/2})]\,\Gamma\,(1/2, z^2)$.

For the calculation of the amount of material diffusing in the medium for a certain interval of space and time, rel (1.5.12) can be integrated for the specific limits (t and x). The integration of error function is given by the formula:

$$\int\operatorname{erf}(z)\,dz = z\,\operatorname{erf}(z) + \frac{1}{\sqrt{\pi}}\exp(-z^2) + \text{const.} \qquad (1.5.13)$$

The graphs for erf (z) and of $\int_0^z\operatorname{erf}(z)\,dz$ are shown in Figure 1.5.3a and 1.5.3b, respectively.

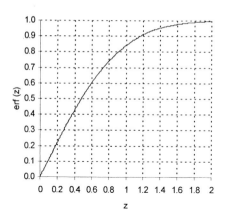

FIGURE 1.5.3a. *Graph of erf (z).*   FIGURE 1.5.3b. *Graph of* $\int_0^z erf\ (z)\ dz.$

With the expression (1.5.12) used for erf (z), the changes of c/c_0 as a function of

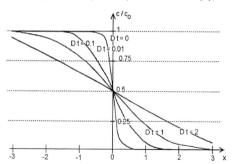

distance for different D t values when the boundary limits are given by rel. (1.5.10) can be obtained, and they are shown in Figure 1.5.4. For $x < 0$ and at large distance from $x = 0$, the concentration is closer to c_0, and for $x > 0$ at large distance from $x = 0$, the concentration remains $c = 0$. At $x = 0$, the concentration $c = c_0/2$. The values for c tend toward $c_0/2$ at wider intervals around $x = 0$ when t is larger. At infinite time, rel. (1.5.11) shows that a uniform concentration $c = c_0/2$ is obtained for the whole system.

FIGURE 1.5.4. *Changes of c/c_0 for different D t values for a diffusion process starting at the initial moment t = 0 with the concentration of the diffusing species $c = c_0$ for x < 0, and c = 0 for x > 0.*

One additional question regarding the diffusion through a channel is related to the value for z corresponding to a constant flow of material. As seen in Figure 1.5.3a, the value of erf(z) attains about 95% of its value for $z \approx 1.5$. At values $z > 1.5$, the variation of c is effectively very small. This shows that 95% of the equilibrium concentration is attained for $z = 1.5$. The relation between t and x obtained for $z = 1.5$ can be written as follows:

$$t_{95\%} = \frac{x^2}{9D} \tag{1.5.14}$$

Another case when the diffusion equation (1.5.2) can be solved is that when the condition $c = c_0$ for $x \leq 0$ at any time t and $c = 0$ for $x > 0$. These boundary conditions are written as follows:

$$c\big|_{x \leq 0} = c_0 \qquad c\big|_{t=0, x>0} = 0 \qquad\qquad (1.5.15)$$

The solution for this case is given by the expression:

$$c = c_0 \, \text{erfc} \left(\frac{x}{2\sqrt{Dt}} \right) \qquad\qquad (1.5.16)$$

The changes of c/c_0 for different D t values for the solution expressed by rel. (1.5.14)

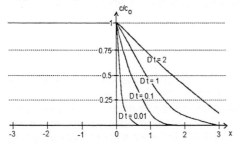

are shown in Figure 1.5.5. The c/c_0 increases with D t values and also c/c_0 is higher when x is closer to x = 0. The solution (1.5.14) gives the expression of the concentration of the diffusing species for x > 0 and at any time t assuming a continuous constant source at the boundary x = 0. When t is very large, D t is also very large, and the concentration for x > 0 tends to c_0.

FIGURE 1.5.5. *Changes of c/c_0 for different D t values for a diffusion process with $c = c_0$ constant at x = 0, and starting at the initial moment t = 0 with c = 0 for x > 0.*

- Mass transfer by diffusion and convection

The equation governing convective diffusion has the general form:

$$\frac{\partial c}{\partial t} = D\nabla^2 c - (\vec{u} \cdot \text{grad})c \qquad\qquad (1.5.17)$$

where $\vec{u}$ is the solution velocity (in a specific direction). The first problem related to the solution of equation (1.5.17) is the type of flow. With assumption that the moving fluid is a Newtonian fluid (has a constant viscosity at a given temperature), the flow can be laminar or turbulent depending on the Reynolds number Re given by the expression:

$$Re = \rho u d / \eta \qquad\qquad (1.5.18)$$

where u is the linear fluid velocity (cm/sec), ρ is the fluid density (g/cm^3), d is the tube diameter (cm), and η is the fluid viscosity (poise). At low Re numbers (below 2100) the flow of the liquid is laminar. In laminar flow the fluid elements remain in streamlines, and for the flow through a circular tube of radius r, the velocity u_z of a stream path at the radial position r_z is given by the expression (parabolic profile):

$$u_z = 2 u [1 - (r_z / r)^2] \qquad\qquad (1.5.19)$$

where u is the mean fluid velocity, and the maximum velocity $u_{max} = 2u$. In order to solve equation (1.5.16) for a laminar flow in direction z, the equation must be written in cylindrical coordinates [25]. Assuming no angular diffusion, the equation (1.5.17) takes the form:

$$\frac{\partial c}{\partial t} = D[\frac{1}{r}\frac{\partial}{\partial r}(r\frac{\partial c}{\partial r}) + \frac{\partial^2 c}{\partial z^2}] - u_z \frac{\partial c}{\partial z} \qquad (1.5.20)$$

The solution of equation (1.5.20) is not simple, and various approximations or simplifications are used to find the solution. For example, under steady state conditions, the term containing the derivative with respect to time will vanish. Considering the flow in direction z, diffusion only in direction x, and neglecting dimension y, which would correspond to the flow through a narrow rectangular channel, it is convenient to express rel. (1.5.20) in Cartesian coordinates, and the equation (1.5.20) becomes:

$$D\frac{\partial^2 c}{\partial x^2} - 6u\frac{x}{b}(1-\frac{x}{b})\frac{\partial c}{\partial z} = 0 \qquad (1.5.21)$$

where b is the height of the channel. The solution of equation (1.5.21) can be obtained for various boundary conditions corresponding to the concentrations at the wall of the channel [26], [27]. However, the solution is expressed as a series of exponential terms multiplied by parabolic cylinder functions. Because of the complicated form of the solutions for equations of the type 1.5.13, the solutions were obtained only by using different approximations (see e.g. [20]) or were calculated numerically.

- *Diffusion rates in various media*

The theory previously developed for the diffusion makes no assumptions regarding the nature of the diffusing species or the medium of diffusion. The diffusion coefficient depends very much on the physical state of the diffusing substance and of the material in which diffusion takes place. For gases, kinetic theory shows that diffusion coefficient of two gases "A and "B" is given by the expression:

$$D_{A,B} = \frac{1}{3}\lambda_A \bar{c}_A x_A + \frac{1}{3}\lambda_B \bar{c}_B x_B \qquad (1.5.22)$$

where λ is the length of the mean free path of the molecule, $\bar{c}$ is the average speed of the molecule, and x is the mole fraction. The average speed $\bar{c}$ and λ can be estimated from the expressions:

$$\bar{c} = (\frac{8RT}{\pi M})^{1/2} \qquad\qquad \lambda = \frac{3}{\rho}\frac{\eta}{\bar{c}}$$

$$(1.5.22)$$

where M is the molecular weight of the gas, ρ is its density and η is the viscosity (expressed in g cm^{-1} s^{-1}). Rel. (1.5.22) shows that a simple relation exists between diffusion coefficient and viscosity of gases. Empirical relations were also developed for the estimation of diffusion coefficients in gases [15]. The diffusion coefficients for a few gases are indicated in Table 1.5.1.

TABLE 1.5.1. *Some diffusion coefficients for gases at 1 atm. [15].*

Gas A	Gas B	Temp. °K	D_{AB} (cm²/s)
hydrogen	nitrogen	273.15	0.401
hydrogen	nitrogen	400	1.270
hydrogen	benzene	311	0.404
hydrogen	n-hexane	289	0.290
air	benzene	298	0.096
air	chlorobenzene	299	0.079

Diffusion coefficients for liquids are much lower than those for gases. Different theories and also empirical relations were developed for the estimation of diffusion coefficients. For example, Stokes theory empirically modified for better prediction gives the following formula for the diffusion coefficient for nonelectrolytes 'A" in liquids "B":

$$D_{A,B} = 7.4 \cdot 10^{-8} \frac{(\psi_B M_B)^{0.5} T}{\eta \overline{V}_A^{0.6}} \tag{1.5.23}$$

where $\overline{V}_A$ is the molecular diffusion volume of solute "A" (in cm³ mole⁻¹), M_B the molecular weight of solvent B, T temperature in Kelvin degrees, η the viscosity of the solution (in 10^{-4} g cm⁻¹ s⁻¹, or centipoise), ψ_B an "association" factor for the solvent (ψ_B is 1 for nonpolar solvents, 1.5 for ethanol, 1.9 for methanol, 2.6 for water). Values of $\overline{V}_A$ for different compounds are given in literature [15]. Some viscosity values for a number of common solvents are given in Table 1.5.2.

TABLE 1.5.2. *Some viscosity values for common solvents at 20° C [15].*

Solvent	Viscosity cP, 20° C	Solvent	Viscosity cP, 20° C
diethyl ether	0.23	chloroform	0.57
pentane	0.23	toluene	0.59
petroleum ether	0.3	methanol	0.60
acetone	0.32	benzene	0.65
isopropyl chloride	0.33	nitromethane	0.67
acetonitrile	0.34	ethylene chloride	0.79
n-propyl chloride	0.35	pyridine	0.94
methyl acetate	0.37	carbon tetrachloride	0.97
diisopropyl ether	0.37	cyclohexane	1.00
carbon disulfide	0.37	ethanol	1.20
diethylamine	0.38	dioxane	1.54
methylene chloride	0.44	dimethylsulfoxide	2.24
ethyl acetate	0.45	n-propanol	2.3
diethyl sulfide	0.45	isopropanol	2.3
cyclopentane	0.47		

The diffusivity in solutions of electrolytes is influenced by the electrolytic dissociation and the different migration of the anion and cation. The diffusion coefficients for a few liquids are indicated in Table 1.5.3.

TABLE 1.5.3. *Some diffusion coefficients for liquids at 20° C [15].*

Solute A	Solvent B	D_{AB} (cm²/s)
ethanol	water	$1.00\ 10^{-5}$
phenol	water	$0.84\ 10^{-5}$
sucrose	water	$0.45\ 10^{-5}$
NaCl	water	$1.35\ 10^{-5}$
phenol	ethanol	$0.8\ 10^{-5}$
phenol	benzene	$1.54\ 10^{-5}$
acetic acid	benzene	$1.92\ 10^{-5}$

The diffusion of large molecules is strongly influenced by the increase in the molecular diffusion volume $\overline{V}_A$, which depends on the molecular weight of the solute. A typical variation for D (for an arbitrary solvent) as a function of log (MW) of the solute is shown in Figure 1.5.6.

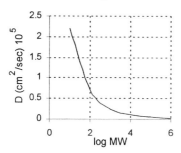

FIGURE 1.5.6. *Variation of D as a function of MW for solutions of nonelectrolytes.*

Diffusion rates are even lower in solids than in liquids. Diffusion coefficients of solids in solids are typically lower than 10^{-8} cm²/s. The diffusion coefficients of liquids in solids are also very low. However, the penetration of liquids in solids occurs in many instances due to the porosity of the solid medium. If it assumed that the pores are filled with solution, the diffusion is similar to that into a solution but with a much reduced available cross-section for diffusion. The diffusion process into porous particles is also affected by the geometry of the porous structure and by the interaction of the solute molecule with the pore walls. The tortuosity of the pores leads to an increased path length for the diffusion. Narrow pores also slow down the diffusion process. Adsorption of the diffusing species on the walls of the pores diminishes further the rate of diffusion. The value for the diffusion coefficient D in a material showing low adsorption properties can be estimated by the formula:

$$D_{eff} = \frac{\phi D}{\gamma (K + 1)} \qquad (1.5.24)$$

where ϕ is the fraction of space available for diffusion, K is the ratio of solute concentration in a unit of solid volume to the concentration in a unit of volume of pore fluid, D is the diffusion coefficient in solution (in the absence of the porous medium), and γ is a tortuosity factor (not larger than 1.73). When the diffusion process is in a medium with very narrow pores (smaller diameter than the length of the mean free path of the molecule), or when the adsorption on the pore walls is important, rel. (1.5.24) cannot be applied. Also when ionic interactions take place during the diffusion process, such as in an ion exchange resin, another expression of the diffusion coefficient must be applied. This is based on Nernst equation:

$$D = \frac{RT}{F} U_s \qquad (1.5.25)$$

where F is Faraday's constant and U_s is the ionic mobility of the ionized solute.

1.6.   CHARACTERIZATION OF CHROMATOGRAPHIC SEPARATIONS

Chromatographic separations are applied in both sample preparation and core analytical separation/measurement. When used in sample preparation, the chromatography is usually simpler, leading to a less detailed separation, and it is not followed by measurement. Regardless of its use, the characterization of the chromatographic process must be done for the evaluation of its attributes such as separation efficiency, sample capacity, etc. The purpose of this section is to describe the parameters used for the characterization of a chromatographic separation. These parameters are also useful in comparing analytical procedures.

A variety of chromatographic techniques are known (see Section 4.1). Among these, gas chromatography (GC) and high performance liquid chromatography (HPLC) are the most common. For this reason, the discussion on peak characterization will focus mainly on these techniques.

- *Distribution process in gas chromatography*

Once the sample reaches the chromatographic column, the distribution of the analytes between the mobile phase and the stationary phase begins. In gas-liquid chromatography, the retention of the analyte in the liquid stationary phase is characterized by its *partition coefficient* K_i given by rel. (1.4.25). With the index "s" for the stationary phase A, and "g" for the mobile phase, the equilibrium can be written as follows:

$$i_g \rightleftharpoons i_s$$

and the expression (1.4.25) for the constant K_i can be written using molar concentration accordingly:

$$K_i = c_{i,s} / c_{i,g} \tag{1.6.1}$$

In a chromatographic process based on the adsorption on a gas-solid interface, the equivalent of partition coefficient is the *adsorption constant* K^A_i given by rel. (1.4.36). Formally the parameters for the characterization of the chromatographic process are not different for the two types of separation processes, and they will not be considered separately.

As shown in Section 1.4, the graph representing $c_{i,s}$ (where $c_{i,s}$ is the concentration of the component "i" in the stationary phase) as a function of $c_{i,g}$ (the concentration in the gas phase) at a given temperature is called an isotherm, and its equation is

$$c_{i,s} = K_i \, c_{i,g} \tag{1.6.2}$$

(For gas-solid adsorption K_i is replaced by K^A_i.) Depending on the isotherm type (linear or nonlinear), the chromatographic process can be classified as linear or nonlinear (see Section 1.4).

An ideal chromatographic process would take place with (thermodynamically) reversible exchange of the analytes between the two phases and with instantaneous equilibrium. However, this case is not met in practice. Whereas the isotherms commonly encountered in gas and liquid chromatography are linear, the process is never ideal. When the sample is injected as a very narrow zone and when the chromatography is linear-nonideal, diffusion effects and nonequilibrium generate broadening of the elution zones. The broadened zones of the eluting components have a Gaussian distribution of the concentrations because they are generated by a random process. This is translated into a Gaussian shape of the chromatographic peak, which is the response of the detector to an eluting analyte. For nonlinear isotherms the shape of the eluting peaks is not Gaussian.

The time necessary for a component "i" injected into the chromatographic column to elute is called the *absolute retention time* t_R (the index "i" is omitted). The chromatographic separation is based on different retention times of the components of the mixture. These retention times are different because the partition of each analyte between the two phases, the gas phase in motion and the stationary phase, are different. The retention time t_R of the analyte "i" depends on a series of factors such as the nature of the compound and of the stationary phase, the gas flow in the column, the temperature, column construction, etc. Because the eluted analyte appears as a broadened zone, the retention time t_R must be measured when the maximum of the Gaussian elution peak is attained. The absolute retention time t_R of the analyte is the sum of the (average) time spent by the analyte in the gaseous mobile phase (t_g) and the (average) time spent in the stationary phase (t'_R):

$$t_R = t_g + t'_R \qquad (1.6.3)$$

The time spent by a component in the stationary phase t'_R is compound dependent (the index "i" is again omitted), but the time spent in the mobile phase is the same for any component and depends on the *linear flow rate* u of the gas (velocity) and the *column length* L where

$$t_g = L / u \qquad (1.6.4)$$

In practice, the value of t_g can be estimated from the time necessary to elute an unretained component (when $t'_R = 0$).

In parallel to the absolute retention time t_R, a *retention volume* V_R can be defined as the volume of mobile phase that flows through the column for the elution of a given component. Also, a volume V_g corresponding to t_g is defined as the elution volume for an unretained component. The volumes are related with the times, using the *volumetric flow rate* U, by the formulas:

$$t_R = V_R / U \quad \text{and} \quad t_g = V_g / U \qquad (1.6.5)$$

The retention time t_R is dependent of the distribution constant $K_i = c_{i,s} / c_{i,g}$ for a given analyte. To show this dependence, it is convenient to introduce the parameter R (or R_i), which is the fraction of moles (molecules) of solute "i" in the mobile phase:

$$R = n_{i,g} / (n_{i,g} + n_{i,s}) \qquad (1.6.6)$$

where $n_{i,s}$ and $n_{i,g}$ are the number of moles of component "i" in the stationary phase "s" and in the gas, respectively. The concentrations $c_{i,s}$ and $c_{i,g}$ can be expressed as a ratio between the fraction of moles in the phase and the volume of the phase, which gives

$c_{i,g} = R / V_g$ and $c_{i,s} = (1 - R) / V_s$, where V_s is the volume of the stationary phase. By use of these relations, the value for the partition coefficient K_i becomes

$$K_i = \left(\frac{1-R}{V_s}\right) \Big/ \left(\frac{R}{V_g}\right) = \left(\frac{1-R}{R}\right) \left(\frac{V_g}{V_s}\right)$$

(1.6.7)

On the other hand, the relation between the partition coefficient K_i and the *capacity factor* k_i given by rel. (1.4.28) can be written as follows:

$$K_i = k_i (V_g / V_s)$$

(1.6.8)

which leads to relation:

$$k_i = (1 - R) / R$$

(1.6.9)

and also:

$$R = 1 / (1 + k_i) = 1 / [1 + K_i (V_s / V_g)]$$

(1.6.10)

As shown in Section 2.1, k_i describes the molar distribution of component "i" between the two phases. The molar distribution of component "i" between the two phases should be equal to the ratio between the time spent by a component "i" in the stationary phase versus the time spent in the gas phase and therefore

$$k_i = t'_R / t_g = (t_R - t_g) / t_g$$

(1.6.11)

Because all analytes spend the same amount of time in the mobile phase t_g, k_i is a good measure of the actual retention of a specific analyte on the stationary phase. The values for k_i covers a wide range, but common values are between 0 and 100. An absolute capacity factor $k_i' = t'_R / t_R = t'_R / (t'_R + t_g)$ is also used in literature. By application of rel. (1.6.10) and (1.6.11), the value for R can be written in the form:

$$R = \frac{t_g}{t'_R + t_g} = \frac{t_g}{t_R}$$

(1.6.12)

This relation shows that R can be seen as the ratio of the time spent by the analyte in the mobile phase and the absolute retention time t_R. The larger is R, the more rapidly the analyte moves through the column. A similar expression for R can be obtained using the retention volumes V_R and V_g. From rel. (1.6.12) it is easily obtained

$$R = V_g / V_R$$

(1.6.13)

Based on rel. (1.6.13) and (1.6.10), the expression for K_i can be written:

$$K_i = (V_R - V_g) / V_s$$

(1.6.14)

Also, K_i can be expressed using rel. (1.6.8) as follows:

$$K_i = k_i (V_g / V_s) = k_i \beta = (t'_R / t_g) \beta$$

(1.6.15)

where β is called the *phase ratio* and has the expression:

$$\beta = V_g / V_s \qquad (1.6.16)$$

For capillary columns (see Section 4.2) with the internal radius r and the film thickness of the stationary phase d_f, the phase ratio β can be approximated by the formula:

$$\beta = r / (2\ d_f) \qquad (1.6.17)$$

and its typical values are between 50 and 500 (usually the columns have a diameter between 0.2 mm to 0.5 mm and the film thickness between 0.1μ to 1μ). Common values for t_g range from 20–30 s to several minutes, and k values may range from 1 to 100 or more. Analytes with high boiling points and/or high polarity (lower vapor pressure) have in general higher elution times and higher values for K_i. This can be shown starting with rel. (1.6.1) for K_i and using $c_{i,A} = n_{i,A} / V_s$ and $c_{i,g} = n_{i,g} / V_g$. For N_s moles of stationary phase and $n_{i,A}$ moles of solute, the mole fraction of solute "i" in the stationary phase is $x_{i,A} \approx n_{i,A} / N_s$. Using rel. (1.3.18), which generalizes Raoult's law, it can be obtained

$$n_{i,A} / V_s = x_{i,A}\ N_s / V_s = p_i\ N_s / (\gamma'_{i,A}\ p^o_i\ V_s) \qquad (1.6.18)$$

where p_i is the partial vapor pressure of the component "i" in the gas phase and p^o_i is the vapor pressure above the pure component. For 0° C as a reference temperature, it is obtained

$$n_{i,g} / V_g = p_i / (R\ 273.15) \qquad (1.6.19)$$

From rel. (1.6.1), (1.6.18), and (1.6.19) it can be obtained

$$K_i = N_s\ R\ 273.15 / (\gamma'_{i,A}\ p^o_i\ V_s) \qquad (1.6.20)$$

By use of rel. (1.6.14) and (1.6.20), it can be obtained

$$V_R = V_g + N_s\ R\ 273.15 / (\gamma'_{i,A}\ p^o_i) \qquad (1.6.21)$$

or

$$t_R = t_g + N_s\ R\ 273.15 / (\gamma'_{i,A}\ p^o_i\ U) \qquad (1.6.22)$$

This relation shows that a larger retention time is obtained when a larger amount of stationary phase is present (larger N_s), when the flow rate U is lower, and when the solute has lower vapor pressure p^o_i. It also shows that t_R depends on both solute "i" and stationary phase "A" (through the coefficient $\gamma'_{i,A}$).

- Distribution process in HPLC and TLC

The stationary phase in HPLC and TLC can be practically considered always covered by a monolayer of molecules of solute "i" or of mobile phase "A". With the index (s) for the stationary phase and (m) for the mobile phase "A", the distribution of "i" between mobile phase and stationary phase can be expressed by the equilibrium:

$$i_m + n\, A_s \rightleftharpoons i_s + n\, A_m$$

where it is assumed that the solute "i" can displace in the stationary phase n molecules of adsorbed phase "A". Using the equality of chemical potentials for the components of the two sides of the equilibrium, the thermodynamic distribution constant can be written as follows (see Section 1.4):

$$K_i = (a_{i,s} / a_{i,m})\, (a_{A,m} / a_{A,s})^n = \exp\left[- (\mu^o_{i,m} + n\,\mu^o_{A,s} - \mu^o_{i,s} - n\,\mu^o_{A,m}) / (RT)\right] \quad (1.6.23)$$

In rel. (1.6.10a) the regular notations (a) are used for activities and (μ) for standard chemical potentials. The two values for the standard chemical potentials in the mobile phase $\mu^o_{i,m}$ and $n\,\mu^o_{A,m}$ are usually equal and cancel each other. This leads to the relation:

$$K_i = \exp\left[- (n\,\mu^o_{A,s} - \mu^o_{i,s}) / (RT)\right] \quad (1.6.24)$$

As shown in Section 1.3 (see rel. 1.3.22), the activities can be expressed as a function of concentrations, and because the same liquid "A" forms both phases and $c_{A,m} = 1$, rel (1.6.24) becomes

$$K_i = (\gamma_{i,s} / \gamma_{i,m})\, (c_{i,s} / c_{i,m})\, [1 / (\gamma_{A,s}\, c_{A,s})]^n = \exp\left[- (n\,\mu^o_{A,s} - \mu^o_{i,s}) / (RT)\right] \quad (1.6.25)$$

where $c_{i,s}$ is the molar concentration of the component "i" in the liquid stationary phase, $c_{i,m}$ is the molar concentration in the mobile phase, and $c_{A,s}$ is the molar concentration of the molecules of the mobile phase "A" in the liquid stationary phase. The concentration of solvent "A" in the mobile phase is not affected by the partition process, therefore the equilibrium can be considered only for the solute "i" between the stationary and mobile phases:

$$i_m \rightleftharpoons i_s$$

This equilibrium is characterized as shown in Section 1.4 by its partition coefficient:

$$K_i = c_{i,s} / c_{i,m} \quad (1.6.26)$$

With rel. (1.6.26) and (1.6.25) and the assumption $\gamma_{i,s} / \gamma_{i,m} = 1$ and $\gamma_{A,s} = 1$, the expression for K_i becomes:

$$K_i = \exp\left[- (n\,\mu^o_{A,s} - \mu^o_{i,s}) / (RT)\right] c_{A,s}{}^n \quad (1.6.27)$$

This formula can be expressed in a different form by replacing the potentials $\mu^o_{i,s}$ and $\mu^o_{A,s}$ with specific experimental parameters as follows:

$$\mu^o_{i,s} = 2.302\, RT\, \alpha'\, S^o \quad (1.6.28a)$$

and

$$\mu^o_{A,s} = 2.302\, RT\, \alpha'\, \varepsilon^o\, A_m \quad (1.6.28b)$$

In these expressions, α' is a characteristic property of the solid phase that measures the ability of a unit of adsorbent surface to bind adsorbed molecules [15], and S^o is a measure of adsorption energy of "i" onto a standard adsorbent surface (defined for $\alpha' = 1$) from a standard mobile phase (for which $\varepsilon° = 0$). Elutropic strength $\varepsilon°$ is the adsorption energy of solvent "A" per unit area of standard adsorbent surface ($\alpha' = 1$) and characterizes the solvent "strength." Some values of $\varepsilon°$ for different solvents and different stationary phases are given in Table 4.3.4. The parameter A_m is the surface area occupied on the adsorbent by a solvent molecule. By use of rel. (1.6.28a) and (1.6.28b) in rel (1.6.27) and replacement of the area occupied by the solvent with A_i, the area occupied by the solute, where $A_i = n A_m$, the expression for log K_i (in solvent "A") can be written as follows:

$$\log K_i = n \log c_{A,s} + \alpha' (S° - \varepsilon° A_i) \tag{1.6.29}$$

The value for $c_{A,s}$ varies little for different solvents and in practice is proportional with the adsorbent surface area. For this reason $c_{A,s}$ is commonly treated as a stationary phase parameter [15]. Therefore, expression (1.6.29) for the partition coefficient indicates that distribution of a solute "i" between the mobile and stationary phase is determined by the adsorption process of the solute and mobile phase, while the interactions in the solution are not very important.

The evaluation of the distribution constants K_i in HPLC also can be done directly from rel. (1.4.16), which is expressed in this case in the form:

$$\log K_i = V_i [(\delta_i - \delta_m)^2 - (\delta_i - \delta_s)^2] / (RT) \tag{1.6.30}$$

where δ_m is the solubility parameter for the mobile phase, while δ_s is that of the stationary phase liquid. In the case of chromatographic columns where a true liquid is "stationary" on the surface of a solid phase, the value for δ_s is that of the liquid. In cases where the stationary phase is a bonded compound (such as in a C18 column), an approximate value can be obtained for δ_s.

Other parameters used for the characterization of the separations of liquid chromatography are very similar to those used in partition GC. For example, the capacity factor k_i (see rel. 1.6.8) is given by the expression:

$$k_i = K_i V_s / V_m \tag{1.6.31a}$$

In rel. (1.6.31a) V_s is the volume of the stationary phase and V_m that of the mobile phase used for eluting an unretained compound (V_m replaces V_g and t_m replaces t_g from gas chromatography). Typical values for k_i in HPLC are usually lower than in GC. and values between 0 and 10 are common. An absolute capacity factor defined by the expression:

$$k_i' = t'_R / t_R = t'_R / (t'_R + t_g) \tag{1.6.31b}$$

is also used in HPLC similarly to GC. Also, the value of R given by rel. (1.6.12) is the same in HPLC and GC, except that for TLC chromatography this value is noted R_f. On a TLC plate, the R_f is measured as the ratio of the distance from the start to the spot of the analyte and the distance from the start to the eluent front. For a constant flow of the eluent, these distances are proportional to t_m and t_R, respectively. Therefore, the R_f value can be seen as the relative migration rate of the analyte vs. the eluent.

In liquid-liquid partition separations, the relation between R and K_i is similar to rel. (1.6.7) for gas chromatography. By rearrangement of rel. (1.6.7) it can be obtained

$$R = \frac{V_m}{V_m + K_i V_s}$$

(1.6.32)

where V_g from rel. (1.6.7) is replaced with V_m. This relation is used to estimate the migration rate of an analyte vs. the eluent as a function of the distribution constant K_i. Also, rel. (1.6.14) for gas chromatography remains valid for liquid chromatography (V_g being substituted with V_m), and the expression for the retention volume in HPLC can be written in the form:

$$V_R = V_m + K_i V_s = V_m (1 + k_i)$$

(1.6.33)

The retention volume in separations using a liquid mobile phase also has implications in the evaluation of parameters used in solid phase sample preparation (see Section 11.1).

- *Peak broadening in chromatography*

Considering a chromatographic process where ideally the only diffusion process is ordinary diffusion, the peak broadening can be studied based on Fick's laws. For a separation where the mobile phase has a linear flow rate u, the distance from the origin to the center of the moving zone is $\xi = u \, t_R$. Therefore, by use of rel. (1.5.5) for a given ξ, the resulting σ is given by the expression:

$$\sigma^2 = 2 \, D \, \xi \, / \, u$$

(1.6.34)

By use of rel. (1.5.6b) for c / m as a function of x, a chromatogram measuring c / m shows peak broadening for an eluting analyte as illustrated in Figure 1.6.1.

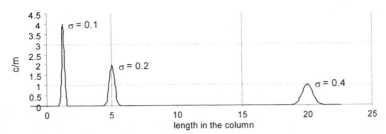

FIGURE 1.6.1. *Peak broadening of an eluting analyte along a chromatographic column.* In this ideal chromatogram, the broadening for the three peaks correspond to $\sigma = 0.1$, $\sigma = 0.2$, and $\sigma = 0.4$, which can be obtained, for example, for D = 1, u = 250 and $\xi = 1.25$, $\xi = 5$, and $\xi = 20$, respectively (arbitrary units).

In reality, besides ordinary diffusion, other diffusion types also occur in chromatography. The more important ones are the eddy diffusion due to the flow along longer or shorter paths in packed columns (eddy diffusion is absent in capillary columns) and transversal diffusion due to differences in the flow rate of the mobile phase along the chromatographic column walls compared to the center of the column. Also, peak broadening can be induced by nonequilibrium effects (see e.g. [20], [27a]). The shape of the chromatographic peaks remains Gaussian when the partition or adsorption process is described by linear isotherms (see also Section 1.4). For this reason, peak broadening for a linear-nonideal process can be studied based on the assumption that the peak shape is Gaussian.

For a common chromatographic process, the concentration c in rel. (1.5.6b) must be replaced with the peak height h, as generated by the response to concentration of a chromatographic detector. The variation of h as a function of distance x from start is described in this case by the equation:

$$h(x) = A (2\pi \sigma^2)^{-1/2} \exp [-(x - \xi)^2 / 2\sigma^2] \qquad (1.6.35)$$

where A is the total peak area, ξ is the middle of the zone (and the maximum of the Gaussian curve), and σ determines the extent of peak broadening. Rel. (1.6.35) is valid for any linear-nonideal process, but the value for σ is not determined only by the simple diffusion, and more factors besides longitudinal diffusion contribute to peak broadening.

Because in most chromatographic processes the measured parameter is the retention time and not the length of the path in the column, σ can be replaced with $\sigma_t = \sigma / R u$ in rel. (1.6.35), and h is expressed as a function of time t as follows:

$$h(t) = A (2\pi \sigma_t^2)^{-1/2} \exp [-(t - t_R)^2 / 2\sigma_t^2] \qquad (1.6.36)$$

In practice peak broadening can be measured on the chromatograms using the peak width W_h (in time) at the half height of the Gaussian curve or the peak width at the baseline W_b (in time) where $W_b = 2 W_i$ (in time). Figure 1.6.2 shows the measurements of t_R, W_i, W_h, and W_b on a recorded chromatographic peak.

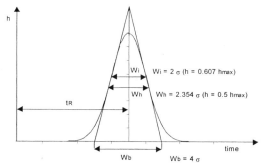

FIGURE 1.6.2. *Measurements of retention time t_R and peak broadening W_i, W_b and W_h on a recorded chromatogram.*

From rel. (1.5.8) or (1.5.9) it results that σ_t can be calculated from W_h or W_b using the expressions:

$$\sigma_t = (8 \, Ln \, 2 \,)^{-1/2} W_h \qquad (1.6.37a)$$

$$\sigma_t = 0.25 \, W_b \qquad (1.6.37b)$$

The value of σ^2 is related to another parameter used to characterize zone spreading, namely the *height equivalent to a theoretical plate* H which is defined as

$$H = \sigma^2 / L \tag{1.6.38}$$

This parameter is very useful in chromatography for the characterization of peak broadening per unit length of the column. In addition to H, the peak broadening characterization in a column can be done using the *theoretical plate number* n. For a column of length L, n is defined as

$$n = L / H \tag{1.6.39}$$

Rel. (1.6.40) indicates that n is proportional to the column length L and inversely proportional to H. The theoretical plate number n can be expressed as a function of length by a simple substitution of rel. (1.6.38) in rel. (1.6.39):

$$n = L^2 / \sigma^2 \tag{1.6.40}$$

Since $\sigma = \sigma_t\, R\, u$ and $L = t_R\, R\, u$, n can be expressed from rel. (1.6.40) as a function of time by the expression:

$$n = 16\, t_R^2 / W_b^2 \tag{1.6.41}$$

In addition to the theoretical plate number, an *effective plate number* N is defined by substituting t_R in rel. (1.6.41) with t'_R. The formula for N will be

$$N = 16\, t'^2_R / W_b^2 \tag{1.6.42}$$

Rel. (1.6.42) shows how N depends on chromatographic retention time t'_R, and since t'_R is compound related (index "i" omitted), it also shows that N (as well as n) are compound dependent. Both rel. (1.6.41) and (1.6.42) can be used to measure the theoretical plate number or effective plate number based on experimental data obtained with a given column. This measurement is useful in practice to select columns (higher n gives lower peak broadening) and also to assess the loss in performance of a column after a certain period of usage.

Since n is a function of t_R and N is function of t'_R, it is useful to note the following relation between t_R and t'_R:

$$t_R / t'_R = k / (k + 1) \tag{1.6.43}$$

(where the index "i" for k was omitted). Using rel. (1.6.43), the following relation can be obtained between n and N:

$$N = n\, [k / (1+ k)]^2 \tag{1.6.44}$$

The plate number of a chromatographic column depends on a number of factors. Important are the nature of the stationary phase and construction characteristics of the column (length, diameter, particle size for packed columns, void spaces, etc.), the nature of the mobile phase and of the solutes, the flow rate of the mobile phase, etc. A number of mechanisms contribute to peak broadening and some are discussed below.

Longitudinal diffusion, as shown previously, contributes to peak broadening (or to the plate height), and by combination of rel. (1.6.34) and (1.6.38) this contribution to the height of a theoretical plate can be written in the form:

$$H_L = 2\, D_g\, /\, u \qquad\qquad (1.6.45a)$$

where the index g was added to the diffusion coefficient for gas phase. Rel. (1.6.45a) must be modified when a linear diffusion is hindered by a packing material and rel. (1.6.45a) becomes:

$$H_L = 2\,\Phi\, D_g\, /\, u \qquad\qquad (1.6.45b)$$

where Φ =1 for gas capillary columns but is about 0.6 for packed columns.

In a packed material, the flow occurs through a tortuous channel system with various path lengths, and additional factors influence peak broadening. One such broadening is referred to as the eddy diffusion and is produced by different paths of molecules from the same solute. The contribution to the plate height of the eddy diffusion is given by the expression:

$$H_E = \Lambda\, d_p \qquad\qquad (1.6.46)$$

where d_p is the average diameter of a particle (in the packed column), and Λ is a parameter depending on column packing.

Lateral movement of material due to convection also occurs in a packed column. The contribution to the plate height of this type of diffusion is given by the expression:

$$H_D = \frac{\Omega d_p^2 u}{D_M} \qquad\qquad (1.6.47)$$

where Ω is a parameter depending on column packing and D_M is the diffusion coefficient of the solute in the mobile phase.

Another contributing factor to peak broadening is caused by nonequilibrium effects regarding the transfer into and out the stationary phase. This rate is controlled by the rate of diffusion in the liquid stationary phase or by the adsorption-desorbtion kinetics in the case of adsorption processes. It can be shown with a random walk model [e.g. 27b] that for a distribution process the contribution to the plate height due to nonequilibrium effects can be expressed by the formula:

$$H_S = \frac{\Theta k d_f^2 u}{D_s (1+k)^2} \qquad\qquad (1.6.48a)$$

where Θ is a factor depending on the shape of the film, d_f is the film thickness, and D_s is the solute diffusivity in the stationary phase. For capillary columns $\Theta = 2/3$, while for packed columns $\Theta = 1/30$.

In size exclusion, there are no sorption effects (ideally) to affect band broadening. However, a contribution to band broadening comes from the stagnant mobile phase in the porous material. For spherical porous particles it has been demonstrated that this contribution is given by an increase of the plate height with the quantity:

$$H_S = \frac{kd_p^2 u}{30D_M(1+k)^2} \tag{1.6.48b}$$

For gas chromatography, the resistance to mass transfer into the gas phase brings an additional factor. The broadening due to this factor is expressed by the formula:

$$H_g = \frac{1+6k+11k^2}{24(1+k)^2} \cdot \frac{r^2 u}{D_M} \tag{1.6.49}$$

where r is the radius of the capillary column.

Combining all the factors contributing to the increase of the plate height, the total result can be written in the form:

$$H = H_L + H_E + H_D + H_S + H_g \tag{1.6.50}$$

Rel. (1.6.50) can be written in the general form known as van Deemter equation:

$$H = A + B/u + Cu \tag{1.6.51}$$

where A is the eddy diffusion term, B is the ordinary diffusion term, and C the non-equilibrium mass-transfer term plus the lateral diffusion term. For GC using capillary columns, the term A is not present in rel. (1.6.51). This expression gives explicitly the plate height in the gas chromatographic process as a function of linear gas velocity u, although the detailed expression for each term shows that H depends on a number of parameters describing the chromatographic system. A graphic representation of van Deemter equation for GC is given in Figure 1.6.3. The variation of plate height for capillary columns is further discussed in Section 4.2.

In HPLC, the theoretical plate height H is expressed by a similar expression to that for a GC on packed columns. The main difference between the HPLC and GC with packed columns regarding peak broadening arises from 10^4 to 10^6 times higher diffusivity in gases compared to that in liquids. For this reason, the term B/u is much smaller in liquid chromatography then in gas chromatography, while the term H_D is significantly higher in HPLC. Other equations also can be applied for describing the height of the theoretical plate in HPLC, such as the expression:

$$H = E\,u^{0.4} \tag{1.6.52}$$

where E is a constant. The variation of H following rel. (1.6.52) is shown in Figure 1.6.4. This type of variation seems to follow more closely the experimental observations.

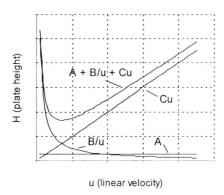

FIGURE 1.6.3. *Graph of van Deemter equation showing dependence of plate height H as a function of linear gas velocity u in GC.*

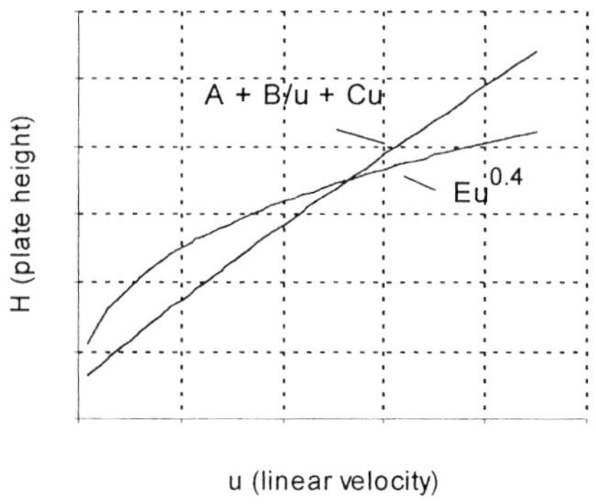

FIGURE 1.6.4. *Graph of van Deemter equation and an approximation for H as a function of liquid velocity in HPLC.*

The plate height in HPLC, as commonly applied, is lower than in GC. This allows the use of much shorter columns in HPLC as compared to packed GC columns or even capillary GC columns.

For non-Gaussian peaks, an asymmetry parameter is also defined for peak characterization. This is determined by taking a perpendicular from the peak maximum and a parallel to the baseline at 10% peak height. The asymmetry is defined as the ratio of the rear (r) to front (f) segments cut on this parallel by the chromatographic peak and the perpendicular, as shown in Figure 1.6.5a. The asymmetry is generated in chromatography by various factors such as column overloading. In this case, due to the excessive amount of sample, the stationary phase is covered or saturated with the sample, and the distribution process is far from ideal, leading to peak "fronting" (r / f <1). When more than one type of interaction of an analyte with the stationary phase takes place, for example due to "active spots" in the path of the analyte, peak "tailing" takes place (r / f >1). The deviation from the linear isotherm of the overall distribution process is shown in Figure 1.6.5b, where curve "b" corresponds to a linear process, curve "a" shows a negative, and curve "c" a positive deviation. As indicated in Figure 1.6.5b, a given concentration of the analyte in the gas phase $C_{i,g}$ corresponds to a concentration $C^a_{i,s}$ for curve "a", a concentration $C^b_{i,s}$ for curve "b", and a concentration $C^c_{i,s}$ for curve "c". Therefore, in the chosen point $C_{i,g}$, depending on the type of isotherm, the distribution constant can be K^a_i, K^b_i, or K^c_i, where:

$$K^a_i = C^a_{i,s} / C_{i,g} \qquad K^b_i = C^b_{i,s} / C_{i,g} \qquad K^c_i = C^c_{i,s} / C_{i,g} \qquad (1.6.53)$$

and $K^a_i < K^b_i < K^c_i$. Using rel. (1.6.14) the retention volume (or the retention time) can be calculated by the formula $V_R = V_g + K_i V_s$. The result is the difference in the retention time depending on the isotherm with $V^a_R < V^b_R < V^c_R$. This is illustrated in Figure 1.6.5a.

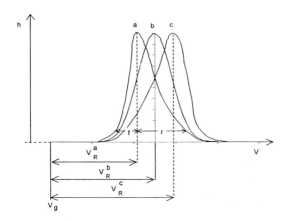
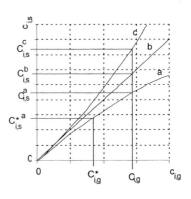

FIGURE 1.6.5a. *Variation of the peak shape as a function of isotherm (b linear, a, c nonlinear).*

FIGURE 1.6.5b. *Linear (b) and nonlinear (a, c) isotherms.*

The nonlinearity of the variation of $c_{i,s}$ as a function of $c_{i,g}$ makes the peak shape non-Gaussian. For a selected concentration $C^*_{i,g}$ the corresponding concentration in the stationary phase is $C^*{}^a_{i,s}$ and $K^*{}^a_i = C^*{}^a_{i,s} / C^*_{i,g}$. It can be seen that $K^*{}^a_i > K^a_i$ and not a constant as in the case of a linear isotherm. The peak asymmetry characterized by the rear (r) and front (f) segments, shown in Figure 1.6.5a for the peak "a", results from the fact that rel. (1.6.14) is not valid when K_i is not a constant [15].

- Peak separation in chromatography

The peak separation will depend on the nature of the two components to be separated. The more different are the distribution constants K_i and K_j for the components, the more different are their retention times t'_R, as seen in rel. (1.6.15), considering t_g and β as constants. The *separation factor* α defined in Section 1.4 (see rel. 1.4.29) is used to characterize the chromatographic separation, where:

$$\alpha = t'_{Ri} / t'_{Rj} = K_i / K_j = k_i / k_j \qquad (1.6.54a)$$

A good separation is attained for a large separation factor (by convention $\alpha > 1$) when the product $k_i\, k_j$ is as close as possible to unity. Using rel. (1.6.54a) and (1.6.20) it can be seen that:

$$\alpha = (\gamma_{j,A}\, p^o_j) / (\gamma_{i,A}\, p^o_i) \qquad (1.6.54b)$$

This relation shows that the separation in gas chromatography depends on two ratios, $\gamma_{j,A} / \gamma_{i,A}$ and p^o_j / p^o_i. The ratio $\gamma_{j,A} / \gamma_{i,A}$ depends on the interaction of the two components to be separated with the stationary phase. The ratio p^o_j / p^o_i shows that, for relatively close values for $\gamma_{j,A}$ and $\gamma_{i,A}$, the more different are the vapor pressures of the pure analytes, the better is the separation. Separation of compounds with $\alpha > 1.01$ or even lower are typically possible in GC.

In liquid chromatography the separation factor α is directly obtained from rel. (1.4.30), and for the separation of two components "i" and "j" has the form:

$$\alpha = K_i / K_j = (c_{i,s} \, c_{j,m}) / (c_{i,s} \, c_{j,m}) \qquad (1.6.54c)$$

where "s" indicates stationary and "m" mobile phase.

An evaluation of how well two peaks are separated can be obtained experimentally from the chromatogram by the formula:

$$R_s = (2 \, \Delta \, t_R) / (W_{b1} + W_{b2}) \qquad (1.6.55)$$

where R_s is called *resolution* and $\Delta \, t_R = t'_{R1} - t'_{R2}$. The values $\Delta \, t_R$, W_{b1} and W_{b2} are measured from the chromatogram as shown in Figure 1.6.6.

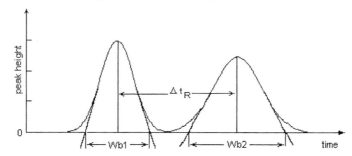

FIGURE 1.6.6. *Measurement of $\Delta \, t_R$, W_{b1} and W_{b2} on a chromatogram for the calculation of resolution.*

The two peaks are considered base-to-base separated when $R_s > 1$, which corresponds to $\Delta \, t_R = W_b$ (when W_{b1} is assumed equal to W_{b2}). Therefore, a good separation will be achieved when

$$\Delta \, t_R > 4 \, \sigma \qquad (1.6.56)$$

The resolution R_s depends on the capacity factors k_i, separation factor α, and the number of theoretical plates n. If it is assumed that $W_{b1} = W_{b2} = W_b$ and rel. (1.6.54a) is written in the form:

$$\Delta \, t_R = (\alpha - 1) \, \alpha^{-1} \, t'_R \qquad (1.6.57)$$

then rel. (1.6.55) can be written:

$$R_s = (\alpha - 1) \, \alpha^{-1} \, t'_R / W_b \qquad (1.6.58a)$$

This expression can be further written using rel (1.6.43) and (1.6.45) in the form:

$$R_s = 0.25 \, (n)^{1/2} \, (\alpha - 1) \, \alpha^{-1} \, k \, (k + 1)^{-1} \qquad (1.6.58b)$$

where the index "i" was neglected for k and for t'_R (k refers to the larger of the two k_i, k_j). Rel (1.6.58b) shows that the separation of two components on a chromatographic column is better (higher R_s) when n for the specific column is higher. Rel. (1.6.40) for n shows that for a constant height of a theoretical plate H, the resolution is proportional to the square root of the column length ($L^{1/2}$).

Rel. (1.6.58b) can be used in the choice of a GC column with a certain value for the number of theoretical plates n when α and k are known and the separation R_s is set higher than unity. This number should be

$$n > 16 \frac{(1+k)^2}{k^2} \frac{\alpha^2}{(\alpha - 1)^2} \qquad (1.6.59)$$

In many practical applications, the separation factor α between an analyte and other components of a specific matrix may be too close to unity. The increase in the number of theoretical plates of the column can be helpful in these cases, although n cannot be higher then a specific limit (common GC columns 60 m long with 0.32 mm i.d. may have $n \approx 300000$). Other procedures such as changes in the temperature gradient for the GC oven, which may change the retention time of some components, change of the GC column with a column having a different stationary phase, or sample preparation can be applied for improving the results of the analysis. The use of sample preparation simplifies the mixture of components on which the chromatographic separation must be performed (see also Chapter 4) or modifies the nature of the analyte by derivatization and therefore modifies the separation factor.

Separation in HPLC is achieved not only by selecting a column with a high number of theoretical plates. The typically lower number of theoretical plates for HPLC columns (around 10,000), requires larger values for α than in GC (usually $\alpha > 1.05$). The use of gradient elution allows enhanced separation. This is based on the modification of the equilibrium constant when the nature of the mobile phase is changed. For example, by assumption that the mobile phase is water, the variation of the distribution constant at the addition of an organic modifier in the mobile phase is ideally expressed by the formula:

$$\ln K_i (c_M) = \ln K_i (c_M = 0) - \varphi_{i,M} c_M \qquad (1.6.60)$$

where $K_i (c_M)$ is the distribution coefficient for the analyte "i" in the presence of an organic modifier "M" in the mobile phase, $K_i (c_M = 0)$ is the distribution constant for water as a mobile phase, $\varphi_{i,M}$ is a constant for a given analyte-solvent-stationary phase combination, and c_M is the molar concentration of the organic solvent. Rel. (1.6.60) shows that by modification of the composition of the mobile phase, the value of K_i changes differently for different analytes "i", and specific separations can be achieved.

- Sample capacity in gas chromatography

In addition to different column characteristics, one more factor must be taken into consideration in gas chromatography, namely the sample capacity. This is defined as the maximum permissible sample size that can be injected into a column without more than 10% loss of efficiency, and it is expressed as:

$$A = a \ v_{eff} \ (n)^{1/2} \qquad (1.6.61a)$$

where A is the maximum permissible volume of vaporized sample (exclusive of the carrier gas), *a* is a constant depending on the system, n is the number of theoretical plates, and v_{eff} is given by [28]:

$$v_{eff} = V_g / n + K V_s / n \qquad\qquad (1.6.62)$$

The parameter v_{eff} is called the effective volume of one plate. For a capillary column $V_g = \pi r^2 L$ and V_s can be obtained from rel. (1.6.8). With these values rel (1.6.61a) can be written in the form:

$$A = a \ (\pi r^2 L) \ (1 + k) \ (n)^{-1/2} \qquad\qquad (1.6.61b)$$

The reported values for *a* are on the order of 10^{-2}, and common values for A for capillary columns are less than 0.1 mL (gas in normal conditions).

This limitation indicated by the theory regarding the sample volume injected in a GC system imposes a serious problem for analysis of traces in a given sample. The detectors used in GC have limited sensitivity (see Section 4.2), and an amount of sample below a certain limit cannot be detected. Therefore, a compromise should be chosen such that the sample should be small enough to be accommodated by the chromatographic column but sufficiently large for the detector sensitivity.

An alternative for achieving a lower column load and enough analyte in the detector is to perform an additional separation before the analytes reach the analytical column. In this separation, part of the sample that is not of interest can be eliminated, and at the same time the important analytes can be kept. This preliminary separation can be done with bidimensional chromatography, but simpler techniques are also reported, such as programmed temperature vaporization (PTV) injection, etc. Sample preparation that allows the simplification of the matrix for the analytes or the concentration of the analyte such that a smaller amount of sample must be injected in the GC is a common procedure for achieving lower column load.

- Isothermal and programmed temperature gas chromatography

As indicated previously, the retention time of an analyte is temperature dependent. For this reason, the gas chromatographic separations are always performed at controlled temperatures. The temperature can be kept constant (isotherm separation), can be modified at a given rate (gradient separation), or may consist of a sequence of isotherm and gradient portions. The temperature program is commonly chosen to achieve two main practical purposes: an acceptable separation of the components of the sample and a reasonable time span for the whole chromatographic process. While as a rule the separation is better at lower temperatures, isothermal chromatography is not practical for complex samples. The elution times in a gas chromatographic separation depend on the boiling points of the constituents of the sample (similarly to a distillation process). Because of that, when the isothermal conditions are set significantly lower than the boiling point of a given compound, its elution can take a very long time. The light fractions of a sample may elute very fast and with poor or no separation if the temperature is set too high. When the components of a sample cover a wide range of volatilities, a good separation cannot be done using isothermal conditions.

The gas chromatographic separation in temperature gradient is affected by more than one factor [28]. Simultaneously there is variation in the gas flow, variation in the distribution constants, and variation in peak broadening.

In isothermal conditions a constant inlet gas pressure would automatically maintain a constant volumetric and linear flow rate. When the temperature of the column is programmed while the inlet gas pressure is kept constant, the flow rate will change considerably, becoming lower because of the modifications of the column geometry, expansion of the gas, and modifications of the gas viscosity, which increases with $(T)^{1/2}$. Besides isobaric operation, which is common, constant flow at programmed temperature capability is also available for certain instruments.

The distribution constant is another parameter that varies with the temperature. Being an equilibrium constant, it is expected that its variation follows the formula:

$$\Delta G^0 = -RT \ln K_i \qquad (1.6.63)$$

where ΔG^0 is the standard free enthalpy for the solute to be transferred from the mobile to the stationary phase. Rel (1.6.63) can be written in the form:

$$k_i \, \beta = \exp[-\Delta G^0 / (RT)] \qquad (1.6.64)$$

The phase ratio β does not vary with the temperature, the parameter that varies being the capacity factor k_i. For a hypothetical system with $\Delta G^0 = -20,000$ J mol^{-1} and $\beta = 100$, the variation with the temperature of k (where the index i is omitted) is shown in Figure 1.6.7. As seen from rel. (1.6.64) and shown in Figure 1.6.7, the value of k β tends to 1 when the temperature increases. From rel. (1.6.11) it can be seen that in this case t'_R tends to become equal to t_g. The resolution R_s given by rel. (1.6.58) and depending on the ratio k / (k+1) will decrease as temperature increases. The variation of the ratio k / (k+1) with the temperature for the same hypothetical system with $\Delta G^0 = -20,000$ J mol^{-1} and $\beta = 100$ is shown in Figure 1.6.8.

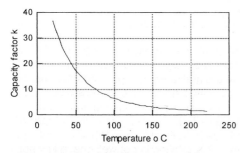

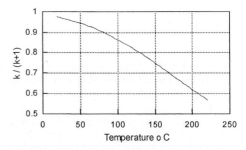

FIGURE 1.6.7. *The variation of k with temperature for a system with* $\Delta G^0 = -20,000$ *J mol* $^{-1}$ *and* $\beta = 100$.

FIGURE 1.6.8. *The variation of k/(k+1) with* T^0 *C for a system with* $\Delta G^0 = -20,000$ *J mol* $^{-1}$ *and* $\beta = 100$.

Similar considerations can be applied to the variation of the separation factor α with temperature. The general formula for its dependence is

$$\log \alpha = a \, (1 / T) + b \qquad (1.6.65)$$

which indicates that α decreases when T increases. More elaborate studies were done to precisely calculate the variation of t'_R or of other chromatographic parameters with temperature [29], [30]. However, in practice these calculations do not have significant applicability. As a general rule, higher temperatures are needed for the analysis of polar and/or high boiling point constituents. Usually, the main limiting factor to this temperature increase is not the excessive decrease in the values for k and α but the decomposition (bleeding) of the stationary phase used as a column coating.

- Retention index in gas chromatography

The qualitative information in a chromatogram is related to the retention time of the peaks. Although no structural information for an analyte can be obtained from a chromatogram generated with a FID (or similar type of detector), chromatograms are widely used for qualitative analysis. One possibility for qualitative analysis is to match the retention times of standards identical to the compounds that are analyzed. The retention time of each compound has a specific value in given chromatographic conditions (although the danger always exists that different compounds may have the same retention time) and for a specific stationary phase. One problem with this approach is that by changing the chromatographic conditions (like carrier gas head pressure, column dimensions, etc.), these retention times vary. One procedure to avoid this variability is to replace the retention time of a compound with a *retention index* I_j [31], [32]. Kováts retention index is a measure of the relative retention in isotherm conditions with the normal alkanes used as a standard reference. In isotherm conditions, the retention times of normal alkanes increase exponentially with the carbon number of the alkane. The measurement of the adjusted retention times t'_R (in isotherm conditions) for a series of normal alkanes allows a scale to be generated. Kováts retention index for a given compound "j" showing a peak between the alkane with n carbons and the alkane with $(n + a)$ carbons is defined as:

$$I_j = 100\,n + 100\,a\left(\frac{\log t_{R_j'} - \log t_{R'_n}}{\log t_{R'_{(n+a)}} - \log t_{R'_n}}\right)$$

(1.6.66)

where t'_R are the corresponding adjusted retention times. Kováts retention index for the alkanes with n carbons as shown in rel. (1.6.66) will be 100 n. The retention index is dependent on temperature and on the nature of the stationary phase. The dependence of the retention index on temperature can be approximated with the formula [33]:

$$\Delta I\,/\,10^\circ\,C = (I_{T1} - I_{T2})\,/\,[10\,(T_2 - T_1)]$$

(1.6.67)

where T_1 and T_2 are two different temperatures. This relation is not well obeyed by polar compounds or when the difference between T_1 and T_2 is larger than 50° C. Extended lists of standard retention indexes (LTPRI indexes) are available in literature [32].

Extensions of the definition of a retention index were done by the use of gradient temperature conditions [33] or methyl esters as standards instead of normal hydrocarbons. Tables for retention indexes of certain substances and stationary phases have been published [34]. The retention index system has limited utility for the GC analysis of complex samples.

- ***Estimation of the number of peaks separated in a chromatogram***

The complex nature of certain samples may lead to a considerable number of visible chromatographic peaks. However, the number of peaks that can be separated in a chromatogram may still be lower than the number of compounds in the mixture. The relation between the number p of visible peaks in a chromatogram and the real number of components m can be written as follows [35], [35a]:

$$p = m \exp(-m / n_c) \qquad (1.6.68)$$

where n_c is the *peak capacity* defined as the maximum number of peaks that can be put side by side (with acceptable resolution) into the available separation space in a chromatographic separation. The value for n_c in rel. (1.6.68) should be chosen as the value corresponding to a critical resolution. In isotherm conditions, n_c can be estimated with the formula:

$$n_c = 0.5 \, (n)^{1/2} \qquad (1.6.69)$$

where n is the number of theoretical plates of the column. This shows that even for a good column (n=300,000), n_c has values between 250 and 300. For a mixture with $m = n_c$, rel. (1.6.68) indicates that only 36.8% of the expected number of peaks are observed, the rest of them being merged with the others. This represents approximately 110 peaks that can be seen in a chromatogram. The number of single component peaks s is even lower, and it is estimated by the formula [35]:

$$s = m \exp(-2m / n_c) \qquad (1.6.70)$$

which shows that only 13.5% of the total number of components appear as single component peaks. In practice, therefore it is impossible to completely separate a mixture with more than 30 to 40 components using one chromatographic column. However, the number of compounds in a mixture may very well exceed this number. Various procedures are utilized to circumvent this problem. Among these are instrumental procedures such as the use of selective detectors, bidimensional GC, etc. Mass spectrometry detection with various procedures such as extracted ions, single ion monitoring, MS/MS detection, and peak deconvolution [36] are widely utilized for allowing the determination of specific analytes in a simpler chromatogram. However, the most important procedure for allowing or improving the chromatographic analysis in a complex matrix remains sample preparation.

1.7. GENERAL CHARACTERISTICS OF SEPARATIONS

Most cleanup and concentration techniques are based on separations. Separations have a number of characteristics that are important for their applicability to a specific problem. Some of the general characteristics are [15] a) fractionation capacity, b) load capacity, c) adaptability to analyte volatility, d) type of selectivity, and e) speed and convenience.

Fractionation capacity (fraction capacity) is defined as the maximum number of components that can be separated in a single operation. Some separation techniques have a fractionation capacity equal to two. Examples of such techniques are crystallization, dialysis, single stage extraction, etc. For a chromatographic process, fractionation capacity is equivalent to *peak capacity,* defined in Section 1.4 as the maximum number of peaks that can be put side by side (with acceptable resolution) into the available separation space in a chromatographic separation. Similarly to the definition for chromatography, fractionation capacity can be defined for electrophoresis. Fractionation capacity is a parameter that cannot be assigned with a precise value for many separations where the change in conditions can modify significantly the number. For example, in GC, the fractionation capacity depends on the column used for the separation, the GC conditions, etc.; in HPLC it depends on the column, type of gradient, detector cell volume, etc.

Load capacity is the maximum quantity of a mixture that can be separated in the process. Load capacity depends on the equipment utilized and the manner in which the separation procedure is applied. Only a range for the load capacity can be assigned to a technique. For example, distillation and crystallization can be performed on large quantities without significant problems. On the other hand, gas chromatography is limited to a small column load even when packed columns are used. Preparative LC may have a considerable sample load, but this requires special equipment, while analytical HPLC instruments have a very limited load capacity.

Adaptability to analyte volatility is defined as the ability of a separation technique to be used with particular types of samples regarding their volatility. Compounds can be classified as volatile, semivolatile, or nonvolatile. Some techniques are applicable only to volatile compounds. The volatility can be displayed at ambient temperatures or at higher temperatures with the condition that the compound can generate vapors without decomposition. Other compounds are semivolatile. These compounds do have some volatility, but their partial vapor pressure is low in normal conditions. At higher temperatures above a certain value, these compounds tend to decompose. Semivolatile compounds still can be separated by some techniques such as distillation under reduced pressure. Nonvolatile compounds are in fact of different types. Some are not stable when heated, and others, although thermally stable, are extremely difficult to volatilize. Many organic compounds decompose upon heating, and ionic compounds, for example, cannot be changed into gas phase except at unusually high temperatures. Macromolecules not only are not volatile but may also decompose upon mild heating. Due to the wide utilization of separation techniques of gas phase compounds (distillation, gas chromatography, etc.), the limitation in the use of a technique to semivolatile or only to volatile compounds is important.

The type of selectivity refers to the physico-chemical property distinguished by the separation. Among physico-chemical properties toward which a separation shows selectivity are a) boiling point, b) solubility in various solvents, c) partition coefficient, d) acid/base dissociation constant e) chemical properties f) diffusion coefficient, g) dipole moment, h) molecular shape i) isomerism, j) biochemical character, k) molecular weight, l) electrophoretic mobility, and m) density. A given separation may depend on more than one parameter, such as gel permeation that depends on molecular weight and size of the molecule. Some parameters are not independent from each other, such as boiling point, which depends on molecular weight and polarity, etc. However, the separation is essentially based on a specific property. For example, differences in boiling point allow the separation by distillation and play a role in gas chromatographic separations. Solubility differences allow separations by crystallization/precipitation and play a role in liquid chromatography. Partition coefficient is important in various chromatographic separations. Acid/base dissociation constant is important in ion exchange. Other chemical properties such as capability of a metal ion to form complexes also can be used in separations. Diffusion coefficient plays an important role in several separation techniques based on diffusion, such as membrane filtration, dialysis, etc. Dipole moment and other related physical characteristics such as polarization play a role in various techniques such as chromatographic separations. Molecular shape is also important, for example in gel filtration. Isomerism can be of various types (see Section 11.7). Chromatographic separation on chiral stationary phases is done based on the differences in the chirality of the analytes. Biochemical character is used in separations such as precipitations based on antigen-antibody interactions or in a number of chromatographic separations where the column contains an immobilized biopolymer. Numerous other separations are based on biochemical properties, but a discussion of these is beyond the subject of this book. Molecular weight is directly and indirectly implicated in many separations. Electrophoretic mobility is involved in electrophoretic separations. Density is also important for example in centrifugation, etc.

Speed and convenience of a separation are other features of interest in sample preparation. Some techniques are fast and can be done in a few minutes. For example, solid phase extraction (SPE) is done by passing a solution that contains the sample through a small column of solid phase (usually in the form of a cartridge). If the analytes have a high affinity for the solid phase, the column is washed with an appropriate solvent that carries at least part of the undesired matrix components. After this, the analytes are eluted from the solid phase and obtained as a cleaner solution. The process can be done in a few minutes, and the technique is classified as fast. Other techniques such as dialysis separate, for example, large molecules from small molecules, based on the property of small molecules to penetrate through a special membrane material from solution. The process of separation requires a long time, and the technique can be classified as slow. Slow techniques are not preferred as sample preparation techniques unless no other separation in available. Related to speed and convenience, another factor to be considered is the ease of a separation to be automated. Some methods, although not fast, can be easily automated, which highly increases the convenience of their use.

Some techniques do not require complex instrumentation, one example being SPE. Other techniques such as ultracentrifugation require dedicated instrumentation that can be rather complex. The speed and the instrumentation complexity can be important factors in the selection of a specific separation technique for sample cleanup or concentration.

A number of separation techniques frequently utilized in sample cleanup and concentration are shown in Table 1.7.1 together with some of their properties [15].

The detailed description of each technique and its application in sample preparation are further discussed in Chapters 7 through 15.

TABLE 1.7.1. *Separation techniques frequently utilized in sample cleanup and concentration.*

Technique	Fraction-ation capacity	Load capacity (max.)	Adaptability to analyte volatility	Selectivity	Speed and convenience	Instrumen-tation
centrifugation/ ultracentrifugation	2/5	kg/mg	semi/nonvolatile	density	slow	simple/ complex
crystallization	2	kg	semi/nonvolatile	solubility	medium	simple
dialysis	2	g	nonvolatile	molecular weight	slow	simple
distillation	10	kg	volatiles	boiling point	medium	simple
exclusion chromatography	5-10	mg	semi/nonvolatile	molecular shape	medium	complex
electrophoresis	50-100	μg/mg	semi/nonvolatile	electrophoretic mobility	medium	complex
extraction	2	kg	all volatilities	partition coeff.	fast	simple
filtration/ultrafiltration	2	kg/g	semi/nonvolatile	particle size	medium	simple
gas chromatography	100-300	μg/mg	volatiles	partition coeff.	medium	complex
gas desorption	2-3	mg	volatiles	partition coeff.	fast	simple
ion exchange	10	g	nonvolatile	acid/base character	medium/fast	simple
liquid chromatography (HPLC, TLC, paper, etc.)	20-30	mg/g	semi/nonvolatile	partition coeff.	medium	simple/ complex
osmosis	2	kg	nonvolatile	molecular weight	slow	simple
precipitation	2	kg	semi/nonvolatile	solubility	fast	simple
solid phase extraction	2-4	mg	semi/nonvolatile	partition coeff.	fast	simple

- *Operational aspects of separations*

The properties of various separations indicated in Table 1.7.1 refer to general characteristics without including the operational aspects for the procedure. The basic operation modes for separations can be classified using different criteria. One classification distinguishes static and dynamic procedures. In a static procedure, the fractions separated remain in contact during and after the separation, but usually as distinct phases. Because the fractions are in contact, they are ideally in equilibrium. Examples of static procedures are distillation, crystallization, the static headspace separation of volatiles from a solid or liquid matrix, static liquid-liquid extraction, etc. In some cases the time necessary for achieving the equilibrium is long and the separation is carried only close to equilibrium conditions. In some static separations the change of composition of the phases is continuous during the process, such as in distillation, while in others it is done in a rapid step such as in precipitation. Repetition of the same static separation for the same sample is also possible, these being sometimes indicated as multiple stage static separations. Dynamic procedures imply moving phases, such as continuous extraction, dialysis, and dynamic headspace separations. In these techniques, the phases carrying the separated fractions have a dynamic contact, and the equilibrium may not be reached. This is not the case for chromatography, which can be considered a dynamic type of separation, but one where the equilibrium is attained. Dynamic procedures have the advantage of being efficient and able to produce exhaustive separation although not always reaching equilibrium, in contrast to static separation where the equilibrium is obtained, but the separation is limited to this equilibrium. Static separations are preferred for systems with high rates of establishing equilibrium and with good separation factors. In general, static separations are simple to perform, not requiring special instrumentation. On the other hand, dynamic processes are not limited in the separation of a fraction by the value of an equilibrium constant. A typical example is dynamic headspace extraction where the sample is placed in a flow of gas in which the volatile compounds are transferred. Because the gas passing over (or through) the sample is continuously refreshed, no "saturation" of the gas stream is possible.

The classification as static or dynamic does not capture enough details for separations, and a more detailed classification can be done in the following groups [15]: a) single contact, also indicated as co-current or static, b) differential processes, c) crosscurrent processes, d) countercurrent processes (dynamic), e) countercurrent with reflux (dynamic), f) differential migrating processes, etc. More than one operation mode can be used for some separation types. A typical example is distillation, which can be done as a differential process, countercurrent with reflux, etc. Also, liquid extraction can be done in a single stage, in countercurrent, crosscurrent, etc. Other separations such as chromatography are done in a single operational mode.

Single contact separations typically are done in one stage (or unique operation) where two fractions are obtained after the separation is performed. This type of separation usually is done in equilibrium conditions. Typical examples of this type of operation are filtration, single batch liquid-liquid extraction, and static headspace separation of volatiles from a sample. This type of separation usually is applied for systems where the separation factor between the two fractions is large. It has the advantage of simplicity and is commonly applied in sample preparation. One of its disadvantages is the inability to separate fractions with low separation factor. For some separations the time required for attaining equilibrium can be long, reducing the separation efficiency.

Differential processes are characterized by a single stage operation where one of the separated fractions is continuously removed from the mixture. The typical example is simple distillation where the composition of both fractions is changed continuously. The change in the composition of the two separating phases during differential separation is exemplified for distillation in Section 8.1. Rel. (8.1.8) relates the final composition of the fraction remaining undistilled with the initial composition and the separation factor for the process. The same formulas can be used for other differential processes.

Crosscurrent processes are exemplified by repeated batch liquid-liquid extractions, Soxhlet extraction, etc. In this process portions from a specific component of the same sample are repeatedly separated from the sample. For example, the same water sample containing organic contaminants is extracted repeatedly with fresh portions of solvent in several stages. This type of process is discussed in detail for the example of liquid-liquid extraction in Section 10.1. The dependence of the recovery efficiency on the number of repetitions of the process is given for crosscurrent solvent extraction by rel. (10.1.3). In crosscurrent separations equilibrium is usually assumed at every stage of the process, and these can be considered repeated single contact separations.

Countercurrent separations are typical dynamic separations, including two streams flowing in opposite directions. This type of process is applied, for example, in countercurrent extraction and dialysis. Similar with countercurrent separations are other dynamic separation processes where one phase is stationary and the other is moving. This type of separation is discussed for dialysis in Section 13.2. Countercurrent techniques can be modeled based on a process with a number of discrete stages. This type of model is discussed in Section 10.1 for solvent extraction in a continuous flow operation with stages [15].

Countercurrent-with-reflux separations are typically exemplified by fractional distillation. The operation has been thoroughly studied for the fractional distillation applied for industrial purposes. A short discussion on fractional distillation with reflux, where a cascade of theoretical single stage distillation units are interconnected such that the vapors and liquid phases move in countercurrent, can be found in Section 8.1.

Differential migrating processes are exemplified by the chromatographic techniques. This is a dynamic technique where one phase is stationary and the other is mobile. The special characteristic for this type of separation is that equilibrium between the analyte concentration in the mobile and stationary phase is established quickly. Nonequilibrium processes are encountered in some chromatographic separations, but this is not the ideal case. The separation in chromatography is discussed in more detail in Section 1.6. The separation of this type of process is good even for components with low separation factors. The chromatographic peak broadening discussed in Section 1.6 is the main limiting factor in the separation in a chromatographic process. Different types of chromatography are discussed in Chapters 4 and 12.

1.8.   ELEMENTARY DISCUSSION ON THERMODYNAMIC AND KINETIC FACTORS IN CHEMICAL REACTIONS

For any isolated (closed) system, spontaneous transformations take place with an increase in the entropy ΔS of the system (see e.g. [37]). For a nonisolated system in isothermal conditions (see rel. 1.2.3), spontaneous processes take place with a negative variation of free enthalpy ΔG where:

$$\Delta G = \Delta H - T \, \Delta S + \sum_{j=1}^{r} n_j \mu_j \qquad (1.8.1)$$

and where ΔH is the variation in enthalpy of the system and T is the absolute temperature (measured in Kelvin degrees $^{\circ}$K, where 0° K = - 273.15° C), n_j is the number of moles, and μ_j is the chemical potential of component "j". From rel. (1.8.1) for ΔG and based on the fact that a spontaneous process takes place with a negative variation of free enthalpy ΔG, it can be seen that in a spontaneous process a system will have the tendency to lower its enthalpy (or at constant volume its energy $\Delta E = \Delta H - p \, \Delta V$) and to increase its entropy. For a transformation at constant pressure p and temperature T, equilibrium will correspond to a minimum of the free enthalpy. Otherwise, the process would spontaneously continue. For equilibrium, the following expression can be written for the closed system:

$$\delta G = 0 \quad \text{or} \quad \Delta G = 0 \qquad (1.8.2)$$

Now by consideration of a chemical reaction with the form:

$$aA + bB + ... \, \rightleftharpoons \, dD + eE.... \qquad (1.8.3)$$

where "A", "B",… are different molecular species, at the equilibrium the relation between the molar concentrations of each component is given by the formula:

$$K = \frac{[D]^d [E]^e \cdots}{[A]^a [B]^b \cdots} \qquad (1.8.4)$$

where K is the equilibrium constant (activities should in fact be used in this expression instead of concentrations). For gases, partial pressures replace the molar concentrations. On the other hand, the total free enthalpy for a system with r components at constant temperature and pressure is defined as the sum of the chemical potentials and

$$\Delta G = \sum_{i=1}^{r} n_i \mu_i \qquad (1.8.5)$$

The expression for μ_i as a function of standard chemical potential is given by the expression:

$$\mu_i = \mu_i^{\circ} + RT \ln a_i \qquad (1.8.6)$$

By introduction of rel. (1.8.6) in rel. (1.8.5) and replacement of the activities with molar concentrations, the expression for the free enthalpy variation for the chemical equilibrium given by rel (1.8.3) can be written in the form:

$$\Delta G - \Delta G^{\circ} = RT \ln \frac{[D]^d [E]^e \cdots}{[A]^a [B]^b \cdots} \qquad (1.8.7)$$

where ΔG° is the standard free enthalpy (noted $^{\circ}$) at constant pressure of 1 atm. On the other hand, ΔG at equilibrium is equal to zero, and therefore the equilibrium constant is related to the standard free enthalpy by the expression:

$$\Delta G^{\circ} = - RT \ln K \qquad (1.8.8)$$

The free enthalpy accompanying the reaction is the difference in the standard free enthalpies of the products and the reactants and is given by the expression:

$$\Delta G^{\circ} = \Sigma \, \Delta G^{\circ}_{products} - \Sigma \, \Delta G^{\circ}_{reactants} \qquad (1.8.9)$$

(Similar expressions are true for the enthalpy and the entropy of the system.) Because ΔG°, ΔH°, and ΔS° are temperature dependent, they must be calculated at a certain temperature. Values of thermodynamic functions at 298.15° K (or 25° C) for many chemical compounds are tabulated. For a large number of organic compounds, the decomposition reactions at standard temperature of 25° C have negative values for the standard free enthalpy ΔG°. These reactions should, therefore, occur spontaneously. However, their reaction rates are in most cases slow enough such that this assures the chemical stability of numerous organic compounds. This also points out the importance of the kinetic factors over the thermodynamic ones in many reactions.

Expression (1.8.8) for the relation between standard enthalpy and equilibrium constant can be extended to the evaluation of partition coefficients K_i for a solute between two phases. Use of rel. (1.8.6) for the chemical potential of a component "i" in a solvent "A" or in a solvent "B", and replacement of the activities with concentrations, rel. (1.8.5) leads to the expression:

$$\Delta G - \Delta G^\circ = RT \ln (c_{i,A} / c_{i,B}) \tag{1.8.10}$$

Because ΔG for a system at equilibrium is equal to zero, the distribution constant K_i is related to the standard free enthalpy by an expression similar to rel. (1.8.8):

$$\Delta G^0 = - RT \ln K_i \tag{1.8.11}$$

The *reaction rate* of a chemical process where R is a reactant and P is a product is defined as the variation of the concentration of R or of P versus time. The reaction rate can be expressed by the relation:

$$- \frac{d[R]}{dt} = \frac{d[P]}{dt} \tag{1.8.12}$$

where [R] and [P] are the (molar) concentrations at any time during the reaction. If in a chemical reaction of the type:

$$A + B + C + ... \longrightarrow D + E + F + ...$$

the rate of reaction depends linearly on the concentration of one reactant A, then

$$- \frac{d[A]}{dt} = k[A] \tag{1.8.13}$$

This type of reaction is of the first order. The constant k is the *rate constant*. Concentrations are commonly expressed in mol/L, and k is expressed in s^{-1}. The rate constant k is temperature dependent, and it is a constant only in isothermal conditions.

It is possible that the rate of reaction depends on the concentrations of the A and B species simultaneously. In this case the dependence is given by

$$- \frac{d[A]}{dt} = - \frac{d[B]}{dt} = k[A][B] \tag{1.8.14}$$

and the reaction is of the second order. The reaction rate constant k for second order kinetics is expressed in $mol^{-1} s^{-1}$ (different units from those used for the constant of first order kinetics).

Some chemical reactions have a reaction rate of the form:

$$- \frac{d[A]}{dt} = k[A]^n \tag{1.8.15}$$

where n is the reaction order. The value of. n can be an integer, or for certain chemical reactions it can be a fractional number.

In order to understand how the constant k depends on temperature, it was assumed that the chemical reactions may take place only when the molecules collide. Following this collision, an intermediate state called an activated complex is formed. The reaction rate will depend on the difference between the energy of the reactants and the energy of the activated complex. This energy $E^{\#}$ is called activation energy (other notation E^{a}). The reaction rate will also depend on the frequency of collisions. Based on these assumptions, it was shown (e.g. [38]) that k has the following expression (Arrhenius reaction rate equation):

$$k = A \exp\left(- \frac{E^{\#}}{RT}\right)$$

(1.8.16)

where A is a parameter related to the collision number, and it is called *frequency factor*. Rel. (1.8.16) indicates the explicit dependence of the rate constant k on temperature (expressed in $^{\circ}K$).

1.9. BRIEF SURVEY ON MOLECULAR INTERACTIONS

One theoretical approach for predicting the behavior of the analyte in a separation process is based on the calculation of the interactions between the analyte molecules and those with the molecules of a solvent or those with a stationary phase. Ideally, these techniques can predict the behavior of sample components in a specific separation. However, the success of these theoretical techniques is not always very good because of the complexity of the separation processes and because of the approximations used to evaluate molecular interactions and properties [39]. The application of statistical techniques helps in obtaining the desired information. For example, a common procedure for expressing a given chromatographic parameter P_r as a function of physico-chemical or molecular parameters x_i uses the formula:

$$P_r = f\left(a_1 x_1 + a_2 x_2 + \ldots a_n x_n\right)$$

(1.9.1)

where $a_1, a_2 \ldots a_n$ are calculated by multiple (linear) regression. Among the molecular parameters x_i are one or more structural additive parameters (carbon number, molecular mass, parachor, molar volume, molar refractivity, polarizability), physico-chemical parameters (boiling point, dipole moment, ionization potential, formal charges in the molecule, n-octanol-water partition coefficient), topological parameters related to the molecule shape, etc.

Because a chromatographic separation is the result of the interactions between the analyte and the molecules of stationary phases, the evaluation of these interactions can be informative for understanding the chromatographic process. As shown in Section 1.4, the separation process can be predicted when the distribution constants K_i for a component "i" between two phases "A and "B" are known, or when the chemical potentials $\mu^{\circ}_{i,B}$ and $\mu^{\circ}_{i,A}$ are known. The chemical potentials are related to the partial molar enthalpy of a component in a given phase. From rel. (1.2.1) it results

$$H = TS + G$$

(1.9.2)

and for constant temperature it can be obtained for the partial molar quantities (see rel. 1.2.11 and 1.2.10) the following relation:

$$H_i = TS_i + \mu \tag{1.9.3}$$

and for the standard state:

$$\Delta H^o_i = T\Delta S^o_i + \Delta\mu^o \tag{1.9.4}$$

where the partial molar enthalpy is related to the energies between "i" and the molecules of the two phases (see Section 1.4).

Other physical properties of a molecule that are relevant for the separation process also can be estimated from molecular interactions. For example, the stronger the intermolecular attractions in a liquid, the lower its vapor pressure at any temperature, and the higher the boiling point of that liquid. Based on the increase (or decrease) in the boiling point, guidance also can be obtained regarding the proper chromatographic conditions that must be applied for the separation.

Common contributors to the molecular interactions in gas-liquid chromatography are van der Waals forces, hydrogen bonds, and hydrophobic interactions. For a solid surface stationary phase, the main contributors to the chromatographic process remain van der Waals forces and hydrogen bonds between the molecules of the analyte and stationary phase. The interactions with active solid surfaces that may occur uncontrolled and perturb the chromatographic process also depend on van der Waals forces and hydrogen bonds. In GC, the molecular interactions with the carrier gas can be neglected. However, the adsorption-desorption mechanism in a liquid type stationary phase (common case) is still a very complex process and difficult to estimate theoretically.

The value of van der Waals energy alone is not a good predictor for any specific chromatographic parameter, but it represents a component of the interactions between the molecules of the analyte and stationary phase. Therefore these interactions may provide some insight regarding the interactions in the chromatographic process. Van der Waals energy depends on the dipole moment μ, polarizability α, and the ionization potential I of each interacting molecule. (The notation μ is used for chemical potential and μ for dipole moment.) The variation of these parameters during derivatization shows the direction in which van der Waals forces are modified.

Molecular interactions depend on certain physical properties of molecules, which can be calculated theoretically. Commonly known interactions between molecules include: the ion-dipole interactions, interactions between permanent dipoles, inductive forces, dispersion forces, hydrogen bonds, electron pair donor-electron pair acceptor interactions, solvophobic interactions [39], and possibly "finite interactions" [40], [41]. This section provides a short discussion on these interactions.

The ion-dipole interactions requiring the presence of an ion are not common in processes where organic molecules interact with organic nonionic phases. However, inorganic stationary phases may be utilized, for example, in columns using molecular sieves or diatomite as stationary phase. For a stationary phase that contains ions in gas chromatography or for other types of separations where ions are involved in the separation process, this type of interaction may have a certain contribution. The potential energy of an ion-dipole interaction (E_{i-d}) is given by the expression:

$$E_{i-d} = - \frac{1}{16 \pi^2 \varepsilon_0^2} \frac{z \mu \cos\theta}{\varepsilon r^2}$$

(1.9.5)

where z is the charge of the ion, μ is the dipole moment of the neutral molecule, θ is the dipole angle relative to the line joining the ion and the center of the dipole, ε is the electric permittivity of the medium (relative), and r is the distance from the ion to the center of the dipole, all expressed in SI units [12], where vacuum electric permittivity is $\varepsilon_0 = 8.854 \ 10^{-12}$ F m^{-1}, and the coefficient $1 / 16 \ \pi^2 \ \varepsilon_0^2$ is needed for appropriate calculation in SI units.

- Van der Waals forces

The permanent dipole interactions, the inductive forces, and the dispersion forces are known as van der Waals forces, and they play a significant role in molecular interactions. When the separation takes place for a gas phase (such as in GC), it is expected that these forces are important for the behavior of a molecule in the separation process. The separation in a liquid phase (extractions, HPLC, TLC, etc.) would certainly involve additional contribution from the interactions in condensed phase.

The dipole-dipole interactions between two molecules with permanent dipole moments μ_1 and μ_2 (Keesom effect) have the potential energy given by the formula:

$$E_{d-d} = - \frac{1}{16 \pi^2 \varepsilon_0^2} \frac{2}{3 k_B T} \frac{\mu_1^2 \mu_2^2}{\varepsilon r^6}$$

(1.9.6)

where k_B is Boltzmann constant ($k_B = 1.380658 \ 10^{-23}$ J K^{-1}) and T is the absolute temperature. The decrease (in absolute value) of the calculated E_{d-d} energy for two molecules of decanol as a function of temperature is shown, as an example, in Figure 1.9.1. The calculations were based on the value $\mu = 1.5$ D and assuming an intermolecular distance of 3 Å.

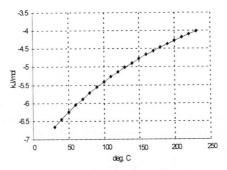

FIGURE 1.9.1. *The temperature dependence of dipole-dipole interaction energy E_{d-d} for two molecules of decanol at 3 Å distance.*

Besides the permanent dipoles, molecular interactions may also take place due to the dipoles induced in a molecule by a neighboring molecule with a permanent dipole moment. The induced dipole moment is characterized by the polarizability, which can be defined as the proportionality coefficient between the induced dipole moment in a molecule by the field $\mathcal{E}$ as shown in the formula:

$$\mu_i = \varepsilon_0 \, \alpha \, \mathcal{E} \tag{1.9.7}$$

The polarizability is in fact a tensor, but for practical purposes an average value is used. The potential energy between permanent dipoles and induced dipoles due to the polarization of molecules (Debye effect) can be written in SI units as follows:

$$E_{d\text{-}id} = - \frac{1}{16 \, \pi^2 \, \varepsilon_0^2} \; \frac{\alpha_1 \mu_2^2 + \alpha_2 \mu_1^2}{\varepsilon \, r^6} \tag{1.9.8}$$

where the units for polarizability in rel. (1.9.8) are $J^{-1} \, C^2 \, m^2$.

Both the Keesom and Debye interactions described by rel. (1.9.6) and (1.9.8) are calculated based on overall molecular parameters α and μ. In reality, adjacent molecules in liquids tend to interact based on individual bond dipoles and polarizabilities. The bond dipole effect is more pronounced for compounds with polar substituents. For some compounds with two or more polar substituents, the calculation of the net dipole moment of the molecule can lead to relatively low values for μ due to the compensation of the charges in the calculation. In reality, the interactions are more dependent on the group polarity and not the entire molecule polarity. For this reason the dipole moment is not a useful measure of polarity for many compounds. The lack of correlation between dipole moment and boiling point is shown in Figure 1.9.2, as an example, for difluorobenzenes. One additional type of interaction caused by so-called London dispersive forces may take place between molecules. The potential energy of dispersion interactions for two molecules is approximated by the expression:

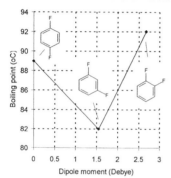

FIGURE 1.9.2. *The plot of boiling point as a function of dipole moment for difluorobenzene isomers.*

$$E_d = - \frac{1}{16 \, \pi^2 \, \varepsilon_0^2} \; \frac{3}{2} \; \frac{I_1 \, I_2}{I_1 + I_2} \; \frac{\alpha_1 \alpha_2}{\varepsilon \, r^6} \tag{1.9.9}$$

where I_1 and I_2 are the ionization potentials of the interacting molecules (expressed in J). The total dispersion energy E_d for a polyatomic molecule composed from "j" atoms interacting with a surrounding phase composed from "k" atoms is obtained as a sum of all possible interactions:

$$E_d = \sum_j \sum_k (E_{j,k})_d \tag{1.9.10}$$

The summation over all interacting atoms can be replaced with the integration over interacting volume elements, and the polarizabilities α_j, α_k can be replaced with electron polarizabilities per unit volume $(\alpha_e^V)_j$ and $(\alpha_e^V)_k$. This leads to the expression:

$$E_d = C_{const} \, V_j \, (\alpha_e^{\,V})_j \, (\alpha_e^{\,V})_k. \tag{1.9.11}$$

where C_{const} is approximately constant for similar molecules, V_j is the molar volume of compound "j", and the electron polarizabilities per unit volume $(\alpha_e^{\,V})_j$ can be calculated using Lorenz-Lorentz equation:

$$\alpha_e^{\,V} = 3/4 \, \pi \, N \, (n^2 - 1) \, / \, (n^2 + 2) \tag{1.9.12}$$

In rel. (1.9.12) N is Avogadro's number ($N = 6.0221367 \; 10^{23} \; mol^{-1}$), and n is the refractive index of the compound.

The overall value of the energy for van der Waals interactions (considered in an ideal gas) can be evaluated from the sum:

$$E_W = E_{d-d} + E_{d-id} + E_d \tag{1.9.13}$$

where each energy is given by one of the rel. (1.9.6), (1.9.8), or (1.9.9). It is common for most interactions that the E_d component is the main contributor to the intermolecular energy, followed by the E_{d-d} component, while the E_{d-id} component is less important. As an example, for decanol at $50°$ C and with a distance of 3 Å between the two decanol molecules, $E_d = -19.1$ kJ/mol, $E_{d-d} = -6.2$ kJ/mol, and $E_{d-id} = -0.8$ kJ/mol (1 cal = 4.1868 J), E_d representing 73% from the total van der Waals energy.

As seen from the expression of each van der Waals interaction, the energy varies with the intermolecular distance r as a function of $1 / r^6$. This dependence is exemplified for two molecules of decanol at $50°$ C in Figure 1.9.3. The calculations were done based on the following values: $\mu = 1.5$ D, $\alpha = 4.2 \; Å^3$, I = 10.9 eV.

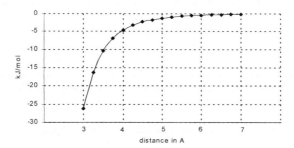

FIGURE 1.9.3. *The dependence of van der Waals energy E_W with the distance for two molecules of decanol at $50°$ C.*

The dependence of the intermolecular energy as a function of distance is described in fact by Lenard-Jones equation, which includes a repulsive term in addition to the negative attraction term given by rel. (1.9.13). Because the attraction term E_W is a function of $1 / r^6$, the total energy can be written in the form:

$$E(r) = - \frac{a}{r^6} + \frac{b}{r^{12}} \tag{1.9.14}$$

For this case the graph showing the energy between two decanol molecules as a function of distance is given in Figure 1.9.4.

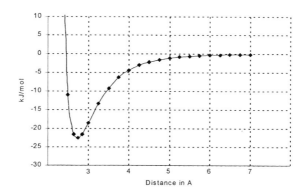

FIGURE 1.9.4. *The dependence of van der Waals energy E_W with the distance for two molecules of decanol at 50°, including a repulsion term.*

The energy of van der Waals interactions depends, as shown from rel. (1.9.6), (1.9.8), and (1.9.9), on the distance r and also on the dipole moment μ, polarizability α, and the ionization potential I.

- Hydrogen bonds and charge transfer interactions

Another important type of molecular interaction is caused by the hydrogen bonds. The hydrogen bond appears to be partly electrostatic and partly covalent. In this bond, a hydrogen atom covalently bonded by atom A is attracted by an atom containing a free electron pair :B. This leads to a strong polarization of the A-H bond and to electrostatic interactions between $H^{\delta+}$ and :B. The average energy of a hydrogen bond from an OH group is 20–25 kJ/mol [39] and is 8–12 kJ/mol for NH_2 groups. In certain systems, this energy can be much higher, approaching 150 kJ/mol, which is comparable to that of a covalent bond (210–420 kJ/mol). Some values for the energy of different types of interactions are given in Table 1.9.1 [41a].

TABLE 1.9.1. *Typical values for different interaction energies [40].*

Interaction type	Energy (kcal/mol)
dispersion	2 - 7
dipole – induced dipole	1 – 2
dipole – dipole	1 – 3
hydrogen bonding	5 – 10
ionic	50 – 200
covalent	50 – 100

The groups involved in a hydrogen bond may have neutral, acidic, or basic character, and both donor and acceptor atoms can be involved in hydrogen bonding. In the hydrogen bond involving oxygen atoms, for example, the bonds can be of the form $O^{+} \cdots H \cdots O$, $O \cdots H \cdots O$, or $O \cdots H \cdots O$. These are in general strong bonds, and the classification as acidic or basic hydrogen bonds is sometimes used in characterizing solute/solvent interactions.

Similar to hydrogen bonding is the type of interaction occurring between electron pair donor and electron pair acceptor compounds (EPD-EPA or charge transfer interaction). This interaction leads to complexes with a mesomeric structure between one formed by two noninteracting molecules (except for van der Waals forces) D and A, and one with two components with ionic interactions D^+A^-. The electron donor molecules (Lewis bases) and the electron acceptor molecules (Lewis acids) can be n, σ, or π donors or respectively acceptors. Their interacting energy can be as low as about 10 kJ/mol but can reach up to 180 kJ/mol. An example of this type of interaction is that between tetracyanoethylene and an alkylbenzene, the formula of the complex being written as a resonance between two structures:

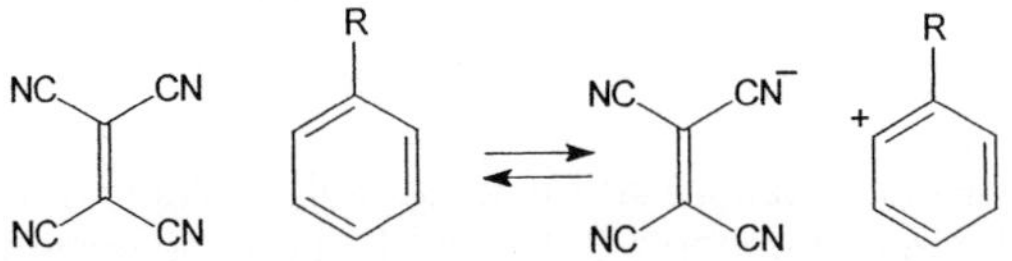

The addition of tetracyanoethylene to the stationary phase for the improvement of a gas chromatographic separation of alkylbenzenes has been reported [42]. Strong Lewis acids and strong Lewis bases have the tendency to form bonds mainly based on electrostatic interactions as shown for tetracyanoethylene and alkylbenzenes. The donor acceptor type of bond can also take place between compounds with high polarizability, although they are weak Lewis acids and bases.

- Solvophobic and other interactions

The solvophobic interactions are commonly invoked for explaining the higher solubility of a nonpolar compound in a nonpolar solvent than in a polar one, although the interactions at the molecular level may indicate that the energy between a nonpolar molecule with another nonpolar molecule is smaller (in absolute value) than between a nonpolar molecule and a polar one. In order to dissolve a nonpolar molecule in a polar solvent, a nonpolar $\longleftrightarrow$ polar interaction must replace a polar $\longleftrightarrow$ polar one between the solvent molecules. Therefore, it is favored energetically to have polar $\longleftrightarrow$ polar and nonpolar $\longleftrightarrow$ nonpolar interactions instead of nonpolar $\longleftrightarrow$ polar ones. The formalism of solvophobic interactions [43] is important mainly for HPLC separations. A measure of the interaction between a molecule S_{sol} in solution and a hydrophobic stationary phase L can be obtained [44] from the equilibrium constant K for the process:

$$S_{sol} + L \rightleftharpoons SL + sol$$

In this equilibrium, the stationary phase is viewed as a hydrophobic ligand L, while the molecules of the solvent X are indicated only as bulk (sol). The equilibrium constant of the process, $K = [SL] / [S_{sol}] [L]$, can be evaluated based on the classical expression:

$$\ln K = -\frac{\Delta F^0}{RT}$$

(1.9.15)

where ΔF^0 is the free energy of the process (which replaces ΔG^0 for equilibria that take place at constant volume), R is the gas constant ($R = 8.31451$ J deg^{-1} mol^{-1}), and T is the temperature (in deg. Kelvin). The free energy ΔF^0 of the interaction between S_{sol} and L can be conceptually separated in two components [45]. One component is the interaction between S and L molecules in an ideal gas phase. The other component comes from the interactions of individual molecules S, L and SL with the bulk solvent. The first component ΔF^0_W for the interaction in gas phase can be estimated from van der Waals forces using $\Delta F^0_W \approx E_W$ (see rel. 1.9.13). The free energy change due to the interaction with the solvent for a molecule A (where A can be S, L, or SL) consists of three terms. One term accounts for van der Waals interactions (in gas phase) $\Delta F_{X,A}$ between the molecule X of the solvent and the molecule A, another term $\Delta F_{cav, A}$ accounts for the creation of the cavity needed to accommodate the molecule A in the solvent, and a third term accounts for the change in the free volume of the system. Therefore, the term for the interaction of a molecule A with the solvent has the expression:

$$\Delta F^0_A = \Delta F_{cav, A} + \Delta F_{X,A} + RT \ln (RT / p_0 V) \tag{1.9.16}$$

where p_0 is the atmospheric pressure and V is the molar volume of the solvent. By use of rel. (1.9.16) for all three components S, L and SL, it can be written:

$$\Delta F^0 = \Delta F^0_W - (\Delta F_{cav, S} + \Delta F_{X,S}) - (\Delta F_{cav, L} + \Delta F_{X,L}) + (\Delta F_{cav, SL} + \Delta F_{X,SL}) - RT \ln (RT / p_0 V)$$

Based on this relation, it was calculated [44] that the value for $RT \ln K$ has the form:

$$RT \ln K = - \Delta F^0_W + \Delta F_{X,S} - N (\lambda - 1) \mu_S^2 D \mathcal{P} / (2\lambda V_S) + N \Delta A \gamma + 4.836 N^{1/3} (\kappa - 1) V^{2/3} \gamma +$$
$$+ RT \ln (RT / p_0 V) \tag{1.9.17}$$

where N is Avogadro's number, λ is a proportionality factor between the molecular volume of the SL complex v_{SL} and the molecular volume of S the solute v_S ($v_{SL} = \lambda v_S$), μ_S is the dipole moment of the solute, and D and $\mathcal{P}$ are given by the expressions:

$$D = \frac{2(\varepsilon - 1)}{2\varepsilon + 1} \qquad\qquad \mathcal{P} = \frac{1}{4 \pi \varepsilon_0 (1 - D \alpha_S / v_S)} \tag{1.9.18}$$

In rel. (1.9.18) α_S is the polarizability of the solute, and ε is the dielectric constant of the solvent X. The value for ΔA in rel. (1.9.17) is given by the difference in the surface areas of the species S plus L and that of the complex SL:

$$\Delta A = A_S + A_L - A_{SL} \tag{1.9.19}$$

Also, γ is the surface tension of the solvent (at a given temperature T), and κ is a parameter related to the energy required for the formation of a cavity in the solvent [43]. The value of κ depends on the heat of vaporization of the solvent ΔE_{vap}, and is given by:

$$\kappa = \frac{N^{1/3} \Delta E_{vap}}{V^{2/3} \gamma (1 - \dfrac{d \ln \gamma}{d \ln T} - 2/3 \mathcal{A} T)} \tag{1.9.20}$$

where $\mathcal{A}$ is the coefficient of liquid thermal expansion.

The theory of solvophobic interaction as given by previous relations was proven valid for different solvents and solutes. Several useful results were obtained based on this theory, such as the proof that the surface tension of the solvent plays an important role in reverse phase HPLC. The capacity factor k (index "i" omitted) given by rel (1.4.28) is related to K by the expression:

$$\log k = \log K + \log (V_s / V_m) \tag{1.9.21}$$

and the capacity factor k (see Section 1.4) can be related to the surface tension of the solvent by the expression:

$$\log k = a + b\,\gamma \tag{1.9.22}$$

In rel. (1.9.22) **a** is independent of γ, and **b** is constant when κ remains constant.

The finite interaction between molecules is not a new type of force but a technique that attempts to account for the energy between a macromolecular structure and "N" small molecules without direct calculation of all local interactions. However, this type of calculation may be able to account for contributions to the total energy that are neglected when only molecule-to-molecule interactions are considered [41]. The total interaction energy with a macromolecular structure for an average distance r between a small molecule and the macromolecule is expressed by the formula:

$$E\,(r) = 2\,\pi\,\frac{N}{V}\left(\frac{a}{12\,r^3} - \frac{b}{90\,r^9}\right) \tag{1.9.23}$$

where a and b are the parameters used in rel. (1.9.14), and **V** is the volume of the considered macromolecular structure.

1.10. EMPIRICAL AND ESTIMATED PARAMETERS FOR SAMPLE PREPARATION PROBLEMS

Sample preparation problems require knowledge of a significant number of parameters, some intrinsic to a specific molecule and some molecular parameters depending on the chemical nature of two or more compounds. The dependence of a parameter on more than one molecular species is, for example, the case for partition coefficients, adsorption constants, solubility parameters, etc. The liquid-liquid partition process, involving a solute and two solvents, is characterized by a partition coefficient depending on three substances. Gas-liquid partition coefficients depend on two compounds, which are the distributed compound and the liquid phase, no influence coming from the chemical nature of the gas. Solubility is determined by the nature of the solvent and that of the solute. Temperature and pressure, which also influence the values of equilibrium constants, add supplementary complexity in the characterization of a chromatographic process. Chemical parameters such as equilibrium constants or reaction rate constants are another group of constants very important for predicting the chemical behavior of a specific analyte.

- Estimation of various parameters for sample preparation problems

A large number of physico-chemical parameters that are difficult to directly measure can be estimated with relatively good accuracy. The estimation of properties particularly important in separations and in sample preparation can be very useful when the experimental values are not available. A list of some parameters pertinent for sample preparations is given in Table 1.10.1.

TABLE 1.10.1. *Parameters of interest in sample preparation estimated by various techniques.*

Parameter	Estimation technique	Reference
solvent characterization	correlations[1]	[46], [47]
parachor	additive parameters	[48], [49]
molar refraction	additive parameters	[48], [49], [50]
boiling point	additive parameters, correlations[1]	[51], [51a]
solubility in water	structural information, correlations[1]	[52]
solubility of a liquid in liquid	thermodynamics	[53]
octanol / water partition	fragment constants, correlations	[54]
solubility of a solid in liquid	thermodynamics	[48]
soil / water partition	correlations[1]	[49]
solubility of a gas in liquid	thermodynamics	[48]
air water partition	semiempirical	[49]
acid dissociation constant	correlations[1]	[54]
rate of hydrolysis	correlations[1]	[48]
activity coefficient	thermodynamics, additive parameters	[55], [56]
heat (enthalpy) of vaporization	semiempirical, Clausius-Clapeyron equation (Section 1.3)	[57], [58]
vapor pressure	thermodynamics, Clausius-Clapeyron equation	[48], [49]
volatilization from water	semiempirical	[59]
diffusion coefficients	semiempirical	[60], [48]
surface tension	semiempirical	[61]
heat capacity	additive parameters	[62], [63], [64]
dipole moment	semiempirical	[65]

[1] Note: "correlations" indicate correlation with other physical parameters.

The estimation using thermodynamic relations is one of the most common procedures. These relations allow the determination of an unknown value based on some other parameters with known values. For example, the heat of vaporization can be estimated based on Trouton's rule. This rule states that most nonpolar liquids have about the same molar entropy of vaporization at their normal boiling points. This is caused by the fact that the increase in disorder in changing a mole of a liquid into a gas is about the same for every substance. Trouton's rule can be written as follows:

$$\Delta S_{vap} = 88 \pm 5 \text{ J deg}^{-1} \text{ mol}^{-1} \tag{1.10.1}$$

Based on the relation between the entropy and the enthalpy of vaporization, the formula for the calculation of the heat of vaporization can be written as follows:

$$\Delta H_{vap} = T_b \, \Delta S_{vap} = 88 \, T_b \text{ J mol}^{-1} \tag{1.10.2}$$

where T_b is the boiling point of the substance (in Kelvin deg.). For benzene, for example, the boiling point (bp) is 80.1° C = 353.25° K. The calculated ΔH_{vap} is 31.09 kJ/mol compared to 30.8 kJ/mol. However, the results for polar liquids such as water are larger than those predicted by Trouton's rule. It can be seen in this example, that thermodynamic relations cannot be applied without restrictions.

One other alternative for obtaining estimations is the use of extrathermodynamic methods [48], [39], [49]. Extrathermodynamic estimations apply a variety of procedures that do not have always a theoretical base. These procedures can be grouped as follows: a) procedures that correlate two physical properties, one known and the other to be predicted, b) procedures that correlate a molecular structural information with the physical property to be predicted, c) procedures using molecular fragments constants, d) semiempirical methods that use calculations with a correct theoretical background, but adding arbitrary adjustments, and e) procedures using thermodynamic relations but including modified parameters for a best fit of the expected property.

Multiple linear regression on a known set of compounds is commonly used to establish the "best fit" parameters in a given predicting technique. Extension of a known property in a group of isomers or a homolog series leads to useful applications. Group contribution models, various similarity models, etc. also are used for predicting physical properties.

The extrathermodynamic predicting procedures frequently use additive structural parameters such as the carbon number for an organic molecule, the molecular mass, the parachor, the molar volume, van der Waals volume, molar refractivity, and polarizability. These additive parameters describe structural or thermodynamic properties. Besides the additive parameters, some structural or topological indices are commonly needed to predict a certain physical property. Some of these indices are related to the molecule structural formula, such as the connectivity indices, or depend on the functional groups in the molecule.

For example, for a compound with the molecular weight M, the parachor Pc can be calculated from atomic and bond contributions using a relation of the type:

$$Pc = \Sigma \text{ (atomic contributions)} + \Sigma \text{ (bond contributions)} \qquad (1.10.3)$$

The parachor is defined by relation:

$$Pc = M \, \gamma^{1/4} / (\rho_l - \rho_g) \qquad (1.10.4)$$

where γ is the surface tension of the liquid, and ρ_l and ρ_g are the densities of the liquid and of the vapors in equilibrium at the given temperature. The molar refraction Rm, which is defined by the relation:

$$Rm = \frac{n^2 - 1}{n^2 + 2} \frac{M}{\rho}$$

$$(1.10.5)$$

where n is the refractive index and d the density of the compound, can be obtained using the expression:

$$Rm = \Sigma \text{ (atomic contributions)} + \Sigma \text{ (bond contributions)} \qquad (1.10.6)$$

Some atomic and bond contributions for the calculation of the parachor and molar refraction are given in Table 1.10.2.

TABLE 1.10.2. *Atomic and bond contributions for parachor and molar refraction.*

Atomic Contribution	Parachor Pc	Molar Refraction Rm	Bond Contribution	Parachor Pc	Molar Refraction Rm
C (single bond)	9.2	2.418	3-member ring	12.5	0
H	15.4	1.1	4-member ring	6	0
O (hydroxyl)	20	1.525	5-member ring	3	0
O (ether)	20	1.643	6-member ring	0.8	0
O (carbonyl, double bond)	39	2.211	7-member ring	-4	0
2 O (esters)	54.8	3.736	double bond	19	1.733
F	25.5	0.95	polar double bond	0	1.733
Cl	55	5.967	triple bond	38	2.398
Br	69	8.865			
I	90	13.9			
N (primary amine)	17.5	2.322			
N (secondary amine)	17.5	2.502			
N (*tert* amine, or in ring)	17.5	2.84			
N (nitrile, triple bond)	55.5	5.516			
S	50	7.69			
P	40.5	various			
Si	70	6.1			

The parachor in addition to other parameters is used in the calculation of various physical properties. For example the boiling point T_b (bp) of a liquid can be calculated using the relation:

$$T_b \text{ (in deg. C)} = (637 \, Rm^{1.47} + B) / Pc - 273.15 \qquad (1.10.7)$$

where *B* is a structural parameter and has the values given in Table 1.10.3. The accuracy of the results in the boiling point calculation depends significantly on the compound structure.

TABLE 1.10.3. *The values of parameter B for the calculation of boiling point.*

Group Contribution	B	Group Contribution	B	Group Contribution	B
carboxyl	28000	esters	15000	aromatic hydrocarbons	-2500
hydroxyl	16500	ethers and mercaptans	-2000	olefins	-4500
amino (primary)	6500	silyl ether	-5000	carbonyl	15000
amino (secondary)	2000	alkanes	-2500	monochlorinated alkanes	4000
amino (tertiary)	-3000	acetylenic hydrocarbons	-500	nitriles	20000

The boiling point is an important parameter for sample preparation because it is related to distillation and also to the calculation of partial vapor pressure of a compound.

One other parameter important for the choice of a sample preparation procedure is the diffusion coefficient $D_{A,B}$. For gases, this parameter can be estimated using the expression:

$$D_{A,B} = \frac{10^{-3} T^{1.75}}{p(\overline{V}_A^{1/3} + \overline{V}_B^{1/3})^2} (\frac{1}{M_A} + \frac{1}{M_B})^{1/2} \qquad (1.10.8)$$

where $\overline{V}_A$ and $\overline{V}_B$ are the molecular diffusion volumes (in $cm^3 \, mole^{-1}$), M_A and M_B the molecular weights of the two gases, T temperature in Kelvin degrees, and p the pressure. Tables for the calculation of the molecular diffusion volumes $\overline{V}$ are available in literature [15], [48], [60]. For liquids, there are also various formulas for the calculation of the diffusion coefficient, such as rel. (1.5.23). For diffusion in water the following expression gives results within 5.9% relative error:

$$D_{A,w} = \frac{1.4 \cdot 10^{-5}}{\eta_w^{1.1} \overline{V}_A^{0.6}} \qquad (1.10.9)$$

where η_w is water viscosity (in centipoise) at a desired temperature [48].

Another parameter that is frequently evaluated using extrathermodynamic estimation procedures is the octanol/water partition coefficient K_{ow}. The experimental values for K_{ow} are also known for many compounds, and extensive tables are available [66], [66a]. However, a significant effort was made in developing a procedure for its estimation. This parameter is relevant in drug bioavailability problems, and a number of other parameters such as solubility in various solvents and distribution coefficients can be estimated from K_{ow} values. One well-known procedure for this purpose uses the fragment methodology [54], [66], [66a]. In the fragment procedure, the calculation of log K_{ow} uses the relation:

$$\log K_{ow} = \Sigma\, a_n\, f_n + \Sigma\, b_n\, F_n \qquad (1.10.10a)$$

where a_n is the number of occurrences of a fragment, f_n is a constant for the particular fragment, b_n is the number of occurrences when a correction factor is needed, and F_n is a correction factor for a structural feature in the molecule.

The estimation of K_{ow} using rel. (1.10.10a) begins with the identification of fragments in the molecule. The fragments are usually classified as single atoms, group of atoms that form common fragments such as -OH, -NH$_2$, -COOH, etc., larger fragments, and special fragments that are involved in hydrogen bonding. The values for f_n depend, besides the fragment itself, on the atoms or structures to which the fragment is connected. The connection to an aromatic ring (noted ϕ), two aromatic rings (noted $\phi\phi$), or groups such as vinyl, benzyl, or special electron-withdrawing groups lead to different f_n values for the same fragment. Some values for f_n for a number of fragments are given in Tables 1.10.4 to 1.10.7 [48]. These tables do not include the values for parameter f_n when the connection occurs with groups such as vinyl, benzyl, and special electron-withdrawing groups, and these values can be found in the dedicated literature [48].

TABLE 1.10.4. *The values of the parameter[1] f_n for fragments not containing C and H.*

Fragment	f	f^ϕ	$f^{\phi\phi}$	Fragment	f	f^ϕ	$f^{\phi\phi}$
-F	-0.38	0.37		-SP(S)O$_2$<	-2.89		
-Cl	0.06	0.94		-SO$_2$ F		0.30	
-Br	0.20	1.09		-SO$_2$N<		-2.09	
-I	0.59	1.35		-S(O)-	-3.01	-2.12	-1.62
-N<	-2.18	-0.93	-0.50	-SO$_2$ -	-2.67	-2.17	-1.28
-O-	-1.82	-0.61	0.53	-SO$_2$O-	-2.11	-2.06	-0.62
-S-	-0.79	-0.03	0.77	-SF$_5$		1.45	
-NO		0.11		-SO$_2$O$^-$	-5.87	-4.53	
-NO$_2$	-1.16	-0.03		-OSO$_3^-$	-5.23		
-ONO$_2$	-0.36			-N=N-			0.14
-IO$_2$		-3.23		-NNN-		0.69	
-OP(O)O$_2$<	-2.29	-1.71		-N=NN<		-0.85	
-P(O)O$_2$<		-2.33		>NNO	-2.40	-0.84	
-OP(S)O$_2$<,		-0.30		-O$^-$		-3.64	
>NP(S)(N<)$_2$	-3.37			-Si≡	-0.09	0.65	

[1] Note: f indicates an aliphatic structural attachment, f^ϕ attachment to an aromatic ring, and $f^{\phi\phi}$ attachment to two aromatic rings for bivalent fragments.

TABLE 1.10.5. *The values of the parameter[1] f_n for fragments not containing C or not containing H.*

Fragment	f	f^ϕ	$f^{\phi\phi}$	Fragment	f	f^ϕ	$f^{\phi\phi}$
-H	0.23	0.23		-As(OH)$_2$O-		-1.84	
-NH-	-2.15	-1.03	-0.09	-As(O)(OH)$_2$		-1.90	
-NH$_2$	-1.54	-1.00		-B(OH)$_2$		-0.32	
-OH	-1.64	-0.44		>C<	0.20		
-SH	-0.23	0.62		-CF$_3$		1.11	
-SO$_2$NH-		-1.75	-1.10	-CN	-1.27	-0.34	
-SO$_2$(NH$_2$)		-1.59		-C(O)N<	-3.04	-2.80	-1.93
-SO$_2$NH(NH$_2$)	-2.04			-SCN	-0.48	0.64	
-NHSO$_2$ (NH$_2$)		-1.50		-C(O)-	-1.90	-1.09	-0.50
-NH(OH)		-1.11		-C(O)O-	-1.49	-0.56	-0.09
-NHNH-			-0.74	-C(O)O⁻	-5.19	-4.13	
-NH(NH$_2$)		-0.65		-N=CCl$_2$		0.64	
-SP(O)(O-)NH-	-2.18			-OC(O)N<	-2.54	-1.84	
-SP(O)(NH$_2$)O-	-2.50			-C(=S)O-	-1.11		

[1] Note: f indicates an aliphatic structural attachment, f^ϕ attachment to an aromatic ring, and $f^{\phi\phi}$ attachment to two aromatic rings for bivalent fragments.

TABLE 1.10.6. *The values of the parameter[1] f_n for fragments with C and H.*

Fragment	f	f^ϕ	$f^{\phi\phi}$	Fragment	f	f^ϕ	$f^{\phi\phi}$
-CH$_3$	0.89	0.89		-OC(O)H	-1.14	-0.64	
-C$_6$H$_5$	1.90			-NHC(O)H		-0.64	
-C(O)H	-1.10	-0.42		-C=NOH(OH)		-1.64	
-C(O)OH	-1.11	-0.03		-C(=S)NH$_2$		-0.41	
-C(O)NH-	-2.71	-1.81	-1.06	-N(C(O)NH$_2$)-		-2.25	-2.07
-C(O)NH$_2$	-2.18	-1.26		-SO$_2$NHN=CH-			
-OC(O)NH-	-1.79	-1.46		-NHC(=S)NH-		-1.79	
-OC(O)NH$_2$	-1.58	-0.82		-NNO(C(O)NH$_2$)	-0.95		
-CH=N-		-1.03	0.08	-C(O)NHNH$_2$		-1.69	
-CH=NOH	-1.02	-0.15		-NHC(=S)NH$_2$	-1.29	-1.17	
-CH=NNH-	-2.75			-CNH$_2$(=NH· HCl)		-3.49	
-NHC(O)NH-	-2.18	-1.57	-0.82	-NHC=NH(NH$_2$)	-5.65		
-NHC(O)NH$_2$	-2.18	-1.07		-C(O)C(O)-	-3.00		-0.30
>NC(O)NH$_2$		-2.25	-2.15	-C(O)NHC(O)-	-3.31		-3.00
>C=NH			-1.29	-C(O)NHC(O)H	-2.84		
>NC(O)H	-2.67	-1.59		-C(O)NHN=CH-			-1.12
-OC(O)NH-	-1.79	-1.45		-C(O)NHC(O)NH$_2$	-1.91		
-C(=S)NH-	-2.00			-CH(NH$_2$)C(O)OH	-3.97		
-NHCN		-0.03		-CH=NNHC(O)NH$_2$	-0.63	-0.66	
-CH=NN<		-1.71		-CH=NNHC(=S)NH$_2$		-0.05	
-NHC(O)N<		-2.29		-CH=NNHC(O)NHNH$_2$		-1.09	
-NNO(C(O)NH-)	-1.50			-C(O)NHC(O)NHC(O)-	-2.38		

[1] Note: f indicates an aliphatic structural attachment, f^ϕ attachment to an aromatic ring, and $f^{\phi\phi}$ attachment to two aromatic rings for bivalent fragments.

TABLE 1.10.7. *The values of the parameter[1] f_n for fragments in an aromatic ring.*

Fragment (no C)	$\underline{f}^\phi$	Fragment (with C)	$\underline{f}^\phi$
-N=	-1.12	C	0.13
-N<	-1.10	C (ring fusion C)	0.22
ϕ-N<	-0.56	C (ring fusion hetero)	0.44
-N=N-	-2.14	CH	0.355
>N-O-	-3.46	-C(O)-	-0.59
-O-	-0.08	-OC(O)-	-1.40
-S-	0.36	-CH=NNH-	-0.47
>S=O	-2.08	-N=CHNH-	-0.79
-Se-	0.45	-NHC(O)-	-2.00
-NH-	-0.65	-N=CH-O-	-0.71
-NHN=N-	-0.86	-N=CH-S-	-0.29
		-CH=N-O-	-0.63
		-N=CHN=	-1.46
		-NHC(O)NH-	-1.18
		-C(O)NHC(O)-	-1.08
		-C(O)NHC(O)NH-	-1.78
		-C(O)NHC(O)NHN=	-1.36

[1] Note: $\underline{f}^\phi$ indicates fragments in the ring.

The correction factors F_n are necessary for geometric feature, unsaturation, multiple halogenation, involvement in H-polar proximity, or involvement in intramolecular hydrogen bonding. Some values for F_n are given in Table 1.10.8 [48].

TABLE 1.10.8. *The values of the correction factor[1] F_n.*

	Geometric feature (and symbol)	F_n		Unsaturation (and symbol)	F_n
F_b	bond in chain ($b_n = n - 1$)	-0.12	$F_=$	double bond	-0.55
F_{rb}	bond in ring ($b_n = n - 1$)	-0.09	$F_\equiv$	triple bond	-1.42
F_{cBr}	branching alkane chain	-0.13	$F_=^\phi$	double bond conjugate to ϕ	-0.42
F_{HBr}	branching H-polar fragment (e.g. OH)	-0.22	$F_=^{\phi\phi}$	double bond conjugate to $\phi\phi$	0.0
F_{rCl}	branching ring cluster	-0.45	$F_=^=$	double bond conjugate to a second double bond in chain	-0.38
F_{bN}	branching amine	-0.20	$F_\equiv^{\phi\phi}$	triple bond conjugate to $\phi\phi$	0.0
F_{bP}	branching phosphorous ester	-0.31			

	Multiple halogenation (and symbol)	F_n		Two H-polar fragments proximity (and symbol)	F_n
F_{h2}	on the same carbon n=2	0.30	F_{P1}	chain 1 carbon separation	-0.42 ($f_1 + f_2$)
F_{h3}	on the same carbon n=3	0.53	F_{P2}	chain 2 carbon separation	-0.26 ($f_1 + f_2$)
F_{h4}	on the same carbon n=4	0.72	F_{P3}	chain 3 carbon separation	-0.10 ($f_1 + f_2$)
F_{hv}	on adjacent carbons ($b_n = n-1$)	0.28	F_{rP1}^c	aliphatic ring 1 carbon sep.	-0.32 ($f_1 + f_2$)
			F_{rP2}^c	aliphatic ring 2 carbon sep.	-0.20 ($f_1 + f_2$)
	Intramolecular H-bond (and symbol)	F_n	F_{P1}^ϕ	aromatic ring 1 carbon sep.	-0.16 ($f_1 + f_2$)
			F_{P2}^ϕ	aromatic ring 2 carbon sep.	-0.08 ($f_1 + f_2$)
F_{HN}	for nitrogen	0.60			
F_{HO}	for oxygen	1.0			

[1] Note: bonds to hydrogen of any atom are not counted.

The calculation of log K_{ow} starts with identification of the fragments, and if parameters for a fragment are known, the fragment is not calculated from components. For example, the fragment -C(O)O- in esters should not be computed but simply used from the tables. After the fragments are identified, the structural correction factors are identified based on the molecule structure. All parameters are added as indicated by rel. (1.10.10a). The calculation for three compounds is exemplified in Table 1.10.9 [48].

TABLE 1.10.9. *Exemplification of the calculation of K_{ow} [48].*

H₃C—C(=O)—O—C₂H₅	log K_{ow} = 3 f_C + 8 f_H + $f_{C(O)O}$ + (3-1) F_b = 3 (0.20) + 8 (0.23) + (-1.49) + 2 (-0.12) = 0.69 Experimental log K_{ow} = 0.73
(CH₃)₂CH—CH(CH₃)₂ structure	log K_{ow} = 6 f_C + 14 f_H + (5-1) F_b + 2 F_{cBr} = 6 (0.20) + 14 (0.23) + 4 (-0.12) + 2 (-0.13) = 3.68 Experimental log K_{ow} = 3.85
benzodioxole structure	log K_{ow} = 4 f^ϕ_{CH} + 2 f^ϕ_{C} ring fussion hetero + 2 f_H + 2 f^ϕ_O + f_C + (4-1) F_{rb} + F_{rP1} + $F_{P2}{}^\phi$ = 4 (0.355) + 2 (0.44) + 2 (0.23) + 2 (-0.61) + 0.20 + 3 (-0.09) + (-0.32) (-0.61 - 0.61) + (-0.08) (-0.61 - 0.61) = 1.96 Experimental log K_{ow} = 2.08

The calculation of log K_{ow} for a compound similar to another that has a K_{ow} value already known can be calculated using the expression:

$$\log K_{ow} \text{(new)} = \log K_{ow} \text{(known)} \pm a_n f_n \pm b_n F_n \qquad (1.10.10b)$$

where the fragments and the corrections are subtracted and/or added appropriately to change the known compound into the new one.

The success of this type of calculation for K_{ow} is demonstrated in Figure 1.10.1, which shows the variation of calculated K_{ow} as a function of experimental K_{ow} for 76 compounds.

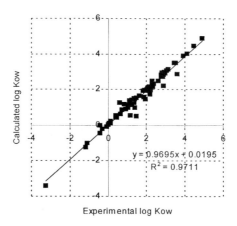

FIGURE 1.10.1. *Variation of calculated K_{ow} as a function of experimental K_{ow} for 76 compounds.*

Dedicated computer programs are available for the calculation of K_{ow} from structural data, some following the fragment procedure described in this section [49].

- Parameters for the characterization of solubility and solvents

The dependence between K_{ow} and solubility in water of various solutes has been frequently investigated. A number of regression equations between K_{ow} and **S** (solubility in water) are reported in literature, the most commonly proposed expression being of the form:

$$\log (1 / S) = a \log K_{ow} + b \qquad (1.10.11)$$

Table 1.10.10 shows some a and b values used in equation (1.10.11) for different classes of compounds.

TABLE 1.10.10. *The values of parameters a and b in regression equation 1.10.11 for different classes of compounds [48].*

Class of compounds	a	b	units for S
aromatic and chlorinated hydrocarbons, etc.	1.37	-7.26	µmol/L
mixed classes	0.922	-4.184	mg/L
mixed classes	1.49	-7.46	µmol/L
alcohols	1.113	-0.926	mol/L
ketones	1.229	-0.72	mol/L
esters	1.013	-0.520	mol/L
ethers	1.182	-0.935	mol/L
alkyl halides	1.221	-0.832	mol/L
aromatics	0.996	-0.339	mol/L
alkanes	1.237	0.248	mol/L
mixed classes	1.214	-0.85	mol/L
mixed classes	1.339	-0.978	mol/L

More complicated than solubility is the problem of extraction of one compound "i" from a solvent into another solvent. The main difficulty comes for the fact that there are three components in such a system (solute "i", solvent "A", and solvent "B"). Considerable utility can be found for a simplified parameter that would characterize the behavior of a certain compound in a separation, regardless the second partner. One such parameter is the solvent polarity P'. This parameter is frequently used for describing the behavior of a solvent in liquid-liquid chromatography [46], [47]. For a given solvent S, the calculation of polarity P' starts with the experimental distribution coefficient for four solutes, n-octane, ethanol, dioxane, and nitromethane. With a slightly modified notation for the distribution coefficient, rel. (1.4.26) can be written as follows:

$$K_S = c_{i,S} / c_{i,g} \qquad (1.10.12)$$

Values for K_S were reported for a variety of systems involving liquid substances [21], [22]. Experimentally, the distribution constants can be measured by adding a volume of test solution (5 µL mixture n-octane, ethanol, etc.) in 2 mL solvent S placed in a closed vial of a specific volume (13.4 mL). After equilibration, the composition of the gas phase can be measured and K_S values calculated [21]. These K_S values are used in the calculation of a modified constant that is intended to eliminate the effect of the solvent molecular weight. The modification is done by use of the solvent molar volume V_S (mL/mole) in the expression:

$$K'_S = K_S V_S \qquad (1.10.13)$$

The values K'_S are then used to calculate the coefficients K''_S, which are obtained with further correction of K'_S values for the molecular weight of the solute. This is done with the relation:

$$K''_S = K'_S / K'_v \tag{1.10.14}$$

where K'_v is the estimated K' value of an n-alkane whose molar volume is the same as that of the solute. The values of K'_v are calculated using the expression:

$$\log K'_v = (V_i / 163) \log K_{octane,S} \tag{1.10.15}$$

In rel. (1.10.15) V_i is the molar volume of the solute (ethanol, dioxane, or nitromethane), and $K_{octane,S}$ is the experimental distribution coefficient of n-octane in the evaluated solvent. The constants K_S'' are further corrected to have zero value for n-hexane as a solvent. The resulting constants K_S'' are used to measure the excess retention of a solute relative to an n-alkane of equivalent molar volume. A polarity parameter P' is then defined by the expression:

$$P' = \log K''_{ethanol,S} + \log K''_{dioxane,S} + \log K''_{nitromethane,S} \tag{1.10.16}$$

Larger values for P' indicate a polar solvent (such as alcohol or water), and values close to zero show nonpolar solvents such as hexane, cyclohexane, etc. Solvent polarity is used in selecting solvents in LC separations and can be a measure of the behavior of a given volatile compound in GC separations. However, parameter P' is not always sufficient for the characterization of solvent properties. The types of interactions that dominate solvent behavior can be quite different between solvents with the same P'. For example, a polar solvent and a solvent forming hydrogen bonds, although they may have identical P', may not act in the same manner toward different solutes. An additional parameter x_i was developed for solvent characterization, defined by the formula:

$$x_i = \log K''_{i,S} / P' \tag{1.10.17}$$

where "i" can be ethanol (x_e), dioxane (x_d), nitromethane (x_n), toluene (x_t), or methyl ethyl ketone (x_m). It can be assumed that the larger is x_i value for a specific compound, the higher is the similarity with the comparing solvent. However, the value of x_i also depends on P', and relatively large K_S'' do not necessarily lead to large x_i values. For this reason, the values for x_i were used to group the solvents in nine main groups, the solvents in the same group having similar properties. These groups are a) solvents with very low P' values (nonpolar), b) aliphatic ethers, tetramethylguanidine, hexamethyl-phosphoric acid amide, c) aliphatic alcohols, d) pyridine derivatives, tetrahydrofuran, amides, glycol ethers, sulfoxides, e) glycols, benzyl alcohol, acetic acid, formamide, f) methylene chloride, ethylene chloride, g) tricresyl phosphate, aliphatic ketones and esters, dioxane, h) aromatic hydrocarbons, halo-substituted aromatic hydrocarbons, nitro compounds, aromatic ethers, and i) fluoroalkanols, m-cresol, water, chloroform. Values for P' for some solvents classified in these groups are given in Tables 1.10.11 to 1.10.19 [47].

TABLE 1.10.11. *Parameters for solvent characterization in the group of nonpolar solvents [47].*

Compound	P'	Compound	P'
n-hexane	0.1	n-decane	0.4
iso-octane	0.1	squalane	1.2
cyclohexane	0.2	carbon tetrachloride	1.6
carbon disulfide	0.3		

TABLE 1.10.12. *Parameters for solvent characterization in the group of aliphatic ethers, tetramethylguanidine, and hexamethylphosphoric acid amide [47].*

Compound	P'	Compound	P'
triethyl amine	1.9	ethyl ether	2.8
butyl ether	2.1	tetramethyl guanidine	6.1
i-propyl ether	2.4	hexamethyl phosphoric acid triamide	7.4

TABLE 1.10.13. *Parameters for solvent characterization in the group of aliphatic alcohols [47].*

Compound	P'	Compound	P'
n-octanol	3.4	n-propanol	4
i-pentanol	3.7	tert-butanol	4.1
n-butanol	3.9	ethanol	4.3
i-propanol	3.9	methanol	5.1

TABLE 1.10.14. *Parameters for solvent characterization in the group of pyridine derivatives, tetrahydrofuran, amides, glycol ethers, and sulfoxides [47].*

Compound	P'	Compound	P'
tetrahydrofuran	4	triethylene glycol	5.6
2,6-lutidine	4.5	tetramethyl urea	6
2-picoline	4.9	methyl formamide	6
quinoline	5	dimethyl formamide	6.4
diethylene glycol	5.2	N,N-dimethyl acetamide	6.5
pyridine	5.3	N-methyl-2-pyrrolidone	6.7
methoxy ethanol	5.5	dimethyl sulfoxide	7.2

TABLE 1.10.15. *Parameters for solvent characterization in the group of glycols, benzyl alcohol, acetic acid, and formamide [47].*

Compound	P'	Compound	P'
benzyl alcohol	5.7	ethylene glycol	6.9
acetic acid	6	formamide	9.6

TABLE 1.10.16. *Parameters for solvent characterization in the group of methylene chlorid and ethylene chloride [47].*

Compound	P'	Compound	P'
methylene chloride	3.1	ethylene chloride	3.5

TABLE 1.10.17. *Parameters for solvent characterization in the group of tricresyl phosphate, aliphatic ketones and esters, and dioxane [47].*

Compound	P'	Compound	P'
ethyl acetate	4.4	cyanomorpholine	5.5
bis-(2-ethoxy ethyl) ether	4.6	acetonitrile	5.8
tricresyl phosphate	4.6	propylene carbonate	6.1
methyl ethyl ketone	4.7	aniline	6.3
cyclohexanone	4.7	formyl morpholine	6.4
dioxane	4.8	butyrolactone	6.5
benzonitrile	4.8	tris-cyanoethoxypropane	6.6
acetophenone	4.8	oxydipropionitrile	6.8
acetone	5.1	tetrahydrothiophene-1,1-dioxide	6.9

TABLE 1.10.18. *Parameters for solvent characterization in the group of aromatic hydrocarbons, halo-substituted aromatic hydrocarbons, nitro compounds, and aromatic ethers [47].*

Compound	P'	Compound	P'
toluene	2.4	ethoxybenzene	3.3
p-xylene	2.5	phenyl ether	3.4
benzene	2.7	anisole	3.8
chlorobenzene	2.7	benzyl ether	4.1
bromobenzene	2.7	nitrobenzene	4.4
iodobenzene	2.8	nitroethane	5.2
fluorobenzene	3.2	nitromethane	6

TABLE 1.10.19. *Parameters for solvent characterization in the group of fluoroalkanols, m-cresol, water, and chloroform [47].*

Compound	P'	Compound	P'
chloroform	4.1	dodecafluoroheptanol	8.8
m-cresol	7.4	water	10.2
tetrafluoropropanol	8.6		

The classification of solvents is useful in different utilizations in sample preparation and in solvent selection for chromatographic separations. Further discussion on solvent properties is included in Sections 4.3 and 8.2.

- *Parameters for the characterization of distribution constants in gas chromatography*

A value frequently necessary for the estimation of the behavior of a compound in gas chromatography or in techniques of sample preparation such as headspace is the vapor pressure of a pure liquid. A common procedure for the calculation of vapor pressures is based on Clausius-Clapeyron equation, which can be written in integrated form:

$$\ln (p^{o}_{2} / p^{o}_{1}) = - \Delta H_{vap} (1 / T_{2} - 1 / T_{1}) / R. \qquad (1.10.18)$$

At the boiling point of a solvent, the pressure of the pure liquid is equal with atmospheric pressure, and ΔH_{vap} can be estimated from rel. (1.10.2). Therefore, the vapor pressure of a pure liquid at a temperature T (in Kelvin deg.) can be estimated with the expression:

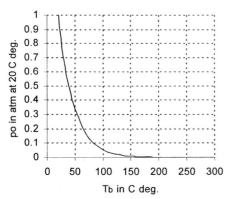

$$\ln p^{\circ} = \Delta S_{vap} (1 - T_b / T) / R \quad (1.10.19)$$

where with Trouton's rule, $\Delta S_{vap} = 88 \pm 5$ J deg^{-1} mol^{-1}. The variation of p° at 20° C as a function of boiling point of the liquid following the approximation given by rel. (1.10.19) is shown in Figure 1.10.2. The agreement of results using rel. (1.10.19) and the experimental data for vapor pressure are relatively modest. A number of improved methods of estimation are

FIGURE 1.10.2. *The variation of p° at 20° C as a function of boiling point of the liquid.*

available [48]. One of these uses a corrected Clausius-Clapeyron equation of the form:

$$\frac{d \ln p}{dT} = \frac{\Delta H_{vap}}{\Delta Z \, RT^{2}} \qquad (1.10.20)$$

where ΔZ is a compressibility factor that is typically taken $\Delta Z = 0.97$. More precise calculation of ΔH_{vap} must take into consideration its variation with temperature. One such equation has the form:

$$\Delta H_{vap} = K_F (8.75 + R \ln T_b) T_b \qquad (1.10.21)$$

where K_F is an empirical parameter with values shown in Table 1.10.20 [48].

TABLE 1.10.20. *The values of the parameter K_F for the calculation of ΔH_{vap}.*

Compound class	K_F for total number of carbon atoms		
	C_2	C_5	C_{10}
hydrocarbons	1.0	1.0	1.0
halides	1.04	1.03	1.02
ketones	-	1.06	1.03
esters	1.09	1.07	1.03
aldehydes	1.09	1.07	1.03
primary amines	1.13	1.1	1.07
secondary amines	1.09	1.07	1.04
tertiary amines	-	1.01	1.01
nitriles	1.05	1.06	1.03
nitro compounds	1.07	1.06	1.03
alcohols	1.31	1.31	1.20
diols	1.33	1.33	-
ethers	1.03	1.02	1.01
thiols	1.03	1.01	1.01

A better approximation for p^o is obtained by use of the expression:

$$\ln p^o = \frac{\Delta H_{vap}}{\Delta Z\, RT_b^2} \frac{(T_b - C_2)^2}{1} [\frac{1}{(T_b - C_2)} - \frac{1}{(T - C_2)}]$$

(1.10.22a)

where $C_2 = -18 + 0.19\, T_b$ and ΔH_{vap} is calculated with rel. (1.10.21). Another estimation used for the calculation of p^o for liquids and also for solid compounds is given by the equation:

$$\ln p^o = \frac{\Delta H_{vap}}{\Delta Z\, RT_b}[1 - \frac{(3 - 2\, T/T_b)^m}{T/T_b} - 2\, m\, (3 - 2\, T/T_b)^{m-1} \ln T/T_b]$$

(1.10.22b)

where $m = 0.19$ for liquids, and $m = 0.36$ for solids with $T/T_b > 0.6$, $m = 0.8$ for solids with $0.6 > T/T_b > 0.5$, and $m = 1.19$ for solids with $T/T_b < 0.5$. The values for ΔH_{vap} and ΔZ are the same as for equation 1.10.22a.

The calculation of vapor pressure of a compound at a specific temperature is useful for the estimation of Henry's constant k. When Henry's law is written in the form:

$$p = k\, c$$

(1.10.23)

where p is the vapor pressure of a compound above the solution and c the molar concentration of the compound in the solution (at equilibrium), the value of k can be calculated knowing p and c. For example, for partition between gas and water, Henry's law takes the form:

$$k = p^o / S$$

(1.10.24)

where S is the solubility in water (expressed in mol/L, with k resulting in atm. L /mol). By use of estimated p^o values as given by rel. (1.10.22a) or (1.10.22b) and estimated values for S as given by rel. (1.10.11), the values for k for a certain system can be calculated. Henry's constant is related to the equilibrium constant K_i between a liquid and a gas for a compound "i" by the expression:

$$K_i = RT / k_i$$

(1.10.25)

By use of rel. (1.10.25), the partition coefficient can be calculated for a given system. The value obtained in this way is only an estimation, which can be quite different from the experimental values. However, for sample preparation purposes these numbers may be sufficient for guidance.

- Parameters for the characterization of distribution constants in liquid chromatography

It was found [67] that the partition coefficients K_{iA} and K_{iB} for a compound "i" in two systems, a) solvent A / water and b) solvent B / water, are related by the expression:

$$\log K_{iA} = a \log K_{iB} + b \qquad (1.10.26)$$

where a and b are constants that are characteristic for the two solvents that are utilized. Rel. (1.10.26) gives excellent results for similar systems such as octanol/water and hexanol/water, but the correlation becomes weaker for systems that are very different such as octanol/water and petroleum ether/water.

The correlation between K_{ow} for a compound "i" and the capacity factor k_i for liquid chromatography in a system with water as a mobile phase follows a relation similar to rel. (1.10.26). In this case, the dependence is written in the form:

$$\log k_i = a \log K_{ow}(i) + b \qquad (1.10.27)$$

The coefficients a and b can be obtained using a regression line. As an example, the dependence of $\log k_i$ vs. $\log K_{ow}$ for 72 mono and disubstituted aromatic compounds,

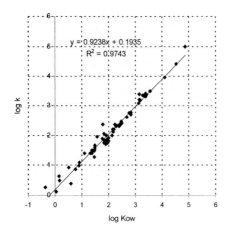

with k_i values obtained for a C18 stationary phase using water as a mobile phase [39], is described by an expression of the form (1.10.27) with a = 0.9238 and b = 0.1935 with R^2 = 0.9743 as shown in Figure 1.10.3 [39]. The same type of correlation between distribution constant k_i and K_{ow} was found for other stationary phases such as polydimethylsiloxane (PDMS) with water as a mobile phase [67a].

FIGURE 1.10.3. *Dependence of log k_i vs. log K_{ow} for 72 mono and disubstituted aromatic compounds with k_i values obtained for a C18 stationary phase with water as a mobile phase [39].*

The use of pure water as a mobile phase in HPLC is not common, and mixtures of solvents are usually utilized in separations. The modification of the mobile phase leads, as expected, to significant variation of k_i. As shown in rel. (1.6.60), the expression that can be used for transforming k_i values for water into k_i values for water + organic solvent can be written in the form:

$$\log k_i (c_M) = \log k_i (c_M = 0) - \varphi'_{i,M} c_M \qquad (1.10.28)$$

where $\varphi'_{i,M}$ is a parameter that depends on the solvent, solute, and stationary phase. However, for the same class of compounds, a dependence between the distribution constants in water and a partially organic solvent can be written in the form:

$$\log k_i (c_M) = a \log k_i (c_M = 0) - b c_M \qquad (1.10.29)$$

where the parameters a and b do not depend on the solute "i" and are constant for a specific solvent. As an example the dependence of k_i in water/methanol (50:50) as a

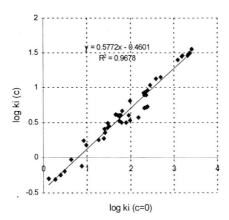

FIGURE 1.10.4. *Dependence of k_i in water/methanol (50:50) as a function of k_i in water on a C18 stationary phase for 48 aromatic compounds [39].*

function of k_i in water on a C18 stationary phase for 48 aromatic compounds is shown in Figure 1.10.4.

A number of other chromatographic parameters were estimated using extra-thermodynamic relations. For example, it has been shown that the logarithm of the aqueous molar solubility can be well correlated with the logarithm of the capacity factor k_i for analytes isolated on a macroporous acrylic resin [67b], [68] from an aqueous solution. The dependence was found of the form $\log k_i = 1.77 - 0.52 \log S_i$. This method ignores stationary phase contributions to retention and is limited by the availability of solubility data. A large number of solubility related phenomena and transfer properties have been characterized using linear solvation energy relations and cavity model approach [69], [70], [71]. In such a model, a free energy related property P is related to a number of parameters through a relation of the form [72]:

$$P = c + m\frac{V_i}{100} + rR_2 + s\pi_2^H + a\alpha_2^H + b\beta_2^H \qquad (1.10.30)$$

where V_i is a parameter characteristic of the size of the solute , R_2 is the solute's excess molar refraction, π_2 is a measure of the solute's ability to stabilize a neighboring dipole by its capacity for orientation and induction interactions, and α and β are parameters characterizing the solute's hydrogen-bond acidity and hydrogen-bond basicity, respectively. V_i can be calculated as the sum of atomic volumes using an expression of the form:

$$V_i = \sum_A n_A V_A \qquad (1.10.31)$$

where n_A is the number of atoms "A" in the molecule, and V_A is the atomic volume of the atom "A". A few atomic volumes are given in Table 1.10.21. Some values for the parameters R_2, π_2^H, α_2^H, and β_2^H are given in Table 1.10.22.

TABLE 1.10.21. *The values of several atomic volumes [72].*

Atom	Atomic volume (cm³/mol)	Atom	Atomic volume (cm³/mol)
carbon	16.35	phosphorus	24.87
hydrogen	8.71	fluorine	10.48
oxygen	12.43	chlorine	20.95
nitrogen	14.39	bromine	26.21
sulfur	22.91	iodine	34.53

TABLE 1.10.22. *The values of R_2, π_2^H, α_2^H, and β_2^H for several compounds [72].*

Compound	R_2	π_2^H	α_2^H	β_2^H
naphthalene	1.340	0.92		0.20
benzene	0.610	0.52		0.14
chlorobenzene	0.718	0.65		0.07
bromobenzene	0.882	0.73		0.09
iodobenzene	1.188	0.82		0.12
1,2-dichlorobenzene	0.872	0.78		0.04
benzaldehyde	0.820	1.00		0.39
hexanal	0.146	0.65		0.45
benzonitrile	0.742	1.11		0.33
anisole	0.708	0.75		0.29
1,4-dioxane	0.329	0.75		0.64
acetophenone	0.818	1.01		0.49
nitrobenzene	0.871	1.11		0.28
3-nitrotoluene	0.874	1.10		0.25
methylbenzoate	0.733	0.85		0.48
cyclohexanone	0.403	0.86		0.56
phenol	0.805	0.89	0.60	0.31
3-cresol	0.820	0.88	0.57	0.34
1-naphthol	1.520	1.05	0.61	0.37
2-chlorophenol	0.853	0.88	0.32	0.31
2-nitroaniline	1.180	1.37	0.30	0.36
benzamide	0.990	1.50	0.49	0.67
acetanilide	0.870	1.40	0.50	0.67

The system constants c, m, r, s, a, and b are solute independent and are characteristic of the stationary phase and the solvent. These parameters are evaluated using multiple linear regressions by determining the property P. Once established, the property P can be estimated for any solute in the same sampling system for which the solute parameters are known or can be estimated from empirical combining rules. For example, the breakthrough volume V_B (see Section 11.1) for particle loaded disks, with styrene-divinylbenzene (SDVB) or with C18 silica-bound polymers and using water with 1% methanol, can be obtained using rel. (1.10.30) and the parameters in Table 1.10.23.

TABLE 1.10.23. *The values of parameters c, m, r, s, a, and b for the calculation of breakthrough volume V_B on styrene-divinylbenzene or C18 silica-bound polymers using water with 1% methanol [73].*

Parameter	SDVB	Silica-C18
c	-0.77	-1.23
m	5.16	5.14
r	0.81	0.0
s	-0.65	-0.92
a	-1.85	-1.05
b	-2.93	-2.24

This type of correlation allows the prediction of separation properties of classes of compounds. However, the success of these predictions is still limited.

Besides various books and papers in the scientific literature, several computer packages are available for the estimation of physico-chemical properties of organic compounds (see e.g. [74]), and their use is beneficial in sample preparation practice.

1.11. INTRODUCTORY DISCUSSION ON PRECISION, ACCURACY, SENSITIVITY, AND LIMIT OF DETECTION IN CHEMICAL ANALYSIS

A large amount of information exists regarding data evaluation in chemical analyses, and more details about this subject can be found in the dedicated literature (e.g. [75], [2], [76]). Also, computer packages that perform statistical data analysis are available (e.g. [77]). Only some basic concepts are discussed in this section, some other aspects of data evaluation being presented in Chapter 5.

It is common in a quantitative analytical determination to attempt the measurement of the amount or concentration of an analyte using a specific property or signal "y" dependent on the amount or the concentration "x" of the analyte, where

$$y = \mathcal{F}(x) \tag{1.11.1a}$$

The amount or the concentration x is obtained from rel. (1.11.1a) using a number of procedures, the most common being that of a calibration curve. The resulting values for x are typically scattered around a specific value, all types of measurements being affected by errors. In a survey conducted by LC-GC magazine [14], the sources of errors in a chemical analysis were ranked based on the percentage of respondents selecting the specific cause. The results are given in Figure 1.11.1.

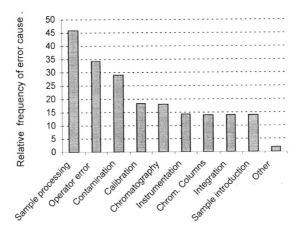

FIGURE 1.11.1. *Relative frequency of the error cause encountered in chemical analysis obtained as the percentage of respondents [14].*

The errors of measurement can be classified as systematic (determinate) or random [1]. Systematic errors are generated by a specific cause, and the main types are the constant errors and the proportional errors. Constant errors are independent of the true measured value. The proportional errors depend on the measured value. The source of constant errors can be a) insufficient selectivity and measurement of signals from other components together with the analyte, b) interference of the matrix, c) inadequate blank corrections, d) contamination, e) problems with the analytical instrumentation, etc. The source of proportional errors can be a) incorrect slope of the calibration line, b) incorrect assumption of linearity, c) changes in time of the sensitivity of the analytical instrumentation, etc. The comparison of the results from different analytical methods may indicate when a procedure has a systematic error. Sample preparation is an important source of systematic errors, and these should be determined and avoided.

Random errors do not have an assigned cause, and they affect the precision of the measurement. Precision refers to the reproducibility of measurement within a set, indicating the scatter or dispersion of a set about its central value [78]. It can be assumed that random errors are scattered within a continuous range of values. Therefore, the measurement of one variable x can generate any values in this continuous range. Any obtained set of measurements $\{x_1, x_2...x_n\}$ is defined in statistics as a *sample* of this continuous range of values. Only all possible measurements (which must be infinite in number to cover the whole range) would generate the real set, which is called *population*. (The statistical term *sample* can easily be confused in analytical chemistry with the term sample = specimen, as the term *population* may also have a different meaning. To avoid this confusion, the statistical terms *sample* and *population* are italicized in this book.)

Repeated determination of the amount or concentration of an analyte generates for each measurement "j" a value x_j. If the number of measurements is n, they will generate the *sample* $\{x_1, x_2...x_n\}$. The *average* (or the *mean*) of these measurements and the *standard deviation* (SD) that shows the distribution of measurements about the mean are given by the formulas:

$$m = \frac{\sum_{j=1}^{n} x_j}{n} \qquad s = \sqrt{\frac{\sum_{j=1}^{n} (x_j - m)^2}{n - 1}} \qquad (1.11.2)$$

The mean and the standard deviation are the most common values to express analytical results. Standard deviation is the measure characterizing *precision* (of a set of measurements), a small standard deviation indicating good precision for the set of measurements. Both m and s are expressed in the same units. The value s^2 (variance) is sometimes used instead of s. The values m and s describe the *sample* $\{x_1, x_2...x_n\}$, but they are not necessarily the same for the *population*. In a *population*, μ will replace the average m, and σ will replace the standard deviation s. For n → ∞, the limit of mean m is μ, and the limit of standard deviation s is σ. The value μ is the mean of the population, and m is an experimental estimate for μ; σ is the standard deviation for the population, while s is its experimental estimate. Relative standard deviation s/m is also used to describe precision.

The value of μ as represented by m (for a large enough n) may or may not be the same as the true value μ_o of the measured amount or concentration. The absolute difference between μ and μ_o is called the *bias*. The differences between μ and μ_o are caused by the systematic errors in the measurement of m that approximates μ. In most experiments, neither μ, nor μ_o are known (μ_o may be considered known when taken as a standard to be used for analysis). For this reason μ is replaced with the mean m, and μ_o is replaced with a value m_o, which is accepted as the true or correct result of the measurement. The difference between these known values is considered *accuracy* and is given by the expression:

$$a = m - m_o \qquad (1.11.3)$$

The accuracy can be considered as an experimental value that approximates the bias. For one measurement x_i the difference $x_i - m_o$ is a combination of the systematic error and random error and can be considered a total error of the measurement [79].

If a number of independent *sample* sets, each containing n data, are taken randomly from a *population*, the mean of each set will also show some scatter. The set of means are values of another random variable **m**, which will show less and less scatter as n increases. The total mean can be calculated as the average of all data in all *samples* or as the average of the means of each *sample* (they are equal due to the associativity of addition). The standard deviation of the mean, called *standard error* of the mean s_m, is given by the formula:

$$s_m = s / n^{1/2} \qquad (1.11.4)$$

The standard error of the mean s_m does not evaluate the precision, and it is only a measure of the confidence in a result indicated by the mean.

One important question regarding the distribution of measurements about their mean is the expected frequency of occurrence of an error as a function of the error magnitude. The most commonly utilized function, which describes well the *relative frequency of occurrence of random errors* in large sets of measurements, is given by Gauss formula:

$$f(x) = (2\pi \sigma^2)^{-1/2} \exp [- (x - \mu)^2 / 2\sigma^2] \qquad (1.11.5)$$

This frequency function (density function) shows that the point of maximum frequency is obtained for the mean (when $x = \mu$), the distribution of positive and negative errors is symmetrical, and as the magnitude of the deviation from the mean increases, an exponential decrease in the frequency takes place. The errors with the relative frequency of occurrence given by rel. (1.11.5) have a so-called *normal distribution* $N(\mu,\sigma)$. With the substitution:

$$z = (x - \mu) / \sigma \qquad (1.11.6)$$

and with the assumption that the variable x has a normal distribution $N(\mu,\sigma)$, z has a distribution $N(0,1)$. The value $(x - \mu)$ is so-called *mean centered value*, and by division with σ, it is expressed in σ units (or it is *standardized*). Mean-centered standardized variables (*standardized variates*) are commonly used in statistical data processing. The area under the curve f(x) for $x < \alpha$ will give a *cumulative frequency distribution* expressed by the formula:

$$F(\alpha) = \int_{-\infty}^{\alpha} f(x) \, dx \qquad (1.11.7)$$

where f(x) is the distribution function given by rel. (1.11.5). This cumulative frequency distribution is equal to the probability P for x to have a value below α in any measurement. The integral of f(x) over the whole space gives P = 1. The values of function $F(\alpha)$ are known and tabulated (e.g. [2]) or given in computer statistical packages.

The mean of a *sample* of n values $\{x_1, x_2...x_n\}$, as indicated previously, is itself a value of a random variable. Assuming that x has a normal distribution $N(\mu, \sigma)$, the mean **m** takes a continuous range of values with a normal distribution $N(\mu, \sigma/n^{1/2})$, such that the variable:

$$z = (m - \mu) / (\sigma/n^{1/2}) \qquad (1.11.8)$$

has an $N(0,1)$ distribution. It is possible now to evaluate how close the values of μ and m are for a certain *population* and an experimental m. For the variable z given by rel. (1.11.8), two values $z_{\alpha/2}$ and $z_{1-\alpha/2}$ can be found such that the probability for z of being outside the interval $(z_{\alpha/2}, z_{1-\alpha/2})$ is equal to α (darkened area under the curve in Figure 1.11.2). Therefore, the value for probability will be $P = 1 - \alpha$ when

$$z_{\alpha/2} < (m - \mu) / (\sigma/n^{1/2}) < z_{1-\alpha/2} \qquad (1.11.9)$$

Because $z_{\alpha/2} = - z_{1-\alpha/2}$, rel. (1.11.9) is equivalent to

$$\mu - z_{1-\alpha/2} (\sigma/n^{1/2}) < m < \mu + z_{1-\alpha/2} (\sigma/n^{1/2}) \qquad (1.11.10)$$

With rel. (1.11.9) it is easy to obtain the maximum possible differences between m and μ. Figure 1.11.2 shows the curve $N(0,1)$ with two values $z_{\alpha/2}$ and $z_{1-\alpha/2}$ such that the probability for z of being outside the interval $(z_{\alpha/2}, z_{1-\alpha/2})$ is equal to α (area under the curve). The larger α is, the smaller is P and the smaller is the value for $z_{1-\alpha/2}$.

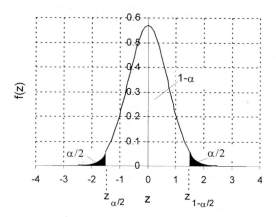

FIGURE 1.11.2. *Gaussian curve $N(0,1)$, showing two values $z_{\alpha/2}$ and $z_{1-\alpha/2}$ such that the probability for z of being outside the interval $(z_{\alpha/2}, z_{1-\alpha/2})$ is equal to α (and area under the curve is $1-\alpha$).*

For small sets of measurements it was found that the relative frequency of occurrence of random errors is well described by a frequency function (density function) named "t" or *Student function*, $f(t, \nu)$ (where $\nu = n - 1$ represents the degree of freedom of the *sample*) with a known expression [2]:

$$f(t, \nu) = \frac{1}{\sqrt{\nu \pi}} \frac{\Gamma [(\nu +1) / 2]}{\Gamma (\nu / 2)} (1 + t^2/ \nu)^{-(\nu+1)/2}$$

$$\qquad (1.11.11)$$

where

$$\Gamma\,(z) = \int\limits_{0}^{\infty} t^{(z-1)} \exp\,(-z)\,dt$$

For the Student distribution, the values of cumulative frequency $F(t, \nu)$ are also known and tabulated [75] or present in computer packages. When the number of degrees of freedom increases, the density function for Student distribution tends to the Gaussian distribution. The variable t is equivalent with variable z for Gaussian distribution. For a selected probability P (and $\alpha = 1 - P$) chosen for the decision, two values $t_{\alpha/2,\nu}$ and $t_{1-\alpha/2,\nu}$ (with $t_{\alpha/2,\nu} = - t_{1-\alpha/2,\nu}$) can be found such that the probability for t of being outside the interval $(t_{\alpha/2,\nu}, t_{1-\alpha/2,\nu})$ is equal to α. The variation of $t_{1-\alpha/2,\nu}$ values for several ν and at three different values for P is shown in Figure 1.11.3. The graphs indicate that for any specified probability, an increase in the number of measurements $n = \nu + 1$ leads to a decrease in the value of $t_{1-\alpha/2,\nu}$. Using for Student distribution an expression similar

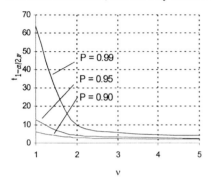

FIGURE 1.11.3. *Variation of $t_{1-\alpha/2,\nu}$ as a function of ν for a Student distribution.*

to rel. (1.11.10) to evaluate the maximum possible difference between m and μ, it can be seen from the variation shown in Figure 1.11.3 that a larger interval is expected for a small number of determinations.

Several other frequency (density) functions are also known and utilized for describing the occurrence of random errors. Among these, binomial distribution, χ^2 distribution, and F distribution are used in data processing in analytical chemistry [2].

- *Comparison of two procedures regarding accuracy*

The selection of an analytical method must take into consideration a number of desirable characteristics, such as simplicity, fast analysis time, use of specific instrumentation, etc. Adopting a method for analysis also requires proof that the method is accurate. This can be done by analyzing standards or reference materials for which the amount or concentration of the analyte is known with high accuracy and precision. The difference between the known true value and the experimental mean obtained by the new method is due to the sum of the method bias and the random error. If the differences are mainly due to random errors, the method can be considered accurate. This decision can be made using the "t-test." For a set of n identical sub-samples with a true concentration m_o, which are measured leading to n results with the mean m and the standard deviation s, the variable:

$$t = \frac{m - m_o}{s} \sqrt{n}$$

(1.11.12)

has a Student distribution. The larger t is, the higher is the probability that the difference between m and m_o is due to a systematic error. In the same manner as for the evaluation of how close are the values of μ and m (see rel. (1.11.9)), the values m and m_o can be compared. For a desired probability P chosen for the decision (such as P = 0.95 or 95%), a value α = 1 - P is obtained, and then an interval $(t_{\alpha/2,v}, t_{1-\alpha/2,v})$ is found from the tables containing F(t, v) and t values for the particular v = n - 1. The errors can be considered random if

$$t_{\alpha/2,v} < t < t_{1-\alpha/2,v}$$ (1.11.13)

When the comparison is not done for a known true concentration m_o but between a standard (accepted) method and a new method, the difference between the two methods also can be determined as caused by a bias or by random errors. This comparison can be done for the same sample, with the assumption that the standard method generates the set of results $\{x_1, x_2...x_n\}$ and the new method the results $\{x_1', x_2'...x_k'\}$. For the comparison, the first step is to calculate the two averages m and m' and the two standard deviations s and s'. It can be shown that if the *samples* are large enough, the variable:

$$H = \frac{m - m'}{\sqrt{s^2/n + s'^2/k}}$$ (1.11.14)

has an $N(0,1)$ distribution. In the same manner as for the evaluation of how close are the values of μ and m (see rel. (1.11.9)) and with the selection of a desired probability P for this decision, a value α = 1 - P is obtained, and then an interval $(z_{\alpha/2}, z_{1-\alpha/2})$ is found from the tables containing F(z) and z for Gaussian distribution. The two methods are not biased if

$$z_{\alpha/2} < H < z_{1-\alpha/2}$$ (1.11.15)

In the case of *samples* with a low number of measurements, one with n determinations and the other with k determinations and the standard deviations known to be equal, instead of a Gaussian distribution, a Student distribution can be assumed for a variable T, where:

$$T = \frac{m - m'}{s'' \sqrt{1/n + 1/k}} \quad , \quad s'' = \sqrt{\frac{(n - 1)s^2 + (k - 1)s'^2}{n + k - 2}}$$ (1.11.16a)

For the desired probability P chosen for the decision, a value α = 1 - P is obtained, and then an interval $(t_{\alpha/2,v}, t_{1-\alpha/2,v})$ is found from the tables for the particular v = n + k - 2. The two methods are not different if

$$t_{\alpha/2,v} < T < t_{1-\alpha/2,v}$$ (1.11.17)

In case of problems where the standard deviations are not known to be equal, the variable:

$$t = \frac{m - m'}{\sqrt{\frac{s^2}{n} + \frac{s'^2}{k}}}$$ (1.11.16b)

has a Student distribution with v degrees of freedom, where v is given by the expression:

$$v = \frac{1}{[c^2 / (n-1)] + [(1-c)^2 / (k-1)]} \quad \text{and} \quad c = \frac{s^2 / n}{s^2 / n + s'^2 / k} \qquad (1.11.16c)$$

One obvious problem with the previous comparisons is that they are done for one sample. A more general conclusion can be obtained by performing the determinations on n samples. The differences $d_i = x_i - x_i'$ for n samples analyzed by the two methods should be as close as possible to zero. The t-test can be used to verify if d_i are generated by systematic or random errors. For this purpose, an average d and a standard deviation s_d are calculated, and the t-test is applied to the variable:

$$t = \frac{d - 0}{s_d} \sqrt{n} = \frac{d}{s_d} \sqrt{n}$$

$$(1.11.18)$$

Some erroneous results can be obtained using this test when a systematic error is present only for a few samples and the systematic error is not detected, or when only one sample is affected by a large systematic error, and instead of evaluating the cause of the problem sample, the whole method is considered biased.

The comparison of two procedures also can be done using least-squares fitting. For this purpose, the two methods are applied to the same set of samples. The results of the first method are noted $\{y_1, y_2...y_n\}$ with the mean m_y, and the results for the second method are noted $\{x_1, x_2...x_n\}$ with the mean m_x. The pairs $\{y_i, x_i\}$ can be seen as the values of a dependence $y = x$, and the comparison will try to verify how well the pairs fit this equation. The comparison starts with the calculation of a and b for the linear parametric function:

$$y = a + b x \qquad (1.11.19)$$

such that to obtain y_i as close as possible to $a + b x_i$ for all "i". Using the notation $r_i = y_i - (a + b x_i)$ where r_i are called the residuals, the required minimization is written as follows:

$$\sum_{i=1}^{n} (y_i - a - bx_i)^2 = \sum_{i=1}^{n} r_i^2 = \text{minimum} \qquad (1.11.20)$$

The minimization is done by setting to zero the partial derivative with respect to a and b for the expression (1.11.20). Besides the values for the intercept a and for the slope b, several other parameters can be calculated, including the standard deviation for the residuals s_r, the standard deviation for the slope s_b, and the standard deviation for the intercept s_a. Using the notations:

$$S_{xx} = \sum_{j=1}^{n} (x_j - m_x)^2, \quad S_{yy} = \sum_{j=1}^{n} (y_j - m_y)^2, \quad S_{xy} = \sum_{j=1}^{n} (x_j - m_x)(y_j - m_y) \qquad (1.11.21)$$

the values for b and a can be written as follows:

$$b = S_{xy} / S_{xx} \qquad a = m_y - b\, m_x \qquad (1.11.22)$$

and the expressions for s_r^2, s_b^2, and s_a^2 can be written as follows:

$$s_r^2 = (S_{yy} - b^2 S_{xx}) / (n - 2) = (S_{yy} - b S_{xy}) / (n - 2) \qquad (1.11.23)$$

$$s_b^2 = s_r^2 / S_{xx} = [(S_{yy} - b^2 S_{xx}) / (n - 2)] / S_{xx} \qquad (1.11.24)$$

$$s_a^2 = (s_r^2 \sum_{j=1}^{n} x_j^2) / (n \, S_{xx}) \qquad (1.11.25)$$

The comparison of the two methods must test that the differences between b and 1 and between a and 0 are random and not systematic. For this purpose, it can be shown [80] that a Student distribution with n - 2 degrees of freedom is followed by the variable:

$$t = \frac{b - 1}{\sqrt{1 - R^2}} \sqrt{n - 2}$$

$$(1.11.26)$$

where R is called correlation coefficient and has the expression:

$$R = [S_{xy}^2 / (S_{xx} S_{yy})]^{1/2} \qquad (1.11.27)$$

Also, a Student distribution with n - 2 degrees of freedom is followed by the variable [81]:

$$t = a / [\sum_{j=1}^{n} r_j^2 \, (1/n + m_x^2 / S_{xx}) / (n - 2)] \qquad (1.11.28)$$

With the selection of a desired probability P for the decision, a value $\alpha = 1 - P$ is obtained, and then an interval $(t_{\alpha/2,v}, t_{1-\alpha/2,v})$ is found from the tables for the particular $v = n - 2$. The two methods are not different if

$$t_{\alpha/2,v} < t < t_{1-\alpha/2,v} \qquad (1.11.29)$$

for t given by rel. (1.11.26) and t given by rel. (1.11.28). The comparison may lead to erroneous results when the range of values chosen for comparison is too small, or if the values $\{y_i\}_i$ are in a large number in one narrow region and only one or a few in another region.

The technique of least-squares fitting has a number of other applications, for example, for finding the optimum calibration curves from a set of pairs concentration and signal (x_i, y_i). Also, not only linear dependencies can be fit using least-squares fitting (see e.g. [2]).

- **Comparison of two procedures regarding precision**

The problem of determining whether the variability of one set of data is significantly different from the variability of another set is also common when attempts are made to compare analytical results. If the distribution of errors is assumed to be Gaussian, and the set of data are independent, the procedure to evaluate the precision is based on the fact that the variable $s^2\sigma'^2/s'^2\sigma^2$ follows a $F(\nu,\kappa)$ distribution [82] where $\nu = n - 1$, $\kappa = k -$ 1n, and k the number of measurements in each set. If it is assumed that there is no difference in the variances σ^2 and σ'^2 of the two populations, the variable s^2/s'^2 must follow an $F(\nu,\kappa)$ distribution. For testing the hypothesis $\sigma^2 = \sigma'^2$, a desired probability P is selected, and a corresponding $\alpha = 1 - P$ is obtained. With the tabulated values used for $F_{\alpha/2}(\nu,\kappa)$, if the sample value s^2/s'^2 is not above this value, the hypothesis $\sigma^2 = \sigma'^2$ is accepted, and therefore the precision of the two sets of measurements do not have a statistical difference.

- Sensitivity and limit of detection

The sensitivity of a quantitative analytical method can be defined as the slope of the curve that is obtained when the result of a series of measurements is plotted against the amount (or concentration) that is to be determined. For the dependence described by rel (1.11.1a), the sensitivity is defined as the first derivative of the function $\mathcal{F}(x)$ or

$$S = d \, \mathcal{F}(x) \, / \, dx \qquad (1.11.30)$$

When the dependence described by rel (1.11.1a) is a linear function and has the expression:

$$y = a + bx \qquad (1.11.1b)$$

the sensitivity is equal to the constant b. Sensitivity can therefore be determined from the calibration curve for the method. For nonlinear dependencies the definition still can be applied, but the sensitivity is not constant for all concentrations $(S(x) = d \, \mathcal{F}(x) \, / \, dx$ for the concentration x_j). Many methods have only a range of linear dependence, with an upper region and a lower region that are not linear. In the nonlinear regions, the sensitivity varies, and for this reason sensitivity is not necessarily a convenient way to characterize an analytical method. Two examples are given in Figures 1.11.4a and 1.11.4b. Curve A in Figure 1.11.4a shows a linear dependence with the slope 0.4, and curve B shows a linear range with the slope 0.75. However, for concentrations below 0.2 or higher than 0.8, the curve B is not linear. At a low concentration, the slope indicated by C shows lower sensitivity than for curve A.

It is common that concentrations below a certain value show a sensitivity decrease. The deviation from linearity at lower concentrations in chromatographic analysis is probably due to a combined effect of the loss of a small amount of sample that is decomposed or adsorbed irreversibly and the overall modification of the chromatographic process. The loss of sample in itself is expressed in the dependence $y = a + bx$ by the negative value of a. The losses may depend on concentration and are smaller at lower concentrations. This changes the slope of the calibration curve as shown in Figure 1.11.4b.

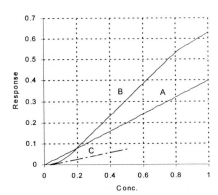

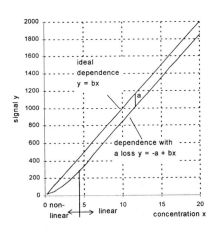

FIGURE 1.11.4a. *Calibration curves, one showing constant sensitivity (A) and the other (B) with constant sensitivity for a limited interval (0.2-0.8 arbitrary units).* Curve C shows the slope at a low concentration for curve B.

FIGURE 1.11.4b. *The calibration line for an ideal linear dependence and one for a dependence that is linear only for a specific concentration interval*

In general, detection limit is described as the limit below which detection is not possible. The detection limit can be discussed in terms of signal and transformed in terms of amount or concentration using the calibration function. The signal y for an analytical measurement is in fact the difference between the signal for a sample y_s and that of a blank y_b, both affected by errors. The signals for the blank (or the noise) are assumed to have a mean μ_b and the signals from the sample to have a mean μ_s. The standard deviations for both the blank and the sample can be considered equal to the same value $\sigma_y = \sigma$. A specific value of the signal can be selected and considered as not generated by the noise. This value is known as decision limit L [83]. Assuming that the errors in the signal of the blank and sample have a normal distribution, the probability to obtain signals from the blank higher than the decision limit L is given by the expression:

$$\alpha = \int_L^\infty f(y_b)\, dy_b \qquad (1.11.31)$$

where $f(y_b)$ is given by rel (1.11.5) with $\mu = \mu_b$. The decision limit L can be expressed in terms of signals using the expression:

$$L = \mu_b + k\,\sigma \qquad (1.11.32)$$

The probability to consider a noise as being signal from the analyte for a value higher than L is given by $P = \alpha$. For $k = 2.33$ the resulting value for the one sided normal distribution gives $\alpha = 0.01$ (or 1% if expressed in percent). Therefore if the signal is higher than $\mu_b + 2.33\ \sigma$, the probability of false positives is 1%. A signal y with $\mu_b = L$ has, however, the problem of generating false negatives. The probability of false negatives is given by the expression:

$$\beta = \int_{-\infty}^{L} f(y_s)dy_s \qquad (1.11.33)$$

where $f(y_s)$ is given by rel (1.11.5) with $\mu = \mu_s$. This probability is P = 0.5 (or 50%) because the normal curve is symmetrical around the mean, as shown in Figure 1.11.5. Therefore the possibility of false negative is very high, and 50% of the true positives will be considered noise.

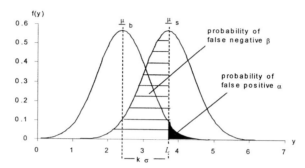

FIGURE 1.11.5. *Graph showing on the horizontal axis the signal y, the values μ_b and μ_s as well as the decision limit L. The vertical axis gives the frequencies f(y) where f is given by rel (1.11.5) and describes a normal distribution.*

A higher signal (higher μ_s) will continue to diminish the chances for false positives and also will diminished the false negatives. For a chosen probability P = 0.01 of obtaining a false negative, the corresponding value of the signal can be calculated. This value noted by *D* has the expression:

$$D = \mu_s - k'\,\sigma \qquad (1.11.34)$$

For the probability β = 0.01, the resulting value for the one-sided normal distribution is k' = 2.33. The level previously considered to generate true negative was y < *L*, in other words the signal from the analyte lower than *L*, which indicates that in fact *D = L*, as also shown in Figure 1.11.6.

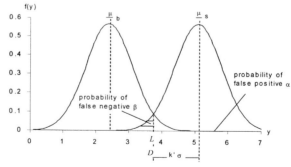

FIGURE 1.11.6. *Graph showing on the horizontal axis the signal y, the values μ_b and μ_s as well as the limit D corresponding to a probability of 1% to have the analyte signal below L. The vertical axis gives the frequencies f(y) where f is given by rel. (1.11.5) and describes a normal distribution.*

The subtraction of rel. (1.11.34) from rel. (1.11.32) together with the use of relation $D = L$ leads to the expression:

$$\mu_s = \mu_b + (k + k') \sigma \qquad (1.11.35)$$

where $k + k' = 4.66$ and corresponds to a probability of 1% for a false negative signal. The probability to generate a false positive is lower than 0.01%.

The value of the average signal μ_s that is with 4.66 σ higher than average of the blank μ_b corresponds (through rel. 1.11.1a) to a concentration that is defined as *detection limit* [84], [85], [86], [87]. This definition of the detection limit assumes that the value for σ is known. The value of the concentration that represents the detection limit can be obtained assuming a linear dependence of the form y = a + bc between the signal y and the concentration c. With this assumption, the expression for the detection limit is the following:

$$c_s = (\mu_s - a) / b = \frac{1}{b} (\mu_b - a + 4.66 \sigma) \qquad (1.11.36a)$$

Some analytical procedures have the signal depending on the amount of material introduced into the detector and not on the concentration. For these procedures, and assuming a linear dependence of the form y = a + bq between the signal y and the amount q, the expression for the limit of detection can be written as follows:

$$q_s = (\mu_s - a) / b = \frac{1}{b} (\mu_b - a + 4.66 \sigma) \qquad (1.11.36b)$$

Rel. (1.11.36a) and (1.11.36b) relate the low detection limit with the high sensitivity of an analytical measurement. Assuming a linear dependence between the signal and the concentration (or amount), a high value for b will lead to a lower value for the detection limit expressed in concentration or amount.

It is common in analytical chemistry to use the rel. (1.11.35) by considering $\mu_s - \mu_b$ as the "signal" (S) and the value of 3σ as "noise" (N). The expression from rel. (1.11.35) can be written in the form:

$$S/N \approx 2 \qquad (1.11.37)$$

The *detection limit* in this case is the amount or the concentration of the analyte that leads to a value S/N = 2. The noise is assumed to be measured on a blank, at the same point where the signal is measured for the sample. However, it is common to measure the noise as the width of the oscillation for the baseline at a flat region of the chromatogram close to the peak of the analyte and the signal as the peak height of the analyte, as shown in Figure 1.11.7.

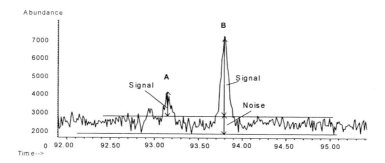

FIGURE 1.11.7. *Graph showing the measurement of signal to noise ratio, for a chromatographic peak.* Signal A is below detection limit, while signal B is above detection limit.

Other definitions of noise are also known. More correctly, the magnitude of drift and noise as a process taking place in time is characterized by its variance σ_n^2, which is given by the expression:

$$\sigma_n^2 = \lim_{T \to \infty}(1/T) \int_0^T (\Delta y(t))^2 dt \approx \lim_{n \to \infty}(1/n) \sum_{j=1}^n (\Delta y(t_j))^2 \qquad (1.11.38)$$

where $\Delta y(t)$ is the difference between the signal at time t minus the average signal, and n is the number of discrete measurements at times t_i. Also, for deciding the detection limit, values S/N = 3 or even S/N = 5 are used.

The definition of detection limit as developed previously is expressed as a function of the mean of the blank μ_b, that of the analyte μ_s, and of the standard deviation σ. However, the values of μ_b, μ_s, and σ are usually only estimated by the measured values for a *sample* $\{y_1, y_2...y_n\}$. Using the notations m_b for the mean values of the blank measurements, m_s for the sample measurements, and s for the standard deviation, rel. (1.11.35) is replaced in practice by:

$$m_s = m_b + 4.66\,s \qquad (1.11.39)$$

Also, in order to have for a selected probability P a small possible difference between μ_s and m_s, a larger number of measurements is recommended, which is specified in various protocols. In practice, it is also common to notice that the value for s is not the same when measuring low concentrations of analyte or high concentrations. For this reason, when the determination of the detection limit is not based on the S/N ratio but on the value of s and rel. (1.11.39), it is recommended to make measurements for establishing s at levels of the analyte close to an estimated value for the detection limit.

When a number of independent *sample* sets each containing n data are taken randomly from a *population*, the mean of all the measurements in the sets will generate in this case the values m_s and m_b, while s will be replaced by s_m given by rel. (1.11.4). In this case, rel. (1.11.39) is replaced by the expression:

$$m_s = m_b + 4.66 \; \frac{s}{\sqrt{n}} \qquad\qquad (1.11.40)$$

A *determination limit (or quantitation limit)* is also defined for an analytical process as the limit at which a given procedure is sufficiently precise to yield a satisfactory quantitative estimate of the unknown amount or concentration [88]. The value of a signal m_d can be written similarly to that for detection limit as follows:

$$m_d = m_b + k'' \, s \qquad\qquad (1.11.41)$$

where k" depends on a chosen probability of false negative, being chosen usually between 5 and 10 [89].

1.12. VALIDATION OF AN ANALYTICAL METHOD

The validation process describes the measures taken to ensure that an analytical process provides reproducible and secure results that are suitable for the application intended. The formal validation also includes documented evidence indicating that the analytical process generates a consistent result. The validation involves internal confirmation or external confirmation by other laboratories, the use of other methods, and the use of reference materials in order to evaluate the suitability of the chosen methodology [88]. Validation issues have been addressed by several public and private organizations such as International Standardization Organization (ISO), U.S. Food and Drug Administration (FDA), U.S. Environmental Protection Agency (EPA), Association of Official Analytical Chemists (AOAC), etc. Detailed information on this subject is given in a considerable number of publications, which include [90], [1], [91], [92], [93], [94]. Only an outline of the subject is presented here.

The validation covers the qualification of the sample and sampling (sample collection, chain of custody, etc.), certification of chemical standards, qualification of instrumentation (operational qualification, performance qualification, calibration specifications, necessary documentation, etc.), validation of the analytical method, data validation, scheduling of audits, training and operator qualifications, etc. There are various levels of formalized validation.

The validation of an analytical method is the only subject of interest here. Method validation is typically performed based on the verification of a number of parameters regarding the method. These parameters are summarized in Table 1.12.1.

The validation of the analytical method can be part of a formal validation or only needed to ensure that the analytical process is adequate for the intended application. For sample preparation, the same general scheme of method validation is followed, the subject being further discussed in Section 3.9.

TABLE 1.12.1. *Parameters used for the validation of an analytical method.*

Parameter	Short description
Specificity/selectivity	Specificity refers to a method that produces response for a single analyte in the presence of other components in the matrix. Selectivity refers to a method responding to a limited number of chemical compounds.
Precision	Precision refers to the reproducibility of measurement within a set, indicating the scatter or dispersion of the set about its central value (mean). The scatter is characterized by the standard deviation (see Section 1.11).
Reproducibility, intermediate precision, and repeatability	Reproducibility is typically considered the precision between different laboratories. Intermediate precision refers to long-term variability within a single laboratory. Repeatability refers to precision obtained over a short period of time with the same equipment (in the same lab.) when using different matrices (at least 3) and different concentrations of the analyte.
Accuracy	Accuracy can be considered as an experimental value that approximates the bias. Bias is the difference between an accepted (or true) value for an amount or a concentration analyzed and the result of the analysis (see Section 1.11).
Linearity	Linearity indicates the linear dependence between the signal and the concentration or amount and is characterized by the standard deviation for the slope s_b, and the standard deviation for the intercept s_a, given by rel. (1.11.24) and (1.11.25).
Range, linear range	The range is the interval between the upper and lower levels that have been demonstrated to be determined with precision and accuracy. The range with linear response is the linear range.
Limit of detection	The concentration (or amount) corresponding to the average signal that is with 4.66 standard deviations higher than the average of the blank signal (see Section 1.11).
Limit of quantitation (or of determination)	The minimum concentration (or amount) that produces quantitative measurements with acceptable precision and accuracy (signal usually about 10 times higher than the blank).
Recovery	Recovery is the ratio (in percent) between a known added amount of an analyte and the measured amount.
Robustness	Robustness refers to the quality of an analysis to not be influenced by small experimental modifications during the performance of the process.
Ruggedness	Degree of reproducibility under a variety of conditions such as different laboratories, analyses, or instruments.
Stability	Stability indicates that the same results are obtained in time and under different conditions.

REFERENCES 1

1. F. M. Garfield et al., *Quality Assurance Principles for Analytical Laboratories*, AOAC International, Gaithersburg, 2000.
2. C. Liteanu, I. Rica, *Teoria si Metodologia Statistica a Analizei Urmelor*, Ed. Scrisul Romanesc, Craiova, 1979, or its English translation, *Statistical Theory and Methodology of Trace Analysis*, E. Horwood, Chichester, 1980.
3. V. David, A. Medvedovici, J. Chem. Inf. Comput. Sci., 40 (2000) 976.
4. M. Valcarcel, A. Rios, Anal. Chem., 65 (1993) 781A.
5. J. Brignell, N. White, *Intelligent Sensor Systems*, Institute of Physics Publishing, Bristol, 1994.
6. W. Wegscheider, *Proper Sampling: A Precondition for Accurate Analysis,* in H. Gunzler (ed.), *Accreditation and Quality Assurance in Analytical Chemistry*, Springer, Berlin, p.99, 1996.
7. M. Stoeppler, *Sampling and Sample Preparation*, Springer, Berlin, 1997.

8. V. David, D. Ciomartan, Rev. Chim. (Bucharest), 36 (1985) 954.
9. C. Liteanu et al., *Separatologie Analitica*, Ed. Dacia, Cluj-Napoca, 1981.
10. Anal. Chem. Fundamental Reviews, 72 (2000) 37R, 137R.
11. E. R. Tufte, *The Visual Display of Quantitative Information*, Graphics Press, Cheshire, 1983.
12. I. Mills et al. (eds.), *Quantities, Units and Symbols in Physical Chemistry*, Blackwell Science, Oxford, 1993.
13. R. Panico et al., *A Guide to IUPAC Nomenclature of Organic Compounds*, Blackwell Science, Oxford, 1993.
14. R. E. Majors, LC-GC, 14 (1996) 754.
15. B. L. Karger et al., *An Introduction to Separation Science*, J. Wiley, New York, 1973.
16. R. S. Berry et al., *Physical Chemistry*, Oxford Univ. Press, Oxford, 2000.
17. W. J. Moore, Physical Chemistry, Prentice-Hall, Englewood Cliffs, 1955.
18. A. Medvedovici et al., Eur. J. Drug Metabol. Pharmacokinetics, 25 (2000) 91.
19. J. H. Hildebrand, R. L. Scott, *The Solubility of Nonelectrolytes*, Dover, New York, 1964.
20. F. Dondi, G. Guiochon, *Theoretical Advancements in Chromatography and Related Separation Techniques*, NATO ASI Ser. vol. 383, Kluwer, Dordrecht, 1992.
21. L. Rohrschneider, Anal. Chem., 45 (1973) 1241.
22. W. O. Reynolds, J. Chromatogr. Sci., 8 (1970) 685.
23. V. I. Smirnov, *Kurs Vischei Matematiki*, vol. 2, Gosudarstvennoe Izd., Moscow, 1952.
24. M. Abramowitz, I. A. Stegun, *Handbook of Mathematical Functions*, Dover, New York, 1972.
25. G. B. Arfken, H. J. Weber, *Mathematical Methods for Physicists*, Academic Press, New York, 1995.
26. S. Moldoveanu, J. L. Anderson, J. Electroanal. Chem., 175 (1984) 67.
27. S. Moldoveanu et al., J. Electroanal. Chem., 179 (1984) 119.
27a. G. J. Kennedy, J. H. Knox, J. Chromatogr. Sci., 10 (1972) 549.
27b. F. Reif, *Statistical Physics*, McGraw Hill Book Co., New York, 1967.
28. L. S. Ettre, *Open Tubular Columns in Gas Chromatography*, Plenum, New York, 1965.
29. J. C. Giddings, J. Chromatogr., 4 (1960) 11.
30. J. C. Giddings, in N. Brenner et al. (eds.), *Gas Chromatography*, Academic Press, New York, 1962.
31. E. Kováts, Helv. Chim. Acta, 41 (1958) 1915.
32. The Sadtler Capillary GC Standard Retention Index Library and Data, Base, Sadtler Res. Lab., Philadelphia.
33. L. S. Ettre, Anal. Chem., 36 (1964) 31A.
34. W. Jennings, T. Shibamoto, *Qualitative Analysis of Flavor and Fragrance Volatiles by Glass Capillary Gas Chromatography*, Academic Press, New York, 1980.
35. H. J. Cortes, *Multidimensional Chromatography*, M. Dekker, New York, 1990.
35a. A. Felinger, M. C. Pietrogrande, Anal. Chem., 73 (2001) 619A.
36. S. E. Stein, J. Am. Soc. Mass. Spec., 10 (1999) 770.
37. I. M. Klotz, R. M. Rosenberg, *Chemical Thermodynamics*, J. Wiley, New York, 1994.
38. E. G. Tibor, S. Geza, *Chimie Fizica Teoretica*, vol. 2, Ed. Technica, Bucuresti, 1958.

39. R. Kaliszan, *Quantitative Structure–Chromatographic Retention Relationship*, J. Wiley, New York, 1987.
40. C. Mircioiu et al., Rev. Roum. Biochim., 19 (1982) 37.
41. V. A. Voicu, C. Mircioiu, *Mecanisme Farmacologice la Interfete Membranare*, Ed. Academiei, Bucuresti, 1994.
41a. M. C. Hennion, J. Chromatogr., A, 856 (1999) 3.
42. D. E. Martire, Anal. Chem., 48 (1976) 398.
43. T. Halicoglu, O. Sinanoglu, Ann. N. Y. Acad. Sci., 158 (1969) 308.
44. Cs. Horvath et al., J. Chromatogr., 125 (1976) 129.
45. O. Sinanoglu, Theor. Chim. Acta, 33 (1974) 279.
46. L. R. Snyder, J. Chromatogr., 92 (1974) 223.
47. L. R. Snyder, J. Chromatogr., Sci., 16 (1978) 223.
48. W. J. Lyman et al., *Handbook of Chemical Property Estimation Methods*, ACS, Washington, 1990.
49. M. Reinhard, A. Drefahl, *Handbook for Estimating Physicochemical Properties of Organic Compounds*, J. Wiley, New York, 1999.
50. F. Eisenlohr, Z. Physik. Chem., 75 (1910) 585.
51. H. P. Meissner, Chem. Eng. Prog., 45 (1949) 149.
51a. J. C. Forman, G. Thodos, AIChE J., 6 (1960) 206.
52. F. Irmann, Chem. Eng. Tech., 37 (1965) 789.
53. J. H. Hildebrand et al., *Regular and Related Solutions*, Van Nostrand, New York, 1970.
54. C. Hanch, A. J. Leo, *Substituent Constants for Correlation Analysis in Chemistry and Biology*, J. Wiley, New York, 1979.
55. G. Nicolaides, C. Eckart, Ind. Eng. Chem. Fundam., 17 (1978) 331].
56. G. M. Wilson, J. Am. Chem. Soc., 86 (1964) 127.
57. V. A. Kistiakovskii, J. Russ, Phys. Chem. Soc., 53 (I) (1921) 256.
58. D. S. Viswanath, N. R. Kuloor, J. Chem. Eng. Data, 11 (1966) 69.
59. P. S. Liss, P. G. Slater, Nature, 247 (1974) 181.
60. E. N. Fuller et al., Ind. Eng. Chem., 58 (1966) 19.
61. S. Sugden, J. Chem. Soc., 125 (1924) 1167.
62. A. I. Johnson, C.-J. Huang, Can. J. Technol., 33 (1955) 421.
63. D. N. Rihani, L. K. Doraiswamy, Ind. Eng. Chem. Fundam., 4 (1965) 17.
64. R. C. Reid et al., *The Properties of Gases and Liquids*, McGraw-Hill, New York, 1973.
65. S. H. Fishtine, Ind. Eng. Chem., 55 (1963) 47.
66. C. Hansch, A. Leo, *Exploring QSAR, Fundamentals and Applications in Chemistry and Biology*, ACS Washington, 1995.
66a. C. Hansch et al., *Exploring QSAR, Hydrophobic, Electronic and Steric Constants*, ACS Washington, 1995.
67. R. Collander, Acta Chem. Scand., 5 (1951) 774.
67a. E. Baltussen et al., J. Microcol. Sep., 11 (1999) 737.
67b. E. M. Thurman et al., Anal. Chem., 50 (1978) 775.
68. C. M. Josefson et al., Anal. Chem., 56 (1984) 764.
69. M. H. Abraham, Chem. Soc. Rev., 22 (1993) 73.
70. C. F. Poole et al., Chromatographia, 34 (1992) 281.
71. R. W. Taft et al., J. Solution Chem., 14 (1985) 153.
72. N. J. K. Simpson (ed.), *Solid-phase Extraction, Principles, Techniques, and Applications*, M. Dekker, New York, 2000.
73. S. K. Poole, C. F. Poole, Analyst, 120 (1995) 1733.

74. P. Baricic, M. Mackov, *Toolkit for Estimating Physicochemical Properties of Organic Compounds,* based on M. Reinhard, A. Drefahl, *Handbook for Estimating Physicochemical Properties of Organic Compounds,* J. Wiley, New York, 1999.
75. D. L. Massart et al., *Evaluation and Optimization of Laboratory Methods and Analytical Procedures,* Elsevier, Amsterdam, 1978.
76. A. Papoulis, *Probability, Random Variables, and Stochastic Processes,* WCB/McGraw-Hill, Boston, 1991.
77. Statistica, '98 Ed., StatSoft Inc., Tulsa, 1998.
78. Guide for Use of Terms in Reporting Data, Anal. Chem., 47 (1975) 2527.
79. E. F. McFarren et al., Anal. Chem., 42 (1970) 358.
80. M. R. Spiegel, *Theory and Problems of Statistics,* McGraw-Hill, New York, 1972.
81. F. Grémy, D. Salmon, *Bases Statistiques,* Dunod, Paris, 1969.
82. A. J. Duncan, *Quality Control and Industrial Statistics,* Irwin McGraw-Hill, Boston,1986.
83. L. A. Currie, Anal. Chem., 40 (1968) 586.
84. H. Kaiser, Fresenius' Z. Ann. Chem., 209 (1965) 1.
85. C. Liteanu, I. Rica, Mikrochim. Acta, (1973) 745.
86. C. Liteanu, I. Rica, Mikrochim. Acta, (1975) 311.
97. C. Liteanu et al., Anal. Chem., 48 (1976) 2013.
88. M. Otto, *Statistics and Computer Application in Analytical Chemistry,* Wiley-VCH, Weinheim, 1999.
89. EURACHEM Guidance Document, Document No. 1/WELAC Guidance Document No. WGD 2. *Guidance on the Interpretation of the EN 45000 Series of Standards and ISO/IEC Guide 25,* Teddington, 1993.
90. L. Huber, *Validation and Qualification in Analytical Laboratories,* Interpharm Press, Inc., Buffalo Grove, 1999.
91. U.S. FDA, *Technical Review Guide: Validation of Chromatographic Methods,* Center for Drug Evaluation and Research (CDER), Rockville, 1993.
92. U.S. EPA, *Guidance for Methods Development and Methods Validation for the Resource Conservation and Recovery Act (RCRA) Program,* Washington, 1995.
93. AOAC *Peer-Verified Methods Program, Manual on Policies and Procedures,* Arlington, 1993.
94. U.S. EPA, *Federal Insecticide, Fungicide and Rodenticide Act (FIFRA): Good Laboratory Practice Standards, Federal Register* 48 (230):53946-53969; Nov.29, 1983; effective May 2, 1984.

CHAPTER 2

Sampling and Its Connection with Sample Preparation

2.1. RELATION OF SAMPLING TO SAMPLE PREPARATION

Sample collection is a complex subject. Particular aspects of sample collection are encountered in many fields where chemical analyses are required, such as processes characterization, biological sample analysis, food and beverage analysis, and environmental problems. Only some general aspects regarding sampling are discussed here for understanding of the relation of sampling to sample preparation. The subject is fully described in dedicated books and original publications (see e.g. [1], [2], [3], [4]).

The choice of the sampling technique is highly dependent on the physical nature of the sample. The physical nature of the sample can be determined either by its matrix, by the analytes, or by both, depending on the proportion of the analytes in the matrix. The purpose of analysis and the required precision and accuracy of the results are also important in choosing a specific type of sampling. Other specific operations during the analysis such as the type of sample preparation, choice of separation, etc., have less importance for the choice of a specific sampling procedure. The amount of required sample depends on the level of analytes in the sample, the nature of the matrix, but also on the chosen analytical method.

Sampling can be done randomly, based on sample availability, based on a specific experimental design (see e.g. [5]), or can be performed following an established protocol. However, sampling is not always done by the analyst, and in many instances the analysis must be performed on the sample as provided, unless major problems are seen in the sampling process and the analysis should not be performed.

The upper and lower bounds for the amount of sample needed in an analysis must accommodate the sample preparation capability and the sensitivity of the selected analytical method. The minimum amount of sample must provide enough analyte for performing the analysis. A less sensitive analytical measurement requires a larger amount of analyte and therefore more sample. Also, for samples that decompose, the speed of an analysis may influence the timing of sample collection as well as the number of samples that are collected at a given time. The choice of a specific sample preparation may influence sampling, again through the amount of sample collected. A large amount of sample that exceeds the sample preparation capability cannot be processed, and it is either resampled or processed in stages that may introduce additional problems. The nature of the matrix containing the analytes also influences the amount of sample. Chromatographic separations and measurements through their sensitivity also influence sampling decisions.

On the other hand, sampling may influence the choice of other steps in the analysis. For example, for analysis of a trace impurity in a gas when sampling is done in a gas bag, direct GC analysis may be performed with no sample preparation step. If the impurity in the gas is collected by passing the gas through a solid phase adsorbing cartridge, the analysis continues with the cartridge desorption or some other elution procedure of the analyte and not directly with the GC analysis. Some sampling techniques dependent on the nature of the sample are described in Section 2.3.

Sampling can have as a purpose the analysis of a specific material or of a specific process. A number of special problems are related with sampling of a process, such as the need for real time sampling, continuous sampling, sampling at specific or random intervals of time, sampling period, etc. These characteristics must be decided in connection with the purpose of analysis. A special statistical evaluation of the results is typically needed in process sampling.

In some cases the sampling step can be combined with a sample preparation process. For example, gas collection can be done using techniques that concentrate the analyte or even separate the analytes from a gas mixture (see Section 2.3). One of the gas collection procedures consists of the collection of specific analytes from a gas by freezing the flowing gas at a specific low temperature. Various components condense on the walls of the condensing trap while others may remain in gas form and are eliminated. This procedure can be applied for example for the collection of various vapors from air. The procedure serves as sample collection and sample preparation at the same time. The subject will be further discussed in Section 2.3. Sampling of solids also can be done using a dissolution process. The dissolution of the solid sample can be selective, and a sample preparation step is imbedded in the sampling step by selectively separating only a part of the initial sample.

2.2. REPRESENTATIVE SAMPLES

Sampling or *sample collection* can be seen in most cases as a mass reduction process applied to the *bulk sample* that is the object of analysis [6]. For special cases such as surface analysis, analysis of inclusions, etc., this definition of sampling does not apply. However, the following discussion will focus on sampling for quantitative analysis applied on a small representative portion of a larger amount of material. Sampling can be necessary for the analysis of a specific material, but also can be related to the characterization of a process or the characterization of the behavior of a material in a specific process.

A number of different styles of sampling are utilized [7], [8], and their choice depends on the nature of the bulk material, the system investigated, and the purpose of analysis. Random sampling is the most common sampling procedure. It assumes that the bulk material is made from a large number of equally sized and discrete portions, a random sample resulting if each portion has an equal chance of being selected. If performed correctly, a random sampling procedure will lead to a system of unbiased representative samples. Systematic sampling is achieved by taking portions of the bulk material on a regular basis in space or time. For instance, a material moving steadily on a conveyer belt or in a pipeline may be sampled systematically by removing equal amounts of material at fixed time intervals. Stratified sampling is executed when the bulk material is distributed in zones of priority, either in reality (as when the material is put in a number of containers) or schematically (as when a system is subdivided into a number of areas). Sampling is then accomplished either randomly or systematically in each zone.

Convenience sampling is performed without any regard to representativeness for the bulk. This is, for example, the case of analysis of inclusions, and most of the time this type of sampling is performed without the intention of quantitation. Sampling can be done following a specified experimental design (see e.g. [5]). The experimental design is typically performed to answer, within imposed restrictions, two kinds of questions: what is the optimal value of a parameter, and what are the best circumstances for an experiment. The optimization of specific sampling parameters is typically done using statistical techniques.

Proper sampling is critical for any analysis because the errors made during sampling are usually impossible to correct. Sampling must be reproducible, timely, economical, and safe. In addition, sampling should not disturb the analyzed system, a series of circumstantial criteria being important in specific analyses, for example, in analyses on specimens of human tissues, art and historic objects, etc. In specific instances, sampling can be just a well-defined protocol that includes a certain level of arbitrariness. One such example is the characterization of cigarette smoke where the smoking conditions and smoke collection protocol can change significantly the outcome of the analysis [9], [10], [11], [12], [13].

For trace analysis, or for samples that are not readily available, the minimum amount of sample is usually conditioned by the minimum amount of analyte that can be measured in a specific method. It must be assured that at least 3–5 times more than the minimum amount that can be detected is present in the sample (see also Section 3.1). Considering the need for replicates, the amount of sample must accommodate this requirement of any analysis. In addition to this requirement, the amount of sample must be done in such a way to assure a minimum error for the analysis. The quantitative estimation of sampling errors is commonly done using statistics [14] and treating sampling errors as random errors (see Section 1.10).

The quality of sampling is described by its representativeness. In a simple case, it can be assumed that only one measurement per sample is done and the analytical error is absent. In this case, the representativeness of an analytical sample may be quantified by the degree and the reliability of approximation of the obtained mean value m as compared to the true mean μ of the parent *population*. The confidence interval (m - d) < μ < (m + d), which characterizes representativeness, depends on the value d that can be derived for a small number of measurements n following Student's "t" distribution, with the formula:

$$d = \frac{t_{n-1,P}\, s}{\sqrt{n}} \qquad (2.2.1)$$

In rel. (2.2.1), $t_{n-1,P}$ is a tabulated variable for Student's "t" distribution and depends on the number of measurements n and the chosen statistical certainty P, and s is the standard deviation obtained from the results (see e.g. [15] and Section 1.11). Representativeness decreases with the increase in the confidence interval and vice versa. For a large number of samples, the variable $t_{n-1,P}$ can be replaced with the variable z_P for a Gaussian distribution.

The representativeness can be calculated when sampling is done by collection of a specific amount at different locations or at different times. Spatial representativeness (site, area) and temporal representativeness (time, duration, frequency of collection) are important in specific instances. However, the most common type remains the representativeness of a certain amount of sample. For example, the value for d_1 can be calculated for n samples of the same weight w_1 and then compared with the value d_2 for n samples of weight w_2. This comparison can establish whether or not modification of the amount of sample modifies representativeness. For location representativeness, the values for d are calculated for n samples collected at n random sampling points in a location and compared to n samples collected at n random sampling points at a different location. Representativeness can be diminished or even annihilated by incorrect collection of the sample, by sample contamination, etc.

In the evaluation of standard deviation s, it should be noted that its value can be obtained only after the analyses are performed, and therefore it includes sampling errors together with analytical errors. With the typical relation for the standard deviation of a sum of results each one with its own error, standard deviation can be written:

$$s^2 = s_a{}^2 + s_b{}^2$$

(2.2.2)

where s_a can be considered the standard deviation resulting from the random errors in the analysis, and s_b the standard deviation resulting from the errors in sampling. For the comparison of sampling representativeness in rel. (2.2.1), only s_b should be used. However, independent values for s_a and s_b cannot be obtained. For this reason, either it can be assumed that s_a is much smaller than s_b, such as in the case of highly nonhomogeneous samples, or that s_a is constant for practical purposes in comparison of sampling procedures. In both cases, s will be used instead of s_b for evaluating sampling representativeness.

A formula that allows the estimation of the required number of samples for a desired confidence interval for the results and a chosen statistical certainty P can easily be obtained from rel. (2.2.1). Rearrangement of rel. (2.2.1) gives

$$n = (\frac{t_{n-1,P}\, s}{d})^2$$

(2.2.3)

In this case, $t_{n-1,P}$ and s have the same meaning as in rel. (2.2.1), and d is the tolerable error in estimation of the mean m (amount, concentration, etc.) [16]. For a large number of samples, the variable $t_{n-1,P}$ can be replaced with the variable z_P for a Gaussian distribution.

The application of statistics by experimental optimization for sampling can be done, for example, with representativeness as an optimization parameter. By use of this parameter, sample collection can be seen as an experiment with a number of variables. The variables can be modified one by one with univariate optimization, factorial design, Latin squares, or complete randomization of conditions [17] to obtain optimum representativeness.

The statistical procedure previously described for the characterization of sampling has the disadvantage of being able to estimate the errors only after the samples are collected and analyzed. Also, it is applicable only when the sampling is not affected by systematic errors. Systematic errors in sampling can be avoided only based on observations or protocols that may vary from problem to problem. However, the statistical procedure is able to determine the sampling reproducibility.

The minimum amount needed of a sample for a required relative error of sampling can be evaluated with no analysis for the simple case of a sample made of equal particles, only some containing the analyte. The selection of one particle during the sampling can be viewed as a *binomial experiment* (see e.g. [18] with two possible outcomes: particle with analyte (+) or particle with no analyte (-). If the analyte is present only in a fraction p of particles and n particles are sampled, the average number of particles with the analyte is

$$m = n\,p \qquad (2.2.4)$$

In an ideal sampling (no sampling error), multiple samples with n particles must have the same number m of particles containing the analyte. However, multiple sampling will lead to variations in the number m that will have a standard deviation given by the formula [18]:

$$s = [n\,p\,(1-p)]^{1/2} \qquad (2.2.5)$$

The value for s given by rel. (2.2.5) can be used to calculate d (using rel. (2.2.1)) and thereby the confidence interval (m - d) < μ < (m + d), which characterizes representativeness. For this model, no analysis is needed to estimate sampling error. The use of rel. (2.2.4) and (2.2.5) also allows the calculation of relative standard deviation s/m for various numbers n of sampled particles and for various fractions p of particles containing the analyte in a bulk material. The dependence of relative standard deviation s/m % as a function of the number of sampled particles for a bulk material with p = 0.1 is shown in Figure 2.2.1a. Figure 2.2.1b shows the variation of relative standard deviation s/m % as a function of p for three different numbers n of sampled particles.

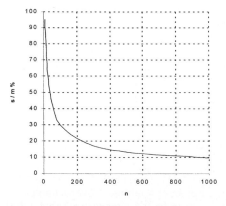
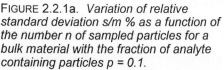
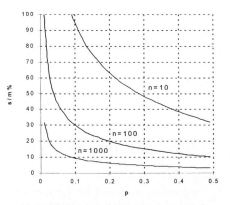

FIGURE 2.2.1a. *Variation of relative standard deviation s/m % as a function of the number n of sampled particles for a bulk material with the fraction of analyte containing particles p = 0.1.*

FIGURE 2.2.1b. *Variation of relative standard deviation s/m % as a function of p for three values of the number n of sampled particles.*

The results of this model show that the relative standard deviation of sampling decreases with the increased number of sampled particles, and also that for a material with a small fraction of particles containing the analyte (small p), the relative standard deviation is larger compared to a material with larger p. When the (average) weight of a single particle is known, the calculation of the number of particles based on rel. (2.2.3) can easily give the required weight of the sample.

In practice, extremely nonhomogeneous samples show very large values for the relative standard deviation s/m after analysis. As shown also for the model presented in Figures 2.2.1a and 2.2.1b, the problem of representative sampling is solved by collecting a large number of samples. The limitations to this procedure are usually the cost of the analyses and insufficient capacity of a laboratory to handle the large number of samples. The quantity of sample collected is another parameter that can be optimized. Larger samples from nonhomogeneous materials usually lead to better representativeness. However, this is limited by the inadequate capability to handle a large amount of material in performing a specific analysis.

For truly liquid or gaseous samples (not containing particles), the homogeneity is not usually a problem. The critical aspects can be the points of sampling, time of sampling, frequency of sampling, and maintenance of the integrity of the sample prior to the analysis. Specific protocols for sampling, which depend on the nature of the sample, are described in literature [19], [20], [21], [22].

One common procedure in sampling that reduces the number of required analyses is the use of composite samples. Composite samples can be obtained by mixing various samples collected at different points of the bulk sample, different locations, or different times. A composite sample is expected to have a better representativeness, being more homogeneous and describing better the bulk sample. The problem of making a composite sample is that the amount of final sample is larger than individual samples. For composites easy to homogenize such as liquid or gaseous samples, the composite can be easily "resampled" by taking for analysis only a portion of the composite. For solid samples, the composite can be ground, further homogenized and resampled. This is in many cases a common procedure but not always a straightforward task. As an example, grinding of a gold ore does not lead to a more homogeneous material because only the sterile is ground, while the gold is not and remains separated in certain spots. Depending on the nature of the bulk material, composite samples are either recommended or not considered adequate.

2.3   TECHNIQUES FOR SAMPLE COLLECTION

The choice of sample collection technique depends on several factors, the main ones being the physical nature of the sample and the purpose of analysis. Several sample characteristics to consider for sampling are shown in Figure 2.3.1 (see also Table 1.1.1).

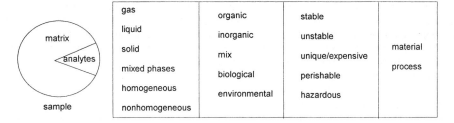

FIGURE 2.3.1. *Several sample characteristics to consider in sampling.*

Based on the overall appearance of physical nature, the samples are classified as gas, liquid, or solid. It is common that samples are not homogeneous, and under the appearance of one phase, the sample may consist of mixed phases. Gas sampling, for example, is applied for gases and also for vapors, liquid aerosols, as well as solid particles dispersed in a gas. Liquid sampling is applied for homogeneous liquids, but also for nonmiscible mixtures of liquids, for liquids containing suspensions, or for liquids containing dissolved gases. Solids can be nonhomogeneous as a mixture of solids, but also may contain liquids (such as water) or even trapped gases. A special case is the sampling of biological materials, which may be available only at very limited amounts or pose problems of perishability, hazard, etc. [19]. Mixed phase samples, where it is not possible to define it as a solid or a liquid, also are known.

The physical nature of the analyte may be different from that of the sample, for example, when most of sample material is a solid matrix and the analyte is a liquid or a gas. Specific sampling procedures must be applied when sampling is done with the purpose of measuring an analyte that may be lost from the matrix during sampling, during the storage of the sample, or even during sample preparation. Similar precautions must be taken when the matrix can be modified during sampling. For example, water is a common constituent of plant materials, food samples, etc. When the result of the analysis is reported on "as collected" basis, the preservation of the initial amount of water in the sample is important. The variability in the water content can be disregarded by reporting the analytical result to a "dry basis," which requires the elimination of water and measurement of the weight of a dry sample.

Usually, sampling precedes sample preparation, but in certain cases the sampling technique is associated with a preconcentration and/or a separation process. In these cases, sampling and sample preparation are concurrent. Special techniques for sample collection that aid preferentially the selection of the analyte and help the analyte measurement, such as solid phase microextraction (SPME), stir bar sorptive extraction (SBSE), headspace sample collection, are discussed in Part 2 of this book.

For environmental problems, it is common that the substances of interest (contaminants) are present at very low levels. For this reason, the direct analysis of the sample as collected is not appropriate [23]. The collection must be associated with a concentration step. All substances that are not expected in atmosphere and water or that exceed their background levels are considered pollutants [24]. Among the typical analytes for an air or water sample are a number of volatile organic compounds (VOCs) and numerous semivolatiles and nonvolatiles that exhibit health hazards or act as precursors to ozone formation in the atmosphere [25]. Extensive lists and details about the threshold levels of these compounds are available (see e.g. [26], [27], [28], [29], [30]). Inorganic constituents of various samples are also important, and contamination during the sampling can be a significant problem when the sampling is not done using procedures that avoid metal contamination. Some elementary aspects of sample collection based on the physical nature of the sample are discussed in this section.

- Techniques for gas collection

For most quantitative analyses on gases, the result is expressed as a concentration, usually as the amount or the volume of the analyte in a specified amount or volume of sample (at a specified pressure and temperature). When the concentration is expressed using volumes, the equivalent amount must be easily obtainable. For gases, the mass is dependent on volume, pressure, and temperature. For this reason, gas sample collection usually requires the measurement of the volume, the pressure, and the temperature at the same time. In addition, when flowing gases must be sampled, the gas possesses both static pressure and pressure due to the velocity of the fluid molecules. In these specific measurements the gas flow rate also must be measured.

Another problem related to gas sampling is related to the collection of vapors and particles as part of the gas sample. The presence of vapors and particles can make the gas sample unstable in the sense that vapors may condense or particles may deposit in the collection system. Special precautions must be taken in these cases for measurment of the modifications in the sample initially collected.

A gas can be collected as is, but the collection also can be done into a liquid or solid medium, either without chemical modification or following a chemical reaction with a collecting reagent. The gas collection techniques can be grouped in static sampling, dynamic sampling, and sampling based on diffusion (another classification recognizes active and passive sampling [31]). Several procedures are used for the determination of the amount of gas. One type of procedure consists of the measurement of a static volume, the pressure, and temperature. Other procedures consist of the measurement of the mass flow rate of the gas and of the collection time. Also the measurement of the linear velocity of the gas flow through calibrated tubes may lead to the determination of the amount of gas. In practice, combinations of different collection techniques and measurement procedures are applied.

There is no single sampling technique adequate for all kinds of analytes measured in gases. The choice of any of the collection procedures is determined by the purpose of analysis, type of analyte, as well as by the available capability. In specific cases, continuous analysis is necessary, and special instrumentation can be devised for this purpose [32], [33], [34].

In a static sampling procedure, a specific volume of sample is collected without altering its chemical composition. In these techniques, it is common to measure the volume using measuring devices such as spirometres, displacement bottles, bubble meters, dry gas meters, Roots meters, etc. In this type of measurement, the volume of the container of the collected gas is usually known (by previous calibration). The gas collection can be done using a) a collapsible gas bag [35], b) displacement of an inert liquid from a flask or a container of a known volume with the mixture to be analyzed [36], c) evacuation of the air from a calibrated flask or canister using a vacuum pump to a very low and known pressure and then filling it with the gaseous sample [37], or d) displacement of the air from a calibrated container by passing the gaseous mixture through it for such a long time that no modification of the composition occurs.

Static gas collection usually is not done for flowing sources except when the gases to be analyzed are homogeneous. In this case, the results can be reported relative to the mass of the gas already collected. However, various factors can influence the composition of the gases sampled by the static methods [38]. Among these are a) leaks of analytes in the valves or needles of the vessels, b) chemical reaction or adsorption of the analytes on the wall of the recipients, c) desorption of the unremoved contaminants from previous usage of the vessel generating so-called wall memory effect, and d) stratification process for the analytes when high density vapors may separate to a certain extent from lighter gases.

Another source of errors is related to the calibration standards used for the analysis. For the sampling done by a static method, it is recommended to make the calibration necessary for analytical measurements using gaseous standard mixtures obtained by the same method. Static gaseous mixtures are obtained when a known amount of one or several analytes are introduced into a known volume of a diluting gas. The static methods can be used for the preparation of relatively small amounts of standard gaseous mixtures containing the analyte in the concentration range from less than 1 ppm to several tens of a percent. When the calibration standard is kept in a container with constant volume V_c, it is common to maintain also a constant pressure. For this purpose, when a volume V (mL or μL) is removed from the container, it is followed by the introduction of the same volume V of diluting gas. The concentration of the analyte after n usages, denoted by C_n, is given by the formula [39]:

$$C_n = C_o (1 - V / V_c)^n \qquad (2.3.1)$$

When no gas replacement is performed, the concentrations remain the same, but the pressure is modified. Possible modifications due to chemical reactions or adsorptions must be monitored.

The dynamic type of gas sampling consists of the collection of a stream of gas sample. With a gas volumetric flow rate U (usually measured in mL/min.), a collection time t (usually measured in min.), and collection efficiency E_f ($E_f = 1$ for complete recovery and $E_f = 0$ for no recovery), the volume of the gas sample (V) collected is given by the relation:

$$V = U t E_f \qquad (2.3.2)$$

The collection can be done using various procedures depending on the nature of the analyte. The diagram of a simple dynamic collection setup for air samples is shown in Figure 2.3.2.

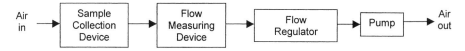

FIGURE 2.3.2. *Configuration of a dynamic sampling procedure.*

The gas flow rate can be measured with orifice meters, Venturi meters, rotameters, mass flow meters, etc. When the sampling is done from a stream of flowing gases, such as for smoke stacks or dryer exhausts, the velocity of the sampled gas must be measured. The gas velocity measurement can be done with Pitot tubes or electronic velocity meters [1]. From the gas velocity and the cross-section area of the pipe in which the gas is flowing, the flow rate of the sampled gas can be obtained. During the dynamic sample collection, after a particular flow rate has been optimized for sampling, it is necessary to maintain it at exactly that value. Sources of pressure drop in the system must be considered for proper evaluation of the flow rate at the collecting device [40].

The use of a dynamic sampling procedure from a flowing stream of gas poses special problems. For heterogeneous gas samples the solids and liquids have momentum vectors dependent on both their mass and their velocity, and these are very different from single gas molecules. For this reason the sampling at flows lower than the velocity of the source for streams containing particles may result in poorly representative samples. In these instances, isokinetic sampling is applied. In isokinetic sampling, the sample is withdrawn from the source such that the flow into the sampling probe has the same linear velocity as the flow of the sample through the source. The probe should be positioned so that its entrance is pointed directly into the direction of flow (the path of the molecules that pass from the stream and into the probe nozzle is straight). Sampling a source using a velocity higher than the velocity of the source is also undesirable since the sampling would perturb the flow dynamics of the stream being sampled, resulting in a poorly representative sample.

In order to achieve isokinetic sampling, the collection of the sample must be subject to mechanical controls. In general, a vacuum pump is used, but in cases where the stream being sampled is under high pressure (i.e., more than a few Torr above ambient), pressure-reducing devices (e.g., a fixed orifice) must be used to reduce the flow rate of the withdrawn sample.

Sampling from flowing streams of gases that do not have a constant composition (for example, when a batch process generates the gases) is an even more complex problem. Proportional sampling [1] may yield more useful results in these cases, especially if the species of interest are in the gas phase at the conditions sampled.

One of the most commonly analyzed gases is air. The analyses can be performed for various reasons such as the determination of organic pollutants, inorganic pollutants, dust particles, etc. The major components of the atmosphere are N_2 78% (in volume), O_2 20.93%, and Ar 0.93%. Also, CO_2 and H_2O vapors are minor components. When air

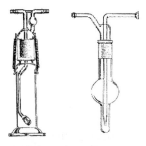

FIGURE 2.3.3. *Gas absorption bottles.*

or other diluting gases are sampled for trace components, the trace components must be collected, but only the volume (or amount) must be known for the diluting gas that acts as a matrix for the analytes. In some procedures, the whole gas is collected, but it is also common to use selective sampling by passing the sample through an absorbing solid or liquid medium where the collection is done only for the specific analytes. This type of collection can be done using the dissolution or reaction of the analytes with a reagent. Other procedures use impactors where the gas stream hits a surface, condensing the analyte. Condensation of the analytes by freezing is also a common collection procedure. For collection of dust, mist, and fumes, the gases are passed through special filters, electrostatic precipitators, etc

In selective sampling the composition of the fraction collected is different from the composition of the initial gas sample, because the absorption of the analytes is selective. Therefore, selective sampling may act in specific cases as a preconcentration technique and be viewed as a sample preparation procedure. The collection efficiency E_f can be defined for a specific volume of gas as the ratio between the amount of analyte recovered from the gas and the initial amount of analyte in the gas. Several mechanisms can be involved in gas absorption when selective sampling is utilized. The more common types of collection mechanisms are summarized below.

Dissolution of the gas in a liquid medium without any chemical reaction involved is a common procedure in dynamic gas collection. It can be applied using impingers containing a solvent that does not react with the analytes and provides an appropriate medium for dissolving them. The yield of recovery in this case can be approximated with the equilibrium constants of the gas liquid partition process (see Section 1.4). The absorption of a gas when no chemical reaction is involved obeys Henry's law given by rel. (1.4.20), which describes the solubility of the gaseous analyte "i" in a liquid medium. Henry's law is written as follows:

$$k' = c_{i,A} / p_i \qquad\qquad (2.3.3)$$

where $c_{i,A}$ is the molar concentration of "i" in the liquid medium and p_i is its partial pressure in gas phase. Rel. (2.3.3) shows that increased pressure increases the analyte concentration in the solvent. However, the temperature and other factors such as ionic strength (through activity coefficient) of the liquid medium also influence the equilibrium. With increase in temperature, the solubility of the compound "i" in the liquid medium decreases. The influence of the ionic strength of the solvent depends on the nature of the analyte. Henry's law is not valid for gases that interact chemically with the solvent (see Section 1.4).

In order to use rel. (2.3.3) to estimate gas absorption, the assumption must be made that the analytes are at the equilibrium concentration in the gas and in the liquid, which is not always the case. When the contact between the gases and the dissolving solution is not efficient, or when the contact time is too short, this theory is not applicable. For this purpose high shear mixing devices were developed [41], which allow a better contact between the liquid and the gas, such that a more rapid equilibrium is attained.

A parameter related to k' describing the dissolution of "i" in a liquid at a given pressure and temperature is the "solubility." Solubility is expressed as the volume (usually in liters) of a gaseous compound dissolved in one liter of solvent. The solubilities in water of a few common gases are given in Table 2.3.1 [42].

TABLE 2.3.1. *Solubilities of some gases in water (L gas/1 L water) at 1 atm. and two temperatures.*

Temperature	H_2	N_2	O_2	CO_2	NH_3
0°C	0.017	0.023	0.049	1.70	58.2
30°C	0.006	0.013	0.026	0.66	8.5 (at 100°C)

The value of constant k' in rel. (2.3.3) (and of solubility) depends on the nature of the gas and that of the liquid, as shown in Section 1.4. Empirical rules based on structural similarity ("like-to-like") between analyte and the solvent were developed for predicting solubility (see Section 8.2).

Adsorption of the gas sample on a solid surface with high affinity for specific analytes is another collection procedure. This procedure can be applied for adsorbing gases and vapors using cartridges containing specific stationary phases (Tenax, active carbon, various stationary phases also used in chromatography, etc.). Similarly to this procedure of gas sampling, a special sample introduction technique, dynamic headspace, was developed for GC analysis (see Part 2). The selectivity toward specific analytes may be used to improve the analysis, but when it is unknown may lead to erroneous results. For example, Tenax does not adsorb water and adsorbs methanol very poorly, but it has a high affinity toward aromatic compounds. The elimination of water is very useful in GC analysis, where water is not desirable as a sample component.

Chemical absorption is a collection procedure based on the absorption in a medium involving a chemical reaction with the analytes from the gaseous sample. Impingers with a reagent solution are commonly used for collecting the analytes. For the same purpose can be used special cartridges that adsorb specific analytes through a chemisorption process or denuder tubes with the inside wall coated with a specific reagent that absorbs the vapor phase analytes [43], [44]. The gaseous sample is passed through the tube using a vacuum pump. The advantages of denuders over impingers include a higher sampling flow rate (up to several L/min.) and a small volume of the absorbing medium that leads to a more concentrated analyte in the sample. The reaction kinetics influences the recovery of the analytes from the gaseous mixtures. Some absorbing solutions used for collection of pollutants from air followed by IC determination are indicated in Table 2.3.2.

TABLE 2.3.2. *Some absorbing solutions used for collection of pollutants from air followed by IC determination [45].*

Component	Absorbing solution	Species analyzed by IC	Reference
SO_2	0.6% H_2O_2 + 0.06 mM HCl;	SO_4^{2-}	[46]
	80% propan-2-ol + 3% H_2O_2	SO_4^{2-}	[47]
SO_2, NO_2, CO_2	0.25 M $KMnO_4$ + 1.25 M NaOH	SO_4^{2-}, $S_2O_6^{2-}$, NO_3^-, CO_3^{2-}	[48]
NH_3	0.1% N H_2SO_4 + 3% H_2O_2	NH_4^+	[47]
HCl	0.1 N NaOH + 3% H_2O_2	Cl^-	[47]
HCN	0.2 N NaOH	$HCOO^-$	[49]
$CH_3\text{-}SO_3H$	5 mM KOH	$CH_3\text{-}SO_3^-$	[50]
smoke	3 mM $NaHCO_3$ + 2.5 mM Na_2CO_3	Cl^-, PO_4^{3-}, Br^-, NO_3^-, SO_4^{2-}	[51]
lower aldehydes	80% propan-2-ol + H_2O_2	$HCOO^-$, CH_3COO^-, etc.	[52]
lower carboxylic acids	deionized H_2O	$HCOO^-$, CH_3COO^-, etc.	[53]
Cl_2	12 mM Na_2SO_3 (pH = 8.7)	Cl^-	[54]
Cl_2, ClO_2	10 mM KI in phosphate buffer	Cl^-, ClO_2^-	[55]

Condensation of the analytes at low temperature from the flowing gas is another common collection technique. This kind of sampling of gaseous mixtures is known also as cryogenic collection, and it is used for the collection of volatile and semivolatile compounds that can condense at a certain low temperature. The method is based on passing the gaseous sample through a collection space (trap) set at a very low temperature. In this trap, the components from the gaseous sample with the boiling point equal to or higher than the trap temperature will condense. Very often a U-shaped or spiral tube is used as condensing vessel. Two parameters should be taken into account in order to optimize the yield of collection. The first is the temperature, which can be controlled by using different cold liquids. For example, a mixture of ice and salt reaches a temperature of -15° C, a mixture of dry ice + acetone or methyl-cellosolve reaches -80° C. Good yields can be obtained with liquid oxygen (bp -183° C) or liquid nitrogen (bp -195.8° C). The other parameter is the flow rate of the gas stream passing through the trap. A low flow rate involves high yield of condensation, but the sampling time is long, mainly when the analytes are in low concentrations. Both parameters can be optimized, depending on the level of concentration of analytes from the gaseous mixture. Cryogenic collection has several advantages to other collecting procedures, allowing relatively high flow rates of passing gases, and does not add significant backpressure to the flow. By properly setting the condensing temperature, the cryogenic collection may also act as a preconcentration procedure. For example, the collection of various VOCs from atmosphere by this procedure allows separation of VOCs from air main constituents (oxygen, nitrogen, etc.). For this reason, cryogenic collection has been widely used for the analysis of VOCs in air, as summarized in Table 2.3.3, which lists several reported methods.

A combination of cryogenic collection and dissolution of gases in a liquid medium is also utilized. For example, vapor phase of cigarette smoke can be collected in a series of methanol traps cooled with dry ice in isopropanol [62]. The procedure can be very useful for the collection of specific analytes, but some gases dissolved in the cool solvent may be lost when the solvent is brought to room temperature [35].

TABLE 2.3.3. *Analysis of VOCs and other analytes in air with cryogenic collection and chromatographic analysis.*

Analyte	Analytical procedure	Reference
VOC	GC FID	[56]
sulfur compounds	GC with sulfur chemiluminescence detector	[57]
C_2 - C_{10} VOC	GC FID	[58]
C_2 - C_{12} VOC	GC FTIR-MS, GC FID, HPLC	[59]
VOC	GC	[60]
hydrocarbons	GC	[61]

Filtration using special filter media is typically used for the sampling of aerosols. The filters can be made from various materials such as glass fiber, porous nylon, polypropylene, polyurethane, Teflon (PTFE), etc. [63], [64]. The filter medium must exhibit a high efficiency and should not be hydroscopic. Various procedures regarding aerosol collection are reported in literature. Some examples for the analysis of aerosols in atmosphere are given in Table 2.3.4.

TABLE 2.3.4. *Examples of filter media for collection of gases, aerosols and solid particles from atmosphere [45].*

Sample	Filter medium	Extractant	Species absorbed	Reference
aerosols	quartz	0.05 mM $HClO_4$	SO_4^{2-}, NO_3^-, Na^+	[65]
aerosols	PTFE	1 mM phthalate	Cl^-, NO_3^-, SO_4^{2-}	[66]
paint aerosols	PVC	2% NaOH 3% Na_2CO_3	CrO_4^{2-}	[67], [68]
particles	paper	H_2O (t° C)	NH_4^+, Na^+, K^+, Cl^-, NO_3^-, SO_4^{2-}	[69]
particles	PTFE	deionized water	Cl^-, NO_2^-, NO_3^-, SO_4^{2-}	[70]
acidic gases	paper coated with 1M KOH	deionized water	Cl^-, NO_2^-, NO_3^-, SO_4^{2-}	[70]
sulfur dioxide	paper coated with Na_2CO_3	H_2O_2 solution	SO_4^{2-}	[71]
ammonia	membrane coated with 10% H_3PO_4	distilled water	NH_4^+	[72]

Other procedures for analyte collection from a flowing gas sample are also known. Gas sample collection can be done using electrostatic precipitation, impactors, etc. Specialized literature is available for detailed description of such procedures [1].

When dynamic collection of gases is used, it is recommended that the standards be made using a dynamic technique. For this purpose, the standards are obtained from a gas with known composition of the analyte by passing it through the absorbing medium for a specific amount of time. Various gases with known composition can be used sequentially or premixed. A dynamic mixture can be obtained from two or more gas streams using known constant flow rates. It is also possible to inject the analyte gas into a diluting gas stream using a syringe or a syringe pump.

Diffusive gas collection is another common sampling technique. A diffusive sampler is a device that is capable of taking samples of gases or vapors at a rate controlled by a physical process, such as diffusion through a static air layer and permeation through a membrane. This procedure does not involve any active movement of the gas through the sampler [73]. The material penetrating the membrane reaches an adsorbing medium that must bound physically or chemically to the analyte. Figure 2.3.3 shows schematically a diffusive sampler.

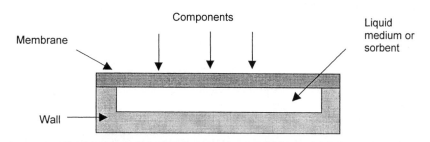

FIGURE 2.3.3. *The principle of diffusive sample technique.*

The diffusion process is governed by Fick's laws (see rel. 1.5.1b and 1.5.2), which can describe the distribution in time and space of the amount of substance that diffuses through the membrane. The efficiency of diffusing sampling can be estimated based on the theory of diffusion from a continuous source. In an ideal diffusion process where no convection takes place and the concentration remains constant at the surface of the sampler, the diffusion is described by rel. (1.5.2) with boundary conditions (1.5.15). The real situation is more complicated, because the membrane is relatively thin and is in contact with the liquid or sorbent medium, which behaves differently from the membrane. In spite of this, the use of rel. (1.5.16) as a solution shows that the amount of material in the sampler is proportional with the concentration c_o from outside the sampler and is larger for larger D t values, which correspond to more permeable membrane and to longer exposure times (see Figure 1.5.5). Because the value of D depends on temperature, this is also a factor influencing the diffusion process. For gases, the partial pressure $p_i = RT\, c_{i,g}$ can replace the concentration c in rel. (1.5.16).

The new sample obtained by diffusive sampling is further analyzed. The ability of a diffusive sampler to respond accurately to transient changes in concentration has been questioned [74]. However, a relatively accurate result can be obtained over a sufficiently long sampling time (30 min. or more). This sampling time is also dependent on sampler geometry. Diffusive samplers can be broadly divided into two classes [75]: badges and tubes. Many commercially available diffusive samplers use activated charcoal as the collecting medium. Decomposition, polymerization, or irreversible adsorption can take place with reactive and high molecular weight compounds when sampled on activated charcoal. For moderately reactive compounds, such as certain organic monomers, ketones, phenols and chloroaromatics, a porous organic polymer such as Tenax, Amberlite XAD, Chromosorb 106, Carboxen 569, Tenax GR, and Carbosieve S-III can be used instead of charcoal [75], [76], [77], [78]. There are some applications using zeolites in diffusive sampling, such as in carbon monoxide collection, followed by thermal desorption and analysis [79].

Similar to the diffusive sampler is the reactive filter sampler, where the analytes are collected on a reagent-coated filter [80]. One of the difficult problems related to the use of the filter sampler is a back-diffusion process that occurs in time. Aldehydes can be sampled using a 2,4-dinitrophenyl-hydrazine coated filter and amines using a filter coated with 1-naphthylisothiocyanate. The derivatives from such reactions can be determined subsequently by HPLC with UV detection. From the occupational hygiene point of view, five aldehydes are of special interest: formaldehyde, acrolein, glutaraldehyde, acetaldehyde, and furfural. They are reactive, excepting furfural, which is stable enough to be sampled on an uncoated organic porous polymer such as Amberlite XAD-2. Both aliphatic and aromatic amines are reactive compounds. Only tertiary aliphatic amines are sampled on uncoated solid sorbents. For the more reactive primary and secondary amines chemisorption can be used. The simplest way is to use acid-coated sorbents or filters followed by GC or HPLC analysis.

One other procedure where the diffusion process is involved is the use of an adsorbing material that is kept for a certain period of time in a confined volume together with the gas sample to be analyzed. This procedure is related to special sample introduction techniques used in GC analysis such as static headspace and SPME. The gases from the sample headspace are selectively adsorbed on a specific material, which can be Tenax, polymeric silicones, etc., and then desorbed and analyzed. This procedure leads to a selective adsorption of some compounds and can be used to concentrate specific analytes. The subject will be further discussed in Part 2 of this book.

- Liquid sample collection

Sampling of liquids is in many respects similar to sampling for gases. The sampling of a uniform liquid is not complicated unless special problems are present. Among these can be chemical instability, volatility, high viscosity, hazard related problems, adsorption of water from atmosphere, etc. The sampling can be done by measuring either the volume or the weight of the liquid.

Among the liquids typically sampled are rain, fresh water, and seawater. Special sampling procedures are used in the oil industry for sampling mixtures of hydrocarbons. Problems related to sampling strategy including area and sampling time period, which are important for generating the appropriate information [19]. Also, special sampling equipment has been developed for sampling rain and snow [81] or fresh and seawater [82]. Similarly to gas analysis, sampling of environmental liquid samples must avoid contamination, mainly when the level of analytes is in the low ppb level.

Some liquid samples may have homogeneity problems. Homogeneity problems may be present for static liquids and also for liquids moving through pipes or channels. Proper techniques must be applied for obtaining a representative sample, considering either homogenization of the liquids or proportional sampling of all the components in the mixture. Sampling from moving liquids must take into account whether the flow is laminar or turbulent. In a laminar flow, poor mixing takes place within the stream. Also, heavier liquids or solid particles may be part of the flow and not distributed randomly, but be accumulated at the lower part of the pipe or channel. Some nonhomogeneous liquid samples that are collected as one phase may separate in time in different layers based on their hydrophobic-hydrophilic interactions. A nonhomogeneous subsample may lead to wrong results if the separation is not considered. This type of sample must either be homogenized again when a subsample is collected, or each portion must be analyzed separately. The use of an extraction procedure may be applied when only a small portion of the sample is segregated from the bulk.

The sampling of liquids can be done similarly to gases, with static, dynamic and diffusive techniques. Static sampling is common and can be done with a variety of measuring devices. Depending on the nature of the sample, these can be volumetric measuring devices or weighing devices. Special static collection devices can be used, for example, for rain sampling. Dynamic sampling is done mainly for moving liquids [83]. Diffusive sampling can be applied for liquids, using for example a stir bar coated with a specific sorbent [84]. This technique known as SBSE is used for the enrichment in specific analytes from the sample and is discussed separately as a sample preparation technique.

- *Techniques for solid sample collection*

The problems related to solid sample collection are mainly those of representativeness and homogeneity. The amount of sample collected should be large enough to be representative, as discussed in Section 2.2. Larger samples can be resampled after a homogenization step. Sample homogenization is typically done using various types of particle reduction procedures (see Section 3.3). A simple procedure used for sample homogenization is coning. Coning is used mainly when dealing with large samples. The bulk material is shaped into a cone shape by pouring the material on the apex of the cone. Then the cone is flattened. The flattened material is reformed into a cone shape, which is again flattened. The procedure is repeated a number of times. The sample reduction can be done using quartering. This consists of flattening the bulk material into a circular layer. This layer is quartered and the alternate quarters are discarded. The remaining material is reshaped into a cone, flattened and the process can be repeated until a sample of suitable size is obtained. For a 3D bulk material, a better sampling can be made by converting the bulk into a flattened narrow band or into a flattened object. The sampling of the band can be done by cutting equal segments that are selected randomly or using a specific protocol. The flattened material can be sampled using a grid with random or systematic selection of the grid elements.

A number of sample preparation techniques including extractions, melting followed by centrifugation, etc, may be applied as a combination of sampling and sample preparation. However, these procedures are seldom used without a preliminary sampling step, and are discussed in Part 2 of this book as sample preparation techniques.

Soil is a special type of solid sample that poses special collection problems. It consists of solid matter of mineral and organic origin, but also contains numerous pores that can be filled with water and with air. In addition, soils contain a biotic part including bacteria, fungi, worms, and insects. Depending on the analysis purpose, various sampling protocols are applied for soil collection [85], [86].

A challenging sample problem is that of solid waste. Waste is examined for establishing the suitability for recycling, defining the appropriate disposal site, finding calorific value, evaluating potential of environmental contamination, etc. Solid waste is very nonhomogeneous, containing portions with extreme values in different components. The evaluation of sample representativeness including necessary amount of sample, place (location and number of points) of collection, and time (timing and period) is important and sometimes difficult to establish. The answer to the lack of representativeness for this type of sample is the proper presentation of data, indicating precisely the conditions for the data collection, and evaluation of the results as a scattered data set, possible for a large number of sampling points and over a longer period of time.

- Sampling of mixed phases

Some samples are made of mixed phases, usually liquid and solid. Examples are various types of food (preserves with seeds, canned vegetables, etc.), special soils (mud, oil shales, etc.), sewage sludge, etc. These samples are in general difficult to sample with good sample representativeness. Special sampling protocols are designed for these types of samples, either for establishing the minimum amount necessary for a representative sample or the number of points or the necessary frequency of sampling. Separate sampling for the different phases and the evaluation of the ratio of each phase may also be chosen for obtaining the correct information on the sample.

2.4.   HANDLING, PRESERVATION, AND STORAGE OF SAMPLES

Sample preservation, until the sample preparation starts, can be an important factor for correct analytical results. A number of problems may arise when the samples are analyzed a long time after collection (long holding time). The handling (e.g. transportation) and the storage must avoid any contamination, alteration, or loss of analytes. Alteration of the sample content can occur because of physical and chemical changes in the sample.

Among physical effects that can modify the sample are adsorption on the container walls, volatilization, or precipitation of the analytes. For example, it is known that in water monitoring, samples containing pollutants that are hydrophobic cannot be kept in plastic containers. The best choice in this case is glass, possibly adding a small volume of a solvent towards which the pollutants have a good solubility (e.g. addition of acetonitrile in water samples prevents the adsorption of polycyclic aromatic hydrocarbons (PAHs) on the walls of glass containers [87]). Losses of volatiles during transportation can be estimated, for example, using a test sample with known initial composition (trip blank) that after analysis indicates the losses.

Chemical processes (photodecomposition, thermal decomposition, microbial action, oxidation or reduction processes, etc.) can occur during handling operations and storage. It is also possible for a laboratory to have only a limited capacity of analysis. In this case, samples must be stored over a period of time in special conditions, depending on the nature of samples: dark, low temperature, addition of preservatives, antioxidants, or adjustment of pH value. Biological samples (human or animal tissues, food samples), which can change their composition, should be stored in frozen state under liquid N_2.

Samples designated for trace analysis have special requirements for handling and storage. It is preferable to store the samples in a separate physical location from analytical standards or any other material that may contain a high concentration of the analyte to be investigated. Also it is recommended to use a max./min. thermometer to check for temperature fluctuations during storage of temperature sensitive samples [88]. Freezing the samples is a common preservation procedure. Samples that cannot be frozen or that do not need to be frozen (with nonvolatile matrix and analytes, or stable at ambient temperature) are usually stored at 0–5°C. Several conditions for sample storage are given in Table 2.4.1.

TABLE 2.4.1. *Some common storage conditions for samples.*

Storage conditions	Appropriate sample types	Inappropriate sample types
deep freeze (-18°C)	samples with high enzymatic activity (e.g. liver), most sample types, less stable analytes	fresh fruit and vegetables, samples liquefying on thawing, aqueous samples
refrigerator (4°C)	soils, minerals, fresh fruits and vegetables, aqueous samples	samples with possible enzymatic activity
room temperature (dark)	dry powders and granules, minerals, stable analytes	fresh foods
desiccator	hydroscopic samples	samples more hydroscopic than the desiccant

Other sample preservation techniques besides freezing are indicated in Table 2.4.2.

TABLE 2.4.2. *Some physical and chemical methods used for sample preservation.*

Method	Examples	Critical aspects
freeze drying	food samples, aqueous samples	unsuitable for volatile analytes
irradiation	aqueous samples, biological samples	stability of analyte must be established
electromagnetic radiation	biological samples	stability of analyte must be established
chemical sterilization	addition of preservatives	may alter the sample or may introduce contaminants
addition of antioxidants	liquids and solutions	stability of analyte must be established; check specific interference effects
addition of anticoagulants	blood and clinical samples	check specific interference effects
autoclaving	sterilizing body fluids	stability of analyte must be established
ultrafiltration	use of membranes to separate microorganisms	applied only in special cases
preservation at constant relative humidity (RH)	food samples. biological samples	temperature control needed

A common operation used for preservation of biological samples is sterilization. The aim of a sterilization process is to destroy or eliminate microorganisms that may be present on or in the sample. The sterilization may have as an objective the preservation of the sample, but also to assure that the sample is free from infection hazards. The accepted performance target for a sterilization process is to have a probability of finding a nonsterile unit of less than one in one million. The process including production, storage, shipment, etc. will provide a Sterility Assurance Level (SAL) equal to or better than 10^{-6} [89].

Some samples must be stored under a specific relative humidity (RH). Relative humidity is temperature dependent, and a specific RH value can be achieved using saturated salt solutions in a confined space. Several RH values at 25° C for the commonly used salts are the following: NaCl (RH = 75.3), KI (RH = 68.9), potassium citrate (RH = 62.4), NaBr (RH = 57.6), $Mg(NO_3)_2$ (RH = 52.9), K_2CO_3 (RH = 42.16), $MgCl_2$ (RH = 32.8), and $KOOCCH_3$ (RH = 22.51) [90]. Sample drying is another step sometimes necessary between sample collection and further sample preparation. Sample drying is discussed in Section 8.1 as a sample preparation procedure. Dried samples can be stored with less chance of decomposition, biological activity, etc.

Contamination of the sample can occur from the containers or exposure to the environment. The containers used for storage and handling are the main source of contamination. Glass containers are suitable for most samples. A variety of other materials are also available. The use of a specific material must be proven to not generate contamination, for example, with volatile compounds or plasticizers, or to produce losses due to permeability [91]. In trace metal analysis, contamination is the main source of errors [45]. While glass is preferred for organic analytes, for metal analysis containers made from polyethylene or other inert material must be used. The washing of these containers must be done with nonionic detergents (sulfate-free) followed by thorough rinsing [92]. Standard solutions used for the calibration procedure must be stored in inert recipients. It was proved that even brief exposure of aqueous solutions to conventional laboratory glassware may produce significant contamination, particularly with sodium and chloride [93]. Also, aqueous samples are readily contaminated with bicarbonate, ammonium, and nitrate ions by the absorption of CO_2, NH_3 and NO_2 from the air [94]. Reagents used as additives also may introduce detectable levels of contaminant ions [95], [96]. The level of contamination can be estimated using a blank sample processed in the same way as the real samples.

REFERENCES 2

1. D. D. Wight, *Fundamentals of Air Sampling*, Lewis Pub., Boca Raton, 1994.
2. G. E. Baiulescu et al., *Sampling*, E. Horwood, Chichester, 1991.
3. K. G. Carr-Brion, J. R. P. Clarke, *Sampling Systems for Process Analysers*, Butterworth-Heinemann, Oxford, 1996.
4. H. Maarse, *Volatile Compounds in Food and Beverages*, M. Dekker, New York, 1991.
5. D. L. Massart et al., *Evaluation and Optimization of Laboratory Methods and Analytical Procedures*, Elsevier, Amsterdam, 1978.
6. P. M. Gy, Mikrochim. Acta, II (1991) 457.
7. M. Thomson, M. H. Ramsey, Analyst (London), 120 (1995) 261.
8. L. H. Keith (ed.), *Principles of Environmental Sampling*, ACS Professional Reference Book, ACS, 1988.
9. H. C. Pillsbury et al., J. Assoc. Off. Anal. Chem., 52 (1969) 458.
10. Routine Analytical Cigarette Smoking Machine-Definitions and Standard Conditions; ISO 3308:1991(E), International Standardization Organization, Genéve, 1991.

11. Tobacco and Tobacco Products - Atmosphere for Conditioning and Testing; ISO 3402:1991(E), International Standardization Organization, Genéve, 1991.
12. Cigarettes - Determination of Total and Nicotine Free Dry Particulate Matter using a Routine Analytical Smoking Machine; ISO 4387:1991 (E), International Standardization Organization, Genéve, 1991.
13. G. N. Connoly, H. Saxner, Memorandum, The Commonwealth of Massachusetts, Exec. Office of Health and Human Services, Aug. 19, 1997.
14. A. Rios, M. Valcarcel, Analyst (London), 120 (1994) 109.
15. C. Liteanu, I. Rica, *Teoria si Metodologia Statistica a Analizei Urmelor*, Ed. Scrisul Romanesc, Craiova, 1979, or its English translation, *Statistical Theory and Methodology of Trace Analysis*, E. Horwood, Chichester, 1980.
16. L. H. Keith et al., Anal. Chem., 55 (1983) 2210.
17. E. L. Crow et al., *Statistics Manual*, Dover, New York, 1960.
18. W. Mendenhall, *Introduction to Probability and Statistics*, Duxbury Press, Belmont, 1971.
19. M. Stoeppler, *Sampling and Sample Preparation*, Springer, Berlin, 1997.
20. *Annual Book of ASTM Standards, Section 11: Water and Environmental Technology,* p.95, 1988.
21. R. Soniassy et al., *Water Analysis. Organic Micropollutants*, Hewlett-Packard, p.191, 1994.
22. M. L. Wilson et al., *Atmospheric Sampling. Student Manual*, United States Environmental Protection Agency, 1983.
23. F. Bruner, *Chromatographic Environmental Analysis. Principles, Techniques, Instrumentation*, VCH Publishers, Cambridge, 1993.
24. E. Meszaros, *Atmospheric Chemistry. Fundamental Aspects*, Akad. Kiado, Budapest, 1981.
25. S. E. Manahan, *Environmental Chemistry*, Lewis Pub., Chelsea, 1991.
26. United States Environmental Protection Agency (US EPA), *Health Effects Notebook for Hazardous Air Pollutants*, www.epa.gov/ttn/atw/hapindex.html.
27. United States Environmental Protection Agency (US EPA), *Drinking Water Standards and Health Advisories*, EPA 822-B-00-001.
28. United States Environmental Protection Agency (US EPA), *National Recommended Water Quality Criteria-Correction*, EPA 822-Z-99-001.
29. U.S. Dept of Health and Human Services *NIOSH Pocket Guide to Chemical Hazards*, J. J. Keller & Assoc., Neenah, 1997.
30. *Threshold Limit Values (TLVs) and Biological Exposure Indices (BEIs) for Chemical Substances and Physical Agents*, ACGIH, Cincinnati, 1999.
31. M. R. Guerin et al., *The Chemistry of Envirionmental Tobacco Smoke*, Lewis Pub., Boca Raton, 1992
32. S. Mitra et al., J. Microcolumn, 8 (1996) 21.
33. S. Mitra et al., J. Chromatogr., A, 727 (1996) 111.
34. N. H. Zhu et al., J. Microcolumn, 10 (1998) 393.
35. J.-Z. Dong et al., J. Microcolumn Sep., 12 (2000) 142.
36. W. K. Harrison, J. S. Nader, Am. Ind. Hyg. Assoc., 21 (1960) 115.
37. B. B. Hicks et al., in L. H. Keith (ed.), *Principles of Environmental Sampling*, American Chemical Society, p.229, 1993.
38. J. Namienik, J. Chromatogr., 300 (1984) 79.
39. V. David et al., Rev. Chim. (Bucharest), 40 (1989) 528.
40. J.-Z. Dong et al., Beiträge Tabak. Int. 19 (2000) 33.

41. J. L. Harris, L. E. Hayes, Tob. Sci., 21 (1977) 58.
42. C. D. Nenitescu, *Chimie Generala*, Ed. Didactica si Pedagogica, Bucuresti, 1972.
43. M. J. Joseph, E. W. Cochran, 53rd Tobacco Science Research Conference, Sept. 12-15, 1999, Montreal, Canada.
44. H. Frank et al., J. High Resolut. Chromatogr., 18 (1995) 83.
45. P. R. Haddad, P. E. Jackson, *Ion Chromatography. Principles and Applications*, Elsevier, Amsterdam, 1990.
46. J. D. Mulik et al., in E. Sawicki et al. (eds.), *Ion Chromatographic Analysis of Environmental Pollutants*, vol. 1, Ann Arbor Sci. Publ., Ann Arbor, 1978.
47. B. Dellinger et al., Environ. Sci. Technol., 14 (1980) 1244.
48. J. H. Margeson et al., Anal. Chem., 57 (1985) 1586.
49. T. W. Dolzine et al., Anal. Chem., 54 (1982) 470.
50. D. Grosjean, J. D. Nies, Anal. Lett., 17 (1984) 89.
51. R. D. Holm, S. A. Barkdale, in E. Sawicki et al. (eds.), *Ion Chromatographic Analysis of Environmental Pollutants*, vol. 1, Ann Arbor Sci. Publ., Ann Arbor, 1978.
52. J. M. Lorain et al., Anal. Chem., 53 (1981) 1302.
53. T. Yasuoka et al., Buseki Kagaku, 32 (1983) 580.
54. W. C. Askew, S. J. Morisani, J. Chromatogr. Sci., 27 (1989) 42.
55. J. Braunstein, G. D. Robbins, J. Chem. Educ., 48 (1971) 52.
56. P. Aragon et al., Anal. Chem., 30 (2000) 121.
57. Y. C. Chen, J. G. Lo, Anal. Sci., 13 (1997) 199.
58. V. Giarrocco et al., Tec. Lab., (1995) 546.
59. B. Zielinska et al., Atmos. Environ., 30 (1996) 2269.
60. C. C. Huang et al., J. Chromatogr., A, 731 (1996) 217.
61. N. Moschonas, S. Glavas, J. Chromatogr., A., 733 (1997) 117.
62. G. D. Byrd et al., J. Chromatogr., 503 (1990) 359.
63. B. C. Turner, D. E. Glotfelty, Anal. Chem., 49 (1977) 7.
64. R. G. Lewis et al., Anal. Chem., 49 (1977) 1668.
65. R. A. Wetzel et al., Chem. Anal. (N.Y.), 78 (1985) 355.
66. P. J. Schoenmakers, *Optimization of Chromatographic Selectivity*, Journal of Chromatographic Library, Vol. 35, Elsevier, Amsterdam, 1986.
67. J. C. Berridge, Analyst (London), 107 (1984) 291.
68. T. Sunden et al., Anal. Chem., 55 (1983) 2.
69. M. L. Balconi, F. Sigon, Anal. Chim. Acta, 191 (1986) 299.
70. P. R. Haddad, C. E. Cowie, J. Chromatogr., 303 (1984) 321.
71. D. R. Jenke, Anal. Chem., 56 (1984) 2674.
72. J. Slanina et al., in E. Sawicki, J. D. Mulik (eds.), *Ion Chromatographic Analysis of Environmental Pollutants*, vol.2, Ann Arbor Sci. Publ., Ann Arbor, 1979.
73. R. H. Brown, M. D. Wright, Analyst, 119 (1994) 75.
74. M. Harper, C. J. Purnell, Am. Ind. Hyg. Assoc. J., 48 (1987) 214.
75. J. O. Levin, L. Carleborg, Ann. Occup. Hyg., 31 (1987) 31.
76. J. O. Levin et al., J. Chromatogr., 454 (1988) 121.
77. M. Harper, Analyst, 119 (1994) 65.
78. R. J. B. Peters, H. A. Bakkeren, Analyst, 119 (1994) 71.
79. K. Lee et al., Environ. Sci. Technol., 26 (1992) 697.
80. J. O. Levin, R. Lindahl, Analyst, 119 (1994) 79.
81. P. Winkler et al., Fresenius J. Anal. Chem., 340 (1991) 575.
82. B. Market (ed.), *Environmental Sampling for Trace Analysis*, Verlag Chemie, Weinheim, 1994.

83. K. E. Creasy, T. W. Francisco, in R. A. Meyers (ed.), *Encyclopedia of Analytical Chemistry: Applications, Theory and Instrumentation*, J. Wiley, Chichester, 2000.
84. E. Baltussen et al., J. Microcolumn Sep., 11 (1999) 737.
85. A. L. Pape et al. (eds.), *Methods of Soil Analysis*, ASA-SSA, Madison, 1982.
86. W. Salomons, W. M. Stigliani, *Biogeodynamics of Pollutants in Soils and Sediments*, Springer, Berlin 1995.
87. A. V. Medvedovici et al., Chem. Anal. (Warsaw), 43 (1998) 47.
88. E. Prichard et al., *Trace Analysis: A Structured Approach to Obtain Reliable Results*, The Royal Society of Chemistry, London, 1996.
89. B. Garfinkle, M. Henley, in R. Remington (ed.), *The Science and Practice of Pharmacy*, 19th Edition, vol. II, Mack Pub., Easton, 1995.
90. L. Greenspan, J. Res. Natl. Bur. Stand. Sect. A, 81A (1977) 89.
91. J. R. Moody, Philos. Trans. Royal Soc. London, 305 (1982) 669.
92. M. A. Fulmer et al., in E. Sawicki, J. D. Mulik (eds.), *Ion Chromatographic Analysis of Environmental Pollutants* , vol. 2, Ann Arbor Sci. Publ., Ann Arbor, 1979.
93. S. Saigne et al., Anal. Chim. Acta, 203 (1987) 11.
94. M. Legrand et al., Anal. Chim. Acta, 156 (1984) 181.
95. R. M. Cassidy, S. Elchuk, J. Chromatogr. Sci., 19 (1981) 503.
96. L. W. Green, J. R. Woods, Anal. Chem., 53 (1981) 2187.

CHAPTER 3

Overview on Sample Preparation for Chromatography

3.1   THE ROLE OF SAMPLE PREPARATION

Sample preparation consists of one or more operations necessary to modify the sample for delivering the analytes in a convenient form for the analysis by a chosen chromatographic procedure. Sample preparation, being an intermediate step between sample collection and the chromatographic analysis, is influenced by both sides of the process. The end result of sample preparation is a processed sample that can be delivered to the chromatographic instrument.

Chromatographic techniques are initially considered as candidates for the core step in an analytical method when they satisfy two conditions: a) they are appropriate for the analysis of the pure compounds of interest, and b) they have the sensitivity necessary for the determination of the lowest quantity of analyte estimated to be transferred in a processed sample. Further selection from a number of possible techniques is done based on the chances to achieve a good separation of the analytes and other sample components (see Section 4.1). Once the potential capability of a chromatographic technique to analyze the sample is established, a sample preparation is selected such that the matrix of the sample is simplified or eliminated and does not produce interferences in the chosen chromatographic analysis. Then the sample preparation is designed to assure that the analytes from the initial sample are transferred to the chromatographic instrument with the appropriate concentration relative to the limit of detection of the chromatographic method. This process may involve a concentration of the sample. Some concentration operations can be done to increase selectively only the concentration of the analyte. Only the use of special techniques such as selective adsorption followed by desorption or specific membrane separations can achieve a selective concentration. Common concentration techniques may increase at the same time the concentration of the analyte and that of the matrix containing interfering species. When this type of concentration takes place, a separate cleanup may be necessary either previous to the concentration or after it. The selection of the chromatographic process and of the sample preparation can be summarized by the diagram in Figure 3.1.1.

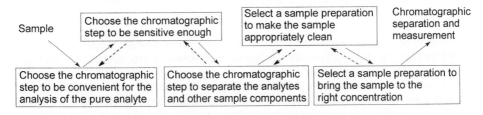

FIGURE 3.1.1. *Diagram summarizing the selection of a chromatographic core step and of an appropriate sample preparation.*

Although the scheme in Figure 3.1.1 shows a simple sequence of selections starting with the sample and ending with the chromatographic process, iterations are sometimes included in the process. Also, a feedback flow of information is present during this selection. For example, the selection of a sample preparation that brings the sample to the right concentration but increases the interferences from the matrix to unacceptable limits requires a modification of the cleanup operation and the process is repeated. The selection of chromatography also can be reconsidered based on the sample preparation selection. For these reasons, the whole process of choosing a sample preparation procedure can be iterative.

A common reason to perform sample preparation is the dissolution of the sample (required mainly when the sample is solid). Also, sample preparation attempts to solve the two most common problems encountered when performing an analysis, interferences from the matrix (or among the analytes) and the lack of sensitivity of the detection technique. The first problem is solved by the elimination of compounds generating analytical signals similar to the analytes, modification of a matrix that interferes nonspecifically in the analysis, or fractionation of the sample to generate simpler samples for analysis. The second problem is solved by enhancing the properties of the analyte such that the detection technique can measure the analytes. The reduction or even the elimination of interferences is done using the cleanup or fractionation of the sample as well as derivatization. For this purpose, the role of cleanup is to reduce the matrix of the sample; the fractionation selects from the sample a portion that contains the analytes, and the derivatization modifies only the analytes or only the matrix and may improve the separation. For improving the sensitivity, concentration and derivatization are used.

The number of reported sample preparation procedures is very large and can include various operations with quite different purposes. For obtaining a specific result, both physical and chemical operations can be used. It is common that sample preparation modifies the matrix of the sample by eliminating unwanted components, concentrating the analytes, etc. The preparation procedures not involving chemical reactions are applied mainly for cleanup and concentration of the sample. After this stage in sample preparation, the analyte may be chemically modified, for example, for further fractionation of for better detectability. In practice, more than one operation can be used for the same purpose. For example, sample cleanup can be achieved by precipitation of a part of the matrix, extraction with an organic solvent of the analyte, reextraction of the organic layer with water, etc. In many methods, the resulting processed sample is a solution cleaned from interfering compounds, possibly more concentrated, and suitable for the core analytical procedure.

As shown in Section 1.2, depending on the use of a solvent, sample preparation procedures can be classified as using solvents or solvent-free procedures. In solvent-free procedures, a processed sample in the form of a solution containing the analyte is not generated. The analytes are just desorbed from the initial sample or a partially processed sample on a solid phase material. The analytes are then transferred into the analytical instrument, for example by desorption in the GC injection port. The same solid phase can be reused after releasing the analytes into the chromatographic instrument. This type of sample preparation will be discussed in more detail in Chapters 9 and 11.

The simplified diagram of a procedure that uses dissolution, cleanup, concentration, and derivatization as operations that lead to a processed sample in the form of a solution containing the analytes is shown in Figure 3.1.2a. The diagram for a solvent-free procedure is shown in Figure 3.1.2b.

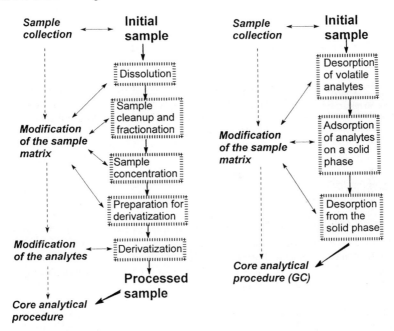

FIGURE 3.1.2a. *Diagram of a sample preparation involving dissolution, cleanup, fractionation, concentration, and derivatization.*

FIGURE 3.1.2b. *Diagram of a solvent-free sample preparation procedure.*

The selection and the order of different operations are dictated only by the need of the analysis. Cleanup and concentration operations, for example, can be done even after derivatization. An example of a complex sample preparation scheme is shown in Figure 3.1.3. The procedure has been used for the analysis of pregnenolone in rat brain using derivatization with 1-antroylnitrile and with the internal standard (I.S.) 3β-hydroxy-16-methylpregna-5,16-dien-20-one [1]. In this procedure, the analytical separation was done by HPLC on an ODS column, using CH_3CN/water (90:10) as eluent and fluorescence detection with excitation (ex.) 370 nm, emission (em.) 470 nm. The derivatization step was followed by additional sample preparation using a cleanup step on a silica column. The order of each operation is important, following a logical sequence determined by the goal of sample preparation. In the example shown in Figure 3.1.3, the goal is to provide for the HPLC analysis a derivatized amount of pregnenolone at a concentration proportional to that from the analyzed rat brain and appropriate for the sensitivity of detection of the HPLC technique, in this case the fluorescence detector. In addition, the processed solution was free of proteins and other interfering compounds.

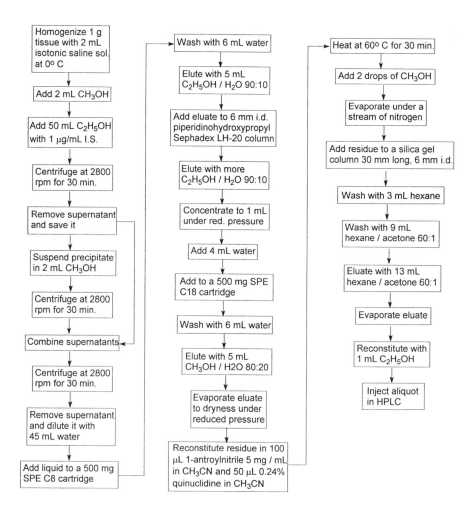

FIGURE 3.1.3. *Diagram for sample preparation for the analysis of pregnenolone in rat brain using derivatization with 1-antroylnitrile [1].*

As seen in Figure 3.1.3, sample preparation must be viewed as a complex process, formed from a chain of operations each one with its own role, in a specific sequence necessary to achieve the goal of the analysis. The operations can be independent, each with its own role, but also some operations can be merged in a more elaborate procedure. The presentation of individual operations used in sample preparation is in Part 2 and 3 of this book. This presentation cannot capture properly the possible interconnections between different operations, and only the description of an entire method can give this view. Also, new sample preparation procedures are continuously developed and reported. For these reasons, the search of original literature related to a particular subject is frequently necessary for selecting the appropriate sample preparation method.

3.2. THE USE OF PRELIMINARY INFORMATION FOR THE CHOICE OF A SAMPLE PREPARATION PROCEDURE

In the selection of a particular sample preparation procedure, a number of criteria must be taken into consideration. The use of information about the intended analysis, immediate purpose of analysis, nature of the sample, and required quality of results, as given in Table 1.1.1, can be used for this decision. Starting with this information, a number of suggestions regarding sample preparation are given in Tables 3.2.1 through 3.2.4. These suggestions are based on the information related to the type of intended analysis, the immediate purpose of analysis, the nature of sample, the required quality of results, etc.

TABLE 3.2.1. *The relation between the type of intended analysis and the external information.*

Information	Recommendation
Is the analysis needed for a known type of sample or for a new type?	For samples of a known type, it is preferable to use a previously established and reported (in literature) sample preparation technique. Sample cleanup and concentration can be done using a procedure not necessarily related with a specific derivatization. Improvements may be attempted by varying the existent possibilities.
Is a new analytical method needed or can a reported one be used?	For a totally new type of sample, a new method may be needed. Similarities with known procedures must be identified and evaluated. Full knowledge of reported methods may provide useful guidance.
Is the analysis a "one time" need or a routine analysis?	For "one time analyses," a search for an outside laboratory to perform the analysis is highly recommended. Full development of an analytical method also can be avoided by applying a "scan type" technique such as GC-MS or LC-MS using multipurpose columns for the separation.
What is the required turnaround time (rapid, or long term analysis)?	Analyses requiring a short turnaround time must avoid extensive sample preparation. Specialized selective detectors may work well, eliminating the need for sample cleanup or concentration.
A specific protocol must be followed, or no regulations are imposed?	The need to follow a specific protocol reduces the options in selecting a different analytical method.
Is the analysis done for a process evaluation?	Process evaluation analyses must carefully evaluate the optimization of sampling and the relevance of the results.
What instrumentation is available?	The lack of certain instruments may limit the choice of a specific analytical method. The use of a derivatization technique either for GC or for LC may expand analytical possibilities. Also, extensive sample preparation may eliminate the need for a very sensitive instrument.
What are the cost, manpower, and funding?	Insufficient manpower and funding are common. The estimation of the level where the problem becomes prohibitive is important.
What expertise is available?	Lack of expertise can be compensated in part with self education and outside training.
How certain is the info. about the sample?	Uncertain information can be misleading, and the choice between further research and use of available information must be evaluated.

TABLE 3.2.2. *Suggestions based on the immediate purpose of analysis.*

Information	Recommendation
Is the analysis done only to determine elemental composition?	Nonchromatographic techniques such as ICP, AA, X-ray fluorescence, or combustion with H, C, N, O analysis are typically utilized.
Is the analysis a complete or a selective qualitative or semiquantitative measurement?	Selective qualitative analyses can be done using dedicated tests, some of them not using chromatographic techniques. Only in complex mixtures and when other procedures fail, must chromatography be used. TLC may offer an excellent alternative for qualitative analysis.
Is the analysis qualitative, quantitative, or both?	Quantitative analysis poses significantly more problems than qualitative ones. A "scan type" technique using GC-MS or LC-MS may be sufficient for the identification of compounds even in a complex mixture. A derivatization such as silylation may expand the range of identifiable compounds by GC. Quantitative methods typically require sample preparation procedures associated with the chromatographic analysis.
Is quantitation needed for all or for specific components?	Most quantitations need a complete chromatographic analytical method, with sample preparation. Only some compounds may be analyzed using dedicated instruments such as specific electrodes (CN, sulfide, glucose, etc.) or other nonchromatographic techniques.
Is any structural analysis, isomer analysis, MW of macromolecules needed?	Structural analyses typically require specialized techniques. Depending on the problem, dedicated methodology and instrumentation (such as NMR) may be needed for structure elucidations. Isomers may be analyzed as any other analytes except for special chromatographic separations. The analysis of macromolecules may have different requirements than usual analyses.
Is the sequence of a polymer or a weak interaction compound determined ?	Polymer analysis and weak interaction compounds typically require specialized analytical methodology.
Is the analysis destructive or must be nondestructive?	Spectroscopic techniques are usually applied when nondestructive analysis is required.
Is the whole sample analyzed or only a specific part (surface, etc.)?	If only specific parts of the sample need analysis, a resampling may be recommended or the use of specialized techniques (for example for surface analysis).
Do other analyses follow?	Concern about preserving part of the sample appears if other analyses follow. Also, proper communication of the results between analyses is recommended.
Are there preparative purposes?	Nondestructive procedures must be adopted when preparative purposes follow a specific analytical method.

TABLE 3.2.3. *Suggestions based on the nature of sample including the analytes and the matrix.*

Information	Recommendation
Is the sample solid, liquid (solution, colloid), gas or mixed phases? Is the sample homogeneous?	Nonhomogeneous samples typically need careful sampling followed by the separation of phases. Gases and the compounds that are thermally stable at their boiling point are preferably analyzed by GC. Labile compounds or those impossible to volatilize are typically analyzed by LC. GC-MS offers excellent sensitivity and capability for quantitation with simultaneous positive identification.
Is the sample inorganic, organic, mix, of biological origin, environmental?	Inorganic samples can be analyzed using chromatographic methods such as IC, but techniques such as ICP, AA, etc. are more frequently applied. The analysis of complex samples typically requires sample preparation. Chromatography is the technique preferred in many organic, biological and environmental analyses. Sample preparation is typically unavoidable.
What is the number of samples and the number of analytes to measure?	For large number of samples, automation must be seriously considered. Chromatographic techniques with sample preparation are common for samples with multiple analytes.
Is the matrix complex or simple? Are interferences expected?	Complex matrices typically require extensive sample preparation including cleanup, concentration, and possible derivatization. These steps are more likely to avoid the effects of interferences.
Has the sample small molecules, ionic compounds, polymers, or is it a mix?	Samples with small molecules are typically analyzed using GC (with or without sample preparation). LC also may be used sometimes using derivatization for enhanced sensitivity. Ion compounds can be analyzed using IC. Polymers can be separated using GPC and analyzed unmodified, or using hydrolysis or pyrolysis.
What is the estimated level of the analyte (major, trace, etc.)?	Major analytes are easier to analyze and typically require less elaborate sample preparation. Trace analytes may need extensive sample preparation, concentration, and possible derivatization for enhancing sensitivity.
What is the solubility of the sample? Is there a need for digestion?	Soluble samples are less likely to be altered during sample preparation. Digestion techniques may destroy the organic compounds.
What is the thermal stability, perishability of the sample?	Thermally labile samples may need special storage and handling. LC type techniques are more frequently used for analysis. Cold on-column GC may also be applicable for some analyses. Perishability may be modified using preservatives.
What are the availability and the value of the sample?	Samples that are not available in sufficient quantity require sensitive analytical methods (or nondestructive techniques). Special micro sample preparation procedures may be needed.
Are there hazard problems?	Appropriate measures must be taken if hazard problems are present.

TABLE 3.2.4. *Suggestions based on the required quality of results.*

Information	Recommendation
Is the evaluation of qualitative results needed?	Qualitative results are typically evaluated based on the extent they provide the necessary information.
Is high sensitivity required (known detection limit)?	High sensitivity typically required in trace analysis implies sensitive instrumentation and sample preparation that perform appropriate cleanup and concentration of the analytes.
What precision and accuracy are required?	High precision and accuracy imply the use of analytical methods well verified and compared with other methods. Higher number of replicates increases the probability of obtaining a desired precision and accuracy.
Is statistical analysis on various sets of samples required?	A carefully designed sampling protocol is recommended when statistical analysis of the data is required.
Is a comparison with info. in a data base or other laboratories necessary?	Comparison of the data require careful consideration of drift in the analytical results, conditions of analysis, the use of the same or of different methods, etc.
What is the desired robustness of the analysis?	Robust analytical procedures are always preferred. Transferability from person to person and from laboratory to laboratory is important.
What is the desired ruggedness of the analysis?	Rugged analytical procedures are always preferred. Factors affecting easily the results of an analytical method are highly undesired. Personnel qualification, instrument reliability, etc. must be evaluated when a method is not stable.
What is the utility of the results?	The utility of the results is the most important factor that must be considered when selecting a sampling procedure, a sample preparation technique, and a core analytical method.

Tables 3.2.1 to 3.2.4 provide guidance only regarding the general strategy applied for the choice of an analytical method in general and a sample preparation procedure in particular. From a number of criteria for choosing a specific sample preparation procedure, the main one is to be part of a method capable of achieving the purpose of analysis. In addition, a number of circumstantial conditions are important in the selection of a specific sample preparation procedure, including the availability of proper equipment or materials required to perform the analysis and the time (when and how long) required to perform the analysis. These restrictions must be known from the initial information about the sample and the required analysis.

The initial information may be crucial in deciding if an analysis is possible using a specific method. The knowledge about the matrix may indicate if interferences are likely and how significant may be their role. Also, the knowledge about the level of the analyte in the sample is extremely important. Entirely different methods are chosen when the analytes are in high concentrations compared to when they are in trace amounts. Another important piece of information regarding the sample is its availability. For a sample available in large amount and with the analytes at low concentrations, a sample preparation procedure leading to the concentration of the analyte may be the best alternative. However, if the same sample is available only in a small amount, a modification of the properties of the analytes, such as a derivatization, and the use of a very sensitive detection technique may be the correct path. For a group of chosen methods, all able to achieve the purpose of analysis, the other selection criteria are used only for optimization.

A simplified diagram that can be applied for the choice of a specific path in sample preparation is shown in Figure 3.2.1 [2].

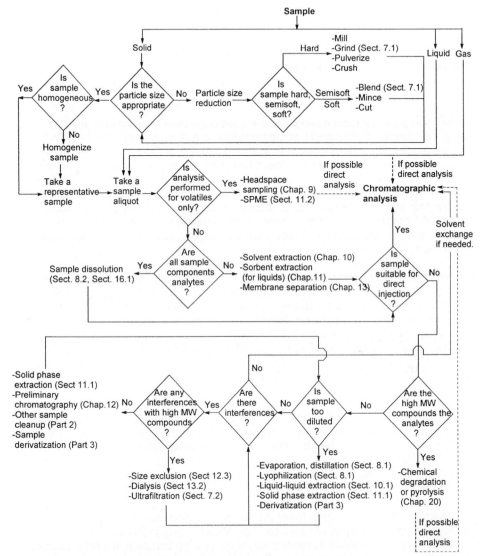

FIGURE 3.2.1. *Simplified diagram of the sample preparation path [2].*

The details about the technical aspects of each sample preparation operation are given in Part 2 and Part 3 of this book.

3.3　SAMPLE HOMOGENIZATION AND DISSOLUTION

Sample homogenization and dissolution are frequently the first operations in sample preparation. Samples can be classified based on the overall appearance of their physical nature as gas, liquid, or solid (see Section 2.3). Most samples are not homogeneous, although they may be large enough to be representative (see Section 2.2). For nonhomogeneous samples a decision must be made whether or not they should be homogenized before analysis, or if specific components must be separated and analyzed independently. Homogenization is always applied when resampling is necessary to use in the analysis only a small part of the initial sample.

For solid samples, homogenization is sometimes done after grinding and possibly sieving the sample. Solid sample homogenization is further discussed in Section 7.1. A common procedure for homogenization is the dissolution of the sample. Dissolution is usually applied to solid samples, but liquid and even gaseous samples can be dissolved in a specific solvent. Dissolution is discussed in more detail in Section 8.2. Besides dissolution in an appropriate solvent, some samples require special chemical reactions to be transferred into solution. These chemical reactions may be applied to modify the analytes, the matrix, or the entire sample. Some aspects of the use of chemical reactions with the sole purpose of taking the sample into solution are discussed in Section 16.1.

Homogenization is not always used or necessary for a sample. In some instances, the lack of homogeneity is used in favor of easier analysis. For example, for gas samples containing liquid or solid aerosols, the separation of physically distinct parts of the sample can be done using filtration, followed by the analysis of the gas, of the residue on the filter when the analytes are not in the gas, or of both gas and particles. Other separations can be applied such as electrostatic precipitation, adsorption in a liquid medium, etc. Stability in time of the gas sample can be an issue when the sample is not homogeneous, because if the particles can be separated by filtration, it is likely that the sample as a gas is not stable in time (see Section 2.3). Gas samples containing liquid aerosols may deposit some of the aerosols on specific filters, but some of the material may remain in vapor phase if the partial vapor pressure of the liquid is not negligible. This leads to erroneous results when only the deposit on the filter is analyzed. Various alternatives are used to recover the whole sample, such as collection in impingers, in denuders, or cryo collection of the gas emerging from the filter. Adsorbing materials for the gases or vapors also can be utilized for retaining the analytes of interest. These aspects were briefly discussed in Section 2.2.

Liquid samples pose, in general, fewer problems regarding resampling or homogenization. Most liquid samples can be handled easily in GC or HPLC. Liquids can be dissolved more easily in specific solvents, insuring that no precipitation occurs. The proper dissolution of liquids is usually only a matter of selecting properly a specific solvent. When the liquid sample is not homogeneous and contains particles, these can be separated using filtration, centrifugation, etc. Specific care must be taken for the possibility of having the analyte distributed in two phases. Solids present in a liquid may adsorb selectively part of the analyte and modify the result of an analysis if the analytes are not redissolved. Liquids containing gases are analyzed using specific procedures that are able to properly evaluate the amount of gases in the sample. The analysis of gases and volatile analytes from a solution can be done using a number of methods such as sparging of the liquid with a stream of gas followed by adsorption of the analytes on a solid phase material. Some of the methods used to separate the volatiles from a liquid are described in Chapter 9 and Section 11.2. Other analytes in solution can be separated by extraction with liquids, with various sorbents, by preliminary chromatography, using membranes, etc. A number of technical details regarding the concentration and cleanup of the analytes in solution are given in Part 2.

Some solid samples do not need homogenization, and a selective separation is performed on a part of the sample that contains the analytes. The samples may contain the analytes as solid compounds that can be either volatile or nonvolatile, but possibly as liquids or even gases. When the analytes are gases or volatile compounds, a number of headspace techniques are available for analysis. These are discussed in detail in Chapter 9. For the analytes present as liquids or solids, the sample is typically subjected to dissolution or to the extraction of a part of the sample containing the analytes. Simple extraction in water can be sufficient, for example, for the analysis of soluble ionic species from solid samples followed by IC analysis [3]. A number of techniques are available for extracting solid samples with specific solvents, for removing volatile compounds of interest from the solid matrix and collecting them from the sample headspace, etc. These techniques are discussed in Chapters 9 to 11.

3.4 SAMPLE CLEANUP, FRACTIONATION, AND CONCENTRATION

Interferences are a major reason why some analyses are not possible. One common purpose of sample preparation operations is the elimination of some (and sometimes all) of the compounds that form the matrix and interfere with the analyte measurement. The interference can be caused by some specific compounds that are difficult to separate in the core analytical step or by the matrix that interferes in an unselective manner. Not only the separation or the detection of the analyte may suffer because of a certain matrix, but also problems may arise with solubility, adsorption, depositions in the injection port, deterioration of the chromatographic column, etc. that are caused by the compounds in the matrix. The process of elimination of undesired compounds from the matrix is commonly known as *sample cleanup*. A variety of sample cleanup techniques and procedures are reported in literature (e.g. [4], [5]).

Most cleanup procedures are based on separation techniques. The separations using the differences in the partition between two phases are very common. Among these are solid phase extraction (SPE), headspace desorption, supercritical fluid extraction, accelerated solvent extraction, microwave assisted extraction, solid phase microextraction (SPME), matrix solid phase dispersion, and chromatography itself (such as preparative chromatography). Separations based on transport rate such as various types of electrophoresis, ultracentrifugation, etc. and mechanical separations such as filtration, sedimentation, sieving, membrane filtration, gel permeation, and many others also are utilized. A detailed discussion of each technique is given in Part 2 of this book. Some cleanup and fractionation operations are not based on physical separations. The typical example is that of a sample containing an acidic and a basic fraction. Chemical treatment of the sample can be used in this case for sample fractionation. These types of operations are discussed in Part 3 of the book.

The cleanup process is not always able to eliminate completely the undesired matrix components. These may include compounds interfering with the analytes or other materials that affect negatively the analysis. This calls for further sample preparation operations such as derivatizations. Also, many compounds may remain in the matrix after cleanup, and these are separated by the core chromatographic process. However, when the matrix is simpler ("cleaner processed sample"), the results of the core chromatographic analysis are usually better.

An alternative to the elimination of the undesirable compounds from the matrix is the separation of the sample into a number of fractions, such that the analyte is found in only one of these fractions. When dealing with more analytes, it is also possible to have the analytes separated in groups, each one in a separate fraction. This process of separation of the sample into a number of groups of subsamples is known as *fractionation*. A clear distinction between sample cleanup and fractionation is difficult to make, because fractionation, by separating parts of the matrix from the initial sample, can be considered just a specific cleanup process. The distinction is still made, mainly when fractions containing analytes are isolated from a sample, as opposed to other ways of reduction of some interfering matrix component.

After sample preparation, the analytes are typically present in a carrier, such as a solvent appropriate to be injected into the chromatographic system. This solvent is not considered a new matrix, because it does not interfere with the analysis and can be easily reduced or eliminated. The nature of the carrier (solvent) of the analytes can be chosen during the sample preparation to satisfy specific requirements such as volatility in GC or miscibility with the mobile phase in HPLC.

A clear differentiation between concentration and cleanup of the sample is, again, difficult to make. *Concentration* without cleanup is done only when the analytes are in a carrier solvent and this is eliminated (in part or completely). The concentration done with respect to the matrix of the sample can be considered a cleanup operation because it reduces the proportion of the matrix in the sample. When part of the matrix is eliminated, the partially processed sample may be subject to a concentration process if, for example, the sample is in a solution. This step may increase the concentration of both the analytes and of the remaining compounds from the matrix. If the analysis of the concentrated sample is adversely affected by the matrix, an additional cleanup operation may be necessary.

Specific samples such as gas or water samples from the environment have air or water as matrix. The elimination of water, for example, from this type of water sample should be considered sample cleanup if the definition is followed strictly. However, it is common in these cases to consider the elimination of water a concentration step. Some other operations that eliminate a major component from the matrix are sometimes indicated as concentrations. For this reason, no effort has been made to differentiate cleanup from concentration procedures, and the two terms are frequently used interchangeably.

In an attempt to eliminate the undesired matrix, sample cleanup procedures may lose part of the analyte. The sample preparation process is usually designed such that the amount of analyte from the initial sample is transferred as efficiently as possible into the processed sample. However, during sample preparation it is possible that some portion of the analyte is lost. The transfer efficiency is described by a parameter known as analyte recovery R. (In order to differentiate recovery R from the fraction of moles of solute "i" in the mobile phase also noted with R, the recovery is usually expressed as %.) When the processed sample is isolated as an intermediate material to be analyzed, analyte recovery is usually expressed in percent and is defined by the relation:

$$R = \frac{q_p}{q_i} \quad (R\% = 100\ R) \tag{3.4.1}$$

where q_p is the amount of analyte in the processed sample and q_i the amount of analyte in the initial sample. Good analyte recovery is considered when $R\% > 90\%$. The measurement of analyte recovery in an analytical method is typically done by adding a known amount of analyte in a blank sample and measuring it in the processed sample. This procedure requires the availability of a blank sample. When a blank sample is not available, the standard addition method is typically used to find the amount of the unknown (see also Section 3.8). The measurement of an analyte in the processed sample by the core analytical procedure is considered in this step as always generating correct results. However, it must be verified that the processed sample is clean enough and does not interfere with the analytical measurement having deviations from the calibration line (the values for the quantities of analyte are obtained using the measured concentration in the processed sample).

Analyses with low recoveries may still be utilized for qualitative analysis and even for quantitative analysis. Quantitative analysis can be performed using internal standards that have the same recovery as the analyte. When both the analyte and the standard have the same recovery, their ratio still can be used for quantitation. However, it is not always possible to have standards with the same recovery as the analyte, even when using labeled compounds as standards. The problem of appropriate choice of the standards is further discussed in Section 3.8.

Another parameter that must be considered in sample preparation is the factor of concentration change F (enrichment factor) between the processed sample and the initial sample. This factor is defined by the expression:

$$F = \frac{c_p}{c_i} \tag{3.4.2}$$

where c_p is the concentration in the processed sample and c_i the concentration in the initial sample. Because c_i is not known for an unknown sample, only an estimate of F can be initially obtained. However, the value for F must be determined for every analytical method. In some instances only a simple calculation is necessary to obtain F, but in other cases a calibration with known standards must be performed. The factor F is used to change the concentration c_p measured by the analytical instrument into the concentration c_i for the initial sample. The higher is the factor F, the higher has been the increase in the concentration of the analyte during sample preparation. When the calibration of an analytical instrument is not done based on the processed sample (using standards directly injected into the chromatographic instrument), but based on the initial sample (for example using the standard addition technique), the calculation of the concentration factor is not necessary. However, the information on the concentration factor for the material introduced into the chromatograph is always beneficial.

All analytical methods, including chromatography, require a minimum volume V_p^{min} and a minimum concentration c_p^{min} for the processed sample. These values depend on the type of chromatographic technique and are either known or can be readily estimated or measured using standards. For example, GC methods typically require a minimum of 0.1–0.2 mL of processed sample, although the injection volume in a GC system can be as low as 0.1 μL. The volumes needed for HPLC are usually larger. The minimum concentration of the sample highly depends on the sensitivity of the detection system, but concentrations between 0.1–10 μg/mL are typically required for GC. A larger range of concentrations is used in HPLC (see Section 4.3). Even for sample preparation techniques that transfer a number of analytes from the initial sample to a solid phase material and further to a GC system without preparing a processed sample, the amount of adsorbed analytes should be estimated.

The choice of the core chromatographic technique, of the sample preparation procedure, and of the amount of initial sample must be made such that they fulfill the requirements for the minimum volume of processed sample and of the minimum concentration required for successful chromatographic determination. With a known c_p^{min} and the minimum concentration of the analyte in the sample estimated as c_i^{min}, the sample preparation procedure must provide a concentration factor $F^{min} = c_p^{min} / c_i^{min}$. The value for F^{min} is not always predicted accurately, and trial and error with increasing concentration factors may be needed to achieve the c_p^{min} for a given sample.

The minimum required amount of processed sample q_p^{min} can be estimated based on the minimum necessary volume of processed sample V_p^{min} and minimum concentration of the processed sample ($q_p^{min} = V_p^{min} c_p^{min}$). With a known q_p^{min} and recovery R%, the necessary amount of analyte q_i^{min} in the initial sample can be evaluated using rel. (3.4.1) ($q_i^{min} = q_p^{min} 100 / R\%$). From this, the necessary volume of initial sample V_i^{min} can be calculated based on the expression:

$$V_i^{min} = \frac{V_p^{min} 100 \ F}{R\%}$$

(3.4.3)

The initial sample may not be a liquid, and in this case the volume of the initial sample must be changed to the amount of initial sample $Q = V_i^{min} \rho$ where ρ is the initial sample density. Rel. (3.4.3) indicates that the amount of initial sample for one analysis is higher when the needed volume of the processed sample is higher, when the recovery is lower, or when the concentration factor is higher.

The evaluation of the necessary amount for the initial sample is a very useful step in the decision regarding the choice of a specific sample preparation procedure. If the sample is not available in sufficient quantity, it is obvious that the chosen analytical procedure is not appropriate for the specific analysis. The alternative is to select a different measurement technique with higher sensitivity (lower detection limit), requiring less analyte and therefore less initial sample.

One advantage brought by the sample concentration is a lower detection limit for the overall method in the analysis of the initial sample. The detection limit is defined, as shown in Section 1.11, by the concentration corresponding to the value of the average signal that is higher with 4.66 s than the average of the blank sample (s is the standard deviation of the signal). The increase in the concentration of the analytes in the processed sample will give a lower detection limit for the analysis of the compounds of interest. Assuming that the analytical procedure has a linear dependence $y = a + bc$ between the signal y and the concentration c, the detection limit for the processed sample is given by rel. (1.11.36a). Using in rel. (1.11.36a) mean values of the measurements and the calculated standard deviation s, the expression for the detection limit for the concentration of the processed sample becomes:

$$c_{s \text{ (processed sample)}} = \frac{m_s - a}{b} = \frac{1}{b} (m_b + 4.66 \text{ s} - a) \qquad (3.4.4)$$

where m_b is the mean value of the blank measurements (see Section 1.11). The detection limit in the initial sample is given by:

$$c_{s \text{ (initial sample)}} = (1 / F) \, c_{s \text{ (processed sample)}} \qquad (3.4.5)$$

where F is the factor for concentration change between the initial sample and the processed sample given by rel. (3.4.2). The increase in the concentration of the analytes during sample processing leads to an increase in the overall value for F. Rel. (3.4.5) shows that the higher is the factor F for concentration change, the lower is the detection limit of an analyte in the initial sample. Therefore, a sample preparation technique that increases the concentration of the analytes has the effect of decreasing the detection limit relative to the initial sample, such that the measurements can be made at larger S/N ratios.

The process of concentrating the analytes during sample preparation may have additional benefits besides the decrease in the detection limit for the analyte in the initial sample. One such benefit may be related to the increase in sensitivity. As indicated in Section 1.11, sensitivity is defined as the slope of the curve that is obtained when the result of a series of measurements is plotted against the amount that is to be determined. For a truly linear dependence no benefit regarding sensitivity is obtained, therefore, by concentrating the sample. However, it is common for analytical chromatography to have a decrease in sensitivity toward lower concentrations (or amount) of sample. A sample preparation that increases the concentration in the processed sample may allow the measurements to be done in the linear portion of the calibration curve. External addition of a specific amount of analyte can be done in order to increase its concentration and bring it in the linear range of measurement. However, this procedure may add undesirable errors and should not be applied without careful evaluation of the results.

3.5   CHEMICAL MODIFICATIONS OF THE SAMPLE

Chemical modifications of the sample are done for different purposes. Some chemical transformations are used for dissolution as shown in Section 3.3, and others are done for separations or to remove interferences. A chemical change of the analytes (and sometimes of the matrix) may lead to a better separation or may eliminate significant interferences. The separation process may utilize differences in the chemical properties of the compounds to be separated or can be based on differences in the reaction rate of the same type of reaction. Among the separation techniques based on chemical reactions are ion exchange, electrodeposition, differences in the acid/base character of certain sample components, etc. However, the most common use of chemical modifications is to improve the sensitivity of detection. In HPLC, for example, labeling the analyte molecule with groups that have strong UV absorbance or have fluorescent properties can increase significantly the sensitivity (limit of detection) of an analytical method. The chemical modifications done to change the structure of the analyte and obtain new compounds with better analytical properties can be classified as derivatizations, polymer degradations, or pyrolytic reactions.

Derivatizations use reagents that chemically interact with specific components of the sample to form new compounds having the desired properties for analysis. They are the most common chemical modifications used for analytical purposes on analytes with small molecules. Typically, the derivatizations are done with the purpose of modifying the analytes and not the matrix. Polymer chain fragmentations are considered separately from derivatizations. These are reactions performed in the presence of a reagent with the intention to generate smaller molecules from a polymer. The fragments are easier to analyze with standard analytical procedures. The degradations can be performed for structural elucidation but also for quantitation of the polymer. Protein chemical modifications and labeling without chain fragmentation are also used for analytical purposes, but the subject is only tangentially addressed in this material. Pyrolytic reactions are chemical transformations performed by heating the sample at temperatures significantly higher than ambient [6], and no reagent is necessary in the pyrolytic step. Applied to polymers, pyrolysis commonly generates smaller molecules that are easier to analyze. Further derivatization may be applied after polymer chain fragmentation or pyrolysis.

A detailed description of various chemical modifications and their uses in chromatography are given in Part 3 of this book. The purposes and benefits of derivatization in chromatographic analysis are presented in Chapter 17. The chemistry of different derivatization reactions is presented in Chapter 18, and the reagents tailored for specific functionalities of the analyte are discussed in Chapter 19. Chapter 19 begins with the derivatization of monovalent functionalities such as alcohols, phenols, thiols, amines, aldehydes, ketones, acids, and derivatives of carboxylic acid. The last part of Chapter 19 describes the derivatization for compounds with more than one functionality type such as amino alcohols, carbohydrates, hydroxy acids, amino acids, etc. Several polymer chain fragmentations using reagents or pyrolysis are discussed in Chapter 20.

3.6   OTHER TYPES OF SAMPLE MODIFICATIONS

A number of other problems encountered with the samples may be addressed by sample preparation. A group of operations related to modifications required for easier handling of the sample do not involve any complicated process. Among these can be considered the mixing (vortexing), taking of sample aliquots, dilutions, selecting and adding internal standards (see Section 3.8), etc. However, these types of operations are common in laboratory work and are not specifically addressed in this book.

Among other operations common in sample preparation is the pH change. This is done very frequently for aqueous samples or for extractions involving an aqueous phase. Another change sometimes necessary in sample preparation is that of ionic strength of sample solutions. This operation is used, for example, when handling biological samples that must be maintained at a specific ion strength.

One very common type of sample modification during sample preparation is the exchange of solvents (reconstitution). This operation is typically applied to the processed samples at different points along the sample preparation chain. The exchange of solvent can be necessary for a separation step such as solid phase extraction (SPE), for derivatization purposes, or for accommodating the chromatographic process. For example, after an extraction process from the initial matrix, the analytes are present in the organic solvent "A". For further cleanup the sample must be passed through a SPE cartridge, but the analytes are not retained on the SPE phase when present in solvent "A". The common procedure is to evaporate solvent "A" and redissolve the residue in another solvent such that the SPE separation process can be done. Special solvents may be needed for sample derivatization, and some aspects regarding the preparation for sample derivatization are discussed in Section 17.2.

Solvent exchange frequently must be done before the sample is introduced into a chromatographic system. A number of solvents (such as water) are not recommended to be injected in a GC instrument (although direct GC injection of water solutions has been reported), and other solvents may generate interfering peaks or long tailing peaks (such as methanol on certain columns). In HPLC, the solvent of the sample must be soluble in the mobile phase. When such situations are encountered, a solvent change is applied. Solvent change can be done in various ways, the simplest being distillation/evaporation of the first solvent (see Section 8.1) and reconstitution in a second solvent (see Section 8.2). Other procedures may use solvent extraction (see Section 10.1) or other more elaborate sample processing techniques for solvent exchange. Solvent reconstitution is sometimes done with the same solvent. In some concentration procedures, instead of evaporating the solvent up to a desired level, it is preferable to eliminate the solvent and redissolve the analytes in a precise (small) volume of the same solvent.

Another type of problem addressed in sample preparation is related to sample instability. Sample instability can be of physical or chemical nature. Samples may lose their initial state due to physical modifications that occur in time. Some of these modifications may take place during storage, and the problem must be addressed immediately after the sample collection (see also Section 2.4). However, some samples may change their physical state during the analysis. For example, a sample of gases collected in a solvent at low temperature may remain stable during storage but lose some components during the injection in a GC when introduced in a warm syringe. This type of sample requires different sample processing, such as direct gas analysis or adsorption on an SPME fiber and analysis, instead of using liquid injection.

Special attention must be paid to samples that form colloidal solutions. Some substances such as proteins, polysaccharides, and Maillard browning polymers [6] may form stable colloidal solutions with the appearance of a true solution. However, the colloid may adsorb irreversibly on a stationary phase of the column for example when using gel permeation for separation. The filtration of the sample that is commonly done on HPLC samples does not remove the colloids. This type of problem is a common cause of column damage in HPLC. The use of guard columns may protect the chromatographic column but does not solve the problem of a complete analysis of the sample when some components are never recovered from the column. Sample preparation is necessary to eliminate this type of problem, which may go easily unnoticed. This can be done for example by precipitating the colloid before the analysis using pH modifications, dialysis, etc.

Special sample modifications are required for samples that are reactive or that can degrade (by chemical modifications or biological processes). The addition of preservatives in biological samples is a common practice for preserving them. Samples that are too basic or too acidic also can be processed immediately after collection for eliminating the chances of getting contaminated for example by reaction with the containers. A number of reports discuss in detail the sample preparation requirements for the preservation and analysis of samples with reactivity problems [7], [8]. Samples that present health risks or environmental problems also must be prepared using special procedures. A discussion on the requirements related to the processing of hazardous samples is beyond the purpose of this book.

3.7   DIFFERENCES IN SAMPLE PREPARATION FOR DIFFERENT CHROMATOGRAPHIC TECHNIQUES

As shown in Section 1.1, sample preparation is a step done after sample collection, and before the chromatographic analysis. A simple diagram showing the selection of the chromatography and of a sample preparation was given in Figure 3.2.1. The selection typically starts with the chromatography and is followed by the selection of a sample preparation. Sample preparation is frequently formed from a number of operations. The nature of the sample determines some of these operations. Other operations are, however, determined by the type of the chromatographic separation. Feedback from the selection of sample preparation is also utilized in adjusting the initial choice of chromatography.

A discussion of the more common chromatographic procedures is given in Chapter 4. The main characteristics to consider for the chromatography are the required volatility of the analytes, the separation power, and the sensitivity.

The required volatility is important for choosing a gas or a liquid chromatographic technique. A shown in Section 1.7, adaptability to analyte volatility is an important characteristic of a separation method. Gas chromatography is a very good technique for the analysis of volatile compounds, and the use of mass spectrometry as detection for GC makes the GC-MS an almost ideal method for qualitative and quantitative analysis. For this reason, GC is highly preferred when its use is possible. The lack of volatility or thermal instability of many analytes is the main limiting factor for the use of this technique. Sample preparation for GC is for this reason frequently geared toward isolating the volatile compounds from a matrix and toward generating volatile compounds from nonvolatile analytes when chemical modifications allow it. The discussion of various techniques able to separate the volatile compounds from a sample is given in Chapters 9 and 11. Procedures to chemically modify the analytes to increase their volatility and thermal stability are discussed in Chapter 18, and a number of examples are given in Chapter 19.

When GC analysis is chosen for solving an analytical problem, the selection of a sample preparation also must consider the separation power of the GC technique and the type of detector utilized. Although GC is a powerful separation technique, it is invariably influenced negatively by a complex matrix ("dirty sample"). The elimination of compounds interfering in the chromatography, of as much as possible from the nonvolatile matrix, and of the components that decompose when heated is a common purpose of sample preparation for GC. Various procedures to achieve this task are discussed in Part 2 and Part 3 of this book. The improvement of the separation in general also can be achieved using physical or chemical modifications of the sample (see Section 17.3).

The sensitivity of the GC is another issue when performing sample preparation. This sensitivity depends on the detector. In addition to the selection of a good sensitivity of the GC system, sample preparation must be done such that an appropriate concentration of the analyte reaches the detector. A number of procedures are applied for this purpose. A group of techniques allowing the increase in the concentration of the analyte are general cleanup/concentration operations discussed in Part 2. A different strategy is chemical modification of the analytes for making them appropriate for detection using special, more sensitive detectors. For example, the use of electron capture detection or of negative chemical ionization MS (NCI-MS) leads to excellent sensitivity for compounds with high electron affinity such as chlorinated or fluorinated compounds. For this reason, a number of derivatization procedures are applied to introduce in the molecule of the analyte new groups or atoms that lead to high electron affinity. These aspects are further discussed in Section 17.4.

When using a liquid chromatography technique, the volatility issue discussed for GC does not exist. The roles of sample preparation to clean the sample and to improve sensitivity are also applicable in liquid chromatography. The most common liquid chromatography technique is HPLC. Sample cleanup in HPLC can be done using various techniques. The principles of some separation techniques and chemical modifications used for sample cleanup are given in Parts 2 and 3 of this book. Details on many other procedures can be found in the literature dedicated to HPLC analytical methods (see e.g. [9]).

The methods for sensitivity increase in HPLC also form the subject of a significant amount of published work. Besides procedures for sample concentration (described in Part 2), various derivatization procedures are applied for improving the detection in HPLC (see Section 17.6). Selected examples of using sample preparation for increasing the analytical sensitivity of HPLC methods are also found in Part 3.

3.8 QUANTITATION TECHNIQUES IN CHROMATOGRAPHY AND THEIR RELATION TO SAMPLE PREPARATION

The common detectors used in gas chromatographs such as FID (and its variations) and MS respond to the instantaneous amount of mass of analyte that is introduced into the detector. For this reason, quantitative analysis is based on the area under the chromatographic peak. This area is proportional to the amount of the component generating that peak and is independent of the rate of introduction of the analyte into the detector. For very narrow peaks and for those with the shape close to a Gaussian curve, the peak height can also be used for quantitation. However, the peak height is proportional with the rate of introduction of the analyte into the detector. Therefore, changes in temperature and carrier gas flow rate will lead to changes in peak height but not to changes in the area of the peak. Other detectors such as the thermal conductivity detector respond to the instantaneous concentration in the analyte. For these detectors, the peak height is independent of the gas flow rate, while the area is inversely proportional to the carrier gas flow rate. For a constant temperature, the peak height will be more suitable for measuring concentrations using these detectors, although the peak area is still usable for quantitation. For HPLC detectors, the response depends on the instantaneous concentration of analyte that is introduced into the detector, and for a constant flow, the peak areas are used for quantitation.

Ideally, for a given volume of sample injected into the chromatograph, the peak area is linearly dependent on the concentration of the sample. The procedure requires a calibration curve obtained with the compound to be analyzed. The concentrations c of interest can be determined from the areas using the relation:

$$c = b A \qquad (3.8.1)$$

where A is the peak area and b is the slope of the calibration curve for compound x (sensitivity). The value b of the slope for the calibration curve may be different for different compounds. For this reason, the generation of calibration curves is usually necessary for each analyte that must be quantitated. For generating the calibration curve, it is possible to use solutions of different concentration made using the pure compound to be analyzed considered as a *calibration standard*. The calibrations are done independently of the sample, and the calibration standard can be considered in this case an *external calibration standard*. (An external standard is analyzed in a different run from the sample, while an internal standard is added and analyzed together with the sample.) In many practical applications it is preferable to make the calibrations by adding different levels of the calibration compound to a blank sample that does not contain the analyte. This procedure makes the analysis of the samples containing the calibration standards as close as possible to the analysis of a real sample and allows the subtraction of the overall influence of the matrix in the analysis. For compounds that have similar structures, sometimes the calibration curve for only one of the compounds is utilized, and different compounds are quantitated based on the same calibration. However, this is not a recommended practice and should be used only when the calibration standards of all compounds are not available.

Most detectors used in chromatography, such as FID, NPD, and MS, have a linear response to the amount (within a certain interval) of material introduced in the detector. The injection of a constant volume of sample makes these detectors have a linear response to the concentration of the sample. The RI and UV-Vis detections in HPLC also provide a linear dependence of the peak areas with the sample concentration (at constant flow). Nonlinearity may be due to overloading of detectors that otherwise are expected to have a linear response. Also very low levels of analyte may lead to nonlinear response (see Section 1.11). For this reason, linearity must be verified for a whole range of concentrations, and particular attention must be paid to very low and very high concentrations. The linearity of fluorescence detection is valid only for low concentrations. A nonlinear calibration curve is sometimes necessary for fluorescence and chemiluminescence detectors. This still allows the calculation of the concentration directly from the nonlinear calibration curve or using a corresponding relation. However, a linear calibration curve is preferable to other cases. Some linear calibrations do not pass through the origin, and the calculation of the peak area must be done using a relation of the form:

$$c = a + b\,A \qquad\qquad (3.8.2)$$

This type of dependence may indicate some problems with the particular analytical method, such as sample decomposition, loss of sample in the chromatographic process due to selective adsorption, interfering signal from the blank sample, etc. A negative value for the parameter a indicates in general a loss of analyte, while a positive value indicates background or interferences.

In many quantitative techniques an *internal standard* also is introduced with every chromatographic run. Internal standards are compounds absent in the real samples, which are added in a constant amount at a chosen point during the analysis for verifying the reproducibility, accounting for sample losses, etc. The internal standard must be chosen in such a way as to behave in the analytical process as close as possible to the analytes, to not interfere with the analyte determination, to give a chromatographic peak convenient to integrate, etc.

Besides the internal standards that are added in the sample such that they go through the sample preparation process, it is sometimes useful to add a *chromatographic standard* in the processed sample. This standard is a type of internal standard used only for verifying that the chromatographic process works properly, and it is introduced in the processed sample that is ready to be injected in the chromatograph. In a long chain of operations, and when the processed sample injected in the chromatographic system is still a complex material, it is not uncommon to encounter problems after a certain number of samples have been analyzed successfully. The cause of such problems is sometimes difficult to locate. The sample preparation may have been executed differently, or the chromatographic system may have problems, for example due to column degradation. An internal standard undergoing the sample preparation process is not a good indicator for the differentiation between a sample preparation problem and a chromatography problem. A chromatographic standard added to the processed sample is affected only by a chromatography problem. This second type of standard (chromatographic standard) does not need to be as close as possible to the analytes as the standards used for quantitation or as the internal standard. *External chromatographic standards* also can be used, these being run independently of any other sample and used exclusively to check the chromatographic process.

A different quantitation technique besides the external calibration is that of standard addition. Standard addition method can be used to analyze an unknown sample of concentration c_x without the use of a calibration curve obtained in separate runs. It must be assumed, however, that the dependence of the peak area of the concentration follows rel. (3.8.1). A set of known amounts of analyte $\{q_i\}_{j=0,1,2...n}$ with $q_0 = 0$ are added to the unknown sample, leading to the concentrations $c_i = (q_x + q_i) / (V_x + V_i)$ where V_x is the known volume of the sample to be analyzed, V_i is the volume of the added solution with the "i" standard, and $c_x = c_0 = q_x/V_x$. The relation between the concentration c_i and the signal (peak areas A_i) is in this case given by the relation:

$$c_i = b \, A_i \tag{3.8.3}$$

The values for c_0 and b (as parameters) can be obtained from the added amounts and peak area measurements $\{q_j, A_j\}_{j=0,1...N}$ using, for example, least-squares fitting (see Section 1.11, rel. 1.11.19 and 1.11.22). The standard addition method can be used even with a single added amount to the unknown sample. If one single addition q_1 is made to the sample, and the unknown sample is considered as having $q_0 = 0$, two peak areas A_0 and A_1 are generated corresponding to c_0 and c_1. The two equations of the form (3.8.3) for c_0 and c_1 are $q_x / V_x = b \, A_0$ and $(q_x + q_1) / (V_x + V_1) = b \, A_1$, and they lead to the result:

$$\frac{(q_x + q_1)V_x}{(V_x + V_1)q_x} = \frac{A_1}{A_0} \tag{3.8.4}$$

This relation can be easily rearranged to give

$$c_x = \frac{q_1 A_0}{(V_x + V_1)A_1 - V_x A_0} \tag{3.8.5}$$

When the addition of the standard does not dilute the sample ($V_1 = 0$), rel. (3.8.5) can be written in the form:

$$c_x = \frac{c_1 A_0}{A_1 - A_0} \tag{3.8.6}$$

In order to use the standard addition technique in this form, it must be assumed that the calibration curve passes through the origin (rel. (3.8.1)). In the case that the calibration line is not given by rel. (3.8.1) and is given by rel. (3.8.2) (with $a \neq 0$), each concentration is affected by a constant that must be known. Experimentally this is equivalent with the need to have a calibration obtained by standard additions to a blank sample and to have a measurement of the signal y for $x = 0$, which will allow determination of the value of a.

Other procedures also can be used for quantitation [10]. One of them is based solely on the peak area ratios for two compounds. For this procedure, a response factor F_x must be obtained initially. This response factor using an internal standard is calculated from the peak area A_{is} of the internal standard and the peak area A_x of the compound to be analyzed, both added to a blank sample at equal amounts (concentration). The ratio of the two areas, usually obtained as an average of several measurements, gives the response factor:

$$F_x = A_{is} / A_x \qquad (3.8.7)$$

Ideally, the value for F_x remains constant for an interval of values for the pair of concentrations of the standard and the sample. The concentration of the unknown is then obtained by measuring in the same run the peak area of the compound to be analyzed (at unknown concentration) and peak area of the standard using the formula:

$$c_x = F_x (A_x / A_{is}) \, c_{is} \qquad (3.8.8)$$

where A_x is the area of the compound x at unknown concentration, A_{is} is the area of the standard at the concentration c_{is}, and F_x is the response factor. In order to achieve a constant value for the response factor F_x in a range of concentrations, it is recommended that the two compounds, the internal standard and the analyte, be chemically similar or even identical except for use of a labeled compound for the standard. The use of labeled standards requires that detection be done using a mass spectrometer for differentiating the analyte from its labeled pair, which may have an identical retention time in the GC (or HPLC) separation. The response factors for similar compounds are usually expected to be very close. However, even for labeled compounds it is possible to obtain quite different response factors, mainly when the sample processing involves a longer chain of operations.

The quantitation based on peak area ratio also can be applied using areas of the analyte and that of an external standard. In this procedure, peak area A_{ext} of the external standard and peak area A_x of the compound to be analyzed are obtained in separate runs. An expression similar to rel. (3.8.7) is used to calculate a response factor. The quantitation can be further performed using the same run for measuring the peak of the analyte and that of the standard added this time on the sample (as internal standard) using rel. (3.8.8) or again from separate runs.

3.9 METHOD DEVELOPMENT IN SAMPLE PREPARATION

Method development in sample preparation is frequently needed. Even for a method well documented in literature, some method development is needed when the analysis is a new implementation in the laboratory. The method development is always necessary when a known method is applied to a new type of sample or when a new analytical method is needed.

Due to the strong interaction of different steps in an analytical method, it is convenient to discuss in short the whole method development outline, although the discussion will emphasize the sample preparation part. The first stage in a method development is usually the choice of the core chromatographic technique, as indicated in Section 3.1 and shown in Figure 3.1.1. For the chosen chromatography it is common to obtain a calibration curve with the pure analytes to verify that the compounds expected to remain in the processed sample after sample preparation are separated from the analytes and that the analytes are separated among themselves. This stage also allows the evaluation of the detection limits for the analytes (in the absence of a matrix) and also the identification of potential internal standards. This application of a chromatography to some standards is also very useful in setting up the chromatographic conditions for the analysis. The chromatography may need further optimization when working with the real sample, but this will start with conditions established for the pure compounds.

The process of evaluating the chromatographic behavior of pure compounds and of selecting internal standards is very useful for further quantitative analysis. When the analysis is done for qualitative purposes or when the chemical composition of the analytes is not known, the initial sample may be directly subjected to the chromatographic process for estimation of the results and evaluation of possible paths for improvement. The use of the initial sample in a preliminary chromatographic process implies that the sample is in a convenient form to be analyzed. In some cases, a dissolution operation is necessary. Both chromatography of the unprocessed sample and calibration with pure compounds of interest can be applied in the preliminary assessment of a given analytical problem.

Based on the preliminary information about the analytical problem, as well as from the results of a chromatography of an unprocessed sample and from the calibration with pure compounds of interest, a sample preparation procedure is decided. This sample preparation includes:

a) The potential need for dissolution or the use of a solvent-free technique,

b) The need for sample cleanup when the matrix is complex ("dirty sample"), when interferences which cannot be resolved by the core chromatographic process are expected, when the matrix may destroy the chromatographic column or affect negatively the chromatography, etc.,

c) The need for concentration (or dilution) based on the evaluation of the detection limit of the core chromatographic step and of the expected level of analytes in the initial sample, or

d) The need for derivatization for the elimination of interferences, the increase in sensitivity, etc.

Each process may consist of one or more operations. Also, some iterations may be necessary for achieving the desired result. A simplified diagram of this strategy is shown in Figure 3.9.1.

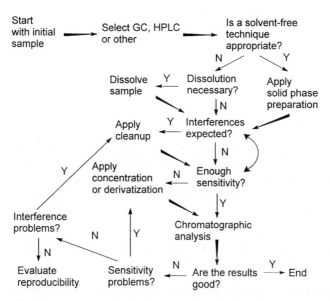

FIGURE 3.9.1. *Diagram summarizing a common strategy applied for sample preparation.*

In the diagram from Figure 3.9.1, the process of cleanup of interferences and the process of concentration to eliminate a sensitivity problem can be interchanged. This diagram presents only a general view on sample preparation, and consequently the information regarding details is very weak. In practice, after establishment of a specific path, it is common to perform optimization of each process. For example, if a dissolution step is necessary, a series of solvents are evaluated and some are experimentally tested until the sample is completely dissolved. The choice of an appropriate solvent is not limited to being able to dissolve the sample. It must be suitable for further operations and should not pose volatility problems or problems related to hazard to the worker or the environment. As in every optimization process, the best choice may require several iterations until the result is good. Details about the optimization of various sample preparation operations are given in Part 2 and 3, where more information is given for each operation.

The graphic design shown in Figure 3.9.1 allows a simple explanation regarding the application of a specific process but leaves unanswered the criteria of deciding when a Y/N conclusion is reached along the diagram. In order to answer questions such as "enough sensitivity?" or "are the results good?" in method development of a sample preparation, it is necessary to have the capability to check the results (final and intermediate) and to have a set of well defined criteria for the evaluation of the method.

Some differences exist between solvent-free sample preparation procedures and those involving solvents. The solvent-free procedures typically use a solid phase material to carry the sample into the chromatographic instrument, and this can be considered similar to a solvent (details regarding solvent-free procedures are discussed in Chapter

9 and 11). Based on this similarity, further discussion for sample preparation procedures using solvents can be extended easily to solvent-free techniques.

Depending on the type of sample available for analysis, a number of paths can be selected for evaluating the success of a sample preparation procedure. Ideally, a blank sample containing the same matrix as other samples but not containing the analytes is available. In this case, a given amount of pure analyte is added to the blank sample, and the sample preparation is performed. The analysis of the processed sample must generate a result identical with the amount added to the blank (or indicate the correct concentration of the added analyte). The calculation of the analytical result can be done using various procedures, some of which are described in Section 3.8. When a blank sample is not available, other procedures are applicable, such as the use of standard addition or the use of a surrogate analyte. In the standard addition procedure the analyte is measured in the unknown sample, and then a known amount of pure analyte is added to the sample, which is analyzed again. The peak ratios of the analyte in the initial sample and that of the analyte with the addition are used for quantitation. The surrogate analyte is a pure compound unlikely to be found in the sample but similar to the real analyte. For the surrogate analyte the real sample can be considered a blank and processed accordingly [11].

Once a specific sample preparation procedure is established, a validation process must be performed, following in principle the steps described in Section 1.12. The first aspect that must be verified in a method development procedure is the improvement in the selectivity of the analytical method as a whole. From the overall aspect of the chromatograms, the characteristics related to selectivity that can be evaluated include a) the separation resolution, b) the baseline drift, and c) the baseline noise. If interferences are still possible or likely, the method still needs improvements. The interferences can be seen using qualitative information about the chromatographic peaks, such as mass spectra, IR spectra, etc., or based on the existence of systematic errors in quantitative measurements. Sometimes, additional separation, the use of high resolution MS, or the use of MS/MS detection is necessary for distinguishing a possible interference.

After the analysis is performed and the results are obtained using calibration with a blank sample, with standard addition, with surrogate analytes, etc., it is common to evaluate the precision of the method. The calculation of precision is based on the calculation of standard deviation of a set of measurements (see Section 1.11). However, various protocols are described in literature for the conditions of choosing the samples for which the precision must be estimated. These protocols depend on the type of sample, purpose of analysis, type of laboratory, etc.

Besides the analytical precision, the main parameter to be considered is the method accuracy. The accuracy can be estimated based on the deviation of the mean of a number of measurements from the results obtained on a sample with known composition, by comparison of the results with values reported in literature, or by comparison of the results with the values obtained by other methods or in other laboratories. Again, depending on the type of sample, purpose of analysis, type of laboratory, etc., various protocols are available for the conditions for which the accuracy should be estimated.

In addition to precision and accuracy, a number of other parameters must be evaluated. From the results of a number of samples it can be evaluated a) the range of linearity of the method, b) the limit of detection, c) the sensitivity, d) the percent recovery of the analyte from the initial sample, e) the factor of concentration change between the initial sample and the processed sample, f) the stability in the measurement of the internal standard, etc.

In addition to those measured parameters, some other aspects of the analytical method in general and of the sample preparation step must be evaluated. Among these are the robustness and ruggedness of the method, the degree of difficulty in performing different operations, the cost, and the analysis duration. The time factor also can be very important if the analytical results vary when the analysis is not done within a specified length of time. For this reason, in many analytical methods the analytical measurement must be done within a specified time frame. Time variability can be influenced by various factors such as instability in time of a certain derivatized analyte, possibility for hydrolysis of the analyte, adsorption on the walls of the containers in the analysis of traces, etc.

The adaptability for automation is another important factor to consider, mainly when the number of samples to be analyzed is high. Automation of sample preparation is an important means of reducing the manpower required by an analysis and also of standardizing various operations in sample preparation. Some aspects related to automation are further discussed in Chapter 6.

REFERENCES 3

1. K. Shimada, T. Nakagi, J. Liq. Chromatogr. & Rel. Technol., 19 (1996) 2593.
2. Guide for Sample Preparation, LC-GC, Aug. 2000.
3. P. R. Haddad, P. E. Jackson, *Ion Chromatography, Principles and Applications*, Elsevier, Amsterdam, p.417, 1990.
4. Current Trends and Developments in Sample Preparation, LC-GC, Supplement, May 1998.
5. A. J. Handley (ed.), *Extraction Methods in Organic Chemistry*, Sheffield Academic Press, Sheffield, 1999.
6. S. C. Moldoveanu, *Analytical Pyrolysis of Natural Organic Polymers*, Elsevier, Amsterdam, 1998.
7. J. Versieck, R. Cornelius, *Trace Elements in Human Plasma or Serum*, CRC Press, Boca Raton, 1989.
8. M. Stoeppler (ed.), *Sampling and Sample Preparation*, Springer, Berlin 1996.
9. G. Lunn, L. C. Hellwig, *Handbook of Derivatization Reactions for HPLC*, J. Wiley New York, 1998.
10. J. Drozd, *Chemical Derivatization in Gas Chromatography*, Elsevier, Amsterdam, 1981.
11. United States Environmental Protection Agency, *Methods for the Determination of Organic Compounds in Drinking Water*, EPA-600/4-88/039, PB91-231480; National Technical Information Service, Springfield, VA, 1991.

CHAPTER 4

Chromatography as the Core Step in an Analytical Process

4.1   THE CHOICE OF THE CORE ANALYTICAL TECHNIQUE

The elements determining the choice of the core analytical technique are similar to those determining the choice of a whole analytical method (see Section 1.1). The nature of the sample, a specific sample preparation procedure, and the required accuracy and precision of the results are important factors in this choice. Regarding the nature of the sample, this includes sample physical properties, sample complexity, the matrix composition, the number of analytes, expected level of analytes, sample availability, and the number of samples to be analyzed. External circumstances also play a role in the choice of the core analytical technique. These include available instrumentation, automation capability, manpower, expertise available in the laboratory, the cost of analysis, funding, etc. The correlation between all these elements is not rigid, the same goal being achievable through different paths.

Chromatography as a separation technique, as shown in Section 1.7, has an excellent fractionation capacity. The adaptability to the analyte volatility determines the choice between GC and a liquid chromatography type. Also, chromatography has a low load capacity that in this case is an advantage because it requires a small amount of sample. For these reasons, the chromatographic technique is chosen as a core analytical method when the sample to be analyzed has a complex matrix and requires detailed separation, even after a cleanup and concentration step. Even when a separation is not absolutely needed, the sensitivity, the need for a small amount of sample for analysis, and the capability to make compound identifications make chromatography the technique of choice in many analyses. Other techniques such as various kinds of spectroscopy or certain electrochemical techniques may also have the ability to analyze mixtures. However, when the number of the components in the sample to be measured is larger, the utility of nonchromatographic techniques tends to be more limited. When advanced separations are the only ones able to single out the analytes, chromatography is the technique of choice. Even single component samples that in fact do not need separation are sometimes analyzed with chromatographic techniques for convenience. Besides chromatography, capillary electrophoresis (CE) and other electrophoretic techniques having advanced separation capabilities also are used as core analytical techniques. More recently, enough progress has been made in the instrumentation of CE such that this technique is becoming extensively utilized for purposes similar to the use of HPLC. In fact, the similarity of CE to HPLC can go so far that certain analyses have identical sample preparation, differing only in the choice of HPLC or CE as a core analytical method.

Chromatography has the capability to separate compounds better than any other common separation procedure. The separation capability of GC has been estimated to be about 110 compounds in one sample. However, with the help of selective detectors or with MS identification, up to 300 compounds can be identified and measured in a GC chromatogram (see Section 1.6). For HPLC, the number of separated peaks is lower.

The number of theoretical plates for an HPLC column is in the range of 10,000 compared to 300,000 for GC. However, HPLC can have excellent separation capability by the use of selective eluents and the use of solvent gradient separations. The selectivity during detection can also enhance the separation capability of HPLC. The sensitivity of GC depends on the detector and the analyzed compound. The lower limit of detection for the analyte amount injected in a GC is estimated around 10^{-13} mg for ECD detection or 10^{-11} mg for sensitive MS systems. Fluorescence and luminescence detectors in HPLC can go to even lower limits of detection. The need for a small amount of sample is another reason that chromatography is a very attractive technique. The injection volume in a GC system is typically around 1 μL (except for large volume injections) for a liquid sample, or 0.1–1 mg solids for pyrolysis-GC. The HPLC injection volume is typically 5–50 μL.

Chromatography is also an excellent choice when compound identification is necessary. The connection of gas chromatography with mass spectrometry can be done very conveniently. The use of electron ionization (EI+) technique in GC-MS leads to MS spectra that can be compared with standard spectra from the available libraries containing several hundred thousands of compounds, or can be interpreted based on specific rules (see e.g. [1]). Also the coupling of an IR detector (IRD) with a GC system can be done successfully. Liquid chromatography can be coupled with MS with different types of interfaces such as electrospray (ES), atmospheric pressure chemical ionization (APCI), particle beam (PB), continuous flow FAB (fast atom bombardment), and thermospray (TS). Although the spectra in LC-MS typically show less fragmentation and are less informative than EI+ spectra, the technique is successfully used in compound identification. For enhancing the fragmentation, LC-MS-MS systems can be applied, leading to excellent identification capability [2]. In HPLC, the use of UV-Vis or fluorescence detection also can provide qualitative information based on the spectra of the analyte. For all chromatographic techniques, the retention times can be applied for compound identification, although this requires preliminary use of standards (see Section 5.1), and the identification is not unequivocal.

Both GC and HPLC systems have the capability of using more than one detector. Some of the detectors in GC are not destructive (such as the IRD or the photoionization detector PID) and can be connected in series with other detectors. For other detectors, the exit of the chromatographic column can be split and connected to more than one detector (parallel connection). Combinations of a flame ionization with a nitrogen phosphorus detector (FID + NPD), FID + MS, FID + atomic emission (AE), and MS + IRD are not uncommon. These combinations provide enhanced qualitative and quantitative results from the same separation [3]. Combinations of detectors are also possible for HPLC. The use of UV and MS in series is possible when special high backpressure cells are used in the UV system, leading to excellent results.

Another advantage of chromatographic methods is the ease with which they can have some parameters changed. For GC methods, it is very simple to modify oven temperature program, injection port temperature, injection volume, split ratio, etc. The modification of these parameters allows adjustment of an available method to suit better a new type of sample. For HPLC separations, also it is simple to modify a solvent gradient and obtain a better separation. Other modifications such as the change of the chromatographic column (for both GC or HPLC) can be done without difficulty, but in this case a new optimization of the separation is necessary.

Sample preparation plays an important role in the choice of the core analytical technique. If the analytes become "enough clean" after sample preparation, chromatography is not indispensable, and other analytical techniques can be chosen for measurement. However, in many cases the sample preparation would become too complicated to generate a material easy to analyze without further separation, and chromatography is frequently chosen as a core analytical technique.

Between the choice of the core chromatographic technique and the sample preparation process is a strong dependence. Sample preparation can be done with the purpose of modifying the matrix of the sample, the analytes, or both. When the analytes are not chemically modified during sample preparation, the physical nature and properties of the analytes are very important in the choice of the type of chromatography used as a core step in the analysis. However, derivatization or other chemical modifications of the analytes may generate new compounds that can be analyzed with chromatographic techniques that are not appropriate for the initial analyte. As an example, monosaccharides cannot be analyzed directly using GC, but after silylation the GC analysis is perfectly applicable for monosaccharide analysis (see Section 17.3). Another example is formaldehyde, which is difficult to analyze using HPLC, but after derivatization with 2,4-dinitrophenylhydrazine, an HPLC analysis using UV detection gives very good results (see Section 13.5). Within the same type of chromatography (such as GC or HPLC), the choice of the separation column or the choice of the detection technique is also dependent on the type of sample preparation applied to the sample. Because derivatization may modify significantly the initial properties of the analyte, the choice of the core chromatographic technique may vary depending on other criteria, such as the type of sample preparation selected, the available instrumentation, the limit of detection required in the analysis, etc.

The large variety of chromatographic techniques represents an attractive feature for choosing one of them as a core analytical procedure. A number of classifications are available for distinguishing different types of chromatography. Criteria for such classifications are the following:

a) The physical state of stationary and mobile phase. The stationary phase can be liquid or solid. Based on the mobile phase, three types of chromatography are common: namely gas chromatography (GC), supercritical fluid chromatography (SFC), and liquid chromatography (LC). In supercritical fluid chromatography most stationary phases are liquid or liquid-like. In GC and LC, the stationary phase can be either liquid (or liquid-like) or solid. For this reason the chromatography is usually classified as gas-liquid, gas-solid, liquid-liquid, liquid-solid, or supercritical fluid.

b) The mechanism of the separation process. Based on this criterion, various types of processes can be distinguished, including partition, adsorption, ion exchange, ion exclusion, ligand exchange, size exclusion, chemical affinity [4], hydrophobic interactions, ion pair formation, etc.

c) The physical shape of the medium on which the separation takes place. This distinguishes column chromatography and planar chromatography. Planar chromatography includes thin layer chromatography, paper chromatography, and several electrochromatographic techniques. Column chromatography is more common than planar, and the type of column can further be used in the classification.

d) Type of chromatographic column (for column chromatography). For gas chromatography and supercritical fluid chromatography (SFC) this criterion distinguishes packed columns and capillary columns. For liquid chromatography, low pressure chromatography and high performance liquid chromatography (HPLC) are distinguished. Among the chromatographic techniques, GC and HPLC are the most commonly utilized. Some unusual chromatographic techniques are also described in literature [5].

e) The polarity of the stationary and mobile phase. This criterion is applied only to liquid chromatography. Based on the difference in the polarity of the stationary phase and mobile phase, it distinguishes normal (or direct) phase chromatography where the stationary phase is more polar than the mobile phase, and reverse phase where the mobile phase is more polar than the stationary phase.

Various types of chromatography can be chosen based on the criteria indicated above, although not all combinations are possible, some being very common and others very seldom utilized. A very common type is gas-liquid chromatography (GC) with the separation based on gas-liquid partition and performed on a capillary column. Also very common is liquid-liquid chromatography based on partition, using high performance separation (HPLC) on a reverse stationary phase.

Chromatographic techniques can use several types of detectors. These include detectors specific for certain chromatographic techniques such as flame ionization (FID) or thermal conductivity detector (TCD) that are specific for gas chromatography. Other detectors are mass spectrometers (MS), UV-Vis spectrophotometers, infrared instruments, various electrochemical detectors, etc. that can be used with various types of chromatography. It is therefore possible to combine the capability of a spectroscopic technique with that of an advance separation. Some of the detectors used in chromatography are highly sensitive. This leads to exceptional capabilities for qualitative and quantitative analysis of complex mixtures when techniques such as gas chromatography-mass spectrometry (GC-MS) are used, or to extremely high sensitivity when HPLC with a fluorescence detector is used.

The physical nature of the material introduced in a GC can be gas or liquid (typically a solution of a gas, liquid, or solid material), but also can be solid when using special sample introduction techniques. In pyrolysis, for example, the sample can be solid, and the pyrolyzer is interfaced with the GC or GC-MS system. However, GC analysis separates the analytes in gas form, and therefore the analytes must be able to exist in gas (vapor) form without decomposition. The material introduced for analysis in an HPLC instrument is typically a liquid. Solutions of liquids or solids can be used for analysis. The analysis of solutions of gases by HPLC, although possible, is not common. The solution introduced for analysis by HPLC must be miscible with the mobile phase used for separation. HPLC is typically used for a variety of compounds including large molecules, molecules with high polarity, or those that are labile at higher temperatures.

Some criteria for the selection of a chromatographic technique as a function of specific features of the analytes after sample preparation are detailed in Table 4.1.1. The initial sample can be significantly modified using sample preparation operations such as derivatization. For this reason the selection of a specific chromatographic technique must consider the properties of the sample after the sample preparation step.

TABLE 4.1.1. *Criteria for the selection of a chromatographic technique after sample preparation.*

Physical state of the sample	Volatility of the analyte	Thermal stability of analyte	Minimum quantity of sample	Quantity of analyte	Purpose of analysis	Type of chromatography
solid, liquid, gas	bp. below 300 - 400° C	good	about 1 μL solution	1 ppb - 0.1%	qualitative, quantitative	Gas with capillary col.
solid, liquid, gas	bp. below 300° C	good	a few μL solution	1 ppm - 1%	qualitative, quantitative	Gas with packed col.
solid, liquid	not necessary	not important	a few μL solution	1 ppb - 1%	qualitative, quantitative	HPLC
solid, liquid	not necessary	not important	about 1 μL solution	1 ppb - 0.1%	qualitative, quantitative	SFC
solid, liquid	low	not important	about 1 μL	a few ppb - 1%	mainly qualitative	TLC, paper
solid, liquid	low	not important	a few μL solution	10 ppm - 1%	mainly quantitative	LC low pressure
solid (ions)	not necessary	not important	a few μL solution	1 ppb - 1%	mainly quantitative	Ion exchange
solid (polymers)	not necessary	not important	a few μL solution	a few ppm - 1%	qualitative, quantitative	Size exclusion (GPC), HPLC

As seen in Table 4.1.1, for the samples that are more volatile and thermally stable, GC is the technique of choice. Less volatile samples, including various biological materials such as proteins or nucleic acids, must be analyzed using HPLC. Size exclusion chromatography having the capability of separating the components based on their molecular weight (at least in the first approximation) is used mainly for polymer separations. Ions are well separated using ion exchange chromatography. When qualitative analyses are needed, mainly for liquid and solid samples, TLC can be extremely useful. However, in order to better understand the selection of a specific type of chromatography as a core analytical technique, more details about the chromatographic procedures must be considered. Some of these details are discussed in Sections 4.2 through 4.4.

4.2 GAS CHROMATOGRAPHY AS A CORE STEP IN THE ANALYTICAL PROCESS

Gas chromatography (GC) is one of the techniques where sample preparation is frequently necessary. Several types of procedures are used for this purpose. Besides general sample cleanup and concentration, derivatization is frequently used for modifying the properties of the analytes, as described in Chapter 12 and 13. In development of GC instrumentation, various procedures have been applied for improving the analysis results. Among these procedures are special sample introduction techniques such as pyrolysis, programmed-temperature injection techniques, etc. Pyrolysis is applied mainly in the case of polymeric materials, and this subject will be discussed in Chapter 20. Certain special sample introduction techniques may enhance the analytical results and eliminate part of the sample matrix. Some of these procedures are discussed in more detail in Part 2. For a better understanding of how these sample preparation procedures are related with the GC analysis itself, a few details about gas chromatography are necessary. The description of gas chromatographic instrumentation and of the theory of chromatography is available in a wealth of dedicated literature (e.g. [6], [7], [8], [9], [10]. Only for self-consistency of this book, some basic concepts in chromatography are presented in this section (see also Section 1.6).

- The injection system of a gas chromatograph

The gas chromatograph (GC) has in principle three parts: the injection port (inlet system), the oven with the analytical column, and the detector. A measured volume of sample that is a gas or a solution containing the analytes is introduced in the injection port. The injection port allows the sample to be placed at the beginning of the chromatographic column in a narrow zone loaded with the analytes. Also, the carrier gas, which can be helium, hydrogen or less frequently nitrogen, flows into the injection port. The pressure of the carrier gas (column head pressure) is utilized to control the gas flow in the analytical column. Some GC systems can maintain a constant flow in the chromatographic column. The constant flow in the column requires special control, because the increase in the temperature of the column leads to increased viscosity of the gas carrier and a decrease in the flow rate if a constant head pressure is maintained. A schematic diagram of a GC system (not to scale) showing some details of the injection system is given in Figure 4.2.1.

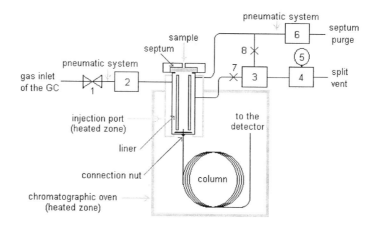

FIGURE 4.2.1. *Simplified diagram of a GC system (not to scale).* The injection port of the GC is shown in more detail. The carrier gas is controlled by a pneumatic system that consists of (1) a mass flow controller, (2) an electronic flow sensor, (3) a solenoid valve, (4) a backpressure regulator, (5) a pressure gauge, and (6) a septum purge controller. Connection (7) is closed and connection (8) is open by the solenoid valve (3) when the GC works in splitless mode (purge off), and (8) is closed and (7) is open when the GC works in split mode (purge on). No details on the GC oven or on the detector are given.

The injection port (typically containing a glass or silica liner) is usually set at a specific temperature where the sample is flash vaporized before entering the column. In the "split/splitless" type injection port, the temperature is set higher than the boiling point of the solvent carrying the sample and usually lower than the maximum temperature attained by the chromatographic column. Once a certain volume of sample has been injected, the amount of vapors going into the column is further adjusted by choosing the split or splitless mode of the GC and by varying the split ratio (the ratio of the total gas flow and the flow in the analytical column) using the pneumatic system of the GC. A number of specific constructions for the liner of the split/splitless injection port allow some discrimination based on volatility for the compounds that actually go into the chromatographic column. Split/splitless injection port can eliminate, for example, polymeric materials present in the sample.

Other models of injection ports are also known. A common injection port is the "on column" type. In "on column" injection port, the sample is directly injected at the beginning of the chromatographic column. This technique is used when temperature sensitive samples are analyzed, and the hot injection port must be avoided. Using special types of injection systems, significant discrimination of the sample components can be done, also based on the differences in volatility of sample components. This discrimination may allow the measurement of specific analytes, avoiding the interference of the matrix of the sample. The choice of the injection parameters is related to the sample preparation procedure, and the injection volume must be chosen such that the sample capacity of the column is not exceeded. The minimum amount of sample needed to reach the column is determined by the sensitivity of the detector, but except for some special injection port systems (PTV), the injection volume is limited to 2-3 µL. The main types of injection are summarized in Table 4.2.1.

TABLE 4.2.1. *Types of injection techniques.*

Injection technique	Description	Injection volume (µL)	Percentage in column
direct	The sample passes directly from the syringe into a hot inlet, where it vaporizes. Entire sample enters the GC column.	0.1 - 2	~100
split	The sample passes from the syringe into a hot inlet, where it vaporizes. Only a fraction of sample enters the GC column.	0.1 - 2	0.01 - 10
splitless	The sample passes from the syringe into a hot inlet, where it vaporizes, and the bulk of sample enters the column for a duration of 0.3–2 minutes.	0.1 - 2	80 - 95
cold on-column	The sample passes from the syringe into the column or its extension as a liquid. The portion of column or its extension accepting the sample is kept relatively cool during injection.	0.1 - 1	100
programmed-temperature split	The sample passes as a liquid from the syringe into a cooled inlet that is subsequently heated to vaporize the sample. Only a fraction of sample enters the column.	0.1 - 2	0.01 - 10
programmed-temperature splitless	The sample passes as a liquid from the syringe into a cooled inlet that is subsequently heated to vaporize the sample. The bulk of sample enters the column for a duration of 0.5–1 min.	0.1 - 2	80 - 95
solvent elimination without splitting	The sample passes as a liquid from the syringe into a cooled inlet. The solvent is allowed to evaporate through a vent leaving a nonvolatile residue behind. The inlet is subsequently heated to vaporize the residue, which enters the column as in splitless injection.	1 - 100	80 - 95

A special type of injection system for GC is the programmed temperature, large volume sample introduction system (PTV). The use of PTV allows a preconcentration of the sample and also some separation of low temperature boiling compounds. In addition, it overcomes the 2–3 µL constraint of the injection volume of other injection systems being capable of using up to 1 mL sample for one injection, and usually in the range of 50–100 μL. The PTV inlet resembles in principle the classical split/splitless inlet (shown in Figure 4.2.1), having a solenoid valve that allows the switching between split and splitless operation. The PTV inlet is equipped with a rapid temperature controller that can maintain a constant temperature and also is capable of rapidly heating or cooling the liner.

The sample is injected in the PTV system having the column head pressure reduced to ambient (stop flow) and the temperature typically adjusted slightly below the solvent boiling point. The sample is injected at a slow constant rate such that the solvent, having a high vapor pressure, has time to vaporize and be swept from the injector. Solute molecules with higher boiling points are expected to remain in the injector. After the injection is done, for a short period of time the conditions are kept unchanged for further purging of the remaining solvent. The column head pressure is then restored, the split valve is switched to splitless mode, and the inlet is rapidly heated to transfer the sample into the chromatographic column. The system works well when the vapor pressures of the analytes are significantly lower than that of the solvent. The optimization of the speed of injection, temperatures, flow rates, and length of time for injection and purging are necessary for obtaining reproducible results. Depending on the injected volume, the system may achieve significant increase in sensitivity. Various improvements of the PTV system are possible. For example the injection port may be packed with an adsorbing material able to act as a short chromatographic column, or may be made from a wide bore tube coated with a thick stationary phase layer. Associated with the capability to cool and heat rapidly the injection system, a separation of the solvent from the rest of the analytes can be achieved in the injection port. Such a system can have a system for monitoring the solvent elution using the UV absorption of the purging gases for a better timing of the process. This type of inlet adds in fact a simplified chromatography operation (with separation and detection) in front of the main chromatographic process. A rudimentary chromatographic separation also can be performed using on-column injection and a retention gap long enough to accommodate the liquid sample [11] and to perform some separation that differentiates enough the solvent peak. With this procedure, the solvent (or most of it) can be eliminated using an exit valve. A system with a retention gap is schematically shown in Figure 4.2.2. Variations of this concept and corresponding instrumentation are available, offering systems with special injection systems, precolumn, and intermediate valves that are able to eliminate the solvent from the sample. Significant effort was made to improve such systems due to the benefits of less sample handling, increased sensitivity of the chromatographic determination, and possible increase in reproducibility [12], [13], [14], [15]. The alternative to large volume injection

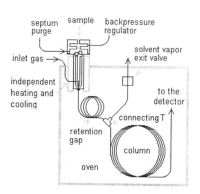

FIGURE 4.2.2. *Diagram of a large volume injection system with retention gap.*

is an off-line concentration step. This can be done using a number of sample preparation procedures, such as evaporation, solvent extraction, solid phase extraction, etc. Depending on the instrumentation available, one system or another may be more convenient. For processed samples containing analytes with a high boiling point compared to the solvent, the use of PTV or of a retention gap with a solvent vapor exit valve is a convenient procedure to save additional sample processing. However, for volatile analytes the use of automatic systems using solvent elimination must be done only after careful optimization of the process.

The solvent elimination and specific segregation of sample components in PTV systems or use of retention gaps with solvent elimination valves may be viewed as sample preparation procedures. However, these systems are typically considered as part of the GC instrumentation.

- *The oven and the column in a gas chromatograph*

The oven of the GC provides a controlled temperature for the chromatographic column. In most GC systems, the set temperature can be kept within $0.1°$ C, and a range between $-100°$ C to $400°$ C can be achieved using either a cryogenic agent (liquid N_2 or CO_2) or electric heating. Also, the GC ovens are commonly able to provide temperature gradients such that a sequence of isotherm and gradient portions (usually, three or four ramps) are available. Heating rates for the GC oven may vary, but standard ovens provide a heating rate as high as $30°$ C/min. with good temperature control. Faster heating capabilities are also available, useful mainly for fast (high speed) chromatographic procedures [16], [17], [18], [19], [20]. The chromatographic column is placed in the GC oven, and it is assumed to have the same temperature as the oven. Capillary chromatographic columns can be attached directly to the injection port of the GC or connected through a piece of uncoated (deactivated) capillary. This piece of uncoated capillary (also known as retention gap) accommodates and focuses the initial sample before the chromatographic process starts. The temperature of the GC oven can be set below the boiling point of the solvent carrying the sample, and although vaporized in the injection port, the solvent and the sample will recondense in the retention gap. Only when the temperature of the oven reaches a certain temperature, the material in the retention gap becomes vapors and the separation begins. A number of compounds that have low volatility or are formed due to some decomposition processes can be permanently retained in the gap. Therefore, the retention gap may serve the purpose of cleaning the sample that reaches the analytical column. The diminishing performance of a chromatographic system that has been used for a certain period of time for the analysis of "dirty" samples and has a retention gap can be sometimes improved by replacing the gap. The role of the retention gap has been thoroughly reviewed in literature [21].

In gas chromatography the most common process used for separation is based on the different partition of each analyte between two phases, the gas phase and the stationary phase that is a liquid or a liquid-like polymer (gas-liquid chromatography). Gas chromatography based on the differences in the adsorption on a solid surface is also known but less common (gas-solid chromatography).

From the construction point of view, two basic types of columns are known: a) packed columns and b) open-tubular columns (capillary columns). The packed columns contain solid particles of certain dimensions that can be coated with the stationary phase, while the capillary columns contain the stationary phase as a film on the inner wall. The stationary phase can be a liquid, a cross-linked material, or a bonded phase. Some special columns such as capillary columns containing fine solid particles on the wall are also known. The use of regular packed columns is limited to specific applications, while the capillary columns are widely utilized. The capillary columns, with the stationary phase as a film on the inner wall, are commonly made from silica and have an outer coating (polyimide, aluminum, etc.) that improves their mechanical resistance. The column internal diameter (i.d.) can be chosen in a range between 50–100 μm to 0.6 mm or even wider. The columns are classified as microbore for i.d. < 0.1 mm, minibore for i.d. = 0.18 mm, narrow bore for i.d. = 0.25 mm, regular for i.d. = 0.32 mm, and megabore for i.d. = 0.53 mm. The stationary phase film thickness on the inner wall of the column is a very important parameter. The film thickness may be chosen between 0.1 μm to 5 μm, a frequently utilized film thickness being 0.25 μm. However, narrower columns have thinner films. The length L of the capillary column is another important parameter, lengths between 5 m and 100 m being common. As seen in Section 1.6, these parameters determine the phase ratio β (see rel. 1.6.16) and the number of theoretical plates n (see rel. 1.6.40) for the chromatographic column. As shown in Section 1.6, the number of theoretical plates n for a column depends on its construction. Narrow bore columns may have a theoretical plate number for 1 m of length between 5000 to 10,000 plates or even higher for special columns. However these columns have low sample loading capacity. Conventional columns (minibore, narrow bore) have n/L between 3000 and 5000, and wide bore columns typically have n/L between 1500 and 3000. However, the value of n depends as shown in Section 1.6 on the linear flow rate u and the nature of the carrier gas. The minimum value for H can be obtained by taking the value for u where the derivative of rel. (1.6.51) is null. This corresponds to

$$u = (B / C)^{1/2} \tag{4.2.1}$$

The best H_{min} that can be obtained for a given column is given by the expression:

$$H_{min} = A + 2 (B\ C)^{1/2} \tag{4.2.2}$$

For a capillary column rel. (4.2.2) can be estimated using the expressions for A, B, and C calculated in Section 1.6. The eddy diffusion for capillary columns does not exist, and therefore A = 0. The value for B can obtained from rel. (1.6.45a) with the expression:

$$B = 2\ D_g \tag{4.2.3}$$

The values for C are composed of two terms H_g given by rel. (1.6.49) and H_s given by rel. (1.6.48a) with the film thickness d_f. These two contributions can be written in the form:

$$C = H_g + H_s \tag{4.2.4}$$

Using these expressions in rel. (4.2.2), the value for H_{min} for a capillary column can be written as follows:

$$H_{min} = r\ (\mathcal{F}_1 + \mathcal{F}_2)^{1/2} \tag{4.2.5}$$

where

$$\mathcal{F}_1 = (1 + 6k + 11k^2) / [3 (1 + k)^2] \tag{4.2.6}$$

and

$$\mathcal{F}_2 = (D_g / D_s) (1 / \beta^2) \{(4\ k / [3 (1 + k)^2]\} \tag{4.2.7}$$

For capillary columns with very thin coating, β is relatively large and $\mathcal{F}_2$ is small. Neglecting $\mathcal{F}_2$ the expression for H_{min} becomes

$$H_{min} = r\,(\mathcal{F}_1)^{1/2} \qquad\qquad (4.2.8)$$

Figure 4.2.3 shows the variation of $\mathcal{F}_1$ as a function of k (with k between 1 and 20).

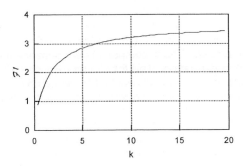

FIGURE 4.2.3. *Variation of $\mathcal{F}_1$ as a function of k.*

Using an upper limit for $\mathcal{F}_1$, rel. (4.2.8) shows that for large β (thin film), the narrower is the capillary column (small r) the smaller are the values for H_{min} that can be obtained.

When the values for β are relatively low (thick film capillary columns), because the ratio D_g / D_s can be quite large, the term $\mathcal{F}_2$ must be taken into consideration in rel. (4.2.5). The variation of the factor depending on k in the expression of $\mathcal{F}_2$ is shown in Figure 4.2.4.

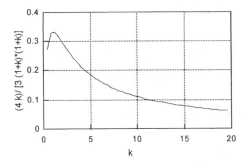

FIGURE 4.2.4. *The factor depending on k in the expression of $\mathcal{F}_2$.*

The decrease in $\mathcal{F}_2$ for k increasing above about 10 is not significant, as shown in Figure 4.2.4. Therefore, for capillary columns the construction basically determines the value of H_{min}, and through rel. (1.6.39) the value for n.

Numerous stationary phase materials have been utilized for the gas chromatographic columns (see e.g. [22]). The phases are evaluated and selected based on a variety of criteria, such as thermal stability, lowest temperature at which the column acts as a liquid phase, and lack of volatile compounds generated from the stationary phase during heating (bleeding). An important criteria for choosing a column is its polarity. Polarity of a stationary phase is usually characterized using McReynolds constants [23], which are obtained with the help of Kováts retention indexes. The constants are defined as the difference between the Kováts index for a reference compound "j" in the phase to be evaluated and in a reference stationary phase:

$$\Delta I_j = I_j - I_j^\circ \tag{4.2.9}$$

where I_j is the Kováts index for the phase to be evaluated and I_j° is the Kováts index for the reference phase. Squalane is used as the reference stationary phase, and five compounds are used as reference. These compounds are benzene, butanol, 2-pentanone, nitropropane, and pyridine. This list can be extended with five more compounds [24], [25]. An average "polarity" constant P can be obtained from McReynolds constants [23] for the individual reference compounds as follows:

$$P = \left(\sum_{j=1}^{j=5} \Delta I_j \right) / 5 \tag{4.2.10}$$

Using McReynolds index, a classification of stationary phases has been done, from the least polar, which is squalane, to the most polar, which is 1,2,3-tris(2-cyanoethoxy) propane (TCEP). Also it was shown that only 12 stationary phases are able to cover the whole range of polarities of the known materials used as stationary phases in GC [26], [27]. A thermodynamic justification for the polarity concept of stationary phases also was developed [28]. The variation of parameter P for several liquid stationary phases is shown in Figure 4.2.5 [26].

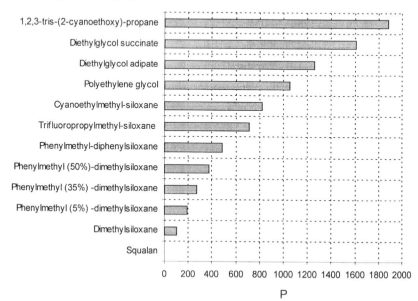

FIGURE 4.2.5. *The variation of P for several liquid stationary phases.*

In GC separations, the capillary columns are by far more utilized than other types of columns. Various models of capillary columns are available, and they are produced by different manufacturers. Certain types of stationary phase are very frequently utilized because of their ability to separate a variety of compounds. Such phases are a) polyethylene glycol, which is a polar material and is used for the separation of the analytes based on their differences in polarity and tendency to form hydrogen bonds, b) dimethylpolysiloxane, which is nonpolar and is used for separations based mainly on the boiling point of the analytes, and c) dimethylpolysiloxane copolymer with a certain amount of diphenylpolysiloxane (such as 5%, 35%, or 50%), which has some polarity. In these phases, the separation is based on the different partition of each analyte between the gas phase and the liquid (or liquid-like) phase.

Many other choices are available for the stationary phase of the capillary columns. Some are modifications of stationary phases indicated above, such as polysilyl-phenylenemethylphenylsiloxane, which can be used as a replacement for the combination dimethylpolysiloxane + diphenylpolysiloxane, or polyethylenglycol ester replacing polyethylene glycol phases. Others are more special phases, such as cyanopropyl(methyl)polysiloxane, which is a polar phase with higher temperature resistance than polyethylene glycol, or carboranedimethylsiloxane phase, which is nonpolar and very resistant to high temperatures. Several other stationary phases made from different proportions of typical phases (methyl, phenyl, cyanopropyl, cyanopropylphenyl) or from special compounds such as polytrifluoropropylsiloxane are available. Combinations of two columns are also applicable for obtaining better separation capabilities [29]. The structures of some of these phases are indicated below:

dimethyl polysiloxane

dimethyl polysiloxane + x% phenyl

dimethyl polysiloxane + x% phenylcyanopropyl

cyanopropyl polysiloxane

polyethylene glycol

carborane polysiloxane

Columns such as PLOT (porous layer open tubular), columns coated with a modified graphitized carbon or with a silicone based polymer with chiral groups incorporated into the polymeric chain, columns coated with derivatized cyclodextrins (for the separation of chiral compounds), etc. also are utilized for specific applications. For the packed columns, the variety of stationary phases is also large. Among common stationary phases for packed columns are both liquid or liquid-like materials as well as solids (where the separation process is based on adsorption) such as porous polymers (see Section 9.2). Some commercially available stationary phases that can be used for packed chromatographic columns in GC as well as for other adsorption purposes are given in Table 4.2.2. Some of these materials are available in solution and can be applied on the walls of a capillary column or an an inert support.

TABLE 4.2.2. *Bulk chromatographic stationary phases for GC* .

Stationary phase	Min[1] °C	Max[1] °C	Stationary phase	Min[1] °C	Max[1] °C
Apiezon L	50	300	Halocarbon	20	100
Bentone	0	180	Halocarbon K-352	0	250
bis(2-ethoxyethyl)adipate	0	150	Halocarbon wax	50	150
bis(2-ethoxyethyl)phthalate	0	150	1,2,3,4,5,6-hexakis (2-cyanoethoxy-cyclohexane)	125	150
bis(2-methoxyethyl)adipate	0	150	Igepal CO-880 (Nonoxynol)	100	200
bis(p-methoxylbenzylidene)-α,α'-bi-p-toluidine (BMBT)	189	225	Igepal CO-990	100	200
butandiol succinate	50	225	N-n-lauryl-N-L-valine-*tert*-butylamide (SP-300)	60	140
Carbowax 20M	60	225	neopentyl glycol adipate	50	225
Carbowax 20M-terephthalic acid	60	225	neopentyl glycol sebacate	50	225
Carbowax 400	10	100	neopentyl glycol succinate	50	225
Carbowax 600	30	125	OV-1 (vinylmethyl silicone)	100	350
Carbowax 1000	40	150	OV-1 (dimethyl silicone, gum)	100	350
Carbowax 1540	50	175	OV-3 (phenylmethyldimethyl silicone, 10% phenyl)	0	350
Carbowax 4000	60	120	OV-7 (phenylmethyldimethyl silicone, 20% phenyl)	0	350
cyclohexanedimethanol succinate (CHDMS)	100	250	OV-11 (phenylmethyldimethyl silicone, 35% phenyl)	0	350
Dexsil 300, carborane/methyl silicone	50	450	OV-17 (phenylmethyl silicone, 50% phenyl)	0	375
Dexsil 400, carborane/methyl phenyl silicone	50	400	OV-22 (phenylmethyldiphenyl silicone, 65% phenyl)	0	350
Dexsil 410, carborane/methyl cyanoethyl silicone	50	375	OV-25 (phenylmethyldiphenyl silicone, 75% phenyl)	0	350
dibutyl maleate	0	50	OV-61 (diphenylmethyl silicone, 33% phenyl)	0	350
di-n-decyl phthalate	10	175	OV-73 (5.5% diphenylmethyl silicone)	0	325
di(2-ethylhexyl)sebacate	0	125	OV-101 (dimethyl silicone, fluid)	0	350
diethylene glycol adipate (DEGA)	0	200	OV-105 (cyanopropylmethyl silicone)	0	275
diethylene glycol succinate (DEGS)	20	200	OV-202 (trifluoropropyl silicone, fluid)	0	275
diglycerol	20	100	OV-210 (trifluoropropyl silicone, fluid)	0	275
diisodecyl phthalate	0	175	OV-215 (trifluoropropyl, gum)	0	275
2,4-dimethylsulfolane	0	50	OV-225 (cyanopropylmethyl-phenylmethyl)	0	265
dinonyl phthalate	20	150	OV-275 (dicyanoallyl)	25	250
dioctyl sebacate	0	125	OV-330 silicone - Carbowax	0	250
ethyl N,N-dimethyloxamate (EDO-1)		40	β,β-oxydipropionitrile	0	75
ethylene glycol adipate (EGA)	100	225	phenyldiethanolamine succinate	0	230
ethylene glycol phthalate	100	200	polyethylene glycol adipate (EGA)	0	225
ethylene glycol succinate	100	200	polyethyleneimine	0	175
ethylene glycol tetrachlorophthalate	120	200	polyphenyl ether (5 rings) OS-124	0	200
free fatty acid phase (FFAP)	50	250	polyphenyl ether(6 rings) OS-138	0	225

TABLE 4.2.2 (continued). *Bulk chromatographic stationary phases for GC.*

Stationary phase	Min[1] °C	Max[1] °C	Stationary phase	Min[1] °C	Max[1] °C
polypropylene glycol	0	150	SP-2331 (methyl silicone)	25	275
polypropyleneimine	0	200	SP-2340 (poly(90% biscyanopropyl / 1 0% cyanopropylphenyl siloxane))	25	275
polypropylene carbonate	0	50	SP-2380 (poly(biscyanopropyl siloxane))	25	275
Quadrol	0	150	SP-2401 (methyl silicone, trifluoropropyl)	0	275
SE-30 (methyl silicone,	50	300	squalane	20	100
SE-52 (methyl silicone)	50	300	sucrose acetate isobutyrate (SAIB)	0	200
SE-54 (methyl silicone: 5% phenyl, 1 % vinyl silicone)	50	300	tetracyanoethylated pentaerythritol	30	175
SF-96 (methyl silicone)	0	250	tetraethylene glycol dimethyl ether		80
Silar 5 CP (methyl silicone)	0	250	1,2,3,4-tetrakis (2-cyanoethoxy)butane	110	200
Silar 10 CP (methyl silicone)	0	250	tetraethylenepentamine	0	125
sorbitol		150	tetraethylenepentamine	0	125
SP-300 (N-n-lauroyi-N-L-valine-T-butylamide)	60	140	tetrahydroxyethylethylenediamine (THEED)	0	125
SP-400, chlorophenyl methyl silicone	0	350	β,β'-thiodipropionitrile (TDPN)		100
SP-2100 (methyl silicone)	0	350	tricresyl phosphate	20	125
SP-2250 (methylphenyl silicone, 50% phenyl)	0	375	tris(2 cyanoethyl) nitromethane (TCENM)	20	140
SP-2300 (poly(cyanopropylphenyl siloxane))	20	275	1,2,3-tris(2-cyanoethoxy)propane (TCEP)	0	175
SP-2310 (poly(50% biscyanopropyl / 50% cyanopropylphenyl siloxane))	25	275	Triton X-100	0	200
SP-2330 (poly(80% biscyanopropyl / 20% cyanopropylphenyl siloxane))	25	275	Triton X-305	0	200

[1] Min and Max indicate minimum and maximum temperatures of applicability.

The choice of the chemical nature of the stationary phase in a chromatographic column determines at least in part the range of compounds that can be analyzed. This chemical nature is selected based on the composition of the sample (polarity, volatility, etc.) and offers an important practical procedure for obtaining a better separation. The choice of a specific column is frequently related to a specific sample processing procedure, as it will be further shown in this book. Various tests were devised for the choice of optimal column, such as by comparing the results in the separation and in the peak shape of specific compounds [30], [31]. The columns are usually kept filled with an inert gas (He) for storage or use, and after a number of utilizations, the performance of the column diminishes. This process is determined by various factors including the presence of air and water flowing together with the carrier gas, but more frequently by the matrix and the solvent of the analyzed samples. One important role of sample preparation is the elimination or at least reduction of sample components that affect the column lifetime.

- The detectors in gas chromatography

The detector of a GC is an important part of the instrument, as gas chromatography is used as an analytical technique and not only as a separation procedure. The detector senses the presence of a component different from the carrier gas and generates an electrical signal preferably proportional with the amount of the analyte. Various sensitive detectors are utilized. Some detectors are nonselective and do not have the capability of qualitative identification of the eluting compounds. Some detectors are element specific and can determine if the eluting compounds contain, for example, nitrogen or sulfur. Instruments such as a mass spectrometer or an infrared spectrophotometer can also be used as detectors for the GC, offering the capability of qualitative identification of the eluting compounds. Elaborate descriptions of different detectors can be found in literature (e.g. [32]). Some of the detectors known in gas chromatography and their characteristics are indicated in Table 4.2.3.

TABLE 4.2.3. *Main types of GC detectors, their sensitivity and limit of detection.*

Detector type	Abbrev.	Sensitivity[1]	Limit of detection in mg analyte	Linear range	Noise	Type of selectivity
thermal conductivity	TCD	10 mV mL/mg	$2 \cdot 10^{-5}$ to 10^{-8} (50 mL gas/min.)	10^6	0.01 mV	nonselective
flame ionization	FID	0.01 C/g	$2 \cdot 10^{-8}$ to 10^{-11}	10^7	10^{-14} A	nonselective
nitrogen phosphorus other thermoionic	NPD		10^{-8} to 10^{-12}			nitrogen, phosphorus specific
electron capture	ECD	40 A mL/g	10^{-10} to 10^{-13}	10^4	$2 \cdot 10^{-12}$ A	halogen, some carbonyl
flame photometry	FPD	$4 \cdot 10^{-10}$ A	10^{-6} to 10^{-10}	10^3	$2 \cdot 10^{-12}$ g/sec	some specificity
photoionization	PID		10^{-5} to 10^{-12}	10^7		nonselective (with some exceptions)
electrolytic conductivity	Hall		10^{-6} to 10^{-11}			halogen, sulfur, nitrogen
sulfur chemiluminescence			10^{-6} to 10^{-14}			sulfur
nitrogen chemiluminescence			10^{-5} to 10^{-13}			nitrogen, NO
aroyl-luminescence [33]	ALD		very low			
atomic emission	AED					specific elements
helium ionization	HID	100 C/g	$4 \cdot 10^{-10}$	$5 \cdot 10^3$	$3 \cdot 10^{-14}$ A	
mass selective	MSD		instrument dependent $10^{-9} - 10^{-11}$	10^5		qualitative and quantitative
infrared	IRD		instrument dependent 10^{-6} 10^{-5}	10^3		qualitative and quantitative

[1] The sensitivity units are dependent on the nature of detector. TCD detector responds to changes in concentration, while FID, NPD, etc. respond to mass of material entering the detector per unit of time. Also, the flows and the parameters in the detector may strongly influence sensitivity [34]. Typical flows are 20 mL/ min He for TCD and FID, 35 mL/min. N_2 for ECD, 80 mL/min. N_2 for FPD, 60 mL/min. He for HID.

There are several important characteristics of a good detector: sensitivity, dynamic range, stability, and for specific ones selectivity. Sensitivity in fact should be characterized by two parameters, the ratio of the detector response to the amount of sample (sensitivity slope) and the minimum detectable level of a given compound (commonly measured for a signal to noise ratio of 3). The dynamic range is the range of sample amount for which the detector can be calibrated to provide accurate concentration in quantitative measurement. Stability refers to the capability to generate the same response for the same amount of sample.

Among the GC detectors the most frequently utilized is probably the FID. A wide variety of technical developments were done for improving FID performances regarding construction, the flow of the gases, and electronic detection of the signal. For example, both analog (continuous) and digital monitoring of the signal were developed, but the signal sampling with a certain frequency (such as 60 times in a sec.) is the more common procedure in modern instruments.

Another common detector for GC is the mass spectrometer. This detector (which was in fact developed independently as a stand-alone instrument) offers the capability of compound identification. Extensive literature is available regarding gas chromatography/mass spectrometry (GC-MS) analysis of organic molecules (e.g. [35], [36].

A number of mass spectral instruments can be used as GC detectors, and these include magnetic sector (usually indicated as high resolution instruments or HR-MS because they can operate at resolutions $m/\Delta m$ up to 10,000 or even higher) quadrupols, ion trap (IT), or time of flight (TOF) instruments. Magnetic sector instruments can have excellent sensitivities (10^{-7} C μg^{-1}), among the other systems the ion trap being also very sensitive.

The MS detector typically generates a total ion chromatogram (TIC), which is a plot of the total ion count (detected and processed by the data system) as a function of time. The single ion chromatogram (extracted from the TIC or obtained using single ion monitoring or SIM) plots the intensity of one ion (m/z value) as a function of time. These chromatograms have a discrete structure being made from scans (the scan number is linearly dependent of time). When the points of the chromatogram are close to each other, this gives a continuous aspect of the graph. Each scan has an associated spectrum for the TIC, and therefore the TIC has in fact a tridimensional structure. Tridimensional plots are available with certain data systems, although their practical utility is rather limited. Most data systems display the TIC and spectra for chosen scan numbers.

A variety of other detectors are used such as NPD, which is practically a modified FID, PID and ECD, which are very sensitive detectors for specific classes of compounds, and AED [37]. With an AED detector, it is possible for example to monitor in parallel eluting compounds that contain carbon, hydrogen, oxygen, nitrogen, and sulfur. Gas chromatography–Fourier transform infrared spectrometry (GC-FTIR) in combination with MS detection is another detection alternative that enhances the information in GC analysis. The two detectors provide complementary information. Mass spectra of structural isomers are sometimes similar, and their identification is difficult. On the other hand, FTIR is less proficient in detecting homolog compounds and in general provides less structural information than MS. However, structural isomers are identifiable with FTIR spectrometry, and mass spectra of homologs exhibit mass differences, which can be used in their identification. Thus, it is advantageous to use both detectors. Detection limits are better for MS than for light-pipe FTIR [38]. The sensitivity in GC-FTIR can be increased with a system that collects the analytes by cryotrapping the GC effluent on a ZnSe window at low temperature (80° K) followed by IR detection [39], [40]. The IR measurement on a specific spot on the ZnSe window can be done using repeated scans, which increase the sensitivity of the technique.

- The need for sample preparation in GC

The need for sample preparation in GC is summarized in Table 4.2.4.

TABLE 4.2.4. *Main types of problems in GC analysis and the application of sample preparation to solve them.*

Problem	Solution
Analytes in a form not adequate for GC analysis.	Solubilization that takes the analytes in solution, or use of solventless procedures (P&T, SPME) for adsorbing the analytes on a solid phase material followed by desorption for GC analysis.
Analytes affected by specific interferences.	Fractionation, derivatization of specific analytes, chemical separation.
Analytes in a complex matrix generating global interferences.	Cleanup, derivatization
Sample decomposition in the injection port.	Derivatization.
Analytes in a concentration too low for detection.	Sample concentration, derivatization.
Overall poor chromatography.	Derivatization, sample cleanup, concentration.
Short lifetime of the chromatographic column.	Sample cleanup, derivatization.
Enantiomers coelution.	Derivatization with chiral reagents to form diastereoisomers.

4.3 CONVENTIONAL HIGH PERFORMANCE LIQUID CHROMATOGRAPHY AS A CORE ANALYTICAL TECHNIQUE

High performance liquid chromatography (HPLC) is the most common separation technique, being about four times more frequently applied than GC. HPLC is classified as a separate technique that uses a liquid mobile phase and is based on the high performance separation capability of the chromatographic column (see Section 4.1). Because various processes may be involved in the separation, a variety of HPLC types may be identified. The usual process of separation in HPLC is liquid-liquid partition. Liquid-solid adsorption is also used in various separations, but less frequently. For the separation based on size exclusion (SEC), the technique is commonly identified as high performance gel permeation chromatography (GPC) when applied for compounds soluble in organic solvents, or gel filtration chromatography (GFC) when applied for water soluble compounds. The high performance separation capabilities also may be based on an ion exchange process (used in both ion chromatography and for the separation of certain nonionic species) or on "ion mediated" separation. This diversity is achieved with specific choices for the chemical nature of the stationary and mobile phases. A diagram of the more common HPLC chromatography types and their uses depending on sample solubility and sample molecular weight is given in Figure 4.3.1.

Regardless the specific separation process, most of the instrumental parts in these techniques are similar. Some details on size exclusion chromatography, ion chromatography, and ion pair chromatography are given in Section 4.4.

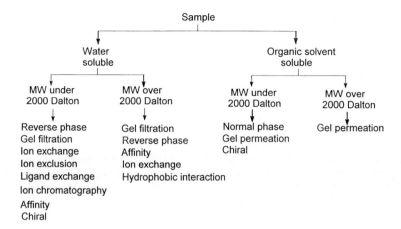

FIGURE 4.3.1. *Diagram showing several types of HPLC.*

A typical block diagram of a high performance liquid chromatography (HPLC) system is shown in Figure 4.3.2. A variety of instrumentation is available for performing HPLC, and the specialty literature is abundant with information regarding this field (e.g. [41], [42], [43], [44]).

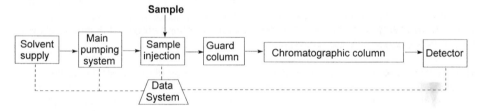

FIGURE 4.3.2. *Block diagram of a simple HPLC system.*

The solvent supply plays the role of delivering the solvent(s) necessary as a mobile phase for the HPLC. The separation can be achieved using a solvent with a constant composition (isocratic separation) or using a solvent with a variable composition (gradient separation). The variable composition can be achieved by mixing the mobile phase constituents before the main pumping system (low pressure mixing) or after it when each mobile phase component requires a separate pump (high pressure mixing). The main pumping system consists of pump(s) able to deliver a constant flow of solvent through the injector, through the chromatographic column (in certain cases having also a precolumn), and through the detector. Most pumps are capable of delivering between 0.1 mL/min to 10 mL/min and generate pressures up to 5000 psi or higher (1 psi = 6.89476 kPa = 6.89476 10^{-2} bar = 6.80460 10^{-2} atm.). The high pressure is needed to overcome the resistance to flow of the chromatographic column, which usually contains very fine particles. Numerous other details about the solvent delivery systems and injectors can be found in the dedicated literature (see e.g. [45]). From the point of view of sample preparation, only some of the instrument components used in HPLC are of interest. These are the chromatographic column and the detector.

The choice of the chromatographic column and of the mobile phase in HPLC is critical in any chromatographic separation. This choice also has impact on the type of sample preparation used in the applied analytical method. Also, the detector used in an HPLC system is very important for the sample preparation, especially when chemical modifications are necessary for the utilization of a specific type of detection. The choice of a stationary phase is based on the nature of sample and the main scope of the analysis. For nonionic nonpolymeric organic compounds direct phase or reverse phase chromatography is commonly chosen. Reverse phase columns where the mobile phase is more polar than the stationary phase are in general preferred because the separations show better reproducibility. The solubility of the analytes in the mobile phase also plays an important role in the column choice. For compounds soluble in water or partially aqueous solvents, the selection of a reverse phase HPLC is not a problem, but for compounds insoluble in polar solvents, a direct phase column is usually necessary. The evaluation of the polarity of the sample, that of the stationary phase, and that of the mobile phase are also important in the choice of the column.

A large variety of columns is available for HPLC, the nature of the stationary phase being a determining factor for the type of chromatography, as indicated in Section 4.1. The stationary phase is made from a finely divided porous solid material or "base material" on which a layer of solvent is adsorbed or on which specific organic fragments are chemically bonded (bonded phases). The base material is usually made from porous silica that can be shaped in irregular particles or spherical particles with an i.d. typically varying between 3 and 10 μm and with a pore size varying between 50 to 100 Å. Both particle size and pore size must be as homogeneous as possible for better quality of the column. In addition to silica, zirconia and also organic polymers such as acrylates or polystyrene-divinylbenzene are used as a base material. Silica is by far more common. However, the other materials may have some advantages for specific applications. Zirconia for example, is very stable in a wide pH range and may be a better choice than silica regarding stability [46]. A variety of physical characteristics for the column can be chosen. Some typical column physical characteristics are given in Table 4.3.1. The columns are typically made from stainless steel or special plastics, such as polyetheretherketone (PEEK), and have adaptors (fittings) to the HPLC lines. The ends of the columns typically contain special filter frits.

TABLE 4.3.1. *Common physical characteristics of HPLC columns.*

Column type	Length mm	i.d. mm	Particle size μm	Theoretical plate No.	Flow rate mL/min.
packed analytical	200 - 300	4 -10	3 - 10	10^4	1 - 2
minibore	10 - 200	1 - 2	3 - 5	$5\ 10^4$	0.01 -1
microcolumn	500 - 3000	0.04 - 0.3	3 - 5	10^5	0.001 - 0.1
microcolumn in nano LC	~150	75 μm	3	-	200 nL/min

For normal phase chromatography one common stationary phase is bare silica. The silanol groups and the adsorbed water are responsible for the solute retention. As shown in Section 1.6, parameters such as $c_{A,s}$ (which is the molar concentration of the molecules of the mobile phase "A" in the liquid stationary phase but is treated as a stationary phase parameter [47] because it varies little for different solvents) and α' (which measures the ability of a unit of adsorbent surface to bind adsorbed molecules) depend on the adsorbent surface area and water content. The variability in the amount of water adsorbed that can be removed by certain solvents during the use of the column makes this column less reproducible. However, silica columns are frequently used in HPLC. Much more reproducible properties have bonded phases that have been developed for both normal phase and reverse phase HPLC. Bonded phases are obtained by chemical modifications of silica surface in reactions that attach various groups replacing the active silanols. Several bonded phases common in HPLC are indicated in Table 4.3.2.

TABLE 4.3.2. *Bonded phases common in HPLC.*

Phase	Name	Structure	Properties
C1 - C6	alkyl depending on R with R = CH_3, C_2H_5, C_3H_7, C_4H_9, C_5H_{11}, C_6H_{13}	$-Si-R$	Reverse phase material with retention increasing from C1 to C6. Various other applications such as in ion pairing chromatography.
C8	RP-8, octyl, etc.	$-Si-C_8H_{17}$	Reverse phase material with wide applicability, similar to C18
C18	ODS, RP-18, octadecyl	$-Si-C_{18}H_{37}$	Reverse phase material with wide applicability. Most retentive for nonpolar compounds. Used in other applications such as ion pairing.
C_6H_5	phenyl	$-Si-(CH_2)_3-$ ⬡	Reverse phase material used in the analysis of aromatic compounds
CN	cyano, CN, cyanopropyl, nitrile	$-Si-(CH_2)_3-CN$	Stationary phase used in both normal and reverse phase.
NH_2	amino, amino propyl silyl, NH_2	$-Si-(CH_2)_3-NH_2$	Stationary phase used in both normal and reverse phase or even a weak anion exchange. Used in many separations including carbohydrates.
OH	diol	$-Si-(CH_2)_3-O-CH_2-CH-CH_2$ with OH OH	Stationary phase used in both normal and reverse phase, sometimes in gel filtration.
SAX	quaternary amine, strong base, anion exchange	$-Si-(CH_2)_3-\overset{+}{N}(CH_3)_3$	Anion exchange material, strongly basic.
SCX	sulfonic acid, strong acid, cation exchange	$-Si-(CH_2)_3-SO_3H$	Cation exchange material, strongly acidic.

The bonded phases on silica are obtained by two general procedures (see other reactions for the modification of silica surface in Section 11.1). In one procedure the silica particles are treated with a monochloro or a monoalkoxy silane (R_3SiX), which reacts with the silanol groups on silica. In the other procedure the silica particles are treated with trifunctional silanes ($RSiCl_3$), which react with the silanol groups on silica and also with each other. This treatment leads to higher reticulation and lower content of free silanol groups. The two types of treatment generate different loading with organic material. For a C18 column carbon loading is usually between 7 and 30%. The increase in the carbon loading % leads to higher k values for compounds containing hydrophobic moieties. A characterization of hydrophobicity of the HPLC column can be made using a hydrophobicity index (HI), defined as k value for phenylheptane with 75% CH_3CN / 25% H_2O mobile phase. The variation of k for several types of columns is shown in Figure 4.3.3. As seen in Figure 4.3.3, a wide range of k values is possible for the same bonded phase type and is particularly wide for C18 columns. This variation in the k values depends on the stationary phase structure, which can be different from one manufacturer to another. One problem with silica base reverse phase column is that only some of the silanol groups react and generate an organic phase (7–8 $\mu M/m^2$ of silanol are present on silica, and only about 4 $\mu M/m^2$ bonded silane is achievable). Also, some silanol groups may be generated during column utilization. The presence of these silanol groups contributes to the retention process and leads to a mixed retention mechanism, frequently with undesired results. This effect is reduced, for example, by an

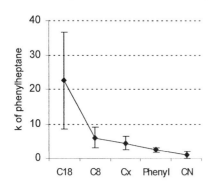

FIGURE 4.3.3. *Variation of k (phenyl-heptane) for different stationary phases (x = 6, 2, 1).* The wide range of values is shown by the error bar.

additional treatment of the column material with $(CH_3)_3SiCl$, which reduces the number of free silanol groups (endcapping). The extent of endcapping is not well related to k for phenylheptane and is usually tested with a mixture pyridine/phenol at pH 7. Well deactivated sorbents generate two well separated peaks, while active silanol groups lead to chromatograms with two merged peaks for pyridine and phenol.

Many other phases are available for HPLC in addition to those indicated in Table 4.3.2. (See also Section 12.3 for a discussion of size exclusion stationary phases.) Various types of phases available in HPLC include other bonded phases, polymer coated materials, bonded affinity solid phases, adsorbed biomolecules on solid supports, etc. Among these are phases containing groups such as nitro, weak basic, or acidic groups, various groups used for affinity chromatography or for chiral separations, and GPC type columns. In affinity chromatography, for example, a biomolecule (often a protein) able to have a selective interaction with another molecule is firmly adsorbed on a solid phase and further used as a biospecific stationary phase. The solid support capable of adsorbing the first biomolecule contains groups such as OH and NH_2 in structures as shown below:

$$HO-\overset{\overset{\displaystyle O}{|}}{\underset{\underset{\displaystyle O}{|}}{Si}}-(CH_2)_3-O-CH_2-\overset{}{\underset{\underset{\displaystyle OH}{|}}{CH}}-CH_2-O-CH_2-(CH_2)_6-CH_3$$

Attempts were made to generate a stationary phase that elutes proteins in the excluded volume and retains and separates smaller molecules. This type of stationary phase has been used successfully for the direct analysis of small molecules in biological samples, eliminating a sample preparation step [48]. This type of stationary phase is known as internal-surface reversed-phase (ISRP) or restricted access materials (RAM) [49]. One such packing material can be obtained, for example, starting with a porous silica that has glycyl-L-phenylalanyl-L-phenylalanine bonded on the surface. By exposing the material to a specific enzyme such as carboxypeptidase A, the phenylalanine is removed from the outer surface, creating a hydrophilic surface (diol-glycine). Because the enzyme cannot penetrate inside the pore, the packing retains its inner hydrophobic surface characteristics. When a serum sample is injected, proteins and other large biomolecules are excluded from the packing by repulsion from the hydrophilic surface group, and small molecules will diffuse into the pores and interact with the hydrophilic surfaces by a reversed-phase mechanism [50]. Columns with external diol and internal C18 pore surface are also constructed. This type of column combines a size exclusion character with reverse phase properties of the inner core of the stationary phase.

The performance of HPLC columns modifies in time due to the accumulation of materials that do not elute from the column either because of precipitation or irreversible adsorption. Also the performance may decrease due to the destruction of the stationary phase during the column use in acidic or basic conditions. The columns are commercially available in a specific solvent. The drying of the column also can damage the stationary phase, modifying the solid phase or creating channels. Sample preparation and periodic flushing of the HPLC column with appropriate solvents are important for maintaining the column performance. The use of precolumns (or guard columns) that can be replaced more frequently is a common practice for protecting the column performance.

The use of a precolumn in HPLC is similar to a sample cleanup step performed on-line in HPLC. The material used for precolumns (or guard columns) is usually the same as the one used in the main analytical column. A large number of precolumn materials are commercially available. Specific requirements are imposed to the precolumn. They must have a minimal dead volume for avoiding additional band broadening. Also, they must have no detrimental effect on separation, and the same particle size material as the one used in the column is recommended. However, the use of the same particle size in the guard column is not always practiced. The protection of silica based columns from too basic or acidic samples can be done with special guard columns (saturator guard columns). Trace enrichment and sample cleanup on a precolumn are also successfully applied. One typical example is the determination of drugs and their metabolites in biological fluids by HPLC. The main difficulty in this type of analysis comes from the necessity to remove the proteins from the plasma before injecting it in a reverse phase type column. The proteins are absorbed irreversibly on the stationary phase or precipitate in the column, significantly shortening the useful lifetime of the analytical column. For this reason it is convenient to use a precolumn, which acts as a guard column as well as a preseparation operation of the main analytes from the complex fluid matrix. In this way, a large number of injections can be performed with an analytical column before a significant loss of performances occurs [51], [52].

Another important choice in HPLC is that of the mobile phase. Pure or mixed solvents as well as solvents with modifiers that are solids have been used as mobile phase. The solubility in water of several solvents used in HPLC is given in Table 4.3.3.

TABLE 4.3.3. *Solubility in water of various solvents used in HPLC.*

Solvent	Solvent % in water	Solvent	Solvent % in water
water	100	isopropanol	100
formamide	100	butyl acetate	7.81
dimethylsulfoxide	100	1,2-dichloroethane	0.81
dimethylformamide	100	methylene chloride	1.6
acetic acid	100	diethyl ether	6.89
acetonitrile	100	benzene	0.18
ethanol	100	methyl *tert*-butyl ether	4.8
methanol	100	xylene	0.018
acetone	100	toluene	0.051
dioxane	100	di-isopropyl ether	?
methyl ethyl ketone	24	carbon tetrachloride	0.08
ethyl acetate	8.7	trichloroethylene	0.11
chloroform	0.815	cyclohexane	0.01
tetrahydrofuran	100	pentane	0.004
n-butanol	0.43	hexane	0.001
n-propanol	100	heptane	0.0003

The solvents in LC are frequently classified based on a polarity scale as described in Section 1.10 [53], [54]. The polarity P' of various solvents used in HPLC is given in Tables 1.10.11 to 1.10.19. Besides polarity, some other parameters were developed for the characterization of the behavior of a certain solvent towards a specific stationary phase. One such parameter is the elutropic strength ε° [47] described in Section 1.6.

The equilibrium constant K_i of a specific compound "i" on a specific stationary phase and using a specific solvent is given as shown with rel. (1.6.29) by the expression:

$$\log K_i = n \log c_{A,s} + \alpha' (S^\circ - \varepsilon^\circ \mathcal{A}_i) \qquad (4.3.1)$$

where n is given by $n = \mathcal{A}_i / \mathcal{A}_m$ with $\mathcal{A}_i$ and $\mathcal{A}_m$ the areas occupied on the adsorbent by the solute and the solvent, respectively, $c_{A,s}$ is a parameter proportional with the adsorbent surface area, α' is characteristic for each solid phase, S° is a measure of adsorption energy of "i" onto a standard adsorbent surface, ε° is the elutropic strength. As seen in expression (4.3.1), a number of factors are included, and its use for the calculation of K_i is limited. The problems arise from the difficulty of capturing in one expression the variation in the structures of the analyte, solvent, and stationary phase, as well as their interactions. However, for specific values for $c_{A,s}$, α' and $\mathcal{A}_i$ (see e.g. [55]), it is useful to understand the direction of variation for K_i for different solvents and for different analytes. Some values for the elutropic strength ε° for solvents on silica, alumina, Florisil, MgO, and C18 are given in Table 4.3.4.

For a different solid phase, the elutropic strength of the same solvent is not the same as seen in Table 4.3.4. However, the trend remains similar on polar phases and, for example, $\varepsilon^\circ (\equiv SiOH) \approx 0.77\ \varepsilon^\circ (Al(OH)_3)$. On the other hand, on C18 phase the elutropic strength is significantly different. In addition to pure solvents, it is common in HPLC to use solvent mixtures. The resulting elutropic strength is in between the ε° of the two solvents, and specific rules are established to estimate the value [56].

TABLE 4.3.4. *Elutropic strength on silica, alumina, Florisil, MgO, and C18 stationary phases of various solvents used in HPLC.*

Solvent	ε^o $\geqslant$SiOH	ε^o Al(OH)$_3$	ε^o Florisil	ε^o MgO	ε^o C18 (relative to ε^o =1 for methanol)
fluoroalkanes		-0.25			
pentane	0.00	0.00	0.00	0.00	
hexane	0.00-0.01	0.00-0.01			
iso-octane	0.01	0.01			
petroleum ether		0.01			
cyclohexane	0.03	0.04			
cyclopentane		0.05			
carbon disulfide		0.15			
carbon tetrachloride	0.11	0.17-0.18	0.04	0.10	
1-chlorobutane	0.20	0.26-0.30			
xylene		0.26			
toluene	0.22	0.20-0.30			
chlorobenzene	0.23	0.30-0.31			
diisopropyl ether		0.28			
isopropyl chloride		0.29			
n-propyl chloride		0.29			
benzene	0.25	0.32	0.17	0.22	
diethyl ether	0.38-0.43	0.38			
dichloromethane	0.30-0.32	0.36-0.42	0.23	0.26	
chloroform	0.26	0.36-0.40	0.19	0.26	
diethyl sulfide		0.38			
1,2-dichloroethane		0.44-0.49			
methyl ethyl ketone		0.51			
acetone	0.47-0.53	0.56-0.58			8.8
dioxane	0.49-0.51	0.56-0.61			11.7
1-pentanol		0.61			
tetrahydrofuran	0.53	0.45-0.62			3.7
methyl *tert*-butyl ether	0.48	0.3-0.62			
ethyl acetate	0.38-0.48	0.58-0.62			
methyl acetate		0.60			
dimethyl sulfoxide		0.62-0.75			
diethylamine		0.63			
nitromethane		0.64			
acetonitrile	0.50-0.52	0.52-0.65			3.1
1-butanol		0.70			
pyridine		0.71			
2-methoxyethanol		0.74			
n-propyl alcohol		0.78-0.82			10.1
isopropyl alcohol	0.60	0.78-0.82			8.3
ethanol		0.88			3.1
methanol	0.70-0.73	0.95			1.0
ethylene glycol		1.1			
dimethyl formamide					7.6

In reverse phase chromatography it is common to chose a solvent with the polarity higher than that of the most polar analyte. For gradient HPLC the polarity of the solvent is slowly decreased to a solvent with polarity lower than that of the least polar component of the sample. Detailed descriptions of the separation procedures are always given for a particular HPLC analysis.

The variation of parameter $S°$ in rel. 4.3.1 can be estimated making the assumption that $S°$ is an additive property depending on the stationary phase and the group of atoms in the molecule to be separated. This allows the calculation of $S°$ using the expression:

$$S° = \sum_m E_m \qquad (4.3.2)$$

where E_m is the free energy of adsorption for a group of atoms on a specific sorbent. Some values for E_m are given in Table 4.3.5. The values are not totally independent of the neighbor or the structure of the rest of the molecule. Monovalent groups "m" have only one neighbor "A" (A-m), while other groups may have two neighbors, "A" and "a" (A-m-a).

TABLE 4.3.5. *Values for the additive parameter E_m for several common organic groups.*

Group	A-aliphatic, a-aliphatic SiO_2	Al_2O_3	A-aliphatic, a-aromatic SiO_2	Al_2O_3	A-aromatic, a-aromatic SiO_2	Al_2O_3
methyl (-CH$_3$)	0.07	-0.03	-	-	0.11	0.06
methylene (-CH$_2$-)	-0.05	0.02	0.01	0.07	0.07	0.12
aromatic C (-CH=)	0.25	0.31	0.25	0.31	0.25	0.31
fluoro (-F)	1.54	1.64	-	-	-0.15	0.11
chloro (-Cl)	1.74	1.82	-	-	-0.20	0.20
bromo (-Br)	1.94	2.00	-	-	-0.17	0.33
iodo (-I)	1.94	2.00	-	-	-0.15	0.51
disulfide (-S$_2$-)	1.90	2.70	0.94	1.10	-	-
sulfide (-S-)	2.94	2.65	1.29	1.32	0.48	0.76
ether O (-O-)	3.61	3.50	1.83	1.77	0.87	1.04
tertiary amine (-N<)	5.80	4.40	2.52	2.48	-	-
aldehyde (-CHO)	4.97	4.73	-	-	3.48	3.35
nitro (-NO$_2$)	5.71	5.40	-	-	2.77	2.75
nitrile (-CN)	5.27	5.00	-	-	3.33	3.25
ester (-COO-)	5.27	5.00	3.45	3.40	4.18	4.02
ketone (>CO)	5.27	5.00	4.69	3.74	4.56	4.36
hydroxyl (-OH)	5.60	6.50	-	-	4.20	7.40
imine (>C=N-)	-	6.00	-	4.46	-	4.14
primary amine (-NH$_2$)	8.00	6.24	-	-	5.10	4.41
sulfoxide (>SO)	7.20	6.70	4.20	4.00	-	-
acid (-COOH)	7.60	21	-	-	6.10	19
amide (-CO-NH$_2$)	9.60	8.90	-	-	6.60	6.20

The effect of temperature in HPLC has been studied both experimentally and theoretically. As shown in Section 1.6, the partition coefficients K_i are expected to vary with temperature (see rel. 1.6.64, which can be applied to any equilibrium process). However, the range of temperatures applicable during HPLC is limited, and the separation in HPLC is influenced less dramatically by the changes in temperature compared to GC. Some analytical methods using HPLC indicate, however, the necessity to maintain a specific temperature for achieving the desired separation.

Various techniques are used for detection in HPLC. The more common ones are based on the measurement of UV-Vis absorption, refractive index (RI), fluorescence (FL), chemiluminescence, various electrochemical parameters, or mass spectra. Other special detectors were also developed such as the evaporative light scattering detector (ELS), IR or Raman detectors, etc. Depending on the detection type, various paths can be selected for the sample preparation. The selection of a specific derivatization that is used frequently as a part of a sample preparation procedure is strongly dependent on the choice of the detector used in HPLC. Also, other elements in sample preparation such as the solvent used to carry the analytes from the sample preparation step to the core analytical step are influenced by the choice of a specific detector. Several aspects regarding detection in HPLC are further discussed in this section.

- Refractive index detection in HPLC

Refractive index detection (RI) is very common in HPLC. Due to the universal modification of the refractive index of a solution as a function of the concentration of the solute, RI can be used for the quantitation of a variety of analytes. This type of detection can be applied without the need for chromophore groups, fluorescence bearing groups, or other specific properties in the molecule of the analyte. In many cases the sensitivity of RI detection is, however, below that of other types of detection. Also, there are problems with using elution with a concentration (composition) gradient for the mobile phase in HPLC.

- UV-Vis detection in HPLC

As shown in classical UV-Vis spectrophotometry, two related quantities, transmittance T and absorbance A, are used to measure the absorption of light passing through the solution of the analyte. Transmittance is defined as follows:

$$T = I_1 / I_o \qquad\qquad (4.3.3)$$

where I_o is the intensity of the radiant energy incoming to the sample and I_1 is the intensity of the emerging light (T can also be expressed as percent). As expected, T is a function of the frequency ν or of wavelength $\lambda = c_{light} / \nu$ of the radiation (c_{light} is the speed of light). Absorbance is defined as follows:

$$A = \log_{10} (1 / T) \qquad\qquad (4.3.4)$$

Absorbance is related to the molar concentration c_M of the absorbing species by Lambert-Beer law:

$$A_\lambda = \varepsilon_\lambda\, c_M\, L \qquad\qquad (4.3.5)$$

where ε_λ is the molar absorption coefficient at the specific wavelength λ, and L is the path length of the light through the sample. The absorption is commonly used in UV-Vis quantitations because it is proportional with the concentration. In HPLC, the absorption of the liquid eluting from the separation system is measured, generating the chromatographic peaks as the absorption A_λ changes. The wavelength of absorption is selected for best results, and usually it is set at the maximum absorption for the analyte. Also, some determinations that are done in UV measure the absorbance at 254 nm. This wavelength corresponds to the maximum emission (253.7 nm) of a mercury lamp that is commonly used as a UV source in simpler spectrophotometers. The area under the chromatographic peak of the analyte being proportional with its concentration, the quantitation can be done using a (linear) calibration curve or a response factor between peak area and the analyte concentration. Various factors may influence this linearity. In practice a specific range of linearity is indicated for each analytical method. The peak height also can be used for quantitation assuming equal peak widths, but this procedure is less common in HPLC because the peaks are wider than in GC. If the entire UV-Vis spectrum is measured in different points across a chromatographic peak, this can be used for the evaluation of peak purity and also can be a guide for qualitative identification of the analyte, although UV-Vis spectra are very seldom sufficient for compound identification.

The UV region of spectrum starts at 200 nm, but the common range of practical utility in UV spectrophotometric measurements in HPLC with UV detection starts at about 220 nm or higher. At lower values than this wavelength, a strong light absorption usually takes place because of the solvent. The wavelength cut-off of various solvents can be found in tables e.g. [56].

UV and Vis spectra are generated by the electronic energy transitions in the molecule, as shown in the simplified scheme given in Figure 4.3.4. Electronic levels are classified as bonding, nonbonding, and antibonding depending on their energy (e.g. [57]), and each electronic level is associated with vibrational levels. Due to transitions involving a number of vibrational levels, the electronic spectra appear as broad bands.

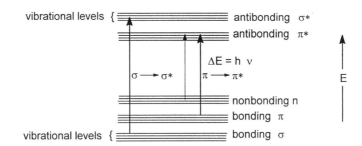

FIGURE 4.3.4. *Simplified scheme of electronic transitions during UV or Vis light absorption.*

Depending on the molecule, the number of electronic energy levels can be quite large. For a closed shell molecule, for example, the number of bonding levels is usually equal to half of the number of electrons in the molecule. These energy levels are commonly classified based on the symmetry of the molecule [58]. For molecules belonging to the point groups $C_{\infty v}$ and $D_{\infty h}$ (see e.g. [58]), the molecular orbitals corresponding to a given energy (and accordingly the energy levels) can be classified as σ, π, δ,.... This classification is also extended to planar molecules but cannot be applied to molecules with different symmetry. Besides the transitions seen in Figure 4.3.4 exemplified for a planar molecule, such as n —> π^* generating the R bands, π —> π^* generating K bands, σ —> σ^* transitions, or n —> σ^* transitions, the molecules also may have transitions that generate so-called B bands, E bands, etc. The B bands are typical for aromatic compounds, and the E bands are typical for compounds with ethylenic double bonds (involving other orbital symmetry classifications).

The energies ΔE involved in σ —> σ^* or in n —> σ^* transitions are usually quite large and correspond to short wavelengths λ where $\lambda = h\, c_{light} / \Delta E$ (h is Planck constant).

These absorption bands commonly fall in the far UV region (below 220 nm). The energy involved in n —> π* or π —> π* transitions may also correspond to UV absorptions but at higher wavelength values or fall in the visible region (above 360 nm).

The use of UV-Vis absorption for the detection in HPLC is very common. Detectors with constant wavelength absorption, with variable wavelength absorption, or able to collect simultaneously a range of wavelength in discrete intervals (diode array detectors DAD) are commercially available. Depending on the nature of the analyzed material, the detection limit of the UV-Vis detection in HPLC can be 0.1–1.0 ng, with a linear range of five orders of magnitude. With an appropriate solvent that does not absorb in the range of UV-Vis measurements, the use of elution gradient can be applied for separation.

- Fluorescence detection in HPLC

Fluorescence (FL) is the process of emission of light by a molecule after absorbing an initial radiation (excitation light). A molecule M goes from a lower energetic state (commonly ground state) to an excited state M* by absorbing energy. The emission process may take place by the molecule bouncing back to the initial state without the change in the wavelength of the absorbed light. In this case the process is difficult to use for analytical applications. However, it is possible that part of the energy of the excited molecule M* is lost by nonradiative processes such as collisions with other molecules. In this case, the electron may go to another excited electronic state with lower energy and then, emitting a photon, reaches the ground state. It is also possible that no intermediate electronic state is present, but the molecule acquires a lower vibrational energetic level and jumps to the ground state by emitting a photon of lower energy than the absorbed one. In both these cases, the fluorescence radiation has a lower frequency than the excitation radiation, as shown in Figures 4.3.5a and 4.3.5b. Fluorescence by emission of radiation at higher frequency than the absorbed one is also possible (anti-Stokes radiation) but it is uncommon.

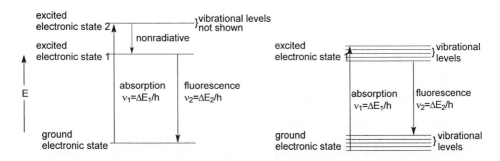

FIGURE 4.3.5a. *Simplified scheme of electronic transitions during fluorescence with two distinct excited elelctronic states. Vibrational energetic levels not shown.*

FIGURE 4.3.5b. *Simplified scheme of fluorescence with the difference between absorbtion and emission caused by differences in vibrational levels.*

The average lifetime of an excited state of a molecule M* undisturbed by collisions is about 10^{-8} s, and fluorescence can take place within this length of time. For some special compounds, the molecules can remain for a longer time in a metastable excited state. In this case fluorescence can be observed long after the initial radiation is interrupted. This type of fluorescence is commonly called phosphorescence.

Fluorescence is less frequently observed than expected based on the processes shown in Figures 4.3.5a and 4.3.5b because it is very common that the nonradiative loss of energy takes place as low as the ground state M. In gas phase, for example, at one atm. pressure, about 100 collisions occur in 10^{-8} s (lifetime of an excited state). The loss of energy by collisions is called quenching. The two processes, fluorescence and quenching by collision with a molecule X, take place in fact in parallel:

$$M^* \xrightarrow{k_1} M + h\nu$$

$$M^* + X \xrightarrow{k_2} M + X + \text{kinetic energy}$$

The intensity of fluorescence F can be expressed as a function of the absorbed light, which is the difference between the intensity I_o of the radiant energy incoming into the sample and the intensity of the emerging light I_1. The expression of F is the following:

$$F = (I_o - I_1)\frac{k_1\,[M^*]}{k_1\,[M^*] + k_2\,[M^*]\,[X]} = (I_o - I_1)\frac{1}{1 + (k_2/k_1)\,[X]} = (I_o - I_1)\,\Phi \qquad (4.3.6)$$

where Φ is the (quantum) fluorescence efficiency of the process. Based on the Beer's law, it can be seen that $I_o - I_1 = I_o\,[1 - \exp(\varepsilon_\lambda\,c_M\,L)]$, and using rel. (4.3.6) the expression for F can be written as follows:

$$F = I_o\,[1 - \exp(\varepsilon_\lambda\,c_M\,L)]\,\Phi \qquad (4.3.7)$$

The intensity of fluorescence is in fact lower than shown by rel. (4.3.7) due to the fact that only a part of the fluorescence is measured in an analytical instrument. For low concentrations $[1 - \exp(\varepsilon_\lambda\,c_M\,L)] \approx \varepsilon_\lambda\,c_M\,L$, and the intensity of measured fluorescence F' is related to the concentration c_M by the approximation relation:

$$F'_{\lambda 1} = k\,I_{o\,\lambda 2}\,c_M \qquad (4.3.8)$$

where λ_1 is the emission wavelength, λ_2 is the excitation wavelength, and k is a constant coefficient. Measurement of fluorescence intensity (usually at the maximum of the emission band) is the base of quantitation of the fluorescent species. The technique can be extremely sensitive, and amounts as low as a few fmol of analyte can be analyzed in HPLC with fluorescence (FL) detection. The detection in fluorescence methods encounters several difficulties because of nonlinearity of fluorescence due to self-absorption effects, difficulty in discriminating between overlapping broad spectra of interfering molecules, quenching produced by oxygen dissolved with the sample, etc.

Because the intensity of fluorescence increases linearly with the intensity of the initial radiation, laser induced fluorescence (LIF) detection is a successful technique applied in HPLC. For HPLC, lasers are a convenient excitation source because they have intense light focused into a small volume, they are highly monochromatic, and the associated Raman light has a well-defined wavelength that can be avoided with the monochromator used for observing fluorescence. However, laser induced fluorescence is still affected by background interference commonly arising from the Raman effect in the blank (molecular scattering) or from low level of solid impurities in the solvent producing Rayleigh light scattering.

The use of fluorimetric detection in HPLC is also common. Detectors with constant excitation wavelength and variable absorption or with variable wavelength excitation and absorption are commercially available. Depending on the nature of the analyzed material, the detection limit using fluorimetric detection can be 10^{-2} to 10^{-3} ng, with a linear range of four orders of magnitude. When appropriately selected, the use of elution gradient can be applied for separation without interfering with the fluorescence.

- *Chemiluminescence detection in HPLC*

Chemiluminescence (CL) is the emission of light as a result of a chemical reaction. Certain compounds achieve excited energy states in specific chemical reactions and emit light following a transition to ground state. The wavelength of the light emitted by a molecule in chemiluminescence is the same as in its fluorescence, the energy levels of the molecules being the same. The difference comes from a different excitation process. If the energy of the chemical reaction is lower than required for attaining the excited state, the chemiluminescence does not occur. Also, the deactivation of the excited molecule by nonradiative processes such as collisions with other molecules takes place for chemiluminescence similarly to fluorescence. A process efficiency Φ_{CL} similar to quantum efficiency Φ from fluorescence is also defined. For a molecule able to fluoresce with fluorescence efficiency Φ, the efficiency Φ_{CL} increases with the efficiency of the chemical reaction producing the excitation (such as the oxidation process). Higher energies required by molecules to achieve the excited state diminish Φ_{CL}. Chemiluminescence efficiency Φ_{CL} can be expressed as a "quantum yield," which is defined as the proportion of analyte molecules that emit a photon. In analytical uses of chemiluminescence, one more factor that must be taken into account is the time frame of the light emission. Certain chemiluminescent systems, although with very good Φ_{CL}, may emit the light for a period of 40–50 min. [59]. Much shorter times can be achieved using a catalyst.

Because no excitation light is needed in chemiluminescence, the interfering light from Raman effect or light scattering by trace particles is nonexistent. In addition, the development of detectors virtually able to detect single photons makes the technique highly sensitive. Levels as low as a few hundred amol were detected using chemiluminescence for certain analytes [60]. However, the luminescent molecules are not very common.

- *Electrochemical detection in HPLC*

Electrochemical analytical techniques are commonly used for the analysis in bulk solution (see e.g. [61]), but they can be easily adapted for HPLC detection. From the techniques with analytical utility that include amperometric, coulometric, potentiometric and conductometric methods, the ones more commonly applied in HPLC are amperometric and to a lesser extent coulometric procedures [62]. These techniques can have very high sensitivity, and the price of the detectors is relatively low.

In amperometric techniques, the current intensity is measured in an electrochemical cell when a specific potential is applied between two electrodes. Usually, only the reaction at one electrode is of interest, and a cell can be composed of a working electrode coupled with a nonpolarizable electrode (one that does not modify its potential upon passing of a current). This is known as the reference electrode, and examples are the saturated calomel electrode (SCE) and Ag/AgCl electrode. More frequently, a three-electrode cell arrangement is used. In this arrangement, the current is passed between a working electrode (made for example from glassy carbon) and an auxiliary electrode, while the potential of the working electrode is measured relative to a separate reference electrode. The two types of cells are shown in Figures 4.3.6a and 4.3.6b.

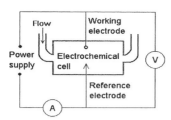

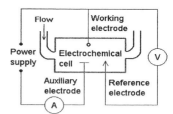

FIGURE 4.3.6a. *Schematics of a two-electrode flow-cell.*

FIGURE 4.3.6b. *Schematics of a three-electrode flow-cell.*

Any overall cell reaction comprises two independent half-reactions, and the cell potential can be broken into two individual half-cell potentials. The half-reaction of interest that takes place at the working electrode surface can be either an oxidation or a reduction. A simple reduction reaction is written as follows:

$$Ox + ne^- \rightleftharpoons Red$$

The electrode potential E for this half-reaction is reported to the potential of a reference standard hydrogen electrode (NHE), which is taken as zero. Experimental measurements are commonly done with a SCE or Ag/AgCl reference electrode. The potential of a SCE electrode vs. NHE is +0.242 V, and the potential of a Ag/AgCl electrode is +0.197 V vs. NHE. The reaction takes place spontaneously in a cell with a positive resulting potential.

Considering a reversible reduction that has a very rapid electron transfer, and assuming that both Ox and Red are soluble species, the molar concentrations c_{Ox} and c_{Red} at the electrode surface (x = 0) are governed by Nernst equation,

$$E = E^0 + \frac{RT}{nF} \ln \frac{c_{Ox\,(x=0)}}{c_{Red\,(x=0)}}$$

(4.3.9)

where E^0 is the standard electrochemical potential of the half cell, R is the gas constant (in SI units $R = 8.31451$ J deg^{-1} mol^{-1}), T is the temperature (in K deg.), n is the number of electrons involved in the electrochemical reaction, and F is Faraday constant (9.6485309×10^4 C mol^{-1}). Tables of standard electrochemical potentials in specific media are available in literature [63], and the half-reactions are expressed as reductions. For example, the reduction of $[Ru(bpy)_3]^{3+}$ in acetonitrile and 0.1 M tetraethylammonium tetrafluoroborate takes place at E = 1.32 V (vs. aq. SCE).

For the very rapid electron transfer at the electrode surface, the rate "v" of the reaction is given by the rate at which the electroactive species is brought to the surface of the electrode by the mass transfer in solution. This can be written in the following form:

$$v \text{ (mol-1 s-1 cm-2)} = \frac{i}{nFA} = v_{mass \ trransfer}$$

(4.3.10)

where A is the electrode area and i the current intensity. In the absence of convection (for bulk solutions) and of migration under the influence of the electric field, diffusion is the only mechanism for mass transfer, and the rate $v_{mass \ transfer}$ can be approximated by the following expression:

$$v_{mass \ transfer} = m_{Ox} \ [c_{Ox}{}^* - c_{Ox}(x=0)]$$

(4.3.11)

where m_{Ox} is a proportionality coefficient called mass transfer coefficient, and $c_{Ox}{}^*$ is the concentration of Ox in the bulk solution. The largest rate of mass transfer for Ox occurs when $c_{Ox}(x=0) = 0$. The value of the current in these conditions is called the limiting current i_l, and its value is given by the following rel.:

$$i_l = nFA \ m_{Ox} \ c_{Ox}{}^*$$

(4.3.12)

The expression for $c_{Ox}(x=0)$ can be written now from rel. (4.3.11), (4.3.10), and (4.3.12) as follows:

$$c_{Ox} (x = 0) = \frac{i_l - i}{n \ m_{Ox} \ FA}$$

(4.3.13)

When the reducing species Red is absent in the bulk solution, $c_{Red}{}^* = 0$, and using a relation similar to rel. (4.3.11) for the reduced species, the expression for c_{Red} can be written as follows:

$$c_{Red} (x = 0) = \frac{i}{n \ m_{Red} \ FA}$$

(4.3.14)

Using rel. (4.3.13) and (4.3.14) in Nernst equation (4.3.9), an i-E relation can be obtained as follows:

$$E = E^o - \frac{RT}{nF} \ \ln \frac{m_{Ox}}{m_{Red}} + \frac{RT}{nF} \ \ln \frac{i_l - i}{i}$$

(4.3.15)

For the current intensity equal to half of the limiting current, $i = 1/2 \ i_l$, the last term in rel. (4.3.15) is null, and the corresponding potential $E_{1/2}$ is independent of the concentrations of the oxidant or reduced species. The name "half wave potential" is used for $E_{1/2}$. Rel. (4.3.15) can be written in this case as follows:

$$E = E_{1/2} + \frac{RT}{nF} \ \ln \frac{i_l - i}{i}$$

(4.3.16)

The current-potential curve of a reversible reaction with a very rapid electron transfer (Nernstian reaction) involving two soluble species and only oxidant present initially is shown in Figure 4.3.7. The equation of this curve can easily be obtained from rel. (4.3.16). In most cases, electrochemical processes are much more complex than the case presented above. One critical assumption, that the process is Nernstian, is not true in numerous cases, and the electron transfer can be slow. In this case, the rate of the process is no longer controlled by the rate of the mass transfer. The departure of the electrode potential from the Nernstian behavior upon the passage of a Faradaic current is called polarization. The extent of polarization is measured by the overpotential η [61]. Another factor adding complexity is that the reaction products (Red for reductions or Ox for oxidations) may undergo further reactions after the electron transfer. An example of such case is the oxidation of p-aminophenol in aqueous solutions:

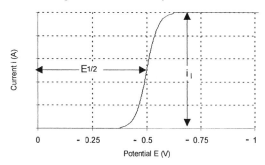

FIGURE 4.3.7. *The current-potential curve of a Nernstian reaction involving two soluble species and only oxidant present initially.* In this example, $E_{1/2} = -0.5$ V.

A special type of reaction following the electron transfer is the catalytic reaction in which the product of an electrochemical process reacts with a nonelectroactive species that is present in excess in solution and regenerates the initial compound. The process takes place as follows:

$$Ox + ne^- \rightleftharpoons Red$$
$$Red + A \rightleftharpoons Ox + B$$

In this reaction, because the compound A is able to oxidize Red, it would be expected that it is also possible to be reduced electrochemically. However, the rate of the reduction process at the electrode surface may be very slow for some compounds that otherwise can be reduced in solution. The associated chemical reaction can affect the measured parameters of the electrochemical process, such as the characteristic potential of the reaction or the current intensity.

For the case of the electrochemically active species flowing over the surface of an electrode, which is the case of electrochemical detection in HPLC, the current-potential dependence is the function of a convective diffusion process. This makes, for example, the limiting current intensity for a Nernstian process dependent on the mobile phase flow rate and on channel and electrode geometry. For a rectangular channel flow electrode in steady state laminar flow with the working electrode at one wall, the limiting current intensity is given by the relation:

$$i_l = 1.467 \ n \ F \ c^* \ (A \ D \ / \ b)^{2/3} \ U^{1/3} \tag{4.3.17}$$

where c^* is the bulk concentration of the analyte, A is the electrode area, D is the diffusion coefficient, b is the channel height, and U is the volumetric flow rate [64], [65], [66]. For different channel and electrode shapes, the expression for the current intensity is different [67], [68].

In amperometric detection, the current passing through the cell is measured at a fixed potential E, commonly chosen higher in absolute value than $E_{1/2}$ specific for the analyte. In these conditions, the desired electrochemical process takes place, but also all other species present in solution and having $E_{1/2}$ lower (in absolute value) than the chosen E value can become electrochemically active species. This may include even the solvent if the working potential E is very high. For eliminating this type of interference, compounds with low electrochemical potentials (in absolute value) are preferred for electrochemical detection. In HPLC, amperometric detection is frequently used for oxidation reactions. The quantitation can be done by calibration of the measured current vs. different concentrations of analyte while maintaining strictly controlled flow conditions. Also, instead of a constant oxidation potential, a pulse amperometric detection (PAD) can be used, alternating the oxidation analytical potential with a reducing pulse used for cleaning the electrode [69], [70]. The application of different working potentials is done at specific time intervals, and the measurement is made only when the active species are oxidized.

- Other detection types used in HPLC

Liquid chromatography-mass spectrometry (LC-MS) is a technique that is becoming more and more utilized due to the advancements in the instrumentation [71]. A wide variety of procedures are reported for coupling liquid chromatography to mass spectrometry [72]. The role of the LC-MS interface is to convert the dissolved analyte from the mobile phase of the LC into gas phase ions at reduced pressure that are further analyzed with the mass spectrometer. Among the techniques used for interfacing the LC with the MS are the following:

a) The electrospray (ES), which consists of a hollow needle set at a high electrical potential through which the effluent flows (at 1–10 µL/ min.) generating a spray of charged droplets. The evaporation of this spray generates the ions, while some of the solvent is removed in successive vacuum chambers toward the mass spectrometer.

b) Atmospheric pressure chemical ionization (APCI). In APCI the effluent is sprayed into a heated chamber at atmospheric pressure where the solvent and the analytes are evaporated. The gases are ionized using a separate source of electrons, and the excess of solvent is removed similarly to ES.

c) Particle beam (PB), which consists of an aerosol generator from the LC flow (at 0.1–1 mL/min.) followed by a desolvation chamber and a separator that directs the aerosols through a series of apertures separating the volatile compounds including the solvent from the solid aerosols. The beam of solute particles is directed to a hot surface, and the vapors generated are ionized by a conventional electron bombardment. The technique is successful only for compounds that have some volatility and do not decompose when heated.

d) Continuous flow FAB, where the effluent is introduced directly into a vacuum region (with a flow rate of 5–10 μL/min.) mixed with a matrix material such as glycerol, and the ionization is achieved using a beam of ions at 5–8 keV.

e) Thermospray (TS). In TS the effluent is introduced in the MS through a heated capillary that produces a spray. The vaporization process itself generates ions when, for example, an ammonium salt is added to the effluent. An additional electron bombardment may increase the ion generation efficiency. The allowed flow rates in thermospray are higher then ES or APCI (up to 1 mL/min.), but the effluent must contain some water.

Except for PB the spectra generated in LC-MS do not show the typical fragmentation seen in EI+ spectra because the ionization techniques are too mild, similar to a chemical ionization. The lack of fragmentation does not help to the same extent in compound identification as in typical EI+ spectra. This explains the need and the successful use of LC-MS-MS as a detection technique in HPLC.

Other detectors are also used in HPLC analysis such as evaporative light-scattering detector (ELSD) [73], [74], direct deposition FTIR detector [75], [76], NMR [77], ICP-MS [78], circular dicroism detector, etc.

- *The need for sample preparation in HPLC*

The need for sample preparation in HPLC is similar to that for GC and is summarized in Table 4.3.6.

TABLE 4.3.6. *Main types of problems in HPLC analysis and the application of sample preparation to solve them.*

Problem	Solution
Analytes in a form not adequate for HPLC analysis.	Solubilization that takes the analytes in solution with an appropriate solvent to be soluble in the mobile phase.
Analytes affected by specific interferences.	Fractionation, derivatization of specific analytes, chemical separation.
Analytes in a complex matrix generating global interferences.	Cleanup, derivatization
Analytes in a concentration too low for detection.	Sample concentration, derivatization
Overall poor chromatography.	Derivatization, sample cleanup, concentration.
Short lifetime of the chromatographic column.	Sample cleanup, derivatization.
Enantiomers coelution.	Derivatization with chiral reagents to form diastereoisomers.

4.4 OTHER CHROMATOGRAPHY TYPES USED AS A CORE ANALYTICAL TECHNIQUE

Besides GC and standard HPLC, some other types of chromatography have rather frequent applications. Among these are some special types of HPLC such as size exclusion, ion chromatography, and ion-pair chromatography. Also chromatography types such as planar chromatography, including the common thin layer chromatography (TLC), and supercritical fluid chromatography (SFC) are used as core analytical procedures.

- *Size exclusion chromatography*

Size exclusion as a principle for separation is used in many HPLC applications. Some other names are used for this type of separation, such as gel filtration (when used to separate molecules in aqueous systems) or gel permeation (GPC) (when used to separate organic polymers in nonaqueous systems), the names sometimes being used interchangeably. The separation uses a column packed with a porous separating medium. The separation in size exclusion depends on the hydrodynamic volume or "effective" molecular size of the solute, the smaller molecules being retained longer than the larger molecules. This process is based on the fact that small molecules can enter freely the pores (interstices) of the gel and are retained longer, while the larger molecules cannot enter the pores and are flushed by the eluent. Among the first materials used as stationary phases in size exclusion was dextran treated with 1-chloro-2,3-epoxypropane (epichlorhydrin), known as Sephadex. Numerous other materials are currently available including polyhydroxymethacrylate, polyvinyl acetate, various dextrans, porous silica, acrylamide-methylene bisacrylamide, agarose, polymethylmethacrylate, polystyrene, styrene-divinylbenzene copolymer, etc. The solvents used for these phases can be organic, organic/aqueous, or aqueous [79].

All the parameters used for the formal characterization of the separation process that are used in HPLC are also applicable in GPC. For example, the variation of the retention volume V_R with the distribution constant as expressed by rel. (1.6.33) is valid for size exclusion, and:

$$V_R = V_m + K V_s \qquad (4.4.1)$$

where the distribution constant K depends on the molecular weight of the sample and the pore size of the stationary phase. Band broadening in HPLC baaed on size exclusion mechanism is given by an expression of the form (1.6.51). The component H_S in this expression is given by rel (1.6.48b), which indicates increased band broadening for compounds with low diffusion in the liquid.

The retention volume V_R (in GPC typically called elution volume) is larger for small molecules and much smaller for large molecules. A typical relation between V_R and the molecular weight is given by the expression:

$$V_R = A + B \log (MW) \qquad (4.4.2)$$

This expression is valid only for a certain range of MW values and Figure 4.4.1 shows four types of curves common for GPC. As seen in this Figure 4.4.1, only certain ranges

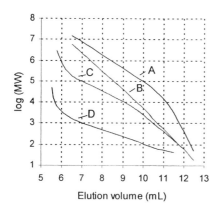

Elution volume (mL)

FIGURE 4.4.1. *Calibration curves of MW vs. elution volume in GPC.*

show linearity, the deviation appearing at the end of the calibration interval (curve A), at the beginning of the calibration interval (curve D), or at both ends (curve C). The linearity between log MW and the elution volume is very frequently utilized for the characterization of molecular weight distribution in both synthetic and natural polymers. Due to possible deviations from linearity, as shown in Figure 4.4.1, the dependence between MW and the elution volume is typically calibrated in the molecular range of interest using polymer standards with known molecular weight. The application of size exclusion chromatography is very common for the separation of biopolymers including polysaccharides, proteins, nucleic acids, etc. In sample preparation, size exclusion is used mainly for the separation of low molecular weight compounds from high molecular weight ones. The subject is further discussed in Section 12.3.

The steric exclusion as a unique mechanism for the separation of polymers is sometimes difficult to achieve, because other types of interactions may be present simultaneously with size exclusion. Although no sorption is assumed between the stationary phase and the analytes, this restriction is frequently difficult to achieve. When specific interactions take place between the stationary phase and the solutes, deviations from the expected values of the elution volume may appear. For example, the separation of Maillard polymers that have numerous polar and hydrogen bonding groups, may lead to erroneous results regarding the assignment of MW even on polystyrene-divinylbenzene gels.

In comparison of the instrumentation used in GPC and that used in conventional HPLC, there are many similarities. Only a few construction details are different for GPC instrumentation, where the control of the column temperature is more important than in other techniques. The gel permeation process is more sensitive to temperature changes than standard sorption HPLC. Constant temperature is necessary for obtaining good reproducibility of the calibration log MW vs. V_R. Also some polymers are soluble in specific solvents only at temperatures above ambient, and an increased temperature during separation is necessary to maintain the solutes in solution.

- Ion chromatography

Another special type of HPLC technique is ion chromatography (IC) (see e.g. [80]). In ion chromatography the analyte is, in principle, ionizable. Among the compounds analyzed by IC are metal ions, acids, and many other ionizable organic compounds. Several separation mechanisms can be used in IC, including ion exchange, separation of ion pairs, and ion exclusion. The typical detection system for IC uses a measurement of the electrical conductance of the solution eluting from the chromatographic column after the solution passes through a suppressor element that reduces the background conductance. In addition to conductance measurement, spectrophotometric and pulse amperometric detection are used for IC.

The IC with ion exchange separation is applied for both organic and inorganic ions. The separation of anions including those of organic acids is performed using anion exchange resins such as polystyrene/divinylbenzene copolymers with quaternary ammonium groups. The separation of cations is typically performed using polystyrene/divinylbenzene copolymers with sulfonic or carboxylic acid groups.

The suppressor used in IC reduces the conductivity background and simultaneously converts the sample components to acids (for anions) or hydroxides (for cations). The background conductivity is typically generated by strong bases such as NaOH, which is used as eluent for anion exchange columns, or generated by strong acids such as HCl, which is used as eluent for cation exchange columns. Membrane suppressors that are typically used for this purpose contain a microporous ion exchange membrane. This membrane is capable of exchanging ions from the effluent leaving the separation column and of containing the analytes and the excess of eluent. At the same time the membrane can be regenerated with a continuous flow of a regenerant solution (in countercurrent with the effluent). In a separation containing for example anions in an eluent containing NaOH, the role of the suppressor is to eliminate NaOH and generate H_2O and to transform the salts of the analytes into the corresponding acids, as shown schematically in Figure 4.4.2. The regenerating solution for the suppressor membrane in anion analysis is typically H_2SO_4.

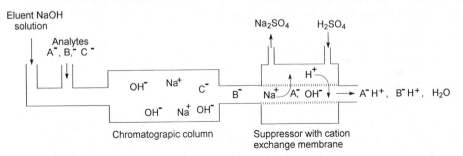

FIGURE 4.4.2. *Schematic diagram of an IC system showing the role of a membrane suppressor for anions.*

In cation exchange chromatography where the eluent is likely to contain a strong acid such as HCl, the regenerating solution for the suppressor membrane is typically tetrabutylammonium hydroxide.

Other parts of the IC instrumentation are usually similar to those used in conventional HPLC, except for replacement of the metal parts that come in direct contact with the liquid flow with plastic components for reducing the chance of ion formation. Size exclusion applied to ionic compounds such as weak organic acids is known as ion exclusion chromatography [81].

- Ion pair chromatography

Ion pair chromatography (IPC) is based on the separation of analyte ions using in the mobile phase a specific modifier having ions of opposite charge to the analyte and a lipophilic moiety. The ions interact compensating the ionic character of the analyte, and the separation is typically done using a reverse phase HPLC column. The procedure has many applications and can be used with standard HPLC equipment [82]. Among the applications of IPC are included the separation of organic and inorganic ions and of organic ionizable compounds (amines, acids, etc.). In IPC technique, several variables such as the type of pairing ion, pairing ion concentration, ionic strength, pH, and mobile phase composition can be used to control the solute retention and separation selectivity. The broad choice and combination of these variables allow the separation of complex sample mixtures containing ionic/ionizable and even some neutral analytes.

The main process in IPC is the formation of the ion pairs, and two types of models (stoichiometric and nonstoichiometric) are proposed to describe it [83]. Stoichiometric models are based on the assumption that solute ions and pairing ions form stoichiometric complexes, either in the mobile phase (ion-pair model) or at the stationary phase (dynamic ion-exchange model). The ion-pairing adsorption model assumes the formation of an ion pair in the polar mobile phase followed by the adsorption of this complex on the hydrophobic stationary phase. The dynamic ion-exchange model assumes that the ion-pairing reagent molecules are adsorbed on the stationary phase and cause the column to behave as a dynamically generated ion exchanger. The retention of analyte ions is in this case due to the ion exchange process with the inorganic counter ions of the ion-pairing reagent [84].

Nonstoichiometric models describe the retention of ionic species without the formation of chemical complexes. These models assume that the retention of solute ions is partly determined by their interaction with the electric field created by the adsorbed pairing ions. Therefore, the effect of the pairing is assumed to be indirect and acts through establishing a certain electrostatic surface potential. In this theory, the pairing ion is adsorbed on the stationary phase surface forming a primary ion layer. The electrolytic (inorganic) counter ions form a secondary ion layer between the charged surface and the bulk eluent. The analyte ions are attracted or repelled by the primary ion layer (replacing part of the secondary layer) depending on the sign of their charge and that of the layer. The development of nonstoichiometric models was largely stimulated by experimental evidence of the adsorption of ion pairing reagent on the hydrophobic stationary phase [85], [86], [87], [88].

Similar to other HPLC type separations, an equilibrium takes place between a mobile and a stationary phase where the notation "X" is used to indicate the analyte:

$$X_m \rightleftharpoons X_s$$

and $K_X = c_{X,s} / c_{X,m}$. A theoretical examination of the factors influencing the equilibrium process in ion pair chromatography [89] can begin with the expression (1.8.8) for the distribution constant. This can be written for the distribution process with ion pair formation in the form:

$$K_X = \exp\left(-\frac{\Delta G_{t,X}^0}{RT}\right)$$

(4.4.3)

where $\Delta G_{t,X}^0$ represents the work needed to transfer the molecule X from the bulk of the mobile phase onto the stationary phase surface. This work has two components expressed by the formula:

$$\Delta G_{t,X}^0 = \Delta G_X^0 + z_X F \Delta\Psi$$

(4.4.4)

The first component (ΔG_X^0) is determined by the adsorption of the analyte in the absence of ion-pairing reagent. The second component defines the electrostatic contribution of the process, and z_X is the charge of the analyte "X", F is Faraday's constant, and $\Delta\Psi$ is the difference in electrostatic potential between the bulk of the mobile phase and the stationary phase surface induced by the adsorption of the ion-pairing reagent. From rel. (4.4.4) it can be seen that the distribution constant in ion pair chromatography can be written in the form:

$$K_X = K_X^0 \exp(-z_X F \Delta\Psi)$$

(4.4.5)

where K_X^0 indicates the distribution coefficient in the absence of the ion-pair reagent. Based on the theory of the electrical double layer, the expression of $\Delta\Psi$ can be estimated and the result for the expression of K_X can be expressed as follows:

$$\ln K_X = \ln K_X^0 - \frac{z_P z_X}{z_P^2 + 1}[\ln n_0 K_P + \ln c_P] + \zeta$$

(4.4.6)

where z_P is the charge of the ion-pairing reagent, n_0 is the monolayer capacity of the stationary phase, K_P is the adsorption constant for the ion-pair reagent, c_P is the molar concentration of the ion pair reagent in the mobile phase, and ζ is a constant factor depending on temperature, dielectric constant of the mobile phase, and the charges and total ion concentration in the mobile phase.

As seen from rel. (4.4.6), only when the charges of the analyte and that of the pairing ion are opposite, is the distribution coefficient for the stationary phase higher than in the absence of the pairing ion. The effect is higher when the concentration of ion pair reagent c_P is higher and when the term $n_0 K_P$ is higher. This term is related to the hydrophobicity of the pairing ion. The retention of oppositely charged analytes increases with increasing hydrophobicity of the pairing ions. The retention change can be attributed to the higher adsorption of more hydrophobic pairing ions and a corresponding higher electrostatic surface potential at a constant value of the ionic strength of the solution and other chromatographic variables. In terms of the electrostatic model, the hydrophobicity of the pairing ion influences the free energy of adsorption (ΔG_P^0 and K_P), and the size of the pairing ion affects the monolayer capacity (n_0). The more hydrophobic is the pairing ion, the larger is the retention increase (for oppositely charged analytes) or decrease (for similarly charged analytes).

In both conventional reverse phase HPLC and in ion-pair chromatography, it is common that the mobile phase contains water. In addition to water an organic solvent, such as methanol, acetonitrile, etc., is frequently added to the mobile phase. This reduces the polar character of water and diminishes the distribution constant between the reverse phase column and the mobile phase for lipophilic solutes. This is expressed by rel. (1.6.54), which can be written in the form:

$$\ln K_X^0 (c_M) = \ln K_X^0 (c_M = 0) - \varphi_X c_M \tag{4.4.7}$$

where $K_X^0 (c_M)$ is the distribution coefficient for the analyte "X" in the presence of an organic modifier "M" in the mobile phase, $K_X^0 (c_M = 0)$ is the distribution constant for water as a mobile phase, φ_X is a constant for a given analyte-solvent combination, and c_M is the molar concentration of the organic solvent. From thermodynamic point of view, rel. (4.4.7) expresses the fact that the free energy of adsorption is a linear function of the organic modifier concentration. For an uncharged analyte ($z_X = 0$) there is no electrostatic term in the total free energy of adsorption, and based on expression (1.8.8) for the distribution constant, rel. (4.4.7) can be written in the form:

$$\frac{\Delta G_X^0 (c_M)}{RT} = \frac{\Delta G_X^0 (c_M = 0)}{RT} - \varphi_X c_M \tag{4.4.8}$$

A similar relation to (4.4.8) can be written for the adsorption of the ion pair complex, and therefore the modification of K_P in the presence of an organic solvent will follow an expression similar to rel. (4.4.7). By introduction of these expressions in rel. (4.4.6), the variation of K_X as a function of c_M can be written as follows:

$$\ln K_X = \ln K_X^0 (c_M = 0) - (\varphi_X - \varphi_P \frac{z_P z_X}{z_P^2 + 1}) c_M - \frac{z_P z_X}{z_P^2 + 1} [\ln n_0 K_P (c_M = 0) + \ln c_P] + \zeta \tag{4.4.9}$$

According to rel. (4.4.9), the reverse-phase retention of ion pairs when using an organic modifier is diminished when c_M increases. When the analyte ion is oppositely charged to the pairing ion, the decrease is larger. In practice, when higher organic modifier concentrations are used, more hydrophobic pairing ions are needed to reach a high enough retention.

Another factor influencing the formation and retention on the stationary phase of the ion pairs is pH. Many analytes and pairing reagents are weak acids and bases. For example, compounds such as alkanesulfonic acids or quaternary ammonium compounds are typically used as modifiers. Therefore, in the solution the interacting ionic species are also subject to equilibria of the type:

$$AH \rightleftharpoons A^- + H^+ \quad \text{with} \quad K_a = \frac{[A^-][H^+]}{[AH]} \tag{4.4.10}$$

or of the type:

$$BH^+ \rightleftharpoons B + H^+ \quad \text{with} \quad K_a = \frac{[B][H^+]}{[BH^+]} \tag{4.4.11}$$

These equilibria influence significantly the concentration of the free ions of the analyte and/or of the pairing reagent. The choice of an appropriate pH where most of the analyte or of the reagent is predominantly in ionic form is typically selected for successful use of ion-pair separation.

- Immunoaffinity chromatography

Antibodies immobilized on a solid support can be used as stationary phase. The chromatography performed on such materials is known as immunoaffinity chromatography [4]. The technique is usually applied for the analysis of biological samples for which the antibodies can be efficiently generated. Each antibody has high affinity for the compound that generated it, and a selective retention is easily achieved. However, a fast elution of the compound of interest without damaging the antibody column may be a problem. It is possible to use eluents that dissociate the antigen-antibody complex, but if these are gentle enough to protect the column, they do not produce in general a sharp peak of the analyte. For this reason, the immunoaffinity column can be used together with a second standard column. After retention of the desired antigen and elimination of interferences, the antigen is eluted as a broad band and collected on the second column. Changing the solvent for the second column, the antigen can be eluted as a sharp peak. Immunoaffinity chromatography has numerous applications in the analysis of biological samples, including immunoaffinity chromatography with the direct detection of the eluate, use of competitive binding immunoassay, sandwich immunoassay, etc. The subject of immunoaffinity chromatography is beyond the purpose of this book, the subject being presented in the dedicated literature [90], [91], [92].

- Thin layer chromatography

In thin-layer chromatography (TLC), the stationary phase is coated (0.5–2 mm) on an inert plate of glass or plastic usually 10–15 cm long [93]. A variety of stationary phases are used in TLC, including silica, alumina, C18 bound on silica, etc. The samples are applied as a spot close to one end of the plate. The end containing the sample(s) is then immersed in the mobile phase, which migrates based on capillarity toward the top of the plate. The migrating solvent acts as the mobile phase, and the separation is characterized by the ratio of the distance from the start to the spot of the analyte and the distance from the start to the eluent front, ratio known as R_f (see section 1.6). The choice of the solvent is very important in TLC, and different studies have been done for the appropriate choice depending on the analyte [93]. The detection in TLC is done either visually or with instrumentation such as densitometers. The analytes are not always visible, and various procedures are used to allow detection. Some derivatization procedures of the analytes can be performed, but much more commonly after the separation the analytes are treated with (or exposed to) a reagent that produces color (such as iodine vapors) or fluorescence to indicate their positions on the plate. Typically the sample preparation in TLC is not elaborate. In addition to one-dimensional TLC, bidimensional TLC is also practiced. In bidimensional TLC the plates are usually wider (20 x 20 cm) and the sample is applied in only one corner of the plate. After obtaining migration in one direction with the use of a specific solvent, the plate is usually dried and a second migration is performed using a different solvent and on a perpendicular direction compared to the first one. This procedure allows better separations and has been applied successfully for the analysis of complex mixtures.

TLC can be a very useful technique for qualitative analysis of a mixture. One interesting advantage in TLC is the possibility to see any material that remains at the start after the chromatographic process is finished. In most column chromatographic systems, only the material exiting the column can be detected, and the substances that do not leave the chromatographic column are never detected. This advantage of TLC makes it a valuable tool for assessing the presence of unidentifiable polymeric materials in samples.

- Supercritical fluid chromatography

One other technique successfully used in separations is SFC [94]. SFC is a form of chromatography in which the mobile phase is at a pressure and temperature that correspond to the supercritical fluid state (see Section 1.3). Based on the column type, SFC can be classified as open tubular or packed. The properties of supercritical fluids are intermediate between gases and liquids and can be varied by small changes in the temperature and pressure (see also Section 10.6). The changes in these properties also modify the solvating properties of the mobile phase and influence the separation. The most commonly used fluid in SFC is CO_2. The critical parameters, which are critical temperature T_c and critical pressure p_c, must be situated in a convenient range of temperatures ($30–300°C$) and pressures ($30–300$ atm.) for practical applications.

The solvating power of a supercritical fluid is found to be, in a first approximation, proportional to its density. The change in density by the variation of pressure and temperature, allows modification of solvating properties and therefore the separation. For supercritical CO_2, the density can be varied almost by an order of magnitude, and this variation is used to modify the separation. However, the range of polarity of supercritical fluids still remains in a relatively narrow range. The hydrocarbons and CO_2 are situated in the low polarity range. Other compounds that can form conveniently supercritical fluids, such as NH_3 and SO_2, are very polar, but due to their relatively high reactivity are seldom used in analytical practice.

The solvating power of a mobile phase also can be changed using modifiers. The modifiers are organic compounds miscible with the mobile phase that have different polarity compared to the pure compound used as a main material for the mobile phase. The use of modifiers significantly extends the solubility of many compounds in the mobile phase, allowing more flexibility in the properties of the mobile phase and higher sample load.

A number of stationary phases can be used in SFC. For packed columns, silica and modified silica based materials can be used. For capillary (open tubular) columns, bonded phase materials such as polymethylsiloxanes similar to those used in GC are common. However, the capillary SFC columns are usually microbore columns ($50–100$ μm i.d. and $5–10$ m long).

The detection in SFC can be done with flame based detectors such as FID and NPD or based on UV light absorption or fluorescence. Due to the small samples typically used in SFC, the sensitivity of the detector is an important element for successful SFC analysis. The use of MS as a detector, although not uncommon, encounters some problems due to the relatively large amount of carrier fluid compared to the amount of sample [95] and requires some type of jet separator for successful operation.

- Other types of liquid chromatography

A number of other types of chromatographic techniques are known, such as countercurrent chromatography, electrochromatography, microchromatography performed on a silicon chip [96], [97] , etc. Also, a variety of new types of stationary phases are available such as continuous bed stationary phase made of a porous polymer rod and not from finely divided particles. In electrochromatography, for example, the separation is performed in a capillary column containing a finely divided stationary phase. The migration of the sample is based on electro-osmotic flow, similar to electrophoresis, but the separation is based on both differential migration in the electric field and differential partition between stationary and mobile phase [98], [99]. Sample preparation for these types of chromatography is basically the same as for other types of chromatography.

REFERENCES 4

1. F. W. McLafferty, *Interpretation of Mass Spectra*, Univ. Science Books, Mill Valley, 1980.
2. K. L. Bush et al., *Mass Spectrometry/Mass Spectrometry: Techniques and Applications of Tandem Mass Spectrometry*, VCH, New York, 1988.
3. S. Moldoveanu, C. A. Rapson, Anal. Chem., 59 (1987) 1207.
4. D. S. Hage, M. A. Nelson, Anal. Chem., 73 (2001) 199A.
5. W. R. LaCourse, Anal. Chem., 72 (2000) 37R.
6. J. C. Giddings, *Unified Separation Science*, J. Wiley, New York, 1991.
7. D. Rood, *A Practical Guide to the Care, Maintenance, and Troubleshooting of Capillary Gas Chromatographic Systems*, Hüthig Buch, Heidelberg, 1991.
8. R. L. Grob, *Modern Practice of Gas Chromatography*, J. Wiley, New York, 1995.
9. *Alltech Chromatography Catalog 350*, Alltech Assoc. Inc., Deerfield, 1995.
10. G. A. Eiceman et al., Anal. Chem., 72 (2000) 137R.
11. A. Kaufmann, J. High Resolut. Chromatogr., 21 (1998) 258.
12. A. Termonia, M. Termonia, J. High Resolut. Chromatogr., 20 (1997) 447.
13. E. Boselli et al., J. High Resolut. Chromatogr., 21 (1998) 355.
14. B. Grolimund et al., J. High Resolut. Chromatogr., 21 (1998) 378.
15. E. Boselli et al., J. High Resolut. Chromatogr., 22 (1999) 327.
16. C. A. Cramers et al., J. Chromatogr., A 856 (1999) 315.
17. L. M. Blumberg, J. High Resolut. Chromatogr., 20 (1997) 597.
18. L. M. Blumberg, J. High Resolut. Chromatogr., 20 (1997) 697.
19. L. M. Blumberg, J. High Resolut. Chromatogr., 22 (1999) 403.
20. L. M. Blumberg, J. High Resolut. Chromatogr., 22 (1999) 501.
21. K. Grob, *On-Column Injection in Capillary Chromatography*, A. Hüthig, Heidelberg, 1987.
22. Phenomenex for Chromatography, Catalog 2000/2001.

23. W. O. McReynolds, J. Chromatogr. Sci., 8 (1970) 685.
24. J. R. Mann, S. T. Preston, J. Chromatogr. Sci., 11 (1973) 216.
25. C. Liteanu et al., *Separatologie Analitica*, Ed. Dacia, Cluj-Napoca, 1981.
26. J. J. Leary et al., J. Chromatogr. Sci., 11 (1973) 201.
27. G. E. Baiulescu, V. A. Ilie, *Stationary Phases in Gas Chromatography*, Pergamon, New York, 1975.
28. J. Novak et al., Anal. Chem., 45 (1973) 1365.
29. J. V. Hinshaw, L. S. Ettre, Chromatographia, 21 (1986) 561.
30. M. Donike, Chromatographia, 6 (1973) 190.
31. W. Blum., J. High Resolut. Chromatogr., 8 (1985) 719.
32. R. P. W. Scott, *Introduction to Analytical Gas Chromatography*, M. Dekker, New York, 1997.
33. Y.-Z. Tang, W. A. Aue, J. Chromatogr., 409 (1987) 243.
34. C. H. Hartmann, Anal. Chem., 43 (1971) 113A.
35. R. Buffington, M. K. Wilson, *Detectors for Gas Chromatography*, Hewlett-Packard, 1987.
36. J. H. Beynon, *Mass Spectrometry and its Applications to Organic Chemistry*, Elsevier, Amsterdam, 1960.
37. R. Oguchi et al., J. High Resolut. Chromatogr., 14 (1991) 412.
38. P. R. Griffith, Appl. Spectrosc., 31 (1977) 284.
39. P. Jackson et al., J. High Resolut. Chromatogr., 16 (1993) 515.
40. T. Hankemeier et al., J. High Resolut. Chromatogr., 21 (1998) 341.
41. L. R. Snyder, J. J. Kirkland, *Introduction to Modern Liquid Chromatography*, J. Wiley, New York, 1979.
42. G. Lunn, L. C. Hellwig, *Handbook of Derivatization Reactions for HPLC*, J. Wiley, New York, 1998.
43. G. Szepesi, *How to Use Reverse Phase HPLC*, VCH, New York, 1992.
44. L. R. Snyder, Anal. Chem., 72 (2000) 413A.
45. J. J. Kirkland, *Modern Practice of Liquid Chromatography*, J. Wiley, New York, 1971.
46. C. J. Dunlap et al., Anal. Chem., 73 (2001), 599A.
47. B. L. Karger et al., *An Introduction to Separation Science*, J. Wiley, New York, 1973.
48. D. L. Gisch et al., J. Chromatogr., 433 (1988) 264.
49. K. S. Books, C. H. Grimm, TrAC, 18 (1999) 175.
50. R. E. Majors, LC-GC Int., 8 (1995) 128.
51. C. F. Poole, S. K. Poole, Anal. Chim. Acta, 216 (1989) 109.
52. K. K. Unger et al., J. High Resolut. Chromatogr., 23 (2000) 259.
53. L. R. Snyder, J. Chromatogr., 92 (1974) 223.
54. L. R. Snyder, J. Chromatogr. Sci., 16 (1978) 223.
55. L. R. Snyder, *Principles of Adsorption Chromatography*, M. Dekker, New York, 1968.
56. P. G. Sadek, *The HPLC Solvent Guide*, J. Wiley, New York, 1996.
57. S. Moldoveanu, A. Savin, *Aplicatii in Chimie ale Metodelor Semiempirice de Orbitali Moleculari*, Ed. Academiei., Bucuresti, 1980.
58. S. Moldoveanu, *Aplicatile Teoriei Grupurilor in Chimie*, Ed. Sti. Enc., Bucuresti, 1975.
59. M. Stigbrand et al., Anal. Chem., 66 (1994) 1766.
60. T. J. Novak, M. L. Grayeski, Microchemical J., 50 (1994) 151.

61. A. J. Bard, L. R. Faulkner, *Electrochemical Methods, Fundamentals and Applications*, J. Wiley, New York, 1980.
62. B. E. Erickson, Anal. Chem., 72 (2000) 353A.
63. B. E. Conway, *Electrochemical Data*, Elsevier, Amsterdam, 1952.
64. H. Matsuda, J. Electroanal. Chem., 15 (1967) 325.
65. S. Moldoveanu, J. L. Anderson, J. Electroanal. Chem., 175 (1984) 67.
66. S. Moldoveanu et al., J. Electroanal. Chem., 179 (1984) 119.
67. J. Yamada, H. Matsuda, J. Electroanal. Chem., 44 (1973) 189.
68. S. Moldoveanu, J. L. Anderson, J. Electroanal. Chem., 185 (1985) 239.
69. P. Edwards, K. Haak, Amer. Lab., April, (1983) 26.
70. H. J. Lee, H. H. Girault, Amer. Lab. News Ed., 33 (15) (2001) 8.
71. W. M. A. Niessen, J. Chromatogr., 856 (1999) 179.
72. R. Willoughby et al., *A Global View of LC-MS*, Global View Pub., Pittsburgh, 1998.
73. H. Bünger et al., J. Chromatogr., A, 870 (2000) 363.
74. L. Voress (ed.), *Instrumentation in Analytical Chemistry 1988-1991*, ACS, Washington, 1992.
75. S. Bourne, Amer. Lab., August 1998.
76. G. W. Somsen et al., J. Chromatogr., 856 (1999) 213.
77. K. Albert, J. Chromatogr., 856 (1999) 199.
78. L. K. Sutton, J. A. Caruso, J. Chromatogr., 856 (1999) 243.
79. W. Chi-San (ed.), *Handbook of Size Exclusion Chromatography*, M. Dekker, New York, 1995.
80. J. Chromatogr., A, 920 (2001) issues 1-2.
81. J. Weiss, *Ion Chromatography*, VCH, Weinheim, 1995.
82. C. Wolf et al., Anal. Chem., 72 (2000) 5466.
83. M. T. W. Hearn (ed.), *Ion-Pair Chromatography–Theory and Biological and Pharmaceutical Applications*, M. Dekker, New York, 1985.
84. J. H. Knox, R. A. Hartwick, J. Chromatogr., 204 (1981) 3.
85. B. A. Bidlingmeyer, J. Chromatogr. Sci., 18 (1980) 525.
86. F. F. Cantwell, J. Pharm. Biomed. Anal., 2 (1984) 153.
87. H. Liu, F. F. Cantwell, Anal. Chem., 63 (1991) 993.
88. H. Liu, F. F. Cantwell, Anal. Chem., 63 (1991) 2032.
89. A. Bartha, J. Stahlberg, J. Chromatogr., A, 668 (1994) 255.
90. C. P. Price, D. J. Newman, *Principle and Practice of Immunoassay*, Stockton, New York, 1997.
91. P. Bailon et al. (eds.), *Affinity Chromatography: Methods and Protocols*, Humana, Ottawa, 2000.
92. H. Schott, *Affinity Chromatography*, M. Dekker, New York, 1984.
93. E. Stahl (ed.), *Thin Layer Chromatography*, Springer, New York, 1969.
94. M. L. Lee, K. E. Markides (eds.), *Analytical Supercritical Fluid Chromatography and Extraction*, Chromatography Conference Inc., Provo, 1990.
95. J. F. J. Todd et al., Rapid Commun. Mass Spectrom. Ion Proc., 60 (1984) 85.
96. B. He, F. Regnier, J. Pharm. Biomed. Anal., 17 (1998) 925.
97. B. He et al., Anal. Chem., 70 (1998) 3790.
98. R. Stevenson et al., Amer. Lab., 30 (16) (1998) 16A.
99. L. Schweitz et al., Anal. Chem., 69 (1997) 1179.

CHAPTER 5

Data Processing in Chromatographic Methods

5.1 INFORMATION GAIN FROM CHROMATOGRAPHIC DATA

The data processing in chromatographic methods is not necessarily related to a specific sample preparation procedure. However, the results for an analytical method depend on the type of data processing, and the success of a specific sample preparation procedure can be decided based on the type of information obtained from the method. Qualitative information may show if a sample preparation is adequate or does not clean enough the sample. Also, quantitative information can be severely affected by a deficient sample preparation step. For these reasons, a short review on data processing in chromatographic methods is useful.

A formal approach to estimate the relation between the initial information about the sample (input) and the analytical results (output) is to view it as the relation between the information source and the reception in an information transmission system. Similarly to an information transmission system, the degree of incertitude in analytical information can be characterized by the informational entropy H. For a system consisting of n independent events each with the probability p_i, the informational entropy is defined by the relation:

$$H = - \sum_{i=1}^{n} p_i \log p_i \qquad (5.1.1)$$

(the units for H are bits when the logarithm base is 2). For a continuous range of values for a random variable in the interval [a,b] with the density function p(x), the informational entropy is defined by the expression:

$$H = - \int_{a}^{b} p(x) \log p(x) dx \qquad (5.1.2)$$

The informational entropy characterizes the level of uncertainty associated with the specific information. The amount of information I_{nf} generated by performing the operations in an analytical system can be described by the difference:

$$I_{nf} = H_0 - H_1 \qquad (5.1.3)$$

where H_0 is the initial informational entropy and H_1 is the informational entropy after the experiment has been performed [1]. The amount of information is positive when $H_0 > H_1$, and the chemical analyses are selected to lead to positive values for I_{nf}. The maximum information corresponds to $I_{nf} = H_0$, when the uncertainty of the sample is eliminated.

As an example, a sample containing three components has from informational point of view the following possible states: 000, 100, 010, 001, 110, 101, 011, 111, where the presence of a component is noted with "1" and its absence with "0". Assuming equal probabilities for each state, they are $p_i = 1/8$. This leads to $H_0 = \log_2 2^3 = 3$ bits. In general, an analysis requiring detection of n components has $H_0 = n$ bits. Assuming that all analytes are determined in the analysis (uncertainty eliminated), $\{p_i = 1\}_{i = 1, n}$, $H_1 = 0$, and the information $I_{nf} = H_0$.

For a quantitative measurement of one analyte, the evaluation of the informational entropy can be done starting with the assumption that the analyte can be present in the sample in the concentration interval between c' and c", and that knowledge of the correct concentration within an interval Δc is accurate and precise enough. The concentration can have in this case $n = (c' - c') / \Delta c$ discrete values, and the informational entropy is given by the expression:

$$H_0 = \log n = \log \frac{c'' - c'}{\Delta c} \tag{5.1.4}$$

If the analysis is not subject to systematic errors, and the analysis errors have a normal (Gaussian) distribution with a standard deviation s, H_1 can be calculated based on rel. (5.1.2). Using for the probability distribution function the expression of a Gaussian given by rel. (1.11.5), the informational entropy for the measurements has the expression:

$$H_1 = -\frac{1}{s\sqrt{2\pi}} \int_0^{\Delta c} \exp(\frac{-x^2}{2s^2}) \log[\frac{1}{s\sqrt{2\pi}} \exp(\frac{-x^2}{2s^2})] dx = \log (\frac{s\sqrt{2\pi e}}{\Delta c}) \tag{5.1.5}$$

The information for the analysis is in this case

$$I_{nf} = \log \frac{c'' - c'}{s\sqrt{2\pi e}} \tag{5.1.6}$$

Rel. (5.1.6) shows that a more precise determination provides more information, and that information also depends on the preliminary knowledge on the possible concentration interval for the sample.

The application of information theory to analytical systems offers the possibility of evaluating accurately the information obtained following an analysis that leads to the identification of a given number of analytes, as well as the information obtained during the quantitation with a specific error of an analyte. However, this formal procedure does not have significant practical applications, and the information capability of an analytical system is usually not formalized.

5.2   QUALITATIVE AND QUANTITATIVE DATA FROM THE CHROMATOGRAPHIC PROCESS

Gas chromatographic analysis generates both qualitative and quantitative information regarding the sample. Depending on the detector used in the chromatographic analysis, the qualitative information can be limited to the retention times of the chromatographic peaks or, for example, when a MS is coupled as a detector with the GC, full qualitative information can be obtained based on the mass spectra of each peak. The development of large libraries with standard spectra and of algorithms for automatic library searches makes GC-MS the best technique for the identification of the compounds volatile enough to be analyzed by GC. IR detectors also may provide significant qualitative information [2], [3]. However, due to the lower sensitivity of the IR detectors compared to MS, a larger amount of a specific compound in the chromatogram is necessary for successful identifications.

HPLC data are more commonly used as a source of quantitative information. HPLC using a refractive index detector or evaporative light-scattering detector (ELSD) does not provide, besides retention times, additional qualitative information about the sample. However, MS, FTIR, and NMR detection techniques are more informative regarding compound identification (see Section 4.4). The most common HPLC detectors, UV-Vis and fluorescence, provide limited qualitative information because the UV-Vis spectra as well as fluorescence emission spectra are compound dependent but are not diagnostic for compound identification.

The use of multiple detectors can lead to enhanced qualitative information in chromatography. Typical examples are the use of IRD and MS in series as GC detectors, and of UV-Vis and MS in series for HPLC. Numerous other examples of combinations of detectors are known, the combined information from more than one detector being useful in the identification of unknown compounds [4].

Besides the suitability for use in connection to a mass spectrometric detector, chromatography can be applied in compound identification based on the specificity of retention time or of Kováts retention indexes (although the estimation of retention indexes for a chromatographic separation with temperature gradient is a problem in itself). As discussed in Section 1.6, the retention times (and Kováts retention indexes) in a chromatographic separation are compound specific. The retention time can be used for compound identification only when other parameters in the chromatographic process are kept constant. For GC analysis, besides the same chromatographic column, parameters such as eluting gas flow rate and oven temperature program must be kept constant. In any identification it is necessary to use standards and determine the retention time for the compound to be analyzed or to make a previous identification using a mass spectrometer. The use of the retention time is applicable, however, only to infer the nature of the unknown, because more than one compound can have the same retention time. Also, retention times may vary randomly as much as 0.1–0.2 min. (for a 0.32 mm i.d. column). Typically, the retention time (or the retention index) must be associated with a mass spectral detection and the presence of at least three specific ions in the mass spectrum for a positive identification of a given compound. In HPLC, the mobile phase must have the same flow rate, and the composition gradient must be the same for standards and the analyzed material.

In addition to the standard information provided by the retention time (index), global qualitative comparisons between chromatograms can be done based on the profile of the chromatogram. Global profile comparisons are useful for evaluating complex samples. A simple global comparison can be obtained using a similarity index. There are various ways to calculate such an index. One procedure [5], which is aimed to qualitatively compare two chromatograms A and B, uses the relation:

$$S\% = 100 \, N_c / (N_c + N_u^{(A)} + N_u^{(B)}) \qquad (5.2.1)$$

where N_c is the number of peaks that are common in the two chromatograms, $N_u^{(A)}$ is the number of unique peaks in chromatogram A, and $N_u^{(B)}$ the number of unique peaks in chromatogram B. The peaks can be tabulated with electronic integrators and their number is readily obtainable. The pairing of peaks is based on their retention times, which must be equal or within a small interval. This comparison does not take into account the peak areas.

Another method that takes into account the areas of a series of peaks from the two chromatograms starts with the values for the peak areas A_1, A_2, A_3...A_n in the first chromatogram and B_1, B_2, B_3...B_n in the second chromatogram. Pairs of these areas are used to calculate the ratios:

$$R_i = A_i / B_i \text{ or } B_i / A_i \text{ such that } R_i < 1 \qquad (5.2.2)$$

and $R_i = 0$ for each missing corresponding peak in any of the chromatograms. The fit factor can be calculated using the formula:

$$SI = (\Sigma \, R_i) / n \qquad (i = 1, 2, 3...n) \qquad (5.2.3)$$

The SI value can be expressed in percentage (by multiplying SI with 100).

The absolute peak areas in rel. (5.2.2) may be replaced by the peak areas normalized by an internal standard, by the sum of all peak areas, or by the concentrations of the corresponding compounds 1, 2, 3...n (if these concentrations are known). Due to random errors in the chromatographic process, for chromatograms with 30–100 peaks, a similarity higher than 90% typically indicates identical chromatograms.

- Qualitative analysis using MS as a detector

Mass spectrometers are frequently used as detectors in gas chromatography and offer excellent capability of compound identification. The qualitative information is obtained from the fragmentation pattern seen in the mass spectra. The mass spectrum can be used as a fingerprint leading to the identification of the molecular species that generated it. The fragmentation (when done in standard EI+ conditions) generates typical patterns that allow the identification of each compound, either based on interpretation rules or by matching it with standard spectra found in mass spectral libraries. The development of large libraries with standard spectra (over 275,000) and of algorithms for automatic library searches made the use of these tools for spectra interpretation routine. The algorithms for automatic library searches use several criteria for evaluating the quality of a match between the unknown spectrum and a reference spectrum. One well-known algorithm is the Probability Based Matching (PBM) system. This algorithm compares the unknown with all the spectra in the library and has two characteristic features: it weights the mass peaks and performs a reverse search. The weighting of the mass peaks refers to the fact that the mass peaks corresponding to higher masses are given more importance in the search than the peaks with lower mass. This action is necessary because the larger molecular fragments have in general a lower probability of occurrence than lower mass peaks. The reverse search refers to the fact that the algorithm checks whether a peak from the reference spectrum is present in the unknown (and in the appropriate abundance) and not the other way around. In this way, the reverse search ignores peaks in the unknown that are not present in the reference. The mass library searches are in fact more elaborate and provide at the end a list of possible matches for the analyzed compound and for each possible finding a calculated percentage match.

Spectra interpretation for CI+ is less standardized than for EI+ spectra. CI spectra are commonly utilized for the confirmation of a specific compound. The CI MS spectra interpretation is more difficult because of the lack of fragmentation and because the reproducibility of CI spectra is affected significantly by the experimental conditions in which they are generated. CI spectra provide valuable information regarding the molecular mass, and this can be useful in combination with EI spectral information.

Computer programs known as target compound software are available, combining the stability of retention times in a given separation with the identification of specific compounds based on several characteristic mass ions. These software packages have the general purpose of quantitation of specific compounds in GC-MS chromatograms.

The use of a mass spectrometer as a detector in HPLC gives, in general, less qualitative information compared to GC-MS. The reason is the lack of typical fragmentation in LC-MS spectra and the variability of the spectra, which depend on the type of ionization and on the experimental conditions utilized in the HPLC method. However, LC-MS is a valuable tool for obtaining qualitative information, the most common being the molecular mass of the analyte (see also Section 4.4).

- Quantitative results in chromatographic analysis

The quantitative information in a chromatogram is related to the peak areas or peak heights. The signal from the chromatographic detector is processed electronically by an "integrator." Integrators can tabulate the information regarding each chromatographic peak such as retention time, peak area, relative peak area (to the total area of all detected peaks), peak height, relative peak height, type of baseline in the region of each peak, etc. Peak area measurement raises a series of problems such as the choice of the baseline for a given portion of a chromatogram, integration of peaks which are merged and are not separated at the baseline, or integration of peaks appearing as shoulders on other peaks [6]. The answers to these kinds of problems are frequently detailed in the manuals associated with specific integrators and described in a number of books on chromatography (e.g. [7], [8], [9]). The quantitation techniques used for chromatography are discussed in Section 3.8.

A last part of the data processing regarding the quantitative information consists of the evaluation of means, standard deviations, elimination of unusual results, comparison of data between different laboratories, etc. These operations are done for chromatographic data using the typical statistical procedures for any type of analytical results (see also Section 1.11). Extensive discussion on this subject can be found in the dedicated literature (see e.g. [1], [10], [11]).

REFERENCES 5

1. C. Liteanu, I. Rica, *Teoria si Metodologia Statistica a Analizei Urmelor*, Ed. Scrisul Romanesc, Craiova, 1979, or its English translation, *Statistical Theory and Methodology of Trace Analysis*, E. Horwood, Chichester, 1980.
2. N. Ragunathan et al., J. Chromatogr., 856 (1999) 349.
3. R. J. Leibrand (ed.), *Basics of GC/IRD and GC/IRD/MS*, Hewlett Packard, 1993.
4. H.-J. de Geus et al., Trends in Anal. Chem., 15 (1996) 168.
5. P. G. Vincent, M. M. Kulik, Appl. Microbiol., 20 (1970) 957.
6. V. David, S. Moldoveanu, Bull. Inst. Polytech. Bucuresti, 15 (1979) 23.
7. R. L. Grob, *Modern Practice of Gas Chromatography*, J. Wiley, New York, 1995.
8. K. J. Hyver (ed.), *High Resolution Gas Chromatography*, Hewlett-Packard, 1989.
9. N. Dyson, *Chromatographic Integration Methods*, RSC, Cambridge, 1990.
10. P. C. Meier, R. E. Zünd, *Statistical Methods in Analytical Chemistry*, J. Wiley, New York, 1993.
11. D. L. Massart et al., *Evaluation and Optimization of Laboratory Methods and Analytical Procedures*, Elsevier, Amsterdam, 1978.

CHAPTER 6

Automation in Sample Preparation

6.1 AUTOMATION OF THE INSTRUMENTATION USED IN SAMPLE PREPARATION

The term automation is defined as "the use of combinations of mechanical and instrumental devices to replace, refine, extend, or supplement human effort and facilities in the performance of a given process, in which at least one major operation is controlled without human intervention, by a feed-back system" [1]. Similarly defined is mechanization, which is "the use of mechanical devices to replace, refine, extend, or supplement human effort" [2]. Automation in general is a common practice in chemical analysis. Chromatographic instruments are controlled using computers, the detection systems frequently have sophisticated data processing capabilities, and autoinjectors are very common with all chromatographic instruments. Automation is also extended to sample preparation. However, sample preparation automation is in many instances a challenge because of the extreme differences in the operations applied for sample preparation. The advantages of automation are reduced manpower and typically higher repeatability of the operation. This leads in many applications to higher precision of the analysis. The cost of automation is usually recovered from the reduction of manpower, and for large numbers of samples automation can be very economical.

A sample preparation procedure is typically assembled from a number of unit operations. The automation may act on specific operations or on the entire process and may operate on one sample or on a number of samples simultaneously. From this operational point of view, automated sample processing can be classified as follows [3]:

a) Sequential processing, where a single sample follows one by one the steps of an automatic procedure. As soon as this sample is processed, the next one starts the program. The time necessary to process a set of samples is equal to the sum of the times for each unit operation, multiplied by the number of samples.

b) Batch processing where each unit operation from the procedure is carried out on all samples one after the other before the next unit operation is performed. It is a convenient path in case of manual processing, when the same operation is performed on a set of samples. The entire time taken to process a set of samples is equal to the sum of times for each unit operation, multiplied by the number of samples in the batch. Although apparently there is no time difference from sequential processing, the completion of one unit operation on all samples before starting the next unit operation is in general less time consuming than changing the operations for each sample. However, in batch processing the time interval between starting the sample procedure and starting the chromatographic analysis cannot be the same for all samples from the batch. This can be a drawback in cases where a constant time for preparing a sample is required (such as in a derivatization reaction with a specific kinetics).

c) Concurrent sequential processing in which more than one sample enters the chain of the operations, so that multiple identical operations take place simultaneously. The maximum number of samples that are treated concurrently is equal to the number of operations running concurrently. The reduction in the entire processing time is proportional to the number of operations running concurrently.

d) Parallel batch processing in which batches of samples are treated in parallel. Parallel processing reduces the entire process time according to the number of samples processed simultaneously.

The automation of sample preparation can be done on the whole process or only on specific operations. The transfer between two automated segments can be done on-line without human intervention or off-line requiring human intervention. The transfer of the sample from the sample preparation instrument to the chromatographic instrument also can be done on-line or off-line [4].

In off-line sample transfer to the chromatographic systems, sample processing is entirely separated from the chromatograph. In off-line systems the sample is manually transferred to an autosampler or injected by hand in the chromatograph. In case of on-line configuration, the automated sample preparation instrument and the chromatographic system are linked, and there is no human intervention during the run. Several parameters of the on-line system can be programmed to make the instrument work independently on the intervention of the analyst.

The instrumentation used for automation is complex and can be classified in several groups, based on the degree of automation, on-line or off-line connection between modules, etc. A possible instrument classification is the following [5]:

a) Workstations that process automatically a limited number of unit operations (possibly only one). This type of automation is quite frequent in sample preparation. Dedicated instruments, such as sample collectors, solid phase extractors, evaporators, etc., can be very useful for processing uniformly a large number of samples in a limited number of operations.

b) On-line instrumentation that allows convenient sample transfers from operation to operation. This type of automation refers usually to column switching devices or SPE switching devices, which allow the use of various sorbents for the separations of sample constituents, direct transfer of samples to a chromatographic instrument, etc. On-line connections usually switch the flow of liquid or gaseous samples from one path to another.

c) xyz-Handlers that are used to perform operations that require movements of samples or labware [6]. This type of automation is designed to simulate human movements to handle samples and allow simple operations such as filtration, use of SPE cartridges, addition of solvents for dilution, etc.

d) Robotic workstations that have more complex capabilities combining unit operations, such as a simple workstation and possibly xyz movements of the sample.

e) Fully robotic systems that are designed to completely automate the entire analysis starting with sample weighing or the measurement of sample volume, through the transfer to the chromatographic instrument.

The application of workstations and of on-line instrumentation is very common and successful in sample preparation. A large number of instruments of this type are available and used in sample preparation. xyz-Handlers are also successful in some applications, and for large number of samples is a convenient way to reduce the necessary manpower and increase repeatability. Robotic and fully robotic systems are less common in sample preparation. The fully robotic systems typically offer good flexibility in performing different operations, but at the same time must be used for a large number of samples or for an extended period of time doing the same operation in order to be economical. For this reason, in spite of the advantages offered by such systems, the cost of the instruments and the time required for the implementation of a specific set of operations is in many instances prohibitive.

6.2 COMPUTER ASSISTED METHOD DEVELOPMENT

Part of automation can be considered the computer assisted (expert system) development of a separation method [7]. Dedicated programs are available for the optimization of GC or HPLC separations [8]. The DryLab program, for example, uses retention data from two or three initial experimental runs on a specific column to model LC or GC separation conditions. Several separation parameters can be modeled such as percent organic solvent for reverse phase isocratic HPLC, gradient conditions for HPLC, pH of the eluent of the isocratic HPLC, etc. For GC, the program allows optimization of GC temperature gradient. This type of computer assisted method development can be extended to SPE separations. Also, a number of parameters used in sample preparation, such as partition coefficient K_{ow}, boiling point, vapor pressure of pure compounds, water solubility, etc., can be obtained using estimating programs [9] (see Section 1.10). Combining direct experience with the information from such programs, the efficiency in developing an analytical method can be significantly improved.

REFERENCES 6

1. Commission on Analytical Nomenclature of Analytical Chemistry, Division of the International Union of Pure and Applied Chemistry.
2. J. K. Foreman, P. B. Stockwell, *Automatic Chemical Analysis*, E. Horwood, Chichester, 1975.
3. R. D. McDowall, *Automated Sample Preparation,* in *Encyclopedia of Analytical Chemistry*, vol. 1, Academic Press, London, 1995.
4. G. D. Nichols, *On-line Process Analyzers*, J. Wiley, New York, 1988.
5. G. A. Smith, L. L. Lloyd, LC-GC Supplement, May (1998) S22.
6. S. de Koning et al., J. Chromatogr., A, 922 (2001) 391.
7. D. Brown, S. Hardy, *Expert Systems*, J. Wiley, New York, 1985.
8. LC Resources Inc., DryLab 2000, Walnut Creek, 2000, www.lcresources.com.
9. P. Baricic, M. Mackov, *Toolkit for Estimating Physicochemical Properties of Organic Compounds,* based on M. Reinhard, A. Drefahl, *Handbook for Estimating Physicochemical Properties of Organic Compounds*, J. Wiley, New York, 1999.

PART 2

Sample Preparation Techniques Using Physical

Processes

CHAPTER 7

Mechanical Processing in Sample Preparation

7.1 GRINDING, SIEVING, AND BLENDING

Many samples taken for analysis are solid. Solid samples are usually homogenized before analysis. The process typically starts with a particle size reduction. This can be achieved with milling [1], grinding, etc., which can be done for the material as collected or after a cryogenic treatment with liquid CO_2 or N_2. The cryogenic treatment makes many materials brittle enough to be ground easily. For example, substances that contain a certain proportion of water and have some elasticity at room temperature can be ground only after cryogenic treatment. Sieving and repeated particle size reduction can be applied when a certain particle size is desired. Particle size reduction can be beneficial for further sample preparation such as dissolution, extraction, etc. (see Section 8.2) and for sample homogenization. The integration of size reduction in a sample preparation chain is schematically shown in Figure 3.2.1.

Other operations for size reduction in sample preparation can be used. Special size reduction is necessary, for example, for samples that are semi-rigid (rubber, plastics) when cutting is applied. Biological samples may be processed either involving the rupture of the cell walls and intracellular structures or maintaining the cell structure. The use of procedures that involve cell wall rupture, such as sonication, may lead to mixing of enzymes from the inside of the cell with substrates that were previously separated by the cell membrane. This effect must be taken into consideration when results are reported for the composition of extracellular fluid composition.

Sieving is a common method for separating particles according to size. Sieving is typically used for particles above 38 μm and can be extended to even lower dimensions using special micromesh sieves. One screen produces only two fractions, but by use of a series of screens with mesh sizes in decreasing order in the direction of material movement, a number of fractions can be obtained. The screen specifications described by its aperture in mm or by the number of meshes per linear inch are selected for a desired particle size. However, the particle dimensions allowed to pass the sieve are not well limited by the screen aperture due to clogging or particles sticking to each other. Wet screening is used for some applications, and the particle association is diminished. The sieving operation implies also a mechanical movement of the screen. This movement is done manually or with a mechanical shaker with specific vibration frequency [2]. Also, sonic sifters and air jet sifters are commercially available. The use of appropriate sieving labware is necessary for obtaining reproducible results, which can influence the precision in chemical analysis, mainly when some sample discrimination occurs during sieving [3].

The homogenization can be achieved after particle size reduction by mixing (blending) of powders. Mixing may pose specific problems depending on whether the powder is free flowing or cohesive [4]. Typically, free flowing powders have a particle size above 50 μm. Free flowing powders may get unmixed by settling differentially upon movement, etc. [5]. Cohesive powders lead to stable mixtures, but the mixing is more difficult and requires adequate shearing and inclusion of the entire sample. Also, separation of liquids (such as oils from vegetal materials) should be avoided, or the separated material must be properly included in the sample. An evaluation of the homogenization process is desirable when the success of homogenization is questionable. Food analysis may have special sampling problems for nonhomogeneous materials such as meat with bones, cereal mix, etc. Solid sample homogenization must avoid inadvertent changes in the sample composition. These changes may occur due to the loss of fine dust, modifications in the moisture content, or contamination.

7.2   FILTRATION, MICROFILTRATION, AND ULTRAFILTRATION

Filtration is one of several separation processes based on a barrier that prevents mass movement but allows restricted and/or regulated passage of one or more species through it [6]. The separation process through a barrier can be driven by several forces, such as a) mechanical pressure used in filtration, microfiltration, ultrafiltration, and reverse osmosis, b) difference in the chemical potential used in gas diffusion, reverse osmosis, and dialysis, and c) migration in an electrical field used in electrodialysis, etc. The separation by osmosis and dialysis are further discussed in Chapter 13, and electrodialysis is discussed in Chapter 14. This distinction between the different types of separation processes using membranes is somewhat arbitrary because in each separation more than one single mechanism is present. The approximate range of molecular sizes separated in the barrier type techniques is indicated in Table 7.2.1.

TABLE 7.2.1. *Particle dimension ranges separated in different barrier type techniques.*

Separation driving force	10^{-4}–10^{-3} μm	10^{-3}–10^{-2} μm	10^{-2}–10^{-1} μm	10^{-1}–1 μm	1–10 μm	10–100 μm	higher than 100 μm
mechanical pressure	reverse osmosis	ultrafiltration, reverse osmosis	ultrafiltration, reverse osmosis	micro-filtration	micro-filtration	filtration	filtration
osmotic pressure	osmosis, dialysis	osmosis, dialysis	osmosis, dialysis				
electric field	electro-dialysis	electro-dialysis	electro-dialysis				

Barriers used for the separation are identified by terms such as filter, membrane, or other porous media. The term filter should indicate a simple fiber filter but is also used as a generic term for membranes and porous media. The membranes are of several types, their classification based on criteria such as a) the structure of the membrane as being porous or nonporous, b) the morphological characteristics as being solid or liquid, c) the application of the membrane for the separation of two gases, a gas and a liquid, a liquid and a liquid, a liquid and a solid, etc., d) the mechanism of membrane action as nonselective, adsorptive, diffusive, ion-exchange, or osmotic, and e) electrically conductive or nonconductive. Other porous media are made from finely divided particles forming a filtering bed. The materials used for this purpose must be inert in order to avoid in the filtered material discriminations based on properties other than particle size. Porous media may be used for the separation of molecules by a different mechanism known as size exclusion. Size exclusion is discussed in Section 12.3.

The common use of the term filtration is applied for the separation of heterogeneous samples such as a solid from a fluid medium or of aerosols from a gas, using a filter. Adsorption or even chemical reactions retaining aerosols are sometimes indicated as filtration. The filtration of gases is typically considered separately from that of liquids. The main concern in gas filtration is the complete collection of the particles from the gas. In addition to filtration, other techniques are used for collecting particles from a flowing gas. Several procedures used for this purpose are discussed in Section 2.3. The filters used for collecting particles from a gas can be made from various materials such as metallic silver, glass fiber, high purity quartz fibers, cellulose esters on a cellulose pad, polytetrafluoroethylene (PTFE), nylon, etc. The materials for gas filtration must satisfy specific requirements regarding purity. The filtration is frequently done with the goal of analyzing the collected material. For example, the metallic silver screen does not bring any organic impurities, is solvent resistant, and can be reutilized. The nature and the purity of the adsorbing media must be adjusted to the analysis needs.

The filtration of liquids is classified, as shown in Table 7.2.1, in three groups, filtration, microfiltration, and ultrafiltration, this classification being based on convention. Simple filtration is designed to separate suspended particles larger than 10 μm. Microfiltration is designed to retain suspended particles in the range of 0.02 μm to about 10 μm. Ultrafiltration separates molecules with a diameter below 0.02 μm. A microfilter can be regarded simply as a mechanical sieve with uniformly located pores or capillaries, functioning on the principle of surface separation. Tortuous capillary paths from the pores of the membrane provide additional filtration (depth versus surface filtration) through the principles of mechanical entrapment and random adsorption [7].

In ultrafiltration, the filters are usually made of a thin, dense layer with a substructure of progressively larger pores that are open to the permeate side of the membrane. Hydrostatic pressure is applied to the solution at one side of the membrane, and the solvent and small molecules of solutes pass through the membrane while the larger molecules of solutes are retained. Depending on the purpose of analysis, the analytes of interest can be collected from the feed solution (donor) or from the downstream side (named permeate or acceptor). It is common in ultrafiltration to describe the process using the molecular weight cutoff instead of particle size of the retained molecules. Depending on the pore diameter, ultrafiltration separates particles or molecules that range from about 1000 to about 1,000,000 Daltons. This is in average equivalent to a range of particles of about 0.001–0.02 mm. The characteristics of the filter or membrane control which component permeates and which component is retained. The relation between the molecular weight cutoff rating (based on a globular protein molecular weight) and the approximate pore size in nm is shown in Figure 7.2.1 [8]. The filtration being based on particle size, the logarithm of molecular weight of macromolecules is proportional to the size with a unique slope only for polymers with similar shape. For example, globular proteins, branched polysaccharides, and linear synthetic

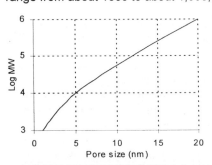

Figure 7.2.1. *Approximate dependence of MW cutoff on filter pore size.*

polymers show a different slope for the curve indicated in Figure 7.2.1. Molecule shape is a factor important to consider when separating molecules in a narrow range of molecular weight.

The molecular size range where the ultrafiltration is applied is overlapping with reverse osmosis, which uses pressure as a driving force for the separation. Reverse osmosis (or hyperfiltration) is designed to retain all components in a fluid other than the solvent and in general is applied to solutes with even smaller molecules than those encountered in ultrafiltration. Reverse osmosis is discussed in Section 13.2.

The theory of simple filtration can be based on the model of a fluid flowing through a tube and the assumption that the fluid at the walls of the tube is stagnant. According to Poiseuille's law, the relation between the flow rate U of the fluid through the tube and the physical parameters involved in the process is the following:

$$U = \frac{dV}{dt} = \frac{\pi \Delta p r^4}{8 \eta L} \qquad (7.2.1)$$

where V is the volume of fluid, t is time, r is the tube radius considered to be large compared with the mean free path of a molecule in the fluid, Δp is the difference in the pressure from one end to the other of the tube, η is the viscosity of the fluid, and L is the length of the tube. This expression is modified for a filter to the expression:

$$U = \frac{dV}{dt} = \frac{\varepsilon \pi \Delta p r^4}{8 \eta L} \qquad (7.2.2)$$

where ε is the porosity of the material. This expression assumes uniformly distributed evenly sized pores, negligible concentration polarization at the surface of the filter (see further), laminar flow through pores, incompressible fluid, etc. The porosity ε depends on the number of pores for the whole filter surface. The pressure on the flowing fluid can be obtained from rel. (7.2.2) for a given flow rate [9], [10]. The filtration is not in fact only a mechanical process. Hydration effects occurring in the pore of a filter, adsorption phenomena, or electrokinetic effects should also be taken into consideration. The characterization of filter pore size may be based on rel. (7.2.2) using instead of pore size the bubble point or water breakthrough. Bubble test measures the air pressure necessary to form bubbles at a membrane surface covered with fluid, and the water breakthrough measures the pressure required to transmit water through the largest pore of a dry hydrophobic material.

Other models are also applied for describing the flow of material, especially in the ultrafiltration process [11]. One such model assumes that the solute is brought to the membrane surface by convective transport with the flux J_f, and the permeate flux J_p is given by the expression:

$$J_p = J_f / c_f \qquad (7.2.3)$$

where c_f is the concentration of the rejected solute (U = J A where A is the membrane area). The rate of back transport into the solution from the membrane surface is given by the expression:

$$J_f = D \frac{dc}{dx} \tag{7.2.4}$$

where D is the diffusion coefficient of the solute and dc/dx is the variation of concentration over a differential element at the boundary layer. When equilibrium is established between the convective transport and the back diffusion into the solution, from rel. (7.2.3) and (7.2.4) it can be obtained after integration:

$$J_p = \frac{D}{\delta} \ln \frac{c_s}{c_f} = k \ln \frac{c_s}{c_f} \tag{7.2.5}$$

where δ is the thickness of the boundary layer, c_s is the concentration at the membrane surface, and the coefficient k is known as a mass transfer coefficient. Rel. (7.2.5) shows that the flux can be improved by selecting a membrane with larger k for a given c_f and for c_s determined by the nature of the solution and the compounds to be separated.

The models previously discussed do not include explicitly the dependence of filtration parameters on temperature. However, parameters such as viscosity η or the diffusion coefficient D are temperature dependent. In general, the increase in temperature has the effect of increasing the flux (or the volumetric flow).

A membrane filter may become more efficient during an application after an accommodation time during which smaller particles are trapped or absorbed within the pores of the filter. The filter captures the larger particles that may have escaped through previously larger pores that now are constricted mechanically by the trapped particles. Therefore the values of ε and r in rel. (7.2.2) may change during the filtration process. This phenomenon can be useful when filtering particles with sizes similar to the pore-size rating of the membrane, not allowing larger particles to penetrate the filter. On the other hand, the process of pore plugging may be very detrimental by diminishing the flow rate through the membrane.

Filtering characteristics of a specific filter can be sometimes changed using additives into the feed solution. For example, the use of detergents may lower the cut-off value of a membrane. During filtration the depletion of the detergent can allow larger particles to pass through the membrane. With appropriate detergent concentration the membrane characteristics can be manipulated to a certain degree. The presence of other solutes may influence the ultrafiltration process. Small molecules do not affect in general the permeability, although the pH and the ionic strength are important parameters to control when the separation is intended for proteins. Large molecules such as a secondary polymers may influence the separation due to concentration polarization.

- Filter and membrane properties

Pore dimension is the most important property of a filter or membrane. However, the pores are not all of the same dimensions. A distribution of pore size in a filter or membrane always is present and is an important characteristic. Each filtering material (fiber filter, membrane, or other porous medium) has a specific range of pore sizes. The mean pore size determines the minimum molecular weight (or cutoff value) of a particle that cannot penetrate the membrane. Due to the manufacturing process of most fibrous and polymeric membrane filters, the capillary channels can vary in size. The pore size of each membrane is expressed as a nominal value, which is equal to a given size of particles retained by the membrane. However, the retention is not complete, and the rating indicates the particle size that is retained in a certain proportion. This can vary for different manufacturers between 98% and 60%. An absolute rating can be indicated, and this is based in ultrafiltration on the penetration of a specific microorganism (with known dimensions) through the membrane. The rating is given by the dimension which does not allow more than a single microorganism in the permeate. The cutoff, although frequently expressed as a value, is for this reason a range. In ultrafiltration, this property is described by the rejection characteristic, which is given by the formula:

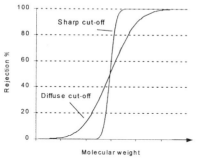

Figure 7.2.2. *Rejection characteristics of a membrane.*

$$R_{rej} = (1 - C_p/C_f)\ 100 \qquad\qquad (7.2.6)$$

where C_p is the concentration in the product and C_f the concentration in the feed solution. The variation R_{rej} is shown in Figure 7.2.2 for two membranes, one with a sharp cutoff and another with a diffuse cutoff.

Solute removal efficiency is dependent on the solute size and shape, the shape of the pores and physico-chemical interactions between solute molecules and membrane surface, pH of the solute, solute aggregation, and the degree of concentration polarization at the membrane surface [12]. Concentration polarization is a local concentration of the solutes at the membrane surface as compared to the feed (or bulk) solution when the solute is brought to the membrane surface by convective transport. A portion of the solvent is removed from the feed, and the large molecules are retained and do not diffuse back into the bulk. The variation of concentration toward the membrane surface is shown in Figure 7.2.3. The concentration polarization is responsible for the difference in the flux between water and a solution of polymers and also is responsible for the modification in the rejection characteristics of a membrane from an individual polymer to a mixture of polymers. The retention of polymers is more efficient from a mixture than if they are present individually. Due to concentration polarization, a secondary dynamic membrane is formed [11] on the surface of the membrane.

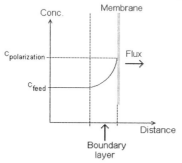

Figure 7.2.3. *Concentration polarization during ultrafiltration.*

Besides filter porosity and pore distribution, some other physical characteristics of the filter are important. The dimension of the filter must be adjusted to accommodate the required solution volume and flow rate. Also, the hold-up volume is another important characteristic to consider in sample preparation. This is the volume of solution retained in the filter when no additional pressure is applied on the filter. Therefore, if recovery of the whole sample after filtration is not complete, the results cannot be accurate. The solution can be eliminated from the filter using pressure, vacuum, or centrifugation. Also the filter can be washed with pure solvent. The washing operation of the retentate also can be necessary to remove quantitatively the analytes found in precipitate matrix. However, this involves a dilution of the permeate, and for situations where the amount of solvent must be as low as possible, the hold-up volume of the filter must be minimized.

Filters and membranes may have different structures depending on their applications for prefiltration, microfiltration, or ultrafiltration. Filters can be simple screen filters or "depth" filters. Depth filters are made from a matrix of randomly oriented fibers bonded together. Screen filters separate the particles on their surface, and most membrane filters are screen filters. Screen filters can be further classified as microporous and asymmetric. Microporous filters have the pores of the same size through the membrane, while asymmetric membranes have small pores at one surface and larger on the other. The microporous membranes reject particles larger then the pore, but the particles with dimensions similar to the pores may penetrate the pore but remain retained in the membrane, blocking the pores. This process is avoided with asymmetric membranes where, once the particle penetrates the membrane, it is typically found in the permeate due to the wider opening on the opposite side.

- Filter and membrane materials

The most common filtration material for liquids is probably cellulose (paper filters). Cellulose filters of different porosities are available. Table 7.2.2 gives some common characteristics of paper filters.

TABLE 7.2.2. *Common characteristics of filter papers [13].*

Filter type	Coarse	Medium	Medium-fine	Fine
particle retention	>20 μm	5–10 μm	2–5 μm	1–3 μm
filtration speed[1]	55 s (Herzberg)	230 s (Herzberg)	730 s (Herzberg)	1400 s (Herzberg)
flow rate[2]	160 mL/min.	60 mL/min	25 mL/min.	5 mL/min.
wet strength	limited	limited	limited	limited
applications	retains gelatinous and large particle precipitates	general purpose	general purpose	retains fine particles

[1] Time for 100 mL to pass through 10 cm^2 filter paper with 10 cm water head pressure (Herzberg). A different filtering time protocol is recommended by ASTM, which is the time for 100 mL filtered water to pass through 15 cm quadrant-folded paper.

[2] Rate of flow for distilled water through a 5 cm diameter filter with constant 5 cm water head pressure.

Paper filters of other qualities are also available. For example, Whatman filter papers are graded in six porosities from coarse to fine. The filter papers are commercially available in different dimensions and shapes such as circles or squares, including smooth, creped or pleated surface, etc. For some applications the quality of the cellulose is evaluated based on the amount of ash generated by combustion. The ash content for special quantitative paper can be less than 0.01% of its weight. Ashless paper is also graded based on characteristics such as particle retention, filtration speed, etc. Another characteristic of the paper filter that can be selected is the wet strength. This can be controlled by the length of the cellulose fiber or by chemical modifications. Papers with high and low wet strength are commercially available.

Other filter materials besides cellulose are commonly used in practice. Filters from glass fiber (usually inert borosilicate glass) are also used as general purpose filters, mainly for prefiltration. These filters can be made with binders (such as acrylic resin) or with no organic binders and chemically and biologically inert. The filters with binders have better mechanical properties. Binder-free filters can be used for various purposes including corrosive reagents, biological fluids, etc. Various porosities are available, starting with efficiency for particles larger than 0.2–0.6 μm, or for more rapid filtration with larger particle retention. Glass fiber filters can be heated in autoclave for sterilization, which makes them suitable for many applications such as filtration of clinical specimens. The glass fiber filters are also used as prefilters for submicron further filtration.

A number of properties besides pore size and distribution are important for the filter material. These include filter inertness toward the solvent and solute, filter material purity, etc. The filter material may be soluble or decompose in specific solvents. Also, there is a required compatibility between the solvent and the nature of the membrane filter. For example, for aqueous solutions the membrane type can be one of the following materials: nylon, poly(vinylidene fluoride) (PVDF), poly(vinyl chloride), polypropylene, cellulose acetate, regular cellulose, etc. For biological liquids the preferred membrane materials are polyethersulfone (PES) or poly(vinylidene fluoride), which are characterized by low protein binding properties (4 μg / cm^2 protein retention). Mixed cellulose esters (MCE), polypropylene, or regular cellulose also can be used in biological fluid filtration. Partially organic/aqueous buffered solutions can be filtered with membranes made from nylon, poly(vinylidene fluoride), polypropylene, or regular cellulose, while organic solutions with high solvating capability can be filtered with polytetrafluoroethylene (PTFE) or polypropylene. Other properties such as pyrogenicity, sterile character, surfactant free character, plasticizer free character, and lack of cytotoxicity are are also important for a filtering material. Some common materials used for filters are indicated in Table 7.2.3.

TABLE 7.2.3. *Common materials used for filters.*

Filter material	Solution type	Pore size μm	Applications[1]
cellulose	water, organic	1, 5, 10, 20	prefiltration
cellulose acetate (CA)	water	0.22, 0.45, 0.80	biological fluids, MF, UF
cellulose triacetate	water	0.22, 0.45, 0.80	biological fluids, MF, UF
cellulose nitrate (CN)	water	0.2, 0.45, 0.8	biological fluids, MF
glass microfiber	any	various	prefiltration
mixed cellulose ester (MCE)	water	0.22, 0.44	biological fluids, MF
nylon (NYL)	water, organic	0.22, 0.45	water or solvent solutions, MF, UF
polyacrylonitrile	water, organic		water, solvents, UF
polycarbonate	water, etc.	0.2, 0.4, etc.	MF
polyesters	organic, water		MF
polyethersulfone (PES)	water	0.1, 0.22, 0.45	biological fluids, MF, UF
polyimides	water		biological fluids, MF
polypropylene filaments	organic, water	1, 5, 10, 20	water, solvents, MF
polypropylene hydrophilic			MF
polytetrafluoroethylene (PTFE) hydrophobic	organic	0.45, 0.50	aggressive solvents, MF
polytetrafluoroethylene (PTFE) hydrophobic (bonded with polyethylene)	organic	0.45, 0.50	aggressive solvents, MF
polytetrafluoroethylene (PTFE) hydrophilic	organic, water	0.45, 0.50	water or solvent solutions, MF
polyvinyl chloride	water	0.45, etc.	water solutions, MF, UF
polyvinylidene fluoride (PVDF)	water	0.22, 0.44	biological fluids, MF, UF
porcelain			water or solvent solutions, MF
regenerated cellulose	water, organic	1, 5, 10, 20	prefiltration
SiO$_2$, Cellite, diatomaceous earth, Celatom, Fuller's earth	organic, water	various particle sizes	prefiltration, water or solvent solutions
surfactant-free cellulose acetate (SFCA)	water	0.22, 0.45, 0.80	biological fluids, MF, UF

[1] MF = microfiltration, UF = ultrafiltration

The chemical structures of some membrane materials are shown below:

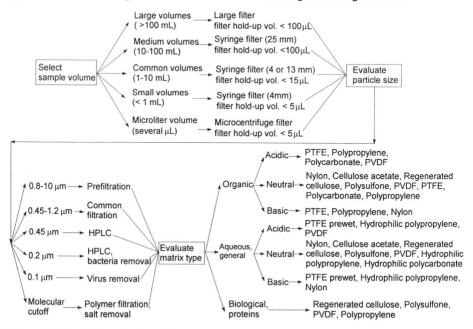

Other materials or composite materials are also used for filtration. For example, hydrophilic filters with hydrophobic edges are available for preventing solution intrusion under the filter holder. The filtering material must be selected in accordance with the type of filtration. A diagram useful for filter selection is given in Figure 7.2.4.

Figure 7.2.4. *Filter selection diagram.*

The diagram shown in Figure 7.2.4 can be used as a guide for a filtration operation. In the course of a sample preparation sequence, several filtrations with different requirements may be necessary.

- *Filtering devices*

Separation by use of filtration is achieved by placing a liquid sample on one side of the filter or membrane (feed side) and applying a pressure difference to drive all molecules of appropriate size, including the solvent molecules, through the membrane pores to the other side (permeate side). Various types of labware are designed for assisting filtration. For the common filtration (prefiltration) the most frequently used glassware is the filtering funnel. Models differing in top diameter, bowl angle (usually 60°), stem length, and special ridges are commercially available. In this type of funnel the pressure necessary for performing the filtration is obtained by the gravity flow of the water column. To speed the filtration it is also common to use vacuum. Labware used with vacuum are the Büchner funnel or other filtration assemblies that allow, besides the use of vacuum, the separation of the filter without significant loss of the retained material. Some common filtering labware is shown in Figure 7.2.5. Prefiltration or common filtration

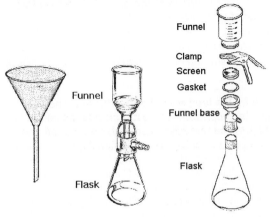

Figure 7.2.5. *Common filtering labware.*

is frequently recommended before microfiltration or ultrafiltration. The process removes larger particles that affect the flow rate of micro or ultrafiltration.

Microfiltration and ultrafiltration are performed in various filtering devices. The material of the filter holder or of the labware necessary to contain the sample can be glass, polypropylene, nylon, solid polytetrafluoroethylene, etc. This material must have specific chemical and physical properties to not affect sample composition. Microfiltration devices used in chromatography are flat disk membranes used in glass vacuum filtration units for solvent preparation, encapsulated membranes in syringe filter holders, disposable syringe filter units, etc. In off-line applications the driving force for moving the permeate through the filter is pressure using the syringe, vacuum, or a centrifugal force. The syringe filters are among the most common and are available in different diameters and filter thicknesses. One typical syringe filter is shown in Figure 7.2.6. Filter holders made from stainless steel are also available, as shown in Figure 7.2.7. Other types of filtration devices are available. These include systems containing a syringe filter and a syringe incorporated in a single device, filters incorporated into a sample vial, filters incorporated into a centrifuge tube, etc. The filter incorporated into a sample vial consists of two pieces, one being a plastic cylindrical cup that contains the

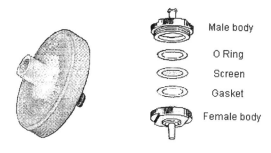

Male body

O Ring

Screen

Gasket

Female body

Figure 7.2.6. *Common*   Figure 7.2.7. *Stainless*
syringe filter.   *steel filter holder*

sample and the other a (crimped top) vial that fits as a piston the cylindrical cup and has a filter at the bottom. As the vial is introduced in the cup containing the sample, the permeate fills the vial and the whole assembly is used for sample injection. Centrifuge filters have also become more popular for different chromatographic applications. These devices are especially well suited for the filtration of small volumes and viscous samples. The driving force in this kind of filtration is given by the force of gravity generated by a centrifuge. Sample volume ranges from 50–200 µL. By using a multi-place rotor within a centrifuge, these devices can process many samples simultaneously. Although microfiltration membranes are also used in this format, ultrafiltration and molecular weight cutoff filters are used more for centrifugation [8].

In on-line filtration, the sample is pumped on the feed side of a membrane. The filtrate flow is directed for example, into an injection loop, from where it is introduced into an analytical system. In the direct interfacing between a filtration membrane and an analytical system, it is possible that only a part of the filtrate is transferred to an injection loop and injected into the analytical system. This procedure cannot be applied if the sample requires further processing. On-line filtration of the HPLC solutions before they enter the HPLC column is common and uses cartridges or filters with special filter holders made for example from stainless steel, similar to that shown in Figure 7.2.6, having the appropriate fittings and with very small dead volume.

- Applications of filtration, microfiltration, and ultrafiltration

Filtration is probably the most common operation in sample preparation (see Figure 1.2.1). Filtration and microfiltration are procedures for the separation of suspended particles from dissolved substances in a feed stream, provided the particles meet the size requirements for the specific filter or membrane. The procedure has numerous applications including the filtration of samples, partially processed samples, filtration before introducing a sample in the LC column, etc. The selection of a filter for HPLC is particularly important for maintaining the column performances. The columns are commonly protected by a porous frit that may have 2 µm pore diameter. The typical spaces between the particles of a HPLC column are about 1/3 of the particle diameter and can be lower than 1 µm. The frit is able to filter particles that may accumulate in the HPLC column. The filtration of the HPLC solutions and samples is necessary for preventing the blockage of the frit and blockage of the column void spaces. For this purpose, filtration using 0.45 µm is common for the sample and for the solvents. However, the use of 0.22 µm filters is preferable for the sample, if this is not precluded by the sample viscosity or the long time necessary for filtration.

Ultrafiltration is a method for purifying, concentrating, and fractionating macromolecules or fine colloidal suspensions. The flow chart for the application of ultrafiltration in a typical sample preparation of a biological sample is shown in Figure 7.2.8 [14].

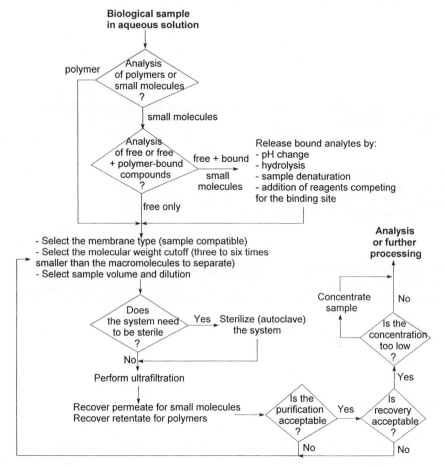

Figure 7.2.8. *Flow chart for the application of ultrafiltration in the processing of a biological sample* [14].

Further integration of ultrafiltration operations in a sample preparation procedure is shown in Figure 3.2.1.

Among other applications for ultrafiltration, the indirect analysis of small molecules is also reported. By use of the equilibrium in the reaction between a metallic ion and a specific water-soluble polyelectrolyte polymer and between the metallic ion and a small complexing molecule, the analysis of the small molecule is possible. For this purpose the reaction is followed by the retention of the polymer/metallic ion combination by ultrafiltration and analysis in the permeate of the complex formed by the metallic ion and the small molecule [15].

7.3 CENTRIFUGATION

Centrifugation is another technique used for the separation of heterogeneous samples. During centrifugation two kinds of forces act on the particles from a heterogeneous sample, the gravitational field and a centrifugation force, applied by rotation at a high speed. The gravitational force is homogeneous over the bulk of sample and in general much smaller than centrifugation force. For these reasons, gravitational force is usually neglected in centrifugation processes.

The centrifugal force Φ depends on the radius r to a point in the rotating sample (distance from the rotation axis) and of angular velocity ω (expressed in radians per time unit) by the expression:

$$\Phi = (1 - \rho / \rho_p)\, m\, \omega^2\, r \tag{7.3.1}$$

where m is the mass of the particle, ρ is the density of the solvent, and ρ_p is the density of the particle. The mass of the rotating particle is affected by buoyancy, and the effective mass affected by the centrifugation is $(1 - \rho / \rho_p)\, m$. The centrifugal acceleration a_c is given by the expression:

$$a_c = \omega^2\, r = (2\pi\, n)^2\, r \tag{7.3.2}$$

where n is number of rotations per time unit. This acceleration is usually expressed in units of acceleration of gravity ($g_n = 9.80665$ m s^{-2}), and the relative centrifugal field (RCF) is given by the ratio between the centrifugal and gravitational forces:

$$RCF = \frac{a_c}{g_n} = \frac{4\pi^2 n^2 r}{g_n} = 1.11825 \cdot 10^{-6}\, r \cdot n^2 \tag{7.3.3}$$

where r is expressed in mm and n in rotations per min. (rpm). Depending on the values of RCF, the centrifuges can be classified in the three groups, a) common laboratory centrifuges with RCF < 3000, b) supercentrifuges with RCF 3 10^3–10^5, and c) ultracentrifuges with RCF 10^5–10^8.

The use of a centrifugal force increases the rate of separation of the particles in a solution, which in a gravitational field usually could occur only with a low rate. The main operations that are based on centrifugation in sample preparation are sedimentation, formation of a meniscus of materials lighter than the solvent, separation of heterogeneous liquid systems, and the use of centrifugation in connection with filtration.

Sedimentation of the particles heavier than the solvating medium has been studied in detail for small spherical particles moving in a laminar motion under the influence of gravity. The force F exerted on these particles at the equilibrium velocity is given by the expression:

$$F = \frac{\pi}{6} d^3 g_n (\rho_p - \rho) \tag{7.3.4}$$

where d is the diameter of the particle. The resistance to motion of the particle due to the viscosity of the fluid is given by the expression:

$$R = 3\pi d \cdot u \cdot \eta \tag{7.3.5}$$

where η is the viscosity of the fluid and u is the particle velocity. The condition $R = F$ leads to Stokes law that gives the final settling velocity:

$$u_s = \frac{d^2 g_n (\rho_p - \rho)}{18\eta} \tag{7.3.6}$$

For sedimentation during centrifugation, g_n in rel. (7.3.6) is replaced by a_c given by rel. (7.3.2) and becomes

$$u_s = \frac{2\pi^2 d^2 (\rho_p - \rho) \cdot n^2 r}{9\eta} \tag{7.3.7}$$

Rel. (7.3.7) shows that particles of different diameters have different settling velocities. This velocity depends on the angular velocity (or the number of rotations per min.), the viscosity of the solvent, the density of the solid particles, and also on the distance to the rotation axis r of the particle in the centrifuge tube. The sedimentation velocity for solid particles having the size of a few μm is reached in several microseconds, while larger particles (about 100 μm size) can reach the same sedimentation speed in several milliseconds. It is assumed that 0.1 seconds are sufficient for all dispersed particles to reach the sedimentation speed [16]. This makes the sedimentation process with centrifugation much faster than that due to gravity. If the density of the solute is lower than that of the solvent, such as in the case of lipoproteins and lipids in aqueous solutions, the separation takes place because solute phase floats towards the meniscus at the top of the solution. Centrifugation also can be used for the separation of heterogeneous liquid systems (emulsion, extraction). For example, a liquid-liquid extraction emulsion is frequently formed, and centrifugation can be used to separate the two nonmiscible phases. For some complex aqueous phases, such as blood samples, the extraction procedures may lead to a gel type phase that can be separated in its two nonmiscible phases only by centrifugation [17].

Centrifugation is widely used in many simple separations of heterogeneous samples and is frequently used instead of filtration for the separation of solids from a liquid [18]. In typical laboratory practice, the material to be separated is introduced into an appropriate cenrtifuge tube and centrifugated. Values of rpm are 3000–5000

A special application of separation by centrifugation was developed using rpm values higher than 72,000. This procedure, known as ultracentrifugation, can be applied for the separation of macromolecules from a solution, and the separation takes place differentiating macromolecules by their molecular weight [19]. For a homodisperse solute in ideal conditions, the relation between the molecular weight M, the solute concentrations C_1 and C_2 at two radial positions r_1 and r_2, and the angular velocity ω of the centrifuge, the following expression can be written:

$$M = \frac{2\,RT\,\ln\,(C_1\,/\,C_2)}{\omega^2\,(1 - \rho\,/\,\rho_p)\,(r_1{}^2 - r_2{}^2)} \tag{7.3.8}$$

In order to perform analytical ultracentrifugation, the stirring of the ultracentrifuge cell must be eliminated, the angular velocity must be high, the temperature of the process must be kept constant, etc. [20]. Special optical systems are designed for the measurement of polymer molecules movement in the centrifuge tube. Analytical centrifugation cannot be considered a sample preparation technique, being used mainly for molecular weight determinations of polymers. However, the technique can be used for the separation of sub-cellular particles from a solution. For the separation of cell size particles, a medium-speed centrifuge is sufficient. Such centrifuges are designed to handle a large number of samples and are commonly used in clinical laboratories [21].

- The use of centrifugation in connection with filtration and ultrafiltration

Centrifugation can be applied to provide the necessary pressure difference in filtration (see rel. 7.2.3) or for the separation of the layers in liquid-liquid extraction. For example, a device that can be applied in solvent extraction is the centrifugal separator [22], [23].

The efficiency of filtration is not influenced by the operating pressure, which in theory affects only the rate of filtration. The necessary pressure is usually mechanical force in a syringe, generated by a gas pressure or use of vacuum. However, the use of centrifugal acceleration can be successfully used for the same purpose. Special centrifuge tubes having a double bottom, the upper one consisting of a filter, are commercially available. The material to filter is introduced in the filtration/centrifugation tube and subjected to centrifugation, and the permeate is collected at the bottom of the tube. The method is suitable for both microfiltration and ultrafiltration. For example, the separation of protein solutions from the ionic and low molecular weight molecules can be done by use of centrifugation/ultrafiltration. Special ultrafilter cones can be fitted into a special centrifuge tube. Swinging-head centrifuges are most suitable in these applications. The centrifugation/ultrafiltration can be used for the concentration of protein solutions. For instance, 5 mL of 0.01% albumin solution can be concentrated to a volume of 1 mL with this procedure. The ultrafilter retains components with a molecular weight of $5 \cdot 10^4$ but has no significant retention for molecular weights lower than $2 \cdot 10^4$. In current applications, artificial membranes with well-defined permeabilities are frequently used. Such membranes can be obtained from cellulose nitrate, cellulose acetate, regenerated cellulose, polyvinyl chloride, polyamide, polyethersulfone, etc. [18]. The procedure also can be applied for the separation of small molecules from particles and polymeric components. For example, in clinical analysis, the preparation of fecal samples for assay of volatile fatty acids can be done with steam distillation or centrifugation/ultrafiltration. For centrifugation/ultrafiltration the samples are mixed with H_2SO_4 followed by the centrifugation of the mixture through a membrane with a molecular weight cut-off of 3000. The solution is collected and analyzed by GC or HPLC. The results obtained by the two procedures are very similar, although the centrifugation/filtration procedure is significantly simpler [24].

REFERENCES 7

1. M. J. Lichon, J. Assoc. Off. Anal. Chem., 73 (1990) 820.
2. H.-J. Meyer, American Lab. News Ed., 33 (2001) 28.
3. B. L. Karger et al., *An Introduction to Separation Science*, J. Wiley, New York, 1973.
4. M. J. Lichon, J. Chromatogr., 624 (1992) 3.
5. N. Harnby et al., *Mixing in the Process Industries*, Butterworth-Heinemann, Oxford, 1992.
6. N. Lakshminarayanaiah, *Equations in Membrane Biophysics*, Academic Press, New York, 1984.
7. O. I. Nachinkin, *Polymeric Microfilters*, E. Horwood, New York, 1991.
8. R. Lombardi, LC-GC Suppl., S46 May 1998.
9. N. C. van de Merbel, Trends Anal. Chem., 16 (1997) 162.
10. N. C. van de Merbel et al., Anal. Chim. Acta, 279 (1993) 39.
11. M. Cheryan, *Ultrafiltration Handbook*, Technomic, Lancaster, PA, 1986.
12. J. Murkes, C. G. Carlsson, *Crossflow Filtration*, J. Wiley, New York, 1988.
13. Fischer Catalog 2001, Fischer Scientific, Pittsburgh, 2001.
14. Guide for Sample Preparation, LC-GC, Aug. 2000.
15. I. Moreno-Villoslada et al., Anal. Chem., 73 (2001) 5468.
16. G. Jinescu, *Procese Hidrodinamice si Utilaje Specifice in Industria Chimica si Petrochimica*, Ed. Didactica si Pedagagica, Bucuresti, 1983.
17. A. Medvedovici et al., Anal. Lett., 33 (2000) 2219.
18. L. Gaspar, in I. Kerese (ed.), *Methods of Protein Analysis*, Akademiai Kiado, Budapest, 1984
19. T. Svedberg et al., *The Ultracentrifuge*, Clarendon Press, Oxford, 1940.
20. G. Ralston, *Ultracentrifugation*, in *Encyclopedia of Analytical Chemistry*, Vol. 1 (A-Che), Academic Press, London, 1995, p.578.
21. P. Spielman, International Laboratory, Nov./Dec. (1977) 69.
22. D. G. Vallis, U. K. Patent Application, 14964/67 (1967).
23. J. K. Foreman, P. B. Stockwell, *Automatic Chemical Analysis*, E. Horwood, Chichester, 1975.
24. H. M. Chen, C. H. Lifschitz, Clin. Chem., 35 (1989) 74.

CHAPTER 8

Phase Transfer Separations Applied in Sample Preparations

8.1 DISTILLATION, VAPORIZATION, AND DRYING

Distillation is a separation technique with many applications, including industrial as well as laboratory scale separations. Distillation is used for both preparative and analytical purposes, and as a sample preparation technique it is not limited to analyses where a chromatographic core step is utilized. During distillation, a liquid solution is boiled and the vapors are condensed. When the vapor pressure of the liquid equals the atmospheric pressure, the liquid boils, and the vapors are generated from the whole liquid and not only at the surface separating the phases. In a simple distillation, the vapors generated by boiling are sent directly into a condenser, and the condensed liquid (distillate or overhead product) is collected. A diagram of a setup for simple distillation is shown in Figure 8.1.1a. A different procedure includes between the boiling liquid and the condenser a column where the temperature decreases upward (either controlled or by heat loss into environment), and some of the vapors condense in this column and return to the distillation unit as a reflux. A reflux-distillate ratio-controlling device can also be incorporated into the distillation apparatus. The distillation can be carried out batchwise or continuously. This technique is known as fractional distillation, and a simple diagram is shown in Figure 8.1.1b.

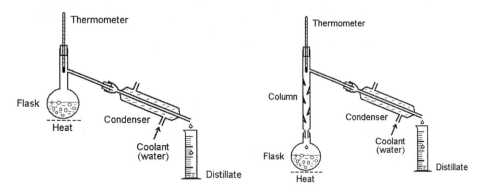

FIGURE 8.1.1a. *Diagram of a simple distillation apparatus.*

FIGURE 8.1.1b. *Diagram of a fractional distillation apparatus*

The theory of simple distillation shows that the separation is based on the differences in the composition of the liquid and of its vapors at equilibrium (see Section 1.3). For a system with two components, Gibbs phase rule (see rel. 1.3.1) indicates that the mole fraction, pressure, and temperature can be used as variables. Keeping one variable constant, a phase diagram can be obtained allowing the study of the dependence of the other two variables on each other. A diagram showing the variation of vapor pressure as a function of mole fraction for a two-component solution at constant temperature is given in Figure 1.3.2a (see Section 1.3). For a constant pressure (e.g. at 1 atm.), the phase diagram shows the variation of boiling temperature of the mixture as a function of composition. One such diagram is shown in Figure 8.1.2. For a composition x_i^L of the liquid, below the temperature T_2 the liquid does not boil. At the temperature T_1 all the liquid with the composition x_i^L is transformed into vapors. In between these

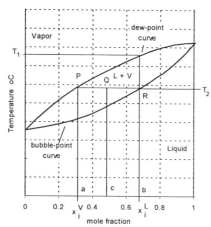

temperatures, both liquid and vapors (L + V) coexist. The difference in the mole fractions x_i^L and x_i^V shows the difference in the composition of vapors and of the liquid for the same temperature (except for the start and end points). If the solution is boiled until all the liquid is transformed into vapors and these are condensed, the collected condensate is obviously the same as the original liquid. The difference from a pure liquid is that boiling takes place in a range of temperatures and not at a single temperature. If the collection of the condensed vapors is stopped before all the liquid is gone, the condensed liquid (distillate) is richer in the component with a lower boiling point, and the liquid left in the boiling flask (residue or bottom) is richer in the less volatile component. For a point Q in the two-phase region, at a temperature T_2, the total composition (gas and liquid) expressed in mole fraction of "i" equals c. The line PR is

FIGURE 8.1.2. *Diagram showing the composition of vapors x_i^V above an ideal solution as being different from the composition of the liquid x_i^L at a constant pressure (set at 1 atm.).*

called *tie line*. The relative amount of liquid and vapors for the point Q can be calculated based on the material balance. The mole fraction of the vapors in P being a, and for the liquid in R being b, the following relation can be written:

$$c\,(n^V + n^L) = a\,n^V + b\,n^L \qquad (8.1.1)$$

where n^V is the total number of moles of material ("i" and "A") in vapor phase, and n^L the total number of moles of material in liquid phase. Form rel. (8.1.1) the following expression can be obtained:

$$\frac{n^V}{n^L} = \frac{b - c}{c - a} \qquad (8.1.2)$$

Rel. (8.1.2) is known as the lever rule, and shows the ratio of the amount of vapors vs. liquid in the two-phase region.

In practice, not only binary solutions are subject to distillation. Solutions with multiple components can be distilled, and the separation capability between any pair is still described by the separation factor α. For nonideal solutions, instead of rel. (1.3.16), the expression for α must include an activity coefficient and has the form:

$$\alpha = \gamma_i'\, p_i^{\circ} / \gamma_A'\, p_A^{\circ} \qquad (8.1.3)$$

This factor is dependent on the volatility of the two components "i" and "A" in pure state.

The distillation process can be performed in batches or continuously (see Section 1.7). In batch distillation, the initial material to be distilled is not replenished. The continuous distillation is not common in sample preparation for analytical purposes but is very common in the industrial applications of distillation.

For a simple distillation of a binary mixture containing n_i^L moles of a compound "i" and n_A^L moles of a compound "A" in the liquid phase, the total amount of mixture is $W = n_i^L + n_A^L$. During distillation when a differential amount dW of material is evaporated and is removed from liquid, the material balance for the compound "i" can be written in the form:

$$x_i^V \, dW = d(W \, x_i^L) = W \, dx_i^L + x_i^L \, dW \tag{8.1.4}$$

where the common notations for the mole fraction of compound "i" in the liquid phase x_i^L and in vapor phase x_i^V were used. From rel. (8.1.4) the following expression can be written:

$$\frac{W dx_i^L}{dW} = x_i^L - x_i^V \tag{8.1.5}$$

Rearranging rel. (8.1.5) and integrating it for the initial state of the liquid (1) and for the final state, after the distillation is performed (2), the result is the following:

$$\ln \frac{W(1)}{W(2)} = \int_{x_i^L(1)}^{x_i^L(2)} \frac{dx_i^L}{x_i^V - x_i^L} \tag{8.1.6}$$

Using rel. (1.3.17) written in the form:

$$\alpha = \frac{x_i^V / x_i^L}{(1 - x_i^V) / (1 - x_i^L)} \tag{8.1.7}$$

the variable x_i^V can be expressed as a function of α and x_i^L. Assuming that α is constant during distillation, rel. (8.1.6) can be completely integrated giving the formula:

$$\ln \frac{W(1)}{W(2)} = \frac{1}{\alpha - 1} \ln \frac{x_i^L(2)[1 - x_i^L(1)]}{x_i^L(1)[1 - x_i^L(2)]} + \ln \frac{1 - x_i^L(1)}{1 - x_i^L(2)} \tag{8.1.8}$$

The expression given by rel. (8.1.8) relates the initial amount and final amount of undistilled liquid and the initial and final composition of the fraction remaining undistilled. With known values for $W(1)$, $W(2)$, α, and $x_i^L(1)$, this expression allows the calculation of $x_i^L(2)$.

Simple distillation is used when an efficient separation is not required, such as for the separation of simple mixtures with components having large differences between their boiling points, or for obtaining product cuts with defined boiling point ranges (for example, in petroleum products) with the individual fractions further suitable for analysis by GC. Also distillation is frequently used for removing a specific solvent from a solution by boiling of the solution. Together with vaporization of solvents from a diluted sample, simple distillation plays an important role in sample concentration. Even in simple distillation, some applications are not accomplished without problems. Due to the increasing boiling point of solutions as a function of the concentration of the solutes, the initial boiling temperature changes during distillation. For some solutions, the boiling point increases beyond the point where the sample is stable or can be distilled using simple equipment. Other practical problems such as uneven boiling, mainly for samples with higher viscosity, should be avoided for successful use of distillation in sample preparation.

In fractional distillation the column can be viewed as a cascade of single stage distillation units interconnected such that the vapors and liquid phases move in countercurrent (see Section 1.7). In each stage the liquid and vapor phases are in intimate contact, and a fraction of liquid evaporates while a fraction of vapors condenses. The vapors that flow to the upper stage are enriched in more volatile component, and the liquid flowing to the lower stage is enriched in the less volatile component. In the upper stages a distillate is collected, while at the lower stages a bottom fraction is collected. This model can be practically implemented as a batch fractional distillation, as opposed to the continuous fractional distillation where the separation in stages is only a model. Fractional distillation is frequently used for the separation or enrichment of multicomponent solutions. An elaborate theory of efficiency for fractional distillation is available in literature [1]. The theory relates the column efficiency with the amount of reflux, establishes a theoretical plate number for a column in specific reflux conditions, etc. The theoretical plate for distillation is equivalent to the efficiency of a stage with thermodynamic equilibrium between the two phases leaving the stage (vapors and reflux). In practice, stages have less than one theoretical plate. For a column with n theoretical plates, the relation between the composition of distillates and bottoms is given by the expression:

$$(x_i^V / x_A^V)_{Distillate} = \alpha^n (x_i^L / x_A^L)_{Bottom} \qquad (8.1.9)$$

Fractional distillation is used when a more efficient separation is required. Even the separation of hydrocarbons differing in the boiling point with $0.5°$ C can be done using an efficient column (4–4.5 m tall) and a long distillation time (1 week).

A large variety of distillation systems are available for industrial or laboratory uses (see e.g. [2]). Among these are various systems of flasks, condensers, distillation adapters, columns, condensers, column heads, and collecting vials. For example, the flask used to boil the sample can be a simple round bottom flask, a rotating flask that forms a film of liquid for faster evaporation, etc. A number of procedures are used to avoid sudden violent boiling (stirrers, boilerizers). Systems where the distillation flask is stationary but contains inside rotating discs are also utilized. The film that ensures efficient evaporation is formed and continuously renewed on the discs. The columns for fractionation can be a simple empty tube, a tube with indentations (Vigreux column), a tube packed with an inert material (glass, ceramics, etc.) shaped in spirals, helixes, spheres, etc., a column with a spinning band that rotates inside, etc. The purpose of the packing material is to provide adequate contact between the rising vapors and the returning condensed liquid. Special systems for controlling the amount of reflux are also available. Also, condensers and systems allowing the collection of different fractions without interrupting the distillation are common. Although for heat-stable compounds, atmospheric pressure distillation is usually applied, some artifacts may be formed during distillation. These may be generated due to reactions between sample components, between the solvent and some of the analytes, or by oxidation reactions. In the flavor industry where distillation is frequently utilized for the isolation of flavor compounds (see e.g. [3]), the formation of artifacts during distillation has been a subject of considerable interest. For example, the comparison of cold-pressed and distilled lime oils shows that distillation causes the almost total disappearance of α-phellandrene and α- and β-terpineols [4], [5]. Artifact formation has been noticed mainly during the distillation of aldehydes, unsaturated fatty acids, and terpenes.

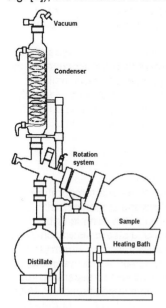

FIGURE 8.1.3. *Diagram of a distillation system using vacuum and rotation of the sample flask.*

In order to avoid higher temperatures required for the distillation of compounds with high boiling points or to avoid possible decomposition at atmospheric pressure, distillation under reduced pressure (vacuum distillation) is quite common. Boiling under vacuum takes place at lower temperatures than the standard boiling point of each compound. This makes vacuum distillation a common procedure in sample preparation, mainly for concentration purposes. A schematic drawing of a distillation system using vacuum with simultaneous rotation of the sample flask is shown in Figure 8.1.3. The separation factor α increases as absolute pressure decreases, and vacuum distillation may also provide better separation than conventional distillation. The pressure used for vacuum distillation may vary between values close to atmospheric pressure to as low as 1 Torr. Typically values between 20 to 200 Torr are used.

- *Azeotropic distillation*

Azeotropic distillation uses the addition of a solvent capable of forming an azeotrope with one of the sample components that must be separated (azeotropes of ternary type are also known). Azeotropes are solutions having identical composition of the liquid and vapor phase at the boiling temperature. In the majority of cases the azeotropes boil at a temperature lower than that of the components (minimum-boiling azeotrope), and therefore most of the added substance distills and is found in the distillate, while the other components remain as bottom product. Azeotropic distillation can be used as a simple distillation but more commonly is used as a fractional distillation.

Azeotropic distillation can be used for the separation of binary mixtures of close-boiling point components. By the addition of a particular solvent, it is possible to produce azeotropes with sufficiently different boiling points so that in a distillation process one azeotrope will be the overhead and the other the bottom product. An example is the separation of a mixture of aromatic and paraffinic hydrocarbons by addition of ethyl alcohol. The alcohol forms an azeotrope with both hydrocarbon types, but these are sufficiently different in boiling point that they can be separated by fractional distillation. Further separation of the components of an azeotrope cannot be done by distillation, but it is possible that the two components of the azeotrope are not miscible as liquids and form two liquid layers upon condensation.

In the selection of a compound to form azeotropes (azeotrope agent) with a sample component, it is desirable to have a lower boiling point compared to the sample components, it should show a positive deviation from ideal, and it should be separable from the component forming the azeotrope. The composition of the product changes as the distillation proceeds when the distillation of more than one azeotrope takes place. Azeotrope formation can be used for different purposes in sample preparation, typically for the elimination of a solvent (such as water) that is present as a matrix or is added for different purposes during sample preparation and must be eliminated.

- Steam distillation

Steam distillation is a technique allowing the separation of mixtures at a temperature lower than the normal boiling points of their constituents. The vaporization of the mixture is achieved either by continuously blowing steam through the sample or by boiling water and the sample together. A schematic diagram of a steam distillation apparatus with an external source of steam is shown in Figure 8.1.4. The steam is injected continuously into the bottom of the flask, assuring good contact between the sample and the steam. The temperature of the sample can be adjusted separately (by heating) or may be heated only by the passing steam when some condensation of water may occur in the sample flask. The mixed steam and vaporized material is condensed similarly to conventional distillation. Typically, the liquid distillate and the water are not miscible and separate in distinct layers. Nonvolatile impurities remain in the distilling flask.

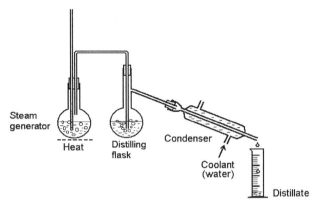

FIGURE 8.1.4. *Diagram of a steam distillation apparatus with external steam generation.*

The steam passing through the sample carries the material as vapors in the molecular ratio of the partial pressures, up to a maximum partial pressure equal to the vapor pressure of the material at the operating temperature. Not only steam can be used in this technique, but also any stream of gas (carrier) when at the same time the sample is heated to a temperature that does not affect it negatively. If equilibrium is established between the steam and the vaporized compounds, their partial pressure equals their vapor pressure. For a binary mixture of steam "w" and an organic compound "i", the amount of condensate of "i" can be calculated from the relation:

$$F_i = F_w \frac{p_i M_i}{p_w M_w} \qquad (8.1.10)$$

where F_i is the mass flow rate of "i", F_w the mass flow rate of steam, p_i the partial vapor pressure of "i", p_w the partial vapor pressure of water (or other carrier), M_i the molecular weight of the compound "i", and M_w the molecular weight of water (or other carrier). It is possible to adjust the mass flow rate of steam F_w and by changing the temperature to modify p_i and p_w. These parameters allow different steam distillation regimes. As the value of p_i depends on the molar fraction of "i" in the solution (see Henry's law rel. 1.4.21), the mass flow rate of the analyte decreases during the steam distillation process. The recovery of the organic volatile compounds can be increased by simultaneous gas sparging, by addition of a large amount of an inorganic salt (salting-out effect), pH adjustment, the addition of a co-distiller such as benzene or toluene, or performance of the operation at a reduced pressure. An approximation of the recovery (see rel. 3.4.1) of an analyte during steam distillation can be obtained using the expression:

$$R = [1 - \exp(-\frac{F_w K_w \gamma_{i,w} t}{W})] \qquad (8.1.11)$$

where K_w is the ratio of the concentration of the analyte in the condensed vapors to the concentration of the solution, $\gamma_{i,w}$ the activity coefficient of the analyte in water, t the time, and W the volume of the water + sample.

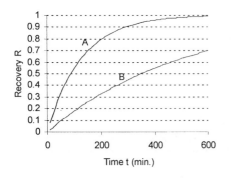

FIGURE 8.1.5. *Recovery in a steam distillation system with F_w = 10 mL/min, W = 500 mL, K_w = 0.4 (curve A), and K_w = 0.1 (curve B).*

The recovery for a hypothetical system at two K_w values is shown in Figure 8.1.5. Simultaneous distillation and extraction (SDE) is another special steam distillation technique with frequent applications in sample preparation. The subject is discussed in more detail in Section 10.5.

A significant number of applications of steam distillation as a sample preparation for chromatography are reported in literature. Some of these applications are associated with steam distillation and simultaneous extraction, and others use steam distillation followed by additional cleanup of the distillate using operations such as solid phase extraction (SPE). Other methods extract the distillate with an appropriate solvent and analyze it by GC or HPLC. These procedures are applied successfully mainly when the analytes are significantly different from the matrix, which represents a large part of the sample. This is the case of pesticides or other pollutants in water samples, soil, sediments, food samples, plant materials, etc.

Steam distillation has been applied, for example, for the analysis of organochlorine pesticides in vegetables [6]. Macerated vegetables were mixed with water and boiled vigorously for 1–9 h at a reflux rate of 6 ml/min., and the distillate was collected in iso-octane. When the extraction was completed, the organic phase was separated and dried over anhydrous Na_2SO_4, pentachlorobenzene was added (internal standard), and the solution was diluted with H_2O. A portion of the resulting solution was analyzed by GC-ECD. Sixteen pesticides were studied, and the recoveries ranged from 80–100% for γ-HCH, α-HCH, aldrin, heptachlor epoxide, endosulfan I, dieldrin, 4,4'-DDE, 4,4'-DDD and 4,4-DDT and 30–60% for five other pesticides. The recovery of the water soluble pesticide aldrin aldehyde was about 20%. The procedure was applied for the analysis of tomatos, potatos, bananas, pineapples, and cocoa beans.

Another example is the analysis of residues of organochlorine pesticides, PCBs, and chlorophenols in fish using steam distillation followed by further fractionation on silica gel and analysis by GC-ECD [7]. Other analyses include that of γ-hexachlorcyclo-hexane, heptachlor, aldrin, heptachlor epoxide, and dieldrin from slices of sugar beet. This can be done using steam distillation followed by the extraction of the distillate with hexane and analysis by GC-ECD. The recoveries of the procedure, except for heptachlor, are from 80 to 96%, and the relative standard deviation (RSD) is less than 4% [8]. The recoveries of aldrin, o,p'-DDT and p,p'-DDT can be enhanced by addition of a solution of oxalic acid with the sample during the steam distillation operation [9]. Various PCBs can be analyzed in the 0.1 to 10 ppb range with recoveries from 72 to 104% in soil and food samples [10]. The determination of the herbicides frenock (sodium 2,2,3,3-tetrafluoropropanoate) and dalapon (sodium 2,2-dichloropropanoate) in soil can be done using steam distillation followed by GC-MS analysis with recoveries higher than 92% and RSD of less than 5% [11].

The determination of chlorophenol levels in urine can be performed using a sample preparation starting with a steam distillation. The urine is acidified with concentrated H_2SO_4 and subjected to steam distillation. The distillate is collected in HCl, and the solution is further passed through a cation-exchange column and a reversed-phase column (RP-18) connected in series. The chlorophenols are eluted with CH_2Cl_2 / heptane and derivatized with pentafluorobenzoylchloride. The derivatives are analyzed by GC-MS [12]. Phenols from various aqueous and solid samples can be first separated by steam distillation, followed by a cleanup step on a XAD-4 column and analyzed by HPLC [13]. The reported detection limits are 0.1–0.4 ng. The recoveries of 4-chlorophenol, 4-chloro-2-methylphenol and 2,4-dichlorophenol are 78, 99, and 102%, respectively; but the recovery for 4-nitrophenol is poor.

Several essential oils also have been analyzed using steam distillation in the sample preparation step. For example, the essential oils from Colombian ylang-ylang are extracted from flowers by steam distillation in a NaCl solution, followed by extraction of the condensate with ethyl ether, drying with Na_2SO_4, and separation of the solvent by rotary evaporation. The extracts are analyzed by GC-FID [14]. The essential oil of *Origanum majorana* can be isolated from air-dried plant material by steam distillation for 5 h. The oil is extracted with pentane, and the extract is analyzed by GC-FID on a Carbowax [15]. Similarly, the analysis of the essential oils in *Illicium* fruit [16], certain flavor components in apples [17], or even of flour samples [18] have been done using steam distillation in the sample preparation step.

Some other analyses not related to pesticides or flavor analysis have been reported to use steam distillation in the sample preparation step, such as the determination of volatile fatty acids in anaerobic waste digesters [19] or the determination of N,α-dimethyl-N-2-propynylbenzene-ethanamine (selegiline) in plasma [20].

- Other distillation procedures

Some other types of distillation are applied in practice. A special type of distillation is molecular distillation, which is used for the separation of heat-sensitive, high molecular weight, and low-volatility compounds. Thermal decomposition can be avoided or reduced to a minimum using this technique. Molecular distillation can be considered a version of vacuum distillation that takes place at pressures lower than 10^{-3} Torr. The equipment used for this operation is designed such that the distance between the evaporation and condensation surfaces is of the same order of magnitude as the mean path of the molecules. The free motion of the molecules is not hindered mechanically between the two surfaces. The liquids are evaporated without boiling in this technique, where the temperatures of the evaporation surface and condensing surfaces are strictly controlled [21], [22]. The evaporation is done using various systems including falling-film stills, wiped-film stills, centrifugal stills, etc. Among other types of distillation is flash distillation, which uses an instantaneous vaporization of a specific light fraction of the sample.

The utilization of distillation in laboratory practice is widespread, mainly related to preparative procedures. For sample preparation in chromatography, the separation using distillation is not extensively utilized. Distillation uses large amounts of sample, while in chromatographic separation the amount of sample tends to be small. Also, the separation by distillation is not always complete, and some of the analyte may be lost or not efficiently recovered. However, distillation is still used as a sample preparation technique in fields such as essential oils [23], petroleum products [24], [25], and soil analysis.

- Vaporization

Vaporization is also a process of changing the physical state of a substance in vapor phase and is one of a number of techniques involving phase equilibria. The vaporization (or evaporation) term typically refers to changing the state into vapor phase below boiling temperature. However, the term also can indicate the operation of eliminating a solvent, regardless the temperature utilized. Vaporization is a common sample preparation technique used for sample concentration or when the solvent of a processed sample must be eliminated or changed (see Section 3.6). The reduction of the sample volume by elimination of most or all of the solvent results in the concentration of the analytes.

The vaporization of a liquid is associated with a significant increase in the distance between molecules and decrease in their interactions. The average molar enthalpy of vaporization ΔH_{vap} for nonpolar compounds is given by the empirical Hildebrand's rule $\Delta H_{vap} \approx 11 \ RT_{vap}$ [26], where T_{vap} is the vaporization temperature. Also, Clausius-Clapeyron equation (see Section 1.3) provides a means for the estimation of ΔH_{vap}. Vaporization taking place with absorption of energy requires external heating or occurs with a decrease in temperature. The variation in temperature can be estimated based on Clausius-Clapeyron formula:

$$\frac{1}{T_2} - \frac{1}{T_1} = -R \ \frac{\ln(p_2^\circ / p_1^\circ)}{\Delta H} \tag{8.1.11}$$

When a liquid reaches equilibrium with its vapors (see Section 1.3), the net transfer of material stops. The evaporation can be continued only by removing the vapors present above the liquid. This is usually done using a stream of gas. This gas can be air or, depending on the sample requirement, can be a stream of nitrogen, dry air, dry nitrogen, etc. Special equipment is available for evaporation in sample preparation (see e.g. [27] using a helical gas jet to remove the vapors above the sample. The diagram of such an instrument is shown in Figure 8.1.6. Automation can be added to control the sample temperature, the level of liquid left in the vial (optical sensor), etc. The vaporization of the solvent from processed samples at different stages in the sample preparation process is a very common operation (as can be seen in Figure 1.2.1). The process can be done at room temperature (with mild heating) or even at more elevated temperatures. Vaporization from a precise starting volume ending with a precise final volume is sometimes applied in sample preparation. However, the addition of an internal standard may be used for the calculation of the concentration ratio generated by the elimination of some of the solvent and necessary in quantitative measurements. Because the control of the final volume during the evaporation is sometimes imprecise, complete evaporation of the solvent is sometimes preferred, followed by the reconstitution of the sample in a precise smaller volume of solvent.

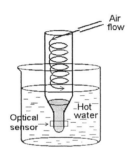

FIGURE 8.1.6. *Diagram of an evaporative concentrator.*

A common apparatus used in the laboratory for the concentration of small amounts of sample dissolved in a volatile organic solvent is Kuderna Danish (KD) evaporative concentrator. The apparatus consists of a pear-shaped flask containing a lower vial, adapted with a condenser and placed over a vigorously boiling water bath, as shown in Figure 8.1.7. The flask is filled with the volatile solution of the sample 30-40% of its volume. The level of the boiling water in the bath is maintained below the joint of the concentrator tube and the flask. The final concentrated sample remains in the lower tube, and the volume of the sample usually can be measured. The upper part of the

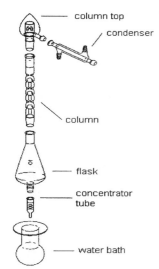

- column top
- condenser
- column
- flask
- concentrator tube
- water bath

FIGURE 8.1.7. *Diagram of a Kuderna Danish evaporative concentrator.*

apparatus is used to recover the solvent. The temperature used in the KD concentrator is not necessarily below the boiling point of the solvent. The vaporization process can be used in various other ways to eliminate the solvent and concentrate the sample (see e.g. [28]). Some precautions must be taken during solvent evaporation, including the use of temperatures that do not decompose the sample, the verification that the analytes are not evaporated together with the solvent, etc. The use of evaporation in an inert atmosphere is frequently used to avoid oxidations that may occur in the process. Large volume injection systems are available for gas chromatographs that are designed to accommodate a large sample (100–200 μL) and eliminate a solvent evaporation operation (see Section 4.2). Besides evaporation at atmospheric pressure, evaporative systems using vacuum are utilized, mainly for the evaporation of solvents with low vapor pressure or for sensitive samples that must avoid excessive temperatures.

The vaporization may take place directly from a solid, and the process is known as sublimation. Sublimation can be used as a sample preparation for specific compounds such as iodine.

- Drying

Sample drying is a vaporization process where the compound eliminated from the sample is water. Sample drying can be applied to gaseous, liquid, or solid samples and is the subject of numerous studies, including both industrial and laboratory applications [29]. Many quantitative results are reported on a dry weight basis (DWB) for materials that contain water. This assumes a thorough elimination of the water present in the sample, followed by a correct measurement of the weight of material (which may be hygroscopic).

For gases, a typical procedure for water vapor removal is the adsorption on solid desiccants. The choice of the desiccant is done such that it can be easily handled, stored, and possibly regenerated. It is also useful to have an indicator in the desiccant, which can change its color when the desiccant cannot adsorb more water. Among the criteria for choosing a desiccant are the relative humidity (RH) produced by the desiccant, stability in time, temperature dependency, capacity, possibility of reaction with moisture giving rise to other gases, etc. The most used drying adsorbents are probably silica gel, calcium sulfate, and anhydrous magnesium perchlorate. Several characteristics of these materials and of other water adsorbents are listed in Table 8.1.1.

TABLE 8.1.1. *Properties of different desiccants used in a drying gases* [30].

Desiccant	Initial composition	Regeneration requirements		Efficiency[1] (mg/L)	Relative capacity[2] (L)
		Drying time (hours)	Temp. (°C)		
calcium oxide	CaO	6	500, 900	0.656	51
magnesium oxide	MgO	6	800	0.753	22
phosphorus pentoxide	P_2O_5	-	-	$3.5 \cdot 10^{-3}$	566
activated alumina	Al_2O_3	6 - 8	175, 400	$2.9 \cdot 10^{-3}$	263
silica gel	SiO_2	12	116 - 127	0.07	317
molecular sieve 5A	$Me_2(I)O \cdot Me(II)O \cdot Al_2O_3 \cdot n\ SiO_2$	-	-	$3.9 \cdot 10^{-3}$	215
barium oxide	96.2% BaO	-	1000	$2.8 \cdot 10^{-3}$	244
ascarite	91% NaOH	-	-	0.093	44
anhydrous barium perchlorate	$Ba(ClO_4)_2$	16	127	0.599	28
anhydrous magnesium perchlorate	$Mg(ClO_4)_2 \cdot 0.12\ H_2O$	48	245	$2 \cdot 10^{-4}$	1169
anhydrone	$Mg(ClO_4)_2 \cdot 1.48\ H_2O$	-	240	$1.5 \cdot 10^{-3}$	1157
anhydrous lithium perchlorate	$LiClO_4$	12	70, 100	0.013	267
anhydrous calcium chloride	$CaCl_2 \cdot 0.13\ H_2O$	16	127	0.067	232
Drierite	$CaSO_4 \cdot 0.02\ H_2O$	1 - 2	200 - 225	0.067	232
anhydrocel	$CaSO_4 \cdot 0.21\ H_2O$	1 - 2	200 - 225	0.207	683

[1] The average amount of water remaining in a N_2 stream after it was dried to equilibrium.
[2] The average maximum volume of N_2 dried at a specified efficiency for a given volume of desiccant.

Drying gases by condensation (cooling) is another good alternative. The procedure is quite simple since it requires only a low temperature to condense the water vapors. The gas sample is passed through a vessel at a temperature below the dew point where the water vapors condensate on the inner walls and thus is removed from the gas stream. The vessel can be put into a cold bath. Different temperatures of the bath are obtained using specific cooling materials such as ice and water (0° C), dry ice and a solvent (-78.5° C), organic solvent slashes (-23° C–160° C depending on the solvent), liquid nitrogen (-196° C), etc.

Liquid samples also require drying in many sample preparation procedures. The drying of the liquid samples is frequently done using drying materials such as inorganic salts that can adsorb water (Na_2SO_4, $MgSO_4$, $Mg(ClO_4)_2$, etc.), special molecular sieves of various dimensions and properties (molecular sieve 3A to 5A, 13X, AW-300, AW-500), or some chemicals that can react with water. Common compounds reacting with water are metallic Na, CaC_2, $(CH_3)_2C(OCH_3)_2$ (2,2-dimethoxypropane or acetone dimethyl acetal), $SOCl_2$ (thionyl chloride), etc. It is important that the compound reacting with water does not affect unfavorably other properties of the sample. The drying of liquid samples can be done by addition of the drying agent in the liquid followed by filtration or by passing the liquid through filters or a small column containing the desiccant. Even cellulose filters can be used for drying, these adsorbing water and allowing the organic liquids to pass with little adsorption. Other techniques for drying liquid samples are also known. One of these is distillation of water or of the liquid sample. In some cases it is also possible to find an azeotrope agent for water and distill the water after its addition.

Solid samples such as plant materials and also sludges, gels, and animal or human tissues may require drying. These samples can be dried using several procedures such as evaporation with heat, evaporation under a stream of gas (such as dry N_2), exposure to a dry atmosphere, vacuum, lyophilization (freeze-drying) [31], critical point drying [32],

 or replacement of water with other solvents. The dry atmosphere can be generated in a confined space by the presence of a chemical desiccant such as $MgSO_4$, $CaSO_4$, $CaCl_2$, $Mg(ClO_4)_2$, Na_2SO_4, or other drying agents, some indicated in Table 2.4.2 [33]. The confined space is typically obtained in a desiccator (see Figure 8.1.8). Special attention should be given to the possibility of interaction of volatiles from the sample with the desiccant. Another common way of drying solids (and also liquids) is by using heating in an oven.

FIGURE 8.1.8. *Common glass desiccator.*

Various types of ovens are commercially available (see e.g. [34]), including ovens with gravity convection, mechanical convection, vacuum, temperature programming, etc. The ovens are used for drying of thermally stable compounds at temperatures between 105° C and 135° C and exposure times from 30 min. to several hours, depending on the nature of the sample and the tendency to adsorb water. It is important during drying in an oven to ensure that the sample does not undergo unwanted chemical changes and does not lose volatile compounds of interest. Drying with the loss of volatiles can be intentionally applied when the volatiles are not analytes. An alternative to drying is the use of controlled moisture content of the sample. Lower temperatures may also be used for drying in an oven. However, temperatures above 60° C favor the acceleration of reactions such as Maillard condensation, which may occur in food or plant materials. Samples sensitive to temperatures above 50–60° C must by dried using other techniques.

Freeze-drying is another common procedure for water removal. It has a number of advantages including keeping the sample at low temperatures and almost complete removal of water. Freeze-drying is accomplished by freezing the material to be dried below its melting point and then providing part of the latent heat of sublimation, usually under reduced pressure. The transformation of water into vapors adsorbing the latent heat of sublimation is usually sufficient to maintain the samples in frozen state. Precise control of heating and of the vacuum allows drying from frozen state without melting (which must be avoided). Water solutions must be sufficiently diluted (about 10% for sugars) to obtain a good dried sample after freeze-drying and to avoid formation of sticky liquids.

8.2 DISSOLUTION, CRYSTALLIZATION, AND PRECIPITATION

Dissolution, crystallization, and precipitation can be used in separations for various purposes, including sample preparation for chromatography. However, these techniques have multiple other applications. Dissolution, crystallization, and precipitation are simple operations, and they are frequently included in other more complex sample preparation processes.

- Dissolution

Dissolution is a generic term for the process of taking a solid, a gas, or even another liquid in solution (the *solution* is a phase containing more than one component). The term dissolution is applied more frequently for the formation of liquid solutions. However, gases are miscible in all proportions, and they may be seen as forming solutions. Also, solid crystalline solutions can be formed when the ions or atoms in a crystal are replaced with ions or atoms of a different kind. Solutions of noncrystaline solids are also known, and their structure can be complicated (such as the structure of glass), the subject being presented in the dedicated literature [35].

The process of dissolution may involve physical interactions or chemical reactions. For example, the dissolution of solid naphthalene in hexane does not involve chemical changes and is equivalent with the formation of a liquid homogeneous mixture of the two types of molecules. The process of solution formation can be much more complicated than it appears, and various types of interactions can take place between the liquid (solvent) and the solute. For example, the solutes can undergo electrolytic dissociation upon dissolution in water or other highly polar solvents, and this leads to the differentiation of solutes into electrolytes and nonelectrolytes. For this reason, dissolution in water is typically treated differently from that in organic solvents. The term dissolution is also utilized for some chemical reactions. In these reactions an insoluble substance is transformed into a soluble compound, which is simultaneously transferred into a solution, such as the dissolution of a metal in an acid. This type of dissolution is in fact a combination of two very different processes. The dissolution of gases depends on the nature of the gas and the solvent, on temperature, and on the pressure of the gas. The dissolution of gases in a liquid can be viewed as a change in physical state from gas to liquid and follows the principles discussed in Section 1.3 (see also [36]).

A solution where equilibrium exists between the solid substance and its dissolved form is a *saturated solution*. The equilibrium can be written as follows:

$$A_{solid} \rightleftharpoons A_{solution}$$

and considering $[A_{solid}] = 1$, the equilibrium constant can be written in the form:

$$[A_{solution}] = K \qquad\qquad (8.2.1)$$

where [A] indicates molar concentration. Expression (8.2.1) shows that the molar concentration of a saturated solution is constant (for a constant temperature). In a saturated solution in the presence of solid, by addition of solvent more solid dissolves (until the solution is not saturated anymore). The elimination of solvent leads to the formation of more solid. The *solubility* is defined as the maximum amount of solute that dissolves in a fixed volume of solvent at a given temperature and can be expressed as moles / L of solute, mole fraction, etc. The solubility is dependent on the interactions between solute molecules and the surrounding medium. The simple principle of "'like-to-like" describes in the best way the possibility of dissolving the compounds in liquid solvents. The presence of other compounds may influence the solubility either through other interactions or even through chemical reactions.

The dissolution process of nonelectrolytes can be viewed hypothetically as formed from two steps. The first step is melting of the compound "i" to form a supercooled liquid (see Section 1.3). The second step is the mixing of this liquid with the solvent "A". For a transformation at constant pressure and temperature at equilibrium, the free enthalpy $\Delta G = 0$, and this holds true for melting at fusion temperature T_f. Therefore, the free enthalpy of melting (fusion) is $\Delta G_f = 0$ and $\Delta H_f = T_f \Delta S_f$, where ΔH_f is the heat of fusion and ΔS_f the entropy of fusion. At a given temperature T, the free energy of melting will be given by the expression:

$$\Delta G_f = \Delta H_f - T\,\Delta S_f = \Delta H_f - (T/T_f)\,\Delta H_f \qquad (8.2.2)$$

The free energy of mixing for ideal solutions does not take place with any heat change, and therefore

$$\Delta G_m = -\,T\,\Delta S_m \qquad (8.2.3)$$

The variation in the entropy of mixing ΔS_m of ideal solutions is given by the expression [26]:

$$\Delta S_m = -\,R \sum_j n_j \ln x_j \quad (j = \text{each component}) \qquad (8.2.4)$$

For the dissolution process, it can be assumed that $x_A = 1$ for the solvent, and therefore $\Delta G_m = -\,RT\,n_i \ln x_i$, (the index A from $x_{i,A}$ and $n_{i,A}$ is omitted), or for $n_i = 1$ the expression for ΔG_m becomes

$$\Delta G_m = -\,RT \ln x_i \qquad (8.2.5)$$

The total free enthalpy of dissolution is $\Delta G = \Delta G_f + \Delta G_m$, and at the equilibrium obeys the condition $\Delta G = 0$. The addition of rel. (8.2.2) and (8.2.3) leads to the expression $(1 - T/T_f)\,\Delta H_f + RT \ln x_i = 0$, which can be rearranged in the form:

$$\ln x_i = [\Delta H_f (1/T_f - 1/T)]\,/\,R \qquad (8.2.6)$$

Rel. (8.2.6) gives the expression for solubility (expressed as the maximum mole fraction) for the formation of an ideal solution. As shown by rel. (8.2.6), the solubility increases with the temperature T. This is true for many compounds including both nonelectrolytes and electrolytes. However, there are exceptions such as for the solubility of $CaSO_4$ and $Ce_2(SO_4)_3$ in water.

Changing in rel. (8.2.6) the mole fraction x_i into molar concentration c_i (using rel. (1.3.21)) and also including an activity coefficient γ_i to correct for the deviation from ideal solutions, rel. (8.2.6) can be written in the form [37]:

$$\ln [\gamma_i\, c_i\, M_A\, /\, (1000\,\rho_A)] = \frac{\Delta H_f}{RT} \frac{T - T_f}{T_f} \qquad (8.2.7)$$

From rel. (8.2.7), the solubility for nonelectrolytes can be estimated based on solubility parameters δ, which are introduced through rel. (1.4.15) expressing γ_i. Rearranging rel. (8.2.7) and using rel (1.4.15), the expression for ln c_i is given by the formula:

$$\ln c_i = \frac{\Delta H_f}{RT} \frac{T - T_f}{T_f} - \frac{V_i}{RT} (\delta_i - \delta_A)^2 + \ln \frac{1000 \, \rho_A}{M_A}$$

(8.2.8)

Considering $\Delta H_f \approx T_f \Delta S_f$, and taking $\Delta S_f \approx 13$ cal mole^{-1} deg^{-1} (the fusion entropy being relatively constant for many compounds), rel. (8.2.8) can be estimated by the formula:

$$\ln c_i = 6.54 \frac{T - T_f}{T} - \frac{V_i}{RT} (\delta_i - \delta_A)^2 + \ln \frac{1000 \, \rho_A}{M_A}$$

(8.2.9)

The molar volumes V_i, and the solubility parameters δ are given in Table 8.2.1 for a number of compounds [38], [39], [40], [41].

TABLE 8.2.1 *Molar volumes and solubility parameters for some common compounds.*

Compound	V (in mL)	δ	Compound	V (in mL)	δ
perfluoroalkanes		6.0	chlorobenzene	102	9.6
CFCl$_2$-CF$_3$		6.2	anisole		9.7
isooctane	165.1	7.0	1,2-dichloroethane	64.4	9.7
isopropyl ether	142.0	7.0	methyl benzoate		9.8
diisopropyl ether		7.0	dioxane		9.8
n-pentane		7.1	methyl iodide		9.9
CCl$_3$-CF$_3$		7.!	bromobenzene	105.3	9.9
n-hexane	131.1	7.3	CS$_2$	60.6	10.0
n-heptane		7.4	propanol	75.1	10.2
diethyl ether	104.4	7.4	octanol	158.0	10.3
triethylamine		7.5	cyclohexanone	103.8	10.4
cyclopentane		8.1	pyridine	80.6	10.4
cyclohexane	108.4	8.2	benzonitrile	103.3	10.7
propyl chloride		8.3	nitromethane		11.0
CCl$_4$	96.9	8.6	nitrobenzene		11.1
diethyl sulfide		8.6	ethanol	58.6	11.2
ethyl acetate	98.1	8.6	phenol		11.4
propyl amine		8.7	dimethylformamide	77.3	11.5
ethyl bromide	75	8.8	acetonitrile	52.7	11.8
p-xylene	132.6	8.8	methylene iodide		11.9
rn-xylene		8.8	acetic acid	71.3	12.4
toluene	106.6	8.9	dimethylsulfoxide	70.9	12.8
CHCl$_3$	80.4	9.1	methanol	40.6	12.9
tetrahydrofuran		9.9	1,3-dicyanopropane		13.0
methyl acetate		9.2	propylene carbonate		13.3
benzene	89.2	9.2	ethanolamine		13.5
perchloroethylene		9.3	ethylene glycol	55.8	14.7
acetone	73.8	9.4	formamide	39.7	17.9
methyl ethyl ketone	89.9	9.5	water	18.02	21
CH$_2$Cl$_2$	64.4	9.6			

From rel. (8.2.9) the variation in ln c_i for three different compounds as a function of the value δ_A for the solvent is shown in Figure 8.2.1. The solution temperature was chosen 25° C, $M_A = 100$, and $\rho_A = 1$. Compound A was chosen with $\delta_i = 7$, $V_i = 400$ and $T_f = 380\,°K$, compound B was chosen with $\delta_i = 11$, $V_i = 200$ and $T_f = 400\,°K$, and compound C was chosen with $\delta_i = 14$, $V_i = 300$ and $T_f = 500\,°K$. The figure shows that the maximum solubility is obtained when the solubility parameter of the solvent is equal to that of the solute, which proves the principle of "'like-to-like". Compounds with higher molar volume have a narrower range of solubility, and the increase in the fusion temperature diminishes the solubility. At the same time, the increase in ambient temperature increases the solubility. Also, rel (8.2.9) and Figure 8.2.1 show that considerable solubility can extend for a relatively wide range of values for δ_A. This observation is equally applicable when considering the variation of ln c as a function of δ_i. The solubility of a range of compounds in a solvent with a given δ_A value shows similar curves as in Figure 8.2.1. Therefore, in a given solvent are soluble compounds with a wide range of δ_i values. This result shows that solubilization is not an efficient procedure for separation.

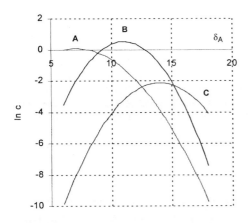

FIGURE 8.2.1. *Diagram showing solubility variation as a function of solvent parameter δ.* (See text for various choices of other parameters in rel. (8.2.9) to generate curves A, B and C).

Another parameter used for solvent characterization is the polarity P' discussed in Sections 1.10 and 4.3 (see Tables 1.10.11 through 1.10.19) [41]. A positive correlation between the values of parameter δ and the polarity P' is shown in Figure 8.2.2. The classification of solvents based on the parameter P' takes into consideration the type of interactions likely to be exhibited in the solvent. Based on the type of molecular interactions, solvents can be classified in several groups. An initial classification distinguishes nine groups (see Section 1.10). For practical purposes a less detailed classification distinguishes three groups. The first group is that of solvents with "nonlocalized" interactions including nonpolar to intermediate polarity solvents. These solvents do not have a specific site of the molecule where the interactions take place. To this first group belong solvents such as benzene,

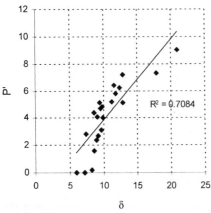

FIGURE 8.2.2. *The variation of P' as a function of δ for a number of solvents.*

chloroform, and methylene chloride. The second group includes dipolar solvents with interactions that are localized but nonsolvent specific. This group contains solvents such as acetone, acetonitrile, ether, etc. The third group is that of basic solvents with localized interactions that are solvent specific. Among these are dimethylsulfoxide, 2-propanol, water, etc. The "like-to-like" principle is well verified for solvents in each group. Nonpolar compounds are better dissolved in solvents of the first group, polar compounds that do not have specific interacting groups are better dissolved in those from the second group, and the compounds with polar groups able to form hydrogen bonds or to participate in acid/base interactions are better dissolved in the third.

An expansion of the concept of solubility parameters δ also has been done [42] to take into account the type of interactions in the solvent , such as dispersion, (d), dipole-dipole orientation (o), proton donor interactions (a), or proton acceptor interactions (b). For each type of solvent having predominantly one type of interaction, four different δ parameters are assigned [38], these describing more accurately the solvent properties.

For the evaluation of the solubility of nonelectrolytes in water, another estimating procedure is to use the correlation between the molar solubility S in pure water and the water/octanol partition coefficient K_{ow}. The values of K_{ow} can be obtained experimentally and are tabulated, or they can be estimated semiempirically (see Section 1.10). The solubility is related to K_{ow} by rel. (1.10.11), given here again:

$$\log (1 / S) = a \log K_{ow} + b \qquad (8.2.10)$$

where a and b are empirical parameters (see Table 1.10.10).

The solubility of an organic compound in water is typically decreased in the presence of large concentrations of inorganic compounds present in the same solution. The decrease is given by the expression [43]:

$$\log (S / S^*) = K_s \, c_s \qquad (8.2.11)$$

where S^* is the molar solubility in a salt solution, K_s is an empirical salting parameter, and c_s is the molar salt concentration. This effect of decreasing the solubility is frequently applied as "salting out effect" for increasing the elimination (by extraction or volatilization) of organic compounds from water solutions.

The solubility of sparingly soluble ionic compounds (which typically undergo electrolytic dissociation) is usually approached differently and is obtained based on the equilibrium established when the pure solid ionic salts are put in water. For a salt of the form A_nB_m, this equilibrium can be written as follows:

$$A_nB_m \rightleftharpoons n \, A^{+m} + m \, B^{-n}$$

and the equilibrium constant (taking $[A_nB_m] = 1$) for the reaction is written in the form:

$$[A^{+m}]^n \, [B^{-n}]^m = K_{sp} \qquad (8.2.12)$$

The constant K_{sp} is known as the *solubility product* of the substance A_nB_m. The solubility **S** expressed in moles per 1 L solvent at a given temperature can be obtained from the solubility product using the relation:

$$S = n^{-n} \, m^{-m} \, K_{sp}^{1/(m+n)} \qquad (8.2.13)$$

The solubility of sparingly soluble ionic compounds can be affected by the common ion, pH, chemical complexation, etc.

The dependence of solubility on temperature of nonelectrolytes and also of electrolytes does not follow closely rel. (8.2.6). As an example, the graph of solubility for CH_3COONa in water as a function of temperature is given in Figure 8.2.3.

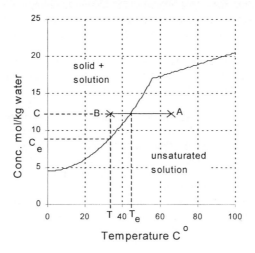

FIGURE 8.2.3. *Solubility of CH_3COONa in water at different temperatures.*

It is possible that the concentration of a solution exceeds the equilibrium value for the amount of dissolved solid, and the solution becomes supersaturated. Supersaturation can be obtained for the solutions of some compounds that have increased solubility for increased temperature by decreasing the temperature. In Figure 8.2.3 point A corresponds to a solution that did not reach the saturation. If temperature is decreased, a temperature T_e is reached and crystals are expected to start forming. However, it is possible that a point B must be reached where the concentration of the solution is above C_e (the equilibrium concentration for the given temperature) for the crystallization to start. The supersaturation can be expressed as the difference $\Delta C = C - C_e$ or as the supercooling temperature $\Delta T = T_e - T$.

The dissolution (and crystallization) process frequently comes to equilibrium slowly, because it takes time to transfer material across the phase boundary. This delay in attaining the equilibrium is a factor to consider in sample preparation using dissolution. Various procedures are used to accelerate the dissolution such as heating, agitation, sonication [44], size reduction of the particles to be dissolved in order to increase the contact surface between the solid and the liquid, etc. Sonication, for example, uses vibrations with frequencies higher than 16 kHz to produce mechanical stress, heating, and cavitation, all contributing to the acceleration of the dissolution process.

The most common type of processed sample introduced into the analytical instrument is in liquid form, but many samples are solids and their analysis requires at least one dissolution operation. For this reason, dissolution is a very common operation, frequently part of a more elaborate process. The typical problem is the selection of the appropriate solvent for dissolution. This selection may need to consider solubilities and sometimes other factors such as the reactivity of the solvent. The solubilities of a large number of compounds are indicated in tables (see e.g. [45]). The preliminary estimation of the solubility in water for nonelectrolytes can be done using estimation procedures, for example based on water/octanol partition coefficient [46], using rel. (8.2.10), etc. Tables with solubility products are available for sparingly soluble electrolytes. The reactivity of a solvent must be considered in relation to possible hydrolysis or interference in further derivatization and also regarding the interaction with the chromatographic separation. Some solvents do not have a desirable behavior in the GC because they are not volatile, decompose, or lead to tailing chromatographic peaks (such as CH_3OH). For HPLC, the solubility of the solvent used for the sample in the liquid mobile phase is critical.

The dissolution process may need in some applications to generate solutions of nonionic compounds in polar solvents or of ionic compounds in nonpolar solvents. As the theory of solubility is previously discussed, this is not a likely possibility. However, there are various procedures attempting to perform this kind of task. One such procedure is to form molecular associations that help the formation of true solution or of colloidal solutions. For example, crown ethers can form complex compounds with inorganic cations (such as K^+ or Na^+) by binding the ion in the center of the crown ether molecule [47] and increase the solubility in nonprotic solvents of various salts of organic acids. Solubilization of anions can be done by replacing the small inorganic cations with large tetraalkylammonium ions. The use of solvent mixtures is another important possibility to enhance solubility. A synergistic effect has been obtained in some cases of solvent mixtures, the solubility being higher for the mixture than for each individual solvent (see rel. 10.1.20). Mixtures of solvents partially aqueous may be used for the dissolution of organic compounds with polar character.

- *Crystallization*

The term *crystallization* is also used with several meanings indicating several processes in which a solid is formed from a solution. This includes a phase change from a melted material into a solid (solidification) or separation of solids from a solution due to the increase in concentration beyond the solubility limit. A short discussion on the formation of crystals from a bicomponent melt has been given in Section 1.3. *Precipitation* is a form of rapid crystallization and may involve a chemical reaction that precedes the deposition of a newly formed solid compound. For example, the addition of a solution of $NaBr$ in a solution of $AgNO_3$ precipitates $AgBr$. The term *precipitation* does not always indicate a chemical reaction. For example, the change in a solvent may lead to the precipitation of a solid without any chemical reaction, such as when alcohol is added in a water solution of glucose leading to the precipitation of the solid substance. In some instances, the precipitation is done with some structural modification of the solute without a chemical reaction. For example, biological samples containing proteins can be treated with 5% CCl_3COOH for the precipitation of proteins. In this reaction, the proteins are degraded without undergoing a precise chemical reaction. Colloid solutions can also form precipitates as a result of the aggregation of colloidal particles into larger ones. Other types of solid formation are also known. Among these are the crystallization following *sublimation* where a solid is transformed directly in its vapor phase and then back to crystals, *chemical vapor deposition* where gaseous components react to form solids, *electrodeposition* where a solid is formed as a result of an electrochemical reaction, etc.

Crystallization from solutions (not including chemical precipitation) can be achieved in several ways, such as a) removal of the solvent, b) change of the solvent by mixing it with other liquids, c) change in the temperature of the solution, d) induction of the precipitation using seeding, or e) modifying the ionic strength of the solution. The crystallization process starts with a nucleation. Nucleation can be homogeneous or heterogeneous. In homogeneous nucleation, a small number of molecules come together to form a submicroscopic solid. To this submicroscopic solid, molecules are added and subtracted randomly, and some of the particles continue to grow. At a certain point, a critical size is reached beyond which the free energy decreases with the size increase due to the reduction of surface energy (relative to the volume). At this point, a nucleus is formed and crystallization continues. The nucleation rate depends on the critical size that is temperature dependent. Heterogeneous nucleation is caused by foreign particles and the walls of the container. Foreign particles are sometimes added intentionally to promote nucleation. These particles can be a different material or crystals of the same compound that is to be crystallized. Nucleation also can be induced using friction, shocks, electric fields, etc. Details about crystal formation can be found in the dedicated literature [48].

Crystallization that allows the formation of a solid material from a liquid can be followed by a mechanical separation of the solid from the liquid. Because only the compound that crystallizes is assumed to have reached the saturation concentration, the crystal formation is also a purification procedure. The compound that forms the crystals is selectively removed from the solution, and the crystals should not contain other compounds present in solution. However, in practice the crystals contain inclusions from the solution, and some of the compound that crystallizes remains in solution. The inclusion formation can be explained by the trapping process. The crystal growth takes place with a mass transfer process from the bulk of solution to the surface of the crystal. Due to this mass transfer, a concentration gradient is present around the crystal, and the edges and corners projecting further into the solution experience a slightly higher solute concentration. This difference makes the growth of the crystal nonuniform, with the corners growing faster than the faces. The new layers formed toward the corners and not at the center of the face may create steps, new faces and the possibility to include solvent or impurities into the crystal. Besides trapping, various adsorption processes also may take place and include impurities into the crystal. Large crystals and amorphous solids have a higher tendency to trap impurities than smaller well defined crystals. Larger crystals also can grow at the expense of small crystals. In this process, depending on the impurities in solution, the larger crystals may have higher purity.

- *Precipitation*

Precipitation from solutions is a type of crystallization that involves a momentary production of an enormous supersaturation [49]. This can be achieved, for example, with chemical reactions that produce an insoluble product. Also, precipitation can be induced by changing the solvent. The precipitation of insoluble salts is frequently applied in inorganic chemical analysis and typically takes place in water. Precipitation of organic compounds is also common. Precipitation is frequently used in sample preparation for the elimination of an undesirable matrix component or for the separation of specific analytes. The formation of crystalline precipitates is desirable because they include less foreign material from the solution. Amorphous precipitates having a large surface may include or adsorb significant amounts of other compounds. For this reason, some cleanup procedures involving precipitation require the redissolving of the precipitate and reprecipitation. The adsorption and trapping in precipitates also can be used to the advantage of concentrating compounds from a diluted solution. The procedure is particularly applied for inorganic compounds.

The precipitation process can be enhanced by the same procedures as crystallization. In addition to chemical reactions that form precipitates, the change in solvent by adding another solvent, or evaporation, the precipitation can be enhanced by decreasing the solution temperature, stirring, addition of a salt, etc.

A typical example of the use of precipitation in sample preparation is the removal of proteins from animal or human plasma samples for the determination of analytes such as metabolites, drugs, toxins, etc. This process is known as deproteinization and is based on the denaturation of the protein. This is done mainly by modifying the tertiary and quaternary structure of the protein. In practice, deproteinization can be performed by adding acids (such as trichloroacetic acid, which is commonly used), bases, inorganic salts, organic compounds such as urea, guanidine, short chain alcohols, acetonitrile, mixtures of organic compounds, heat, etc. The structure change takes place mainly by the alteration of numerous hydrogen bonds, but also may affect ionic interactions, sulfide bonds, etc. In the denaturation process, most proteins having a specific physiological activity, such as enzymes, hormones, antibodies, etc., lose their activities. In most cases, the process is irreversible, and it is associated by small variation of enthalpy (about 0.3 kcal/residue of amino acid mole) and high variation of the entropy of the protein molecules.

Precipitation also can be applied to concentrate the analytes. For trace analysis, the precipitation of the analyte itself is not common. However, for other analyses it can be used as a convenient separation procedure. For example, soluble peptides can be precipitated using alginic acid. The procedure can be applied to separate peptides from a hemofiltrate by addition of alginic acid followed by the separation of the precipitate. This is washed with ethanol and 0.005 M HCl, and then the peptide is eluted with 0.2 M HCl and further analyzed [50].

REFERENCES 8

1. J. H. Perry, *Chemical Engineers' Handbook*, McGraw-Hill, New York, 1963.
2. Chemglass, Catalog 2000, www.chemglass.com.
3. S. Arctander, *Perfume and Flavor Chemicals*, S. Arctander, Montclair, 1982.
4. M. A. Azzous et al., J. Food Sci., 41 (1976) 324.
5. J. A. Pickett et al., Chem. Ind. (London), 13 (1975) 571.
6. M. Parreno et al., J. Trace Microprobe Tech., 11 (1993) 133.
7. A. Hollstein, Nahrung, 35 (1991) 1029.
8. A. G. Ober et al., Bull. Environ. Contam. Toxicol., 38 (1987) 404.
9. Y. I. Chang, M. Sampath, Bull. Environ. Contam. Toxicol., 32 (1984) 657.
10. C. Hemmerling et al., Nahrung, 35 (1991) 711.
11. T. Tsukioka et al., Analyst, 110 (1985) 39.
12. J. Angerer et al., Fresenius' J. Anal. Chem., 342 (1992) 433.
13. Y. Ishii et al., Nippon Noyaku Gakkaishi, 15 (1990) 445.
14. E. Stashenko et al., J. High Resolut. Chromatogr., 16 (1993) 441.
15. M. E. Komaitis et al., Food Chem., 45 (1992) 117.
16. C. S. Yang et al., Zhongguo Yaoxue Zazhi, 27 (1992) 206.
17. I. Yajima et al., Agric. Biol. Chem., 48 (1984) 849.
18. J. A. Ronalds et al., J. Cereal Sci., 21 (984) 25.
19. P. Abbaticchio et al., Environ. Technol. Lett., 4 (1983) 179.
20. Z. Juvancz et al., J. Chromatogr., 286 (1984) 363.
21. P. R. Watt, *Molecular Stills*, Chapman and Hall, London, 1963.
22. J. Hollo et al, *The Application of Molecular Distillation*, Akademiai Kiado, Budapest, 1971.
23. G. Kretchmar, J. Pictet, Chem. Eng. Techn., 29 (1957) 16.
24. D. Severin, Analyst, 117 (1992) 305.
25. W. Holstein et. al., Fresenius' Z. Anal. Chem., 319 (1984) 421.
26. J. H. Hildebrand, R. I. Scott, *The Solubility of Non-electrolytes*, Dover, New York, 1964.
27. Zymark Corp. TurboVap Evaporator Operator's Manual, Hopkinton, 1990.
28. C. F. Poole, S. K. Poole, *Chromatography Today*, Elsevier, Amsterdam, 1991.
29. A. S. Majumdar, T. Kudra, Drying Technol., 14 (1996) 1.
30. M. L. Wilson et al., *Atmospheric Sampling. Student Manual*, US EPA, 1983.
31. G. Durand et al., Chromatographia, 28 (1989) 597.
32. V. Neuhoff (ed.), *Micromethods in Molecular Biology*, Springer, Berlin, 1973.
33. D. R. Gere, E. M. Derrico, LC-GC, 7 (1994) 370.
34. Barnstead/Thermoline Lab. Products Catalog, 2001, www.barnsteadthermoline.com.
35. H. Bach, D. Krause (eds.), *Analysis of the Composition and Structure of Glass and Glass Ceramics*, Springer, Berlin, 1999.
36. G. R. Cysewski, J. M. Prausnitz, Ind. Eng. Chem. Fundam., 15 (1976) 304.
37. J. G. Gmehling et al., Ind. Eng. Chem. Fundam., 17 (1978) 269.
38. B. L. Karger et al., *An Introduction to Separation Science*, J. Wiley, New York, 1973.
39. A. F. Barton, Chem. Rev., 75 (1975) 731.
40. K. Shinoda, *Principles of Solution and Solubility*, M. Dekker, New York, 1978.
41. L. R. Snyder, J. Chromatogr. Sci., 16 (1978) 223.

42. J. J. Kirkland (ed.), *Modern Practice in Liquid Chromatography*, Wiley Interscience, New York, 1971.
43. R. P. Eganhouse, J. A. Calder, Geochim. Cosmochim. Acta, 40 (1976) 555.
44. E. Dimitriu et al., *Ultrasunetele, Posibilitati de Utilizare in Industria Alimentara si Biologie*, Ceres, Bucuresti, 1990.
45. D. R. Linde (ed.), *Handbook of Chemistry and Physics*, CRC Press, Boca Raton, 1990.
46. W. J. Lyman et al., *Handbook of Chemical Property Estimation Methods*, ACS, Washington, 1990.
47. C. D. Pederson, H. K. Frensdorff, Angew. Chem. Int. Ed. Eng., 11 (1972) 16.
48. R. F. Strickland-Constable, *Kinetics and Mechanism of Crystallization*, Academic Press, New York, 1969.
49. A. G. Walton, *The Formation and Properties of Precipitates*, J. Wiley, New York, 1967.
50. R. Richter et al., J. Chromatogr., B, 726 (1999) 25.

CHAPTER 9

Headspace Sampling Techniques in Gas Chromatography

9.1 STATIC HEADSPACE TECHNIQUES

The determination of the nature and quantity of volatile organic compounds (VOCs) and semivolatile organic compounds in various matrices is a common type of chemical analysis. A number of solvent-less procedures that collect the analytes from the headspace of the sample were developed for the analysis of VOCs. These techniques can be classified into two general groups, a) static headspace and b) dynamic headspace. Static headspace (SHS) techniques are those that collect the analytes (and possibly part of the matrix) from a closed vessel where the sample is assumed to be in equilibrium with its vapors at a specific temperature and pressure [1]. Dynamic headspace techniques remove the analytes (and possibly part of the matrix) from the sample in a stream of gas flowing over or through the sample and processes them further for analysis (see Sections 9.2 and 11.2). Headspace type separations of the analytes from the sample significantly simplify the matrix and are common cleanup procedures selected for the analysis of volatile compounds.

The simplest SHS technique consists of collecting an aliquot of the headspace of the sample introduced in a closed vial. The collection can be done using a gas-tight syringe and is followed by injection in a GC instrument for analysis. A number of parameters influence the transfer of analytes from the sample into the headspace, and for a compound "i" in the sample (index "s") and in headspace (index "h"), the following equilibrium occurs:

$$i_s \xrightleftharpoons{K_{hs}} i_h$$

This equilibrium is described by the liquid-to-gas partition coefficient given by the expression:

$$K_{hs} = c_{i,h} / c_{i,s} \tag{9.1.1}$$

where $c_{i,s}$ is the concentration of the analyte in the sample, and $c_{i,h}$ the concentration in the headspace. By comparison of rel. (9.1.1) for the analyte "i" with the expression (1.4.26), which gives for the partition coefficient K_i in gas-to-liquid equilibrium, it can be seen that $K_{hs} = 1 / K_i$. For solid samples, the constant K_i must be replaced with the adsorption constant K^A_i (see rel. 1.4.36), which characterizes the adsorption process at a constant temperature on a gas-solid interface. In this case, the constant K_{hs} is given by $K_{hs} = 1 / K^A_i$. Formally, the transfer between sample and headspace can be studied in the same way, regardless the adsorption/desorption mechanism.

Considering the initial concentration in the sample c_o, the mass balance for the system can be written as follows:

$$c_o V_s = c_h V_h + c_s V_s \tag{9.1.2}$$

where V_s is the volume of the sample and V_h the volume of the headspace. The concentration c_s can be expressed as a function of c_h as follows (neglecting the index "i"):

$$c_s = c_h / K_{hs} \tag{9.1.3}$$

Substituting rel. (9.1.3) in rel. (9.1.2), an expression that relates c_h with c_o can be easily obtained and is written as follows:

$$c_h = \frac{K_{hs} c_o V_s}{K_{hs} V_h + V_s} \tag{9.1.4}$$

The quantity of material in the headspace is given by $q_h = c_h V_h$. Typically, only a fraction f from the volume V_f is transferred to a GC system, the injected quantity being given by the expression:

$$q_c = \frac{f K_{hs} c_o V_s V_h}{K_{hs} V_h + V_s} \tag{9.1.5}$$

Rel. (9.1.5) indicates that the amount of material injected in the analytical system is proportional with the fraction f of volume collected from the headspace and with the initial concentration of the sample. The ratio of the quantity q_c given by rel. (9.1.5) and the initial amount of sample $q_o = c_o V_s$ gives a recovery of the analytes from the sample and headspace, expressed as follows:

$$R = q_c / q_o = \frac{f K_{hs} V_h}{K_{hs} V_h + V_s} \tag{9.1.6}$$

This expression shows that R increases with the increase of K_{hs}. (R also can be expressed in percent as R% = 100 R.) The increase in R% with the increase in K_{hs} is shown in Figure 9.1.1 for a sample with $V_h / V_s = 10$ and $f = 1$ (all sample injected). As seen from this figure, the recovery of compounds with higher K_{hs} is very good. Some K_{hs} values for different organic compounds volatilized from a water sample are given in Table 11.2.2.

Like any equilibrium constant, K_{hs} is temperature dependent. The variation of K_{hs} with the temperature can be expressed by the typical expression (see rel 1.6.63):

$$K_{hs} = \exp\frac{\Delta S^0_{hs}}{R} \exp\frac{-\Delta H^0_{hs}}{RT} \tag{9.1.7}$$

where besides the explicit variation of K_{hs} with T, ΔS^0 and ΔH^0 are also temperature dependent. The estimation of the entropy and enthalpy in rel. (9.1.7) can be done for an organic volatile compound on a solid sample using the expressions $\Delta S^0_{hs} = \Delta S_{vap}$ and $\Delta H^0_{hs} = \Delta H_{vap}$. Selecting benzene as an example of analyte and applying Trouton's rule with $\Delta S_{vap} = 88$ J deg^{-1} mol^{-1} and using $\Delta H_{vap} = 30.8$ kJ/mol, the variation of K_{hs} with temperature is shown in Figure 9.1.2 (assuming that ΔS^0_{hs} and ΔH^0_{hs} do not vary considerably with temperature). As seen in this figure, K_{hs} increases when the temperature increases.

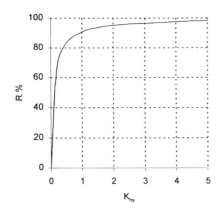

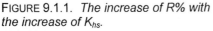

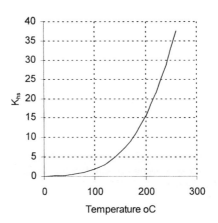

FIGURE 9.1.1. *The increase of R% with the increase of K_{hs}.*

FIGURE 9.1.2. *Variation of K_{hs} with temperature for benzene on a solid sample.*

The theoretical approach on SHS indicates that the technique is applicable for the analysis of volatile compounds, leading to excellent recoveries. The increase in temperature may help the emission of volatile compounds from the matrix, and maintaining a constant temperature is important because K_{hs} is temperature dependent.

The removal of a small amount of the headspace from a sample (to be injected in a GC) does not affect significantly the analysis, however removing a large volume of the headspace modifies the pressures enough to represent a parameter that must be taken into consideration. Also, from the same headspace it is frequently necessary to take more than one aliquot volume for analysis. For an aliquot taken in the syringe from the headspace, the resulting pressure can be calculated using the formula $p\,V = n\,RT$. The decrease in the amount of material is reflected in the decrease in n. This has the effect of reduction of p because the volume V is the constant volume of the vial. The decrease in the pressure is not desirable. It is difficult to stop the air from penetrating into the vial diluting inadvertently the headspace. Also, the equilibrium described by rel (9.1.1) is not maintained properly. It is common for this reason to replace the syringe volume in the headspace with a diluting gas. The addition of the replacement gas to the headspace can be done after the aliquot is removed or before it. Pressurization of the vial having the headspace to be analyzed followed by removal of the appropriate volume with a syringe is a common procedure. Both manual and automatic systems are available for performing this operation. Dedicated instruments to collect the headspace of samples maintained at controlled temperature are commercially available [2].

Quantitation using SHS can be done using standards. However, the use of internal standards is typically a problem. The addition of another volatile compound in the sample modifies the partial pressures of the analytes (see rel. 1.3.12). Therefore, assuming the total pressure of the gases in the sample to be constant, less analyte will be present in the headspace. Addition of a solvent carrying the standard modifies further the equilibrium. It is frequently necessary to add such a solvent because the amount of internal standard added in the sample is typically low, and a carrying solvent is utilized. The difficulties with the internal standard may be avoided in the quantitation using an external standard, but this procedure has its own problems. Calibration with an external standard may not represent the sample properly (see Section 3.8) and may lead to inaccurate results.

Better results in quantitation using SHS are usually obtained using standard addition technique or multiple headspace extractions. The standard addition must avoid the use of a solvent when performing the addition of the analyte, and this requirement may create some difficulties. The multiple headspace extraction attempts to remove completely the analyte from the sample, by use of repeated pressurizations followed by the aliquot removal and analysis. Assuming that the recovery R for a number of extractions remains constant, after n extractions the amount of analyte will be given by the expression:

$$q_n = q_0 (1 - R)^n \qquad\qquad (9.1.8)$$

which can be written in the form:

$$\ln q_n = \ln q_0 + n \ln (1 - R) \qquad\qquad (9.1.10)$$

By measuring experimentally q_n for a number of analyses of the same headspace, the unknown values for R and q_0 can be easily calculated.

A different procedure to collect the static headspace from a sample is the use of an adsorbent material. The procedure consists of allowing the adsorbent to stay in the headspace of the sample for a specific period of time and at a constant temperature. A convenient procedure is to put the adsorbent in a small bag made from clean tissue paper (teabag) suspended in the headspace of the sample, as shown in Figure 9.1.3 [3]. A number of solid phase materials can be used as adsorbent. Among these are solid supports with a specific coating, very similar to the packing for gas chromatographic columns. Other materials are porous polymers such as Tenax® (polymer of 2,6-diphenyl-p-phenylene oxide) or other polymeric sorbents (see Section 9.2). After the equilibration of the sample (several hours), an aliquot of the solid sorbent is transferred into a desorber. Special desorption devices are commercially available, including some with autosampling capability. The simplest procedure, however, is to load a weighed quantity of sorbent in an injector liner, which can be quickly introduced in the injection port of the GC. The heating of the sorbent at a certain temperature (e.g. 275° C) desorbs the compounds adsorbed from the sample headspace. The desorption is not instantaneous, and for obtaining sharp chromatographic peaks it is usually necessary to use cryofocusing. This can be done using special attachments to the chromatographic system. Various cryofocusing systems are commercially available, and their principle is to cool at very low temperature a small portion of a precolumn. Some systems use as a cryogenic agent liquid nitrogen, and others use liquid CO_2. The cryofocusing attachment may have its own heating system that allows a very rapid heating of the previously cooled portion of the capillary with the purpose of

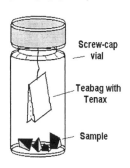

Screw-cap vial

Teabag with Tenax

Sample

FIGURE 9.1.3. *Adsorption of volatiles from the headspace.*

providing a narrow injection. Immersing a loop from the precolumn into liquid nitrogen contained in a small polystyrene cup may serve as a cryofocusing system, with the heating resulting from the chromatographic oven after the removal of the liquid nitrogen.

At equilibrium, the amount of analyte adsorbed on the solid sorbent can be calculated following the same expressions as those developed for solid phase microextraction (SPME) [4], a similar tecnique, which utilizes a special adsorbent device (see Section 11.2). Using the index "s" for the sample, "h" for headspace, and "a" for the adsorbing material, the equilibrium that takes place can be written as follows:

$$i_s \overset{K_{hs}}{\rightleftharpoons} i_h \overset{K_{ah}}{\rightleftharpoons} i_a$$

The equilibrium between the analyte in the adsorbing material and in the headspace is described by the constant K_{ah}, where $K_{ah} = c_a / c_h$ and c_a is the equilibrium (molar) concentration of the analyte "i" in the adsorbing material. The adsorbing material can be a stationary phase (liquid or liquid like) on a solid support. For this case, the constant K_{ah} is given by $K_{ah} = K^A_i$. (see rel. 1.4.36). The concentration c_a indicates either the concentration in the adsorbing layer or as an average in the whole sorbent. The volume V_a of the stationary phase must be considered accordingly, either that of the adsorbing layer or of the whole sorbent. The mass balance for the system can be written as follows:

$$c_o V_s = c_a V_a + c_h V_h + c_s V_s \tag{9.1.11}$$

The concentrations c_h and c_s can be expressed as a function of c_a as follows:

$$c_h = c_a / K_{ah} \qquad\qquad c_s = c_h / K_{hs} = c_a / (K_{hs} K_{ah}) \tag{9.1.12}$$

By subsitution of rel. (9.1.12) in rel. (9.1.11), an expression that relates c_a with c_o can be obtained. This expression can be written in the form:

$$q_a = c_a V_a = \frac{K_{ah} K_{hs} V_a c_o V_s}{K_{ah} K_{hs} V_a + K_{hs} V_h + V_s} \tag{9.1.13}$$

where q_a is the amount of adsorbed material. Rel. (9.1.13) shows that the amount of adsorbed material in the solid sorbent is higher for a larger initial concentration and a larger volume of adsorbent and is lower for a larger headspace volume. A recovery of the analytes from the sample (taking f =1) can be defined similar to rel. (9.1.6), as follows:

$$R = q_a / q_o = \frac{K_{ah} K_{hs} V_a}{K_{ah} K_{hs} V_a + K_{hs} V_h + V_s} \tag{9.1.14}$$

By comparison of this technique with SPME, several advantages and disadvantages can be noticed. Both techniques have a recovery dependent on the partition coefficients K_{ah} and K_{hs}. Large values for these constants assure a high recovery. Also the recovery depends on the volume of the adsorbing phase. Because the volume of the adsorbing phase is large in a teabag technique compared to SPME (see Section 11.2), the value of R is typically significantly high for SHS with adsorbing phase. On the other hand, SHS with the adsorbing phase may require a special desorber and cryofocusing, while

SPME does not. This is a significant practical advantage of SPME, and for analyses where the extremely high sensitivity is not an issue, SPME is preferred.

9.2 DYNAMIC HEADSPACE

A procedure applied to separate the volatile analytes from various matrices consists of transferring them into a stream of gas flowing over or through the sample. Inert gases such as N_2 or He can be used for desorbing volatiles, but also air or other gases can be utilized. This procedure is known as dynamic headspace. Dynamic headspace compared to static headspace has the advantage of replenishing the headspace with pure carrier gas such that the volatilization is not limited by the equilibrium at a value determined by the partition coefficient liquid-to-gas. For a liquid sample, the gas flow can be done through the sample, the procedure being known as sparging. The typical glassware used for sparging a liquid is shown in Figure 9.2.1. For solid samples, various types of glassware are available for containing the sample and allowing the gas to pass through. The transfer of the analytes from the sample into the purging gas can be increased by heating the sample. The heating must be done only if artifacts are not generated and influence negatively the analysis results. The volume of gas passing through the sample can be relatively large, and therefore the concentration of the analytes in the purging gas is low. For this reason, a concentrating procedure is necessary after the purging step. The concentration can be done using a trapping procedure. This is the reason the dynamic headspace is also known as purge and trap (P&T). The trapping can be done using several procedures: a) dissolution in a solvent, b) collection in a cryo trap, c) dissolution in a solvent at low temperature, which combines the previous two procedures, and d) adsorption in a solid phase sorbent trap. After trapping, the analytes must be further released from the trap and injected in a chromatographic system for analysis. The release of the analytes from the trap can be done using a solvent or by thermal desorption. After trap desorption the analytes are transferred into a chromatographic instrument. The block diagram of the purge and trap process is shown in Figure 9.2.2 [5].

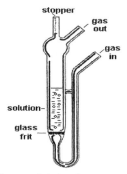

FIGURE 9.2.1. *Glassware used for sparging.*

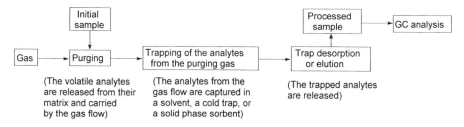

FIGURE 9.2.2. *Block diagram of the purge and trap process*

A number of parameters are important regarding the purging process. These include the selection of the amount of sample, sample temperature, selection of the purging gas, volume of the gas, and purging time. For the optimization of the purging, it is useful to inspect the mass balance during the process. For this purpose, it is convenient to consider a liquid sample of volume V_s containing an analyte "i" in contact with a volume of gas V_g. The equilibrium distribution of the compound "i" between gas and liquid is described by the partition constant K_i, which can be obtained from rel. (1.4.26), and can be written in the form:

$$K_i = c_{i,s} / c_{i,g} \tag{9.2.1}$$

where $c_{i,s}$ and $c_{i,g}$ are the molar concentrations in sample and in gas phase. If the flow of gas in the purging cell is u_p, the mass balance at time t is given by the expression:

$$V_g \frac{dc_{i,g}(t)}{dt} = -V_s \frac{dc_{i,s}(t)}{dt} - u_p c_{i,g}(t) \tag{9.2.2}$$

In rel. (9.2.2) it is indicated that the concentration of the analyte in the two phases is a function of time. However, rel. (9.2.1) is still valid for equilibrium conditions, and, therefore, $c_{i,g}$ can be expressed as a function of $c_{i,s}$ and substituted in rel. (9.2.2.). Assuming u_p constant, rel. (9.2.2) can be integrated, the result being the expression:

$$c_{i,s}(t) = c_{i,s}(0) \exp\left(\frac{-u_p t}{V_g + K_i V_s}\right) \tag{9.2.3}$$

A similar relation is valid for the concentration of the analyte "i" in the gas volume in contact with the sample, and:

$$c_{i,g}(t) = c_{i,g}(0) \exp\left(\frac{-u_p t}{V_g + K_i V_s}\right) \tag{9.2.4}$$

In rel. (9.2.3) and (9.2.4), K_i can be replaced with $1/K_{hs}$ as defined in Section 9.1.

The recovery R(t) of the analyte "i" defined as the ratio of the amount of analyte extracted from the sample $q_{i,s}(0) - q_{i,s}(t)$ and the amount initially present in the liquid sample $q_{i,s}(0)$ is given by the expression:

$$R(t) = 1 - q_{i,s}(t) / q_{i,s}(0) \tag{9.2.5}$$

The amount q can be obtained from the concentration multiplied with the volume, where the volume of the solution is constant, and R is given by the expression:

$$R(t) = 1 - c_{i,s}(t) / c_{i,s}(0) \tag{9.2.6}$$

By replacement of $c_{i,s}(t)$ with its expression given by rel. (9.2.3), the result can be written as follows:

$$R(t) = 1 - \exp\left(\frac{-u_p t}{V_g + K_i V_s}\right) \tag{9.2.7}$$

For volatile nonpolar analytes, K_i is small, and for small volumes of samples, the term $K_i V_s$ can be neglected such that the expression for R can be written in the form:

$$R\ (t) = 1 - \exp \left(\frac{-u_p t}{V_g} \right) \qquad (9.2.8a)$$

This relation indicates an equal recovery for all nonpolar compounds, irrespective of the temperature or the sample volume. This result is not in fact more than a rough approximation, because the term containing the value of K_i is neglected in rel. (9.2.7). A different expression is obtained if K_i is replaced with $1/K_{hs}$. In this case, the term $K_{hs} V_g$ becomes small, and the approximation of R can be obtained from the expression:

$$R\ (t) = 1 - \exp \left(\frac{-K_{hs} u_p t}{V_s} \right) \qquad (9.2.8b)$$

Rel. (9.2.7) shows that the recovery increases in time and is higher when K_{hs} is larger. The illustration of the variation of R with time for a small K_i and $u_p/V_g = 0.25$ min^{-1} is shown in Figure 9.2.3. This figure shows that regardless the value of K_i, the recovery can be complete if the time of purging is long enough. Also, from rel. (9.2.8a) it can be seen that for a large volume of V_g the recovery of all volatiles can be quantitative. In practice, the compounds with large K_{hs} values are close to completely recovered. These include mainly hydrophobic compounds recovered from aqueous solutions. Because K_{hs} increases with temperature (see rel. 9.1.7 and Figure 9.1.2), the temperature increase favors the transfer of volatile compounds in the gas phase. The heating is limited because of possible artifact formation and possibly due to the volatilization of part of the matrix, which is not desirable in the vapor phase. Mild heating at temperatures around 50–60° C are common for many samples. For water samples, the increase in temperature may lead to a higher amount of water to be carried together with the analytes, but water can be eliminated in the further steps of the process.

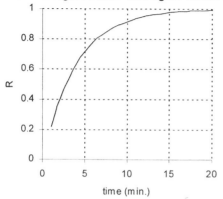

FIGURE 9.2.3. *Variation of R with time for K_i negligible and $u_p/V_g = 0.25$ min^{-1}.*

Other procedures can be used for desorbing the volatile compounds from the sample. One such procedure is salting out aqueous solutions for decreasing the partition in water of the organic volatiles (see rel. 8.2.11). Recoveries of volatiles can be significantly improved using this procedure [6], [7]. The use of vacuum extraction of analytes from the sample also has been applied [8].

- Trapping step in dynamic headspace

Once the volatile compounds are transferred into the vapor phase, the second step is the trapping of the analytes from the purging gas. Trapping in solvents by use of cryofocusing or solid phase adsorbents is frequently practiced. A short discussion on trapping of volatile compounds from gases is also given in Section 2.3.

The adsorption based on the solubility of the analytes uses a certain volume of a convenient solvent to collect the analytes. This collection can be done in an impinger and is followed by the injection of this solvent in a chromatographic instrument. Because the analytes are typically volatile, gas chromatography is more commonly used than other techniques as a core analytical technique. However, if a derivatization reaction is necessary for enhancing sensitivity, the gases can be trapped in a reagent solution and further analyzed using either GC or HPLC techniques.

Condensation of the analytes at low temperature from the flowing gas, known as cryogenic or cryofocusing collection, is based on passing of the carrier gas through a collection trap set at a very low temperature (as low as -190° C for some applications). Very often a U-shaped or spiral tube is used as condensing vessel. The collection is influenced by the trap temperature and the contact time with the cold surface, which is further determined by the flow rate of the gases, vessel surface, and type of flow (turbulent, laminar). The temperature is controlled by the cooling agent, several cooling agents being discussed in Section 2.3. A secondary solvent can be used in the cold trap, a combination solvent/cooling being an efficient way of trapping the analytes. Also, the analytes from the traps with cool walls can be further recovered using a small amount of an appropriate solvent. The cold traps also collect the water vapors that deposit as ice [9], [10]. This can be a disadvantage because the sample becomes diluted with water. Cold traps can be used for the cryogenic removal of water in the determination of noncondensing analytes [11].

The use of solvents, cool traps, and cool traps with solvents has a number of advantages and disadvantages. Among the advantages are the good efficiency of collection, capability for addition of an internal standard that can be very useful for quantitation, relatively large capacity of collecting analytes when these are in higher concentrations, and the possibility to use the same sample for more than one GC or LC injection. Among the disadvantages of these systems are the presence of an additional solvent added as carrier for the analytes and potential dilution beyond desirable levels. The solvent may elute in the GC analysis at the beginning of the chromatogram and very close to many volatile compounds. In such cases, trace analytes may be covered by the solvent peak in the chromatogram. The use of cryo systems may condense a large amount of water, which is undesirable in the GC analysis. Also, the work with cryo traps must avoid "sample degasing" when the collected material is brought at room temperature and part of the analytes may be lost. Cold solvents may need to be injected, keeping the sample at low temperatures. However, formation of bubbles in the syringe of the GC system may occur, which leads to nonreproducible injections.

A very common trapping procedure is based on the use of a solid phase sorbent. A large number of solid phase sorbents are used as trapping materials, including stationary phases similar to those used in packed GC columns. The equilibrium distribution of the compound "i" between gas and the sorbent is again described by the partition constant K_i or in the case of porous polymers by the adsorption constant K^A_i (see rel. 1.4.36), which characterizes the adsorption process at the gas-solid interface. The sorbent must be selected to have large K_i (or K^A_i) values for the analytes of interest. The adsorption on the trapping sorbent is the reverse process from the sample desorption. The use of room temperatures for adsorption (lower than those used for desorption) increases the efficiency of analyte trapping. The trapping material is usually contained in a special chamber that also allows rapid heating of the sorbent. In some instruments, it is possible to take only an aliquot of sorbent for further desorption. However most instruments use the same amount of sorbent for trapping and further desorption.

Among the sorbents used for trapping are the following groups: a) charcoal sorbents, b) silica, molecular sieves, alumina, c) porous polymers, and d) bulk chromatographic phases. Solid phase sorbents are characterized by their partition coefficient values (for the class of analytes of interest), loading capacity (the maximum amount of material that can be retained in the sorbent as mg analyte / g sorbent), average surface area (m^2 / g), pore size, breakthrough volume, temperature stability, insolubility in common solvents, etc.

The breakthrough volume is defined as the volume of a sample with a concentration low enough not to overload the sorbent, which can be passed through a specific amount of sorbent until the analyte starts appearing at the outlet or shows at the outlet above a certain threshold concentration (see also Section 11.1). In the definition of the breakthrough volume, it is assumed that the retention of an analyte is similar to a chromatographic process. After passing enough gas through the sorbent when the analyte is injected as a spike, the analyte will elute from the sorbent. The process is similar when injecting the sample continuously (but without overloading the sorbent). After passing the gas for a period of time that is long enough, the analyte will show at the outlet above a certain threshold concentration. This volume of gas (or liquid for a liquid sample in SPE) is defined as breakthrough volume. The breakthrough volume V_B depends on the nature of the analyte, the nature of the sorbent material, and the carrier used in the process. The relation between the value of the breakthrough volume V_B in mL gas and the capacity factor k_i for the specific analyte and sorbent can be estimated with the expression:

$$V_B \approx 200 \; k_i \; V_{bed} \qquad\qquad (9.2.9)$$

where V_{bed} is the volume of the sorbent bed in cm^3 [12].

Charcoal sorbents are very common. Various procedures are used to prepare and activate charcoal sorbents, such as high temperature steam treatment of ordinary charcoal. The average surface area of charcoal sorbents can be around 10–15 m^2/g for lower surface area charcoals and 100–120 m^2/g for higher surface area charcoals. The charcoals are suitable for nonspecific adsorption and highly inert. Charcoal may adsorb some compounds too strongly and further desorption can be difficult. Charcoal adsorption is still influenced by the presence of water in the gas stream, although a low level of water is retained on this material. A number of commercial types of charcoal are available such as graphitized carbon (Carbopack®), carbon molecular sieves (Carboxen®), etc.

A number of inorganic sorbents are also used for trapping analytes. Among these are silica gel, diatomites, several molecular sieves or zeolites, hydrated alumina, and Florisil. Silica gel is a polar sorbent used for collecting amines or other polar analytes.

Compounds with low polarity can be displaced on silica gel by compounds with higher polarity, and this may be a problem during absorption of analytes from a gas stream with high humidity, because the water may replace some of the analytes. Diatomites are natural silica materials, commercialized under various names such as Chromosorb® G, P or W. Depending on the moisture and number of OH polar groups in these materials, they are available at different polarities. Alumina and Florisil are not frequently used for trapping analytes from gases. Molecular sieves are more frequently used. These are synthetic and natural zeolites or aluminosilicates with the molecular structure characterized by a tridimensional array of AlO_4 and SiO_4 tetrahedra, with oxygen atoms shared by two tetrahedra. The formula of a zeolite can be written $Me_{x/n}[AlO_2]_x (SiO_2)_y]$ m H_2O, where Me is a monovalent or divalent cation such as Na^+, K^+, or Ca^{2+}, x, y and n are integers, and m is the

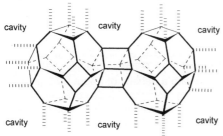

Cubo-octahedron idealized
structure of zeolites

number of water molecules. A number of basic crystalline structures are possible for zeolites, such as the cubo-octahedrons connected by cubes or by hexagonal prisms. In the cubo-octahedron, the Al or Si atoms are placed in the vertices while the sides consist of –O– bonds. The cubo-octahedrons are also connected through –O– bonds forming a structure with specific pore sizes (cavities). The neutrality of the molecule is maintained by the Me ions. Varying the ratio Al/Si and the nature of Me ion, the dimensions of the pores can be modified. For example, molecular sieve 4A has the formula $Na_{12}[(AlO_2)_{12}(SiO_2)_{12}]$ 27 H_2O and a pore diameter of 4 Å. The replacement of Na^+ with K^+ gives the molecular sieve 3A with the pore diameter of about 3 Å. The replacement of the Na^+ with Ca^{2+} gives the molecular sieve 5A with the pore diameter of about 5 Å. Molecular sieve 3A can trap H_2O, and molecular sieve 4A can trap H_2O, CO_2, NH_3, and CH_4, C_2H_6, C_2H_4, C_3H_6. Larger pores for molecular sieve 5A allow trapping of small molecules but also of n-olefins, alcohols, amines, etc. The materials are commercially available as granules of different sizes (mesh number).

Porous polymers represent a common type of sorbent frequently used in trapping of volatile organic compounds from a stream of gas. These materials are used in dynamic headspace analytical procedures and also in collection of analytes from gaseous samples such as those related to environmental analysis. The porous polymers are based on a few typical structures, such as poly-2,6-diphenyl-p-phenylene oxide (Tenax®) polystyrene, and polyacrylate. The structures of these polymers are shown below:

poly-2,6-diphenyl-p-phenylene oxide   polystyrene   polyacrylate

The crosslinking of the polymeric chains can be done using divinylbenzene, ethyleneglycol dimethacrylate, etc.

| divinylbenzene | ethyleneglycol dimethacrylate |

Tenax is available as a pure polymer with specific surface area around 35 m^2/g, average pore size 200 nm, and the density 0.25 g/cm^3 (Tenax-TA®). Tenax containing 23% graphite (Tenax-GR®) is also commercially available with a density of about 0.55 g/cm^3. Tenax is very stable up to 350° C, and the desorption of the compounds collected from a gas stream can be done easily using heating. The breakthrough volume (of gas) for a number of analytes on Tenax-TA and Tenax-GR are given in Table 9.2.1.

TABLE 9.2.1. *Breakthrough volumes V_B on Tenax, as mL gas (for 1 g sorbent) at 20° C for several compounds.*

Compound	Tenax-TA	Tenax-GR
methane	5.7	2.2
ethane	19.7	14.4
water	55	47
methyl chloride	190	182
benzene	44000	33800
1,1,2-trichloroethane	110000	105000

The large V_B value indicates high retention of the specific analyte on the sorbent.

Other porous polymers and copolymers include materials known as Chromosorb® and Porapak®. Table 9.2.2 indicates the general composition of Chromosorb type polymers, and Table 9.2.3 indicates the general composition of Porapak type polymers. Other porous polymers are also available such as HayeSep® [13], [14], Amberlite XAD polymers (polyaromatic materials), and Dowex type adsorbents. The selection of any of these porous polymers is determined by the chemical composition of the analytes and the need to retain or eliminate specific volatile components. For example, Tenax-TA is nonpolar, and the retention of water is much lower that that of benzene or 1,1,2-trichloroethane (as shown in Table 9.2.1). For this reason, the polymer can be used for adsorbing organic compounds even from a wet stream of gas.

TABLE 9.2.2. *Cromosorb specifications.*

Series	Composition	Surface area m^2/g	Polarity	Maximum temperature °C [1]
101	poly-DVB/styrene	>50	nonpolar	275/325
102	poly-DVB/Styrene	300–400	slightly polar	250/300
103	crosslinked polystyrene	15-25	nonpolar	275/300
105	polyaromatic	600-700	moderately polar	250/275
106	crosslinked polystyrene	700-800	nonpolar	250/275
107	crosslinked acrylic ester	400-500	polar	250/275
108	crosslinked acrylic	100-200	polar	250/275

[1] Temperatures are expressed for constant and for short gradient heating.

TABLE 9.2.3. *Porapak specifications.*

Series	Composition	Surface area m²/g	Polarity	Maximum temperature °C
Q	poly-DVB/ethylvinylbenzene	500-600	slightly polar	250
P	poly-DVB/styrene	100-200	nonpolar	250
R	poly-DVB/vinyl pyrrolidone	450-600	moderately polar	250
S	poly-DVB/vinyl pyridine	300-450	moderately polar	250
T	poly-ethylene glycol dimethacrylate	225-350	polar	190
N	poly-DVB/vinyl pyrrolidone	250-350	very polar	190

A variety of bulk chromatographic phases also can be used as trapping materials. Some of these materials and the range of temperatures of their utilization are given in Table 4.2.2.

The selection of the appropriate solid phase material as a sorbent depends on the nature of the sample, the purpose of analysis, and the sensitivity of the instrumentation used as a core analytical step. The amount of sorbent also depends on the amount of sample, amount of analytes, etc. The solid phase material may come as bulk, and a selected amount can be used for loading a trap (e. g. 1–5 g), or special traps can be available for specific operations. It is very common for the solid phase materials utilized for adsorption to require previous to utilization a conditioning step at a specific temperature (usually the same temperature necessary for desorbing the analytes). For this purpose, the sorbents must be kept for a specific period of time (1–2 hours) at the elevated temperature until no impurity or interference is present in the sorbent. Excessive temperature for conditioning must be avoided. Low temperatures or insufficient heating time may lead to incorrect results due to carry-over problems from sample to sample. The same sorbent usually can be used for a number of adsorption-desorption cycles. However, nonvolatile materials or decomposition effects can be noticed after a period of utilization, and the trap material must be replaced.

- Water elimination from purging gases

Many samples subject to analysis using dynamic headspace contain a significant amount of water. These samples may include food, environmental samples including soil, various types of water samples, biological materials, etc. The increased temperature sometimes used for the removal of volatile compounds from the sample may increase the amount of water in the purging gases. For this reason, water removal in dynamic headspace is a common problem. One method to remove the moisture from gas flows is the use of desiccants. Dehydrated inorganic salts can be used to dry efficiently the gaseous mixture prior to the injection in the chromatographic column. Several sulfates (Na, Ca, Mg, or Cu) can be used for this purpose, as well as $Mg(ClO_4)_2$, $Ba(ClO_4)_2$, or K_2CO_3. Some problems related to the use of salts are related to significant losses of aromatic compounds (other than benzene and toluene). This effect is observed mainly when K_2CO_3 is used [15]. Methanol and acetone also can be absorbed by K_2CO_3 [16]. Another technique for water removal is the use of CaC_2, where no losses are observed for many volatile halocarbons [17]. The use of this method is restricted to the isolation of volatile nonpolar compounds; the polar volatile and semivolatile compounds exhibiting recovery losses. Also, the use of desiccants requires regeneration or replacement after a few operations.

Another procedure for water elimination is the selective permeation of water from a gas flow through perfluorosulfonic acid polymeric membranes. Commercially available under the names Nafion (Du Pont de Nemours Co.) and Perma Pure (Perma Pure, Inc.), these membranes can be successfully applied for the analyses of halocarbons in gases [18], SO_2 in humid air [19], hydrocarbons, ethers and esters in air [20], etc. However, incomplete recovery of different alcohols, ketones, and aldehydes has been observed, and even the formation of artifact compounds is possible, as it was established for monoterpenes [21].

The application of pre-columns in order to dry the gas streams is based on the selective chromatographic separation of water and organic compounds. On hydrophobic porous polymers like Tenax or Porapak as well as on charcoal, water has only a small retention volume and can be vented before the analytes of interest break through. However, some volatile compounds can pass the sorbent bed concurrently with the water, and they cannot be trapped quantitatively. The pre-separation of these compounds can be achieved on packed hydrophilic column, when many compounds like terpenes, pyrans and furans elute before the water [21].

- Desorption of the analytes from the trap

The analytes adsorbed into a trap must be further transferred into the chromatographic instrument. The volatiles collected into a solvent are typically injected without further problems. For the materials collected in cryotraps, the transfer may be a problem if the analytes are lost when heated at room temperature. The most common trapping procedure uses solid phase sorbents. The analytes can be removed from the sorbent using specific solvents or more frequently using heat. The on-line desorption with the direct transfer of the analytes into the chromatographic column is a common procedure. A simplified diagram of an on-line P&T system, which adsorbs the analytes in a Tenax trap followed by desorption is shown in Figure 9.2.4.

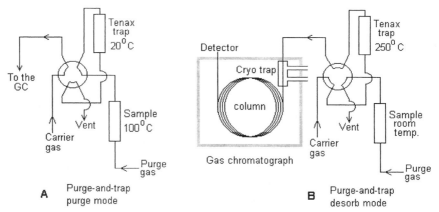

FIGURE 9.2.4. *Diagram of an on-line P&T system connected to a GC.*

In Figure 9.2.4, configuration A indicates the purging of the analytes from the sample and the adsorption in a trap. The temperature for the sample desorption is decided based on each type of sample, and the temperature of the trap necessary for the adsorption of the analytes is typically close to room temperature. An intermediate step (pre-desorption) is used in some analyses, with the purpose of eliminating some of the undesirable compounds collected in the trap, such as water. In the predesorption step, the instrument is still kept in configuration A, but the temperature of the trap is increased, for example to 110° C for a short period of time. After this step, the instrument is switched to configuration B, and the trap is heated (e.g. at 250° C), the analytes being desorbed. Some desorption temperatures and necessary desorption times for a number of analytes from a XAD-4 sorbent are given in Table 9.2.4.

Because the desorption process is not instantaneous, it is frequently necessary to add a cryofocusing device to the chromatographic instrument. Otherwise, the peaks in the analytical separation are too broad. To this broadening contributes also the use of high flow rates (30–50 mL/min.) required at the thermal desorption process. Cryofocusing uses low temperatures to focus the sample into a plug at or near the head of the separation column in order to improve the peak shape. Two types of cryofocusing can be used, namely on-column and external. For on-column cryofocusing, volatile compounds are trapped as a narrow band on the column or pre-column in the chromatographic oven. External cryofocusing traps the volatile compounds as a narrow band outside the oven compartment, either in the injector or completely outside the gas chromatograph [23].

TABLE 9.2.4. *Some desorption characteristics for a XAD-4 trap.*

Analyte	Desorption time (min.)	Desorption temperature (°C)	Recovery (%)
toluene	15	210	88
ethylbenzene	15	210	79
indene	10	180-210	96
naphthalene	15	210	90
1-methylnaphthalene	10	180-210	95
hexane	8	175	88
chloroform	4	175	93
dibromomethane	4	175	88
cyclohexanol	5	200	98
n-heptyl alcohol	6	200	100
benzyl alcohol	13	220	83
methyl iso-butyl ketone	8	220	100
amyl iso-propyl ketone	8	200	102
methyl nonyl ketone	13	220	96
p-methylacetophenone	13	220	99
ethyl heptanoate	10	200	96
octyl acetate	10	200	61
bromobenzene	10	200	106
o-dichlorobenzene	10	200	102

A relatively unexplored desorbing procedure is the use of microwave heating. If water is present in the sample, its thermal energy can be transferred to other volatile compounds and enhance their volatilization. The process can be used for desorbing the trapping material but also can be used for desorbing the volatiles from the sample itself using a static or a dynamic setup [24].

- *Practice of purge and trap analysis*

Purge and trap analysis is commonly applied for the analysis of volatile compounds in a variety of samples. These include environmental samples, food, plant materials, flavors, etc. Even the scent of live plants and flowers can be analyzed with this technique (as shown in Figure 9.2.5). The application of P&T must consider a number of problems related to this technique. One of them is the carry-over problem. During sample desorption, some semivolatile compounds may be transferred to the vapor phase, mainly when heating is applied to enhance volatilization. These compounds may condense on transfer tubes and any colder parts of P&T instrumentation. This requires heating of various zones of the P&T instruments with the purpose of avoiding condensation spots. Complete elimination of these compounds from sample to sample may be a problem. As an example, in the analysis of trace flavors from cigarettes, triacetin (glycerin triacetate) present in cigarette filters is also volatilized. Triacetin carry-over can be noticed in some P&T instruments even after four or five runs of different samples containing no triacetin. Also, the contamination and the formation of artifacts can be a problem when applying P&T. Many organic sorbents have a polymeric structure based on monomers such as styrene, divinylbenzene, etc. Traces of these monomers are sometimes present in the polymeric material or may be generated due to some decomposition when the polymer is heated to desorb the analytes. When the carrier gas is not pure N_2 or He, and the experimental conditions require the use of air, the contact with reactive compounds such as traces of O_3 and NO_x normally present in ambient outdoor air may enhance the release of monomeric compounds [25], [26], [27]. For Tenax, degradation reactions have been extensively studied, including reactions with ozone, chlorine [28], [29], nitrogen dioxide [30], nitrogen oxide, sulfur dioxide, and sulfuric acid [31]. Potential contaminants generated from Tenax include phenol, benzaldehyde, acetophenone, decanal, dibutylphthalate, 2,6-diphenyl-p-quinone (DPQ), and 2,6-diphenylhydro-p-quinone [32].

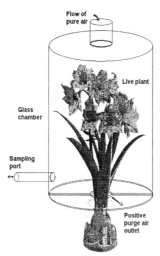

FIGURE 9.2.5. *P&T applied to a live flower.*

Artifact formation during the desorbing of analytes at temperatures reaching $250°$ C (or even higher) may affect some compounds. This process must be evaluated for each type of sample. Artifact forming reactions may also occur between the adsorbed analytes and the reactive species from the gaseous phase. Some compounds are more sensitive than others, the reaction of terpenes and ozone [25], [26], [27] being among the best characterized.

The application of a P&T analysis requires a number of optimization steps, which are summarized in Figure 9.2.6.

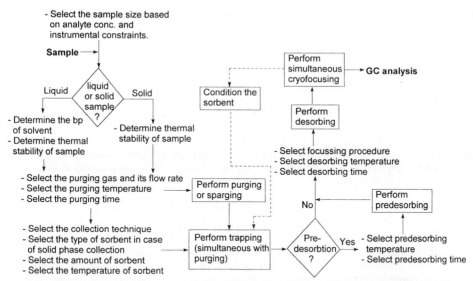

FIGURE 9.2.6. *Diagram showing steps requiring optimization in P&T applications.*

Many analyses using P&T are performed for quantitative purposes, and because the technique can be highly sensitive, it is widely applied in trace analysis. The quantitation using dynamic headspace poses several problems. The first is the relatively lower reproducibility of the results in P&T when a chain of operations is performed on the sample, as shown in Figure 9.2.6. Assuming that all the reproducibility problems are solved, and that no artifacts, carry-over, or contamination problems occur, the calibration for quantitation is the next problem. The choice of internal standards and the calibration procedures are similar to those from the static headspace. The use of stable isotope standards (deuterated or containing ^{13}C, ^{15}N, etc.) for calibration is highly recommended, these substances having distribution constants and bp temperatures close to natural isotope compounds. Standard addition technique is also commonly used for calibration in P&T. The high potential for concentrating the analytes from the whole sample for only one GC injection makes P&T a very sensitive method [33]. Because by use of P&T as sample processing device, the higher molecular weight compounds are absent from the compounds injected in the GC, the use of mass spectrometry for detection enhances further the advantages of this type of analysis.

9.3 OTHER HEADSPACE TECHNIQUES

Several other headspace techniques are developed for the analysis of volatile compounds. These techniques include static and trapped headspace (S&THS) [34], short path thermal desorption [35], closed-loop stripping [36], spray and trap [37], etc.

Static and trapped headspace uses a large syringe volume to remove the static headspace gases from the sample, but instead of injecting them into a GC system, passes the gases through a sorbent. Once adsorbed, the sorbent material can be transferred into a desorber and injected into a GC. In short path thermal desorption the sample is placed directly in a thermal desorber connected to the GC system. Instrumentation capable of performing desorption under a flow of gas and injection of the gases in a GC system is available. A sorbent on which the sample was previously collected also can be used in the desorber. Cryofocusing is frequently needed with short pass desorption. However, the procedure cannot handle samples with high levels of moisture, because the water entering the GC column either affects the chromatography or in case of cryofocusing may block the capillary column [38]. Closed-loop stripping [39] is in many respects similar to a P&T system, where the gas passing through the sample is circulated using a pump through the sample and the trap. A common problem with this technique is the contamination due to the passing of the gases through the pump.

Spray and trap technique uses a spray nozzle forming very small droplets of the aqueous solution to be analyzed in an extraction chamber. The large total interface area between the liquid and gas phase helps to reach quickly the partition equilibrium of dissolved compounds between the liquid and the gas phase. Compounds extracted into the gas phase are preconcentrated in a sorption tube gas-sampling device. A thermal desorption following the preconcentration allows the collected compounds to enter the GC column. The spray-and-trap method is very sensitive and can be applied to sampling of very low concentrations samples (down to 10–30 ng/L) and with short sampling times. The transfer into the gas phase is enhanced by the use of small droplets not only because of the large interfacial area, but also because the partition equilibrium is shifted to the advantage of the gaseous phase. The partial vapor pressure of liquids in the form of droplets ($p_{droplets}$) increases compared to the vapor pressure in bulk (p_{bulk}) with decreasing radius of curvature of the particles according to the equation

$$p_{droplets} = p_{bulk} \exp(\frac{2\,\gamma\,V_m}{r\,RT})$$

(9.3.1)

where γ is the surface tension of the solution, V_m is the molar volume of the liquid (in liters), r is the radius of the droplet, R is the gas constant, and T is the temperature. The total amount of the compounds transferred into the gas phase increases when droplets are formed from the bulk liquid. However, in water this effect becomes significant only for droplets with radii less than 1 μm.

Spray extraction of VOCs has been developed to circumvent bubble formation problems associated with the purge and trap procedure of isolating different analytes from solutions containing high concentrations of surfactants. The analytes in gas phase are concentrated on Tenax and thermally desorbed into a GC or GC-MS system. This technique has been used to detect benzene, toluene, *tert*-butylbenzene, tetrachloroethene, p-dichlorobenzene, and naphthalene at 10 to 30 ppb levels, using a sampling time of 2 minutes and 900 mL water. When water is cycled only once through the spray extraction system, the extraction efficiencies are 10 to 15% better than those of the conventional purge and trap procedures. Cycling the water sample through the system several times further improves extraction recovery [40].

REFERENCES 9

1. B. Kolb, L. Ettre (eds.), *Static Headspace Gas Chromatography*, Wiley-VCH, New York, 1997.
2. B. V. Burger et al., J. Chromatogr., 402 (1987) 95.
3. E. D. Alford, unpublished results.
4. J. Pawliszyn, *Solid Phase Microextraction, Theory and Practice*, Wiley-VCH, New York, 1997.
5. Y. Seto, J. Chromatogr., 674 (1994) 25.
6. D. de la Calle Garcia et al., J. High Resolut. Chromatogr., 21 (1998) 373.
7. B. McGillavray, J. Pawliszyn, J. Chromatogr. Sci., 32 (1994) 317.
8. P. Werkhoff et al., J. Agric. Food Chem., 46 (1998) 1076.
9. T. H. Wang et al., Chromatographia, 17 (1983) 411.
10. A. Cert, J. Bahima, J. Chromatogr. Sci., 22 (1984) 7.
11. H. T. Badings et al., J. High Resolut. Chromatogr., 8 (1985) 755.
12. E. M. Thurman et al., Anal. Chem., 50 (1978) 775.
13. O. L. Hollis, Anal. Chem., 38 (1966) 309.
14. O. L. Hollis, W. V. Hayes, J. Gas. Chromatogr., 4 (1966) 235.
15. R. D. Cox, R. F. Earp, Anal. Chem., 54 (1982) 2265.
16. D. A. Heatherbell, J. Agric. Food Chem., 19 (1971) 1069.
17. R. Boos et al., J. Chromatogr., 328 (1985) 233.
18. W. A. McClenny et al., Anal. Chem., 56 (1984) 2497.
19. D. C. Thorne et al., Anal. Chem., 58 (1986) 2688.
20. R. A. Rasmussen, A. K. Khalil, J. Geophys. Res., 86 (1981) 9826.
21. W. F. Burns et al., J. Chromatgr., 269 (1983) 1.
22. B. J. Tyson, G. C. Carle, Anal. Chem., 46 (1974) 610.
23. T. A. Bettell, R. L. Grob, Int. Lab., April (1986) 30.
24. R. E. Majors, LC-GC Int., 8 (1995) 128.
25. F. Juttner, J. Chromatogr., 442 (1988) 157.
26. R. J. B. Peters et al., Atmos. Environ., 28 (1994) 2413.
27. A. Calogirou et al., Anal. Chem., 68 (1996) 1499.
28. E. D. Pellizarri et al., Anal. Chem., 56 (1984) 793.
29. X. Cao et al., Environ. Sci. Technol., 28 (1994) 757.
30. R. L. Hanson et al., Environ. Sci. Technol., 15 (1981) 701.
31. M. B. Neher et al., Anal. Chem., 49 (1977) 512.
32. P. M. Clausen et al., Atmos. Environ., 31 (1997) 715.
33. B. V. Ioffe, A. G. Vitenberg, *Head-space Analysis and Related Methods in Gas Chromatography*, J. Wiley, New York, 1984.
34. A. Chaintreau et al., J. High Resolut. Chromatogr., 11 (1988) 830.
35. J. J. Manura, T. G. Hartman, American Lab., May (1992) 11.
36. R. G. Westendorf, American Lab., Dec. (1982) 2.
37. G. Baykut, A. Voigt, Anal. Chem., 64 (1992) 677.
38. R. Van Wijk, J. Chromatogr. Sci., 8 (1970) 418.
39. K. Grob, J. Chromatogr., 84 (1973) 225.
40. C. J .Koester, R. E. Clement, Crit. Rev. Anal. Chem., 24 (1993) 263.

CHAPTER 10

Solvent Extraction

10.1 LIQUID-LIQUID EXTRACTION

Liquid-liquid extraction (LLE) is a solvent extraction technique applied to liquids (liquid samples or samples in solution). LLE is practiced commonly in both industry and laboratories. LLE is performed for achieving the isolation of one or more compounds from a liquid or for concentration purposes. A schematic diagram of the extraction process is shown in Figure 10.1.1.

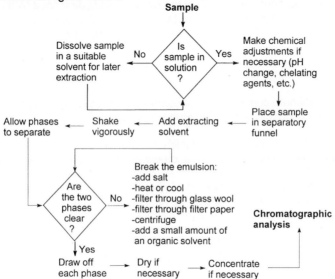

FIGURE 10.1.1. *Schematic diagram of LLE process.*

The separation is obtained by selective extraction of the analytes from the sample or of components from the sample matrix that must be eliminated. The concentration is achieved by extracting in a small volume of solvent specific compounds that were previously in a larger volume of liquid.

- *Batch extraction*

The simplest extraction procedure and the technique frequently applied in laboratory for analytical separations involves the use of a given volume of solution that contains the analytes. This solution is put in intimate contact with a given volume of a nonmiscible solvent. After a short period of shaking (frequently using mechanical shakers), the two layers of liquid are allowed to separate. The layer of interest is then taken aside and, if necessary, the extraction is repeated after addition of fresh solvent. Usually, the

FIGURE 10.1.2.
Separatory funnel.

operation is done using a separatory funnel (see Figure 10.1.2). This batch extraction process provides fast and simple separations and offers many advantages, especially when the extraction efficiency of the solute of interest is large and the process involves only a few extraction operations. Ideally, the separations are quantitative, but a perfect separation is never possible. Various methods for increasing the extraction efficiency as well as the selectivity of an extraction can be applied.

The theory of liquid-liquid extraction is based on the partition of a component "i" between two liquid phases "A" and "B" that are not miscible. This theory was discussed in Section 1.4. The ratio of the concentrations of "i" in the two phases is described by the partition coefficient (or distribution constant) K_i given by rel. (1.4.6). Considering solvent "A" an organic solvent and solvent "B" water, rel. (1.4.6) can be written in the form:

$$K_i = \frac{c_o}{c_w} \qquad (10.1.1)$$

where c_o is the equilibrium concentration of compound "i" in the organic phase "o", and c_w is the equilibrium concentration in the aqueous phase "w".

The value of K_i can be estimated from solubility parameters based on rel. (1.4.16). In general, the partition of an organic compound between water and a solvent is closely related to its solubility in water, because this solubility can be considered as partition of a substance between water and its own liquid phase [1]. This allows the calculation of partition coefficients from solubility values [2], [3], [4]. Various other procedures are proposed for the estimation of partition coefficients [5], some being discussed in Section 1.10. However, the choice of an optimum organic solvent for an extraction of particular interest is usually based on empirical evaluations. The partition coefficient for octanol/water system is well studied, and extensive data are available for this system [6].

The fraction from the total amount of analyte "i" extracted in the organic phase is given by the extracted fraction E_i (equivalent to recovery R for the component "i"):

$$E_i = \frac{q_o}{q_t} = \frac{c_o V_o}{c_o V_o + c_w V_w} = \frac{K_i \beta}{1 + K_i \beta} = 1 - \frac{1}{1 + K_i \beta} \qquad (10.1.2)$$

where q_o is the amount in the organic phase and q_t the total amount of analyte, V_o and V_w are the volumes of organic and aqueous phase, respectively, and β is the phase ratio with $\beta = V_o / V_w$. The variation of E_i as a function of $K \beta$ is shown in Figure 10.1.3 for a single extraction (curve with n = 1). This figure shows that the increase in K_i and in β lead to a higher extracted fraction (which also can be expressed in %). For equal volumes of organic and aqueous phase $\beta = 1$, the graph in Figure 10.1.3 shows that a large value for K is needed for an extracted fraction close to 100% ($E_i = 1$). The improvement of the extraction yield can be obtained using repeated extractions, for example using portions of pure solvent "o" to extract the same water solution a number of times. The extracted fraction $E(n)_i$ from the total amount, assuming that the volume V_w remains constant after n extractions each with the volume V_o of fresh organic solvent, is given by the expression (see Section 1.7):

$$E(n)_i = 1 - \left(\frac{1}{1 + K_i\beta}\right)^n \tag{10.1.3}$$

Figure 10.1.3. shows the effect of two and of three extractions for different $K\beta$ values.

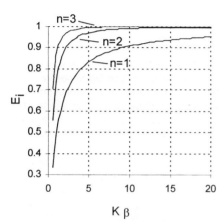

FIGURE 10.1.3. *The variation of distribution ratio E_i as function of $K\beta$ for single extraction (n = 1) and for repeated extractions (n = 2, n = 3).*

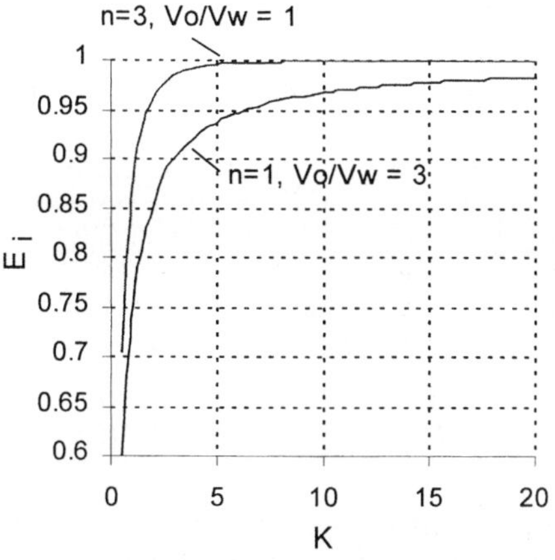

FIGURE 10.1.4. *The variation of distribution ratio E_i for three extractions and $V_o/V_w = 1$ compared to one extractions and $V_o/V_w = 3$, for various K values.*

The extracted fraction increases significantly when the number of extractions increases. For the same volume of organic phase used in one step or divided in more steps, higher efficiency is obtained when more steps are used, as can be seen by comparing rel. (10.1.3) and (10.1.2). This is exemplified in Figure 10.1.4 for n = 3 and $V_o/V_w = 1$, compared to n = 1 and $V_o/V_w = 3$ for various K values.

The selectivity of the extraction process can be characterized by a separation factor α. This factor can be defined by an expression analogous to rel. (1.4.30), as the ratio:

$$\alpha = E_i / E_j \tag{10.1.4}$$

for two different species "i" and "j", where E_i and E_j are the extracted fractions for the two compounds. The concentrations remaining in the aqueous phase can be characterized by the ratio $(1 - E_i) / (1 - E_j)$. Larger values for α correspond to a better separation between the two species. As shown in Figure 10.1.5, for the same ratio K_i / K_j, the value for α is variable, increasing as both the $K\beta$ values are smaller. However, very low values for $K\beta$ lead to low extraction efficiency of both components, and a compromise must be made between low K and β values and high values for the separation factor α. The ideal conditions are those with a low extraction of one component and high of the other, such that one component has the tendency to stay in the organic phase, and the other in the aqueous phase. This is achieved more easily when the chemical nature of the compounds is significantly different. Only gradual differences in the extraction

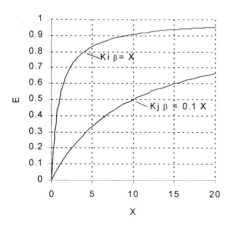

FIGURE 10.1.5. *Variation of E for two values of Kβ at a constant ratio $K_i / K_j = 10$.*

fractions are usually present in compounds with similar chemical structures. A significant change in the partition coefficients can be obtained by changing a substance from ionic into nonionic form. For this reason, organic compounds that contain acidic or basic functional groups, such as R-COOH, R_3N, R_2NH, RNH_2, can be changed into ionic or nonionic forms by pH change when their extraction properties are significantly modified. In this way, many extractions can be controlled by pH modifications. As an example, the flow diagram of a cleanup procedure used for sample preparation applied for the analysis of heterocyclic amines in particulate phase of cigarette smoke is shown Figure 10.1.6 [7].

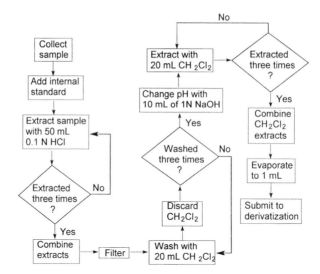

FIGURE 10.1.6. *Diagram for sample preparation before derivatization for the analysis of heterocyclic amines in particulate phase of cigarette smoke.*

As seen in Figure 10.1.6, the cleanup steps consist of washing the acidic sample of amines with CH_2Cl_2 (which is discarded) followed by changing the pH and reextracting the analytes in an organic phase (again CH_2Cl_2). The amines salts are not extractable in the organic phase, while the free bases are efficiently extracted.

In general it cannot be expected that solvent extraction provides a high separation factor for all the sample components, having a fractionation capacity of 2 (see Table 1.7.1). However, LLE is used mainly as a cleanup procedure, where it may be extremely useful, and not as an advanced fractionation technique.

The use of an internal standard is frequently necessary during the extraction process when quantitative measurements are intended. Precise measurement of the volumes of the two liquids are sometimes sufficient, but the accidental loss of solvents influences negatively the accuracy of quantitations.

- Other extraction procedures

Besides batch extraction, continuous extraction is also applied in sample preparation. The procedure is used typically when the extraction efficiency is relatively small and a large number of batch extractions are necessary to perform a quantitative separation. There are various designs for continuous separation devices. One type of continuous extractor operates on the principle of extracting the sample followed by the distillation of the extracting solvent, which is then condensed and passed again through the solution being extracted. The extracting liquid is recycled. The extracted solutes are expected to remain in the evaporation flask. Some restrictions regarding the separated material are imposed in this procedure, including low volatility and thermal stability. On the other hand, the solvent must be easily distilled. Two continuous liquid-liquid extractors, one for a solvent heavier than water and the other for a solvent lighter than water, are shown in Figure 10.1.7. The efficiency of the continuous extraction procedure depends on various factors: the partition coefficient K_i, the mixing efficiency of the two phases, the relative volumes of the two phases, etc. One practical method of improving the efficiency is to ensure as high an area of contact as possible between the two liquids. As the extracting solvent passes through the solution, fritted-glass discs, small orifices, baffles, and stirrers may be used to bring the two nonmiscible layers in closer contact.

Countercurrent systems also have been designed for extraction. In principle, counter-current distribution can be performed in a series of separatory funnels (stages), each filled with the same volume of lower phase. The sample is introduced in the upper phase of the first separatory funnel and transferred to the second funnel after attaining equilibrium. The process is repeated many times, with each upper phase transferred to the next separatory funnel. With automated countercurrent distribution equipment, this process can be performed with several hundred transfers. The countercurrent distribution process closely resembles low-resolution column chromatography, and an automated countercurrent extractor was developed as a model for partition chromatography (Craig apparatus [8]). The countercurrent distribution process is mainly useful in large-scale preparative separations [9]. The theory of countercurrent extraction can be modeled based on a process with many discrete stages [10]. In each stage, the equilibrium between the upper stage and

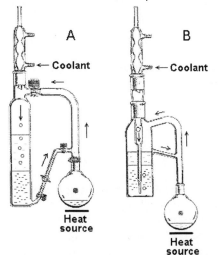

FIGURE 10.1.7. *Two continuous liquid-liquid extractors, one for a solvent heavier than water (A) and the other for a solvent lighter than water (B).*

lower stage is described by the constant $K = c^L / c^U$. When the volume dV of the upper stage is moved from stage i-1 to stage i, it carries the amount c^U_{i-1} dV to stage i. Also, when the volume dV of the upper stage i is moved to stage i+1, it carries the amount c^U_i dV to this stage. The process is shown schematically in Figure 10.1.8.

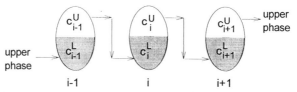

FIGURE 10.1.8. *Schematics of continuous flow operation with stages [10].*

At the same time, in the stage i the change in concentration is given in the lower phase by dc^L_i and in the upper phase by dc^U_i. The mass balance for stage i is given by the expression:

$$c^U_{i-1} \, dV - c^U_i \, dV = V^U \, dc^U_i + V^L \, dc^L_i \tag{10.1.5}$$

where V^U is the volume of the upper phase and V^L that of the lower phase. Expressing c^L_i as a function of c^U_i, rel. (10.1.5) can be written in the form:

$$\frac{dc^U_i}{dV} = \frac{c^U_{i-1} - c^U_i}{V^U + KV^L} \tag{10.1.6}$$

For the initial concentration of an analyte in the first stage c^U_0, the solution of the equation (10.1.6) is given by the formula:

$$c^U_i = c^U_0 \frac{\exp(-F) \; F^i}{i!} \quad \text{with} \quad F = \frac{V^T}{V^U + KV^L} \tag{10.1.7}$$

where V^T is the total volume of upper phase passed through stage i. A graph showing the variation of c^U_i / c^U_0 as a function of K and taking as an example $V^U = V^L$ and $V^T = 3\,V^U$ is shown in Figure 10.1.9 for two values of i. The variation of c^U_i / c^U_0 as a function of F is shown in Figure 10.1.10 for two values of i.

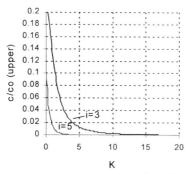

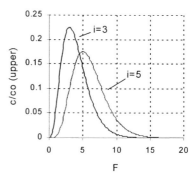

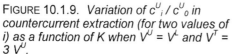

FIGURE 10.1.9. *Variation of c^U_i / c^U_0 in countercurrent extraction (for two values of i) as a function of K when $V^U = V^L$ and $V^T = 3\,V^U$.*

FIGURE 10.1.10. *Variation of c^U_i / c^U_0 in countercurrent extraction (for two values of i) as a function of F.*

Figure 10.1.9 shows that in the countercurrent procedure the extracted substance diminishes rapidly due to the depletion of material in the L phase as K increases. The increase in the number of stages gives a higher value for F. Figure 10.1.10 shows the formation of the elution peak during the extraction as a function of the F value [11].

Microextraction is another form of LLE in which extractions are performed with phase ratio values in the range 0.001–0.01. Compared with conventional LLE that uses a small volume of organic solvent, microextraction provides less analyte recovery, but the analyte concentration in the organic phase is significantly enhanced. In addition, solvent usage is reduced. This type of extraction can be performed in a volumetric flask using organic solvents that have lower densities than the aqueous sample, such that the small volume of organic solvent accumulates in the narrow neck of the flask for convenient withdrawal. It is highly recommended in this procedure to use internal standards for controlling the extraction [8]. Examples of microextraction include US-EPA Methods 504 and 505, which use only 2 mL of hexane to extract analytes from 35 mL of water samples, and US-EPA Method 551, which uses 2 mL of methyl-t-butyl ether to extract chlorinated solvents and chlorinated by-products of disinfectants from 35 mL of water. Microextraction with 2 mL of methyl-t-butyl ether to 30 mL of water has been used to extract haloacetic acids and chlorophenols from water samples.

The incorporation of LLE units into continuous-flow systems is another procedure to perform extractions. This procedure provides several advantages over more time-consuming manual procedures, because it is automatic and has good sample throughput. Continuous-flow LLE assemblies typically use three basic units, a segmentor, an extraction coil and a phase separator. The correct operation of the segmentor and of the separator is important for the efficiency of the separation [12].

On-line LLE with a GC instrument also has been reported [13], [14]. The operation can be performed by periodic injection of the aqueous sample into a stream of organic solvent. Because it takes some time before the extraction reaches equilibrium, long lines (2–3 m) of PTFE or fused silica tubing are used, in which a segmented flow is formed. After extraction, phase separation can be achieved using a semi-permeable membrane [15] or a sandwich-type phase separator [16]. The organic phase is then led through an interface, and a suitable portion is transferred into the GC system. The GC is typically equipped with a large volume sample injector and a portion of 100–500 µL is transferred into the injection port. The total on-line setup has been used successfully in several applications [17].

On-line extractions with simultaneous derivatization of the sample are also reported. For example, a continuous extraction system with two-phase derivatization has been developed for the determination of phenols [18] and for the analysis of carbamate pesticides [19]. Another application is the determination of organic acids and chlorinated anilines in aqueous samples [20]. In this analysis, the organic acids and chlorinated anilines are alkylated and acylated, respectively, in a two-phase reaction. Continuous LLE systems including a derivatization module coupled on-line with the GC have been used in several analyses [12], [21]. A disadvantage of the on-line system is that the excess of the derivatizations reagent in the organic phase is not removed and appears as a very large peak in the GC chromatogram.

Automation in LLE plays an important role because classical LLE requires manual handling. Several instrument manufacturers have developed instrumentation that can automate all or a portion of the extraction and concentration process. A number of autosamplers and workstations for HPLC and GC can perform LLE. Most of these systems use their liquid dispensing and mixing capabilities to perform LLE in sample vials or microvials [9]. Some systems mix the layers by alternately loading the solvents into the autosampler needle and dispensing the contents back into the sample vial [22]. By controlling the depth of the needle, either the top layer or the bottom layer can be removed for injection or further sample preparation. Most of these devices use small volumes of sample (in the mL range) and have limited capability to perform LLE. In other cases, the units use vortex mixing to spin the vial at a high speed for enhancing the separation [23], [24]. Large volumes, such as 1 L of water samples required in certain methods can be handled only by using specific robotic systems.

- *Extraction of ionic species and of metal complexes*

As a general rule, ionic species are not easily extractable with organic phases, while nonionic ones are significantly more easily extracted. Considering a compound AH that can dissociate in the form:

$$AH \rightleftharpoons A^- + H^+$$

the distribution constant for the species [AH] when the substance is extracted in a LLE system is given by the expression:

$$K = \frac{[AH]_o}{[AH]_w} \tag{10.1.8}$$

However, due to the dissociation, a part of AH is present in A^- form, and for analytical purposes both forms AH and A^- are usually of interest. The characterization of the distribution of both forms is done using a total distribution constant K^T defined by the expression:

$$K^T = \frac{[AH]_o}{[AH]_w + [A^-]_w} \tag{10.1.9}$$

The concentration of A^- in the organic phase is neglected in rel. (10.1.9), being assumed very low. Using the dissociation constant for the acid AH defined by the expression:

$$K_a = \frac{[H^+]_w [A^-]_w}{[AH]_w} \tag{10.1.10}$$

the value for $[A^-]_w$ obtained from rel. (10.1.10) and the value for $[AH]_o$ obtained from rel. (10.1.8) can be substituted in rel. (10.1.9), leading to the expression:

$$K^T = \frac{K[AH]_w}{[AH]_w + K_a[AH]_w /[H^+]_w} = \frac{K}{1 + K_a /[H^+]_w} = \frac{K[H^+]_w}{[H^+]_w + K_a} \tag{10.1.11}$$

Rel. (10.1.11) shows that $K^T \approx K$ when [H$^+$] is significantly higher than K_a. However, always a small amount of material is present in A^- form, and in general $K^T < K$. The increase in pH of the aqueous phase favors the extraction in the organic phase. For polyprotic acids, the expression for K^T can be obtained easily following the same procedure. For example, for an acid of the form AH_3, the expression for K^T is given by the expression:

$$K^T = \frac{K[H^+]_w^3}{[H^+]_w^3 + K_{a1}[H^+]_w^2 + K_{a1}K_{a2}[H^+]_w + K_{a1}K_{a2}K_{a3}} \qquad (10.1.12)$$

where

$$K_{a1} = \frac{[H^+][AH_2^-]}{[AH_3]} \qquad K_{a2} = \frac{[H^+][AH^{2-}]}{[AH_2^-]} \qquad K_{a3} = \frac{[H^+][A^{3-}]}{[AH^{2-}]} \qquad (10.1.13)$$

Solvent extraction applied to liquid phases has been frequently applied for the extraction of metal chelates. A number of chromatographic techniques are available for the analysis of metal chelates (see Section 19.8). The extraction of a metallic ion M^{n+} with a chelate agent HA dissolved in an inert organic solvent can be generally described by the following equation [25]:

$$M^{n+} + nAH \rightleftarrows MA_n + nH^+$$

This reaction leads to an uncharged species MA_n, which is extractable in the organic solvent. The distribution constant K for the extraction, assuming that no other species are formed in the reaction, is defined by the expression:

$$K = \frac{c_{M,o}}{c_{M,w}} = \frac{[MA_n]_o}{[M^{n+}]_w} \qquad (10.1.14)$$

where $c_{M,o}$ is the concentration of the metal present in the form of a chelate in the organic phase, and $c_{M,w}$ is the concentration of the metal present in the aqueous phase as an ion.

The formation of the chelate is typically done starting with an organic reagent having chelating properties in the organic phase and with the metal in the aqueous phase. The equilibria taking place in the system are schematically indicated in Figure 10.1.11.

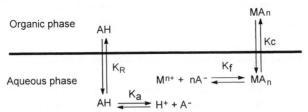

FIGURE 10.1.11. *Equilibria between an organic and an aqueous phase in a system of an organic chelating reagent and an inorganic ion.*

The constants K_a and K_f for the chemical equilibria are in general known for aqueous solution, and the distribution constants K_R and K_c are known for specific solvents and compounds. The expressions for these constants are given below:

$$K_a = \frac{[H^+][A^-]}{[AH]} \qquad K_f = \frac{[MA_n]}{[M^{n+}][A^-]^n} \qquad K_R = \frac{[AH]_o}{[AH]_w} \qquad K_c = \frac{[MA_n]_o}{[MA_n]_w} \qquad (10.1.15)$$

By the use of a sequence of replacements in rel. (10.1.14) with the values given in rel. (10.1.15), the expression for K can be written as follows:

$$K = \frac{K_c K_f K_a^n \, [AH]_o^n}{K_R^n \quad [H^+]_w^n} \tag{10.1.16}$$

Rel. (10.1.16) indicates that the constant K depends on the concentration of the reagent AH in the organic phase and the pH of the aqueous phase, as well as on the constants given in rel. (10.1.15). From rel. (10.1.16) it can be easily obtained

$$\log K = \log K_{ex} + n \, pH + n \, \log [HA]_o \tag{10.1.17}$$

where $K_{ex} = K_c \, K_f \, K_a^n / K_R^n$ and can be expressed as:

$$K_{ex} = \frac{[MA_n]_o [H^+]_w^n}{[M^{n+}]_w [AH]_o^n} \tag{10.1.18}$$

In most cases, the metallic ion is present in the aqueous phase not as a unique ion but as different complex species, such as hydroxo-complexes $M(OH)_m^{n-m}$, intermediate complexes with the main chelating anion (A$^-$) of the form MA_i^{n-i}, or complexes ML_j^{n-j} with other ligands (L$^-$). In these cases, the formula giving K for a unique complex must be modified. The distribution constant will have in this case the expression:

$$K = \frac{c_{M,o}}{c_{M,w}} = \frac{[MA_n]_o}{[M^{n+}]_w} \delta_M \tag{10.1.19}$$

where δ_M represents the fraction of all metal containing species in the aqueous phase that is in ionic form. The expression for δ_M can be obtained based on the equilibria present for each particular system [26].

Ion pair formation is also a procedure to form uncharged species extractable in organic solvents. As shown in Section 1.4, for a simple system the distribution of a single ion is given by rel. (1.4.40). However, the formation of dimers and trimers in the organic phase, as well as the presence of several ionic species in the aqueous solution for the same analyte, complicate the picture of ion pair extraction [10].

- Optimization of the factors influencing the extraction process

A number of factors are important when choosing LLE as a sample preparation technique. These factors address on one hand the modifications of the analyte or of the compounds to be extracted, such that the partition into the organic phase or in water takes place with a high partition coefficient. On the other hand are the nonchemical elements of the extraction, which include a) the choice of the extraction technique, b) the choice of the solvent, c) the choice of solvent and aqueous phase volumes, d) the time of extraction, e) the drying of the solvent, f) the solvent evaporation procedure, etc.

In the first group of factors must be included the solution pH, the concentration of the reagents such as chelating agents added to the solution, addition of other chemical modifiers such as cryptands used to solvate ions for extraction in organic solvents, the formation of ion pair complexes, etc. The influence of these factors affecting the extraction can be inferred from the theory presented in this section. The use of buffers to control the pH of the solutions is frequently required. A considerable number of buffer solutions are described in literature (see e.g. [27]). One other procedure to influence the partition coefficient is the addition of salts (known as salting-out effect: see rel. 8.2.11), which increases the distribution coefficient of different species.

The other group of factors is also important for a successful utilization of LLE in sample preparation. The choice of the extraction procedure (batch or continuous), the number of extractions when using the batch procedure, etc., are important for achieving the appropriate extraction efficiency. As shown in Figure 10.1.3 and using rel. (10.1.3), the extraction efficiency can be calculated if the distribution coefficient is known or at least estimated. An extraction higher than 99% is typically desirable. However, using appropriate internal standards, the extraction with lower efficiency can be used. In this case it is important to verify for quantitative determinations that the internal standard has a very close distribution coefficient with the analytes. Otherwise, the normalization based on the internal standard may lead to proportional systematic errors.

The choice of solvent with a low boiling point or the ease of removing the solvent after extraction for the concentration of the analytes are only two of the criteria used in the selection of a solvent when the possibility of choice exists. The solvents with very low boiling point, however, may give some problems with controlling the organic phase volume, which can be a source of errors. The solvents having high boiling points are usually avoided because they are difficult to remove when a concentration is required after extraction. Solvent extraction often requires the removal of solvents by evaporation. This operation can be time consuming and may lead to loss through evaporation or decomposition of unstable compounds, particularly when heating is required. The degree of miscibility of the two phases, the relative specific densities, viscosity, and tendency to form emulsions also should be taken into consideration in the choice of a solvent. The safety, the toxicity, and flammability of the organic solvent must also be considered when making a choice. Some of the main physical properties of several solvents used in a LLE procedures are given in Table 10.1.1 (see also Tables 1.10.11 to 1.10.19, 4.3.3, 17.2.1 and 17.2.2).

TABLE 10.1.1. *Physical properties of some common solvents used in extractions.*

Solvent	Dielectric constant	Dipole moment (D)	Boiling point (°C)	Density (g/mL)	Solubility in water (% weight)	Solubility of water in solvent (% weight)
hexane	1.89	0	68.7	0.66	0.01	-
benzene	2.28	0	80.1	0.88	0.08	0.06
cyclohexane	2..02	0	80.72	0.78	-	0.01
toluene	2.33	0.31	110.62	0.87	0.05	0.06
m-xylene	2.4	0.4	139	0.86	0.01	0.04
chloroform	4.81	1.04	61.15	1.49	0.81	0.97
CCl4	2.2	0	77	1.58	0.08	0.01
chlorobenzene	5.5	1.6	132	1.11	0.05	0.05
diethyl ether	4.4	1.2	35	0.71	6.9	1.26
diisopropyl ether	3.9	1.2	68	0.72	0.9	0.6
dibutyl ether	3.1	1.2	142	0.77	0.3	0.19
diamyl eter	3.1	-	187	0.78	-	-
ethyl acetate	6.4	1.8	77	0.9	7.94	3.01
methyl isobutyl ketone	13.1	-	116	0.8	1.7	1.9
cyclohexanone	18.2	2.8	157	0.95	2.3	8

Sometimes a mixed-solvent system is used to obtain the desired characteristics of the solvent. A synergistic effect has been obtained in some cases of solvent mixtures, the partition coefficients being better for the mixture than for each individual solvent. For example, if a water solution is extracted with a solvent containing x_1 mole fraction solvent "A" and $1-x_1$ mole fraction solvent "B", it would be expected to obtain a distribution coefficient $K = x_1 K_A + (1-x_1) K_B$. If a different distribution constant K_{A+B} is obtained, a synergy coefficient can be defined with the expression:

$$S = \log \frac{K_{A+B}}{K_A + K_B} \qquad (10.1.20)$$

Another parameter in LLE is the ratio between the volumes of the solvent and the aqueous solution. Larger solvent volumes favor partitioning of the analyte into the solvent layer. Serial solvent extraction is a better approach for obtaining high recoveries of the analytes, as shown previously. The amount of sample taken for processing must contain enough analyte to accommodate the sensitivity of the core analytical chromatography. The volume of solvent must be adjusted to compromise between efficient extraction and convenient liquid volumes to handle and possibly to evaporate when the concentration in the organic layer is too low. LLE of large volumes of water also can be performed using continuous extractors. Continuous LLE with CH_2Cl_2 has been used for example to extract 10 to 100 L of water samples for pesticide analysis. A total of 35 pesticides, including organochlorines, organophosphates and triazines, at concentrations of approximately 15 ppb were extracted by this procedure [28], [29].

The extraction time is another parameter to consider. Although for the extraction of liquids the equilibrium is achieved faster than for solid samples, the extraction time and the shaking efficiency must be chosen to assure that the extraction is close to equilibrium. In order to achieve equilibrium, longer time and more intensive mixing are typically required than usually practiced. For this reason, an "effective" extraction parameter K_i^{ef} is more appropriate to describe practical extractions. This extraction parameter is only a fraction of thermodynamic distribution constant, and $K_i^{ef} = \zeta K_i$. The coefficient ζ is determined by the duration of extraction, the volume of extraction solvent, and other extraction conditions [30].

Frequently, the organic solvent is separated after extraction, dried, and evaporated (in part or completely) [30a]. Several procedures for drying and solvent evaporation are described in Section 8.1. Conditions such as use of an appropriate drying agent (or procedure) or the choice of an evaporation temperature that does not lead to sample losses are very important for appropriate transfer of the analytes into the processed sample (see Section 3.4).

Optimization of an LLE procedure is usually focused on the selectivity of the process for achieving appropriate cleanup and on the extraction yield (or recovery). However, time and cost of the operation are also elements to be taken into consideration. The selectivity of the LLE may influence the choice of the chromatographic separation, and the LLE recovery influences the quantitation limit and the precision and accuracy of the entire process. A stable value of the recovery over a wide interval of concentration of the analyte in a sample to be proceeded by LLE is also important for obtaining a high precision of the analytical data.

One of the main problems in solvent extraction is the formation of emulsions. Emulsion formation is a drawback, particularly if a solvent of intermediate polarity (such as isoamyl alcohol) is used as the extracting solvent, or if the volume ratio of organic solvent and aqueous phase is not sufficiently large. Emulsions may be avoided by changing the phase ratio and may be broken by filtration over glass wool or filter paper, centrifugation refrigeration, addition of salt, addition of a less polar organic solvent, etc. However, these approaches are not always satisfactory, and a solid phase extraction procedure must be considered.

Other limitations of LLE sample preparation are known. For example, polar compounds that can form hydrogen bonds with water molecules are difficult to extract from aqueous samples. Analytes with carboxyl or hydroxyl groups are difficult to remove from water, even using acidification of the sample (pH < 2), which may improve the extraction. Trace contaminants are generally present in solvents used for extraction, some enough to interfere in the determination of the analytes of interest [31]. Preservatives that are purposely added to solvents also may be a source of analytical artifacts. Unexpected reactions such as that of cyclohexene with bromide and chloride ions may raise the levels of chlorinated or brominated derivatives [32]. Many artifacts formed during LLE using US-EPA 625 method have been identified by GC-MS, including artifacts formed by oxidation, halogenation, or nitration of phenolic compounds, and halogenation and autooxidation of cyclohexene [33]. Another problem associated with solvent extraction can be the adsorption on glassware, which can occur at the extraction or solvent evaporation stage and can influence the analysis of traces. Also, the contamination of the sample with grease and components from the grease that is frequently used on stopcocks and glass stoppers may be a serious problem. The total elimination of grease and its replacement with polyethylene sleeves is preferable.

- *Examples of applications*

As shown in Figure 1.2.1, LLE is one of the most common operations in sample preparation, being applied in almost 50% of all sample preparation methods. A variety of samples are processed with methods that include LLE. Only a few selected examples are discussed in this section, including analysis of biological samples, analysis of semivolatile pollutants in water, and analysis of inorganic ions.

Many biological samples such as plasma, animal tissues, plant materials, etc. are processed by LLE. For analysis of plasma samples, for example, LLE is used to remove the large part of plasma matrix, which interferes in the chromatographic process, and also to concentrate the analyte in the organic phase. The concentration is achieved either by use of a high ratio of plasma/solvent volumes or by evaporation of the organic solvent after the extraction. Many compounds used as drugs and their metabolites in plasma are analyzed after extraction in an organic solvent. Metabolites are usually more polar than the original compound, and the isolation procedure should be capable of extracting all the compounds of interest. Selective separation can be achieved by extraction with a number of solvents of increasing polarity and/or by extraction at different pH values. Another possibility for extracting the more polar metabolites is to extract the analytes into a solvent by addition of a salt to the aqueous phase, which shifts the partition such that the metabolites are better extracted into the organic phase (salting-out effect). Common solvents are frequently used in separation of drugs and their metabolites from plasma samples. The choice of the solvent is dependent on the extraction yield, volatility, and possibility of avoiding the formation of emulsion. Also the solvent must be selected to avoid the extraction of interfering compounds or plasma components that are undesired in the chromatographic separation. An example of an HPLC chromatogram for an extract in methyl-*tert*-butyl ether of a plasma sample is given in Figure 10.1.12. The separation is done on a C8 column using a mobile phase 20% mixture of methanol:acetonitrile = 1:1 and 80% phosphate buffer with pH = 7.3, a flow-rate of 1 mL/min., and detection at λ = 220 nm. As seen in this figure, compounds coextracting from plasma are present up to about 10 min. in the chromatogram. Determination of traces of drug metabolites eluting in this time interval cannot be done unless a higher wavelength of detection is possible.

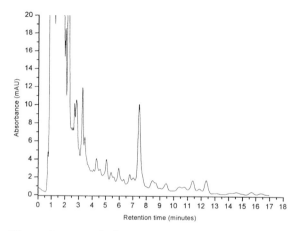

FIGURE 10.1.12. *Chromatogram of a human plasma extract in methyl-tert-butyl ether, monitored at λ = 220 nm.*

LLE is also a common procedure for extracting semivolatile organic compounds (SVOCs) from water samples. After extraction, the organic phase is separated from the water and is often dried with Na_2SO_4. The volume of the extract is then reduced, further sample cleanup may be performed, and the sample is analyzed. Some common LLE procedures used in the US-EPA methods for drinking water analysis are given in Table 10.1.2. Some other examples of LLE sample preparations used in various chromatographic analyses are given in Tables 10.1.3 and 10.1.4.

TABLE 10.1.2. *Liquid-liquid extraction methods used by US-EPA for drinking water analyses* [34].

US-EPA Method	Analytes	Procedure
504	1,2-dibromoethane, 1,2-dibromo-3-chloropropane	2 mL hexane to 35 mL sample; 6 g NaCl added before extraction; GC-ECD (detection limit: 10 ppt).
505	organohalide pesticides, PCBs	2 mL hexane to 35 mL sample; 6 g NaCl added before extraction; GC-ECD (3 ppt to 15 ppb).
506	phthalate and adipate esters	60 mL CH_2Cl_2 (twice) and 40 mL hexane to 1 L sample; 50 g NaCl added before extraction; GC-PID (0.84 to 11.5 ppb).
507	nitrogen and phosphorous pesticides	60 mL CH_2Cl_2 to 1 L sample (twice); pH = 7 with phosphate buffer and 100 g NaCl added before extraction; GC-NPD (75ppt to 5 ppb).
508	chlorinated pesticides	60 mL CH_2Cl_2 to 1 L sample (twice); pH = 7 with phosphate buffer and 100 g NaCl added before extraction; GC-ECD (1.5 ppt to 5 ppb).
508A	PCB screening	60 mL CH_2Cl_2 to 1 L sample (twice); GC-ECD; (> 0.5 ppb).
513	2,3,7,8-tetrachlorodibenzo-p-dioxin	60 mL CH_2Cl_2 to 1 L sample (three times); GC-MS (detection limits matrix dependent: 20 ppq to 2 ppt).
515	chlorinated acids	250 g NaCl added to 1 L sample; pH = 12 and sample washed with 60 mL CH_2Cl_2 (twice); pH = 2; sample extracted with 120 mL ethyl ether; acids derivatized to methyl esters with diazomethane; GC-ECD (20 ppt to 1 ppb).
550	PAHs (polycyclic aromatic hydrocarbons)	60 mL CH_2Cl_2 to 1 L sample (twice); HPLC-DAD-FLD (2 ppt to 3 ppb).
551	chlorinated by-products of disinfectants, chlorinated acids	2 mL methyl-*tert*-butyl ether to 35 mL sample; GC-ECD (2 to 100 ppt).
552	haloacetic acids	pH = 11.5; 100 mL sample washed with 30 mL methyl-*tert*-butyl ether; pH < 0.5; extraction with 15 mL methyl-*tert*-butyl ether (twice); GC-ECD (7.5 ppt to 0.3 ppb).
553	benzidines, N-containing pesticides	pH = 7; add 100 NaCl; extraction with 60 mL CH_2Cl_2 (twice).

TABLE 10.1.3. *LLE-GC applications for the analysis of various aqueous samples.*

Analyte	Matrix	Interface	Detector	Quantification limit	Reference
halocarbons	sea water	loop-type interface (LTI)	ECD	0.5 ng/L	[15]
aromatic hydrocarbons	municipal sewage water	LTI	FID	0.1 µg/L	[13]
hexachloro-cyclohexanes	ground water	LTI	ECD	0.1µg/L	[14]
carboxylic acids	water standard	LTI	FID	0.1 µg/L	[20]
chlorinated anilines	water standard	LTI	ECD	0.01 µg/L	[20]
phenolic compounds	water standard	splitless injection	FID	0.2 mg/L	[8]
carbamates	water standard	splitless injection	FID	0.2 mg/L	[19]
carbamates	milk	splitless injection	ECD	2 mg/L	[35]

TABLE 10.1.4. *Other applications of LLE in sample preparation for chromatography.*

Analytes[1]	Sample	Extraction solvent	Chromatographic analysis	Reference
humulene epoxides	beer	CH_2Cl_2	GC-MS	[36]
clavulanic acid	fermentation broth	PEG		[37]
pesticides	water	methyl-t-butyl ether	CGC-PD	[38]
nitrosamines	water	CH_2Cl_2	GC-NPD	[39]
phenols	water	CH_2Cl_2	HPLC, UV detection	[40]
pesticides	freeze-dried water	CH_2Cl_2	GC-NPD	[41]
antihistaminic H1 drug	human urine	$(C_2H_5)_2O$	HPLC	[42]
antidisrhythmic agent (UK 68798)	human urine	methyl-t-butyl ether	HPLC-DAD	[43]
remifentamil	human blood	CH_2Cl_2	GC-MS	[44]
dimyristoylphosphatidyl-glycerol	human serum	methyl-t-butyl ether	-	[45]
2-(4-t-butylphenoxy)-7-(4-chlorophenyl) heptanoic acid	blood	$(C_2H_5)_2O$	HPLC	[46]

[1] For information and chemical structure regarding various drugs see e.g. M. O'Neil et al. (eds.), *The Merck Index*, Merck & Co., Inc., 13th ed., Whitehouse Station, 2001.

Another field where solvent extraction has proven to be very useful is that of extraction of chelate complexes of inorganic ions. The procedure was initially used in connection with preparatory or other analytical purposes. More recently chromatographic methods were developed for the analysis of a number of inorganic ions (see Section 19.8). Several examples of reagents used for LLE extraction and chromatographic analysis of metals are given in Table 10.1.5.

For appropriate use in LLE, the complexes must be fairly well soluble in nonpolar organic solvents, which provide high distribution coefficients and concentration factors of metallic ions during extraction. The possibility of controlling the complex formation by pH-variation or by adding other complexing agents for modifying the extraction of specific ions can increase the selectivity of solvent extraction. A typical problem with the use of solvent extraction of chelates is the simultaneous extraction of a large amount of reagent. Some reagents are, however, acidic enough, and their distribution constant is not extremely high in nonpolar solvents. For these compounds, the excess of the reagent is not excessively high in the organic solvent, and the chromatographic separation is not complicated by its presence. Some extractions of the inorganic ions are performed with solvents that can act simultaneously as a complexing reagent. Such solvents include carbonyl compounds with weak basic character and also neutral organophosphorus compounds (basic phosphates, phosphonates, phosphinates and phosphine oxides). Some molecules containing C=O or P=O groups are able to coordinate to a central inorganic ion, their oxygen atom generally displacing a water molecule from the coordination sphere of the metallic ion or in some cases forming a coordination compound through a water molecule bridge. Tributyl phosphate (TBP) frequently has been used for this purpose either in pure form or diluted with inert solvents.

TABLE 10.1.5. *HPLC methods using LLE as sample preparation in the analysis of metals with the formation of chelating complexes.*

Reagent	Metallic ion[1]	Matrix	Detection	Detection limit	Reference
diethyldithio-carbamate	Se^{4+}	drinking and waste water	UV (254 nm)	3.6 µg/L	[47]
	Pt^{4+}	biological	UV (254 nm)	0.75 µg/L	[48]
	Pt^{4+}	biological	UV (254 nm)	25 µg/L	[49]
	Bi^{3+}, Co^{2+}, Cu^{2+}, Hg^{2+}, Mn^{2+}, Ni^{2+}, Pb^{2+}	steel, waste water	UV (280 nm)	25 – 250 ng	[50]
	Cr^{3+}, Cr^{6+}	water	UV (254 nm)	-	[51]
	Co^{2+}, Ni^{2+}, Cu^{2+}	natural water	UV (325 nm), and VIS (440 nm)	0.2 – 10 µg/L	[52]
	Cu^{2+}, Pb^{2+}, Zn^{2+}	river water	UV (265 nm)	0.2 – 0.5 ng	[53]
	Co^{2+}, Ni^{2+}, Cu^{2+}, Cr^{3+}, Hg^{2+}, Se, Te, Pb^{2+}	galvanic solutions	UV (254 nm)	0.5 – 18 ng	[54]
	Cd^{2+}, Hg^{2+}, Pb^{2+}	industrial wastes	UV (350 nm)	-	[55]
	Pd^{2+}, Rh	alloys	UV (245 and 297 nm)	0.05 ng	[56]
	Cd^{2+}, Hg^{2+}, Cu^{2+}, Ni^{2+}, Zn^{2+}	drinking water	UV (254 nm)	0.04 – 0.5	[57]
pyrrolidine-dithiocarbamate	Bi^{3+}, Co^{2+}, Cu^{2+}, Hg^{2+}, Ni^{2+}, Pb^{2+}	river water	UV (254 nm)	0.2 – 2 ng	[58]
	Cu^{2+}, Ni^{2+}, Pb^{2+}	citrus leaves	UV (254 nm)	-	[59]
	Cd^{2+}, Co^{2+}, Cu^{2+}, Hg^{2+}, Ni^{2+}, Pb^{2+}	electrolytes	UV and electro-chemical	0.1 – 1 µg/mL	[60]
hexamethylene-dithiocarbamate	Bi^{3+}, Co^{2+}, Cu^{2+}, Cd^{2+}, Zn^{2+}, Hg^{2+}, Ni^{2+}, Pb^{2+}	electrolytes	UV (260 nm)	0.16 µg/mL	[61]
	Bi^{3+}, Co^{2+}, Cu^{2+}, Cd^{2+}, Hg^{2+}, Ni^{2+}	river water	UV (260 nm)	0.045 – 0.3 ng	[62]
	Bi^{3+}, Co^{2+}, Cu^{2+}, Cd^{2+}, Hg^{2+}, Ni^{2+}, Zn^{2+}, Pb^{2+}	biological	UV	0.5 µg/mL	[63]
8-quinolinol	Co^{2+}, Mo^{2+}, Cr^{3+}	steel	UV (254 nm)	0.5 – 1.5 ng	[64]
	Mn^{2+}, Mn^{3+}	river water, soil	UV-VIS (394 and 404 nm)	-	[65]
	Al^{3+}, Cu^{2+}, Fe^{3+}, Mn^{2+}	biological	UV or electro-chemical	5 – 10 ng, 1 – 2 ng	[66]
dithizone (diphenylthio-carbazone)	Bi^{3+}, Co^{2+}, Cu^{2+}, Hg^{2+}, Ni^{2+}, Pb^{2+}, Mn^{2+}	river and waste water, biological matrices	UV (280 and 285 nm)	5 – 200 ng	[67]
	Cd^{2+}, Co^{2+}, Cu^{2+}, Hg^{2+}, Ni^{2+}, Pb^{2+}	biological	UV	0.1 – 5 ng	[68]
tetraphenyl-porphyrine	Cu^{2+}, Ni^{2+}, Zn^{2+}	tap water	Vis (412 nm)	-	[69]
tributyl phosphate	various ions	industrial wastes, industrial materials	UV	traces and % quantities	[70], [71], [72]

[1] No charge indicates various forms of the ion.

10.2 CONVENTIONAL LIQUID-SOLID EXTRACTION

Liquid-solid extraction (LSE) is the process of using a solvent for extracting compounds from a solid sample. The compounds to be extracted can be the analytes that must be separated from the matrix or, less often, specific matrix compounds that must be eliminated from the sample.

During LSE the compounds that are taken into solution can be adsorbed on the solid matrix or simply present in various forms as a mixture in the solid sample. In most cases, it is rather difficult to distinguish between the extraction of adsorbed compounds and the dissolution process of some constituents of a solid sample. True LSE could be considered as taking place only when the analytes from a solid matrix are physically interacting with it and are desorbed and partitioned in the solution. In practice, LSE refers to the extraction of compounds from any sample that does not allow simple dissolution. This includes in addition to the extraction of the adsorbed compounds, the extraction from solid samples of the compounds trapped or encapsulated in the solid materials. Among these are vegetal or animal tissues made from cells that do not allow immediate release of specific components, heterogeneous solid samples such as coal, bituminous shales, polymers of various kinds analyzed for the presence of small molecules (residual monomers, plasticizers, antioxidants), and char from pyrolysates, etc., with inclusions of substances of interest. For extraction, a specific amount of sample is measured (weighed) and extracted with the intention to completely remove the analytes (or other compounds of interest). Quantitative measurements are possible using internal standards even when the extraction is not complete. However, if the internal standard is added on the surface of the sample, it can be extracted very differently than a compound included in the matrix, and it is always preferable to have higher yields for the extraction of analytes during liquid-solid extraction.

For solid samples with the analytes trapped in the solid matrix, in addition to LSE, other operations are needed to help the extraction. Among these are cell disruption, sonication [73], grinding, or other means of particle size reduction. Special extraction procedures for solid samples improve the extraction yield by increasing the temperature and pressure of the solvent, by microwave exposure, etc. Some of these procedures are discussed in more detail in Sections 10.3 to 10.6. For solids that can be solubilized, a procedure of disrupting the matrix is the solubilization followed by precipitation and reextraction with a selective solvent, which dissolves only the compounds of interest.

The LSE applied to extract the compounds adsorbed on a solid surface is in fact the reverse process of adsorbing compounds on a solid material from a solution. This is the basis of a common sample preparation procedure known as solid-phase extraction (see Section 11.1). More details about the adsorption/desorption process on solid surfaces are discussed in Section 11.1.

In general, the extraction of solids requires the choice of a specific solvent, which depends on the chemical nature of the analytes and also on the nature of the matrix. The nature of the solvent is chosen based on solubility criteria (as discussed in Section 8.2) and on the diffusion coefficient of the solvent in a specific matrix that allows the penetration and removal of the solvent from the matrix. The dissolution rate can be approximated based on Fiks's law (see rel. 1.5.1b) by the expression:

$$\frac{dq}{dt} = \frac{AD}{d}(C_s - C_b)$$ (10.2.1)

where q is the quantity of solute, A is the surface area of the solid sample, D is the diffusion coefficient of the solute in the sample, d is the diffusion layer thickness, C_s is the concentration of the dissolving solute in the sample, and C_b is the concentration in the dissolving solution [73a]. Rel. 10.2.1. shows that a solvent with a high diffusion coefficient is preferable to one with low diffusivity. As diffusivity is related to solvent viscosity as shown in rel. (1.5.23), a more viscous liquid has in general a lower diffusion coefficient. The viscosities of several common solvents are given in Table 1.5.2. Also, a larger access area A of the solvent into the sample gives better dissolution rate, which indicates that a sample with low particle size is more easily extracted than one with large pieces (the division of the sample increases its area).

The compounds to be extracted should have good solubility in the selected solvent. The choice of solvent for LSE must also consider the type of separation expected when this technique is used. When a selective extraction is the purpose of the process, the compounds to be eliminated must be insoluble in the solvent. The amount of solvent must be larger than the amount of sample. A minimum of four times more solvent than sample is usually necessary, although much larger volumes of solvent can be used for specific applications.

Insufficient solvent may lead to incomplete extraction, difficulties to recover enough solvent after the extraction mainly when the sample adsorbs the solvent, etc. A lower volume of solvent is sometimes useful for achieving higher concentrations in the analytes. However, low recoveries during the extraction can be more damaging than low concentrations of the analyte. The extraction process can be done by placing the sample and the solvent in a closed vial, which is further subjected to mechanical shaking. The particle size of the sample is typically reduced by various mechanical procedures (see Section 7.1). The shaking can be done manually but much more frequently is done using mechanical shakers. The shakers are of various types, such as platforms with gyratory movement, roller bottles, wrist action, 3D-type, vibrational, etc. During the extraction, parameters such as temperature, shaking range and speed, etc. may need to be controlled. Other extractions are done by simple stirring of the solid sample with a certain volume of solvent. A version of this procedure, quite frequently applied, uses a boiling solvent with reflux as shown in Figure 10.2.1. This procedure has the advantage of performing the extraction at the temperature of the boiling point of the solvent. Due to the increase in solubility with the increase in the solvent temperature, this procedure may lead to better extraction. Closed vials with a tight cap can be used for extraction by heating the sample with the solvent at temperatures even slightly higher than the solvent boiling temperature. Sonication is another common procedure for increasing the rate of dissolution and of solvent penetration in the solid matrix.

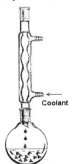

Coolant ←

FIGURE 10.2.1. *Extraction with solvent reflux.*

After the extraction, the solvent containing the compounds of interest must be separated from the remaining solid material. This can be done using filtration, decantation, or centrifugation (see Section 7.3). The solution remaining soaked in the solid material must be recovered as much as possible because it may retain a significant amount of analytes. Internal standards can be used (typically added in the solvent) to account for the loss of analytes when the solution is not completely recovered from the solid sample. Repeated washes may solve the problem of analyte loss in the solid material. After extraction, it is not uncommon that the extracted solution must be filtered. Losses in the filter must be avoided or accounted for using internal standards.

The optimization of the process of extraction of certain analytes from a matrix is not simple, since usually the analyte-matrix association is poorly understood. More information about the nature of interactions between analytes and matrix would allow a more appropriate choice of the extraction parameters based on models where the characteristics of the sample and the matrix are known [73b]. However, the subject is typically approached in an empirical manner, taking a number of parameters to be modified such as the choice of solvent, time of contact, mechanical shaking, extraction temperature, etc. The amount of solvent used during LSE depends in principle on the solubility of analytes. One problem with extraction optimization is the choice of an appropriate reference. The addition of standards and the study of their extraction does not give good results, since the standards can be added only on the surface of the sample and typically are easily extractable. Repeated extractions of the same sample until no material of interest is further extracted can be utilized for determining the extraction time. For some samples, the needed extraction time can be surprisingly long (6 to 12 hours or even longer), and the tendency to use shorter extraction time than needed is common. Comparison with a continuous extraction process such as Soxhlet extraction can be used for evaluating the optimum extraction time.

- *Soxhlet extraction*

Soxhlet extraction is a continuous extraction procedure that takes place in a special apparatus (see Figure 10.2.2). The procedure is a very common technique for

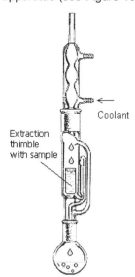

FIGURE 10.2.2. *Soxhlet apparatus.*

extracting organic compounds into a solvent and can be applied to solid and semi-solid samples. For extraction, the solid material to be extracted is placed in a thimble or in a fritted crucible and introduced in the middle extracting part of the Soxhlet. The thimble is usually made from cellulose and is permeable to the solvent. It is recommended that the samples be chopped, ground, or pulverized (see Section 7.1) to produce a fine particulate matter with a larger surface area prior to Soxhlet extraction. Some nonrigid samples can be better pulverized if the sample is first frozen and then chopped in a high speed blender or ground in a mortar and pestle together with dry ice [74]. The amount of sample used for extraction (typically grams) depends on the level of analytes and the sensitivity of the chromatographic analysis following the extraction. Several dimensions for the apparatus and for the sample thimble are commercially available. Many samples selected for Soxhlet extraction may contain water. The extraction with organic solvents can be in this case impeded by the presence of water and the inability of the solvent to wet the sample. For this reason, either the sample is dried before extraction (without losing analytes), or a desiccant can be added together with the sample in the thimble. Addition of an internal standard together with the sample in the Soxhlet apparatus is also recommended in many analyses.

The solvent used for extraction is added in the middle part of the Soxhlet to a level able to siphon into the lower round bottom flask. From practice, it is recommended for the solvent amount to be approximately 70 mL + 1.1 times the sample volume. After this, the solvent is distilled and allowed to condense back into the middle part of the apparatus. The temperature of the condenser must be low enough to avoid any loss of solvent. The process is repeated in a number of extraction cycles. This number depends on the solubility of the analytes and the capacity of the solvent to penetrate the sample matrix. The choice of the extracting solvent must be done depending on the nature of the sample. Besides the extraction efficiency and the insolubility of the matrix, this choice must consider other aspects. The requirements for thermal stability of the analytes at the boiling point of the solvent must be verified in order to avoid decompositions in the bottom flask. Also, the volatility of the extracting solvent must be selected such that excessive heating is avoided. Solvent flammability is also a factor to be considered. Ethyl ether frequently used for Soxhlet extractions has the disadvantage of being very flammable [75]. After the extraction is stopped, in some instances a filtration of the extract is necessary. Also, the solvent containing the extracted material can be further evaporated to increase the concentration of analytes.

The time required for the extracting solvent to diffuse into and out of the sample matrix is strongly influenced by the size of the particles (and the nature of the solvent). For samples easy to permeate, Soxhlet extraction times can be around 30 min., and for more compact samples about 1 h. However, for some samples the optimum extraction time can be significantly longer. Because the extraction once begun does not require constant tending, long extraction time is not necessarily a major problem with this technique. Also, it is common in laboratories applying Soxhlet extraction routinely to have batteries of several extractors processing multiple samples in parallel.

The extraction efficiency using the Soxhlet apparatus can be very good (close to 100%), and the procedure can be successfuly used in quantitative analyses with good recoveries. This efficiency was compared, for example, in the extraction of organic compounds from sediments using Soxhlet extraction with CH_2Cl_2 and the ultrasonic extraction with either CH_2Cl_2 or with CH_2Cl_2 / methanol (19:1). Soxhlet extraction gave the highest recoveries for nonpolar compounds, but the differences were marginal for compounds of medium polarity. Recoveries of 92% for aromatic compounds and 96% for aliphatic compounds and PCBs were obtained [76]. For most polar compounds the ultrasonic extraction with CH_2Cl_2 / methanol gave better recoveries. The addition of methanol to CH_2Cl_2 improved the efficiency of ultrasonic extraction [77]. Comparison of ultrasonic extraction and Soxhlet extraction has been discussed for many other applications. Among these are extractions for polycyclic aromatic hydrocarbons from sediments and air particulate material [78], PCBs in sand and soil [79], polychlorinated dibenzo-p-dioxins and dibenzofurans in fly ash [80], pentachlorophenol in water and soils [81], priority pollutants in soil [82], phthalate esters in sewage sludge [83], alcohols and phenols from airborne particulate matter [84], etc.

Data from an inter-laboratory study of the determination of PCBs in soils were used to examine the contribution of factors such as extraction equipment, solvents, and cleanup procedures to the variance and bias in the mean of the analytical results. Five soils spiked with Aroclor 1260 (mixture of polychlorinated biphenyls with the average number of Cl/molecule equal to 6.30) at concentrations covering four orders of magnitude were employed. The results obtained using Soxhlet extraction were more accurate than those obtained using sonication, particularly at higher PCB concentrations, but had equal precision. The use of nonpolar solvents gave analytical results of lower accuracy than those obtained with more polar solvents [85].

Comparison of methods for the extraction of flavonoids from birch leaves (*Betula pendula*) was carried out using HPLC as a technique of analysis. The compared extraction methods were a) short extraction (3 min.) with unheated solvent, b) refluxing with solvent for 30 min., and c) Soxhlet extraction for 18 h. The tests were performed in each instance with acetone, ethanol, and methanol as solvents either in the anhydrous form or as 1:4 mixtures with H_2O (H_2O:solvent). The extracts were evaporated to dryness, and the residues were dissolved in 50% methanol/water and analyzed by HPLC. Soxhlet extraction and refluxing were the most efficient methods of extraction of flavonoids using several solvents. Refluxing extraction with anhydrous methanol was the most convenient exhaustive extraction procedure for flavonoid glycosides [86].

In addition to the common Soxhlet extractor, a high-pressure Soxhlet extractor is also described in literature [74], [87], [87a]. This consists of a Soxhlet enclosed in a high-pressure chamber such that a standard glass apparatus can be used with low boiling point extracting liquids (CCl_2F_2, NH_3, CO_2, N_2O, pentane). The conventional solvents are replaced with substances that can be liquefied at temperatures of 0 to 20°C at pressures up to 100 atm. At the end of the extraction, the chamber is depressurized, the solvents evaporated, and the analytes are recovered from the bottom flask with a small amount of another solvent [88]. High recoveries were obtained by this technique, for example, in the analysis of morphine from blood and of caffeine from kola nuts [89].

During Soxhlet extraction the temperature of the solvent is below its boiling point. As the solubility increases with the solvent temperature, attempts were made to perform an extraction similar to Soxhlet, but maintaining the temperature of the solvent at the boiling point. This can be achieved in a Soxtec system [90]. A Soxtec system has the sample in a thimble introduced in a sample holder that is immersed in the boiling solvent. Fresh condensing solvent flushes the sample. After extraction, the sample and holder are removed from the boiling solvent and rinsed again with fresh solvent.

The continuous heating of the solvent in a conventional Soxhlet extractor may lead to the formation of artifacts, mainly when this heating is extended for a long time (such as a few days) or when the boiling temperature of the solvent is relatively high [91]. For example, the use of acetone – hexane in the Soxhlet extraction of solid environmental samples was proved to produce artifacts due to the reactivity of acetone. A more inert solvent such as CH_2Cl_2 led to fewer problems [91a]. The formation of trace levels of nitrosamines was noticed during Soxhlet extraction with CH_2Cl_2 of rubber nipples for baby-food bottles. Using morpholine as a marker amine, between 9 and 80 ng of nitrosomorpholine was formed per analysis, depending on the sample type and the brand of CH_2Cl_2 used. The problem was minimized by incorporating propyl gallate as a N-nitrosation inhibitor in the extraction process [92].

- *Examples of applications*

Solvent extraction is a useful and common technique in many solid sample preparations. The frequency of use of LSE and of Soxhlet extraction in particular compared to other sample preparation techniques is considerable, as shown in Figure 1.1.1. Many extractions are done on the initial sample. However, the extraction also can be done on traps or filter materials after a particular sample collection. For example, the filters used for sampling environmental pollutants retain the analytes very efficiently, and a Soxhlet extraction is necessary to release them. This procedure is applied to analyze urban air samples collected on glass fiber filters (particulate) and polyurethane foam plugs (semivolatiles). These filters are subjected to Soxhlet extraction with benzene for the recovery of the analytes, followed by a cleanup step and analysis [93]. Traps such as XAD-2 also can be extracted using a Soxhlet apparatus. For example, in a sampling system for the determination of polychlorinated dibenzo-p-dioxins, dibenzofurans (PCDDs and PCDFs), PCBs, hexachlorobenzene (HCB), and PAHs in ambient air, two glass fiber filters and two XAD-2 traps were used for collection. After the air was passed through the collection device, the glass-fiber filters and XAD-2 traps were subjected to Soxhlet extraction with toluene for 24 h, followed by sample cleanup and GC-MS analysis [94]. A number of applications using LSE as a sample preparation procedure followed by a chromatographic analysis are indicated in Table 10.2.1.

TABLE 10.2.1. *Some applications of LSE extraction in connection with chromatographic analyses.*

Sample	Analytes[1]	Solvent	Extraction Conditions	Technique	Ref.
propolis	26 phenolic components	petroleum ether	Soxhlet, 40 – 60°C, 3 h.	HPLC, UV detection	[95]
legumes (peas, beans, soybeans)	saponins	methanol	Soxhlet, 50 h.	TLC	[96]
river sediments	organo-chlorine pesticides	hexane/acetone (1 : 1)	Soxhlet, 10 h	GC-ECD	[97]
soil	pyridine	CH_2Cl_2		GC-MS	[98]
emissions	dioxins, dibenzofurans	toluene	collect on XAD-2, Soxhlet, 24 h	GC-MS	[99]
pigments	dioxins, dibenzofurans	CH_3Cl/toluene (19 : 1)	24 h	GC-MS	[100]
fish	arsenobetaine arsenocholine	$CHCl_3$ and CH_3OH	-	HPLC with off-line AAS	[101]
Chelidonium majus	alkaloids	$CHCl_3$	Soxhlet, 2 h	HPLC, UV detection	[102]
fish	PCBs (polychlorinated biphenyls)	hexane/CH_2Cl_2 (1:1)	Soxhlet, 3 h	GC-MS	[103]
roots of *Plumbago zeylanica*	plumbagin	$CHCl_3$	Soxhlet, 4 h	HPLC, UV detection	[104]
pecan nuts	tocopherol	hexane	-	HPLC-FLD	[105]
stalk, leaves	flavonoids	95% ethanol	-	TLC, UV detection	[106]
seeds	carbaryl	acetone, hexane or methanol	Soxhlet, 4 h	HPLC, UV detection	[107]
cocoa beans	triglyceride	petroleum ether (bp 40-60°C) – ethyl ether (1:1)	4 h	HPLC, refractive index detection	[108]
soil	2,4,6-trinitrotoluene	CH_3OH	-	HPLC, UV detection	[109]
wine	carbamate	ethyl ether	5 – 6 h	GC-MS	[110]
cigarette filter	benz[c]acridine	chlorobenzene	-	HPLC-FLD	[111]

TABLE 10.2.1 (continued). *Some applications of LSE extraction in connection with chromatographic analyses.*

Sample	Analytes[1]	Solvent	Extraction Conditions	Technique	Ref.
soil	fluazifop	acetone	18 h	HPLC, UV detection	[112]
air	PAHs (polycyclic aromatic hydrocarbons	CH_2Cl_2	50° C	GC-MS	[113]
air	PAHs (polycyclic aromatic)	cyclohexane	Soxhlet	HPLC, UV detection	[114]
air	polychlorinated dibenzofurans and dibenzo-p-dioxins	toluene		GC-MS	[115]
air	2-nitrofluoranthene and 2-nitropyrene	CH_2Cl_2	Soxhlet, 17 h	GC-MS	[116]
air	nitro-PAHs	benzene / methanol (4:1)	8 h	GC-MS	[117]

[1] For information and chemical structure regarding various drugs see e.g. M. O'Neil et al. (eds.), *The Merck Index*, Merck & Co., Inc., 13th ed., Whitehouse Station, 2001.

10.3   ACCELERATED SOLVENT EXTRACTION

Accelerated solvent extraction (ASE) is an extraction technique for solids and semi-solid samples that takes advantage of the increase in solubility with temperature increase and of the capability of liquids to penetrate better various matrices when they are at higher pressure (the name pressurized fluid extraction is also used for this technique). The pressure increase allows the use of temperatures above the boiling point at atmospheric pressure of the solvents. The increase of extraction efficiency in ASE with temperature from a soil sample contaminated with petroleum products is shown, as an example, in Figure 10.3.1 [8]. Due to higher extraction efficiency, the extraction times are also reduced in ASE. In extreme cases, this can be as significant as from hours to just several minutes [118]. The extraction of compounds adsorbed on the surface of specific matrices as well as that of analytes encapsulated in the matrix are more efficient at higher temperature. The increase in extraction efficiency allows

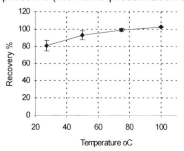

FIGURE 10.3.1. *Increase with temperature of extraction efficiency in ASE of petroleum products from a soil sample [8].*

in many instances the use of a lower volume of solvent than with conventional LSE techniques. Extraction using a heated solvent and at a pressure that keeps the solvent liquid even above its boiling temperature at normal pressure can be performed with a simple well-closed reaction vial that is loaded with the sample and the solvent and then heated by using a common heating block. However, accelerated solvent extraction usually refers to an automated system. A block diagram of an accelerated solvent extraction system is shown in Figure 10.3.2.

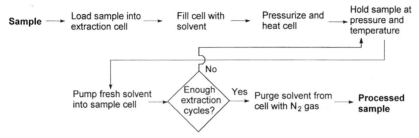

FIGURE 10.3.2. *Operations involved in an ASE procedure.*

The diagram shown in Figure 10.3.2. indicates that for the same sample the cell can be refilled with fresh solvent. This step can be repeated a number of times (cycles) and is similar to the cycles in a Soxhlet extraction, assuring a high recovery for the analytes. Automated instruments are capable of holding the solvent for a specified time period in the extraction cell (static extraction) and also of having a solvent flow (at a specified flow rate) through the cell. The final purge step (typically done with nitrogen) has the role of recovering the solvent of the last cycle from the sample. These operations are usually performed in commercially available instruments that have the schematic diagram shown in Figure 10.3.3.

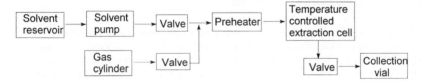

FIGURE 10.3.3. *Block diagram of an ASE instrument.*

In many respects the extraction using ASE is similar to other extraction procedures, except with a higher extraction efficiency. For example, the extraction selectivity cannot be expected to be improved in ASE compared to other procedures. In fact, a potential disadvantage of ASE comes from the fact that it tends to be exhaustive and leads to nonselective extractions, which require additional cleanup [119]. A comparison of several solvent extraction procedures showing the advantages of ASE for solid samples is shown in Table 10.3.1 [120]:

TABLE 10.3.1. *Comparison of several extraction techniques of solid samples [120].*

Extraction technique	Typical volume of solvent (mL)	Typical extraction time
Soxhlet	300	4–48 hours
sonication	200	30–60 min.
supercritical fluid	8-50	30–120 min.
accelerated solvent	15-40	12–18 min.
microwave	25-50	30–60 min.

Besides a number of advantages, ASE also has limitations. One important limitation is related to sample modifications due to the heating. Samples sensitive to heating and those that may generate artifacts during heating cannot be extracted using this technique. Another problem is related to possible precipitation of solids when the sample leaves the heated extraction cell. The temperature of the extraction fluid typically lowers significantly at the exit valve and on its way to the collection vial. During this transfer, part of the analyte may be lost and precipitation of solutes may occur. This may cause carry-over of material from sample to sample. For this reason, ASE is less useful for the extraction of samples where an efficient extraction is required only because of the solubility limitations in normal conditions. The main applications of ASE are related to the extraction of analytes that are adsorbed in the sample matrix and are not easily extracted unless additional energy helps the process.

- *Optimization of the extraction process in ASE*

The parameters that can be used to optimize the extraction process in ASE are a) the nature of the solvent (or mixture of solvents), b) sample amount and preliminary processing, c) solvent temperature, d) solvent volume, and e) contact time of the sample and the solvent. The choice of the extraction solvent is made based on the same criteria as for other liquid-solid extraction techniques. If a solvent is known from another extraction procedure to work well with a specific type of sample, it can be chosen in ASE. Other aspects such as the compatibility with the postextraction chromatographic procedure, solvent volatility, cost, and environmental factors are also considered in choosing the solvent. Typical solvents include water, buffered aqueous solutions, hexane, acetone, methylene chloride, etc.

The sample amount is selected to accommodate the sensitivity of the analytical instrument. Larger samples are typically needed when the analytes are at low concentration in the sample. The need for a representative sample and practical considerations regarding handling usually set the limitations regarding the smallest acceptable amount of sample. In general, it is preferable to use samples that are not at the limit of the smallest or the largest amount. Commercial instruments for ASE are usually equipped with different extraction cell sizes (e.g. 11, 22, and 33 mL) to accommodate small to large amounts of sample. Previous to extraction it is recommended to grind the sample. Also, the aggregation of sample particles can be avoided by mixing of the sample with an inert material. Samples that are extracted for organic compounds and contain a certain amount of water are usually better extracted if they are dried before or during the extraction. Drying materials, such as anhydrous Na_2SO_4 , diatomaceous earth, etc., can be added together with the sample in the extraction chamber. Also, adsorbing solid phase materials may be added together with the sample for selectively retaining specific components from the sample during the extraction process and eliminating some interferences [121].

Solvent temperature is selected depending on the nature of the sample and that of the solvent utilized. The high-pressure ASE allows the solvent to be heated above its normal boiling point, typically in the range of 50 to 150°C and in special applications up to 200° C. The high temperature accelerates the diffusion process through the sample and consequently improves the extraction rate. A higher temperature is therefore preferable to a lower temperature. The concerns about sample degradation and formation of artifacts are limiting factors for the temperature increase.

The pressure is applied in ASE mainly for maintaining the solvent in liquid state at temperatures above its boiling point (at atmospheric pressure). Gases do not have good solvating capability (see Section 10.6), and the solvent must be kept in liquid phase. Typical pressures used in ASE are between 1000 and 2000 psi, which are significantly higher than the threshold required to keep the solvents in liquid form. The pressure is also necessary to move the liquid through the system.

The solvent volume is typically reduced in ASE compared to other extraction procedures such as Soxhlet extraction [120]. The optimum volume must be established such that the recovery of the analytes from the sample is as close to complete as possible. Also, the filtration at the end of the extraction usually is not necessary. Typical extraction time with ASE systems is 10 to 20 minutes per sample. However, the time can be adjusted depending on the recovery efficiency [122].

- Example of applications

Various applications are reported for automatic ASE, such as extraction of small molecules from synthetic polymers, analyses of foods, natural products, soils, and pharmaceuticals [8]. These applications can be related to the extraction of major components such as oils from oilseeds or of trace components such as various contaminants [123], [124]. For example, a typical extraction of oil from oilseeds such as canola, soybeans, corn, flax, or cotton [125] is done starting with 4 g of seeds that are ground and kept in an oven for 2 hours at 130° C. The seeds are then extracted under reflux with petroleum ether for 4 h. After this extraction, the solvent is separated and the seeds are reground for 7 min. and reextracted for another 2 h. The total volume of solvent used is 150–250 mL, and the total time of extraction 8–10 h. Using ASE with petroleum ether as a solvent, a pressure of 6.9 MPa ($\approx$1000 psi), a temperature of 105° C, 100% flush volume and 3 cycles, the operation can be done in about 45 min. with about 35 mL solvent. Some other applications of ASE are given in Table 10.3.2.

TABLE 10.3.2. *Some examples of ASE applications.*

Analytes	Sample matrix	ASE conditions	Reference
PCBs, PAHs, pesticides	marine sediments, urban dust.	acetone-hexane (1:1, v/v) or dichloromethane-acetone (1:1), 100°C, 1,500 - 2,000 psi, 5 min. equil., 5 min. static, pretreatment with HCl	[118], [126], [127], [128]
PAHs, chlorinated pesticides, dioxins and dibenzofurans	soil, scrubber dust, fly ash	acetone-hexane (1:1) at 100°C, or toluene at 175°C/200°C, or dichloromethane-acetone (1:1) at 100°C, 10/14 MPa.	[128]
dioxins	soil, sediment, dust, fly ash	toluene, 180°C, 2000 psi, 9 min. equil.,5 min. static.	[130]
organophosphorus pesticides	clay, loam, sand	dichloromethane-acetone (1:1), 100°C, 2,000 psi, 5 min. equil., 5 min. static.	[131]
organophosphorus pesticides	foods	cyclohexane-acetone (1:1), or dichloromethane-acetone (1:1), or ethyl acetate-acetone (1:1), 100°C, 1,500 psi	[132]
chlorophenoxy acid herbicides	clay soil, loam soil, sand	dichloromethane-acetone (1:2), with 4% H_3PO_4-H_2O (1:1), 100°C, 2,000 psi, 5 min. equil., 5 min. static.	[131]
phenols	soil	dichloromethane, 30-70°C, 600-3,000 psi, time: 5-25 min.	[128]
diflufenican	soil	acetonitrile, 100°C, 2,000 psi, 5 min. equil., 4 min. static, 60 s. purge.	[133]
PCBs	fish tissue	hexane, 100°C,1450 psi, 5 min. equil., 5 min. static, 90 s purge.	[121], [134]

10.4 MICROWAVE-ASSISTED SOLVENT EXTRACTION

Microwaves can be used for heating materials based on radiation adsorption followed by dissipation as heat of the accumulated energy. Microwave sample preparation was developed initially as a digestion procedure (wet ashing) for eliminating the interference of organic constituents from samples such as animal tissues, botanical materials, etc. This was necessary for the analysis of metals in these samples by atomic absorption or various types of emission spectroscopy [135]. The procedures used for wet ashing are applied to aqueous solution of strong acids such as hydrochloric, nitric, perchloric, sulfuric, with microwave radiation as the heat source. Microwave assisted extraction (MAE) was developed from these initial techniques by replacement of the solutions of strong acids with solvents and by application of the procedure as an extraction technique and not for digestion [136]. Because microwave radiation is not significantly absorbed by nonpolar materials, the solvents must contain a polar component such as water, methanol, acetone, acetonitrile, etc. [137], [138], [139]. MAE has proved to be an efficient extraction procedure that may lead to shorter extraction times and minimal solvent usage in comparison with Soxhlet or sonication procedures.

The microwaves used for industrial and laboratory heating are limited to four frequency bands at 915, 2450, 5800, and 22125 MHz (with specific band widths). The power output of microwave units is typically between 300 and 700 W. The microwaves are partially absorbed by the sample, similarly to absorption of light (see Section 4.3). The process is characterized by a dissipation factor defined as

$$\tan (\delta) = \varepsilon" / \varepsilon \qquad (10.4.1)$$

where ε is the dielectric constant of the medium, and $\varepsilon"$ is a loss factor that describes the energy lost from the incident radiation and absorbed by the sample. A high dissipation factor leads to a high absorption of the microwave energy and significant heating. Depending on the frequency of the microwaves, some materials have very low dissipation factor being transparent to the microwaves, while others have high dissipation factors. Another way of characterizing the microwave absorption is the use of penetration depth. A half-power penetration depth is the distance in the path of microwaves into the sample where the power of the microwaves is reduced to half that of incident radiation.

The absorption of energy takes place mainly by two mechanisms, dipole rotation and ionic conduction. The dipole rotation is caused by the alignment of molecules with permanent or induced dipole moment in the electric field. The oscillation of the electromagnetic field intensity produces alignment in one direction at increased electric field, followed by the return to disorder at zero field, and then orientation in the opposite direction as the direction of the electric field vector goes in the opposite direction. At 2450 MHz for example, the alignment of molecules and their return to disorder should take place $4.9 \; 10^9$ times per second. The absorbed energy during the orientation process is dissipated as heat, and the process results in rapid heating. However, the dipole orientation and relaxation does not depend solely on the frequency of the oscillating electromagnetic field. The dielectric relaxation time τ of the molecules is another factor that influences energy absorption. The dielectric relaxation time is defined as the time required for 63% of the molecules of the sample to return to disorder. The maximum energy absorption of the molecules occurs for a frequency given by the expression [140]:

$$\nu = \frac{1}{2\pi\tau} \qquad (10.4.2)$$

The dielectric relaxation time varies with the temperature, and therefore, as the sample is heated, the microwave absorption efficiency is modified. Also, a higher absorption may have the effect of a shorter half power penetration depth, and this may be advantageous in some samples and not in others. Microwave frequency and power are experimental parameters of the microwave generating system, while the dissipation factor and the dielectric relaxation time depend on the nature of the material irradiated with microwaves. In general, the larger is the dipole moment of the solvent, the faster the solvent will heat.

Ionic conduction is the process of migration of the ions present in a solution in the applied electromagnetic field. This ion migration is equivalent with a current flow that has as a result production of heat following Joule's law:

$$Q = I^2R \qquad (10.4.3)$$

where I is the electric current intensity and R the resistance to ion flow. Heating by ionic conductance mechanism is more prevalent in aqueous ionic solutions. In addition to the parameters of the microwave generating system, parameters affecting heat generation by this mechanism are ion concentration, ion mobility, solution temperature, etc.

MAE is typically performed on solid samples, which are introduced in a common organic or organic/aqueous solvent and irradiated for a short period of time (0.5 to 10 min.). After cooling, the supernatant solution is decanted and the matrix is rinsed several times with the solvent. The combined extracts are further filtered or centrifuged if necessary, concentrated, or further cleaned prior to chromatographic analysis. The solvents with very low dipole moments such as hexane do not absorb efficiently microwave energy, and therefore the addition of at least 10% of a polar solvent is generally required. The absorption of energy by the solvent results in the disruption of weak hydrogen bonds and improved solvent penetration [141], [142].

Extraction using MAE can be done at atmospheric pressure at the boiling point of the solvent, possibly using a condenser to prevent solvent loss, or can be done at elevated pressure, similarly to ASE. In a system at atmospheric pressure (open cell), the microwave radiation is focused on the vessel containing the extraction solvent. For this case, the radiation power of the system is less important. The use for extraction of organic solvents such as CH_2Cl_2 has been proven possible, with the condition that the sample contains enough moisture or that some water is added in the sample and solvent mixture. Quantitative recoveries were obtained, for example, for extraction of PAHs from soil and sediments during 10 min. process, using CH_2Cl_2 and 30% water as solvent [143].

Another MAE approach uses a closed vessel. In this case, in addition to the transfer of energy to the solvent/sample using microwaves, an increase in extraction temperature above the normal solvent boiling point is possible because of the higher pressure in the extraction vessel. For a number of solvents (acetone, acetone-hexane, dichloromethane-acetone, etc.), the temperature inside the vessel can be up to three times the boiling point of the solvent at atmospheric pressure.

The equipment used for closed vessel MAE consists of an oven (adapted with a magnetron tube) where the individual extraction (closed) vessels are placed on a turntable. Monitoring devices for temperature and pressure as well as safety features such as rupture membranes are usually attached to the cells. The vessels are typically made of microwave transparent materials such as ceramics, quartz, TFM (poly-tetrafluoromethoxyl), Teflon, etc. This equipment used in this technique frequently has the capability to extract multiple samples simultaneously.

In the selection of a solvent for MAE, besides the microwave absorbing properties of the solvent, some other factors must be considered, such as the solubility of the analytes in the selected solvent, the compatibility between the extraction solvent and the chromatographic technique used as the core step, the extraction selectivity of the solvent, etc. [122]. A cleanup step is frequently necessary after performing MAE because nonselective extractions are common in this technique.

- *Examples of applications*

Most applications of MAE are related to exhaustive extraction of samples where trace analytes are adsorbed to the matrix material. On the other hand, the matrix is usually not soluble in the extracting solvent such that the high efficiency of the extraction does not result in a complete sample dissolution. These characteristics make MAE an adequate procedure, for example, for the analysis of pollutants in soils and sediments. Some examples of this type of analysis are given in Table 10.4.1.

TABLE 10.4.1. *Some applications of MAE to the extraction of soil and sediments for pollutant analysis.*

Pollutants	Matrix	Solvent	MAE conditions	Reference
PAHs	soil	acetone, dichloromethane	29 min. at 120°C in closed vessel	[144]
PAHs	marine sediments, mussel tissue, air particles	dichloromethane, hexane-acetone (1:1)	5-15 min. at 115, closed vessel	[145]
PAHs	reference marine sediments	hexane-acetone (1:1)	5-10 min. at 35°C in closed vessel	[145]
PAHs	fly ash	hexane-acetone (9:1)	70°C in closed vessel	[146]
phenols, organochlorine pesticides	topsoil, clay, sand	hexane-acetone (1:1)	closed vessel extraction at 115°C for 10 min.	[147]
PCBs	municipal sewage sludge	hexane-acetone (1:1)	10 min., 30 W, open vessel	[148]
PCBs	river sediments	hexane-acetone (1:1)	15 min., closed vessel	[149]
herbicides	soil	methanol	screw-capped vials	[150]
imidazolinone herbicides	soil	0.1 M buffer CH$_3$COONH$_4$/NH$_4$OH (pH = 9-10)	3-10 irradiation at 125°C in closed vessel	[151]
organotin compounds	sediments	0.5 M CH$_3$COOH in CH$_3$OH	3 min., in open vessel, at different temperatures	[152], [153]

Besides the analysis of soils and sediments, MAE is applied as sample preparation in a number of other analyses such as the determination of PCBs and hexachlorobenzene in seal fat, pork fat, and in cod liver [154], [155], analysis of aromatic hydrocarbons in H_2O samples [156], [139], etc.

10.5 SIMULTANEOUS DISTILLATION AND EXTRACTION

Simultaneous distillation and (solvent) extraction (SDE) is a combined technique that uses steam distillation for removing the analytes from a sample having a complicated matrix and then a liquid/liquid extraction for transferring the analytes from the water solution into an organic solvent. A simplified scheme of the process is shown in Figure 10.5.1. The process takes place in a dedicated apparatus shown schematically in Figure 10.5.2 [157], [158], [159] with two versions, one for solvents with higher density than water (drawing A in Figure 10.5.2) and another for solvents with lower density (drawing B). The sample is introduced together with water (Solvent 1 in Figures 10.5.2) in a flask and the mixture is brought to boiling. At the same time a different solvent that is nonmiscible in water (Solvent 2 in Figures 10.5.2) is boiled from the second flask. The vapors from both flasks are condensed together such that an intimate mixture is produced. The separation of two layers of solvents allows each solvent to return in its own flask. The extraction process may take a number of hours to achieve efficient extraction. However, similarly to Soxhlet extraction, once started the process does not require constant tending. For samples with long extraction time, it is also useful to process more samples in a battery of extractors working in parallel. Different other constructions of an SDE apparatus are reported in literature, such as a system that uses an external source of steam [160], [161]. The cooling efficiency of the apparatus is very important for avoiding losses of more volatile analytes and of the solvent. It is common that cooling is done with coolant mixtures circulated in the condenser.

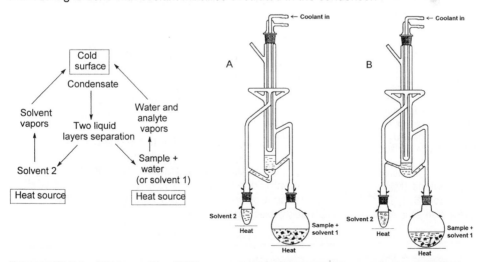

FIGURE 10.5.1. *Diagram of an SDE apparatus.*

FIGURE 10.5.2. *SDE apparatus for solvent 2 heavier than water (A) and for solvent 2 lighter than water (B) [159].*

A number of solvents are used in SDE sample preparations, CH_2Cl_2 being probably the most common. Other solvents include ethyl acetate, pentane, hexane, diethylether, etc. [162], [163]. The decrease of the water solubility for various organic compounds can be achieved by salting out the water solution. This procedure has been shown to increase recovery efficiency [164].

The efficiency of the process is mainly controlled by the efficiency of the steam distillation step [165]. Various theoretical models have been proposed for the calculation of SDE recoveries [165], [166], [167]. The best agreement with experimental data are obtained using the following expression for the recovery R of the analyte:

$$R = \frac{KK_wV_s}{KK_wV_s + K_sV_w}\{1 - \exp[\frac{(KK_wV_s + K_sV_w)U_sU_w}{V_wV_s(U_w + KU_s)}t]\} \qquad (10.5.1)$$

where U_s is the volumetric flow of solvent distillation, U_w is the volumetric flow of steam distillation, K_w is air-to-water partition coefficient, K_s is air-to-solvent partition coefficient, K is solvent-to-water partition coefficient, V_s is the volume of the solvent in the flask, V_w is the volume of the water in the flask, and t is time (K_s and K_w are expressed as the ratio of the concentration of the condensed vapors to the concentration of the solution). The recovery as a function of time for two values of K and three values of K_w are shown

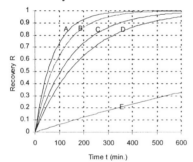

FIGURE 10.5.3. *SDE calculated recovery as a function of time for K=10, K_w=2 (trace A), K=5, K_w=2 (trace B), K=10, K_w=1 (trace C), K=5, K_w=1 (trace D), and K=10, K_w=0.1 (trace E). The* other parameters are the same, and K_s = 0, U_s = 1 mL/min., U_w= 5 mL/min., V_w = 500 mL, and V_s = 50 mL.

in Figure 10.5.3. The value of K_w influences significantly the recovery process, as seen in Figure 10.5.3 by comparison of traces A, C, and E where K is the same and K_w varies. The value of K_w can be obtained using empirical estimation methods [168]. Water soluble compounds or compounds with low vapor pressure do not have a good transfer during the SDE process. As an example, a comparison regarding the extraction efficiency of a number of PAHs from cigarette smoke condensate with SDE and with solid phase extraction (SPE) on two different cartridge materials (C18 and cyclohexyl CH) is shown in Figure 10.5.4 [169]. The SDE extraction was done using CH_2Cl_2 as a solvent with 5 h of extraction in an apparatus similar to the one shown in Figure 10.1.2 [170]. As seen in Figure 10.5.4, the extraction efficiency for naphthalene (MW = 128) or $^2H_8$-naphthalene is close to 100% for SDE and SPE (CH) and about 65% for SPE (C18). As the molecular weight of PAHs increases, the extraction efficiency using SDE decreases significantly.

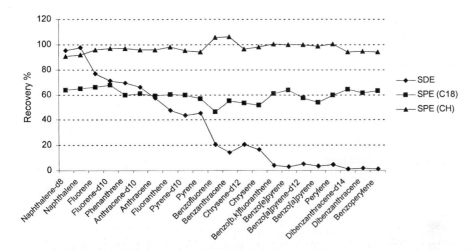

FIGURE 10.5.4. *The variation of recovery % for a number of PAHs from cigarette smoke condensate using SDE and using solid phase extraction (SPE) on two different cartridge materials (C18 and cyclohexyl CH).*

The quantitation using SDE as a sample preparation procedure gives very good reproducibility when the extraction efficiency is very high. When the recovery is low, the use of appropriate standards such as deuterated homologs still allows quantitation with acceptable precision and accuracy [171].

SDE has some advantages and disadvantages in comparison with other extraction techniques such as supercritical fluid extraction (SFE) and Soxhlet extraction. A comparison regarding the analysis of hexachlorobenzene (HCB) in soils or in fatty sample materials reported in literature [160] is shown in Table 10.5.1.

TABLE 10.5.1. *Comparison of the processing parameters using SDE, SFE, and Soxhlet extraction for the isolation of hexachlorobenzene (HCB) in soils or fatty sample materials [160].*

Parameters	SDE	SFE	Soxhlet extraction
Processing time	60 min	25 min	4 h
Maximum sample amount (g):			
- soil	100	5	100
- fatty matrix	10	0.5	10
Recovery (η%) from:			
- soil	100	98	55
- fatty matrix	85	90	92
Quantification limit (ppb HCB):			
- in soil	0.05	0.1	0.5
- in fatty matrix	2	2	2
Further cleanup of extracts:			
- soil samples	not required	not required	SPE
- fatty matrix	not required	SPE	SPE
Overall analysis time (h)	1.5	0.5 – 1.3	6

Besides SDE performed at atmospheric pressure, vacuum SDE is also reported in literature [172], [173]. The main objective of vacuum SDE is to perform the process at lower temperatures to avoid any artifact formation [174], [175].

Similarly to other distillation procedures, some artifacts may be present in SDE sample preparates. The sources of artifacts are the contamination with grease from glass joints, possible oxidation for compounds such as benzaldehyde, terpenes, etc. [176], [177], and thermal reactions such as Maillard reactions, ester hydrolysis, and sugar degradations [178].

- Examples of applications

Steam distillation extraction procedures are applied in a variety of analyses, mainly when the matrix is complicated but with very different physical properties from the analytes. Also, because the extraction is dependent on the partial vapor pressure of the analytes (see rel. 10.5.1), the procedure is applied predominately to the analysis of volatile compounds such as flavors, but also to pesticides [160] and other analytes. Typical samples analyzed using SDE are soils, food, plant materials, animal tissues, but also water, essential oils, etc. For example, aroma concentrates were obtained from cooked, ground meat samples by the continuous steam distillation extraction using pentane as a solvent. After extraction the solvent was dried over Na_2SO_4 and concentrated. The concentrates were then analyzed by GC-MS on a DB-5 capillary column and MS detection [179]. Other types of samples of food or vegetables were analyzed using SDE for sample preparation. For example, peeled ripe bananas were homogenized with H_2O and extracted using SDE with CH_2Cl_2 as the organic solvent. The extract was mixed with methyl octanoate (internal standard) and analyzed by GC-MS [180].

The volatile compounds of water-boiled duck meat, duck fatty tissue, and Cantonese style roasted duck were isolated by SDE (3 h) in pentane-ethyl ether (1:1). A C_{17} alkane was added as chromatographic standard after extraction. The volatile extracts were dried with anhydrous Na_2SO_4, concentrated to a minimum volume and determined by GC-FID or GC-MS [181]. Optimization of a SDE procedure has been described for pre-concentration of volatile components in foods [162]. Extracted volatiles were analyzed by capillary GC-FID. Volatile components in blue crab (*Callinectes sapidus*) meat and processing by-products were isolated by SDE and analyzed by GC-MS and GC-FTIR [182].

The analysis of tetrachlorodibenzo-p-dioxin (TCDD) in biological samples also has been done using SDE as a sample preparation operation [183]. Rat liver was homogenized with H_2O, and the homogenate was adjusted to pH = 1 with concentrated HCl. Internal standard (^{14}C-labeled TCDD) was added, the solvent was evaporated, and the residue was subjected to steam distillation extraction with hexane. After addition of octanol, the hexane extract was evaporated to dryness, and the residue was redissolved in acetonitrile and cleaned up by HPLC on a LiChrosorb RP18 column, with aqueous 85% acetonitrile as mobile phase. A 2-min. fraction containing TCDD was collected and extracted with hexane. After evaporation under N_2, the residue was analyzed by GC-MS and liquid scintillation counting [184].

Bromophenol distribution in salmon and selected seafoods of fresh and salt water origin was investigated using SDE and GC-MS (in SIM mode) procedure [185]. Samples were homogenized, the pH adjusted to 2, and pentachloroanisole (internal standard) was added. The homogenate was extracted for 4.5 h using a mixture of pentane-ethyl ether-toluene. The extract was dried over sodium sulfate and evaporated in a stream of purified N_2. Analysis by GC was conducted on a DB-5 column with He as carrier gas.

Water samples were also analyzed using SDE. The initial samples were treated with KCl and were subjected to SDE with ethyl ether as extractant for triazine pollutants. Portions of the extract were subjected to GC analysis using photoionization detection. The recovery of seven 1,3,5-triazines added to water ranged from 69 to 89%, except for simazine (<10%). The best recoveries were obtained with 3 h of extraction at pH=7 [184].

Pentachlorophenol was extracted from waste water, soils, and sediments by steam distillation end extraction with hexane at pH 1 for 4, 16 and 16 h, respectively, the extract was evaporated to dryness, and the residue was redissolved in acetonitrile. The resulting solution was derivatized by N-methyl-t-butyldimethylsilytrifluoroacetamide to *tert*-butyldimethylsilyl derivative, spiked with ^{13}C-pentachlorophenol and analyzed by GC on a column coated with DB5 and MS detection [186].

SDE and dynamic headspace methods in the GC analysis of cheese volatiles were compared [187]. Cheese was mixed with H_2O and placed in a simultaneous steam distillation-extraction apparatus. Isobutylthiazole (internal standard) in CH_2Cl_2 was placed in the extractor, the separation chamber was filled with CH_2Cl_2 and H_2O, and steam distillation was performed for 2 h at 110° C. The aqueous distillate was simultaneously extracted with CH_2Cl_2 in a water bath at 50° C. For dynamic headspace sampling, ground cheese was mixed with methyl salicylate (internal standard) and placed in an Erlenmeyer flask at 40°C with purified He for 25 min at 60 ml/min, the volatiles were adsorbed in a Tenax GC trap, and the adsorbent was flushed with He to remove moisture. The volatiles were thermally desorbed at 300°C for 12 min under He at 10 ml/min, trapped in a silica capillary at -100° C, and the trap heated to 200° C. The volatiles were analyzed by GC on a DB-WAX column with temperature programming, He as carrier gas, and MS detection. The headspace technique permitted the identification of 110 compounds, while about 50 compounds were isolated by SDE procedure.

SDE and GC-MS have been applied to the determination of volatile composition of vinegars. To determine minor volatile components of the vinegars, the samples were concentrated by use of a Chrompack micro steam distillation-extraction apparatus in the high-density solvent configuration, with CH_2Cl_2 as extractant and an extraction time of 1.5 h. The extracts, with methyl octanoate added as internal standard, were analyzed on a fused-silica capillary column coated with cross-linked BP-21, temperature programmed, with He as carrier gas, and quadrupole MS detection. Major volatile components were determined by direct analysis of the vinegars, with pentan-3-ol added as internal standard, on a column, with Carbowax-bis-(2-ethylhexyl) sebacate (23:2) mixture, coated on desilanized Volaspher A-2, with H_2 as carrier gas and with FID detection [188].

10.6 SUPERCRITICAL SOLVENT EXTRACTION

Supercritical solvent (fluid) extraction (SFE) uses a supercritical fluid as the extraction solvent. SFE is typically applied on solid or semisolid samples with a problem matrix such as gummy or tar-like materials, solids in which analytes are deep in the bulk of the sample or strongly adsorbed on the surface, materials of vegetal or animal origins containing encapsulated analytes, etc. SFE is also applied when the analytes of interest are temperature sensitive. Similarly to other liquid-solid extraction operations, SFE requires two basic steps, which are the contact of the supercritical fluid with the sample and the separation of the fluid containing the extracted components from the rest of the sample. In addition to this, SFE includes a fluid exchange step or a precipitation step where the extracted compounds are recovered from the supercritical fluid not usable directly in a chromatographic system. Supercritical fluid extraction is performed in dedicated instrumentation necessary for generating a supercritical fluid and also for controlling the fluid movement. The instrumentation for SFE is developed mainly for the use of supercritical CO_2 because CO_2 has significant advantages regarding range of temperatures (from ambient up to 100–120° C) and pressures utilized (from tank pressure about 870 psi up to 5000 psi), lack of toxicity, lack of reactivity, etc.

As shown in Section 1.3, a fluid becomes supercritical when its temperature and pressure are above the critical point (critical temperature T_c and critical pressure p_c). The values for T_c and p_c are characteristic for every compound. Some critical values for compounds that can be used in SFC are given in Table 10.6.1.

TABLE 10.6.1. *Critical properties of some extraction fluids.*

Solvent	T_c (°C)	p_c (atm.)
CO_2	31.1	72.9
N_2O	36.5	71.7
$CHClF_2$ (Freon-22)	96	48.5
CHF_3 (Freon-23)	25.9	46.9
CH_3OH	240	78.5
H_2O	374.1	218.3

Among the compounds that can be used as supercritical fluids for extraction, CO_2 is by far the most common. In a supercritical fluid, intermolecular forces are lower than the molecular kinetic energy and therefore are insufficient for the condensation of the compound into a liquid. However, the number of molecules per unit volume (consequently the density) can be that of a liquid while the tendency to expand as the volume increases is gas-like. The density, viscosity, and diffusion coefficient of supercritical fluids are intermediate between liquids and gases, as it can be seen in Table 10.6.2 [189], [190].

TABLE 10.6.2. *Approximate ranges for density, viscosity, and diffusion coefficient of liquids, supercritical fluids, and gases.*

Fluid	Density (g cm^{-3})	Viscosity (g cm^{-1} s^{-1})	Diffusion coefficient (cm^2 s^{-1})
gas	$0.6 \times 10^{-3} - 2 \times 10^{-3}$	$1 \times 10^{-4} - 3 \times 10^{-4}$	0.1–1
supercritical fluid	0.2 – 0.9	$1 \times 10^{-3} - 3 \times 10^{-3}$	$0.1 - 5 \times 10^{-4}$
liquid	0.6 – 1.6 (except metals)	$2 \times 10^{-3} - 3 \times 10^{-2}$	$0.2 \times 10^{-5} - 3 \times 10^{-5}$

The most important characteristic for the utilization of any solvent in an extraction procedure is its capability to dissolve the solutes (solvation capability). Below the critical temperature, the gas and liquid phases coexist, the solvent power for the gas being minimal and only the liquid being used as a solvent. For supercritical fluids the solvation capability is highly related to the density of the fluid, showing a significant increase as the density increases toward the critical density. The critical density for CO_2 is $\rho_c = 0.468$ g/mL, and critical volume $V_c = 94$ mL mol^{-1}. Very little solute is dissolved in a gas-like supercritical fluid. Only when the properties of a fluid start to become liquid-like, does the solubility start to increase. The liquid-like behavior for CO_2, for example, starts only when the fluid density ρ is 0.2–0.3 times that of the supercritical fluid (reduced density $\rho_r > 0.2$). This is illustrated by the variation in the solubility of naphthalene in CO_2 at 45° C as a function of reduced density ρ_r of the fluid, which is shown in Figure 10.6.1 (it is common to express the properties of a supercritical fluid using reduced parameters such as reduced temperature $T_r = T/T_c$, reduced pressure $p_r = p/p_c$, reduced volume $V_r = V/V_c$, and reduced density $\rho_r = \rho/\rho_c$). Above the critical

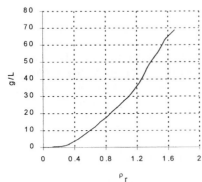

FIGURE 10.6.1. *The variation of naphthalene solubility (g/L) in CO_2 as a function of reduced density ρ_r of the fluid.*

temperature, the solvent power of a fluid can be adjustable by modifying the density. The density of supercritical fluids can be varied in a relatively wide range by the variation of pressure. This variation is shown for CO_2 for different temperatures (expressed as reduced temperatures) in Figure 10.6.2. As the pressure increases (expressed as reduced pressure p_r), the density also increases. However, for $T_r = 0.8$, the formation of liquid is possible and for the branch corresponding to the variation of the liquid density, a large increase in the pressure does not affect significantly the density (liquids are not compressible). For $T_r = 1$, a significant increase in the density is seen for a small increase in pressure above the value of p_c (log $p_r + 1 = 1$ corresponds to $p_r = 1$ and $p = p_c$). A number of theoretical models were developed to provide guidance regarding the solubility of a compound in supercritical fluids [191], [192], [193].

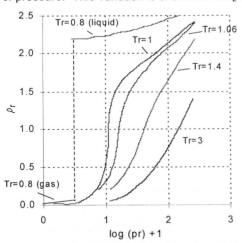

FIGURE 10.6.2. *The variation of ρ_r for CO_2 as a function of p_r for different T_r values.*

The variation in the density of a supercritical fluid influences directly the solubility for a given compound, but also modifies the solvent polarity. It has been shown [194] that the solubility parameter δ varies with the fluid density ρ following the relation:

$$\delta = 1.25 \sqrt{p_c}\; \rho/\rho_{liq} \qquad\qquad (10.6.1)$$

Rel. (10.6.1) shows that with the increase in the fluid density the solubility parameter δ increases proportionally. Consequently curves similar to those shown in Figure 10.6.2 are obtained when the variation of solubility parameter is plotted with the pressure (log $p_r + 1$). Based on rel. (10.6.1), the resulting value of δ for liquid CO_2 is around 10, showing solubility properties similar to CS_2. The polarity P' of CS_2 is 0.3 [195], and liquid CO_2 is also nonpolar. This characteristic has been proven experimentally for liquid CO_2, and when a more polar extracting fluid is necessary, either a different supercritical fluid is utilized (such as supercritical acetone, methanol, or ammonia), or modifiers are added to the CO_2 to increase its polarity. Because SFE is done with liquid-like agents having adjustable solvating properties, and because supercritical fluids have more gas-like transport properties (viscosity and diffusion) allowing efficient sample penetration (see rel. 10.2.1), SFE is a very attractive solvent extraction technique.

- Operating parameters in SFE

SFE sample preparation technique is in fact a chain of operations consisting of a) sample extraction, b) collection of the extract in a trap, and c) rinsing of the trap to generate a processed sample that is further utilized for a chromatographic analysis or for other sample preparation steps. Other extraction procedures also require more steps besides the extraction such as the separation of the solvent containing the extracted materials from the remaining matrix. However, in SFE, all operations are performed in one instrument, and specific parameters from one step to the other are interdependent. The main functions of the SFE process are indicated in the diagram from Figure 10.6.3.

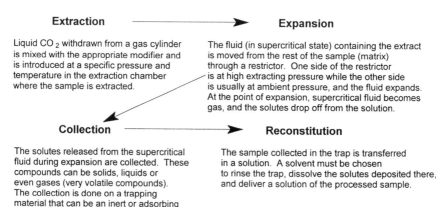

Extraction ⟶ **Expansion**

Liquid CO_2 withdrawn from a gas cylinder is mixed with the appropriate modifier and is introduced at a specific pressure and temperature in the extraction chamber where the sample is extracted.

The fluid (in supercritical state) containing the extract is moved from the rest of the sample (matrix) through a restrictor. One side of the restrictor is at high extracting pressure while the other side is usually at ambient pressure, and the fluid expands. At the point of expansion, supercritical fluid becomes gas, and the solutes drop off from the solution.

Collection ⟶ **Reconstitution**

The solutes released from the supercritical fluid during expansion are collected. These compounds can be solids, liquids or even gases (very volatile compounds). The collection is done on a trapping material that can be an inert or adsorbing solid (silica, charcoal, Tenax) or a liquid.

The sample collected in the trap is transferred in a solution. A solvent must be chosen to rinse the trap, dissolve the solutes deposited there, and deliver a solution of the processed sample.

FIGURE 10.6.3. *Main functions performed in the SFE process.*

A number of SFE instruments are commercially available. Although there are various differences in their construction, they accomplish the same basic operations. A flow diagram of a common SFE instrument is shown in Figure 10.6.4.

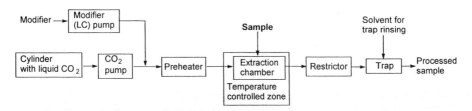

FIGURE 10.6.4. *Flow diagram of the SFE process.*

The initial choice of parameters and their optimization in a SFE sample preparation process are discussed in more detail below.

- Optimization of the extraction process

The key parameters selected for the extraction when CO_2 is used as a supercritical fluid are the pressure of the fluid and the nature and concentration of the modifier. A "threshold density" is defined for the density of the solvent as the minimum density at which the supercritical fluid can dissolve a detectable level of solute. This value is not strictly defined, but curves similar to that shown in Figure 10.6.1 indicate that the solubility shows a significant increase starting at a certain density value of the fluid. The threshold density for a solute increases as its melting point increases and as its molecular weight increases. The pressure for any extraction is selected such that the density of the fluid is above the threshold density. In many instruments that use computer control, instead of selecting a pressure for the extraction it is possible to select directly a certain density, the pressure being automatically adjusted accordingly.

The limit in solvation power of pure CO_2 and the limit in increasing the pressure of CO_2 are usually overcome using chemical modifiers added to the solvent. The nature of the modifier used in SFE with CO_2 must be chosen based on the solvent type interaction with the sample. Compounds expected to have nonlocalized interactions with the solvent such as nonpolar compounds (see Section 8.2) are expected to dissolve well in supercritical CO_2, and the improvements in the extraction must be done by the modification of the solvent density and temperature. For compounds with polar molecules without specific functional groups, polar modifiers such as dioxane, ether, etc. are advantageous. For compounds with localized interactions solvent specific, the modifiers are typically chosen to match the solute character, acidic modifiers being appropriate for basic solutes and basic modifiers for acidic solutes. Alcohols (such as methanol) or even water are frequently used as modifiers, increasing both the polarity and the capability to form hydrogen bonds of the solvent. The amount of modifier is also an important element to consider during extraction. The variation in solubility with the addition of a modifier is not linear, and the amount is usually established experimentally. For example, up to 20–30% methanol can be added to supercritical fluid CO_2, although the molar fraction of the modifier is usually recommended at maximum 0.15 in order to avoid liquid-like diffusivity and other properties too close to a liquid.

In addition to density and the modifier, the solvent temperature is also important for determining the solvating power of the supercritical fluid solvents. The influence of temperature on solubility in general was discussed in Section 8.2, showing that an increase in temperature increases solubility. Also, melted materials dissolve better than solids, and if the temperature is above the melting point of the substance to be extracted (without producing chemical modifications), the dissolution typically takes place very well. However, for supercritical solvents, the temperature also affects the density of the solvent, and temperature influence is more complicated. In some cases, the increase in temperature affects favorably the dissolution process. In other cases, a solute can be dissolved in a supercritical solvent at a given density, and with increase in temperature the density is lowered such that the solute may precipitate from the fluid. The increase in pressure to maintain the density constant when temperature increases favors the dissolution process. For solutes with MW > 350, it is common that besides the choice of an appropriate supercritical fluid modifier, it is necessary to increase temperature for better solubility.

Increased temperatures in the extraction process affect not only the solubility of the analytes, but also the efficiency of extracting the analytes from a compact matrix. This can be applied successfully mainly for volatile and semivolatile compounds that can have significantly better extractions at increased temperatures. The increased temperatures also increase the diffusivity of the fluid. This effect can be beneficial for extracting samples with encapsulating matrices. The increase in temperature, although favorable for sample dissolution, must be maintained below a value that can cause artifact formation or even advanced decomposition of the sample. Some guidance in the choice of extraction parameters is given in Table 10.6.3 [196].

TABLE 10.6.3. *Guide for selecting extraction parameters in SFE.*

Intermediate volatility (vapor pressure $p^0 \approx 1$ mm for T = 35-145° C)	Nonvolatile ($p^0 < 1$ mm for T=35-145° C, nonpolar, lower T_f, MW<300)	Nonvolatile, nonpolar, higher T_f, MW > 300	Nonvolatile polar
- Low density CO_2. - Temp. is dominant.	- Density can be dominant. - Increased temp. may help. (Verify artifact formation). - Nonpolar modifier.	- High density CO_2. - Increased temp. (Verify artifact formation). - Nonpolar modifier.	- High density CO_2. - Increased temp. (Verify artifact formation). - Polar modifier.

Several operational aspects of the extraction in SFE are also important for the success of this process. As indicated in Figure 10.6.4, the sample is loaded (after appropriate particle size reduction) into an extracting thimble, which is introduced in an extraction chamber. The choice of the amount of sample depends on the level of analytes to be analyzed in the sample, the amount of sample, and the volume of the thimble available for a specific instrument model. The thimble can be filled completely with the sample, and if not enough sample, with an inert material (sand, silica, diatomaceous earth, sodium sulfate, alumina, HPLC or GC column packing, or even stainless steel rods) to take up the volume not occupied by the sample itself. In this way the consumption of carbon dioxide is low [197]. It is also possible to leave the thimble partially filled. This approach has the disadvantage of higher CO_2 consumption, but the risk of contamination is low and the flow is more dispersed. Some instruments also have the capability of automation using a number of thimbles that can be loaded with different samples. These are usually extracted sequentially.

If pure CO_2 is used as an SFE extracting solvent, this is transferred with a pump into the extracting chamber. If a modifier is added to the solvent, there are several procedures for adding it. Some modifiers are provided directly in the gas cylinders (at a constant concentration). A desired fluid composition can be obtained by mixing in the transfer line pure CO_2 and CO_2 with the modifier. However, formation of layers with different composition of modifier in the cylinders has been noticed, mainly during storage at low temperatures, and this may bring some irreproducibility to the extraction process. A different addition of the modifier is the use of a separate pump (an LC pump) continuously adding the pure modifier to the CO_2 flow. This procedure also allows the addition of a mixture of solvents as modifiers by employing binary or ternary pumps. A manual application of the modifier is also common. This type of application is done directly on the sample before extraction. The modifier is frequently extracted by the supercritical fluid, and the sample is exposed to a variable solvent composition. However, the simplicity of the procedure makes it a common practice. Variable composition of the extracting solvent, even when a constant composition of the incoming fluid is used, also can be generated by condensation of the modifier on the sample during the extraction step. This is done by modification of the pressure and temperature during the extraction. This operation may be done intentionally for wetting the sample with the modifier and achieving an initial higher concentration than in the incoming fluid.

Another operational aspect is the regimen of adding the solvent to the sample. The addition can be done in static mode, in dynamic mode, or a combination of the two. Also, the volume of the fluid and the extraction time are important parameters for the extraction. A typical extraction starts with filling the extraction chamber with the fluid at the desired pressure, composition, and temperature. Once the filling to set point is complete, the sample is held statically at these conditions for a period of time indicated as "equilibration time." The equilibration time is chosen depending on the sample, and a time of 2–5 min. is typical. At the completion of the equilibration time, the sample is exposed to a continuous flow of the extraction fluid maintained at the extraction density and temperature. The amount of fresh fluid needed for sweeping the sample (expressed as mass flow of the CO_2 at the pump or number of times fresh solvent is swept through the thimble) depends on the extraction efficiency. Times between 5 and 10 min. and flows around 2 g/min. (at the CO_2 pump) are typical for a 7 mL thimble. However, depending on the density of the fluid, the number of thimble volumes swept by the fluid varies considerably for the same mass flow rate. The number of thimble volumes swept is a calculated parameter that depends not only on the mass flow of the CO_2 and its density, but also on the actual free volume left in the thimble after the addition of the sample.

- *Optimization of the expansion and collection process*

After performing the extraction, the supercritical fluid flows through a restrictor into a trap. One end of the restrictor is at the chosen pressure for extraction, and the other end is typically at atmospheric pressure. The restrictor (nozzle) has a special geometry such that the expansion takes place within a very narrow region. From different manufacturers, a number of constructions for the restrictors are available, including capillaries, frits, variable openings, etc. During the expansion, CO_2 becomes gas, and the solutes are isolated from the fluid and carried into the collection trap. Since the solutes can be gases, liquids, or solids, the isolation process takes place differently, the solids being deposited, liquids deposited or vaporized, and gases remaining in gas form. The restrictor temperature plays an important role in the way the material is deposited and in controlling the process. It is common that SFE instruments have the capability of restrictor temperature control. The expansion process is associated with a cooling process due to the evaporation of CO_2. A low restrictor temperature may lead to plugging of the restrictor due to the ice formation. A temperature that is too high also is undesirable due to potential decompositions and due to rapid depositions of solids that can plug the restrictor. An appropriate selection of the restrictor temperature may allow proper deposition of the solutes. By adjusting the restrictor temperature slightly below the boiling point of the modifier, it is also possible to obtain some solvent condensation at the restrictor exit such that the solutes are dissolved and carried properly into the trap. In order to avoid deposition of the solutes in the restrictor, some other procedures have been applied such as the mixing of expanding supercritical effluent with an overheated organic solvent [198].

The role of the trap is to retain the solutes emerging from the restrictor. A number of parameters can be adjusted for the trap, including a) the trap content (packing), b) the trap temperature during collection, c) the solvent used for extracting the analytes in the trap, and d) the volume and the flow of the solvent used to extract the trap. As a simple rule, if the compounds extracted from the sample are volatiles, it is recommended to have a trap with high adsorbing capability and not at elevated temperature. The trap content can be a solid phase material similar to that used in chromatographic columns such as Tenax, or ODS, silica, etc. One example of packing consists of ODS spheres 40 μm diameter. For gases or highly volatile compounds the trap acts as a packed chromatographic column. Trap temperature can be as low as -30° C for collecting compounds with high volatility. The trap temperature cannot be, however, much lower than the restrictor temperature, and low temperatures of the trap may lead to problems such as formation of ice plugs at the restrictor. An alternative to adsorbing materials for trap packing is an inert material such as stainless steel balls and a solvent.

For compounds with lower volatility, a number of materials can be used for packing the trap. A common material is stainless steel balls. One example of packing consists of stainless steel balls 400 μm diameter. The walls of an empty vial also can be used in some instruments for collecting the nonvolatile materials. The trap temperature for collecting nonvolatile solutes can be chosen to accommodate the rinse conditions. A guide for the choice of restrictor and trapping parameters is given in Table 10.6.4 [196]:

TABLE 10.6.4. *Guide for choice of restrictor (nozzle) and trapping parameters in SFE.*

Intermediate volatility ($p^0 \approx 1$ mm for T = 35-145° C)	Nonvolatile ($p^0 < 1$ mm for T=35-145° C, nonpolar, lower T_f, MW<300)	Nonvolatile, nonpolar, higher T_f, MW > 300	Nonvolatile polar
Restrictor conditions			
- Restrictor at chamber temp. ± 5° C	- Restrictor at extracting chamber temp. ± 5° C	- Restrictor at extracting chamber temp. or higher. - Modifier at boiling point	- Restrictor at extracting chamber temp. or higher. - Modifier at boiling point
Trapping conditions			
- Adsorbing trap (ODS) -Cold trap (5-25° C)	- Inert trap (stainless steel balls). -Cold to chamber temp., warm for oils and fats, or at modifier bp.	- Inert trap (stainless steel balls). - Warm for oils and fats, or at modifier bp.	- Inert trap (stainless steel balls). - At modifier bp.

- *Optimization of the trap rinsing*

The extraction of the solutes deposited in the trap (rinse cycle) is practically a reconstitution operation that generates a processed sample in a desired solvent. For this purpose, the solvent(s) must be chosen to dissolve the trap content. Because some irregular depositions may occur at the restrictor (nozzle) exit, these also must be dissolved during the extraction step, and many SFE instruments allow the solvent to pass through the restrictor. Also, more than one solvent can be chosen in some instruments for the rinse steps, not only for the complete recovery of the analytes but also to avoid carryover from one sample to another. From 15 to 20 rinse cycles are used in some methods using SFE for sample preparation. The parameters to optimize during the rinse operation are a) number of solvents used for rinse, in case that more than one solvent is used, b) the nature of each solvent used in a rinse cycle, c) the volume of rinse solvent, d) trap temperature during rinsing, e) flow rate of the rinsing solvent, and f) number of vials filled with rinse solvent. A larger volume of the rinsing solvent is preferable for achieving a complete extraction of the trap. However, the decrease in the concentration of the analytes is not usually desirable. For this reason the rinsing solvent is collected in separate vials, and only a limited number of vials containing detectable levels of analyte are collected as sample.

Trap rinsing also can be seen as an elution process from an LC or SPE column when the trap contains a retaining material. Therefore some fractionation of the analytes can be achieved by appropriate selection of the solvent and flow rate. Each fraction that may have a specific composition must be collected in a different vial. Some criteria for the selection of rinse cycle parameters are shown in Table 10.6.5 [196]:

TABLE 10.6.5. *Guide for the choice of rinse parameters.*

Intermediate volatility ($p^0 \approx 1$ mm for T = 35-145° C)	Nonvolatile ($p^0 < 1$ mm for T=35-145° C, nonpolar, lower T_f, MW<300)	Nonvolatile, nonpolar, higher T_f, MW > 300	Nonvolatile polar
- Shorter extracting time for volatiles - Longer for less soluble compounds. - Low volume of solvent	- Longer extracting time. - Larger solvent volume	- Longer extracting time. - Higher temperatures - Larger solvent volume	- Longer extracting time. - Polar solvents - Larger solvent volume

In SFE, the process usually starts with a set of selected parameters. The process is evaluated only after the whole chain of operations is finished. A number of alternatives may need to be evaluated before selecting optimum (or appropriate) conditions for a successful sample preparation. The flow diagram of a complete SFE process is given in Figure 10.6.5.

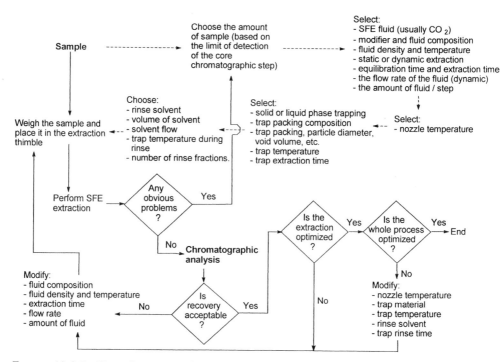

FIGURE 10.6.5. *Flow diagram and parameters to be adjusted during the SFE process.*
(Dotted arrows indicate a selection process rather than operations.)

- SFE on-line with other chromatographic techniques

Sample preparation using SFE can be followed off line by other sample preparation
operations or by any core chromatographic step such as GC, LC, or SFC. On-line
coupling is also possible and has several advantages compared to off-line SFE sample
preparation. Such systems reduce the sample handling, diminish losses allowing
quantitative extraction, diminish the chances for contamination, eliminate the need for a
rinsing solvent, etc. However, a number of problems are still related to the direct
connection.

The connection of the SFE extractor can be done with a GC, HPLC or an SFC system.
For SFE-GC coupling, the analytes in supercritical CO_2 exiting the SFE extraction
chamber must be first depressurized, which requires a special interface. The SFE
effluent can be depressurized and the analytes can be collected in an external cold trap.
From this trap, the extracts can be transferred to the GC using heating and a sweeping
carrier gas [199]. Another possibility is to depressurize the SFE extract directly into a
retention gap at the head of the GC column [200]. It is also possible to use the GC
injection port and depressurize the SFE extract inside the GC injection port (either
split/splitless or on-column injector) [201], [202]. The use of an external cold trap for
collecting the SFE is similar to the use of a trap and no solvent in an SFE extractor.
This procedure has several problems related to the separation of volatiles from CO_2 and
to transferring less volatile compounds into the GC. The direct injection is more

convenient, but a number of other difficulties are related to this operation. The introduction of a large volume of CO_2 into the GC may lead to serious problems regarding both the separation and the detection. For this reason, the use of a retention gap and on-column injection, although leading to good results mainly in the analysis of trace compounds, must be done such that the retention gap is able to accommodate the (large) sample. Split/splitless injection has the advantage of eliminating part of the CO_2, but losses of analytes are possible. For these reasons, the direct connection SFE-GC should be done considering the sample type and the analysis needs. A considerable number of applications using on-line SFE-GC coupling are reported in literature [203].

On-line coupling of SFE with LC can be done successfully using various schemes. A simple procedure consists of passing the supercritical CO_2 extract through a standard trap as previously described and connecting the trap to a HPLC system that elutes the analytes from the trap and sends them for an HPLC separation. Systems capable of loading only a small portion of the extract or focusing the analytes are preferable for enhancing the range of applicability [204]. A diagram of a SFE-HPLC on-line system is shown in Figure 10.6.7. In stage A the extracted sample is transferred into the SFE trap. The trap is further washed with a solvent from an HPLC pump or with the rinsing solvent from the SFE system and collected (completely or in part) in the precolumn. After collection, the system is shifted to stage B. Solvent B removes the analytes from the precolumn and sends them into the analytical column. If the solvent B is appropriate for the chromatographic separation, the process is continued. Otherwise, the system is shifted back to stage A, and the elution is done with solvent C (the system works in this case in A-B-A mode). The valve after the trap may allow elimination of part of the extract from the trap if not all the extract is intended to be transferred into the precolumn.

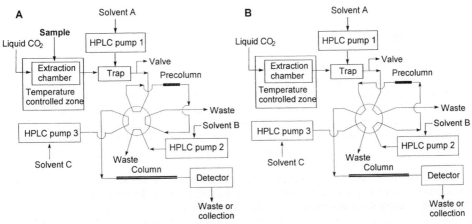

FIGURE 10.6.7. *Diagram of a SFE-HPLC system working in A-B or in A-B-A mode.*

The generation of a solution in a selected solvent from SFE does not impose specific restriction for further sample operations, and the sample can be analyzed, for example, by off-line SFC as core analytical step. On-line connection of an SFE extraction with an SFC core analytical step is also common. A variety of approaches for on-line SFE-SFC are reported using simpler or more complex interfaces [205]. One possibility is shown in Figure 10.6.8.

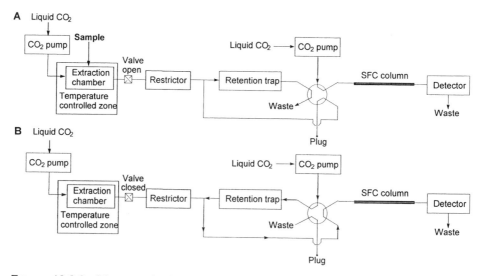

FIGURE 10.6.8. *Diagram of a SFE-SFC system working in A-B mode.*

In the SFE-SFC coupling shown in Figure 10.6.8, the restrictor is directly connected to a retention trap that allows the collection of the analytes. This can be done using cooling or specific adsorbent materials. After the adsorption of an appropriate amount of analyte, the system is switched to mode B of operation. The solvent used for the SFC instrument is passed through the retention trap and further into the SFC column.

- Applications

A large number of applications are reported using SFE as a sample preparation procedure. Some of these applications are associated with the use of SFC as a core analytical technique. The use of SFE on line to an SFC or to a GC such that the SFE replaces the injection port of the chromatographic instrument has various advantages regarding automation. However, this technique imposes several limitations to the SFE procedure, such as the sample size. Many applications use off-line SFE. This procedure does not exclude the use of SFC as a core chromatographic step, although, more frequently uses either GC or HPLC as the core analytical method. Similarly to other extraction procedures, SFE extraction can be done on the initial sample or on the traps or filter materials after a particular sample collection. For example, the filters used for sampling environmental pollutants can be extracted using SFE. Some practical applications using SFE are given in Table 10.6.6.

TABLE 10.6.6. *Example of applications for SFE.*

Sample	Analytes	Reference
air	PAHs, alkanes	[206], [207], [208], [209]
air	PAHs	[210]
soil and sediments	PAHs	[197], [211], [212]
soil	Pesticides	[213], [214]
vehicle exhaust	PAHs, PCBs	[215]
smoke from wood	guaiacol, syringol and derivatives	[200]
cigarette smoke	nicotine, phenols, tobacco specific nitrosamines (TSNA]	[197], [216], [217]
wood	terpenes	[200]
tobacco	nicotine, menthol	[200]
lemon peels	terpenes, other constituents	[218]
paper	anthraquinone	[219]
meat	pesticides	[220]
milk powder, infant formula	vitamin K_1	[221]
various samples	cannabinoids	[222], [223]

Besides standard SFE extraction, a number of modifications designed to improve the selectivity of the procedure have been developed. For example, the sample can be mixed with a solid phase material that adsorbs specifically some sample components, improving the separation [224], [225]. The analysis of polychlorinated dibenzo-p-dioxins (PCDDs) as target analytes was investigated using this procedure in the presence of Tenax, Florisil, alumina, carbon, and chemically modified silicas (C18 and CN). A 15-min extraction with supercritical CO_2 at 3000 psi removed over 75% of the chlorinated benzenes and PCBs from Florisil. Full recoveries of the PCDDs were subsequently obtained only by extraction with nitrous oxide (N_2O) for 90 min. at 6000 psi. This fractionation process was used to clean up municipal incinerator fly ash extracts prior to quantitation of PCDDs. Differences in the basic properties between CO_2 and N_2O are used to explain the differences in extraction recoveries from Florisil. The affinity of Florisil toward PCDDs is associated with the presence of magnesium ions in the silica lattice. Specific metals in the lattice, which are present in many common environmental matrices such as fly ash, result in chemisorption of PCDDs [226].

Another approach that has been successfully applied to the extraction of polar compounds in soil samples is based on SFE coupled with in situ derivatization. This technique involves the addition of common derivatization reagents directly into the extraction chamber, where the acid herbicides (2,4-D, 2,4,5-T, dicamba, silvex, trichloropyr, bentazone) are derivatized to extractable esters or ethers [227].

The application of SFE is not always the best choice, some samples being better extracted using a different procedure. For example, Soxhlet extraction can be replaced by supercritical fluid extraction or accelerated solvent extraction (ASE) for the removal of dioxins from municipal waste incinerator fly ash. SFE showed higher percent recoveries versus Soxhlet extraction for low-carbon-level fly ash, but only a few percent of dioxins can be extracted from high-carbon fly ash. The addition of large quantities of toluene in the extraction cell prior to SFE extraction of high-carbon fly ash improves the recovery of the lowest chlorinated dioxins (about 90%), but only a maximum of 20% of octachloro-dibenzo-p-dioxins can be extracted compared to Soxhlet procedure. Since large quantities of toluene are needed to improve the recoveries, SFE is not the extraction of choice. ASE with toluene for 2 hours at 80°C can be used, leading to recoveries similar to Soxhlet extraction. Increasing the temperature to 150°C improves the extraction rate and yields recoveries of 110–160% compared to 48 h Soxhlet extraction for both low and high-carbon fly ashes. These results question the choice of Soxhlet extraction as a reference method for dioxin determination [228].

REFERENCES 10

1. I. Akaza, *Correlation Between Extraction Chromatography and Liquid-Liquid Extraction*, in T. Braun, G. Ghessini (eds.), *Extraction Chromatography,* Akademiai Kiado, Budapest, 1975, p. 8.
2. C. McAuliffe, J. Phys. Chem., 70 (1966) 1267.
3. A. D. Nelson et al., Rec. Trav. Chim. Pays Bas, 87 (1968) 528.
4. H. S. Frank, M. W. Evans, J. Chem. Phys., 13 (1957) 133.
5. W. J. Lyman et al., *Handbook of Chemical Property Estimation Methods,* ACS, Washington, 1990.
6. C. Hansch et al., *Exploring QSAR, Hydrophobic, Electronic and Steric Constants,* ACS, Washington, 1995.
7. T. A. Sasaki et al., Anal. Lett., 34 (2001) 1749.
8. A. J. Handley (ed.), *Extraction Methods in Organic Analysis,* CRC Press, Boca Raton, 1999.
9. R. E. Majors, LC-GC, 10 (1997) 93.
10. B. L. Karger et al., *An Introduction to Separation Science,* J. Wiley, New York, 1973.
11. G. H. Morrison, H. Freiser, *Solvent Extraction,* in C. L. Wilson, D. W. Wilson (eds.), *Comprehensive Analytical Chemistry,* Elsevier, Amsterdam, Vol. IA, 1959, p.147.
12. M. Agudo et al., Anal. Chem., 65 (1995) 2941.
13. J. Roeraade, J. Chromatogr., 330 (1985) 263.
14. E. C. Goosens et al., J. High Resolut. Chromatogr., 13 (1990) 438.
15. E. Fogelqvist et al., Anal. Chem., 58 (1986) 1516.
16. C. de Ruiter et al., Anal. Chim. Acta, 192 (1987) 267.
17. J. J. Vreuls et al., J. Chromatogr., A, 856 (1999) 279.
18. E. Ballesteros et al., Anal. Chem., 62 (1990) 1587.
19. E. Ballesteros et al., J. Chromatogr., 633 (1993) 169.
20. E. C. Goosens et al., J. High Resolut. Chromatogr., 15 (1992) 242.
21. M. Valcarcel et al., Trends Anal. Chem., 13 (1994) 68.
22. J. B. Bell et al., Amer. Lab., 18 (1986) 94.
23. V. A. Nau, L. B. White, Amer. Lab., 24 (1992) 64.
24. R. E. Majors, K. D. Fogelmann, Amer. Lab., 25 (1993) 40W.
25. Y. Marcus, A. S. Kertes, *Ion Exchange and Solvent Extraction of Metal Complexes,* Wiley-Interscience, London, 1969.
26. F. Sebesta, *Chelating Agents as Stationary Phase,* in T. Braun, G. Ghessini (eds.), *Extraction Chromatography,* Akademiai Kiado, Budapest, p. 304 (1975).
27. T. J. Bruno, P. D. N. Svoronos, *Handbook of Basic Tables for Chemical Analysis,* CRC Press, Boca Raton, 1989.
28. G. D. Foster et al., Int. J. Environ. Anal. Chem., 41 (1990) 105.
29. G. D. Foster et al., J. Agric. Food Chem., 39 (1991) 1618.
30. A. Medvedovici et al., European J. Drug Metabolism and Pharmacokinetics, 25 (2000) 91.

30a. R. Soniassy et al., *Water Analysis*, edited by Hewlett- Packard, 1994.
31. B. S. Middleditch, *Analytical Artifacts*, Elsevier, Amsterdam, 1989.
32. D. T. Wang et al. , Anal. Lett., 21 (1988) 1749.
33. P. H. Chen et al., Chemosphere, 26 (1993) 1743.
34. C. J. Koester, R. E. Clement, Crit. Rev. Anal. Chem., 24 (1993) 263.
35. E. Ballesteros et al., Anal. Chem., 65 (1993) 1773.
36. X. Yang et al., J. Agric. Food Chem., 41 (1993) 1300.
37. M. Videira et al., J. Chromatogr., A, 668 (1994) 237.
38. G. R. Van der Hoff et al., J. Chromatogr., 644 (1993) 367.
39. M. De-Queiros et al., Analusis, 20 (1992) 12.
40. G. Lamprecht et al., J. Chromatogr., A, 667 (1994) 47.
41. R. Alzaga et al., Chromatographia, 38 (1994) 502.
42. B. Malavasi et al., Chromatographia, 36 (1993) 337.
43. D. K. Walker et al., J. Chromatogr., B, 568 (1991) 475.
44. C. M. Grosse et al., J. Pharm. Biomed. Anal., 12 (1994) 195.
45. K. M. Wasan et al., J. Pharm. Biomed. Anal., 12 (1994) 851.
46. R. Neidlein et al., J. Chromatogr., B, 616 (1993) 129.
47. G. Schwedt, Fresenius' Z. Anal. Chem., 288 (1977) 50.
48. S. J. Bannister et al., J. Chromatogr., 173 (1979) 333.
49. R. F. Borch et al., Anal. Lett., 12 (1979) 917.
50. E. B. Edward-Inatami et al., Anal. Proc., 17 (1980) 40.
51. T. Tande et al., Chromatographia, 13 (1980) 607.
52. N. Haring et al., Talanta, 27 (1980) 873.
53. M. Yamazaki et al., Bunseki Kagaku, 30 (1981) 40.
54. S. R. Hutchins et al., J. Chromatogr., 252 (1982) 185.
55. R. M. Smith et al., Analyst, 110 (1985) 35.
56. B. J. Mueller et al., Anal. Chem., 57 (1985) 2693.
57. B. Wenclawiak et al., Mikrochim. Acta, 2 (1984) 251.
58. S. Ichinoki et al., Bunseki Kagaku, 31 (1983) E319.
59. S. Ichinoki et al., Bunseki Kagaku, 32 (1983) 285.
60. A. M. Bond et al., J. Liq. Chromatogr., 6 (1983) 1799.
61. S. Ichinoki et al., Anal. Chem., 57 (1985) 2219.
62. S. Ichinoki et al., J. Liq. Chromatogr., 6 (1983) 2079.
63. S. Ichinoki et al., J. Liq. Chromatogr., 7 (1984) 2467.
64. B. Wenclawiak, Fresenius' Z. Anal. Chem., 310 (1982) 144.
65. B. W. Hoffman et al., J. High Res. Chromatogr., 5 (1982) 439.
66. A. M. Bond et al., Anal. Chim. Acta, 178 (1985) 197.
67. E. B. Edward-Inatimi, J. Chromatogr., 256 (1983) 253.
68. G. Drasch et al., Fresenius' Z. Anal. Chem., 317 (1984) 468.
69. K. Saitoh et al., Anal. Chim. Acta, 178 (1985) 169.
70. I. Akaza et al., Bull. Chem. Soc. Japan, 43 (1970) 2063.
71. I. S. Bark et al., Analyst, 92 (1967) 347.
72. I. Akaza et al., Bull. Chem. Soc. Japan, 46 (1973) 1199.
73. E. Dimitriu et al., *Ultrasunetele- Posibilitati de Utilizare in Industria Alimentara si Biologie*, Ceres, Bucuresti, 1990.
73a. T. Veress, J. Chromatogr., A., 668 (1994) 285.
73b. J. Pawliszyn, Trends Anal. Chem., 14 (1995) 113.
74. W. G. Jennings, A. Rapp, *Sample Preparation for Gas Chromatographic Analysis*, Huthig, Heidelberg, 1983.
75. N. L. Firth et al., J. Assoc. Off. Anal. Chem., 68 (1985) 1228.

76. J. Grimalt et al., Int. J. Environ. Anal. Chem., 18 (1984) 183.
77. S. Sporstoel et al., Anal. Chim. Acta, 151 (1983) 231.
78. C. H. Marvin et al., Int. J. Environ. Anal. Chem., 49 (1992) 221.
79. E. G. Van der Velde et al., J. Chromatogr., 626 (1992) 135.
80. A. Beard et al., J. Chromatogr., 589 (1992) 265.
81. A. J. Wall, G. W. Stratton, Chemosphere, 22 (1991) 99.
82. M. Richards, R. M. Campbell, LC-GC, 9 (1991) 358.
83. T. Zurmuehl, Analyst (London), 115 (1990) 1171.
84. A. R. Gholson et al., J. Assoc. Off. Anal. Chem., 70 (1987) 897.
85. D. E. Kimbrough et al., Analyst (London), 119 (1994) 1283.
86. M. Keinanen, J. Agric. Food Chem., 41 (1993) 1986.
87. W. G. Jennings, J. High Resolut. Chromatogr., 2 (1979) 221.
87a. J. L. Bernal et al., Chromatographia, 34 (1992) 468.
88. T. S. Reighard, S. V. Elesik, Crit. Rev. Anal. Chem., 26 (1996) 61.
89. N. P. Ndiomu, C. F. Simpson, Anal. Chim. Acta, 213 (1988) 237.
90. B. B. Sitholé et al., TAPPI J., 74 (1991) 187.
91. A. L. Lafleur, N. Pangaro, Anal. Lett., 14 (1981) 1613.
91a. K. Hunchak, I. H. Suffet, J. Chromatogr., 392 (1987) 185.
92. N. P. Sen et al., Analyst (London), 111 (1986) 139.
93. U. L. Nilsson, C. E. Ostman, Environ. Sci. Technol., 27 (1993) 1826.
94. M. Hippelein et al., Chemosphere, 26 (1993) 2255.
95. J. Serra-Bonhevi, J. Am. Oil Chem. Soc., 71 (1994) 529.
96. A. H. Khalil, T. A. El-Adawy, Food Chem., 50 (1994) 197.
97. G. H. Tan et al., Pestic. Sci., 40 (1994) 121.
98. R. J. B. Peters et al., Fresenius' J. Anal. Chem., 348 (1994) 249.
99. L. Bonfant et al., Fresenius' J. Anal. Chem., 348 (1994) 136.
100. K. S. Brenner, Fresenius' J. Anal. Chem., 348 (1994) 56.
101. J. F. Lawrence et al., J. Agric. Food Chem., 34 (1986) 315.
102. J. P. Rey et al, J. Chromatogr., 641 (1993) 180.
103. M. T. Galceran et al., Chemosphere, 27 (1993) 1183.
104. M. M. Gupta et al., J. Chromatogr., 637 (1993) 209.
105. F. Yao et al., J. Food Sci., 57 (1992) 1194.
106. Z. Zian et al., Shenyang Yaoxueyuan Xuebao, 10 (1993) 24.
107. J. G. Brayan et al., Pestic. Sci., 34 (1992) 215.
108. B. Hernandez et al., Food Chem., 41 (1991) 269.
109. S. D. Harvey et al., J. Chromatogr., 518 (1990) 361.
110. C. Fauhl, A. Wittkowski, J. High Resolut. Chromatogr., 15 (1992) 203.
111. K. Kamata et al., J. Liq. Chromatogr., 15 (1992) 1907.
112. M. Zanco et al., Fresenius' J. Anal. Chem., 344 (1992) 39.
113. K. M. Hart et al., Environ. Sci. Technol., 26 (1992) 1048.
114. S. O. Baek et al., Environ. Technol., 12 (1991) 107.
115. J. Koenig et al., Chemosphere, 26 (1993) 851.
116. J. N. Pitts et al., Atmos. Environ., 19 (1985) 1601.
117. P. Ciccioli et al., Ann. Chim. (Roma), 85 (1995) 455.
118. B. E. Richter et al., Anal. Chem., 68 (1996) 1033.
119. T. S. Reighard, S. V. Olesik, Crit. Rev. Anal. Chem., 26 (1996) 61.
120. K. Bartle et al., LC-GC, May (1998) 32.
121. E. Bjorklund et al., J. Chromatogr., A, 83 (1999) 285.
122. V. Lopez-Avila, Crit. Rev. Anal. Chem., 29 (1999) 195.
123. B. E. Richter, J. Chromatogr., A, 847 (2000) 217.

124. B. E. Richter, L. Covino, LC-GC, 18 (2000) 1068.
125. American Oil Chemist Society, Official Method AM 2-93.
126. M. M. Schantz et al., Anal. Chem., 69 (1997) 4210.
127. J. A. Fisher et al., Environ. Sci. Technol., 31 (1997) 1120.
128. J. R. Dean, Anal. Commun., 33 (1996) 191.
129. P. Popp et al., J. Chromatogr., A, 774 (1997) 203.
130. B. E. Richter et al., Chemosphere, 34 (1997) 975.
131. J. L. Ezzel et al., LC-GC, 13 (1995) 390.
132. H. Obana, Analyst, 122 (1997) 217.
133. E. Conte et al., J. Chromatogr., A, 765 (1997) 121.
134. O. Zuloaga et al., Trends Anal. Chem., 17 (1998) 642.
135. K. M. Kingston, L. B. Jassie (eds.), *Introduction to Microwave Sample Preparation*, ACS, Washington, 1988.
136. J. R. Pare, J. Lapointe, U.S. Patent 5,002,784, 1991.
137. V. Lopez-Avila et al., Anal. Chem., 66 (1994) 1097.
138. O. F. X. Donard et al., Anal. Chem., 67 (1995) 4250.
139. J. R. Pare et al., J. Microcol. Sep., 7 (1995) 37.
140. A. R. Von Hippel, *Dielectric Materials and Applications*, J. Wiley, New York, 1954.
141. K. Ganzler et al., J. Chromatogr., 520 (1990) 257.
142. T. S. Reighard, S. V. Olesik, Crit. Rev. Anal. Chem., 26 (1996) 61.
143. H. Budinski et al., J. Chromatogr., A, 837 (1999) 187.
144. I. J. Barnabas et al., Analyst, 120 (1995) 1897.
145. K. K. Che et al., J. Chromatogr., A, 723 (1996) 259.
146. T. B. Hsu et al., Organohalogen Compd., 27 (1996) 450.
147. V. Lopez-Avila et al., J. Chromatogr. Sci., 33 (1995) 481.
148. G. Dupont et al., Analyst (London), 124 (1999) 453.
149. Z. C. Kodba et al., Chromatographia, 49 (1999) 21.
150. F. Hernandez et al., Anal. Chem., 72 (2000) 2313.
151. S. J. Stout et al., Anal. Chem., 68 (1996) 653.
152. B. Lalere et al., Analyst, 120 (1995) 2665.
153. I. Rodriguez et al., J. Chromatogr., A, 774 (1997) 379.
154. K. Hummert et al., Chromatographia, 42 (1996) 300.
155. W. Vetter et al., Chemosphere, 37 (1998) 2439.
156. T. S. Reighard, S. V. Olesik, Crit. Rev. Anal. Chem., 26 (1996) 61.
157. S. T. Likens, G. B. Nickerson, Proc. Am. Soc. Brew. Chem., (1964) 5.
158. G. B. Nikerson, S. T. Likens, J. Chromatogr., 21 (1966) 1.
159. M. Godefroot et al., J. Chromatogr., 203 (1981) 325.
160. V. Seidel, W. Lindner, Anal. Chem., 65 (1993) 3677.
161. Z. Sichum et al., Envir. Monitor. Asses., 44 (1997) 536.
162. G. P. Blanch et al., J. Chromatogr., 628 (1993) 261.
163. A. Jayatilaka et al., Anal. Chim. Acta, 302 (1995) 147.
164. A. Bartsch, F. J. Hammerschmidt, Perfum. Flavor., 5/6 (1993) 41.
165. J. Rijks et al., J. Chromatogr., 279 (1983) 395.
166. V. Janda, B. Dolezal, Coll. Czech. Chem. Commun., 50 (1985) 2115.
167. P. Polien, A. Chaintreau, Anal. Chem., 69 (1997) 3285.
168. W. J. Lyman et al., *Handbook of Chemical Property Estimation Methods*, ACS, Washington 1990.
169. Q. Zha et al., 55th TSRC, Greensboro, 2001.
170. J. B. Forehand et al., J. Chromatogr., A, 898 (2000) 111.
171. A. Chaintreau, Flavour Fragr. J., 16 (2001) 136.

172. R. A. Flath, R. R. Forrey, J. Agric. Food Chem., 25 (1977) 103.
173. T. H. Schultz et al., J. Agric. Food Chem., 25 (1977) 446.
174. S. M. Picardi, P. Issenberg, J. Agric. Food Chem., 21 (1973) 959.
175. L. Maignial et al., J. Chromatogr., 606 (1992) 87.
176. A. S. McGill, R. Hardy, J. Sci. Food Agric., 28 (1999) 89.
177. R. G. Clark, H. E. Nursten, J. Sci. Food Agric., 27 (1999) 713.
178. P. Sandra, C. Bicchi, *Essential Oil Analysis*, Hüthig, Heidelberg, 1987.
179. N. Ramarathnam et al., J. Agric. Food Chem., 39 (1991) 1839.
180. G. P. Blanch et al., J. Chromatogr., A,, 655 (1993) 141.
181. C. M. Wu, S. E. Liou, J. Agric. Food Chem., 40 (1992) 838.
182. H. Y. Chung, K. R. Cadwallader, J. Food Sci., 58 (1993) 1203, 1211.
183. W. Zwickenpflug, E. Richter, Chemosphere, 17 (1988) 647.
184. V. Janda et al., J. Chromatogr., 329 (1985) 186.
185. J. L. Boyle et al., J. Food Sci., 57 (1992) 918.
186. E. H. Jenkins, P. J. Baugh, Anal. Proc. (London), 30 (1993) 441.
187. M. Careri et al., Chromatographia, 38 (1994) 386.
188. G. P. Blanch et al., J. Agric. Food Chem., 40 (1992) 1046.
189. G. M. Schneider, Pure Appl. Chem., 55 (1983) 479.
190. K. P. Johnston et al., Ind. Eng. Chem. Res., 28 (1989) 1115.
191. K. D. Bartle et al., J. Supercrit. Fluids, 2 (1989) 30.
192. T. S. Storvick, S. I. Sandler, *Phase Equilibria and Fluid Properties in the Chemical Industry*, ACS, Washington 1977.
193. J. W. King, J. Chromatogr. Sci., 27 (1989) 355.
194. J. C. Giddings et al., Science, 162 (1968) 67.
195. L. R. Snyder, J. Chromatogr. Sci., 16 (1978) 223.
196. C. R. Knipe et al., *Designing a Sample Preparation Method That Employs Supercritical Fluid Extraction*, Hewlett-Packard Co., Wilmington, 1993.
197. D. R. Gere, E. M. Derrico, LC-GC International, 7 (1994) 370.
198. J. Vejrosta et al., Anal. Chem., 71 (1999) 905.
199. S. A. Liebman et al., J. Chromatogr. Sci., 27 (1989) 371.
200. B. W. Wright et al., Anal. Chem., 59 (1987) 640.
201. S. B. Hawthorne et al., J. Chromatogr. Sci., 27 (1989) 347.
202. J. M. Levy et al., J. Chromatogr. Sci., 27 (1989) 341.
203. M. L. Lee, K. E. Markides (eds.), *Analytical Supercritical Fluid Chromatography and Extraction*, Chromatography Conf. Inc., Provo, 1990.
204. C. Maugin et al., J. High Resolut. Chromatogr., 19 (1996) 700.
205. M. E. P. McNally, J. R. Wheeler, J. Chromatogr., 435 (1988) 63.
206. S. B. Hawthorne, D. J. Miller, J. Chromatogr., 403 (1987) 63.
207. S. B. Hawthorne et al., J. Chromatogr. Sci., 27 (1989) 347.
208. J. M. Levy et al., J. Chromatogr. Sci., 27 (1989) 341.
209. S. B. Hawthorne et al., Fresenius Z. Anal. Chem., 330 (1988) 211.
210. B. W. Wright et al., Anal. Chem., 59 (1987) 38.
211. M. R. Anderson et al., J. Chromatogr. Sci., 27 (1989) 371.
212. F. I. Onuska, K. A. Terry, J. High Resolut. Chromatogr., 12 (1989) 357.
213. J. R. Wheeler, M. E. McNally, J. Chromatogr. Sci., 27 (1989) 534.
214. P. Capriel et al., J. Agric. Food Chem., 34 (1986) 70.
215. S. B. Hawthorne, D. J. Miller, J. Chromatogr. Sci., 24 (1986) 258.
216. C. A. Rouse, unpublished results.
217. S. B. Hawthorne et al., Anal. Chem., 61 (1989) 736.
218. K. Sugyiama, M. Saito, J. Chromatogr., 442 (1988) 121.

219. J. M. Wong, K. P. Johnston, Biotech. Prog. 2 (1986) 29.
220. J. W. King et al., J. Agric. Food Chem., 37 (1989) 951.
221. M. A. Schneiderman et al., J. Assoc. Off. Anal. Chem., 71 (1988) 815.
222. T. Veress, LG-GC International, 10 (1997) 114.
223. T. Lehmann, R. Bresseisen, J. Liq. Chromatogr., 18 (1995) 689.
224. K. Anton, C. Berger (eds.), *Supercritical Fluid Chromatography with Packed Columns. Techniques and Applications*, Science ser., Vol. 75, M. Dekker, New York, 1997.
225. C. D. Thiebault (ed.), *Practical Supercritical Fluid Chromatography and Extraction*, Horwood Academic Pub., Amsterdam, 1999.
226. N. Alexandrou et al., Anal. Chem., 64 (1992) 301.
227. M. S. David et al., Anal. Chem., 72 (2000) 3665.
228. I. Windal et al., Anal. Chem., 72 (2000) 3916.

CHAPTER 11

Sorbent Extraction

11.1 SOLID PHASE EXTRACTION PROCEDURES

Solid phase extraction (SPE) is a common sample preparation procedure that uses a solid material for retaining specific compounds from a solution. The retention takes place by passing the solution to be processed through a specific amount of finely divided porous solid phase (50 mg up to 10 g) usually contained in a small column, a cartridge, or a disk. The retained compounds can be later released, using an eluent. This process is frequently used for sample cleanup as well as for sample concentration. The cleanup process can be done in several ways. One basic procedure consists of the selective retention on the solid phase of the interfering compounds while the analytes are eluted and collected for analysis, as illustrated in Figure 11.1.1. Another basic procedure consists of retention of the analytes in the solid phase while the interfering compounds from the matrix are rinsed away. When the analytes are retained on the solid phase, it is common that after the interfering components are eliminated, the analytes are eluted with a different solvent and collected. This procedure is schematically shown in Figure 11.1.2.

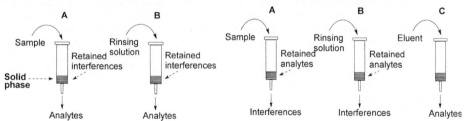

FIGURE 11.1.1. *Diagram of SPE cleanup that retains the interferences.*

FIGURE 11.1.2. *Diagram of SPE cleanup with analyte retention followed by elution.*

The two main procedures can be combined in retention of the analyte and some matrix components, followed by rinsing while retaining the analyte and some other matrix components, and continuing with elution of the analytes while part of the matrix is still retained in the solid phase. This procedure is shown schematically in Figure 11.1.3.

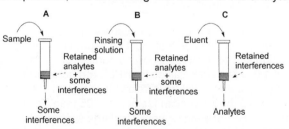

FIGURE 11.1.3. *Diagram of SPE cleanup with analyte retention followed by selective elution.*

The concentration using SPE is typically done using a scheme similar to the one shown in Figure 11.1.2. The analytes present in the sample at low concentration are retained in the solid phase by passing the solution through the solid phase, followed by elution with another eluent using a small volume compared to the volume of initial solution. Large volumes of solutions (several liters) containing traces of analytes can be processed by this procedure, leading to excellent concentration ratios.

The retention and elution in SPE can be viewed as a distribution process between a mobile phase and a stationary phase similar to a LC separation performed on a very short column, with a low number of theoretical plates, but involving compounds with very different distribution coefficients K_i. For example, the number of theoretical plates of an HPLC analytical column can be around 10,000, while that of an SPE column is around 70 (for SPE disks can be 10–20). At the same time, in HPLC it is possible to separate compounds with α as low as 1.05. This is not feasible in SPE. The adsorbing materials in SPE are made in a finely divided solid phase with the dimensions of the particles around 60–100 μm in diameter. This is significantly larger than those used in HPLC (which are 3–5 μm, see Section 4.3). This makes a significant difference in the pressure necessary for passing the fluid through an HPLC column and an SPE device. For these reasons, SPE is seen as a different technique where the separation is not based on the gradual elution of the analytes along a chromatographic column. In SPE the separation is based on a complete retention of the compounds of interest, followed by a step change of solvent and rapid elution from the sorbent. Elution similar to isocratic chromatography (see Section 1.6) based on the difference in the retention volume V_R between the components to be separated is also possible in SPE but is less common than other separation procedures. Also, it is possible to use SPE without elution of the analytes with a solvent but the release by the use of increased temperatures. For this purpose, the analyte is adsorbed on the SPE material, and the matrix (or part of it) is eliminated. After drying the solvent that initially contained the sample or the rinsing solvent, the sorbent can be transferred in a desorbing device and with increased temperatures transferred to a GC system. The desorption using increased temperatures is discussed in Section 9.1.

- *Retention process in SPE*

Depending on the nature of solid phase, a number of mechanisms of solutes retention take place [1], [2], [3], [4]. The most common retention mechanism is liquid/liquid partition (although the name of the technique indicates solid phase). Liquid/solid adsorption as well as ion exchange are also used in various separations.

The efficiency of SPE process can be evaluated based on its capability of retaining the desired type of compounds while the other compounds that must be separated are not retained. The SPE separation, although similar to HPLC, shows a number of differences. As discussed in Section 1.6, common k_i values for HPLC are between 0 and 100, while k_i in SPE for retained compounds must have values higher than 100. For separation in HPLC, it is necessary to have the separation factor $\alpha > 1.05$ (see Section 1.6), while in SPE it is necessary that $\alpha > 4$–5 or even higher. The large separation factor is achieved in SPE by the selection of special stationary phases (sorbent) and by the selection of specific solvents during the retention and elution of the compounds of interest. The choice of sorbents and of rinsing and elution solvents in SPE are discussed further in this section. Another aspect in SPE different from HPLC is the relation between the sample amount and the loading capacity of the stationary phase. While in HPLC it is common to use the analytes in much lower quantity than necessary to exceed the loading column capacity, this is not always the case in SPE. The maximum amount of material that can be retained in the SPE device is characterized by the maximum retention capacity (mg analyte / g sorbent).

A specific parameter used to characterize the retention process in an SPE device is the breakthrough volume V_B. Assuming that a solution of solute "i" with the concentration C_0 is percolated through a SPE cartridge, a frontal or breakthrough curve giving the concentration of the solution exiting the SPE device can be observed. This type of curve is shown in Figure 11.1.4. The analyte concentration is chosen to avoid the overloading of the sorbent capacity. The effluent does not contain at the beginning any measurable levels of the solute. For a volume of effluent denoted by V_B, the concentration in the solute begins to increase. This is due to the elution of the analyte in the SPE device as more solution is passed through the sorbent (and not due to overloading the sorbent). The corresponding concentration for V_B is usually taken as 1% of the initial concentration [5]. As the solution continues to pass through the sorbent, the maximum value of the

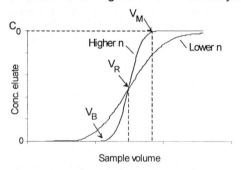

FIGURE 11.1.4. *Breakthrough curve for a SPE device.*

eluate concentration is that of the initial solution. The volume of collected eluent with a concentration equal to 99% of the initial concentration value is denoted by V_M. Under ideal conditions, the breakthrough curve has a bilogarithmic shape, having an inflection point situated at the value of the effluent volume denoted by V_R. This corresponds to the retention volume V_R in a standard chromatographic elution for an injection of a compound at a low concentration through the SPE device (see rel. 1.6.5). Because the point used for establishing V_B is arbitrarily chosen at 1% for the ratio of the effluent concentration vs. the initial concentration, other values can be selected, depending on the purpose of SPE separation [6], [7], [8], [9], [10]. The breakthrough volume V_B can be defined as the volume of a sample assumed to have a constant concentration, low enough not to overload the sorbent, which can be passed through the SPE device until the concentration of the analyte at the outlet C_E reaches a certain value $b = C_E/C_0$. The breakthrough volume V_B depends on the nature of the analyte, the nature of the sorbent material, and of the solvent used in the process.

In the definition of the breakthrough volume, it is assumed that in the whole process the capacity of the SPE device is never exceeded. However, the SPE capacity depends on the number of available sorption sites for the analyte and retained matrix components. Saturation of these sites leads to lack of retention or distortion of the sorption isotherm, resulting in lower retention as the surface concentration of sample on the sorbent surface increases. This produces a mass overload, which becomes important when the concentration of analyte and/or retained matrix is high. Typical chemically bonded sorbents have a capacity of about 1–10% of their weight. This limit is unlikely to be exceeded during trace enrichment of samples such as surface water, biological fluids, etc. However, higher sample concentration can break through the SPE device because the retention capacity is exceeded, and this is not related to V_B as defined previously.

The measurement of the breakthrough volume for a number of components present in a sample is not practical. An estimation for V_B can be based on the similarity of SPE with the LC separation. As shown in Section 1.6, the peak broadening in LC is characterized by the peak width at the baseline W_b (expressed in time) or at the half height W_h. The breakthrough volume can be estimated using the expression:

$$V_B = V_R - (W_b/2)\ U \qquad (11.1.1)$$

where the retention volume V_R is identical to that in an LC procedure, and U is the volumetric flow rate of the solution, necessary in rel. (11.1.1) to change peak width W_b from time units to volume units. This expression indicates that the peak broadening effect is the cause of the increase in the solute concentration before V_R.

By the use of rel. (1.6.41) for the number of theoretical plates, rel. (11.1.1) can be written in the form:

$$V_B = V_R - \frac{2 V_R}{\sqrt{n}} = V_R (1 - \frac{2}{\sqrt{n}}) \qquad (11.1.2)$$

The value of V_R can be expressed as a function of the volume of the mobile phase in the sorbent V_m based on rel. (1.6.33) such that rel. (11.1.2) can be written in the form:

$$V_B = V_m (1 + k_i)(1 - \frac{2}{\sqrt{n}}) \qquad (11.1.3)$$

The value of V_m can be estimated as being the interparticle volume of the sorbent bed (void volume), n is the number of theoretical plates for the SPE device, and k_i is the capacity factor for the solute "i". As shown in Figure 11.1.3, V_B has a lower value for a SPE column with lower n. Also, when a compound is strongly retained by the sorbent, k_i is higher, and the breakthrough volume is larger. The calculation of the void volume V_m can be done using the expression:

$$V_m = U\ t_o \qquad (11.1.4)$$

where t_o is the retention time of a totally unretained compound. The value of V_m is estimated for the C18 silica sorbent as 0.12 ± 0.01 mL per 100 mg of sorbent [11]. For a SPE cartridge packed with 500 mg C18-silica, the average number of theoretical plates for a flow-rate of 5 mL/min is about 20 [12]. Various reports describe the estimation of the breakthrough volume for SPE devices [8], [13], [14], [15].

The validation of the estimation for V_B as previously described was done in a number of experimental measurements for several analytes using extraction disks with C18-silica or poly(styrene-divinylbenzene) (PS-DVB) sorbents [16]. The k_i values were measured by LC with water as mobile phase or from measurements in water-methanol mobile phase and extrapolation of the results in an expression similar to rel. (1.10.28) or (1.10.29).

Other approximating expressions were proposed for the breakthrough volume V_B (expressed in mL liquid) One simple expression is the following:

$$V_B = 1/6 \; k_i \; V_{bed} \qquad (11.1.5)$$

where V_{bed} is the volume of the sorbent bed in cm^3 [17]. Another approximation takes into consideration the level set for b % for the calculation of V_B and is given by rel. (11.1.6). The values for parameters a_0, a_1 and a_2 are given in Table 11.1.1 [8], [18]:

$$V_B = V_R (a_0 + \frac{a_1}{n} + \frac{a_2}{n^2})^{-\frac{1}{2}} \qquad (11.1.6)$$

TABLE 11.1.1. *The values of parameters a_0, a_1 and a_2 for the calculation of breakthrough volume.*

b (%)	a_0	a_1	a_2
0.5	0.990025	17.92	26.74
1	0.9801	13.59	17.60
2	0.9604	9.686	10.69
5	0.9025	5.36	4.60
10	0.81	2.787	1.94

The information about breakthrough volume is very important for practical applications. If the analytes are not properly retained in the SPE device, recovery R% of the sample preparation procedure is low and frequently not reproducible. It is not possible in this case to use internal standard, even very similar to the analytes, to account for the losses because the losses are not proportional to the analyte concentration. The retention of the analytes on the SPE device must be done such that the volume of solution to be analyzed is below the breakthrough volume. The variation of breakthrough volume with the flow rate of the solvent also must be considered for an SPE process. Figure 11.1.5 shows a typical variation of the recovery R% of an analyte as a function of the sample solution flow rate.

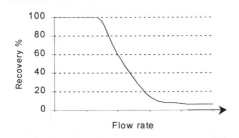
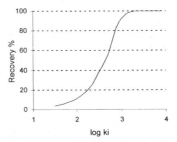

FIGURE 11.1.5. *Recovery variation with the increase in the solution flow rate in SPE.*

FIGURE 11.1.6. *Recovery variation with the increase in the k_i for the same sample volume and solution flow rate.*

The decrease in the number of theoretical plates of the column leads to a decrease in the breakthrough volume, and losses of the sample occur. The large particle size of sorbents in SPE is unfavorable to recoveries at large flow rates. For this reason, it is common to adjust the flow through an SPE device for obtaining appropriate recoveries.

On the other hand, the increase in k_i for a specific analyte increases the recovery, as shown in Figure 11.1.6. Results for the data in Figure 11.1.6 were obtained for a C18 SPE disk with n = 20, the amount of sorbent 450 mg, and a volume of 500 mL solution.

- Materials used as solid phases

A large number of stationary phases (sorbents, solid phases) are used in SPE. These solid phases are made similarly to those used in LC from particles of a porous substance. Some natural materials (such as porous silica or alumina) have specific properties allowing the adsorption process. Other materials are synthetic and are made from a "base material" on which specific organic fragments are chemically bonded (bonded phases). The most common base material is porous silica. Organic polymers such as polystyrene-divinylbenzene or acrylates are also used as sorbents.

The solid phase materials (sorbents) used in SPE are classified as nonpolar, polar, and ion exchange types. Sometimes the materials having a synthetic organic polymer as the backbone material are classified separately. A significant number of sorbents have a porous silica backbone, which is derivatized with specific reagents for obtaining a bonded phase with the desired properties. The silica used for this purpose is typically an amorphous material with a surface area of 50–500 m^2/g and pore (mesopore) diameters of 50–500 Å. Depending on the silica source, the apparent pH varies between 3.8 and 9.8, although most materials have an acidic character. The silica surface contains a number of silanol groups, which can be single silanol, geminal silanol, or associated silanol groups, some having hydrogen bonds to water molecules as shown below:

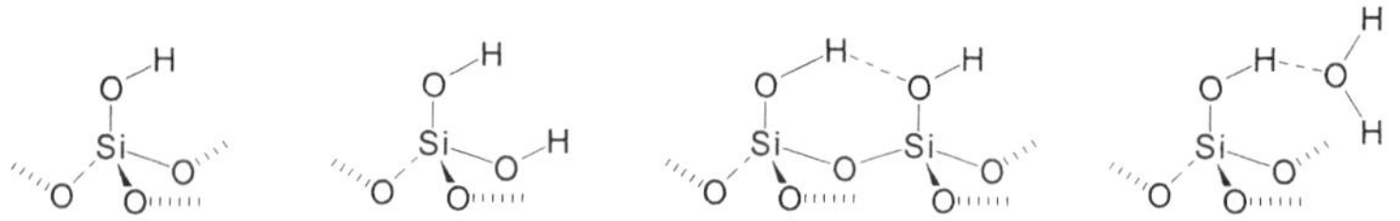

A silica type with more uniform structure is also available (Type B silica). This silica contains single silanol groups, associated by hydrogen bonds. Silica without any modification can be used as a SPE polar sorbent. However, similarly to the use of silica as a stationary phase in HPLC or TLC, the bonds between the backbone and the water on the surface of silica are not very strong, and some of the water can be removed by different eluents. This leads to modifications in the nature of the silica surface and therefore of the polarity of this material. On the other hand, the silanol groups from the silica surface can be chemically modified, leading to materials of different polarities that are widely used as stationary phases. There are several reactions applied for obtaining stationary phases with a silica backbone. One such reaction uses a reactive organosilane as shown below:

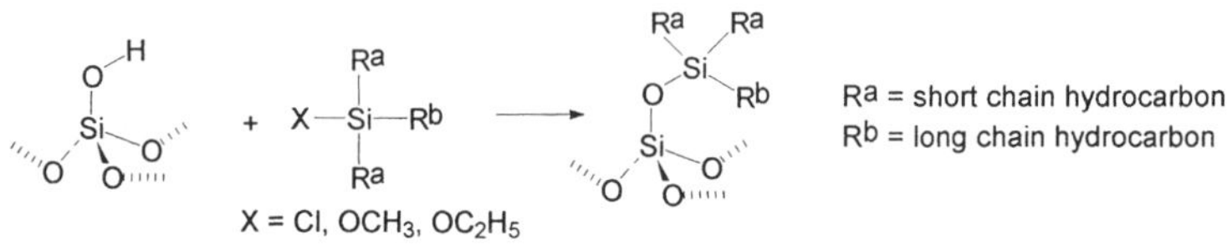

Ra = short chain hydrocarbon
Rb = long chain hydrocarbon

X = Cl, OCH3, OC2H5

Better results are obtained using as a reagent a silane with three reactive functionalities, and the reaction can be written as follows:

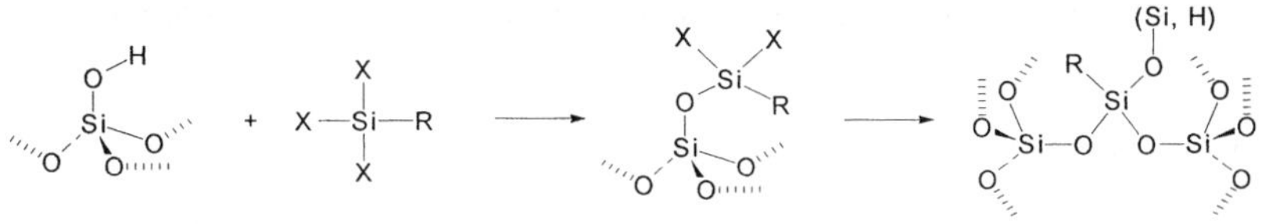

This type of reaction generates cross-linking between more OH active groups, and two adjacent silanols react as follows:

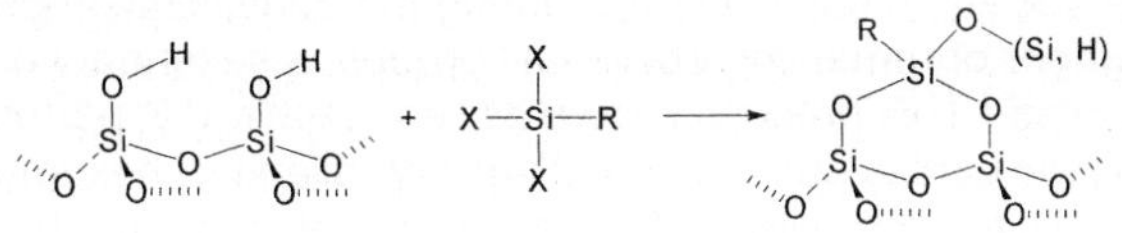

Some other reactions are used to replace the active OH groups on the silica surface. Among these are the reaction with alcohols and elimination of water from acidic OH groups, reaction with $SOCl_2$ that replaces OH with Cl followed by reaction with a Grignard reagent or organolithium compound, silanization followed by reaction with a reactive alkene, etc. The idealized structures of some of the materials obtained using a silica base are shown below:

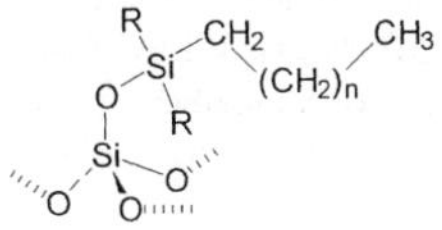

n = 0 (C2), 6 (C8), 16 (C18) 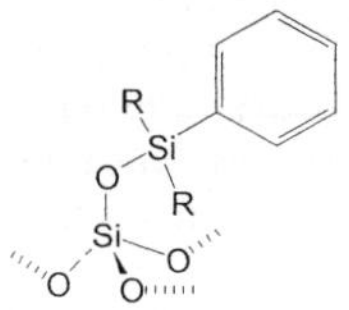   PH 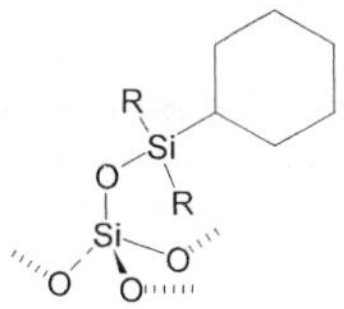   CH 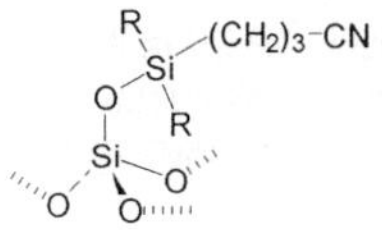   CN

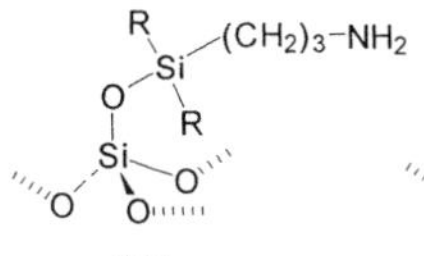

NH$_2$ 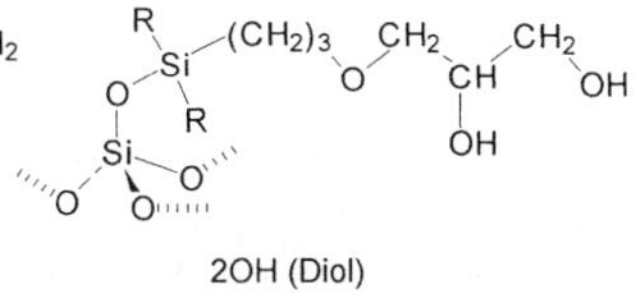   2OH (Diol) 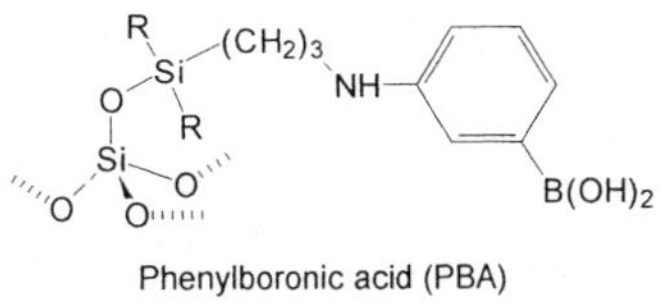   Phenylboronic acid (PBA)

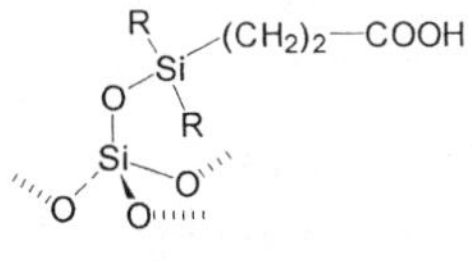

CBA 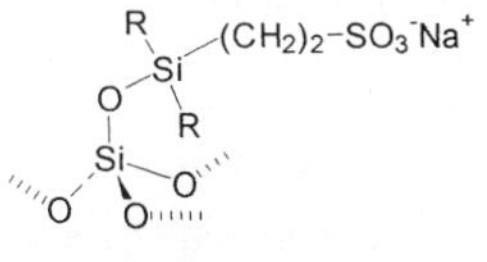   Propyl sulfonic (PRS) 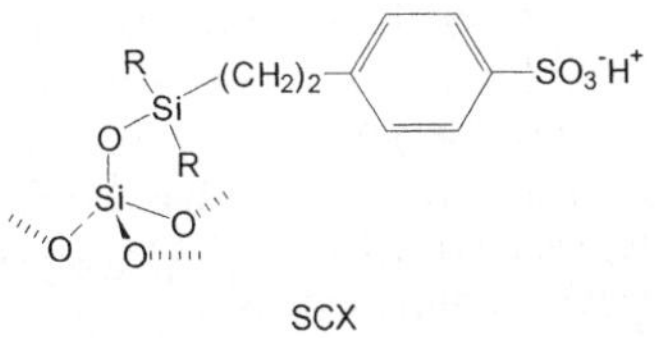   SCX

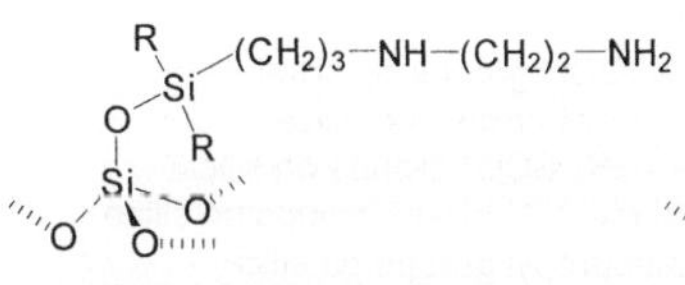

Ethylenediamine-N-propyl 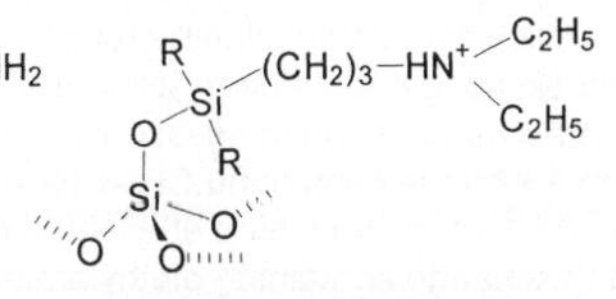   Diethylaminopropyl 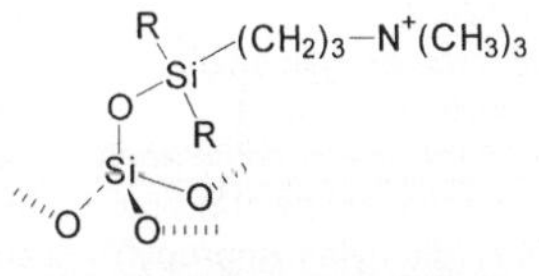   SAX

Depending on the derivatization procedure, the R groups can be CH_3, $Si(CH_3)_3$, or for SPE materials obtained using silanes with three reactive groups (such as trichlorosilanes), R can be the oxygen bound to another Si atom from the silica base material.

Although silica surface derivatization is conducted with the intention to eliminate the active OH groups, a number of these groups always remain free after the reaction. Besides the mesopores present in silica particles, some very small pores (micropores) are also present, and their surface contains active OH groups. In addition to the OH groups present in micropores, the attachment of long hydrocarbon chains such as C_{18} groups hinders the access to some silanols during the derivatization process. In order to minimize the number of remaining active OH groups, a secondary derivatization process can be applied. This process known as "endcapping" is performed with reagents with small molecules such as trimethylchlorosilane or hexamethyldisilazane. The small hydrophobic trimethylsilyl (TMS) groups diminish the number of free OH groups, although a complete endcapping is not possible. The polarity of the resulting solid phase is homogeneous nonpolar. For phases that are not endcapped, both nonpolar long hydrocarbon chains and some polar silanol groups are present and "seen" by the analytes. The result in HPLC is a decrease in column performance, mainly for basic compounds that show peaks with tail. The same effect is seen in SPE. Different types of polarity obtained for SPE materials on a silica backbone are indicated in Table 11.1.2.

Besides the polarity and the endcapping process, another parameter is controlled during synthesis, which is the amount of organic material loaded on the silica base. This parameter is expressed as C% load and can vary from a few % up to about 18%.

TABLE 11.1.2. *Different types of polarity obtained for SPE materials on a silica backbone.*

Stationary phase (in order of increasing polarity)	Description
C18 (octadecyl)	nonpolar, C_{18} chains on a silica support
C8 (octyl)	nonpolar, C_8 chains on a silica support
C2 (ethyl)	nonpolar, C_2 chains on a silica support
PH (phenyl)	nonpolar, phenyl groups on a silica support
CH (cyclohexyl)	nonpolar, cyclohexyl groups on a silica support
CN-E (cyanopropyl endcapped)	nonpolar, C_3H_6-CN on a silica support (E indicates endcapped)
C18/OH (not endcapped)	nonpolar, C_{18} chains on a silica support but low hydrocarbon load and some free silanol groups present
NH_2 (aminopropyl)	polar, C_3H_6-NH_2 on a silica support
CN-U (cyanopropyl not endcapped)	polar, C_3H_6-CN on a silica support (U indicates no endcapping)
2OH (diol)	polar, C_3H_6-O-CH_2-CHOH-CH_2OH on silica
PBA (phenylboronic acid)	polar N-propyl aminophenylboronic acid on silica
SI (silica)	polar, silica containing silanol groups
CBA (carboxylic acid)	cation exchange containing CH_2-CH_2-COOH groups on silica
propylsulfonic acid	cation exchange containing CH_2-CH_2-SO_3H groups on silica
SCX (benzensulfonic acid)	cation exchange containing CH_2-CH_2-C_6H_4-SO_3H groups on silica
ethylenediamine-N-propyl	anion exchange containing ethylenediamine-N-propyl groups on silica
DEA (diethylaminopropyl)	anion exchange containing diethylaminopropyl groups on silica
SAX (quaternary amine)	strong anion exchange containing propyltrimethyl ammonium groups on silica

Alumina, $MgSiO_3$ (Florisil), cellulose, hydroxyapatite, and graphitized carbon are also materials used as sorbents. Alumina has been used for a long time as a stationary phase in TLC, and different states of hydration and acidic or basic characters can be obtained depending on the treatment of this material. Alumina, Florisil, cellulose and graphitized carbon have a number of applications as SPE phases [19]. Similar properties with alumina can be found in other oxides. For example, magnesium oxide and zirconium oxide were used as sorbents for separation purposes. Some sorbent materials using a modified zirconia backbone have been evaluated for HPLC use.

Another type of material frequently used as a sorbent as well as for base to add functional groups that confer specific desired properties to the sorbent is polystyrene/divinylbenzene (SDVB or PS-DVB). Divinylbenzene units having two reactive groups determine the degree of cross-linking in the copolymer and influence the properties of the material. Contents around 8–12% DVB in styrene are common in commercial products. Other cross-linking substances can be used besides divinylbenzene, such as ethylenedimethacrylate. Also, SDVB can be further modified by chloromethylation (with HCl and formaldehyde in the presence of $ZnCl_2$) followed by reaction with a tertiary amine. This derivatization leads to a strong anion exchange material. Sulfonation of SDVB leads to a strong cation exchanger. The idealized structure of SDVB and of the anion and cation exchangers obtained from this material are shown below:

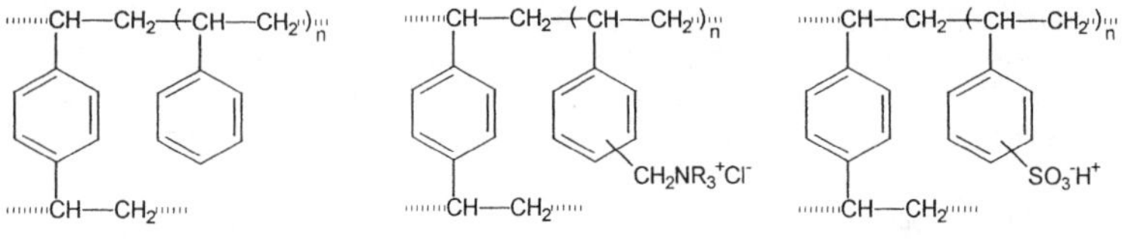

SDVB   Anion exchanger   Cation exchanger

Other groups can be attached to the SDVB backbone. These may include C18 chains, pyridine groups, N-vinylpyrrolidone, or iminodiacetic groups as shown below:

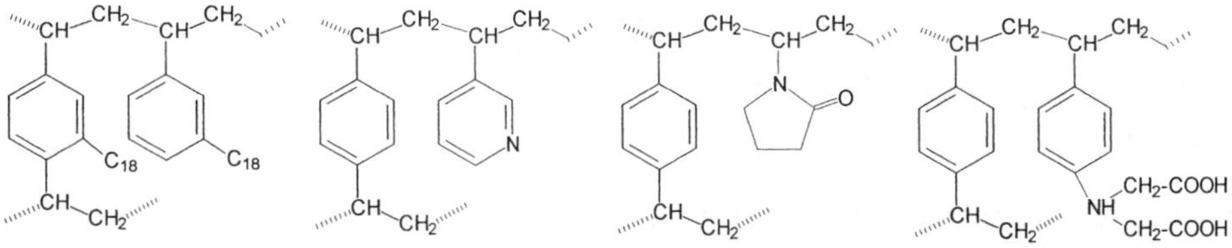

C18   Vinylpyridine   N-Vinylpyrrolidone   Iminodiacetic

Some few other polymers and copolymers are also used as sorbents for SPE applications. Examples are the copolymer SDVB/methyl methacrylate, the copolymer PS-DVB/N-vinylpyrrolidone (PS-DVB-NVP), and the copolymer SDVB/ethylvinylbenzene (PS-DVB-EVB). Depending on the chemical nature of the organic radicals attached to the backbone either of silica or of an organic polymer structure, a variety of polarities can be obtained for the resulting SPE material. Some polymers contain a more complex backbone, such as a copolymer PS-DVB/N-vinylpyrrolidone with additional functional groups such as $-SO_3H^-$ or $-CH_2N^+(CH_3)_3$. The vinylpyrrolidone groups, being less hydrophobic compared to the benzene rings, allow better access of aqueous solvents to the polymer.

One very important characteristic of the sorbents used in SPE is the purity of the material. Silica may contain specific metal ions, and the derivatization process may leave traces of the reagent or of the reaction products of the reagent with silica or with water. Traces of solvents also may be present in the sorbent. These impurities may affect negatively the core chromatographic analysis, introducing unexpected interferences. The quality of SPE materials is continuously improving, but in specific trace analysis the purity of the stationary phase may still remain an issue.

A number of physical characteristics of the solid phase material are also important for the performance of the sorbent [13], [20]. Particle dimensions, porosity, and surface area were mentioned previously. The measurement of porosity P_t can be done by use of an unretained compound in a solution and measurement of its retention time t_0 at a specific volumetric flow rate U. The value for P_t is obtained from the expression:

$$P_t = U t_0 / V_c \qquad (11.1.7)$$

where V_c is the volume of the sorbent. These parameters contribute to the pressure drop across the particle bed Δp. From Δp a specific permeability B_0 is defined for a specific sorbent bed length L and bed area A:

$$B_0 = \frac{U \eta L}{A \Delta p} \qquad (11.1.8)$$

where η is the fluid viscosity. The larger is B_0 (expressed in m^2), the more permeable is the bed. Standard SPE columns and cartridges have a specific permeability around 25 10^{-14} m^2. The permeability may change during the SPE device utilization, and particles present in the sample may decrease B_0 below acceptable limits of utilization. Particle dimensions also influence significantly Δp, and although small particles may be preferred for better adsorption properties and lower plate height, they lead to the increase in Δp. This increase may influence unfavorably the SPE functionality, requiring longer processing time, higher pressure (or vacuum) for passing the solutions through the sorbent, and in extreme cases leading to device plugging. Some performance criteria for SPE devices are summarized in Table 11.1.3.

Another aspect that may differentiate SPE sorbents from one manufacturer to another is the homogeneity of the material from batch to batch, leading to good or questionable reproducibility in the chemical analysis. The quality of a specific SPE material is typically described in the manufacturer's catalogs (see. e.g. [21], [22], etc).

The cost of SPE devices is especially important because these devices are usually disposable. Some devices are used only once, and some may be reutilized a few number of times, but a large number of utilizations like with HPLC columns is not practiced.

TABLE 11.1.3. *Some performance criteria for SPE devices [19].*

Property	Influenced by:	Affected result
Pore diameters (50–500 Å) and volume	Nature of base material, type of surface derivatization, crosslinking in synthetic polymers, water removal	Permeability, binding capacity, recovery of some analytes
Surface area 50–600 m^2/g	Nature of base material, crosslinking in synthetic polymers, water removal	Capacity, recovery
Particle shape	Irregular or spherical shape of initial material	Permeability, may influence the number of theoretical plates
Particle dimensions (30-150 μm diameter d_p)	Selection of base material	Permeability, number of theoretical plates n (or effective plate N), flow rate through the sorbent bed
Particle size distribution (uniform or large variability)	Selection of base material or polymer	Permeability, reproducibility by influencing channeling
Specific permeability B_o (20-30 10^{-14} m^2)	Pore size, particle size, and shape distribution	Maximum flow rate through the bed
Capacity / g sorbent (50 mg/g silica base, 250 mg/g synthetic polymers)	Sorbent type, carbon load, pore and particle physical properties	Maximum amount absorbed / g of sorbent
Recovery (R%)	Nature of sorbent, nature of eluents, sorbent homogeneity.	Recovery of analyte, success of cleanup process
Physical stability	Particle rigidity and particle size distribution	Loss of sample due to channeling, formation of colloids
Chemical stability	Nature of phase, quality of manufacturing process	Change in the nature of sorbent during analysis, irreproducibility
Biological stability	Nature of sorbent	Limits on storage conditions
Cleanliness	Purity of construction materials, homogeneity of the derivatization	Interferences in analysis
Reproducibility	Homogeneity of material, physical and chemical stability of sorbent	Precision of results, accuracy of measurements
Selectivity	Nature of stationary phase, choice of eluent, and procedure in SPE use	Success of cleanup, lack of interferences in core process
Cost	Various factors	Overall cost of analysis

- Nonpolar SPE sorbents

A quite large number of SPE materials are commercially available, and if they are of the same type, although of different brands, some have very similar properties. The most utilized SPE sorbent is probably octadecyl (C18). This sorbent has high retentive properties for nonpolar compounds. C18 is generally regarded as the least selective silica-based sorbent, since it retains most organic analytes from aqueous matrices, which is often a benefit when the compounds of interest vary widely in structure. The potential for polar interactions between analytes and sorbent is less significant with C18 than with any other sorbent because of the predominant effect of the long hydrocarbon chain. For analysis of small to intermediate molecules, C18 can also be utilized for desalting aqueous matrices prior to ion exchange because salts pass through it unretained. The stability of this phase is usually between pH 2 and 9. This is one of the limitations of silica-based sorbents. At low pH, the alkyl-bonded phases are susceptible to hydrolysis, and silica dissolution occurs at high pH. The stability of silica base sorbents within this interval is in general good, although the specific pH range of utilization must be verified before use. Besides C18 with the highest possible carbon load and high degree of endcapping, a number of other C18 materials with a decreased carbon load and some other characteristics such as high porosity, various particle sizes, etc. are commercially available. These sorbents show a slightly lower interaction with nonpolar analytes and may be useful for a more convenient elution.

C18/OH is a non-endcapped low-load version of the octadecyl bonded phases that enables the silanol groups on the surface to be more active. This C18 SPE has controlled silanol activity. The silanol activity permits fractionation of metabolites and enhanced retention of basic compounds compared with endcapped C18. In addition, the 150 Å intermediate pore size works well for intermediate molecular weight compounds that may not be retained on a standard 60 Å material. Also, some applications require the retention of compounds with a wide range of polarities. This may require two types of SPE material and therefore additional sample preparation processes. Some phases, having more than one type of group such as C18 and silanol may be useful for processing a wider range of compounds.

C8 is very similar in property to C18, but is not as retentive for nonpolar compounds due to its shorter hydrocarbon chain. C8 can be used as a replacement for C18 when analytes are too strongly retained on C18 for effective elution. The potential of C8 for polar interactions with analytes is somewhat higher than for C18 because the shorter hydrocarbon chain does not mask the silica surface as effectively, but polar interactions are still not a significant property of C8. The C8 sorbent has been successfully utilized in the simultaneous extraction of fat and water soluble vitamins from human plasma samples. Some of the main characteristics of C18 as well as C8 sorbents that are commercially available are given in Table 11.1.4.

TABLE 11.1.4. *Characteristics of some commercial SPE C18 and C8 silica gel base sorbents.*

Characteristic property	C18	C8
porosity (Å)	55–125	55–125
particle diameter d_p (μm)	40–100, some uniform diameters some in a range (37–55 range)	40–100, some uniform diameters some in a range (37–55 range)
particle shape	spherical or irregular	spherical or irregular
silane function in synthesis	mono, tri	mono, tri
endcapping	some yes, some no	some yes, some no
carbon content (%)	4.7–18	8–12.5

C2 shows some polar character because of the short chain length of the functional groups, which exposes the silica surface. C2 is used as a replacement for C18 and C8 when molecules are retained too strongly by these phases. Polarity of ethyl-silica is slightly lower than a CN for polar interactions. This SPE is useful for the extraction of drugs and their metabolites from plasma and urine.

Due to the single carbon functional group, C1 is the least retentive of all alkyl group bonded phases for nonpolar compounds. However, because the sorbent is endcapped, masking the polar silanol activity, easier retention and elution of polar and multifunctional compounds can be achieved.

Cyclohexyl phase also shows some polarity and exhibits unique selectivity for certain solutes such as PAHs [23]. When employed as a nonpolar sorbent, cyclohexyl-silica has the approximate polarity of a C2 sorbent. The polar subsurface is not an important factor for cyclohexyl-silica properties. Because of its unique selectivity, CH is often a good choice when other nonpolar sorbents such as C18, C8, C2, and PH do not provide the desired selectivity.

Phenyl phase is most commonly employed for nonpolar extractions, with a similar polarity to a C8 sorbent. Like cyclohexyl, phenyl-silica exhibits a different selectivity from other nonpolar sorbents. This added selectivity results from the specific interactions with the aromatic ring. Retention of planar, conjugated organic molecules is enhanced compared to the aliphatic bonded phases.

Synthetic polymers are used to make a significant number of nonpolar sorbents. Polymeric sorbents are characterized by a higher hydrophobicity than modified silica materials, providing better retention in many applications. Also, polymeric materials are stable in a wider pH range. For polystyrene-divinylbenzene polymers (SDVB or PS-DVB) various cross-linking degrees achieved by controlling the amount of divinylbenzene and the polymerization conditions lead to materials of various porosities and surface areas. Divinylbenzene-based polymers provide a high stability within an extended pH range of 0–14. This can be an advantage when using extreme pH values for development of procedures for basic or acidic solutes. Another advantage of polymeric sorbents is given by the unique type of interaction only through hydrophobic forces. The lack of silanol groups eliminates the polar interactions that complicate the retention mechanism. Methacrylate copolymers also can be used as nonpolar sorbents. They are more stable than silica-based sorbents, although hydrolysis of the ester group can occur at high pH. Different other groups can be attached to the benzene ring, and copolymers with ethylvinylbenzene (EVB) or N-vinylpyrrolidone (NVP) are used as sorbents. Some characteristics of nonpolar synthetic polymer sorbents commercially available are given in Table 11.1.5.

TABLE 11.1.5. *Characteristics of some commercial nonpolar synthetic polymer sorbents.*

Characteristic property	SDVB	PS-DVB-Methacrylate	PS-DVB-EVB	PS-DVB-NVP
porosity (Å)	80–450		300	55–82
particle diameter d_p (μm)	40–125 (some products as low as 5)	70	40-120	30–120
surface area (m²/g)	500-1200		1060	550–800

One other nonpolar sorbent is graphitized carbon. This is a hydrophobic material, although the carbon surface may have some functionalities (such as OH, C=O or COOH). Poor recoveries are usually noticed for larger or strongly aromatic compounds when this sorbent is used. However, basic compounds with nonpolar moieties can be adsorbed from neutral media and eluted using low pH eluents that generate positive charges on the analyte, followed by easier elution with polar solvents.

- *Polar SPE sorbents*

A medium polarity sorbent with many uses is cyanopropyl (CN). This phase is ideal for applications in which extremely nonpolar compounds would irreversibly retain on nonpolar sorbents, such as C18 or C8. CN phase can be used for the adsorption of polar analytes, but many applications are for unsaturated compounds. Besides polar and nonpolar interactions between this adsorbent and solutes, special π-π interactions can be involved between solutes and the cyano group in the adsorbent. CN sorbents are available in endcapped or nonendcapped forms. The endcapped version of the cyano sorbent is typically recommended for extraction of analytes from an aqueous matrix. The nonendcapped material has a higher polarity, being less retentive than silica gel or 2OH (Diol) for very polar analytes that might be retained irreversibly on these sorbents. The nonendcapped cyanopropyl phase is suited for the extraction of polar compounds from a nonpolar matrix, such as hexane or oils. Typical carbon loading of cyanopropyl phases is 8–9%, and the hydrophobic interactions are still considerable. This allows this phase to be used also as reversed-phase sorbent.

Aminopropyl (NH_2) is a polar sorbent, which can interact through hydrogen bonding and anion exchange properties. Since the pK_a of the NH_2 sorbent is about 9.8, at any pH below this value, the majority of the functional groups are positively charged. NH_2 is a weaker anion exchanger than sorbents like SAX (a quaternary amine sorbent that is always charged). This makes it a better sorbent for the retention of strong anions such as sulfonic acids, which may be retained irreversibly on SAX. Because a propyl group is connected to NH_2 functionality, this phase also can be used for the isolation of nonpolar compounds from polar samples. Like 2OH and silica, NH_2 sorbents are useful for the separation of structural isomers. Aminopropyl silica can be used with aqueous matrices, especially when the sample volume is low. This sorbent is commonly used for cleanup of biological samples. For example, in the analysis of vitamin D_3 metabolites in human plasma, aminopropyl silica proved to give better detection and quantification limits by chromatographic analysis in comparison with the samples processed with C18 SPE [24].

Diol (2OH) is a polar sorbent, typically used for polar extractions from nonpolar solvents. 2OH resembles unmodified silica in its tendency to form strong hydrogen bonds with the analytes. It also shares silica's ability to discriminate between compounds of high similarity such as structural isomers. Prostaglandins, for example, can be separated into classes by use of diol sorbent by slight variation of the solvent polarity. 2OH is a more predictable sorbent than silica gel for extraction procedures. Silica polarity is due in part to adsorbed water, which can be removed by some anhydrous solvents, and because of this, silica polarity changes easily in different applications. In addition to its usefulness as a polar sorbent, 2OH also is used in nonpolar mode because of the hydrocarbon moiety, which provides enough nonpolar character for retention of some hydrophobic solutes such as tetrahydrocannabinols (THC) from polar matrices like urine.

Silica is generally regarded as the most polar sorbent available, and -Si-O-Si- groups, although hydrophobic, have low activity. Its character is determined by the free Si-OH groups, and the density of silanols on chromatographic grade silica is 7–9 $\mu mol/m^2$, which is higher than the maximum possible concentration of alkyl groups on a bonded phase (4–5 $\mu mol/m^2$). Unbonded, activated silica is typically fairly acidic. This sorbent is one of the best sorbents available for selective separation of compounds with a very similar structure. For this purpose, the analytes are applied on the SPE material in a nonpolar solvent, followed by slow increase of the solvent polarity by addition of a polar modifier such as tetrahydrofuran (THF), ethyl acetate, etc. The hydrogen bonds and dipole-dipole interactions between silica and polar analytes are strong, especially when the analyte has functional groups like hydroxyl, carboxyl, or amino. In the use of silica, the adsorption of polar solutes from solutions is usually high, but the desorption yield of the same solute may be low. Strong interactions are expected mainly when the silanol groups are ionized and the analytes are positively charged. The dissociation of silanol groups can be controlled by the pH of the analyte solution, and at pH = 2 the silanol groups are not ionized. As pH increases, more dissociation takes place, and above pH = 4 the silica surface becomes charged with negative ionic groups [25], [26], [27].

Some other polar SPE materials not having a silica base material are also commercially available. Florisil ($MgSiO_3$) is an extremely polar material ideal for the isolation of polar compounds from nonpolar matrices. The larger particle size of the Florisil material enables fast flow of large volume samples and is an alternative to silica for viscous samples. Florisil is recommended in a number of AOAC and EPA regulated methods.

Alumina (Al_2O_3 with free OH groups), like silica, is an extremely polar sorbent. The alumina surface tends to be slightly more stable under high pH conditions than unfunctionalized silica. The smaller particle size ensures high extraction efficiency, and small sorbent beds are therefore commonly used. This sorbent is usually prepared at a pH that ensures electrically neutral surface (Alumina N with pH ≈ 7.5). However, acidic alumina (Alumina A with pH ≈ 4.5) as well as basic alumina (Alumina B with pH ≈ 10.0) are commercially available. Neutral alumina is a strongly polar sorbent, which shows good retention of compounds such as aromatic species and aliphatic amines. It also favors retention of compounds containing electronegative groups (for example, functional groups with oxygen, phosphorus, or sulfur atoms). Alumina can be used for extracting nonpolar compounds from aqueous solutions or polar species from aqueous samples. Acidic alumina has a slightly cationic nature, while basic alumina a slightly anionic nature. Chemically modified alumina has been proposed for SPE material, but is much less common than modified silica materials.

Cellulose powder is another known SPE material. Microgranular particles (made usually from cotton) of a controlled size and with a high α-cellulose content ensure good column packing properties. The cellulose SPE is very stable over a wide pH range and contains an extremely low metal content. Cellulose has a polar nature due to its OH groups and is able to have high loading of polar substances from aqueous and organic phases. The binding property of the free hydroxyl groups is associated with slight reducing properties due to the presence of a small number of residual aldehydes. Also, cellulose has a weak ion-exchange character. Microgranular cellulose has been used as a prefilter for suspended matter (e.g. cell debris and suspended fats), as a pre-fractionation step prior to reverse phase HPLC, in separation of biomolecules, and in affinity chromatography [21]. Large-scale industrial applications are also known for this material.

- Ion exchange SPE sorbents

Based on a silica substrate, a number of ion exchange SPE materials are also available. CBA is a medium polarity sorbent based on carboxypropyl functionalized silica. Its main characteristic is the weak cation-exchange properties. However, this material can display either polar or nonpolar properties depending on the matrix or the solvent utilized. Because of its pK$_a$ value of 4.8, CBA does not require the extreme basic conditions for elution that are required by other cation exchange sorbents. Above a pH 4.8, the majority of functional groups carry a negative charge that can be used for retaining cationic compounds. A pH below 2.8 effectively neutralizes the surface of CBA material, allowing convenient elution of the analytes retained by the sorbent. For this reason CBA is often the best choice for cationic exchange applications, especially when dealing with strong cations (i.e. cations with a high pK$_a$). In addition, CBA shows a wider range of selectivity towards counter-ions than other cation exchange phases, providing more options for elution. Typical samples that can be processed by this sorbent are aqueous samples and biological fluids.

A strong cation-exchange sorbent that is also very polar contains propylsulfonic acid groups (PRS). PRS does not exhibit any appreciable degree of nonpolar interactions. In nonpolar solvents, PRS is also capable of polar and hydrogen bonding interactions. The pK$_a$ of PRS is low, and usually cationic compounds must be eluted either with high ionic strength solutions or by neutralization of the charge on the cationic isolate. Therefore, PRS is most effective for weaker cations, such as pyridinium compounds. The relative lack of nonpolar secondary interactions, combined with its low capacity, gives PRS an unique set of selectivity properties.

Similar to PRS as an ion-exchange sorbent, SCX is a strong cation exchanger with a very low pK$_a$. The SCX sorbent, compared to PRS, has a higher potential for nonpolar interactions due to the presence of the benzene ring on its surface. This nonpolar character should be taken into consideration when the sorbent is used as an ion-exchange for aqueous solvent systems. This dual nature is useful with compounds exhibiting both cationic and nonpolar character. The overall result of the dual character is superior cleanup capability in specific applications. Organic polymers with sulfonic groups are also used as SCX sorbents.

Materials with anion exchange properties are also obtained from a silica base material. PSA, for example, contains ethylenediamine-N-propyl bonded functional groups. This sorbent is similar to NH$_2$, but has two amino groups that offer higher ionic capacity than NH$_2$ SPE material (1.4 times higher). The pK$_a$'s for PSA (10.1 and 10.9, respectively) are also higher than for NH$_2$ material. PSA is therefore a stronger anion exchanger than NH$_2$. The PSA functional group is a very good bidentate ligand, making this sorbent an excellent candidate for chelation, as shown in the structure connecting a Cu^{2+} ion. Its higher carbon content compared to NH$_2$ makes it also exhibit nonpolar properties. This sorbent is in general a better choice than NH$_2$ for very polar compounds that are retained too strongly on NH$_2$ sorbent.

PSA Cu^{2+} complex

Similar to PSA is DEA sorbent, which has bonded diethylaminopropyl groups. DEA has some resemblance to NH$_2$ properties. It has slightly lower capacity as an anion exchange sorbent (1.0 meq/g) and a more nonpolar character due to the additional carbon chain on the functional group. The carbon chains make DEA a medium polarity sorbent in spite of its amino functionality. DEA is somewhat more polar than C8, but less than C2 or CN. It is usually applied to water samples, biological fluids, and nonpolar extracts.

The strongest anion exchange sorbent available is considered SAX, which contains quaternary ammonium groups. Due to these groups, the sorbent is typically present in ionic form. SAX offers minimal nonpolar interactions, because any effect of the carbon atoms in its structure is masked by the quaternary ammonium group. Because SAX is such a strong anion exchanger, it is a good sorbent for the retention of weaker anions, such as carboxylic acids that may not be retained strong enough on weaker anion exchange phases. It is supplied in the chloride counter-ion form, and the user may modify its selectivity by conditioning it with appropriate buffers. Organic synthetic polymers with quaternary ammonium groups also are used as SAX sorbents.

- Moisture and particulate removal SPE

The removal of water and of particulate materials from a sample or from a partially processed sample is a common operation in sample preparation. A number of procedures are applicable for removal of water from samples, some being described in Section 8.1. The removal of particulate materials from samples is usually done with filtration or centrifugation. These were discussed in Sections 7.2 and 7.3. However, SPE with appropriate sorbents can be used as a convenient technique for drying samples or for removal of particles. For example, for drying, cartridges containing granular anhydrous Na_2SO_4 are available. Also, many other materials are available for drying, such as a hydrophilic molecular sieve applied on a hydrophobic phase separation membrane. The filtration materials are designed to be used with samples that contain particulate materials, such as precipitated proteins, suspensions from food analysis, etc. Depth filters containing inert porous materials such as diatomaceous earth are used for this purpose. The efficiency of cartridges and column depth filters can be better than simpler filtration procedures, although the process is only a filtration. Filtration and water removal also are sometimes needed before a SPE separation. The small cross-sectional areas of typical columns or cartridges may result in slow sample processing rates and a low tolerance to blocking by particulate materials and adsorbed matrix components. Also, channeling may occur, which reduces the capacity of the sorbent bed to retain the analytes (decreases the breakthrough volume). Combinations of two types of sorbent, one for filtration and the next for a specific cleanup, are used for this reason in some SPE devices.

- Affinity, immunoaffinity, and molecular imprinted sorbents

The possibility to bind specific molecules on a polymeric backbone has been widely applied for the separation of biological samples. These can be selected by the user or can be general purpose immobilized compounds. For this purpose, compounds such as heparin [28], lectins [29], or nucleotides are bound to a solid support and used as selective sorbents. The support can be agarose (the neutral gelling fraction of the complex natural polysaccharide agar), cross-linked agarose, cross-linked dextrans (sepharose, sephacryl, etc.), or cellulose. The linking process is done using an activating reagent such as cyanogen bromide. The molecule to be bound must contain a free primary amine, sulfhydryl, or hydroxyl groups for attachment. On the obtained sorbent, proteins (or other molecules) with specific binding capability can be retained. Immobilized heparin, for example, acts with a specific binding site to retain certain proteins, lectin resins can be used for the purification of glycoproteins from other glycoconjugate molecules, and nucleotide resins are used for the purification of specific proteins. Heparin resins, lectin resins, and others are commercially available. Specific immunoproteins also can be bound on a solid support and used as sorbents. The binding can be done, for example, on agarose activated with cyanogen bromide or with other activation reagents such as 6-aminohexanoic acid, carbonyldiimidazole, or thiol [30]. These types of materials have a very high specificity for the specific antigen that generates the immunoprotein. Affinity resins containing immobilized sugars and sugar derivatives and resins with immobilized biotin or avidin are also available. Several activation reagents for agarose and cross-linked dextrans are indicated in Table 11.1.6.

TABLE 11.1.6. *Reagents used to make activated resins able to bind proteins.*

Activating reagent	Linkage to resin	Available reactive group	Specificity of group	Reaction conditions	Bond type to ligand	Stability
6-aminohexanoic acid	isourea	carboxyl	amine, with carbodiimide coupler	pH 4.5-6.0	amide	good
6-aminohexanoic acid N-hydroxy-succinimide ester	isourea	succinimidyl ester	amine	pH 6.0-8.0	amide	good
carbonyldiimidazole	carbamate	imidazolyl carbamate	amine	pH 8.0-10.0	carbamate	good below pH 10
cyanogen bromide	ester	cyanate	amine	pH 8.0-9.5	isourea	moderate
epoxy	ether	epoxy	SH>NH	pH 7-8 SH pH 8-11 NH_2	SH: thioether, NH_2 amino ether	very good
N-hydroxy-succinimide ester	isourea	succinimidyl ester	amine	pH 6.0-8.0	amide	good
periodate	oxidizes agarose, saccharides	aldehyde	amine	pH 4.0-10.0	reductive animation with $NaBH_3CN$	very good
thiol	isourea	disulfide	sulfhydryl	pH 6.0-8.0	disulfide	good in nonreducing conditions

The affinity and immunoaffinity type sorbents have excellent selectivity and work well in aqueous solutions, but each material must be developed for a specific analyte, are unstable with organic solvents, and may be stable only in a narrow pH range. This type of sorbent is used for many biological applications. However, a detailed discussion on the subject of affinity materials for bioanalytical purposes is beyond the purpose of this book.

The high selectivity of affinity and immunoaffinity sorbents also can be achieved with molecular imprinted polymers (MIP). An imprinted polymer for antiprotozoal drug 4,4'-[1,5-pentandi-yl-bis(oxy)]bis-benzenecarboximidamide (pentamidine) can be prepared, for example, by the polymerization in solution water/2-propanol (1.3 mL / 2.8 mL) of a mixture of ethylene glycol dimethacrylate and methacrylic acid in the presence of pentamidine (used as template) [31]. The structures of the reagents and their relative amounts are shown below:

12 mmol

0.5 mmol

0.125 mmol

By the use of an initiator such as azobis(isobutyronitrile 12 mg in 0.5 mL 2-propanol) and the solution being kept at 60° C for 24 hours, a polymeric material is obtained. This material is eluted with a CH_3CN/potassium phosphate buffer to completely eliminate the pentamidine template. The material can be further used for selectively retaining pentamidine from samples to be analyzed at pH 5–7. At pH 2 the analyte pentamidine can be eluted from the column. Several other imprinted polymers were synthesized and reported in literature [32], [33], [34]. The materials can be used either for LC columns or as SPE sorbents [35], [36]. The use of MIPs has the advantage of high selectivity, they work well in various solvents (organic or aqueous), and they are stable in a wide pH range [37]. However, this type of material is still rather difficult to develop, and the quality of the polymer may raise problems such as difficulty to completely remove the template from the sorbent.

- *Other SPE type sorbents*

PBA is a unique sorbent containing phenylboronic acid, which is able to form covalent bonds with specific analytes. Because the covalent retention involves interaction energies 10–100 times stronger than other extraction mechanisms, the retained compounds are strongly bound to the sorbent surface, resulting in superior cleanup. The boronate group has a high specificity for cis-diols like catechols, nucleic acids, low molecular weight proteins, and carbohydrates. PBA has proven to be especially effective in the isolation of catecholamines from biological fluids and for the separation of RNA from DNA. The elution of the retained compounds requires hydrolysis of the covalent bond, which can be done, for example, using acidic methanol.

Another type of surface modification based on sorbents made from derivatized silica can be used to obtain an SPE material. For this purpose, cationic surfactants such as cetyltrimethylammonium bromide are absorbed on the surface of C18 bonded silica, and the ionized functional group of the surfactant can then act as an ion-exchange site. This type of SPE can be used, for example, for the analysis of phenolic compounds from aqueous samples [38].

A special type of synthetic organic polymers is used for the retention of some ionic or highly polar solutes. These may be insufficiently retained on silica-based sorbents, which results in reduced recoveries because of the analyte breakthrough during the sample-loading step. For such polar solutes polymeric resins may provide higher recoveries. Amberlite XAD-4, XAD-7, XAD-16, and XAD-1180 resins are used to extract trace organic compounds from aqueous samples. These materials are nonionic macroreticular resins, which can retain selectively organic compounds and weak organic acids and bases. Strongly dissociated compounds are not retained, while the recovery of weak acid-basic compounds is strongly dependent on the pH-value. The retention efficiency is higher for aromatic compounds and lower for aliphatic compounds with low molecular weight. Within a homologous series, the retention efficiency varies with the molecular weight of the solute.

In some applications, the cleanup process requires passing the sample through more than one type of SPE material. In order to simplify this type of analysis, mixed-mode sorbents were developed. They combine two or more functional groups into a single cartridge, allowing multiple retention interactions to occur between the sorbent and the analytes. The results are superior cleanup, improved reproducibility, and high recovery. Mixed beds may contain a mixture of two materials such as C8 and SCX phases. Also, it is possible to use synthetic polymers containing two different active groups on the same polymeric backbone as shown below:

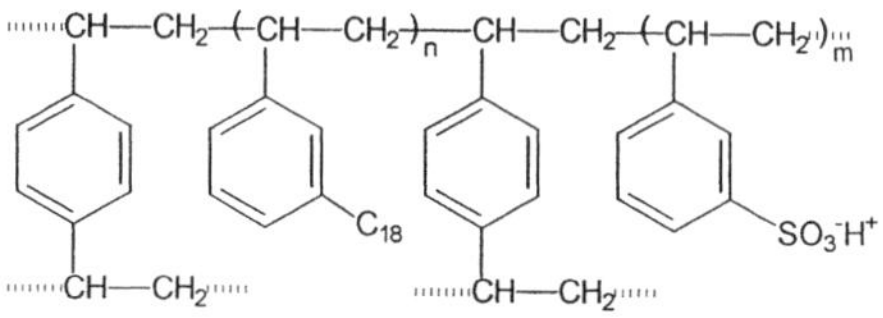

Also, layered sorbents beds are commercially available. These contain multiple sorbents usually separated by a polyethylene frit. This type of sorbent is utilized when multiple sorbents are required for the extraction method. For example, organochlorine pesticides show good recovery rates on C18 phases, but these recovery rates are much lower for more polar compounds, such as triazine derivatives. On the other hand, polystyrene-divinylbenzene based SPE phases are very well suited for triazines, but not for organochlorine pesticides.

Some other special sorbents are commercially available. Examples are phases specifically designed to be used in a given application such as the analysis of polychlorinated biphenyls (PCBs), herbicides (such as Bond Elut Atrazine), specific pollutants or particular drugs, sorbents made from restricted access materials (RAM) with diol groups on the surface and C18 on the internal pores [39], etc.

- *Conditioning, eluting, and rinsing solvents in SPE*

Similar to LC separations, the nature of the mobile phase (eluent) is very important in SPE. The two operation modes "retain interferences" or "retain analyte" require the choice of the solvent to fit the purpose of the SPE operation. In the case with the retained analyte, further elution is expected, and a change in solvent usually allows this operation.

The choice of solvents in SPE has a number of similarities with the choice of a solvent for HPLC. One of these criteria is that the solvent must be compatible (miscible or at least soluble, not chemically interacting) with other solvents used before it in the SPE process. The solvent must satisfy the requirement for which it is used, either to elute the analyte without eluting the matrix (in retain interferences mode) or to elute the matrix without affecting the analyte.

The use of solvents in SPE may be necessary for a number of purposes such as a) conditioning of the SPE sorbent, b) dissolving the sample to be processed and acting as a proper carrier for the solutes (allowing adsorption on the sorbent of desired compounds), c) rinsing the material that must be eliminated from the sorbent without eluting undesirable compounds, and d) eluting the material to be analyzed from the sorbent bed.

Conditioning of the SPE sorbent before use is a common operation necessary to obtain the expected performance from a SPE device. The materials used as sorbents may come dry or with a specific moisture content. It is possible that the solvent of the sample will act as a conditioning agent, but it is not always the case. In many situations, a specific solvent or even two or three solvents are necessary to condition the column. For C18 SPE sorbents, for example, often the solvent used for the sample is water. Water does not wet properly the C18 material and is excluded from access to the sorbent surface because of surface tension effects. The surface C18 chains also have the tendency to aggregate among themselves to decrease the surface exposed to water. This requires the use of a conditioning step before the use of the sorbent, which can be done using an organic solvent such as methanol, isopropanol, or acetonitrile. This allows better access of the solutes from the water solution to the hydrophobic surface, and also arranges the sorbent to have maximum interaction with the solute. The performance of the C18 unconditioned material in contact with water is additionally diminished due to the exposure of the hydrophilic remaining silanol groups to the water solution, which is schematically indicated in Figure 11.1.7.

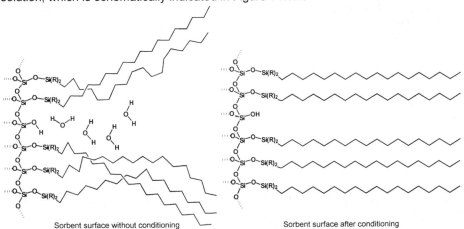

Sorbent surface without conditioning   Sorbent surface after conditioning

FIGURE 11.1.7. *Models of the orientation of octadecyl chains from the surface of a C18 bonded silica before and after conditioning with a nonpolar solvent.*

The problem regarding wetting of the hydrophobic surfaces of the sorbent may not be sufficiently addressed by conditioning. The addition of a proportion of organic solvent (up to 50%) in the sample before the SPE operation may be beneficial for better adsorption. It was also observed that the contact between some polar analytes and a hydrophobic C18 silica material during the SPE processing of water samples is better when no endcapping has been performed to the silica material, or when C18 silica material was prepared using a monofunctional silane.

Conditioning is also needed when using ion exchange SPE sorbents. The use of buffers that condition the adsorbent to the desired form and pH are typically needed to assure a proper ion exchange process.

The choice of a specific solvent for the retention on the sorbent of the compounds from the solution is based on the same rules as for HPLC (see Section 4.3). The "like-to-like" principle applied for solubility is a good guidance for determining where a specific solute is likely to stay. A polar solvent such as water + methanol or water + acetonitrile containing as solutes organic compounds with large hydrophobic moieties and a nonpolar stationary phase (C18, C8 etc.) will have the result of retaining the solutes on the sorbent. The elution with hexane or other hydrophobic solvents will tend to elute the retained compounds. The opposite is applicable for compounds with polar groups dissolved in an organic solvent. These can be retained on silica or alumina and eluted with methanol + water. Some physical properties for a number of solvents commonly used in SPE are given in Table 11.1.7.

The polarity P' of various solvents is discussed in Section 1.10, and P' values for a number of solvents are given in Tables 1.10.11 to 1.10.19. Also, the elutropic strength ε° can be used as a guidance regarding the behavior of a specific solvent on a specific sorbent. The values for elutropic strength ε° for a number of solvents and some polar sorbents are given in Table 4.3.4. These tables indicate solvent polarity and can be used to determine which solvent may elute too much or too little from the sample. The polarity values P' shown in Table 11.1.7 are slightly different from those shown in Tables 1.10.11 to 1.10.19, because the calculation of P' was done considering the corrected distribution constants K" for all five solvents, ethanol, dioxane, nitromethane, toluene, and methyl ethyl ketone (see rel. 1.10.16).

TABLE 11.1.7. *Common solvents used in SPE and some of their physical properties.*

Solvent	Polarity P'	Dielectric constant ε (rel.)	Viscosity (at 20°C) mPa s	Boiling point (°C)	Refractive index	UV cut-off (nm)
acetone	5.40	20.7	0.32	57	1.395	330
acetonitrile	6.20	35.7	0.37	82	1.344	210
amyl alcohol	0.61	13.9	4.10	138	1.410	210
benzene	3.00	2.28	0.65	80	1.501	280
n-butanol	3.90	17.8	2.95	117	1.347	210
n-butyl acetate	4.00	5.01	-	126	1.390	254
carbon disulfide	0.15	2.64	0.37	46	1.626	380
carbon tetrachloride	1.60	2.24	0.97	77	1.466	265
chloroform	4.40	4.81	0.57	61	1.443	245
cyclohexane	0.10	2.02	1.00	81	1.427	210
cyclopentane	0.20	2.00	0.47	49	1.406	210
n-decane	0.04	1.99	0.92	174	1.412	210
dimethylformamide	6.40	37.6	0.92	153	1.427	270
dimethylsulfoxide	7.20	46.2	2.24	189	1.476	268

TABLE 11.1.7 (continued). *Common solvents used in SPE and some of their physical properties.*

Solvent	Polarity P'	Dielectric constant ε (rel.)	Viscosity (at 20°C) mPa s	Boiling point (°C)	Refractive index	UV cut-off (nm)
dioxane	4.80	2.21	1.54	102	1.422	220
ethanol	4.30	24.3	1.20	79	1.361	210
ethyl acetate	4.30	6.02	0.45	77	1.370	260
ethyl ether	2.90	4.34	0.23	35	1.353	220
ethylene glycol	6.90	37.7	19.9	197	1.427	210
n-heptane	0.20	1.97	0.41	98	1.385	200
n-hexane	0.06	1.89	0.33	69	1.375	210
isobutyl alcohol	3.00	16.68	4.70	108	1.384	220
isooctane	0.10	1.94	0.53	99	1.404	210
methanol	6.60	32.8	0.60	65	1.329	210
methyl acetate	-	6.68	0.37	57	1.362	260
methyl ethyl ketone	4.50	18.5	0.43	80	1.381	330
methyl isobutyl ketone	4.20	13.11	-	119	1.394	330
methylene chloride	3.40	9.08	0.44	40	1.424	245
morpholine	-	-	7.42	128	-	285
nitromethane	6.00	35.9	0.67	101	1.394	380
iso-pentane	0.00	2.10	-	30	1.371	-
n-pentane	0.00	1.84	0.23	36	1.358	210
petroleum ether	0.01	-	0.30	30-60	1.365	210
isopropanol	4.30	18.3	2.37	82	1.375	210
n-propanol	4.00	20.1	2.27	98	1.383	210
Iso-propyl ether	2.40	3.88	0.37	68	1.368	220
tetrahydrofuran	4.20	7.58	0.55	66	1.408	220
toluene	2.40	2.37	0.59	111	1.496	285
trimethylpentane	0.10	1.94	0.47	99	1.389	215
water	10.2	78.5	1.00	100	1.323	-
p-xylene	2.50	2.27	0.65	138	1.439	290

One aspect that also must be considered in SPE is the possible unwanted mixing of solvents when changing from adsorption to elution. For eliminating this possibility, some SPE procedures recommend a complete drying of the cartridge prior to the liquid desorption. If this operation is not completely performed, the effect will be a "wet' extract transferred to the GC column. This problem can affect negatively the results or the stability of the GC columns when, for example, water is constantly present in the prepared sample. The incorporation of a cartridge packed with a drying material has been proposed for elimination of this effect [40]. Though this approach is likely to extend the GC column lifetime, it also offers a new source of errors by analyte loss due to the fact that polar compounds are easily adsorbed on the drying material. Drying can be needed not only to remove water from the cartridge. When the solvents are changed during the SPE process, in order to avoid mixing of two solvents, a intermediate drying step can be necessary in specific applications.

The final operation on a SPE is usually the passing of an appropriate solvent through the cartridge, which is specifically chosen to disrupt the analyte-sorbent interactions and assure the elution of the analytes. It is common that this is an organic solvent that is able to disrupt the interaction between analytes and the sorbent. Recovery of an analyte depends on its properties and specific interaction with the sorbent. A high affinity of the analyte to the sorbent surface involves a high adsorption yield but can influence negatively the elution and lead to a low value of the final recovery. For example, high hydrophobicity for the analyte favors good adsorption on the surface of a C18 bonded phase, and on the other hand makes the desorption process more difficult.

A useful procedure for eluting solutes with acidic or basic properties is the pH change of the eluting solution. In nonionic state, solutes have a lower hydrophilic character, while in ionic form this character is significantly increased. Rel. (1.4.45) shows the variation of the distribution constant for the solvent extraction with pH for a weak acid. The expression can be immediately applied to SPE separations, replacing the distribution constants with the corresponding capacity factors for the distribution process.

For ion exchange SPE, the choice of pH and of the ionic strength of the solution is a critical factor for use of specific SPE materials, allowing the retention or elution of specific analytes. The equilibrium for a cation exchange can be written in the simplified form:

$$R\text{-}X^-H^+ + M^+ \rightleftharpoons R\text{-}X^-M^+ + H^+$$

where R is a polymeric structure containing usually covalently connected the group X^- and a counterion, usually H^+. In the adsorption process the counterion is replaced (in part) with the cation M+. The equilibrium of this reaction is described by the equilibrium constant:

$$K_{H,M} = \frac{[M_r^+]}{[M_s^+]} \cdot \frac{[H_s^+]}{[H_r^+]} \tag{11.1.9}$$

For anion exchange sorbents, a typical equilibrium can be written in the form:

$$R\text{-}X^+Cl^- + Y^- \rightleftharpoons R\text{-}X^+Y^- + Cl^-$$

where the group X^+ has a counterion, usually Cl^- or OH^- that is replaced in the exchange process. The equilibrium of this reaction is described by the equilibrium constant:

$$K_{H,M} = \frac{[Y_r^-]}{[Y_s^-]} \cdot \frac{[Cl_s^-]}{[Cl_r^-]} \tag{11.1.10}$$

The pH is important for the equilibria involving H^+ or OH^-, and the concentration of different ions are important when these are involved in the equilibrium. In addition to concentrations, specific affinities for the resin are characteristic for different ions. More discussion on this subject is found in Section 15.1.

In addition to the role of releasing or eluting the sorbent, some other properties of the solvents used in SPE must be considered. Among these properties is the volatility of the solvent, which can be high or low depending on the solvent. Low volatilities have the disadvantage of being more difficult to remove in case a concentration step with solvent evaporation is necessary after the SPE operation. Solvents with low volatility also may have problems with losses and concentration modification. The use of internal standards during SPE process may compensate for this problem. Also, reactivity, toxicity, environmental problems, available purity, and cost represent factors to consider when selecting a specific solvent for SPE use.

- *Columns, cartridges, disks, and accessories used in SPE*

The material used as a sorbent in SPE is contained in columns, cartridges, or disks. A number of parameters important for the sample preparation process depend on a) the amount of sorbent, b) the dimensions of the sorbent column (bed), c) the volume of solvent used in the separation, d) the operational aspect of the SPE, etc. Common shapes of columns, cartridges, and disks used in SPE are shown in Figure 11.1.8.

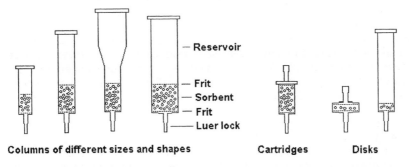

| Columns of different sizes and shapes | Cartridges | Disks |

FIGURE 11.1.8. *Common shapes of columns, cartridges, and disks used in SPE.*

The material used to make the reservoir (syringe barrel) is usually polypropylene, although glass or other materials also can be used. Some of the dimensions used for the columns are indicated in Table 11.1.8.

TABLE 11.1.8. Some of the dimensions used for SPE columns and cartridges.

Mass of sorbent[1]	Type	Reservoir volume	Maximum retention capacity for silica base	Maximum retention capacity for synthetic polymers	Minimum elution volume[2]
500 mg	cartridge	-	25 mg	125 mg	1.2 mL
1000 mg	cartridge	-	50 mg	250 mg	2.4 mL
50 mg	column	1 mL	2.5 mg	12.5 mg	125 µL
100 mg	column	1 mL	5 mg	25 mg	250 µL
200 mg	column	3 mL	10 mg	50 mg	500 µL
500 mg	column	3 mL	25 mg	125 mg	1.25 mL
1 g	column	6 mL	50 mg	250 mg	2.5 mL
2 g	column	12 mL	100 mg	500 mg	5 mL
5 g	column	20 mL	250 mg	1.25 g	12 mL
10 g	column	60 mL	500 mg	2.5 g	24 mL

[1] Bed volume $\approx$ 120 µL / 100 mg sorbent.
[2] Minimum elution volume is $\approx$ 2 bed volumes.

The design of the columns and cartridges may vary in order to meet different requirements regarding sample volume, solvent delivery systems (syringes), or automation [41]. The reservoirs of the columns can be adapted in order to increase the sample volume. In addition to "clean" sorbents, high-purity materials with low extractable contents using medical-grade polypropylene and polyethylene for the reservoir and frits are used.

Limitations of packed SPE conventional columns and cartridges include restricted flow-rates and plugging of the top frit when handling water containing suspended solids, such as surface water or waste water. Therefore, the percolation of larger volume samples, such as 500 mL can take a long time. In order to avoid previous filtration and clogging, various approaches have been investigated to overcome the flow limitation. Depth filters containing diatomaceous earth or other materials [5] are used for filtration. These filters can be integrated in the SPE cartridges.

The passing of the sample through the SPE columns is done using free flow, or more frequently using a syringe piston system to push the liquid sample through the sorbent, or a vacuum system to produce suction and accelerate the percolation process. Positive and negative pressure manifolds were developed for processing a number of SPE columns simultaneously. Another development uses columns specifically shaped to be put into a centrifuge and accelerate the percolation process through centrifugation. The flow rate through the SPE material must be limited due to the possibility of breakthrough when the flows are too high. Automatic systems are also available for processing SPE operations (see Section 6.1).

An alternative to the use of columns and cartridges in SPE is the use of SPE disks (also named improperly membranes). The disks (such as Empore disks) contain sorbent particles tightly bound within an inert matrix, such as polytetrafluoroethylene (PTFE) (90% sorbent, 10% PTFE) or glass fiber. The amount of sorbent contained in the disks is 10 mg or less, and several diameters such as 7 mm, 10 mm, or 20 mm disks are commercially available. The fibrils of the matrix do not interfere with activity of the sorbent. The disks allow higher flow-rates, without channeling effects, due to their large cross-sectional area and thin bed. The sample capacity of the disks is lower than that of standard cartridges because they contain lower amount of sorbent. The sample volume must be adjusted appropriately. However, the disks are ideal for processing small samples. Some of the same limitations as for columns and cartridges exist for disks when the samples contain suspended matter [42]. The disks may be obstructed by the suspended matter in the sample. Continuous improvements are made to obtain better materials for the SPE disks [43]. For example, disks with a depth filter above the membrane material have been developed [44], [45].

The demand for high-throughput applications (combinatorial chemistry, drug screening) has been at the origin of the development of more small-volume SPE systems. One such system using the disk technology is the 96-well plate shown in Figure 11.1.9, which can be used for simultaneous rapid sample preparation of a large number of samples [46]. Each well in the plate can be eluted with a small volume of solvent (100–200 μL). The SPE pipette is an alternative to the 96-well plates and uses a conventional pipette with the tip fitted with a SPE disk. The SPE pipettes are designed to be handled quickly by automated liquid-handling systems. Also, more special solid phase devices are being developed. For example, on-chip solid phase extraction has been demonstrated using

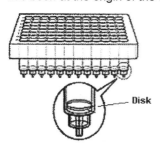

FIGURE 11.1.9. *96-Well plate.*

a porous polymer prepared within the channels of a microfluidic device [47]. SPE sorbents also can be used in HPLC type precolumns. The solution is passed through these precolumns using an HPLC pump, and the precolumn can be considered as part of sample preparation for the HPLC. However, the adsorption/desorption can be done similarly to SPE techniques.

- The practice of SPE operation

The decision to use SPE is typically based on the need to provide a cleaner more concentrated sample for the core chromatographic process. The procedure can provide very good results, and the application of SPE is in general simple. Polarity interplay between analyte, sorbent, and eluting solvent must be chosen in such a way to eliminate as many interferences as possible. The diagram in Figure 11.1.10a illustrates this balance for a polar analyte.

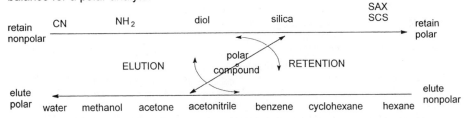

FIGURE 11.1.10a. *The selection of a sorbent and of a solvent for a polar analyte*

The bi-directional arrow can rotate and the sorbent/solvent pair on the left side of the arrow line indicates elution of the compound, while the pair at the right indicates retention. For example, a silica SPE retains the compound from acetonitrile, but methanol or water elutes the compound. The same compound is retained on a NH_2 sorbent only if the compound is present, for example, in cyclohexane or hexane. The diagram in Figure 11.1.10b illustrates the same balance for a nonpolar analyte.

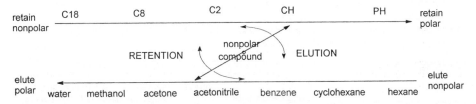

FIGURE 11.1.10b. *The selection of a sorbent and of a solvent for a nonpolar analyte.*

A nonpolar compound is retained, for example, from acetonitrile on a C8 or C18 sorbent, but hexane, for example, elutes the compound from a C8 column. This type of diagram provides guidance regarding the sorbent and the solvent necessary for retaining or eluting a specific compound. However, further experimental fine tuning is necessary for the optimization of a cleanup process where compounds of different polarities are present in the sample.

For some samples, cleanup preliminary operations are needed before SPE. This is particularly needed for extremely "dirty" samples containing materials that can block the SPE device. A diagram providing guidance regarding the operations required before the use of SPE is indicated in Figure 11.1.11.

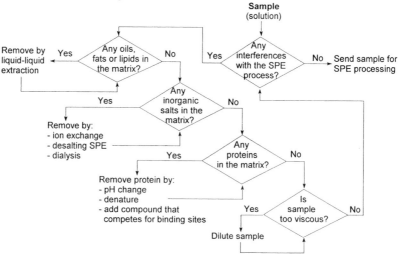

FIGURE 11.1.11. *Diagram of the cleanup process necessary before the use of SPE.*

The choice of the appropriate SPE process is very important for the success of many chemical analyses. Part of the optimization for a sample preparation procedure using SPE is selection of the SPE sorbent and of the solvents for rinsing and elution. Typical choices depending on the nature of the analyte and sample are given in Table 11.1.9.

TABLE 11.1.9. *Selection of solid phase sorbent, rinsing solvent, and eluent.*

Analyte type	Matrix type	Sorbent	Extraction mechanism	Rinsing solvent	Eluent
Compounds with long alkyl chains, aromatic rings, low polarity	aqueous, biological fluids	C18, C8, C2, phenyl, cyclohexyl, cyanopropyl, polymeric SDVB	distribution (reversed phase)	water, methanol, other polar solvents	hexane, chloroform, ethyl acetate
Hydrophylic groups, hydroxyls, amines, heteroatoms	nonpolar, lipids	CN, 2OH-Diol, silica, aminopropyl, Florisil, alumina	distribution (direct phase)	hexane, CH_2Cl_2, other nonpolar solvents	polar solvents, methanol/water, methanol, etc.
Positively charged groups, such as amine cations	aqueous, low ionic strength, biological fluids	strong (benzene or propyl sulfonic groups) or weak (carboxylic acids)	cation exchange	water, methanol, other polar solvents, low ionic strength	alkaline buffers, acetate, citrate, phosphate, high ionic strength
Negatively charged groups, ionized organic acids	aqueous, low ionic strength, biological fluids	strong (tetraalkyl-ammonium), weak (DEA, amino)	anion exchange	water, methanol, other polar solvents low ionic strength	acidic buffers, high ionic strength
Vicinal diols	aqueous, biological fluids	PBA (phenyl boronic)	covalent bonds	water, methanol	acidic methanol
Specific analytes	water, biological fluids	specific, tailored for the analyte	usually distribution		

Following the preliminary cleanup process, the sample is further processed by SPE. One important decision at this point is whether the "retain analyte" or "retain interferences" path is selected. After this selection, it must be assured that the selected compound a) binds to the SPE cartridge, b) is retained during rinsing, and c) elutes completely during elution. The application of SPE for sample cleanup and concentration can be guided following the diagram shown in Figure 11.1.12.

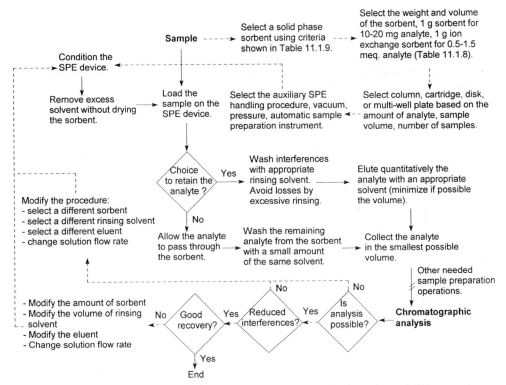

FIGURE 11.1.12. *Diagram of the path for the selection and validation of an SPE preparation of the sample.*

Only the basic SPE procedures have been discussed so far in this section The practice of SPE can use a number of variations of these basic procedures including combinations of adsorptions and desorptions, use of large volumes of samples, even use of adsorption from liquid samples followed by drying and further thermal desorption of the analytes for direct GC analysis. These particular procedures related to adsorption of the analytes from a solution on a solid phase followed by thermal desorption such as SBSE (stir bar solute extraction) are further discussed in Section 11.4. Many other interesting procedures for SPE use are presented in literature [19].

Excellent automation is also available for sample preparation using SPE (e.g. RapidTrace instrument from Zymark). Automatic systems having the possibilities to use more elaborate SPE sample preparation, for example using more then one device with automatic switching of cartridges, are available. Also, systems using on-line SPE with HPLC systems or other chromatographic analytical instruments are available (see Section 6.1). An example of an experimental setup for on-line use of SPE-HPLC is shown in Figure 11.1.13. The diagram shows a system that works in three stages A-B-A. In the first stage A, the SPE cartridge is conditioned with the solvent from pump 1, and the sample (e.g. large volume) is injected. The cartridge and the solvents are selected to collect the sample and eliminate excess of solvent and interferences. In stage B, the compounds retained in the DPE cartridge 1 are eluted with the solvent from pump 2 and collected in a sample loop. Then the system returns to stage A when the analytes retained in the sample loop are washed by the solvent in HPLC pump and sent into the analytical column. This arrangement allows the use of low-pressure pumps 1 and 2 for washing the cartridge and a HPLC pump only for the HPLC column and the loop.

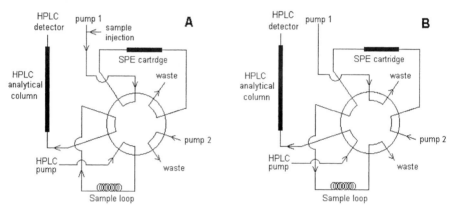

FIGURE 11.1.13. *Diagram of a SPE-HPLC system that works in three stages A-B-A, using an SPE cartridge and a HPLC precolumn.*

The same diagram works when the sample loop is replaced with a HPLC precolumn that is able to retain the analytes that are eluted from the SPE cartridge in stage B. The solvent in pump 2 and the sorbent in the precolumn can be selected to have good retention of analytes but be able to eliminate further interferences. The solvent in the HPLC pump must be able to elute the analytes from the precolumn and separate them in the analytical column. Systems that inject the SPME eluate directly into a GC system are also reported in literature [48].

- *Examples of applications*

SPE is a very common operation in sample preparation, as seen in Figure 1.2.1 by comparison of the frequency of different operations used for preparing samples for analysis. Among these can be included analysis of drugs and drug metabolites [49], various biological samples, environmental samples [50], etc. The usage of different phases in applications of SPE has been estimated in a survey regarding trends in sample preparation conducted by *LC-GC* magazine [41], the results being shown in Figure 11.1.14.

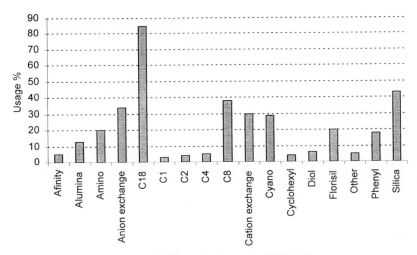

FIGURE 11.1.14. *The usage of different phases in SPE [41].*

As seen in Figure 11.1.14, the most common phases are C18 and silica. However, other phases also have significant usage. The number of applications of SPE is very high and still increasing. This is due to numerous advantages of this technique compared to other cleanup procedures. A comparison of the results obtained using SPE with simultaneous distillation and extraction (SDE) in the analysis of polycyclic aromatic hydrocarbons (PAHs) from mainstream cigarette smoke condensate [51] is shown in Figure 10.5.4. The diagram of the SPE procedure used for this analysis is shown in Figure 11.1.15.

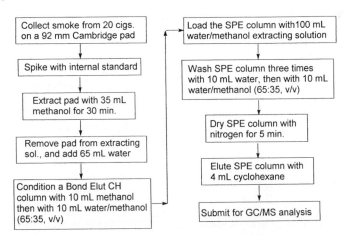

FIGURE 11.1.15. *Diagram for the analysis of PAHs in mainstream cigarette smoke [51].*

After the SPE cleanup, the analysis is performed on an Agilent 6890/5973 GC/MSsystem. The GC is equipped with a 30 m long, 0.25 mm i.d. and 0.25 µm film thickness ZB-5 capillary column. The oven is programmed at an initial temperature of 45°C for 7 min., a heating rate of 8.5°C/min. to 360°C, and a final time of 5 min. The mass spectrometer operates in SIM (selected ion monitoring) mode. Since all PAHs generated the molecular ion as the base peak in their mass spectra, the molecular ions can be used for SIM. A typical chromatogram for the smoke from a Kentucky reference cigarette (1R4F) is shown in Figure 11.1.15. The quantitation of each analyte is done with the use of the area ratio for the peaks of the compound and of its corresponding deuterated standard.

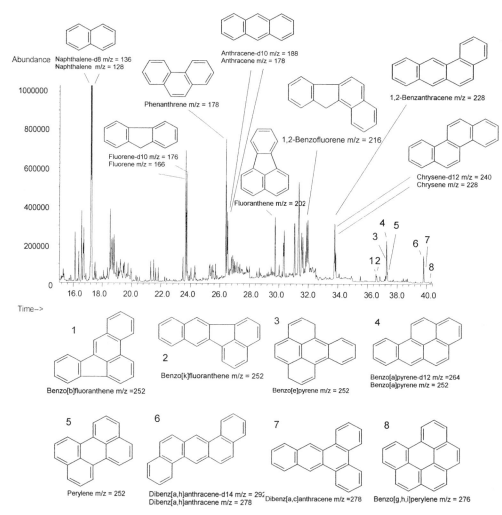

FIGURE 11.1.15. *Chromatogram (SIM signal) for the analysis of PAHs in mainstream cigarette smoke of a 1R4F cigarette.*

Besides advantages regarding separation efficiency, SPE also has other benefits. One advantage is the decreased usage of solvents, which in case of liquid-liquid separations (LLS) can be significant. Problems such as formation of emulsions common in case of LLS are not encountered in SPE. Also, SPE separations are automated more easily than LLS techniques. This explains the increased number of officially recommended methods, for example by the United States Environmental Protection Agency (EPA). Some examples showing EPA methods used for the analysis of pollutants from water are given in Table 11.1.10 [41].

TABLE 11.1.10. *Examples of EPA methods using SPE cleanup [41].*

EPA Method	Nature of analytes	Matrix	SPE phase
506	phthalates and adipate esters	drinking water	C18
513	tetrachlorodibenzo-p-dioxin	drinking water	C18
508.1	chlorinated pesticides, herbicides, and organohalides	drinking water	C18
515.2	chlorinated carboxylic acids	drinking water	PS-DVB = poly(styrene-divinylbenzene)
525.1	organic compounds (extractable)	drinking water	C18
525.2	organic compound	drinking water	C18
548	endothall[1]	drinking water	C18, anion exchange
549.1	diquat[1] and paraquat[1]	drinking water	C8
550.1	polycyclic aromatic hydrocarbons	drinking water	C18
552.1	haloacetic acids and dalapon	drinking water	anion exchange
553	benzidines and nitrogen-containing pesticides	drinking water	C18 or PS-DVB
554	carbonyl compounds	drinking water	C18
555	chlorinated acids	drinking water	C18
1656	organohalide pesticides	waste water	C18
1657	organophosphorus pesticides	waste water	C18
1658	phenoxy-acid herbicides[1]	waste water	C18
3535	organochloropesticides, phthalate esters, TCLP[1] leachates	aqueous samples	C18 or PS-DVB
3600	organochlorine pesticides and polychlorinated biphenyls	waste water	Florisil, alumina, silica gel
8325	benzidines and nitrogen-containing pesticides	water, waste water	C18
8440	total recoverable petroleum hydrocarbons	sediments, soil, sludge	silica gel

[1] For information and chemical structure, see e.g. D. Hartley, H. Kidd (eds.), *The Agrochemicals Handbook*, 2nd ed., Royal Society of Chemistry, University of Nottingham, England, 1987.

11.2 SOLID PHASE MICROEXTRACTION

Solid phase microextraction (SPME) is a technique used for extracting the analytes from a sample and introducing them directly into an analytical instrument. In SPME a stationary phase is exposed to the compounds in the headspace of a sample or to a sample solution and accumulates the analytes. After that, the stationary phase is desorbed and the analytes determined by a chromatographic technique. The ingenious

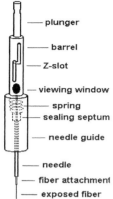

— plunger

— barrel

— Z-slot

— viewing window

— spring

— sealing septum

— needle guide

— needle

— fiber attachment

— exposed fiber

FIGURE 11.2.1.
Diagram of a SPME
sample holder.

use of a stationary phase material, which coats a fiber that can be retracted in a syringe needle, made this technique very useful and popular [52], [53]. The coated fiber is kept inside the needle during puncture of the septum of the vial containing the sample or during the injection in the chromatographic instrument. During the process of analyte extraction and during the desorption in a GC injection port or in a HPLC, the fiber is exposed by pushing it outside the needle. SPME can be compared with the use of an inside-out piece of a GC chromatographic column that collects the analytes, followed by desorption and analysis. A diagram of a SPME fiber holder is shown in Figure 11.2.1 [54].

The analyte can be in the gas phase or in a liquid phase. The fiber coating is typically a bound polymer acting as a liquid-like material, and the extraction process is based on a partition equilibrium and not on a true solid phase adsorption (see Section 1.4). Nevertheless, the name solid phase is still kept for the coating (which in fact is liquid-like), and the extraction process is indicated as adsorption (with the larger meaning of concentrating a solute into a stationary phase), although it is a partition involving a liquid-like stationary phase. Sample desorption that takes place in the injection port of a GC system is again a distribution process between the analytes from the stationary phase and the analytes liberated in the injection port in gas form at the elevated temperature. The schematics

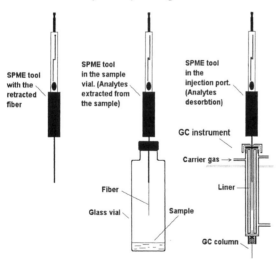

SPME tool
with the
retracted
fiber

SPME tool
in the sample
vial. (Analytes
extracted from
the sample)

SPME tool
in the
injection port.
(Analytes
desorbtion)

GC instrument

Carrier gas →

Fiber

Liner

Glass vial

Sample

GC column

of SPME use with sampling from the headspace and followed by desorption for GC analysis is shown in Figure 11.2.2. The use of SPME in sample preparation is typically performed in the following steps: a) the fiber is kept for a period of time (0.5–1 hour) at a relatively high temperature (e.g. 250° C) for the desorption of any residual compounds from a previous use or contamination, b) the fiber is exposed to the sample (headspace of the sample or in a solution of the sample) for a period of 1 to 30 min. at room temperature, c) the fiber is desorbed in the injection port of a GC for 1 min. By this procedure, the compounds with high affinity for the fiber are extracted from

FIGURE 11.2.2. *Diagram of SPME sampling of the*
headspace followed by desorption in a GC.

the sample into the fiber. Then, due to the high temperature of the GC, the analytes are transferred into the analytical instrument, SPME being a solventless procedure. As seen from the description of the use of SPME, the procedure has the double role of extracting selected compounds from the sample (cleanup function) and also to concentrate the analytes such that all the accumulated material from a large sample is injected into the chromatographic system (concentration function).

The adsorption from a solution simply eliminates the headspace, and the SPME fiber is directly introduced into the solution. The headspace of the vial containing the solution must be in this case as small as possible in order to avoid some of the volatile compounds being lost in the headspace. In case of direct adsorption from the solution, the liquid can be kept stationary or can be stirred during the adsorption process. Stirring accelerates the adsorption process, but otherwise is not supposed to affect the quantity of the analyte collected on the fiber. Water solutions of organic materials are easily analyzed by this procedure. The choice between headspace analysis and solution analysis depends on a number of factors, the physical state of the sample being very important. Solid samples are frequently analyzed using the headspace technique. Examples include the analysis of volatile materials from food or packaging materials. Some samples are better analyzed from solution (usually water solutions), mainly when the sample solubilization is not a problem. When both solution and headspace adsorption can be utilized, an optimization of the process is recommended to choose the best procedure. For these cases, the amount of material adsorbed in the headspace procedure and in direct solution procedure typically do not have significant differences.

The desorption of the fiber can be done not only into a GC at the elevated temperature of the injection port. Re-extraction with a solvent can be achieved in a system using a HPLC. The schematic of SPME use with HPLC is shown in Figure 11.2.3 [55]. In step A, the sample passes through the fiber (at very low flow rate) and is adsorbed. In step B the sample is eluted from the sample and sent to the HPLC chromatographic column.

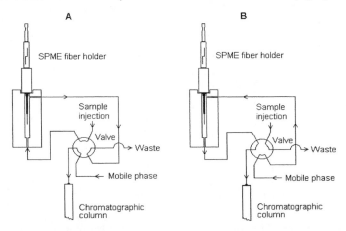

FIGURE 11.2.3. *Diagram of SPME use with a HPLC system.* In phase A the fiber is loaded with the sample. In phase B the valve is switched and sample is extracted from the fiber and introduced into the HPLC system.

The diagrams from Figure 11.2.3 show both adsorption and desorption of the fiber performed in the same setup. The sample adsorption also can be done off-line, and only the desorption is done in the HPLC system.

After the desorption, the fiber can be utilized for another analysis. However, although the desorption is assumed to be quantitative for a specific analyte, the SPME fiber must be conditioned before the next analysis. This conditioning step consists of a lengthy desorption (1–2 hours) at elevated temperature (around 250° C, depending on the fiber).

The chemical nature of the stationary phase of the SPME fiber is a key element in this technique. There are various fibers utilized in SPME. A short list and some descriptions of the more common SPME fiber coatings are given in Table 11.2.1. The typical length of the fiber is 1 cm, and the diameter of the silica rod covered by the active phase is 0.05 mm.

TABLE 11.2.1. *Solid phase microextraction common fibers.*

Type of coating	Film thickness	Description	Color code	Polarity	Use
polydimethyl-siloxane (PDMS)	100 μm	nonbonded	red / plain	low	GC/HPLC
	30 μm	nonbonded	yellow / plain	low	GC/HPLC
	7 μm	bonded	green / plain	low	GC/HPLC
polydimethyl-siloxane/divinyl-benzene (PDMS/DVB)	65 μm	partially crosslinked	blue / plain	low	GC
	60 μm	partially crosslinked	brown / notched	low	HPLC
	StableFlex 65 μm	highly crosslinked	pink / plain	low	GC
polyacrylate	85 μm	partially crosslinked	white / plain	high	GC/HPLC
Carboxen/polydimethyl-siloxane (CAR/PDMS)	75 μm	partially crosslinked	black / plain	medium	GC
	StableFlex 85 μm	highly crosslinked	lt. blue / plain	medium	GC
Carbowax/divinyl-benzene (CW/DVB)	65 μm	partially crosslinked	orange / plain	high	GC
	StableFlex 70 μm	highly crosslinked	yellow-green / plain	high	GC
Carbowax/Templated Resin (CW/TPR)	50 μm	partially crosslinked	purple / notched	high	HPLC
StableFlex divinylbenzene/Carboxen/PDMS (DVB/CAR/PDMS)	60/30 μm	highly crosslinked	gray / plain	medium	GC
	50/30 μm	highly crosslinked	gray / notched	medium	GC

The chemical structures of the materials for some stationary phases are similar to those of the phases used as coating for capillary columns. Among these are polydimethyl-siloxane (PDMS) and polyethylene glycol (Carbowax). Other phases are similar to those used in SPE, such as polydivinylbenzene or polyacrylate. A special material is Carboxen, which is a synthetic porous carbon. Glassy carbon also is used as adsorbing phase [56]. Other materials are used for the fiber coating, such as templated resins (hollow spherical DVB coated with another phase such as Carbowax [57]).

For each application the choice of a specific fiber can be made based on the evaluation of absorption efficiency for the analytes of interest. The adsorption process is discussed in more detail further in this section.

Besides simple fibers, a fiber doped with a reagent for specific analytes can be used in SPME. The reaction between the analyte and the reagent also can be performed after the extraction. The adsorption process is influenced by the presence of the reagent in the fiber, while the derivatization after the adsorption is performed only for improving the chromatographic behavior or increasing sensitivity of detection of the analyte. The derivatization on a solid support is further discussed in Section 18.8.

- Distribution equilibrium in SPME

SPME has two steps, sample extraction and sample desorption. The first process to be studied is that of the extraction of an analyte "i" from the headspace of a sample containing the analyte at the initial (molar) concentration c_o. The goal is to determine the amount (mass) of the analyte found in the fiber coating when the system reached equilibrium. The extraction mechanism of the analyte "i" is based on partition equilibria between three phases, the sample (s), the headspace (h), and the fiber (f). This equilibrium can be written as follows (see also Section 9.1):

$$i_s \overset{K_{hs}}{\rightleftharpoons} i_h \overset{K_{fh}}{\rightleftharpoons} i_f$$

The partition constant K_{hs} describes the equilibrium between the analyte in the headspace and in the sample, having $K_{hs} = c_h / c_s$ where c_h is the equilibrium (molar) concentration of the analyte "i" in the headspace of the sample, and c_s the equilibrium (molar) concentration in the sample (comparison of the expression for K_{hs} with rel. 1.4.26 for a compound "i" shows that $K_{hs} = 1 / K_i$). The equilibrium between the analyte in the fiber coating and in the headspace is described by K_{fh}, having $K_{fh} = c_f / c_h$ where c_f is the equilibrium (molar) concentration of the analyte "i" in the fiber coating (comparison of the expression of K_{fh} with rel. 1.4.26 for a compound "j" shows that $K_{fh} = K_j$).

The mass balance for the system can be obtained as shown in Section 9.1 and can be written as follows:

$$c_o V_s = c_f V_f + c_h V_h + c_s V_s \qquad (11.2.1)$$

where V_s is the volume of the sample, V_f the volume of the coating on the fiber, and V_h the volume of the headspace. The concentrations c_h and c_s can be expressed as a function of c_f as follows:

$$c_h = c_f / K_{fh} \qquad\qquad c_s = c_h / K_{hs} = c_f / (K_{hs} K_{fh}) \qquad (11.2.2)$$

By substitution of rel. (11.2.2) in rel. (11.2.1), an expression that relates c_f with c_o can be obtained. This expression can be written in the form:

$$c_f V_f = \frac{K_{fh} K_{hs} V_f c_o V_s}{K_{fh} K_{hs} V_f + K_{hs} V_h + V_s} \qquad (11.2.3)$$

The mass of analyte extracted in the fiber coating is $q_f = c_f V_f$, and therefore rel. (11.2.3) gives the amount of analyte extracted in the fiber assuming that equilibrium is attained. The equilibrium constants in rel. (11.2.3) can be related to Henry's law constants. For this purpose Henry's law (see rel. 1.4.21) for the equilibria headspace-sample and fiber-headspace can be written (assuming the solutions diluted enough) as follows:

$$p_h = \pmb{k'}_{fh}\, c_f \quad \text{and} \quad p_h = \pmb{k'}_{hs}\, c_s \qquad (11.2.4)$$

where p_h is the partial vapor pressure of the analyte in the headspace, and $\pmb{k'}_{fh}$ and $\pmb{k'}_{fh}$ are Henry's law constants. Using the expression $p_h = c_h RT$, the equilibrium constants K_{fh} and K_{hs} can be written as follows:

$$K_{fh} = RT\, /\, \pmb{k'}_{fh} \qquad \text{and} \qquad K_{hs} = \pmb{k'}_{hs}\, /\, RT \qquad (11.2.5)$$

By multiplying K_{fh} and K_{hs} and from rel. (11.2.5), it can be noticed that $K_{fh}\, K_{hs} = c_f\, /\, c_s = \pmb{k'}_{hs}\, /\, \pmb{k'}_{fh}$. Using the notation $K_{fs} = c_f\, /\, c_s$ for a new constant, expression (11.2.3) can be written in the form:

$$q_f = \frac{K_{fs} V_f c_o V_s}{K_{fs} V_f + K_{hs} V_h + V_s} \qquad (11.2.6)$$

A number of useful conclusions can be obtained based on rel. (11.2.6). The first is the proportionality between the amount of material in the fiber and the initial concentration c_o of the analyte in the sample. This proportionality indicates that ideally SPME leads to a linear relation between the concentration of the analyte and the peak area in the chromatogram. Another conclusion is that a large headspace volume V_h diminishes the amount of analyte collected in the fiber. Also, the increase in the volume of the coating material V_f leads to an increase of q_f, because the volume of the headspace V_h and that of the sample V_s are significantly larger than V_f, and the term $K_{fs} V_f$ is not important in the denominator of rel. (11.2.6). However, the increase of V_f has other implications related to the kinetics of the SPME process, which are discussed later in this section. The end result is that larger V_f requires longer extraction and desorption times for reaching equilibrium, and this may be detrimental for the success of the analysis.

For the use of SPME in solution, rel. (11.2.6) takes the form:

$$q_f = q_0 \frac{K_{fs}\, /\, \beta}{1 + K_{fs}\, /\, \beta} \qquad (11.2.7)$$

where $\beta = V_s/V_f$ and q_0 is the initial amount of analyte with $q_0 = c_o V_s$. This expression shows that for solutions, a large V_s is not favorable for extraction. As the volume for the stationary phase from a fiber is about 0.5 μL, for a solution volume of 10 mL the value for V_s/V_f is $\beta \approx 2\ 10^4$. This indicates that values for $q_f\, /\, q_0 > 0.95$ are obtained only for K_{fs} $> 3.8\ 10^5$. Approximating K_{fs} for a polydimethyl-siloxane fiber as roughly being equal to K_{ow} value for the organic compound (see Section 1.10 and Figure 1.10.2), it is obvious that only a limited number of compounds satisfy this requirement. In conclusion, the extraction of the analytes from a sample using SPME in equilibrium conditions does not lead in most cases to an exhaustive transfer of the analytes to the fiber. The repeated extraction from the same sample can be used in practice for sampling the same material

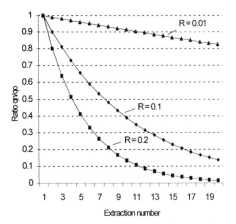

FIGURE 11.2.4. *Depletion of the analyte upon repeated extractions for three R values.*

a number of times. The extraction efficiency after n extractions can be characterized by the ratio q_n/q_0, where q_n is the amount of material left in the sample. After n extractions, the ratio q_n/q_0 is given by the expression [58], [59]:

$$q_n/q_0 = (1 - R)^n \qquad (11.2.8)$$

where $R = q_f/q_0$ is the efficiency of one extraction and where q_f can be obtained from rel. (11.2.6) or rel. (11.2.7). Upon repeated extractions the depletion of the analyte occurs as is shown in Figure 11.2.4. The figure shows that repeated extractions are acceptable only when the extraction efficiency is not very high.

Several limitations must be considered regarding the application of rel (11.2.3) and related formulas. Rel. (11.2.3) and (11.2.6) are written for only one analyte and only one phase of the sample. However, it is very common that SPME is applied for the analysis of multiple component samples. Assuming ideal conditions, the amount of each analyte in the fiber is expressed by a relation of the type (11.2.3) or (11.2.6) where the constants K must be replaced with K_i and i = 1,2,3...j for each analyte. For heterogeneous samples comprised of more phases, the use of an expression of the form (11.2.3) or (11.2.6) is not straightforward. Each compound may have a different constant for the equilibrium between a specific phase and the fiber. At the same time the analyte is in equilibrium between different phases of the sample (see e.g. [53]. For a simple system with two phases "A" and "B," with the volumes V_A and V_B, and containing the analyte in concentrations c_A and c_B, respectively, the equivalent of rel. (11.2.3) can be written as follows:

$$q_f = \frac{K_{fh}K_{hA}K_{hB}V_f(c_A V_A + c_B V_B)}{K_{fh}K_{hA}K_{hB}V_f + K_{hA}K_{hB}V_h + K_{hA}V_B + K_{hB}V_A} \qquad (11.2.9)$$

where $K_{hA} = c_h / c_A$ and $K_{hB} = c_h / c_B$. The comparison of rel. (11.2.9) with rel. (11.2.3) indicates that significant differences in the amount q_f may appear when two samples contain the same amount of analyte but in different matrices. With the assumption that $c_A V_A + c_B V_B = c_0 V_s$ and $K_{hA} = K_{hB}$, rel (11.2.9) becomes identical with (11.2.3). However, the equilibrium constants are different from material to material, and in general $K_{hA} \neq K_{hB}$. In case of multicomponent matrices, the value of q_f may differ significantly from matrix to matrix.

Another problem with the use of rel. (11.2.3) for estimating the amount of material adsorbed in the fiber is related to the assumptions that the system is ideal or that the expression of the equilibrium constants is a function of concentrations and not of activities. As $K_{hs} = a_h / a_s = (\gamma_h c_h) / (\gamma_s c_s)$, and $K_{fh} = a_f / a_h = (\gamma_f c_f) / (\gamma_h c_h)$ (where a indicates activity and γ the activity coefficients), and because the activity coefficients are dependent on the other constituents of the system, further dependence of the value q_f on the matrix and on other analytes is shown.

The assumption that the system attains equilibrium also is not fulfilled for many practical applications. A simplified study of the kinetics of SPME analyte distribution is discussed later in this section.

For SPME applied by immersion of the fiber into the solution of the analyte, the headspace does not exist between the fiber and the sample. In this case, the equilibrium taking place during the extraction step can be written as follows:

$$i_s \xrightleftharpoons{K_{fs}} i_f$$

where $K_{fs} = c_f / c_s$ and rel. (11.2.3) is reduced to the expression:

$$q_f = \frac{K_{fs} V_f c_o V_s}{K_{fs} V_f + V_s} \tag{11.2.10}$$

When the volume of the solution containing the sample is very large compared to the volume of the fiber, the term $K_{fs} V_f$ can be neglected, and rel. (11.2.10) takes the form:

$$q_f = K_{fs} V_f c_o \tag{11.2.11}$$

Various procedures are reported for the estimation of the distribution constants K_{fs} and K_{hs} [53]. The values for K_{hs} for nonpolar compounds in solution samples (such as organic volatiles in water samples) can be obtained from the tables for Henry's constants. The values for K_{fh} for a specific fiber coating can be estimated based on Kováts retention indexes or on the linear temperature programmed retention indexes (LTPRI) [60] obtained for a chromatographic stationary phase identical to the fiber coating. The relation between K_{fh} and LTPRI is given by the formula:

$$K_{fh} = 0.00415 \text{ LTPRI} - 0.188 \tag{11.2.12}$$

Some values for partition constants K_{hs} where the sample matrix is water $K_{hs} = K_{hw}$ and some K_{fh} and K_{fs} values where $K_{fs} = K_{fw} = K_{fh} K_{hs}$ for a number of hydrocarbons in a PDMS fiber are given in Table 11.2.2 [53] (see also Section 1.10).

The values for the partition constant K_{fh}, although expected to be independent of the film thickness, may vary with this parameter. Some values of K_{fh} for a PDMS coating at two different thickness values are given in Table 11.2.3.

TABLE 11.2.2. *Some values for partition constants K_{hs} where the sample matrix is water (K_{hw}) and some K_{fh} and K_{fs} values where $K_{fs} = K_{fw} = K_{fh} K_{hw}$ for a number of hydrocarbons in a PDMS fiber (50 μm film).*

Compound	K_{hw}	K_{fh} (25°C)	K_{fw} (22°C)
3-methylpentane	21	159	3270
2-methylhexane	26	387	10202
2-methylheptane	26	993	25806
4-methylbeptane	26	1060	27274
3-methylheptane	29	1090	31856
2-methyloctane	17	2600	45267
3-methyloctane	23	2890	66682
3,3-diethylpentane	24.5	2610	63718
2,2-dimethyloctane	19	4320	82430
benzene	0.19	301	58
toluene	0.23	818	189
ethylbenzene	0.27	2070	566
m-xylene	0.26	2090	533
p-xylene	0.23	2500	564
o-xylene	0.17	2900	485
isopropylbenzene	0.36	3880	1412
n-propylbenzene	0.33	5040	1664
isobutylbenzene	0.50	8360	4197
sec-butylbenzene	0.47	8590	4011
1-methyl-3-isopropylbenzene	0.33	10100	3284
1,4-dimethyl-2-ethylbenzene	0.23	15900	3628
1,2-dimethyl-4-ethylbenzene	0.23	17400	3984
1,3-dimethyl-2-ethylbenzene	0.24	18100	4345
2-methylbutylbenzene	0.38	24100	9099
n-pentylbenzene	0.24	34500	8195
tert-1-butyl -3,5 -dimethylbenzene	0.40	45600	18260
1,3,5-triethylbenzene	0.28	67300	18517
1,2,4-triethylbenzene	0.21	75600	16253

TABLE 11.2.3. *Some values of K_{fh} for a PDMS fiber at two different thickness values (T = 25°C).*

Analyte	K_{fh} for 30 μm PDMS	K_{fh} for 100 μm PDMS
benzene	260	300
toluene	710	980
ethylbenzene	2000	2100
o-xylene	3100	3100
p-xylene	2300	2400

The values for the SPME partition constants are not always available. In practice, the choice of a specific fiber for a group of analytes is usually done based on a "polarity match" between the fiber and analytes.

In case of the use of fibers doped with specific reagents, the adsorption process is affected by the stability constant of the derivatized analyte with the reagent, and by the partition constant for the derivatized analyte.

The application of SPME is not limited to the analysis of static samples. A nonequilibrium sampling also can be used, for example for the quantitation of volatiles in an air stream [61], [62]. In this case, the amount of analyte accumulated in the fiber can be described by the expression:

$$q_f(t) = c_g \, K_{fg} \, V_f \, (1 - \exp \frac{- AUt}{K_{fg} V_f})$$ (11.2.13)

In rel. (11.2.13), c_g is the concentration of the analyte in the gas stream, $K_{fg} = K_{fh}$ being the equilibrium constant between the analyte in the fiber and in the gas ($K_{fh} = c_f / c_g$), U is the volumetric gas flow, and t the time. Parameter A represents the fraction of analyte in the gas stream that has time to interact with the fiber. The values of A depend on the compound and on the residence time of the gas in contact with the fiber, these values being determined experimentally for a specific sampling setup [63].

- Kinetics of the distribution process in SPME

The rate of the distribution process in SPME is controlled by the diffusion of the adsorbed species from the sample into the headspace and from the headspace into the fiber (for the headspace adsorption). In case of solution adsorption, the diffusion from the sample takes place directly into the fiber. The diffusion is described by Fick's law (see Section 1.5). For headspace adsorption, the diffusion takes place in two different stages, a) from the sample into the headspace and b) from the headspace into the fiber, as shown in the diagram from Figure 11.2.5. The simplified model of the diffusion from the sample is similar to the diffusion in a tube or a rectangular channel from a medium where at the initial moment t =0, the concentration of the diffusing species is c = c_o for x ≤ 0, and c = 0 for x > 0, the limit between the sample and headspace occurring at x = 0. These conditions (boundary conditions) are written as follows:

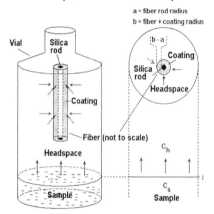

FIGURE 11.2.5. *Diagram of the diffusion process during the headspace SPME adsorption.*

$$c\big|_{t=0,x\leq0} = c_o \qquad c\big|_{t=0,x>0} = 0$$ (11.2.14)

and the solution c(x,t) of this problem is discussed in Section 1.5, being given by rel. (1.5.11). Figure 1.5.4 shows the variation in time of the ratio c/c_o for different values D·t, where t is the time and D the diffusion coefficient. The situation for the diffusion from the sample into the headspace in SPME is complicated by fact that the diffusion coefficient in the sample (x ≤ 0) is different from that in the headspace and by the condition that at the equilibrium the concentrations in the sample are not equal to those in the headspace, obeying the expression $K_{hs} = c_h / c_s$. This leads to two diffusion equations:

$$\frac{dc_s}{dt} = D_s \frac{\partial^2 c_s}{\partial x^2} \qquad\qquad \frac{dc_s}{dt} = D_h \frac{\partial^2 c_h}{\partial x^2} \qquad (11.2.15)$$

where D_s is the diffusion coefficient of the analyte in the sample, and D_h is the diffusion coefficient of the analyte in the headspace. The equations (11.2.15) are subject to the limit (boundary) conditions:

$$c_s\big|_{t=0,x\leq 0} = c_o \qquad c_h\big|_{t=0,x>0} = 0 \qquad (11.2.16a)$$

and:

$$D_s \frac{\partial c_s}{\partial x}\bigg|_{x=0} = D_h \frac{\partial c_h}{\partial x}\bigg|_{x=0} \qquad c_h = K_{hs}\, c_s \qquad (11.2.16b)$$

The solution of this system is more complicated. Conditions such as sample perfectly stationary or perfectly stirred must be imposed to obtain the solution. When the equilibrium is attained ($t = \infty$), the remaining concentration in the sample c_{so} and the concentration in the headspace c_{ho} will be given by the expressions:

$$c_{so} = \frac{L_s c_o}{K_{hs} L_h + L_s} \qquad \text{and} \qquad c_{ho} = K_{hs}\, c_{so} \qquad (11.2.17)$$

where L_s is the sample depth and L_h the headspace length, and are proportional to V_s and V_h, respectively. The expressions of c_{so} and c_{ho} given by rel. (11.2.17) can be obtained from expressions similar to rel. (11.2.10) and (11.2.11) applied for the sample/headspace system.

The analytes in the headspace are further adsorbed into the fiber. The diffusion process in the fiber follows an equation of the form:

$$\frac{\partial c_f}{\partial t} = D_f \nabla^2 c_f \qquad (11.2.18)$$

where D_f is the diffusion coefficient in the fiber. Assuming no angular diffusion and no diffusion along the fiber, rel. (11.2.18) can be simplified to the form:

$$\frac{\partial c_f}{\partial t} = D_f \frac{1}{r} \frac{\partial}{\partial r}\left(r \frac{\partial c_f}{\partial r}\right) \qquad (11.2.19)$$

Equation (11.2.19) also describes the diffusion process in the fiber if the analytes are coming from a solution. The boundary conditions for the differential equation (11.2.19) can be chosen using two models for the diffusion processes, each of them describing a limiting case. In the first model it is assumed a perfectly agitated solution (or headspace). In this situation, the migration of the analyte into the organic coating is not affected by the diffusion in the liquid phase or in the headspace. The second model assumes an unstirred solution or headspace, where the analyte must penetrate a layer of liquid or gas surrounding the fiber coating before it can diffuse into the coating. Neither of these models describes attainable experimental conditions, since it is neither possible to ensure perfect agitation of the solution nor possible to eliminate mixing through convection. The solution of equation (11.2.19) using different approximations has been reported in literature [53].

The diffusion in the headspace and in the fiber also can be modeled using a unidimensional approximation. In this case, the sample diffusion is assumed to take place in one single direction in an unique channel, as shown in Figure 11.2.6. The diffusion into the headspace is described by equations (11.2.14) with conditions (11.2.16a) and (11.2.16b), but the fiber coating is treated as a flat membrane at the end of the diffusion channel. In order to account for the difference between the membrane surface and the vial cross-section (diffusion channel), a factor $F = A_f / A_s$ must be introduced. Taking the membrane thickness $L_f = b - a$ (see Figure 11.2.5), the diffusion problem is described by a set of three differential equations that can be written in the form:

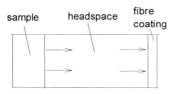

FIGURE 11.2.6. *Diagram of the unidimensional channel.*

$$\frac{dc_i}{dt} = D_i \frac{\partial^2 c_i}{\partial x^2} \qquad \text{with } i = s, h, \text{ and } f \qquad (11.2.20)$$

Assume equilibrium between the sample and the headspace (see rel. 11.2.16), and the boundary conditions are written as follows:

$$c_s\big|_{t=0, x\leq 0} = c_{so} \qquad c_h\big|_{t=0, x>0} = c_{ho} \qquad c_f\big|_{t=0, x>Lh} = 0 \qquad (11.2.21a)$$

$$D_s \frac{\partial c_s}{\partial x}\bigg|_{x=0} = D_h \frac{\partial c_h}{\partial x}\bigg|_{x=0} \qquad D_h \frac{\partial c_h}{\partial x}\bigg|_{x=Lh} = D_f \frac{\partial c_f}{\partial x}\bigg|_{x=Lh} \qquad \frac{\partial c_f}{\partial x}\bigg|_{t>0, x=Lh+Lf} = 0 \qquad (11.2.21b)$$

$$c_h = K_{hs}\, c_s \qquad c_f = K_{fh}\, c_h \qquad (11.2.21c)$$

The analytical solution of the equation (11.2.20) can be obtained [53], [64], and it is graphically pictured in Figure 11.2.7 for three moments in time A < B < C after the diffusion process starts. As shown in Figure 11.2.7, the concentration in the sample is the highest at the moment A and the lowest at the moment C, while the concentration in the fiber is in the opposite order.

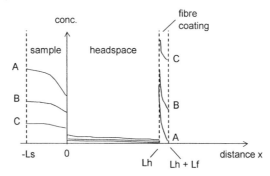

FIGURE 11.2.7. *Diagram for the simplified unidimensional diffusion process in SPME with headspace adsorption.* Concentration profiles (arbitrary units) are given at three moments in time A < B < C.

The average concentration of material in the fiber can be calculated using the integral over the fiber depth as follows:

$$c_f(t) = 1/L_f \int_{Lh}^{Lh+Lf} c_f(x,t)dx \qquad (11.2.22)$$

A solution obtained assuming the sample and the headspace perfectly stirred with the sample at constant concentration all the time c_{so} corresponds to the solution for the extraction from a gaseous sample similar to that described in Section 1.5. for the diffusion equation (1.5.2) with the boundary conditions given by rel. (1.5.15). The concentration in the fiber is given in this case by the formula:

$$c_f(t) = K_{fh} \, c_{so} [1 - \exp(T) \, \mathrm{erfc}(\sqrt{T})] \quad \text{with} \quad T = D_h t / (K_{fh} L_f F)^2 \quad (11.2.23)$$

The equilibration time for this system is considered attained when 95% of the ideal amount of analyte is extracted from the sample and is given by the expression:

$$t_e = t_{95\%} = 100 \, (K_{fh} \, L_f \, F)^2 / D_h \qquad (11.2.24a)$$

This expression neglects the diffusion time of the analyte in the coating and is valid only when $L_h > 15 \, K_{fh} \, F \, L_f$. For $L_h < 15 \, K_{fh} \, F \, L_f$, the expression for the equilibration time is given by the formula [53]:

$$t_e = t_{95\%} = [2.9 / D_h] \, (K_{fh} \, F \, L_f \, L_h + L_h^2 / 10) \qquad (11.2.24b)$$

The second term in rel. (11.2.24b) is similar to rel. (1.5.14). Both rel. (11.2.24a) and (11.2.24b) show that the equilibration time increases when the fiber volume and the distribution constant K_{fh} increase. Particularly the increase in the fiber thickness, which can be chosen in the analysis, should not be larger than necessary for accommodating the amount of analytes in the sample, because it affects unfavorably the adsorption time. The time decreases as the diffusion coefficient of the analyte increases. Rel. (11.2.24b) also shows that longer path for the headspace increases the equilibration time. Because D_h increases with temperature, the diffusion process is accelerated by heating. The calculation of the equilibration time for a compound with $K_{fh} = 100$, in a fiber with $L_f = 50$ μm, $F = 1$, and $D_h = 0.1$ cm^2/s, rel. (11.2.24a) predicts a time $t_e = 250$ s. Rel (11.2.24b) for $L_h = 2$ cm predicts a time $t_e = 40.6$ s and a time of about 25 min. for $K_{fh} = 5000$ (when also more material is accumulated in the fiber).

For the adsorption from a perfectly stirred solution, the equilibrium time depends only on the diffusion in the coating and is given by the formula:

$$t_e = t_{95\%} = L_f^2 / (2D_f) = \frac{(b-a)^2}{2D_f} \qquad (11.2.25)$$

This formula shows that for a fiber of 100 μm thickness, in a perfectly stirred solution the equilibration time for a molecule such as benzene ($D_f = 2.8 \cdot 10^{-6}$ cm^2 s^{-1}) is about 20 sec.

Assuming for the diffusion from the sample an equilibration time given by a formula similar to rel. (11.2.25), where the thickness of the diffusing medium is much larger, the equilibration time between the sample and the headspace becomes very long (the values for D_s are typically of the order of 10^{-5} cm^2 s^{-1}). For this reason, sample heating at temperatures that do not create artifacts and extraction times of the order of 1/2 hour are typically recommended for achieving an efficient diffusion of the analytes in the headspace [65].

The extraction times in practice vary between a few min. to 1/2 hour. Rel. (11.2.24a) and rel. (11.2.25) for the sample show, however, that longer times may be needed for attaining 95% equilibrium. Because the equilibration time is dependent on the partition constants, the equilibration time can be quite different for different components of a sample. This fact adds to the limitations of the application of SPME for quantitation.

- Desorption of analytes and conditioning of the SPME fiber

The desorption process in SPME can be done in the injection port of a GC system or in a HPLC. The use of SPME for the direct transfer of the adsorbed compounds into the GC injection port and the selective adsorption capacity of the fiber coating that prevents oxygen and moisture from getting into the GC column are important features of this technique. These features in the headspace SPME provide major advantages over the other known headspace techniques [66].

For the GC desorption, two different parameters affect desorption: a) the elevated temperature of the GC port, and b) the fast flow of pure gas around the fiber. The effect of temperature on the adsorption can be approximated based on temperature dependence of an equilibrium constant given by rel. (1.8.8). The expression of the equilibrium constant K_{fh} can be written in the form:

$$K_{fh} = \exp \frac{\Delta S^0_{fh}}{R} \exp \frac{-\Delta H^0_{fh}}{RT}$$

(11.2.26)

In rel. (11.2.26) besides the explicit variation in the function of T, ΔS^0 and ΔH^0 are also temperature dependent. For adsorption processes with $K_{fh} > 1$, the adsorption should be an exothermic process, and, therefore, K_{fh} decreases when the temperature increases. A model of the variation of K_{fh} for benzene, with the assumption that ΔS^0_{fh} and ΔH^0_{fh} do not vary considerably with temperature for the given interval, is shown in Figure 11.2.8.

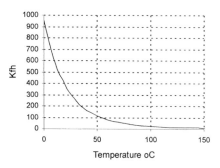

The quantity of material q_g desorbed from the fiber in a volume V_g of gas at equilibrium can be estimated by the expression:

FIGURE 11.2.8. *Model of variation of K_{fs} for benzene, assuming ΔS^0_{fs} and ΔH^0_{fs} do not vary with temperature for the given interval.*

$$q_g = \frac{q_f V_g}{K_{fg} V_f + V_g}$$

(11.2.27)

The lower is the value of K_{fg}, the closer is the amount of material desorbed from the fiber to total sample load q_f. Also, thinner fibers will release the load more completely compared to thicker ones. Considering that the desorption is done in flowing gas, the desorption is very efficient. Regarding the desorption kinetics, the desorption process can be seen as the reverse of the adsorption into the fiber, and the equilibration time is described by rel. (11.2.25). Therefore, the desorption is fast, and 1 min. desorption time into a GC oven is typically sufficient. Longer desorption times are not recommended unless some type of cryofocusing at the beginning of the chromatographic column is applied. With desorption times of 1 min. or less, and using only the solvent effect for focusing the analytes at the beginning of the chromatographic column, SPME can be used without any peak broadening in the chromatograms. The desorption temperature is limited by the type of fiber utilized in the specific method. Some recommended temperatures for desorption are given in Table 12.2.4 [67].

Before any fiber utilization, it is recommended that the fiber should be conditioned. Fiber conditioning is in fact a thorough fiber desorption. In many instances, some compounds will not desorb completely within 1 min. [65]. This is particularly true for compounds with large K_{fg} values (see rel. 11.2.24), and these may remain in part in the fiber and interfere when this is used the second time. Also, left unutilized for a certain period of time, the fiber may absorb trace compounds from the atmosphere. The recommended conditions for the fiber conditioning also are given in Table 11.2.4.

The desorption in HPLC also can be done very efficiently in a flow of pure solvent flushing the fiber. The desorption must be done in a special setup as shown in Figure 11.2.3. In this setup, sealing the fiber assembly is the critical step due to the high pressure of the eluent in HPLC. Also, a modification of an external lop injection valve of a packed column supercritical fluid chromatograph (PC-SFC) has been reported for the insertion of the fiber and desorption of the retained analytes by supercritical CO_2. This approach was applied to the determination of low ppb levels of alkylphenols in environmental samples. [68]. Other desorption procedures were also reported.

TABLE 11.2.4. *Desorption, conditioning, and maximum temperature recommended for several SPME fibers.*

Coating	Film thickness in μm	Maximum temperature	Desorption temperature	Conditioning temperature	Conditioning time in hours	Applications
	100	280°C	200-270°C	250°C	1	VOCs
PDMS	30	280°C	200-270°C	250°C	1	
	7	340°C	220-320°C	320°C	2–4	PAHs
PDMS/DVB	65	270°C	200-270°C	260°C	0.5	amines, nitroaromatics
polyacrylate	85	320°C	220-310°C	300°C	2	pesticides, phenols
CAR	50	320°C	200-250°C	300°C	1	surfactants
CAR/PDMS	75	320°C	240-300°C	280°C	0.5	alcohols, polar compounds
CW/DVB	65	265°C	200-260°C	250°C	0.5	polar compounds
DVB/CAR/	50	270°C	230-270°C	270°C	4	
PDMS	30	270°C	230-270°C	270°C	4	

- *Method development and quantitation in SPME*

The goal of SPME method development is the optimization of a number of parameters that influence the selective adsorption and the desorption of the analytes. Among these parameters are a) the type of fiber to be utilized, including chemical nature of the phase and film thickness, b) the choice between solution adsorption or headspace adsorption, c) adsorption time, d) adsorption temperature, e) salting effect and the presence of organic solvents in the sample, f) the pH of the sample, g) desorption temperature (for GC analysis), h) desorption time, i) selection of quantitation technique, etc.

The choice of the SPME fiber is usually done to achieve a high adsorption of the analytes in the fiber. For this reason fibers for which K_{fs} is high are typically selected. However, when this value is not available, the polarity match is utilized (some polarities are indicated in Table 11.2.1). Compounds with low polarity typically show very high K_{sf} values for nonpolar fibers. Trial and error is also frequently used for fiber selection, mainly when the analysis is done for a whole group of analytes that may have different behavior toward a specific fiber. Besides the polarity, a specific film thickness of the fiber must be chosen. Depending on the amount of sample, a thinner fiber is preferred when trace levels of the analyte must be determined. Thicker fibers have a higher "load capacity," but the adsorption and desorption processes are slower.

The choice between solution and headspace adsorption also is based on the nature of the sample. For compounds with higher solubility in water or for compounds with low volatility, the solution adsorption is usually preferred. These compounds have low K_{hs} values, and even when using polar fibers (with high K_{fh} for the particular compounds), rel. (11.2.3) shows that low K_{hs} values lead to low values for q_f. The compounds with high volatility and low water solubility are typically analyzed from the sample headspace. The sample can be kept in solid form or in a water solution or slurry. Another factor sometimes considered in the choice between solution and headspace is that fiber degradation after a number of analyses is usually higher for solution adsorption. Also, for very dirty samples, the adsorption from the solution must be avoided if possible.

An important element that must be considered when choosing the fiber and selecting solution or headspace adsorption is that during SPME adsorption, not only the analytes are adsorbed into the fiber, but also other compounds from the sample matrix. When undesired matrix components are adsorbed in the SPME fiber, the ratio analyte/matrix adsorption can be more important than the actual amount of material absorbed into the fiber. The sensitivity of the analytical instrumentation is also important in this decision. When ample sensitivity is offered by the chromatographic instrument, the choice of the fiber and the adsorption method is based on achieving the best "cleaning" of the sample and not necessarily on the highest amount of analyte collected from the system.

Optimization of the adsorption time depends on sample nature, sample amount, type of fiber, etc. It is common that adsorption times are not optimized to achieve equilibrium but to obtain sufficient amount of the analytes in the fiber such that the chromatographic procedure works well. For this reason the quantitation using SPME requires strictly maintained adsorption times. The change in sample matrix may affect the diffusion, and therefore the quantitation of the same amount of a given analyte can lead to different results in different matrices if the calibration is not performed for the same matrix.

The optimization of the adsorption temperature is also important, because the process of desorption of the analytes from the sample and the adsorption into the fiber are affected by temperature. This indicates that the temperature during the adsorption process from headspace or solution must be kept constant. As shown by rel. (11.2.26) and the graph from Figure 11.2.8, for exothermic adsorption the increase in temperature does not favor the adsorption into the fiber. However, higher temperatures favor the desorption of the sample into a headspace, and for this reason heating of the sample is frequently utilized in SPME adsorption. Systems using sample heating and simultaneous fiber cooling were experimented, but they are not popular.

In order to modify the distribution constants when using water samples, the addition of an inorganic salt such as NaCl has been proven useful in a number of cases by increasing the desorption of some organic compounds. The addition of salt can be done for both solution extraction and for headspace extraction. The variation in K_{fs} with the addition of salt can be described by the expression:

$$K_{fs}^{salt} = K_{fs} \exp (k_{salt}\, c_{salt}) \tag{11.2.28}$$

where K_{fs}^{salt} is the partition coefficient after the addition of salt, k_{salt} is a parameter depending on the system, and c_{salt} is salt concentration.

While the increase in salinity favors the adsorption of organic compounds from aqueous solutions, the presence of organic solvents in water has an opposite effect. The formula providing a rough estimation of the variation of K_{fs} compared with K_{fw} when adding a solvent is the following:

$$K_{fs} = K_{fw} \exp [(c\%/100)\, (P'_{solvent} - P'_{water})] \tag{11.2.29}$$

where c% is the percent concentration of the solvent in water, and P' is the polarity of the solvent as given in Tables 1.10.11 to 1.10.19. Rel. (1.12.29) shows that the addition of a solvent of low polarity to the sample may decrease significantly the partition into the fiber for a nonpolar analyte. For example, 2% of such a solvent can decrease the partition coefficient to about 20%. This effect can be seen in practice when adding to an aqueous sample an internal standard dissolved into an organic solvent.

The pH modification also can be very important for compounds that may exist in solution in both ionic or nonionic forms, such as organic acids or amines. For a substance subject to the equilibrium:

$$AH^+ \underset{}{\overset{K_1}{\rightleftharpoons}} A + H^+$$

where K_1 is the acidity constant, the concentration of the nonionic compound at equilibrium is given by $[A] = K_1 [AH^+] / [H^+]$. In the case of amines, for example, a high pH will favor the existence of nonionic form and, therefore, higher adsorbance in the fiber. The acids will show better extraction at lower pH values. For a substance subject to the equilibrium:

$$AH \underset{}{\overset{K_1}{\rightleftharpoons}} A^- + H^+$$

the concentration of the nonionic compound is given by [AH] = [AH⁻] [H⁺] / K_1, and a low pH will favor the existence of nonionic form with higher adsorbance in the fiber.

The desorption of the SPME fiber is typically done for a short period of time (about 1 min.) at temperatures recommended in literature for the specific fiber (see Table 12.2.4). Modifications of these parameters do not provide, in general, a large variation in the results using SPME analysis.

Selection of an appropriate quantitation procedure when SPME is used as a sample preparation procedure is important because a number of factors may influence the accuracy of the determination. The method of external calibration using a blank sample such that calibration standards have the same matrix as real samples is typically successful. However, addition of analyte on the blank sample done using an organic solution may influence the results [69]. Use of an internal standard in quantitative SPME determinations may also encounter problems because adsorption of the internal standard into the fiber can be affected by the analyte concentration. Up to 65% reduction in the level of an internal standard (50 ng/g d_3-methyl *tert*-butyl ether) was noticed for a CAR/PDMS fiber used in the determination of benzene and alkylbenzenes when the level of the analytes increased from 0.18 µg/g to 1.8 µg/g in the sample [70].

The application of standard addition method for quantitation in SPME also can be affected by some inaccuracies because of the nonlinearity of the calibration. As shown in Section 3.8, the application of one-step standard addition method assumes a linear dependence on the analytical response and the sample concentration and also that for no analyte there is no analytical signal. This is not always true for SPME, and the quantitation must be verified with known surrogate analytes.

Automation in SPME is a common means to improve sample throughput and reproducibility. An SPME autosampler is able to locate the sample vial, the heater position (in case the sample adsorption is not done at room temperature), and the injection port position. Also, it must control a) the plunger movement so that the fiber is sheathed inside the needle when vial or injector septum is pierced and outside when the sample is adsorbed or desorbed, b) the depth of the needle and the fiber in the vial so that the exposed volume is controlled, c) the absorption time in the vial, d) the depth of the fiber in the injector, and e) the desorption time in the injector [71]. On-line SPME-GC techniques were also reported in literature [72].

- Examples of applications

SPME technique has proved to be versatile and extremely useful in the analysis of environmental, food, forensic, and many other types of samples. The capability to extract the analytes of interest from a complex matrix and the concentration of the analyte for a single injection into the chromatographic system are important qualities of SPME. Although the factor of concentration of the analyte (see Section 3.5) for SPME cannot always reach that of standard purge and trap techniques that are estimated to be up to 100 times more sensitive, SPME is applied more and more for analyses where purge and trap was used in the past [73]. The headspace SPME technique can be used to extract a wide range of organic compounds, volatile or semivolatile, from various matrices, such as air, water, and soil. Direct extraction SPME from a solution can be used to extract organic compounds from drinking water, waste water, sludge, soil, and other complex matrices. Quantification limits and good precision have been achieved when samples are collected at room temperature from drinking water and waste water. For example, a 15μm of PDMS fiber SPME can be used to extract semivolatile compounds such as PAHs and PCBs from aqueous samples. For sampling of phenols in water by use of a polyacrylate coating; the detection limit, linear range, and precision were reported to be better than or equivalent to a EPA method using purge and trap [74]. SPME also can be used to efficiently extract herbicides and pesticides from aqueous matrices. Several applications of SPME in sample preparation of environmental samples are summarized in Table 11.2.5.

TABLE 11.2.5. *Some SPME-based procedures for the analysis of several pollutants in environmental samples.*

Pollutants	SPME fiber	Sample volume (mL)	Chromatographic technique	Quantification limit (ng/L)	Reference
volatile organic compounds	PDMS, 100 μm	50	GC-ITD	1	[75]
PAHs, PCBs	PDMS, 15 μm	40	GC-ITD	< 1	[76]
N-herbicides	acrylate, 85 μm	4	GC-FID / GC-NPD / GC-ITD	200 / 10 / 10	[77]
N,P-pesticides	PDMS, 100 μm	4	GC-NPD / GC-ITD	20 / 10	[78]
Cl-pesticides	PDMS, 100 μm	35	GC-FID / GC-ECD / GC-MS	2000 / 50 / 20	[79]
P-insecticides	acrylate, 85 μm	4	GC-FID / GC-NPD / GC-ITD	200 / 10 / 2	[80]
P-pesticides	PDMS, 100 μm / acrylate, 85 μm	3	GC-NPD / GC-NPD	2 / 1	[81]
triazine, herbicides	acrylate, 85 μm	3	GC-NPD	10	[81]
2,6-dinitroaniline herbicides	acrylate, 85 μm	3	GC-NPD	8	[81]
anilines	PDMS/DVB	5	GC-FID	180	[82]
phenolic compounds	acrylate, 85 μm	40	GC-FID / GC-MS	600 / 10	[74]

The continuous monitoring of levels of organic compounds in process streams and wastewater effluents are of particular importance to industry and environmental monitoring. This can be accomplished by using specific sensors, but this approach is not always possible because of the complexity of such samples. The rapid sampling and concentration response of SPME facilitate the monitoring of organic compounds in a flowing stream. It was determined that in addition to convection due to the flow in the system, an efficient agitation is sometimes required to achieve efficient extractions. Three agitation techniques are more common: magnetic mixing, intrusive mixing, and sonication. While magnetic stirring is inexpensive and easy to automate, it exhibits low mixing efficiency. Intrusive mixing allows efficient agitation but may cause heating and is not easily amenable to on-line analysis. Sonication proved to be an efficient and convenient means of agitation for sampling from flowing streams. However, different studies have demonstrated that prolonged sonication may affect the sample composition, mainly in the analysis of organic trace components in water, possibly due to the formation of traces of hydrogen peroxide and molecular hydrogen.

Analysis of soil and sludge samples can be more difficult than water samples because of possible chemisorption of analytes in the solid matrices, which limits their recovery [83]. This problem can be solved by using a combination of several extraction procedures. In the headspace SPME method, volatile organic compounds that are physically adsorbed by the solid sample can be sampled from the headspace. The less volatile compounds or those with strong interaction with the matrix can be desorbed by heating the sample to an elevated temperature and then sampled by SPME. The increase in temperature can overcome the energy barrier of the chemisorption of analytes on the solid matrix, releasing these compounds in the solution [84].

An interesting application of SPME is in the determination of various analytes involving derivatization. The derivatization reaction can be done directly in the fiber using a fiber doped with a specific reagent. This subject is discussed in Section 18.8. Numerous other applications involve a derivatization to make the analytes more volatile, followed by SPME sample processing. Several examples of such applications for analysis of organometallic pollutants in the environment are indicated in Table 11.2.6 [85].

TABLE 11.2.6. *Examples of the application of SPME sample preparation after derivatization in the analysis of organometallic pollutants [85].*

Pollutant	Method	Detection limit	Reference
alkyl-Pb in water	derivatization with $NaB(C_2H_5)_4$, SPME, GC-MS	100 ng/L	[86]
butyl- and phenyl-Sn in water and sediments	derivatization with $NaB(C_2H_5)_4$, SPME, GC-FPD	0.006-0.6 ng/L	[87]
butyl-Sn in water and sediments	derivatization with $NaB(C_2H_5)_4$, SPME, GC-FID	30	[88]
methyl-Hg in spiked soil	water extraction, derivatization with $NaB(C_2H_5)_4$, SPME, GC-MS.		[89]
Se in tap and river water	derivatization with 4,5-dichloro-1,2-phenyldiamine, SPME, GC-MS	6 ng/L	[90]
Sn, Hg, Pb in surface water and sediments	derivatization with $NaB(C_2H_5)_4$, SPME, GC-ICP-MS	0.1-4 ng/L	[91]
tetraethyl-Pb in water	derivatization with $NaB(C_2H_5)_4$, SPME, GC-MS	400 ng/L	[92]
triphenyl-Sn in water, potatoes, mussels	extraction with KOH-ethanol, derivatization with $NaB(C_2H_5)_4$, SPME, GC-ICP-MS	0.1	[93]

Various other samples can be analyzed qualitatively and quantitatively using the SPME technique. For example, mainstream cigarette smoke can be analyzed by collection of the smoke in water followed by immersion of a StableFlex DVB/CAR/PDMS 50 μm film fiber in the solution for 20 min. followed by desorption for 1 min in the injection port at 250° C of a GC/MS system equipped with a Chrompack PoraPLOT Q column (25 m x 0.25 mm i.d., 8 μm film). The separation is performed with a gradient temperature from -60° C to 250° C. The resulting chromatogram is shown in Figure 11.2.9, and the identification of peaks from the chromatogram in Figure 11.2.9 is given in Table 11.2.7.

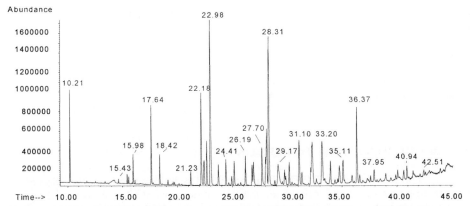

FIGURE 11.2.9. *GC/MS total ion chromatogram (TIC) of liquid phase SPME of an aqueous solution of smoke from 1R4F Kentucky reference cigarette.*

TABLE 11.2.7. *The identification of peaks from the chromatogram shown in Figure 11.2.9, obtained using SPME sample preparation for a solution of cigarette smoke.*

Peak No.	Ret. time	Compound	MW	Peak No.	Ret. time	Compound	MW
1	10.21	carbon dioxide	44	33	30.19	2-butenal	70
2	14.64	carbon oxide sulfide	60	34	30.46	1-methyl-1,3-cyclopentadiene	80
3	15.43	hydrogen cyanide	27	35	30.70	thiophene	84
4	15.57	propene	42	36	31.10	benzene	78
5	15.98	chloromethane	50	37	31.34	2-butenenitrile	67
6	16.15	2-propadiene	40	38	31.98	2-ethylacrolein	84
7	17.64	acetaldehyde	44	39	32.16	3-methyl-3-buten-2-one	84
8	18.42	methanethiol	48	40	32.30	3-methylbutanal	86
9	19.16	formic acid methyl ester	60	41	32.68	1-penten-3-one	84
10	21.23	acetonitrile	41	42	33.20	2-pentanone	86
11	22.18	propenal	56	43	33.43	pentanal	86
12	22.43	furan	68	44	33.96	1-methyl-1H-pyrrole	81
13	22.68	propanal	58	45	34.38	dimethyldisulfide	94
14	22.98	propanone	58	46	34.63	2-methyl-2-butenal	84
15	23.73	2-propenenitrile	53	47	34.77	2-methyl-butanenitrile + pyridine	83 + 79
16	24.41	acetic acid methyl ester	74	48	35.11	3-methyl-butanenitrile	83
17	24.65	ethylene sulfide (thiirane)	60	49	35.24	2-pentenal	84
18	24.93	1,3-cyclopentadiene	66	50	35.91	cyclopentanone	84
19	25.16	1,2-pentadiene	68	51	36.15	4-methyl-2,3-dihydrofuran	84
20	25.73	2,3-dihydrofuran	70	52	36.37	methylbenzene	92
21	26.19	propanenitrile	55	53	36.63	4-methyl-2-pentanone	100
22	26.79	2,5-dihydrofuran	70	54	36.90	3-methyl-2-pentanone	100
23	26.92	2-methylpropanal	72	55	37.30	furancarboxaldehyde	96
24	27.70	3-buten-2-one	70	56	37.62	3-hexanone	100
25	28.03	2-methylfuran	82	57	37.95	2-hexanone	100
26	28.12	3-methyl-2-butanone	86	58	38.99	2-methylcyclopentanone	98
27	28.31	2-butanone	72	59	39.90	hexanenitrile	97
28	28.87	2-propenoic acid methyl ester	86	60	40.08	4-methyl-pentanenitrile	97
29	29.17	2-methyl-2-methoxypropane (I.S.)	88	61	40.64	ethylbenzene	106
30	29.56	propanoic acid methyl ester	88	62	40.94	1,4-dimethylbenzene	106
31	29.76	2-methyl-propanenitrile	69	63	41.51	styrene	104
32	29.87	2-butenenitrile	67	64	42.51	phenol	94

Numerous other applications are described for SPME sample preparation. For example, SPME can be used for the analysis of pyrolysates by pyrolysis in a closed vial and not in a GC pyrolysis interface (see Section 15.1), followed by SPME-GC-MS analysis of the pyrolysate in the vial [94]. Other analyses include analysis of flavors [95], various natural products, pharmaceutical products, biological materials, etc. Various monographs reporting SPME applications as well as lists of published SPME applications are available [53], [57], [73].

11.3 MATRIX SOLID PHASE DISPERSION

Matrix solid phase dispersion (MSPD) is an analytical technique used for extraction of analytes from semi-solid and viscous samples. The principle of this technique is based on the use of the same bonded-phase solid supports as in SPE, which also are used as grinding material for producing the disruption of sample matrix. During this procedure, the bonded-phase support acts as an abrasive, and the sample disperses over the surface of the support. The classic methods used for sample disruption such as mincing, shredding, grinding, pulverizing, and pressuring are avoided in this procedure [96]. The technique has many applications to the processing of samples of biological origin (animal tissues, plant materials, fats, etc.). A schematic of MSPD preparation is given in Figure 11.3.1.

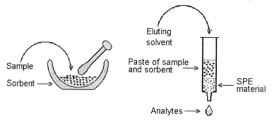

The sample is placed in a mortar containing the sample and a bonded phase material (with or without any solvent). The mixture is then crushed with a pestle. During this operation, the bonded phase and its support

FIGURE 11.3.1. *Schematics of a MSPD preparation [19].*

serve several functions including a) is an abrasive that promotes mechanical disruption of the sample structure, b) assists in sample disruption and lysis of cell membranes similar to a solvent, and c) adsorbs the analytes or other compounds of interest from the sample. After this step, the material containing the sample, the solid sorbent (and possibly an internal standard and a solvent) are transferred into a SPE column. This SPE column may contain the same or a different sorbent from the one blended with the sample. Empty columns containing only the frit also can be used in certain applications. By the use of solvents, the analytes can be eluted from the unique column material made from two layers (one in some cases). This allows a new degree of sample fractionation [97], [98], [99]. The selection of sorbent to be mixed with the sample depends on the nature of the material to be analyzed. Principles similar to those used for the selection in standard SPE are utilized in MSPD. The amount of sample is adjusted depending on the concentration of analytes and on the sensitivity of the analytical determination. Typical values are 0.1–0.5 g or mL material. The amount of solid phase added depends on the amount of sample, and an excess of 4–6 times higher than sample is typical. The dimensions of the particles may play a specific role in MSPD, having the double function of abrasive and adsorbing material. Particles that are small may have better adsorption, but they may generate problems with the rinsing and elution, leading to very low flow rates or plugging the SPE column. For example, particles with 3–20 μm d_p lead to extended solvent elution times and possible plugging. Particles with 40 μm d_p have been used extensively and successfully [100]. The transfer of the sample mixed with the solid phase into the SPE column must be quantitative, or a known aliquot must be used. Also, the sorbent in the SPE cartridge is selected according to the purpose of analysis and may serve as further cleaning material or for removing water or particulate material. The selection of the solvents used for rinsing and eluting the analytes also follow the same rules as in SPE.

- Examples of applications

A number of applications of MSPD are reported in literature, the procedure having good applications on samples that require cell disruption, animal tissue homogeneization, etc. [101]. Some examples of MSPD applications are given in Table 11.3.1.

TABLE 11.3.1. *Some applications of MSPD reported in literature.*

Sample	Analytes	Carrying sorbent	SPE sorbent	Elution	Recovery %	Ref.
salmon muscle	five sulfonamides[2]	C18 silica base	none	CH_2Cl_2	66–82	[102]
bovine tissues	aminoglycoside antibiotics	C18 silica base	none			[103]
bovine liver	ivermectin	C18 silica base	none			[104]
meat and milk	sulphonamides	C18 silica base	none			[105]
liver tissue	triglycerides, phospholipids	C18 silica base	none	CH_2Cl_2	98	[101]
pork muscle	five benzimidazole anthelmintics[2]	18% ODS silica (40 μm)	alumina			[106]
pork meat	seven sulfa drugs	C18 silica base	none	CH_2Cl_2	70–96	[107]
catfish (*Ictalurus punctatus*)	sulpha-dimethoxine[2]	C18 silica base	none	CH_2Cl_2		[108]
pork muscle	furazolidone[2]	C18 silica base	none			[109]
cow milk	chloramphenicol[2]	C18 silica base	none		61–79	[110]
milk	oxytetracycline, tetracycline, chlortetracycline[2]	C18 silica base, Na$_2$EDTA, and oxalic acid	none	$CH_3CO_2C_2H_5/$ CH_3CN (1:3)	63–93	[111]
beef liver	benzimidazole, antihelmintics[2]	C18 silica base	none followed by alumina	CH_3CN		[112]
catfish (*Ictalurus punctatus*)	oxytetracycline	C18 silica base	none	CH_3CN / CH_3OH (1:1)		[113]
grass carp	polychlorinated biphenyls	ODS silica (40 μm)	none	C_6H_{12}	89–93	[114]
oysters, catfish	14 chlorinated pesticides	C18 silica base	Florisil	CH_3CN / CH_3OH (9:1)		[115] [116]
beef fat	chlorinated pesticides	Florisil	none			[117]
liver	clenbuterol[1]	C18 silica base (40 μm)	filter paper			[118]
milk	clorsulon[1]	C18 silica base	none followed by Florisil	$C_2H_5OC_2H_5$	89–96	[119]

[1] For information and chemical structure of some pesticides , see e.g. D. Hartley, H. Kidd (eds.), *The Agrochemicals Handbook*, 2nd ed., Royal Society of Chemistry, University of Nottingham, England, 1987.
[2] For information and chemical structure regarding various drugs, see e.g. M. O'Neil et al. (eds.), *The Merck Index*, Merck & Co., Inc., 13th ed., Whitehouse Station, 2001.

11.4   OTHER SOLID PHASE SAMPLE PREPARATION TECHNIQUES

The use of solid phase sample preparation techniques offers a number of advantages that explain their common use. In addition to standard procedures (discussed in Section 11.1), other related techniques were developed to assist in sample preparation for analysis, such as SPME (see Section 11.2), matrix solid phase dispersion, etc. One such technique is stir bar sorptive extraction (SBSE), and another is open tubular trapping (OTT).

- Stir bar sorptive extraction

This technique uses the partition of the analytes into an extracting medium covering a small glass-coated magnetic stir bar. The dimension of the stir bar may vary, a typical one being 1 cm long and 3.2 mm in diameter. A common material immobilized on the surface of the stir bar is polydimethylsiloxane in a layer of 1 mm. The stir bar is typically conditioned before use by heating at temperatures around 300° C for a few hours. The stir bar is then introduced in the sample solution or slurry [120]. The polydimethyl-siloxane (PDMS) stationary phase extracts most organic compounds, and the common use of the stir bar is in aqueous media. The sample is stirred for a certain time (using a magnetic stirrer), and then the stir bar is removed from the sample, washed with a small volume of pure water, dried with filter paper, and introduced in a thermal desorption instrument. The desorption is typically performed at temperatures between 250° C and 300° C for 2–4 min. in a flow of an inert gas, usually He. The desorbing equipment is on-line connected to a GC or GC-MS. Due to relatively long desorption time, for obtaining narrow chromatographic peaks a cryofocusing step is necessary in this technique. Cryofocusing can be done using dedicated devices and low temperatures.

The adsorption on polydimethylsiloxane from a water solution can be estimated at equilibrium as similar to partition between water and octanol (see Section 1.10). Using the expression:

$$K_{ow} \approx K_{PDMSw} \qquad (11.4.1)$$

the concentration ratio of a specific analyte in the stir bar and in water can be estimated with the expression:

$$c_{SBSE} / c_w = K_{ow} \qquad (11.4.2)$$

The mass of the analyte in water and in the stir bar material can be evaluated starting with the amount q_0 of material in a given water solution of volume V_w and using the ratio $\beta = V_w/V_{SBSE}$ (where V_{SBSE} is the volume of the stationary phase on the stir bar). For this purpose, from rel. (11.4.2) it can be written for the equilibrium:

$$q_{SBSE} / q_w = q_{SBSE} / (q_0 - q_{SBSE}) = K_{ow} / (V_w/V_{SBSE}) = K_{ow} / \beta \qquad (11.4.3)$$

Rel. (11.4.3) can be written in the form:

$$q_{SBSE} = \frac{q_0 (K_{ow} / \beta)}{1 + (K_{ow} / \beta)} \qquad (11.4.4)$$

The variation of q_{SBSE} / q_0, which represents the recovery of an analyte from the aqueous solution as a function of K_{ow} / β, is shown in Figure 11.4.1 (recovery expressed as %). Figure 11.4.1 shows that large volumes of water are not favorable for extraction. In practice, the volume of PDMS coating is about 55 µL (for 1 cm long commercially available stir bar). With a 10 mL solution, values of $\beta \approx 180$ can be obtained. Practically, the compounds with $K_{ow} > 2 \cdot 10^3$ are completely adsorbed on the SBSE coating. Many organic compounds are covered in this range of values for K_{ow}, which is a significant advantage, for example, when the analysis of trace compounds in a very diluted water solution is intended and a complete recovery is important. This makes SBSE particularly useful in the analysis of various environmental aqueous samples. The analysis of compounds with a "dirty" matrix, such as plant extract, plasma, etc., may be less convenient using SBSE.

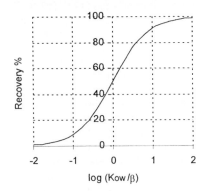

FIGURE 11.4.1. *Recovery in SBSE as a function of log (K_{ow} / β).*

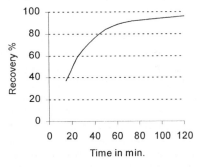

FIGURE 11.4.2. *Recovery in SBSE as a function of stirring time for various PAHs from 60 mL water with 5% methanol and 55 µL PDMS [120].*

One procedure that can be applied for partial removal of specific compounds adsorbed on SBSE coating that are not intended to be analyzed and may act as interference is the use of a solvent wash. Selective re-extractions can be achieved for the removal of partially polar compounds that are adsorbed in the SBSE coating, not due to their high K_{ow} but due to their high concentration in the sample.

The kinetics of SBSE partition process depends on the diffusion of the analyte in the PDMS coating. The recovery in SBSE in a 55 µL PDMS coating as a function of stirring time for various PAHs from 60 mL water with 5% methanol is shown in Figure 11.4.2. This figure shows that full equilibration in SBSE may require a relatively lengthy period of time.

This technique can be used for the enrichment of volatile and semivolatile compounds from aqueous samples, using stir bars coated with PDMS or any other gum used as stationary phase in capillary gas chromatography. Due to the possibility of increase in the quantity of PDMS coated on the stir bar compared to SPME, the quantification limit can be increased up to 500-fold for a time of stirring of 30–60 minutes. This concentration process allows determination at ppt level of concentration when GC-MS is used further for the analysis. Additionally, as drying of the stir bars is not required, volatile compounds can be conveniently handled. Environmental, biomedical, food and drinks, or other samples can be prepared by this technique. Aroma profiles from coffee and tea are two examples of the first applications of SBSE. Also, fatty matrices (milk, fresh cheese, and yogurt) have been processed by this technique. The method is adequate for various samples such as surface water samples for the analysis of benzene, toluene, ethylbenzene, xylenes, and dichlorobenzenes at ng/mL levels [18].

- *Open tubular trapping*

Open tubular traps (OTT) made from tubes with the interior coated with adsorbent particles or with reagent solutions are used, for example, for gas collection as denuders. This type of trapping was later applied using sorbent-coated capillaries [121], [122]. For this purpose, short capillary traps coated with materials similar to those used as stationary phase in GC (such as PDMS) are used and can be further desorbed directly in the injection system of a gas chromatograph (such as a programmed-temperature vaporizer or split-splitless injector). The traps made from a short piece of a capillary GC column were evaluated for direct coupling to a GC capillary system [123], [124], [125], [126]. Not only gas samples can be processed using OTT. A liquid sample can be passed through the capillary for the collection of the analytes. For example, aromatic hydrocarbon contaminants present in an aqueous sample can be trapped on a 2 mm x 0.32 mm i.d. OTT coated with a 5-μm thick stationary phase, the water being further removed by a slow flow of nitrogen. The desorption can be carried out with an organic solvent and transferred to the GC system via a PTV injector as interface [127]. The adsorption process in OTT must be conducted to ensure the complete adsorption. The breakthrough volume (V_b) of an analyte can be approximated by the following relation:

$$V_b = K_D V_S (1 - 0.9 \cdot \sqrt{\frac{U}{D_m L}})$$ (11.4.5)

where K_D is the distribution constant of the analyte between the sorbent and mobile phase, V_s is the volume of stationary phase, U is the flow-rate of the aqueous sample, D_m is the diffusion coefficient of the analyte in the solvent (water), and L is the length of the trap. According to rel. (11.4.5), the flow rate should not exceed a certain threshold to avoid analyte breakthrough. In practice the maximum value of U is about 0.2 mL/min. for 2 m x 0.32 mm i.d. OTT. The use of a thick-film stationary phase will increase the breakthrough volume, as seen in the expression of V_b given by rel. (11.4.5) [128]. However, the diffusion in the thick film is slower, and rel. (11.4.5) does not capture the kinetic aspect of the process. Also, the type of flow in the capillary influences the adsorption process. Coiled columns allow a higher sample flow rate (up to 4 mL/min.) because in deformed capillaries a turbulent flow enhances the radial dispersion of the analyte [129]. Other techniques using solid phase sorbents for the analysis of volatile compounds are discussed in Chapter 9.

REFERENCES 11

1. V. Pichon et al., Int. J. Environ. Anal. Chem., 65 (1996) 11.
2. L. K. Tan, A. J. Liem, Anal. Chem., 70 (1998) 231.
3. M. J. C. Rozemeijer et al., J. Chromatogr., A, 761 (1997) 219.
4. H. Carlsson, O. Ostman, J. Chromatogr., A, 790 (1997) 73.
5. M.-C. Hennion, J. Chromatogr., A, 856 (1999) 3.
6. M. L. Larrivee, C. F. Poole, Anal. Chem., 66 (1994) 139.
7. C. M. Josefson et al., Anal. Chem., 56 (1984) 764.
8. P. Lövkvist, J. A. Jönsson, Anal. Chem., 59 (1987) 818.
9. H. M. J. Mol et al., J. Chromatogr., 630 (1993) 201.
10. R. W. Frei, K. Zech, *Selective Sample Handling and Detection in High-performance Liquid Chromatography*, part A, Elsevier, Amsterdam, 1988.
11. P. Subra et al., J. Chromatogr., 456 (1988) 121.
12. K. G. Miller et al., J. High Resolut. Chromatogr., 176 (1994) 125.
13. M. L. Larrivee, C. F. Poole, Anal. Chem., 66 (1994) 139.
14. C. F. Poole et al., J. Chromatogr., B, 689 (1997) 245.
15. D. Seibert, C. F. Poole, Chromatographia, 41 (1995) 51.
16. M.-C. Hennion et al., J. Chromatogr., A, 823 (1998) 147.
17. E. M. Thurman et al, Anal. Chem., 50 (1978) 775.
18. H. A. Baltussen, *New Concepts in Sorption Based Preparation for Chromatography*, Cip-Data Library, Technische Universiteit Eindhoven, 2000.
19. N. J. K. Simpson (ed.), *Solid-Phase Extraction, Principles, Techniques and Applications,* M. Dekker, New York, 2000.
20. M.-C. Hennion, Trends Anal. Chem., 10 (1991) 317.
21. Varian, *Chromatography & Spectroscopy Supplies*, Walnut Creek, 2001 or www.varianinc.com.
22. *Phenomenex for Chromatography*, Torrance, 2001.
23. Q. Zha et al., J. Chromatogr. Sci., submitted for publication.
24. F. Ortiz Boyer et al., Chromatographia, 47 (1998) 367.
25. L. C. Sander et al., Crit. Rev. Anal. Chem., 18 (1987) 299.
26. J. Nawrocki, Chromatographia, 31 (1991) 177, 193.
27. J. G. Dorsey et al., Anal. Chem., 66, (1994) 857A.
28. H. Sasaki et al., J. Chromatogr., 400 (1987) 123.
29. K. Yamamoto et al., Meth. Mol. Biol., 14 (1993) 17.
30. G. T. Hermanson et al., *Immobilized Affinity Ligand Techniques*, Academic Press, New York, 1992.
31. B. Sellergren, Anal. Chem., 66 (1994) 1578.
32. G. Brambilla et al., J. Chromatogr., B, 759 (2001) 27.
33. V. T., Remcho, Z. J. Tan, Anal. Chem., 71 (1999) 248A.
34. I. Ferrer et al., Anal. Chem., 72 (2000) 3934.
35. K. Ensing et al., LC/GC, 19 (2001) 942.
36. C. T. F. Fleischer, K.-S. Boos, American Lab., 33 (2001) 20.
37. A. Bereczki et al., J. Chromatogr., 930 (2001) 31.
38. F. Abu-Shammala, Anal. Lett., 32 (1999) 3083.
39. K. S. Books, C. H. Grimm, TrAC, 18 (1999) 175.
40. T. Hankemeier et al., J. High Resolut. Chromatogr., 21 (1998) 450.
41. R. J. Majors, LC-GC Suppl., May 1998, S8.
42. D. D. Blevins, D. O. Hall, LC-GC Suppl., May 1998, S16.

43. I. Urbe, J. Ruana, J. Chromatogr., A, 778 (1998) 337.
44. D. F. Hagen et al., Anal. Chim. Acta, 236 (1990) 157.
45. V. Pichon et al., J. Chromatogr., A, 795 (1998) 83.
46. D. D. Blevins, D. O. Hall, LC-GC Int., Sept. (1998) 17.
47. C. Yu et al., Anal. Chem., 73 (2001) 5088.
48. T. Hankemeier et al., J. High Resolut. Chromatogr., 21 (1998) 341.
49. X. H. Chen et al., J. Anal. Toxicol., 16 (1992) 351.
50. D. Barcelo (ed.), *Sample Handling and Trace Analysis of Pollutants Techniques, Applications and Quality Assurance*, Elsevier, Amsterdam, 2000.
51. Q. Zha et al., 55 TSRC, Paper 24, Greensboro, 2001.
52. C. L. Arthur, J. Pawliszyn, Anal. Chem., 62 (1990) 2145.
53. J. Pawliszyn, *Solid Phase Microextraction, Theory and Practice*, Wiley-VCH, New York, 1997.
54. Z. Zhang, J. Pawliszyn, J. High Resolut. Chromatogr., 19 (1996) 155.
55. J. Chien, J. Pawliszyn, Anal. Chem., 67 (1995) 2530.
56. M. Giardina, S. V. Olesik, Anal. Chem., 73 (2001) 5841.
57. S. A. Scheppers-Wercinski (ed.), *Solid Phase Microextraction*, M. Dekker, New York, 1999.
58. L. Urruty, M. Montury, J. Chromatogr. Sci., 37 (1999) 277.
59. L. Tuduri et al., J. Microcol. Sep., 12 (2000) 550.
60. The Sadtler Capillary GC Standard Retention Index Library and Data Base, Sadtler Res. Lab., Philadelphia.
61. R. J. Bartelt, Anal. Chem., 69 (1997) 364.
62. P. A. Martos et al., Anal. Chem., 69 (1997) 402.
63. R. J. Bartelt, B. W. Zilkowski, Anal. Chem., 72 (2000) 3949.
64. H. S. Carslaw, J. C. Jaeger, *Conduction of Heat in Solids*, Clarion Press, Oxford, 1959.
65. D. Louch et al., Anal. Chem., 64 (1992) 1187.
66. Z. Zhang, J. Pawliszyn, Anal. Chem., 65 (1993) 1843.
67. T. Gorecki, J. Pawliszyn, Anal. Chem., 67 (1995) 3265.
68. A. Medvedovici et al., J. High Resolut. Chromatogr., 20 (1997) 619.
69. T. Gorecki et al., Analyst (1999) 124.
70. L. Black, D. Fine, Environ. Sci. Technol., 35 (2001) 3190.
71. C. L. Arthur et al., Anal. Chem., 64 (1992) 1960.
72. S. Motlagh, J. Pawliszyn, Anal. Chim. Acta, 284 (1993) 265.
73. Supelco, *SPME Application Guide*, Sigma-Aldrich Co., 1999.
74. K. D. Bucholz, J. Pawliszyn, Anal. Chem., 66 (1994) 160.
75. C .L. Arthur et al., J. High Resolut. Chromatogr., 15 (1992) 741.
76. D. W. Potter, J. Pawliszyn, Environ. Sci. Technol., 28 (1994) 298.
77. A. A. Boyd-Boland, J. Pawliszyn, J. Chromatogr. A, 723 (1995) 163.
78. T. K. Choudhury et al., Environ. Sci. Technol., 30 (1996) 3259.
79. S. Magdic, J. Pawliszyn, J. Chromatogr. A, 723 (1996) 111.
80. S. Magdic et al., J. Chromatogr. A, 736 (1996) 219.
81. R. Eisert, K. Levsen, Fresenius' J. Anal. Chem., 351 (1995) 555.
82. D. S. Debruin et al., Anal. Chem., 70 (1998) 1986.
83. N. Alexandrou et al., Anal. Chem., 64 (1992) 301.
84. A. J. Robbat et al., Anal. Chem., 64 (1992) 1477.
85. R. E. Clement, P. W. Yang, Anal. Chem., 73 (2001) 2761.
86. X. Yu, J. Pawliszyn, Anal. Chem., 72 (2000) 1788.

87. S. Aguerre et al., Analyst, 125 (2000) 263.
88. E. Millán, J. Pawliszyn, J. Chromatogr., A, 873 (2000) 63.
89. A. Beichert et al., Appl. Organomet. Chem., 14 (2000) 493.
90. M. Guidotti et al., J. High Resolut. Chromatogr., 22 (1999) 414.
91. T. De Smaele et al., Mikrochim. Acta, 130 (1999) 241.
92. M. Sandra Fragueiro et al., J. Anal. At. Spectrom., 15 (2000) 705.
93. J. Vercauteren et al., J. Anal. At. Spectrom., 15 (2000) 651.
94. S. C. Moldoveanu, J. Microcol. Sep., 13 (2001) 102.
95. J. Song et al., J. Agric. Food Chem., 46 (1998) 3721.
96. S. A. Barker et al., J. Chromatogr., A, 475 (1989) 353.
97. S. A. Barker, Chemtech., 23 (1993) 42.
98. S. A. Barker, R. Haley, Int. Lab., 46 (1992) 16.
99. S. A. Barker, A. R. Long, J. Liq. Chromatogr., 15 (1992) 2071.
100. S. A. Barker, Supplement to LC-GC, May 1998, 37.
101. S. A. Baker, J. Chromatogr., 629 (1993) 23.
102. G. J. Reimer, A. J. Suarez, J. Assoc. Off. Anal. Chem. Int., 75 (1992) 979.
103. L. G. McLaughlin, J. D. Henion, J. Chromatogr., 591 (1992) 195.
104. F. J . Schenck et al., J. Assoc. Off. Anal. Chem. Int., 75 (1992) 655.
105. L. S. G. Van-Poucke et al., J. Chromatogr. Sci., 29 (1991) 423.
106. A. R. Long et al., J. Food Compos. Anal., 3 (1990) 20.
107. A. R. Long et al., J. Agric. Food Chem., 38 (1990) 423.
108. A. R. Long et al., J. Assoc. Off. Anal. Chem., 73 (1990) 868.
109. A. R. Long et al., J. Assoc. Off. Anal. Chem., 74 (1991) 292.
110. A. R. Long et al., J. Agric. Food Chem., 38 (1990) 427.
111. A. R. Long et al., J. Assoc. Off. Anal. Chem., 73 (1990) 379.
112. A. R. Long et al., J. Assoc. Off. Anal. Chem., 73 (1990) 860.
113. A. R. Long et al., J. Assoc. Off. Anal. Chem., 73 (1990) 864.
114. Y. C. Ling et al., J. Chromatogr. A, 669 (1994) 119.
115. H. M. Lott, S. A. Barker, J. Assoc. Off. Anal. Chem. Int., 76 (1993) 67.
116. A. R. Long et al., J. Assoc. Off. Anal. Chem., 74 (1991) 667.
117. A. R. Long et al., J. Assoc. Off. Anal. Chem., 74 (1991) 493.
118. D. Boyd et al., Anal. Chim. Acta, 275 (1993) 221.
119. F. J. Schenck et al., J. Liq. Chromatogr., 14 (1991) 2827.
120. E. Baltussen et al., J. Microcol. Sep., 11 (1999) 737.
121. B. V. Burger, Z. M. Munro, J. Chromatogr., 402 (1987) 95.
122. K. Grob et al., J. High Resolut. Chromatogr., 13 (1990) 257.
123. K. Grob et al., J. High Resolut. Chromatogr., Chromatogr. Commun., 8 (1985) 726.
124. R. E. Kaiser, R. Rieder, J. Chromatogr., 477 (1989) 49.
125. G. Goretti et al., J. High Resolut. Chromatogr., 15 (1992) 51.
126. H. G. J. Mol et al., J. High Resolut. Chromatogr., 16 (1993) 413.
127. J. Stanieswski et al., J. Microcol. Sep., 4 (1992) 331.
128. H. G. J. Mol et al., J. Chromatogr., 630 (1993) 201.
129. H. G. J. Mol et al., J. Microcol. Sep., 7 (1995) 247.

CHAPTER 12

Chromatographic Procedures as Preliminary Separation

12.1 GAS CHROMATOGRAPHY AS A SAMPLE PREPARATION TECHNIQUE AND BIDIMENSIONAL GC

Chromatographic separations can be used successfully as sample preparations for further chromatographic analysis. The sample is typically separated in specific fractions, and the fraction(s) of interest are further submitted for the core chromatographic analysis. The utilization of chromatography for sample preparation more frequently involves various types of liquid chromatography, such as classical column chromatography or preparative HPLC. The sample load in these types of chromatography can be relatively high, and a larger amount of sample can be processed. On the other hand, the sample load in GC is usually low, and only small amounts of sample can be processed in a GC. This leads to very low amount of material in the fractions that can be collected in a GC separation and consequently to less frequent utilization of GC as a sample preparation technique. One additional problem is that GC can be applied only to compounds that have at least some volatility. For most samples that can be introduced in a GC instrument, the separation power of the GC is usually sufficient for the separation of the analytes, which are further detected, without the need of an additional preparative GC step. However, even the very high separation power of a GC (with columns having up to 300,000 theoretical plates) is not always sufficient for some separations. In such cases, a GC-GC system can be necessary. The problem of a collection of very small fraction after a preliminary GC separation is solved using dedicated instrumentation known as bidimensional GC. Bidimensional GC (or GC-GC) is usually viewed as a stand-alone core chromatographic technique, the first chromatographic stage not being considered a sample preparation step. However, a short description of the procedure is given in this section.

In bidimensional gas chromatography the sample is injected in a GC and sent to a first chromatographic column (Column 1) where the separation takes place. The system allows a small portion of the eluting material to be redirected to a second chromatographic column for additional separation [1]. This process can be accomplished using different instrumental setups. One such system, allowing bidimensional chromatographic separation with one GC oven, is shown in Figure 12.1.1.

In this bidimensional system the eluted compounds and the carrier gas from the first column are sent into a mid-point restrictor where they can take two different paths. The first possible path does not use the second column, and the eluted compounds are diverted through a restrictor loop to the detector (Detector 1). In this case, the valve to Detector 1 is open. At the same time enough make-up gas is sent to the mid-point restrictor in order to have flow in the second column (Column 2) and through Detector 2. Gas chromatographic columns must have a flow of inert gas when heated to avoid the degradation of the stationary phase. No bidimensional separation takes place when this path is activated.

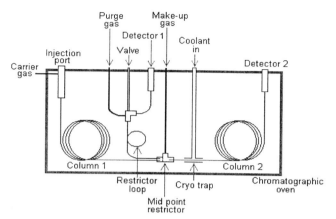

FIGURE 12.1.1. *Bidimensional chromatographic system with one GC oven.*

The second possible path uses the second column and leads to bidimensional separation. In this case, the valve to Detector 1 is closed, and the flow is forced to the second column. A so-called heart-cut is taken from the first separation by sending the eluted compounds to the second column. The make-up gas is stopped or reduced in this case. Because the peak broadening will be too large if the material from the first column goes directly to the second separation, a cryo-focusing step is needed. A small portion of the beginning of the second column passes through a cryo-trap, cooled at a low temperature (-60° C to -100° C, or even lower) with liquid CO_2 or N_2. The heart-cut materials are accumulated in a narrow zone as long as the cryo-trap is cold, and the heart-cut is taken. The cryo-trap is maintained cold as long as the system is not ready for the second separation.

In practice, a preliminary chromatographic run is done in single dimension mode, using no heart-cut, with a temperature oven program that is suitable for the separation in the first column. This run shows where the heart-cuts must be taken. After this run, one or more narrow regions of interest are selected from the single dimension chromatogram for taking heart-cuts. In a following run with a heart-cut, the cryo-trap is set cold a few minutes before the heart-cut is taken. Then, at the precise established run-time the valve to Detector 1 is closed and the mid-point restrictor is set for heart-cut mode. The analytes of interest are collected in the cryo-trap. After the heart-cut is taken, the oven temperature program may be continued for eluting the rest of material from Column 1, or the material left in the first column can be purged by reversing the flow in the first column (back-flushing). The GC oven is then reset for the new program needed for the separation in the second column. When the second separation starts, either the cryo-trap is heated or the cooling is stopped and the entire column takes the temperature of the GC oven, evaporating the heart-cut. The separation in the second column proceeds with the second temperature program for the oven.

It is common to use different stationary phases for Column 1 and Column 2. A variety of possibilities are available, but frequently one column is polar while the other is nonpolar. In this way, one separation is done mainly depending on the polarity of the analytes, while the other is based mainly on boiling point differences. Both polar/nonpolar and nonpolar/polar pairs have been used depending on the specific need for the separation. Other bidimensional configurations are described in the dedicated literature [1].

Bidimensional GC can lead to excellent separations, it being possible to select a very narrow heart-cut if necessary. As shown in Section 1.6, in isotherm conditions, the maximum number of peaks n_c that can be put side by side into an available separation space is given by rel. (1.6.69) and is proportional with the square root of the number of theoretical plates n. This shows that, for example, by doubling the length of a column, the value of n_c increases with the factor $\sqrt{2}$. In bidimensional GC, separating each peak in a heart-cut, the maximum number of peaks for the system is virtually n_c^2 [2], although in reality lower numbers are obtained [3].

As an example of increased peak capacity, a small portion from a chromatogram of a cellulose pyrolysate separated on a 30 m x 0.32 mm Carbowax column with 0.5 μm film thickness is shown in Figure 12.1.2. The pyrolysis of the sample was performed at 600° C, and the GC oven temperature was initially kept at 35° C for 2.5 min., then heated with 30° C/min. at 55° C, and further heated with 3° C/min. to 240° C. The detector for the chromatogram in Column 1 was a MS system. The peak identifications for the time interval shown in Figure 12.1.2 are given in Table 12.1.1.

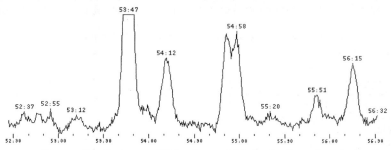

FIGURE 12.1.2. *The portion between 52.3 min. and 56.3 min. from the chromatogram of a cellulose pyrolysate.*

TABLE 12.1.1. *Identification of peaks shown in Figure 12.1.2.*

Retention time	Compound name	Formula	MW
53:47	hexanal	C6H12O	100
54:12	tetrahydro-5-methyl-furfuryl alcohol	C6H12O2	116
54:58	dihydro-4-methyl-2(3H)-furanone	C5H8O2	100
55:00	tetrahydro-2H-pyran-2-one ? or isomer of dihydro-4-methyl-2(3H)-furanone	C5H8O2	100
55.51	unidentified		
56:16	3-buten-1,2-diol or 3-hydroxytetrahydrofuran ?	C4H8O2	88
56:30	propylcarboxaldehyde	C8H10O2	138

A heart-cut of the interval 52.3 to 56.3 min. was taken in the GC bidimensional system, and the separation was performed on a 30 m x 0.32 mm DB5 column with 0.5 μm film thickness. The oven temperature for the second separation was initially kept at 45° C for 5 min. and then heated with a ramp of 3° C/min. to 240° C. The detector used for the second dimension was also a MS system. The chromatographic results are shown in Figure 12.1.3.

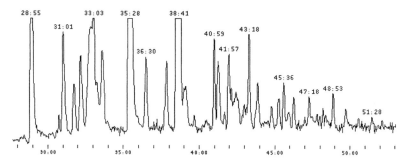

FIGURE 12.1.3. *The chromatogram of the heart-cut taken between 52.3 min. and 56.3 min. in the second dimension separation from the chromatogram of cellulose pyrolysate shown in Figure* 12.1.2. The peaks at 38:33 and 38:41 min., corresponding to the peaks at 54.12 and 54.58 in Figure 12.1.2, are only partially separated (not at the baseline) and appear as one peak due to magnification.

The comparison of Figures 12.1.2 and 12.1.3 shows clearly that a much better separation can be obtained using a bidimensional system for the heart-cut interval. This allows identification of a series of peaks, as indicated in Table 12.1.2. (Note: the peaks previously identified in the first dimension are shown in Table 12.1.2 with black borders.)

TABLE 12.1.2. *Identification of peaks shown in Figure 12.1.3, heart-cut of the run-time interval between 52.3 and 56.3 min. of the first chromatogram of a cellulose pyrolysate.*

Retention time	Compound name	Formula	MW
28:55	3-buten-1,2-diol or 3-hydroxytetrahydrofuran ?	C4H8O2	88
31:01	5-methyl-2(5H)-furanone	C5H6O2	98
31:40	2,3-dihydro-5,6-dimethyl-1,4-dioxin	C6H10O2	114
32:40	tetrahydro-5-methyl-furfuryl alcohol	C6H12O2	116
33:03	1,6-heptadien-ol	C7H12O	112
33:40	3-methylphenol	C7H8O	108
35:28	hexanal	C6H12O	100
36:30	1-(2-furany)-propanone	C7H8O2	124
37:50	cyclopropanecarboxylic acid ?	C4H6O2	86
38:33	dihydro-4-methyl-2(3H)-furanone	C5H8O2	100
38:41	tetrahydro-2H-pyran-2-one ? or isomer of dihydro-4-methyl-2(3H)-furanone	C5H8O2	100
39:08	2,3,5-trimethylphenol ?	C9H12O	138
39:38	2,5-dimethylphenol	C8H10O	122
40:59	4-methyl-4-penten-2-one ?	C6H10O	98
41:25	2,3,6-trimethylphenol ?	C9H12O	138
41:57	4-hydroxybenzaldehyde	C7H6O2	122
43:18	2-coumaranone	C8H6O2	134
43:50	propylcarboxaldehyde	C8H10O2	138
44:56	2,6-dimethylphenol ?	C8H10O	122
45:10	(1,1'-bicyclopentyl)-2-ol	C10H18O	154
45:36	a dihydroxytrimethylbenzene	C9H12O2	152
46:20	2-methyl-4-(1-methylethyl)-2-cyclohexen-1-one	C10H16O	152
47:18	6-methyl-3-(1-methylethyl)-2-cyclohexen-1-one	C10H16O	152
48:55	5-methyl-1(3H)-isobenzofuranone	C9H8O2	148
49:50	a methylisobenzofuranone isomer ?	C9H8O2	148
51:28	4-methyl-1-indanone	C10H10O	146

Besides the system described previously, a series of other instrumental setups are reported or manufactured, including double-oven bidimensional systems, systems allowing the monitoring of the separation in the first dimension even when the heart-cut is taken, etc.

The main problem with bidimensional GC is the fact that the process is time consuming, requiring two GC runs for each heart-cut. The time consuming factor becomes even more significant when a number of heart-cuts are required for a given sample. Bidimensional chromatography can be extremely useful for the analysis of complex samples, such as natural flavors and other volatile plant materials, pyrolysis products, etc. A bidimensional system can be used as a truly preparative instrument by replacing the second column with a piece of (uncoated, deactivated) capillary column immersed in a cooling agent such as liquid nitrogen or solid CO_2/acetone [4]. The first step in the bidimensional GC process is conducted as previously described, and a desired heart-cut is stored in the cryo-trap. After the collection and the elution or the elimination of the material left in the first column, instead of starting the second separation, the cryo-trap is heated and the collected material is sent into the piece of cooled capillary column. The piece of capillary column containing the collected material (at a low temperature) can be further moved to a different instrument such as a GC-MS, mounted similarly to a precolumn, and the collected material analyzed upon removing the cooling agent.

A development of bidimensional GC is comprehensive two-dimensional GC [5]. In this technique, the second chromatographic column allows the performance of fast chromatography. The heart-cuts from the first column are collected from the whole first separation at short intervals, and each heart-cut is separated using a fast chromatographic process in the second column. The collection from the first column and the release of the material in the second column is done using a modulator. A cryogenic modulator works very similarly to the cryo-trap described previously for a standard bidimensional system. The flows and the choice of the second column must be done such that the fast chromatography in the second column takes place with about the same duration as the collection of the next heart-cut. The first column typically works in a temperature gradient, while the second column practically works isothermally. The procedure has been applied for the analysis of complex samples. However, comprehensive chromatography is not viewed as a combination of a chromatographic sample preparation followed by a GC analysis, but as a unique chromatographic technique [6], [7], [8], [9].

The applications of bidimensional GC are numerous. They range from analyses of flavors or fragrances [10], [11], [12], [13] to complex environmental samples [14], plant extracts, other complex samples such as pyrolysates [15], etc. Many specific applications are presented in the dedicated literature (see e.g. [1]). Achievements using multidimensional GC are summarized in Table 12.1.3 [16].

TABLE 12.1.3. *Some benefits of the application of bidimensional GC.*

Application	Comments
Solvent removal	Large amounts of solvents and excess of derivatizing reagents, etc. can be excluded from the main separation column. This allows better resolution and analysis of compounds that elute close to the excess sample component.
Enrichment of trace components	This can be achieved by multiple injections into a precolumn with selective trapping and storage of the fraction of interest; often performed with packed precolumns because of their higher sample capacity and because maximum resolution is not required in the first separation step.
Separation of trace components buried under major peaks	It is possible to vent the major portion of large peaks from the system and transfer the trace analytes of interest to the second column.
Heart-cutting of several sections of the chromatogram	In a sample having a very wide boiling-point range, only the components of interest are transferred to second separation column. High-boiling components can be back-flushed at an elevated temperature.
Detailed analysis of single peaks or selected cuts in a chromatographic run	Single peaks or whole areas of a chromatogram can be reanalyzed in a second column to provide a better separation. With intermediate trapping and columns of different selectivity, two sets of retention index data can be obtained.

12.2   LIQUID CHROMATOGRAPHY AS A SAMPLE PREPARATION TECHNIQUE

Liquid chromatographic techniques represent excellent separation procedures that are used not only as core analytical techniques for separation and detection of analytes, but also in the sample preparation step. Various types of liquid chromatographic procedures are known, and some are discussed in Sections 4.3 and 4.4. Also, many detection techniques used in HPLC are not destructive, and the eluate can be collected and further analyzed. Therefore, in addition to its use as a core analytical technique, HPLC can be applied as a sample preparation step. Some other LC procedures also can be used in sample preparation. The use of LC in sample preparation can be differentiated by a) the procedural setup used for the separation, b) the mechanism of the separation process, c) the physical shape of the medium on which the separation takes place (column or planar chromatography), d) the type of chromatographic column (for column chromatography) distinguishing low pressure chromatography and high performance liquid chromatography (HPLC), e) the polarity of the stationary and mobile phase, etc. Each of these differences may lead to a different type of application that is appropriate for solving a particular analytical problem.

The procedural setup differentiates off-line and on-line techniques. Off-line LC sample preparation is usually independent of the core technique that follows and not necessarily related to a specific instrumentation. In on-line LC sample preparation, the core analytical procedure is directly connected with the LC system. Two general types of techniques are more common, LC-GC and LC-LC. Both LC-GC and LC-LC can be viewed as hyphenated chromatographic techniques and not as a combination sample preparation-chromatographic technique. Both on-line LC-GC and LC-LC are frequently considered bidimensional chromatographic techniques and are not disconnected in a first preparative procedure followed by a core procedure.

- *Off-line LC sample preparation*

A number of LC techniques are applied for sample preparation purposes. The classification of the technique based on the mechanism of the separation process distinguishes a) adsorption or distribution chromatography, b) size exclusion, c) ion exchange, etc. The selection of one mechanism is influenced by the purpose of separation. General purpose separations are usually done on typical adsorption materials such as C18, silica, alumina, etc. This type of separation is related to conventional LC or HPLC, as discussed in Section 4.3. The use of size exclusion is typically applied when the separation is done with the purpose of differentiating the sample components based on their molecular mass. This procedure has many applications, mainly in processing biological samples where the separation of small molecules from large molecules such as proteins or polysaccharides is important. The use of size exclusion as a core chromatographic technique is described in Section 4.4, and further discussion of this procedure for sample preparation is done in Section 12.3. Ion exchange is mainly used for the separation of ionic species. Further discussion of ion exchange utilization in sample preparation can be found in Section 15.1.

Based on the physical shape of the medium and the type of chromatographic column, the type of chromatography can be classified as a) thin layer chromatography, b) low pressure column chromatography, c) high performance chromatography (HPLC), d) counter current chromatography, and others.

Thin layer chromatography on supports such as alumina or silica has been utilized for a long time as a very useful technique in qualitative and semiquantitative analysis (see also Section 4.4). Among the main advantages of TLC are the simplicity of the procedure not requiring expensive instrumentation, the relatively short turnaround time, and the possibility to visualize the entire separation. Compounds that do not migrate on the TLC plate can be identified at the plate start. Various procedures using TLC are developed. The simplest is probably physical removal (scraping off) from the plate of stationary phase from the spot of interest containing the compounds. The material can be re-extracted in an appropriate solvent, filtered, and further analyzed. This simple procedure leads to very good results, but the loss of material is possible, and for quantitation it is necessary to have standards.

Among the disadvantages of TLC in sample preparation is the limited amount of sample that can be processed in this way. Thicker adsorbent layers can be prepared for particular purposes, but these are not always commercially available. Depending on the plate, 10–100 mg sample can be processed by TLC. This can be achieved, for example, by applying the sample in a band instead of a spot, on a wider TLC plate. Because the spots on a TLC separation are usually detected after treatment with a specific reagent such as iodine vapors, this modifies the analytes or may impede the analysis. For this reason, two plates may be processed similarly, one being treated to identify the position of the separated compounds, and the other for preparative purposes. More elaborate TLC separations are also possible such as the application of centrifugal TLC [17].

Column chromatography is another sample preparation procedure applied usually for cleaning samples from undesired matrix components. Column chromatography has

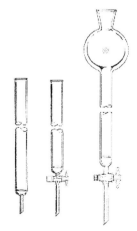

FIGURE 12.2.1. *Common glassware for columns.*

been utilized for many separations, some with preparative purposes not necessarily related to chemical analysis. The technique is applied using glass or plastic columns of various diameters (i.d. between 10 mm and 50 mm are common) and lengths (10 mm to 1 m are used). Some common shapes of columns having a glass frit at the bottom are shown in Figure 12.2.1. Typically a glass wool plug is put over the frit, and over this the stationary phase. A variety of stationary phases are available, including silica, alumina, Florisil, powdered cellulose, as well as modified silica with C18, C8, C4, phenyl, amino, chloropropyl, etc. Numerous organic resins are also available for column (low pressure) chromatography. These include polyaromatic resins such as Amberlite XAD series, various Dowex resins some polyaromatic, some polyacrylic resins, and ion exchange resins. The common particle diameters used in columns are 50–100 μm. The technique is similar to analytical chromatography described in Section 4.3 or solid phase extraction techniques described in Section 11.1.

The main difference from HPLC is the particle size of the stationary phase and the number of theoretical plates, which is several hundreds in column chromatography and around 10,000 for an HPLC analytical column. No pressure or only mild pressure and in some applications vacuum is applied in column chromatography for passing the solution through the stationary phase. The flow with no pressure is caused by gravity, and the flow rate is controlled using a stopcock. Therefore the instrumentation from HPLC is not necessary in this type of separation. The main difference from column chromatography and SPE is that in column chromatography the separation can be done for lower α separation factors. For this reason, elution similar to standard chromatography (see Section 1.6) based on the difference in the retention volume V_R between the components to be separated is common, while in SPE the separation is usually achieved using a step change in solvent. The separation in column chromatography can be done in isocratic conditions and also in gradient.

The selection of stationary phase and of the solvent in column chromatography is based on the same rules as in SPE. Pore size (usually 100 or 300 Å) and diameter distribution for the sorbents is also similar to SPE. The conditioning of the solid phase sorbent is also common and applied for the same reasons as in SPE. The sample application in column chromatography can be done using several procedures. One is simple flowing of the solvent containing the sample through the column. Sample addition may disturb the stationary phase at the top of the column, and a disk of filter paper or another filtering material may be added at the top of the stationary phase bed. A different addition procedure is the separate adsorption of the sample on a small portion of stationary phase, followed by the addition of the whole material on the top of the column. The disturbance of the solid phase when elution solvents are added, as well as channeling in the adsorbing bed, must be avoided during the separation.

Besides simple columns with manual addition of solvents, more automated and better controlled systems are also available. For example, the flow can be controlled using syringe pumps, different gradient delivery systems, and also detectors used for determining the correct time when the elution of compounds of interest takes place. In addition to automation in controlling the separation process, sample collection can be done using a variety of mechanical sample collectors. These can collect the sample based on different volumes, number of fractions in a unit of time, number of drops for a fraction, or they can be triggered by a detector in the flow of the eluent.

Column chromatography has many applications in sample preparation and some advantages compared to SPE and HPLC. The most common applications are related to sample cleanup. The larger sample capacity of column chromatography and the capability to separate compounds without the need for a step change in solvent make the technique very useful in situations where SPE is not efficient enough for the separation. Also, column separation is usually able to handle significantly larger samples compared to HPLC, and the price of the stationary phase is not usually prohibitive. These advantages explain the utility and applicability of column separations in sample preparation of samples such as essential oils, fats, plasma, etc. [18], [19], [20].

Analytical HPLC also can be used as a sample preparation technique. The separation capability of HPLC makes it very useful for specific separations that are difficult to achieve otherwise. The collection of the HPLC eluate followed by further sample preparation or a core analytical separation can be applied successfully for complex samples. Analytical HPLC has, however, a very limited sample capacity, and also the columns can be relatively easily damaged. For these reasons, the application of this technique in sample preparation is limited in spite of the common presence of HPLC instrumentation in laboratories. Also, samples may require extensive preliminary cleanup before using HPLC. The theory of HPLC is developed in Section 1.6, and a presentation of HPLC as an analytical technique is in Sections 4.3 and 4.4. If a sample must be processed using HPLC, but the limit in sample capacity in analytical HPLC is a significant problem, preparative HPLC must be applied.

Preparative HPLC uses dedicated instrumentation and large columns (e.g. 20 mm i.d, 300–500 mm length) with stationary phases similar to that used in analytical HPLC. The use of preparative HPLC is required in special applications when a large amount of sample and a high separation efficiency are necessary. Sample preparation for further analysis using preparative HPLC is only a minor application of this technique. Many stationary phases commonly used in analytical HPLC are also available for preparative columns. Preparative procedures applied for various preparative purposes use besides HPLC medium pressure chromatography, flash chromatography, etc.

Other types of chromatographic separations can be used as a sample preparation step. Among these are several types of counter-current chromatographic techniques such as droplet counter-current chromatography, rotation locular counter-current chromatography, etc. [20]. Droplet counter-current chromatography is a liquid-liquid separation technique based on the partition of a solute between two nonmiscible solvents. A system of tubes filled with one solvent are connected with small-diameter tubes, and the other solvent is pumped as small droplets through all these tubes. The passing droplet has an intimate contact with the stationary solvent, and the distribution process can approach equilibrium. The system can work in ascending mode or descending mode depending on the relative densities of the two solvents. The sample is introduced as solution in one of the solvents, and based on different partition properties the separation takes place. The process is similar to a continuous liquid-liquid extraction, but also similar to a chromatographic process with a truly liquid stationary phase [21].

The automation available in sample collection for various LC separation techniques allows the use of these techniques as convenient sample preparation procedures for further analysis. However, better automation is possible if direct connection between LC and the core analytical step is done. This is done using a number of on-line LC-hyphenated techniques.

- *On-line LC-GC*

The coupling of an HPLC system with a GC (LC-GC) can be necessary for detailed separation of the components in a fraction of the HPLC eluate [22]. Such separation is useful, for example, when the HPLC differentiates classes of compounds and the GC differentiates the compounds among themselves. This type of application is common for analysis of complex mixtures such as biological samples (plant extracts, plasma samples, etc.). Many compounds in biological systems are present in free form and as conjugates, for example with sugars. This is, for example, the case of various drugs and their metabolites or of agrochemicals and their plant metabolites. The separation of the free form and conjugates can be done with different procedures, a convenient one being HPLC. Further separation after the HPLC is frequently necessary, and this can be done for example with GC. Other applications include HPLC separations based on molecular size followed by detailed GC separation, separations based on polarity in the HPLC such as grouping gasoline and diesel fuel fractions in alkanes, olefins, and aromatics with further separation by GC [23], etc. It is not uncommon that before the HPLC operation the sample must be subjected to cleanup operations. The HPLC columns can be relatively easily damaged, and HPLC should not be used to eliminate large amounts of interferences.

Several problems must be solved for a successful connection between a HPLC and a GC system. The first is that the connection must be done such that the HPLC eluate is monitored and an appropriate liquid fraction is collected. The second is related to the capacity of the GC to accommodate the LC fraction. This can be related to the nature of the solvent, because water and salt buffers are not suitable for the GC injection. Besides the nature of the solvent, the solvent volume injected in the GC may be a problem. The flow rate in HPLC using conventional columns is usually 0.5–1.5 mL/min., and an attempt to cut a range of 2 to 3 min. from an HPLC run is equivalent with a 2–4 mL injection. This volume can be injected only with special injectors (such as PTV) or by special adjustments to the GC system to accommodate such a large injection volume and to eliminate most of the solvent.

The simplest connection between an LC and a GC system [24] is based on an automatic injector continuously fed by an HPLC system and connected to a waste reservoir. When the specific portion of the HPLC is eluting, the injector switches to GC injection. Another procedure for injecting a portion of the eluate in the GC is the use of a sampling valve that diverts the effluent into the GC system. The systems with a

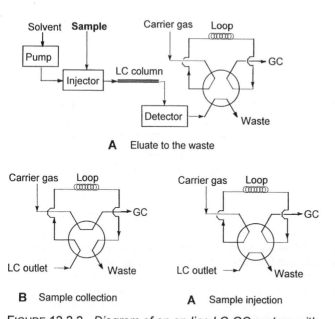

A Eluate to the waste

B Sample collection

A Sample injection

FIGURE 12.2.2. *Diagram of an on-line LC-GC system with the collection of the eluate into an external loop.*

sampling valve may be directly connected to a GC or may have a sampling loop. A system with a sampling loop is shown in Figure 12.2.2. It works in an A-B-A stage cycle, with the sample injected into the HPLC in stage A, and the elution monitored for example using an UV detector (with high pressure flow cell). When the region of interest in the HPLC elution is detected, the system switches to stage B, and the eluate is passed through the loop. The loop volume must be selected to collect the desired sample volume. This volume should also be adjusted based on the GC sample volume capacity. After the loop is filled, the system reverses to stage A, and the carrier gas sweeps the sample into the GC system.

Several other procedures have been devised to transfer the LC eluent into the GC system. For example, if the sample loop has a limited volume but the compounds of interest are eluted in a larger volume of solvent, it is possible to collect for a longer period of time but only a fraction of the eluent, using a flow splitter to load the loop and send the rest of the effluent to the waste. The collection of the eluate from the HPLC is relatively simple compared to the problem of handling a large volume of liquid in the GC.

The loading of the GC with a large volume sample is possible only if the GC is equipped with a large sample injection system (see Section 4.2). Several alternatives are possible for accommodating large samples, including PTV injectors, large retention gaps with solvent vent capabilities, or various other types of concentrators that retain the analyte and eliminate the solvent. Extensive literature is dedicated to the subject [25], [26], [27], [28], [29], [30], [31], [32], [33], [34]. A successful system uses the diagram shown in Figure 12.2.3.

Other instrumental arrangements are possible and used for particular applications [35], [36], [37]. An alternative to the (relatively) large volumes of solvent generated by conventional HPLC is the use of minibore or microbore HPLC. Flow rates around 100 µL/min used for minibore HPLC or as low as a few µL/min used with microbore columns are more easily accommodated in the GC system. Minibore HPLC may still deliver too much solvent for direct injection in a GC system, and special instrumentation using syringe pumps and microcells for the detection must be used with microbore liquid chromatography [38], [39]. However, further development of these techniques is a promising alternative.

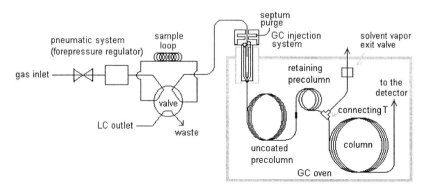

FIGURE 12.2.3. *Diagram of an on-line LC-GC system with a retention gap (uncoated precolumn), a separation precolumn with a solvent vapors exit valve, and an analytical column.*

- On-line LC-LC

Significant effort has been invested in attempts to improve sample preparation for HPLC [40]. Because sample preparation is typically a time-consuming operation, other directions were explored for allowing direct sample injection in an HPLC system. Among these is the use of different types of precolumns or guard columns. For example, for the analysis of biological samples with direct injection in an HPLC, precolumns of styrene-divinylbenzene [41] or butyl modified polymethacrylate [42] can be used. Also, the use of IRSP packing allows direct injection in HPLC of samples containing proteins (see Section 4.3). The packing material used in IRSP is designed to enable large biomolecules to be eluted quickly at or near the void volume of the column and to retain small molecules such as drugs or their metabolites beyond the void volume [43], [44], [45], [46], [47]. However, these practices that protect the analytical column by performing preliminary cleaning or the use of special columns that avoid sample preparation are not always successful. The cleanup process is sometimes carried out using two columns in series but without the capability to select fractions from the first column to be further analyzed or to change the solvents in the two columns.

The use of LC-LC (HPLC-HPLC) systems is a useful alternative in the analysis of complex samples where the separation is not sufficient in one chromatographic step. The first LC process can be identified as a sample cleanup step, and the second separation as a core chromatographic technique. However, LC-LC is in fact a stand-alone chromatographic method, similar to bidimensional GC. Only a short overview on the technique is presented here. The LC-LC is usually applied for the analysis of a sample containing a number of classes of compounds that must be separated, followed by the separation of the components in a specific class. In a bidimensional HPLC the first column also may be used only for cleanup purposes. The selection of columns in LC-LC can be done such that the separation mechanism is very different. Examples are the connection of a GPC column that differentiates the compounds by the molecular mass (volume) and a typical reverse phase column for detailed separation of a specific fraction from the first column eluate. An example of a LC-LC connection is shown in Figure 12.2.4.

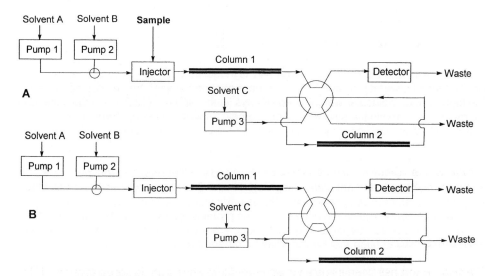

FIGURE 12.2.3. *Diagram of an on-line LC-LC system with two-column switching system.*

The system in Figure 12.2.3 allows column switching from Column 1 to Column 1 + Column 2 and has only one detector. Multiple other possibilities can be imagined using more columns, one detector for monitoring the first column and another detector for the second column, a number of valves and collecting loops for the bidimensional analysis of more than one fraction, etc. [1].

12.3 SIZE EXCLUSION USED IN SAMPLE PREPARATION

Size exclusion (SEC) is commonly named gel permeation chromatography (GPC) when applied for compounds soluble in organic solvents, and gel filtration chromatography (GFC) when applied to water soluble compounds. The principle of separation, briefly described in Section 4.4, is based on the fact that flowing in a column that contains a porous medium, the small molecules can enter freely the pores and are retained longer, while the larger molecules cannot enter the pores and are flushed by the eluent [48]. The gels with SEC separation mechanism are used in HPLC columns, in SPE cartridges or columns, and in (low pressure) column chromatography. The separation using columns with size exclusion gels are based on the variation of the retention volume V_R, which for a specific range of molecular weights (MW) follows relation (4.4.2):

$$V_R = A + B \log (MW) \tag{12.3.1}$$

The range of MW values where the parameters A and B are true constants depends on the nature of the stationary phase and in some instances on the chemical structure of the analyte, as shown in Section 4.4. The dependence of the retention volume on the molecular weight is the basis for the use of SEC in the determination of the molecular weight of the components of a sample using a specific calibration [49]. The calibration must be done with a series of polymers of known MW and similar molecular shape (such as globular). The selection of specific pore size is related to the range of molecular weights where the gel shows linearity to log MW. Larger pores are used for linearity at a higher MW range. However, the separation process in SEC is mainly based on size, and the dependence on MW is in fact indirect, due to the linear dependence between size and MW. The dependence between the molecular size and the logarithm of molecular weight is not always linear, and rel. (12.3.1) is applicable only for molecules with similar structure. For example, globular proteins, branched polysaccharides, and linear synthetic polymers have different behavior.

The application of size exclusion in sample preparation allows separation of large molecules from small molecules. This can be used in SPE, and the molecules that show no retention in the gel can be separated in the first eluate, while other molecules may show some degree of retention. This is particularly useful in analysis of biological samples that contain large molecules, such as proteins, polysaccharides, etc., together with small molecules. Separation of metal ions are also performed on various gels used in SEC, but the retention mechanism may be different from size exclusion [50].

The gels used in SEC must have certain properties, which can be summarized as follows: a) the matrix of the gel must be inert towards the components from sample, b) the interactions with gel must not give rise to any irreversible processes, c) the gel must be stable both mechanically and chemically, such that its separation properties are constant for a long time in normal conditions, d) the gel must have high stability in a wide pH range, usually between 2 and 10, e) organic solvents must not change the gel separation properties, f) the range of variation for the values of particle size of the gel must be limited, and g) the gel must be relatively stable in a range of temperatures. The inertness of the stationary phase regarding sorptive properties is, for example, important in order to avoid modifications in the retention time of certain analytes that are retained by a different mechanism from size exclusion. The inertness is difficult to obtain. For example, the separation of amino acids, peptides, and proteins of low molecular weight on SEC material Sephadex showed that there is a strong dependence between the retention and the composition of the eluent [51], indicating other interactions than simple size exclusion. Various types of van der Waals forces or even hydrogen bonding cannot be avoided between the gel structures and the solute molecules.

A number of materials can be used as stationary phase in size exclusion chromatography. From the composition point of view, three main types of gel filtration materials are available, including silica base materials, synthetic polymers, and natural or chemically modified natural polymers. These materials can be divided into rigid, semi-rigid, and soft gels. Rigid gels are usually based on silica, semi-rigid gels are based on polystyrene cross-linked with divinylbenzene, and soft gels are based on cross-linked dextrans.

Silica-based material contains a silica modified sorbent, which can be used in both aqueous and organic solvents. The coating of the silica backbone can be done with a hydrophilic material for allowing its use in aqueous solutions. Various silica types gels are commercially available.

Among the synthetic polymers, polystyrene-divinylbenzene (SDVB) is a common material. It can be obtained by polymerization in emulsion under conditions that allow the control of the pore size, pore volume, particle size, and degree of cross-linking. The pore size of such material must be selected tightly distributed around a specific value that can be chosen between 50 Å to 10^6 Å. Besides simple gel filtration capabilities, SDVB gels also show adsorptive and partition effects [52], [53]. Polystyrene with 2% divinylbenzene has been applied to the separation of lipids according to their molecular weight and is one of the best materials in separating mixtures of oligophenylenes [54], [55]. Besides SDVB, which is mainly appropriate for the use in organic solvents, sulfonated polystyrene gels are used for SEC in aqueous media.

A number of other gels are available for use in aqueous media. These types of gels must have hydrophilic properties. Among the most common gels in this group are the dextrans cross-linked with epichlorhydrin (Sephadex) [56]. The dextran gels are polymeric carbohydrates consisting of glucose units with more than 90% α-1,6-glycosidic linkages. Dextran is produced during the growth of *Leuconostoc mesenteroides* on sucrose. Dextran contains three hydroxyl groups per glucose unit and is therefore water soluble. The reaction between an alkaline dextran solution and epichlorhydrin leads to a cross-linked gel as shown below:

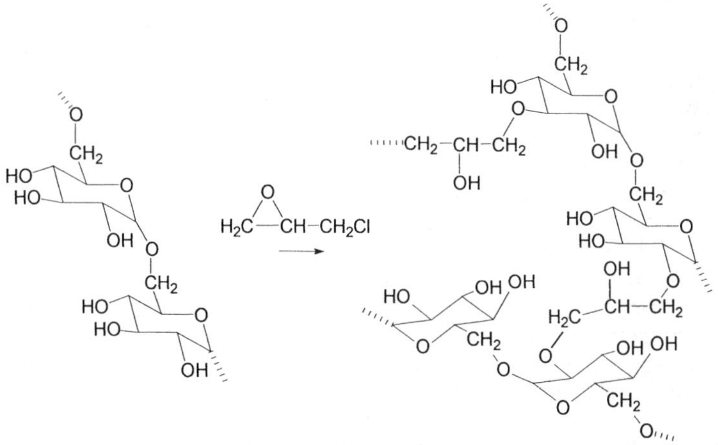

These gels are commercially available as Sephadex and produced as beads of defined sizes. Sephadex G-25 is one of the most common materials used in gel filtration. This gel allows the passing without retention of substances with molecular weights above 50,000. For macromolecules with higher molecular weights, separation is usually performed on Sephadex G-50. The growth of microorganisms that can occur in these gels can be prevented by the addition of 0.02% of sodium azide to the eluent or by saturation with chloroform.

Another common hydrophilic gel is produced by the polymerization of acrylamide in an aqueous solution in the presence of a bifunctional acrylamide, such as N,N'-methylene-bis-acrylamide ($CH_2=CH-CO-NH-CH_2-NH-OC-CH=CH_2$), which produces a water insoluble gel. The presence of the bifunctional reagent results in the formation of cross-linkages between two different chains of the polymer. On the other hand, the presence of amide groups produces polarity and consequently the capability of swelling in water. Gels of allyl dextran copolymer with N,N'-methylenebisacrylamide (Sephacryl) are also used as separation media.

Agar gels are also used as GFC media. They are polysaccharides that are extracted from red seaweed. It was found that agar consists of two main components: agarose, which is a neutral component, and agaropectin, which contains carboxyl and sulfate functional groups [57]. Granulated agar is another GFC. Crosslinked agarose with epichlorhydrin generates the gel commercially known as Sepharose. Other materials used in GFC having hydrophilic character include polyvinyl alcohol gels, cross-linked hyaluronic acid, polyvinyl ethyl carbitol, polyvinyl pyrrolidone, etc. Hydrophobic Sepharose media are also available, obtained by the substitution of the active hydrogens with phenyl, butyl, or octyl groups.

The particles of a gel are suspended in a solvent for use in size exclusion. The total volume of the gel bed V_t consists of three components: a) the solvent volume between the gel grains V_0, b) the solvent volume inside the gel particles V_i, and c) the volume of the gel matrix V_m. The information about these volumes is important for gel characterization and selection. The value of V_t can be measured experimentally, V_0 can be estimated by elution of a solution containing a colored compound with a high molecular weight, and the solvent volume leaving the column between the initial moment and that of elution of the high molecular compound corresponds to the value of V_0. The volumes V_i and V_m are more difficult to estimate. The value of V_i is related to the porosity of the gel and can be obtained using the formula:

$$V_i = \frac{S_r m_m}{\rho_{solvent}} - V_0 \tag{12.3.2}$$

where S_r represents the solvent regain expressed in grams of solvent absorbed by 1 g of dry gel (xerogel) during swelling [58], m_m is the mass of the dry gel, and $\rho_{solvent}$ is solvent density. Several values of the parameters characterizing the gel are given for Sephadex in Table 12.3.1.

TABLE 12.3.1. *Approximate values of various volumes for 1 g of Sephadex after swelling in water.*

Sephadex type	Gel bed volume (V_t, mL)	Outer volume (V_0, mL)	Inner volume (V_i, mL)	Density (swollen, g/mL)
G-10	2	0.8	1	1.24
G-15	3	1.1	1.5	1.19
G-25	5	2	2.5	1.13
G-50	10	4	5	1.07
G-75	13	5	7	1.05
G-100	17	6	10	1.04
G-150	24	8	15	1.03
G-200	30	9	20	1.02

Solid phase materials used for SEC columns must be packed in a controlled manner such that the volume of the solid material, the pore volume, and the interstitial volume must be kept constant. In the use of HPLC separations with size exclusion as a separation mechanism, the backpressure must be carefully monitored not to exceed a specific value (e.g. 1000 psi) for avoiding the modification of gel structure (collapsing the gel). The column in size exclusion must be stored filled with a specific solvent. For SDVB columns, the use of tetrahydrofuran (THF) as a storing solvent is common. Changing solvents for nonaqueous GPC columns must be done only using a sequence of miscible solvents. It must be noted that the degree of swelling of the gel depends on the nature of the solvent. For SDVB gels various swelling characteristics are shown in Table 12.3.2.

TABLE 12.3.2. *Swelling characteristics in a typical SDVB gel.*

70%	60%	50%	30%
toluene	diethyl ether	acetone	acetonitrile
THF	methylene chloride	DMF	cyclohexane
benzene	methyl ethyl ketone	DMSO	hexane
cyclopentane	ethyl acetate	dioxane	isopropanol
pyridine			methanol
o-dichlorobenzene			

The change in the gel volume depending on the solvent indicates that the volume of interstitial solvent and that in the pores of the gel are modified. However, the volume of the bed can change with formation of a void volume, which affects the separation.

- Examples of applications

Numerous applications for size exclusion are known in analytical practice. Only very few examples of such applications are listed in Table 12.3.3.

TABLE 12.3.3. *Several applications of gel filtration.*

Sample	Analytes	Type of gel	Reference
urine	pyridinoline, deoxypyridinoline	Biogel P2	[59]
pharmaceuticals	cephalosporins	Sephadex gel G-10 or G-50	[60]
Listeria monocytogenes, Staphylococcus aureus	listeriolysin O, α-toxin	Butyl-Sepharose, Superose 12HR10/30	[61]
viral lysate	dimeric 16664-base RNA	Spherogel TSK 6000PW	[62]
peripheral blood mononuclear cells	cyclopentyl cytosine 5'-triphosphate	TSK-G2000-SW	[63]
serum and urine	dextran	Sephacryl S-300 SF	[64]
serum	δ-bilirubin		[65]
proteins	tryptophan	Sephadex G10	[66]
mixtures	oligogalacturonic acids	Silica gel Diol 120	[67]
mixtures	glutaraldehyde, aldol condensation products	Bio-Gel SEC-10	[68]
hen egg-white lysozyme	peptides	Asahipak GS320	[69]
coffee	caffeine	TSK-G 3000-SW	[70]
waste water, soils	dimethoate	Bio-Beads SX-3	[71]

REFERENCES 12

1. H. J. Cortes (ed.), *Multidimensional Chromatography*, M. Dekker, New York, 1990.
2. D. H. Freeman, Anal. Chem., 53 (1981) 2.
3. J. M. Davis, J. C. Giddings, Anal. Chem., 57 (1985) 2168.
4. J. R. da Silva, personal communication.
5. J. B. Phillips, J. Beens, J. Chromatogr., A, 856 (1999) 331.
6. R. M. Kinghorn, P. J. Marriott, J. High Resolut. Chromatogr., 21 (1998) 620.
7. R. B. Gaines et al., J. Microcol. Sep., 10 (1998) 597.
8. J. B. Phillips, C. J. Venkatramani, J. Microcol. Sep., 5 (1993) 511.
9. J. Beens et al., J. Chromatogr., A, 882 (1998) 233.
10. A. Bernreuther, P. Scheier, Phytochem. Anal., 2 (1991) 167.
11. P. Kreis, A. Mosandl, Flav. Frag. J., 7 (1992) 199.
12. D. Haring et al., J. High Resolut. Chromatogr., 20 (1997) 351.
13. A. Bartsch et al., Perf. Flav., 18 (1993) 41.
14. E. Benicka et al., J. High Resolut. Chromatogr., 19 (1996) 95.
15. S. C. Moldoveanu, *Analytical Pyrolysis of Natural Organic Polymers*, Elsevier, Amsterdam, 1998.
16. C. F. Poole, S. K. Poole, Anal. Chim. Acta, 216 (1989) 109.
17. Z. Deyl et al., Chromatogr. Rev., 6 (1964) 19.
18. A. B. Alvarado, W. H. Gerwick, J. Nat. Prod., 48 (1985) 132.
19. K.-H. Kubeczka, Chromatographia, 6 (19730 106.
20. K. Hostettmann et al., *Preparative Chromatography Techniques, Applications in Natural Product Isolation*, Springer, Berlin, 1986.
21. K. Hostettmann et al., J. Liq. Chromatogr, 7 (1984) 231.
22. K. Grob, *On-Line Coupling LC-GC*, Hüthig, Heidelberg, 1990.
23. K. J. A. Apffel, H. McNair, J. Chromatogr., 279 (1983) 139.
24. R. E. Major, J. Chromatogr. Sci., 10 (1980) 571.
25. K. Grob, B. Schilling, J. Chromatogr., 391 (1987) 3.
26. A. Trisciani, F. Munari, J. High Resolut. Chromatogr., 17 (1994) 452.
27. F. Modeste et al., J. High Resolut. Chromatogr., 19 (1996) 535.
28. S. Moret et al., J. High Resolut. Chromatogr., 19 (1996) 434.
29. G. A. Jongenotter et al., J. High Resolut. Chromatogr., 22 (1999) 17.
30. E. Grimvall et al., J. High Resolut. Chromatogr., 18 (1995) 685.
31. L. Mondello et al., J. High Resolut. Chromatogr., 17 (1994) 312.
32. I. L. Davies et al., Anal. Chem., 60 (1988) 683A.
33. K. Grob, B. Schilling, J. High Resolut. Chromatogr., Chromatogr. Commun., 8 (1985) 726.
34. T. H. M. Noij, *Trace Analysis by Capillary Gas Chromatography*, Ph.D. Thesis, Eindhoven, 1988.
35. J. P. C. Vissers, J. Chromatogr., A, 856 (1999) 117.
36. A. Lamiri et al., Chem. Anal. (Warsaw), 44 (1999) 977.
37. U. Boderius et al., J. High Resolut. Chromatogr., 18 (1995) 573.
38. F. J. Yang (ed.), *Microbore Column Chromatography: A Unified Approach in Chromatography*, M. Dekker, New York, 1988.
39. M. Krejci, *Trace Analysis with Microcolumn Liquid Chromatography*, M. Dekker, New York, 1992.
40. D. Westerlund, Chromatographia, 24 (1987) 155.
41. G. Tamai et al., Chromatographia, 21 (1986) 519.

42. R. A. Hux et al., Anal. Chem., 54 (1982) 113.
43. D. L. Gisch et al., J. Chromatogr., 433 (1988) 264.
44. L. L. Gluntz et al., J. Liq. Chromatogr., 15 (1992) 1361.
45. K. Kimata et al., J. Chromatogr., 515 (1990) 73.
46. I. H. Hagestam, T. C. Pinkerton, Anal. Chem., 57 (1985) 1757.
47. I. H. Hagestam, T. C. Pinkerton, J. Chromatogr., 351 (1986) 239.
48. J. Porath, P. Flodin, Nature, 183 (1959) 1657.
49. H. Determann, *Gel Chromatography*, Springer, Berlin, 1968.
50. M. Lederer, *Chromatography for Inorganic Chemistry*, J. Wiley, Chichester, 1994.
51. J. Porath, Biochem. Biophys. Acta, 39 (1960) 193.
52. M. F. Vaughan, Nature, 188 (1960) 55.
53. B. Cortis-Jones, Nature, 191 (1961) 272.
54. C. L. Tipton et al., J. Chromatogr., 14 (1964) 486.
55. W. Hertz et al., Makromol. Chem., 98 (1966) 42.
56. J. Leggett Bailey, *Techniques in Protein Chemistry*, Elsevier, Amsterdam, 1967.
57. C. Araki, Bull. Chem. Soc. Japan, 29 (1956) 543.
58. K. W. Pepper et al., J. Chem. Soc., (1952) 3129.
59. P. Arbault et al., J. Liq. Chromatogr., 17 (1994) 1981.
60. C. Q. Hu et al., J. Pharm. Biomed. Anal., 12 (1994) 533.
61. B. Schoel et al., J. Chromatogr. A, 667 (1994) 131.
62. J. Pager, Anal. Biochem., 215 (1993) 231.
63. R. Agbaria et al., Anal. Biochem., 213 (1993) 90.
64. L. Hagel et al., J. Chromatogr., 641 (1993) 63.
65. C. Franzini et al., Ann. Clin. Biochem., 29 (1992) 116.
66. S. Delhaye, J. Landry, Analyst (London), 117 (1992) 1875.
67. J. Naohara, M. Manabe, J. Chromatogr., 603 (1992) 139.
68. L. Holmquist, M. Lewin, J. Biochem. Biophys. Methods, 22 (1991) 321.
69. T. Araki et al., J. Chromatogr., 545 (1991) 183.
70. L. C. Trugo et al., Food Chem., 42 (1991) 81.
71. J. Kjoelholt, Int. J. Environ. Anal. Chem., 20 (1985) 161.

CHAPTER 13

Membrane Separations

13.1 GAS DIFFUSION THROUGH MEMBRANES AND STRIPPING

Several separation processes are based on the differences in the rate of penetrating a specific barrier (see Section 7.2). The compounds to be separated are driven through the barrier by forces such as mechanical pressure, chemical potential, electrical field, etc. Gas diffusion through membranes is a separation technique that belongs to this type of separation. The transport through a membrane can be estimated using Fick's law (see Section 1.5), which in integrated form can be written as follows:

$$J_i = -D_i \frac{\Delta c_i}{\Delta x} \qquad (13.1.1)$$

where J_i is the flux of mass of compound "i" passing in a unit of time through a unit of surface area, D_i is the diffusion coefficient of the medium for compound "i", and c_i is the concentration of the diffusing compound. Assuming for gases that Henry's law (see rel. 1.4.21) is applicable, rel. (13.1.1) can be written in the form:

$$J = -D_i \, k' \, \Delta p_i \, / \, L \qquad (13.1.2)$$

where k' is a constant, L is the barrier (membrane) thickness, and Δp_i the difference in the partial pressure of compound "i" on the two faces of the membrane. A separation between two gases "i" and "j" using a membrane assumes that the membrane shows different diffusion coefficients D_i and D_j for the two gases [1]. The values for the diffusion coefficient of different gases are usually very close to each other, and a separation based on these differences is not efficient. However, a procedure that allows the transfer of volatile compounds from a gas or a solution on one side of a membrane into a carrier gas on the other side of the membrane followed by additional separation has been used in many analytical applications [2]. The diffusion can take place in a static solution, a stirred solution, in the headspace of a solution, or in dynamic mode. The most efficient procedure consists of an extraction of the analytes from the flowing sample into the membrane material on one side and the distribution of the analytes from the membrane into a stripping gas flowing on the other side. A number of membrane materials are available allowing the dissolution for nonpolar compounds. These include membranes made from microporous polypropylene, nonporous silicone rubber, or from composite materials such as polyetherimide(polyester)-silicone. The rate of dissolution in the membrane is controlled by the diffusion coefficient of each specific compound, and large molecules do not permeate the membrane. The procedure is usually applied to nonpolar volatile compounds. The theory of permeation through the membrane is reported in literature [3], [4].

One application of dynamic transport through membranes uses direct mass spectrometric detection of the gases eluted from the membrane and is known as membrane introduction mass spectrometry (MIMS) [5], [6], [7], [8], [9], [10]. The MIMS technique has the significant advantage of being able to provide real time information and therefore to be used for on-line monitoring of specific analytes in a flow. However, for many analytes the MS detection alone does not allow the discrimination. This is the case of the compounds that do not have enough differences in the mass spectra for individual quantitation based on specific ions. For this reason, a chromatographic step following the membrane separation leads to better results and has been used in a technique known as MESI (membrane extraction with a sorbent interface) [11], [12], [13], [14]. This type of method initially used a membrane sheet, but the newer applications have been developed to use a hollow fiber (tubular membrane). The hollow fibers have the shape of a tube with an internal diameter of 100–400 µm and a wall of 100–150 µm thickness. Single fibers or bundles of fibers can be utilized. A schematic diagram of a MESI setup is shown in Figure 13.1.1.

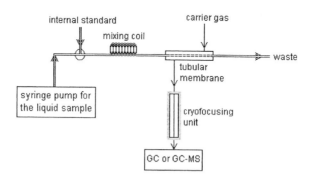

FIGURE 13.1.1. *Schematic diagram of MESI.*

The diffusion process through the membrane is in general slow, and the analytes diffusing through the membrane may have a low concentration. For this reason, either an adsorption step or a cryofocusing unit is necessary after the membrane. The cryofocusing unit has the role of collecting the analytes by condensation at low temperature for a specific period of time that can extend to several minutes. After the collection, a rapid heating of the trap can release the analytes into a chromatographic instrument. A similar role is played by an absorbing material (such as Tenax or activated charcoal), which after collection is desorbed by rapid heating.

Various designs of membrane modules can be used for samples of air, water, or soil. Also, the sample may flow through the tubular membrane or over the membrane. The membrane extraction procedures are limited to the volatile analytes, but some applications to the separation of semivolatile compounds are also reported [15], [16]. The membrane is in direct contact with a sample or its headspace. The use of a hydrophobic membrane prevents moisture from entering the carrier gas. A disadvantage of MESI is the slow response of the membrane to changes in concentration. This results in significant system carry-over, which prevents application of this method to sequential sample analysis. Moreover, the membrane extraction cannot separate polar compounds from a sample, because polar hollow-fiber membranes are not yet commercially available. Variations from the initial method were developed in attempts to correct the method deficiencies. For example, because the water layer on one side of the membrane has a low rate of mass transfer, a technique has been developed to use a gas injection of an aqueous sample for the analysis of trace organic compounds [17]. In this technique, gas injection membrane extraction (GIME), a pulse of sample introduced into a N_2 stream is passed through the membrane, eliminating the diffusion through the water and diminishing the carry-over effects. Also, in order to have the capability for analysis of samples in real time as they flow through the membrane tube, short cycle of sample collection and release from the trap were applied, using fast chromatography for the analysis of each pulse [18].

Typical applications of membrane extraction and gas stripping are reported for the analysis of volatile organic compounds (VOCs) in water [14] such as benzene, toluene, ethylbenzene, xylenes, chlorobenzene, dichlorobenzenes, tetrachloroethane, trihalomethane, etc. However, indirect analysis of larger molecules (such as cyanogenic glycosides) can be done with this type of technique by the measurement of small molecules generated for example during hydrolysis [10].

13.2 REVERSE OSMOSIS AND DIALYSIS

Semipermeable membranes can be used for the separation of small molecules from large molecules based on reverse osmosis or dialysis. A liquid-liquid extraction with the solvent included in a microporous membrane similar to dialysis, however, is used for the separation of small molecules. These techniques can be part of a sample preparation process with applications in chromatography. However, both reverse osmosis and dialysis have numerous other practical applications.

- *Reverse osmosis*

Reverse osmosis consists of application of a pressure on a solution separated by a membrane from a pure solvent, such that the solutes that can penetrate the membrane are sent into the solvent, forming a dilute solution. The spontaneous process takes place with the solvent flowing from the pure solvent side into the more concentrated solution side. The explanation for this process can be given by evaluation of the difference between the value of the chemical potential μ_w of a solvent (such as water) in solution and that of the pure solvent μ_w°. The mole fraction of water in any solution is lower than 1. The chemical potential of water in the solution is given by rel. (1.2.16) and has the expression:

$$\mu_w = \mu_w^\circ + RT \ln x_w \qquad (13.2.1)$$

Because $x_w < 1$, rel. (13.2.1.) indicates that $\mu_w < \mu_w^\circ$. Therefore, in a system separated by a membrane permeable for water, with pure water on one side and solution on the other side, the pure water will flow to the solution side. In this way the chemical potential will decrease (spontaneous processes take place with a negative variation of free enthalpy ΔG, which is defined for a system at constant temperature and pressure by rel. 1.8.5).

The chemical potential difference between the pure water and the solution separated by a membrane is manifested as a specific pressure known as osmotic pressure, created by the pure water penetrating the membrane. With addition of a pressure term to the expression (13.2.1) of the chemical potential, this becomes:

$$\mu_w = \mu_w^{\,o} + RT \ln x_w + \int_{p_o}^{p} \frac{\partial \mu_w}{\partial p}\, dp \qquad\qquad (13.2.2)$$

where p_o is the standard pressure and p is an external pressure. Using rel. (1.2.13) for the partial molar volume V_w, rel. (13.2.2) can be integrated. The value of an externally applied pressure to obtain $\mu_w = \mu_w^{\,o}$, known as osmotic pressure $\Pi = p - p_o$, can be expressed by the formula:

$$\Pi = - \ln x_w\, RT\, /\, V_w \qquad\qquad (13.2.3)$$

The expression (13.2.3) for the osmotic pressure can be approximated with the formula:

$$\Pi = c_i\, RT \qquad\qquad (13.2.4)$$

where c_i is the molal concentration of the solute (van't Hoff equation). This formula can be obtained from rel. (13.2.3) in several steps. Taking $x_w = 1 - x_i$, where x_i is the molar fraction of the solute "i", the value of $\ln x_w$ can be approximated by $\ln x_w = \ln (1 - x_i) \approx - x_i$. Also, using the approximation $x_i = n_i\, /\, (n_i + n_w) \approx n_i\, /\, n_w$ and $n_w\, V_w = V_{solvent}$ where $V_{solvent}$ is the volume of the solvent in the solution of "i", the result is $x_i\, /\, V_w \approx c_i$. (for diluted solutions the molal concentration c_i can be approximated with molar concentration c_i).

When a solution containing low molecular weight solutes is placed on one side of a membrane permeable for small molecules and a pure solvent is placed on the other side of the membrane, the flux of the solvent J_s and the flux of the solute J_i are given by the expressions:

$$J_s = (K_1\, /\, \lambda)\, (\Delta p - \Delta \Pi) \qquad\qquad (13.2.5)$$

$$J_i = (K_2\, /\, \lambda)\, (c_f - c_p) \qquad\qquad (13.2.6)$$

where K_1 and K_2 are the transport coefficients of the membrane for the solvent and solute, λ is the (effective) membrane thickness, Δp is the hydraulic pressure difference between the feed solution and the permeate solution, $\Delta \Pi$ is the difference in the osmotic pressures of the feed solution and the permeate solution, c_f is the concentration of the feed solution, and c_p the solute concentration in the product (permeate). Rel. (13.2.6) is obtained from an expression similar to rel. (13.2.5), where Δp is neglected and $\Delta \Pi \approx RT \cdot (c_p - c_f)$ from rel. (13.2.4). From these expressions, it can be calculated the membrane rejection R%, defined by the expression:

$$R = (1 - c_p\, /\, c_f)\, 100 \qquad\qquad (13.2.7)$$

Considering that the process takes place with the conservation of mass, $c_p = J_i\, /\, J_s$ and therefore:

$$R = \frac{(K_1\, /\, K_2)(\Delta p - \Delta \Pi)}{1 + (K_1\, /\, K_2)(\Delta p - \Delta \Pi)}\, 100 \qquad\qquad (13.2.8)$$

The variation of the rejection with $(K_1/K_2)(\Delta p - \Delta \Pi)$ is shown in Figure 13.2.1. The rejection increases with the increase in Δp, which is the pressure difference across the membrane. Reverse osmosis can be applied for the purification of low molecular compounds of solutions containing polymers and small molecules. Concentration polarization plays an important role in reverse osmosis similar to ultrafiltration (see Section 7.2). The solutes transported by convection to the membrane are rejected and increase the concentration until an equilibrium is established with the back diffusion into the solution. The concentration polarization also leads to higher osmotic pressure difference across the membrane, which decreases the flux J_s and the rejection R.

In addition to the concentration polarization, another factor affecting the separation is the solute loss δ (expressed as a fraction from the initial amount). With the increase in the recovery of the solvent Sr (expressed as volume of solvent removed through the membrane divided by the initial volume), more solute of interest is lost, even for membranes with high rejection values. The solute loss can be expressed by the formula:

$$\delta = 1 - (1 - Sr)^{(1-R/100)} \tag{13.2.9}$$

The variation of δ with the increase in Sr for different rejection values is shown in Figure 13.2.2. This indicates that in reverse osmosis (and also in ultrafiltration) it is important to maintain a large volume of solvent in the initial sample, the concentration of retentate having the effect of losing more material.

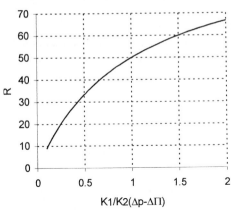

FIGURE 13.2.1. *Variation of rejection R with $(K_1/K_2)(\Delta p - \Delta \Pi)$.*

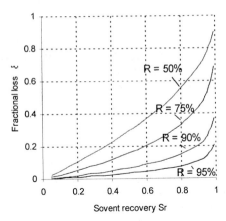

FIGURE 13.2.2. *Variation of fractional loss as a function of solvent recovery.*

The materials used as membranes for reverse osmosis are very similar to those used in ultrafiltration and include cellulose acetates, cellulose triacetate, gelatin, polyimides, polybenzimidazole, polybenzimidazolone, polyacrylic acid + ZrO_2, etc. Reverse osmosis and ultrafiltration have numerous applications in the separation of small molecules from large molecules in samples containing both types (see Figure 3.2.1).

- Dialysis

Dialysis is a separation procedure in which specific solutes are transferred through a membrane from one fluid to another fluid. Dialysis can be viewed as an extracting process that does not take place as an equilibrium between phases. The analyte displacement from one side of the membrane to another is based on the difference in their concentration, and the flux for one component is expressed by rel. (13.2.6). The molecules able to penetrate the membrane are present in the dialysate (or permeate), while the large molecules rejected by the membrane are present in the feed. Most dialysis processes are performed using moving solvents in a dialyzer, the two flows being separated by the membrane. The flow in the dialyzer can be parallel, countercurrent, or mixed. Using the notations U_f and U_p for the volumetric flow rate of the feed and of the dialyzate (or permeate), respectively, the mass balance for a given component leads to the expression [19], [20]:

$$U_f\,(c_{f,i} - c_{f,o}) = U_p\,(c_{p,o} - c_{p,i}) = N \tag{13.2.10}$$

where the index "i" indicates input solution and index "o" output solution. The diagram of a countercurrent dialysis system is shown in Figure 13.2.3. The variable N is the overall solute transport rate through the membrane. This variable can be expressed using a formula similar to rel. (13.2.6):

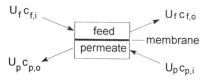

$$N = k\,A\,\Delta c_{ave} \tag{13.2.11}$$

FIGURE 13.2.3. *Diagram of a countercurrent dialyzer.*

where k is a rate transfer constant depending on the properties of the membrane, A is the area of the membrane, and Δc_{ave} is the average of the difference in concentration between the feed and permeate. The efficiency of a dialyzer can be expressed using the "dialysance" $\mathcal{D}$, a variable (with the dimensions of volumetric flow rate) expressed as follows:

$$\mathcal{D} = \frac{N}{c_{f,i} - c_{p,i}} \tag{13.2.12}$$

The efficiency of a dialyzer is usually expressed in terms of $\mathcal{D} / U_f$ (adimensional), and for a parallel flow $\mathcal{D} / U_f$ has the expression:

$$\mathcal{D} / U_f = \{1 - \exp[-Y(1 + Z)]\} / (1 + Z) \tag{13.2.13}$$

and for countercurrent flow has the expression:

$$\mathcal{D} / U_f = \{1 - \exp[Y(1 + Z)]\} / \{Z - \exp[Y(1 + Z)]\} \tag{13.2.14}$$

where $Z = U_f / U_p$ and $Y = k\,A / U_f$. Rel. (13.2.13) and (13.2.14) describe the performance of a dialyzer for specific flow rates. For Z =1, the expression (13.2.14) gives for the countercurrent flow:

$$\mathcal{D} / U_f = Y / (1 + Y) \tag{13.2.15}$$

and the variation of $\mathcal{D} / U_f$ as a function of Y is similar to the variation of rejection R as a function of $(K_1/K_2)(\Delta p - \Delta \Pi)$ as given by rel. (13.2.8).

The flow chart of a dialysis step in the sample preparation chain is similar to that applied in ultrafiltration (see Figure 7.2.8) and is shown in Figure 13.2.4.

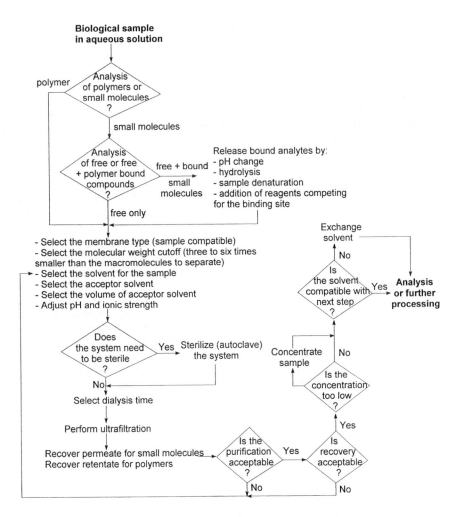

FIGURE 13.2.4. *Flow chart for performing dialysis as a sample preparation operation.*

The membranes used in dialysis are similar to those used in reverse osmosis. The materials used as membranes for dialysis include cellulose acetates (cellophane), various type of silicones, microporous polypropylene, and composite materials such as polyetherimide(polyester)-silicone.

The common laboratory technique of dialysis is used for purifying macromolecules, such as desalting of proteins, or for the separation of small molecules from a matrix containing polymers [21]. [22]. A number of parameters influence the dialysis process, mainly when it is applied for protein purification. Proteins can precipitate or be irreversibly adsorbed on the membrane, and a careful control of the nature of the solvents, solution pH, and solution ionic strength is frequently necessary. Ionic strength of a solution is defined by the expression:

$$s = 0.5 \sum_i c_i\, z_i \qquad\qquad (13.2.16)$$

where c_i are the molalities of the ionic species in the solution, and z_i are the charges of the ions. In some procedures, dialysis is done using iso-osmotic donor and acceptor solutions, which can be achieved by selecting specific concentrations for small ionic molecules such as NaCl added by purpose in the solutions used for dialysis.

A special type of membrane used in dialysis is the liquid membrane. Liquid membranes can be classified into three types: bulk membranes, emulsion liquid membranes, and supported liquid membranes (SLM) [23]. Bulk membranes typically use a U tube with an organic phase heavier than water, which separates the two aqueous phases, or an H tube for a solvent with lower density than water (see Figure 13.2.5). The separation process takes place by extracting a compound from the donor phase into the solvent and from the solvent into the

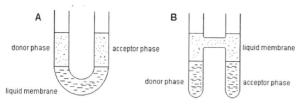

FIGURE 13.2.5. *Bulk membranes separating two aqueous phases with a solvent heavier than water (A) or lighter than water (B).*

acceptor phase [24]. This type of separation is in fact a double extraction, but if the equilibrium is not achieved, it is similar to dialysis. Emulsion liquid membranes are formed as emulsions from two nonmiscible phases. The supported liquid membranes consist of an inert porous support impregnated with a water nonmiscible organic solvent separating two water phases, one being the donor and the other the acceptor phase. This type of membrane can be considered a liquid-liquid extraction with the solvent included in a microporous membrane (MMLLE). Supported liquid membranes (SLM) are more frequently used than others in analytical applications. The driving force of the separation process is the difference in the concentration between the donor and acceptor phases. A typical use of a SLM uses an analyte that is a weak acid or a weak base. For a weak base, for example (such as an amine), on the donor side the pH is maintained relatively high such that the amine is present in free base form and is soluble in the liquid membrane. On the acceptor side, the pH is maintained low, and the amine is extracted as a salt in the aqueous phase. A coupled counter-transport of anions (such as chloride) from the acceptor phase to the donor phase maintains the charge balance in the system. The selectivity and recovery of this process can be controlled by the composition of the liquid membrane and the values of pH in the donor and acceptor phases. This procedure has been applied, for example, for the extraction of amino acids from biological samples. The membrane was made from trioctylmethylammonium chloride (Aliquot 336) supported on commercial porous PTFE membranes with an aqueous donor phase with pH > 11 [25].

The membrane extraction process can be connected on-line with a chromatographic instrument using, for example, the acceptor solution for loading a sample loop of the chromatographic instrument [26]. The coupling can be done either with a GC system [27], [28], [29], an HPLC [30], [31], or even CE [32]. Some of the applications reported in the literature applying membrane separation are included in Table 13.2.1 [33].

TABLE 13.2.1. *Application of on-line membrane extraction for sample preparation* [33].

Analytes	Matrix	Membrane	Reference
amines	urine	SLM	[34]
phenoxy acids	water	SLM	[35], [36]
trace organics	water	silicone	[37]
aromatics and pesticides	water	silicone	[38], [39]
sulphonylurea herbicides	natural water	silicone	[40], [41]
amines	plasma	SLM	[42]
phenols	wasted water	silicone	[43]
amines	air	SLM	[44]
amines	rainwater	SLM	[46]
various herbicides	natural water	SLM	[47]
carboxylic acids	air	SLM	[48], [49]
amperozide	water	SLM	[50]
amperozide	plasma	SLM	[51]
triazines	vegetable oil	silicone	[52]
phenols	natural water	SLM	[53]
aromatic surfactants	natural water	SLM	[54]
bambuterol	plasma	SLM	[32], [55]
chlorinated phenols, sulphonylurea herbicides	ground water	SLM	[53], [56]

REFERENCES 13

1. D. Warren, Anal. Chem., 56 (1984) 1529A.
2. T. Kotiaho et al., Anal. Chem., 63 (1991) 875A.
3. G. Palmai, K. Olah, J. Membr. Sci., 21 (1984) 161.
4. I. N. Beckman, in *Polymeric Gas Separation Membranes*, CRC Press, Boca Raton, 1994.
5. R. C. Johnson et al., Mass Spectrom. Rev., 19 (2000) 1.
6. M. E. Bier, R. G. Cooks, Anal. Chem., 59 (1987) 597.
7. J. S. Brodbelt et al., Anal. Chem., 59 (1987) 454.
8. M. Soni et al., Anal. Chem., 67 (1995) 1409.
9. S. Bauer, D. Solyom, Anal. Chem., 66 (1994) 4422.
10. L. A. B. Moraes et al., Analyst, 125 (2000) 1529.
11. R. D. Blanchard, J. K. Hardy, Anal. Chem., 56 (1984) 1621.
12. K. F. Pratt, J. Pawliszyn, Anal. Chem., 64 (1992) 2101.
13. M. J. Yang, J. Pawliszyn, Anal. Chem., 65 (1993) 1758.
14. M. J. Yang et al., Anal. Chem., 66 (1994) 1339.
15. M. J. Yang, J. Pawliszyn, Anal. Chem., 65 (1993) 2538.
16. C. S. Creaser et al., Anal. Chem., 72 (2000) 2730.
17. D. Kou et al., Anal Chem., 73 (2001) 5462.
18. C. C. Chang, G. R. Her, J. Chromatogr., A, 893 (2000) 169.
19. A. S. Michaels, Trans. Am. Soc. Artificial Internal Organs, 12 (1966) 387.
20. B. L. Karger et al., *An Introduction to Separation Science*, J. Wiley, New York, 1973.
21. U. A. Th. Brinkman (ed.), *Hyphenation: Hype and Fascination*, Elsevier, Amsterdam, 1999.
22. N. C. van de Merbel et al., J. Chromatogr., 634 (1993) 1.
23. J. A. Jonsson et al., Anal. Chim. Acta, 277 (1993) 9.

24. S. Moldoveanu, E. Tudoriu, Rev. Roum. Biochem., 19 (1982) 43.
25. P. Dzygiel et al., Anal. Lett., 31 (1998) 1261.
26. M. Kuntsson et al., J. Chromatogr., A, 754 (1996) 197.
27. G. Audunsson, Anal. Chem., 58 (1986) 299.
28. B. Lindegard et al., J. Chromatogr., 573 (1992) 191.
29. L. Gronberg et al., J. Chromatogr. Sci., 31 (1993) 384.
30. L. Gronberg et al., J. Chromatogr., A, 655 (1993) 207.
31. B. Lindegard et al., Anal. Chem., 66 (1994) 4490.
32. E. Thordarson et al., Anal. Chem., 68 (1996) 2559.
33. N. C. van de Merbel, J. Chromatogr., A, 856 (1999) 55.
34. G. Audunsson, Anal. Chem., 60 (1988) 1340.
35. C. Nilve et al., J. Chromatogr., 471 (1989) 1510.
36. L. Mathiasson et al., Int. J. Environ. Anal. Chem., 45 (1991) 117.
37. R. G. Melcher et al., Process Control Qual., 1 (1990) 63.
38. R. G. Melcher et al., Anal. Chem., 62 (1990) 2183.
39. P. Morabito et al., Process Control Qual., 3 (1992) 35.
40. G. Nilve et al., Chromatographia, 32 (1991) 269.
41. G. Nilve et al., J. Chromatogr., A, 688 (1994) 75.
42. B. Lindegard et al., J. Chromatogr., 573 (1992) 191.
43. R. G. Melcher et al., Anal. Chem., 64 (1992) 2258.
44. L. Gronberg et al., Chromatographia, 33 (1992) 77.
45. L. Gronberg et al., Chemosphere, 24 (1992) 1533.
47. M. Knutsson et al., J. Agric. Food Chem., 40 (1992) 2413.
48. L. Gronberg et al., J. Chromatogr., A, 655 (1993) 207.
49. Y. Chen et al., Anal. Chim. Acta, 292 (1994) 31.
50. J. A. Jonsson et al., J. Chromatogr., A, 665 (1994) 259.
51. B. Lindegard et al., Anal. Chem., 66 (1994) 4490.
52. R. Carabias Martinez et al., Anal. Chim. Acta, 304 (1995) 323.
53. M. Knutsson et al., Chromatographia, 42 (1996) 165.
54. T. Miliotis et al., Anal. Chem., 64 (1996) 35.
55. S. Palmarsdottir et al., Anal. Chem., 69 (1997) 1732.
56. G. Nilve et al., J. Chromatogr., A, 688 (1994) 4490.

CHAPTER 14

Electroseparation Techniques in Sample Preparation

14.1 ELECTROPHORETIC TECHNIQUES

Electrophoresis is an analytical technique widely used for various separations.
However, its application as a sample preparation procedure for chromatography is
minor. Electrophoretic separation can be associated with various detection procedures,
and electrophoresis is typically used as a core analytical technique. The electrophoretic
separation of the sample components is based on the differences in the migration rate
in an electric field through a specific medium. Many molecules can carry charges in a
solution, the typical example being the amino acids. An amino acid molecule can be
charged, depending on the solution pH, with a positive charge in acidic conditions and
with a negative charge in basic conditions, as shown below:

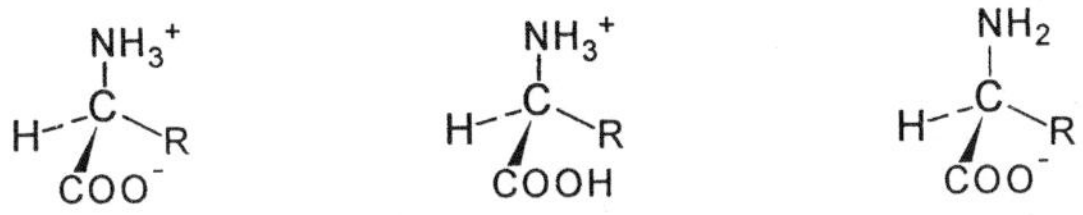

| neutral | in acidic
medium | in basic
medium |

The pH value for which the net charge is zero for an amino acid is known as isoelectric
point. Proteins behave in this respect similar to amino acids, and an isoelectric point
can be determined for each protein. The study of migration in an electric field for amino
acids or proteins shows that the migration is not stopped at the isoelectric point but at a
value close to it known as isoionic point. The explanation of this finding is that in a
solution various ions can be adsorbed on the surface of a neutral molecule or particle
such that it becomes charged. This effect is even more pronounced for charged
molecules or particles. The separation surface between a charged particle with the
charge Q and the solution can be seen as a double layer. A model proposed by
Helmholtz for this layer, which is viewed as a simple parallel-plate capacitor, leads to a
potential difference across the double layer given by the expression:

$$\psi_o = \frac{4\pi\sigma d}{\varepsilon} \qquad (14.1.1)$$

where A is the area of the capacitor, d is the distance separating the oppositely charged
plates, ε is the dielectric constant of the medium, and σ is the surface charge density
with $\sigma = Q / A$ (e.g. coulombs per cm^2). However, this expression for the potential is not
verified by experimental measurements, and the rigid array of charges described by the
Helmholtz model is very likely disturbed by the thermal motion of molecules in solution.
A more complex model for the charged particle in solution is shown schematically in
Figure 14.1.1 [1]. This model (Gouy-Chapman) assumes that the particle is surrounded
by an immobile layer of charged molecules (Stern layer) that does not compensate
completely the molecular charge and continues with a diffuse layer containing molecules

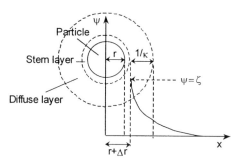

FIGURE 14.1.1. *Diagram of the distribution of charges around a charged particle in solution [1].*

with different charges. As a consequence, for a negatively charged particle the positively charged counterions have a distribution of charges that becomes random only at infinite distance. The potential decreases with the distance x, and for a spherical particle of radius r, this decrease follows an exponential dependence, given by the expression:

$$\psi(x) = \frac{4\pi\sigma r}{\varepsilon} \exp(-\kappa x) \qquad (14.1.2)$$

where κ is Debye-Hückel constant [2] with the expression:

$$\kappa = \left(\frac{8\pi e^2 \sum_{j=1}^{n} c_j z_j^2}{\varepsilon k_B T} \right)^{1/2} \qquad (14.1.3)$$

In rel. (14.1.3), e is the charge of the electron, c_j the concentration, and z_j the charge of ion "j", k_B is Boltzmann constant, and T the temperature in Kelvin deg. At the surface of Stern layer, the electric potential has a specific value ζ known as zeta potential, which is given by the expression:

$$\zeta = \psi(r + \Delta r) \qquad (14.1.4)$$

The charges around the particles are considered to have a distribution different from random up to a distance of $1/\kappa$, this distance being known as the thickness of the double layer. As seen from rel. (14.1.3), in solutions with a high concentration of ions, the diffuse layer is narrower than in solutions with low salts concentration.

In electrophoresis, the particles move in the solution under the influence of an electric field $\mathcal{E}$ (expressed as a vector). Various forces affect the movement of the particle, and the movement takes place with a constant velocity $\bar{v}$ when the vectorial sum of these forces is zero. One such force is the electrophoretic attraction $\bar{F}_1$ of the charged particle in the electric field $\mathcal{E}$ given by the expression

$$\bar{F} = Q\,\mathcal{E} \qquad (14.1.5)$$

Another force is the Stokes friction force $\bar{F}_2$ opposing the particle movement, and is given by the expression:

$$\bar{F}_2 = -6\pi\eta\, r\bar{v} \qquad (14.1.6)$$

where η is the dynamic viscosity of the medium. An additional force that must be considered is created by the action of the field on the ions generated in the double layer. This force is known as electrophoretic retardation and is given by the expression:

$$\vec{F_3} = -(Q - \varepsilon\zeta r)\,\mathcal{E} \tag{14.1.7}$$

A weaker force is also present and is known as the relaxation effect. The relaxation effect is caused by the distortion of the ionic surrounding of the charge due to the differences in the movement of larger charged particles and the counterions around it. The result is that the charged particle is not in the center of its "ionic atmosphere." This force can be neglected. The sum of rel. (14.1.5), (14.1.6) and (14.1.7) leads to the expression:

$$Q\mathcal{E} - 6\pi\eta r\,\vec{v} - (Q - \varepsilon\zeta r)\,\mathcal{E} = 0 \tag{14.1.8}$$

Rel. (14.1.8) allows the evaluation of the velocity of movement of the particle:

$$\vec{v} = \frac{\varepsilon\zeta\,\mathcal{E}}{6\pi\eta} \tag{14.1.9}$$

This expression shows that the movement of a particle or molecule in an electric field is proportional with the field intensity (V/cm), the proportionality constant being known as *electrophoretic mobility* μ, given by the expression:

$$\mu = \frac{\varepsilon\zeta}{6\pi\eta} \tag{14.1.10}$$

The values of μ are expressed in $(cm/s^2)\,/\,(V/cm)$. The movement takes place in the direction of the field, and the vectorial character of the forces, of the velocity, and of the electric field can be neglected. As seen in rel. (14.1.10), electrophoretic mobility depends on the zeta potential of the particle moving in the DC electric field.

Rel. (14.1.10) uses a number of approximations, including the assumption that the electric field is not affected by the presence of the charge Q. More detailed calculations [3] lead to the expression:

$$\mu = \frac{\varepsilon\zeta}{6\pi\eta} f(\kappa r) \tag{14.1.11}$$

where $f(\kappa r)$ is a function with values between 1 and 1.5, which depends on the size of the molecule and the value of κr. For values of κr below 0.1, the value of the function is $f(\kappa r) = 1$, and for values of κr higher than 100, the value of the function is $f(\kappa r) = 1.5$.

The calculation of the migration distance $d = v\,t$ in electrophoresis, where t is the time of the separation, can be done using rel. (14.1.9), and gives the expression:

$$d = \mu\,t\,\mathcal{E} \tag{14.1.12}$$

For weak acids or bases, the apparent mobility μ is also a function of the ionization degree of the molecule. For example, the dependence of the mobility for a weak acid is given by the expression:

$$\mu' = \mu_o \frac{K_a}{[H^+] + K_a} \qquad (14.1.13)$$

where μ_o is the mobility of the completely ionized compound, K_a is the acidity constant of the compound, and μ' is an apparent mobility. For amphoteric compounds such as amino acids or proteins, the dissociation as an acid takes place following two reactions:

$$H_3N^+\text{-R-COOH} \rightleftharpoons H_3N^+\text{-R-COO}^- + H^+$$

$$H_3N^+\text{-R-COO}^- \rightleftharpoons H_2N\text{-R-COO}^- + H^+$$

These reactions lead to two acidity constants (see Table 19.10.2) given by the expressions:

$$K_{a1} = \frac{[H_3N^+ - R - COO^-][H^+]}{[H_3N^+ - R - COOH]} \qquad K_{a2} = \frac{[H_2N - R - COO^-][H^+]}{[H_3N^+ - R - COO^-]} \qquad (14.1.14)$$

At the isoelectric point, $[H_3N^+\text{-R-COOH}] = [H_2N\text{-R-COO}^-]$, and from rel. (14.1.14) the value of pH for the isoelectric point is the following:

$$[H^+]_i = \sqrt{K_{a1}K_{a2}} \quad \text{or} \quad pH_i = 0.5\,(pK_{a1} + pK_{a2}) \qquad (14.1.15)$$

The mobility approaches zero for a pH close to pH_i.

The displacement of a compound applied as a narrow band at the start in the migration medium under the influence of the electric filed is associated with a band broadening (of Gaussian distribution due to the randomness of the process). This band broadening is caused by the ordinary diffusion, field microheterogeneities, turbulence in the flow, and electrosorptive effects [4]. The dispersion (in space) σ^2 of an electrophoretic band can be estimated considering these effects, and for the electrophoresis in a medium containing particles of diameter d_p, the dispersion σ^2 can be calculated with the expression:

$$\sigma^2 = 2D\,t + d_p\,(\mu + \mu_{osm})\,t\,\mathcal{E} \qquad (14.1.16)$$

where D is the diffusion coefficient in the medium and μ_{osm} is the mobility due to an additional type of movement known as electro-osmotic flow. The application of an electrical field to a solution has the effect of a movement of all charged particles. These charged particles are surrounded by water molecules that form a solvation sphere. The movement of charged particles is therefore associated with a flow of solvent, either toward the anode or toward the cathode, depending on the types of ions that have the most pronounced solvation. This solvent flow may be of limited importance, but in some cases it can be an important factor, such in the case of capillary electrophoresis (CE).

For two components with electrophoretic mobilities μ_1 and μ_2, the distances between the centers of the broadening migration zones is given by the expression:

$$\Delta d = (\mu_1 - \mu_2)\,t\,\mathcal{E} \qquad (14.1.17)$$

The resolution obtained during the separation is determined by the value Δd and the broadening W'_{b1} and W'_{b2} of the migration bands that can be defined similarly to the peak broadening (in space) at the peak base in chromatography (see Figure 1.6.2). The expression for *resolution* R_s in electrophoresis is given by the expression:

$$R_s = (2 \Delta d) / (W'_{b1} + W'_{b2}) = 2 (\mu_1 - \mu_2) t \, \mathcal{E} / (W'_{b1} + W'_{b2}) \qquad (14.1.18)$$

The relation between W'_b and dispersion σ^2 (in space) of the band is similar to that in chromatography (see rel. 1.6.37b for W_b and σ^2 in time) and can be written as follows:

$$W'_b = 4 \, \sigma \qquad (14.1.19)$$

Assuming that $W'_{b1} = W'_{b2}$, and using rel. (14.1.16) for σ (with an average value for μ and μ_{osm}), the resolution in electrophoresis is given by the expression:

$$R_s = \frac{(\mu_1 - \mu_2) t \, \mathcal{E}}{2 \sqrt{2Dt + d_p (\mu + \mu_{osm}) t \, \mathcal{E}}}$$

$$(14.1.20)$$

Rel. (14.1.20) shows that both longer time and higher electrical field intensities lead to a decrease in resolution. However the increase in the electric field intensity affects the resolution less than the migration time t.

- Electrophoretic techniques

A number of electrophoretic techniques have been developed, their classification being typically done considering the nature of the migrating medium. The electrophoretic techniques in a free liquid include moving boundary electrophoresis, isotachophoresis, microscopic electrophoresis, etc. Of special interest is capillary electrophoresis, in which the analytes move in a capillary filled with a specific buffer in an electric field applied across the capillary. The technique is a very efficient separation procedure when narrow capillaries with less than 0.1 mm i.d. and high electric field intensity are used. The migration of charged particles under the influence of the electrical field in a support medium (in the shape of a column or a plate) that minimizes convection include zone electrophoresis, isoelectric focusing, electrophoresis in gels with high density, etc. Other classifications are based on electric field intensity (low and high voltage) and type of support (paper, polyacrylamide, etc.). Also, migration in an electric field associated with separation based on differential partition between stationary and mobile phase is used in electrochromatography (see Section 4.4).

Moving boundary electrophoresis is performed in a free solution contained in a large U-shaped tube. It was the first electrophoretic technique and has been applied for the separation of proteins [5]. Isotachophoresis is also performed in free solution but introduces the sample between two electrolyte solutions, a leading electrolyte with a higher mobility than the sample ions and a terminating electrolyte with lower mobility. The resulting electric field generated in the solution by an applied potential is initially stronger in regions with low mobility ions, causing them to move at the same velocity as more mobile ions that are affected by a lower intensity electric field. When the equilibrium is achieved, the sample components move based on their mobility and are separated. Similar to this technique is field amplification, where the ions migrate electrophoretically through a low-conductivity solution such as water into a high-conductivity solution, and the migration slows down dramatically at the boundary of the two solutions. Based on this principle, the sample becomes more concentrated. Stacking gels obtained using differences in pH and in ionic strength between the electrophoresis buffer and the stacking gel are frequently used for narrowing the electrophoretic bands. Isotachophoresis can be coupled with capillary electrophoresis. For this purpose, the separation in zone isotachophoresis is performed with a capillary inlet placed in the leading electrolyte. When the leading electrolyte catches up with the sample zone, the gradient is lost, and CE separation begins. In this technique, the migration mode changes gradually from isotachophoresis to CE.

Electrophoresis in a support medium can be done using a buffer solution impregnated in a plate or a column of paper, cellulose acetate, silica, starch gel, agar/agarose gel, or polyacrylamide gel. For example, a 1% (w/v) agar solution forms a gel with large pore size and low frictional resistance and can be impregnated with a buffer. Polyacrylamide gels can be made with different degrees of crosslinking by using a mixture of acrylamide and N,N'-methylene-bis-acrylamide. Persulfate, N,N,N'N'-tetramethyl-ethylenediamine, or both are used as catalysts for the polymerization of acrylamide gels. These gels are well suited for various separations and are especially used for the analysis of proteins. Electrophoresis has a large number of applications. The experimental conditions may vary regarding the type of sample. The molecules with relatively low molecular weight are usually separated using high voltage electrophoresis, and support medium as paper can be utilized. The heating effect of the high voltage must be eliminated by cooling. A series of chemical reactions can be used for modifying the analytes with the formation of molecular species easier to separate using electrophoresis. For example, the carbonyl groups of reducing sugars can react with $NaHSO_3$ in acidic conditions at mild heating as follows:

$$\begin{matrix} Ra \\ \\ Rb \end{matrix} C{=}O + HSO_3Na \longrightarrow \begin{matrix} Ra OH \\ C \\ Rb SO_3Na \end{matrix}$$

The newly formed α-hydroxysulfonic acids migrate more easily during electrophoresis compared to the neutral sugar.

The addition of sodium dodecyl sulfate (SDS) to proteins is another reaction used in electrophoresis. This reagent binds strongly to proteins, resulting in an overall negative charge of the adduct that travels to anode (about 1.2 g detergent binds to 1 g protein). Using polyacrylamide gels (PAGE) with different crosslinking degrees, the adducts of protein with sodium dodecyl sulfate separate based on the molecular size and not on protein electrokinetic properties (SDS-PAGE). Gradient gels varying in pore size can be used for the separation. Standard proteins with known molecular mass can be added in the separation for calibration and estimation of the MW. Some additional modifications can be done on the protein, such as replacement of disulfide bridges with SH groups by treatment with 2-mercaptoethanol or dithiothreitol (see Section 18.6). The determination of the MW requires calibration with proteins of similar structure to the analyte. The separation using SDS-PAGE can be done after a band sharpening process using isotachophoresis is applied.

Besides the separation in a uniform buffered medium, electrophoresis in a pH gradient is also applicable and is known as electrophoresis with isoelectric focusing. In this technique, the column is made with a variation of pH, for example from low at the anode (+) to high at the cathode (-). The sample is applied in a middle region and, depending on the pH and the K_{a1} and K_{a2} of the protein, this is positively charged in a more acidic region and will migrate toward the cathode and is negatively charged in more basic regions and will migrate toward the anode. The migration will continue until the molecule reaches isoionic point (close to isoelectric point), the separation being determined by the difference in K_{a1} and K_{a2} values of each protein. The diagram of this process is shown in Figure 14.1.2.

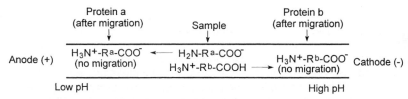

FIGURE 14.1.2. *Schematic description of isoelectric focusing.*

Isoelectric focusing can be performed in free solution but more frequently is applied using a gel support. Methods using preparative isoelectric focusing have been developed for protein separation [6]. Another interesting application of electrophoresis as a sample preparation step is related to the application of western blotting. In this technique, the electrophoresis is performed in a gel, followed by the transfer of proteins into a nitrocellulose sheet by diffusion or electroelution. The electroelution is done by applicaltion of a DC current in the perpendicular direction to the initial migration direction.

Electrophoresis is frequently used for protein separation and can be considered a sample preparation step in immunoelectrophoresis [8]. A typical procedure consists of the electrophoretic separation being performed on a plate covered with a support gel for a protein mixture containing antigens to a specific antiserum (that contains antibodies). After the protein is separated, the antiserum is placed on a longitudinal channel parallel to the direction of migration of proteins. The plate is allowed to incubate for a specific period of time (such as 24 hours). The diffusion of the antigens and of

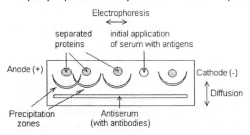

FIGURE 14.1.3. *Schematic representation of immunoelectrophoresis.*

antibodies one toward the other leads to the formation of precipitated complexes that are visible and allow identification. The schematics of this procedure is shown in Figure 14.1.3.

A variety of other techniques based on electrophoretic separation are developed [9]. Some of these techniques may be seen as sample preparation steps before detection, using various procedures to capture the analytes as they approach a specific point in the electrophoretic separation. Among these procedures is the use of a physical barrier such as a dialysis tubing or a hollow fiber used to collect the analytes. Also various types of ligands, free or immobilized in the path of the migrating analytes, can be used for the collection of a specific analyte.

Capillary electrophoresis (CE) is the technique that expanded the electrophoretic analysis to many other analytes besides proteins and amino acids The technique is widely used as a core analytical method, and its description can be found in a number of dedicated books and review articles [10].

14.2 OTHER ELECTROSEPARATION TECHNIQUES

A number of other electroseparation techniques are used in practice, although their application to sample preparation is very limited. Among these are electrolysis, electrodialysis, electrofiltration, electrodecantation, etc. Electrodialysis, for example, is performed in a special cell, where the migration of the analytes is caused not only by the difference in concentration but also under the influence of an electric DC field. The procedure is usually applied for ionic solutions. The dialysis membranes can be selected to have ion exchange properties. A series of cells with one wall made from an anion exchange membrane and one wall from a cation exchange membrane can be used for water desalinization. The diagram of an electrodialysis system showing only three cells is given in Figure 14.2.1. Under the influence of the electric current, the positively charged ions (cations) migrate toward the cathode (-), and the negatively charged ions (anions) migrate toward the anode (+). The migrating cations can pass easily through cation exchange membranes but cannot penetrate the anionic membranes, while anionic membranes are not permeable for cations. With

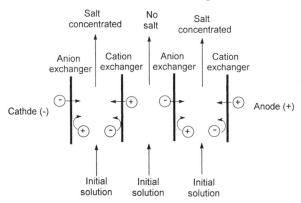

FIGURE 14.2.1. *Diagram of an electrodialysis system.*

this arrangement, considering the middle cell in Figure 14.2.1 and the anions migrating from left to right, the anions initially present in the middle cell will leave through the right (anion exchanger) wall, and the cations will leave through the left (cation exchanger) wall. On the other hand, no anions will enter through the left (cation exchanger) wall, and no cation will enter through the right wall (an anion exchanger). In this way, the middle cell will lose all the ions. The adjacent cells will receive ions through both walls and will concentrate the salts. Electrodialysis has industrial applications, but the laboratory scale use is not common [11].

REFERENCES 14

1. H. A. Abramson, *Electrophoresis of Proteins*, Hafner, New York, 1964.
2. W. J. Moore, Physical Chemistry, Prentice-Hall, Englewood Cliffs, 1955.
3. M. Bier (ed.), *Electrophoresis*, Academic Press, New York, 1967.
4. E. Heftmann (ed.), *Chromatography*, Reinhold, New York, 1967.
5. A. Tiselius, Trans. Faraday Soc., 33 (1937) 524.
6. P. G. Righetti et al., J. Chromatogr., 475 (1989) 293.
7. L. P. Cawley, *Electrophoresis and Immunoelectrophoresis*, Little Brown, Boston, 1969.
9. N. A. Guzman, R. E. Majors, LC-GC Europe, 14 (2001) 292.
10. D. R. Baker, *Capillary Electrophoresis*, J. Wiley, New York, 1995.
11. E. S. Perry (ed.), *Progress in Separation and Purification*, Interscience, New York, 1968.

CHAPTER 15

Other Separation Techniques in Sample Preparation

15.1 ION EXCHANGE IN SAMPLE PREPARATION

An ion exchange material (ion exchanger) consists of a solid matrix containing fixed groups with ionic character and ionically connected counterions that can be exchanged. There are several types of ion exchangers, the most common being the cationic and the anionic ones. The ion exchange process on a cationic ion exchanger can be schematically represented by the reaction:

$$R\text{-}X^- H^+ + M^+ \rightleftharpoons R\text{-}X^- M^+ + H^+$$

where R has a polymeric structure containing the bound group X^- and a counterion (exemplified by H^+). Specific acid groups can be bound on the resin, and various cations can be exchanged in the process, including single or multiple charged species. The solid matrix of an ion exchanger can be an inorganic material or a polymeric organic resin.

The exchange process for an anion exchange material, can be represented by the reaction:

$$R\text{-}X^+ Cl^- + Y^- \rightleftharpoons R\text{-}X^+ Y^- + Cl^-$$

where the bound group X^+ has a counterion exemplified by Cl^-. Instead of Cl^- the anion exchanger can have other groups such as OH^-. These counterions can be replaced with other anions in the exchange process.

- Materials used as ion exchangers

Among the inorganic compounds that have ion exchange properties are several groups of natural silicates such as zeolites. Zeolites are hydrated aluminosilicates with the general formula $Me_{x/n}[AlO_2)_x (SiO_2)_y]$ m H_2O, where Me is Na, K, Ca, etc. (see Section 9.2). These compounds act as selective cation exchangers, the exchange process probably being based on a molecular sieving mechanism. Several hydrated oxides also act as cation exchangers. Among these are hydrated oxides of Si, Al, Zr, Fe, Sn, etc. These hydrated oxides have both adsorbing and ion exchange properties. Other inorganic compounds, such as polymeric hydrated zirconium phosphate, hydroxyapatite, etc., also have ion exchange properties.

More commonly utilized than inorganic ion exchangers are the synthetic organic resins with ion exchange properties. These compounds consist of a polymeric backbone containing covalently bound groups that are able to exist in ionic form. The presence of the ionic groups makes the ion exchange resins act as polyelectrolytes. Some of the groups introduced in synthetic organic ion exchangers are indicated in Table 15.1.1.

TABLE 15.1.1. *Functional groups introduced in synthetic ion exchange resins.*

Cation exchangers		Anion exchangers	
Type	Functional group	Type	Functional group
sulfonic acid	$-SO_3^-H^+$	quaternary amine	$-N(CH_3)_3^+OH^-$
carboxylic acid	$-COO^-H^+$		$-N(CH_3)_2(CH_2CH_2OH)^+OH^-$
phosphonic acid	$-HPO_3^-H^+$	tertiary amine	$-NH(CH_3)_2^+OH^-$
phosphinic acid	$-HPO_2^-H^+$	secondary amine	$-NH_2(CH_3)^+OH^-$
phenolic	$-C_6H_4-O^-H^+$	primary amine	$-NH_3^+OH^-$
arsonic acid	$-HAsO_3^-H^+$	sulfides	$\equiv S^+OH^-$
selenonic acid	$-SeO_3^-H^+$		
phenoxy group	$-C_6H_4-O^-H^+$		

Cation exchange resins are classified as strong acids, which are ionized over a wide pH range, or weak acids, which are ionized only in more basic solutions. From the groups indicated in Table 15.1.1, only the sulfonic groups confer strong acid properties. Anion exchange resins are classified as strong or weak bases. Quaternary amine functional groups form strong base exchangers. The anion exchangers containing trimethylamine groups are more basic than those containing dimethyl-β-hydroxyethylamine. The most common groups in commercially available ion exchangers are sulfonic and carboxylic groups for cation exchangers and tetramethylamino group for anion exchangers.

The polymeric backbone of organic ion exchangers is usually obtained either by polycondensation or by polymerization. A common polycondensation reaction used to obtain polymeric resins is that between phenol and formaldehyde. The reaction can be written as follows:

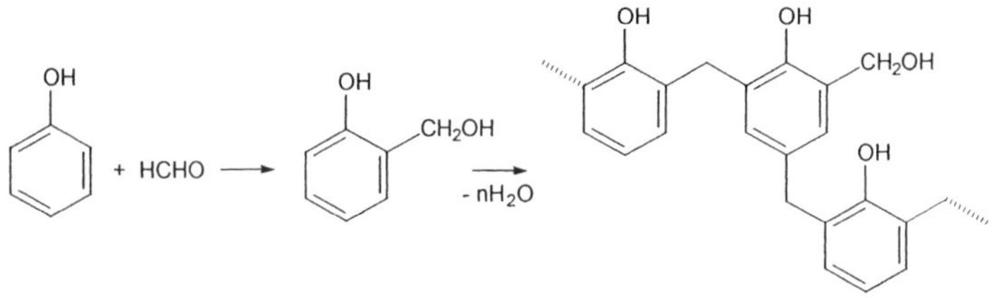

Both acid and basic catalysts can be used in this reaction. In the presence of basic conditions, the reaction takes place more completely and usually leads to a tridimensional thermorigid polymer. The inert resin can be modified by direct sulfonation into a strong cation exchanger.

The polycondensation in the presence of acid catalysts may leave a significant number of free $-CH_2OH$ groups. These can be further derivatized with SO_2Cl, and the $-CH_2OH$ groups are changed into $-CH_2Cl$ groups. Upon treatment with $(CH_3)_3N$, the resin can be changed into a strong anion exchange material as shown below:

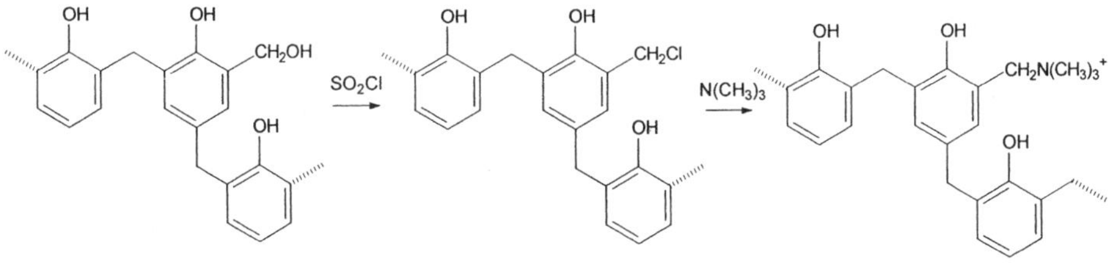

The derivatization of the already polycondensed resin is not the only possibility to generate either cationic or anionic resins. One common procedure starts with the derivatization of the phenol prior to condensation. Sulfonic resins, for example, can be obtained following the reactions schematically shown below:

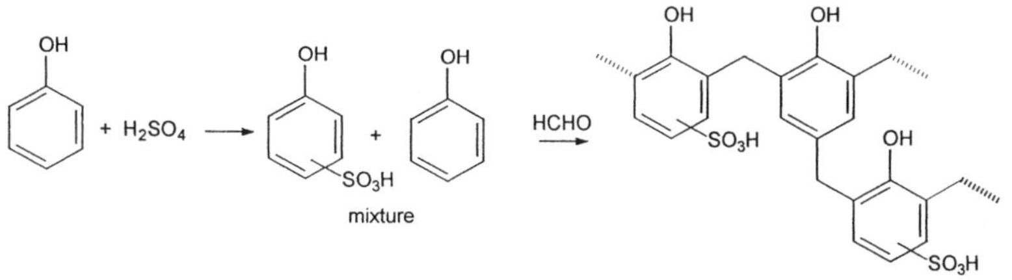

The sulfonation products of phenol can be condensed without separation, and a highly cross-linked material can be obtained in appropriate conditions [1]. Other resins can be prepared similarly to sulfonic resins. For example, starting with salicylic acid and formaldehyde, a resin with carboxylic groups is obtained.

The fixed ionic group also can be generated on the side chain of the resin. For example, condensation of sodium phenolate with Na_2SO_3 and HCHO leads to the formation of a resin with methylenesulfonic acid groups. Also, phenols such as resorcinol, naphthol, phenoxyacetic acid, and other aldehydes are used in the condensation instead of simple phenol and formaldehyde. For example, the condensation of phenoxyacetic acid and formaldehyde leads to a weak acid resin. Other variations of the condensation reaction are utilized, such as condensation of a phenol, a substituted benzaldehyde, and formaldehyde. Also, condensation of aromatic diamines such as m-phenylenediamine with formaldehyde is used to make a polymer with anion exchange properties. Because the amino groups directly connected to the aromatic ring are weak bases, further methylation can be applied to form quaternary amines that are stronger bases. This condensation can be written as follows:

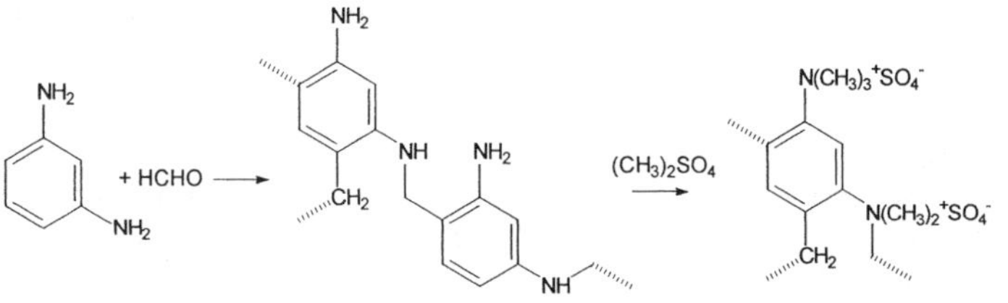

Condensation reactions using aliphatic polyamines, phenol, and formaldehyde also have been applied to generate anionic ion exchangers.

Polymerization of vinyl monomers is another common procedure used to prepare ion exchange resins. Most of these resins are made using styrene/divinylbenzene copolymers. Some procedures to modify the polymer already formed were indicated in Section 11.1. Among these, sulfonated polystyrene (which also may have sulfone bridges) and quaternary amines obtained by chloromethylation of polystyrene/divinylbenzene resins are the most common. Other anionic resins with slightly different properties have longer alkyl chain substituents at the quaternary nitrogen or groups such as trimethylbenzylammonium, dimethylethanolamine, dimethylethanolbenzyl, etc. Weaker anion exchangers contain polyamine groups attached to the polystyrene chain.

Another procedure used to obtain resins with acids or basic groups begins with specific monomers that have the desired group. Sulfonated styrene can be used in a copolymerization reaction with divinylbenzene (usually in the presence of benzoyl peroxide) to form a polymer with sulfonic groups. Copolymerization between acrylic acid or methacrylic acid and divinylbenzene generate a resin with carboxyl groups, as shown below (for methacrylic acid):

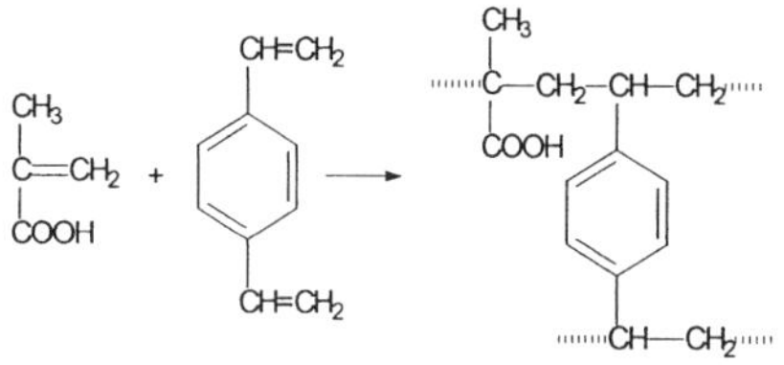

Anionic ion exchangers also can be prepared using specific monomers. For example, acrylonitrile can be polymerized, followed by reduction (with hydrogen and Ni catalyst) followed by the modification of the primary amine to quaternary amine, as shown below:

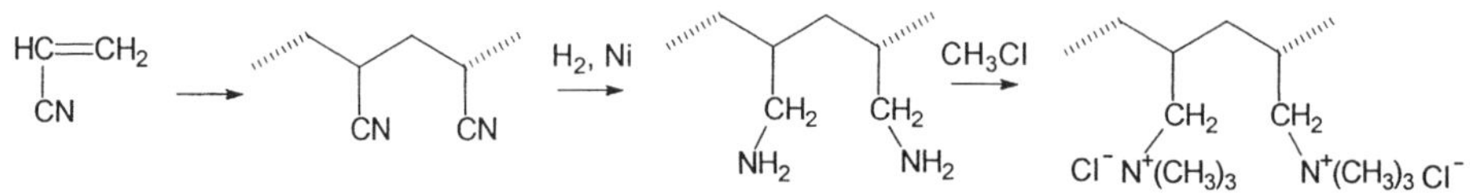

Many other synthetic paths have been either used or only explored for producing ion exchange resins. Some resins are prepared to have more than one type of functional group, others are made to contain unique structures, etc. Amphoteric ion exchangers with both basic and acidic groups, as well as ion exchangers with specific chelating properties are also available for various applications. An amphoteric ion exchanger, for example, can be obtained by copolymerization of styrene, divinylbenzene, and vinyl chloride, followed by amination and sulfonation in a sequence of reactions as shown schematically below:

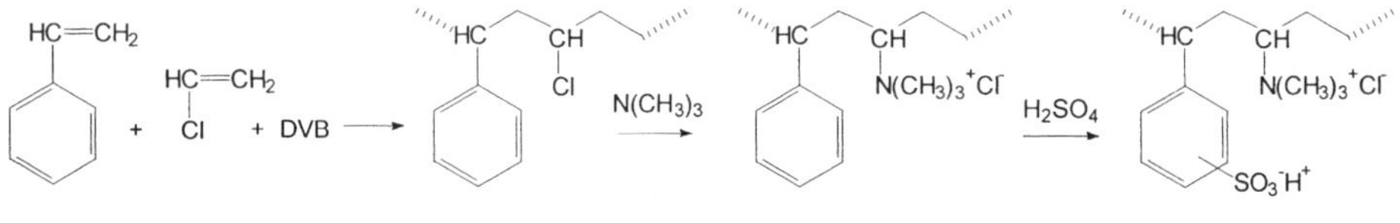

Various commercial names are used for the synthetic resins such as Amberlire, Dowex, Diaion, Duolite, etc. Particular specifications (letters, numbers) indicate the individual type of resin.

Another procedure used to obtain materials with ion exchange properties starts with cellulose. Strong alkali solutions acting on cellulose (at room temperatures) produce alkali cellulose. The studies on the structure of alkali cellulose [2] obtained with 20–40% NaOH solutions indicated that the substance is not a true alcoholate but an addition complex, $R_{Cell}OH:NaOH$. The treatment of alkali cellulose, for example, with chloroacetic acid sodium salt, leads to the formation of carboxymethylcellulose (CMC), following the reaction:

$$R_{Cell}OH:NaOH + ClCH_2COO^- \ Na^+ \rightarrow R_{Cell}O\text{-}CH_2COO^- \ Na^+ + NaCl + H_2O$$

The degree of substitution (D.S.) that can be obtained for this product usually ranges between D.S. = 0.1 to D.S. = 1.2. Pure CMC is commercially available. Carboxymethyl cellulose in itself is a weak acid that can be precipitated from CMC solutions with a mineral acid. The pH of precipitation varies between 6 for low substitution values to 1 for high substitution (D.S. of about 0.9). The material has ion exchange properties.

Ion exchange materials also can be obtained from dextrans. Dextrans are produced by certain bacteria growing on a sucrose substrate, and their structure is characterized by a $(1\rightarrow6)\text{-}\alpha\text{-}D$-glucopyranosyl chain. Branches may occur at $(1\rightarrow3)$-linked points. The dextrans used for ion exchange resins are usually cross-linked using epichlorhydrin, followed by transformation in a cation exchange containing sulfonyl groups or an anion exchange as shown schematically below for the anion exchange preparation:

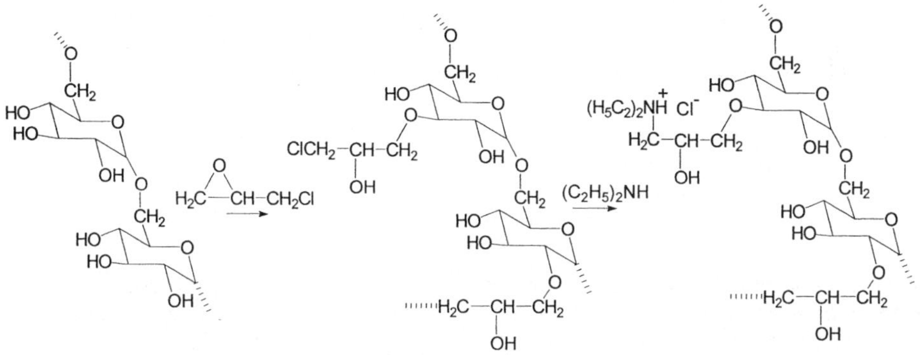

Electron exchange resins are also known, such as one obtained from the condensation of hydroquinone with phenol and formaldehyde.

Besides the chemical structure of the ion exchange material, its physical properties are also important. Many ion exchange resins are used in the form of small particles of different diameters, depending on the application. The polymers are typically obtained in reactions involving water solutions and have the structure of a gel with various water contents. The elimination of this water generates a microporous structure of the polymer. In many applications, it is important to preserve part of this water, and complete drying of the polymer usually is not desired. The polymer is typically used only after swelling in a solvent to be used for the specific application, and complete water elimination may be irreversible, giving a polymer with poor swelling properties. Besides the polymers with a microporous structure, highly porous polymers are useful in some applications. Resins to be used in solvents in which the polymer does not swell, for example, must have a higher porosity to allow better contact with the solvent. Porous polymers are, however, easily compressible and not always useful, for example, as a stationary phase for HPLC. Other types of physical structures for ion exchange resins are also known, such as pellicular ion exchange resins that have an inert core and may be useful in specific applications.

Ion exchange materials are also used in the form of membranes. These membranes can be used as separating barriers inhibiting convection, being at the same time permeable for water and electrolytes. Some membranes can be permeable only to cations or only to anions. The cations are not assumed to enter an anion exchange particle, the anionic membranes being permeable only for anions. Similarly, cationic membranes are not permeable for anions. Three types of ion exchange membranes are common, heterogeneous, interpolymer, and homogeneous. The heterogeneous membranes are produced from finely milled ion exchange granules that are formed into sheets by compression with an inert elastic binder such as polyethylene. In these membranes high molecular weight colloidal particles can penetrate the channels from the membrane matrix and can cause undesirable contamination and precipitation phenomena. Interpolymeric membranes are obtained by casting a film from a homogeneous solution of two polymers, one of which is the polyelectrolyte and the other a soluble filmogenic material. For instance, such membranes can be obtained from polystyrene-sulfonic acid and polyacrylonitrile in dimethylformamide. The polyelectrolyte and the matrix polymer are intricately mixed such that even long immersion in water does not elute the polyelectrolyte. The homogeneous membranes can be obtained from polymerization or polycondensation with production of a polymer that can be made into a thin sheet or film. For example, methacrylic acid and styrene sulfonic acid by copolymerization form a polymer that can be made into a film [3].

A variety of ion exchange resins are commercially available. It is common that the cationic resins are available in H^+ or in Na^+ form (exchangeable cation is H^+ or Na^+), and the anionic resins are available in Cl^- or OH^- form. The exchange of the counterion with another ion can be done by passing a (relatively) concentrated solution containing the desired ions through a resin bed. The reconversion of an ion exchanger into its working form is known as regeneration. For this purpose, different possibilities are used depending on the nature of the exchanger and its acid/base properties. For instance, strong or weak acid cation exchangers are regenerated with HCl or H_2SO_4, and if the desirable form is Na^+, the regeneration of strong cation exchangers should be performed using NaCl, while the weak cation exchangers are regenerated with NaOH. Anion exchangers are regenerated using NaOH, NaCl, NH_4OH, HCl, H_2SO_4, Na_2SO_4, or Na_2CO_3. The operation of regeneration must be complete. Usually, the regeneration solution has a concentration of 1N and is used in a volume more than necessary for stoichiometric conversion. A special precaution must be taken to avoid the formation of precipitates in the resin (such as $CaSO_4$, etc.) during the regeneration process.

- Characterization of ion exchange resins

A number of properties of an ion exchange material are important for an analytical procedure. Among these are its capacity, selectivity, equilibrium behavior, and kinetics of exchange. To these must be added the swelling capability, specific mechanical properties, absence of leakage of organic material in solutions, and resistance to hydrolysis, solvents, temperature, radioactive radiation, etc.

The ion exchange capacity (weight capacity) Q is defined as the maximum number of counterion equivalents exchanged by a specified amount of material (usually at dry basis). The capacity is typically expressed in meq/g and assumes that the ion exchanger is in a specified form such as H^+ for a cation exchanger or Cl^- for an anion exchanger. Other counterions affect the weight of the ion exchanger, and for different counterions different weight exchange capacities are obtained. The capacity for most cross-linked polyacrylic acid polymers is about 9.0 meq/g, for sulfonated polystyrene/

divinylbenzene is about 5.4 meq/g, and for polystyrene/divinylbenzene with methylenetrimethylammonium groups (chloride form) is between 2.5 and 4.0 meq/g. Besides the weight capacity, other ways to express capacity can be utilized. Among these is volume capacity, which is expressed in equivalents exchanged by 1 L of packed bed material. Other ways to express the same property include the concentration of the fixed ionic groups given in meq in a gram of solvent absorbed in the resin.

Capacity, although seemingly a straightforward property, must be defined within rigorous conditions. For example, in addition to ion exchange capacity, resins also may have sorption capacity (silica being a typical example). The sum of sorption capacity and ion exchange capacity give an overall capacity. Even for materials with negligible sorptive capacity, besides weight capacity an apparent weight capacity may be necessary to define. This is the weight capacity in specific conditions that may not allow all the fixed groups with ionic character to be ionized. For example, a weak acid at low pH or a weak base at high pH are not ionizable, and their weight capacity depends on pH. Other factors also influence weight capacity, and an apparent weight capacity may be more important for understanding the property of an ion exchange material.

Cation exchangers in H^+ form can be considered insoluble acids, and the anion exchangers in OH^- form can be considered insoluble bases. Due to their structure (polyelectrolyte gels) they can be titrated with bases (for cationic resins) or acids (for anionic resins). The pH titration curve, gives a good characterization of the ion exchangers. Such titration curves are shown in Figure 15.1.1 for a strong acid resin with capacity 4.5 meq/g and for a weak acid resin with capacity 8 meq/g. The titration of a strong cationic resin with a solution of NaOH is similar to that of a dissolved strong acid, except that the starting pH of the solution is higher because the H^+ ions are in the resin and not in the solution. The titration of a monofunctional weak acid resin is not very different from that of a weak acid, although the buffer solution is not formed in the case of the resin. However, at the beginning of the titration, because the weak acid is not dissociated, the uptake of Na^+ ions is incomplete and a continuous increase in pH is noticed. At the equivalence point, a sharp pH increase is noticed for both weak and strong resins, and this allows the calculation of the weight capacity. The titration of a polyfunctional cation exchanger shows steps similar to the titration of a polybasic acid. Differently from the titration of acids or bases in solution, the titration of ion exchange resins in H^+ or OH^- form is influenced by the presence of salts. The addition of a salt such as NaCl in a solution of an acidic ion exchanger leads to a substitution of the H^+ ions from the resin with Na^+ ions, and the solution becomes more acidic. The shape of the titration curve, although basically similar to the one shown in Figure 15.1.1, has some differences. The resin capacity is not affected by the presence of salts.

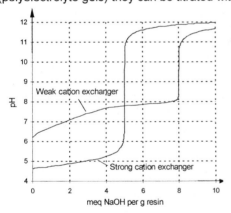

FIGURE 15.1.1. *Titration curves with NaOH of a strong acid resin with capacity 4.5 meq/g and of a weak acid resin with capacity 8 meq/g.*

Weak cation exchangers in H^+ form and weak anionic exchangers in OH^- form can be characterized by their acidity constant K_a (and pK_a) values. Strong acid or basic resins are practically dissociated, and the acidity constant is not important. The acidity constant and the degree of dissociation α (in the resin) are defined by the expressions:

$$K_a = \frac{[RX^-][H_r^+]}{[RXH]} \qquad\qquad \alpha = \frac{[RX^-]}{[RX^-]+[RXH]} \qquad\qquad (15.1.1)$$

(where index r is used for resin). The quantity $[RX^-] + [RXH]$ is related to Q, the weight capacity, except that Q is usually expressed on a dry weight base. For $\alpha = 0.5$ it can be seen that $K_a = [H_r^+]$, or $pK_a = pH_r$. The value for pH_r can be related to that in solution during a titration with NaOH (in the absence of added salts) when it can be assumed that the concentration ratio $[Na_r^+]/[H_r^+]$ is the same as in solution. This gives the expression:

$$[H_r^+] = [H^+]\,[Na_r^+]/[Na^+] \qquad\qquad (15.1.2)$$

For $\alpha = 0.5$, $[Na_r^+] = ([RX^-] + [RXH])/2$, which is half of the concentration of ionogenic groups (which can be easily calculated from the weight resin capacity, taking into account the water content of the resin). From rel. (15.1.2) and with $pK_a = pH_r$ at $\alpha = 0.5$, the value of pK_a can be written as a function of the pH of the solution using the expression:

$$pK_a = pH + \log [Na^+] - \log \frac{[RX^-]+[RXH]}{2} \qquad\qquad (15.1.3)$$

As an example, for a resin with the weight capacity of 8 meq/g, 60% water content, an external pH = 6.5, and $[Na^+] = 0.1$ at the half converted resin, it can be calculated that $pK_a = 6.5 + \log 0.1 - \log [(8{*}40/60) / 2] \approx 5.93$. The value of pK_a for the weak acid or base resin is useful in practical applications involving weak electrolytes, proteins, or enzymes immobilized on ion exchange resins [4].

Among other properties of the resins, the swelling is important for most practical applications. This involves the capability of the resin to uptake specific solvents, and also is related to the modification of the resin volume during utilization. Being a polyelectrolyte, the ions in the resin structure are strongly hydrophilic and attract water molecules. In this process, the inside of the resin becomes equivalent to a concentrated solution that has the tendency to dilute itself when introduced in water. This leads to a specific osmotic pressure of the resin (or swelling pressure), which is defined by the expression:

$$\Pi = p_r - p \qquad\qquad (15.1.4)$$

where p_r is the liquid pressure in the resin, and p the pressure in the external solution. The swelling depends on the nature of the resin and its cross-linking, the nature of the solvent, capacity, nature of the counterion, concentration of the salts in the solution, etc. Polar solvents are in general better swelling agents, and resins with higher capacity have the tendency to swell more.

- *Equilibrium and kinetic factors in ion exchange process*

The study of ion exchange equilibria is usually based on a number of simplifying assumptions. One such assumption is that the maximum uptake of ions by an ion exchange resin is constant and determined by the number of functional groups on the resin matrix [1]. Another assumption is that strict stoichiometric coupling takes place between the different components involved in the ion exchange process. This implies that each ion from the resin phase is replaced by another one from the liquid phase, with an equivalent charge for maintaining electroneutrality. Also, it is assumed that cations do not penetrate an anion exchange particle, and anions do not penetrate a cation exchange particle. These assumptions are useful for the prediction of the ion exchange behavior of simple, dilute, strong electrolyte solutions. However, when more complex systems are involved, such as weak electrolyte solutions or solutions with high solute concentrations exceeding the resin capacity, the simplifying assumptions are not always valid. For example, the uptake of counterions may exceed the total capacity of the resin in case of high concentrations [5].

For a given ion B^+ in solution and a resin in A^+ form, the exchange equilibrium is the following:

$$RX^-A^+ + B^+ \rightleftharpoons RX^-B^+ + A^+$$

The constant for this equilibrium can be written using previous assumptions as follows (see also rel. (11.1.9) and (11.1.10)):

$$K_{A,B} = \frac{[B_r^+]}{[B^+]} / \frac{[A_r^+]}{[A^+]} \tag{15.1.5}$$

(where the index s for solution was omitted and the index A,B indicates substitution in the resin of A with B). The exchange constant $K_{A,B}$ indicates the degree to which an ion "B^+" is preferred in the exchange process compared to the ion "A^+", larger constants for $K_{A,B}$ indicating higher affinity for the resin of species "B^+". The exchange constant can be expressed as a function of a distribution constant between the resin and the solution. The equilibrium described by rel. (15.1.5) can be viewed as equivalent with two independent equilibria:

$$A^+ \rightleftharpoons RX^-A^+ \qquad\qquad B^+ \rightleftharpoons RX^-B^+$$

which are described by the constants:

$$K_{r,A} = \frac{[A_r^+]}{[A^+]}, \qquad K_{r,B} = \frac{[B_r^+]}{[B^+]} \tag{15.1.6}$$

Each solute has a distribution constant (retention constants) for a specific resin, higher constants indicating more affinity for the resin. Using rel. (15.1.6) the equilibrium constants can be expressed in the form:

$$K_{A,B} = K_{r,B} / K_{r,A} \tag{15.1.7}$$

For a multicharge ion the equilibrium taking place between the solution and the resin can be written in the form:

$$zRX^-A^+ + B^{z+} \rightleftharpoons (RX)_z^{-z}B^{z+} + zA^+$$

and the equilibrium constant is

$$K_{A,B} = \frac{[B_r^{z+}]}{[B^{+z}]} / \frac{[A_r^+]^z}{[A^+]^z} \tag{15.1.8}$$

Based on rel. (15.1.5), it can be seen that the concentration of an ion in solution and the value of the equilibrium constant are important factors in the replacement of an ion from solution with another one. In practice, the ion exchange is obtained in a specific form by allowing a solution with a relatively high concentration in one ion to stand in contact with the resin for a period of time (20–30 min.) or by slowly passing the solution through a column containing the resin. For example a solution of 0.1–1 N HCl is passed through a bed containing several grams of swelled cationic resin. This changes the resin into H^+ form (resins in H^+ form are commonly available commercially). The resin is then thoroughly washed with water to eliminate the remaining HCl. When a solution containing the ions M^+, which have a relatively high $K_{r,M}$, is passed through the resin, the ions will attach to the resin, and H^+ ions will be released in solution. The exchange of H^+ ions with M^+ ions can continue until the equilibrium described by rel. (15.1.5) is reached.

The quantitative treatment of ion exchange equilibria can be theoretically approached with the assumption that the electrostatic interactions between fixed charges (functional groups) and mobile charges (ions) in the resin are long-range interactions covering distances much longer than molecular radii. Therefore, the resin phase can be considered as a homogeneous phase instead of a heterogeneous phase. The solvent and solutes, if not size-excluded, are distributed freely over the two phases, but the functional groups are covalently linked to the matrix and cannot leave the resin phase. The stationary boundary can be visualized as a semipermeable membrane, permeable to all species except the functional groups. For these reasons the theory for ion exchange equilibria can be based on the Donnan membrane equilibrium theory. The presence of small pores may cause steric hindrance and create energetic nonhomogeneities, but this is not expected for the small-sized solutes. Physical inhomogeneities may be caused by very large pores. In these, the liquid is not part of the gel and should be regarded as being part of the liquid phase. With these assumptions it can be shown [1] that the Donnan membrane potential for a specific electrolyte is given by the expression:

$$E_{Donnan} = [RT \ln (a_r / a) - \Pi V] / (z F) \tag{15.1.9}$$

where the indexes "A", "B", etc. were omitted for the activity in the resin a_r, for the activity in solution a, for the charge z, and for the partial molar volume V (F is Faraday's constant). At equilibrium, the Donnan potentials of the two species that are exchanged must be equal. This gives for species "A" and "B" the following expression for the equilibrium:

$$RT \ln \left[\left(\frac{a_r(B)}{a(B)} \right)^{z_A} \left(\frac{a(A)}{a_r(A)} \right)^{z_B} \right] = \Pi (z_B V_A - z_A V_B) \tag{15.1.10}$$

(where the charges z_A and z_B must be taken in absolute value). A *thermodynamic equilibrium constant* $K_{A,B}$ defined by the expression:

$$K_{A,B} = \left[\left(\frac{a_r(B)}{a(B)} \right)^{z_A} \left(\frac{a(A)}{a_r(A)} \right)^{z_B} \right] \tag{15.1.11}$$

can be used to describe the exchange between "A" and "B" species, and:

$$RT \ln K_{A,B} = \Pi \, (z_B V_A - z_A V_B) \tag{15.1.12}$$

Higher thermodynamic equilibrium constants show higher affinity for the resin of species "B", and rel. (15.1.12) indicates that species with higher (absolute value) charges are favored by the resin over those with lower charges. The thermodynamic equilibrium constant can be transformed into an equilibrium constant depending on concentrations by transforming the activities into concentrations with the help of activity coefficients γ of each species in the resin and in solution. The expression for the equilibrium constant can be written in this case in the form:

$$RT \ln K_{A,B} = \ln \left[\left(\frac{\gamma_r(A)}{\gamma(A)} \right)^{z_B} \left(\frac{\gamma(B)}{\gamma_r(B)} \right)^{z_A} \right] + \Pi \, (z_B V_A - z_A V_B) \tag{15.1.13}$$

Rel. (15.1.13) shows that the exchange equilibria are influenced by the osmotic term, the charges of the ions, and the activity coefficients of each species. The affinity for the resins of various inorganic ions was found to vary in the order $B^+ < B^{2+} < B^{3+} < B^{4+}$, and $B^- < B^{2-} < B^{3-}$. For the same valence, it was found that the affinity varies in general as follows: $Li^+ < H^+ < Na^+ < NH_4^+ < K^+ < Rb^+ < Ag^+ < Tl^+$, and $Be^{2+} < Mn^{2+} < Mg^{2+} < Zn^{2+} < Co^{2+} < Cu^{2+} < Cd^{2+} < Ni^{2+} < Ca^{2+} < Sr^{2+} < Pb^{2+} < Ba^{2+}$. For anions the order is the following: $OH^- < F^- < CH_3COO^- < HCOO^- < H_2PO_4^- < HCO_3^- < Cl^- < NO_2^- < HSO_3^- < CN^- < Br^- < NO_3^- < HSO_4^- < I^-$. However, inversions are possible due to the nature of the resin, formation of complexes, etc.

The use of various complexing agents in the solution interacting with an ion exchange resin is another procedure used in analytical chemistry for obtaining separations. The equilibrium between the complexing agent (ligand) in solution and the ions to be exchanged reduces the concentration of the free ions available for the exchange process. In this case two simultaneous equilibria take place:

$$RX^-A^+ + B^+ \; \rightleftharpoons \; RX^-B^+ + A^+ \quad \text{and} \quad L^- + B^+ \; \rightleftharpoons \; LB$$

Assuming that the reaction with the ligand is described by the equilibrium constant $K_{LB} = [LB] / ([L^-][B^+])$, the concentration of B^+ in the resin is given by the expression:

$$[B_r^+] = K_{r,B} \frac{[LB]}{[L^-]K_{LB}} \tag{15.1.14}$$

Rel. (15.1.14) shows that a higher concentration of ligand or a high complexation constant diminishes the amount of species B^+ retained in the column. However, complexation can be used to favor retention on the resin. For example, specific ions form negatively charged complexes. Assuming that an ion M^{2+} forms with a ligand L four combinations ML^+, ML_2, ML_3^-, and ML_4^{2-}, the negatively charged complexes and the ligand can be retained on an anion exchange resin, while the positive ions and the neutral molecules are not retained. In a mixture of ions only some having the complexing capability with the formation of negatively charged compounds, an anion exchanger can be used for the separation of the desired species. In this case, considering the constant K_{comp} describing the equilibrium:

$$4 L^- + M^{2+} \rightleftharpoons ML_4^{2-}$$

the concentration of ML_4^{2-} in the resin can be estimated using the expression:

$$[ML_4^{2-}]_r = K_{r,ML4} \, K_{comp} \, [L^-]^4 \, [M^{2+}] \qquad (15.1.15)$$

Each complex ion has its specific distribution constants in the resin, these depending on factors such as bond strength, hydrophobic interactions, steric hindrance, etc. The adsorption of complex ions on resins is a more complicated process, because in addition to the complexation in solution, the ligand adsorbed on the resin may still participate in complex formation. The donor electrons from the ligand retained as counterion are still available for complexation and may further retain M^{2+} ions from solution. The use of ion exchange resin in retaining metal ions as complexes in different pH conditions has been thoroughly studied and reported in literature [6], [7], [8].

In addition to the exchange of ions, ion exchangers are able to retain specific neutral molecules. There are two different processes in which this can take place. The first is related to the retention based on the formation of complexes. Specific ions such as those of transition metals (Cu^{2+}, Zn^{2+}, Co^{3+}, Ni^{2+}, etc.) may be retained on a cation exchange resin and still have the capability to accept lone pair electrons from donor ligands such as amines. Using this mechanism, neutral ligand molecules can be retained on resins already treated with the transitional metal ions.

The second process of retention of neutral molecules is based on adsorption on ion exchange matrix, without involving an ion exchange process. Organic molecules can be adsorbed on the resin. For example, amines can be retained on a strong cation exchanger in K^+ form. The counterion is important in the adsorbing capability of the resin, and the elution is possible by use of solutions of salts at different concentrations. A "salting out" effect is used to modify the adsorption, the variation of the distribution constant in the presence of the salt being described by a formula similar to rel. (1.6.60):

$$\ln K_{r,A}(c) = \ln K_{r,A}(c = 0) + \kappa \, c \qquad (15.1.16)$$

where $K_{r,A}(c)$ is the distribution coefficient for the analyte "A" in the presence of salt, $K_i(c = 0)$ is the distribution constant in water, κ a constant specific for the system, and c is the molar concentration of the salt. As shown in rel. (15.1.16), the increase in salt concentration increases the adsorption (opposite effect compared to addition of organic modifier), and in a chromatographic process that uses salting out, the elution is done by diluting the initial solution of the eluent [9].

Besides the equilibrium aspects necessary to understand the ion exchange behavior, the kinetic factors are also very important. In the dynamic applications such as chromatography or SPE, in addition to the convective process of longitudinal flow of the fluid through the column and the movement of fluid in the void space between the resin particles, other factors are also important. Among these the diffusion of the compound of interest through the solvent immobilized on the resin particles, the diffusion within the gel microchannels, and the kinetics of the exchange process in itself are factors determining the rate of exchange. In static applications, convective factors are of lower importance, although perfectly static applications are uncommon. Due to the complexity of the process, only the estimation of certain kinetic aspects is usually possible. An expression that estimates the time $t_{1/2}$ for half of the complete conversion of a resin from form "A" into form "B", when the limiting factor is the diffusion in the particle, is given by the formula:

$$t_{1/2} = 0.0075 \ d_p^2 \ /D_r \qquad\qquad (15.1.17)$$

where d_p is the particle diameter and D_r is the diffusion coefficient in the resin. However, other expressions were developed to describe the kinetics of the ion exchange process [10].

- Chelating ion exchange resins

The chelating resins are a special type of exchange resin. A chelating resin consists of a polymer matrix in which chelating ligands are attached by covalent bonds. Chelating resins are different from standard ion-exchange resins, having a higher selectivity, stronger bonding ability, and, in general, lower capacity [10], [11]. These resins have special groups able to form chelates with specific inorganic ions. For chelating resins the exchange equilibria involve the formation of complexes with specific ion species. The chelating properties are affected by the nature and number of donor atoms in the chemically active groups, the chemical nature of the polymer matrix, steric factors, the morphology of the resin, and sorption conditions. The interactions involved in the actual complex formation of chelating polymers may be different from those of the corresponding soluble monomeric complexes. Accessibility of the ligands, steric effects, complex stability, and stoichiometry are all factors that play an important role in the ability of a resin to react with the metal ions from a solution, and the kinetic properties are mainly determined by the rate of diffusion of the metal ions into the resins and the rate of complex formation. With a suitable eluent it is also possible to remove the bound metal ions from the resin and reuse the resin in another analysis [12]. A number of resins reported in literature containing chelating groups attached to a polymeric backbone are indicated in Table 15.1.2.

TABLE 15.1.2. *Functional groups introduced in synthetic ion exchange resins.*

Type	Ligand group	Base polymer
amidoxime		polyacrylonitrile-DVB
aminophosphonic acid		polystyrene (PS), phenolic resin, polyacrylic
cryptands		PS-DVB
dithiocarbamic acid		PS-DVB; polyacrylic phenolic resin
ethylenediaminetriacetic acid		PS
8-hydroxyquinoline		polyhydroxyethylmethacrylate
iminodiacetic acid	$-N(CH_2-COOH)_2$	poly(styrene-divinylbenzene) copolymer (PS-DVB), polyacrylic
isothiurea		PS-DVB
phosphonic acid	$-CH_2-PO(OH)_2$	PS-DVB
pyridine		polyvinyl-DVB
thiol	$-SH$	PS-DVB; polyacrylic-DVB, polyhydroxyethyl methacrylate
thiourea		polyacrylonitrile-DVB

- *Utilization of ion exchangers in sample preparation*

Ion exchange resins have many practical applications, sample preparation in analytical chemistry being minor compared to other uses. The uses for ion exchangers in sample preparation can be grouped into two categories, "nonchromatographic" and "chromatographic."

Ion exchangers can be used in a "static" operation, where the sample is simply mixed with the resin and allowed to interact for a certain period of time. This procedure can be used for the removal of specific ions from solution and replacing them with the counterion preexistent in the resin. This procedure can be applied, for example, exchanging undesirable ions with H^+ ions for cations or with OH^- ions for anions. Replacement of H^+ or OH^- ions from solution is also possible, the resin producing a pH change. Also, certain organic species can be retained on an ion exchanger. For example an anion resin in bisulfite form may retain aldehydes from alcohols.

The retention of specific ions from solution is frequently done for analytical applications using the ion exchange in the form of a SPE cartridge or column. The utilization of ion exchange sorbents fits SPE separations very well, the material being able to work efficiently in retaining completely a given species and releasing it with a change in the eluent. Ion exchangers are applicable to organic and inorganic ions and in some instances to nonpolar molecules. Also the capability of ion exchangers to adsorb ions can be used for desalting of specific samples. For this purpose, it is possible to use mixtures of resins, one anionic in OH^- form and the other cationic in H^+ form. Using strong anions and strong cations, it is possible to replace the anions in a solution with OH^- and the cations with H^+, the two ions generating H_2O. The process can be done simultaneously to not modify significantly the pH of the solution. Desalting is successfully applied on sensitive biological samples.

The use of ion exchangers as chromatographic stationary phases is also very common. The application of rel. (1.6.33) for the calculation of the retention volume in a separation on an ion exchange column leads to the expression:

$$V_R = V_m + K_{r,B} V_s \qquad (15.1.18)$$

where $K_{r,B}$ is given by rel. (15.1.5). Based on this expression, it can be seen that the absorption and elution of two components depends on their distribution constant in the resin. Selecting properly the eluent, separation of various components of a sample can be achieved, for example by gradual change of the eluent pH or concentration. Ion separation using ion exchange columns is commonly applied in ion chromatography [13].

- Examples of applications of ion exchange resins in sample preparation

The retention capability of ion exchangers has numerous applications such as in SPE and membrane separations. Columns made of ion exchangers as stationary phase are used in various types of chromatography (see Sections 4.3 and 4.4). Among these, ion chromatography is probably the most common. Ion mediated chromatography is also a useful separation technique done on ion exchangers. In this application, a nonionic solute interacts with the stationary phase of a resin containing ions and specific counterions (such as Na^+, Ag^+, Pb^{2+}), allowing separations of compounds such as carbohydrates, alcohols, etc. Also, ion exchange membranes have applications in sample preparation.

The utilization of ion exchangers as a sorbent in SPE is discussed in Section 11.1. The SPE columns, cartridges, or disks with an ion exchanger as sorbent are ideal in some applications used either for analyte retention followed by elution or for allowing the analytes to pass the column, retaining interfering compounds from the matrix. This can be applied successfully to both inorganic and organic ions. Extraction and purification of amino acids, for example, can be done successfully using this procedure. The analysis of free amino acids from tobacco can be done starting with 2 g sample, which is extracted with 50 mL 0.1 N HCl that contains an internal standard such as norleucine (0.2 mg/mL). The amount of sample and the amount of solution can be scaled for convenience. The extraction is done using mechanical shaking for 1 h. The extract is filtered, and 5 mL solution is passed through a cation exchange column in H$^+$ form (e.g. Poly Prep AG 50 W-X8 Bio-Rad). The column is then washed with 15 mL water. The amino acids are eluted with 15 mL 1N NH$_4$OH. For further analysis, the eluate is evaporated to dryness and derivatized with phenylisothiocyanate or with N-methyl-N-(*tert*-butyldimethylsilyl)-trifluoroacetamide (MTBSTFA) [14].

Ion exchangers being well suited for the separation and purification of ions, many applications of sample preparation using ion exchangers are followed by ion chromatographic analysis. These preparations include sample cleanup, pH adjustment, and elimination of salts that may interfere with the IC analysis. For example, the pH adjustment using ion exchangers is recommended in the IC analysis when the sample neutralization with acids or bases would significantly increase the salt content. Some applications of ion exchangers prior to IC chromatography are given in Table 15.1.2.

TABLE 15.1.2. *Applications of sample cleanup using ion exchangers in IC.*

Sample	Analyte	Ion exchanger	Resin form	Interference removed	Reference
bread	BrO$_3^-$	Dowex 50W-X8-10	Ag$^+$	Cl$^-$	[15]
brine	anions	Dionex ICE suppresser	Ag$^+$	Cl$^-$	[16]
brine	SO$_4^{2-}$	Cation exchanger	H$^+$	Cl$^-$	[17]
water	aldehydes	Dowex 1X8	acetate	Cl$^-$	[18]
water	anions	Bio-Rad X-4, X-8, X-16	Ag$^+$	Cl$^-$	[19]
water	F$^-$, SiO$_3^{2-}$	Dowex 50W-X8	H$^+$	cations	[20]
NaOH	anions	Bio-Rad AG50W-X12	H$^+$	high pH	[21]
Na$_2$CO$_3$ fusion melt	anions	Bio-Rad AG50W-X12	H$^+$	high pH	[22]
urine	Br$^-$	Cation exchanger	CO$_3^{2-}$	anions	[23]
ozonated drinking water	BrO$_3^-$	Dionex OnGuard and Dionex MetPac CC-1	Ag$^+$	Cl$^-$	[24]
food, after microwave digestion	Br$^-$	Sep-Pak C18	H$^+$	high pH	[25]
soil	fluoroacetic acid, formic acid	OmniPac PAX-100G	Ag$^+$	Cl$^-$	[26]

The analysis of a trace analyte in the presence of high levels of interfering ions typically leads to overloading of the IC column and poor recoveries of the target analytes [27]. This problem can be avoided through the use of a matrix elimination with ion exchange resins either as SPE devices or as HPLC precolumns. For example, for the determination of traces of bromate in water containing chloride, the water is passed through an ion exchange column in Ag$^+$ form, which retains Cl$^-$ forming a precipitate. The bromate is further analyzed using conductivity detection with anion membrane suppression and regeneration using a H$_2$SO$_4$ solution. The detection limit of this procedure can reach 0.5 µg/L of bromate, and the recoveries of 5 µg/L can be as good as 95%. Samples containing residual disinfectants must be treated with ethylenediamine to stabilize the bromate concentration [28].

Ion exchange columns also can be used as preconcentration devices. For this purpose, a large volume of a diluted solution is passed through an ion exchange bed that has a high distribution constant in the resin for the selected species. After this step, the collected ion is eluted with an appropriate eluent in much smaller volume than the initial sample solution. Another example is the determination of thiosulfate and polythionates in saline waters using pre-concentration by ion exchange on an anion exchange column (Waters IC-PAK) housed in a guard pre-column module. The elution is done with acetonitrile/H_2O (1:3) containing carbonate buffer solution. For a 6 mL sample of 1:50 diluted sea water, the analytes are pre-concentrated by a factor of 10, and the quantification limits are 1 nM for trithionate and 0.3 nM for tetra- and pentathionate [29]. Various other inorganic and organic anions at trace levels can be concentrated on an anion-exchange column, eluted with toluene-p-sulfonic acid and determined on-line by ion chromatography with spectrophotometric detection [30].

Another example is the analysis of triazine herbicides in water samples. Triazines with basic character (atrazine or 2-chloro-4-ethylamino-6-isopropylamino-1,3,5-triazine, aziprotryne or 2-azido-4-isopropylamino-6-methylthio-1,3,5-triazine) can be retained on strong cation exchange resins and eluted with buffers followed by HPLC analysis [31].

The use of chelating ion-exchange resins for the isolation of metallic species, for example from waste water, allows analyses at very low concentration levels of metals. For example, Al^{3+} at ppb levels can be concentrated on a column of 2-(salicylidene-amino)thiophenol immobilized on glass beads, then eluted with HNO_3, and analyzed.

15.2 OTHER TECHNIQUES IN SAMPLE PREPARATION

A considerable portion of sample preparation techniques can be viewed as conventional procedures. Common techniques making use of the differences in physicochemical properties of the sample components are discussed in Part 2 of this book, and those applying chemical modifications for the sample preparation are discussed in Part 3. However, the variety of procedures and techniques applied for sample processing for chromatographic analysis is very wide. Some procedures are appropriate only for a narrow range of samples, and others use techniques that are not frequently applied in sample preparation. These procedures can be found in the original literature, which offers full information regarding the analysis of a specific type of sample. Also, new procedures are continuously being developed, and they may not be captured in this material.

A number of separation techniques have the potential to be applied as sample preparation procedures. Among these are field flow fractionation [32], [33], ionic flotation [34], formation of clathrates and inclusion complexes [35], ring oven techniques [36], etc. However, these procedures are not applied as typical sample preparation techniques and are not discussed in this book.

REFERENCES 15

1. F. Helfferich, *Ion Exchange*, McGraw-Hill, New York, 1962.
2. E. Ott et al., *Cellulose and Cellulose Derivatives*, Interscience, New York, 1954.
3. K. Dorfner, *Ion Exchangers. Properties and Applications*, Ann Arbor Science Pub. Inc., Ann Arbor, 1973.
4. K. Mosbach (ed.), *Methods in Enzymology*, Vol. 44, Academic Press, New York, 1976.
5. R. H. Perry, D. W. Green (eds.), *Perry's Chemical Engineers' Handbook*, 6th ed., McGraw-Hill, New York, 1984.
6. H. L. Rothbart et al., Talanta, 11 (1964), 33.
7. F. C. Nachod, I. Schubert (eds.), *Ion Exchange Technology*, Academic Press, New York, 1956.
8. K. A. Kraus, F. Nelson, Proc. Intern. Conf. Peaceful Uses Atom. Energ., Geneva, 7 (1956) 113.
9. R. Sargent, W. Rieman, Anal. Chim. Acta, 17 (1957) 408.
10. S. K. Sahni, J. Reedijk, Coord. Chem. Rev., 59 (1984) 1.
11. A. Chakrabarti, M. M. Sharma, Reactive Polymers, 20 (1993) 1.
12. A. R. Timerbaev et al., Fresenius' Z. Anal. Chem., 327 (1987) 87.
13. P. R. Haddad, P. E. Jackson, *Ion Chromatography: Principles and Applications*, Elsevier, Amsterdam, 1990.
14. J. H. Lauterbach, S. C. Moldoveanu, 42nd TCRC, Lexington, 1988.
15. K. Oikawa et al., Bunseki Kagaku, 31 (1982) E251.
16. P. F. Kehr et al., LC-GC, 4 (1986) 1118.
17. F. A. Buytenhuys, J. Chromatogr., 218 (1981) 57.
18. D. L. DuVal et al., Anal. Chem., 57 (1985) 1583.
19. R. Bagchi, R. Haddad, Proc. 9th. Aus. Symp. Anal. Chem., 1987, p. 147.
20. R. Golombek, G. Chwedt, J. Chromatogr., 367 (1986) 69.
21. R. A. Hill, J. High Resolut. Chromatogr., 6 (1983) 275.
22. L. W. Green, J. R. Woods, Anal. Chem., 53 (1981) 2187.
23. H. Moore et al., J. Vet. Res., 2 (1987) 297.
24. R. J. Joyce, H. S. Dhillon, J. Chromatogr., 671 (1994) 165.
25. Miyahara, Y. Saito, J. Agric. Food Chem., 42 (1994) 1126.
26. F. Kingery, H. E. Allen, J. Chromatogr., 671 (1994) 231.
27. P. R. Haddad et al., J. Chromatogr., A, 856 (1999) 145.
28. H. Weinberg, J. Chromatogr., A, 671 (1994) 141.
29. S. I. Weir et al., J. Chromatogr., A, 671 (1994) 197.
30. J. Kobayashi, Bunseki Kagaku, 43 (1994) 727.
31. G. Sacchero et al., J. Chromatogr., A, 671 (1994) 151.
32. J. C. Giddings, J. Chromatogr., 125 (1976) 3.
33. R. A. Meyers (ed.), *Encyclopedia of Analytical Chemistry*, J. Wiley, Chichester, 2000.
34. Y. S. Kim., H. Zeitlin, Anal. Chem., 60 (1971) 1390.
35. D. T. Haworth, R. M. Ziegert, J. Chromatogr., 30 (1967) 637.
36. K. H. Johri, K. Singh, Mikrochim. Acta 1970, 1.

PART 3

Sample Preparation Techniques Involving

Chemical Modifications

CHAPTER 16

Chemical Modifications for Dissolution and Fractionation

16.1. CHEMICAL MODIFICATIONS FOR SAMPLE DISSOLUTION

When solid samples cannot be dissolved in a solvent, some method involving chemical modifications may be the alternative for taking them in solution. Dissolution using chemical reactions can be in some cases convenient even if the sample is soluble in a specific solvent.

For most compounds with small organic molecules, it is possible to select a specific solvent in which the compound is soluble. The dissolution process is presented in Section 8.2. The most common solubilization associated with a chemical modification is that using a pH change. For example, ionic compounds have a significantly better solubility in water compared to that in organic solvents. Many organic compounds are weak acids or bases and have very small dissociation constants. However, weak acids or weak bases can be changed into salts. Salts are usually completely dissociated in aqueous solution. For this purpose, acids can be dissolved in water by being changed into salts at increased pH. By the same procedure, basic compounds such as amines are made more soluble in water at low pH, usually obtained by the addition of a mineral acid that changes the free base into a salt. On the other hand, ionic compounds show a decreased solubility in organic solvents, and if a substance is present in a salt form and can be changed into a weak free acid or a weak free base, these are more soluble in an appropriate organic solvent.

A number of other chemical modification procedures can be done on samples to obtain a better solubility. For example, compounds having active hydrogens can be derivatized (methylated, silylated, etc.) such that they become less polar and soluble in specific organic solvents. Many derivatizations are performed on the initial sample without an initial dissolution. The derivatization process occurs in this case associated with a solubilization. Derivatization procedures are further discussed in this book.

More difficult solubility problems are encountered with two types of compounds. The first are specific inorganic compounds such as oxides, silicates, etc. Some of the inorganic materials are solubilized only after aggressive chemical treatment such as fusion with melted alkali or acidic salts such as $KHSO_4$. Dissolution of inorganic compounds can be followed by ion chromatographic analysis. However, nonchromatographic techniques are more frequently applied for inorganic ion analysis. The second large group of compounds that may be difficult to dissolve is comprised of the polymers. Polymer analysis and their dissolution are further discussed in Chapter 20. Only some general remarks are presented in this section. Particular organic molecules such as some organic dyes (pigments) also can be difficult to dissolve, and their analysis may be done using nonchromatographic techniques or pyrolysis.

The dissolution of insoluble inorganic samples, such as geological or metallurgical materials, may need chemical degradation of the sample. For this purpose, strong acids or bases are usually utilized, the acids targeting the cations with the formation of soluble salts and the bases targeting the anions. Dissolution with the elimination of the organic matrix and with the purpose of the analysis of inorganic components also is done with chemical modification of the sample. In some of these dissolutions, the reaction is performed at elevated temperatures.

Alkali fusion, for example, is based on a melting process of a mixture between the sample and an alkaline compound such as NaOH, Na_2CO_3, $NaHCO_3$, or $Li_2B_4O_7$. The alkali fusion is more frequently applied in the anion analysis. Details regarding melting conditions, such as temperature, reaction time, etc., are highly dependent on the sample and the labware used for the operation [1], [2]. It is common to use an excess of the alkali, 5–10 times larger than the sample. After cooling, the material obtained from the fusion process is dissolved in a suitable solvent and then analyzed, for example, by IC. Halides such as fluoride [3] or chloride [4] can be determined in geological samples using an alkali fusion and IC analysis, and boron and fluoride can be determined in glasses using NaOH for the alkali fusion of these samples [5]. The use of NaOH and Na_2CO_3 + $NaHCO_3$ for alkali fusion have the advantage of being compatible with the type of mobile phase for IC.

Acid digestion is more frequently applied for sample preparation for cation analysis. Classic acid digestion of samples uses closed, chemically inert vessels such as Teflon or special glass tubes that are heated in a conventional oven [6]. Dissolution by fusion with $KHSO_4$ or $K_2S_2O_7$ also can be applied for specific samples such as ceramic oxides like Al_2O_3 or ZrO_2. Dissolution in less volatile acids such as H_2SO_4 or H_3PO_4 has been reported for the analysis of geological samples [7].

A common type of sample to be analyzed for inorganic composition is the biological/botanical one. A variety of digestion procedures (wet ashing) are used, combining the microwave energy, closed vessels withstanding high pressure, strong acids, and oxidative conditions (see also Section 10.4). Some typical procedures applied for the digestion of biological/botanical materials are indicated in Table 16.1.1.

TABLE 16.1.1. *Some procedures for the digestion of biological materials.*

Procedure	Conditions
dry ashing	slow heating in air at 500–700° C
combustion	Parr oxygen bomb, furnace
acid digestion	various acids, common labware
microwave assisted digestion at low pressure	digestion in the presence of various reagents at temperatures lower than reagent boiling point
microwave assisted pressure digestion	digestion in the presence of various reagents at temperatures lower than 250° C
high pressure digestion	temperatures up to 350° C
UV photolysis	special equipment

Microwave-assisted sample preparation is frequently utilized because it can improve the analytical performances of digestion methods [8]. Some acids used in digestion of insoluble samples using microwave heating and pressure vessels are indicated in Table 16.1.2.

TABLE 16.1.2. *Some acids and acid mixtures used in the insoluble sample microwave digestion.*

Acid/mixture	Properties	Application
nitric acid	strong acid, oxidizing agent	biological and botanical matrices
hydrochloric acid	strong acid, volatile	inorganic oxides, sulfates, etc.
hydrofluoric acid	strong acid, reacting with SiO_2, volatile	silica based materials
phosphoric acid	medium acidity, nonvolatile	iron based alloys
tetrafluoroboric acid	reacting with SiO_2	silicates, geological samples
sulfuric acid	strong acid, nonvolatile	inorganic oxides, ores
perchloric acid	strong acid, oxidizing agent	decomposition of organic samples, metals
nitric acid + H_2O_2	strong acid, oxidizing agent	biological and botanical matrices
nitric acid + phosphoric acid	lower volatility than nitric acid	
nitric acid + hydrofluoric acid	strong acid, oxidizing agent	silicates, etc.
nitric acid + hydrochloric acid	generating NOCl (very reactive)	insoluble metals, etc.
tetramethylammonium hydroxide	methylating reagent	biological samples

Organic samples that are subject to wet ashing are practically reduced to their inorganic components, while the organic matter is transformed into CO, CO_2, N_2, etc. Most metals are transformed into their salts, and because the nitrates are water soluble, digestion with HNO_3 is one of the most common digestion procedures. Depending on the sample matrix and the purpose of analysis, a specific digestion system is selected.

When combustion methods are used as a step in sample preparation, the sample is subjected to a combustion process in oxygen or air, followed by the dissolution of the residue or by the collection of the gases resulting from the process. This approach is used for the determination of halides, which are transformed into volatile HF, HCl, HBr, and HI, or for removing S and P from sample as SO_2 and P_2O_5, respectively. Several experimental possibilities [9] are used for sample combustion, such as Schoeniger flask combustion, Parr oxygen bomb combustion, furnace combustion, etc.

Polymer solubilization without the destruction of the structure or without loss of significant information about the polymer structure can be a difficult task. Highly crosslinked polymers are in general not easy to solubilize, and techniques such as hydrolysis or pyrolysis are used in sample preparation for further chromatographic analysis. Nonchromatographic techniques are also common, mainly in the analysis of synthetic polymers. Some procedures applied in sample preparation for polymeric materials are discussed in Chapter 20.

One common problem in the analysis of biological materials is protein solubilization and the keeping of proteins in solution. Simple solubilization of some proteins may affect their structure, and chemical modifications can be done inadvertently. Proteins can be classified based on their solubility, although this type of classification is far from perfect for such a complex group of compounds, mainly due to gradation in solubility. Based on their solubility, proteins are classified as a) albumins, which are soluble in water, dilute salts, acids, and bases, and precipitate with a 2M ammonium sulfate solution, b) globulins, which are insoluble in pure water but dissolve with dilute salts, acids or bases, c) glutelins, which are insoluble in neutral solvents including salt solutions, but dissolve by dilute acids or bases, d) prolamines, which are soluble in alcohol, e) scleroproteins, which are insoluble in most solvents, f) histones, which are basic proteins soluble in water but insoluble in a dilute ammonia solution, and g) protamines, which are similar to histones, but are not coagulated by heat as are other soluble proteins. This classification shows that protein solubilization can be done in some cases by using a specific salt concentration or a specific ionic strength of the solution. Solubilization of proteins followed by their reprecipitation may lead to a modified substance compared to the initial material. In other cases, such as for scleroproteins, the solubilization is very difficult, and protein degradation is necessary for solubilization (see Chapter 20).

16.2. CHANGES IN pH AND CHEMICAL MODIFICATIONS FOR FRACTIONATION PURPOSES

The change in pH (pH adjustment) is a very common operation in sample preparation, either for fractionation purposes or for modifying the sample for further derivatization. The pH is defined as a measure of hydronium ion concentration in an aqueous solution and is given by the expression:

$$pH = -\log_{10} [H_3O^+] = -\lg [H_3O^+] \qquad (16.2.1a)$$

In principle, $[H_3O^+] = [H^+]$, but in an aqueous solution the H^+ ions cannot be free. Their heat of hydration in the reaction $H^+ + H_2O \rightleftharpoons H_3O^+$ is about -200 kcal/mol. Assuming that the variation of the hydration free enthalpy ΔG is not very different from the heat of hydration, from $\Delta G^0 = -RT \ln K$ (see rel. 1.8.8) at room temperature, it results $K \approx 10^{146}$. This value indicates that practically no free H^+ is present in aqueous solutions. The pH is also defined by the expression:

$$pH = -\lg a_{H3O+} \qquad (16.2.1b)$$

where a_{H3O+} is the activity of hydronium ions. The two expressions for the definition of pH give the same result for diluted solutions. For pure water at 25° C, the concentration of $[H_3O^+] = 10^{-7}$, and for water pH = 7 (see also Table 18.1.3). Based on the pH of their water solution, chemical compounds are characterized as acidic or basic, with the acidic ones having the pH < 7. The pH is typically measured using electrochemical procedures with a pH electrode or estimated based on the color change of an indicator compound.

Any compound involved in the reaction of the type:

$$R-H + H_2O \rightleftharpoons R^- + H_3O^+$$

is characterized by the acidity constant:

$$K_a = \frac{[R^-][H_3O^+]}{[RH]} \qquad (16.2.2)$$

and values for K_a for many compounds are available in literature [10], [11], [12]. A number of inorganic acids have very large K_a values and are known as strong acids.

Most organic acids have relatively low K_a values and are considered weak acids. With the analytical concentration c_{RH} of the acid given by the expression:

$$c_{RH} = [RH] + [R^-] \qquad (16.2.3)$$

and based on the electrical neutrality of the solution, which imposes $[R^-] = [H_3O^+]$, the expression of K_a can be written in the form:

$$K_a = \frac{[H_3O^+]^2}{c_{RH} - [H_3O^+]} \qquad (16.2.4)$$

With the assumption that $c_{RH} \gg [H_3O^+]$, rel. (16.2.4) leads to the expression:

$$[H_3O^+] \approx \sqrt{K_a c_{RH}} \qquad (16.2.5)$$

Viewing the dissociation reaction of an acid from right to left, the following equilibrium can be written:

$$R^- + H_2O \rightleftharpoons R\text{–}H + OH^-$$

where R^- is the conjugate base of the acid RH. The constant for this equilibrium, known as basicity constant, is the following:

$$K_b = \frac{[RH][OH^-]}{[R^-]} \qquad (16.2.6)$$

where $K_a K_b = [H_3O^+][OH^-] = K_w$ and K_w is known as the ionic product of water. For a temperature of 25° C, $K_w = 10^{-14}$. The typical reaction for basic compounds is described by the equilibrium:

$$RX + H_2O \rightleftharpoons RXH^+ + OH^-$$

which is characterized by the basicity constant:

$$K_b = \frac{[RXH^+][OH^-]}{[RX]} \qquad (16.2.7)$$

The conjugate acid of the base RX is RXH^+ and has the acidity constant given by the expression $K_a = K_w / K_b$. Compounds with large K_b values are known as strong bases, and those with low K_b values as weak bases. An expression similar to rel. (16.2.5) that relates K_b and $[OH^-]$ can be developed for the $[OH^-]$ concentration in the solution of a weak base. Also, besides acids or bases with only one ionizable group, compounds with more ionizable groups are common. The theory for the acid-base properties of these compounds can be easily developed similarly to that for unique ionizable groups.

The pH of a solution containing a weak acid and its conjugate base (in the form of a dissociated salt) can be easily obtained from rel. (6.2.2) in the form:

$$pH = lg(K_a \frac{[RH]}{[R^-]})$$ (16.2.8)

In rel. (16.2.8) the concentration [RH] can be approximated with the analytical concentration c_{RH} of the free acid, and the concentration [R⁻] can be approximated with the analytical concentration c_{R^-} of the salt present in the solution (salts are assumed totally dissociated). Therefore, rel. (16.2.8) can be written in the form:

$$pH = lg(K_a \frac{c_{RH}}{c_{R^-}})$$ (16.2.9)

Rel. (16.2.8) indicates that the pH of a solution of a weak acid and its salt depends only on the ratio c_{RH} / c_{R^-} and not on the dilution of the solution. This is different from the case of solutions of pure acids (or bases) where the pH varies with the acid concentration, as shown by rel. (16.2.5).

Solutions of weak acids containing a certain amount of its salt also have the capability of maintaining a relatively constant value for the pH at the addition of a small amount of a strong acid or strong base. As an example, the variation of pH calculated for various concentrations in water of a strong acid (completely dissociated) and the variation of pH of the same concentrations of a strong acid in the presence of a buffer solution containing 0.2 M CH₃COOH and 0.2 M CH₃COOH is shown in Figure 16.2.1. The trace showing the variation of the pH for the buffer shows a constant value for a relatively wide range of concentrations of the added acid. Similar behavior can be seen during the addition of a strong base to a buffer. The ability of a buffer to resist pH changes as a result of the addition of strong acids or bases depends on the total concentration of buffering species. A *buffering capacity* is defined as the number of equivalents of strong acid or strong base needed to cause 1.0 L of buffer solution to experience a 1.0-unit change in pH.

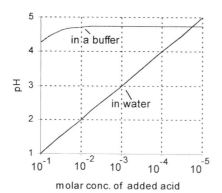

FIGURE 16.2.1. *Variation of pH for various concentrations of a strong acid in water and in the presence of a buffer (0.2 M CH₃COOH and 0.2 M CH₃COOH).*

Besides weak acids and their salts, other mixtures such as solutions of weak bases and their salts are used as buffers. Also, solutions of compounds acting as zwitterions (amphions), such as amino acids with the formula $R-CH(NH_3^+)-COO^-$, can be used as buffers. Detailed recipes for buffers covering various pH ranges are reported in literature [12].

It must be noted that all the equilibria for the definition of pH and for K_a and K_b are assumed to take place in water (and the concentration of water is neglected in rel. 16.2.2 and 16.2.3). However, the pH can be measured not only in aqueous solutions but also in semiaqueous solutions and in emulsions [13], [14]. The meaning of this value must be viewed only as an experimental value and not representative of the theoretical pH. The theory of pH in mixed solvents and nonaqueous media takes into account various factors that affect the activity of the hydronium ions [15]. It is common that the pH change is done in sample preparation only for a practical purpose, and the correct meaning of the measured pH value is not essential.

In sample preparation, the pH change can be achieved using various procedures depending on the nature of the sample and the requirements of the further steps in the analysis. One simple procedure is the addition of strong inorganic acids or bases to neutralize the basic or acidic character, respectively. These reagents may be involved in neutralization reactions of the form:

$$RH + NaOH \longrightarrow RNa + H_2O, \quad \text{or} \quad R\text{-}OH + HX \longrightarrow R\text{-}X + H_2O$$

or may react with acidic or basic compounds forming stable, more neutral salts. Some inorganic salts with basic or acidic character upon hydrolysis also can be used for changing the pH. For example, Na_2CO_3 or $NaHCO_3$ can be used for the neutralization of acidic solution by substituting the acid in the sample with H_2CO_3, which readily decomposes generating CO_2, which is eliminated.

Not only inorganic compounds are added for pH changes. Addition of organic amines, for example, is commonly used for neutralizing acidic samples. Among the most common amines used for this purpose are pyridine, trimethylamine solution in water, and triethylamine. Weak organic acids such as CH_3COOH, etc. can be used for the neutralization of basic samples.

A very common procedure used for the pH change of a sample is the addition of buffer solutions. Among these, acetate buffer, arsenate buffer, borate buffer, citrate buffer, glycine buffer, phosphate buffer, pyrophosphate buffer, 2-amino-2-methyl-1-propanol buffer, and tris(hydroxymethyl)amminomethane buffer are common [16], [17]. It is also possible to use a combination neutralization and buffer addition for achieving a desired solution pH.

Together with the pH change it is important for specific applications to maintain a specific ionic strength of the solution. Ionic strength of a solution is defined by rel. (13.2.16), and the calculation of this parameter can be done considering the concentration of various ions in solution. Both neutralization with acids and bases and the addition of buffer solutions typically add a significant amount of salts in the sample. If addition of salts is not desirable, a third procedure, which can be used for changing the solution pH without significant increase in ionic strength, is the application of ion exchangers. Using for example an ion exchanger in a reaction of the type:

$$R\text{-}X^- Me^+ + H_3O^+ \rightleftharpoons R\text{-}X^- H^+ + H_2O + Me'$$

where R-X represents the resin, allows the replacement of hydronium ions with Me^+ ions like Na^+, K^+, etc., reducing the acidity of the sample. The affinity for the resin of various ions varies in general in the order $Li^+ < H^+ < Na^+ < NH_4^+ < K^+$ (see Section 15.1), and the replacement of H^+ from solution with Na^+ or K^+ requires more resin than the stoichiometric quantity to displace the equilibrium toward Me^+ in solution. The same condition must be achieved when attempting to replace OH^- from a solution with groups such as Cl^- or CH_3COO^-, etc. The affinity of anions for the resin varies in the order $OH^- < F^- < CH_3COO^- < H_2PO_4^- < HCO_3^- < Cl^-$. Some other procedures can be applied for changing the pH of particular samples, such as the elimination by heating of volatile acids or bases, use of specific sorbents, etc.

Chemical modifications are widely used with the purpose of separating the sample components. However, the chemical modification is usually associated with an additional operation, which leads to the fractionation. A simple example is a precipitation reaction where the chemical process is followed by the formation of an insoluble compound that can be separated by filtration or centrifugation. The chemical modification and physicochemical process leading to fractionation (or separation) are discussed as separate subjects for most cases described in this book. Many chemical modifications performed with the purpose of modifying volatility or solubility are discussed in Chapters 18, 19, and 20. The separation techniques were reviewed in Part 2 of this book. There are some fractionations where the chemical reaction is the cause of fractionation, such as some electrochemical separations or the use of enzymes for elimination of a compound from a mixture. For example, pure isomers of amino acids can be obtained from a racemate with enzymatic oxidation with D-amino acid oxidase to remove the D-amino acids or with L-amino acid oxidase to remove L-amino acids [18].

Among the separations where a chemical reaction is the cause of separation can be included ion exchange (see Section 15.1), immunoaffinity chromatography, and other chromatographic techniques with the separation mechanism based on chemical affinity [19], [20], immunoextraction [21], etc. These types of separations are either discussed in connection with other separation procedures or are related to specific separations used in life science, which are beyond the scope of this book.

REFERENCES 16

1. G. Svehla, *Vogel's Textbook of Macro and Semimicro Qualitative Inorganic Analysis*, Longman, London, 1979.
2. F. P. Treadwell, W. T. Hall, *Analytical Chemistry*, vol. II, J. Wiley, New York, 1942.
3. S. A. Wilson, C. A. Gent, Anal. Lett., 15 (1982) 851.
4. S. A. Wilson, C. A. Gent, Anal. Chim. Acta, 148 (1983) 299.
5. C. McCrory-Joy, Anal. Chim. Acta, 52 (1986) 277.
6. M. Stoeppler, *Sampling and Sample Preparation*, Springer, Berlin, 1997.
7. W. T. Kennedy et al., Anal. Lett., 16 (1983) 1133.
8. H. M. Kingston, L. B. Jassie (eds.), *Introduction to Microwave Sample Preparation*, ACS, Washington, 1988.
9. P. R. Haddad, P. E. Jackson, *Ion Chromatography. Principles and Applications*, Elsevier, Amsterdam, 1990.

10. D. R. Lide (ed.), *CRC Handbook of Chemistry and Physics*, CRC Press, Boca Raton, 1990.
11. D. D. Perrin, *Dissociation Constants of Organic Bases in Aqueous Solution*, Butterworths, London, 1965.
12. J. A Dean (ed.), *Lange's Handbook of Chemistry*, McGraw-Hill, New York, 1999.
13. J. Z. Dong et al., Beitrage Tabak. Intern., 19 (2000) 33.
14. V. A. Thorpe, J. Assoc. Off. Anal. Chem., 56 (1973) 154.
15. R. G. Bates, *Determination of pH, Theory and Practice*, J. Wiley, New York, 1973.
16. N. E. Good et al., Biochemistry, 5 (1966) 467.
17. J. W. Ferguson, N. E. Good, Anal. Biochem., 104 (1980) 300.
18. B. L. Karger et al., *An Introduction to Separation Science*, J. Wiley, New York, 1973.
19. D. S. Hage, J. Chromatogr., B, 715 (1998) 3.
20. L. A. van Ginkel, J. Chromatogr., 564 (1991) 363.
21. D. S. Hage, Clin. Chem., 45 (1999) 593.

CHAPTER 17

Purpose of Derivatization in Chromatographic Analysis

17.1 GENERAL ASPECTS REGARDING DERIVATIZATION

From the sample preparation point of view, chemical changes of different types can be classified as derivatizations, polymer degradations, or pyrolytic reactions, as shown in Section 1.2. The derivatizations are reactions with one or more reagents with the purpose of changing the chemical nature of the analyte and making it more suitable for analysis. Polymer chain fragmentations or pyrolytic reactions are mainly applied to polymers to generate smaller molecules easier to analyze. These chemical modifications are considered separately from derivatizations.

The derivatizations typically target the analytes and not the matrix. Some components from the matrix also may be derivatized unintentionally, but the common strategy is to obtain as much chemical difference as possible between the matrix and the analytes. When a chemical derivatization acts only on the analytes while the matrix components are not affected, the chromatographic separation of the analytes from matrix is expected to be better. In polymer degradation, usually the polymer is first obtained in a pure form. In the case of mixtures of polymers, the degradation is typically done selectively on one polymer type. In pyrolysis, if a matrix is present, the whole sample is usually pyrolyzed.

The purpose of derivatization varies depending on the analyte, the matrix of the sample, and the analytical method to be applied. Some derivatizations are used in the sample cleanup or concentration process. Much more frequently, the derivatization is done to change the analyte properties for the core separation (GC, HPLC, etc.). The matrix is typically not affected in these derivatizations. In GC analyses the derivatization can be done to obtain a better behavior of the analyte in the chromatographic column (such as less peak tailing, no decomposition, etc.), to modify the separation, to improve thermal stability in the injection port, or for better detectability. When using GC-MS as the analytical technique, the derivatization may help in spectra identification. For HPLC and other techniques using liquid mobile phase, the derivatization is done mainly to enhance detectability, although improvements in separation also may be intended. The derivatization for CE or electrochromatography is performed in a similar way to the derivatization for HPLC. In CE the detection is similar to that applied in HPLC, and for this reason, similar derivatizations are applied in the two techniques. A variety of other objectives can be achieved using derivatizations, as shown further in this book. For example, the derivatization may be done for protecting some specific groups such as thiol, which can be easily oxidized during the analysis, or a hydrolysis reaction can be used as in the case of carbamates when more reactive groups are generated and can be further derivatized with reagents enhancing detectability. Also, the derivatization process may be combined with simultaneous extraction and concentration of the sample or may be followed by a second preparation step before the chromatographic analysis.

As chemical reactions, derivatizations are chosen to be efficient chemical processes between the analyte and the reagent. These reactions can be classified based on their chemical nature as 1) reactions with formation of alkyl or aryl derivatives, 2) silylation reactions, 3) reactions with formation of acyl derivatives, 4) reactions of addition to carbon-hetero multiple bonds, 5) reactions with formation of cyclic compounds, etc.

Many derivatizations consist of replacing active hydrogens from an analyte Y-H in functional groups such as OH, COOH, SH, NH, CONH. These reactions can be written in a simplified form as follows:

$$Y\!-\!H + R\!-\!X \longrightarrow Y\!-\!R + HX$$

In this reaction, the reagent R–X contains an "active" group X and a group R that carries a desired property (lack of polarity for GC, absorbance in UV for HPLC, etc.). For GC analysis the group R in the reagent is a low molecular mass fragment such as CH_3 or C_2H_5, a short chain fluorinated alkyl in alkylation reactions, $Si(CH_3)_3$ or other silyl groups in silylations, $COCH_3$ or short chain fluorinated acyl groups in acylations, etc. For HPLC, larger groups usually containing chromophores or groups that fluoresce are selected to be added to the analyte molecule. The choice of the appropriate derivatization reagent is not always simple. The type of functional group in the analyte as well as the nature of the rest of the analyte molecule are key factors in choosing a derivatization reagent. The matrix of the sample has a very important role in the choice of a specific derivatization procedure, unless sample preparation steps performed before derivatization changed the initial matrix. Table 17.1.1 gives a simplified view of preferences for the choice of a derivatization reagent for compounds containing active hydrogens.

TABLE 17.1.1. *Derivatization preferences for compounds containing active hydrogens.*

Properties	$K_{A,B} = \dfrac{[B\;;\;]}{[B^+\;]} / \dfrac{[A\;;\;]}{[A^+\;]}$ — increased nucleophile properties →				
COMPOUND	AMINE	AMIDE	ALCOHOL	PHENOL	ACID
first derivatization preference	acylation	acylation	silylation	silylation	alkylation
second derivatization preference	alkylation	alkylation	acylation	acylation	silylation

Besides functionalities with active hydrogens, other functionalities also can be derivatized. Compounds containing carbonyls or containing more than one active group (multifunctional compounds) can be derivatized, for example, using condensation reactions as shown below for a ketone:

$$Ya\!-\!\underset{\underset{O}{\|}}{C}\!-\!Yb + RH_2 \longrightarrow Ya\!-\!\underset{\overset{\|}{}}{\overset{R}{C}}\!-\!Yb + H_2O$$

For performing the appropriate chemical modifications, a number of derivatization reagents containing various reactive groups and groups carrying the desired properties have been developed. In addition, not only single derivatizations but also multiple step derivatizations introducing different substituents can be utilized. These can be applied to compounds with multiple functional groups or when a single derivatization must be followed by a second derivatization to achieve the desired properties of the analyte.

Derivatization is not usually the first step in sample preparation. In many analyses, prior to derivatization there are other preparation steps. Among these can be extractions for preliminary separation of the analytes from a complex matrix, cleanup steps applied to reduce some of the compounds interfering in the separation or with detection, or preconcentrations. The full analytical procedure for the analysis of a given sample is critical for a successful derivatization and for the completion of the analysis. Even for the cases where a direct derivatization is applied to the sample without other sample preparation procedures, it is very common that some other operations are required besides the reaction with the derivatization reagent. These operations are needed for establishing the appropriate medium for the derivatization reaction, such as pH adjustments, addition of proton acceptors or donors, change of the medium (from one solvent to another), etc.

Derivatization can be applied before the core chromatographic process or after it [1]. Precolumn derivatization takes place before the separation and postcolumn derivatization after it. Precolumn derivatization is more common for both gas chromatographic and liquid chromatographic separations. This type of derivatization affects the separation and may also affect the detection. Precolumn derivatization can be done "off line" before the separation process and does not have strict restrictions regarding the length of time required by the derivatization reaction. Because the chromatographic process may separate the reagents from the derivatized analytes, the interference of the reagent in the detector is not usually a problem. For most precolumn derivatizations, simple vessels such as reaction vials can be used for the process. Postcolumn derivatization is performed only for enhancing the detectability of the analytes. In most cases it must be done "on line" and should be completed in the specific time frame needed by the analyte to reach the detector.

Although the purpose of derivatization is an important factor when choosing a specific reagent, the efficiency of the chemical reaction must be considered as the main criterion for choosing a specific process. For a given analyte or group of analytes, the reaction with the derivatization reagent must be complete or at least close to complete, must take place in a length of time that is not prohibitive, and must have very little loss of the analyte with formation of artifacts or decomposition products. Only when these criteria are satisfied, can a specific chosen derivatization be applied successfully.

The application of derivatization in chromatography (and in CE) is the subject of many studies. Numerous derivatizations have been reported in journals (e.g. J. Chromatogr., Chromatographia, J. Liq. Chromatogr., J. High Resolut. Chromatogr., J. Chromatogr. Sci., J. Microcolumn Sep., etc.) and in several excellent books such as D. R. Knapp, *Handbook of Analytical Derivatization Reactions*, J. Wiley, New York, 1979; K. Blau, J. Halket (ed.), *Handbook of Derivatives for Chromatography*, J. Wiley, Chichester, 1993; G. Lunn, L. C. Hellwig, *Handbook of Derivatization Reactions for HPLC*, J. Wiley, New York, 1998; and T. Toyo'oka (ed.), *Modern Derivatization Methods for Separation Sciences*, J. Wiley, Chichester, 1999. Due to the complexity of the topic, compilations such as books or review articles cannot cover all the details, and the examination of the original references is invariably beneficial. This is applicable for both existent analytical techniques that are newly applied and when a new analytical procedure must be developed for a particular analyte or a given matrix and the information from the existent literature is used only as a guide.

17.2 PREPARATION OF THE SAMPLE FOR DERIVATIZATION AND COMMON DERIVATIZATION LABWARE

A set of sample processing operations are sometimes needed and performed after the sample cleanup and/or concentration and before the derivatization reaction. These operations are needed for preparing the best conditions for the derivatization. These preparation steps may include a) an adjustment of the pH, b) a change of the solvent, and c) addition of a specific compound that facilitates the derivatization. In some methods, an internal standard is added at this point of the analysis, although the addition of the internal standard is commonly done before sample preparation starts.

The adjustment of pH is applied when the analyte is present in aqueous solutions or at least partially aqueous solutions (see Section 16.2). A variety of buffers or acid/base adjustments can be used as recommended for each particular method (see e.g. [2], [3]). Tables with different buffer solutions are readily available in literature (see e.g. [4]).

A change in solvent may be needed for providing a specific medium for the derivatization reaction. Certain solvents can influence the efficiency of a specific derivatization. Also, a change in the solvent may be required due to the incompatibility with a derivatization reagent, such as between alcohols as a solvent and a silylating reagent. The change in the solvent also may be needed for reasons not related to derivatization, such as compatibility with the mobile phase in chromatography. Several solvents that do not have active hydrogens and are used as a medium for derivatizations are given in Table 17.2.1.

TABLE 17.2.1. *Solvents used as a medium for derivatization that requires no active hydrogens.*

Solvent	Boiling point °C	Water % solubility in solvent	Dielectric constant (adimensional)	Dipole moment (debyes)	Molecular weight[1]
acetone	56.29	100	21.45	2.69	58.08
acetonitrile	81.60	100	37.5	3.44	41.05
benzene	80.1	0.07	2.28	0.0	78.11
carbon disulfide	46	0.01	2.64	0.0	76.14
chloroform	61.15	0.056	4.81	1.04	119.38
cyclohexane	80.72	0.01	2.02	0.0	84.16
dimethylformamide (DMF)	153	100	36.71	3.86	73.10
dimethylsulfoxide (DMSO)	189	100	46.68	3.96	78.13
dioxane	101.32	100	2.25	0.0	88.11
ethyl acetate	76.5	~2	6.02	1.78	88.11
ethyl ether	34.55	1.26	4.33	1.15	74..12
formamide	210	100	109	3.73	45.04
hexane	68.7	0.01	1.89	0.0	86.18
methylene chloride	39.75	0.24	9.08	1.60	84.93
pyridine	115.25	100	12.4	2.37	79.10
tetrahydrofuran (THF)	66.0	100	7.58	1.75	72.11
toluene	110.62	0.05	2.33	0.31	92.14
triethylamine	88.8	100	2.42	0.66	101.19

[1] These are standard molecular weights, not to be used in mass spectral interpretation.

Solvents with active hydrogens also are used in certain derivatization reactions. In specific conditions, and because in general these are good solvents mainly for highly polar or ionic compounds, they are used as a reaction medium. Probably the most commonly used solvent is water. A few solvents with active hydrogens are given in Table 17.2.2.

TABLE 17.2.2. *Solvents with active hydrogens.*

Solvent	Boiling point °C	Water % solubility in solvent	Dielectric constant (adimensional)	Dipole moment (debyes)	Molecular weight[1]
water	100.0	100	80.1	1.87	18.02
methanol	64.70	100	32.35	2.87	32.04
ethanol	78.0	100	25.3	1.69	46.07
isopropanol	86.26	100	18.62	1.68	60.09
acetic acid	117.9	100	6.15	1.74	60.05

[1] These are standard molecular weights, not to be used in mass spectral interpretation.

Addition of specific compounds that do not react with the analyte but are needed together with the derivatization reagent for better results is also common in many analytical methods. This operation can be considered in certain cases as part of the derivatization itself, but in other cases it is only a step of preparation for derivatization.

- Derivatization labware

Derivatization reactions may be performed in a variety of vials and other types of reaction vessels. Regular crimp top sample vials (Figure 17.2.1) are frequently used for

FIGURE 17.2.1.
Crimp top vial.

performing derivatization reactions when no excessive heating or pressure is involved. Sometimes these vials are tapered or have special inserts that allow the recovery from them small volumes of sample. Another common type is that of vials having a screw cap with a middle hole, a stopper made from silicone rubber, and a Teflon liner that assure a good seal. Also, having thick walls they stand relatively high pressure. These vials are frequently subjected to heating, which should not exceed the temperature where the seal is destroyed. Reaction vials of different volumes and shapes are commercially available. Many other more special vials are available, such those with special valves or with filtration capability built in the vial. Vials with special valves are required when the derivatization must be done in an inert atmosphere. In this case, a special cap can be adapted to the vial allowing a stream of gas such as N_2 to fill the vial. Derivatization reactions can also be conducted in test tubes, centrifuge tubes, special sealable reaction tubes, etc. Special glassware is available for diazomethane generation (see Section 18.1).

The material used for making reaction labware is usually glass, although silica or plastic vials are also common. Both the glassware surface and the plastic surface can be a source of losses due to adsorption. The adsorption becomes important mainly when the analytes are present only in trace amounts. Acidic glass surface has an increased tendency to adsorb mainly basic compounds such as amines. Silanization is a common treatment of the glass surface and has the objective to eliminate the OH groups present on the glass surface. A number of reagents are used for this purpose such as a solution 5–10% of dimethyldichlorosilane (DMCS) in toluene. The treatment is done allowing the solution to cover the glass for 30 min. After that, the solution is removed, the surface is washed with toluene and then with dry methanol, and the solution is dried. Diazomethane solutions in ethyl ether also may be used for glass deactivation [2]. Plastic surfaces easily adsorb traces of hydrophobic organic compounds, and in general it is recommended to avoid plastic labware for analysis of traces of organic compounds.

17.3   DERIVATIZATION FOR IMPROVING GAS CHROMATOGRAPHIC PROCESS

For GC analysis, the effect of derivatization can be beneficial in a variety of circumstances, such as when the peak shape of a compound is not good, when a desired separation is not achievable, when the analyte is not stable in the injection port of the GC, or when the polarity of the analyte is too high and does not elute from the column in the underivatized form (the concept of polarity is used in the sense of solvent polarity as discussed in Section 1.10). The derivatization may change the nature of the analyte such that it becomes more volatile, has better stability when heated and can be analyzed by GC, or generates better chromatographic peak shapes and longer lifetime for the chromatographic column. Also, derivatization may generate more significant differences between the analytes and the matrix. Derivatization also can be associated with sample cleanup and concentration for improving the analysis.

The derivatization for GC attempts to eliminate all possible active hydrogens in an analyte molecule. For this reason multiple step derivatizations may be needed when the analyte has more than one functional group or when the first derivatization does not eliminate all or creates new active hydrogens. Some of the most common rules regarding the results of a derivatization in GC are the following:

a) Derivatization that replaces active hydrogens with nonpolar groups having low molecular mass improves the thermal stability of a compound.

b) Derivatization that replaces active hydrogens improves the behavior of the analyte in the chromatographic column. However, all active hydrogens must be replaced for good results.

c) The increase in the molecular weight by derivatization does not bring a proportional increase in the boiling point, and sometimes the increase can be small. However, a low molecular weight substituent such as CH_3 or $Si(CH_3)_3$ is preferable for GC analysis. Large substituents may increase the boiling point too much and make the compound not acceptable for GC analysis.

d) Condensation reactions may also decrease the boiling point and improve the thermal stability of an analyte. The generation of new active hydrogens must be avoided in condensation reactions or must be followed by a second derivatization.

e) Derivatizing groups containing halogens or nitro functions are likely to improve detectability of the host molecule in systems using electron capture detection (ECD) or negative chemical ionization mass spectrometry (NCI-MS).

- *Some theoretical aspects of the derivatization process*

Several theoretical techniques can be applied in order to predict possible modifications in physical properties after a chemical change. The effects of derivatization on certain molecular parameters are shown below for several examples. Table 17.3.1 gives the values for the dipole moments, the average polarizabilities, and the ionization potentials for decanol and for its derivatized products when some active hydrogens are replaced with methyl, trimethylsilyl, or acetyl groups. The values were calculated using a MOPAC-6 molecular orbital package [5] with specific modification for the calculation of polarizabilities [6].

TABLE 17.3.1. *Dipole moments in debye (D), average polarizabilities α in $4\pi\varepsilon_0$ Å^3 (noted Å^3), and ionization potentials I in eV for decanol and for some of its derivatized products.*[1]

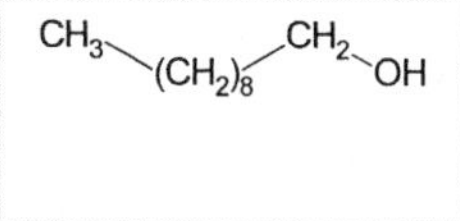 CH$_3$(CH$_2$)$_8$CH$_2$OH	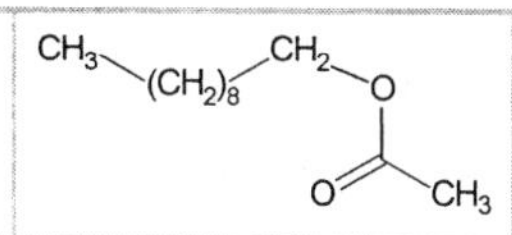 CH$_3$(CH$_2$)$_8$CH$_2$O—C(=O)CH$_3$	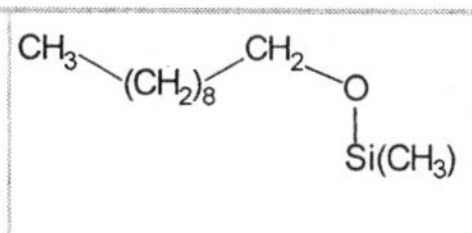 CH$_3$(CH$_2$)$_8$CH$_2$O—Si(CH$_3$)$_3$	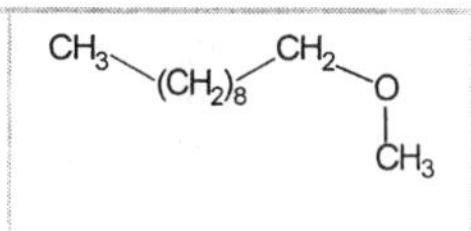 CH$_3$(CH$_2$)$_8$CH$_2$O—CH$_3$
$\mu = 1.5$ D, $\alpha = 4.2$ Å^3	$\mu = 2.3$ D, $\alpha = 5.5$ Å^3	$\mu = 1.5$ D, $\alpha = 6.2$ Å^3	$\mu = 1.3$ D, $\alpha = 4.7$ Å^3
I = 10.9 eV	I = 10.7 eV	I = 10.2 eV	I = 10.5 eV
decanol	acetyl	silyl (TMS)	methyl

[1] The values in SI units are
   1 D = 3.33564×10^{-30} C m,
   $4\pi\varepsilon_0$ Å^3 = 1.11265×10^{-40} J^{-1} C^2 m^2,
   1 eV = 1.60218×10^{-19} J.

Van der Waals energy of interaction between the molecule of an analyte and that of a given stationary phase can be calculated using different models. For example, for a separation performed on a methylsilicone column, the interactions can be calculated assuming that only a small part of the stationary phase interacts with the analyte and no other forces are involved in the process. At distances larger than 7–8 Å, van der Waals forces depending on $1/r^6$ become negligible. The interaction distances are illustrated for a model methyl silicone chain with the formula Si$_{10}$O$_9$(CH$_3$)$_{20}$H$_2$ and a molecule of decanol:

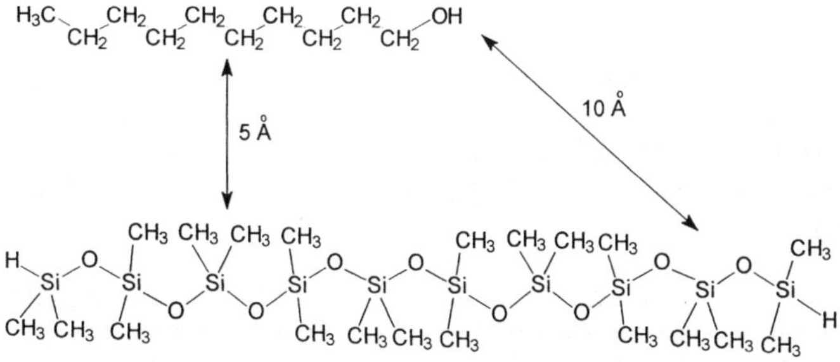

The calculated van der Waals energies (see Section 1.9) using rel. (1.9.13) between the methylsilicone fragment Si$_{10}$O$_9$(CH$_3$)$_{20}$H$_2$ ($\mu = 2.0$ D, $\alpha = 17.6$ Å^3, I = 9.8 eV) and the molecules shown in Table 17.3.1 taken at 5 Å distance are given in Figure 17.3.1 (attractions have negative values).

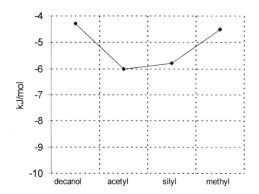

FIGURE 17.3.1. *Calculated van der Waals (attraction) energy for 5 Å distance between a methylsilicone fragment and molecules shown in Table 17.3.1.*

The same types of calculations as those done for decanol were performed for a molecule of p-aminosalicylic acid and for several of its derivatized products. The results are given in Table 17.3.2.

TABLE 17.3.2. *Dipole moments in debye (D), the average polarizabilities α in $4\pi\varepsilon_0$ $\mathring{A}^3$ (noted $\mathring{A}^3$), and ionization potentials I in eV for p-aminosalicylic acid and for some of its derivatized products.*

μ = 3.8 D, α = 4.1 $\mathring{A}^3$ I = 9.0 eV PAS	μ = 3.5 D, α = 5.0 $\mathring{A}^3$ I = 8.9 eV di-CH$_3$	μ = 3.3 D, α = 5.9 $\mathring{A}^3$ I = 9.2 eV tetra-CH$_3$	μ = 3.0 D, α = 6.2 $\mathring{A}^3$ I = 9.5 eV CH$_3$ di-Ac
μ = 2.6 D, α = 6.2 $\mathring{A}^3$ I = 9.5 eV di-Ac	μ = 2.5 D, α = 10.2 $\mathring{A}^3$ I = 8.9 eV tri-TMS	μ = 2.2 D, α = 10.7 $\mathring{A}^3$ I = 9.0 eV CH$_3$ tri-TMS	μ = 2.2 D, α = 6.2 $\mathring{A}^3$ I = 8.8 eV di-CH$_3$ Ac

The calculated van der Waals (attraction) energies for the methylsilicone fragment $Si_{10}O_9(CH_3)_{20}H_2$ (μ = 2.0 D, α = 17.6 $\mathring{A}^3$, I = 9.8 eV) and the molecules shown in Table 17.3.2 taken at 5 Å distance are given in Figure 17.3.2.

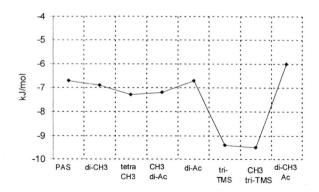

FIGURE 17.3.2. *Calculated van der Waals (attraction) energy for 5 Å distance between a methylsilicone fragment and molecules shown in Table 17.3.2.*

Again, the same types of calculations as those done for decanol and PAS were performed for the molecule of tryptophan and for several of its derivatized products. The results are given in Table 17.3.3.

TABLE 17.3.3. *Dipole moments in debye (D), the average polarizabilities α in $4 \pi \varepsilon_0$ Å^3 (noted Å^3), and ionization potentials I in eV for tryptophan and for some of its derivatized products.*

$\mu = 2.8$ D, $\alpha = 5.5$ Å^3 I = 8.6 eV Tryptophan	$\mu = 3.9$ D, $\alpha = 6.7$ Å^3 I = 10.0 eV Ac	$\mu = 2.7$ D, $\alpha = 6.5$ Å^3 I = 8.5 eV di-CH$_3$	$\mu = 2.7$ D, $\alpha = 6.8$ Å^3 I = 8.3 eV tri-CH$_3$
$\mu = 2.6$ D, $\alpha = 7.3$ Å^3 I = 8.2 eV tetra-CH$_3$	$\mu = 1.9$ D, $\alpha = 11.2$ Å^3 I = 8.2 eV tri-TMS	$\mu = 1.8$ D, $\alpha = 9.6$ Å^3 I = 8.2 eV di-TMD	$\mu = 1.8$ D, $\alpha = 9.9$ Å^3 I = 8.1 eV CH$_3$ di-TMS

The calculated van der Waals (attraction) energies for the methylsilicone fragment $Si_{10}O_9(CH_3)_{20}H_2$ ($\mu = 2.0$ D, $\alpha = 17.6$ Å^3, I = 9.8 eV) and the molecules shown in Table 17.3.3 taken at 5 Å distance are given in Figure 17.3.3.

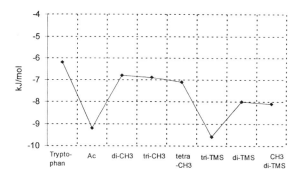

FIGURE 17.3.3. *Calculated van der Waals (attraction) energy for 5 Å distance between a methylsilicone fragment and molecules shown in Table 17.3.3.*

As seen from Tables 17.3.1 to 17.3.3, parameters such as the dipole moment, polarizability, or ionization potential are, as expected, modified upon derivatization. The direction of the modification is in general a decrease in the dipole moment. However, the polarizability does not necessarily decrease. The calculated van der Waals interactions with a methyl silicone fragment simulating a stationary phase that are shown in Figures 17.3.1 to 17.3.3 indicate that, in fact, the overall energy of van der Waals interactions will increase (in absolute value) for many derivatizations. Here, the term polarity refers to polarity P' as presented for solvents in Section 1.10 and not to the modification in van der Waals forces [7]. The polarity concept for stationary phases of GC columns is that obtained by applying McReynolds constants P as discussed in Section 4.2.

The theoretical results regarding the modifications of physical properties after derivatizations show that the interaction with the stationary phase is not always diminished. Although, as shown previously, the permanent dipole moment μ is indeed lowered in many derivatizations, the modification in the molecular polarizability α of the analyte may go in the opposite direction, and van der Waals interactions are increased instead of diminished. This is not in agreement with the generally accepted idea that by derivatization the molecules have weaker interactions with the stationary phase. One explanation for weaker interactions is that by derivatization, besides the modification in the van der Waals interactions, many molecules lose their active hydrogens and the capability of hydrogen bond formation.

The hydrogen bond formation has an important contribution to the chromatographic behavior of a given compound. The lack of hydrogen bonds in the derivatized compounds may be the main cause for the differences in the chromatographic behavior between derivatized and nonderivatized compounds. As compared to the energy of van der Waals interactions, hydrogen bonds may lead to stronger interactions. Therefore, the disappearance of the hydrogen bonds may explain why the increase in the boiling point does not occur as much as expected for some compounds based on the increase in the molecular weight and intermolecular interactions. For example, the molecular weight of a silylated compound is higher than that of the underivatized corresponding substance, and also the polarization α can be larger, indicating strong van der Waals interactions. However, the boiling point is not necessarily higher. This is probably due mainly to the disappearance of the capability to form hydrogen bonds by derivatization.

The solvophobic interactions also may play a role in GC separation, although their role in HPLC is more important. When a free molecule is dissolved in a stationary phase, the interactions between the stationary phase molecules are modified, contributing to the total interaction energy. The theory is not applied for the evaluation of a partition coefficient (or distribution constant) for a compound portioned between a liquid stationary phase and the gas phase.

The use of extrathermodynamic calculations can provide in certain cases information about the direction of change after a specific derivatization. For example, for some compounds such as aliphatic esters [8], a high correlation was reported between the boiling point and the retention (index) in gas chromatography. In general, this type of correlation works better for homolog series, and the change in functionality is not well reflected in the predicted behavior. The difference in the calculated boiling point value vs. the real one and the deviation in the correlation between the two parameters both contribute to poor results. For example, the calculation of the boiling point (T_b or bp) using rel. (1.10.7) gives for decanol the value bp = 207° C (instead of the experimental bp = 231° C). For the acetyl derivative of decanol, the calculated value is bp = 234° C, for the silylated decanol the value is bp = 231° C, and for decyl methyl ether the value is bp = 187° C (instead of the experimental bp = 213° C). On a typical nonpolar column the elution order of the four compounds is decyl methyl ether, silylated decanol, acetyl derivative, and decanol itself.

For derivatization with groups such as CH_3, C_2H_5, several fluorinated alkyl, $Si(CH_3)_3$, $COCH_3$, and several fluorinated acyl, the variation in the boiling point can be only a small increase or even a decrease for particular cases, although the derivatization is associated with an increase in the molecular weight of the initial compound. The calculated value of boiling point is not necessarily related to the elution order on a specific column, but a calculated value may still be useful for example for predicting the behavior of a certain compound in the injection port of the GC.

- *Some applications of derivatization to improve the chromatographic process*

Overall, the chemical modifications brought by the derivatization of the analyte lead to significant changes in the separation. Particularly when the matrix is not derivatized, a significant improvement in the separation can be obtained. Also, regarding the analytes' chromatographic behavior, the improvement of their thermal stability is very beneficial. It may be done to such extent for some compounds that GC analysis becomes applicable while this was not possible otherwise. For other compounds, decompositions in the injection port of the GC may be reduced or eliminated. An additional effect of this increased thermal stability is the possibility to use higher temperatures of the injection port, which may be useful for better transfer of the sample into the chromatographic column. By selection of the appropriate chromatographic columns and conditions, the changes brought by derivatization may be utilized for significantly improving the separation. Among the compounds that cannot be analyzed by GC except after derivatization are mono-, di-, and trisaccharides, as well as polyols with more than three or four OH groups. As an example, myo-inositol cannot be analyzed directly by GC, but after silylation this can be done easily. The reaction can be performed using as a reagent N,O-bis(trimethylsilyl)-trifluoroacetamide (BSTFA) in dimethylformamide (DMF), by heating the mixture with the analyte at 76° C for 30 min.:

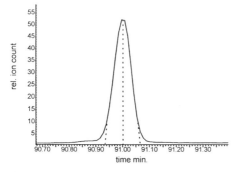

TMS = Si(CH$_3$)$_3$

The GC peak of 1,2,3,4,5,6-hexakis-O-(trimethylsilyl)-myo-inositol (six TMS-myo-inositol) with M = 612 obtained in a separation performed on a dimethylpolysiloxane + 5% diphenylpolysiloxane column (Supelco SPB-5) is shown in Figure 17.3.4a. For this particular GC process the column was 60 m long, 0.32 mm i.d., and with 0.32 µm film thickness. The GC temperature program started at 50° C and was increased to 310° C using a 2° C/min gradient.

Not only can myo-inositol be analyzed after silylation using GC, but the peak shape is better even compared to a compound such as nicotine with M = 162, which was separated on the same column. It can be seen that the peak for nicotine shown in Figure 17.3.4b has at about the same peak height as six-TMS myo-inositol a tail that is not present in the other peak. In addition, the peak width for nicotine is slightly larger than the one for six-TMS myo-inositol, although the retention time for six-TMS myo-inositol is almost double compared to nicotine. (See Section 1.6 for the description of parameters characterizing peak broadening.)

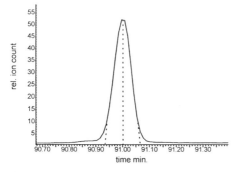 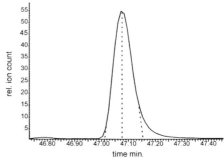

FIGURE 17.3.4a. *Six TMS-myo-inositol peak in a chromatogram obtained on a SPB-5 column.* Another compound elutes in front of the six TMS-myo-inositol peak.

FIGURE 17.3.4b. *Nicotine peak in a chromatogram obtained on a SPB-5 column.*

Another application of derivatization in GC is the analysis of complex mixtures. Plant materials, for example, contain a variety of inorganic and organic compounds including small molecules and polymers. The small organic molecules may consist of organic acids, carbohydrates, amino acids, etc. Because of this variety of molecules, the analysis of specific components in a plant material commonly requires extensive sample preparation. However, by the use of derivatization and simultaneous extraction, this process can be significantly simplified. As an example, Figure 17.3.5 shows a (total ion) chromatogram (TIC) obtained from the direct silylation of a tobacco sample allowing the identification of more than 250 compounds [9]. The sample was prepared by heating at 76° C for 20 min. 50 mg tobacco, 0.5 mL BSTFA containing 1% TMCS (N,O-bis-(trimethylsilyl)-trifluoroacetamide + 1% chlorotrimethylsilane), and 0.3 mL DMF (dimethylformamide) in a reaction vial. The derivatization solution acts also as extracting solvent, and the analysis can be done easily by injecting in the GC 0.5 µL from the supernatant solution. The separation and detection was done in a GC-MS system equipped with a dimethylpolysiloxane + 5% diphenylpolysiloxane column (J&W Sci. DB-5), 60 m long, 0.25 mm i.d., and 0.25 µm film. The GC temperature started at 50° C and ended at 300° C using 2° C/min gradient.

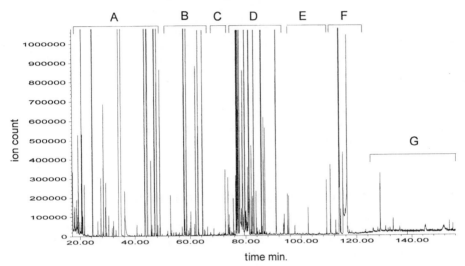

FIGURE 17.3.5. *The chromatogram (TIC) of a silylated tobacco sample, which allows the identification of more than 250 compounds.* The chromatogram includes short chain alcohols, glycolic and lactic acids (group A), other hydroxy acids (group B), citric acid, C_5 sugars (group C), monosaccharides, inositol, C_{16} to C_{20} acids, cembranoids (group E), disaccharides, deoxyfructosazines, fructosazines (group F), chlorogenic acid, and trisaccharides (group G). Besides silylated compounds, some other compounds such as nicotine also are detected in the chromatogram [9].

Another case where complex mixtures are generated is the pyrolysis of polymeric materials. The effect of derivatization may be highly beneficial in such cases for improving the separation. The effects of pyrolysis and further derivatization of pyrolysis products is presented in Chapter 20.

- Separation conditions for different types of derivatizations

The choice of the column dimensions and the nature of the stationary phase vary from one analytical method to another. The choice of a stationary phase is commonly based on its polarity in the sense of the polarity parameters developed by McReynolds [10], [11], [12] (see Section 4.2). However, because of their ability to separate a variety of compounds, some stationary phases are more frequently utilized regardless the use of derivatization or of the type of derivatization that is used. Such phases are polyethyleneglycol, dimethylpolysiloxane, and dimethylpolysiloxane copolymers with a certain amount of diphenylpolysiloxane.

The formation of the alkyl or aryl derivatives usually modifies significantly the behavior of the analyte in the column. However, there is no specific type of column that must be utilized for the separation of alkyl or aryl derivatives. Columns with polar stationary phase such as polyethylene glycol or methylpolysiloxane with 35% trifluoropropyl-polysiloxane can be used successfully for these separations. Also, depending on the desired separation, nonpolar or slightly polar stationary phases such as 100% dimethylpolysiloxane, dimethylpolysiloxane with 5% diphenylpolysiloxane, or polycarboranesiloxane columns can be used.

For silyl derivatives, most GC separations are performed on nonpolar or relatively nonpolar chromatographic columns. The common stationary phases for these columns are bonded phase dimethylpolysiloxane and combinations of dimethylpolysiloxane with different proportions of diphenylpolysiloxane. These columns can be used in a temperature range from about 50° C up to 330° or even up to 350° C. Special high temperature columns using stationary phases such as polycarboranesiloxane can also be used for the separation of silyl derivatives. The choice of the chemical nature of the stationary phase in the chromatographic column determines at least in part the range of compounds that can be analyzed.

17.4 DERIVATIZATION FOR IMPROVING DETECTION IN GC

The derivatization with the purpose of improving detectability in GC is determined by the type of detector utilized. Several types of GC detectors were already discussed in Section 4.2. Most derivatizations are performed precolumn, even if they are applied only with the purpose of improving detection. Some specific postcolumn reactions applied to the analytes are part of certain types of detectors such as chemiluminescence detectors or atomic emission detectors (AED) and are not necessarily classified as derivatization reactions. These types of postcolumn chemical modifications in the detector will be briefly discussed in this section. It is important that the derivatization for improving detection does not deteriorate the separation. Preferably, both the detection and the chromatography are improved by the same derivatization.

For nonselective detectors such as TCD and FID, in most cases no specific derivatization is necessary for achieving good sensitivity. As derivatization is an additional step in sample processing, and supplementary errors may be associated with it, this step is used only for reasons other than improving detection. Particular derivatizations may still be useful for improving detection. For example, for the FID detector, formaldehyde and heavily halogenated compounds give minimal response. This indicates that the analysis of formaldehyde using FID may benefit from derivatization. In other instances, derivatization may have even some adverse results. For example, silylation may have in long term some negative effects because of the deposition of SiO_2 on the jet of the FID. In spite of this, silylation is widely used as derivatization technique in GC because of the important advantages brought for the chromatographic process and for extending the range of analytes amenable to GC analysis. Also, periodical cleaning of the FID detector or the use of silylated reagents containing fluorine results in very few problems with the FID detection because of silylation

Although not very common, derivatization with nitrogenous compounds may lead to higher sensitivity if a nitrogen phosphorus detector (NPD) is used. When such derivatization is performed, it should be considered that different NPD responses can be generated by different nitrogen containing compounds, and in certain cases no real advantage may be gained by derivatization. An adverse result occurs for the NPD detectors when silanization agents are used in the instrumentation or when silylation is performed on the sample. Besides a possible reduction in the NPD response on silylated compounds containing nitrogen, a drastic decrease in the lifetime of the detector may occur, probably due to the excess of silylating reagent that commonly is injected with a derivatized sample and affects the alkali active element of the NPD.

The photoionization detector (PID) is not frequently used in association with a specific derivatization, although its response depends significantly on the nature of the analyte. For a PID, the response depends on the ionization potential of the analyte. Because the photons used for the ionization have a specific energy (such as 10.2 eV), the compounds with higher ionization potential are not sensitive in PID, while those with lower ionization potential may have excellent sensitivity, as low as 10^{-12} g/s of sample. A derivatization resulting in lowering the ionization potential of the analyte may be beneficial for PID detection.

Mass spectrometric detection using EI+ type ionization is a procedure of considerable complexity, and its detailed presentation is beyond the purpose of this book. MS is used for a variety of purposes, some as direct as sample quantitation or identification and some more complex, such as structure determination or detection of sample origin based on isotope distribution. Different uses of derivatization for improving mass spectral identification are discussed further as an independent subject. The use of derivatization for improving sensitivity is not always necessary but still can be applied for example to reduce thermal or catalytic decomposition of the analyte in the ion source. When the quantitation is performed using MS with single ion monitoring (SIM), derivatization can be used to obtain fragmentation with unique intense ions.

Some detectors such as electron capture detectors (ECD) or mass spectrometric detectors using negative chemical ionization (NCI) may benefit highly from certain derivatization types. Both ECD and NCI mass spectrometry (NCI-MS) can be extremely sensitive but they are selective to compounds that are able to form more stable negative ions. ECD, for example, can have sensitivity as low as 10^{-13} g/s of sample, compared to the best sensitivity of FID that can be 10^{-11} g/s of sample. Also, the use of NCI in mass spectrometry for selected compounds can lead to 100 to 1000 times higher sensitivity compared to EI+ ionization (on the same mass spectrometer). Negative EI (EI-) ionization mode in mass spectrometry is less frequently utilized, but the ionization efficiency in this process is also influenced by the electron affinity of the analyte.

Halogenated derivatives in particular show very good response in ECD, EI- and NCI-MS. The efficiency of the process seems to be related to the ease of attaching an electron on the molecule. In ECD this process can be written as follows:

$$A + e^- \longrightarrow A^-$$

With some exceptions, ECD response can be correlated with the electron affinity of the analyte. As an example, the relative ECD response for 5-methyl-2-(methylethyl)phenol (thymol) and some substituted derivatives are represented in Figure 17.4.1 as a function of electron affinity (EA). The relative ECD response for different substituents was measured for identical concentrations of the sample and compared to the response for thymol heptafluorobutanoate, which was considered the unit [13]. The electron affinities were calculated as the energy generated when an electron is added to the molecule in gas phase [6]. The calculations were done using a MOPAC molecular orbital package [5]).

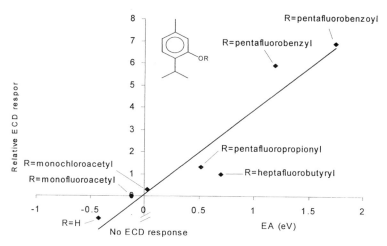

FIGURE 17.4.1. *Relative ECD response for 5-methyl -2-(methylethyl)phenol (thymol) and some substituted derivatives as a function of electron affinity (EA).*

In general, the halogen substituents increase the sensitivity in ECD in the order I > Br > Cl > F. Multiple substitutions seem to have a cumulative effect. Besides halogens, nitro groups seem to have an effect similar to chlorine groups. For aromatic compounds, the substituents affect the sensitivity of the ECD according to their electron withdrawing capability. Strong electron withdrawing groups such as NO_2 increase the sensitivity of the detection, while electron donating groups reduce it.

The higher sensitivity in ECD of heavier halogens compared to the lighter ones can be utilized for lowering the detection limit of certain analytes. For this purpose, a derivatization reaction may be associated with a halogen exchange reaction. As an example, the derivatization of an alcohol with bromodimethylchlorosilane can be performed in the presence of NaI and generates a iodine derivative as shown in the reaction [14]:

$$R-CH_2-OH \;+\; Cl-\underset{\underset{CH_3}{|}}{\overset{\overset{CH_3}{|}}{Si}}-CH_2Br \;\xrightarrow[\text{diethylamine}]{\text{NaI}}\; R-CH_2-O-\underset{\underset{CH_3}{|}}{\overset{\overset{CH_3}{|}}{Si}}-CH_2I \;+\; HCl \;+\; NaBr$$

For NCI-MS the end result of the ionization process is the formation of negative ions. This process is efficient only for molecules with positive electron affinities. The reagent gas used in NCI interacts with the electrons from the filament, lowering their energy but without forming negative reagent gas ions:

$$G + e^-_{(70\ eV)} \longrightarrow G + e^-_{(thermal)}$$

In certain instances, such as when CH_4 is the reagent gas, two low energy electrons may be formed:

$$G + e^-_{(70\ eV)} \longrightarrow G^+ + 2\ e^-_{(thermal)}$$

When a negatively charged species is stable enough, the ionization takes place by electron capture, dissociative electron capture, ion pair formation, or an ion molecule reaction:

$$AB + e^-_{(thermal)} \longrightarrow AB^-$$
$$AB + e^-_{(thermal)} \longrightarrow A^\cdot + B^-$$
$$AB + e^-_{(thermal)} \longrightarrow A^+ + B^- + e^-$$
$$A + B^- \longrightarrow AB^-$$

The sensitivity in NCI-MS is also dependent on the electron affinity of the analyte, similarly to the sensitivity in ECD. This can be shown, as an example, for aniline and some substituted anilines. The calculation of electron affinities for these compounds gave the results shown in Table 17.4.1 [5] [15].

TABLE 17.4.1. *Calculated electron affinities (EA) for aniline and some of its derivatives.*

Compound	EA in eV
phenylamine (aniline)	- 0.64
2-methylphenylamine	- 0.60
3-methylphenylamine	- 0.68
2,3-dimethylphenylamine	- 0.59
2-hydroxyphenylamine	- 0.41
3-hydroxyphenylamine	- 0.54
N-phenyltrifluoroacetamide	0.28
N-phenylpentafluoropropionamide	0.30
N-phenylheptafluorobutyramide	0.74
N-phenyltrichloroacetamide	0.71
N-phenyltribromoacetamide	0.99
2-nitrophenylamine	0.79
3-nitrophenylamine	0.95
4-nitrophenylamine	0.79

As seen in Table 17.4.1, the EA values are negative for aniline and several substituted anilines with electron donating groups. This indicates that negative ionization is not favored. Nitroanilines and the amides with halogenated acyl groups have positive EA values, and they show high sensitivity in detection by ECD or NCI-MS. The correlation between the calculated electron affinities and the sensitivity in NCI-MS may not be applicable, however, to every compound. There are known examples of similar molecules such as benzo[a]pyrene and benzo[e]pyrene where one molecule (benzo[a]pyrene) can be determined with high sensitivity using NCI-MS, while the isomer (benzo[e]pyrene) does not form enough negative ions and cannot be analyzed using this technique. The calculated EA for benzo[a]pyrene and benzo[e]pyrene are both positive, although low [15].

Depending on the derivatization, a variety of substitution groups containing electronegative elements (halogens) or nitro groups can be attached to the analytes. The procedure to attach these groups to the analyte is in most cases the typical substitution of an active hydrogen in the analyte Y–H with a group R from a reagent R–X that has the appropriate active X group. Some groups used for enhancing ECD or NCI-MS sensitivity following an alkylation or aryl derivatization reaction are shown in Table 17.4.2.

TABLE 17.4.2. *Substitution groups used in alkylation and aryl derivatization for enhancing ECD or NCI-MS detectability.*[1]

Group	Formula	Mass of the group	Group	Formula	Mass of the group
p-bromobenzyl	Br—⟨○⟩—CH₂—	169	2,4-dinitrophenyl	O₂N—⟨○⟩— (NO₂)	167
p-bromo-phenacyl	Br—⟨○⟩—C(=O)—CH₂—	197	2,6-dinitrophenyl	⟨○⟩ with NO₂, NO₂	167
3,5-bis-(trifluoromethyl)-benzyl	F₃C—⟨○⟩—CH₂— (F₃C)	227	pentafluoro-benzyl	F,F,F,F—⟨○⟩—CH₂—	181
p-chlorobenzyl	Cl—⟨○⟩—CH₂—	125	pentafluoro-benzylidene	F,F,F,F—⟨○⟩—CH=	180
2-chloroethyl-	Cl-C₂H₄-	63	pentafluoro-propyl	F,F,F,F,F—C—C—CH₂—	133
p-nitrobenzyl	O₂N—⟨○⟩—CH₂—	136	trichloroethyl	$Cl_3C—CH_2—$	131

[1] The masses of molecules or groups can be calculated using two different conventions. One convention considers the natural isotopic abundance of elements. The other convention considers only the masses of the most abundant isotope, which is useful for MS interpretations. For this later case, the resulting mass is rounded to the unit [16]. The convention used throughout this book is that of the most abundant isotope, unless otherwise specified.

Besides alkylation or aryl derivatization, other derivatization techniques used to replace an active hydrogen are applied to introduce into a molecule a substituent containing halogens or nitro groups. Silylation, for example can be used for this purpose. Several silyl groups containing halogens are given in Table 17.4.3.

TABLE 17.4.3. *Substitution groups used in silylation for enhancing ECD or NCI-MS detectability.*

Group	Formula	Mass of the group
bromomethyl-dimethylsilyl-	BrH_2C—$Si(CH_3)_2$—	151
chloromethyl-dimethylsilyl-	ClH_2C—$Si(CH_3)_2$—	107
dimethyl(3,3,4,4,5,5,6,6,6-nonafluorohexyl)silyl	CF_3-$(CF_2)_3$-$(CH_2)_2$--$Si(CH_3)_2$-	305
dimethyltrifluoropropylsilyl-	CF_3-$(CH_2)_2$-$Si(CH_3)_2$	155
iodomethyldimethylsilyl-	IH_2C—$Si(CH_3)_2$—	199
(3,3,4,4,5,5,6,6,6-nonafluorohexyl)-dimethylsilyl-	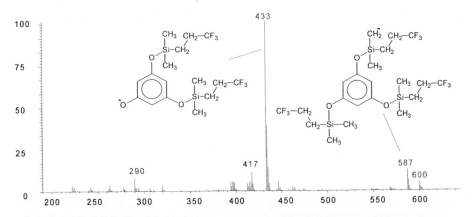	305
pentafluorophenyl-dimethylsilyl		225

An example of an NCI-MS spectrum for 1,3,5-trihydroxybenzene derivatized with 3,3,3-trifluoropropyldimethylchlorosilane is given in Figure 17.4.2.

FIGURE 17.4.2. *Mass spectrum in NCI of 1,3,5-trihydroxybenzene derivatized with 3,3,3-trifluoropropyldimethylchlorosilane.* (Note: the abscissa of a mass spectrum indicates m/z, and the ordinate indicates relative abundance to 100.)

Several acyl groups used to enhance ECD or NCI-MS detection are given in Table 17.4.4, and some other substitution groups introduced by chloroformylation or sulfonation and used for the same purpose are given in Table 17.4.5.

TABLE 17.4.4. *Substitution groups used in acylation for enhancing ECD or NCI-MS detectability.*

Group	Formula	Mass of the group	Group	Formula	Mass of the group
3,5-bis-(trifluoro-methyl)-benzoyl		241	pentafluoro-propionyl	C_2F_5—C— (O)	147
heptafluoro-butyryl	C_3F_7—C— (O)	197	trichloroacetyl	CCl_3—C— (O)	145
pentafluoro-benzoyl		195	trifluoroacetyl	CF_3—C— (O)	97

TABLE 17.4.5. *Other substitution groups used for enhancing ECD or NCI-MS detectability.*

Group	Formula	Mass of the group	Reaction type
pentafluorobenzyloxy-carbonyl		211	chloro-formylation
trichloroethyloxy-carbonyl	Cl_3C—CH_2—O	175	chloro-formylation
2,4-dinitrobenzen-sulfonyl	O_2N—...—S—O— ... NO_2	247	sulfonation

A variety of other reagents can be used for introducing groups containing halogens and enhancing significantly the detectability of the derivatized analytes by ECD or NCI-MS. The type of reaction followed by these reagents may be an alkyl or aryl substitution, a silylation, or an acylation reaction. Some specific groups attached to the analyte for enhancing ECD or NCI-MS detectability will be described together with other specific derivatization reactions in Chapters 18 and 19.

The derivatization with groups containing electronegative elements such as fluorine also has been used for enhancing the detection in LC [17].

- *Detectors in GC involving chemical reactions*

Some specific types of detection used in GC also may be considered derivatization although they are part of the detector. These detectors involve chemical reactions in the detection process. Even the ubiquitous FID, in which a flame is obtained by burning hydrogen in air, involves the combustion or ionization of the compounds eluting from the chromatographic column to generate ions in the flame. Other detectors such as chemiluminescence detectors or AEDs also may use chemical reactions in the detection process. A chemiluminescence nitrogen detector, for example, uses a complete high temperature oxidation of the column eluent in the presence of a catalyst such as Ni. The NO formed during the oxidation process further reacts with ozone, producing NO_2 and emitting light, as shown below:

$$R(N_x)H + O_2 \longrightarrow CO_2 + H_2O + x \, \overset{.}{N}O + \text{other combustion products}$$

$$NO^. + O_3 \longrightarrow NO_2^* + O_2$$

$$NO_2^* \longrightarrow NO_2 + h\nu$$

The oxidation process of the analytes containing nitrogen $R(N_x)H$ can be done at different temperatures and may allow the selection of the detection for specific nitrogenous compounds. For example, by lowering the oxidation temperature to about $500°$ C, it is possible to generate NO only from N-nitroso compounds, while other nitrogen containing compounds do not form NO. This is the principle of a specific detector for nitrosamines.

Similar chemical reactions also are used in a sulfur detector where sulfur-containing compounds are specifically oxidized to SO and other combustion products. SO reacts with ozone, generating SO_2 and emitting light.

More details about this subject can be found in dedicated literature regarding GC detectors [18]. Usually, postcolumn chemical reactions that are intrinsic to the detectors are not classified as analytical derivatizations.

17.5 DERIVATIZATION FOR IMPROVING MASS SPECTRAL IDENTIFICATION

Mass spectroscopy is probably the most powerful tool used for compound identification purposes. The capability of this technique to provide information from very low amounts of material such as that eluting from a chromatographic column adds a significant advantage to this technique. However, the identification of compounds using mass spectroscopy is not a simple process even using the capabilities offered by the electronic searches in the mass spectral libraries. This is particularly true for analysis of complex mixtures or when the analyzed compound is present in traces. Some compounds do not have a very characteristic mass spectrum, or during the chromatographic process the separation is not achieved, and it is difficult to make an identification with a GC/MS and using the mass spectral library search. Also, numerous compounds may have a mass spectrum that matches more than one compound (with a good quality fit) [19]. One common case where the identification may be ambiguous is that of compounds containing OH, COOH, or NH_2 groups on an aliphatic chain longer than four or five carbon atoms. These compounds do not show a good molecular ion in their EI+ mass spectrum, and smaller fragments are the same for different homologs. For example, the library search for the mass spectrum of decanol generates eight other possible compounds (1-undecanol, nonanol, 1-ethyl-2-heptylcyclopropane, 3,7-dimethyl-1-octene, 1-dodecanol, 3,7-dimethyl-1-octanol, formic acid decyl ester, and cyclodecane) having 82% or higher reliability in a PBM (Probability Based Matching) search (see e.g. [16]). In such cases, a derivatization with the purpose of obtaining a compound that forms more informative fragments in the mass spectrum is recommended. For example, the alkylated derivatives of compounds with OH, COOH, or NH_2 groups usually have a mass spectrum where the molecular ion is more intense, and this helps in the identification using mass spectral library searches. For the case of decanol, the methyl ether gives a unique match for a reliability higher than 82%, and the molecular ion is clearly visible in the spectrum. The mass spectra of decanol (MW = 158) and of its methyl ether (MW = 172) are shown in Figures 17.5.1a and 17.5.1b, respectively.

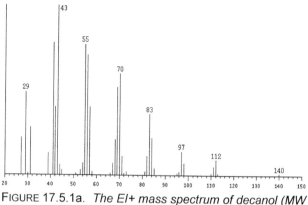

FIGURE 17.5.1a. *The EI+ mass spectrum of decanol (MW =158).*

Derivatizations such as silylation or acylation also can be used to improve mass spectral identifications. Silylated compounds, for example, may have characteristic mass spectra, usually showing loss of a methyl group from the molecular ion (m/z = M - 15). The problem is that for many compounds the derivatives, such as

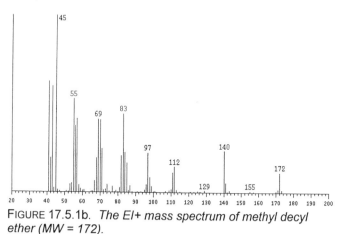

FIGURE 17.5.1b. *The EI+ mass spectrum of methyl decyl ether (MW = 172).*

trimethylsilyl, *tert*-butyldimethylsilyl, or several acyl, are not available in current mass spectral libraries [20], [21], and the use of the spectrum of the derivatized compound is of less value for identification.

In addition to direct improvements regarding spectral searches, there are also other utilizations of derivatization to help with compound identification. One alternative is the use of silylation to differentiate compounds with active hydrogens from those without active hydrogens. An example is the case of 1H-purine and of 7a-hydro-1,2,4-triazolo[1,5-a]pyrazine. Their spectra are shown in Figures 17.5.2 and 17.5.3, respectively. The two spectra have 71% reliability one against the other in a PBM search (see e.g. [16]). The differentiation is obtained immediately using silylation of the purine, while 7a-hydro-1,2,4-triazolo[1,5-a]-pyrazine does not have active hydrogens and cannot be silylated. The silylated mass spectrum of purine is shown in Figure 17.5.4. The utilization of silylation for verification of a specific structure containing active hydrogens, based on the mass spectral identification of the derivatized compound, is very useful in the analysis of complex mixtures. However, not all spectra of the silylated compounds are easier to identify compared to the nonsilylated compound. The dominance of the same ions (such as 73, 147, etc.) in the MS spectra of all silylated compounds makes the search for their identification sometimes difficult.

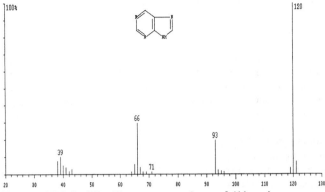

FIGURE 17.5.2. *The EI+ mass spectrum of 1H-purine.*

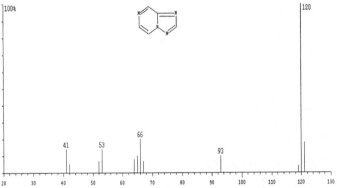

FIGURE 17.5.3. *The EI+ mass spectrum of 7a-hydro-1,2,4-triazolo[1,5-a]pyrazine.*

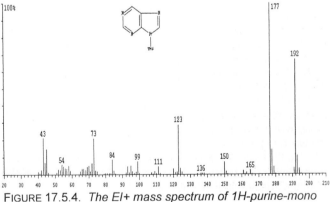

FIGURE 17.5.4. *The EI+ mass spectrum of 1H-purine-mono TMS.*

Another important use of derivatization is that of helping structural elucidations. In this process, a specific derivatization is performed, followed by mass spectral identification of the derivatized compounds. The determination of the structure of a specific compound can be a complicated process, and other techniques such as NMR are usually applied for this task. However, the sensitivity in NMR is much lower than in MS. Also when the compounds to analyze are in a complex mixture, a separation is necessary before the structure determination. The possibility to directly connect an NMR instrument to a GC or HPLC instrument is very limited. Although MS is not the best method for all structural determinations, it is still frequently applied. One typical example is the determination of the position of a double bond in an unsaturated acid. The mass spectra of Z-11-tetradecenoic acid, Z-9- tetradecenoic acid, E-9- tetradecenoic acid, and Z-7-tetradecenoic acid are shown in Figures 17.5.5a through 17.5.5d, respectively. As seen from the spectra shown in Figures 17.5.5a through 17.5.5d, the four isomers have very similar spectra, within 97% similarity in a PBM

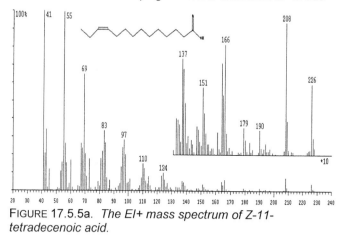

FIGURE 17.5.5a. *The EI+ mass spectrum of Z-11-tetradecenoic acid.*

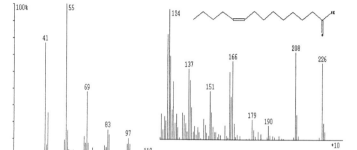

FIGURE 17.5.5b. *The EI+ mass spectrum of Z-9-tetradecenoic acid.*

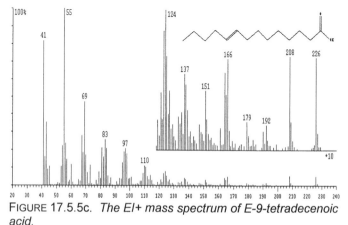

FIGURE 17.5.5c. *The EI+ mass spectrum of E-9-tetradecenoic acid.*

search. Although some details regarding peak intensities are possible to detect, it is very difficult to assess the differences to a specific structural difference or to fluctuation of the intensities in the mass spectrum. The difference between positional isomers can be done using other detection techniques for the GC separation. For example, the use of

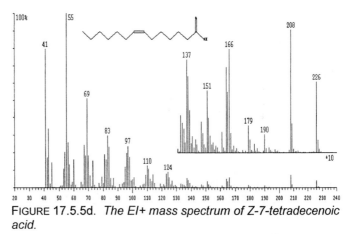

FIGURE 17.5.5d. *The EI+ mass spectrum of Z-7-tetradecenoic acid.*

GC-FTIR may solve some of the problems related to isomer identification (see Section 4.2). However, the determination of the position of the double bond can be done using different derivatizations and MS detection. One possibility is the treatment of the acid with NaIO$_4$ + KMnO4. The double bond is cleaved and the resulting products can be further derivatized for example with ethanol in the presence of BF$_3$. The sequence of reactions is the following:

(1) NaIO$_4$ + KMnO$_4$

(2) C$_2$H$_5$OH / BF$_3$

For this reaction, 0.5 mmol acid is dissolved in 500 mL water containing 1.5 mmol KOH. The solution is treated with 4 mmol NaIO$_4$ (aqueous sol. 1%), and then the pH is adjusted to 8 with a solution containing 1 mL 0.1 M KMnO$_4$ in 400 mL water. The solution is stirred for 2 hours, then the pH is adjusted to 9, and the reaction is allowed to take place for 24 hours. The solution is then acidified and the excess oxidant eliminated with SO$_2$. The solution is made slightly alkaline with KOH and evaporated to near dryness. A few mL ethanol are added to this solution and gas CO$_2$ to eliminate the excess of KOH. The sample is evaporated under vacuum and then treated with 2–4 mL ethanol containing 15% dry HCl (w/w) and 2 mL solution 15% BF$_3$ in ethanol. The mixture is refluxed for 10 min., cooled, and concentrated by removing most of the ethanol. To this mixture, 5–10 mL water saturated with NaHCO$_3$ is added. The solution is saturated with NaCl and extracted twice with 5–10 mL ethyl ether. The extract is washed with water, dried over Na$_2$SO$_4$, and analyzed by GC-MS. The position of the double bonds can be determined easily from the nature of the fragments.

Another possibility to determine the position of the double bond is the use of oxidation with OsO$_4$ to the corresponding glycol (see also Section 18.6), followed by an alkylation or silylation of the glycol. As an example, the methyl ester of 9-octadecenoic acid by oxidation with OsO$_4$ followed by silylation forms 9,10-bis(trimethylsilyloxy)-octadecanoic acid methyl ester. The reaction for this derivatization can be written as follows:

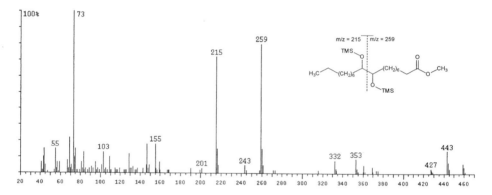

The mass spectrum of the derivatized compound is frequently diagnostic for the position of the double bond, as seen for the spectrum of 9,10-bis(trimethylsilyloxy)-octadecanoic acid methyl ester shown in Figure 17.5.6.

FIGURE 17.5.6. *The EI+ mass spectrum of 9,10-bis(trimethylsilyloxy)octadecanoic acid methyl ester, showing typical ions for the fragmentation that indicates the position of a previous double bond.*

The derivatization for better MS identification in both GC-MS and LC-MS is a rapidly expanding field. Special new techniques using MS detection after a chromatographic separation with chemical modification of the analyte also are evolving, such as the thermal decomposition of biopolymers before MS and after the LC separation [22].

- Use of isotope labeling for enhancing mass spectral identifications

A different procedure that may be utilized for compound identification based on mass spectra is the use of two parallel derivatizations, one of them being done with an isotope labeled reagent. Common labeling isotopes are ^2H (deuterium, d), ^{13}C, ^{15}N, etc. One such isotopic labeling can be done, for example, using silylation with d_{18}-N,O-bis(trimethylsilyl)-trifluoroacetamide (d_{18}-BSTFA) [9] or d_{18}-BSA) [23]. A parallel sample derivatized with regular BSTFA will provide a pairing chromatogram with peaks at retention times that have only small differences from the first, but with spectra differing by a number of units. The comparison of the spectra for corresponding peaks of a given compound allows the calculation of the number of silyl groups attached to that compound. In addition, the fragmentation in the spectra can be better interpreted, allowing easier compound identification.

As an example, two chromatograms obtained by derivatization of a cellulose pyrolysate using MS detection are given in Figures 17.5.7a and 17.5.7b. The chromatogram from Figure 17.5.7a corresponds to the pyrolysate derivatized using BSTFA, and the chromatogram from Figure 17.5.7b corresponds to the pyrolysate derivatized with d_{18}-BSTFA.

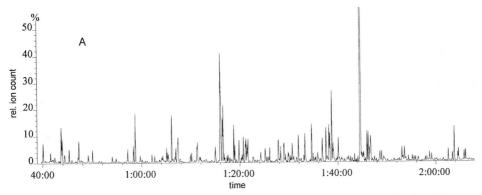

FIGURE 17.5.7a. *Total ion chromatogram of a trimethylsilylated sample obtained from a cellulose pyrolysate using BSTFA.*

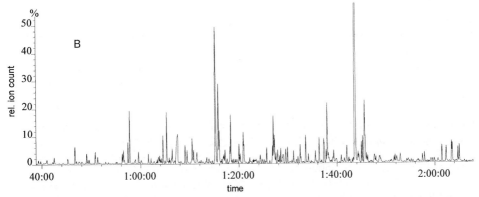

FIGURE 17.5.7b. *Total ion chromatogram of a d_9- trimethylsilylated sample obtained from a cellulose pyrolysate using d_{18}-BSTFA (B).*

It can be seen from Figures 17.5.7a and 17.5.7b that the chromatograms are very similar. Detailed inspection of these chromatograms using electronic data processing capabilities available with mass spectrometers show that the similarity goes even to details in small chromatographic peaks, although the d_9-TMS derivatives may elute slightly faster than the regular TMS derivatives. For every TMS group in a molecule, a difference of 9 mass units is expected between the regular and deuterated compounds.

As an example, the spectra of tri-TMS levoglucosan and tri-(d_9-TMS) levoglucosan are given in Figures 17.5.8a and 17.5.8b respectively.

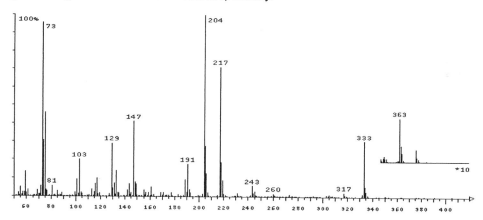

FIGURE 17.5.8a. *Mass spectrum of tri-TMS levoglucosan.* The compound has the molecular mass 378.

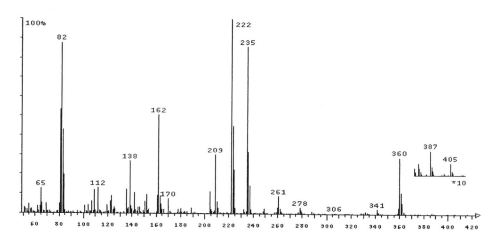

FIGURE 17.5.8b. *Mass spectrum of tri-(d_9-TMS) levoglucosan.* The compound has the molecular mass 405.

The formation of the molecular ion and of a few higher mass fragments for the regular compounds and for the deuterated one (d) are indicated below. The difference of 27 mass units between the regular and deuterated molecular ion indicate three TMS groups substituted in the molecule. The loss of 15 mass units corresponds to the formation of a fragment by the elimination of a CH_3 group, and the loss of 18 mass units corresponds to the formation of a fragment by the elimination of a CD_3 group from the deuterated compound. The elucidation of the structure of specific fragments can be made easily by comparison of the spectra of deuterated and nondeuterated compounds.

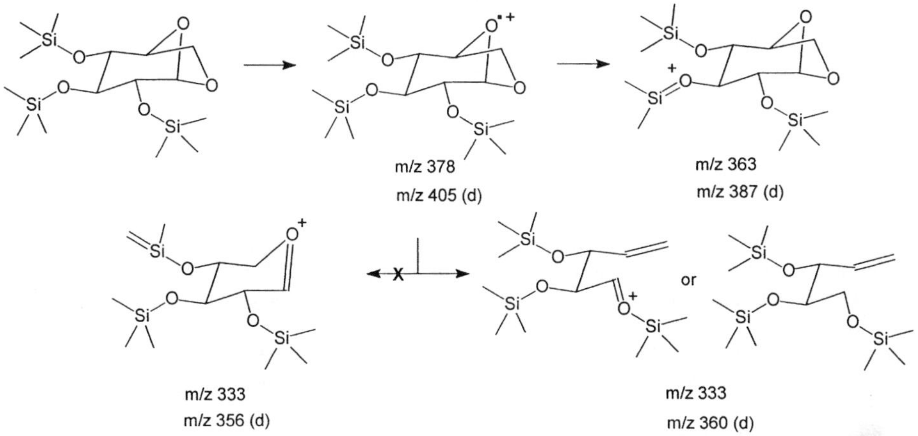

m/z 378
m/z 405 (d)

m/z 363
m/z 387 (d)

m/z 333
m/z 356 (d)

or

m/z 333
m/z 360 (d)

The ions with m/z 333 and 360 correspond to the elimination from the molecular ion (in several steps) of CHO_2 or of C_2H_5O, both with the mass 45. The absence of the ion with the mass 356 from the spectrum of tri-(d_9-TMS) levoglucosan indicates that the fragmentation with the loss of C_2H_5O from the regular molecule and of C_2D_4HO (mass 49) from the deuterated one does not take place.

The structure of other fragments also can be explained based on the same procedure. For example, for the ion 217 from the spectrum of tri-TMS levoglucosan, several structures are possible, as indicated below:

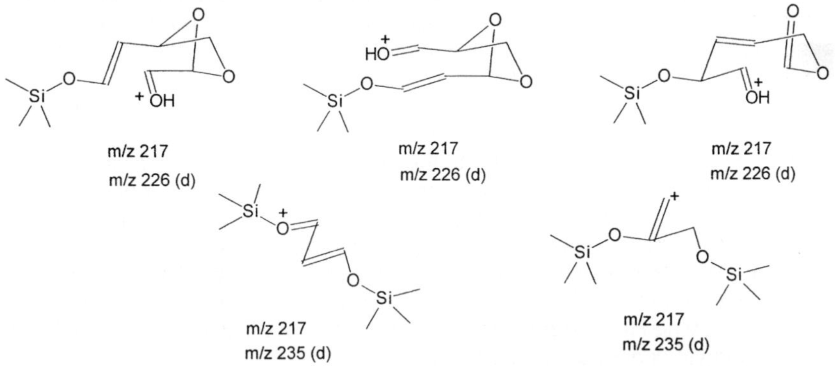

m/z 217
m/z 226 (d)

m/z 217
m/z 226 (d)

m/z 217
m/z 226 (d)

m/z 217
m/z 235 (d)

m/z 217
m/z 235 (d)

The presence in the spectrum of tri-(d_9-TMS) levoglucosan of the fragment with m/z 235 and the absence of the fragment with m/z 226 shows that only the double silylated fragments are formed to generate the ion with m/z = 217.

Isotope labeling for improving mass spectral identification is not uncommon. For example, isotope labeling can be performed in the study of protein structure or DNA sequencing. A significant amount of dedicated literature is available on this subject.

17.6   DERIVATIZATION FOR IMPROVING DETECTION IN HPLC AND TLC

Derivatization reactions are applied in HPLC and thin layer chromatography (TLC) analysis mainly to improve detectability, but modifications in the chromatographic separations are also intended in certain cases. The use of HPLC analysis is extremely common because this technique has no specific restrictions regarding volatility or thermal stability of the analyte. Thus, there are numerous chemical reactions that were developed for improving detectability through specific physical properties that are modified by derivatization. The derivatization can be performed to make the analysis more sensitive, to obtain a linear detection response, to avoid specific interferences, to modify the separation, etc. Based on the physical property that is targeted for enhancement, possible derivatizations can be grouped as follows:

- Modifications of the UV-Vis spectrum of the analyte by generation of new compounds with higher absorption coefficient of light at specific wavelength.

- Modifications of fluorescence of the analyte by fluorescence labeling or by production of molecules that are fluorescent from nonfluorescent analytes.

- Derivatizations related to chemiluminescence properties of the analyte.

- Modifications of electrochemical properties of the analyte.

Details regarding the type of chemical reactions involved in these derivatizations, as well as descriptions of some of the reagents utilized for enhancing detectability in HPLC, TLC, etc., will be given in Chapters 18 and 19. Capillary electrophoresis (CE), although in principle not a chromatographic technique, shares similarities with HPLC. For this reason, some of the derivatizations are identical for the two techniques. Examples of compounds with identical derivatizations and the separation technique being either HPLC or CE are occasionally indicated in Chapter 19.

There are several similarities and also differences between the derivatizations performed for HPLC analysis compared to the derivatization for GC analysis. Both derivatizations seek to generate a unique product from the analyte and ensure that the intended derivatization is complete. Also, in both derivatizations it is common to use the active hydrogens to attach moieties that bring a desired physical property. On the other hand, in HPLC, the elimination of all active hydrogens from the analyte is not usually necessary. Attachment of a single desired chromophore or fluorophore group may be sufficient for improving HPLC detection. Also, in HPLC, the increase in the molecular weight of the analyte is not usually a concern.

- *Postcolumn derivatization in HPLC*

Most derivatizations in HPLC are performed precolumn. Typical injection of the sample followed by separation and detection is done for these cases. Some aspects regarding HPLC analysis were already discussed in Section 4.3. Postcolumn derivatization is also applied in HPLC. This is done more frequently than GC postcolumn derivatization. Postcolumn derivatization is usually performed "on line" and should be completed in the specific time frame required by the flow of the mobile phase with the analyte to reach the detector. Also, the loss of resolution should be as little as possible during the derivatization process. Some other requirements also should be satisfied, such as miscibility of the derivatizing solution and the eluent, no precipitation reactions, and matching of the flow rates of different mixing liquids. Because no separation is done between the reagents and analytes, the reagent should not interfere with the detector. Precolumn derivatization can be done in simple reaction vessels, while specific instrumentation must be used for postcolumn derivatization. A typical scheme of such instrumentation is shown in Figure 17.6.1.

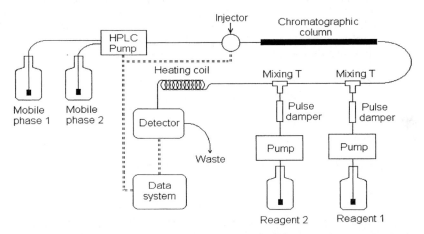

FIGURE 17.6.1. *Typical scheme of a postcolumn derivatization setup in HPLC.*

The diagram shown in Figure 17.6.1 uses HPLC separation, includes the capability of adding two derivatizing reagents, and also has a heating coil in case the reaction needs to be accelerated by heat. Some systems require only one added reagent or no heating. Also, instead of heating, a solid phase reaction may be involved, and a reactor may replace the heating coil. An example of a different postcolumn derivatization setup is shown in Figure 17.6.2. This setup has been applied for the analysis of catecholamines [24]. It uses an Amberlite CG-50 column for the separation and air pressure to force the liquid through the column. A gas-fragmented flow system is applied for postcolumn derivatization. The coils are made from Pyrex tubing with the number of turns shown in Figure 17.6.2. The reagent is a 0.5 M solution of ethylenediamine hydrochloride. A solution 1.5 M of NaOH (also containing 0.1% Triton X-405) is added to control the pH. Potassium hexacyanoferrate 0.3% solution is used as an oxidant, and the excess of the oxidant is removed by reaction with a 10% solution of Na_2SO_3. The fluorescence is measured at 510 nm with excitation at 400 nm. The flow rates of different solutions are shown in the figure.

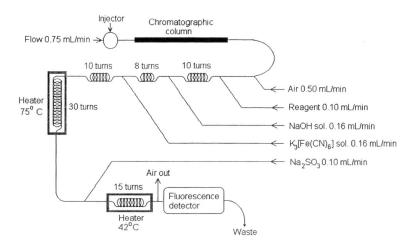

FIGURE 17.6.2. *Scheme of a postcolumn derivatization setup in LC with gas-fragmented flow.*

Several technical problems are related to postcolumn derivatization. Among these is the need to maintain resolution and peak shape, at the same time assuring a thorough mixing of the postcolumn derivatizing reagent with the column effluent. For this purpose, several mixing devices were developed, such as low-volume mixing T's that create local turbulence but keep the volume of the mixing chamber very small [25]. The addition of reagents may be done in a continuous mode or using air segmentation, which is expected to reduce longitudinal diffusion [26], [27]. The heating coil also must preserve resolution by avoiding longitudinal smearing. Laminar flow in tubing appears to generate worse tailing of the chromatographic peaks because the portion of material close to the tubing wall is delayed significantly longer compared to the one far from the wall (deconvolution programs used for data processing may compensate for this tailing). Several designs were applied for the heating coil in an attempt to avoid loss of resolution. When a reactor is needed after reagent addition, the same types of problems must be solved. Some reactors require immobilized reagents (or enzymes), and local mixing should be achieved without altering resolution [26], [28]. More modern devices for postcolumn derivatization have been recently developed, such as microfabricated devices [29], [30], [31], [32], [33].

- Enhancements in UV-Vis detection for HPLC

The enhancement by derivatization of the UV or Vis absorbance of an analyte is a common practice in HPLC, mainly because a variety of excellent spectrophotometric detectors are available for HPLC. The subject has been reviewed in various papers [26], [34], [35], [36].

The groups of atoms responsible in an organic molecule for the absorption of light due to electronic transitions are commonly called chromophores. Many chromophores are unsaturated groups bearing electrons in lone pairs or participating in π bonds such as benzene rings or double bonds. The addition of chromophores to a molecule increases the molar absorption coefficient and possibly shifts the absorption to higher wavelengths (bathocrom effect). This explains why many derivatizing reagents used in chromatography for enhancing UV-Vis detectability have chromophore groups in their molecular structure in addition to the groups able to react with the analyte molecule. When no chromophore is already present in the reagent, it must be generated during the derivatization reaction. Besides the chromophore groups, other groups attached to the molecule may affect the absorbing wavelength or the intensity of absorption. These are commonly called auxochromes. Examples of auxochrome groups are OH, NH_2, Cl, etc. These groups modify the distribution of the energy levels in the molecule and also the values of Einstein coefficients for the probability of transition (e.g. [6]), thus modifying the absorption coefficient at a given wavelength. For example, benzene has a B band at 255 nm with $\varepsilon_\lambda = 215$, while in phenol that has an OH auxochrome, the B band occurs at 270 nm with $\varepsilon_\lambda = 1450$.

Chromophore groups can be added by derivatization to a molecule using the typical procedure of replacing an active hydrogen with a group containing chromophores or by other reactions such as addition to a double bond, and the group property is not significantly influenced by the host structure. The general type of reaction used in these derivatizations can be schematically written as follows:

Property
carrying—Reactive + Reactive—Analyte   →   Property
group   group A   group B   residue   carrying—Analyte + Reactive—Reactive
                                                    group   residue   group A   group B

Other derivatizations change more drastically the structure of the analyte. A typical case of this type of derivatization for enhancing UV detection is the reaction of amino acids with phenylisothiocyanate. This reaction is performed precolumn by adding a solution of phenylisothiocyanate in water/triethylamine/ethanol (1/1/1/7 v/v/v/v) on the sample containing the amino acids and keeping the mixture at room temperature for 30 min. The reaction generates the corresponding 5-substituted 3-phenyl-2-thioxoimidazolidin-4-ones (3-phenyl-2-thiohydantoin):

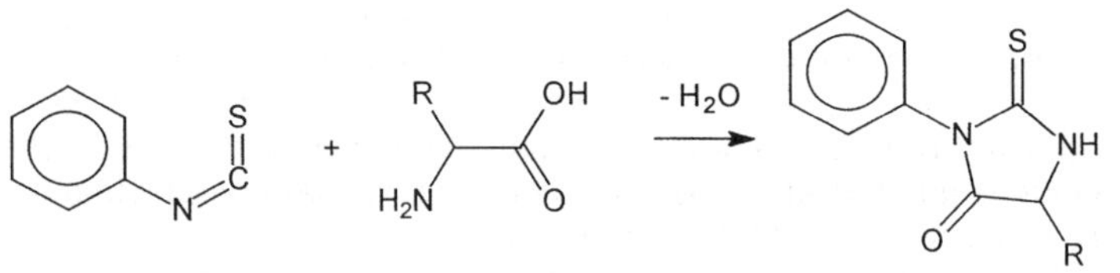

In this reaction, the phenyl group can be considered an attached chromophore, and the formation of 2-thiohydantoin ring generates another chromophore, also enhancing the UV absorption. The maximum of UV absorption for most phenylthiohydantoin derivatives of amino acids is around 245 nm. A chromatogram obtained for the analysis of amino acids from a tobacco sample using derivatization with phenylisothiocyanate is shown in Figure 17.6.3. The derivatization of amino acids using phenylisothiocyanate is furthet discussed in Section 19.10.

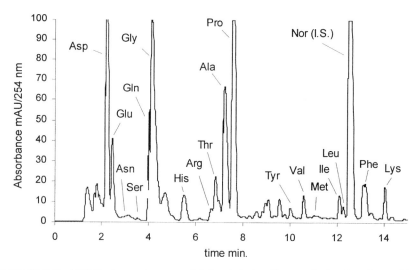

FIGURE 17.6.3. *HPLC chromatogram obtained during the analysis of amino acids from a tobacco sample using UV detection at 254 nm [37].*

For the analysis, the amino acids are extracted from the tobacco sample with HCl, followed by a cleanup step using an ion exchange resin [38]. The HPLC separation was performed using a Waters PICO-TAG column and a gradient elution. The detection was done using a Waters 990 diode array detector set at 254 nm (not at 245 nm where most common amino acids show the maximum absorption). Norleucine was used as an internal standard. For quantitation, the ratios of the peak areas of each individual amino acid and the area of the peak of the internal standard were used. Calibration using standards shows that the molar absorption coefficient ε_λ is about the same for all amino acids analyzed. More reagents used to enhance UV detectability are described in Chapter 19.

- *Enhancements in fluorescence detection for HPLC*

The derivatization for producing fluorescent compounds from nonfluorescent analytes is commonly done using fluorescence labeling reagents. Certain groups that fluoresce can be added by derivatization to a molecule, and the group property is not significantly influenced by the host structure. The number of fluorescence carrying groups (fluorophores) is quite large. Some examples are shown below:

4-(5,6-dimethoxybenzothyazol-2-yl)benzoyl   [2-(5,6-dimethoxybenzothiazol-2-yl)phenyl]-sulfonyl

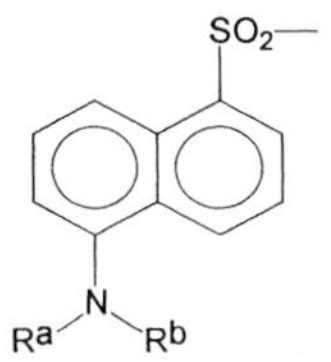

5-(di-R-amino)naphthalene-1-sulfonyl-
R^a = CH$_3$, R^b = CH$_3$ (Dns, dansyl),
R^a = CH$_3$(CH$_2$)$_3$, R^b = CH$_3$(CH$_2$)$_3$ (Bns)

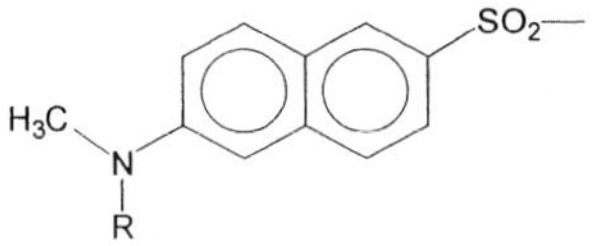

6-N-methyl-R-aminonaphthalene-
2-sulfonyl
R = CH$_3$, R = C$_6$H$_5$

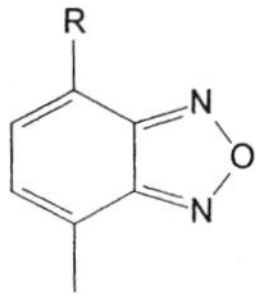

benzofurazans
R = NO$_2$ (NBD),
R = SO$_2$N(CH$_3$)$_2$ (DBD)
R = SO$_2$NH$_2$ (ABD)

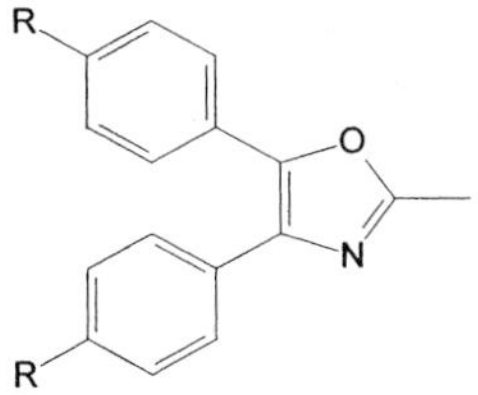

oxazoles
R = H, SO$_2$N(CH$_3$)$_2$

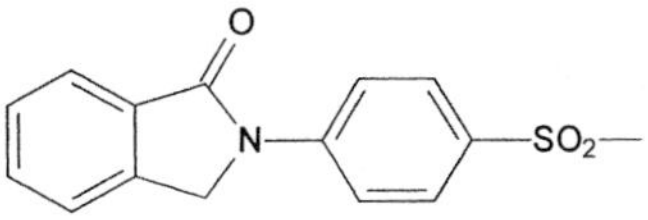

2-[4-(sulfonyl)phenyl]-isoindolin-1-one

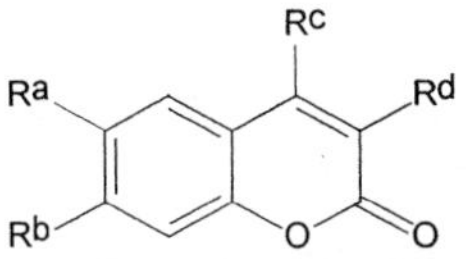

chromen-2-ones (coumarins)
R^a = H, R^b = OCH$_3$, R^c = CH$_2$Br, R^d = H
R^a = OCH$_3$, R^b = OCH$_3$, R^c = CH$_2$Br, R^d = H
R^a = H, R^b = OCOCH$_3$, R^c = CH$_2$Br, R^d = H
R^a = H, R^b = OCH$_3$, R^c = H, R^d = COCH$_2$Br
R^a = O-CH$_2$-O = R^b, R^c = CH$_2$Br, R^d = H
R^a = O-CH$_2$-O = R^b, R^c = H, R^d = COCH$_2$Br

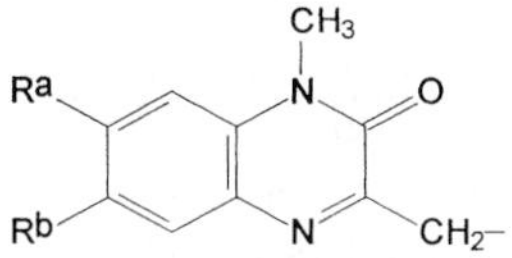

1-methyl-2(1H)quinoxalines
R^a = OCH$_3$, R^b = OCH$_3$
R^a = O-CH$_2$-O = R^b

Some other fluorophores can be used in derivatizations, and they will be described in Chapter 19 related to the derivatization of specific analytes.

The fluorescence-labeling groups are typically present in a reagent that contains a reactive moiety, which allows the formation of a desired compound with the analyte. For example, an active aryl chloride can react with primary and secondary amino groups, OH in phenols and SH in thiols. This can be seen, for example, in 4-chloro-7-nitrobenzo-2-oxa-1,3-diazole (4-chloro-7-nitrobenzofurazan; NBD-Cl), which reacts with an amine as follows:

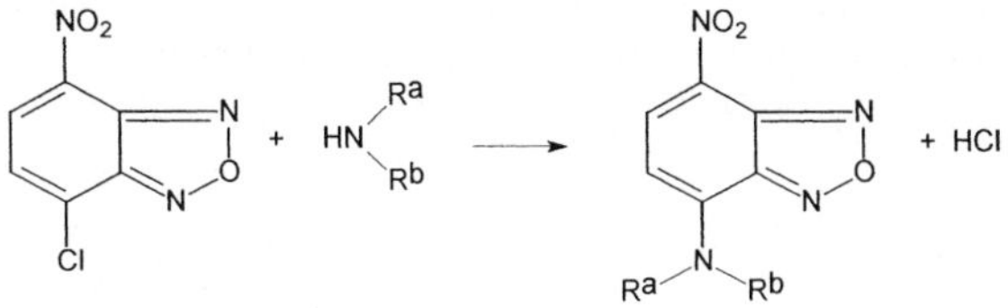

This reaction is favored by basic media and is commonly performed by mixing 0.2–0.5 mL of the solution to be analyzed containing a few µg of analyte with 1–1.5 mL NBD-Cl 0.5 mg/mL in methanol and then adding 50–100 µL of 0.1 M NaHCO$_3$. The reaction mixture is heated at 55° C for one hour and then analyzed [39].

Many other reactive groups are used to achieve derivatization. The nature and the reactivity of the reacting group determine the class of organic compounds that can be derivatized. For example, for reacting with analytes such as amines, phenols, or thiols, the reagents contain groups such as dialdehydes, sulfonyl chlorides, chloroformates, isocyanates, isothiocyanates, etc. For acids, the reagents are amines, alcohols, etc. For aldehydes and ketones, the reagents are substituted hydrazines, hydrazides, etc. The fluorescent moiety can be the same for different reactive groups or may be different. For example, for aldehyde derivatization, instead of NBD-Cl, the reagent that can be used is 4-hydrazino-7-nitro-2,1,3-benzoxadiazole (4-hydrazino-7-nitrobenzofurazan). The reaction takes place as follows:

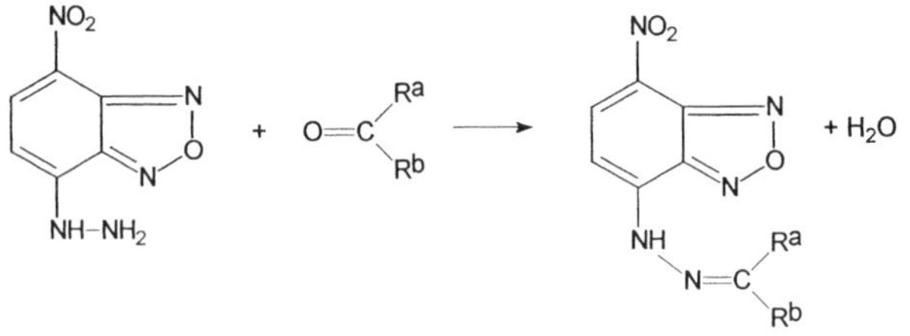

The same strategy is used, for example, for reagents containing the group 5-dimethylamino-naphthalene-1-sulfonyl. Dansyl chloride (Dns-Cl) is used extensively for the analysis of amines and amino acids, phenols, and thiols [26], [34], [3]. The reaction with an amino acid can be written as follows:

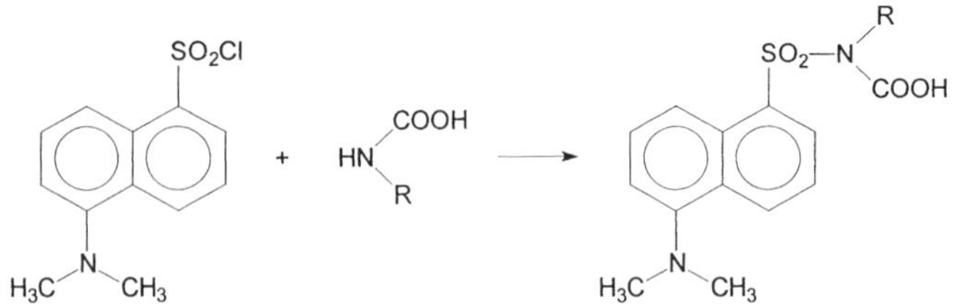

Besides dansyl chloride, dansyl isothiocyanate has been used for amino acid derivatization [40]. In the reaction with an aldehyde or ketone, instead of dansyl chloride, the reagent used is dansyl hydrazine [41]:

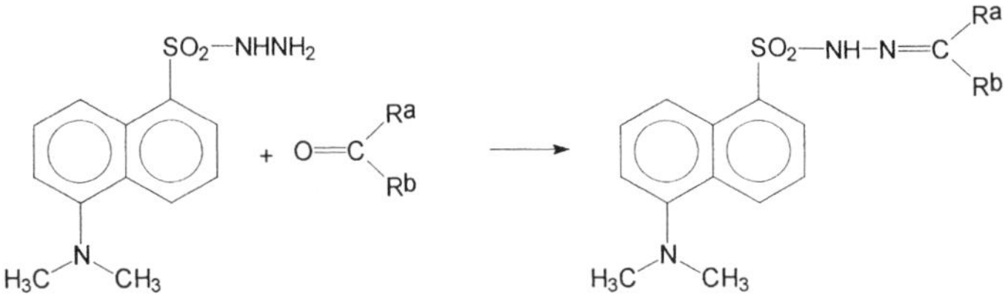

The hydrolysis of Dns-Cl or other Dns derivatives easily generates 5-dimethylamino-naphthalene-1-sulfonic acid, which is also fluorescent, producing interference if it is not separated before detection. Dns-sulfonamides are relatively stable to acid hydrolysis [42] and fairly stable even to hydrolysis with bases [43].

The spectral properties of Dns derivatives depend on the nature of the substituent and also on the solvent, pH, etc. In general, the compounds formed with primary amino groups fluoresce at shorter wavelengths compared to those formed with secondary amino groups, although they have about the same fluorescence efficiency Φ. Dns-amines and Dns-amino acids fluoresce at lower wavelengths than phenols and imidazoles. Phenols and imidazole derivatives also have lower fluorescence efficiencies than amines and amino acids. Fluorescence efficiency Φ also decreases in acidic solutions probably due to the protonation of the dimethylamino group of the Dns fragment. Dns derivatives absorbed on silica gel show the same decrease in fluorescence as in acidic solutions, and spraying silica with a base increases fluorescence [26].

The bonding functionality between the fluorophore and the analyte molecule may play some role in the fluorescence intensity. Among other groups, the hydrazino groups have a diminishing effect on fluorescence intensity. For example, the use of N-methyl-4-hydrazino-7-nitrobenzofurazan in the analysis of aldehydes leads to the same detection limits using UV detection at 470 nm or fluorescence (ex. 470 nm, em. 560 nm). The sensitivity of fluorescence detection is significantly higher than UV detection for amines derivatized with benzofurazan reagents that do not contain a hydrazino group [44].

Another alternative to producing molecules that are fluorescent from nonfluorescent analytes is the use of a nonfluorescent reagent to produce a fluorescent derivative. Fluorescent reagents producing fluorescent derivatives must always be separated in the chromatographic process from the analytes. An example of a nonfluorescent reagent is o-phenylenediamine, which forms with an α-keto acid a quinoxalinol derivative as shown below:

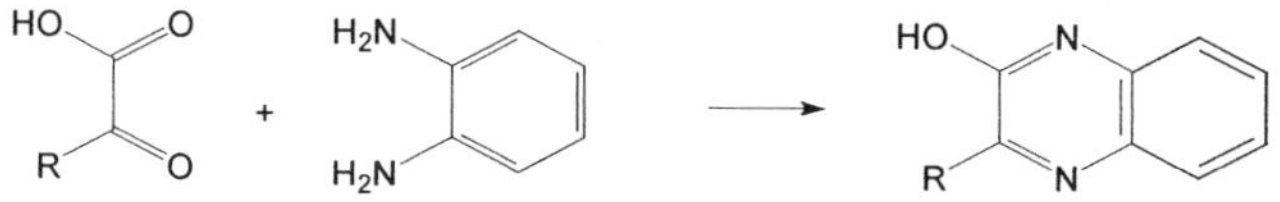

Quinoxalinol derivatives are highly fluorescent. This particular reaction is utilized in precolumn derivatizations. In general, nonfluorescent reagents are preferred in postcolumn derivatizations where the separation of the reagent from the analytes is no longer important.

Numerous other fluorescent moieties have been used in derivatization reagents. Some of them were specifically designed for laser induced fluorescence (LIF), and some compounds such as amines can be analyzed at attomole level using these reagents [45], [46]. More details regarding the type of chemical reactions involved in these derivatizations as well as descriptions of some of the reagents utilized for fluorescence labeling are given in Chapter 18 and 19. Most of these reagents must be used in precolumn derivatizations for a proper separation of the reagent that can be also highly fluorescent.

A different use of fluorescing properties of analytes is selective fluorescence quenching. The technique relies upon the deactivation of a fluorophore by the analyte. The analyte and the fluorophore (in their ground state) may form a complex with different spectral properties than the free fluorophore (static quenching). Also, it is possible that the fluorophore and the analyte interact only in excited state, forming a transient complex (dynamic quenching) that does not fluoresce. In both cases, the analyte is measured from the difference in the fluorescence intensity in the absence and in the presence of the analyte [47]. The technique has several analytical applications, for example in the determination of nitroaromatic compounds. Using pyrene (in cyclohexane) as a fluorophore, ppm levels of 2,4,6-trinitrotoluene and of a number of other nitroaromatic compounds can be measured [48], [49], [50].

- Derivatization in chemiluminescence detection for HPLC

Chemiluminescence detection can be highly sensitive, such that only a few femtomoles of analyte can be measured by this technique. However, the problem is that only a few luminescent molecules are known. For this reason, derivatization can be applied for chemiluminescence (CL) labeling. This is usually achieved using postcolumn derivatization, which involves a combination of molecular species such as the analyte, an oxidant, a catalyst, etc.

Most chemiluminescence reactions are oxidations. As an example, the reaction of luminol (5-amino-2,3-dihydro-1,4-phthalazinedione) and H_2O_2 in the presence of a catalyst such as a metal ion, Fe^{2+}, Co^{2+}, etc., takes place as follows:

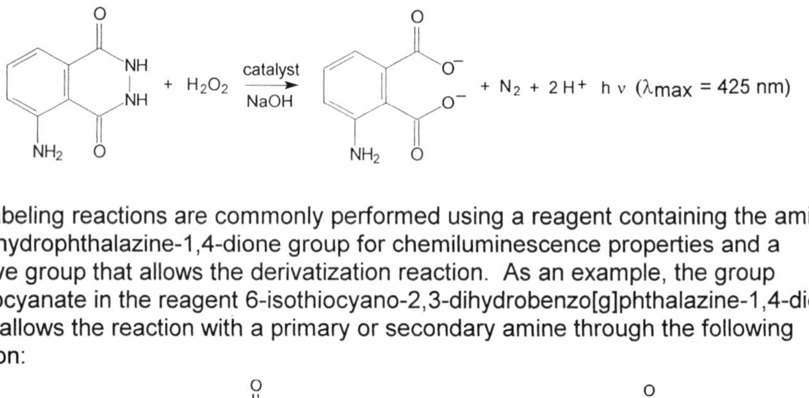

The labeling reactions are commonly performed using a reagent containing the amino-2,3-dihydrophthalazine-1,4-dione group for chemiluminescence properties and a reactive group that allows the derivatization reaction. As an example, the group isothiocyanate in the reagent 6-isothiocyano-2,3-dihydrobenzo[g]phthalazine-1,4-dione (IPO) allows the reaction with a primary or secondary amine through the following reaction:

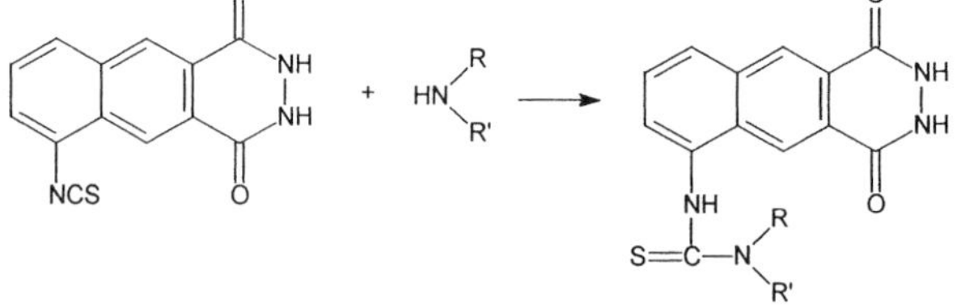

The derivative generated in this reaction can be detected by its CL produced in the reaction with H_2O_2 in the presence of a catalyst such as $K_3[Fe(CN)_6]$. Other oxidants such as $KHSO_5$ can be used instead of H_2O_2 [51]. A variety of labeling reagents will be discussed in Chapter 19.

A different chemiluminescent analytical procedure is based on the observation that certain chemical systems containing an organic oxalate or oxamide, H_2O_2, and a fluorescent compound (fluorophore) produce chemiluminescence, usually in the presence of a catalyst. The reaction for this procedure known as peroxyoxalate chemiluminescence or PO-CL can be written as follows:

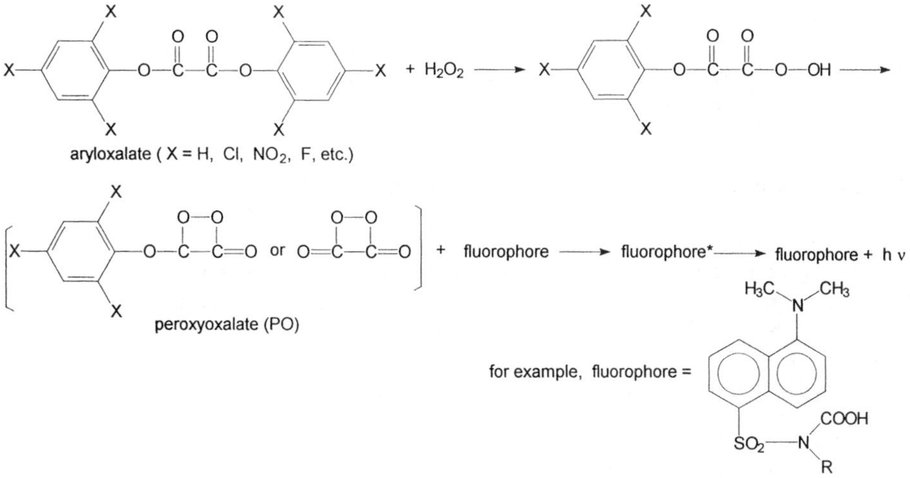

aryloxalate (X = H, Cl, NO$_2$, F, etc.)

peroxyoxalate (PO)

for example, fluorophore =

The chemiluminescence is proportional to the concentration of aryloxalate, hydrogen peroxide, and fluorophore. When performed in the presence of a catalyst, the catalyst concentration also influences the chemiluminescence. The concentration of the fluorophore, for example, can be determined by calibration, measuring the chemiluminescence intensity and keeping other parameters constant. Based on this procedure (PO-CL), many fluorescent compounds obtained by the derivatization of a nonfluorescent analyte can be analyzed. Several aryloxalates can be used to generate the peroxyoxalate, such as bis(2,4,6-trichlorophenyl)-oxalate (TCPO or di 2,4,6-trichlorophenyl ethane-1,2-dioate), bis(2,4,6-dinitrophenyl)oxalate (DNPO), bis(pentafluorophenyl)oxalate (PFPO), bis(2,6-difluorophenyl)oxalate (DFPO), bis[2-(3,6,9-trioxadecyloxycarbonyl)-4-nitrophenyl] oxalate (TDPO), etc. Also a variety of catalysts can be applied in the PO-CL procedures, most of them using general base catalysis. One common catalyst is imidazole, which shortens significantly the lifetime of chemiluminescent species. Besides imidazole as a catalyst, oxalyldiimidazole [52] was proved to act as both reagent and catalyst in a PO-CL reaction of the type:

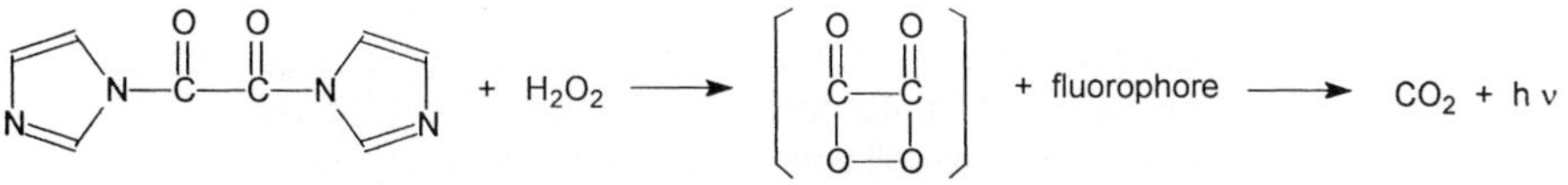

1,2-dioxethandione

An interesting analytical procedure to generate a CL reagent is the use of an electrochemical oxidation for chemiluminescence (ECL). The procedure has been used for the detection of various compounds that can be oxidized with tris(2,2'-bipyridyl)-ruthenium (III) complex [Ru(bpy)$_3$]$^{3+}$, while the complex is generated by the electrochemical oxidation of [Ru(bpy)$_3$]$^{2+}$, which has the following structure:

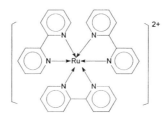

The reducing analytes can be amino acids, Dns-amino acids, amines, organic acids, etc. [53], [54], [55], [56], [57]. The reaction takes place as follows:

$$[Ru(bpy)_3]^{2+} - e^- \longrightarrow [Ru(bpy)_3]^{3+} \text{ (electrochemical process)}$$
$$[Ru(bpy)_3]^{3+} + \text{reducing analyte} \longrightarrow \{[Ru(bpy)_3]^{2+}\}^* + \text{oxidation products}$$
$$\{[Ru(bpy)_3]^{2+}\}^* \longrightarrow [Ru(bpy)_3]^{2+} + h\nu$$

The instrumental setup for this type of analysis is shown in Figures 17.6.4a and 17.6.4b. The reagent $[Ru(bpy)_3]^{3+}$ is generated in the electrochemical flow-cell.

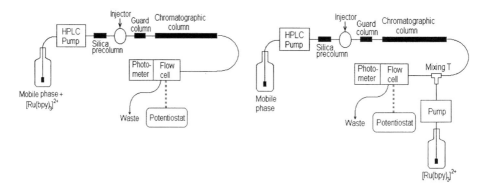

FIGURE 17.6.4a. *Instrumental setup for CL analysis of reducing analytes with* $[Ru(bpy)_3]^{3+}$ *as a reagent.* Ru(bpy)$_3$]$^{2+}$ is added precolumn.

FIGURE 17.6.4b. *Instrumental setup for CL analysis of reducing analytes with* $[Ru(bpy)_3]^{3+}$ *as a reagent.* Ru(bpy)$_3$]$^{2+}$ is added postcolumn.

The addition of the reagent $[Ru(bpy)_3]^{2+}$ can be done either precolumn or postcolumn, as shown in Figures 17.6.4a and 17.6.4b, the concentration being 0.5 mM. The oxidation in the electrochemical flow-cell was done using a working electrode maintained at 1.25 V vs. Ag/AgCl reference electrode. Levels of about 30 µmol of several amino acids were measured by this procedure [55], [58], [59].

- The use of derivatization in other detection techniques for HPLC

Some other detection techniques are available for HPLC, such as electrochemical detection procedures. Derivatization for improving electrochemical detection is applied in the same manner as other derivatizations, using a reagent that contains a reactive moiety capable of reacting with an active hydrogen or a specific functional group in the analyte and also having the rest of the molecule capable of either being reduced or oxidized at a relatively low potential. Compounds containing nitro groups, for example, can be reduced electrochemically and analyzed using amperometric detection. For the oxidation reactions, these being used frequently for analysis, the oxidation potential of the analyte must be lowered following derivatization and set significantly different from the oxidation of the eluent (e.g. acetonitrile). For example, o-phthalaldehyde (OPA) derivatives of amino acids can be analyzed electrochemically using an oxidation potential of 0.6 V (vs. Ag/AgCl). The reaction is done in the presence of 2-mercapto-ethanol and can be applied postcolumn.

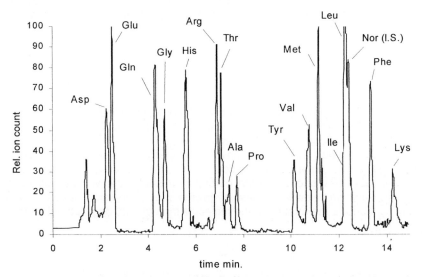

Other derivatizations used for enhancing electrochemical detection in HPLC will be described in Chapter 19.

Derivatization for improving MS detection is less common. However, associated with MS detection, the derivatization can be done mainly to improve the chromatographic separation. As an example, for a better separation, amino acids can be derivatized with phenylisothiocyanate, although the detection is not done in UV but with a MS detector. The same sample analyzed using UV detection and shown in Figure 17.6.3 can be analyzed, for example, using a MS detector with a thermospray source. The TIC obtained with the MS detector is shown in Figure 17.6.5.

FIGURE 17.6.5. *HPLC chromatogram obtained during the analysis of amino acids from a tobacco sample using MS detection with a thermospray source [38].*

Different relative peak intensities can be noticed for the same amino acids by comparison of the UV and the MS chromatograms. The ionization efficiency for thermospray was found to be different for different derivatized amino acids. The MS detection has the advantage of allowing the identification of each amino acid by its spectrum in addition to the corresponding retention time, while the quantitation using UV detection was proven more precise compared to liquid chromatography-mass spectrometry (LC-MS) thermospray detection.

A significant sensitivity improvement in LC-MS has been obtained using electron capture atmospheric pressure ionization (ECAPCI) combined with derivatization using groups containing electronegative elements such as fluorine [60]. Using this procedure the limit of detection for the analysis of estrone is 740 amol, for 2-methoxyestrone is 170 amol, for prostaglandin $F_{1\alpha}$ is 140 amol, for phenylalanine is 610 amol. For example, estrone can be derivatized using pentafluorobenzyl bromide (PFB bromide) by adding to a sample containing 100 ng estrone in 50 μL acetonitrile, 50 μL PFB bromide in acetonitrile (1:19 v/v), followed by 50 μL KOH anhydrous in ethanol (8:1000, w/v). The solution is heated at 60° C for 30 min., evaporated to dryness under N_2 at room temperature and redissolved in about 100 μL solvent. For reversed-phase chromatography the solvent is methanol/water (7:3, v/v), and for normal-phase chromatography the solvent is hexane/ethanol (97:3 v/v). The negative ion for estrone with m/z = 269 corresponds to M - PFP.

17.7 DERIVATIZATION FOR CHIRAL SEPARATIONS

Molecular species having the same composition and bond sequence but with a different arrangement in space of their atoms are called *stereoisomers*. Stereoisomers that are characterized by different interatomic distances between certain atoms that are not bound directly are called *diastereoisomers*. Examples of diastereoisomers are *cis-trans* isomers of compounds containing C = C bonds and *syn-anti* isomers of compounds containing C = N bonds. However, the existence of diastereoisomers is not limited to the presence of a double bond. For example, a molecule with more than one tetrahedral carbon that has different substituents may form diastereoisomers.

A different class of stereoisomers is that of compounds that are mirror images to each other and are not superimposable although the atomic distances are the same in the molecules. These compounds are called *enantiomers*. Molecules that are not superimposable upon their mirror image have the property called *chirality*. Chirality, which is needed for the existence of enantiomers, is commonly caused by the existence in the molecule of at least one tetrahedral carbon atom substituted with groups that are different. This is shown below in the case of glyceraldehyde (the small arrow above the structure at right shows the rotation of the molecule through 120 deg. without changing the geometry):

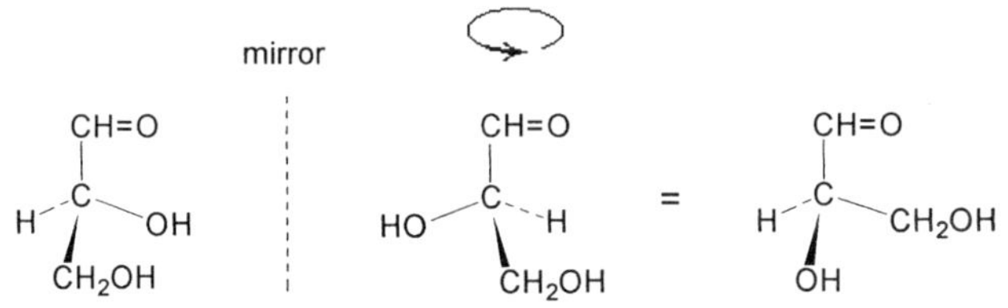

Besides carbon, chiral molecules may be generated with a phosphorus or a sulfur chiral atom. Also, not only a chiral center (such as an asymmetric carbon) generates enantiomers. A chiral axis or a chiral plane also can lead to enantiomers (helicoidal chirality is also known [61]). In the case of a chiral axis, the asymmetry is due to a hindered rotation, as in the example of 2,2',6,6'-substituted biphenyl with the substitution groups having large volumes (NO_2, COOH, Br, etc.) shown below:

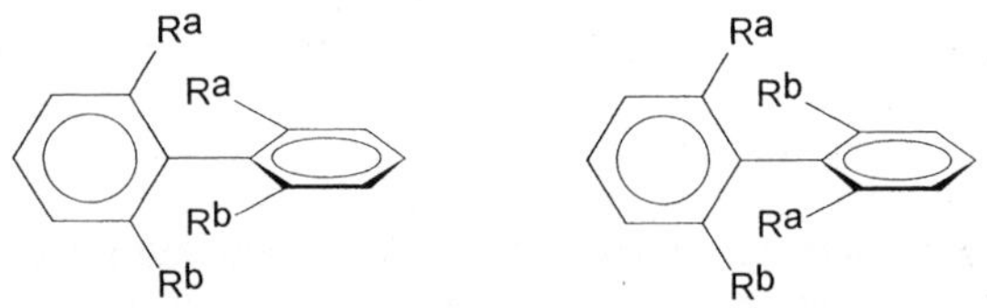

The chirality in an enantiomer is specified using the symbols R and S based on specific rules [61]. For the assignment of a symbol to a chiral carbon, the substituents are arranged in a sequence a > b > c > d. For the four atoms directly attached to the asymmetric carbon, a higher atomic number precedes (outranks) the lower, and a higher atomic mass precedes the lower. For the same atoms directly attached to the asymmetric carbon, the priorities are assigned at the first point of difference. Double and triple bonds are treated by assuming that each such bonded atom is duplicated or triplicated (e.g. C*=O is treated as $C(OH)_2$). After the sequence is established, the molecule is oriented in space with the group "d" of lowest priority behind the asymmetric carbon. When viewed along the C-d bond (from C) and the three substituents a, b, c are oriented clockwise, the compound contains an R asymmetric carbon, and it contains an S asymmetric carbon for counterclockwise arrangement, as indicated below:

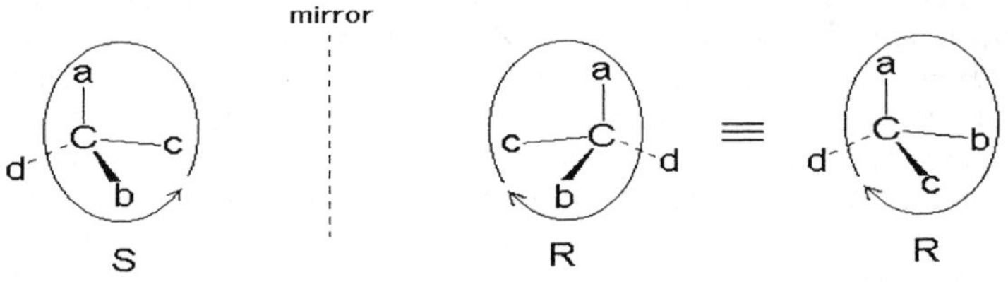

For carbohydrates, an older configurational notation is still in use. For glyceraldehyde, the S enantiomer is also called L, and the R enantiomer is called D. Monosaccharides are classified by convention based on the asymmetry type of the carbon most distant from the carbonyl group in L series and D series when this carbon has the same stereochemistry as in L-glyceraldehyde or in D-glyceraldehyde, respectively.

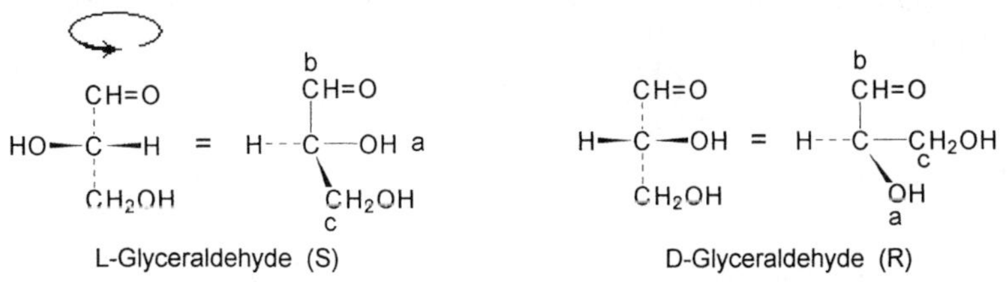

More than one asymmetric carbon can be present in a molecule, as in the case of carbohydrates. The stereoisomers generated by more than one asymmetric carbon can be mirror image one to the other (enantiomers) or may have different steric arrangements, being diastereoisomers. Two diastereoisomers of this type are shown below:

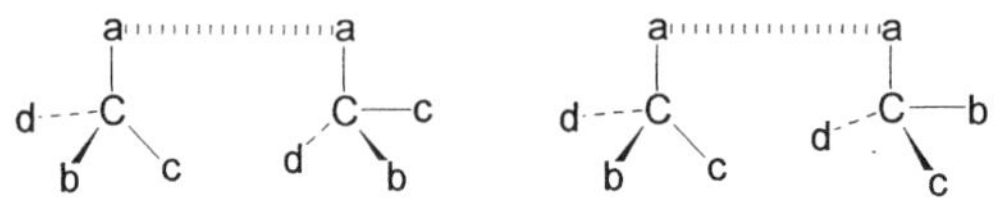

(S,S)-derivative   (S,R)-derivative

For example, C_4 aldoses having two asymmetric atoms and four stereoisomers (the number of stereoisomers being given by 2^n where n is the number of chiral centers) have two pairs of enantiomers. One pair is made from D and L-erithrose and the other from D and L-threose. An erithrose and a threose are diastereoisomers. The structures of the C_4 aldoses are the following:

D-Erithrose	L-Erithrose	D-Threose	L-Threose
(2R,3S)- 2,3,4-	(2S,3R)-2,3,4-	(2S,3S)-2,3,4-	(2R,3R)-2,3,4-
trihydroxybutanal	trihydroxybutanal	trihydroxybutanal	trihydroxybutanal

Stereoisomers can convert one to the other when the free enthalpy of interconversion is lower than about 23 kcal/mol. In this case, the stereoisomers cannot be separated at room temperature. This is the case, for example, of stereoisomers with a chiral axis such as 1-bromopropane. The compound does not have any chiral center, but may be seen as having a chiral axis, and the compounds A and B may be seen as enantiomers, as shown below. The structure B is given in three positions, and the small arrows show rotations of the molecule through 120 deg. without changing the geometry.

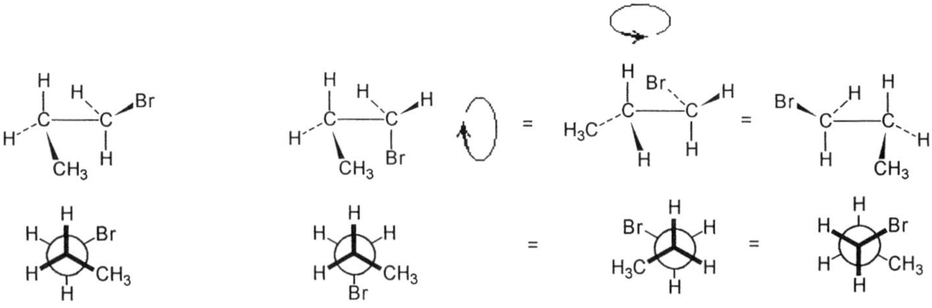

A   B   B   B

However, the free rotation around the C1–C2 bond makes the compounds A and B identical. Such compounds, which can be converted one to the other by rotation of one group around the C–C bond, are commonly called conformers.

Except for the properties that are expressed in chiral conditions and the rotation of the plane-polarized light with opposite (equal) angles, enantiomers have very similar physical and chemical properties. The property of rotating the plane of polarized light is called *optical activity* and is measured by *specific rotation* $[\alpha]^T_\lambda$. Specific rotation depends on the wavelength λ of the polarized light and the temperature T ($^\circ$C) of the optically active medium. For an optically active substance, the specific rotation is obtained from measurement with a polarimeter of the angle α for the rotation of the polarized light. The specific rotation of a solution of concentration c_x (expressed in g/mL solution) with the path length through the optically active medium L (in dm) is given by the relation:

$$[\alpha]^T_\lambda = \alpha / (L\ c_x) \tag{17.7.1}$$

The enantiomer that rotates clockwise the plane of the polarized light (looking in the direction of the source) has a positive $+\alpha$ and is called dextrorotatory (noted + or d). The enantiomer rotating counterclockwise the plane of the polarized light is noted - or l and is called levorotatory. The physical property of a compound of being d (+) or l (-) is not indicative of a specific R or S chiral element (atom, axis, etc.). There are compounds with an R chiral element that are dextrorotatory (d) and some others levorotatory (l). An equimolar mixture of enantiomers does not rotate the plane of polarized light, and this mixture is called a *racemic mixture*.

The presence of asymmetric atoms is very common in numerous compounds, the carbohydrates and amino acids being the best-known examples. Many natural products incorporated in biological systems display chirality, and different stereoisomers may have significant differences in their biological activity. As a result, pharmacological activity of many compounds is different for different enantiomers. However, most organic syntheses generate a racemic mixture of products or at least impurities of the undesired enantiomer. These facts explain the large interest in enantiomer separation.

The separation of enantiomers is difficult, while the separation of other isomers including diastereoisomers can be done much more easily. Common chromatographic techniques can be applied for diastereoisomer separation. The physico-chemical behavior of enantiomer molecules can be different only in a chiral environment. For example, two enantiomers have different behavior toward a chiral stationary phase. For this reason the separation of enantiomers must be done on chiral stationary phases (or using a chiral modifier in the mobile phase for HPLC) [62], [63]. The resolution of a mixture of enantiomers into its components for analytical purposes can be done using various procedures such as:

1) Separation of the underivatized enantiomers (direct resolution) using chiral stationary phases,

2) Derivatization with nonchiral reagents that have the property to enhance the effect of a preexistent chiral element and separation on chiral stationary phases,

3) Derivatization with chiral reagents (in the form of a pure enantiomer) forming diastereoisomers, followed by a separation that can be done on a nonchiral stationary phase, and

4) Separation of the underivatized enantiomers on nonchiral stationary phases but using a chiral modifier in the mobile phase (in the case of HPLC).

Most analytes have specific reactive groups such as OH, SH, NH_2, CH=O, COOH, etc., the derivatization reagents being tailored to react with these groups. For this reason the reagents R–X have the desired property carried by the group R and also must have specific reactive groups X, as shown previously for reagents used in UV-Vis, fluorescence, or chemiluminescence labeling, and must be able to form stable and preferably unique compounds with the analyte. Regardless whether the reagent is chiral or not, it is common that the derivatization for enhancing enantiomer separation is associated with a modification that makes the analyte easier to analyze. For GC analysis this is translated into less polar, thermally more stable compounds. For HPLC the derivatization can be associated with enhancing UV-Vis absorption or with fluorescence labeling, chemiluminescence labeling, etc. This implies that besides the chiral property, the reagent may have chromophores or fluorescent groups.

- Separation of enantiomers on chiral stationary phases

A discussion on the separation of enantiomers without derivatization is beyond the purpose of this book. However, some derivatizations of enantiomers are performed with nonchiral reagents for achieving other desired properties. The resulting derivatized compounds still remain enantiomers. The different behavior of enantiomers does not eliminate the need for better chromatographic properties, better detection capability, or even enhanced separation capability following derivatization. A number of methods and procedures are reported for chiral analyte separation and detection using derivatization with nonchiral reagents, some of these being discussed in Chapter 19.

For both GC and HPLC separations, a variety of chiral phases are available. For GC, two main types are available. One type contains nonbonded derivatized cyclodextrin in a phenyl- or cyanopropylphenyl-polydimethylsiloxane column. The derivatized cyclodextrin can make from 10% to 25% of the weight of stationary phase. Three types of cyclodextrin are common, α -, β - and γ - containing six, seven and eight glucose units, respectively. The cyclodextrin can be used derivatized such as permethylated, as 2,3-di-O-acetyl-6-O-*tert*-butyldimethylsilyl, or as 2,3-di-O-alkyl-6-O-*tert*-butyldimethylsilyl derivatives (where alkyl is methyl, ethyl or propyl). Various contents of phenyl or cyanopropylphenyl were used for the column, varying between 10% and 35%. These columns allow the separation of a wide variety of enantiomers. The second type of chiral GC column uses an amino acid incorporated into the backbone of the silicone-based stationary phase polymer. On this type of column can be separated, for example, amino acids derivatized with nonchiral reagents. The derivatization can be done, for example, using 2-chloropropane followed by the second derivatization with pentafluoropropionyl anhydride, as follows:

The derivatized amino acids can be separated, for example on a Chirasil-L-Val column (*tert*-butylamide linked polydimethylsiloxane) [64].

For HPLC separation, the variety of chiral stationary phases is even larger [65]. The stationary phase must exhibit a number of points of interaction, this number being evaluated to a minimum of three. The interaction can take place through hydrogen bonds, dipole interactions, π electron interactions, etc., and at least one of the interactions must be stereochemically dependent. There are several types of columns such as brush (or Pirkle), ligand exchange, cavity type, and protein type stationary phases. Various Pirkle type columns are available [66], [67]. One such column is made from an optically pure amino acid derivative covalently bonded to a γ-propyl-silanized silica gel. An amide or urea unit is then used for linking a π-electron containing group to the asymmetric carbon of the amino acid. Two examples of such structures, one using (R)-phenylglycine and 3,5-dinitrobenzoic acid with amide linkage and the other using (R)-phenylglycine and 3,5-dinitroaniline with urea linkage, are shown below:

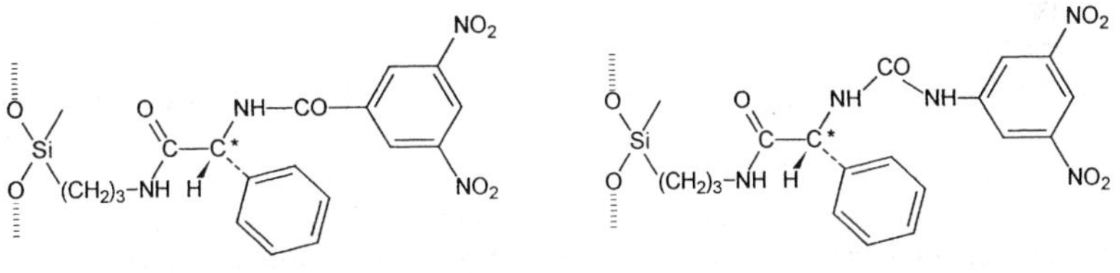

A variety of Pirkle phases are available containing proline, leucine, tartaric acid, etc. as the chiral compound and various π-electron containing fragments [68]. The separations can be done either in normal phase, which is preferred, or in reverse phase. The chiral stationary phases can be used either for enantiomer separations in nonderivatized form or derivatized with nonchiral reagents. One example is the separation of amino acids derivatized with benzyl chloroformate on a Pirkle type column. The derivatization reaction can be written as follows:

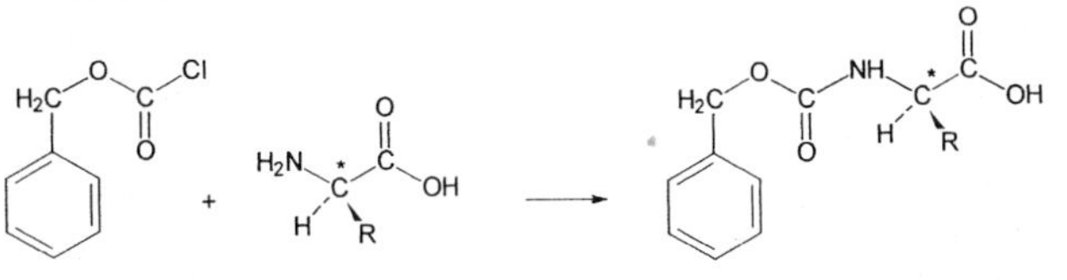

Enantiomeric amino acid derivatives can be separated on a phenylglycine and 3,5-dinitrobenzoic acid with urea linkage Pirkle type column.

The ligand exchange type contains a chiral molecule such as (D)-penicillamine bound to a C18 support [69]. The separation is performed in reverse phase conditions, with the partially aqueous mobile phase containing Cu^{2+} as a modifier. Other ligand exchange columns are also available. A different type contains an optically active Δ-Ru(o-phenantroline)$_3{}^{2+}$ complex in a base clay material.

The cavity type columns contain cyclodextrins, permethylated cyclodextrins, or other modified cyclodextrins as the stationary phase. It is also possible to chemically bond a modified cyclodextrin (such as with carboxymethyl groups) on a polymeric backbone such as poly(methyl methacrylate), as shown in the following structure:

The protein type columns contain a protein such as bovine serum albumin added on a silica base stationary phase.

- Separation of diastereoisomers generated by derivatization with chiral reagents

The separation after derivatization with a pure enantiomer reagent is based on formation of diastereoisomers that can be separated on regular stationary phases. Depending on the nature of the analyte and of the derivatization, different separation techniques can be applied, with GC and HPLC the most commonly utilized. A variety of common columns are used for GC separations. The choice of the column is again dependent on the analyte and the derivatization procedure. For example, α-substituted organic acids, such as α-chloropropionic, α-bromocaproic, etc., can be derivatized with a specific enantiomer of an amino acid ester using benzotriazol-1-yl-oxy-tris(dimethylamino)-phosphonium hexafluorophosphate (BOP) as a peptide coupling reagent in a reaction of the type:

The derivatized acids that are now diastereoisomers can be separated on a common capillary column (OV-1701) [70].

Separations using HPLC or TLC with common stationary phases are also very frequent. As an example, the separation of several chiral amines can be done by derivatization with (S)-(+)-2-(4-chlorophenyl)-α-methyl-5-benzoxazoleacetyl chloride (benoxaprofen chloride). The reaction takes place by heating the amines and the reagent in CH_2Cl_2 at 50° C for 30 min. The separation can be done on a normal-phase silica type column. The reagent brings, besides a chiral center, a fluorescence group, and fluorometric detection is possible [71]. Derivatizations generating diastereoisomers and introducing at the same time a fluorescent group are not uncommon. One other such example is the use of optically pure N-α-9-fluorenyl-methyloxycarbonyl-amino acid-N-carboxy-anhydrides in reactions with amines or amino acids. The reaction with α-methylbenzyl-amine racemate is shown below [72]:

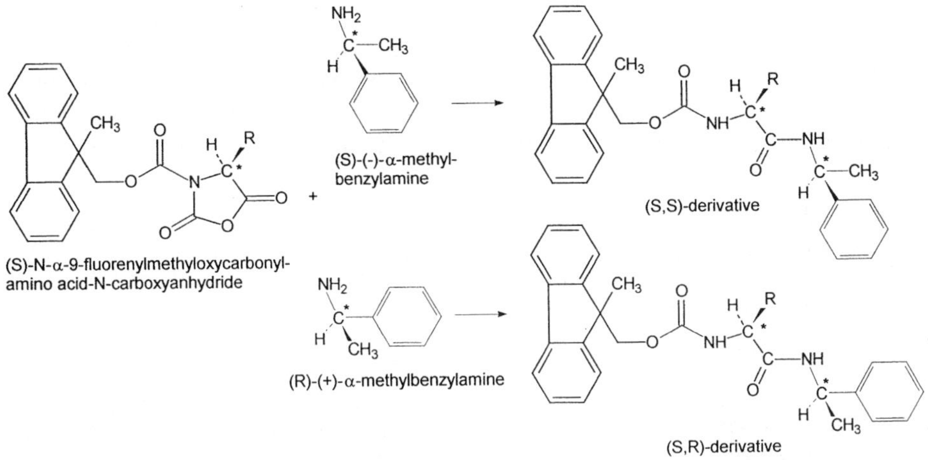

(S)-N-α-9-fluorenylmethyloxycarbonyl-amino acid-N-carboxyanhydride

(S)-(-)-α-methyl-benzylamine

(S,S)-derivative

(R)-(+)-α-methylbenzylamine

(S,R)-derivative

Various groups R corresponding to different amino acids can be used in this reagent [72]. The separation of the derivatized compounds can be done on silica or on ODS (C18) type columns using fluorescence detection.

Not only reagents containing an asymmetric carbon have been used for derivatization. One example of a reagent containing an asymmetric bond is (aS)-2'-methoxy-1,1'-binaphthalene-2-carbonyl cyanide. This reagent can be used for the derivatization of alcohol racemates leading to two diastereoisomers:

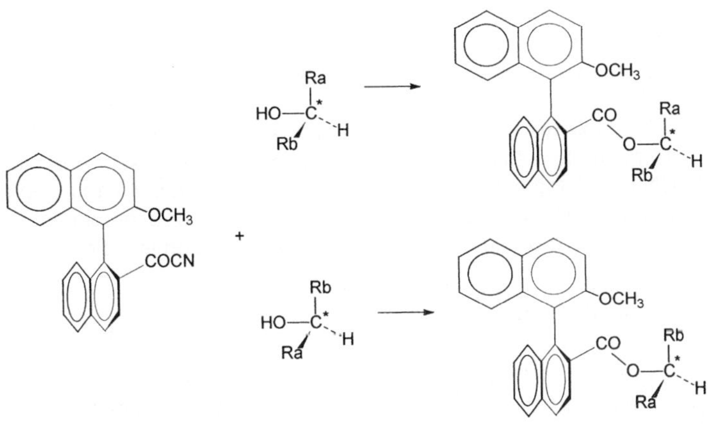

This reaction was applied, for example, for the derivatization of penbutolol, (S)-1-(2-cyclopentylphenoxy)-3-[(*tert*-butyl)amino]-propan-2-ol [73], and the separation was performed on a Cosmosil 5SL column. Many other examples of reagents used for chiral separations are given in Chapter 19.

- *Separation of underivatized enantiomers using a chiral modifier in the mobile phase*

The separation of enantiomers using a chiral modifier in the mobile phase can be achieved using chiral complexing agents or chiral counter ions in ion-pair chromatographic separations. The separation of enantiomers is based on the formation of diastereoisomer ion pairs or complexes [74]. The binding forces in ion pair formation are usually a combination of electrostatic effects and also hydrogen bonding, dipole-dipole, and π-π hydrophobic interactions that make the ion pairs more stable. A "three point rule" indicates that three points of interaction are needed to obtain steroselective retention [75]. Usually, the choice of ion pairs is specifically done for particular analytes [76]. Bulky substituents at the analyte may help in enhancing stereoselectivity [77]. An example of ion pair formation between (1S)-(+)-10-camphorsulfonic acid and (1S,2R)-(+) and (1R,2S)-(-)-ephedrine is shown below:

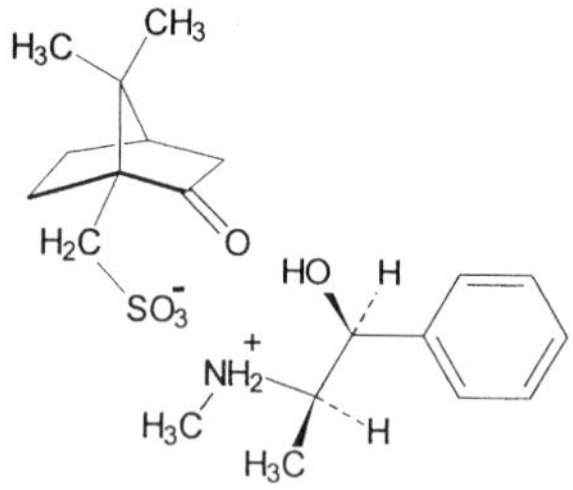

(1S)-(+)-10-camphosulfonic acid   (1S)-(+)-10-camphosulfonic acid

(1S,2R)-(+)-ephedrine   (1R,2S)-(-)-ephedrine

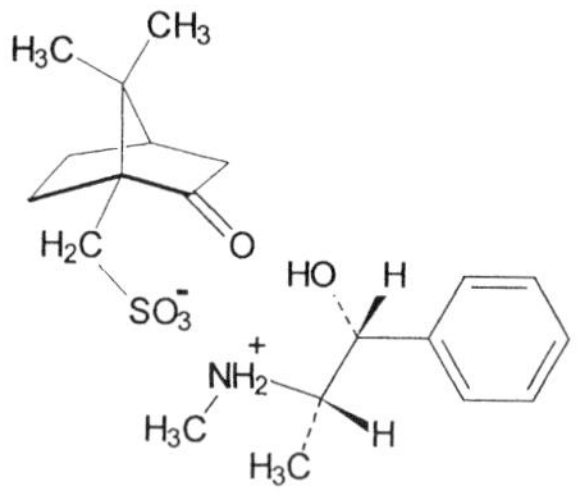

Probably, ion pair formation, hydrogen bonding, and hydrophobic interactions play a role in the stability of adducts in this particular case.

Among other compounds used in the mobile phase for enantiomer separations are also proteins such as albumin, α_1-acid glycoprotein [78], etc.

The formation of complexes with coordinative bonds can also be used in chiral separations. An example is the separation of amino acids in an aqueous mobile phase with 15% CH_3CN by adding in the mobile phase L-proline and Cu^{2+} and using a C18 type stationary phase. The amino acid can be derivatized, for example as dansyl derivatives. The structure of the complexes formed in solution is indicated below:

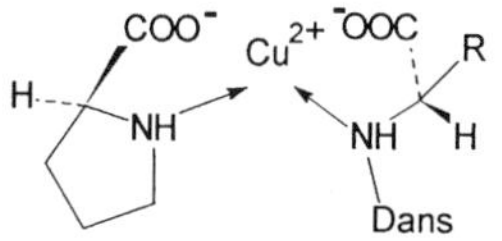

L-proline-Cu^{2+}-dansyl-L-amino acid 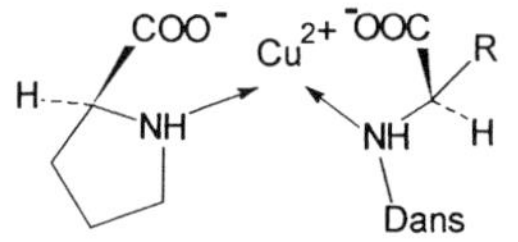   L-proline-Cu^{2+}-dansyl-D-amino acid

Better separations can be obtained using bulkier amino acids in the mobile phase, such as histidine, phenylalanine, or tryptophan [79]. Other derivatizations for the analyte amino also can be applied, following detection with good sensitivity.

REFERENCES 17

1. P. Steinberg, A. Fox, Anal. Chem., 71 (1999) 1914.
2. D. R. Knapp, *Handbook of Analytical Derivatization Reactions*, J. Wiley, New York, 1979.
3. G. Lunn, L. C. Hellwig, *Handbook of Derivatization Reactions for HPLC*, J. Wiley, New York, 1998.
4. J. A Dean (ed.), *Lange's Handbook of Chemistry*, McGraw-Hill, New York, 1999.
5. J. J. P. Stewart, J. Comp. Molec. Design, 4 (1990) 1.
6. S. C. Moldoveanu, A. Savin, *Aplicatii in Chimie ale Metodelor Semiempirice de Orbitali Moleculari*, Ed. Academiei, Bucuresti, 1980.
7. L. R. Snyder, J. Chromatogr., 93 (1974) 223.
8. J. Bermejo, M. D. Guilen, J. Chromatogr., 318 (1985) 187.
9. J. B. Forehand et al., 49th TCRC, Lexington, 1995.
10. W. O. McReynolds, J. Chromatogr. Sci., 8 (1970) 685.
11. J. J. Leary et al., J. Chromatogr. Sci., 11 (1973) 201.
12. G. E. Baiulescu, V. A. Ilie, *Stationary Phases in Gas Chromatography*, Pergamon, New York, 1975.
13. N. K. McCallum, R. J. Armstrong, J. Chromatogr., 78 (1973) 303.
14. J. B. Brooks et al., Anal. Chem., 47 (1975) 1960.
15. S. C. Moldoveanu, unpublished results.
16. F. W. McLafferty, *Interpretation of Mass Spectra,* University Science Books, Mill Valley, 1980.
17. G. Singh et al., Anal. Chem., 72 (2000) 3007.
18. R. Buffington, M. K. Wilson, *Detectors for Gas Chromatography*, Hewlett-Packard, Avondale, 1987.
19. S. E. Stein, J. Am. Soc. Mass. Spec., 10 (1999) 770.
20. Mass spectra library NIST'98, NIST, 1998.
21. Mass spectra library Wiley6, J. Wiley, 1995.
22. P. Lecchi, F. P. Abramson, Anal. Chem., 71 (1999) 2951.
23. J. A. McCloskey et al., Anal. Chem., 40 (1968) 233.
24. T. Seki, J. Chromatogr., 155 (1978) 415.
25. S.-I. Kobayashi, K. Imai, Anal. Chem., 52 (1980) 1548.
26. K. Blau, J. Halket (eds.), *Handbook of Derivatives for Chromatography*, J. Wiley, Chichester, 1993.
27. J. Ruzicka, E. H. Hansen, Anal. Chim. Acta, 78 (1975) 145.
28. I. S. Krull (ed.), *Reaction Detection in Liquid Chromatography*, M. Dekker, New York, 1986.
29. P. R. Brown, E. Grushka (eds.) *Advances in Chromatography*, vol. 33, M. Dekker, New York, 1993.
30. J. M. Ramsey et al., Nat. Med., 1 (1995) 1093.
31. Y. Liu, Anal. Chem., 72 (2000) 4608.
32. S. C. Jacobson et al., Anal. Chem., 66 (1994) 3472.
33. K. Fluri et al., Anal. Chem., 68 (1996) 4285.

34. T. Toyo'oka (ed.), *Modern Derivatization Methods for Separation Sciences*, J. Wiley, Chichester, 1999.
35. H. Lingeman, W. J. M. Underberg (eds.), *Detection Oriented Derivatization Techniques in Liquid Chromatography*, M. Dekker, New York, 1990.
36. R. E. Majors et al. Anal. Chem., 56 (1984) 300R.
37. J. H. Lauterbach, S. C. Moldoveanu, 42nd TCRC, Lexington, 1988.
38. T. H. Burch et al., 44th TCRC, Winston-Salem, 1990.
39. H. J. Klimisch, L. Stadler, J. Chromatogr., 90 (1974) 141.
40. S.-W. Jin et al., FEBS Lett., 198 (1986) 150.
41. R. W. Frei, J. F. Laurence, J. Chromatogr., 83 (1973) 321.
42. W. R. Gray, Methods Enzymol., 25 (1972) 121.
43. N. Seiler, K. Deckardt, J. Chromatogr., 107 (1975) 227.
44. A. Büldt, U. Karst., Anal. Chem., 71 (1999) 1893.
45. H. Higashijima et al., Anal. Chem., 64 (1992) 711.
46. A. J. G. Mank, E. S. Yeng, J. Chromatogr., 708 (1995) 309.
47. R. Bradley, *Fluorescence Spectroscopy*, Plenum, New York, 1983.
48. Y. Rakicioglu et al., Anal. Chim. Acta, 359 (1998) 269.
49. K. J. Albert, D. R. Walt, Anal Chem., 72 (2000) 1947.
50. J. V. Goodpaster, V. L. McGuffin, Anal. Chem., 73 (2001) 2004.
51. J.-M. Lin, M. Yamada, Anal. Chem., 72 (2000) 1148.
52. M. Stigbrand et al., Anal. Chem., 66 (1994) 1766.
53. K. Uchicura, M. Kirisawa, Anal. Sci., 7 (1991) 971.
54. W.-Y. Lee, T. A. Nieman, J. Chromatogr., A, 659 (1994) 111.
55. D. R. Scotty et al., Anal. Chem., 68 (1996) 1530.
56. S. J. Woltman et al., Anal. Chem., 72 (2000) 4928.
57. S. J. Woltman et al., Anal. Chem., 71 (1999) 1504.
58. G. Xu, S. Dong, Anal. Chem., 72 (2000) 5308.
59. I. Rubinstein et al., Anal. Chem., 55 (1983) 1580.
60. G. Singh et al., Anal. Chem., 72 (2000) 3007.
61. F. Badea, F. Kerek, *Stereochimie*, Ed. Stiintifica, Bucuresti, 1974.
62. W. J. Lough (ed.), *Chiral Liquid Chromatography*, Blackie/Chapman, New York, 1989.
63. M. Zief, L. J. Crane (eds.), *Chromatographic Chiral Separations*, vol. 40, M. Dekker, New York, 1988.
64. F. Bruner (ed.), *The Science of Chromatography*, Elsevier, Amsterdam, 1985.
65. D. A. Armstrong, B. Zhang, Anal. Chem., 73 (2001) 557A.
66. W. H. Pirkle et al., J. Chromatogr., 316 (1984) 585.
67. W. H. Pirkle, Y. Liu, J. Chromatogr., A, 749 (1996) 19.
68. F. Gasparrini et al., J. Chromatogr., A, 906 (2001) 35.
69. M. Schlauch et al., J. Chromatogr., A, 897 (2000), 145.
70. H. Mattras et al., J. Chromatogr., A, 803 (1998) 307.
71. H. Weber et al., J. Chromatogr., 307 (1984) 145.
72. M. Pugniere et al., J. Chromatogr., A, 767 (1997) 69.
73. J. Goto et al., Anal. Sci., 7 (1991) 723.
74. C. Pettersson, G. Schill, J. Liq. Chromatogr., 9 (1986) 269.
75. C. E. Dalgliesh, J. Chem. Soc., 137 (1952) 3940.
76. C. Pettersson, J. Chromatogr., 316 (1984) 553.
77. V. Piette et al., J. Chromatogr., A, 894 (2000) 63.
78. J. Hermansson, J. Chromatogr., 316 (1984) 537.
79. S. Lam, A. Karmen, J. Liq. Chromatogr., 9 (1986) 291.

CHAPTER 18

Chemical Reactions Used in Derivatization

18.1   REACTIONS WITH FORMATION OF ALKYL OR ARYL DERIVATIVES

The formation of alkyl or aryl derivatives is applied to replace the active hydrogens from an analyte with an alkyl (R) or aryl (Ar) group. The replacement can be done in functionalities such as OH, COOH, SH, NH, or CONH. The derivatizations can be useful for GC, HPLC, or other type of chromatography. A large number of reagents R-X are known, and in a simplified approach it can be considered that R carries a specific property and X a specific reactivity, although the reactivity of a reagent is influenced by both R and X components of the molecule. Nevertheless, this simplified approach helps to understand the selection process of a reagent for a specific derivatization and has directed the synthesis of many new derivatization reagents.

In most alkylation reactions, the analyte acts as a nucleophile (Y:, Y:H, Y:$^-$) reacting in a substitution (S_N) with the alkylating reagent R-X, which contains a leaving group X and an alkyl group R:

$$Y{:}H + R\!-\!X \; \rightleftharpoons \; Y\!-\!R + X{:}H$$

Alkyl halides, especially alkyl iodides and alkyl bromides, are commonly used for alkylation. Various reagents and conditions are utilized in the derivatizations for analytical purposes. Because some of the derivatizations can be slow and inefficient, depending on the analyte and on the reagent, the reaction rate becomes an important parameter for the analytical applicability. For the preparation of methyl or ethyl substituents, for example, the reaction with an alkyl halide frequently must be performed in the presence of a catalyst and in some instances using a particular solvent. Some catalysts and solvents used in alkylation reactions with alkyl bromide or alkyl iodide as reagents are given in Table 18.1.1.

A typical methylation of a polyol (or monosaccharide) starts with 0.1–100 μmol material that is dissolved in 1 mL dry DMSO. To this solution is added 1 mL methylsulfinyl carbanion solution [9], and the mixture is flushed with N_2. The mixture is allowed to stand for 15–20 min at room temperature, and 0.2 mL CH_3I are added dropwise. After 10 min., 10 mL ethyl ether and 5 mL water are added with mixing, and the organic layer is separated. The aqueous phase is extracted two more times with 5 mL ethyl ether, and the extracts are combined, dried over $MgSO_4$, and evaporated to dryness. The material is redissolved in hexane and analyzed by GC or GC-MS [10].

Methylsulfinylmethanide anion is prepared from dry DMSO and NaH or KH in a reaction as follows:

$$(CH_3)_2SO + NaH \longrightarrow CH_3SOCH_2^- \; Na^+ + H_2$$

TABLE 18.1.1. *Some catalysts and solvents used in alkylation reactions with short chain alkyl bromide or alkyl iodide as reagent.*

Catalyst	Solvent	Analytes	Samples with problems	Reference
Ag$_2$O	alkyl halide	sugars, miscellaneous	easily oxidizable compounds	[1], [2]
Ag$_2$O	DMF	sugars, miscellaneous	peptides with glutamic acid or tryptophan	[3], [4]
Ag$_2$O + (CH$_3$)$_2$S	appropriate	miscellaneous	free sugars may be oxidized	[5]
Ag salt of the acid		acids		[1]
BaO and/or Ba(OH)$_2$	DMF	miscellaneous	compounds sensitive to base catalyzed degradation	[1]
BaO	DMF	miscellaneous	O-acyl migration possible	[1]
NaH	DMF	miscellaneous	compounds sensitive to base catalyzed degradation, O-acyl replacement likely	[1]
NaH	DMSO	miscellaneous	esters, peptides with histidine and arginine	[6]
NaH	ether type	miscellaneous	compounds sensitive to base catalyzed degradation	[1]
NaH	dimethyl-acetamide	peptides	compounds sensitive to base catalyzed degradation	[7]
KOH, or NaOH	DMSO	sugars, phenols, amides, alcohols		[2], [8]
methylsulfinyl carbanion CH$_3$SOCH$_2^-$	DMSO	sugars, amides, etc.	replacement of O-acyl groups with O-alkyl may occur	[9]
HgO and HBF$_4$	CH$_2$Cl$_2$	acids		[4]
K$_2$CO$_3$	DMF	acids	compounds sensitive to base catalyzed degradation	[1]

DMF = dimethylformamide, DMSO = dimethyl sulfoxide

NaH is usually commercially available as a 60–80% suspension in mineral oil. The preparation of methylsulfinylmethanide anion can start, for example, with 8 g suspension NaH that is initially washed with hexane under a stream of N$_2$ and dried with vacuum. To the clean and dry NaH, 150 mL dry DMSO is added. The mixture is heated at 50° C under a stream of N$_2$ until the evolution of H$_2$ is noticed. The reaction can be considered complete in about 1 hour. The reagent can be stored in sealed ampules at -20° C for later use. The reduction of DMSO to prepare the methylsulfinylmethanide also can be done using potassium tert-butoxide or butyllithium [11], [12].

For the alkylation with Ag$_2$O as a catalyst, about 1 part (w) Ag$_2$O is used for about 20 parts alkyl halide (v). The mixture of analyte, reagent (two equivalents for each equivalent of H to be replaced), and Ag$_2$O is stirred in the dark until the reaction is complete. Fresh portions of Ag$_2$O are needed to accelerate the reaction and in some cases the heating of the mixture under reflux in the presence of a desiccant. The alkylation of some compounds that are weak nucleophiles, such as most alcohols, can be slow and inefficient. The OH groups in carbohydrates are easier to alkylate. Even for these compounds where the presence of the polyol groups lead to a slightly higher reactivity, the alkylation reaction with methyl iodide in the presence of Ag$_2$O in DMF requires about 16 hours for completion at room temperature:

The time required for the completion of the reaction is normally established during the method development step, but it can be verified during the derivatization process using for example a TLC procedure. After the completion of the reaction, the solids can be filtered and washed with CH_2Cl_2, adding the solutions together. The remaining silver salts can be removed by washing the organic phase with 5% KCN, followed by further washing with water, drying, and concentrating for GC analysis. For quantitative purposes, an appropriate internal standard is required.

Acidic hydrogens may be replaced with alkyl groups in minutes. As an example, the alkylation of N-(4-hydroxyphenyl)acetamide (acetaminophen) can be done using iodoheptane in the presence of tetramethylammonium hydroxide at room temperature within 10 min. [13]. The reactions can be written as follows:

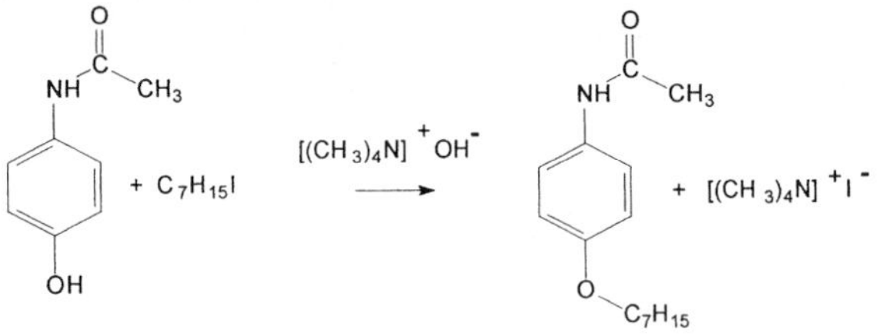

The enhancement of the alkylation efficiency was achieved using several other procedures. For example, specific cryptands can be used to solvate the alkali metal portion of organic acid salts, allowing the anion to be freer and increasing the rate of nucleophilic substitution. These agents were used for the analytical derivatization of carboxylic acids [14], [15].

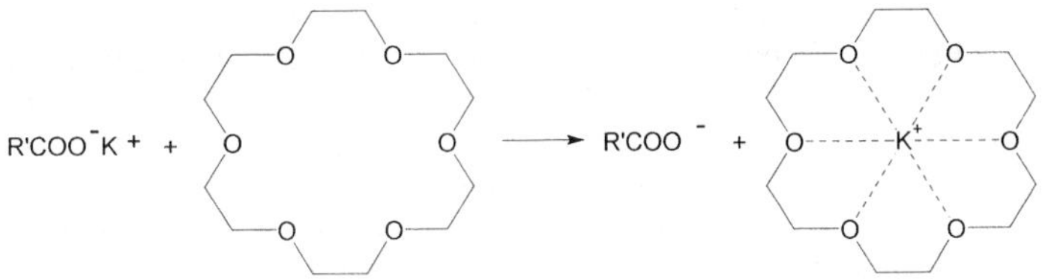

$$R'COO^- + R''{-}X \longrightarrow R'COOR'' + X^-$$

One other approach for enhancing the alkylation efficiency is the use of phase transfer alkylation [16], [17]. This approach is based on the formation of a compound easily extractable in an organic phase and on the displacement of the equilibrium in the direction of the formation of the desired product. For an organic acid, this can be achieved, for example, using a reaction with a tetraalkylammonium salt, following the sequence of reactions indicated below:

$$R'COOH + R_4N^+X^- \text{ (aqueous phase)} \rightleftharpoons R'COO^-R_4N^+ + HX \text{ (aqueous phase)}$$

$$R'COO^-R_4N^+ \text{ (aqueous phase)} \rightleftharpoons R'COO^-R_4N^+ \text{ (organic phase)}$$

$$R'COO^-R_4N^+ + R''X \text{ (organic phase)} \rightleftharpoons R'COOR'' + R_4N^+X^- \text{ (organic phase)}$$

$$R_4N^+X^- \text{ (organic phase)} \rightleftharpoons R_4N^+X^- \text{ (aqueous phase)}$$

The derivatization with short chain alkyl bromides or iodides has numerous analytical applications for compounds such as steroids [18], amino acids [19], catecholamines, sulfonamides, phenols, barbiturates [4], organic acids [20], and mono and oligosaccharides [21], [22]. The derivatization performed for structure elucidations may use deuterated alkyl bromides or iodides, followed by GC-MS for analysis.

One different way of enhancing the alkylation efficiency is the use of different alkylating reagents besides short-chain alkyl bromides or iodides. For a reagent RX, this can be achieved by modifying the R substituent or the leaving group X. One typical example of a halide that is particularly reactive is pentafluorobenzyl bromide. This reagent can be used for the derivatization of a variety of compounds containing active hydrogens [1] in reactions such as the following:

R—OH + C_6F_5—CH$_2$Br $\longrightarrow$ C_6F_5—CH$_2$—OR

R—COOH + C_6F_5—CH$_2$Br $\longrightarrow$ C_6F_5—CH$_2$—O—C(=O)—R

R—SO$_2$—NH$_2$ + C_6F_5—CH$_2$Br $\longrightarrow$ $(C_6F_5CH_2)_2N$—SO$_2$—R

Another reactive halide is 2-bromoacetophenone (phenacyl bromide). This reagent is used mainly for the alkylation of compounds containing more acidic hydrogens such as carboxylic acids:

R—COOH + C_6H_5—C(=O)—CH$_2$—Br $\longrightarrow$ R—COO—CH$_2$—C(=O)—C_6H_5

Reagents containing the group -CH$_2$-Br bound to other activating moieties besides pentafluorophenyl are also used for derivatization in a variety of analytical applications. The modification of the substituent R can be associated with an increase in the molecular weight that may not be desirable for GC applications but may be very useful for enhancing detectability in HPLC analysis. These alkylations can be used for attaching chromophores to the analyte and make possible the detection in HPLC using UV or fluorescence detectors. One such example is the reaction of methanol with 3-bromomethyl-7-methoxy-1,4-benzoxazin-2-one:

+ CH$_3$OH $\longrightarrow$ + HBr

This reaction also has been used to derivatize other alcohols for attaching chromophores with the purpose of HPLC analysis.

The derivatization with the purpose of obtaining aryl derivatives is similar in many respects to the alkylation reaction. The reaction may take place with a variety of compounds containing active hydrogens and has both analytical applications as well as numerous uses in organic synthesis [23]. Simple aryl halides are generally resistant to attack by nucleophiles. This low reactivity can be significantly increased by changes in the structure of aryl halide or in the reaction conditions. The nucleophilic displacement can become very rapid when the aryl halide is substituted with electron attracting groups such as NO_2. As an example, the reaction between a secondary amine and 2,4-dinitrofluorobenzene can be written as follows:

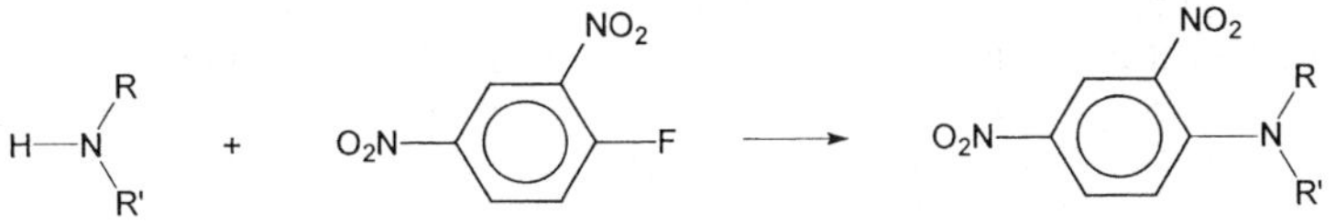

A similar reaction takes place with phenols and other compounds containing acidic hydrogens.

Many reagents generating alkylaryl or aryl derivatives were used for enhancing detectability for HPLC, TLC, or electrochromatographic techniques. A comprehensive presentation for this type of reaction is given in several references [1], [24] (see also Chapter 19).

Besides halides, other alkylating reagents are known (different X in RX), also reacting in a nucleophilic substitution. For example, dimethyl sulfate was one of the first alkylating reagents used for synthetic purposes [25]. Alkylfluoromethylsulfonates are even more reactive than sulfates, and the reaction may take place with the active hydrogen even from alcohols or amines:

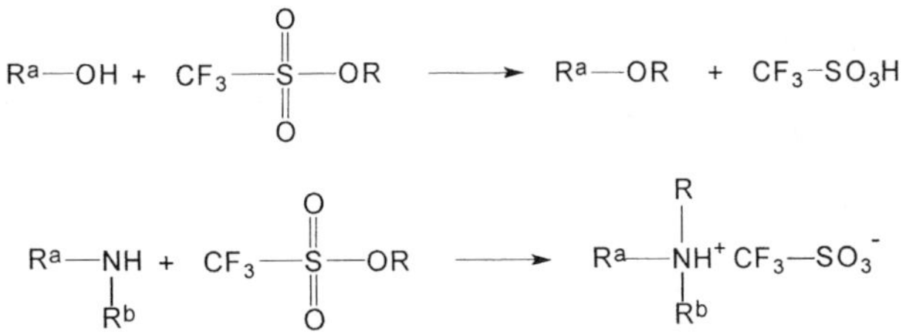

Tertiary amines such as pyridine also react with this type of reagent, forming quaternary ammonium salts:

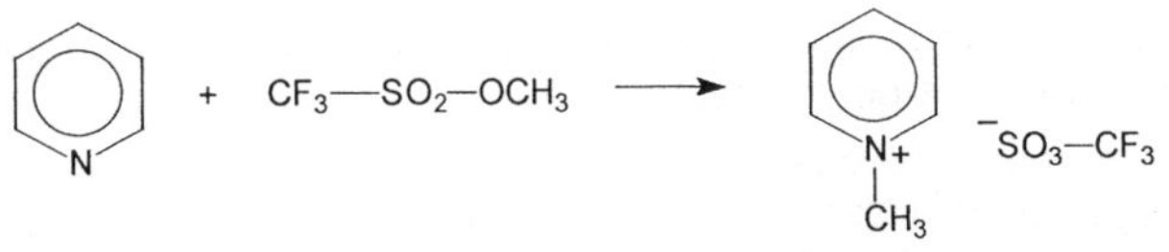

Depending on the reactivity of the analyte, the reaction may take place at room temperature or with heating at 100–120° C for several hours. The alkylation with alkylfluorosulfonates can be catalyzed as other alkylation reactions for increasing the reaction rate. A catalyst that can be used in this reaction is $Hg(CN)_2$ [1].

Other reactive sulfonates may be used, for example, for the alkylation of alcohols, the sulfonic group acting as a good leaving group. One such reaction is shown below, and it is used in HPLC for attaching a chromophore group to alcohols:

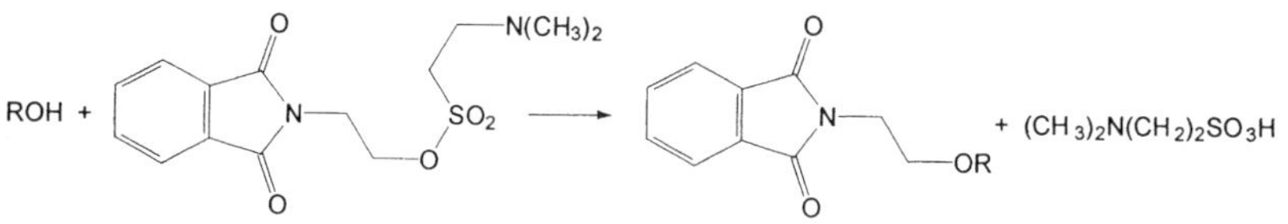

$$ROH + \text{[phthalimide-CH}_2\text{CH}_2\text{-SO}_2\text{-O-CH}_2\text{CH}_2\text{-N(CH}_3)_2] \longrightarrow \text{[phthalimide-CH}_2\text{CH}_2\text{-OR]} + (CH_3)_2N(CH_2)_2SO_3H$$

Sulfonic acids themselves also may be used in some derivatizations, the sulfonic group being in this case the leaving group. For example, the amino group in amino acids can react with sodium 1,2-naphthoquinone-4-sulfonate in the presence of HCl, in a reaction as follows [26]:

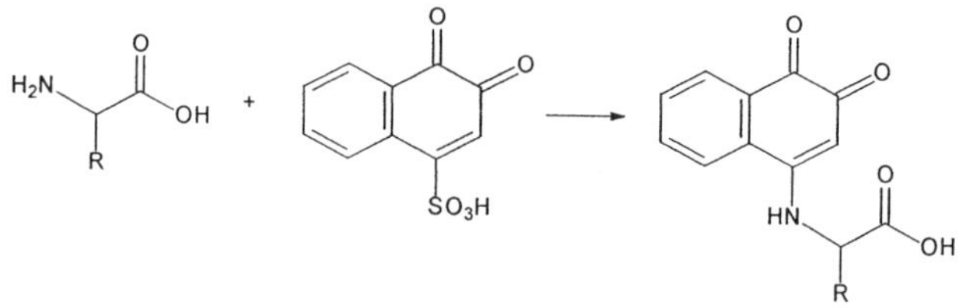

The use of methylsulfinyl carbanion $CH_3SOCH_2^-$ as a catalyst for alkylation using an alkyl halide as a reagent was indicated previously. The compound itself is a strong alkylating reagent and can be used for methylation. Similarly dimethyloxosulfonium methylide $(CH_3)_2SOCH_2^-$ can be used for methylation in strongly basic conditions [23].

Diazomethane is another common alkylating (methylating) reagent. The alkylation using diazomethane can be written as follows:

$$Y:-H \;+\; CH_2{=}\overset{+}{N}{=}\overset{..}{\underset{..}{N}}{:}^- \longrightarrow CH_3{-}\overset{+}{N}{\equiv}N{:} + Y{:}^- \longrightarrow Y{-}CH_3 \;+\; N_2$$

Diazomethane is a gaseous unstable substance, which cannot be stored for long periods of time. It is usually prepared in small quantities and used immediately with or without an intermediate step of dissolution in ether. The preparation can be done from different N-nitroso-N-alkyl compounds in a reaction with a base. The reaction with a N-nitroso-N-alkyl-p-toluenesulfonamide can be written as follows:

$$CH_3{-}\text{[benzene]}{-}SO_2{-}\underset{NO}{\overset{CHR_2}{N}} + NaOH \longrightarrow CH_3{-}\text{[benzene]}{-}SO_3Na + H_2O + CR_2N_2$$

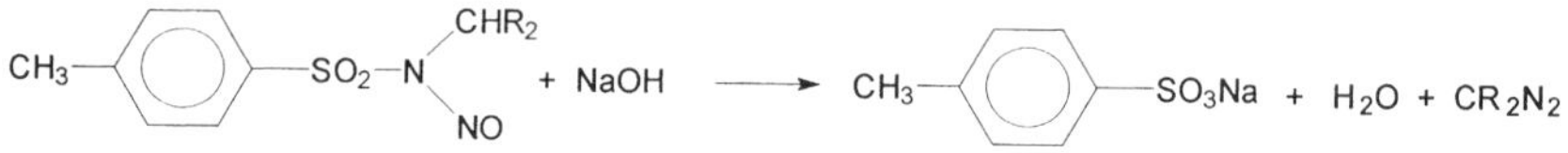

A common procedure for generating diazomethane uses N-nitroso-N-methyl-p-toluenesulfonamide (Diazald) [27], which reacts in a solvent (such as 2-methoxyethyl ether) with a concentrated aqueous solution of NaOH. Other procedures can use the reaction of a strong base with N-nitroso-N-alkylureas, N-nitroso-N-alkylcarbamates, N-nitroso-N-alkylguanidines such as N-methyl-N'-nitro-N-nitrosoguanidine (MNNG), etc. [28], [29], [30], [31]. A commercially available generator for CH_2N_2 for analytical purposes is shown schematically in Figure 18.1.1a, and a generator that can be set up from two reaction vials and a piece of capillary chromatographic column is shown in Figure 18.1.1b. The two parts of the commercial generator are held together and sealed with a pinch-type clamp. The amount of reagent used to generate diazomethane must be less than 30–40 mg, which is mixed with about 0.25 mL organic solvent (diglyme). About 0.25 mL of 5 M solution of NaOH in water is then injected through the septum of the vial cap, and the diazomethane is generated. The analyte in 0.3–0.5 mL of an appropriate solvent (such as ether) is put in the outside tube of the generator (Figure 18.1.1a) or in the second vial (Figure 18.1.1b). Pure ether may be used instead of the analyte solution if further utilization is intended. The generator is usually immersed in an ice bath. The generation of diazomethane is noticed by the effervescence of the reagent when the base solution is added, and the time required for the reaction completion is about 30 min. Because diazomethane is poisonous and can explode, proper safety precautions must be taken during its preparation.

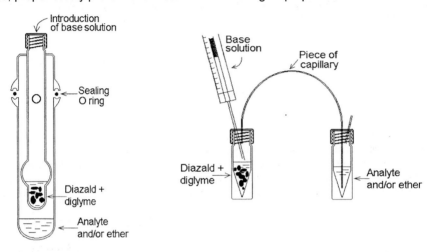

FIGURE 18.1.1a. *A commercially available diazomethane micro-generator (Wheaton).*   FIGURE 18.1.1b. *Simple laboratory setup for diazomethane generation.*

Different diazo compounds besides diazomethane can be used for alkylation, such as diazoethane, phenyldiazomethane [32], or trimethylsilyldiazomethane [33]. Diazo compounds were used even for attaching large fragments such as in the derivatization of organic acids for HPLC analysis using a fluorescence detector. An example is the use of 9-anthryldiazometane for acid derivatization followed by HPLC analysis with fluorescence detection (excitation at 360 nm and emission at 440 nm) [34].

The derivatization with diazo compounds is very efficient for carboxylic acids sulfonic acids, phenols and enol groups, but slower for alcohols and other groups containing active hydrogens. Several Lewis acids were used as catalysts in the alkylation with diazomethane (or other diazoalkanes). The most common catalyst is BF_3, although other Lewis acids were reported to be efficient, such as HBF_4 [4], $SnCl_2$ 2 H_2O [32], or $AlCl_3$. Even methanol was reported to catalyze the methylation of certain hydroxy compounds [35].

The use of diazomethane in the methylation of a variety of carbohydrates using Lewis acids as catalysts was proven to be a very useful technique, mainly because carbohydrates are quite sensitive to oxidation (β-eliminations) in strongly basic conditions. The methylation of partly acetylated sugars and amino sugars using diazomethane and BF_3 in ether leads to the methylation of the free OH groups without the migration or substitution of the existent acyl groups, as shown for a partly acetylated 2-amino-2-deoxy-β-D-glucose [1]:

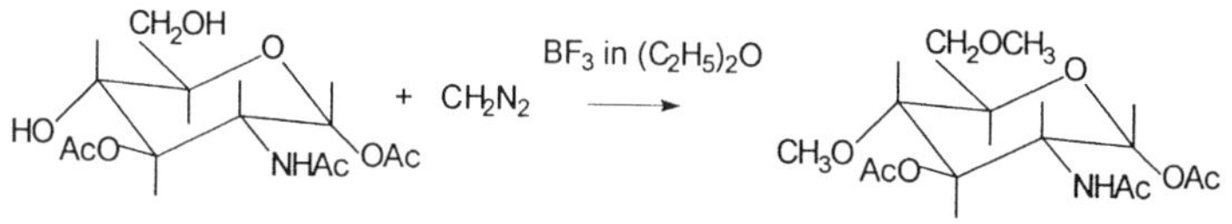

The procedure is important for the analysis of structure of carbohydrates (see Section 20.1).

A common alkylation of acidic analytes such as carboxylic acids, phenols, and thiols [36], [37] is performed using another type of alkylating reagent, namely N,N-dimethylformamide dialkyl acetals. For a compound containing a COOH group, the reaction with this reagent type takes place as follows:

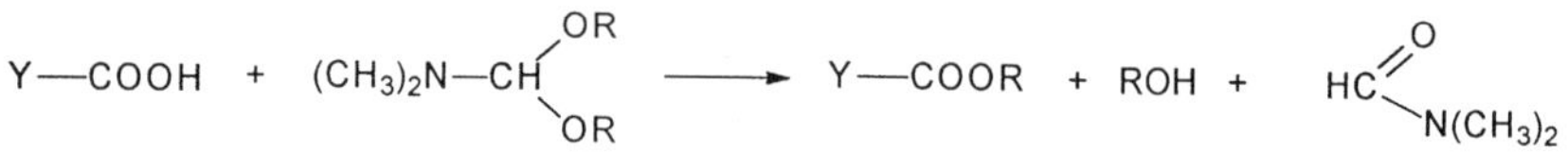

N,N-Dimethylformamide dimethyl acetal is commercially available (Methyl-8® [38]). The reagent is particularly sensitive to moisture, and dry conditions are necessary to avoid hydrolysis. Various solvents were found to be appropriate for this derivatization. The reagent can be used for the alkylation of amines and amides forming N,N-dialkylamino, or N-alkyl derivatives. N,N-Dimethylformamide dialkyl acetals do not react well with aliphatic alcohols unless the OH groups are more acidic.

The acidic groups can also be alkylated (methylated) using trimethyl orthoacetate. The reaction can be written as follows [39]:

$$R—COOH + CH_3C(OCH_3)_3 \longrightarrow R—COOCH_3 + CH_3COOCH_3 + H_2O$$

Alkylation also can be achieved using alkyl-p-tolyltriazenes ($R-NH-N=N-C_6H_4-CH_3$). For example, an alkylation with a benzyl-p-tolyltriazene takes place as follows:

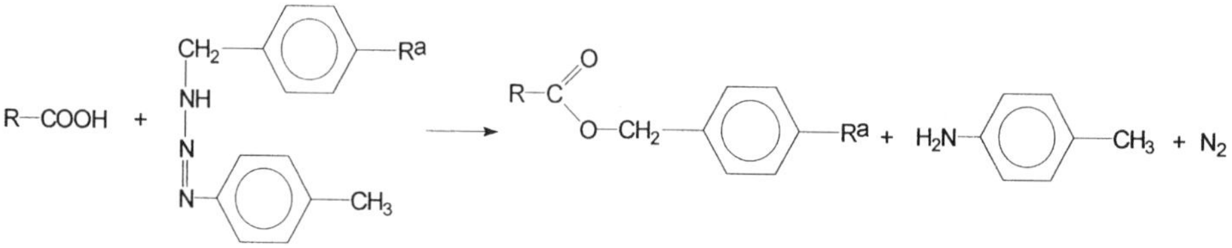

The reaction with less acidic analytes such as phenols or thiols requires heating and longer reaction times.

O-alkyl isoureas also are used for the formation of analytes containing acidic hydrogens. The reaction with a carboxylic acid can be written as follows:

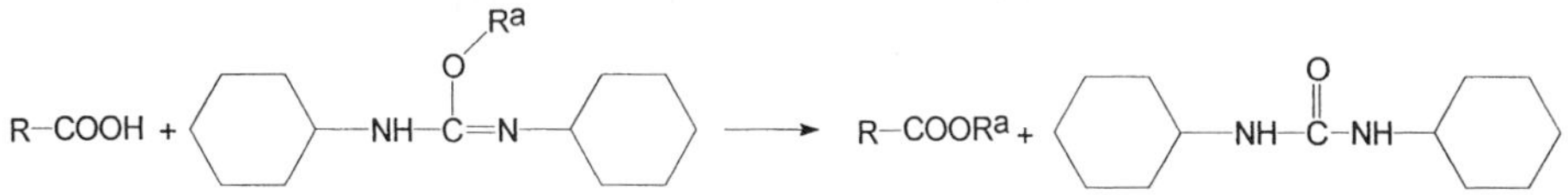

The reaction can be performed in a suitable solvent using heating under reflux [40]. For the same purpose, other isoureas have been used, such as N,N'-diisopropyl-O-(p-nitrobenzyl)isourea [41], [24].

Alkylation also can be done using an imino ester. As an example, alkyltrichloro-acetimidates may react with an alcohol following the reaction:

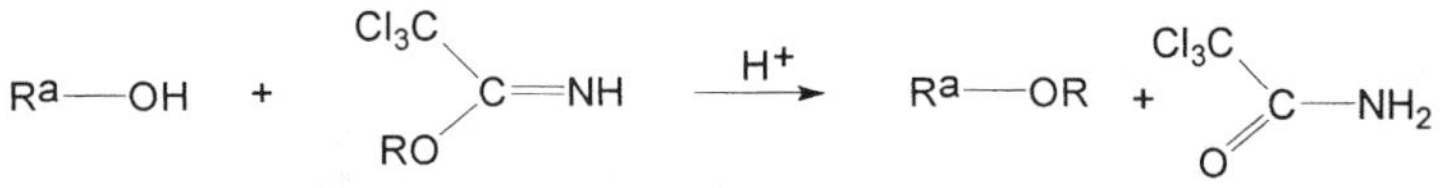

Alcohols themselves can also act as alkylating reagents when the analyte contains a more acidic hydrogen or in the presence of a catalyst such as HCl, BF_3, CF_3-COOH or a cation exchange resin in H^+ form. The addition of HCl can be done as a water solution or as gaseous HCl, which does not bring additional water to the reaction medium. The formation of alkyl or aryl derivatives of acids is a particularly important reaction known as esterification. Derivatization by esterification has been used with acids as the analyte and the alcohol as the reagent, and also with the alcohol as the analyte and the acid the reagent. The esterification can be viewed either as the acid alkylation or as the acylation of the alcohol (see also the esterification mechanism). This reaction can be written as follows:

$$R-COOH \ + \ R^a-OH \ \xrightarrow{H^+} \ R-COOR^a \ + \ H_2O$$

Better efficiency for the esterification is obtained by removing the water using a chemical reagent or distillation when the compounds of interest boil above $100°$ C. Among the materials able to eliminate water are desiccants such as anhydrous $MgSO_4$, molecular sieves, or substances that react with water such as CaC_2, $(CH_3)_2C(OCH_3)_2$ (2,2-dimethoxypropane or acetone dimethyl acetal) and even an appropriately chosen acid anhydride that reacts faster with water than with the reacting alcohol. The derivatization also may be performed in the presence of $SOCl_2$ (thionyl chloride), which reacts with the water assisting in its removal and, when present in excess, may react with the alcohols forming alkyl chlorides or with the acids forming acyl chlorides. Chloride is a better leaving group in a nucleophilic alkylation reaction, and the efficiency of alkylation increases [1]. Acids also can be esterified using a mixture of an alcohol and an acyl halide, as shown below in an example of esterification of alanine with isopropanol in the presence of acetyl chloride:

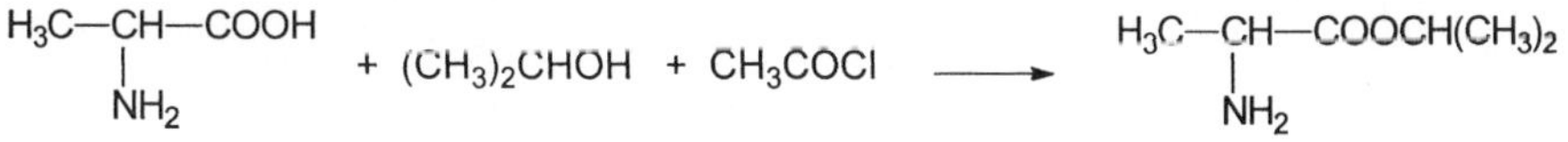

A special procedure for the formation of esters with less active organic acids may utilize an intermediate reaction step that facilitates further reaction with the alcohol. One such procedure applies the addition of dicyclohexylcarbodiimide (DCCI) in the derivatization process. The reaction can be written as follows:

$$R-COOH + \text{(cyclohexyl)}-N{=}C{=}N-\text{(cyclohexyl)} \xrightarrow{H^+} \text{(cyclohexyl)}-N{=}C-NH-\text{(cyclohexyl)} \quad (OOC-R)$$

$$R^a-OH + \text{(cyclohexyl)}-N{=}C-NH-\text{(cyclohexyl)} \ (OOC-R) \longrightarrow R-COOR^a + \text{(cyclohexyl)}-NH-C-NH-\text{(cyclohexyl)} \ (O)$$

The reaction can be performed by adding the appropriate alcohol, usually in a solvent such as pyridine, to the acids that need to be analyzed. To this mixture the DCCI is added in slight molar excess. The reaction may take place at room temperature or using heating at $40°$ C–$80°$ C, in an interval of 30 min. to two hours [42]. Dicyclohexylurea, which is formed in the reaction, is not soluble in pyridine and can be separated. Besides DCCI, other carbodiimides were used in the reaction of acids and alcohols, such as 1-(3-dimethylaminopropyl)-3-ethylcarbodiimide $(CH_3)_2N-(CH_2)_3-N{=}C{=}N-C_2H_5$ [43].

Similarly to DCCI, another reagent that can be used in an intermediate step is N,N'-carbodiimidazole (CDI). The reaction between an acid and an alcohol in the presence of CDI can be written as follows:

$$R-COOH + \text{(imidazole)}N-C(O)-N\text{(imidazole)} \xrightarrow{OH^-} R-C(O)-N\text{(imidazole)} + CO_2 + \text{(imidazole)}NH$$

$$R^a-OH + R-C(O)-N\text{(imidazole)} \longrightarrow R-COOR^a + \text{(imidazole)}NH$$

The procedure has been applied for the analysis of acids from plasma with the formation of methyl esters that were analyzed by GC [44].

Another reagent that can be used in a similar reaction as CDI and DCCI is 6-chloro-1-p-chlorobenzensulfonyloxybenzotriazole (CCBBT) [1].

$$R-COOH + \text{(Cl-benzotriazole)}O-SO_2-\text{(C}_6H_4Cl) \xrightarrow{OH^-} \text{(Cl-benzotriazole)}OCO-R + HSO_3^- + \text{(C}_6H_4Cl)$$

$$R^a-OH + \text{(Cl-benzotriazole)}OCO-R \longrightarrow R-COOR^a + \text{(Cl-benzotriazole)}OH$$

One other example of a coupling reagent for the esterification reaction is 2-chloro-1-methylpyridinium iodide [45]. This compound reacts with the carboxyl groups forming an active intermediate. The use of this coupling reagent is commonly done when the acid is the reagent and the alcohol is the analyte. With naphthalene-1,8-dicarboxylic acid the reaction to form the intermediate takes place as follows:

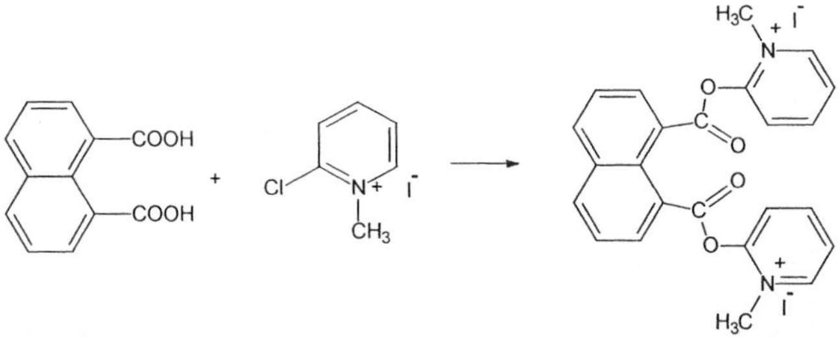

The resulting product reacts easily with an alcohol. Amiprilose (as an alcohol) can be derivatized using this system and analyzed by HPLC with fluorimetric detection.

A list of compounds able to facilitate the reaction between an acid and an alcohol are given in Table 18.1.2

TABLE 18.1.2. *Compounds able to facilitate alcohol-acid esterification reactions.*

Reagent
2,2-dimethoxypropane
thionyl chloride
dicyclohexylcarbodiimide (DCCI)
N,N'-carbodiimidazole (CDI)
1-(3-dimethylaminopropyl)-3-ethylcarbodiimide (EDAC)
6-chloro-1-p-chlorobenzen-sulfonyloxybenzotriazole (CCBBT)
2-chloro-1-methylpyridinium iodide
2,4,6-triisopropylbenzenesulfonyl chloride
2,4,6-trimethylbenzenesulfonyl chloride
2,4,6-trichlorobenzoyl chloride

Short chain alcohols can be used for the alkylation of other alcohols with the formation of an ether. This reaction provides, for example, a procedure for the alkylation of the glycosidic OH group in many sugars leading to the formation of glycosides:

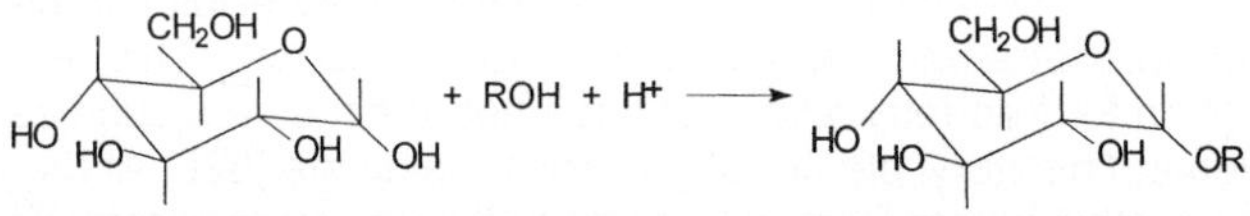

Various experimental conditions are applied for the glycoside formation reactions, depending on the sugar and the alcohol utilized. The most common catalyst in this reaction is HCl. The acid groups in aldonic, uronic, or saccharic acids also are alkylated by this method. The analysis of glycosides using chromatographic techniques is reported in numerous studies [21], [22], etc. The removal of the catalyst is sometimes needed after the derivatization step, and this can be done by different procedures such as evaporation or use of anion exchange resins.

Another class of reagents that are strong alkylating agents is that of oxonium ions, which can react with alcohols, phenols, or carboxylic acids. The reaction for an alcohol can be written as follows:

$$ROH + R^a_3O^+ \longrightarrow ROR^a + R^a_2O + H^+$$

The reaction is usually performed using a trialkyloxonium fluoroborate [46], for example trimethyloxonium or triethyloxonium fluoroborate. These reagents were successfully used for alkylation of a variety of carbohydrates [1].

A special alkylation can be achieved associated with pyrolysis or during the heating in the injection port of a gas chromatograph using tertraalkylammonium hydroxides or alkylarylammonium hydroxides (see e.g. [47]).

Certain amines or even amides can be alkylated using this procedure. For example, after the derivatization with heptyl iodide, acetaminophen can be further derivatized in the GC injection port using trimethylphenyl ammonium hydroxide (trimethylanilinium hydroxide) [13].

Trimethylanilinium hydroxide and tetramethylammonium hydroxide (TMAH) are the most common reagents of this type. Other quaternary N alkyl (or alkyl, aryl) ammonium hydroxides successfully used as derivatization reagents are tetrabutyl-ammonium hydroxide [48], phenyltrimethylammonium hydroxide (or trimethylanilinium hydroxide), and (m-trifluoromethylphenyl)trimethyl ammonium hydroxide (or trimethyltrifluoro-m-tolyl ammonium hydroxide or TMTFTH) [49]. Pyrolytic derivatizations with the formation of ethyl, propyl, hexyl, etc. derivatives also are known [50]. The strong basic character of these reagents puts some limitations to their use. For this reason, the use of phenyltrimethylammonium fluoride or phenyltrimethylammonium acetate as on-injection port methylation reagents without the problems of the basic character is very promising [51]. A differentiation between free acids and esterified fatty acids can be achieved using methylation with tetramethylammonium acetate (TMAAc). This reagent allows pyrolytic methylation for free fatty acids but does not affect esters such as 1-heneicosanyl oleate, colesteryl oleate, or glyceryl tripalmitate [52]. At the same time, TMAH generates methyl esters from both acids and acid esters of fatty acids. The reaction with esters can be considered a transesterification.

Alkylation during the heating in the injection port of a gas chromatograph also can be achieved using trimethylsulfonium hydroxide $(CH_3)_3S$-OH [53], [54], trimethylsulfonium acetate $(CH_3)_3S$-OOCCH$_3$, or cyanide $(CH_3)_3S$-CN [51].

Transesterification is another technique applicable for obtaining certain alkyl derivatives of acids (or acyl derivatives of alcohols). The reaction can be written as follows:

$$R^a\text{—COOR}^b + R^c\text{—OH} \longrightarrow R^a\text{—COOR}^c + R^b\text{—OH}$$

Transesterification can be catalyzed by acids (or Lewis acids) such as HCl, BF3, and H_2SO_4, or by bases such as CH_3OK, CH_3ONa, and C_4H_9ONa. The basic catalysts are commonly used for the methanolysis of triglycerides, followed by the analysis of the fatty acids methyl esters using GC [1], [55]. Trimethyl-(α,α,α-trifluoro-m-tolyl)ammonium hydroxide also can be used for the transmethylation of triglycerides [56]. An interesting reaction is the transformation of a methyl ester of a fatty acid into a pyrrolidide in a reaction with pyrrolidine that takes place as follows [57]:

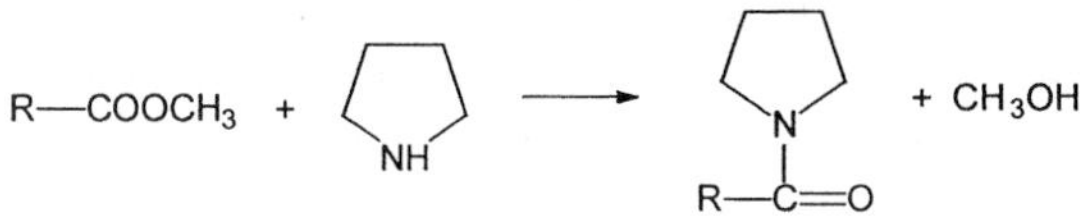

Various alkylation reactions can be performed before the reaction with the main derivatization reagent that leads to the property to be used in the analysis, such as color or fluorescence. This type of reaction is used to make the analyte appropriate for a second derivatization. Some examples are iodoacetic acid and iodomethane, which react with thiols, replacing the active hydrogen and protecting it for other reactions such as oxidation.

Numerous other reactive compounds may be used for replacing active hydrogens in specific compounds. Some other compounds may act as reagents to replace active hydrogens with alkyl or aryl groups, but they do not have analytical applicability. For example, epoxides, aziridines, and episulfides react easily with compounds with active hydrogens, as shown below for an epoxide:

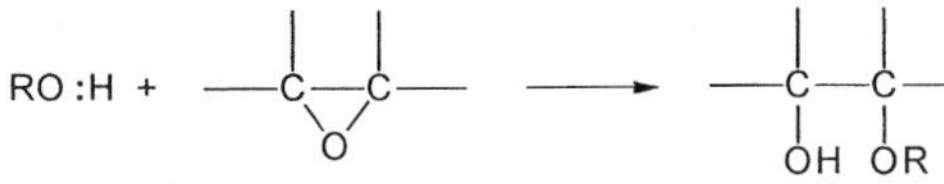

This type of reaction is not always useful for analytical purposes due to the formation of a second group containing an active hydrogen.

- Some aspects regarding alkylation mechanism

Alkylation reactions are commonly reversible, and thermodynamic factors govern the direction in which the equilibrium is displaced. However, many reactions achieve equilibrium rather slowly, and for this reason the reaction rate is an important parameter for their applicability to analytical purposes. It is important to understand both the thermodynamic and the kinetic factors that affect the reaction. As nucleophilic substitutions usually take place in solution and solvation energies are large for polar or ionic species, it is not possible to predict reactivity based on gas phase bond energies. Also, thermochemical data for ionic reactions in solution are very difficult to obtain. This explains why the reactivity for nucleophile substitutions is commonly estimated based on a series of rules and not on thermochemical data.

The nucleophilic substitution may occur following two main mechanisms, each with its own specific kinetics. One is the unimolecular nucleophilic substitution, designated S_N1. The S_N1 mechanism consists of two steps:

$$R-X \xrightarrow{\text{slow}} R^+ + X^-$$

$$Y:-H + R^+ \xrightarrow{\text{fast}} Y-R + H^+$$

The rate for this reaction is given by the relation:

$$d\,[RX] / dt = k\,[RX] \tag{18.1.1}$$

As seen from rel. (18.1.1), the reaction rate for S_N1 reactions is independent of the identity of nucleophile and depends only on the alkylating reagent.

More common than the S_N1 mechanism is the bimolecular mechanism, or S_N2, that takes place in a (concerted) single step with a backside attack (inversion of configuration):

$$Y:-H + \;\overset{\diagup}{\underset{|}{C}}\!-X \longrightarrow Y:--\overset{|}{C}--X \longrightarrow Y-\overset{\diagup}{\underset{|}{C}} + HX$$

The rate of reaction for an S_N2 mechanism is given by the relation

$$d\,[RX] / dt = k\,[RX]\,[YH] \tag{18.1.2}$$

For S_N2 reactions, both the nucleophile Y:H and the nature of the leaving group X will influence the reaction rate. In addition to that, other factors also influence the rate, such as the nature of the alkylating reagent as a whole, the solvent, or the presence of other molecules (catalysts) in the reaction.

Both S_N1 and S_N2 reaction rates are influenced by the nature of the leaving group. The leaving group comes off more easily when it is more stable as a free entity. This is usually inverse to the basicity of the leaving group, and the best leaving groups are the weakest bases. Good leaving groups are -Cl, -Br, -I or $-O-(SO_2)-C_6H_5$, with the group for the strongest corresponding acid being the best (iodine better than the other halogens). Also, as a rule, XH is always a weaker base than X^-. This property can be used to increase the reactivity of certain compounds. For example, alcohols are rather unreactive in S_N reactions because OH^- is a very poor leaving group. It is therefore possible to perform a nucleophilic substitution with an alkoxide previously prepared from the alcohol.

If a strong acid is present, the reactivity of the OH group is significantly increased by the transformation of the R-OH molecule in $R-OH_2^+$, which has H_2O as a leaving group. This can be applied for example in the reaction of ether formation between two alcohols:

$$R—\ddot{O}: \;\; + \;\; —\overset{\overset{\displaystyle H}{|}}{\underset{\underset{\displaystyle H}{|}}{C}}—\overset{+}{O}—H \;\; \longrightarrow \;\; R—O—\overset{\overset{\displaystyle H}{|}}{\underset{|}{C}}— \;\; + \;\; \overset{\displaystyle H}{O}—H \;\; + \;\; H^+$$

A strong Brønsted acid also may act as a catalyst in the esterification of an organic acid when the alcohol acts as a nucleophile and the leaving group in the S_N2 reaction is H_2O:

$$\underset{R}{\overset{O}{\overset{\|}{C}}}—OH \;\; + \;\; H^+ \;\; \longrightarrow \;\; \underset{R}{\overset{O}{\overset{\|}{C}}}—\overset{+}{O}H_2$$

$$\underset{R'}{:\ddot{O}:H} \;\; + \;\; \underset{R}{\overset{O}{\overset{\|}{C}}}—\overset{+}{O}H_2 \;\; \longrightarrow \;\; \underset{R'\;\;R}{HO\text{-}\text{-}\overset{\overset{\displaystyle O^{\delta-}}{|}}{\underset{|}{C}}\text{-}\text{-}\text{-}OH_2} \;\; \longrightarrow \;\; \underset{R'\;\;R}{HO\text{-}\text{-}\overset{O}{\overset{\|}{C}}} \;\; + \;\; H_2O \;\; \longrightarrow \;\; \underset{R'\;\;R}{O\text{-}\text{-}\overset{O}{\overset{\|}{C}}} \;\; + \;\; H_3O^+$$

A strong Lewis acid may play the same role as a Brønsted acid. One such example is the use of BF_3 + CH_3OH as a methylating reagent.

In particular cases, the esterification reaction may take place with a different mechanism. For example, for tertiary alcohols it is possible that in the presence of an acidic catalyst the alcohol R-OH reacts with the proton and eliminates water with the formation of R^+, which further reacts with the acid. It is also possible in special cases that an acylium ion $R\text{-}C(OH)_2{}^+$ is formed from the acid with a strong inorganic acid as catalyst.

The salts of heavy metals such as Ag or Cu catalyze some S_N reactions in the same way as acids. The reaction with a S_N1 mechanism takes place as follows:

$$R—X \;\; \xrightarrow{Ag^+} \;\; RX\text{-}\text{-}\text{-}Ag^+ \;\; \underset{-\;AgX}{\xrightarrow{slow}} \;\; R^+$$

$$Y:—H \;\; + \;\; R^+ \;\; \xrightarrow{fast} \;\; Y—R \;\; + \;\; H^+$$

The structure of the group R plays an important role in determining the rate of nucleophilic substitution. For S_N1 reactions, the rate follows the order (tertiary R) >> (secondary R) >> (primary R), while for S_N2 it is the opposite, (primary R) > (secondary R) >> (tertiary R). Steric hindrance appears to be particularly important in determining S_N2 reaction rates. The influence of voluminous tertiary R groups influence the rate even when they are located one carbon away from the leaving group. Other structural effects such as a double bond, an aromatic ring, or an oxo group β to the leaving group may also increase the S_N2 reactivity, probably due to the stabilization of the transition state by resonance involving the π bonds.

For reactions with S_N2 mechanism, rate also depends on the nature of the nucleophile Y:. This dependence can be summarized in four rules:

a) A nucleophile with a negative charge is always a more powerful nucleophile than its conjugate acid. The alkoxides for example, in contrast to the alcohols, react rapidly with alkyl halides. Also, NH_2^- is a stronger nucleophile than NH_3.

b) In comparing nucleophiles whose attacking atom is in the same row of the periodic table, nucleophilicity is approximately in order of basicity. Some pKa values for several compounds of interest are given in Table 18.1.3.

c) For nucleophiles whose attacking atom is in a higher period, the nucleophilicity increases. For example, the nucleophilicity increases in the order $HS^- > HO^-$; also $PH_3 > NH_3$. In general, larger atoms are better nucleophiles, although the solvation effects may change the order.

d) The freer the nucleophile, the higher the rate. For example, a number of nucleophile agents that are of the type XY where both atoms have unshared electron pairs are very reactive.

TABLE 18.1.3. *Several pKa values for organic compounds.*

Acid	Conjugated base	Approximate pKa (relative to water)[1]	Acid	Conjugated base	Approximate pKa (relative to water)[1]
HI	I^-	-10	ArOH	ArO^-	8 - 11
HCl	Cl^-	-7	R_3NH^+	R_3N	10 - 11
H_3O^+	H_2O	-1.74	RSH	RS^-	10 -11
$Ar_2NH_2^+$	Ar_2NH	1	$R_2NH_2^+$	R_2NH	11
$ArNH_3^+$	$ArNH_2$	3 - 5	CH_3OH	CH_3O^-	15.2
$ArNR_2H^+$	$ArNR_2$	3 - 5	H_2O	OH^-	15.74*
RCOOH	$RCOO^-$	4 - 5	RCH_2OH	RCH_2O^-	16
$HCOCH_2CHO$	$HCOCH^-CHO$	5	RCH_2CHO	RCH^-CHO	16
H_2S	SH^-	7.00	R_2CHOH	R_2CHO^-	16.5
ArSH	ArS^-	6-8	R_3COH	R_3CO^-	17 - 18
HCN	CN^-	9.2	$RCONH_2$	$RCONH^-$	17
NH_4^+	NH_3	9.24	$RCOCH_2R$	$RCOCH^-R$	19-20

[1] The conventionally correct K_a of water $1\ 10^{-14}$ is replaced with $K_a = 1.8\ 10^{-16}$ obtained by treating the water as a solute and using Henry's law standard state [58]. This value expresses better the acidic properties of water when compared with other weak acids.

Based on the data from Table 18.1.3, an approximate order of nucleophilicity is $NH_2^- > RO^- > OH^- > R_2NH > ArO^- > NH_3 > Pyridine > F^- > H_2O$. Another series is $R_3C^- > R_2N^- > RO^- > F^-$. This order is not always followed, as basicity is thermodynamically controlled and nucleophilicity is kinetically controlled. The four rules showing the influence of the nucleophile on the reaction rate are not always followed due to interfering problems such as steric effects.

One other factor with contribution to the rate of nucleophilic substitutions is the reaction medium. The medium may be important just by making the reactants more soluble. This can be achieved by a specific choice of the solvent, by the use of cryptands, or by the formation of specific salts more soluble in organic solvents such as tetraalkylammonium salts of different anions.

Regarding the direct influence of the medium, for neutral substrates and S_N1 mechanism, the more polar the solvent, the faster the reaction. Also, for nonionized substrates the S_N1 reactions are more rapid in protic solvents. For S_N2 reactions, the rate also depends on whether or not the substrate or the reagent is charged. For neutral substrates and reagent, which is the most common case for most analytical applications, polar solvents increase the reaction rate. This is explained by the reduction in the energy of the transition state, which has charges that dissipate in a more polar solvent. The effect is opposite for the charged substrates (such as an alkoxide). The difference between protic and aprotic solvents does not influence significantly the rate of neutral substrates with neutral reagents in S_N2 mechanisms. However, the rate for a charged substrate is decreased in a protic solvent [59].

In addition to the alkylation reactions indicated above, other classes of alkylation reactions are known, but they do not have specific analytical applications. This is for example the case of the well-known Friedel-Crafts alkylation of aromatic rings (which is an aromatic electrophilic substitution with S_E1 mechanism) with numerous applications in organic synthesis.

The reaction mechanism of the formation of aryl derivatives resembles the S_N2 reaction and is noted S_NAr. However, the concerted single step with a backside attack of S_N2 reactions is not likely to occur for the S_NAr mechanism, and the reaction probably takes place in two steps. For an alcohol the reaction can be written as follows:

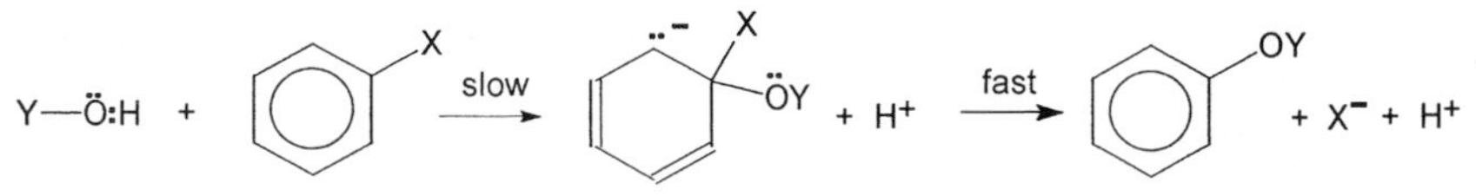

The reaction is not favorable for simple aryl halides because the intermediate molecule loses the aromatic stabilization and also requires the transfer of a negative charge to the ring, which is not very electronegative. Strongly electron-attracting groups in ortho or para positions stabilize the intermediate anion. Substituents in meta position have much less effect on the reactivity of aryl halides.

- Artifact formation in alkylation reactions

Alkylation reactions may generate besides the desired derivatives certain unexpected compounds that can be considered artifacts for the particular analysis. The artifacts may be obtained from unexpected interactions of the reagent with the analyte or may be a result of undesired effects of the catalysts or medium used for derivatization. The control of the alkylation process may be in some cases rather difficult mainly because the derivatization needs to be as efficient as possible to achieve the intended replacement of the active hydrogens but should not modify other groups in the molecule in order to generate an unique derivatized analyte. This requires in many procedures a specific reaction time. Longer or shorter reaction times or intervals between derivatization and analysis may lead to errors, even when an internal standard is used for quantitation.

For the alkylation using short chain alkyl bromides or iodides, one common case of artifact formation occurs during the reaction with compounds containing O-acyl or N-acyl groups, such as previously acylated carbohydrates, glycolipids, or glycoproteins. When the OH groups of different sugars or NH_2 groups of amino sugars were already protected with acyl groups, it was noted that, depending on the catalyst and the chosen medium, these acyl groups can be replaced by alkyl groups, or they may migrate from one position (such as C1) to other positions.

Besides the unexpected reactions during the alkylation with alkyl halides, the catalysts themselves may induce chemical modifications of the analytes, which can be seen as artifacts. For example, the alkylations using strong basic conditions may produce β-eliminations as undesired side reactions:

$$\overset{B^-}{\underset{H}{-\overset{|}{\underset{|}{C}}-\overset{|}{\underset{|}{C}}-X}} \longrightarrow -\overset{|}{C}=\overset{|}{C}- \;+\; X^- \;+\; BH$$

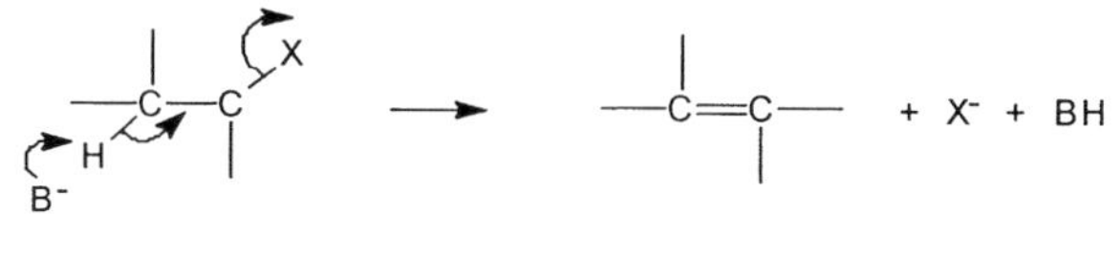

This type of reaction is particularly common in the derivatization of free sugars using alkyl halides as a reagent in the presence of NaH, NaOH, $Ba(OH)_2$, or methylsulfinyl carbanion as catalysts. The same effect can be seen when using dimethylsulfate as a reagent and basic catalysts.

Oxidation is another common side reaction when using Ag_2O as a catalyst. The oxidation effect of Ag_2O can be seen on free sugars as well as when attempting to permethylate peptides. With peptides, Ag_2O may react by producing partial cleavage of the peptide chain or by oxidizing sulfur containing amino acids residues. Sulfhydryl groups are particularly sensitive to oxidation with Ag_2O as a catalyst.

The use of methylsulfinyl carbanion as a catalyst or as a methylating reagent may also produce undesired side reactions with certain esters generating methylsulfinylketones:

$$CH_3-\overset{O}{\overset{\|}{S}}-CH_2^- \;+\; R-\overset{O}{\overset{\|}{C}}-OR' \longrightarrow CH_3-\overset{O}{\overset{\|}{S}}-CH_2-\overset{O}{\overset{\|}{C}}-R \;+\; OR'^-$$

Lewis acid catalysts also may generate undesirable artifacts on substances sensitive to acid catalyzed transformations.

Strong alkylating reagents may produce undesired artifacts by unexpected alkylations. One possible artifact in the alkylation with alkyl fluoromethylsulfonates, methylsulfinyl carbanion, or dimethyloxosulfonium methylide is the methylation of active aromatic hydrogens as shown below:

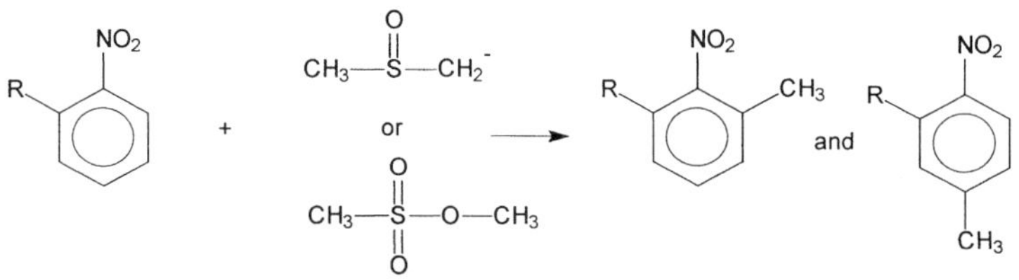

The same strong alkylating reagents may replace already existent alkyl groups with the new alkyl through the cleavage of the old ether bonds [1], as shown in the following reaction:

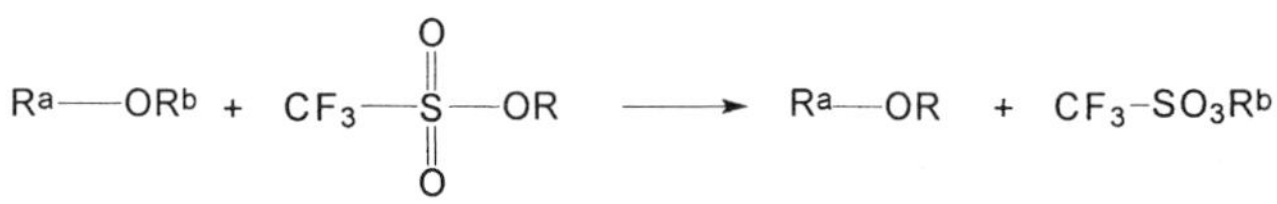

$$R^a-OR^b \;+\; CF_3-\overset{O}{\underset{O}{\overset{\|}{\underset{\|}{S}}}}-OR \longrightarrow R^a-OR \;+\; CF_3-SO_3R^b$$

The use of dimethylformamide acetals also was found to generate a variety of artifacts. Some of the artifacts are unexpected combinations of methylation and condensation reactions. As an example, 2-β-D-ribofuranosyl-1,2,4-triazine-3,5(2H,4H)-dione (6-azauridine) generates an acetal of 3-methylated azauridine at the 2',3' cis diol group [60]:

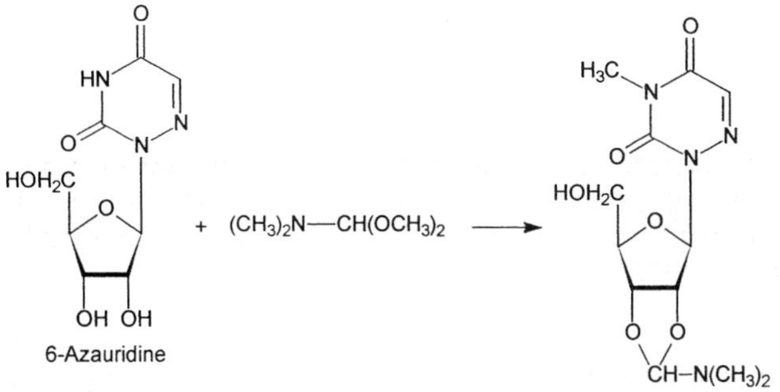

6-Azauridine

The compound can be hydrolyzed with water, generating 3-methylazauridine. Cytidine also forms an acetal at the 2',3' *cis* diol group, but the primary amino group is not methylated:

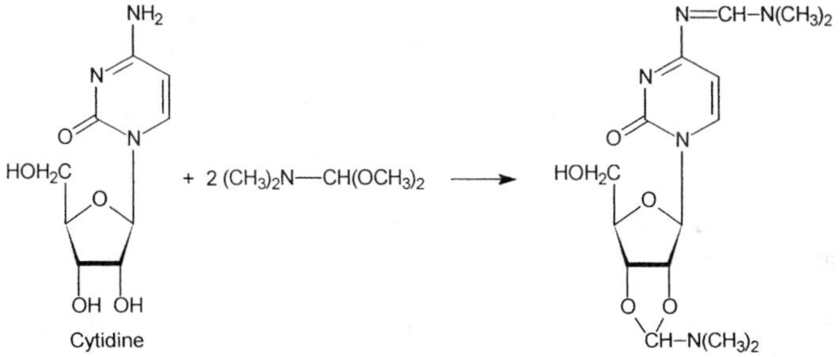

Cytidine

The undesired reaction with primary amines is rather common, taking place as follows:

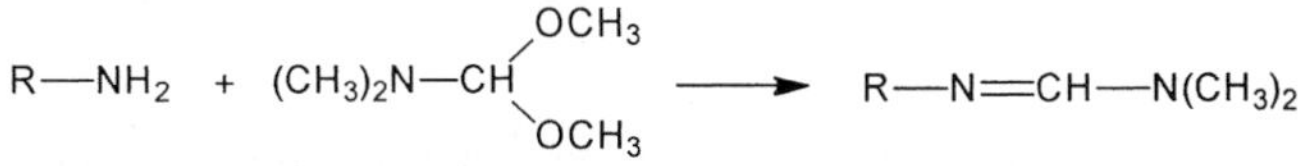

The alkylation using N,N-dimethylformamide dialkyl acetals can generate different artifacts by condensation with other substances containing active methylene groups [1]:

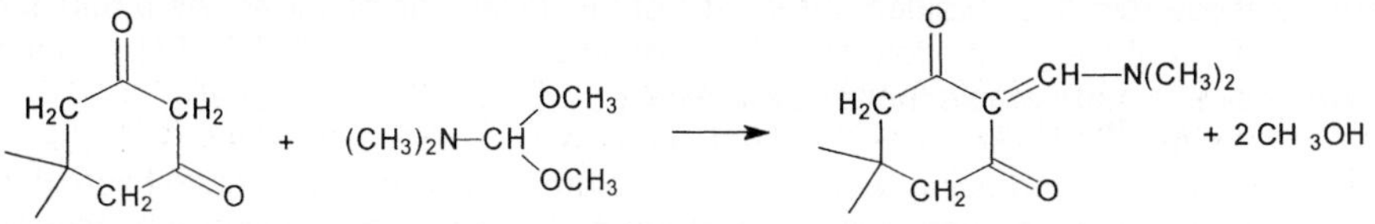

Also, exchange reactions of N,N-dimethylformamide dialkyl acetals leading to the formation of different acetals were noticed [1], as in the reaction:

The reaction using trialkyloxonium fluoroborates, such as trimethyloxonium or triethyloxonium fluoroborate, also may lead to artifacts. The unexpected reactions of trialkyloxonium fluoroborates include replacement of N-acetyl groups, while the O-acetyl groups are not affected, for example, in various acetylated carbohydrates:

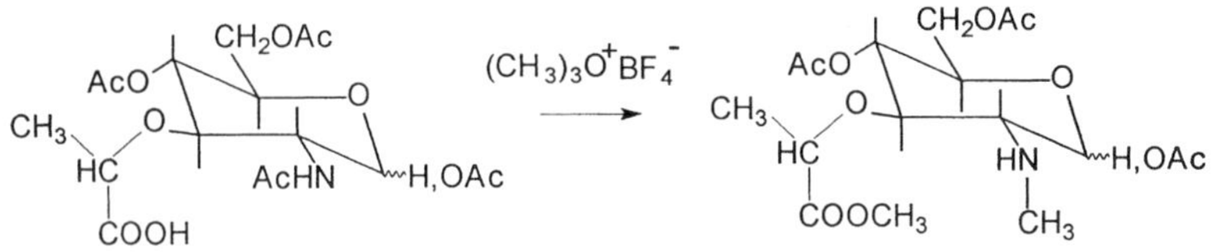

Peracetylated muramic acid

Also, a N-deacylation reaction is possible in proper conditions, leaving unaffected the O-acetyl groups:

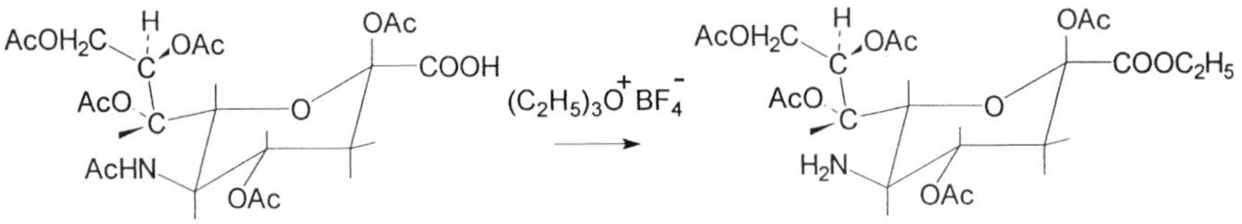

Peracetylated neuraminic acid

Another possible reaction using trialkyloxonium fluoroborates involves compounds that do not have an active hydrogen, such as ethers, sulfides, nitriles, ketones, esters, or amides. In this reaction new -onium fluoroborates are formed as shown, as an example, in the ethylation of 2,3-dihydroindole-2-one:

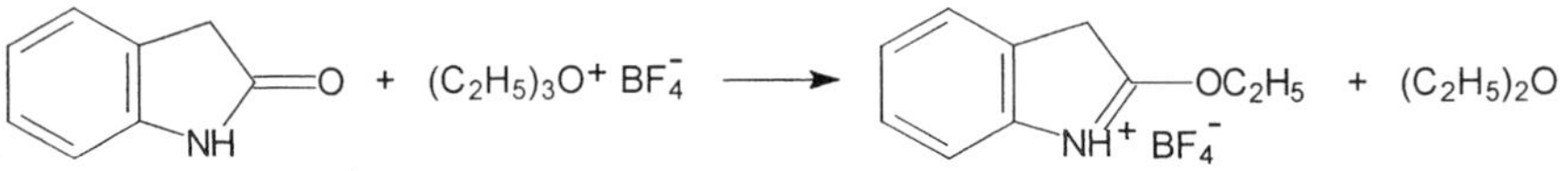

The alkylation using alcohols as reagents also can be affected by artifacts, mainly due to the presence of the acid catalysts used for this reaction. During the alkylation of the OH group in carbohydrates, for example, the equilibrium between the α and β forms as well as between the pyranoside and furanoside forms of the sugar is affected, and a mixture of these forms is obtained regardless of the initial type of sugar. As a rule, the furanoside form of the glycosides tends to be generated more rapidly [1], [61], although the pyranoside is more stable thermodynamically.

In the alkylation achieved using tertraalkylammonium or alkylarylammonium hydroxides by pyrolysis or during the heating in the injection port of a gas chromatograph, several types of artifacts are obtained. For example, in the derivatization of phenobarbital with trimethylanilinium hydroxide (TMAH), together with the N,N-dimethyl derivative, two other compounds are formed [62]:

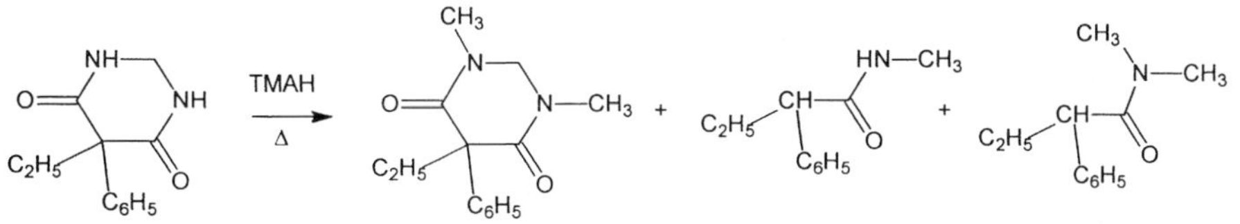

The formation of artifacts is also noted from the reagents themselves. For example, TMAH can generate dimethyl ether, probably as a result of the following reactions:

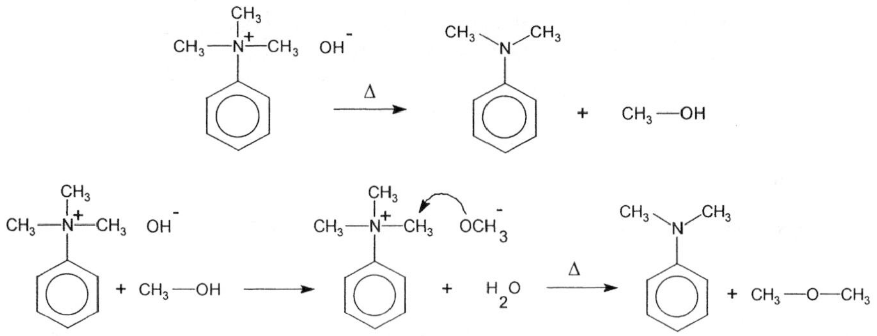

Anisol also can be formed during the decomposition of TMAH. This can be explained by the nucleophilic displacement of the trimethyl ammonium group following the attack of the methoxide ion [47].

When a pyrolytic methylation reaction is intended for a small molecule containing acidic hydrogens, such as an organic acid, and the derivatization is done in the injection port of the GC, it may generate only traces of undesired artifacts. However, true pyrolysis in the presence of a quaternary ammonium hydroxide can be a much more complex chemical reaction. The reaction is not equivalent with the methylation of the compounds generated from pyrolysis, although it may generate some methylated fragments.

18.2 SILYLATION REACTIONS

Silylation is the chemical reaction of replacing a reactive hydrogen in OH, COOH, SH, NH, CONH, POH, SOH, or enolisable carbonyl with a silyl group, usually a trimethylsilyl group (TMS). The purpose of silylation in chromatography is mainly to reduce the polarity of the analyte, increase its stability, and improve the GC behavior. The differences in the mass spectra of the silylated compounds as compared to the initial analyte may also be an advantage for detectability and be a reason for silylation. Other purposes for silylation were already described in Chapter 17. Silylation can be performed on specific analytes or directly on complex samples such as a plant material, and the silylating agent and the solvent can play the double role of extractant and silylating reagent. A series of excellent references are available regarding silylation reactions for analytical purposes [4], [63], [64], etc.

The reaction with the formation of a TMS derivative can be written as follows:

$$Y{:}H \ + \ CH_3{-}\underset{\underset{CH_3}{|}}{\overset{\overset{CH_3}{|}}{Si}}{-}X \ \longrightarrow \ Y{-}\underset{\underset{CH_3}{|}}{\overset{\overset{CH_3}{|}}{Si}}{-}CH_3 \ + \ HX$$

The molecular weight for TMS is 73.047 calculated considering in the elemental composition only the masses of the most abundant isotope. The silylation can be performed with a variety of reagents. Also, a variety of aprotic solvents can be used as medium. For analytical purposes, small amounts of sample and of reagent and solvent can be used. The analysis can be focused on one analyte or on a mixture of analytes that can be quite complex. Various factors contribute to increasing the efficiency and the rate of the silylation reaction. Among these are the silyl donor ability of the reagent and the ease of silylation of different functional groups, but also the solvent (or mixture of solvents) used as a medium and the compounds present or added in the silylation medium, etc. The reagent excess is sometimes important for displacing the equilibrium in the desired direction, and usually an excess up to 10 times larger than stoechiometrically needed is used for silylation. Temperature also increases reaction rate, as expected.

Some reagents used for trimethylsilylation are shown in Table 18.2.1. Only some of these reagents have been applied for analytical purposes, the rest being used only in special syntheses, but all may have a potential for analytical use.

The approximate order of the increasing silyl donor ability for the reagents shown in Table 18.2.1, which are also used frequently for analytical purposes, is HMDS < TMCS < MSA < TMSA < TMSDEA < TMSDMA < MSTFA < BSA < BSTFA < TMSI. This order may be different on particular substrates where other reagents or reagent mixtures may be more reactive.

TABLE 18.2.1. *Some reagents used for trimethylsilylation.*

Reagent	Abbrev.	Analytical use	Structure	Boiling[1] or melting point	Reference
allyltrimethylsilane	Allyl-TMS		$CH_2{=}CH{-}CH_2{-}Si(CH_3)_3$	bp 83-85	[65]
N,O-bis(trimethylsilyl)-acetamide	BSA	Y	$CH_3{-}C({=}N{-}Si(CH_3)_3){-}O{-}Si(CH_3)_3$	bp/35 71-73	[1], [4], [64]
N,O-bis(trimethylsilyl)-benzamide	BSB	Y	$C_6H_5{-}C({=}N{-}Si(CH_3)_3){-}O{-}Si(CH_3)_3$		[64]
N,O-bis(trimethylsilyl)-carbamate	BSC		$(CH_3)_3Si{-}NH{-}C({=}O){-}O{-}Si(CH_3)_3$	mp 77-80	[66]
N,N'-bis(trimethylsilyl)-N,N'-diphenylurea		Y	$(CH_3)_3Si{-}N(C_6H_5){-}C({=}O){-}N(C_6H_5){-}Si(CH_3)_3$		[64]

TABLE 18.2.1 (continued). *Some reagents used for trimethylsilylation.*

Reagent	Abbrev.	Analytical use	Structure	Boiling[1] or melting point	Reference
N,N-bis(trimethylsilyl)-formamide	BSF		H—C=O, (CH₃)₃Si—N—Si(CH₃)₃	bp 158	[67]
N,N-bis(trimethylsilyl)-methylamine		Y	CH₃, (CH₃)₃Si—N—Si(CH₃)₃	bp 144-147	[68]
bis(trimethylsilyl) sulfate	BSS		(CH₃)₃Si—O—S(O)(O)—O—Si(CH₃)₃	mp 41-44 bp/10 99-101	[63]
N,O-bis(trimethylsilyl)-trifluoroacetamide	BSTFA	Y	CF₃—C(O—Si(CH₃)₃)=N—Si(CH₃)₃	bp 145-147	[38], [69]
N,N'-bis(trimethylsilyl)-urea	BSU		(CH₃)₃Si—NH—C(O)—NH—Si(CH₃)₃	mp 219-221	[63]
bromotrimethylsilane	TMBS	Y	(CH₃)₃Si—Br	bp 79	[70]
chlorotrimethylsilane	TMCS	Y	(CH₃)₃Si—Cl	bp 57	[4], [63]
N,N-diethyl-N-trimethyl-silylamine	TMSDEA	Y	(CH₃)₃Si—N(C₂H₅)₂	bp 125-126	[63]
N,N-dimethyl-N-trimethyl-silylamine	TMSDMA	Y	(CH₃)₃Si—N(CH₃)₂	bp 84	[63]
ethylthiotrimethylsilane			CH₃—CH₂—S—Si(CH₃)₃		[71]
ethyl trimethylsilylacetate	ETSA		CH₃—CH₂—O—C(O)—CH₂—Si(CH₃)₃	bp/40 76-77	[72]
hexamethyldisilane			(CH₃)₃Si—Si(CH₃)₃	bp 111-113	[63]
hexamethyldisilazane	HMDS	Y	(CH₃)₃Si—NH—Si(CH₃)₃	bp 124-127	[63], [64]
hexamethyldisiloxane	HMDSO	Y	(CH₃)₃Si—O—Si(CH₃)₃	bp 101	[63]
hexamethyldisilthiane			(CH₃)₃Si—S—Si(CH₃)₃		[73]
iodotrimethylsilane	TMIS	Y	(CH₃)₃Si—I	bp 106-108	[63]
(isopropenyloxy)trimethyl-silane	IPOTMS		CH₃—C(O—Si(CH₃)₃)=CH₂	bp/16 65-67	[74]
1-methoxy-2-methyl-1-trimethylsiloxypropene	MMTSP	Y	(CH₃)₂C=C(O—Si(CH₃)₃)(O—CH₃)	bp 148-150	[75]
N'-methyl -N,N''-bis-(trimethylsilyl)-N,N''-diphenylallofanamide		Y	(CH₃)₃Si—N(C₆H₅)—C(O)—N(CH₃)—C(O)—N(C₆H₅)—Si(CH₃)₃		[64]

TABLE 18.2.1 (continued). *Some reagents used for trimethylsilylation.*

Reagent	Abbrev.	Analytical use	Structure	Boiling[1] or melting point	Reference
(methylthio)trimethyl-silane			CH₃—S—Si(CH₃)₃	bp 110-114	[73]
methyl-3-trimethylsiloxy-2-butenoate				bp/0.3 43-45	[76]
N-methyl-N-trimethylsilyl-acetamide	MSA	Y		bp 159-161	[77]
methyltrimethylsilyl acetate				bp/50 62-64	[63]
N-methyl-N-trimethylsilyl-heptafluorobutyramide	MSHFBA	Y		bp 148	[63]
N-methyl-N-trimethylsilyl-trifluoroacetamide	MSTFA	Y		bp 130-132	[78], [79], [80]
phenylthiotrimethylsilane				bp/12 93-99	[63]
4-trimethylsiloxy-3-pentene-2-one or (trimethylsilyl enol of acetylacetone)	TMSacac	Y		bp/5 61-63	[76]
N-trimethylsilylacetamide	TMSA	Y		mp 38-43	[4]
N-trimethylsilylacetanilide		Y			[81]
trimethylsilyl acetate				bp 102-104	[63]
trimethylsilyl azide		Y	(CH₃)₃Si—N₃	bp 92-95	[63]
trimethylsilylbenzene sulfonate				bp/12 154-156	[82]
trimethylsilyl cyanide	TMSCN		(CH₃)₃Si—CN	bp 114-117	[63]
trimethylsilyl N,N-dimethylcarbamate	DMCTMS	Y		bp/20 74-75	[83], [84]

TABLE 18.2.1 (continued). *Some reagents used for trimethylsilylation.*

Reagent	Abbrev.	Analytical use	Structure	Boiling[1] or melting point	Reference
N-trimethylsilylimidazole	TMSI (TMSIM)	Y	$(CH_3)_3Si$—N⁀N (imidazole ring)	bp/14 93-94	[4], [63]
trimethylsilylmethane sulfonate			CH_3—S(=O)(=O)—O—$Si(CH_3)_3$	bp/12 88-90	[85]
4-(trimethylsilyl)-morpholine			O⟮morpholine⟯N—$Si(CH_3)_3$	bp 160	[86]
3-trimethylsilyl-2-oxazolidinone	TMSO	Y	$(CH_3)_3Si$—N⟮oxazolidinone⟯O (C=O)	bp/6 99-100	[63]
trimethylsilyl perfluoro-1-butane-sulfonate			CF_3—$(CF_2)_3$—S(=O)(=O)—O—$Si(CH_3)_3$	bp/12 69-71	[63]
N-trimethylsilyl-phthalimide		Y	phthalimide N—$Si(CH_3)_3$		[64]
N-trimethylsilylpiperidine	TMSPI	Y	piperidine N—$Si(CH_3)_3$		[4]
N-trimethylsilylpyrrolidine		Y	pyrrolidine N—$Si(CH_3)_3$	bp 139-140	[4]
N-trimethylsilyl-succinimide		Y	succinimide N—$Si(CH_3)_3$		[64]
trimethylsilyltrichloro acetate	TMSTCA		O=C(—O—$Si(CH_3)_3$)—CCl_3	bp/12 70-73	[87]
trimethylsilyltrifluoro acetate			O=C(—O—$Si(CH_3)_3$)—CF_3	bp 88-90	[63]
trimethylsilyltrifluoro-methane sulfonate (triflate)	TMSTF (TMS triflate)		CF_3—S(=O)(=O)—O—$Si(CH_3)_3$	bp/80 77	[63]

[1] The boiling points are given for some compounds at atmospheric pressure and for others at reduced pressure, which is indicated by "/ (pressure in mm Hg)".

Silylation reagents can be used pure or in mixtures of two or even three reagents. Table 18.2.2 gives several common mixtures used for trimethylsilylation, some being commercially available in a premixed form.

Table 18.2.2. *Some silylation reaction mixtures.*

Reagent 1	Proportion	Reagent 2	Proportion	Reagent 3	Proportion	Reference
HMDS	2 vol	TMCS	1 vol			[63], [88]
BSA	3 vol	TMSI	3 vol	TMCS	2 vol	[63]
BSA	1 vol	TMSI	1 vol	TMCS	1 vol	[38]
BSTFA	3 vol	TMSI	3 vol	TMCS	2 vol	[63]
BSA	95%	TMCS	5%			[63]
BSTFA	99%	TMCS	1%			[63]

The reagent mixtures may provide a more efficient silylation for specific compounds. As an example, silylation of 3,4-dimethoxyphenylethylamine takes place with BSA as follows [89]:

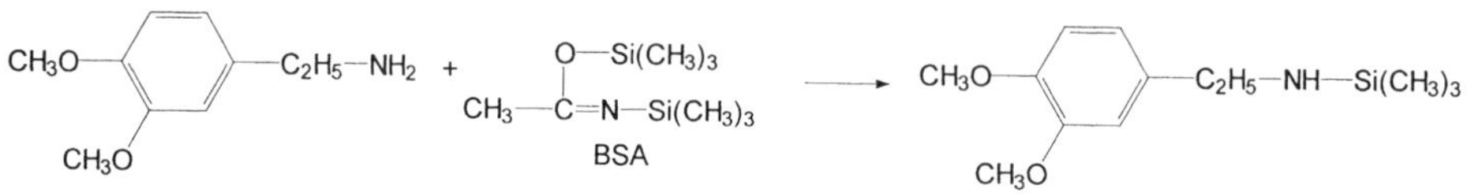

In the presence of TMCS the reaction takes place as follows:

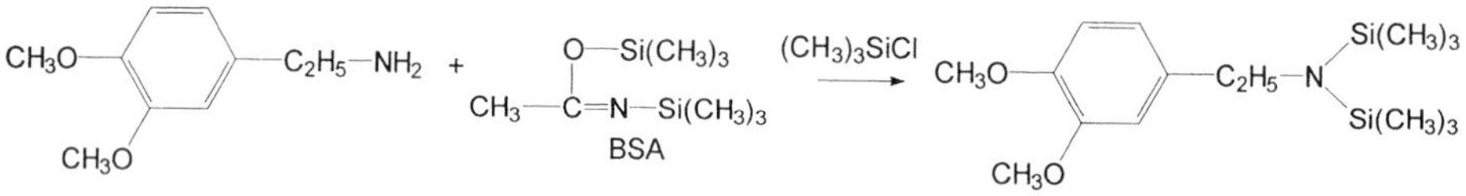

Derivatization using a deuterated silylation reagent also has been applied in connection with MS detection. Better structural identifications can be obtained when the mass spectrum of a compound with d_9-TMS groups is compared with the regular corresponding TMS derivative [90] (see Section 17.5). Deuterated BSTFA (d_{18}-BSTFA) and BSA (d_{18}-BSA) are commercially available.

The nature of the molecule Y:H that is being silylated (the analyte) is also one of the determining factors regarding the silylation efficiency and therefore affecting the choice of the derivatization conditions. It was noticed experimentally [64] that the decreasing ease of silylation follows approximately the order shown in Table 18.2.3.

TABLE 18.2.3. *Several functional groups that can be silylated (listed in the approximate order of decreasing ease of silylation).*

	Compound	Functional group		Compound	Functional group
1	primary alcohol	OH	8	primary amine	NH_2
2	secondary alcohol	OH	9	thiol	SH
3	tertiary alcohol	OH	10	amide	$CONH_2$
4	phenol	OH	11	N-TMS amide	$CONH-Si(CH_3)_3$
5	thiophenol	SH	12	secondary amine	NH
6	aliphatic acid	COOH	13	indole	NH
7	aromatic acid	COOH			

In general, the silylation of OH and COOH groups takes place with better results than that of NH_2, CONH or NH groups.

Besides organic active hydrogens, several inorganic compounds with active hydrogens can also be silylated. Among these are H_2O, H_2O_2, HCl, HNO_3, H_2SO_4, H_2SO_3, H_3BO_3, H_3PO_4, and H_4SiO_4. Also, some salts of the acids may be silylated. For example, the ammonium salts easily generate silylated derivatives [91]. Other salts, such as sodium or potassium salts, may be silylated only in particular conditions, such as in specific solvents.

The silylation reaction commonly takes place in a solvent that does not have active hydrogens. Some solvents with no active hydrogens are given in Table 17.2.1. From those, the most commonly used as a medium for silylation are dimethylformamide (DMF), pyridine, and acetonitrile. The main role of the solvent is to dissolve the analyte and the reagents. However, additional effects influencing the reaction are known for different solvents. In the silylation reaction, the by-product HX can be an acid, a base, or even a neutral compound. As examples, for TMCS the by-product is HCl, for HMDS the by-product is NH_3, for BSTFA the by-product is N-TMS-trifluoroacetamide, and for TMSI the by-product is imidazole. When the silylation reagent generates an acid as a by-product of the reaction, any acid acceptor will promote the silylation. Pyridine, triethylamine, and to a lower extent DMF are used as both solvents and acid acceptors. Mixtures of solvents are commonly used for both enhancing solubility and promoting silylation. For example, formamide in the presence of pyridine may react with an acidic by-product generating an ammonium salt and CO, as follows:

$$ROH + (CH_3)_3SiCl \longrightarrow RO\text{-}Si(CH_3)_3 + HCl$$

$$HCl + C_6H_5N \longrightarrow C_6H_5NH^+Cl^-$$

$$C_6H_5NH^+Cl^- + HCONH_2 \longrightarrow C_6H_5N + NH_4Cl + CO$$

Water plays a special role in silylation. It reacts easily with most silylation reagents, and also can react with the analyte after it is silylated. If the silylated analyte is more sensitive to water than the silylating reagent, anhydrous reaction conditions are needed. Otherwise, the reagent (commonly in excess) will react with the traces of water, and in this case, the amount of water needs to be as low as possible. A common compound resulting from the silylation of water with many silylating reagents is hexamethyldisiloxane $(CH_3)_3SiOSi(CH_3)_3$. (MW = 162). Depending on the reagent, other small molecules, such as trimethylsilanol, also can be formed in the reaction with water, as shown below for BSTFA:

Besides the reagent and the solvent, a variety of added compounds may influence the efficiency of the silylation reaction. For the reactions where the by-product is an acid, basic compounds can be added during the reaction to increase the reaction efficiency. For example, when using TMCS as reagent for the silylation of an alcohol, ammonia gas can be passed through the reacting mixture, and the reaction occurs without the need for heating. Compounds with basic character such as potassium acetate or amines such as triethylamine can be used for the same purpose. As mentioned above, choosing a solvent with a basic character may favor reactions with acidic by-products. For reagents generating basic by-products (NH_3 or amines), acids or substances with acid character can be used for enhancing the reactivity. For example in the silylation using HDMS, the addition of ammonium salts of strong acids or sodium bisulfate has proven useful. Also, addition of a certain amount of TMCS improves the completeness of silylation. In reaction with an alcohol, for example, HDMS generates NH_3 while TMCS generates HCl. Studies regarding the optimum amount of TMCS in this type of reaction indicate that a 1:1 mole ratio HDMS:TMCS is the best mixture for silylation [92]. This shows that, in this case, the role of TMCS is more that of a reagent allowing the elimination of NH_4Cl, which favors the reaction, and not that of a true catalyst. However, TMCS also may act as a catalyst when it is used in mixtures with other reagents such as BSTFA or BSA. Acid catalysts such as trichloroacetic acid are reported to promote the silylation using TMSDEA, mainly in reactions with amines [64].

Different conditions regarding reaction temperature and time are recommended for specific analytes and silylating reagents. These conditions may range from room temperature to heating up to $150°$ C–$170°$ C, and times from a few minutes to 10–12 hours. A typical derivatization uses a sample containing not more than a few hundred μg analyte, 0.25 mL BSTFA (with 1% TMCS), and 0.5 mL DMF. The mixture is heated for 30 min. at about $75°$ C, then allowed to cool, and analyzed by GC or GC-MS. In these conditions, a variety of compounds such as alcohols, phenols, acids, and carbohydrates will be silylated. Besides typical silylation procedures, less conventional ones also are reported, such as vapor phase derivatization [93], derivatization in the injection port of the GC [94], etc. [1].

A summary of the experimental observations made on silylation reactions with different reagents and different substrates can be seen in Table 18.2.5. These results were reported in different references [4], [63], etc., but individual reactions may still have specific characteristics that are not captured in a global description. More details regarding TMS derivatization for specific functional groups will be given in the sections regarding the derivatization of particular functional groups. The choice of specific derivatization reagent may be determined by a variety of reasons, such as purpose of derivatization, silylation efficiency, or convenience of the derivatization process. In addition, other reasons also may play a role, such as reagent availability or behavior of the silylation mixture in the analytical process. For example, the deposition of SiO_2 during FID detection of silylated compounds seems to be diminished for reagents containing fluorine in the molecule. This may be a reason for choosing BSTFA vs. BSA or MSTFA vs. MSA as reagents.

TABLE 18.2.5. *Characteristics of the more common TMS derivatizations [63].*

Reagent	ROH (prim.)	ROH(sec.)	ROH(tert.)	ArOH	R-COOH	Ar-COOH	RNH₂	RSH	ArSH	R-CONH₂	R-CONHTMS	R₂NH	RC=O-CHR₂	Remarks
Allyl-TMS	+	+	+		+	+		+	+					C₃H₆ by-product, need catalyst
BSA	+	+	+	+	+	+	+	+	+	+	+	+		very reactive, acid catalyst, CH₃CN sol.
BSC	+	+	+	+	+	+	-			-	-	-		NH₃ and CO₂ by-product
BSS	+	+	+	+	+	+						+	+	reacts with organic and inorganic salts
BSTFA	+	+	+	+	+	+	+	+	+	+	+	+		highly reactive, neutral by-product, contains fluorine, volatile
BSU	+	+	+		+	+	+	+						highly reactive, neutral by-product, DMSO sol.
ETSA	+	+	+					+	+				+	silylation of ketones
HMDS	+	+												NH₃ by-product, acid catal.
HMDSO	+	+	+	+	+									very poor silylating agent
IPOTMS	+	+	+	+	+	+								neutral, acetone by-product, acid catalyst
MMTSP	+	+	+	+	+	+	+	+	+	+			+	neutral, no catalyst
MSA	+	+	+	+	+	+	+					+		can act as solvent
MSHFBA	+	+	+	+	+	+	+	+	+			+		high fluorine content
MSTFA	+	+	+	+	+	+	+	+	+	+	+	+	+	highly reactive, more volatile than BSTFA, contains fluorine, good solvent
TMBS	+	+	+											reacts with enolisable carbonyls, needs catalyst
TMCS	+	+	+	+	+	+								HCl by-product, used as catalyst in mixtures with other silylating reagents
TMIS	+	+	+										+	highly reactive, also reacts with some carbonyls, used as catalyst
TMSacac	+	+	+	+	+	+	+	+	+	+	+	+		highly reactive, neutral
TMSA	+	+	-	+			+							monosilylation of primary amines
TMSCN	+	+	+	+	+	+								highly reactive, not reacting with amides
TMSDEA	+	+	+		+	+	+	+				+		diethylamine by-product, TMCS, acids catalyst
TMSDMA	+	+	+		+	+	+					+		diethylamine by-product, used for amino acids silylation, TMCS catalyst
DMCTMS	+	+	+	+	+	+								highly reactive, CO₂ and dimethylamine by-products
TMSI	+	+	+	+	+	+	+	+	+	+			-	highly reactive, discriminates amino acids, used in mixtures
TMSO	+	+	+	+	+	+	+	+	+	+				highly reactive in mild conditions
TMSTCA			+	+	+					+				CO₂ and CHCl₃ by-products
TMSTF	+	+	+				+					+	+	highly reactive including for carbonyl compounds

\+ good silylation, method described in literature.
\- no silylation in normal reaction conditions
(empty box) no information available.

- *Silylation for the introduction of groups other than TMS*

Although the TMS derivatives are by far the most commonly used in the derivatization for analytical purposes, other radicals in the silyl group also have been investigated. Some of these groups and the reagents used for their introduction are shown in Tables 18.2.6. and 18.2.7.

TABLE 18.2.6. *Some reagents used for the introduction of alkyl/arylsilyl groups different from TMS.*

Group	Mass of the group[1]	Reagent(s)	Analytical use	Reference
CH_2=CH—CH_2—Si(CH_3)$_2$— allyldimethylsilyl-	99	allyldimethylchlorosilane	Y	[95]
		N,O-bis(allyldimethylsilyl)trifluoro-acetamide (BASTFA)	Y	[96]
BrH_2C—Si(CH_3)$_2$— (bromomethyl)dimethylsilyl-	151	(bromomethyl)dimethylchlorosilane	Y	[63]
CH_3—$\overset{\overset{\displaystyle CH_3}{\mid}}{\underset{\underset{\displaystyle CH_3}{\mid}}{C}}$—Si($CH_3$)$_2$— *tert*-butyldimethylsilyl- (TBDMS)	115	*tert*-butyldimethylchlorosilane (TBDMCS)	Y	[63]
		tert-butyldimethylsilanol		
		4-*tert*-butyldimethylsiloxy-3-penten-2-one (TBDMSacac)		
		1-(*tert*-butyldimethylsilyl)imidazole (TBDMSI)	Y	
		N-methyl-N-(*tert*-butyldimethylsilyl)-trifluoroacetamide (MTBSTFA)	Y	
		tert-butyldimethylsilyl trifluoromethanesulfonate (TBDMS triflate)	Y	
		MTBSTFA + 1% TBDMCS	Y	
		tert-butyldimethylsilylimidazole	Y	[97]
tert-butyldiphenylsilyl-	239	*tert*-butyldiphenylchlorosilane (TBDPCS)	Y	[63], [98]
ClH_2C—Si(CH_3)$_2$— (chloromethyl)dimethylsilyl-	107	1,3-bis(chloromethyl)-1,1,3,3-tetramethyldisilazane (CMTMDS)	Y	[4], [64], [99]
		(chloromethyl)dimethylchlorosilane (CMDMCS)	Y	[64]
CN—C_2H_4—Si(CH_3)$_2$— cyanoethyldimethylsilyl-	112	cyanoethyldimethylchlorosilane	Y	[100]
		2-cyanoethyldimethyl(diethyl)aminosilane (CEDMSDEA)	Y	[101]
cyclotetramethylene-*tert*-butylsilyl-	141	cyclotetramethylene-*tert*-butylchlorosilane (TMTBS-Cl)	Y	[4], [102], [103], [104]
cyclotetramethyleneisopropylsilyl-	127	cyclotetramethyleneisopropylchloro-silane	Y	[102], [103], [104]

TABLE 18.2.6 (continued). *Some reagents used for the introduction of alkyl/arylsilyl groups different from TMS.*

Group	Mass of the group[1]	Reagent(s)	Analytical use	Reference
CH₃—CH₂ N—Si(CH₃)₂— CH₃—CH₂ (diethylamino)dimethylsilyl-	130	bis(diethylamino)dimethylsilane	Y	[99]
CH₃ N—Si(CH₃)₂— CH₃ (dimethylamino)dimethylsilyl-	102	bis(dimethylamino)dimethylsilane	Y	[99]
CH₃ CH₃ CH—C—Si(CH₃)₂— CH₃ CH₃ 2,3-dimethyl(butyl)dimethylsilyl-	143	2,3-dimethyl(butyl)-3-dimethylchlorosilane (*tert*-hexyldimethylchlorosilane)		[63]
		N-[2,3-dimethyl(butyl)-3-dimethylsilyl]-dimethylamine		
		2,3-dimethyl(butyl)-3-dimethylsilyl trifluoromethanesulfonate		
CH₃—CH₂—Si(CH₃)₂— dimethylethylsilyl	87	dimethylethylsilylimidazole	Y	[97]
H₃C CH—Si(CH₃)₂ — H₃C dimethylisopropylsilyl-	101	dimethylisopropylsilylimidazole	Y	[4]
CF₃-(CF₂)₃-(CH₂)₂-Si(CH₃)₂- dimethyl(3,3,4,4,5,5,6,6,6-nonafluorohexyl)silyl	305	dimethyl(3,3,4,4,5,5,6,6,6-nonafluorohexyl)chlorosilane	Y	[107]
—Si(CH₃)₂— dimethylphenylsilyl-	135	dimethylphenylchlorosilane	Y	[105]
		dimethylphenylsilane		
		1,3-diphenyl-1,1,3,3-tetramethyldisilazane	Y	[63]
C₃H₇—Si(CH₃)₂— dimethylpropylsilyl-	101	dimethylpropylchlorosilane	Y	[4]
		dimethylpropylsilylimidazole		[106]
SiH(CH₃)₂— dimethylsilyl-	59	dimethyldichlorosilane (DMCS)	Y	[64]
		1,1,3,3-tertamethyldisilazane (TMDS)	Y	[63]
		bis(dimethylsilyl)acetamide (BDSA)	Y	[108]
C—Si(CH₃)₂— dimethyl(triphenylmethyl)silyl-	301	dimethyl(triphenylmethyl)bromosilane (dimethyltritylbromosilane)	Y	[109]
C₂H₅—Si(CH₃)₂— ethyldimethylsilyl-	87	ethyldimethylchlorosilane	Y	[4]

TABLE 18.2.6 (continued). *Some reagents used for the introduction of alkyl/arylsilyl groups different from TMS.*

Group	Mass of the group[1]	Reagent(s)	Analytical use	Reference
IH₂C—Si(CH₃)₂— (iodomethyl)dimethylsilyl-	199	(iodomethyl)dimethylchlorosilane	Y	[63]
		(bromomethyl)dimethylchlorosilane + NaI	Y	[110]
(CH₃)₂CH—Si(CH₃)₂— isopropyldimethylsilyl-	101.25	isopropyldimethylchlorosilane (DMIPCS)	Y	[111]
methyldiphenylsilyl-	197	methyldiphenylchlorosilane	Y	[63]
		1,3-dimethyl-1,1,3,3-tetraphenyldisilazane (TPDMDS)	Y	[112]
		2-(methyldiphenylsilyl)ethanol		[63]
CF₃—(CF₂)₃—(CH₂)₂—Si(CH₃)₂— (3,3,4,4,5,5,6,6,6-nonafluorohexyl)dimethylsilyl-	305	(3,3,4,4,5,5,6,6,6-nonafluorohexyl)dimethylchlorosilane	Y	[113]
(pentafluorophenyl)dimethylsilyl- (flophemesyl-)	225	(pentafluorophenyl)dimethylchlorosilane	Y	[63], [114]
		(pentafluorophenyl)dimethylaminosilane	Y	[1]
Si(C₄H₉)₃— tri(*n*-butyl)silyl-	199	tributylchlorosilane	Y	[63]
Si(C₂H₅)₃— triethylsilyl-	115	hexaethyldisiloxane		
		triethylchlorosilane	Y	[63]
		triethylsilane triethylsilyl trifluoromethanesulfonate		
Si(C₆H₁₃)₃— trihexylsilyl-	283	trihexylchlorosilane	Y	[99]
Si[(CH₃)₂CH]₃— triisopropylsilyl-	157	triisopropylchlorosilane		[63]
		triisopropylsilane		
		triisopropylsilyl trifluoromethanesulfonate		
Si(C₆H₅)₃— triphenylsilyl-	259	triphenylchlorosilane	Y	[63]
		triphenylsilane		
		triphenylsilylamine (TPSA)	Y	[63]
Si(C₃H₇)₃— tripropylsilyl-	157	tripropylchlorosilane	Y	[63]

[1] The masses of molecules or groups were calculated considering in the elemental composition only the masses of the most abundant isotope. The resulting mass was rounded to the unit [115].

A variety of purposes and advantages (sometimes associated with disadvantages) are intended when the reagents shown in Table 18.2.6 are used. For example, dimethylsilyl derivatives can be used to generate compounds with a lower molecular weight than TMS, which could be more volatile. However, the compounds containing a Si-H group in the molecule are rather reactive and more sensitive to hydrolysis.

The groups heavier than methyl attached to the Si atom have proven beneficial mainly when compounds more resistant to hydrolysis are required. The most common silyl substituent heavier than TMS is *tert*-butyldimethylsilyl (TBDMS), which is frequently used for analytical purposes. The silylation with the formation of TMS groups takes place easier than that with the formation of TBDMS derivatives. Therefore the TMS derivatives are preferred for compounds with steric hindrance or for compounds unstable at higher temperatures. However, TBDMS derivatives are less sensitive to hydrolysis. For example, for the derivatization of amino acids with the formation of TMS derivatives, any trace of water must be eliminated from the derivatization mixture. The use of TBDMS derivatives leads to stable and reproducible results in amino acid analysis (see Section 19.10) [116]. Also, the thermal stability of TBDMS derivatives is better than that of TMS derivatives. For example, it is preferable to use TBDMS derivatives rather than TMS derivatives for the analysis of compounds with relatively high MW, such as long chain fatty acids, that require higher temperatures in the GC injector and chromatographic column.

As seen in Table 18.2.6, other heavier silyl groups are also used, such as ethyldimethylsilyl, chloromethyldimethylsilyl, or even more sterically crowded silyl groups such as cyclotetramethylene-*tert*-butyl and cyclotetramethyleneisopropyl [102], [103], [104]. The compounds silylated with these voluminous groups are stable and resistant to hydrolysis. However, for molecules with more than one active hydrogen, after the first silyl group is introduced, the subsequent substitution may be difficult. Attempts to replace these remaining active hydrogens with TMS may result in interchange of the voluminous silyl groups with TMS.

The analytical use of the derivatization with groups such as *tert*-butyldiphenylsilyl-, dimethyl(triphenylmethyl)silyl-, and methyldiphenylsilyl- is limited mainly to HPLC and TLC utilization or for the silylation of silica and glass capillary columns. Phenyl groups are particularly useful by significantly increasing the UV detectability of the derivatized compounds. The halogenated compounds are useful for ECD detection in GC or for enhancing in mass spectrometry the NCI detectability.

Reagents capable of replacing active hydrogens with alkoxy-dialkylsilyl, and alkoxy-diarylsilyl groups also are reported in literature [63]. Some of these groups and the reagents used for their introduction are shown in Table 18.2.7.

TABLE 18.2.7. *Some reagents used for the introduction of alkoxy-dialkylsilyl groups.*

Group	Mass of the group	Reagent(s)	Analytical use	Reference
tert-butyloxydiphenylsilyl-	255	*tert*-butyloxydiphenylchlorosilane		[63]
tert-butylmethoxyphenylsilyl- (TBMPSi)	193	*tert*-butylmethoxyphenylbromosilane	Y	[117], [118]
dimethoxymethylsilyl-	105	dimethoxymethylchlorosilane	Y	[119]
picolyloxydimethylsilyl-	166	diethylaminodimethylsilyl-3-pyridylmethanol	Y	[120]

The alkoxysilyl ether formation was used for analytical purposes in distinguishing between different hydroxylated species based on the good reactivity of these reagents with primary alcohols but reduced reactivity for secondary alcohols.

A special group of silylation reactions are those generating cyclic silicon derivatives. Cyclic compound formation commonly involves bifunctional molecules. If the silylating reagent contains two reacting sites, it generates cyclic siliconides. This is the case for bis(trialkylsilyl) reagents and for reagents with silylene groups. Table 18.5.1 and Table 18.5.2 (Section 18.5) give lists of reagents containing bis(trialkylsilyl) groups and silylene groups respectively. The reaction of a bis(trialkylsilyl) reagent with a diol takes place as follows:

X = Cl, Br, etc.

The reaction of a silylene reagent such as dimethyldichlorosilane (DMCS) with a diol takes place as follows:

X = Cl, Br, etc.

The efficiency of these reactions depends on the analyte and also on the reagent. Several problems are encountered with isolated OH groups, whcih may react only partially, yielding unstable products. An additional problem is the possibility of reaction with two different molecules forming silyl bridges. Section 18.5 contains more details regarding the subject of cyclic siliconides.

Various other silylation reagents were also tested for derivatization for GC and MS, but their practical applications are very limited [1], [63], [64].

- Some aspects regarding silylation mechanism

Silylation reactions are commonly reversible, and thermodynamic factors govern the direction in which the equilibrium is displaced. Similarly to the alkylation, it is not possible to predict the direction of the equilibrium based on gas phase bond energies. As an example, Table 18.2.8 shows the heat of formation in gas phase for a set of alcohols, amines, and amides and for their silylated correspondent compound. It can be seen that the differences in the heat of formation for each compound and its silylated correspondent are about the same, regardless the nature of the group that is silylated.

TABLE 18.2.8. *Differences in the heat of formation[1] for some compounds and their silylated correspondents.*

Compound	Heat of formation of initial compound kcal/mol	Silylated compound	Heat of formation of silylated compound kcal/mol	Difference in heats of formation kcal/mol
CH_3OH	-48.1	$CH_3OSi(CH_3)_3$	-104.5	-56.4
C_2H_5OH	-56.2	$C_2H_5OSi(CH_3)_3$	-109.2	-53.0
C_3H_7OH	-61.2	$C_3H_7OSi(CH_3)_3$	-115.5	-54.3
CH_3NH_2	-5.5	$CH_3NHSi(CH_3)_3$	-59.6	-54.1
$C_2H_5NH_2$	-11.4	$C_2H_5NHSi(CH_3)_3$	-66.3	-54.9
$C_3H_7NH_2$	-16.8	$C_3H_7NHSi(CH_3)_3$	-72.9	-56.1
$HCONH_2$	-44.5	$HCONHSi(CH_3)_3$	-96.4	-51.9
CH_3CONH_2	-50.5	$CH_3CONHSi(CH_3)_3$	-102.2	-51.7
$C_2H_5CONH_2$	-53.4	$C_2H_5CONHSi(CH_3)_3$	-107.8	-54.4

[1] The heats of formation were calculated using MOPAC molecular orbital package [121], [122].

Once the equilibrium is attained, the displacement of the equilibrium in the desired direction can be achieved for silylation in the same manner as for other reactions, for example, by using a large excess of reagent or by eliminating the by-product of the reaction.

The silylation reaction is not always rapid, and for some systems the reaction can be slow and inefficient. For this reason the reaction rate is an important parameter for the analytical applicability. Also, there are instances where the kinetically determined product is not the more thermodynamically stable one. This stability may be regarding a tautomerization reaction or toward secondary reactions with other components in the silylated medium such as water. This plays an important role in the analytical practice, and in many particular cases the silylated products need to be analyzed within a prescribed time frame to ensure reproducible results.

It is generally accepted that the reaction mechanism for silylation is analogous to S_N2 involving a nucleophilic substitution at the silicon atom being designated S_N2-Si. For an alcohol, for example, the reaction takes place as follows:

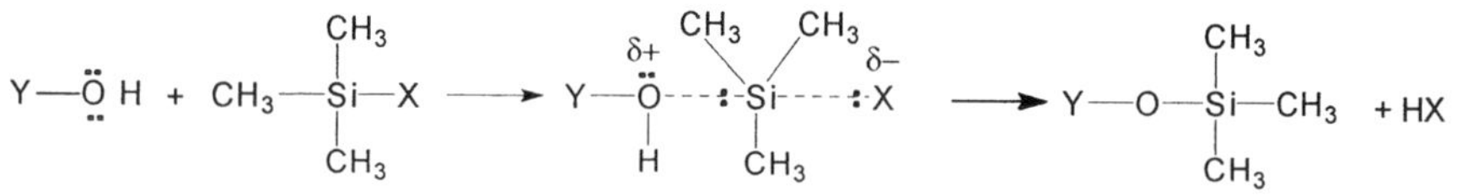

The prediction of silylation efficiency is commonly based on a series of rules and not on thermochemical data, which usually are not available. The nature of the X substituent in the TMS reagent (the leaving group) plays an important role for the silylation efficiency. Similarly to other S_N2 reactions, the leaving group will come off more easily from the TMS reagent when it is more stable as a free entity. This would be expected to be inverse to the basicity of the leaving group, and the best leaving groups are the weakest bases. Indeed, for silylating reagents with acidic leaving groups, the higher the acidity of HX, the better the silyl donor ability of the reagent. However, for increasing the reaction rate in silylation, other factors may play a more important role.

One such factor is the capability of the Si atom to be involved in d orbital bonding. In the transition state, it is possible that the Si atom may form σ (p→d) bonds using $3d_{z^2}$ or $3d_{x^2-y^2}$ orbitals. This may explain the lowering of the activation energy in S_N type reactions for silicon compounds compared to the corresponding carbon compounds, as in the case of the pair $(CH_3)_3SiCl$ and $(CH_3)_3CCl$. The stabilization by a σ (p→d) bond also may explain why silylation reagents with basic X groups can have good silyl donor ability, while following the general rules for a S_N reaction, basic X groups should be poor leaving groups. When such a group is protonated for example by acid catalysis, the (p→d) bonding is eliminated, and the leaving group departs as a neutral molecule with good leaving properties.

Another important factor in understanding the silyl donor ability of a reagent of the type $(CH_3)_3SiX$ is the ability of X substituent to stabilize the transition state of the reaction. For example, the leaving group OCOR has better silyl donor ability than an OR leaving group, and this can be attributed to the more stable transition state formed by OCOR with two possible resonance structures:

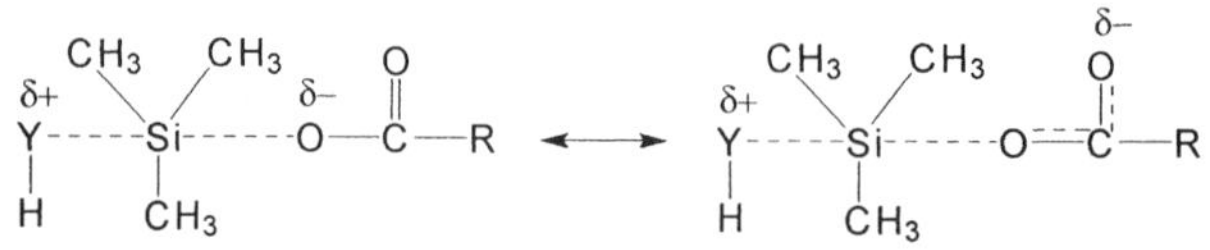

The same concept may explain the better silyl donor ability of N-TMS amides or ureas in comparison with TMS-amines, HMDS, or TMCS. One of the best silyl donor reagents, TMSI has the ability to stabilize the transition state and lower the activation energy by accommodating the negative charge in resonant structures:

In addition to σ (p→d) bonds, Si atom also may form π (p→d) bonds using its $3d_{yz}$, $3d_{xz}$, $3d_{xy}$ orbitals. The formation of π type bonds using the d orbitals of silicon may strengthen the Si-X bond, decreasing the silyl donor ability. Electron withdrawing groups within X may diminish π (p→d) bond formation and increase silyl donor ability. These types of considerations are, however, not easy to verify. The geometry of the activated complex may play an important role in the type of bonds formed. The symmetry of the orbitals in a tetrahedral compound involving the valence s, p, and d orbitals of the central atom and s and p orbitals of the substituents is well known (see e.g. [123]), while the geometry of the activated complex is not. This is an additional factor that makes difficult to understand the electron distribution and the nature of the bonds in the activated complex.

Regarding the ease of silylation, as seen in Table 18.2.3, the compounds with OH and COOH groups are more easily silylated than the compounds with NH or CONH groups. As amines are stronger bases than alcohols or acids, it would be conceivable that they are better nucleophiles. However, oxygen has two electron lone pairs, which may compensate better for the positive partial charge in the transition state, as compared to one pair of electrons on nitrogen. Also, the energy for Si-X bond is higher for Si-O than for Si-N, as shown in Table 18.2.9.

TABLE 18.2.9. *Bond energies for Si-X in kcal/mol [124].*

Bond	Energy kcal/mol	Bond	Energy kcal/mol
Si-O	112	Si-C	69
Si-Cl	93	Si-Si	68
Si-N	75-80	Si-I	59
Si-Br	76	Si-S	54
Si-H	70		

Oxygen has a higher capability to form (p→d) bonds. This may explain the decrease in the energy of the transition state for oxygen-containing compounds compared to nitrogenous ones. The change in the ease of silylation in the order ROH >ArOH > RCOOH > ArCOOH can be explained by the electron withdrawing effects of the double bonds or aromatic rings and lower capability to accommodate the partial positive charge in the transition state. Basic catalysts will promote the reactions for phenols and acids. The same explanation can be given to the change in the **ease** of silylation in the series RNH_2 > $RCONH_2$ > $RCONHSi(CH_3)_3$.

In addition to electronic consideration, the steric factors also should be considered in estimating the reactivity in silylation. In the same manner as for alkylation, steric effects could be a major criteria for the ease of silylation and also for the silyl donor ability of different reagents. The compounds with steric hindrance are more difficult to silylate. However, the sensitivity to hydrolysis is lower for compounds silylated with more voluminous silyl groups such as *tert*-butyldimethylsilyl compounds. This lowered sensitivity to water results in lower hydrolysis rates.

The solvolytic stability of silylated compounds has been thoroughly studied mainly for organic synthesis purposes. In most cases, for a compound with the formula R_3Si-X the solvolysis occurs as a nucleophilic attack at Si and as an electrophilic attack at X, because silicon is more electropositive than most X substituents. Some general rules were established regarding the stability to solvolysis:

a) Stability to solvolysis (hydrolysis) follows the order $R_3Si-O- > R_3Si-N= > R_3Si-S-$.

b) Stability of a $R_3Si-O-Y$ ether to acidic hydrolysis increases and to basic hydrolysis decreases when Y or R is an electron withdrawing group, and vice versa when Y or R is an electron donating group.

c) Stability to hydrolysis increases for compounds with more voluminous Y groups, as well as for more voluminous R groups.

- Specific properties and mass spectra of silyl derivatives

As indicated previously, the silylation reaction replacing an active hydrogen leads to derivatives that are more stable and have better behavior in the chromatographic column. This is attributed mainly to lower hydrogen bonding capability and modification in polarity. Other applications include the use of silyl groups containing halogens for enhancing ECD or NCI-MS sensitivity. The reaction can be applied with success for example for analysis of compounds containing alcohol groups that react easily with silylating reagents. Silylation also has been used for positive identification of compounds with noncharacteristic mass spectra, as shown in Section 17.5.

The mass spectra of TMS derivatives have some specific characteristics, but at the same time the spectra are dependent on the structure of the analyte. Many silylated compounds show in their EI+ mass spectrum the ion M - 15, generated by the loss of a methyl from the TMS group. The ion with m/z =73 is frequently found in the mass spectra of silylated compounds, sometimes being the base peak in the spectrum. The ion with m/z = 147 as well as the ion with m/z = 73 are found in compounds with two (or more silyl groups). These ions are generated by the following fragments:

$(CH_3)_3Si^+$ $HO^+=Si(CH_3)_2$ $(H_3C)_3Si-O^+=Si(CH_3)_2$

m/z = 73   m/z = 73   m/z = 147

Other ions common in the spectra of carbohydrates are 103, 205, and 217. These ions are generated usually from the following fragments:

$$(H_3C)_3Si \overset{+}{\longrightarrow} O = CH_2$$

m/z = 103

$$CH = \overset{+}{O} - Si(CH_3)_3$$
$$|$$
$$CH_2 - O - Si(CH_3)_3$$

m/z = 205

$$\overset{+}{C}H$$
$$||$$
$$C - O - Si(CH_3)_3$$
$$|$$
$$CH_2 - O - Si(CH_3)_3$$

m/z = 217

Rearrangements accompanying fragmentation are common during ion formation of silylated compounds. One such mechanism involves a γ-Si(CH$_3$)$_3$ rearrangement to an unsaturated group (such as =O, etc.) and a β-cleavage (McLafferty rearrangement) [115]. This rearrangement takes place as follows:

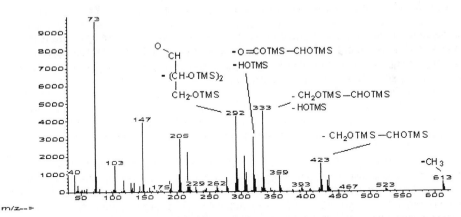

As an example, the spectrum of silylated 2,3,4,5,6-pentahydroxyhexonic acid (gluconic acid 6-TMS) is given in Figure 18.2.1. The figure also shows some neutral fragments that are lost from the initial molecule.

FIGURE 18.2.1. *Spectrum of gluconic acid-5-TMS (EI+ at 70 eV)*. Several neutral fragments lost during the fragmentation process are shown.

The ion with m/z = 292 is formed in the McLafferty rearrangement from the silylated gluconic acid [125] as follows:

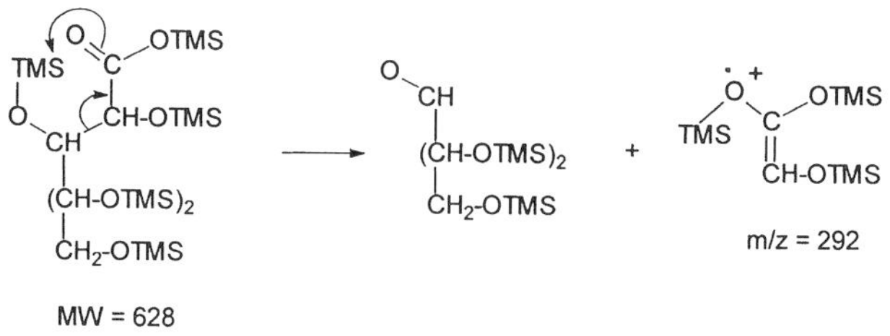

MW = 628   m/z = 292

The study of mass spectra of silylated compounds is beyond the interest of this book, and more information can be found in a variety of reports [126], [127], [116] [128].

- Artifact formation in trimethylsilyl derivatizations

In most situations, only the desired derivatives are formed during silylation. However, the expected silylated compound is not always formed, and either the silylation is not complete, or some compounds such as aldehydes, ketones, or esters with no obvious active hydrogen form silylated compounds. Also, unexpected reactions may occur with amides, carboxylic acids, or phenols during the silylation with particular reagents [129]. The identification of these artifacts is commonly done by MS analysis, as GC-MS is used as the preferred analytical technique.

Incomplete silylation that is the result of inappropriate reaction conditions cannot be considered artifact formation. However, when compounds with multiple functionalities are silylated, it is possible to generate a variety of derivatized compounds, regardless the intention to obtain fully silylated or partly silylated compounds. An example is the silylation of deoxyepinephrine, which generates by silylation with MSTFA in DMF at 76° C for 30 min. the expected O-silylated derivative, but only a limited proportion of the compound is silylated at the secondary amino group:

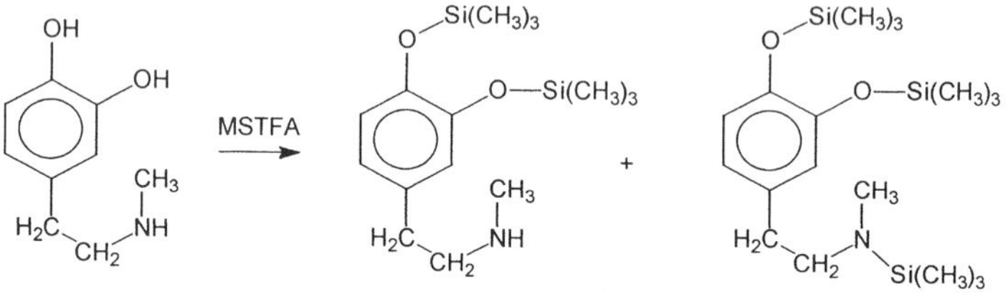

A type of artifact common to all derivatization reactions is the modification of the analyte under the influence of the reagents or catalysts used during derivatization. Similarly to alkylation reactions, when the silylation is done in basic or acidic conditions, the analytes that are sensitive to acidic or basic media may suffer unexpected transformations. This type of transformation also may be the effect of the reagent itself. As an example, the derivatization of β-D-glucopyranose to form a penta-TMS derivative using BSTFA that contains 1% TMCS in DMF as a solvent is associated with the formation of the α anomer and with pyranose-furanose equilibrium. Four peaks can be seen in the chromatogram of the silylated β-D-glucopyranose, as shown in Figure 18.2.2a. All these chromatographic peaks have similar spectra. The spectrum of the chromatographic peak D is shown in Figure 18.2.2.b.

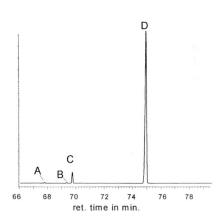

Figure 18.2.2a. *Chromatogram showing four isomers generated from trimethyl-silylation of β-D-glucopyranose using BSTFA (1% TMCS) in DMF.*

Figure 18.2.2b. *Spectrum of chromatographic peak D for trimethylsilylated β-D-glucopyranose.*

The formation of the α anomer and pyranose-furanose equilibrium is significantly stronger when the t-butyldimethylsilyl derivative is obtained. The chromatogram of the silylated β-D-glucopyranose using MTBSTFA with 1% TBDMCS in DMF as solvent has four peaks that have about equal areas, as shown in Figure 18.2.3a. The peaks have similar spectra, the one for peak D being shown in Figure 18.2.3b.

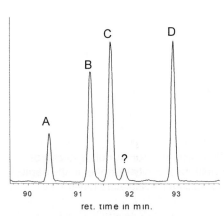

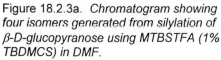

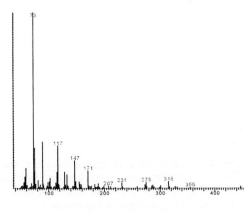

Figure 18.2.3a. *Chromatogram showing four isomers generated from silylation of β-D-glucopyranose using MTBSTFA (1% TBDMCS) in DMF.*

Figure 18.2.3b. *Spectrum of peak D for tert-butyl-dimethylsilylated β-D-glucopyranose.*

The silylation of β-D-glucopyranose using MTBSTFA (1% TBDMCS) in DMF was done in the same conditions as the one used to form TMS derivatives using BSTFA. Both separations were done using a 60 m, 0.32 mm i.d., 0.25 μm film DB5 column.

The most frequent artifacts with compounds not containing obvious active hydrogens occur with aldehydes. Aldehydes are known to be able to undergo two types of chemical reactions with formation of OH groups, namely enolization and acetal formation in the presence of water:

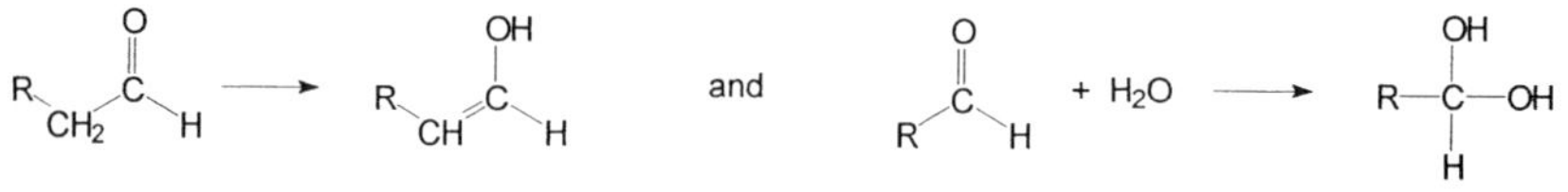

The OH groups formed in this manner react with different silylating reagents and give the corresponding silylated products. Although the enolization or the acetal formation is negligible for the initial aldehyde, the reactions may be significantly displaced toward the formation of the silylated compounds of the enol or of the acetal.

Besides these reactions, aldehydes also may condensate with fragments of the silylating reagent as follows:

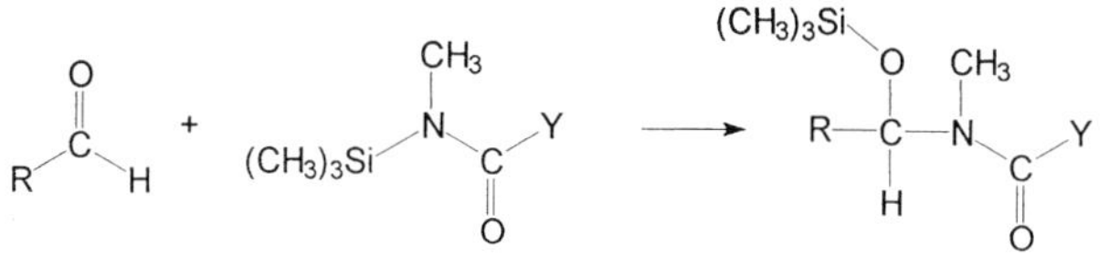

For an aldehyde with the mass M reacting with MSTFA, the ions present in the mass spectrum indicating this type of compound are M + 199 with the loss of H, M + 73 with the loss of $N(CH_3)COCF_3$, and 228 for the ion $(CH_3)_3Si-O^+=CH-N(CH_3)COCF_3$ with the loss of R. Another example [129] is the reaction of chlorobenzaldehyde with an excess of BSA:

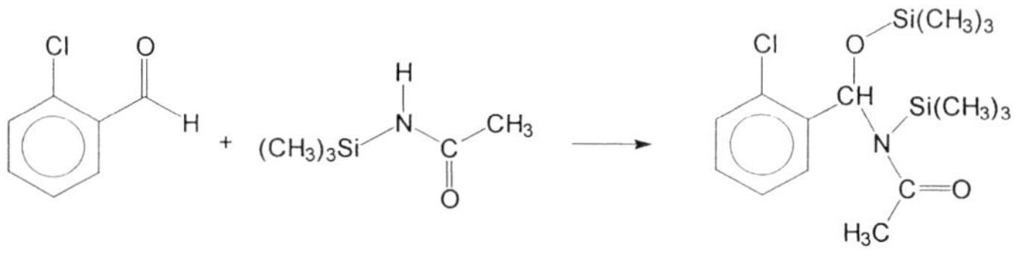

Ketones are less reactive than aldehydes, but they also may form enols that can be silylated. Although the enolization can be negligible for the initial ketone, in the same manner as for aldehydes, the reactions may be significantly displaced toward the formation of the silylated compounds.

The artifacts generated by acids are uncommon. One known example [130] is the formation of a tetrakis-TMS derivative from hydroxymalonic acid, besides the normal tri-TMS derivative:

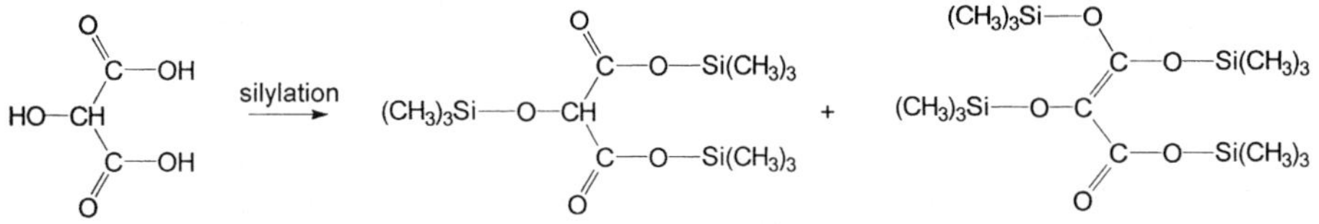

Some acids containing an α-hydrogen also occasionally form artifacts [129]. In the derivatization with BSA, for example, the following reaction may occur:

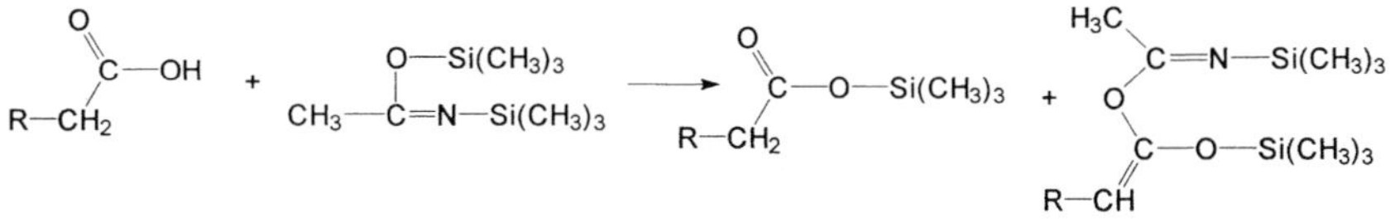

Some active hydrogens that are connected to a carbon also can be acidic enough to be replaced by a TMS group, as in the following example:

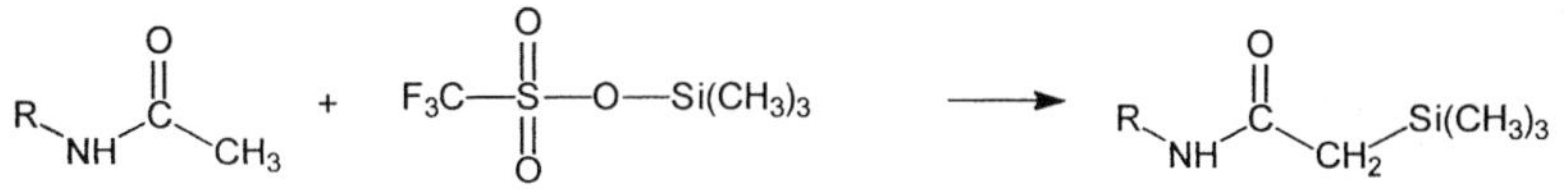

A similar reaction may take place during the silylation of testosterone. Besides the monosilylated compound, several compounds with two TMS groups were obtained when the silylation was done for 72 hours at 60° C using a BSA:TMCS 10:1 mixture. The possible reactions are indicated below [131]:

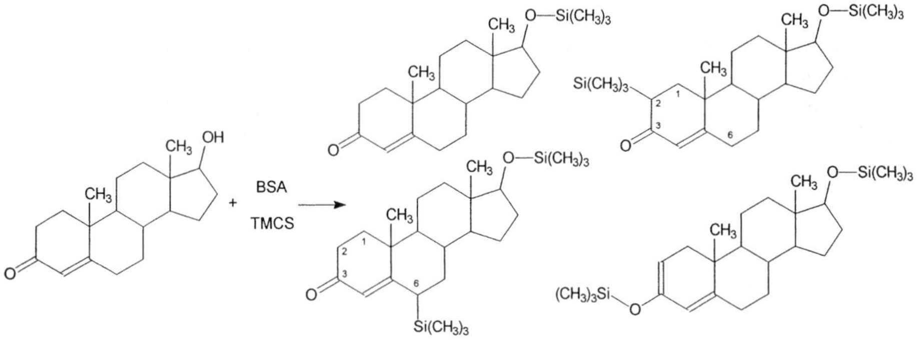

Artifacts also can be generated because of other uncommon reactions with a specific silylation reagent or when the reaction is allowed to continue for an extended period of time. An example of an uncommon reaction is the ring opening of flavanones [129]:

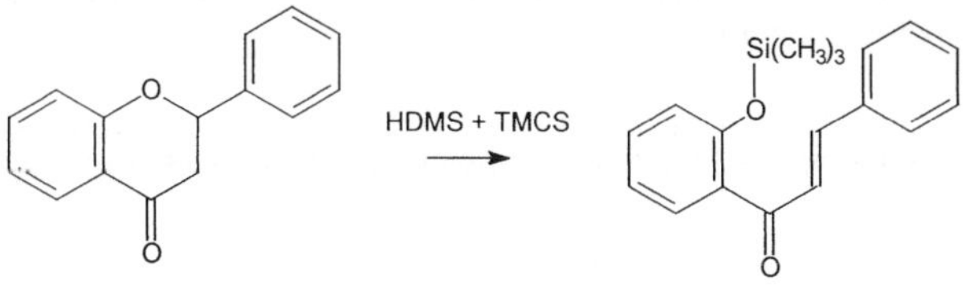

Anthocyanins undergo a ring modification reaction that takes place probably in the hot injection port of the GC [132]:

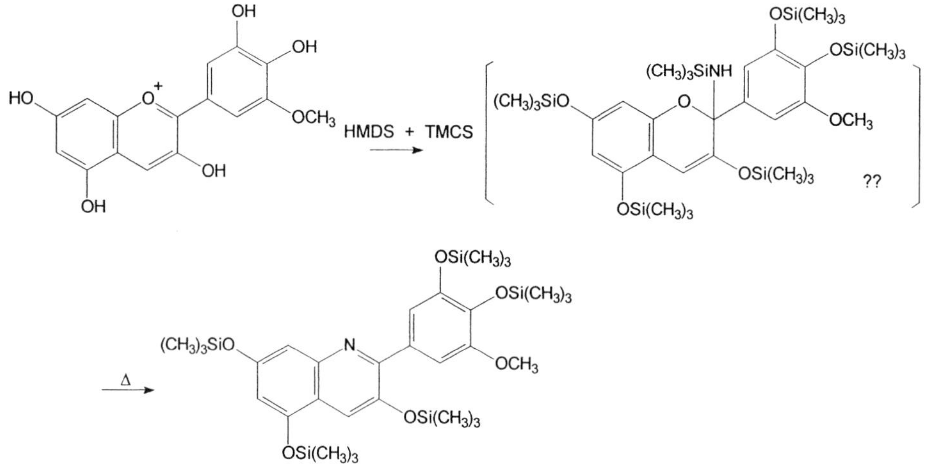

Another example is the silylation of isocyanates generating silylamines:

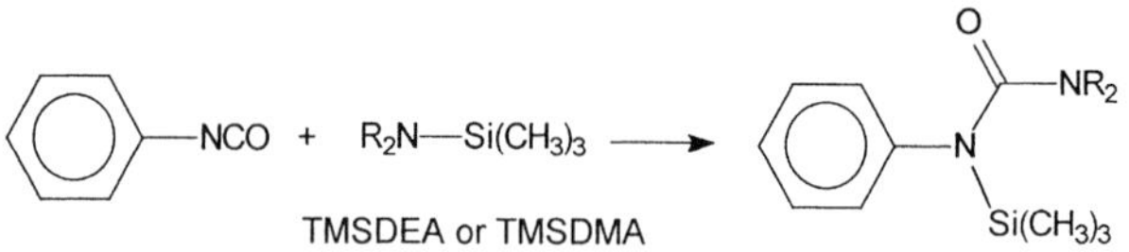

A different type of artifact is generated by an unexpected reaction of the analyte with the solvent used during the silylation. Two such examples, one in which DMF reacts with an amino group and the other where acetone reacts with a phenol are shown below [129]:

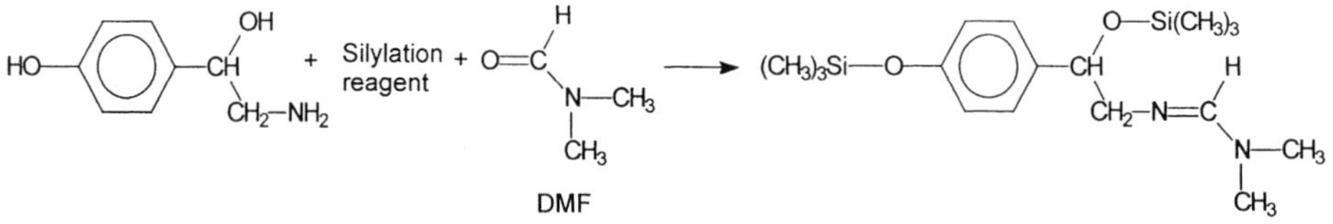

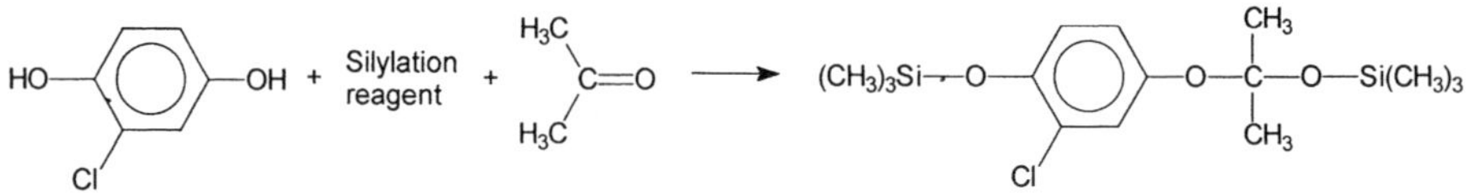

Artifacts can also be generated from the by-products of the silylation reaction [129].

Special silylating reagents used for the formation of derivatives other than TMS also may generate specific artifacts. For example, the formation of allyldimethylsilyl derivatives can be associated with reactions of displacement of the allyl group through nucleophilic attack:

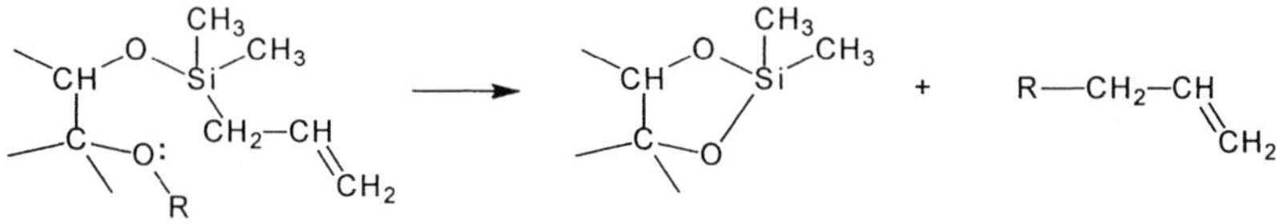

Other special silylating reagents also may have specific problems. For example, the reagents prepared to generate alkylsilylene derivatives that are utilized for bifunctional compounds have the tendency to polymerize. The reagents intended to form chlorinated silyl compounds may produce unexpected condensations, for example, in a reaction with β-hydroxyamines:

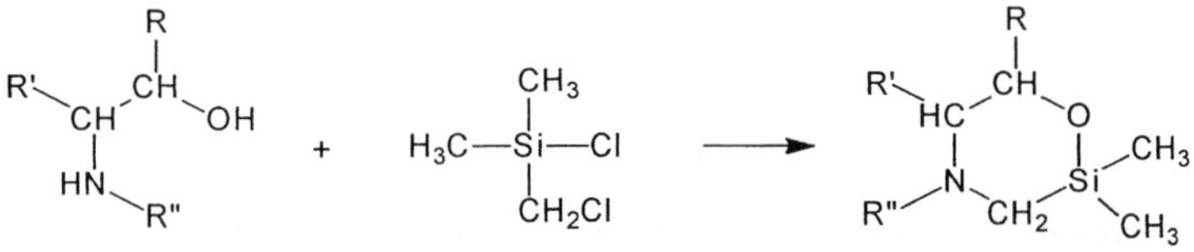

The reagent in the above silylation was a mixture 2/1 v/v of chloromethyldimethyl-chlorosilane and 1,3-bis(chloromethyl)-1,1,3,3-tetramethyldisilazane [133].

The artifact formation in silylation reactions usually can be diminished by not using the silylating reagents that are known to generate artifacts in particular cases and by avoiding unnecessary heating or long reaction times.

18.3 ACYLATION REACTIONS

The formation of acyl derivatives is applied for replacing the active hydrogens from an analyte in functionalities such as OH, SH, NH, CONH, etc. The acylation is used similarly to other derivatizations for reducing polarity and improving the behavior of the analytes in the chromatographic column. A better volatility also may be conferred by acylation, but this is not as marked as for silylation or methylation. Typically, only the derivatization with acetyl groups or with fluorinated acyl groups up to heptafluorobutyryl improve volatility, while other heavier acyl groups are not suitable for this purpose. Acetylation, for example, is used for compounds such as monosaccharides and amino acids to allow their GC analysis. The detectability improvement on the other hand is a very common purpose for acylation, and it has been applied for GC, HPLC, and other chromatographic techniques using a variety of detectors including mass spectrometry. Acylation with fluorinated compounds plays an important role in enhancing detectability in GC with ECD or NCI-MS detection. Other uses of acylation include the enhancement of separation of chiral compounds, etc.

Most acylation reactions are nucleophilic substitutions where the analyte is a nucleophile (Y:, Y:H, Y:⁻) reacting with the acylating reagent RCOX that contains a leaving group X and an acyl group RCO:

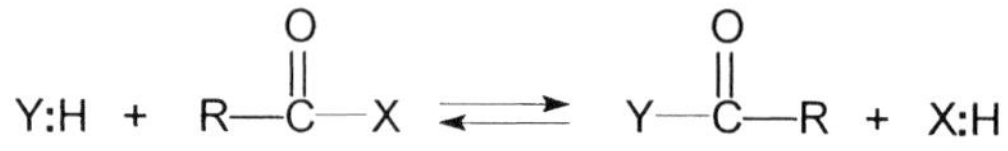

$$Y{:}H \ + \ R{-}\overset{\overset{\displaystyle O}{\|}}{C}{-}X \ \rightleftarrows \ Y{-}\overset{\overset{\displaystyle O}{\|}}{C}{-}R \ + \ X{:}H$$

Acyl halides, commonly chlorides or bromides, are very reactive compounds used for acylation. The reaction of an acyl chloride with an amine, for example, takes place as follows:

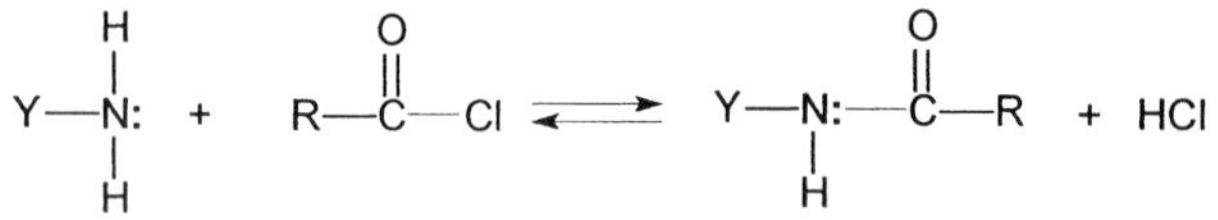

$$\underset{\underset{\displaystyle H}{|}}{\overset{\overset{\displaystyle H}{|}}{Y{-}N}}{:} \ + \ R{-}\overset{\overset{\displaystyle O}{\|}}{C}{-}Cl \ \rightleftarrows \ \underset{\underset{\displaystyle H}{|}}{Y{-}N}{:}{-}\overset{\overset{\displaystyle O}{\|}}{C}{-}R \ + \ HCl$$

The second hydrogen in the amine also may be replaced with an acyl group, although the reactivity of amides is much lower than that of amines. Also, steric hindrance may negatively influence the reaction. The generation of a strong acid such as HCl is a disadvantage in the reaction with acyl halides, and usually the acid should be removed either using a basic compound such as Na_2CO_3 or $MgCO_3$ or by evaporation [1], [134]. Among the acyl halides used more frequently for generating short chain acyl derivatives are those involving fluorinated acyl groups. A few other applications for GC analysis are reported, for example, using acyl chlorides for the formation of propionyl or 2,2-dimethylpropionyl derivatives of amino acids.

The high reactivity of acyl halides is used for the acylation of compounds with less reactive hydrogens. For example, amides can be acylated with 2,3,4,5,6-pentafluoro-benzoyl chloride forming a stable compound [135]. The reaction can be done in hexane as a solvent and in the presence of trimethylamine. The reaction can be completed at room temperature in about two hours.

Much more frequently the acyl halides are used for HPLC derivatization for replacing an active hydrogen with chromophores or fluorescent groups. An example is the use of 4-nitrobenzoyl chloride, which can be used for the derivatization of alcohols, amines, etc. in a reaction of the type:

The reaction can be performed in pyridine for neutralizing the strong inorganic acid that results in this reaction. Other uses of acyl halides as derivatizing reagents for both GC and HPLC analysis will be described in Chapter 19, in connection with the derivatization of specific functional groups.

Certain carbonyl cyanides react similarly to acyl chlorides. For example, pyrene-1-carbonyl cyanide is used as an acylation reagent [136] for alcohols:

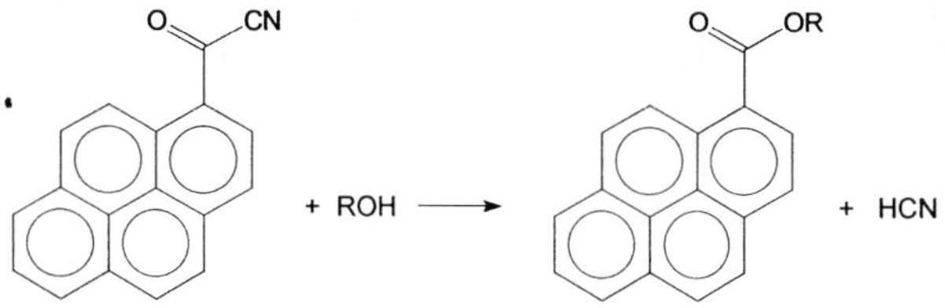

The resulting HCN is a much weaker acid compared to corresponding hydrogen halides.

The disadvantage of generating a strong inorganic acid in the acylation with acyl halides also can be avoided by having, instead of the acyl halide, an anhydride or another reactive acyl derivative as a reagent. The reaction with an anhydride takes place as follows:

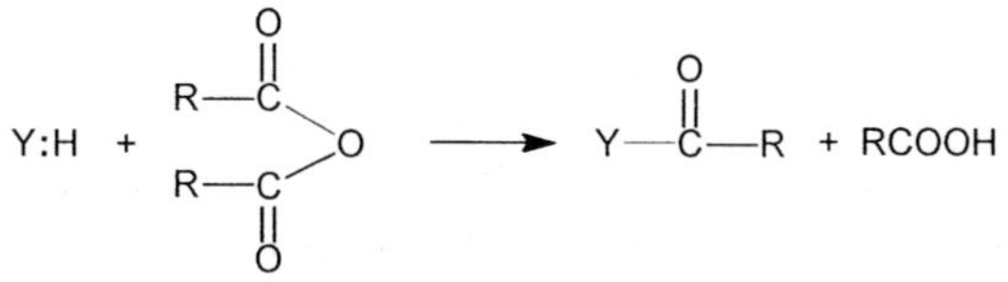

The acid resulting together with the acylated compound is not a strong acid such as HCl or HBr. Using anhydrides, the undesired modifications of the analytes that can be induced by strong acids are not likely to occur. However, the volatility of RCOOH and of the excess anhydride used as a reagent may be lower than for the corresponding acyl halide and hydrogen halide. This may cause some problems regarding the best choice of a reagent with good chromatographic behavior.

The acylation reaction may be strongly influenced by the reaction medium or by the presence of a catalyst. Some typical conditions for the acetylation reaction using acetic anhydride are given in Table 18.3.1.

TABLE 18.3.1. *Catalysts/reacting media used for acylation with acetic anhydride.*

Catalyst	Medium	Conditions	Analytes	Reference
acetic acid	CH_3Cl	2- 16 hours, 50°C	various	[137]
$NaOOC-CH_3$	reagent	1 hour, 100°C	carbohydrates	[138]
pyridine	pyridine	20 min. 100°C or milder	carbohydrates, α-tocopherol, amino acids, etc.	[1]
triethylamine	acetone	30 sec. 60° C	amino acids	[139]
N-methylimidazole	catalyst:reagent (1:1, v/v)	20% molar excess, 10 min. no heating	hydroxy compounds	[140]
	water	addition of $NaHCO_3$	amino, phenols	[1], [141]
methanesulfonic acid	reagent	24 hours, room temperature	glucuronides	[142], [143]
toluenesulfonic acid (TsOH)	reagent		alcohols	[144]
$HClO_4$	ethyl acetate	1 hour 5° C, 5 min. room temperature	alcohols	[145]

The anhydrides of trifluoroacetic (TFA), pentafluoropropionic (PFPA) and heptafluoro-butyric (HFBA) acids commonly are used for derivatization of alcohols, phenols, amines, etc., with the purpose of enhancing detectability and also for improving the chromatographic behavior (higher volatility, better thermal stability, better separation). Other halogenated compounds also are used for enhancing detectability, and the sensitivity in ECD or NCI-MS increases in the order F < Cl < Br < I, as indicated in Section 17.4. However, the volatility of fluorinated compounds being higher, these may be preferred in GC applications [146]. The reactivity of the perfluorinated anhydrides increases in the order HFBA < PFPA < TFA. The differences are not significant. Once formed, the heptafluorobutyrates are more stable to hydrolysis than the trifluoroacetate.

The reactivity of perfluoroacyl anhydrides becomes a problem mainly with less reactive analytes such as compounds with steric hindrance or in the reaction with amides. Primary amines, for example, can be derivatized easily with the substitution of one active proton, forming amides. Further acylation of the second hydrogen may become a problem. Some primary amines can be substituted at both hydrogens. For this purpose longer reaction times are used, as well as the addition of basic compounds such as trimethylamine [135], [147]. As an example, the influence of the concentration of trimethylamine on the yield of acylation of phenylethylamine with trifluoroacetic anhydride is given in Table 18.3.2 [135]. The concentration of the analyte in the experiment was 1.7×10^{-2} M, that of TFA was 3.4×10^{-1} M, the solvent was benzene, and traces of water were present. The reaction was performed at $20°$ C.

TABLE 18.3.2. *The influence of trimethylamine concentration on the derivatization of phenylethylamine with TFA [135].*

	Trimethylamine Concentration						
Reaction Time	0.0 M		4.8×10^{-2} M		7.2×10^{-2} M		9.5×10^{-2} M	
	mono	di	mono	di	mono	di	mono	di
5 min.	96	-	100	1	56	41	-	99
15 min	98	-	99	1	57	42	-	98
30 min.	103	-	102	1	55	43	-	100
60 min.	98	-	94	1	54	43	-	100

In certain derivatizations higher reaction temperatures were found to favor the reaction [148]. In other instances, a different derivatization such as methylation of the second hydrogen was preferred, for example in the case of heterocyclic amines [149].

The reaction with perfluoroanhydrides can be performed in an inert solvent such as CH_2Cl_2 [150], ether, ethyl acetate, acetone, tetrahydrofuran or in CH_3CN [151], [152], etc. Basic compounds such as triethylamine [153], pyridine or even solid $NaHCO_3$ are reported as utilized for neutralizing the acids formed during derivatization [1]. Various anhydrides used for derivatization are discussed in Chapter 19.

Organic acids themselves can be used as acylating reagents. When the nucleophile is an alcohol, the reaction is known as esterification and was discussed in Section 18.1. The acylation with acids can be applied besides alcohols to thiols, phenols, amines, etc. and can be written as follows:

$$Y{:}H + R{-}COOH \longrightarrow R{-}COY + H_2O$$

The reaction can be displaced toward the formation of the acyl derivatives by eliminating the water using distillation or using compounds able to eliminate water such as anhydrous $MgSO_4$, molecular sieve, or substances that react with water such as CaC_2, or $(CH_3)_2C(OCH_3)_2$. Dicyclohexylcarbodiimide (DCCI) also is used for modifying the yield of the desired product. The reaction was described in Section 18.1 for alcohols, but it can be applied in the reaction of alcohols with amines, etc. The reaction with reagents containing a carboxylic acid reactive group also can be done in the presence of 2,4,6-trichlorobenzoyl chloride or with various sulfonyl chlorides such as 2,4,6-triisopropyl-benzenesulfonyl chloride or 2,4,6-trimethyl-benzenesulfonyl chloride.

Formyl derivatives of several steroids were obtained, for example, using direct reaction with formic acid [154]. The formation of formyl derivatives of certain amines also can be achieved using sodium formate in the presence of acetic anhydride [1]. Although the formyl group has a low molecular mass, derivatization with the formation of trifluoroacetyl, pentafluoropropionyl or even heptafluorobutyryl may lead to more volatile compounds.

The reaction of amines with acids can be displaced toward the formation of the amides using a peptide coupling reagent such as benzotriazol-1-yl-oxy-tris(dimethyl-amino)-phosphonium hexafluorophosphate (BOP), diethylcyanophosphonate, O-benzotriazol-1-yl-N,N,N',N'-bis(tetramethylene)uronium hexafluorophosphate, or 2,2'-dipyridyl disulfide + triphenylphosphine. The structures of these reagents are shown below:

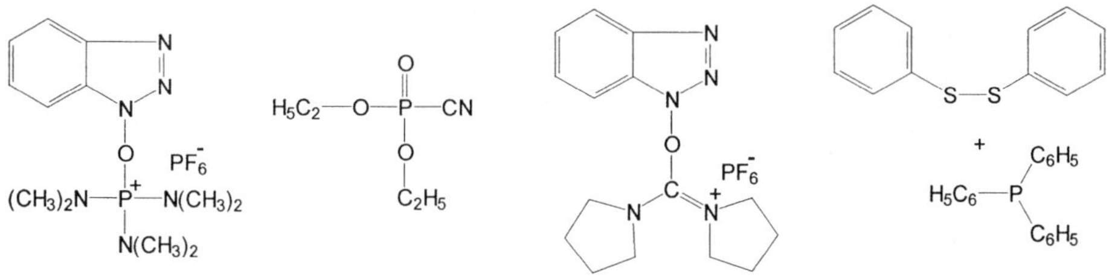

BOP

A list of several peptide coupling reagents is given in Table 18.3.3.

TABLE 18.3.3. *Compounds able to facilitate peptide bond formation.*

Reagent
benzotriazol-1-yl-oxy-tris(dimethyl-amino)-phosphonium hexafluorophosphate (BOP)
diethylcyanophosphonate
O-benzotriazol-1-yl-N,N,N',N'-bis(tetramethylene)uronium hexafluorophosphate
2,2'-dipyridyl disulfide + triphenylphosphine
1-(3-dimethylaminopropyl)-3-ethylcarbodiimide (EDAC)
1-hydroxybenzotriazole hydrate

The reaction of acids with various nucleophiles also can be used in the derivatization of acids for improving their analysis, in a reaction of the form:

$$Y—COOH + R:H \longrightarrow Y—COR + H_2O$$

For this purpose alcohols, amines, hydrazines, etc. are used as reagents. For example, various fatty acids can be analyzed by HPLC with fluorescence detection using derivatization with an amine carrying a fluorophore such as 4-aminomethyl-6,7-dimethylcoumarin in a reaction as follows:

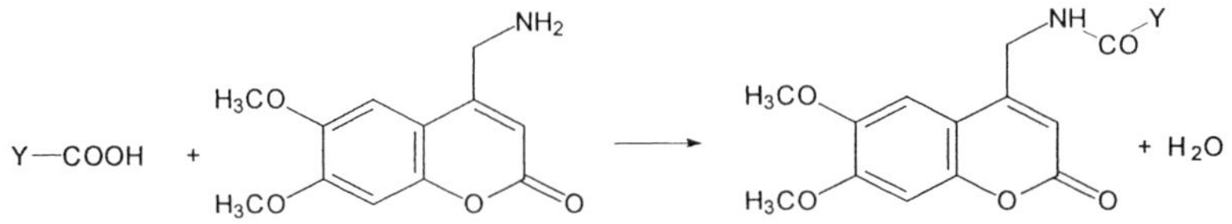

The reaction takes place in the presence of 1-[3-(dimethylamino)propyl]-3-ethyl-carbodiimide [155] leading to a detection limit as low as 20–50 fmol for acids such as palmitic, stearic, arachidonic etc. Some hydrazines react similarly to amines forming hydrazones as shown below:

$$Y—COOH + R-NH-NH_2 \longrightarrow Y—CONH-NH-R + H_2O$$

Various substituted hydrazines are prepared as reagents carrying strong fluorophores and used for HPLC analysis of aldehydes and ketones (see Section 18.4). Some of these labeling reagents are also used for the analysis of carboxylic acids, and the use of peptide coupling reagents or alcohol/acid coupling reagents such as 1-[3-(dimethylamino)propyl]-3-ethylcarbodiimide facilitates the reaction.

Certain amides such as N-methyl-bis(trifluoro-acetamide) or 2,2,2-trifluoro-N-methyl-N-(2,2,2-trifluoroacetyl)acetamide (MBTFA) [156] can be used as acylation reagents, avoiding the formation of water or of a strong acid in the reaction mixture. Bis(trifluoroacetamide) can be used for the same purpose. Acylation of amines takes place at room temperature. Solvents such as CH_3CN, pyridine, DMSO, or THF can be used as a reaction medium:

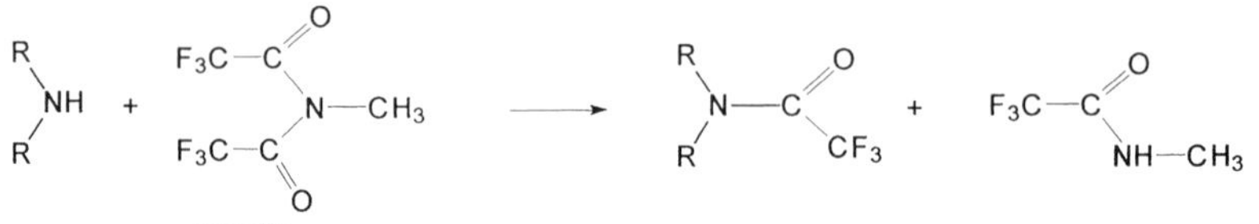

Alcohols and other compounds with active hydrogens also may be acylated with MBTFA, but the reaction takes place with less efficiency and heating may be necessary. An interesting application of this procedure is in a two-step derivatization of compounds such as catecholamines. Silylation with BSTFA of dopamine will generate a silylated derivative. Further reaction with MBTFA replaces the silyl group at the amino function but leaves unmodified the silyl groups at the hydroxyls [157], [158].

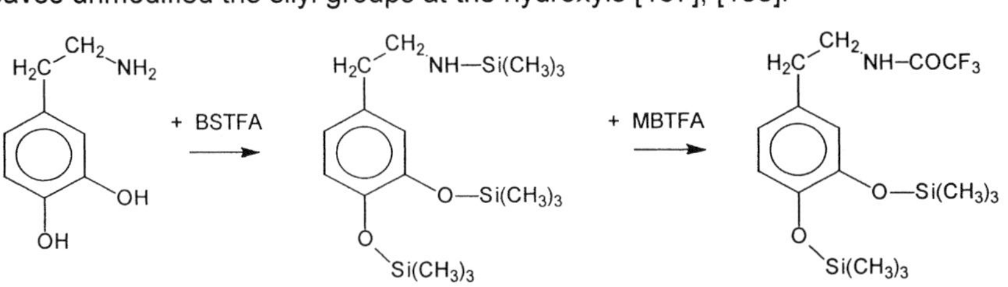

One other procedure successfully applied to obtain acyl derivatives is the use of acyl imidazoles as reagents. This class of compounds reacts with analytes containing alcohol, primary and secondary amino groups, or thiols. The reaction generates as a by-product imidazole:

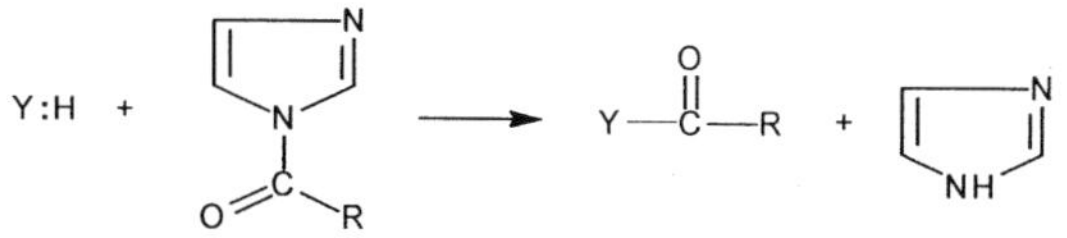

The reagent can be hydrolyzed easily with water generating the acid and imidazole [159]. This allows the removal of the excess reagent using a wash with water [160]. Heptafluorobutyrylimidazoles were used for acylation without adding a solvent or a catalyst, by heating the reagent and the sample for 2–3 hours at 80° C [161]. Trifluoroacetylimidazole (TFAI), pentafluoropropionylimidazole (PFPI), and heptafluorobutyrylimidazole (HFBI) are commercially available compounds used for acylations.

Succinimidyl esters also can be used for acylation purposes. For example, the reaction of an amine with 6-aminoquinolyl-N-hydroxysuccinimidyl carbamate takes place as follows:

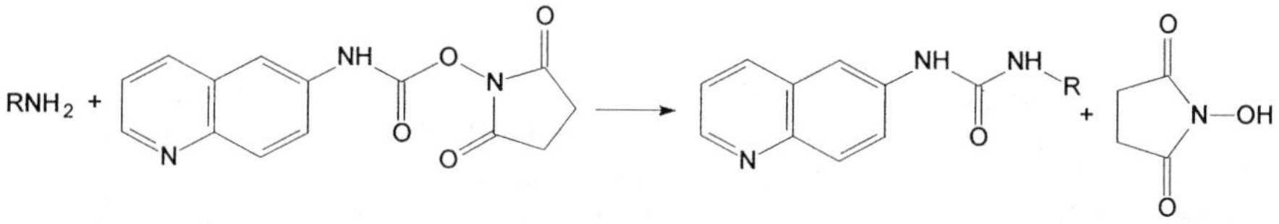

The resulting carbodiamide is highly fluorescent and is used for HPLC determination of amines using fluorescence detection. The reaction takes place in borate buffer at pH 8.8 within 5 min. at room temperature [162]. Alcohols also can be derivatized with this reagent [163].

Amines and the amino group in amino acids also can be acylated using urethane protected α-amino acid-N-carboxyanhydrides or oxycarbonyl-amino acid-N-carboxyanhydrides. The reactions take place as follows:

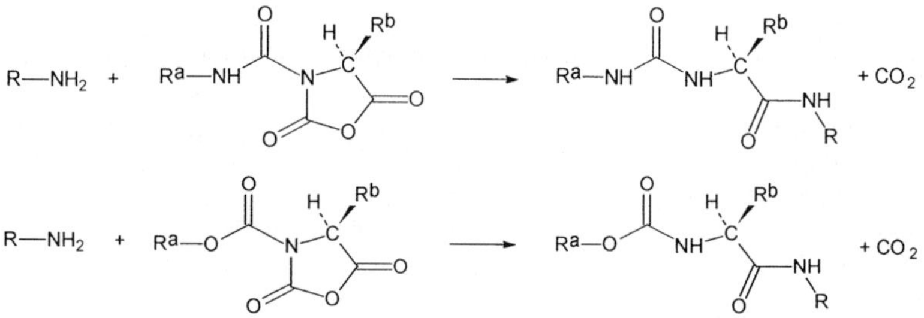

These reactions may take place at room temperature or with mild heating in a short period of time [164], [165].

A summary of some typical acylating reagents used in derivatizations for GC analysis is shown in Table 18.3.4.

TABLE 18.3.4. *Some common groups present in acylating reagents used in derivatizations for GC analysis.*

	Group	Mass of the group	Reagent(s)	Analytes	Reference
formyl		29	formic acid	steroids	[1]
acetyl		43	acetyl chloride	hydroxy group	[166]
			acetic anhydride	alcohols	
trifluoroacetyl		97	N-methyl-bis(trifluoro-acetamide)	alcohols	[156]
			bis(trifluoroacetamide) trifluoroacetic acid		[4]
propionyl		57	propionic anhydride	alcohols	[144]
butyryl		71	butyric anhydride	alcohols	[4]
2,2-dimethyl-propionyl-(pivaloyl)		85	pivaloyl chloride	amino acids	[167]
			pivalic anhydride		
pentafluoro-propionyl		147	pentafluoropropionic anhydride		
heptafluoro-butyryl		197	heptafluorobutyric anhydride		
			heptafluorobutyrylimidazole	amines, alcohols	[1]
trichloroacetyl		145	trichloroacetic anhydride		
pentafluoro-benzoyl		195	pentafluorobenzoyl chloride	alcohols	[168]
			pentafluorobenzoyl-imidazole	amides	[135]
(pentafluoro-phenyl)-acetyl		209	(pentafluorophenyl)-acetyl chloride	alcohols	
(pentafluoro-phenoxy)-acetyl		225	(pentafluorophenoxy)-acetyl chloride	alcohols	

Not only acyl derivatives are used for acylation reactions. Alkylketenes and their dimers may be used for acylation. The reaction with an alkylketene can be written as follows:

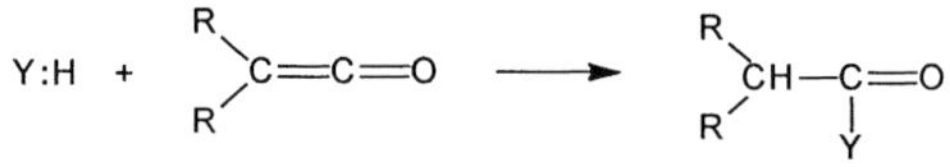

As an example, trimethylsilylketene can react with Δ^9-tetrahydrocannabinol forming the cannabinol trimethylsilyl acetate by allowing the reaction to take place at 25° C for 12 hours [169]:

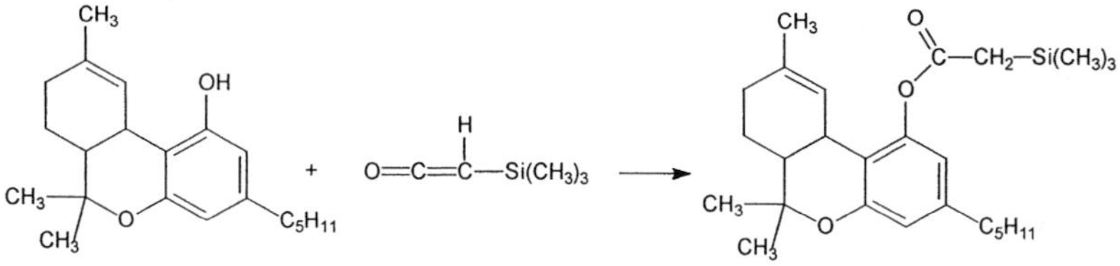

Simple acetylation can be achieved using ketene itself ($H_2C=C=O$). Alkylketene dimers also react with active hydrogens leading to the formation of β-ketoacyl derivatives:

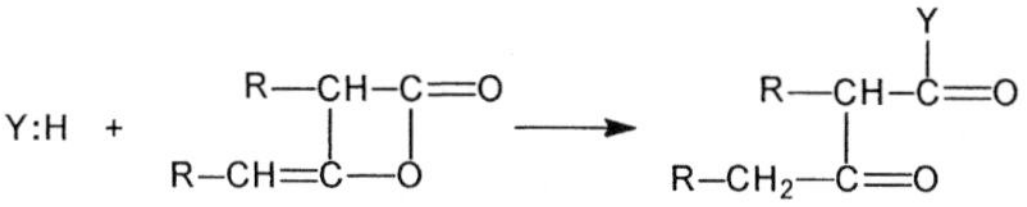

- Derivatization with chloroformates

Carbonic acid, $O=C(OH)_2$, can form amides, esters, halides, etc., due to the presence of two OH groups bonded to the CO group. Carbonic acid ester-halides, also called chloroformates or chloroformate esters, with the formula R–O–C(=O)-X, where R is an alkyl or aryl group, and X is F, Cl, Br, or I, can react with various compounds containing active hydrogens, such as amines and alcohols. Amines, for example, form carbamate esters (urethanes), reacting as follows [170]:

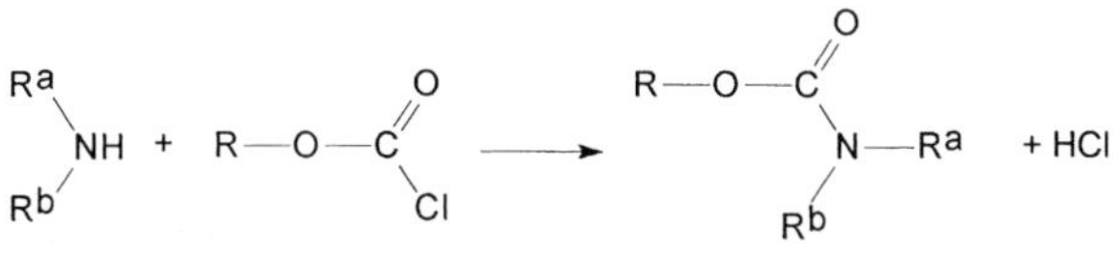

Chloroformates containing in the alkyl or aryl group halogen substituents are particularly reactive. Chloroformates react easily with alcohols, thiols, amines, etc. Even tertiary amines can react with specific chloroformates, such as pentafluorobenzoyl chloroformate (pentafluorobenzoyloxycarbonyl chloride) or trichloroethyl chloroformate, by displacing an alkyl group connected to the nitrogen atom and forming the carbamate ester [171]. Reaction with chloroformates usually takes place in an inert solvent such as heptane, by heating the analyte and the reagent for 30 min–1 hour in the presence of anhydrous Na_2CO_3. Chloroformates may contain in the molecule large groups that bring properties such as fluorescence or strong UV absorption and are rather frequently used

as derivatization reagents. For example, 4-(N-chloroformylmethyl-N-methyl)amino-7-
N,N-dimethylaminosulfonyl-2,1,3-benzoxadiazole can be used to derivatize alcohols,
phenols, amines, and thiols leading to fluorescent derivatives that can be detected at
levels as low as 100 fmoles. The reaction with a thiol can be written as follows:

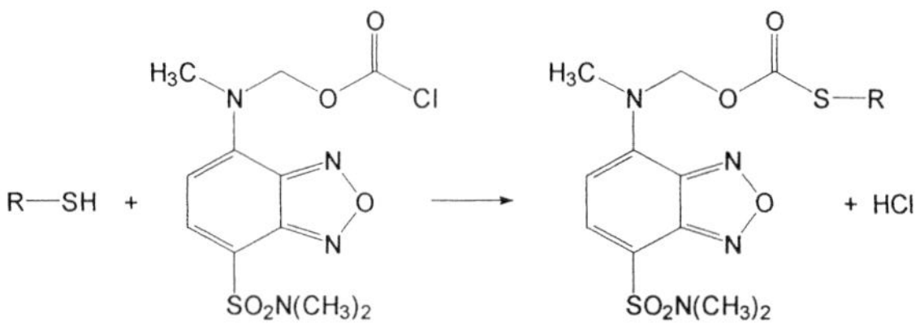

Phosgene (chloroformyl chloride, O=CCl$_2$) reacts similarly to chloroformates. Having
two reactive sites, phosgene is used to form cyclic carbonates in reactions with
polyfunctional analytes such as diols or polyols (carbohydrates). Carbohydrates also
may form cyclic carbonates with chloroformate esters [1] (see also Section 18.5).

- Derivatization with sulfonyl derivatives

The reactions of sulfonyl derivatives R–SO$_2$–X are in many respects similar to that of
acyl derivatives R–CO–X, although sulfonyl halides are in general less reactive than
halides of carboxylic acids. The reaction of a sulfonyl derivative may take place with
alcohols, phenols, amines, etc., as follows:

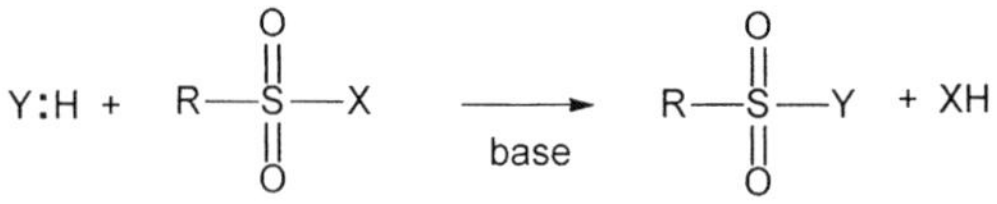

The reactivity toward the sulfonyl sulfur is OH$^-$ > RNH$_2$ > CH$_3$COO$^-$ > H$_2$O > ROH [23].
A typical example of a sulfonyl halide used for derivatization is dansyl chloride 5-
(dimethylaminonaphthalene)-1-sulfonyl chloride (Dns-Cl), which reacts with primary and
secondary amines and with phenolic hydroxyls in a reaction as follows:

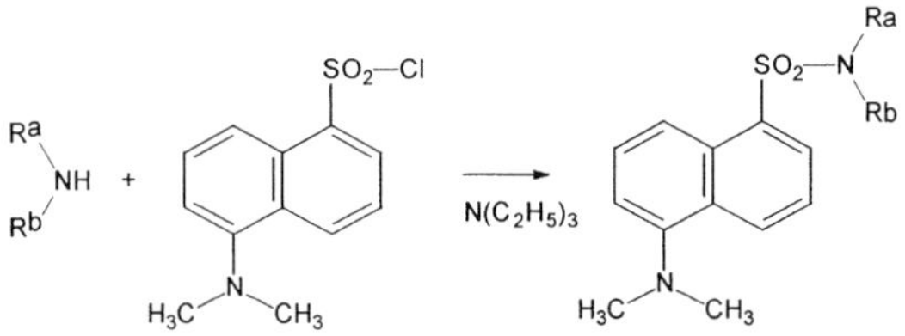

Alkyl hydroxyls react with more difficulty or not at all, and tertiary amines do not react
with Dns-Cl. The reaction can be used for various analyses using HPLC separation with
fluorescence detection [24]. A typical procedure of derivatization consists of adding the
reagent in slight excess to the analyte, in an inert solvent, and in the presence of a basic
buffer or compound. The reaction takes place in 10–20 min., and depending on the
analyte, mild heating may be required.

The fluorescence of the dansyl group can be exploited not only for enhanced fluorescence detection but also for chemiluminescence detection. For the analysis of estradiol, after derivatization with Dns-Cl and HPLC separation, a chemiluminescence detection has been applied using postcolumn reaction with bis(2-nirophenyl)oxalate [172].

Various other sulfonyl halides are used for attaching fluorescent or strongly absorbent UV groups to amines, phenols or thiols.

- *Derivatization with isocyanates, isothiocyanates, carbonyl azides*

A number of other functionalities have high reactivity toward active hydrogens in alcohols, amines, etc. These functionalities include isocyanates, isothiocyanates, carbonyl azides, etc. Formally these reactions can be seen as a replacement of an active hydrogen with a CO-R group or CS-R group, as shown below for amines:

$$Y-NH_2 + R-NCO \longrightarrow Y-NHCO-NH-R$$

$$Y-NH_2 + R-CON_3 \ (R-NCO + N_2) \longrightarrow Y-NHCO-NH-R$$

$$Y-NH_2 + R-NCS \longrightarrow Y-NHCS-NH-R$$

The mechanism of these reactions is in fact based on the addition to the carbon-hetero atom multiple bond and is discussed in Section 18.4.

- *Some aspects regarding acylation mechanism*

The reaction between alcohols, amines, or other compounds containing a nucleophilic atom and an acyl chloride commonly takes place with a typical S_N2 mechanism:

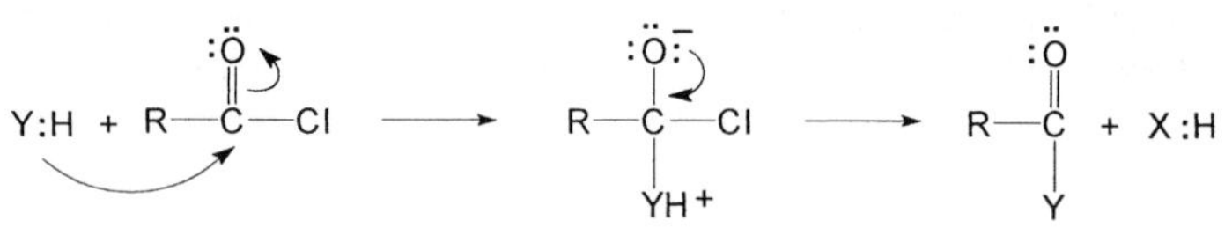

Anhydrides react similarly. The reaction can be catalyzed by acids. For an acid catalysis the reaction takes place as follows:

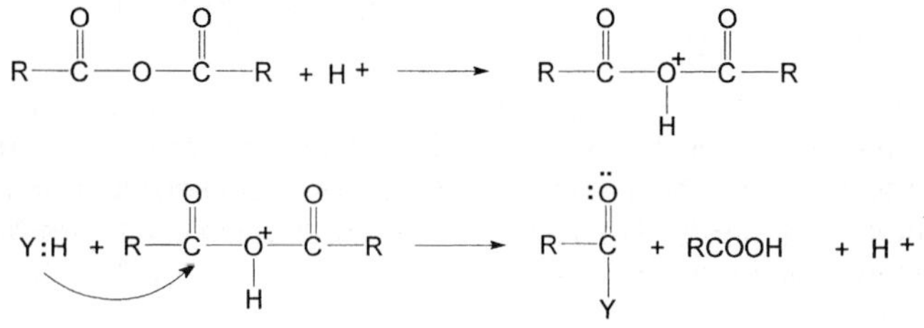

Basic substances such as pyridine frequently are used as a medium (or catalyst) in this reaction. Probably the role of basic compounds such as pyridine in this reaction is to fix the acid resulting from the attack of the nucleophile.

Some other reaction mechanisms such as electrophilic acylations also are known but are much less common.

The mechanism of a reaction with sulfonyl chlorides is similar to S_N2, although the transition state must have five groups on the central atom (sulfur can accommodate 12 electrons in its valence shell). It is likely that the geometry of the transition state is that of a trigonal bipyramid. When the radical in the sulfonyl chloride contains hydrogens ($R-CH_2-SO_2-X$), a sulfene ($R-CH=SO_2$) may act as an intermediate.

- Artifact formation in acylation reactions

Acylation reactions may lead to artifacts due to a variety of reasons. One important factor in artifact formation is related to the acidity of the by-products, such as strong inorganic acids that may be formed during acylation. These acids may lead to undesired or unexpected chemical modification of the analytes.

Another source of problems during acylations consists of unexpected reactions produced by polyfunctional compounds. For example, acylation of amino acids with trifluoroacetic anhydride upon heating of the mixture or the analyte with trifluoroacetic anhydride results in the formation of 1,3-oxazolin-5-one derivatives as shown below:

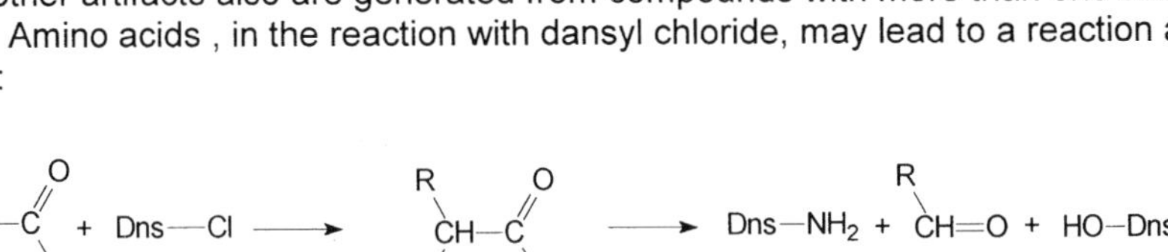

Some other artifacts also are generated from compounds with more than one active group. Amino acids , in the reaction with dansyl chloride, may lead to a reaction as follows:

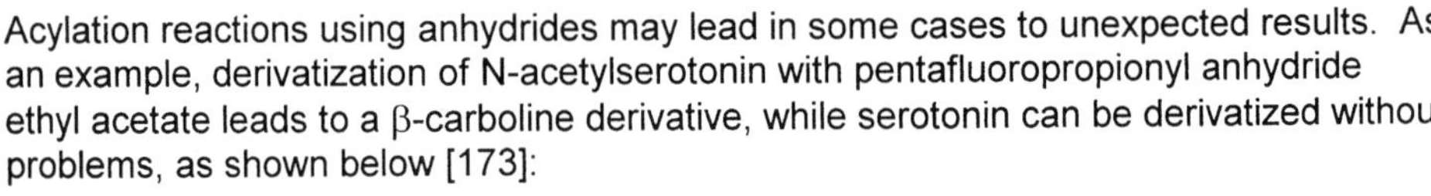

Acylation reactions using anhydrides may lead in some cases to unexpected results. As an example, derivatization of N-acetylserotonin with pentafluoropropionyl anhydride ethyl acetate leads to a β-carboline derivative, while serotonin can be derivatized without problems, as shown below [173]:

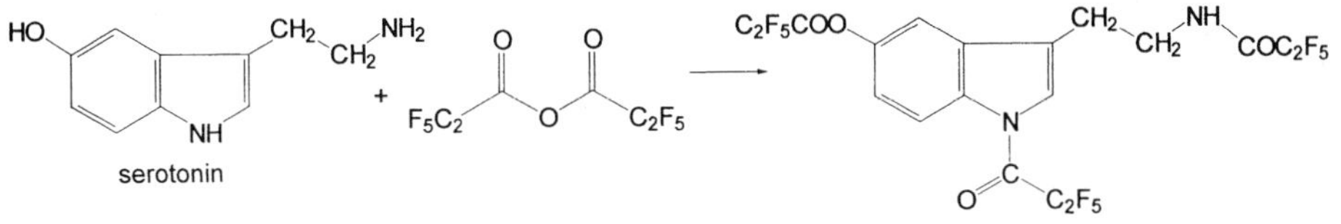

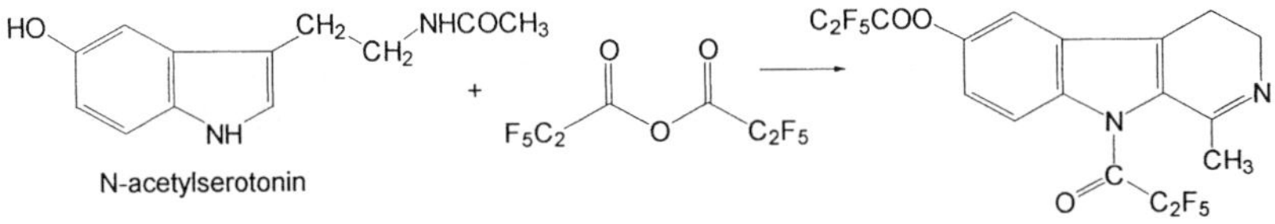

serotonin

N-acetylserotonin

Some other unexpected results of acylation reactions will be indicated in Chapter 19.

18.4 REACTIONS OF ADDITION TO CARBON-HETEROATOM MULTIPLE BONDS USED FOR DERIVATIZATION

A considerable number of derivatization reactions involve addition to hetero multiple bonds in functional groups such as $C=O$, $C=S$, $C=N$ or $C\equiv N$. These reactions are used in two manners. One is the derivatization of analytes with functional groups containing active hydrogens such as OH, NH_2, SH, etc., using reagents containing hetero multiple bonds. The other is the derivatization of analytes with hetero multiple bonds such as aldehydes or ketones using reagents that are able of causing addition reactions. For example, an amine as an analyte may be derivatized with a ketone reagent, or a ketone as an analyte may be derivatized with a primary amine as a reagent.

The reactions can be catalyzed by both acids or bases and, as seen further, the first step in these reactions may be either a nucleophilic attack to the carbon or an electrophilic attack of a proton to the heteroatom. The addition reaction to the multiple bond may be followed by a subsequent elimination reaction. For example, for a carbonyl group the addition reaction may be the final step or may continue with an elimination that generates a new double bond as follows:

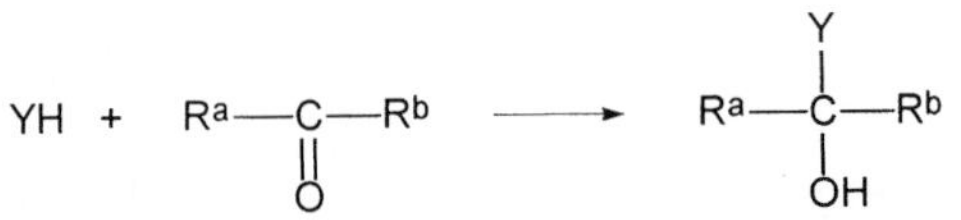

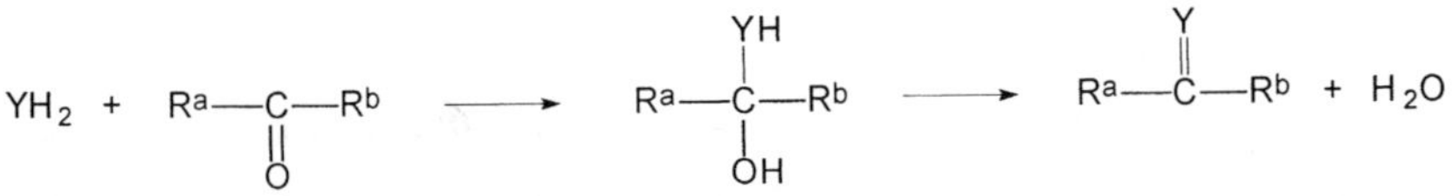

where R^a and R^b is H, R, or Ar in aldehydes and ketones, but also can be OH in acids, OR in esters, NH_2 or NHR in amides, etc.

This section will describe first the reactions for aldehydes and ketones, followed by reactions for isocyanates, isothiocyanates, and for other hetero multiple bonds, regardless whether the compound containing the hetero multiple bond is the reagent or the analyte.

- Reactions at the carbonyl group in aldehydes and ketones

Aldehydes and ketones are able to undergo a number of addition reactions due to the polarity of the C=O group. The reaction can take place even with an active methylene group, as shown below:

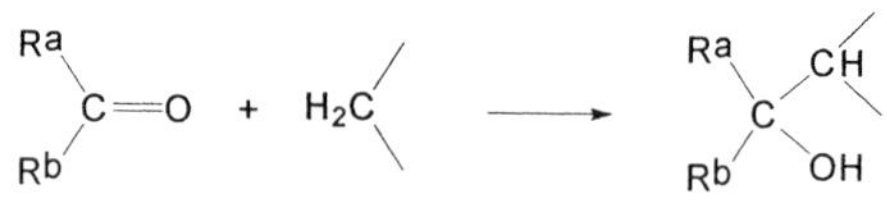

The reaction can be continued as shown previously with water elimination and the formation of a new double bond. An example of a reagent used for aldehyde derivatizations and containing an active methylene is 2-cyanoacetamide. The reaction with this compound takes place as follows:

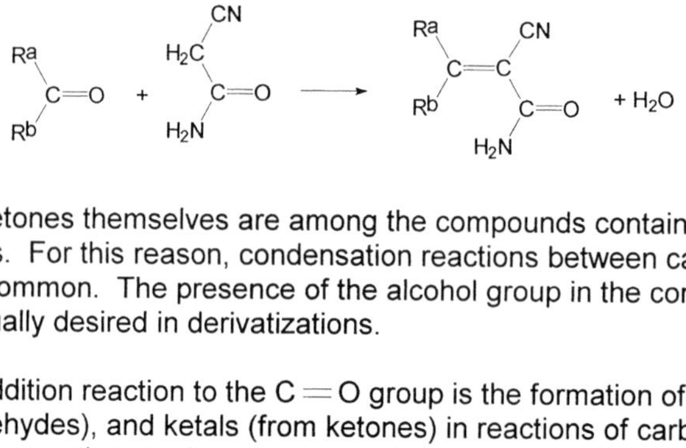

Aldehydes and ketones themselves are among the compounds containing active methylene groups. For this reason, condensation reactions between carbonyl compounds are common. The presence of the alcohol group in the condensation product is not usually desired in derivatizations.

Another typical addition reaction to the C=O group is the formation of hemiacetals, acetals (from aldehydes), and ketals (from ketones) in reactions of carbonyl compounds with alcohols. The reaction can be written as follows:

R^a, R^b = H, CH_3, C_6H_5, etc.

hemiacetal

acetal (ketal)

Most hemiacetals and acetals are not stable compounds, and therefore not suitable for improving stability by derivatization. However, cyclic acetals and ketals may be stable and used for analytical purposes. The reaction with a diol can be written as shown below:

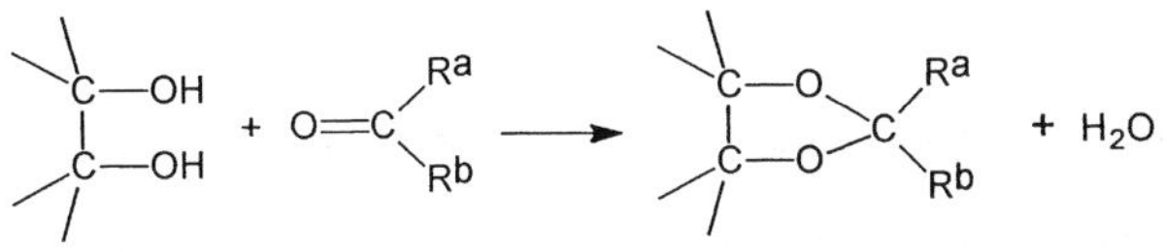

$$\text{>C--OH, >C--OH} + O{=}C<^{Ra}_{Rb} \longrightarrow \text{(cyclic ketal)} + H_2O$$

More details about cyclic acetals and ketals are presented in Section 18.5.

Very common is the reaction of carbonyl compounds with amines. The initial addition reaction usually continues with water elimination forming a substituted imine or a Schiff base:

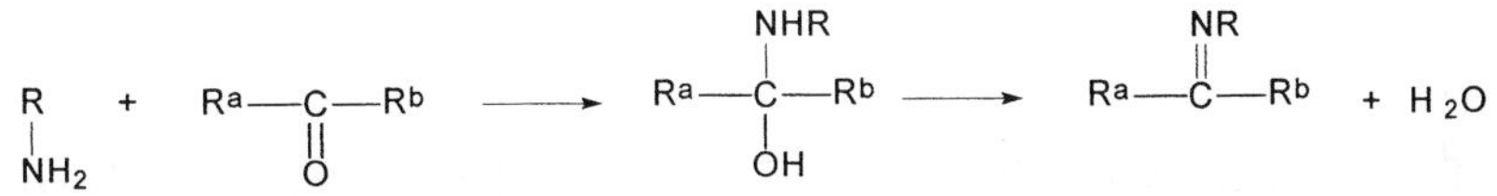

$$R{-}NH_2 + Ra{-}\underset{O}{\overset{}{C}}{-}Rb \longrightarrow Ra{-}\underset{OH}{\overset{NHR}{C}}{-}Rb \longrightarrow Ra{-}\overset{NR}{\underset{}{C}}{-}Rb + H_2O$$

This reaction has been used extensively for primary amine derivatization. For example, alkylamines can be derivatized with 5-(4-pyridyl)-2-thyophenecarbaldehyde [174], the resulting compound being strongly fluorescent and allowing HPLC analysis:

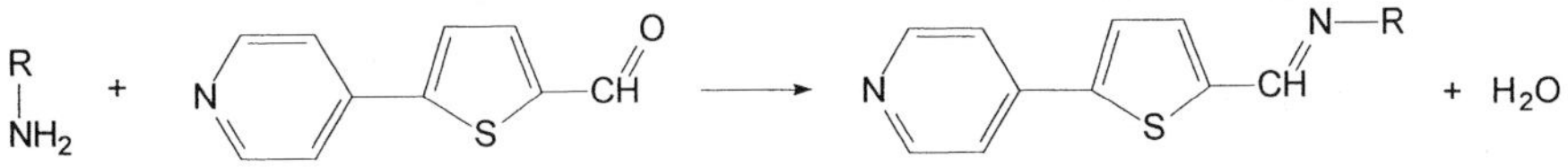

$$R{-}NH_2 + \text{(pyridyl-thiophene-CHO)} \longrightarrow \text{(pyridyl-thiophene-CH=N-R)} + H_2O$$

Although secondary amines react with carbonyl compounds [23], they form unstable hemiaminals that are not used for the analytical derivatization of amines.

The utilization of amines as reagents for the derivatization of carbonyl compounds is also possible. For example, ketones can be derivatized with methylbenzylamine, generating a compound with strong absorbance in UV [175].

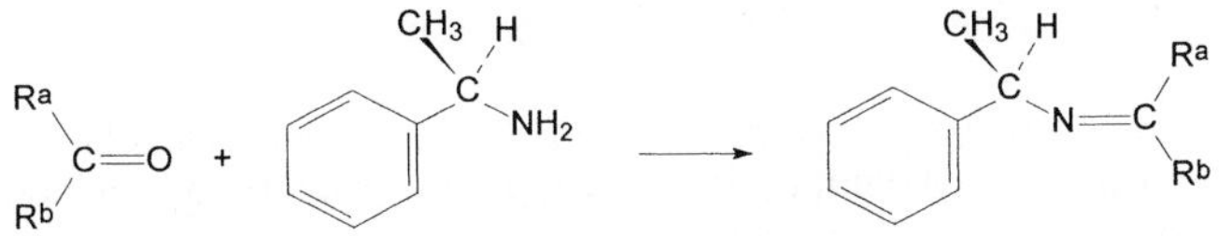

$$\underset{Rb}{\overset{Ra}{C}}{=}O + \text{(PhCH(CH}_3\text{)NH}_2\text{)} \longrightarrow \text{(PhCH(CH}_3\text{)N=C}^{Ra}_{Rb}\text{)}$$

The reaction also is used with a second reaction step consisting of the reduction of the substituted imine with sodium cyanoborohydride [176], [177]:

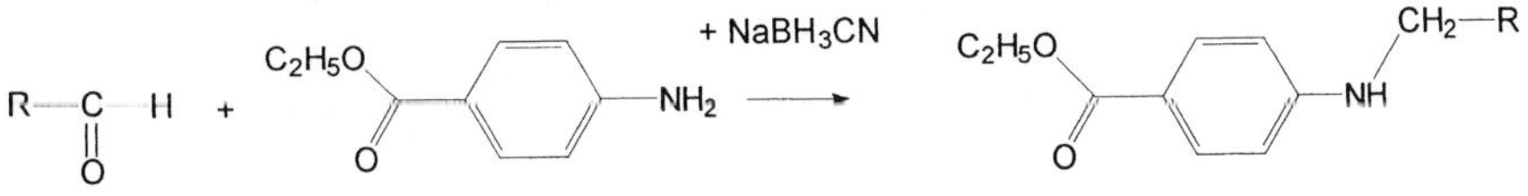

$$R{-}\underset{O}{\overset{}{C}}{-}H + \text{(C}_2\text{H}_5\text{O-C(=O)-C}_6\text{H}_4\text{-NH}_2\text{)} \xrightarrow{+ NaBH_3CN} \text{(C}_2\text{H}_5\text{O-C(=O)-C}_6\text{H}_4\text{-NH-CH}_2\text{-R)}$$

Using this type of reaction, a variety of carbonyl compounds can be analyzed. For example, oligosaccharides derivatized with 2-aminoacridone have been analyzed by HPLC on a C18 column followed by the collection of the derivatized compounds and the analysis using MALDI mass spectrometry [178]. The same derivatization has been used followed by a CE separation. The reactions taking place in this analysis can be written as follows:

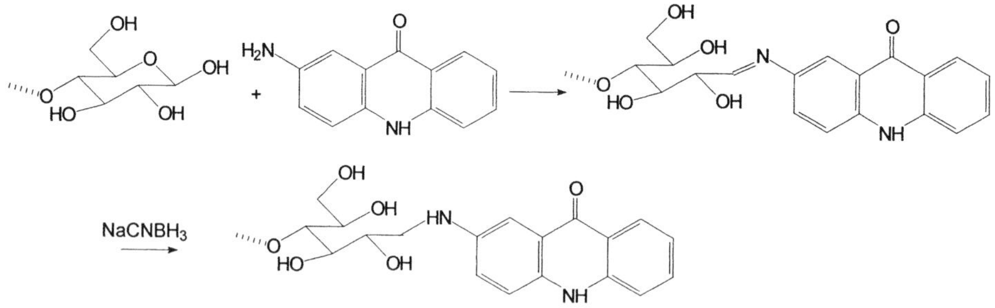

The reaction of amines with carbonyl compounds followed by hydrogenation can be used for amine methylation [23], [179]. The carbonyl compound necessary in this case is formaldehyde, and various hydrogenation reagents are utilized such as $NaBH_4$, $NaBH_3CN$, BH_3-pyridine, formic acid, etc. The reaction can be written as follows:

$$R-NH_2 + HCHO \xrightarrow{NaBH_4} R-\underset{\underset{CH_3}{|}}{N}H + HCHO \xrightarrow{NaBH_4} R-\underset{\underset{CH_3}{|}}{N}-CH_3$$

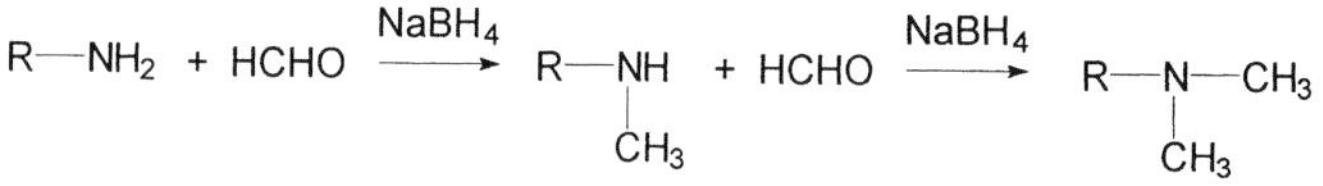

Carbonyl compounds as analytes reacting with NH_3 followed by reduction with $NaBH_3CN$ form primary amines in a reaction of the type:

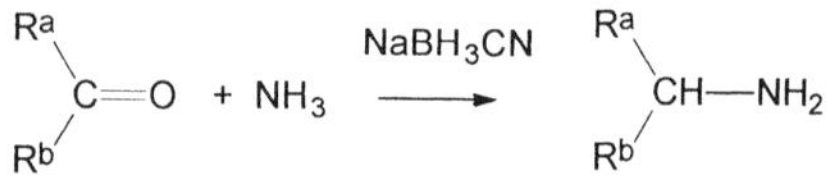

$$\underset{R^b}{\overset{R^a}{>}}C=O + NH_3 \xrightarrow{NaBH_3CN} \underset{R^b}{\overset{R^a}{>}}CH-NH_2$$

The primary amine generated in this way can be further derivatized such that very low detection limits can be achieved using a fluorescence type reagent [180].

The reactivity of the carbonyl group with amines also is used in synthesizing new cycles, usually involving a dicarbonyl compound as a reagent. This type of reaction is described in more detail in Section 18.5.

Hydrazine derivatives are another group of compounds that react with carbonyl-containing molecules forming hydrazones as follows:

$$\underset{R^b}{\overset{R^a}{>}}C=O + H_2N-NH-R \longrightarrow \underset{R^b}{\overset{R^a}{>}}C=N-NH-R + H_2O$$

The reaction is frequently used for attaching chromophore groups to aldehydes or ketones for UV detection in HPLC, for example using as a reagent 2,4-dinitrophenyl-hydrazine [181]. Hydrazine itself may react at both NH_2 groups with certain carbonyl compounds forming azines.

α-Hydroxy aldehydes and ketones in a reaction with arylhydrazines give osazones as shown below:

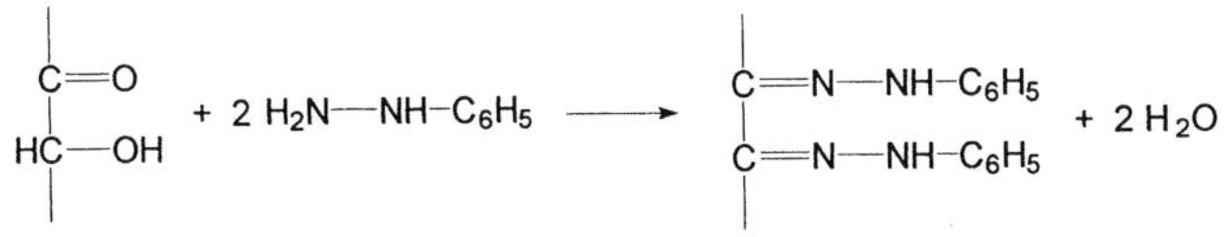

This type of reaction is particularly important in the derivatization of carbohydrates. β-Diketones react differently, generating pyrazole derivatives (see Section 18.5):

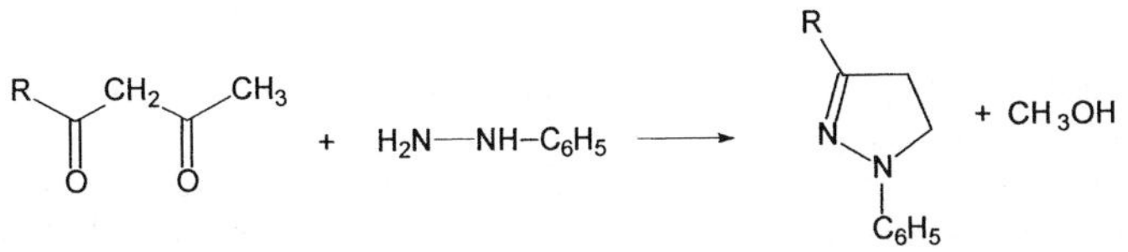

Several other classes of compounds similar to hydrazines react at the carbonyl group. Among these are hydrazones (NH_2—N=CR_2), hydrazides (NH_2NHCOR), and semicarbazide itself ($NH_2NHCONH_2$). For example, N-aminopiperidine can be used for ketone derivatization in the presence of catalytic amounts of acetic acid [4]. The resulting substituted hydrazone can be used in GC analysis.

Various hydrazides also are used as reagents for carbonyl compounds, mainly with the purpose of attaching chromophore groups for UV or fluorescence detection in HPLC analysis:

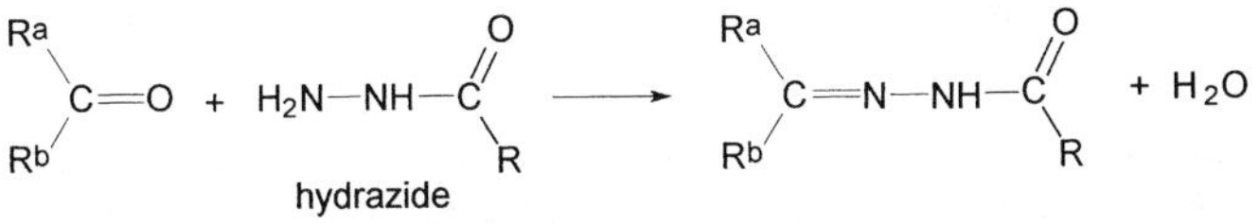

A reagent commonly used for HPLC derivatizations leading to a highly fluorescent compound is 5-dimethylamino-1-naphthalenesulfonic hydrazide (dansyl hydrazine):

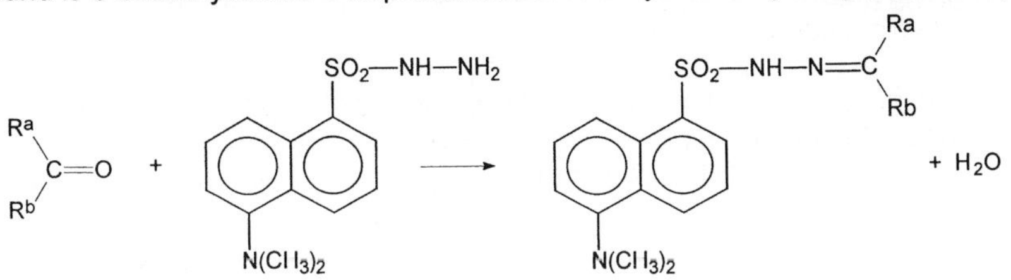

The reaction takes place by heating the aldehyde or ketone for 15 min. at 70° C in methanol containing a twofold molar excess of reagent and a drop of acetic acid. The reaction mixture is then evaporated to dryness, dissolved in benzene, and analyzed [182].

Hydroxylamines also react with carbonyl compounds forming oximes. Hydroxylamine itself, hydroxylamine hydrochloride (STOX® reagent [38]), or derivatives such as H_2NOSO_3H in a solvent like pyridine can be used in the reaction:

Substituted hydroxylamines such as methoxyamine hydrochloride $NH_2OCH_3 \cdot HCl$ (MOX® reagent [38]) and O-(pentafluorobenzyl)-hydroxylamine hydrochloride (FLOROX® reagent [38]) also are used for derivatization purposes as shown below for FLOROX:

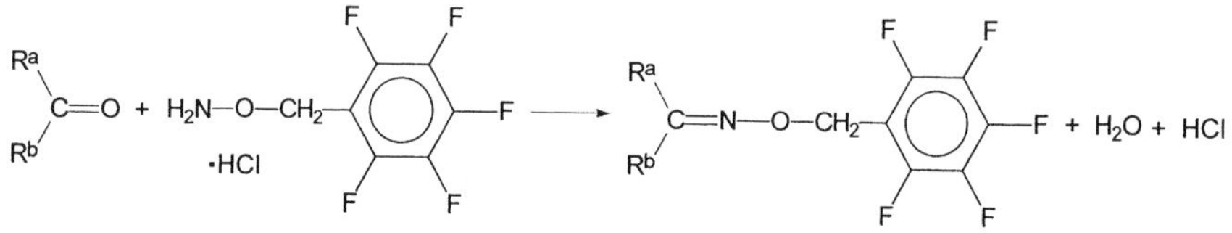

The resulting oximes exist in *sin-* and *anti-* forms, and the two may produce double peaks in chromatographic separations.

When the reaction is performed with hydroxylamine, the generated oxime contains an active hydrogen. This can be further derivatized for example by silylation in a reaction with a common silylation reagent:

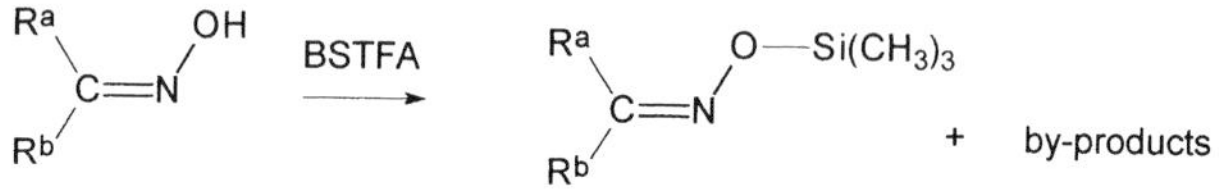

Artifact formation may take place in some reactions of hydroxylamine. For example, in the reaction of a 3-keto acid with hydroxylamine, cyclic compounds may be generated in a reaction of the type:

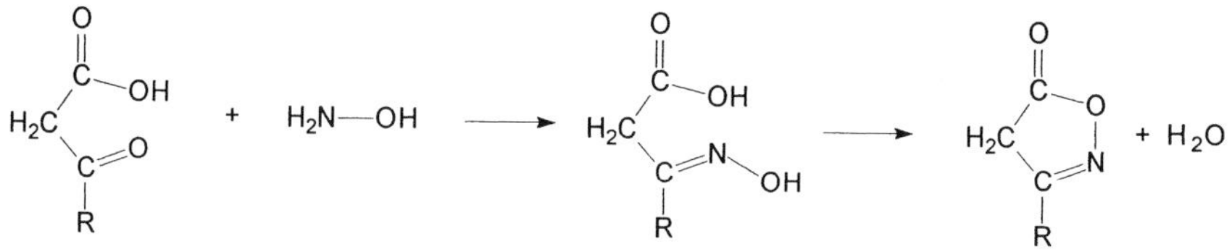

The formation of such compounds possibly complicates even further the chromatogram of oximes that can generate double peaks because of the presence of both *sin*- and *anti*- forms.

Oximes may be converted into nitriles when treated with acetic anhydride in the presence of CH_3COONa. This reaction was used in the derivatization of carbohydrates when a simultaneous acetylation takes place (Wohl degradation [183]). The reaction can be written as follows:

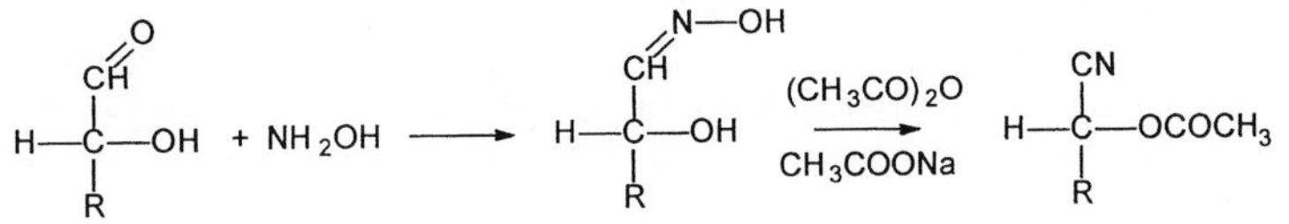

The transformation of the oximes into nitriles generates one single compound from the two (*syn* and *anti*) isomers and can be used to simplify the chromatograms of sugars derivatized as oximes [184], [185].

Hydrocyanic acid can react with carbonyl compounds forming cyanohydrins. Their hydrolysis may lead to hydroxy acids [186], and the reaction takes place as follows:

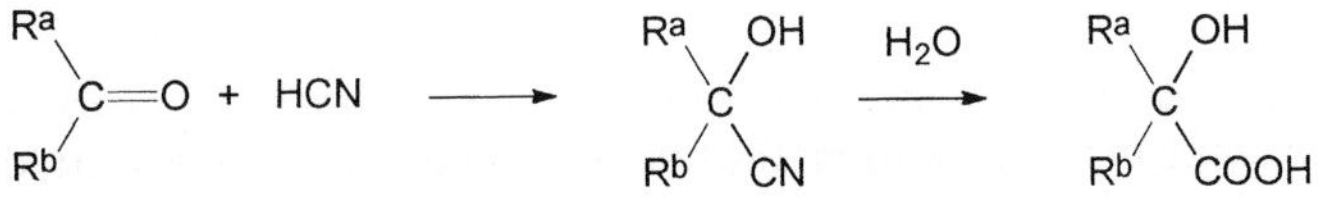

Further derivatization of the hydroxy acids can be used for improving the chromatographic analysis.

- *Reactions at N=C group in isocyanates and isothiocyanates*

Alcohols, amines, and thiols can react at other hetero multiple bonds such as that in isocyanates ($N=C=O$). Numerous applications are known for this reaction with the purpose of attaching chromophores or fluorescent groups to alcohols and amines for HPLC determination. One such example is the reaction of alcohols or amines with 3,5-dinitrophenyl isocyanate [187], [188]:

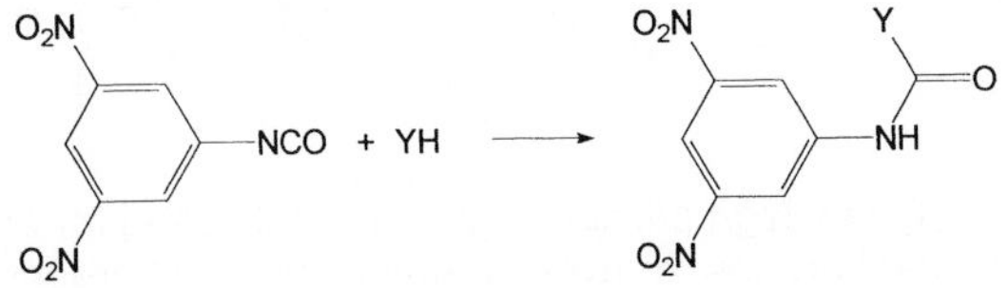

Because the thermal stability of the resulting compound is not necessarily high, the reaction is not commonly used in GC applications but can be used for attaching chromophore or fluorescent groups to compounds with active hydrogens such as alcohols [189]. The presence of an asymmetric carbon also can be used for the derivatization with the purpose of separation of enantiomers (see Section 17.7):

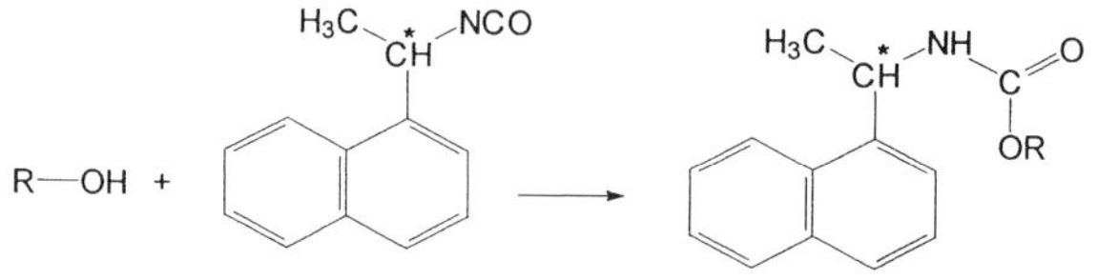

The reaction can be applied using a carbonyl azide that further generates the isocyanate. As an example, N-(p-toluenesulfonyl)pyrrolidinyl carbonyl azide [190] is used for the determination of phenylpropanol. In this example, the isocyanate is generated in situ by refluxing the reagent and the analyte in toluene for 10 min.

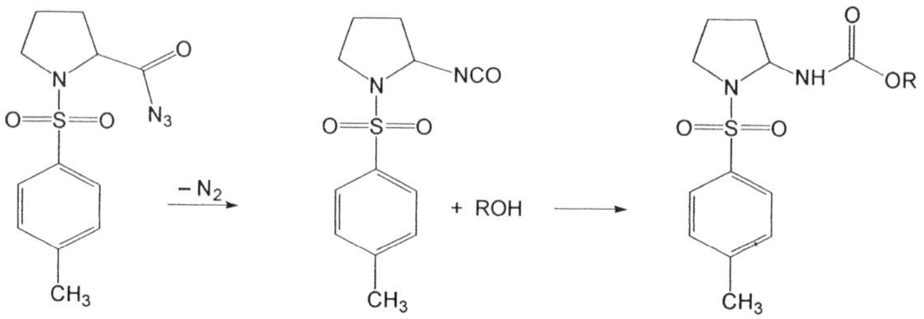

Both primary and secondary amines can react with isocyanates. For example, the reaction of a secondary amine with methylbenzyl isocyanate can be used for attaching a chromophore to an amine and, at the same time, for using a specific stereoisomer of the reagent and achieving an enantiospecific separation [191].

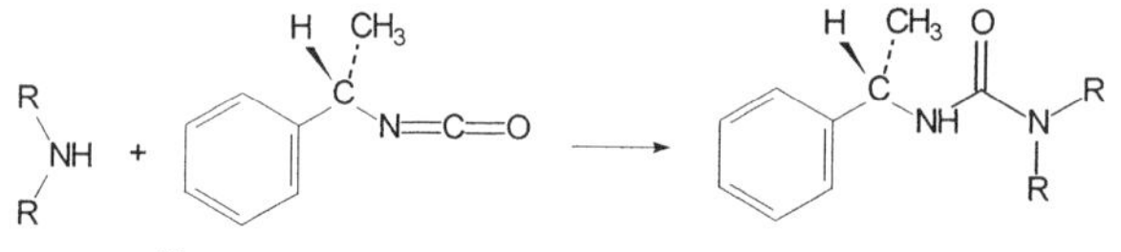

(S) α-methylbenzyl isocyanate

Isothiocyanates react similarly to isocyanates. For example, a secondary amine reacts with benzyl isothiocyanate as follows [192]:

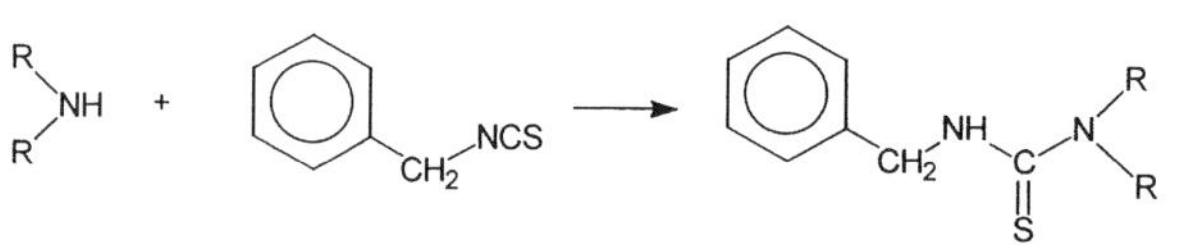

The reaction takes place with high efficiency and is widely used for attaching chromophores, fluorescent groups, or even chemiluminescent groups to amines. Phenyl isothiocyanate reacts similarly. With bifunctional analytes such as amino acids, the reaction of phenyl isothiocyanate is continued with the formation of 5-substituted-3-phenyl-2-thiohydantoins (see also Section 18.5). Thiols react with isocyanates similarly to alcohols forming the dithio-esters. As an example, a thiol reacts with(R)-(-)-4-isothyocyanato-pyrrolidin -1-yl)-7-nitro-2,1,3-benzoxadiazole as follows:

The resulting derivatized thiol can be measured by HPLC with fluorescence detection [24].

- **Other reactions involving addition to a hetero multiple bond**

Various other additions to hetero multiple bonds are known, some of them with important analytical applications. The addition may occur at the C $=$ O group in an amide, at a nitrile, at CS_2 etc. One example is the addition under special conditions, of alcohols to dimethylformamide. The resulting acetals are very reactive and are used themselves as reagents, as shown previously for N,N-dimethylformamide dimethyl acetal (see Section 18.1). Another example is the reaction of CS_2 with alcohols in the presence of a base, leading to the formation of xanthates. Amines also react with CS_2 as follows:

The isothiocyanate can be analyzed using GC [4]. Some of these reactions will be mentioned further in Chapter 19.

Addition to a hetero multiple bond also occurs in the reaction of cyanates with bifunctional compounds such as 2-aminobenzoic acid, forming a new cycle as follows:

The reaction is used for cyanate analysis in blood [24] (see also Section 18.5).

- **Some aspects regarding the reaction mechanism for addition to a hetero multiple bond**

The C=O bond in aldehydes and ketones is rather polar due to the displacement of its π electron system toward the oxygen. The charge distributions for acetaldehyde, acetone, benzaldehyde, and acetophenone are given below as examples showing the negative partial charge on the heteroatom (oxygen) and the positive one on the carbon involved in the double bond:

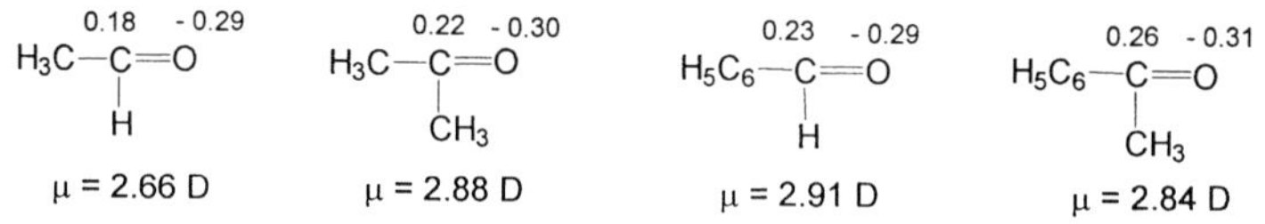

0.18 - 0.29	0.22 - 0.30	0.23 - 0.29	0.26 - 0.31
$H_3C-C=O$	$H_3C-C=O$	$H_5C_6-C=O$	$H_5C_6-C=O$
H	CH_3	H	CH_3
$\mu = 2.66$ D	$\mu = 2.88$ D	$\mu = 2.91$ D	$\mu = 2.84$ D

The values indicated for the charges were calculated using a MOPAC molecular orbital package [121] and may be slightly different from the actual values. As seen from the values of formal partial charges, the C=O bond can be considered about 20% ionic. The addition reaction may start either as a nucleophilic attack to the carbon or an electrophilic attack of a proton to the heteroatom. Seen as a nucleophile attack the process can be written as follows:

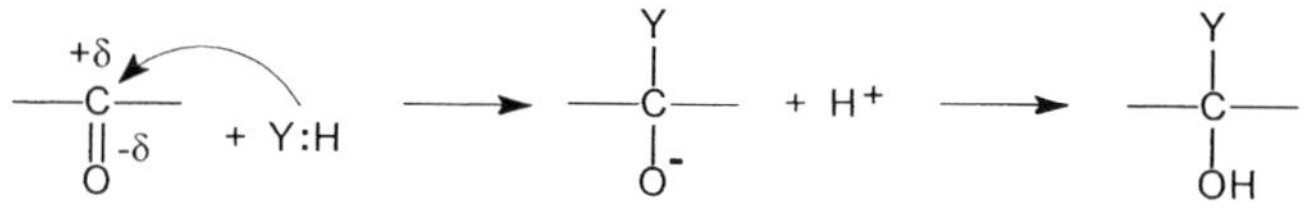

In this reaction, the reformation of the carbonyl group with elimination is not a likely process mainly when the substituents at the carbonyl group are hydrogen, alkyl, or aryl groups. The electrophilic attack of a proton can be written as follows:

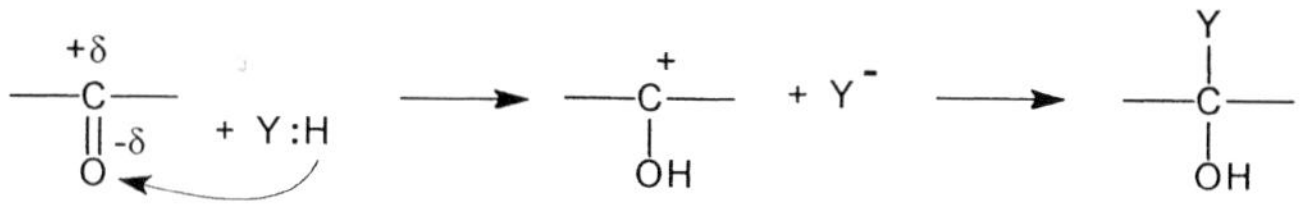

In both cases, the nucleophile step is usually the rate-determining one. The reactions can be catalyzed by both acids and bases [23], in the presence of an acid the mechanism being probably electrophilic.

The reactivity of the carbonyl group is affected by additional factors. For example, aldehydes are more reactive than ketones, and the aromatic carbonyl compounds are less reactive compared to alkyl compounds. The aryl groups stabilize the carbonyl by resonance as shown below:

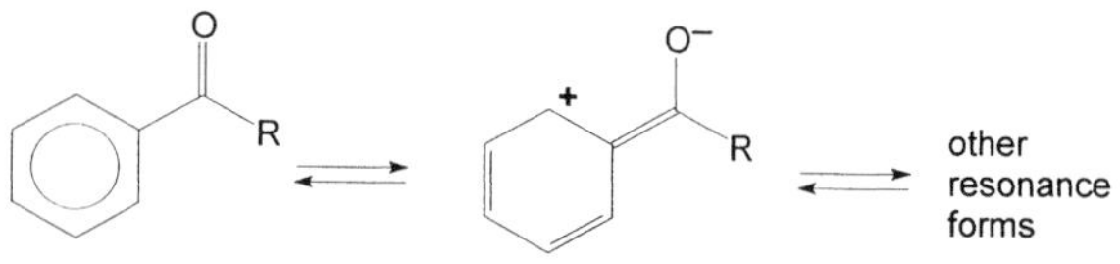

In general, the substituents at the carbon involved in the hetero multiple bond also influence the reactivity. Electron donating groups decrease the reaction rate, while electron attracting ones increase rates. Also, steric factors may play a role in reactivity, and hindered compounds may not be reactive.

The addition to the isocyanates or isothiocyanates takes place similarly to that on the carbonyl group. In many instances, these compounds are even more reactive than the carbonyl compounds due to a higher polarity of the N=C bond.

18.5 DERIVATIZATION REACTIONS WITH FORMATION OF CYCLIC COMPOUNDS

Formation of new cycles from noncyclic compounds or replacement of old cycles with new ones that are more stable or have a desired property is also exploited in sample processing using derivatization. This type of derivatization includes a variety of reactions, which can be classified in the following groups:

- formation of nonaromatic cycles containing oxygen atoms,
- formation of aromatic cycles containing one nitrogen atom,
- formation of azoles and related compounds,
- formation of azines and related compounds,
- formation of cyclic siliconides,
- formation of cyclic phosphothioates, and
- formation of cyclic boronates.

Most reactions with the formation of cyclic compounds involve bifunctional compounds, which can be the analyte, the reagent, or both. Several aspects of cyclic formation reactions are discussed in this section. Reviews on the subject are found in [1], [193].

- Formation of nonaromatic cycles containing oxygen atoms

A three-atom cycle containing an oxygen atom is present in epoxides (IUPAC name oxiranes). Epoxides can be formed in the reaction of a compound with a carbon-carbon double bond and a peroxyacid. Among the peroxyacids more frequently used are peracetic, performic, perbenzoic, trifluoroperacetic, and 3,5-dinitroperoxybenzoic acids. The reaction is not affected, besides the double bond, by the presence of functional groups such as hydroxyl, acid, or ester. Amines may be oxidized and ketones may suffer an oxygen insertion, forming acids [23] in this reaction. The addition maintains the initial stereospecific structure in the sense that a *trans* olefin gives a *trans* epoxide and a *cis* olefin a *cis* epoxide. However, a mixture of enantiomers is formed, as shown below for a cis olefin:

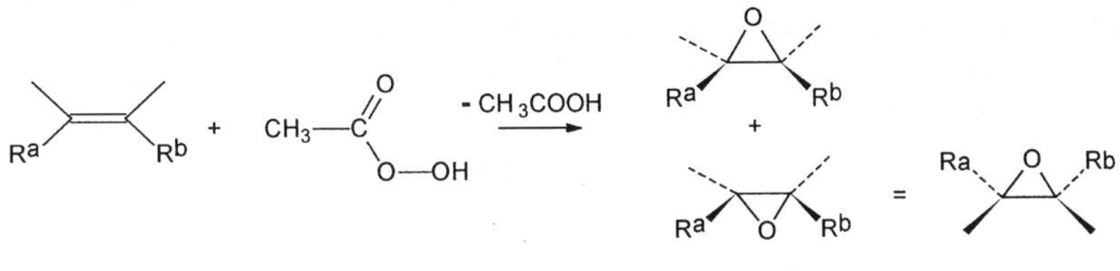

(enantiomers)

A *trans* olefin similarly forms a mixture of enantiomers, maintaining the *trans* position of substituents. For compounds with an OH group in the allylic position, the stereo-specificity of epoxide formation diminishes or disappears, with both *cis* and *trans* isomers giving a product with the incoming oxygen *syn* to the OH group.

A typical derivatization procedure consists of keeping at room temperature for 3 to 6 h the mixture of the analyte containing the carbon-carbon double bond and an excess up to 50 times in weight of reagent. The reaction takes place faster for *cis* olefins [194] than for *trans* olefins.

As the enantiomers are not separated in common chromatographic procedures, the two enantiomers will generate a single chromatographic peak, while the *cis* and *trans* epoxides, being diastereoisomers, are commonly separated. The separation of the epoxides may be easier to achieve as compared to that of olefins, and this type of derivatization has been utilized, for example, for better separation of various *cis* and *trans* fatty esters [194].

Formation of epoxides may lead to certain artifacts. For example, during the derivatization of methyl octadecenoate with peracetic acid, about 1% of analyte is transformed for every hour of reaction into hydroxy-acetoxy methyl octadecanoate. Also, the steric character of the reaction may lead to certain problems. For dienes, which can be derivatized by the same process, two pairs of enantiomers are formed, such that the compounds from one pair are diastereoisomers to the compounds from the other pair. This leads to double peaks in chromatograms. The two pairs of enantiomers are shown below:

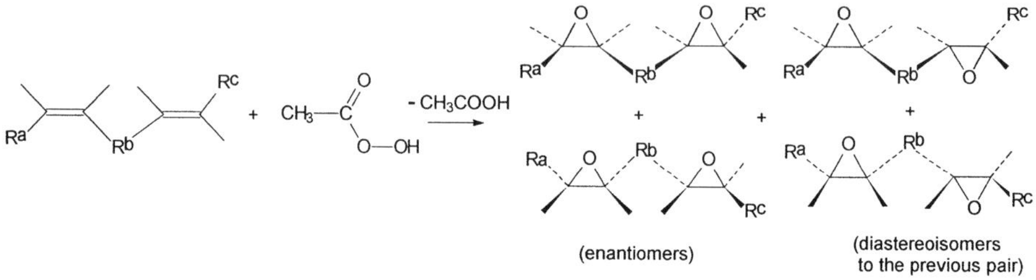

(enantiomers)   (diastereoisomers to the previous pair)

Epoxides can be hydrolyzed in the presence of acids and generate *trans* diols in reactions as follows:

The diols formed from epoxides may be further derivatized for analysis (see Section 19.1). Also, epoxides may isomerize to a ketone as shown below:

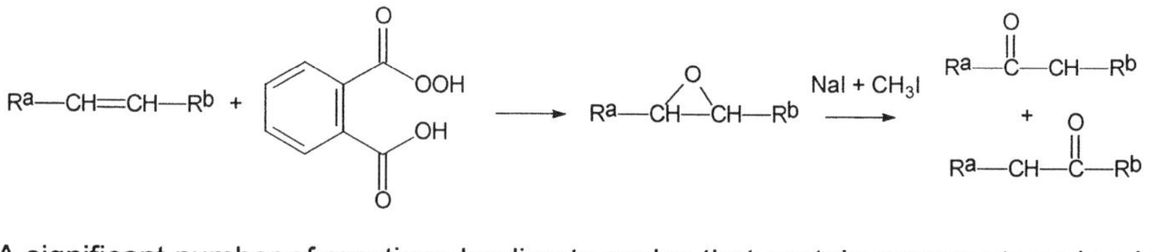

A significant number of reactions leading to cycles that contain oxygen atoms involve the formation of cyclic acetals or ketals. Acetal and ketal formation has been used extensively in synthetic chemistry and also has applications in analytical chemistry. The cyclic acetals are commonly formed in reactions between diols, triols, or polyhydroxy compounds with the diol group on proximal carbons and an aldehyde or ketone. The reaction is commonly catalyzed by acids (strong Brønsted or Lewis acids) and may have one of the following mechanisms [183]:

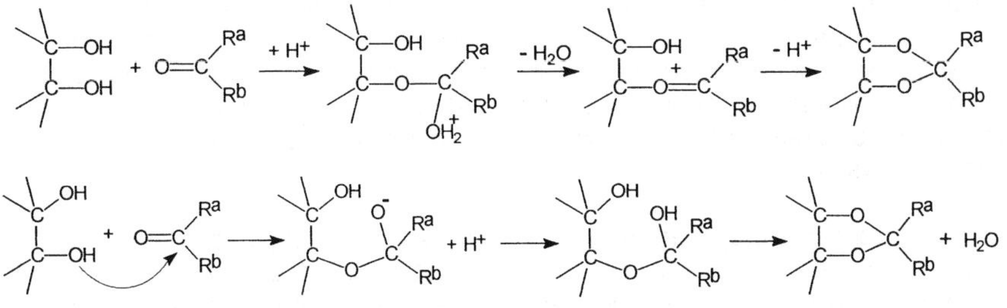

The reaction of the compounds containing more than two OH groups may be more complex. For example, in reactions with aldehydes or ketones, carbohydrates and polyols may form rings of five- (1,3-dioxolane), six- (1,3-dioxane), or seven- (1,3-dioxepan) member cycles. The thermodynamic stability sequence of acetal rings is 6>5>7. The cyclic ketals formed with acetone are particularly common in carbohydrate chemistry, and they are also named acetonides or O-isopropylidene derivatives. For polyols, not only one pair of OH groups can form a cyclic acetal. Two or even three pairs may be involved in the acetal formation, as shown, for example, for the reaction of glucitol with benzaldehyde:

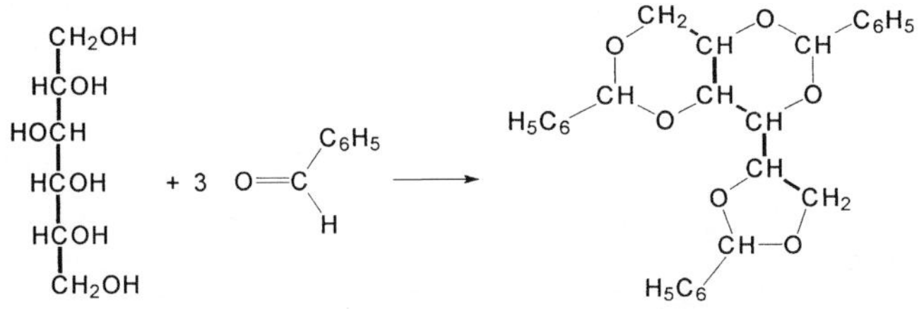

$$CH_2OH$$
$$HCOH$$
$$HOCH$$
$$HCOH \quad + \ 3 \ O{=}C\binom{C_6H_5}{H} \longrightarrow$$
$$HCOH$$
$$CH_2OH$$

When the diol molecule and the carbonyl compound are not symmetrical, the newly formed tetrahedral carbon (originally present in the carbonyl) is chiral. Because the thermodynamic stability of the two forms is comparable, near equimolar mixtures are usually formed. This is shown below for a 1,3-dioxolane ring and a 1,3-dioxane ring:

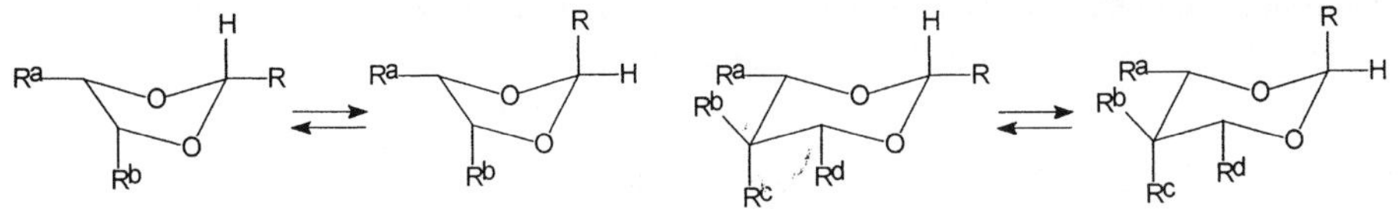

However, for six-membered rings and larger substituents at the aldehyde, the most stable ring has equatorial substituents and higher symmetry [183].

A number of carbonyl compounds are used to form cyclic acetals and ketals [1]. Depending on the nature of the polyol and of the carbonyl compound, different cyclic acetals or ketals are formed [195]. For example, galactose reacts with benzaldehyde forming mainly 4,6-benzylidene acetal with only about 10% 1,2:3,4-diacetal, while acetone forms mainly 1,2:3,4-di-O-isopropylidene galactose. These different acetals or ketals are separable, for example by GC, HPLC, or TLC, and identifiable using different techniques including MS. The acetonides of carbohydrates are used for their identification using TLC separations [193]. Also, certain characteristics in the EI+ mass spectra of the acetonides are important for their identification. These characteristics include a M-15 fragment ion, useful for the molecular mass determination. Also, differentiation between furanose and pyranose isomers can be done in some instances based on acetonide mass spectra, some furanose acetonides showing a characteristic ion with m/z = 101 [193], following a fragmentation of the type:

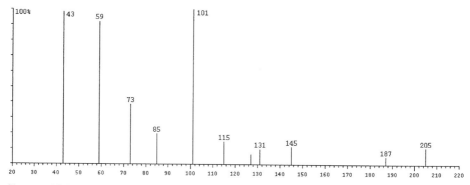

As an example, the EI+ mass spectrum of 5,6-isopropylidene-D-glucofuranose (M = 220) is given in Figure 18.5.1:

FIGURE 18.5.1. *EI+ mass spectrum of 5,6-isopropylidene-D-glucofuranose.*

The chemistry of cyclic acetals of sugars is complex, and the results not always easy to interpret. Glucose, for example, can generate a 1,2:3,5-diisopropylidene-D-glucofuranose, which does not show the ion with m/z=101. The structure of the molecule and the spectrum are shown in Figure 18.5.2.

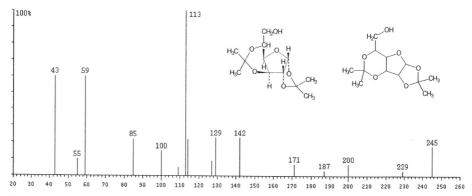

FIGURE 18.5.2. *EI+ mass spectrum of 1,2:3,5-diisopropylidene-D-glucofuranose.*

Another example of complex acetonides is that of D-fructose, which generates with acetone 1,2:4,5- and 2,3:4,5-diisopropylidene-D-fructopyranose (M = 260) and not diisopropylidene furanoses, as shown below:

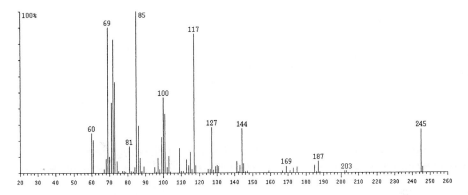

1,2:4,5-di-O-isopropylidene-
β-D-frctopyranose

2,3:4,5-di-O-isopropylidene-
β-D-frctopyranose

The mass spectra of the two diisopropylidene-D-fructopyranoses are shown in Figures 18.5.3 and 18.5.4.

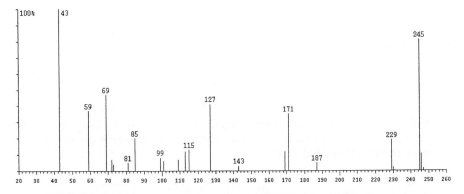

FIGURE 18.5.3. *EI+ mass spectrum of 1,2:4,5-diisopropylidene-D-fructopyranose.*

FIGURE 18.5.4. *EI+ mass spectrum of 2,3:4,5-diisopropylidene-D-fructopyranose.*

As seen in Figures 18.5.1 to 18.5.4, the acetonides show the M-15 ion. The ion with m/z = 101 is present in the 1,2:4,5 diacetonide but probably from a different fragmentation.

An interesting application of acetonide formation is related to the identification of the number and location of double bonds in alkenes or alkene chains. For this purpose, the double bond is first oxidized to a diol, for example, in the presence of OsO_4, which gives a *syn* addition from the less hindered side of the double bond. The diol is then changed into an acetonide. As indicated for carbohydrates, the molecular ion is usually absent, and the ion M-15 can be used for molecular mass determination. From the number of losses of acetone (m/z = 58) from the M-15 ion, it is possible to establish the number of double bonds. In simpler cases (one double bond), from the ions in the mass spectrum generated by α-cleavage, it is possible to locate the double bond [193] as shown below for a double bond between R^a and R^b fragments that was transformed to a diol and then to an acetonide:

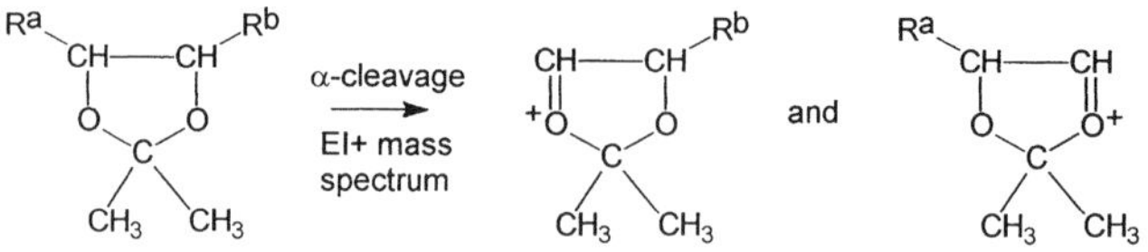

The acetals or ketals can be formed not only from diols but also from epoxides. The reaction usually takes place in the presence of a strong Lewis acid such as BF_3 as shown below:

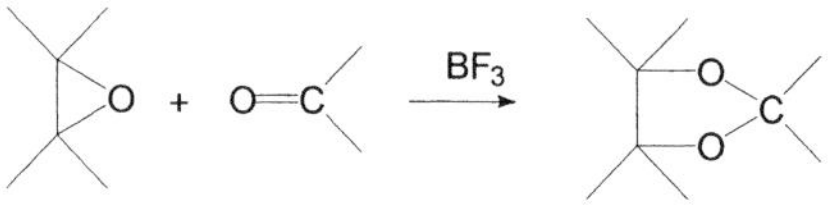

The reaction has been used for the analysis of mono and diepoxyglycerides [196]. Also, certain aminoalcohols may form acetals/aminals, as shown below for ephedrine [197]:

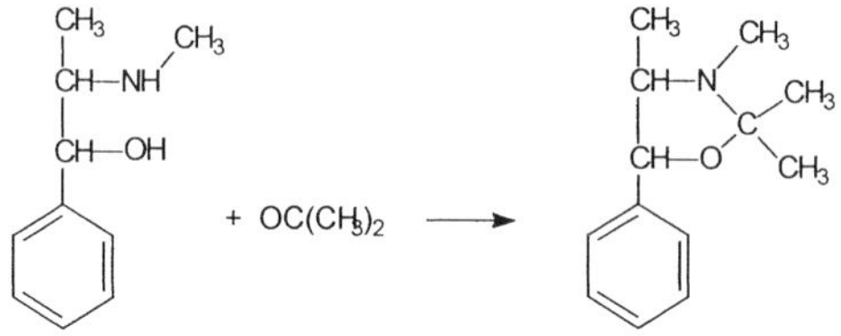

Activated carbonyl groups such as those in hexafluoroacetone or dichlorotetrafluoro acetone may react, besides diols, with other difunctional compounds. For example, α-substituted acids may react with hexafluoroacetone and form a variety of stable nonaromatic cycles, with structures depending on the α substituent. The reactions with α-hydroxy acids, α-mercapto acids, and acids containing an aldehyde group in α-position are shown below [193], [198]:

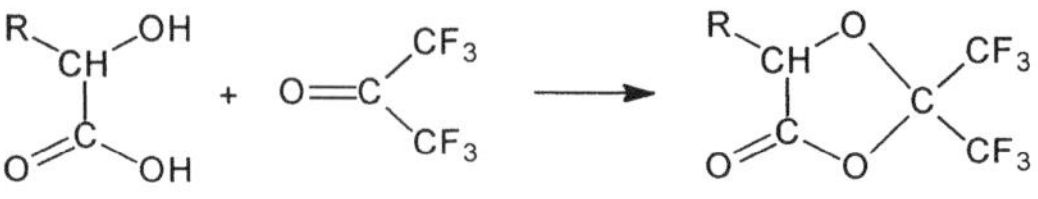

4-substituted-
2,2-bis(trifluoromethyl)-
1,3-dioxolan-5-one

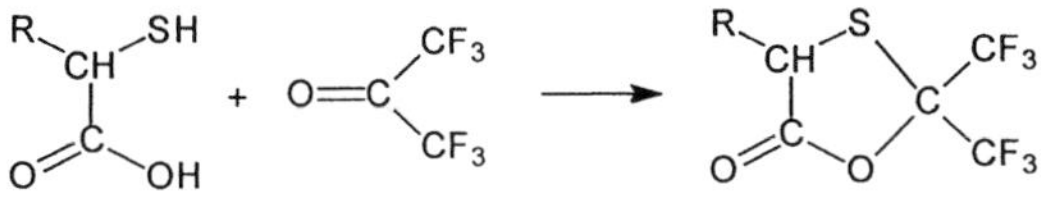

4-substituted-
2,2-bis(trifluoromethyl)-
1,3-oxathiolan-5-one

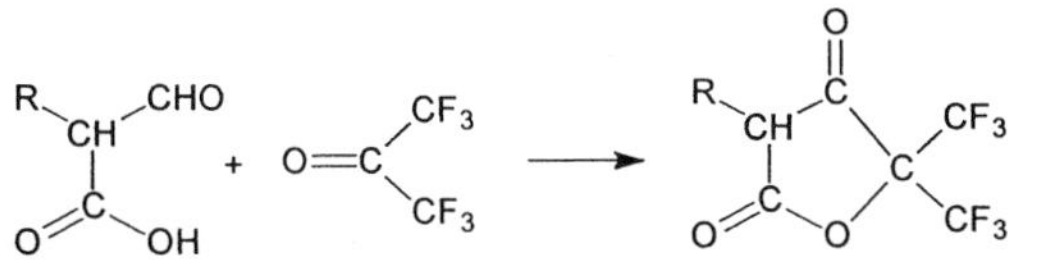

3-substituted-
5,5-bis(trifluoromethyl)-
3,5-dihydrofuran-2,4-dione

Various other analytical applications are reported regarding the formation of cyclic compounds with activated carbonyl groups [193], [199]. Although most carbonyl compounds with simple alkyl groups are not reactive enough to undergo such reactions [200], cyclic isopropylidene derivatives of hydroxy acids can be obtained from 2,2-dimethoxypropane (acetone dimethyl acetal) in methanol and the hydroxy acid [201]. The reaction takes place at room temperature in 24 h, using HCl as catalyst, and the cyclic acetal can be used for GC analysis.

An activated carbonyl can be considered the one reacting in phosgene or in a chloroformate ester. The reaction of phosgene, for example, with a carbohydrate takes place in a reaction with the formation of a cyclic carbonate, as follows:

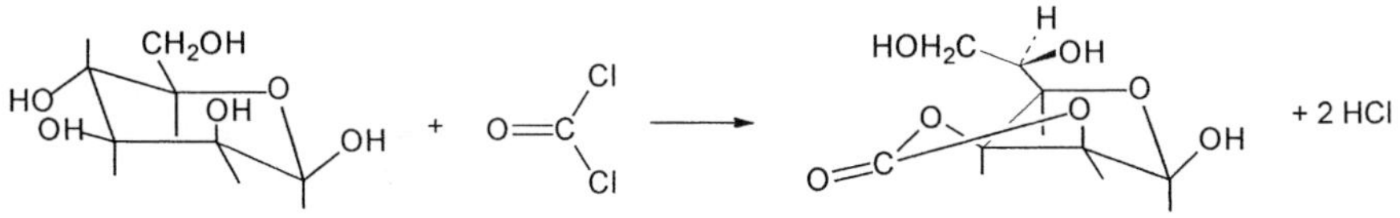

β-D-Mannopyranose   β-D-Mannofuranose-2,3-carbonate

As seen in this example, the pyranose form of the sugar may change into a furanose. The cyclic carbonates obtained from sugars are more stable to acid hydrolysis than the cyclic acetals, although they hydrolyze in basic conditions. These compounds can be used for GC analysis of carbohydrates [202].

- Formation of aromatic cycles containing one nitrogen atom

A variety of aromatic cycles can be formed in reactions involving bifunctional compounds. Addition reactions to hetero multiple bonds in bifunctional molecules frequently lead to cyclic compounds. For example, the reaction of α-dicarbonyl compounds with amines can be used to generate new heterocycles, some with special properties such as fluorescence. For example, a primary amine derivatization using o-phthalaldehyde (OPA) in the presence of a thiol [203] forms a highly fluorescent compound with excitation at 330 nm and emission at 445 nm, the reaction taking place as follows:

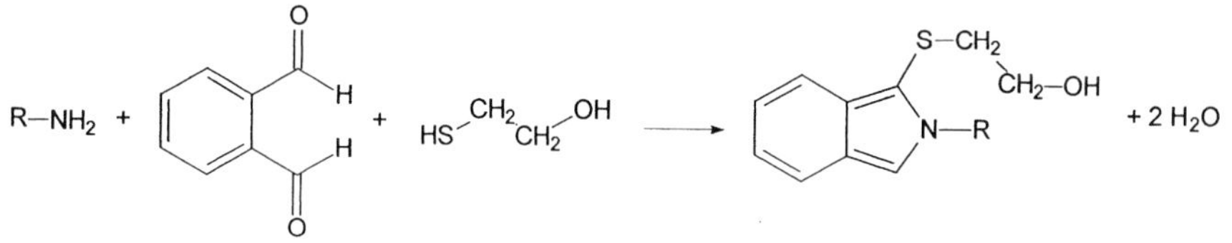

Besides 2-mercaptoethanol, other thiols such as mercaptoacetic acid [204], 3-mercaptopropionic acid [205], thioglucose [206], and ethanethiol [207] can be used in this reaction. Sodium sulfite [208] has been used with OPA, and potassium cyanide has been used with naphthalene-2,3-dicarbaldehyde in amine derivatization in a reaction of the type:

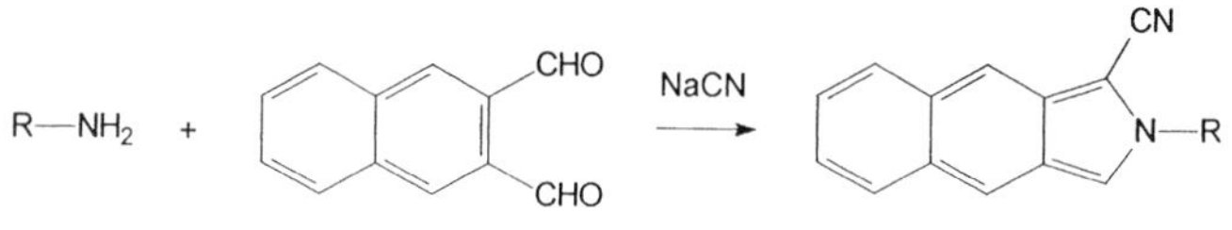

Certain amines may react without the presence of a thiol or cyanide, but undergoing a further condensation. A model reaction of this type is that between peptides containing a N-terminal tryptophan and glyoxal:

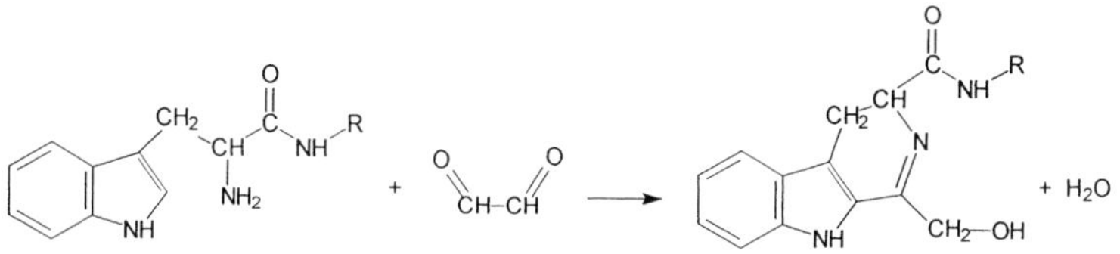

The resulting compound is also highly fluorescent, and a HPLC technique specific for peptides containing a N-terminal tryptophan is based on this reaction [209]. Phenylglyoxal gives a similar reaction [210], [211]. Also, formaldehyde can react with tryptophan or tryptamines generating a β-carboline derivative as follows:

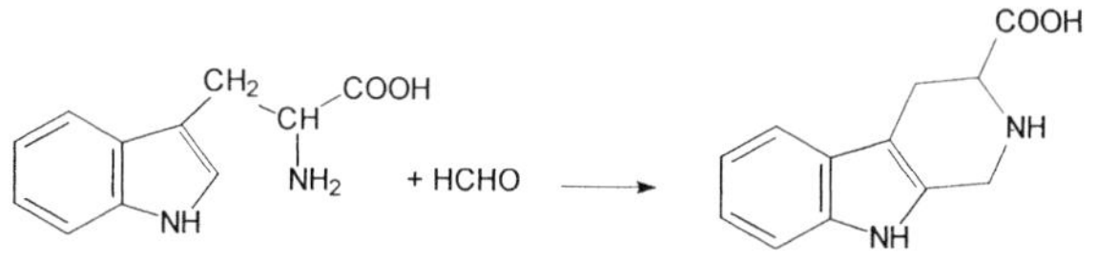

The new compound can be analyzed by GC, usually after further derivatization by silylation [212].

Acetylacetone may react in the presence of formaldehyde with primary amines in a reaction as follows:

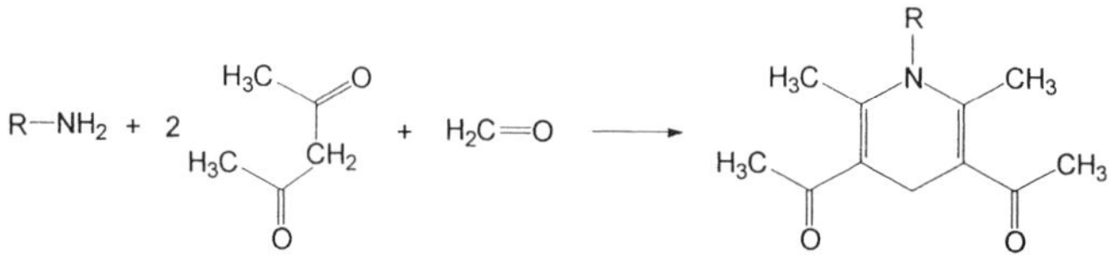

The reaction product is fluorescent. The reaction has been used for the analysis of formaldehyde using ammonia (Nash reagent) as the amine [213]. Also, aldehydes can be derivatized with 1,3-dicarbonyl compounds in the presence of an ammonium salt (ammonium acetate) in a reaction as follows:

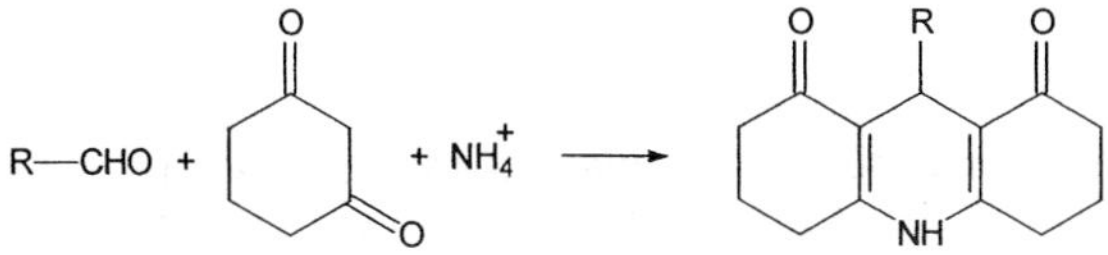

The resulting compound is fluorescent and can be used in HPLC analysis of various aldehydes [214].

Another fluorescence labeling consisting of the replacement of an old cycle with a new one is the reaction of amines with 4-phenylspiro[2-hydrofuran-2,3'-3-hydro-isobenzofuran]-3,6-dione (fluorescamine) [215]:

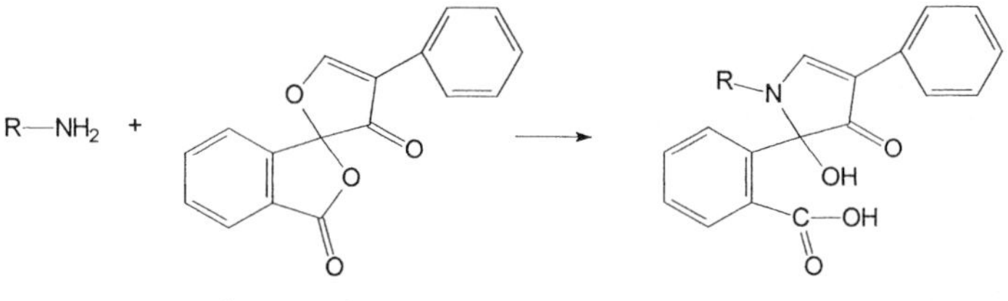

fluorescamine

Similarly, 2-methoxy-2,4-diphenyl-3(2H)-furanone will react with primary amines, but the reaction is associated with the hydrolysis of the methoxy group to a hydroxyl group as shown below:

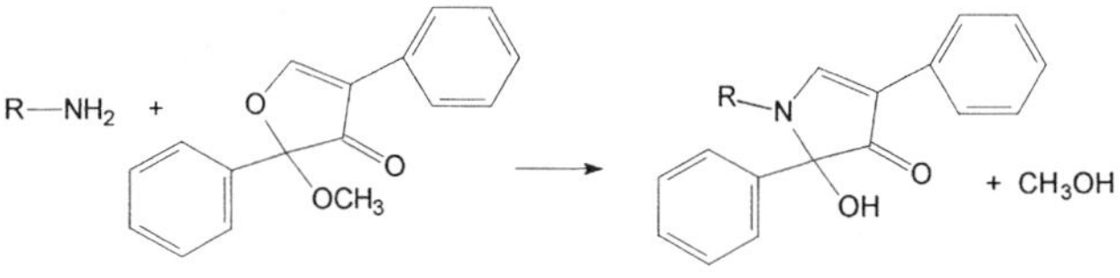

The reaction product is fluorescent. In this reaction, secondary amines form aminodienones that are not fluorescent. However, aminodienones can be converted into fluorescent pyrrolinones by treatment with ethanolamine in a sequence of reactions as follows [216]:

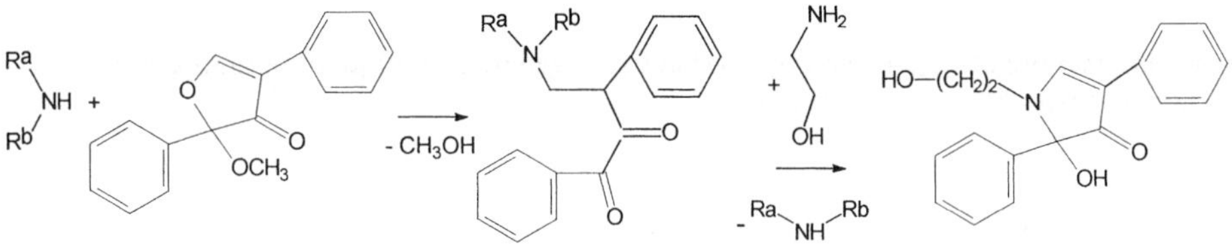

Various amines in biological samples are analyzed using fluorescamine [24]. The reaction occurs very easily. A typical procedure consists of using 50–200 μL of sample containing up to 4–5 µg/mL analyte and adding 100–200 µL buffer with pH 10 (e.g. sodium borate/KCl buffer) followed by vortexing for the separation of insolubles. To this mixture, 200–300 µL solution of 5 mg/mL fluorescamine in acetone is added, with mixing/vortexing between additions. The supernatant solution can be used for HPLC separation with fluorescence detection [24].

- Reactions with the formation of azoles and related compounds

The compounds containing aromatic five-membered heterocycles with a nitrogen and some other hetero atoms (N, O, S) are named azoles. The cycles in the group containing two nitrogen atoms are 1,2-diazole (pyrazole) and 1,3-diazole (imidazole). Besides azoles that are aromatic, various related compounds are formed in derivatization reactions used for analytical purposes. Among these are oxazolinones, oxazolidinones, thioxo-imidazolidinones (thiohydantoins), etc. Many of the reactions with azoles formation involve analytes with two functional groups (or with two active hydrogens on different atoms), reagents with two functional groups, or both.

A typical reaction leading to pyrazoles is the reaction of hydrazines with diketones such as 2,4-pentandione (acetylacetone). For example, the reaction between hydrazine or methylhydrazine and acetylacetone takes place in 1 h at room temperature at pH 6–9 [193]:

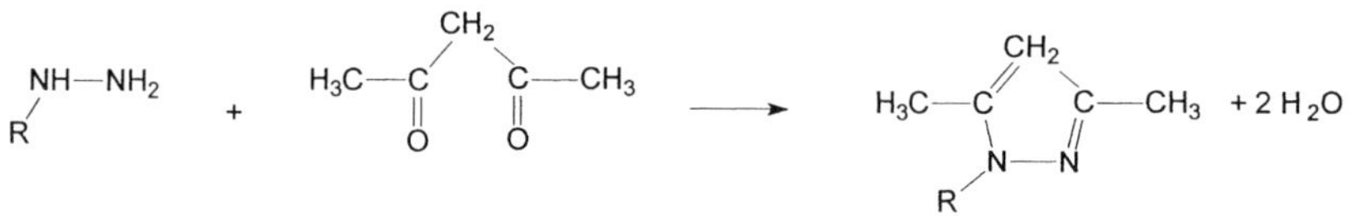

Other hydrazines such as the hypotensive agent hydralazine also react with acetylacetone and can be analyzed by GC [217]:

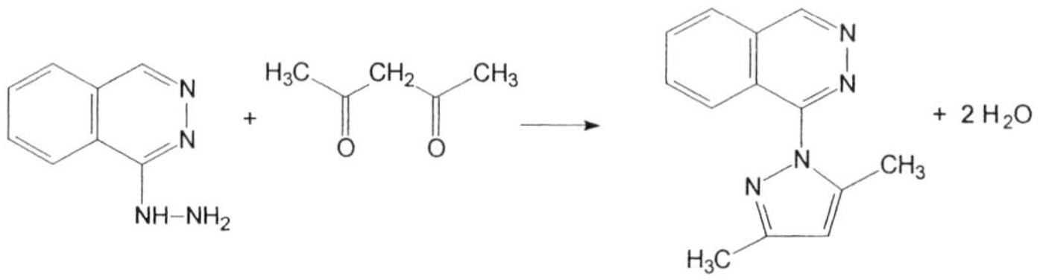

4-Nitrophenylhydrazine has been used for the analysis of 1,3-dicarbonylic compounds [24].

An imidazole derivative can be obtained in the reaction of 9,10-diaminophenanthrene with a carboxylic acid as shown below:

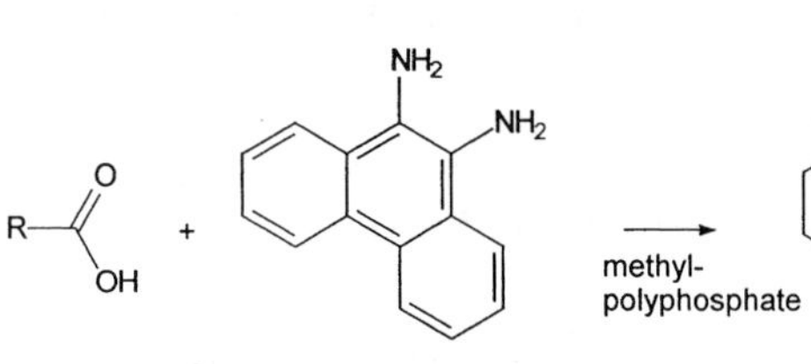

It is difficult to establish the mechanism of this condensation, because as a rule the C=O group in carboxylic acids does not exhibit typical properties of the carbonyl of aldehydes and ketones.

A reaction with the formation of an imidazole takes place between an adenine (adenosine or acid adenosin-3-phosphoric) and chloroacetaldehyde [218]:

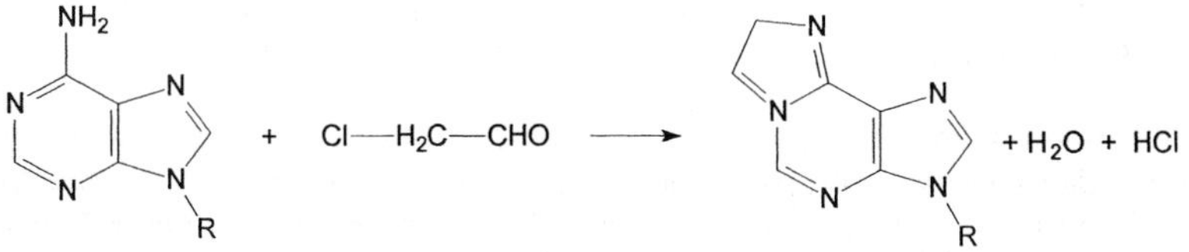

Cytosine containing compounds also can lead to a new imidazole cycle in reaction with a-halogenated carbonyl compounds. As an example, 2-bromoacetophenone reacts as follows:

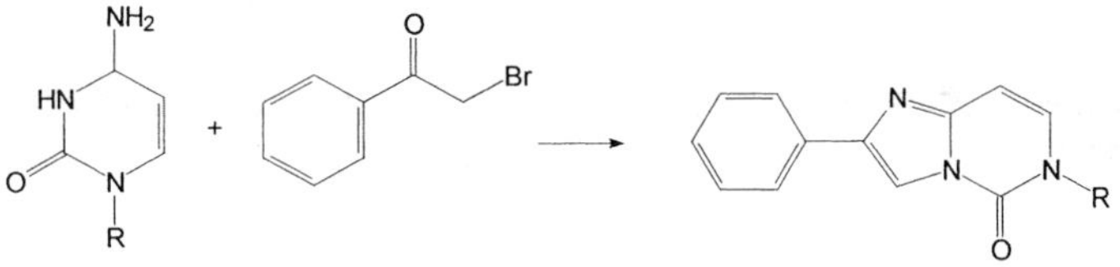

The resulting compound is fluorescent and is used for HPLC determination of cytosine related compounds [219]

Substituted guanidines and benzoin lead to the formation of imidazole derivatives in a reaction that takes place as follows:

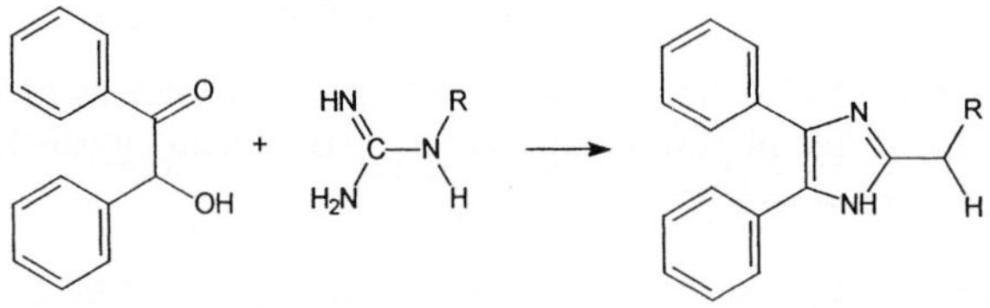

The resulting compounds are highly fluorescent and used for HPLC analysis with fluorescence detection [210], [211].

The formation of oxazoles takes place in the reaction between benzylamine and 5-hydroxyindoles or catecholamines. The reactions take place in the presence of an oxidant such as $K_3[Fe(CN)_6]$ as follows:

catecholamines

The resulting compounds are fluorescent, and the reaction is used for HPLC analysis of various indole derivatives [220]. The catecholamines (R^a = OH, R^b = H norepinephrine, R^a = OH, R^b = CH_3 epinephrine, R^a = H, R^b = H dopamine) also are analyzed with excellent sensitivity by HPLC in biological samples [221].

Activated carbonyl groups such as those in hexafluoroacetone are known to react with difunctional compounds. The reaction may take place with an amino acid as follows:

4-substituted-2,2-bis(trifluoromethyl)-1,3-oxazolidin-5-one

Amino acids can react with an activated anhydride such as trifluoroacetic anhydride (TFAA):

4-substituted-2-trifluoromethyl-1,3-oxazolin-5-one

The reaction takes place by heating the amino acids with an excess of TFAA (10 mg amino acid and 2 mL TFAA) at 150° C for 10 min. The reaction mixture is then dissolved in ethyl acetate and analyzed by GC [222].

A similar condensation may take place between an aldehyde and an aminophenol or aminothiophenol (in the presence of an oxidant). This type of reaction is used for aldehyde derivatization as shown in the following example:

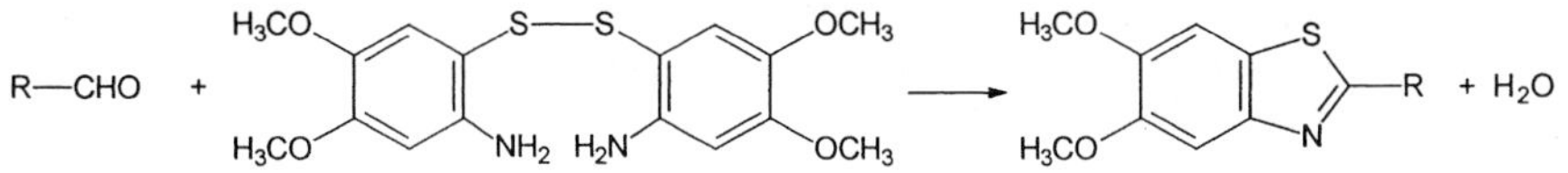

The reaction is done in the presence of NaH_2PO_3, which can generate the thiol from the disulfide. The reagent also generates thiol when reduced by the resulting thiazolidine derivative, which is oxidized to a benzothiazole [223]. Dicyclohexylcarbodiimide (DCCI) can be used for facilitating the reaction [224] in this type of reaction. The reaction takes place in two stages, the first in absence of water and the second in strong acidic conditions (5 - 9 M $HClO4$) as follows:

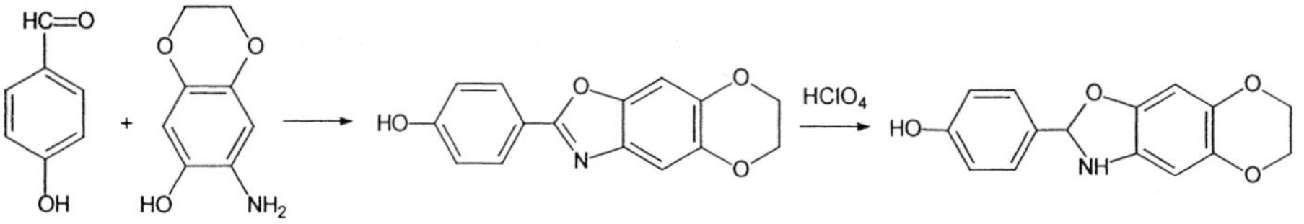

The formation of 5-substituted 2-thioxoimidazolidin-4-ones (2-thiohydantoins), 3-methyl-2-thiohydantoins, and 3-phenyl-2-thiohydantoins is an important reaction used extensively in amino acid and polypeptide analysis [225]. The reaction is performed with SCN^-, CH_3-NCS, or C_6H_5-NCS, respectively. Phenyl isothiocyanate reacts with an amino acid leading first to the formation of an addition product (phenylthiocarbamyl derivative) and is continued with the formation of 5-substituted-3-phenyl-2-thiohydantoins:

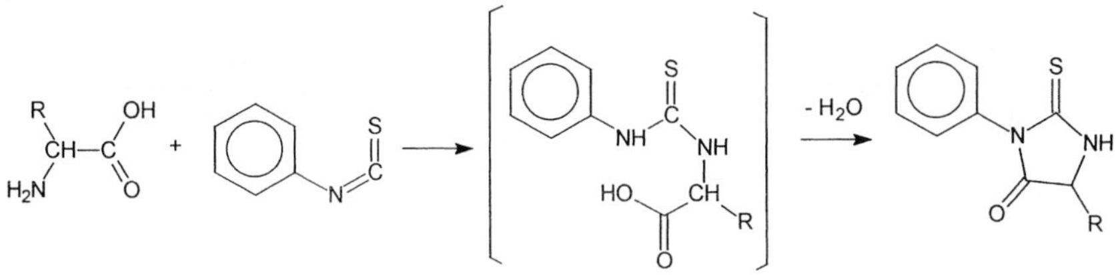

The thiohydantoins are used for amino acid identification with TLC separation [193] and also for GC analysis [226].

Polypeptide analysis can be done using the formation of thiohydantoin derivatives, allowing at the same time a sequential analysis (see Section 20.2). More details about amino acid sequencing using thiohydantoins can be found in dedicated reviews [227], [228].

Some other thiohydantoin derivatives also have been experimented for amino acids and peptide analysis. Methyl and phenyl thiohydantoins are used for amino acids HPLC analysis with UV detection (see also Section 19.10). However, the presence of active hydrogens in thiohydantoins is a problem for obtaining sufficiently volatile compounds for GC analysis. Direct GC analysis is possible only for thiohydantoins of alanine, valine, leucine, isoleucine, and methionine. Other amino acids can be analyzed by GC in thiohydantoin form only after further derivatization by silylation or acylation (see Section 19.10).

An interesting reaction with the formation of cyclic compounds is that of a 1,3-diene with a 1,2,4-triazolin-3,5-dione. The reaction takes place as follows:

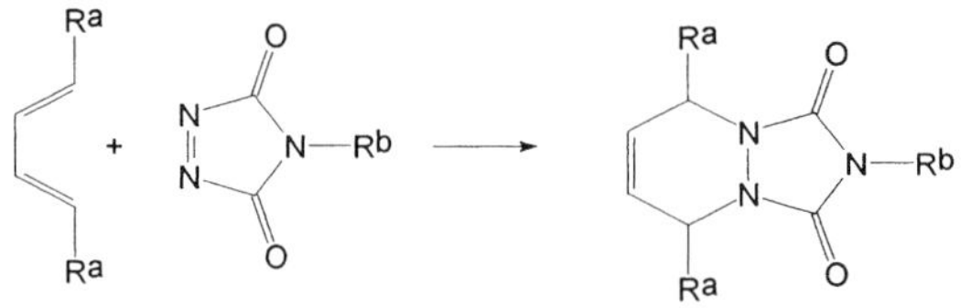

Practical applications were found for this reaction in the analysis of 7-dihydrocholesterol, D_3 vitamin and its metabolites, etc. High sensitivity is achieved with the reagent having the R^b substituent able to carry fluorescence properties through groups such as 1-anthryl, 4-(6-methoxybenzoxazolyl)phenyl, etc.

- Reactions with formation of azines and related compounds

Six-membered heterocycles with more than one nitrogen or one nitrogen and some other heteroatoms are named azines. Diazines contain two N, triazines three N, tetrazines four N, etc., oxazines contain N and O, and thiazines N and S. Various isomers are known for these compounds, and 1,2 diazine (pyridazine), 1,3-diazine (pyrimidine), and 1,4 diazine (pyrazine) are the simplest diazines. In most cases, the formation of cycles from this group requires both an analyte with two functional groups (or with two active hydrogens on different atoms) and a reagent with two functional groups.

A typical reaction with formation of a diazine is the reaction of guanidine derivatives with acetylacetone or hexafluoroacetylacetone [193]. Among the compounds analyzed by GC using this reaction are guanidine $(NH_2)_2C=NH$, methylguanidine, dimethylguanidine, and various compounds of pharmaceutical importance [193]. The reaction takes place as follows:

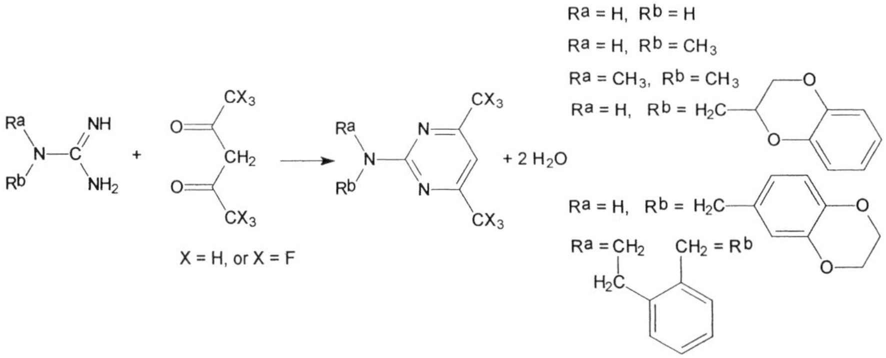

Substituted biguanidines can react with anhydrides and form triazines in a reaction that takes place as follows:

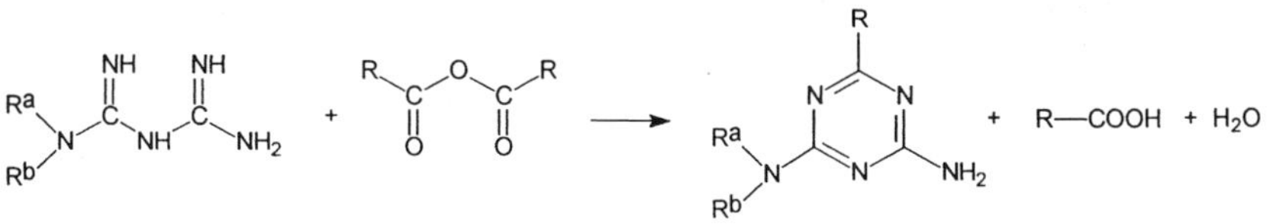

The anhydrides used in this reaction include chlorofluoroacetic, dichlorofluoroacetic, heptafluorobutyric, pentafluoropropionic, and trifluoroacetic anhydride [193]. The biguanides include compounds with R^a = H, R^b = $C_6H_5CH_2CH_2$, R^a = H, R^b = C_4H_9, R^a = CH_3, R^b = CH_3 [229], etc. The resulting compounds are volatile enough for GC analysis. The use of halogen-containing anhydrides allow the use of the highly sensitive ECD detection.

Generation of several fluorescent derivatives [230] involves reactions with the formation of certain azines such as quinoxaline derivatives. One example is the reaction of ethylenediamine with catecholamines using oxidation with $K_3[Fe(CN)_6]$. The reaction can be written as follows:

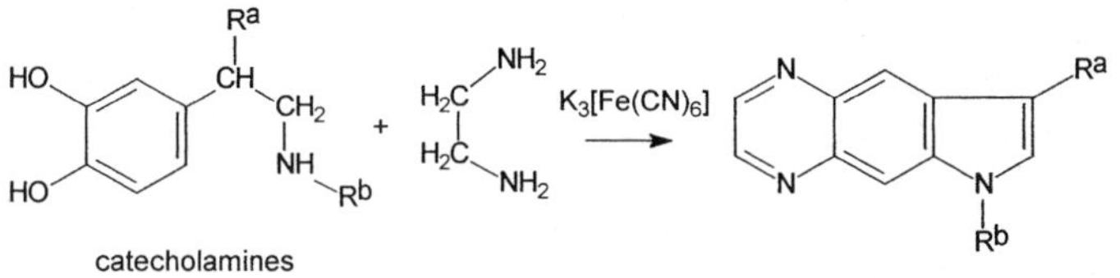

catecholamines

Fluorogenic properties of quinoxaline derivatives are also used in the analysis of various α-keto acids that can react with reagents such as o-phenylenediamine, 4,5-dimethoxybenzene-1,2-diamine (DDB), or 2H-benzo[d]1,3-dioxolane-5,6-diamine (DMB) as follows:

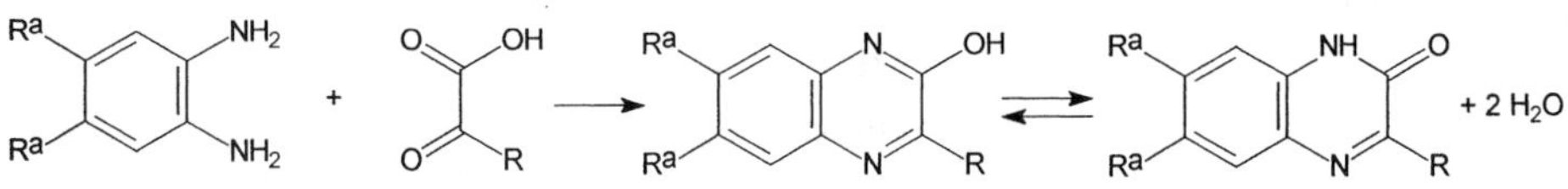

Ra = H, H_3CO, etc.

Besides TLC or HPLC with fluorescence detection, these compounds can be analyzed by GC after blocking the active hydrogen with a TMS group:

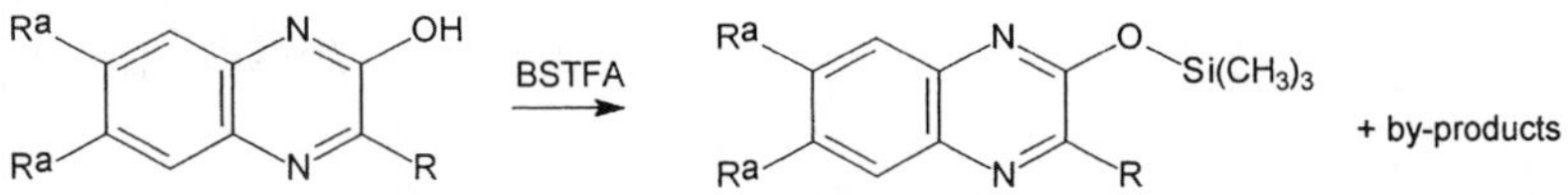

The derivatization with diaminobenzene is commonly carried out in an ethanol-acetic acid solution or in 2–4 N aqueous HCl solution and using heating [231], [232]. Several 1,2-diaminobenzenes substituted with halogens at the aromatic ring and 2,3-diaminonaphthalene are applied for α-keto acids analysis [193].

A similar reaction with that given by keto acids takes place in the presence of an oxidant between a 1,2-diamine and a 1,2-dihydroxybenzene:

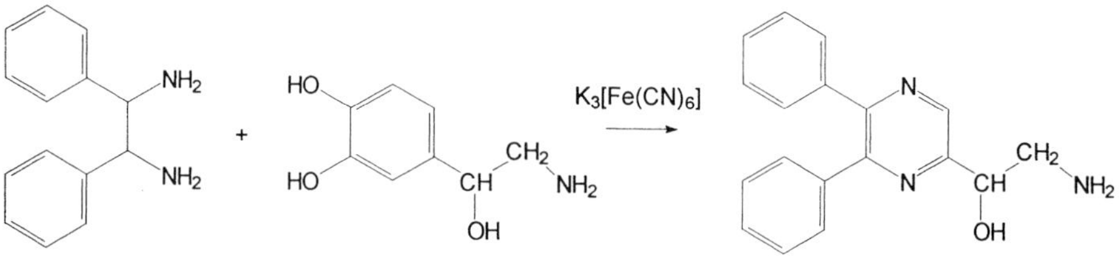

The resulting compounds from various catecholamines can be analyzed by HPLC using fluorescence detection [24].

- Reactions with formation of cyclic siliconides

As indicated previously (see Section 18.2), silylating reagents containing two reacting sites can generate cyclic siliconides. Reagents with bis(trialkylsilyl) reacting groups are used for the silylation of primary amines, diols, etc. generating cyclic compounds. For example, the reaction of a diol in DMF in the presence of imidazole as a base with 1,3-dichloro-1,1,3,3-tetraisopropyldisiloxane takes place as follows [233]:

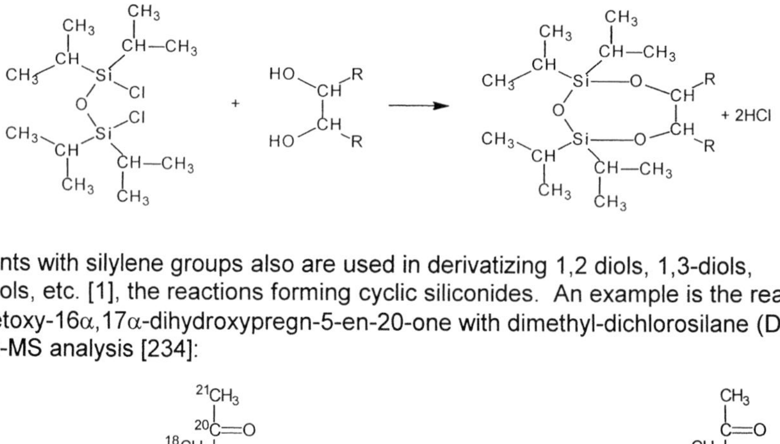

Reagents with silylene groups also are used in derivatizing 1,2 diols, 1,3-diols, catechols, etc. [1], the reactions forming cyclic siliconides. An example is the reaction of 3β-acetoxy-16α,17α-dihydroxypregn-5-en-20-one with dimethyl-dichlorosilane (DMCS) for GC-MS analysis [234]:

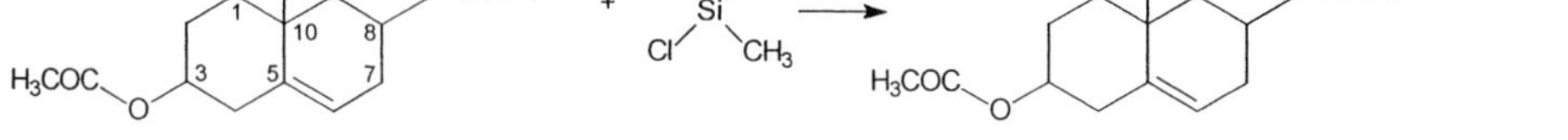

The steroid molecules have several asymmetry centers and numerous possible isomers. The substituents situated above the plane of the rings are indicated with β and those behind the plane with α.

Polyfunctional steroids with isolated functional groups react only partially with DMCS, yielding unstable products. For example, corticosteroids with 11β-hydroxy groups give with DMCS unstable, moisture sensitive products [235]. Attempts to derivatize corticosteroids with a mixture of reagents, one capable of reacting with an isolated OH group and the other reacting simultaneously with two OH groups in vicinal positions, lead to a mixture of products. An additional problem for bifunctional groups is reaction with two different molecules forming silyl bridges. Better results are obtained using dimethyldiacetoxysilane (DMDAS), which in the presence of basic catalysts (triethylamine) reacts with diols or with dihydroxyacetone side chain of steroids forming a cyclic siliconide and with isolated OH groups forming dimethylacetoxy derivatives [236]. However, with cortisol (11β,17,21-trihydroxypregn-4-ene-3,20-dione) DMDAS does not react at 11β-OH, which is more hindered.

A list of reagents containing bis(trialkylsilyl) groups is given in Table 18.5.1, and reagents with silylene groups are given in Table 18.5.2. These reagents are used for analytical purposes and also in organic synthesis for attaching protecting groups.

TABLE 18.5.1. *Some reagents used for the introduction of bis(trialkylsilyl) groups.*

Group	Mass of the group	Reagent(s)	Analytical use	Reference
1,2-bis(dimethylsilyl-)benzene	192	1,2-bis(dimethylsilylchloro)-benzene		
1,1,3,3-tetraisopropyldisiloxane-1,3-diyl-	244	1,3-dichloro-1,1,3,3-tetraisopropyldisiloxane	Y	[237]
1,1,4,4-tetramethyldisilethylene-1,4-diyl-	144	1,2-bis(chlorodimethylsilyl)-ethane		

TABLE 18.5.2. *Some reagents used for the introduction of silylene groups.*

Group	Mass of the group	Reagent(s)	Analytical use	Reference
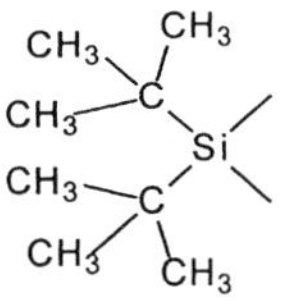 di-*tert*-butylsilylene-	142	di-*tert*-butyldichlorosilane	Y	[238], [239]
		di-*tert*-butylsilyl bis(trifluoromethanesulfonate)	Y	[238]
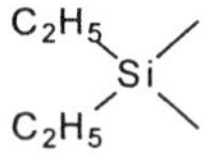 diethylsilylene	86	N,O-bis(diethylhydrogensilyl)trifluoro- acetamide (also diethylhydrogensilyl compounds are formed with mono functional groups)	Y	[240]
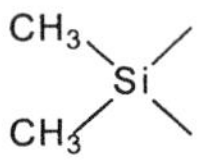		bis(diethylamino)dimethylsilane	Y	[63]
dimethylsilylene-	58	bis(dimethylamino)dimethylsilane		[63]
		dimethyldichlorosilane (DMCS)	Y	[63]
		dimethyldiacetoxysilane		[236]
		2,2,4,4,6,6-hexamethylcyclotrisilazane		
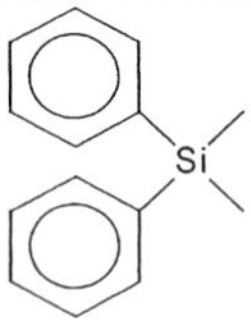 diphenylsilylene-	182	diphenyldichlorosilane		
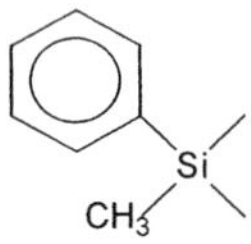 methylphenylsilylene-	120	methylphenyldichlorosilane		

Some cyclic siliconides are generated in reactions where only one bond would be expected to be formed by the silicon atom. For example pentafluorophenyl-dimethylsilylamine is used successfully in steroids analysis [241]. In the reaction with 5β-pregnan-3α,17α-20-triol, the pentafluorophenyl group can be eliminated forming a cyclic siliconide as shown below [193]:

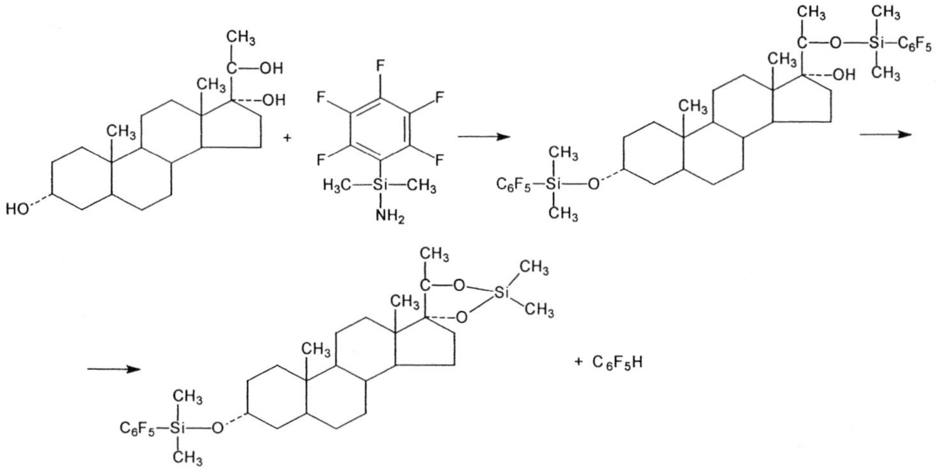

Also, β- and γ-hydroxylated amines may react with silylating reagents containing the chloromethyl group in a silylation reaction that generates cycles. For example, the reaction of 1,3-bis(chloromethyl)-1,1,3,3-tetramethyldisilazane in the presence of chloromethyldimethylchlorosilane (CMDMCS) with a γ-hydroxylated amine takes place as follows:

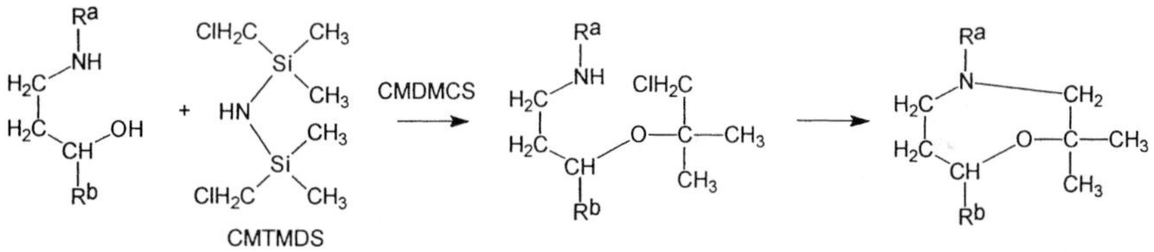

Even tricyclic compounds are reported as generated by silylation when compounds with three reactive groups are involved in reaction. The reaction of organo-trialkoxysilanes with trialkanolamines generates substituted 2,8,9-trioxa-5-aza-1-silatriciclo[3,3,3^{1,5}]-undecanes (silatranes). The compounds were proven to have a transannular N ----→ Si bond [242]. As an example, the reaction of tetraethoxysilane with 1-[bis(2-hydroxyethyl)amino]-propan-2-ol takes place as follows:

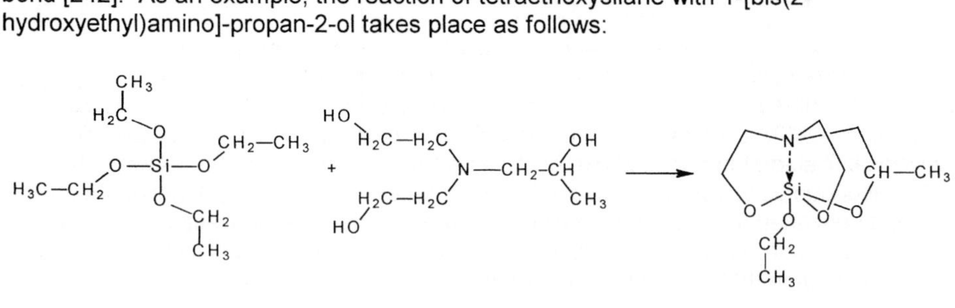

These types of compounds have been analyzed by GC, but their synthesis has not been used for analytical purposes.

- Formation of cyclic phosphonothioates

The reaction of ethylphosphonothioic dichloride with bifunctional compounds containing active hydrogens leads to cyclic derivatives. The active hydrogens can come from groups such as OH, SH, NH_2, and COOH. The reaction takes place in the presence of a basic compound such as $(C_2H_5)_3N$ as follows:

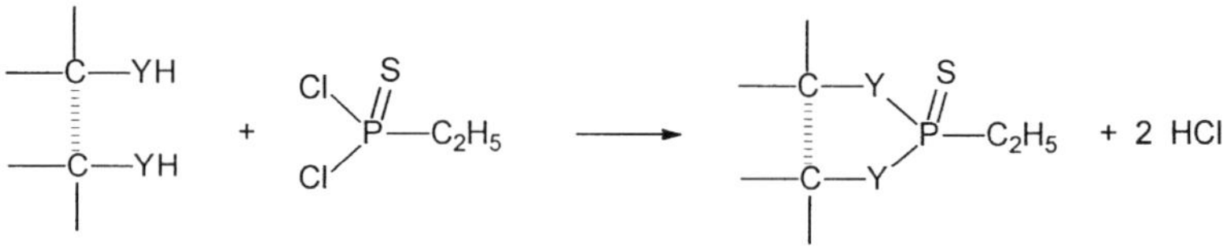

The stability of the cycles is 6 > 5 > 7, and some compounds such as dicarboxylic acids, α-hydroxy acids, and bifunctional aliphatic compounds with the active groups separated by more than two atoms do not form cyclic phosphonothioates. The reaction is reported for ethylene glycol, 1,3-propandiol, 2,3-dimethylbutane-2,3-diol (pinacol), catechol, 2-aminophenol, 2-phenilenediamine, etc. [193]. The reaction takes place by heating up to 10 μmol of analyte with 15 μL reagent and 33 μL triethylamine at 80° C for 30–45 min. [193]. The advantage of this derivatization is the possibility to use an NPD detector in the GC system, which provides excellent sensitivity.

- Formation of cyclic boronates

Another reaction leading to cyclic molecules by derivatization of bifunctional compounds is the formation of cyclic boronates. The reaction takes place between a boronic acid with the formula $R–B(OH)_2$ and bifunctional compounds with the groups in proximity. The reaction is used for 1,2-diols, 1,3-diols, 1,4-diols, 1,2-enediols, 1,2-hydroxyacids, 1,3-hydroxyacids, 1,2-hydroxyamines, 1,3-hydroxyamines, and aromatic compounds with two functional groups such as OH, NH_2 or COOH in 1,2 positions [193]. The compounds reacting with boronic acids can have simple molecules or may have more complicated structures such as carbohydrates, steroids, catecholamines, etc. The reaction between a boronic acid and, for example, a diol takes place as follows:

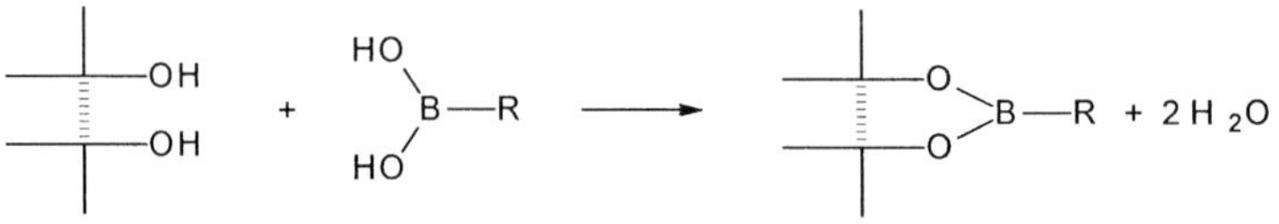

The radical R in the boronic acid can be either an alkyl or an aryl group. Among these, methaneboronic acid, butaneboronic acid, *tert*-butaneboronic acid, cyclohexaneboronic acid, and benzeneboronic acid have all been used for analytical derivatization for GC use. Aromatic boronic acids are stable, while dry alkylboronic acids are slowly oxidized in air, in a reaction that takes place as follows:

$$R–B(OH)_2 + 1/2\ O_2 \longrightarrow R–O–B(OH)_2$$

The alkylborates formed in this reaction can easily hydrolyze to boric acid and are not useful as derivatization reagents. The presence of small amounts of water in the boronic acid inhibits the oxidation of the reagent. In common cases, this water does not interfere with the derivatization, but in cases where it may hydrolyze the analyte, a water scavenger such as 2,2-dimethoxypropane can be used as a solvent. Simpler procedures, such as filtration through dry filter paper, can be used for removing the water. Boronic acids easily form trimeric cyclic anhydrides, as follows:

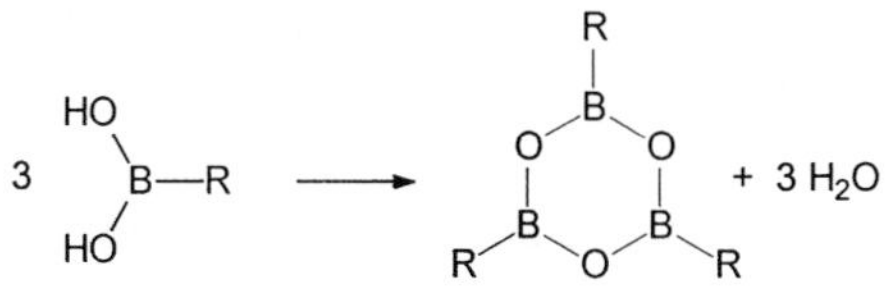

The cyclic anhydrides and the boronic acid react similarly with the analytes. Their formation does not affect negatively the derivatization. The anhydrides also are formed in the injection port of the GC due to the heating and usually are separated in the chromatographic process. Boronic acids react with basic compounds with the formation of salts that are rather stable and much less reactive:

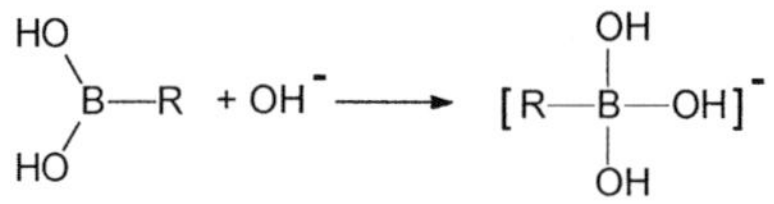

The possible reaction of boronic acids with bases must be considered when choosing the reaction medium for the derivatization with these reagents. Anhydrous inert solvents are commonly utilized as reaction medium.

The reaction with the formation of cyclic boronates takes place in most cases rather easily, at room temperature and within 1 to 30 min. For this purpose, the analyte is mixed with an excess of reagent. The excess water generated in the reaction can be removed using a water scavenger such as 2,2-dimethoxypropane. The most common boronic acid used for derivatization for analytical purposes is butylboronic acid. Except for derivatization of smaller molecules, the cyclohexaneboronates and benzeneboronates are less applicable for GC analysis due to the molecular mass increase. Methyl and *tert*-butylboronates have poor hydrolytic stabilities [193].

Boronate derivatives are analyzed using both GC-MS and LC-MS techniques [243]. The fragmentation in GC-MS for boronates is not strongly influenced by the boronate group. Because the natural isotopes of boron are in the ratio $^{11}B:^{10}B = 4.2:1$, the fragments containing boron generate peaks with m/z and m/z − 1, having the relative intensity in the ratio 4.2:1. The doublet peaks can be seen, for example, in the mass spectrum of 6-deoxy-α-L-galacto-pyranose-1,2:3,4-bis(butylboronate) given in Figure 18.5.5 and of 6-deoxy-α-L-galacto-pyranose-1,2:3,4-bis(phenylboronate) given in Figure 18.5.6.

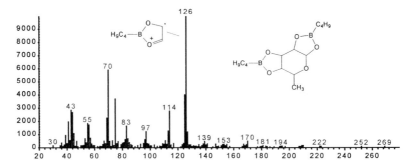

FIGURE 18.5.5. *EI+ mass spectrum of 6-deoxy-α-L-galacto-pyranose-1,2:3,4-bis(butylboronate), MW = 296.* (Note: the abscissa of a mass spectrum indicates m/z and the ordinate indicates relative abundance to 10,000.)

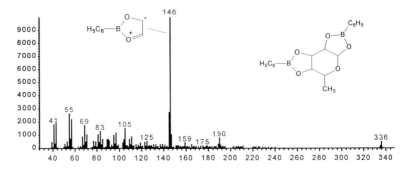

FIGURE 18.5.6. *EI+ mass spectrum of 6-deoxy-α-L-galacto-pyranose-1,2:3,4-bis(phenylboronate) MW = 336.*

The use of boronates as derivatives for GC analysis may be a problem when single groups with active hydrogens remain in the molecule of the analyte. In such cases, the use of a second derivatization such as silylation is possible [244], but the loss of a boronate group also may occur during the second derivatization [193]. The procedure is successfully used for various analyses such as that of 3-methoxy-4-hydroxyphenylene glycol [245] and polyhydroxyalkylpyrazines [246], commonly generated in the reaction of sugars with ammonia [247] or with certain ammonium salts [248].

The formation of cycles containing boron can involve in some reactions formation of B-N coordinative bonds. As an example, the reaction of certain amines (such as ethylenediamine) with salicylaldehyde and diphenylboric anhydride takes place as follows [249]:

18.6 OTHER DERIVATIZATION REACTIONS

A variety of other chemical reactions besides those classified in more general types have been used for derivatization. Although some of these reactions are used to a lesser extent than common reactions such as alkylation or acylation, in specific cases they lead to properties that improve significantly the results of the analysis. This section covers some of these reactions.

- *Reaction of addition to a double bond*

Typical addition reactions to a double bond are used mainly for the derivatization of certain compounds containing reactive double bonds. For example, direct bromination of a double bond is used for derivatization of aflatoxins in a reaction of the type:

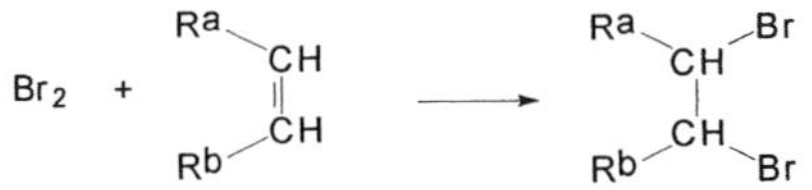

This reaction can be performed postcolumn, with electrochemical generation of bromine (from KBr) and measurement using a fluorescent detector [250]. The addition of bromine to the 8,9- or 9,10-double bond of aflatoxin B_1 or G_1, respectively, is shown below:

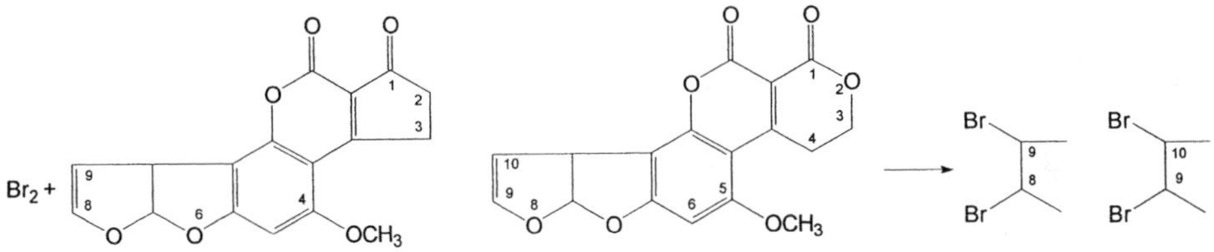

aflatoxin B_1

aflatoxin G_1

Aflatoxins B_1, B_2, G_1, and G_2 are all analyzed by this technique, but it is not determined if aflatoxins B_2 (8,9-dihydro derivative of aflatoxin B_1) and G_2 are also derivatized. In this reaction, pyridinium bromide perbromide is used as an alternative to electrochemically generated bromine [251]. The postcolumn reaction with iodine also is used in aflatoxin HPLC analysis [252]. The double bond in aflatoxins B1, G1 or M (hydroxylated aflatoxin B) also can undergo a hydration reaction to the 8,9- or 9,10-double bond in the presence of trifluoroacetic acid.

The double bond present in olefins can react with a mercuric salt such as $Hg(OOC\text{-}CH_3)_2$, the reaction being typically followed by a hydrolysis. This reaction is written as follows:

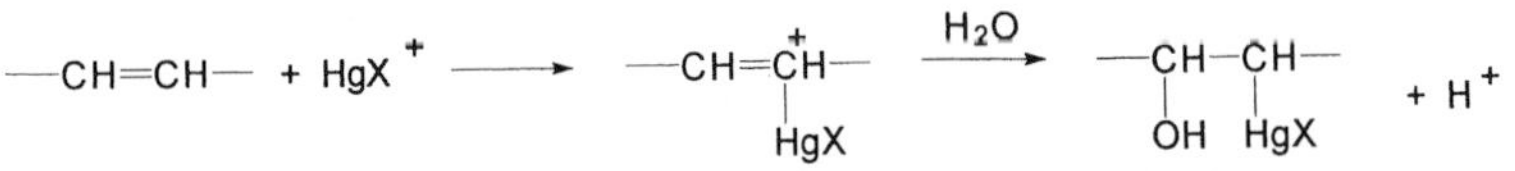

A reactive double bond may be present in an analyte but also in a reagent and used in specific derivatizations. One such case is the reaction between a thiol (analyte) and 1,1-bis-(phenylsulfonyl)ethylene:

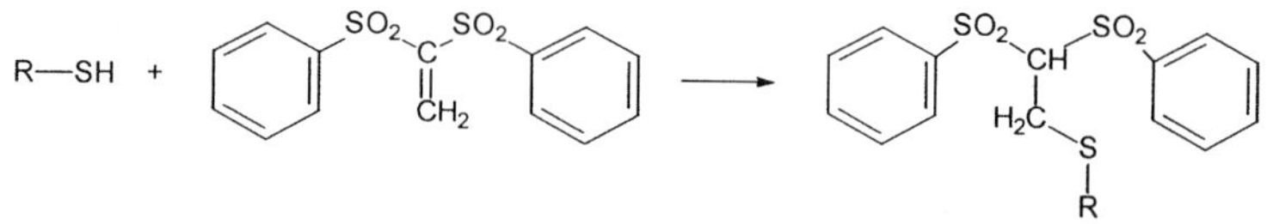

The reaction takes place in a few minutes at room temperature in partially aqueous solutions (pH = 7.5 with borate buffer) and has been used for the analysis of thiol drugs in pharmaceutical formulations [253]. Also, thiols react with maleimides at the 3,4-double bond as follows:

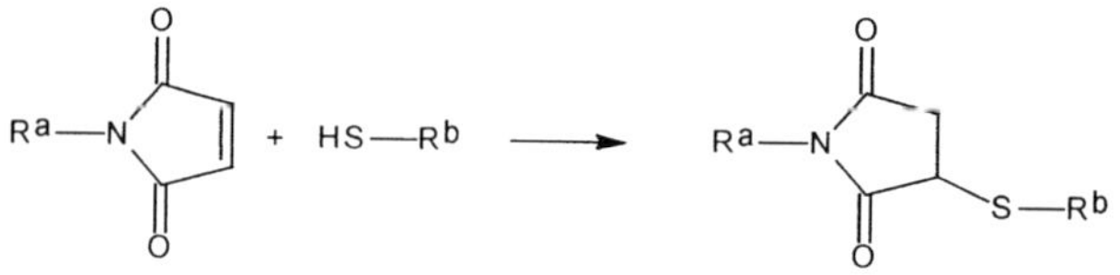

The reaction is applied using various R^a substituents mainly for attaching fluorescent groups to thiols [230]. Double bonds in other α,β-unsaturated carbonyl compounds can participate in the same reaction. As an example, thiols can react with 2-hexenal as follows:

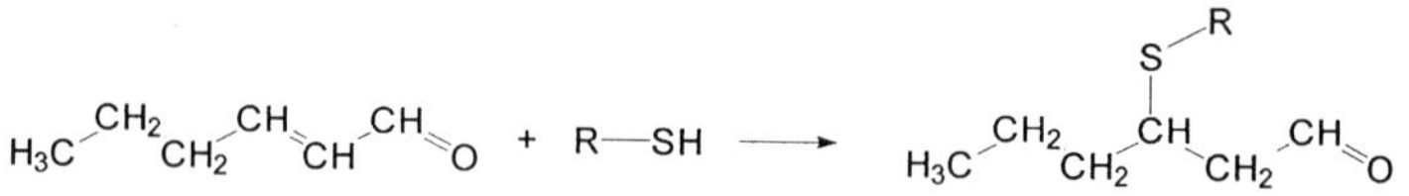

Some other additions to double bonds are known for thiols (see Section 19.3). Even amines may react to specifically active double bonds. As an example, 1-phenylsulfonyl-3,3,3-trifluoropropene can react with putrescine, cadaverine, etc as follows:

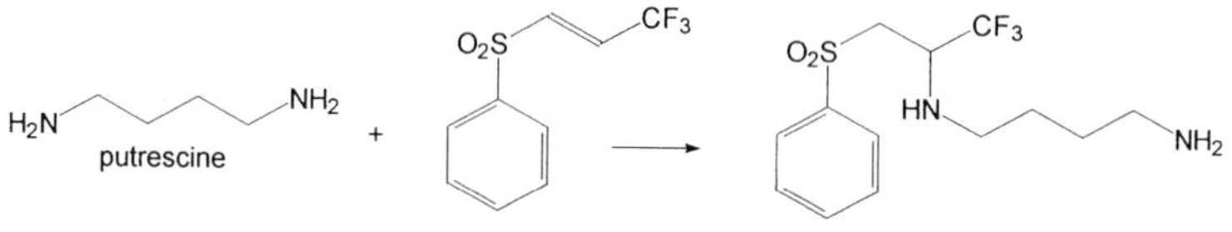

Besides addition to activated double bonds, thiols may react with an aziridine group that is opened with the formation of a thioether. As an example, N-dansylaziridine can be used for thiol derivatization in the presence of $NaBH_4$ in a reaction as follows:

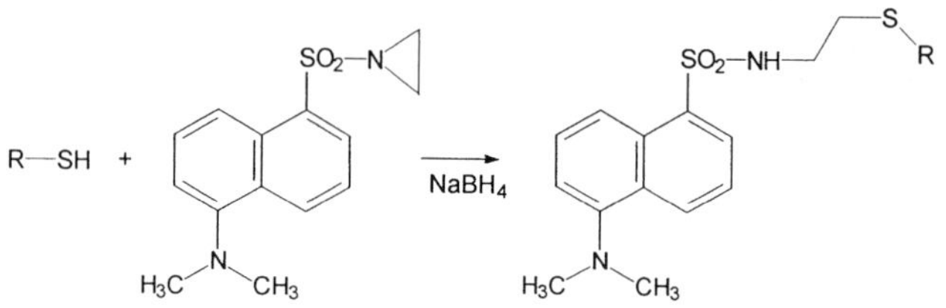

The analysis can be done by HPLC using fluorescence detection.

Diels-Alder reaction can also be considered a case of addition to double bonds. The reaction does not take place easily without the presence of an electron-donating group on the diene and/or of an electron-attracting group on the monoene. However, in the presence of such groups, the reaction takes place with much better yields. For analytical purposes, several 4-substituted 1,2,4-triazolin-3,5-diones are used as activated monoenes (dienophiles) to react in mild conditions with diene analytes. The reaction can be written as follows:

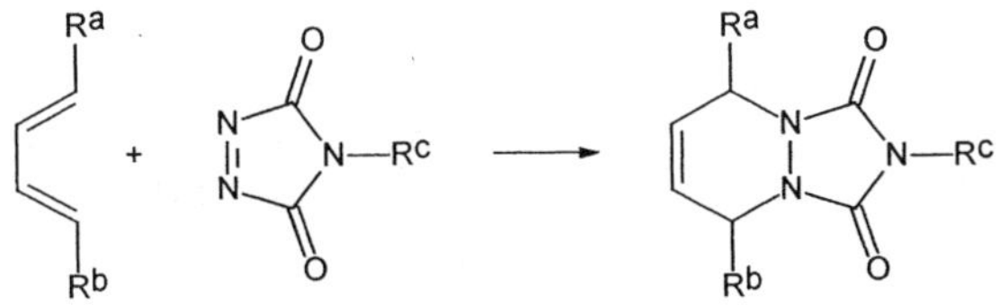

The reagents used for diene derivatizations are designed to have R^c groups that are strong chromophores or fluorophores.

- Oxidations and reductions

Oxidations and reductions are common chemical reactions and some of them have analytical applications. Various oxidations and reductions with analytical applications that are not derivatization reactions take place during electrochemical processes, which were already presented in Section 17.6. Also, the cause of chemiluminescence is commonly an oxidation reaction involving H_2O_2, as shown in Section 17.6.

Derivatization reactions with specific oxidizing or reducing reagents, where the oxidized species have a special physical property such as an enhanced UV-Vis absorption or can fluoresce, also are known. A variety of compounds including alcohols, amines, thiols, etc. can be oxidized, generating compounds that are easier to analyze. For example, certain alcohols can be oxidized to a ketone using pyridinium dichromate in a reaction as follows:

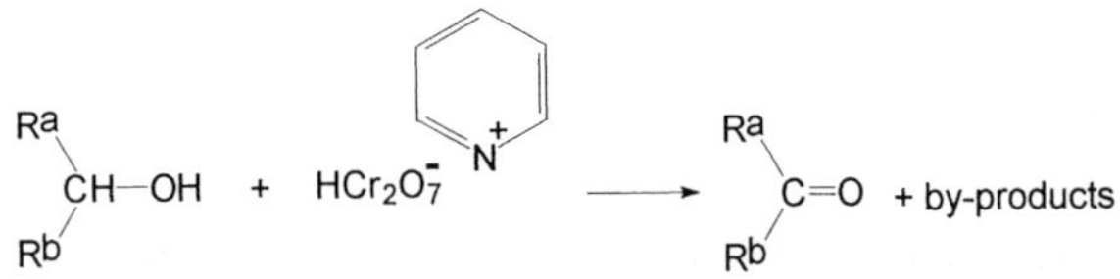

The procedure has been used for prostaglandin analysis in HPLC with UV detection [254]. A variety of other oxidizing reagents are used for analytical purposes, some being inorganic compounds such as $Ce(OH)_3OOH$, V_2O_5 in H_3PO_4, $K_3[Fe(CN)_6]$, etc. For example, thiamine hydrochloride (vitamin B_1 hydrochloride) is oxidized to thiochrome with $K_3[Fe(CN)_6]$ in a postcolumn derivatization reaction as follows:

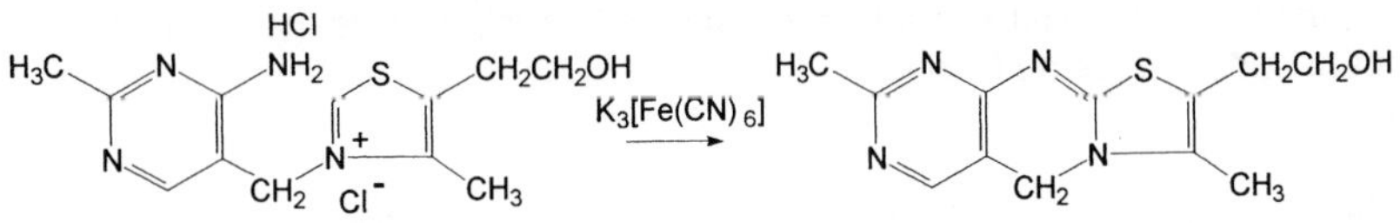

Thiochrome can be detected by its fluorescence, using an excitation light at 390 nm and measuring emission at 475 nm [255], [256].

Phenols and compounds containing a NH_2 group may react in the presence of an oxidant, generating color reactions used for analysis. One such reaction takes place between phenols and 4-aminoantipyrine (Emerson reaction) as indicated below:

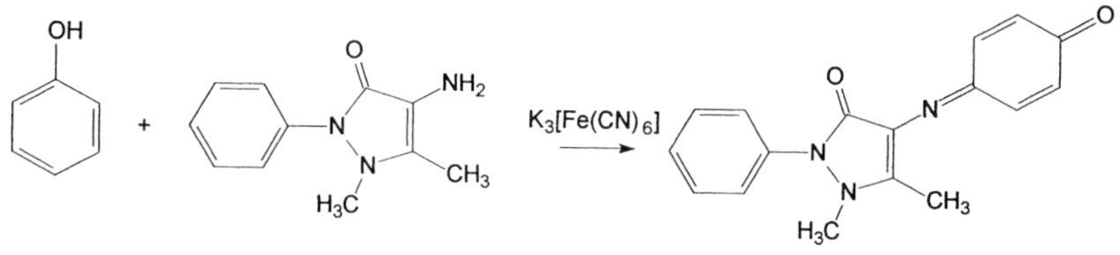

The oxidized product has a strong UV-Vis absorption and has been used for HPLC analysis with detection at various wavelengths for a variety of phenols [24]. Another reaction takes place with 3-methyl-2-benzothiazolinone hydrazone in the presence of ceric ammonium sulfate as oxidant:

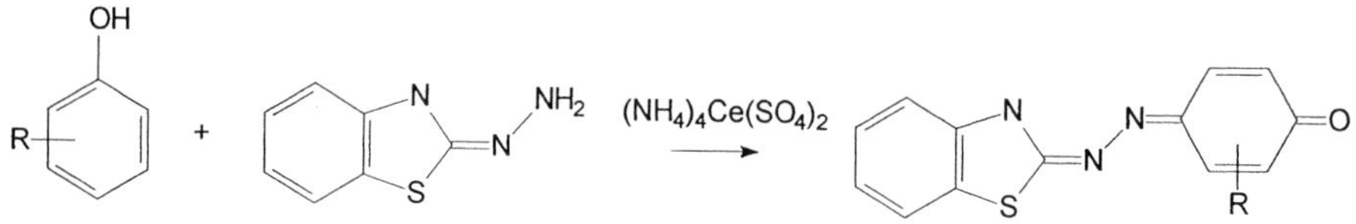

The reactions can be performed postcolumn using UV-Vis detection [257].

The reaction between phenols and ammonia can take place in the presence of an oxidant such as disodiumnitrosylpentacyanoferrate, or trisodiumaquopentacyanoferrate and sodium hypochlorite [257a], and forms 4,4'-dihydroxydiphenylamine (dihydroindophenol) as shown below:

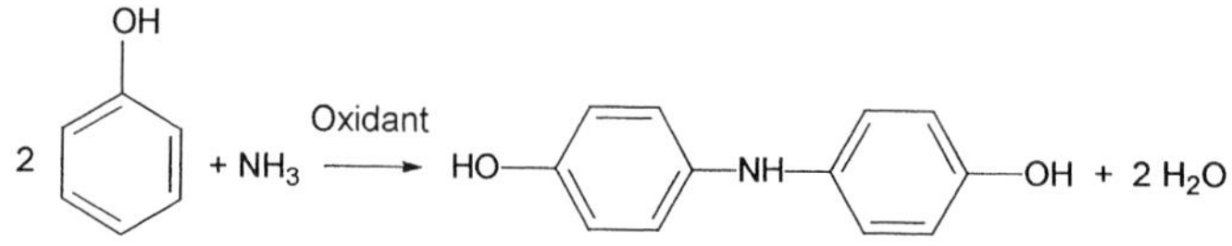

The reaction can be used for the analysis of free ammonia, using GC-MS after further derivatization of the dihydroindophenol with trifluoroacetic anhydride.

Monochlor-sulfonamides $R\text{-}SO_2\text{-}NHCl$ or sodium monochlor-sulfonamides $R\text{-}SO_2\text{-}NClNa$ also are used as oxidation reagents in specific reactions. Sodium monochlor-sulfonamides can be obtained from the reaction of the sulfonamide with NaOCl. For example, thiols can be oxidized with sodium N-chlordansylamide, generating a fluorescent sulfonamide as follows:

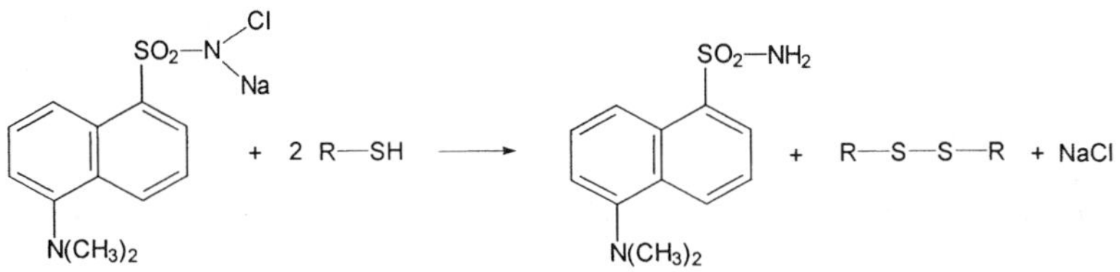

The formation of the fluorescent compound can be used for thiol determination when no other reducing compound is present [258]. Some other oxidation reactions can be used indirectly for analytical purposes. For example, certain proteins are analyzed following reaction with NaOCl, the resulting chlorinated protein being used for the oxidation of thiamine to fluorescent thiochrome [259].

Oxidation reactions also are used for the analysis of specific analytes that are oxidants. For example, peracetic acid can be analyzed following a reaction with 2-([3-{2-[4-amino-2-(methylsulfanyl)phenyl]-1-diazenyl}phenyl]-sulfonyl)-1-ethanol [260]. The reaction can be written as follows:

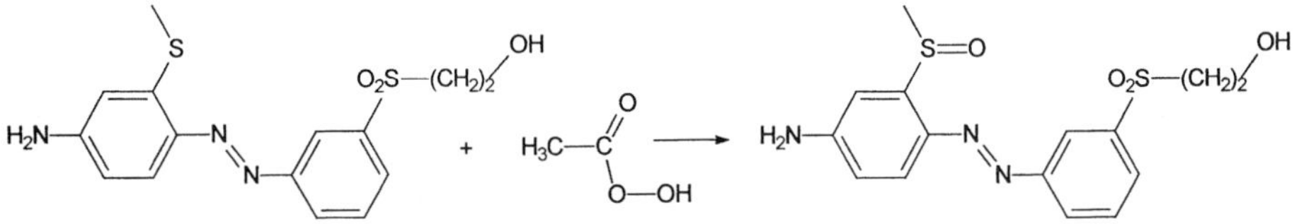

The resulting sulfone allows the detection of peroxyacetic acid at levels as low as 46 ppb without the interference of H_2O_2. The HPLC separation is followed by detection at 410 nm. A reagent that can be used similarly is methyl-p-tolyl sulfide.

Various oxidation and reduction reactions can be performed before a reaction with a second derivatization reagent. For example, the reaction of a thiol with a disulfide with the formation of a new thiol and a new disulfide is rather frequently applied for thiol analysis. The procedure can be used for thiol labeling with groups carrying a desired property. As an example, glutathione can be derivatized using 5,5'-dithiobis(2-nitrobenzoic acid) in a reaction as follows [261]:

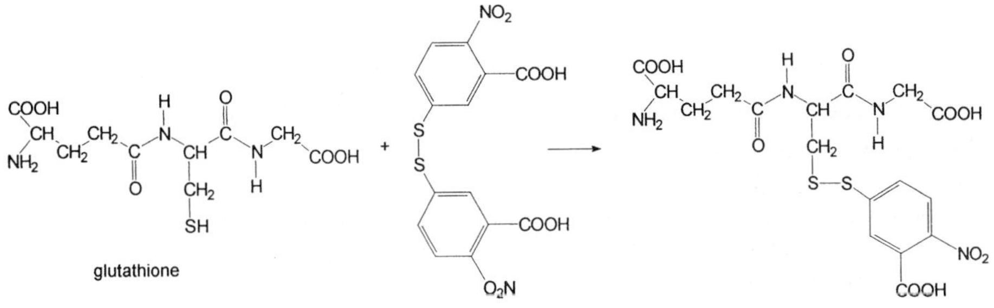

glutathione

The resulting compound can be analyzed by HPLC with UV detection.

Some oxidations or reductions are used with the purpose of modifying the structure of the initial analyte, not necessarily to add specific groups but to generate smaller and more stable molecules that are easier to analyze by GC. One such reaction is the reaction of amitriptyline with $Ce(SO_4)_2$ with formation of anthraquinone [262].

Another oxidation is a reaction that can be used for the cleavage of certain alcohols and can be written as follows:

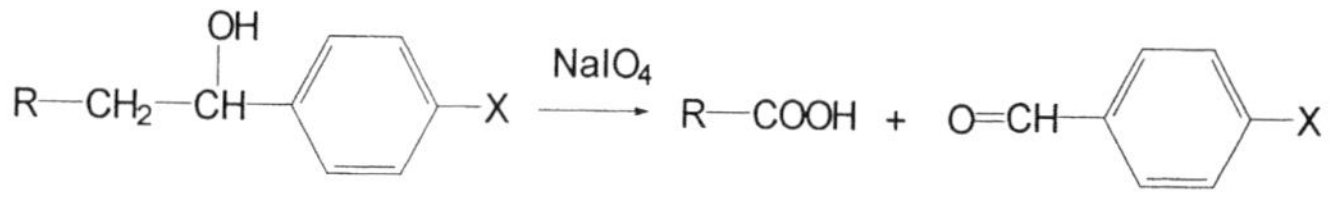

Oxidation reactions also can be applied to compounds containing carbon-carbon double bonds. For example, oxidation with $NaIO_4$ and $KMnO_4$ (neutral pH), with H_2O_2 in the presence of catalytic amounts of OsO_4, or with CrO_3 may lead to the cleavage of C=C bond from aliphatic chains with the formation of carbonyl from a $R_2C=$ group and carboxyl from a RCH= group in a reaction of the type:

The oxidation reaction of the double bond may stop at the formation of a diol in milder reaction conditions. For example, the oxidation with OsO_4 gives a *syn* addition (cis diols) from the less hindered side of the double bond. The same result can be obtained by oxidation with cold $KMnO_4$ at a basic pH. The reactions can be written as follows:

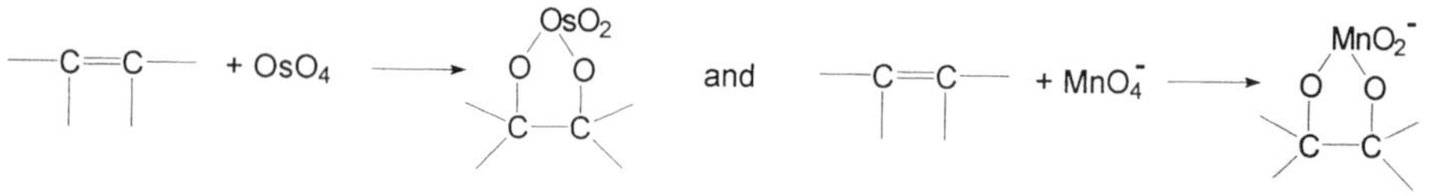

The cyclic esters formed in this reaction are usually hydrolyzed with $NaSO_3$ in ethanol or other reagents generating the diol. *Anti* hydroxylation of the double bond can be achieved with H_2O_2 + formic acid, in a reaction with m-chloroperoxybenzoic acid, or with iodine and silver benzoate followed by hydrolysis. Further derivatization of the OH groups formed in these reactions together with the derivatization of COOH group is common.

Ozone oxidizes the C=C bond in a reaction that forms as a first step an ozonide, as shown below:

The ozonide can be reduced with $NaHSO_3$ to an aldehyde and a ketone. The ozonide formed from monoenoic unsaturated fatty acids can be further derivatized with CH_3OH

and BF_3 leading to the formation of methyl esters of the fragment acids. Other oxidations of the double bond are also known [23]. Oxidation with peracids commonly leads to epoxides, as shown in Section 18.5.

Reduction reactions also are practiced for various double bonds. Carbon-carbon double bond in polyunsaturated fatty acids can be reduced, for example, with hydrazine [263] or $LiAlH_4$ [264]. The C=O double bond in aldehydes and ketones can be reduced with $NaBH_4$ [265], [266]. The nitro group can be reduced to amine using, for example, Zn + a strong acid, followed by the analysis of the amine. Other oxidation and reduction reactions are discussed in Chapter 19.

An interesting oxidation/reduction reaction is probably that of ninhydrin (triketohydrinden hydrate) with amino acids. The reaction can be written as follows:

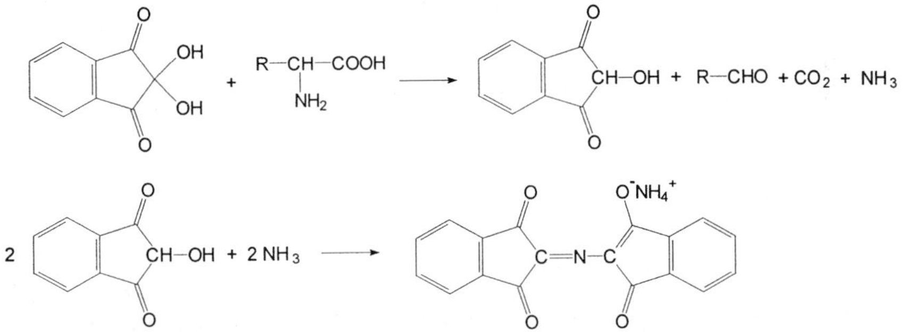

Ninhydrin has been used for qualitative amino acid analysis and in postcolumn HPLC derivatizations [267].

- Hydrolysis

Hydrolysis is a common reaction in the analysis of a number of compounds such as derivatives of organic acids or in the analysis of polymers. Because hydrolysis is a chemical modification of the analyte, it can be seen as a derivatization. Typically hydrolysis is followed by a second derivatization that creates a compound with better analytical properties such as higher detectability or with no active hydrogens. For large molecules such as proteins or polysaccharides, hydrolysis is applied frequently to generate fragments easier to analyze, and the subject is presented in Chapter 20. Certain small molecules such as anhydrides, acyl chlorides, amides, and nitriles can be subjected to hydrolysis before further derivatization or analysis. For example, carbamates can undergo the following reaction:

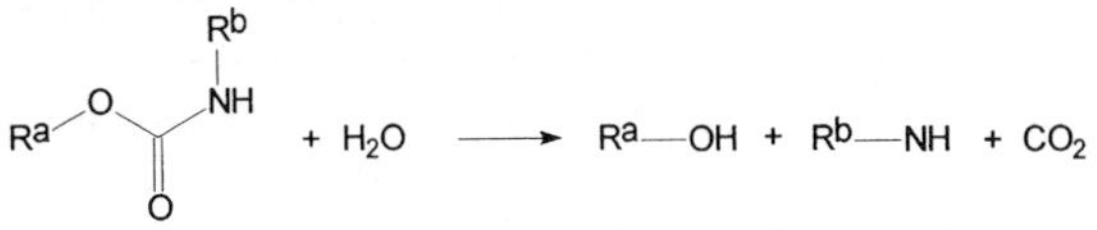

The resulting alcohol and amine can be further derivatized, significantly increasing the sensitivity of the analysis.

- Substitution reactions at the aromatic ring

Electrophilic substitution at the aromatic ring is a common reaction in organic chemistry. Its use in derivatizations is not uncommon for various aromatic compounds, mainly for compounds with activating groups such as O^-, OH, OR, OCOR, NH_2, NHR, NR_2, NHCOR, and SR that contain an unshared pair of electrons on the atom connected to the ring. The groups NR_2, NHR, NH_2, and OH are strongly activating groups. It is interesting that halogens, although they are ortho-para directing groups, deactivate the aromatic ring. Among the electrophilic reactions at the aromatic ring with analytical applications are reactions with halogens used for the ECD detection of the analyte. For example, various amines can be derivatized with iodine (hydroiodic acid and sodium nitrite) [268] in a reaction as follows:

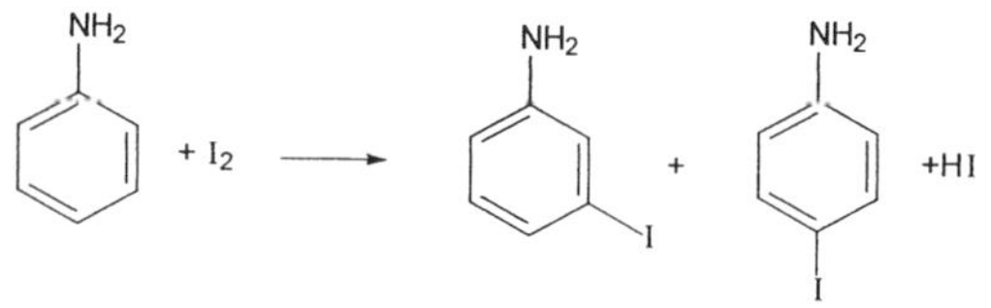

It is usually difficult to predict how much of the product will be ortho isomer and how much para isomer, mainly when various other substituents are present on the aromatic ring.

Diazo coupling is another common reaction used for the preparation of arylazo-derivatives of phenols (naphthols) and aromatic amines. The reaction takes place easily when the acidic solution of an acidic diazonium salt is made alkaline. For a phenol, the reaction can be written as follows:

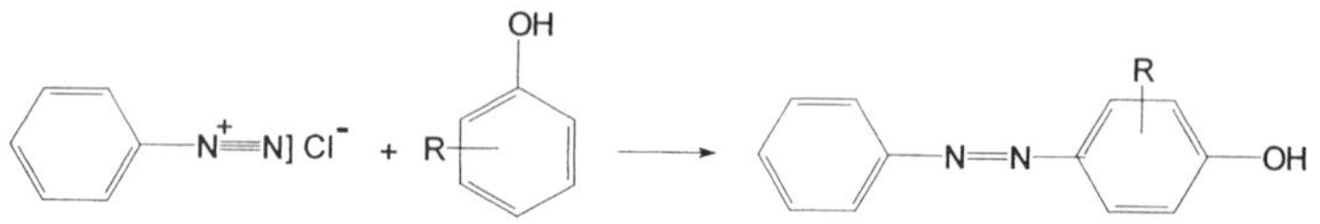

The mechanism of this reaction is electrophilic substitution, and the OH or NH_2 groups are strong *ortho* or *para* directors on the aromatic ring. However, the tendency of *para* substitutions is predominant in most cases. For α-naphthols the coupling tends to occur in position 4 and, if this is occupied, in position 2. For β-naphthols the coupling tends to occur in position 1. Some unexpected reactions may occur during diazo coupling, such as elimination of groups such as COOH when these are in position 4 of a phenol or aromatic amine.

Primary aromatic amines can be transformed into a diazonium salt by treatment with $NaNO_2$ in the presence of an acid. The diazonium salt generated in this way can react with a coupling reagent such as disodium 2-naphthol-3,6-disulfonate, forming a colored compound very sensitive for a UV-Vis detection in HPLC. The sequence of reactions, using as an example 4-amino-biphenyl, is the following:

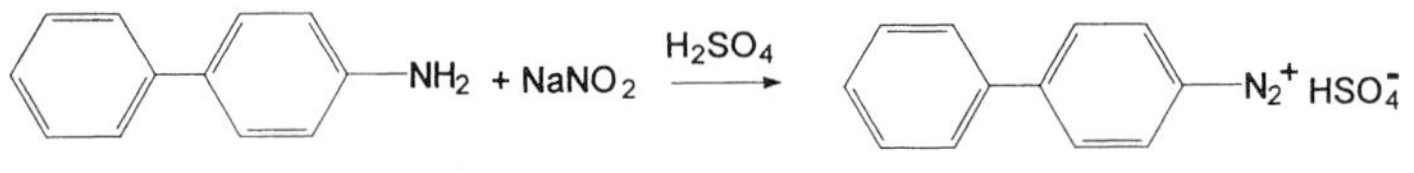

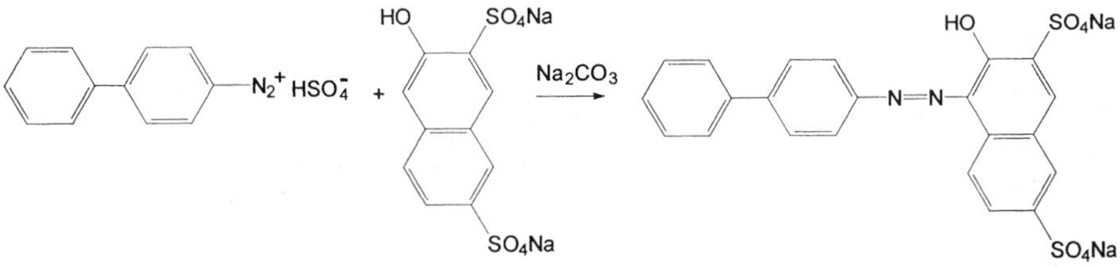

The scheme at the top of the page shows the diazotization and azo coupling reactions:

$$\text{(biphenyl)}-NH_2 + NaNO_2 \xrightarrow{H_2SO_4} \text{(biphenyl)}-N_2^+ \; HSO_4^-$$

$$\text{(biphenyl)}-N_2^+ \; HSO_4^- + \text{(naphthalene with } HO, \; SO_4Na, \; SO_4Na) \xrightarrow{Na_2CO_3} \text{(biphenyl)}-N=N-\text{(naphthalene with } HO, \; SO_4Na, \; SO_4Na)$$

Traces of 4-aminobiphenyl and of other aromatic amines can be detected by this procedure with a limit of detection of 0.15–0.3 ng [269].

- **Complexation and formation of coordinative compounds with metal ions**

Complexation or chelation of metal ions with specific organic molecules has been used extensively in the past for spectrophotometric determination of various metals (see e.g. [270], [271]. Some of these metal complexes or chelates were applied in chromatographic separations, either for the analysis of metals using the organic compounds as reagents or for the analysis of certain organic compounds using metal ions as reagents. For example 1,1,1-trifluoroacetylacetone can be used to form compounds with ions such as Be^{2+}, Al^{3+}, Ga^{3+} etc., which are sufficiently stable and volatile to allow gas chromatographic determination. The reaction leads to the formation of chelates as follows:

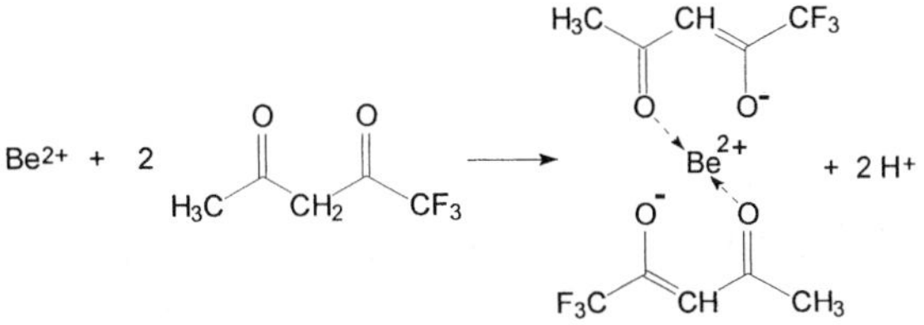

$$Be^{2+} + 2 \; (H_3C-CO-CH_2-CO-CF_3) \rightarrow Be^{2+}(\text{chelate}) + 2 H^+$$

Most derivatizations with a chelating agent are used in HPLC determinations [24]. Many of the reagents used for the derivatization of metal ions have a double role. The first role is to generate a compound that has fluorescent properties or a large extinction coefficient (at a specific wavelength). The second role is to change the ionic form of the metal into a chelate compound leaving practically no free metal ions. For example, arsenazo III or 2,2'-(1,8-dihydroxy-3,6-disulfonaphthalene-2,7-bisazo)bisbenzenearsonic can react with Th(IV), Zr(IV), Sc^{3+}, Y^{3+}, Dy^{3+}, etc., generating a colored compound (absorption at 658 nm with Dy^{3+} ions). The reaction probably involves both ionic and coordinative bonds as shown below:

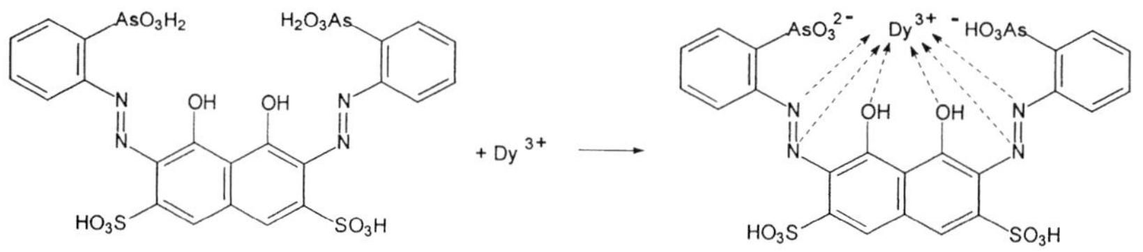

The stability of the chelate compound formed is very high. For example, the reaction $2 La^{3+} + 2 R^{8-} \rightleftharpoons La_2R_2^{10-}$ (where RH_8 is arsenazo III) has an equilibrium constant pK = 81.2, and the reaction $2 Zr^{4+} + 18 H^+ + 2 R^{8-} \rightleftharpoons Zr_2H_{18}R_2^{10+}$ has an equilibrium constant pK = 87.2 [271].

Derivatization with a compound present in the mobile phase also can be applied for HPLC analysis. The compound may assist in the separation or may generate color or a fluorescent compound such as using 8-hydroxyquinoline in the mobile phase for the analysis of Mg^{2+} and Al^{3+} [272]. Some other reactions of this type were used in postcolumn derivatizations in HPLC analyses [24], [273], [274].

- Other reactions

Other reactions are used for derivatization, usually involving special reagents but possibly applied for the determination of common analytes. One example is the derivatization of alcohols with 2-chloro-1,3,2-dioxaphospholane. The reaction, done in the presence of triethylamine for elimination of the formed HCl, takes place as follows:

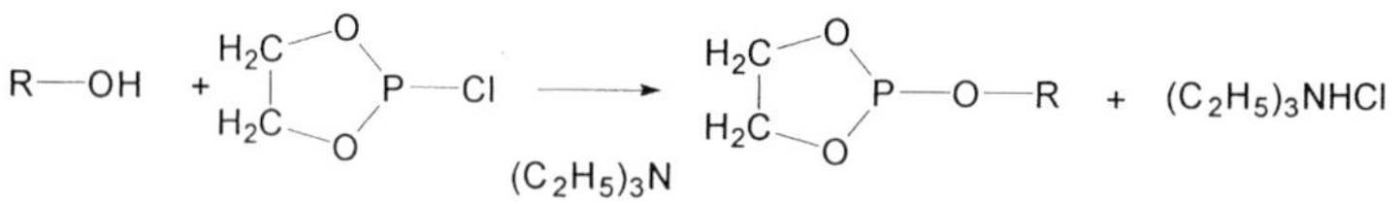

The reaction takes place rapidly. The substituted 4-methyl-2-chloro-1,3,2-dioxa-phospholane as well as 2-chloro-1,3,2-dioxaphosphane also are used in similar reactions [275].

Another reaction sometimes used for the analysis of tertiary amines is the elimination of the amino group using a Hofmann degradation. The reaction takes place for quaternary ammonium hydroxides when they are heated. For this purpose, the tertiary amine is alkylated to a quaternary amine, and this eliminates the amino group during heating, usually in the injection port of the GC. An example of this reaction is the analysis of amitriptyline [276]. First the analyte is methylated using CH_3I and Ag_2O, and then the quaternary amine is decomposed following the reactions:

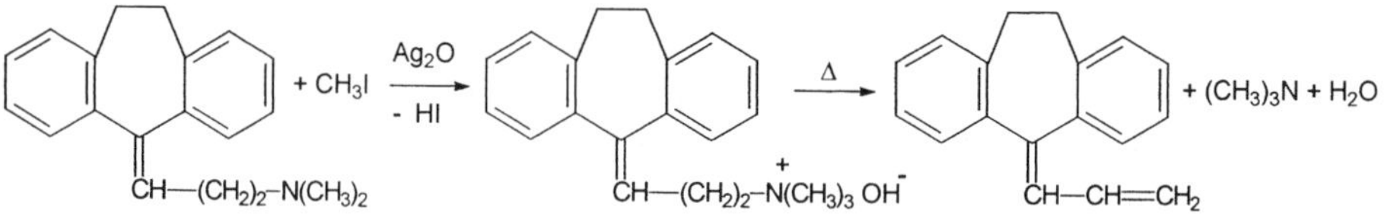

Some other less common reactions that take place between particular analytes with special reagents are given in Chapter 19.

18.7 DERIVATIZATION REACTIONS INVOLVING SOLID PHASE REAGENTS

Solid phase reagents are polymeric materials with specific groups that are reactive and can be transferred to the analyte molecule. These groups may carry fluorescence properties, enhanced light absorbance, etc. For an analyte of the form Y:H, the reaction with a solid phase reagent can be written as follows:

$$Y{:}H \; + \; R\!-\!\text{Polymer} \; \rightleftarrows \; Y\!-\!R \; + \; \text{Polymer:}H$$

Solid phase reagents must work analogously to the corresponding small molecule reagents containing the group R (a tag) and a reactive functionality.

Reagents that are insoluble in certain solvents at high concentrations often can provide a high ratio of analyte/substrate in a polymeric microenvironment that yields a high kinetic rate for the heterogeneous reaction [277], [277a]. In general, immobilized reagents provide a very high ratio of reagent (or tag) to analyte, often at levels of hundreds to one. Because of the microenvironments present in a solid support, reactions are often much more selective and specific when compared to their solution analogs. The polymer plays a direct role in such solid phase reactions, and its pore size, pore diameter, surface area, hydrophobicity, and other physical parameters can all be involved in nucleophilic displacement type reactions [278]. As the derivative tag is immobilized or coated onto the polymeric support, only that portion that actually reacts with the analytes is released into the solution as part of the derivatization. All of the remaining derivatizing reagent remains immobilized, and thus there will be excess or unused reagent [279], [280].

Immobilized reagents can be placed on-line precolumn in HPLC and be used only when needed to perform automated derivatizations, then switched off-line when the analyte conversion is no longer needed. The derivatization also can be performed off-line using disposable cartridges, which can be used several times and then discarded or regenerated. On-line solid-phase reagents can be regenerated overnight in an automated fashion. Both achiral and chiral tags can be placed onto the solid support, using covalent attachments or even adsorption [281].

Mixed-bed polymeric derivatizations allow simultaneous formation of more than a single derivative. The formation of a mixture of reaction products may depend on specific reaction conditions and can be done off-line or on-line in HPLC. This procedure can be applied to biological samples without any sample pretreatment or removal of the analyte from the sample matrix [281] and can be used for the optimization of a complex sample analysis. In conclusion, use of solid phase reagents provides several advantages over solution phase reagents [282]. These advantages include the following [283]:

a) Only the amount of reagent that reacts with the analyte is used, and no reagent excess needs to be further separated as in many other derivatizations. Also, the same amount of polymer can be used a number of times before losing the efficiency of derivatization.

b) The derivatizations of certain traces of analytes can be facilitated by high local concentration of the derivatizing groups, and some reactions take place with higher conversion yield.

c) Some reactions are more selective than the corresponding solution-phase ones, allowing functional group discriminations that are not always possible in solution.

d) Solid phase reagents do not cause problems due to reaction by-products, such as lowering the pH due to the formation of strong acids, etc.

e) More than one solid phase reagent can be used at the same time, allowing the derivatization of a set of analytes that would not be possible in solution-phase due to differences in kinetics or cross-reactivity.

f) Solid phase reagents are more stable over time than many solutions of ordinary reagents.

g) Solid phase reagents can be used easily in on-line derivatization setups.

Various materials are used as solid support for the solid phase reagents, such as modified polystyrene-divinylbenzene, nylon, or silica. Some solid phase reagents have the group carrying the desired properties for derivatization directly connected to the polymeric material. Other solid phase reagents have an anchor group such as nitrophenol or benzotriazolol that reacts with acyl chlorides or chloroformates to form the solid phase reagent.

One type of solid phase reagent is obtained from commercial Amberlite IRA 400 ion exchange resin (chloride form), which can be transformed into polystyrene-divinylbenzene benzoxazole-sulfonate. For making the solid phase reagent, the polymer is treated with sodium benzoxazole-2-sulfonate (1 g resin air-dried and 696 mg sodium benzoxazole-2-sulfonate in 50 mL water at 75° C for 30-45 min.). This solid phase reagent can be used for amines derivatization in a reaction as follows:

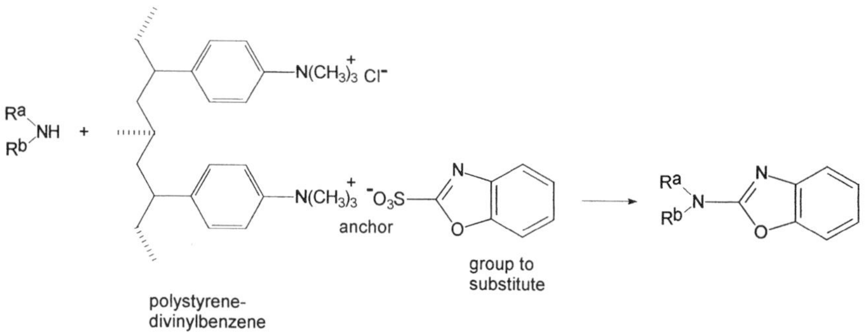

The anchor group in this reagent is based on an ionic bond to the moiety carrying the desired analytical properties. The reagent is in fact a small molecule with multiple functionalities, one of them allowing adsorption into a strong anion exchanger resin.

This type of solid phase reagent can be obtained using different other multifunctional molecules and an ion exchange resin.

Some other solid phase reagents can be prepared easily using available polymers. One such example is trifluoroacetyl nylon 6,6. This solid phase reagent can be obtained from poly(hexamethyleneadipamide) (nylon 6,6) and trifluoroacetyl anhydride [284]. The reagent can be used in amine derivatization in a reaction as follows:

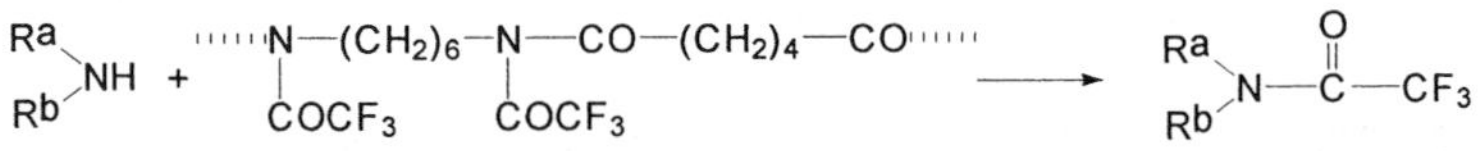

This derivatization of the amine is done by mixing 230 mg solid phase reagent with a solution of 1 mM amine solution in 5 mL CH_3CN. The mixture is stirred for 12 hours at room temperature, and then the solid reagent is separated by centrifugation and washed with two portions of CH_3CN, which is combined with the initial solution. The derivatized analyte is concentrated by evaporating part of the solvent and is analyzed by GC.

The preparation of some other solid phase materials is more elaborate. As an example, polymer bound dimethylamino-pyridinium fluorenylmethyl carbamate can be prepared beginning with chloromethylstyrene-divinylbenzene copolymer as a solid phase. The polymer is subjected to a series of reactions with methylamine, then with 4-chloropyridine hydrochloride, and then with 9-fluorenylmethyl chloroformate [285]. The resulting product can be used for the derivatization of alcohols (in flour) following the reaction:

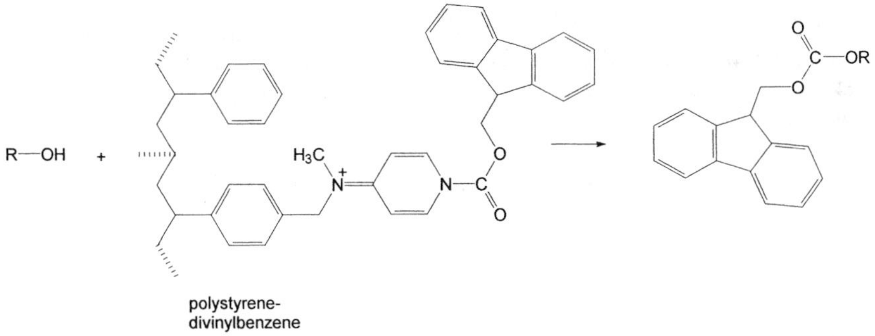

polystyrene-divinylbenzene

The derivatized alcohols can be analyzed by HPLC on a C18 column with a detection limit of 150 ppb with fluorescence detection (ex. 265 nm and em. 320 nm).

By use of a polymer containing benzotriazole groups anchored on a polystyrene type polymer and connected to the 3,5-dinitrobenzoyl group, the following reaction can take place with amines:

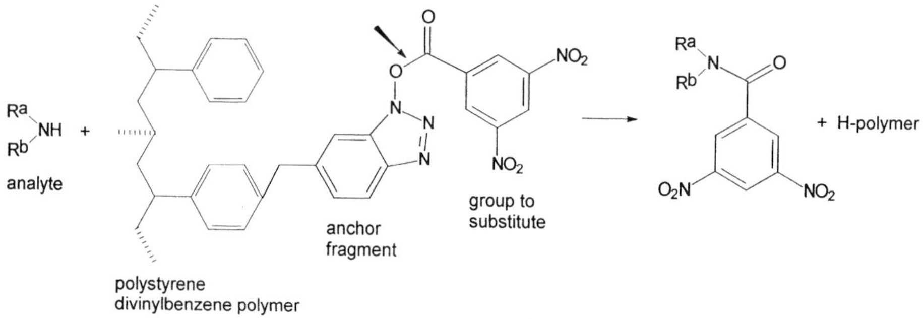

Ra–NH–Rb + [polystyrene divinylbenzene polymer with anchor fragment and group to substitute] → [Ra–Rb–N–C(=O)–benzene–NO₂/NO₂] + H-polymer

analyte

polystyrene
divinylbenzene polymer

anchor
fragment

group to
substitute

The same reaction with a small molecule reagent takes place as follows:

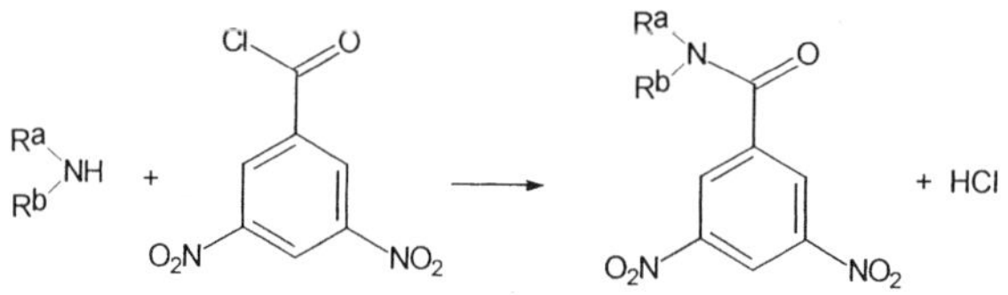

The reactivity of solid phase reagents, including the capability to work with specific analytes and to have desired reaction kinetics, varies significantly from reagent to reagent. The selection of the polymeric moiety and of the anchor fragment strongly influences reactivity. The transferred group may function as acyl, chloroformate, carbonate, etc., and this also may affect reactivity.

Silica also can be used as support, using various attachments to its OH groups. For this purpose, silica dried over a saturated aqueous solution of LiCl is treated with 3-[bis(2-hydroxyethyl)amino]propyl-triethoxysilane in ethanol (e.g. 5 g silica + 8.4 g 62% sol. in ethanol). The mixture is then treated with 40 mL ethanol:pyridine 99.5:0.5 and refluxed with stirring for 6 hours. The result is the following derivatization of silica:

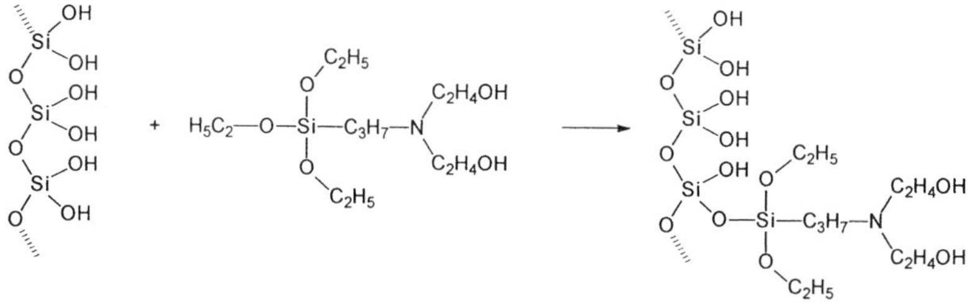

The resulting material is further treated with 4-hydroxy-3-nitrobenzoyl chloride, where the following reaction occurs:

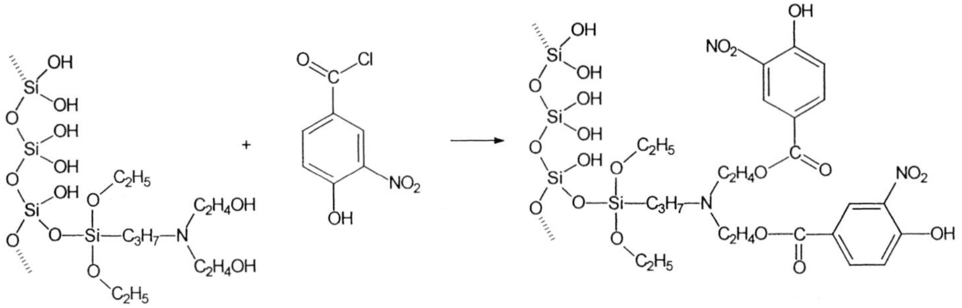

The material can be further modified to attach a fluorescing moiety using, for example, 9-fluoreneacetyl chloride or 3,5-dinitrobenzoyl chloride:

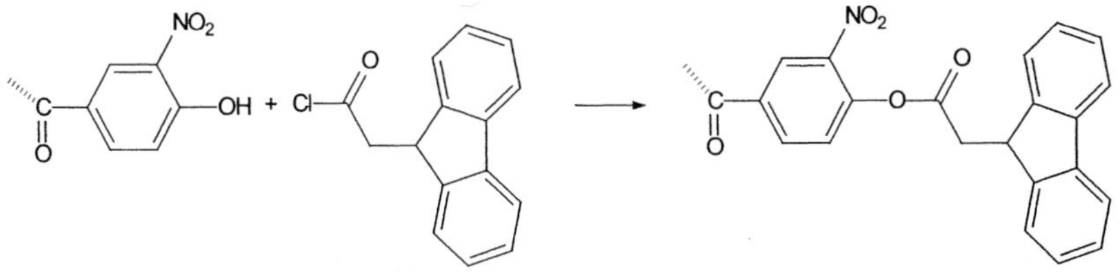

The solid phase reagent can react with amines as follows:

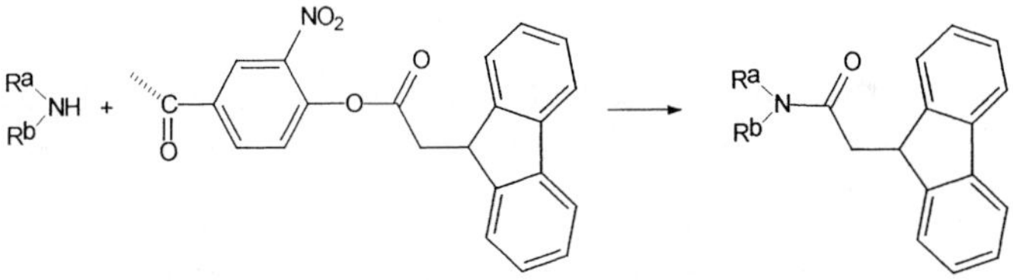

This solid phase reagent was used, for example, for the analysis of cadaverine in urine. For this purpose, the solid phase material (10 mg) was mixed directly with the sample (such as 50 μL urine), heated at 60° C for 10 min., and extracted with 1 mL CH₃CN. An aliquot of this sample is then separated by HPLC on a C18 column, and the fluorescence measured using excitation at 254 nm and emission at 313 nm [24]. The solid phase reagents can be reused after washing with an appropriate solvent such as CH₃CN.

As seen in previous examples, it is preferable that the solid supports have some active groups such as OH, NH, etc. However, polymers such as polystyrene or polystyrene with 10–12% divinylbenzene have been used as starting material for solid phase reagents. One such procedure starts with a reaction of the solid crosslinked polystyrene with 4-chloro-3-nitrobenzyl alcohol in the presence of anhydrous AlCl₃ (at 70° C for three days) as follows:

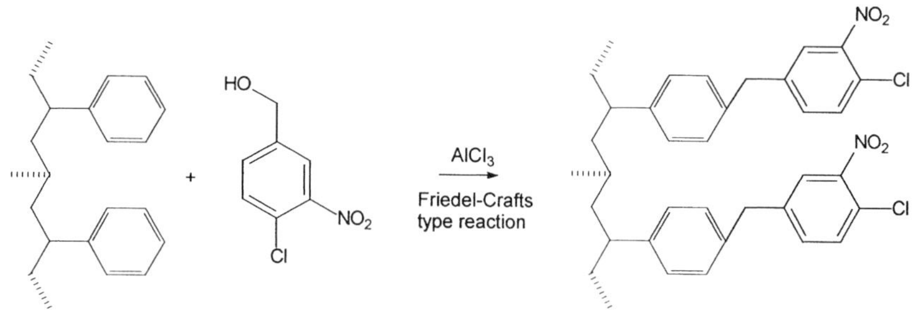

The resulting polymer and hydrazine (hydrazine hydrate in ethoxyethanol 2:3 v/v) are boiled with reflux for 20 hours. The resulting 3-nitro-4-hydrazine-benzylated polymer is then refluxed for 20 hours with HCl in dioxane (1:1 v/v), forming 1-hydroxybenzotriazole-bound polystyrene as shown below [286]:

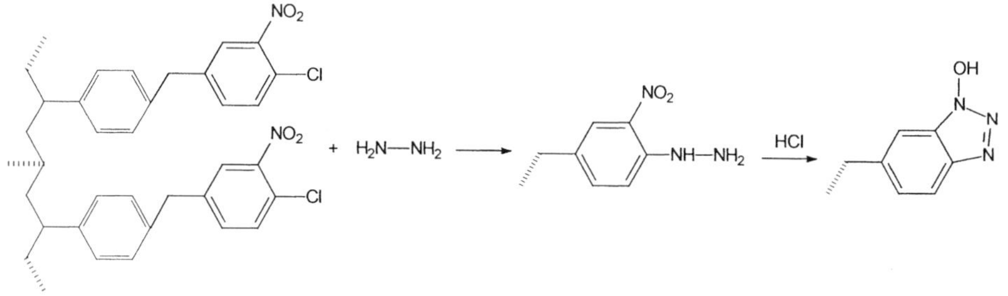

The product 1-hydroxybenzotriazole-bound polystyrene can be further modified to attach the chosen group which carries desired properties and is needed for the derivatization of the analyte. The reaction can be done with o-acetylsalicyloyl chloride, quinoline-6-acyl azide (generating quinoline-6-isocyanate), 3,5-dinitrophenylcarbamate, 3,5-dinitrobenzoyl chloride, 9-fluorenyl-methylchloroformate, etc. [287], [24]. The reaction with 3,5-dinitrobenzoyl chloride is shown below:

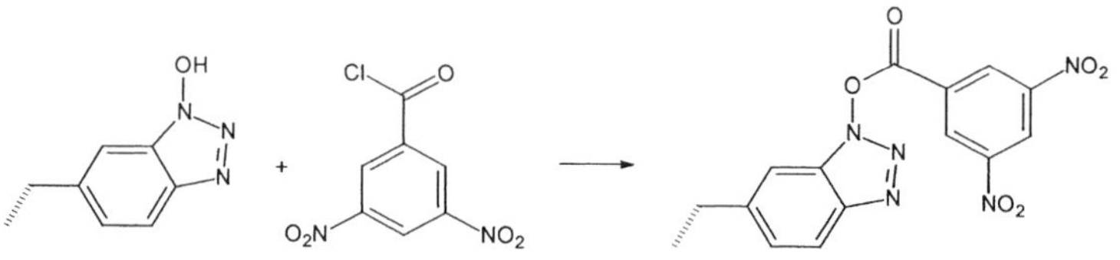

Another possibility is to use polystyrene type polymers in a Friedel-Crafts acylation reaction, for example with 4-methoxy-3-nitro-benzoyl chloride. This reaction can be written as follows:

The modified polymer is further subjected to hydrolysis to form nitrophenol groups that can be used as anchor for fluoreneacetyl, dinitrobenzoyl, or other groups that will be used for derivatization [288]. Other polymer bound compounds also are reported [283], the typical mechanism of making these solid phase reagents being similar to those already described.

18.8 DERIVATIZATION ON A SOLID SUPPORT

In the derivatization on a solid support, the reaction between the reagent and the analyte takes place with the two compounds adsorbed on a solid support (the term adsorption is used here with the larger meaning that indicates the process of concentrating a solute into a stationary phase). It is possible to have the reagent initially adsorbed followed by the adsorption/reaction with the analyte, or it is also possible for the analyte to be adsorbed on the solid support followed by the derivatization. The derivatization on a solid support is usually applied with techniques such as SPE or SPME used for the adsorption step. The main difference in this derivatization procedure from the derivatization using solid phase reagents consists of the nature of the bonds between the solid phase and the analyte or reagent. The adsorbed compounds can be considered as dissolved in the stationary phase and not ionically or covalently bonded as with most solid phase reagents. The interactions between the solid phase and the dissolved compounds are usually weak interactions typical for the partition in a stationary phase.

The reactivity of the compounds in adsorbed form compared to that in solution may show some differences. Because the solvent plays a role in many chemical reactions and adsorption on a solid phase can be viewed as equivalent with a solvent change, the same reaction as in solution is not expected always to occur in the solid phase. Most solid phase materials are typically hydrophobic compounds such as polydimethyl-siloxane PDMS, and the reactivity in the solid phase can be expected to be similar to that in a hydrophobic solvent.

An additional factor to consider for the derivatization on a solid support is the modification in the extraction/derivatization kinetics. For a simple chemical reaction of the analyte A with the reagent R forming the product P, which can be written in the form:

$$A + R \rightleftharpoons P$$

the reaction rate can be assumed of second order, and the variation in the concentration of the analyte can be written in the form (see rel. 1.8.14):

$$-\frac{d[A]}{dt} = k[A][R] \tag{18.8.1}$$

If this reaction takes place for example in a SPME fiber, the concentrations in rel. (18.8.1) refer to the concentrations in the fiber. In this case, rel. (11.2.18), which describes the variation in time of the concentration of an analyte in an SPME fiber, will have an additional term (see Section 11.2). Using the notation $c_f = [A]_f$ for the concentration of the analyte in the fiber, rel. (11.2.18) must be substituted by the expression:

$$\frac{\partial c_f}{\partial t} = D_f \nabla^2 c_f - k[R]c_f \tag{18.8.2}$$

The solution for equation (18.8.2) can be obtained by application of boundary conditions from several models for the diffusion processes (each one describing a limiting case) and with the assumption that either the reaction kinetics is slow compared to diffusion, or that diffusion is the slower process. For slow chemical reactions, and assuming a sample of infinite volume, the kinetics of the formation of the product P in the solid phase is described by the equation:

$$\frac{d[P]}{dt} = k[R]K_{fs}c_o \tag{18.8.3}$$

where c_o is the concentration of the analyte in the sample and K_{fs} is the distribution constant of the analyte between the sample and the fiber. This expression is obtained by neglecting the diffusion term in equation (18.8.2) and using rel. (11.2.11) to express c_f as a function of c_o. Rel. (18.8.3) indicates that the derivatization in the fiber proceeds with a rate dependent on the analyte concentration in the sample, the concentration of the reagent, and the distribution coefficient K_{fs}. If the volume of the sample is limited and the concentration of the analyte decreases during the accumulation in the fiber, an estimation of the equilibration time t_e when 95% of the ideal amount of analyte has reacted can be obtained using the expression [289]:

$$t_e = t_{95\%} = \frac{4.6V_s}{kK_{fs}V_f} \tag{18.8.4}$$

where V_s is the solution volume and V_f is the fiber volume.

A number of applications of the derivatization directly on the fiber are reported in literature [289], [290]. Only two examples are given below. A method has been developed for SPME sampling with in-fiber derivatization for the analysis of airborne formaldehyde (typically, less then 100 ppb). In this method, the derivatization agent is o-(2,3,4,5,6-pentafluorobenzyl)-hydroxylamine hydrochloride, which is loaded onto the SPME fiber coating (PDMS) chosen as a solid sorbent. After exposing the fiber to the air sample, the reaction between the derivatization agent and formaldehyde absorbed onto the coating forms an oxime. The oxime is thermally desorbed in a GC injector port and analyzed by GC with ECD [291]. A similar principle has been used to determine a number of fatty acids ($C_2 - C_{10}$) in aqueous samples. The reagent selected for the derivatization is 1-pyrenyldiazomethane, which requires mild reaction conditions and produces stable reaction products with high yields. The coating sorbent used in this analysis is poly(acrylate), which is polar and has a higher capacity compared to PDMS of retaining polar carboxylic acids. A poly(acrylate) coated fiber doped with the derivatization reagent can be placed in the headspace of the sample, the reaction taking place at room temperature. After the derivatization takes place, the fiber can be desorbed and analyzed [292].

REFERENCES 18

1. K. Blau, J. Halket (eds.), *Handbook of Derivatives for Chromatography*, J. Wiley, Chichester, 1993.
2. I. Ciucanu, F. Kerek, Carbohydr. Res., 13 (1984) 209.
3. H. G. Walker, Jr., et al., J. Org. Chem., 27 (1962) 2100.
4. D. R. Knapp, *Handbook of Analytical Derivatization Reactions*, J. Wiley, New York, 1979.
5. V. Kovácik, P. Kovác, Chem. Zvesti, 27 (1973) 662.
6. H. R. Morris et al., Biochem. J., 125 (1971) 189.
7. H. R. Morris et al., Biochem. J., 125 (1971) 189.
8. R. N. Shah et al., J. Carbohydr. Chem., 6 (1987) 645.
9. S. Hakomori, J. Biochem. (Tokyo), 55 (1964) 205.
10. W. G. Niehaus, Jr., R. Ryhage, Anal. Chem., 40 (1968) 1840.
11. E. J. Corey, M. Chaykovsky, J. Am. Chem. Soc., 84 (1962) 866.
12. J. P. Parente et al., Carbohydr. Res., 141 (1985) 41.
13. W. A. Dechtiaruk et al., Clin. Chem, 22 (1976) 879.
14. C. J. Pedersen, J. Am. Chem. Soc., 89 (1967) 7017.
15. C. J. Pedersen, H. K. Frensdorff, Angew. Chem. Int. Ed. Eng., 11 (1972) 16.
16. E. H. Dehmlow, Angew. Chem. Int. Ed. Eng., 13 (1974) 170.
17. J. Dockx, Synthesis, (1973) 441.
18. K. T. Koshy et al., J. Chromatogr. Sci., 13 (1975) 327.
19. J. P. Thenot, E. C. Horning, Anal. Lett, 5 (1972) 519.
20. P. A. Bond, M. Cagnasso, J. Chromatogr., 109 (1975) 389.
21. C. J. Biemann, G. D. McGinnis (eds.), *Analysis of Carbohydrates by GLC and MS*, CRC Press, Boca Raton, 1989.
22. M. F. Chaplin, J. F. Kennedy (eds.), *Carbohydrate Analysis, A Practical Approach*, IRL Press, Oxford, 1986.
23. J. March, *Advanced Organic Chemistry*, J. Wiley, New York, 1992.
24. G. Lunn, L. C. Hellwig, *Handbook of Derivatization Reactions for HPLC*, J. Wiley New York, 1998.
25. W. N. Haworth, J. Chem. Soc., 107 (1915) 8.
26. J. Saurina, S. Hernandez-Cassou, J. Chromatogr., A, 676 (1994) 311.
27. F. Ngan, M. Toofan, J. Chromatogr. Sci., 29 (1991) 8.
28. H. W. Mueller, J. Chromatogr., B, 679 (1996) 208.
29. W. Dunges, Anal. Chem., 45 (1973) 963.
30. H. W. Mueller, Aldrichim. Acta, 31 (1998) 2.

31. T. H. Black, Aldrichim. Acta, 16 (1983) 3.
32. L. F. Christensen, A. D. Broom, J. Org. Chem., 37 (1972) 3398.
33. S. Kudoh et al., J. Chromatogr., B, 660 (1994) 205.
34. H. Hengy, M. Most, J. Liq. Chromatogr., 11 (1988) 517.
35. A. Schönberg, A. Mustafa, J. Chem. Soc., (1946) 746.
36. H. Vorbrüggen, Angew. Chem. Int. Ed. Eng., 2 (1963) 211.
37. A. Holý, Tetrahedon Lett., 7 (1972) 519.
38. Pierce Catalog 1999, Pierce Chem. Co., Rockford, 1999.
39. A. E. Smith, J. Agric. Food Chem., 37 (1989) 267.
40. H.-P. Klemn et al., J. Chromatogr., 75 (1973) 19.
41. R. Bedoud, G. Pratz, J. Chromatogr., 360 (1986) 119.
42. E. Felder et al., J. Chromatogr., 82 (1973) 291, 390.
43. A. I. Haj-Yehia, L. Z. Benet, J. Chromatogr., A, 724 (1996) 107.
44. H. Ko, M. E. Royer, J. Chromatogr., 88 (1974) 253.
45. S. T. Wu et al., J. Chromatogr., B, 692 (1997) 149.
46. H. Meerwein, Org. Synth., 46 (1966) 113, 120.
47. S. C. Moldoveanu, *Analytical Pyrolysis of Natural Organic Polymers*, Elsevier, Amsterdam, 1998.
48. J. Pecci, T. J. Giovanniello, J. Chromatogr., 109 (1975) 163.
49. K. O. Gerhardt, C. W. Gehrke, J. Chromatogr., 143 (1977) 335.
50. W. C. Kossa et al., J. Chromatogr. Sci., 17 (1979) 177.
51. M. Amijee et al., J. Chromatogr., A, 738 (1996) 43.
52. H.-L. Hardell, N.-O. Nilvebrant, J. Anal. Appl. Pyrol., 52 (1999) 1.
53. W. Butte, J. Chromatogr., 261 (1983) 142.
54. H. Faber, H. F. Scholer, J. Agric. Food Chem., 41 (1993) 217.
55. S.-N. Lin, E. C. Horning, J. Chromatogr., 112 (1975) 483.
56. J. MacGee, K. G. Allen, J. Chromatogr. 110 (1974) 35.
57. B. Anderson, R. Holman, Lipids, 10 (1975) 716.
58. M. L. Campbell, B. A. Waite, J. Chem. Educ., 67 (1990) 386.
59. F. Badea, *Mecanisme de Reactie in Chimia Organica*, Ed. Stiintifica, Bucuresti, 1973.
60. J. Zemlicka, Coll. Czech. Chem. Commun., 28 (1963) 1060.
61. G. G. S. Dutton, Adv. Carbohydr. Chem. Biochem., 30 (1974) 9.
62. R. Osievicz et al., J. Chromatogr., (1974) 157.
63. G. van Look et al., *Silylating Agents*, Fluka Chemie AG, Buchs, 1995.
64. A. E. Pierce, *Silylation of Organic Compounds*, Pierce Chem. Co., Rockford, 1982.
65. T. Morita et al., Tetrahedron Lett., 21 (1983) 835.
66. L. Birkhofer, P. Sommer, J. Organomet. Chem., 99 (1975) C1.
67. K. Kantlehner et al., Chem. Ber., 105 (1972) 2264.
68. M. Dönike, J. Chromatogr., 74 (1972) 121.
69. C. W. Gehrke, A. B. Patel, J. Chromatogr., 130 (1977) 103.
70. L. Aringer at al., Steroids, 17 (19710 377.
71. E. W. Abel, J. Chem. Soc., (1960) 4406.
72. E. Nakamura et al., J. Am. Chem. Soc., 98 (1976) 2346.
73. E. W. Abel, J. Chem. Soc., (1961) 4933.
74. M. Dönike, L. Jaenike, Angew. Chem., 81 (1969) 995.
75. Y. Kita et al., Tetrahedron Lett., 20 (1979) 4311.
76. T. Veysoglu, L. A. Mitscher, Tetrahedron Lett., 22 (1981) 1303.

77. M. Dönike, J. Chromatogr., 103 (1975) 91.
78. M. Dönike, J. Chromatogr., 42 (1969) 103.
79. M. Dönike, J. Chromatogr., 85 (1973) 1.
80. M. Dönike, J. Chromatogr., 115 (1975) 591.
81. R. Piekos et al., J. Chromatogr., 117 (1976) 431.
82. H. H. Hergott, G. Simchen, Liebigs Ann. Chem., (1980) 1718.
83. D. Knausz et al., J. Chromatogr., 365 (1986) 183.
84. E. Csato et al., J. Chromatogr., 348 (1985) 280.
85. H. H. Hergott, G. Simchen, Liebigs Ann. Chem., (1980) 1718.
86. R. Piekos et al., J. Chromatogr., 116 (1976) 315.
87. J. M. Renga, P.-C. Wang, Tetrahedon Lett., 26 (1985) 1175.
88. C. C. Sweeley et al., J. Am. Chem. Soc., 85 (1963) 2497.
89. W. J. A. VandenHeuval, J. Chromatogr., 36 (1968) 354.
90. J. B. Forehand et al., 49th TCRC, Lexington, 1995.
91. W. Butts, Anal. Lett., 3 (1970) 29.
92. S. H. Langer et al., J. Org. Chem., 23 (1958) 50.
93. H. Osman et al., J. Chromatogr., 186 (1979) 273.
94. K. E. Rasmussen, J. Chromatogr., 120 (1976) 491.
95. G. Phillipou, J. Chromatogr., 129 (1976) 384.
96. I. A. Blair, G. Phillipou, J. Chromatogr. Sci., 15 (1977) 478.
97. H. Miyazaki et al., J. Chromatogr., 133 (1977) 311.
98. S. Hanessian, P. Lavallee, Can. J. Chem., 53 (1975) 2975.
99. C. F. Poole, A. Zlatkis, J. Chromatogr. Sci., 17 (1979) 115.
100. M. J. Bertrand et al., J. Chromatogr., 351 (1986) 47.
101. M. J. Bertrand et al., J. Chromatogr., 354 (1986) 331.
102. M. A. Quilliam, J. B. Westmore, Anal. Chem., 50 (1978) 59.
103. M. A. Quilliam, J. B. Westmore, Steroids, 29 (1977) 579.
104. M. A. Quilliam, J. B. Westmore, Steroids, 29 (1977) 613.
105. C. A. White et al., J. Chromatogr., 264 (1983) 99.
106. H. Miyazaki et al., Biomed. Mass Spectrom., 4 (1977) 23.
107. S. C. Moldoveanu, J. B. Forehand, unpublished results.
108. W. J. Richter, D.H. Hunnemann, Helv. Chim. Acta, 57 (1974) 1131.
109. D. J. Ager, I. Fleming, J. Chem. Res., (S) (1977) 6.
110. J. B. Brooks et al., Anal. Chem., 47 (1975) 1960.
111. A. Fukunaga et al., J. Chromatogr., 190 (1980) 339.
112. K. Grob, G. Grob, J. High Res. Chromatogr., 3 (1980) 197.
113. H. Rembold, B. Lackner, J. Chromatogr., 323 (1985) 355.
114. E. D. Morgan, C. F. Poole, J. Chromatogr., 89 (1974) 225.
115. F. W. McLafferty, *Interpretation of Mass Spectra,* University Science Books, Mill Valley, 1980.
116. T. P. Mawhiney et al., J. Chromatogr., 358 (1986) 231.
117. J. W. Gillard et al., J. Org. Chem., 53 (1988) 2602.
118. C. J. W. Brooks et al., J. Chromatogr., 514 (1990) 305.
119. D. J. Harvey, J. Chromatogr., 196 (1980) 156.
120. D. J. Harvey, Spectrosc. Int. J., 8 (1990) 211.
121. J. J. P. Stewart, J. Comp. Molec. Design, 4 (1990) 1.
122. S. C. Moldoveanu, A. Savin, *Aplicalii in Chimie ale Metodelor Semiempirice de Orbitali Moleculari*, Ed. Academiei, Bucuresti, 1980.

123. S. Moldoveanu, *Aplicatiile Teoriei Grupurilor in Chimie*, Ed. St. and Encicl., Bucharest, 1975.
124. O. Bayer, E. Muler (eds.), *Houben-Weil Methoden der Organischen Chemie*, vol. XIII/5, Verlag, Stuttgart (1980), p.14.
125. G. Petersson, Org. Mass Spectrom., 6 (1972) 577.
126. D. C. DeJongh et al., J. Am. Chem. Soc., 91, (1969) 1728.
127. H. Tsuchida et al., Carbohydrate Res., 67 (1978) 549.
128. Zs. F. Katona et al., J. Chromatogr., A, 847 (1999) 91.
129. J. L. Little, J. Chromatogr., 844 (1999) 1.
130. O. A. Mamer, S. S. Toja, Can. Clin. Chem., 19 (1973) 417.
131. E. M. Chambaz et al., Anal. Lett., 1 (1968) 749.
132. E. Bombardelli et al., J. Chromatogr., 139 (1977) 111.
133. C.-G. Hammer, Biomed. Mass Spectrom., 5 (1978) 25.
134. J. E. Staufer, J. Chromatogr. Sci., 7 (1969) 124.
135. H. Ehrsson, H. Brötell, Acta Pharm. Suec., 8 (1971) 591.
136. J. Goto et al., Anal. Sci., 2 (1986) 585.
137. D. G. Saunders, L. E. Vanatta, Anal. Chem., 46 (1974) 1319.
138. C. D. Pfaffenberger et al., Anal. Biochem., 63 (1975) 501.
139. R. F. Adams, J. Chromatogr., 95 (1974) 189.
140. R. Wachowiak, K. A. Connors, Anal. Chem., 51 (1979) 27.
141. M. Belikova, J. Kohlicek, J. Chromatogr., 497 (1989) 159.
142. G. D. Paulson et al., Anal. Chem., 45 (1973) 21.
143. D. P. Schwartz, Anal. Biochem., 71 (1967) 24.
144. D. Zararis et al., Chromatographia, 3 (1970) 180.
145. J. S. Fritz, G. H. Schenk, Anal. Chem., 31 (1959) 1808.
146. G. Scarping et al., J. Chromatogr., 267 (1983) 315.
147. H. Ehrsson, B. Mellström, Acta Pharm. Suec., 9 (1972) 107.
148. P. A. Metz et al., J. Chromatogr., 479 (1989) 107.
149. T. A. Sasaki et al., Anal. Lett., 34 (2001) 1749.
150. F. E. Keiser et al., J. Chromatogr., 94 (1974) 113.
151. G. Gamerith, J. Chromatogr., 256 (1983) 267.
152. G. Gamerith, J. Chromatogr., 256 (1983) 326.
153. A. F. Cockerill et al., J. Chromatogr., 114 (1975) 151.
154. R. W. H. Edwards et al., J. Endocrinol., 30 (1964) 181.
155. K. Sasamoto et al., Anal. Sci., 12 (1996) 189.
156. M. Dönike, J. Chromatogr., 78 (1973) 273.
157. M. Dönike, J. Chromatogr., 103 (1975) 91.
158. G. Schwedt, H. H. Bussemas, J. Chromatogr., 106 (1975) 440.
159. S. Murray, T. A. Baillie, Biomed. Mass Spectrom., 6 (1979) 82.
160. F. Benington et al., J. Chromatogr., 106 (1975) 435.
161. J. Vessman et al., Anal. Lett., 2 (1969) 81.
162. O. Busto et al., J. Chromatogr., A, 737 (1996) 205.
163. J. C. Motte et al., J. Chromatogr., A, 728 (1996) 333.
164. H. Brückner, M. Lüpke, Chromatographia, 40 (1995) 60.
165. M. Pugniere et al., J. Chromatogr., A, 767 (1997) 69.
166. D. P. Rose, P. A. Toseland, Clin. Chim. Acta, 17 (1967) 235.
167. J. E. Staufer, J. Chromatogr. Sci., 7 (1969) 124.
168. A. Zlatkis, B. C. Petit, Chromatographia, 2 (1969) 484.
169. E. E. Knaus et al., J. Chromatogr. Sci., 14 (1976) 525.

170. S. Yamamoto et al., J. Chromatogr., 194 (1980) 399.
171. P. Hartvig, J. Vessman, Anal. Lett., 157 (1974) 223.
172. P. J. M. Kwakman et al., J. Pharm. Biomed. Anal., 9 (1991) 753.
173. F. Cattabeni et al., Science, 178 (1972) 166.
174. R. Nakajima et al., Bull. Chem. Soc. Jap., 63 (1990) 1968.
175. D. K. Slobodzian et al., J. Pharm. Sci., 76 (1987) 169.
176. T. Akiyama, J. Chromatogr., 588 (1991) 53.
177. W. T. Wang et al., Anal. Biochem., 141 (1984) 366.
178. G. Okafo et al., Anal. Chem., 68 (1996) 4424.
179. M. Moore, Org. React., 5 (1949) 301.
180. J. Liu et al., Proc. Natl. Acad. Sci. USA, 88 (1991) 2302.
181. K. Kuwata et al., J. Chromatogr. Sci., 17 (1979) 264.
182. R. W. Frei, J. F. Lawrence, J. Chromatogr., 83 (1973) 321.
183. W. Pigman, D. Horton (eds.), *The Carbohydrates*, Academic Press, New York, 1972.
184. J. Szafranek et al., Anal. Lett., 6 (1973) 479.
185. C. D. Phaffenberger et al., J. Chromatogr., 126 (1976) 585.
186. M. Naoi et al., J. Chromatogr., 434 (1988) 209.
187. J.-N. Zeng et al., J. Chromatogr., B, 654 (1994) 231.
188. W. H. Pirkle et al., J. Liq. Chromatogr., 9 (1986) 443.
189. S. Alessi-Severini et al., J. Chromatogr., 582 (1992) 173.
190. Y. Zhou et al., J. Liq. Chromatogr., 13 (1990) 875.
191. I. Demian, D. F. Gripshover, J. Chromatogr., 466 (1989) 415.
192. D. M. Desai, J. Gal, J. Chromatogr., 579 (1992) 165.
193. C. F. Poole, A. Zlatkis, J. Chromatogr., 184 (1980) 99.
194. E. A. Emken, Lipids, 7 (1972) 459.
195. J. A. Mills, Advan. Carbohydr. Chem., 10 (1955) 1.
196. J. A. Fioriti et al., J. Chromatogr. Sci., 7 (1969) 448.
197. K. Yamasaki et al., Chem. Pharm. Bull., 22 (1974) 2898.
198. P. Husek, J. Chromatogr., 91 (1974) 483.
199. F. Weygand, K. Burger, Chem. Ber., 99 (1966) 2880.
200. H. D. Simmonds, D. W. Wiley, J. Am. Chem. Soc., 82 (1960) 2288.
201. J. L. O'Donnell et al., J. Am. Oil Chem. Soc., 44 (1974) 652.
202. A. S. Perlin, Can. J. Chem., 42 (1964) 1365.
203. O. Busto, J. Chromatogr., A, 718 (1995) 309.
204. A. M. Lovering et al., Antimicrob. Agents Chemother., 26 (1984) 10.
205. G. Forrest et al., J. Chromatogr., B, 681 (1996) 421.
206. D. M. Desai, J. Gal, J. Chromatogr., 629 (1993) 215.
207. N. Bilic, J. Chromatogr., A, 719 (1996) 321.
208. S. Smith, T. Sharp, J. Chromatogr., B, 652 (1994) 228.
209. M. Kai et al., J. Chromatogr., A, 653 (1993) 235.
210. E. Kojima et al., Anal. Chim. Acta, 248 (1991) 213.
211. E. Kojima et al., J. Chromatogr., 612 (1993) 187.
212. B. S. Middledich, J. Chromatogr., 126 (1976) 581.
213. F. Reche et al., J. Chromatogr., A, 896, (2000) 51.
214. M. Matsuoko et al., Chromatographia, 43 (1996) 501.
215. N. F. Swynnerton et al., J. Liq. Chromatogr., 6 (1983) 1523.
216. H. Nakamura et al., Anal Chem., 56 (1984) 919.
217. K. M. Smith et al., J. Chromatogr., 137 (1977) 431.

218. Y. Zhang et al., Am. J. Physiol., 260 (1991) G658.
219. E. J. Eisenberg, K. C. Kundy, J. Chromatogr., B, 679 (1996) 119.
220. J. Ishida et al., Clin. Chem., 39 (1993) 165.
221. M. Yamaguchi et al., Analyst, 123 (1998) 307.
222. G. Nielsen, E. Solheim, J. Chromatogr., 69 (1972) 366.
223. S. Hara et al., Anal. Chim. Acta, 291 (1994) 189.
224. H. Nahota et al., Anal. Chim. Acta, 287 (1994) 223.
225. P. Edman, G. Begg, Eur. J. Biochem., 1 (1967) 80.
226. L. D. Cromwell, G. R. Stark, Biochemistry, 8 (1969) 4753.
227. J. Rosmus, Z. Deyl, J. Chromatogr., 70 (1972) 221.
228. Z. Deyl, J. Chromatogr., 127 (1976) 91.
229. S. B. Martin et al., Anal. Chem., 47 (1975) 545.
230. T. Toyo'oka (ed.), Modern Derivatization Methods for Separation Sciences, J. Wiley, Chichester, 1999.
231. K. H. Nielsen, J. Chromatogr., 10 (1963) 463.
232. U. Langenbeck et al., J. Chromatogr., 143 (1977) 39.
233. T. Ziegler et al., Synthesis, (1992) 1013.
234. R. W. Kelly, Tetrahedron Lett., (1969) 967.
235. R. W. Kelly, Steroids, 13 (1969) 507.
236. R. W. Kelly, J. Chromatogr., 43 (1969) 229.
237. P. Mohr, C. Tamm, Tetrahedron Lett., 28 (1987) 391.
238. C. J. W. Brooks et al., Analyst, 110 (1985) 587.
239. C. J. W. Brooks et al., J. Chromatogr., 315 (1984) 119.
240. H. Miyazaki, Biomed. Mass Spectrom., 11 (1984) 377.
241. I. S. Krull et al., Anal. Lett., 18 (1985) 2619.
242. V. D. Shatz et al., J. Chromatogr., 174 (1979) 83.
243. M. J. Haas et al., Anal. Chem., 71 (1999) 1574.
244. R. W. Kelly, J. Chromatogr., 71 (1972) 337.
245. M. Cagnasso, P. A. Bondi, Anal. Biochem., 71 (1976) 597.
246. H. Tsuchida et al., Carbohydr. Res., 67 (1978) 549.
247. T. Shibamoto, R. A. Bernhard, J. Agric. Food Chem., 25 (1977) 609.
248. W. W. Weeks et al., J. Agric. Food Chem., 41 (1993) 1321.
249. E. Winkler et al., J. Chromatogr., 436 (1988) 447.
250. W. T. Kok et al., J. Chromatogr., 367 (1986) 231.
251. R. C. Garner et al., J. Chromatogr., 648 (1993) 485.
252. M. Holcomb, H. C. Thompson, Jr., J. Agric. Food Chem., 39 (1991) 137.
253. V. Cavrini et al., Chromatographia, 42 (1996) 515.
254. J. Doehl, T. Greibrokk, J. Chromatogr., 529 (1990) 21.
255. J. Bontemps et al., J. Chromatogr., 307 (1984) 283.
256. T. A. Gehring, J. AOAC Int., 78 (1995) 307.
257. O. Fiehn, M. Jekel, J. Chromatogr., A, 769 (1997) 189.
257a. O. Köster, F. Jüttner, J. Microbiol. Meth., 37 (1999) 65.
258. K. Murayama, T. Kinoshita, Anal. Lett., 14 (1981) 1221.
259. T. Yokoyama, T. Konoshita, J. Chromatogr., 518 (1990) 141.
260. S. Effkemann et al., J. Chromatogr., A, 855 (1999) 551.
261. M. A. Raggi et al., Chromatographia, 46 (1997) 17.
262. J. E. Wallace et al., Anal. Chem., 47 (1975) 1516.
263. O. S. Privett, E. C. Nickell, Lipids, 1 (1966) 98.
264. F. D. Gunstone, R. P. Inglis, Chem. Phys. Lipids, 10 (1973) 105.

265. D. H. Shaw, G. W. Moss, J. Chromatogr., 41 (1967) 350.
266. W. F. Lenhardt, R. J. Winzler, J. Chromatogr., 34 (1968) 471.
267. R. L. Cunico, T. Schlabach, J. Chromatogr., 266 (1983) 461.
268. M. Less et al., J. Chromatogr., A, 810 (1998) 173.
269. F. E. Lancaster, J. F. Lawrence, Food Addit. Contam., 9 (1992) 171.
270. G. Popa, S. Moldoveanu, *Analiza Chimica Cantitative cu Reactivi Organici*, Ed. Tehnica, Bucuresti, 1969.
271. G. Popa, S. Moldoveanu, *Reactivii Organici in Chimia Analitica*, Ed. Academiei, Bucuresti, 1976.
272. T. Takeuchi et al., J. Microcolumn Sep., 12 (2000) 450.
273. R. J. McKracken et al., Analyst, 120 (1995) 1763.
274. R. T. Gettar et al., J. Chromatogr., A, 855 (1999) 111.
275. R. Vilceanu, P. Schultz, J. Chromatogr., 82 (1973) 279.
276. H. B. Hucker, J. K. Miller, J. Chromatogr., 32 (1968) 408.
277. S. T. Colgan et al., J. Chromatogr., 333 (1985) 349.
277a. S. T. Colgan et al., Anal. Chem., 58 (1986) 2366
278. T. Y. Chou et al., J. Chromatogr., 454 (1988) 169.
279. C. X. Gao et al., Anal. Chem., 61 (1989) 1538.
280. T. Y. Chou et al., Anal. Chem., 61 (1989) 1548.
281. C. X. Gao, I. S. Krull, J. Pharm. Biomed. Anal., 7 (1989) 1183.
282. L. Nondek et al., J. Chromatogr., 282 (1983) 141.
283. A. J. Bourque, I. S. Krull, J. Chromatogr., 537 (1991) 123.
284. H. Schuttenberg, R. C. Schultz, Angew. Chem. Int. Ed. Eng., 15 (1976) 777.
285. C. X. Gao, I. S. Krull, J. Chromatogr., 515 (1990) 337.
286. C. X. Gao et al., J. Chromatogr. Sci., 26 (1988) 449.
287. J. H. Yu et al., J. Chromatogr., B, 658 (1994) 249.
288. B. J. Cohen et al., J. Org. Chem., 49 (1984) 922.
289. J. Pawliszyn, *Solid Phase Microextraction, Theory and Practice*, J. Wiley-WCH, New York, 1997.
290. Supelco SPME Application Guide, Sigma-Aldrich Co., 1999.
291. H. L. Lord, J. Pawliszyn, LC-GC, May (1998) 41.
292. L. Pan et al., Anal. Chem., 67 (1995) 4396.

CHAPTER 19

Derivatization Reactions for Analytes with Various Functional Groups

19.1 DERIVATIZATION OF COMPOUNDS WITH ALCOHOL, EPOXIDE, OR ETHER GROUPS

Alcohols contain a hydroxyl functional group OH with an active hydrogen. For GC analysis, the formation of hydrogen bonds of the OH may affect the separation process as well as the boiling point of the compound even when the rest of the molecule is not large. Also, water elimination at higher temperatures may be possible in certain alcohols, these cases typically leading to a mixture of compounds. The replacement of the active hydrogen with groups containing halogens also is used to enhance detectability in GC for ECD or NCI-MS detection (see Chapter 17). In HPLC analysis, the importance of the active hydrogen in alcohols is related to the possibility to attach chromophore groups or groups that fluoresce and enhance the detectability. For these reasons, a variety of derivatizations have been developed for replacing the active hydrogen in alcohols.

Seen as Brønsted acids, pure alcohols have the pK_a between 15 and 18. Methanol has $pK_a = 15.2$ and water $pK_a = 15.74$ (see Table 18.1.2). These pK_a values correspond to very low acidity such that the formation of alkoxides from alcohol dissociation does not play an important role in direct alkylations.

The C-O bond in alcohols is polarized, but the formal charges on the carbon atom that is bound to the oxygen are very low. Compared to the formal partial charge of a carbon atom in the corresponding hydrocarbons, they appear shifted toward positive. However, adding the fact that OH^- is a strong base and a very poor leaving group, this carbon is barely susceptible to a nucleophilic attack. In the presence of strong acids, alcohols may accept a proton acting as a base. Protonation of alcohols changes the behavior of alcohols in a nucleophilic attack, the leaving group OH^- being converted into H_2O, which is a good leaving group, and making the carbon atom bound to the oxygen partially positive. This allows reactions of the type:

$$Y:^- + {-}\overset{\overset{\displaystyle H}{|}}{\underset{\displaystyle |}{C}}{-}\overset{+}{O}{-}H \;\rightleftharpoons\; Y{-}\overset{|}{\underset{\displaystyle |}{C}}{-} \;\; + H_2O$$

The electron pairs on the oxygen atom make it basic and nucleophilic. The nucleophile character of the oxygen atom allows alcohols to combine with reagents that have a low electron density on a specific atom. Numerous reactions take place due to this property, such as the reaction with alkenylsulfonyl chlorides:

$$-\overset{|}{\underset{|}{C}}{-}\ddot{O}H \;+\; Cl{-}\overset{\overset{\displaystyle O}{\|}}{\underset{\underset{\displaystyle O}{\|}}{S}}{-}R \;\longrightarrow\; -\overset{|}{\underset{|}{C}}{-}O{-}\overset{\overset{\displaystyle O}{\|}}{\underset{\underset{\displaystyle O}{\|}}{S}}{-}R \;+\; HCl$$

The nucleophile character of oxygen also explains the esterification with S_N2 mechanism or the silylation reaction. Silylation in particular takes place easily and is the derivatization of choice for alcohols in GC analysis. The reaction takes place as a substitution of the active hydrogen with the TMS group:

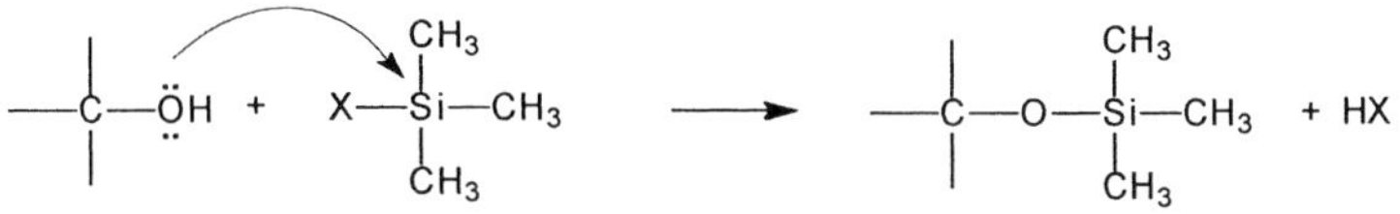

A variety of other chemical reactions can be used for substituting the active hydrogen in alcohols, and those that are more frequently used for analytical purposes are discussed in this section. Some common reactions used for alcohol derivatization are shown below:

Y—OH + R—X ⟶ Y—OR + HX   X = F, Cl, Br, etc.

Y—OH + R—COX ⟶ Y—OCOR + HX   X = Cl, OH, CN, imidazolyl, N-hydroxysuccinimidyl, etc.

Y—OH + R—CO—O—CO—R ⟶ Y—OCOR + RCOOH

Y—OH + R—NCO ⟶ Y—OCO—NH—R

Y—OH + R—CON_3 (R—NCO + N_2) ⟶ Y—OCO—NH—R

Y—OH + R–$Si(CH_3)_2$–X ⟶ Y—O—$Si(CH_3)_2$R + HX

Y—OH + R—OCOCl ⟶ Y—OCOOR + HCl

Y—OH + R—SO_2Cl ⟶ Y—OSO_2R + HCl

These reactions may take place in different conditions such as in inert solvents, in polar solvents, or in the presence of basic compounds such as triethyl amine or dimethylaminopyridine that are needed for the neutralization of the strong acids generated in some of these reactions. The reaction of alcohols with reagents containing a carboxylic acid reactive group is commonly done using the presence of compounds able to facilitate the reaction, as shown in Section 18.1 (see Table 18.1.2).

The sample preparation before derivatization, the precise conditions for the derivatization reaction, as well as the chromatographic conditions used for each separation vary from procedure to procedure, depending on several factors including the nature of the sample and that of the reagent. Details about various analyses can be found in the original reference or in comprehensive review books (see e.g. [1], [2], [3]).

The alcohol groups are found in many organic compounds that also contain other functionalities such as amino, acid, aldehyde, or ketone. Among these compounds particularly important are the carbohydrates that have alcohol groups together with an aldehyde or ketone and steroids that have alcohol groups together with carbonyl or carboxyl. The derivatizations for carbohydrates and for sterols that affect only the alcohol groups are presented in this section. The derivatizations of multifunctional analytes that involve groups other than OH are presented in separate sections (see e.g. Sections 19.9, 19.10, and 19.11).

- Derivatization of simple alcohols for GC analysis

Many compounds containing one hydroxyl group on a saturated carbon are volatile enough for direct GC analysis. However, derivatization in GC analysis is still practiced to reduce polarity and improve GC behavior or to enhance detectability. When more alcohol functional groups are present in the same molecule, derivatization may be needed to enhance volatility. The main types of derivatization for hydroxy compounds in the order of preference are silylation, acylation, and alkylation. Some other reactions also are known for alcohols, and diols and polyols, in addition, can form cyclic derivatives.

The applicability of the alkylation of alcohols for GC analysis is rather limited. There are only a few direct alkylations that work for alcohols, and these are applied only to particular compounds. Most applications are reported for more complex compounds containing OH groups such as carbohydrates. For example, certain sugars can be methylated using methyl iodide in the presence of Ag_2O or in the presence of other basic compounds such as potassium methylsulfinyl carbanion in dimethylsulfoxide (DMSO) [4], [5], [6] (see also Section 18.1).

Alkylation with diazomethane has been reported for a preparative methylation of simple alcohols in the presence of HBF_4 or on silica gel [7], [8]. Diazomethane is used for analytical purposes in the methylation of the hydroxyl group in prostaglandins in the presence of BF_3 in acetone [9]. In this reaction the OH groups as well as the carboxyl groups in prostaglandins are methylated. Ordinary alcohols do not react easily with CH_2N_2.

Formation of trimethylsilyl derivatives (TMS) of alcohols as well as of multifunctional compounds containing OH groups is a very common derivatization procedure used in connection with GC analysis. Alcohols can be derivatized using a variety of silylation reagents (see Section 18.2) such as HMDS [10], MSTFA [11], TMSPI [12], IPTMS [13], N-trimethylsilylacetanilide [14], N,N'-bis(trimethylsilyl)urea [15], BSA [16], [17], d_{18}-BSA [18], BSTFA [19], d_{18}-BSTFA [19], N-trimethylsilylimidazole (TMSI) [20]. A typical reaction with BSTFA is performed using an excess of BSTFA, about five times larger than stoichiometrically required. The procedure can use as a solvent DMF (about 2/3 from the BSTFA volume) or CH_3CN, pyridine, etc. The mixture of sample, solvent, and reagent is heated at a fixed temperature chosen between 75° C and 90° C for 20–30 min. and the sample is ready for injection in the GC system. Excess water in the sample should be avoided, or an even larger excess of BSTFA must be used for compensating the hydrolysis of the reagent (see Section 18.2). Alcohols with no specific hindrance problems are practically completely silylated.

Besides TMS derivatives, other silyl derivatives of alcohols are used for analytical purposes. Among these is the formation of dimethylsilyl derivatives using TMDS [21], BDSA [22]. Also, t-butyldimethylsilyl (TBDMS) derivative of alcohols is used for GC analysis, the derivatization being done with TBDMS-Cl and imidazole in DMF [15]. Silyl derivatives containing a halogen in the molecule can be obtained from alcohols and used for analytical purposes. Among these are iodomethyldimethylsilyl [15], pentafluorophenyldimethylsilyl (flophemesyl) [23], trifluoropropyl-dimethylsilyl [24], etc.

Acylation of alcohols is another derivatization commonly used for analytical purposes (as well as for preparative purposes). Acetic anhydride for example can be used for acetylation in the presence of pyridine [25]. Acetylation with acetic anhydride in pyridine is a very common technique applied to many compounds containing OH groups, and it is not limited to simple alcohols [26]. Acetylation of alcohols with acetic anhydride is also possible in the presence of perchloric acid [27] or in the presence of methanesulfonic acid. The acylating agent when methanesulfonic acid is used is probably acetylmethane sulfonate that is formed as an intermediate in the reaction [28]. In this procedure, K_2HPO_4 is used to remove the excess of reagent and acetic acid.

Acylation with halogenated compounds such as pentafluorobenzoyl chloride [29] is applied for alcohol derivatization in the presence of pyridine. The use of (2,3,4,5,6-pentafluorophenyl)acetyl chloride and (2,3,4,5,6-pentafluorophenoxy)acetyl chloride as acylation reagents show that the relatively high molecular weight of the derivatives makes these reagents quite impractical for GC analysis in the derivatization of larger alcohols [29].

More convenient reagents for alcohol acylation are N-methyl-bis(trifluoroacetamide) or MBTFA and also bis(trifluoroacetamide) [30]. An aprotic solvent such as DMF, tetrahydrofuran (THF), or acetonitrile is useful for completion of this reaction, which leads to volatile trifluoroacetyl derivatives.

Derivatization with isocyanates or carbonyl azides also can be used for derivatization of alcohols for GC analysis [31]. For example, ferrocenoyl azide reacts with various alcohols as follows:

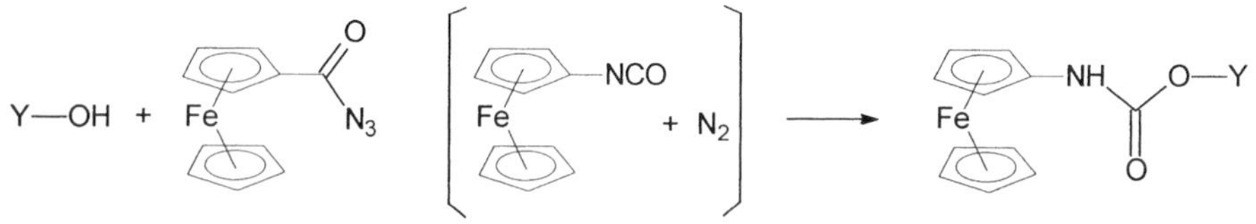

Another reagent reported [15] to be used for the derivatization of alcohols for GC analysis is 2-chloro-1,3,2-dioxaphospholane (see Section 18.6). The reaction takes place rapidly in the presence of triethylamine, which has the role of eliminating the resulting HCl. Similar reagents such as 4-methyl-2-chloro-1,3,2-dioxa-phospholane as well as 2-chloro-1,3,2-dioxaphosphane are used for alcohol derivatization [32] followed by GC analysis. Alcohols also can be derivatized with formylbenzoic acid chloride $O=CH-C_6H_5-COCl$ [33] and analyzed by GC using ALD detection (see Section 4.2).

- Derivatization for chiral GC analysis of alcohols

As indicated in Section 17.7, certain derivatizations are performed with the purpose of generating diastereoisomers from enantiomers that cannot be separated on regular GC columns. Diastereoisomers can be separated on nonchiral columns. Among the reagents used for derivatization of enantiomers of alcohols are acid chlorides, isocyanates and chloroformates. The acid chlorides may be obtained from the corresponding acids by treatment with thionyl chloride [34]. High temperatures or severe derivatization conditions may produce racemization in specific cases. However, the derivatization using isocyanates does not lead to significant racemization, probably because no strong acid is generated in the reaction [2], [35]. Various reagents used for derivatization with the purpose of chiral separations of alcohols are given in Table 19.1.1.

TABLE 19.1.1. *Reagents used for the derivatization of chiral alcohols with GC diastereoisomer separation.*

Reagent	Formula	Reference
3β-acetoxy-Δ⁵-etienic acid chloride		[3]
(S)-acetoxypropionic acid chloride		[3]
carbobenzoyloxy--L-proline acid chloride		[3]
R(+)-*trans*-chrysanthemic acid chloride, or (1R,3R)-*trans*-2,2-dimethyl-3-(2-methyl-1-propenyl)-cyclopropanecarboxylic acid chloride		[34], [36]
drimanoic acid chloride		[2]
D(-) or L(+) mandelic acid chloride		[37]
(-)-menthyloxyacetic acid chloride		[2]
(-)-menthyl chloroformate		[2]
(S)-(-)-α-methoxy-α-(trifluoromethyl)-phenylacetic acid chloride		[38]

TABLE 19.1.1 (continued). *Reagents used for the derivatization of chiral alcohols with GC diastereoisomer separation.*

Reagent	Formula	Reference
(R)(+) or (S)(-)-naphthylethyl-isocyanate		[39]
(R)(+) or (S)(-)-phenylethylisocyanate		[40]
(S)(+)-phenylpropionic acid chloride		[41]
S-tetrahydro-5-oxo-2-furancarboxylic acid chloride		[2]
(S)(-)-N-(trifluoroacetyl)proline acid chloride		[2]

- Derivatization of alcohols for HPLC analysis with UV detection

Derivatization in HPLC is practiced mainly for better detectability (see Section 17.6), although as previously indicated, some benefits for the separation also may be obtained. Only in specific cases when the analyte already has high absorption coefficients in UV or visible, is the derivatization applied only for the modification of the chromatographic behavior. For example, acetic anhydride is used for HPLC derivatization without significantly influencing the absorbance of light by the analyte. Also, some silylation reactions are used for improving HPLC separation. The derivatization for other purposes such as for enhancing chiral separation also is practiced.

Most reagents used for derivatization contain chromophore groups that bring to the derivatized product a high absorption coefficient in UV or even in visible, usually at higher wavelengths than the analyte itself. A reactive group is needed in the reagent such that this is able to react with the active hydrogen from the alcohol group. Among the reactive compounds are alkyl or aryl halides, acyl halides, anhydrides, acids, isocyanates, etc. A list containing reagents reported in literature for the analysis of compounds containing OH groups using UV-Vis detection is given in Table 19.1.2. This list includes analytes that besides the alcohol group may contain other functionalities. However, the derivatization is done at the alcohol and does not affect other parts of the molecule of the analyte.

The sensitivity of the HPLC analyses with UV or visible detection depends on a variety of factors and is usually indicated for each specific method. One of these factors is the absorption coefficient ε_λ of the compound generated after derivatization (see Section 17.6). Therefore the structure of the reagent designed to increase the absorption coefficients is an important contributor to the sensitivity of the analysis. This sensitivity can be as low as about 1 pmol of analyte per injection for a 3/1 S/N ratio. However, as shown in Section 17.6, other aspects of the detection such as absorbing wavelength and linearity of the calibration are equally important in the analysis.

TABLE 19.1.2. *Reagents used in HPLC analysis with UV-Vis detection.*

Reagent		Alcohol type, matrix, sensitivity	Detection λ nm Separation	Reference
acetic anhydride		various samples		[1]
(S)-acetoxypropionic acid chloride		alcohols chiral		[2]
9-anthracenecarboxylic acid		alcohols	254, C8	[1]
9-anthroyl chloride		alcohols 0.5 pg	250, C18	[42]
9-anthroyl nitrile		alcohols	254, ovomucin-conjugated	[1]
benzoic anhydride		saccharides	230, RP-18	[1]
benzoyl chloride		alcohols in blood diols	237, 240, C18, ODS	[1], [43]
bis(trimethylsilyl)-trifluoroacetamide (BSTFA)		hydroxy-vitamins D3	254, silica	[1]
4-biphenylcarboxylic acid + trichlorobenzoyl chloride		alcohols, dihydro-qinghoasu	254, C8	[44]
3-bromomethyl-7-methoxy-1,4-benzoxazin-2-one		aliphatic alcohols 60 μmol	350, Diol	[45]

T<small>ABLE</small> 19.1.2 (continued). *Reagents used in HPLC analysis with UV-Vis detection.*

Reagent		Alcohol type, matrix, sensitivity	Detection λ nm Separation	Reference
tert-butyldimethylsilyl chloride	$(CH_3)_3C-\overset{CH_3}{\underset{CH_3}{Si}}-Cl$	cannabinoids	254, C18	[46]
(1S)-camphanic chloride		urine chiral	275, silica	[1]
carbobenzyloxy-L-proline		blood/warfarin chiral 160 ng (S), 96 ng (R)	313, C18	[1]
N-(carbobenzyloxy)-L-phenylalanine		betamethasone 2.1-4.2 pmol	240, silica	[1]
4-chloro-7-nitrobenzo-2-oxa-1,3-diazole		estrogens 30-50 nmol	380, ODS	[47]
dabsyl chloride, or 4-di-methylaminoazobenzene-4'-sulfonyl chloride, or phenyl-1-azo-(4'-dimethyl-amino-benzene)-4-sulfonyl chloride		alcohols	254, silica	[1]
O,O-dibenzoyl-L-tartaric anhydride		1-methyl-3-pyrrolidinol chiral	254, C8	[1]
dihydrofluorescein diacetate		dihydro-artemisinin[1] 0.1 ng	235, silica	[1]

TABLE 19.1.2 (continued). *Reagents used in HPLC analysis with UV-Vis detection.*

Reagent	Alcohol type, matrix, sensitivity	Detection λ nm Separation	Reference
3,5-dinitrobenzoyl chloride	choline in blood 1 μmol	254, C18, C8	[1]
3,5-dinitrophenyl isocyanate	hydroxy fatty acids	245, chiral phase	[48]
(R,R)-O,O-di-p-toluoyl-tartaric anhydride	chiral	280, silica	[1]
9-fluorenacetic acid + 2,4,6-triisopropylbenzensulfonyl chloride	dihydro-qinghoasu in blood	254, C8	[49]
fluorobenzoyl chloride	alcohols in blood	235, cyano-propyl + silica	[50]
isatoic anhydride	sugars	254	[51]
(1R,2S,5R)-(-) menthoxy-acetyl chloride	chiral	254	[52]
(S)-(-)-α-methoxy-α-(trifluoromethyl)-phenylacetic acid chloride	alcohols chiral		[53]
(-)-menthoxyacetyl chloride	chiral	310, C18	[1]

TABLE 19.1.2 (continued). *Reagents used in HPLC analysis with UV-Vis detection.*

Reagent		Alcohol type, matrix, sensitivity	Detection λ nm Separation	Reference
(-)-1-menthyl chloroformate		blood/warfarin chiral	310, silica	[54]
4-methoxybenzoyl chloride		pentaerithrol	254	[3]
2-[6-methoxy-2-naphthyl]-propionyl chloride		in blood, venlafaxine chiral, 25 ng/mL	229	[1]
6-methoxy-2,5,7,8-tetramethylchromane-2-carboxylic acid (trolox methyl ether)		temazepam[1], oxazepam[1] chiral	254, diol	[55]
4-naphthalene-1-azo-(4'dimethylaminobenzene)-sulfonyl chloride		alcohols	254, RP-18	[56]
naphthoyl chloride		plant extracts	224	[1]
2-(2-naphthyl)propionyl chloride		chiral	254, C18	[1]
2-(2-naphthyl-6-methoxy)propionyl chloride		chiral 0.1% of enantiomer	229, C8	[1]
(S)-(+)-1-(1-naphthyl)ethyl isocyanate		blood, urine, chiral	280	[1]
4-nitrobenzoyl chloride		blood, tissue/ carbohydrates 1-2 ng sugar alcohols	260, silica	[57], [58]

TABLE 19.1.2 (continued). *Reagents used in HPLC analysis with UV-Vis detection.*

Reagent	Alcohol type, matrix, sensitivity	Detection λ nm Separation	Reference
N-1-(2-naphthylsulfonyl)-pyrrolidine-2-carbonyl chloride	diltiazem[1], chiral 0.1% of enantiomer	254, silica	[1]
9-phenanthrenecarboxylic acid	bile	254, C8	[1]
phenyldimethylsilyl chloride	saccharides	254	[59]
phenylisocyanate	milk, blood, sewage 3 ng/mL	240, 230, silica	[1], [60]
1-pyrenesulfonyl chloride	blood estradiol	348, C8	[1]
2-quinoxaloyl chloride	α-hydroxy-carboxylic acids	315, β-cyclo-dextrin	[1]
S-tetrahydro-5-oxo-2-furancarboxylic acid	aromatic alcohols		[2]
N-(p-toluenesulfonyl)-pyrrolidinyl isocyanate (generated from the azide)	phenylpropanol chiral	254, silica	[1]
trifluoroacetic anhydride	blood	223, nitrile	[1]
trimethylacetic anhydride	liothyronine[1], thyroxine	214, C8	[1]

[1] For information and chemical structure regarding various drugs see e.g. M. O'Neil et al. (eds.), *The Merck Index*, Merck & Co., Inc., 13th ed., Whitehouse Station, 2001.

Some other reagents have been used for alcohol analysis in HPLC with UV-Vis detection. For example, the >CH-OH groups in prostaglandins can be oxidized with pyridinium dichromate to ketone groups and measured after HPLC separation at 228 nm and 298 nm [61]. New reagents are continuously developed and reported in literature. Also, it should be noted that some reagents are used for derivatization for HPLC purposes and at the same time for GC analysis. This is for example the case of acetic anhydride, benzoyl chloride, trifluoroacetic anhydride, (S)-acetoxypropionic acid chloride, (S)-tetrahydro-5-oxo-2-furancarboxylic acid, etc. Some of these reagents are also used for chiral separation either in GC or in HPLC.

- Derivatization of alcohols for HPLC analysis with fluorescence detection

Alcohol derivatization for HPLC analysis with fluorescence detection is similar to that used for UV detection. Most of the reagents used for fluorescence detection carry fluorophores, but in some cases either a new fluorescent compound is generated, or the analyte is already fluorescent. The excitation and the emission wavelength depend on both the nature of the analyte and that of the group carrying the fluorescence properties. The sensitivity of an HPLC method using fluorescence detection can be extremely low, in certain cases reaching a few fmol of analyte per injection for a 3/1 S/N ratio. The sensitivity also is affected by factors that depend on the instrumentation (such as type of excitation source, light dispersing system, etc.). A list of reagents used in HPLC analysis with fluorescence detection is given in Table 19.1.3. The list of analytes includes compounds that may contain other groups besides hydroxyl but with the derivatization at the OH group.

TABLE 19.1.3. *Reagents used in HPLC analysis with fluorescence detection.*

Reagent		Alcohol type, matrix, sensitivity	Fluorescence λ Separation	Reference
acetic anhydride		various samples	RP-18, etc.	[1]
4-(N-acetylchloride-N-methyl)amino-7-N',N'-dimethylaminosulfonyl-2,1,3-benzoxadiazole		estrogens 40 fmol	ex 440, em 543, C18, ODS	[1], [62]
6-aminoquinolyl-N-hydroxysuccinimidyl carbamate		alcohols 100 pmol (primary), 250 pmol (sec.)	ex 290, em 345, C18	[63]
9-anthracene-carboxylic acid		blood, dolichols 5 ng	ex 360, em 460, C18	[1]

TABLE 19.1.3 (continued). *Reagents used in HPLC analysis with fluorescence detection.*

Reagent		Alcohol type, matrix, sensitivity	Fluorescence λ Separation	Reference
9-anthroyl chloride		endotoxins	ex 250, em 462	[1]
1-anthroyl nitrile		bile acids in blood, etc. 50 nmol	ex 370, em 470, C18	[64]
9-anthroyl nitrile		carnitine 50 µg/mL	ex 305-395 em 430-470, silica	[1]
2-[2-(azidocarbonyl) ethyl]-3-methyl-1,4-naphthoquinone generating 2-[2-(iso-cyanate)-ethyl]-3-methyl-1,4-naphthoquinone		cholesterol	ex 320, em 430	[1]
(S)-(+)-2-*tert*-butyl-2-methyl-1,3-benzodioxol-4-carbonyl chloride		diacyl glycerols	ex 310, em 370	[65]
2-*tert*-butyl-2-methyl-1,3-benzodioxole-4-carboxylic acid		1-bromo-x-acetylated mono-saccharides a few fmol	various, silica	[66]
carbazol-9-N-acetic acid (in presence of 1-ethyl-3-(3-dimethylamino-propyl)carbodiimide hydrochloride, EDC-HCl as dehydrating agent and 4-dimethylamino-pyridine as catalyst		alcohols 0.1 pg for methanol	ex 355, em 360, C18	[67]
carbazol-9-N-(2-methyl)-acetylbenzene-disulfonate 4-dimethylaminopyridine as catalyst		alcohols	fluorescence	[68]
2-(4-carboxyphenyl)-5,6-dimethyl-benzimidazole		blood, cortico-steroids, estrogens, fatty alcohols 0.6-3 pg/mL	ex 33, em 418	[1]

TABLE 19.1.3 (continued). *Reagents used in HPLC analysis with fluorescence detection.*

Reagent	Alcohol type, matrix, sensitivity	Fluorescence λ Separation	Reference
2-(4-carboxyphenyl)-6-N,N-diethylamino-benzofuran	alcohols 0.2-0.5 pg	ex 387, em 537, C8	[69]
2-(4-carboxyphenyl)-6-methoxybenzofuran	alcohols 0.1-0.5 pg	ex 315, em 390, C8	[70]
7-[(chlorocarbonyl)-methoxy]-4-methylcoumarin	estrogens, bile acids	ex 315, em 389, ODS	[1], [71]
2-(5-chlorocarbonyl-2-oxazolyl)-5,6-methylene-dioxybenzofuran	blood, 2',3'-di-deoxyinosine, nucleotides 1.3 pmol	ex 360, em 460, ODS, C18	[1]
2-(4-chlorophenyl)-α-methyl-5-benzoxazol-acetic acid chloride (benoxaprofen chloride)	alcohols 0.5-1 ng/mL (plasma), 3 ng/mL (urine)	ex 312, em 365, silica, RP-18	[72]
coumarin-3-carbonyl chloride	micotoxins 0.8-2 ng	ex 292, em 425, RP-18	[1], [73]
dansyl chloride or 5-dimethylaminonaphthalene-1-sulfonyl chloride	blood, estradiol, etc. 0.06 ng/mL, etc.	ex 340, em 500, nitrile, C18, silica	[1]
dansylethyl chloroformate	cholesterol, estradiol	ex 342, em 534	[2]
diethylaminocoumarin-3-carbonyl azide	blood acetyl hydrolase 0.5 pmol	ex 400, em 480, C18	[1]
diacetyl-L-tartaric anhydride	propranolol in blood chiral 0.5 ng/mL (R), 1 ng/mL (L)	ex 290, em 335, RP-18	[1]

TABLE 19.1.3 (continued). *Reagents used in HPLC analysis with fluorescence detection.*

Reagent		Alcohol type, matrix, sensitivity	Fluorescence λ Separation	Reference
3,4-dihydro-6,7-dimethoxy-4-methyl-3-oxoquinoxaline-2-carbonyl chloride		alcohols 2-3 fmol	ex 400, em 500, C8	[1]
3,4-dihydro-6,7-dimethoxy-4-methyl-3-oxoquinoxaline-2-carbonyl azide		alcohols 2-45 fmol	ex 360, em 440, C8	[1]
3,4-dihydro-6,7-dimethoxy-4-methyl-3-oxoquinoxaline-2-carbonyl azide		alcohols	ex 360, em 440, C8	[74]
1-dimethylamino-napthyl-4-carbonyl cyanide		bile acids	ex 350, em 530	[75]
1-(ethylthio)-3-(dihalo-1,3,5-triazinyl)-2-propylisoindole		estrogens 1.1 pmol (Cl), 270 fmol (F)	ex. 415 em. 445, size excl., ODS	[76]
1-(ethylthio)-3-(difluoro-1,3,5-triazinyl)-2-propylbenz [f]-isoindole		estrogens 80 fmol	ex. 490, em. 520, size excl., ODS	[77]
(R)-1-(9-fluorenyl)ethyl chloroformate		blood, carnitine, chiral	ex 260, em 310, RP18 CE	[1]
9-fluorenylmethyl chloroformate		C_1-C_4 alcohols 90 pmol	ex 259, em 311, C18	[78]

TABLE 19.1.3 (continued). *Reagents used in HPLC analysis with fluorescence detection.*

Reagent		Alcohol type, matrix, sensitivity	Fluorescence λ Separation	Reference
2-[2-(4-fluorophenyl)-benzoxazol-5-yl]-propanoyl chloride (flunoxaprofen chloride)		blood, urine ciclotropium bromide, chiral 0.5 ng/mL (plasma)	ex 310, em 365	[1]
(-)-(1S,2R,4R)-*endo*-1,4,5,6,7,7-hexachloro-bicyclo[2.2.1]hept-5-ene-2-carboxylic acid (post-column derivatization)		blood, warfarin, chiral 5 ng/mL	ex 313, em 370, C18	[1]
2-[2-(isocyanate)-ethyl]-3-methyl-1,4-naphthoquinone		cholesterol 8-23 fmol	ex 320, em 430	[1]
(aS)-2'-methoxy-1,1'-binaphthalene-2-carbonyl cyanide		penbutolol in blood, chiral 30 pg	ex 290, em 405, silica	[79], [80]
7-methoxycoumarin-3-carbonyl chloride		urine oxosteroids, various alcohols 87-162 ng/mL	ex 355, em 400, size exclusion, ODS	[1]
6-methoxy-2-methyl-sulfonylquinoline-4-carbonyl chloride		various alcohols 0.07-50 pmol	ex 355, em 457, C18	[1]
(-)-2-methyl-1,1'-binaphthalene-2'-carbonyl cyanide		propranolol in blood chiral 100 pg	ex 342, em 420, silica	[81]
3-methylcoumarin-7-carbonyl chloride		alcohols 0.12-0.5 pmol	ex 355, em 400, ODS	[82]
(4-methylcoumarin-7-yl)-methyl imidazole carboxylate		benzyl alcohol 0.8 ng	fluorescence or LC/MS thermo-spray	[1]

TABLE 19.1.3 (continued). *Reagents used in HPLC analysis with fluorescence detection.*

Reagent		Alcohol type, matrix, sensitivity	Fluorescence λ Separation	Reference
1,2-naphthoylene-benz-imidazole-6-sulfonyl chloride		alcohols	ex 365, em 475	[2]
1-naphthoyl chloride		blood digoxin and metabolites 0.25 ng/mL	ex 217, em 340, silica	[1]
naphthoyl imidazole		ouabain[1]	ex 234, em 374, C18	[1]
(S)-(+)-1-(1-naphthyl)-ethyl isocyanate		blood, urine, methocarbamol[1], chiral 10 ng/mL	ex 275, em 336	[1]
1-naphthyl isocyanate		enantiomers of ibutilide in blood 17 pg/mL	ex 290, em 345, C18 + Pirkle	[1]
phenyl isocyanate		blood, stiripentol 8 ng/mL	ex 290, em 355 C18	[1]
7-phenylsulfonyl-4-(2,1,3-benzoxadiazolyl) isocyanate		alcohols 10 fmol	ex 368, em 490, C18	[83]
3-(2-phthalimidyl)-benzoyl azide		blood, alcohols 100-400 fmol	ex 302, em 440, ODS	[1]
propyl isocyanate (post-column)		blood, oxiracetam[1]	ex 340, em 455	[1]
pyrene-1-carbonyl cyanide		carnitine[1] in blood 500 ng/mL	ex 355, em 420, size excl.	[1]

TABLE 19.1.3 (continued). *Reagents used in HPLC analysis with fluorescence detection.*

Reagent		Alcohol type, matrix, sensitivity	Fluorescence λ Separation	Reference
1-pyrenesulfonyl chloride	(structure with SO$_2$Cl)	blood estradiol[1]	ex 350, em 385, ex 325 (Ar laser), C8	[1]
(R)-(+)-4-(pyrrolidin-2-carbonylchloride-1-yl)-7-N',N'-dimethyl-aminosulfonyl-2,1,3-benzoxadiazole	(structure with SO$_2$—N(CH$_3$)$_2$)	alcohols chiral carbohydrates	ex 485, em 530, silica	[84]
(R)-(+)-4-(pyrrolidin-2-carbonylchloride-1-yl)-7-nitro-2,1,3-benzoxadiazole	(structure with NO$_2$)	alcohols chiral	ex 485, em 530, silica	[84]
trimethylsilyldiazo-methane	H_3C—Si—CHN_2 (with three CH$_3$)	7-hydroxy-granisetron[1]	ex 4310, em 420	[1]
veratryl amine (condensation with formation of oxazoles after oxidation with K$_3$[Fe(CN)$_6$]	(structure with H$_3$CO, H$_3$CO, NH$_2$)	5-hydroxy-indoles 450 pmol	ex 345, em 475, C18	[85]

[1] For information and chemical structure regarding various drugs see e.g. M. O'Neil et al. (eds.), *The Merck Index*, Merck & Co., Inc., 13th ed., Whitehouse Station, 2001.

Among other reagents used for alcohol derivatization are solid phase bound compounds such as dimethylpyridinium fluorenylmethyl carbamate. This reagent is used in alcohol derivatization followed by HPLC analysis with fluorescence detection leading to very low detection limit [86].

- *Other derivatizations for HPLC analysis of alcohols*

Chemiluminescence detection is not common in HPLC analysis of alcohols. Detection using laryl chloride (lissamine rhodamine B sulfonyl chloride) in a PO-CL procedure with postcolumn addition of hydrogen peroxide and bis(2-nitrophenyl)oxalate is used for the analysis of estradiol [87]. However, the reaction with laryl chloride probably occurs at the phenolic group of the analyte (see Section 19.2).

Electrochemical detection of alcohol is infrequently used. One procedure applied for the analysis of testosterone and androsterone is the oxidation at a glassy carbon electrode at + 1.0 V of the alcohol esters formed in a reaction with salicyl chloride [88]. Another procedure applied for cholesterol analysis is the reaction with 2-[2-(isocyanate)ethyl]-3-methyl-1,4-naphthoquinone generated from the corresponding azidocarbonyl derivative (see Section 17.6). Electrochemical detection is possible using oxidation at a glassy carbon electrode at + 0.7 V [1]. For derivatization with the purpose of electrochemical detection, the following reagents also can be used: ferrocenyl azide, 3-ferrocenyl-propionyl azide, ferrocenboronic acid, 2-[2-(azidocarbonyl)ethyl]-3-methyl-1,4-naphthoquinone, 2-[2-(isocyanate)ethyl]-3-methyl-1,4-naphthoquinone [2]. The formulas of these reagents are shown below:

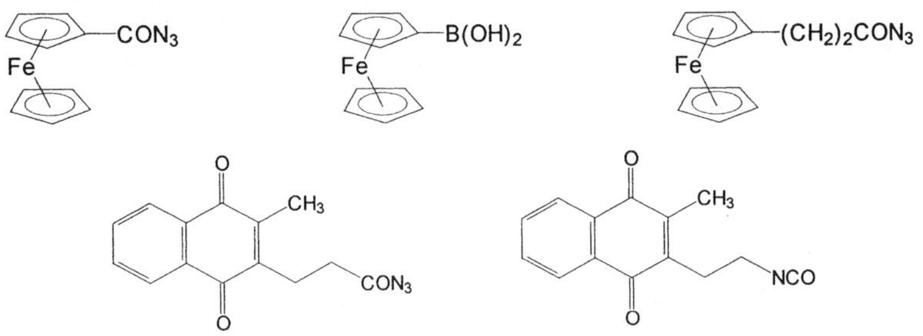

The reaction with alcohols takes place by heating, and a long chain alcohol such as eicosanol is derivatized at 100° C for 15–20 min. These reagents are used for example for the analysis of hydroxysteroids using postcolumn reduction followed by oxidative detection at +0.7 V vs. Ag/AgCl [89].

The progress made in LC/MS instrumentation has led to more analyses by this technique, including the use of derivatization. The derivatization can be performed for better separation or to improve the quality of the mass spectrum. For alcohols, reagents such as 7-[(imidazolemethanoyl)-methoxy]-4-methylcoumarin [90], 4-biphenylcarboxylic acid, 9-fluoreneacetic acid, etc. [91] have been reported as reagents for the derivatization of alcohols with thermospray MS analysis. An example of derivatization for LC-MS analysis is the reaction of antimalarial drug dihydroqinghoasu with 9-fluoreneacetic acid in the presence of 2,4,6-triisopropyl-benzensulfonyl chloride (TIPS-chloride), which takes place as follows:

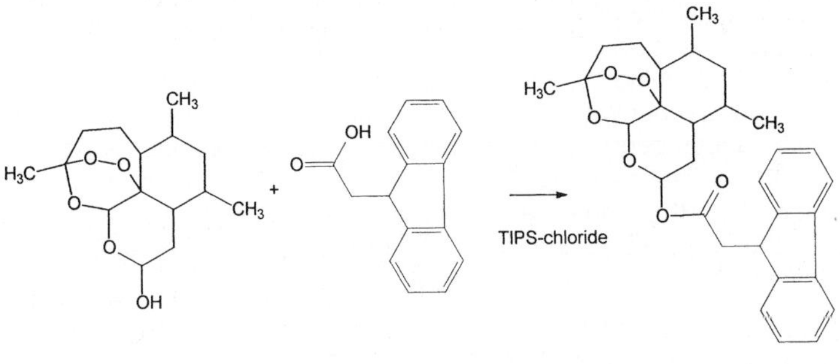

A comparison of the thermospray mass spectra of dihydroqinghoasu (MW = 284) and its derivatized form (MW = 490) are shown in Figures 19.1.1a and 19.1.1b, respectively.

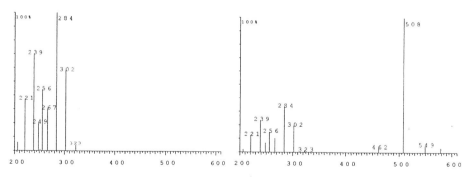

FIGURE 19.1.1a. *Thermospray mass spectrum of dihydroqinghoasu + NH₄⁺ion generated using ammonium acetate/ acetonitrile mobile phase.*

FIGURE 19.1.1b. *Thermospray mass spectrum of derivatized dihydroqinghoasu + NH₄⁺ion generated using ammonium acetate/acetonitrile mobile phase.*

In addition to the molecular ion, the spectra display the ion MW + 18 due to the cluster with NH_4^+.

Many derivatizations of alcohols are reported in literature. Some of these are indirect analyses where the alcohol is transformed first into another functionality and then further derivatized. For example, methanol can be analyzed by oxidation with $KMnO_4$ to formaldehyde, which is further analyzed using 2,4-dinitrophenylhydrazine [92]. Derivatizations of this type are not well captured in this section.

- Derivatization reactions for diols and polyols

The presence of the OH groups in diols and polyols (including sugar alcohols) allows in principle any derivatizations with reagents that are used for alcohols. Silylation and acylation are commonly practiced for the purpose of GC analysis [3], [93]. Reagents such as BSTFA, hexamethyldisilazane + trichlorodimethylsilane (HMDS + TMCS), etc. are successfully used for derivatization. For example, inositols can be easily silylated using BSTFA and DMF (2:1 v/v) in 3–4 times excess to the analyte by heating the mixture for 20 min. at 76° C. The analysis can be done using GC or GC-MS. The acylation can be done with reagents such as trifluoroacetic anhydride (TFA), pentafluoropropionic anhydride (PFPA), acetic anhydride, heptafluorobutyric anhydride (HFBA), or various other acyl chlorides [3]. The reagents used for diols and polyols must be efficient in achieving complete or at least the same degree of derivatization (the same number of groups derivatized in each molecule). The formation of a mixture of compounds with various degrees of derivatization is not acceptable for good analytical procedures. In the derivatization for GC analysis, the elimination of all active groups is usually intended. For HPLC analyses, only the same degree of derivatization may be sufficient, such that the desired chromophores or fluorophores are attached to the analyte molecule in the same number (usually only one).

In addition to reactions typical for all alcohols, the polyols with two OH groups in 1,2 or 1,3 position (sometimes even 1,4 position) may undergo specific derivatization reactions, such as the formation of cyclic boronates or cyclic isopropyliden derivatives (acetonides). For example, long chain aliphatic diols react with acetone in the presence of an acidic catalyst such as p-toluenesulfonic acid (TsOH) [94] to form an acetonide that can be analyzed by GC/MS:

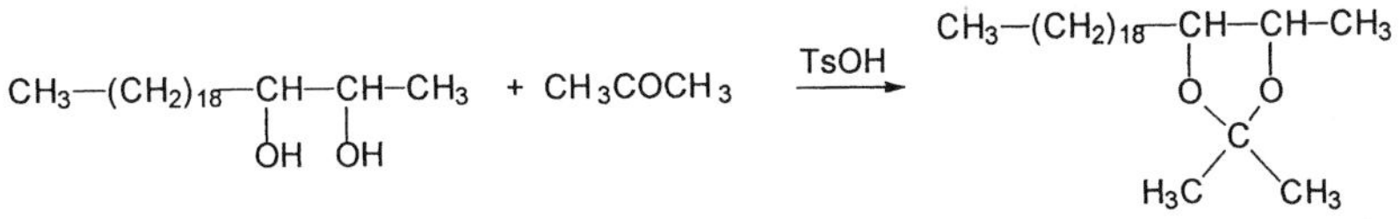

Butylboronic acid is extensively used in diol derivatization (see also Section 18.5). Various glycols, such as 9,10 dihydroxystearates, react by mixing the analyte and the reagent without solvent [95]. In other cases, a solvent is utilized, such as for menthoglycols, which react with butaneboronic acid in pyridine within 3 min. reaction time [96]. Sugar alcohols are analyzed by GC using their butaneboronate derivatives [97]. The butaneboronates of *myo*-inositol, sorbitol, and mannitol are tris-butaneboronates [15].

For polyols with an odd number of OH groups, the reaction with boronic acids can be followed by a second derivatization such as a silylation (see Section 18.5). For example, sequential derivatization can be applied in the analysis of polyhydroxyalkyl-pyrazines. The reaction for 2-D-(arabino-tetrahydroxybutyl)-6-(D-erytro-2,3,4-trihydroxybutyl) pyrazine (deoxyfructosazine) takes place by adding to the analyte in anhydrous pyridine the appropriate amount of butylboronic acid. The mixture is then heated for 10 min. at 100° C in a sealed tube. After the reaction mixture is cooled, the product is silylated with a slight excess of hexamethyldisilazane (HMDS) + chlorotrimethylsilane (TMCS) (2/1 v/v mixture) [98] by evaporating the solvents under reduced pressure. The residue is dissolved in ethyl ether and analyzed. The reaction can be written as follows:

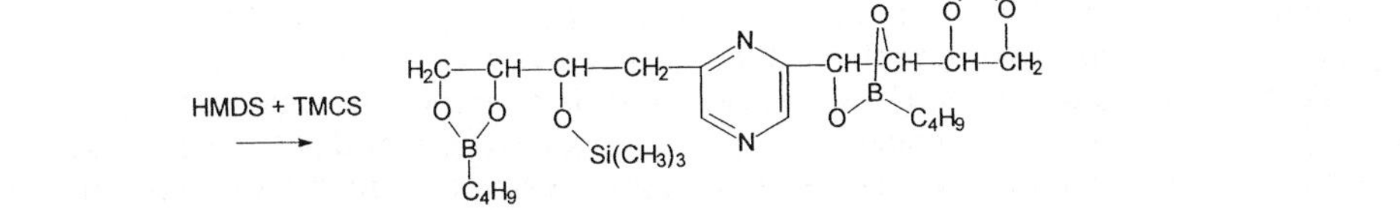

Although relatively large, the molecules obtained by the above derivatization reactions are sufficiently stable and volatile to be analyzed by GC/MS.

The reagents used for HPLC analysis of diols are, as expected, designed to carry groups with specific properties for enhancing detectability. Among the reactive groups are boronic acids or other bifunctional groups. A selection of reagents used for diols HPLC analysis is given in Table 19.1.4.

TABLE 19.1.4. *Reagents used in HPLC analysis of diols and polyols.*

Reagent	Analyte, sensitivity	Detection type, Separation	Reference
1,2-bis(4-methoxyphenyl)-ethylenediamine	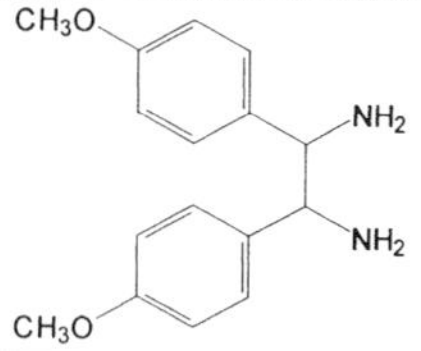 catechol-amines in blood, urine 1-2 nmol (urine), 10-20 pmol (plasma)	fluorescence ex 340, em. 470, size excl.	[1]
3-dansylaminophenyl-boronic acid	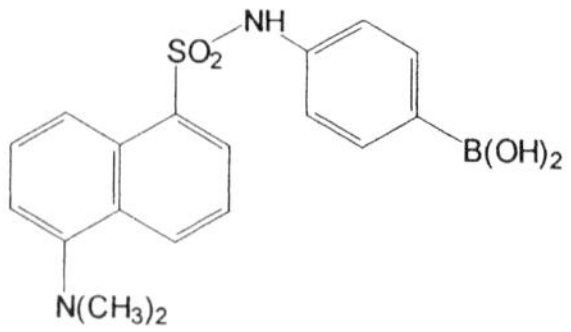 brassinolides, sterols 25 pg	fluorescence ex 345, em. 515, ODS	[99]
3,5-dinitrobenzoyl chloride	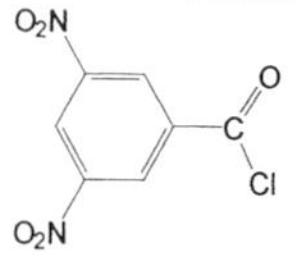 polyols	UV	[1]
1,2-diphenylethylene-diamine	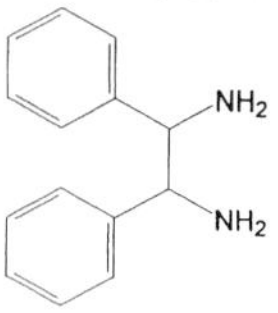 catechol-amines in blood, etc. 7-10 pmol	fluorescence ex 350, em. 480, C18	[1]
ferroceneboronic acid	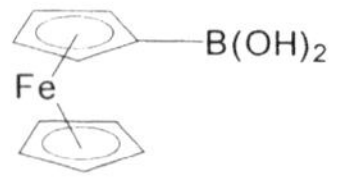 brassinolides, sterols 25 pg	electro-chemical, ODS	[100]
naphthalene-1,8-dicarboxylic acid	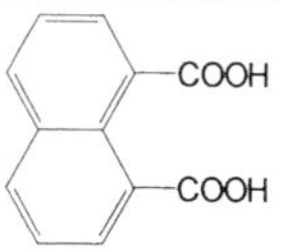 amprilose in plasma 185 ng/mL	fluorescence ex 280, em. 340, ODS	[101]
9-phenantreneboronic acid	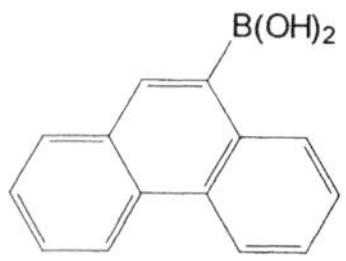 brassinolides, sterols 50 pg	fluorescence ex 305, em. 375, ODS	[102]
phenylboric acid	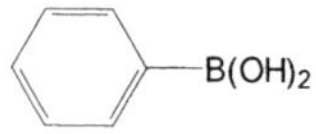 ecdysteroids	UV, silica	[103]
o-phenylenediamine	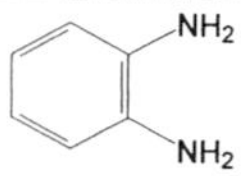 glucans	UV 320, RP-18	[1]

A special group of dihydroxylic (or polyhydroxylic) compounds is that of polyphenols. Some reactions specific for polyphenols are indicated in Section 19.2.

- *Derivatization of the OH groups in carbohydrates*

Carbohydrates (sugars or saccharides) contain in their molecule carbonyl and hydroxyl (aldehyde in aldoses or ketone in ketoses) groups. Carbohydrate derivatization involving both types of functionalities is further discussed in Section 19.9. Also, aspects regarding sample preparation for the analysis of polymeric carbohydrates are discussed in Chapter 20. Carbohydrates commonly exist in their hemiacetal form, and in aqueous solutions the free carbonyls are present only at very low concentrations. For this reason, carbohydrate derivatization is most frequently done at their OH groups, using reagents similar to those for other alcohols.

The most common monomeric saccharides (or monosaccharides) are hexoses and pentoses (containing six or respectively five carbon atoms). For example, glucose is an aldohexose, fructose a ketohexose, and xylose, arabinose and ribose are aldopentoses. The hemiacetal formation from these sugars may lead to a six-membered cycle or a five-membered cycle. For example, in an aqueous solution, D-galactose is about 80% in pyranose form and 20% furanose form. The OH group formed as a result of the cyclization is attached to an asymmetric carbon and leads to two isomers called anomers (noted α and β). The structures of the two anomers for glucopyranose and the structure for α-glucofuranose are shown below:

α-D-Glucopyranose (α-Glup)   β-D-Glucopyranose (β-Glup)   α-D-Glucofuranose (Gluf)

In many chromatographic separations the pyranose and furanose forms as well as the two anomers are separated, leading to multiple peaks (possibly five including the open form). The derivatization at the OH groups of carbohydrates does not imply the modification of the pyranose/furanose form or of the hemiacetal structure. Therefore, the derivatization of sugars as alkyl, silyl, acyl, etc. followed by a chromatographic separation may lead to complex chromatograms, sometimes difficult to interpret or to be used for quantitation based on peak areas. For this reason, various derivatization procedures are applied to block the carbonyl group and eliminate the presence of multiple isomers [104]. A discussion of these possibilities is given in Section 19.9.

All carbohydrates contain more than one OH group. For this reason, some derivatizations are similar to those for polyols. Because of the presence of multiple reacting sites in the molecule, the formation of compounds with various degrees of derivatization is possible and highly undesirable. For GC analysis in particular, the derivatizations are usually done to eliminate all the active hydrogens. For example, silylation is the derivatization of choice in many GC analyses of sugars, as it is for alcohols in general. Monosaccharides, disaccharides, and even trisaccharides can be persilylated using BSTFA + TMCS (3% TMCS in BSTFA) in DMF and analyzed by GC [105]. The high molecular weight of the compounds is the reason why higher oligosaccharides cannot be analyzed by GC. The EI+ mass spectrum of 11-TMS-maltotriose is shown in Figure 19.1.2. The spectrum is very similar to that of 8-TMS-maltose.

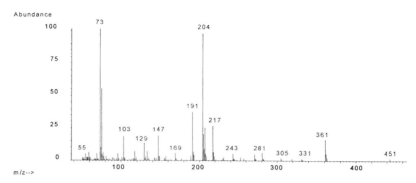

FIGURE 19.1.2. *Mass spectrum (EI+ 70 e.V.) of 11-TMS-maltotriose.*

Mass spectra of silylated sugars have been studied in detail for structure assignments [106]. Silylation of 3-ketoses and 2-heptuloses shows that the open chain form is a large component in the gas chromatograms [107]. The open chain form is presumed to have been produced by opening the hemiacetal during silylation.

A list of several reagents used in carbohydrate analysis with the derivatization of the OH groups followed by GC analysis is given in Table 19.1.5.

TABLE 19.1.5. *Reagents used for the analysis of carbohydrates by GC involving only the OH groups.*

Reagent		Carbohydrate type	React type[1]	Derivative type	Ref.
methyl iodide in the presence of Ag$_2$O in DMF	CH$_3$I	mono-, di-, tri-saccharides	alkyl	permethyl	[15]
methylsulfinylcarbanion/methyl iodide	CH$_3$I + CH$_3$SOCH$_2^-$	carbohydrates	alkyl	permethyl	[6]
methyl iodide in the presence of NaOH in DMSO	CH$_3$I + NaOH	carbohydrates	alkyl	permethyl	[4], [108]
methyl iodide in presence of potassium in liquid ammonia	CH$_3$I + K metal in liq. NH$_3$	carbohydrates	alkyl	permethyl	[15]
(1) (-)-2-butanol + HCl with subsequent addition of Ag$_2$CO$_3$ (2) TMCS + HDMS in pyridine	(1) C$_2$H$_5$CHOHCH$_3$ + HCl (2) [(CH$_3$)$_3$Si]$_2$NH + (CH$_3$)$_3$SiCl	mono-saccharides chiral	alkyl silyl	2-butyl TMS ether	[109]
diazomethane + BF$_3$	CH$_2$NH$_2$ + BF$_3$	carbohydrates	alkyl	permethyl	[3]
acetic anhydride in pyridine	(CH$_3$CO)$_2$O	partially methylated aldoses	acyl	does not replace the methyl	[110]
(1) benzaldehyde + ZnCl$_2$ and (2) acetic anhydride in pyridine	(1) C$_6$H$_5$CHO + ZnCl$_2$, (2) (CH$_3$CO)$_2$O + pyridine	hexo-pyranosides	acetal acyl	acetals, O-acetylated	[111]
(1) butaneboronic acid (2) acetic anhydride	(1) C$_4$H$_9$B(OH)$_2$ (2) (CH$_3$CO)$_2$O	carbohydrates	boron acyl	boronate/acetate	[112]
trifluoroacetic anhydride + sodium trifluoroacetate in formamide or acetonitrile	(CF$_3$CO)$_2$O + CF$_3$COONa	carbohydrates	acyl	trifluoro-acetyl (1-methyl not replaced)	[15], [113]
N-methylbis(trifluoroacetamide) or MBTFA in pyridine	(CF$_3$CO)$_2$NCH$_3$	carbohydrates	acyl	trifluoro-acetyl	[114]
hexamethyldisilazane (HMDS) + trichloromethylsilane (TMCS) in pyridine	[(CH$_3$)$_3$Si]$_2$NH + (CH$_3$)$_3$SiCl in pyridine	carbohydrates	silyl	TMS ether	[15]

TABLE 19.1.5 (continued). *Reagents used for the analysis of carbohydrates by GC involving only the OH groups.*

Reagent		Carbohydrate type	React type[1]	Derivative type	Ref.
trimethylsilylimidazole (TMSI) in pyridine	$(CH_3)_3Si$—N (imidazole ring)	carbohydrates	silyl	TMS ether	[15]
N,O-bis(trimethylsilyl)trifluoro-acetamide (BSTFA) in DMF	CF_3—C(=N—Si(CH_3)_3)—O—Si(CH_3)_3	carbohydrates	silyl	TMS ether	[115]
(1) methaneboronic acid (2) N,O-bis(trimethylsilyl)-trifluoroacetamide (BSTFA)	(1) $CH_3B(OH)_2$ (2) CF_3—C(=N—Si(CH_3)_3)—O—Si(CH_3)_3	carbohydrates	boron silyl	methane-boronate, TMS ether	[116], [117]
1,1,3,3-tetramethyldisilazane + dimethylmonochlor silane in pyridine	$(CH_3)_2HSi$—NH—SiH(CH_3)_2 + $(CH_3)_2HSiCl_2$	aldols and alditols	silyl	dimethyl-silyl	[15]
(1) hypoiodite (or hypobromite) followed by HCl addition (2) hexamethyldisilazane (HMDS) + trichloromethyl-silane (TMCS) in pyridine	(1) HOI or HOBr (2) $[(CH_3)_3Si]_2NH$ + $(CH_3)_3SiCl$ in pyridine	sugars	oxid. silyl	TMS ester of 1,4-aldono-lactones	[15]
acetone + CuSO₄	CH_3COCH_3 + $CuSO_4$	pentoses and hexoses	acetal	O-isopro-pylidene	[118]
acetone + H₂SO₄	CH_3COCH_3 + H_2SO_4	pentoses and hexoses	acetal	O-isopro-pylidene	[119]
2,2-dimethoxypropane + p-toluenesulfonic acid (TsOH)	CH_3O—C(CH_3)(CH_3)—OCH_3 + TsOH	D-glucose	acetal	O-isopro-pylidene	[120]
(1) methylvinyl ether + p-toluenesulfonic acid (TsOH) (2) methyl iodide + Ag₂O	(1) $CH_3OCH=CH_2$ + TsOH (2) CH_3I + Ag_2O	sugars	acetal alkyl	acetal methyl ether	[121]

[1] alkyl = alkylation, silyl = silylation, boron = reaction with boronic acid, oxid. = oxidation, acetal = ketal/acetal formation

Although silylation is common in carbohydrate derivatization, there are still problems related to possible artifacts during this derivatization (see Section 18.2). For example, pyridine may lead to mutarotation of sugars and loss of specific information if this is necessary [122].

As seen in Table 19.1.5, some derivatization reactions for carbohydrates have two steps. For example, for a GC analysis the first step can be the formation of an acetal and the second an acylation. This sequence of reactions is shown below for 1-ethylglucose [123] in a reaction with benzaldehyde and acetic anhydride:

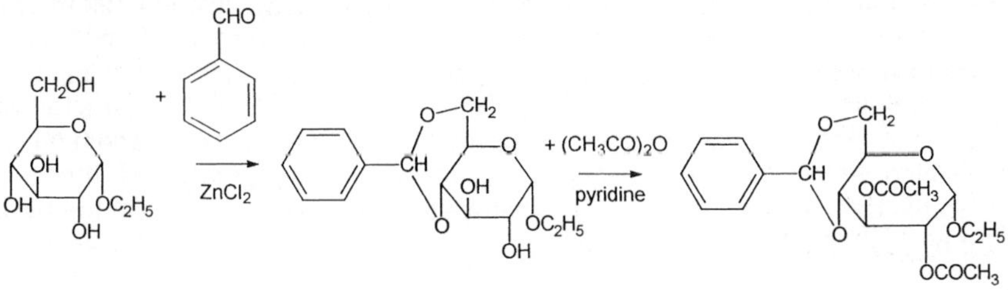

Another example of mixed derivatization is the use of a reaction with butaneboronic acid (see Section 18.5) followed by acylation with acetic anhydride. The sequence of reactions can be written as follows for glucose:

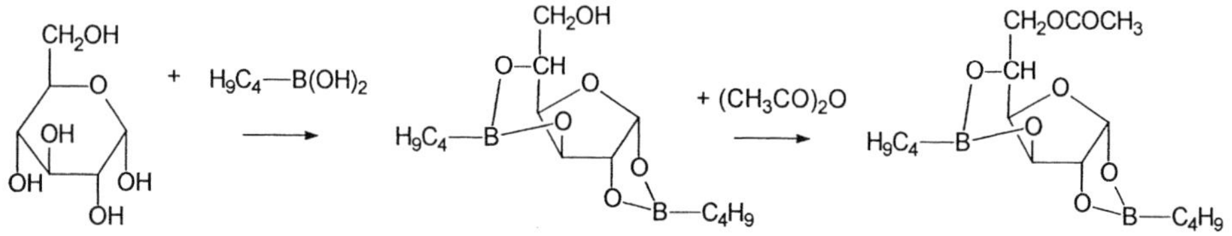

Silylation is frequently used for the derivatization of carbohydrates, and the reaction can be used in combination with other derivatizations. Silylation preceded by a reaction with methaneboronic acid takes place as follows:

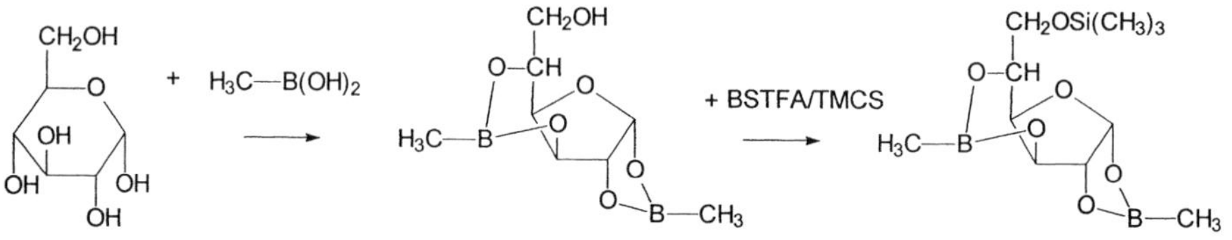

Besides methaneboronates, benzene- and butaneboronates also are used for the analysis of carbohydrates [124]. These reactions are used mainly for the analysis of mixtures of carbohydrates or for structure elucidations. For example, D and L monosaccharides can be derivatized using alkylation with a pure enantiomer alcohol such as (-)-2-butanol in the presence of HCl that leads to the derivatization of one OH group. The remaining OH groups can be further silylated (see Table 19.1.5) and analyzed by GC [125]. Other derivatizations for enantiomer separation can be done using reagents shown in Table 19.1.1.

The derivatization of carbohydrates for HPLC analysis does not require the replacement of all active hydrogens. Usually the attachment of a single chromophore or fluorophore increases sufficiently the detectability. Some of the reagents used for HPLC analysis with the reaction to attach the chromophore or the fluorophore occuring at the OH group were already indicated in Tables 19.1.2 and 19.1.4. However, the formation of hydrazones or reaction involving both carbonyl and hydroxyl groups are more common in HPLC analysis of sugars, and these derivatizations are presented in Section 19.5 and 19.9. Acylation with p-nitrobenzoyl chloride can be used for the derivatization at the OH groups followed by HPLC with UV detection. For the derivatization, 0.5 mg sugar is dissolved in 50 μL pyridine and 150 μL solution of p-nitrobenzoyl chloride 100 mg/mL in pyridine. The mixture is allowed to stand at room temperature for 10 min., and then the solution is evaporated under vacuum at 50° C, and 2 mL 5% Na_2CO_3 containing 5 mg 4-dimethylaminopyridine (catalyst) are added. After 5 min. the material is extracted with $CHCl_3$, and the extract is washed once with 2 mL 5% Na_2CO_3 and twice with 3 mL 0.05 HCl containing 5% NaCl. The reaction takes place at room temperature for mono- di- and trisaccharides, except fructose. The methylated sugars are stable (the methyl is not replaced by acyl) [126].

- Derivatization of the OH groups in steroids

Sterols form an important class of compounds present in nature. They include typical sterols such as cholesterol, β-sitosterol, etc. that contain in molecule only alcohol functional groups, phenolic steroids such as estradiol, ketosteroids such as testosterone that contain both ketone and hydroxyls, bile acids that contain a carboxyl group in addition to hydroxyl, digitalis steroids, etc. (see also Section 19.11 for the discussion on derivatization of other groups in sterols). Due to the presence of the OH groups in all sterols, most derivatization reactions are applied to the OH groups for both GC and HPLC analysis (see also Tables 19.1.2 and 19.1.3). Many of the reported reactions are silylations and acylations.

Extensive research has been done in the analysis of sterols using silylation [15], [127], [128]. The OH groups in various sterols can be present in different positions in the molecule, and some are considerably hindered. For this reason silylation reaction may take place at some OH groups while not affecting others, depending on the position of the OH and on silylating conditions. For example, silylation of β-ecdisone, an insect molting hormone, silylated using trimethylsilylimidazole (TMSI) at 140° C for 20 hours forms a hexa-TMS derivative, while at 100° C for 1 hour forms a mixture of penta- and tetra-TMS derivatives. The reaction below shows only the formation of tetra-TMS derivative:

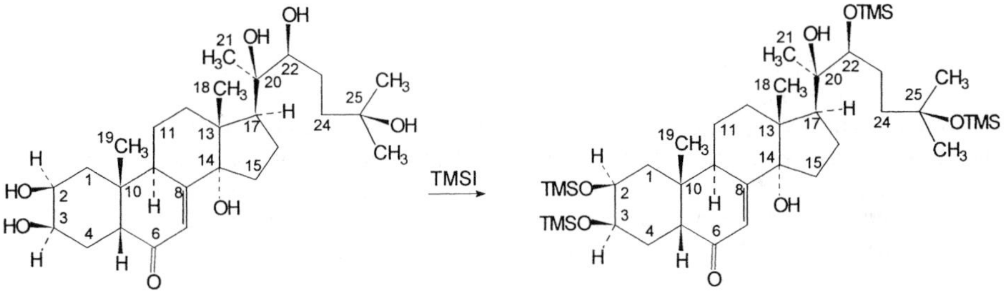

A variety of silylating reagents were evaluated, including HMDS (hexamethyldisilazane), TMCS (chlorotrimethylsilane), BSA (N,O-bis(trimethylsilyl)acetamide), TMSDEA (N,N-diethyl-N-trimethylsilylamine), BSTFA (N,O-bis(trimethylsilyl)-trifluoroacetamide), MSTFA (N-methyl-N-trimethylsilyl-trifluoroacetamide), TMSI (N-trimethylsilylimidazole) and some combinations of reagents such as HMDS + TMCS, BSTFA + TMCS, TMSI + BSA, TMSI + BSA + TMCS, etc. Except for HMDS and TMCS (not in mixture), most reagents lead to the silylation of the OH when it is in positions 3α, 3β, 16α, 20α, 20β, and 21 of the sterol and when the reaction takes place for 24 hours at room temperature. Also, phenolic OH groups are easily silylated. The groups that are more difficult to silylate due to steric hindrance are the OH in positions 11β and 17. Keto sterols with the oxo group in positions 3, 17, and 20 may lead to artifacts when silylation is done with MSTFA or MSTFA in the presence of potassium acetate (see Section 18.2). Silylation of steroids for structure determination is a complex problem, due to the fact that some OH groups may remain free due to steric hindrance, while some keto groups may form enols and be silylated.

A typical silylation procedure for the silylation of a 3-hydroxy group starts with 25–30 mg sterol dissolved in anhydrous tetrahydrofuran (THF) to which 0.5 mL HMDS and 10 μL TMCS are added. The mixture is heated at 55° C for 40 min. and analyzed by GC. TMSI allows persilylation of hydroxy steroids using strong heating and long reaction time [129]. The same result can be obtained using TMSI, BSA, and TMCS (3:3:2, v/v/v) with heating at 60° C for 60 hours [15].

Silylation with different reagents is performed to identify structural features in sterols. As an example, silylation with BSTFA of a hydroxysteroid performed in CH_3CN at 60° C for 1 hour followed by the evaporation to dryness and the addition of 50 μL cyclohexane and further silylated with d_9-TMSI at 60° C was evaluated for various time lengths of second step derivatization [130]. The GC-MS results indicated that BSTFA silylates the steroids at positions 3 and nonhindered 17β, while deuterated TMSI silylates 11β and 17 hindered positions as shown below:

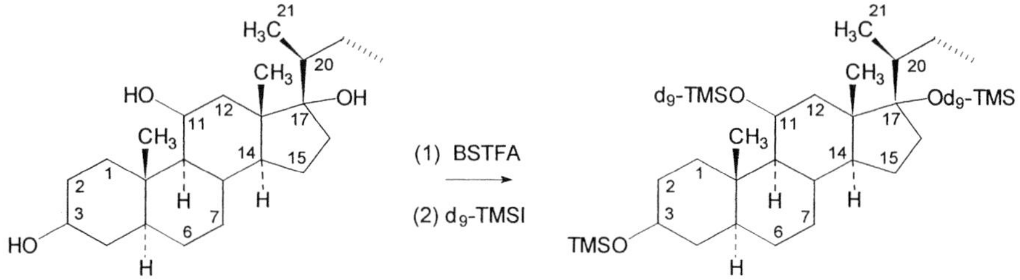

(When atoms 20 and higher are missing, the group in position 17 is TMS and not d_9-TMS.)

Other silylations are performed to generate dimethylsilyl derivatives using tetramethyldisilazane (TMDS) and dimethylmonochlorosilane (DMMCS) [131], dimethylethylsilyl imidazole (DMESI) to generate dimethylethylsilyl derivatives [15], allyldimethylchlorosilane to generate allyldimethylsilyl derivatives [132], [133], tert-butyldimethylsilyl chloride to generate tert-butyldimethylsilyl derivatives [134], etc. This variety of reagents was used because they have different steric structures and have different behavior depending on the sterol steric configuration.

Various silyl derivatives containing halogen in the silyl group are obtained from sterols. Chloromethyldimethylsilyl, bromomethyldimethylsilyl, iodomethyldimethylsilyl, pentafluorophenyldimethylsilyl (flophemesyl) derivatives were obtained from various sterols and used mainly for detection using ECD when the sensitivity was significantly increased [15], [135], [136], [137], [138], [139]. Pentafluorophenyldimethylsilyl derivative obtained using flophemesyl chloride in pyridine at room temperature for 15 min. reacts with nonhindered OH groups, but does not affect enolisable ketones or hindered hydroxyls such as those in positions 17 or 14. This derivatization is also convenient due to the prominent molecular ion in the MS spectrum, as shown in Figure 19.1.3 for cholesterol.

Abundance

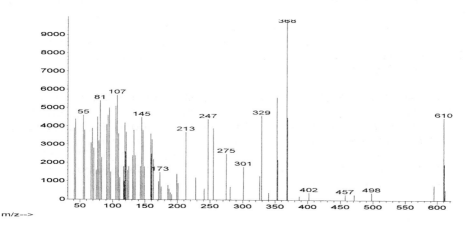

m/z-->

FIGURE 19.1.3. *Mass spectrum (EI+ 70 e.V.) of pentafluorophenyldimethylsilyl ether of cholesterol (MW = 610).*

Silylation of corticosteroids such as cortisone containing 17α,21-dihydroxy group is done using reagents containing sylilene groups to form cyclic siliconides [140].

Other derivatizations for sterols involving the OH groups are the acylations. Acylation of the OH groups in sterols has been done using trifluoroacetic anhydride (TFAA) [141], heptafluorobutyric anhydride (HFBA), [142], pentadecafluorooctanoyl chloride, 9-H-hexadecafluorononanoyl chloride, and chloroacetic anhydride [15], [143]. The reaction of hydroxysteroids with TFAA takes place at room temperature in a few min., using a catalytic amount of pyridine and TFAA several times molar ratio to the sterol. After reaction the mixture is evaporated to dryness under N_2 dissolved in a solvent and analyzed by GC or GC-MS.

Attempts were made to differentiate between equatorial and axial OH groups by using a silylation step followed by an acylation with HFBA. The -OTMS groups in equatorial positions (α) have the tendency to be more easily replaced by F_7C_3COO- groups [144], [145].

Methylation is also reported for various steroids, the reaction being performed using CH_3I and methylsulfinyl carbanion [15], [146]. Also, 17,20,21-trihydroxy steroids can form an acetonide at 20,21-OH groups with acetone and p-toluenesulfonic acid [147]. Steroids containing a 17α,21-dihydroxy group have been derivatized with alkyl boronates [148].

HPLC analysis of sterols by derivatization at an OH group can use reagents typical for other alcohols. Some of these reagents were already indicated in Tables 19.1.2 through 19.1.5. Among these are: 2-[2-(azidocarbonyl)ethyl]-3-methyl-1,4-naphthoquinone, dansylethyl chloroformate, 2-[2-(isocyanate)-ethyl]-3-methyl-1,4-naphthoquinone, 3-dansylaminophenylboronic acid, ferroceneboronic acid, 9-phenantreneboronic acid, etc. Bile acids are usually derivatized at the COOH group.

- Epoxides derivatization

Epoxides can be considered internal ethers of 1,2-diols, and they can hydrolyze with water in the presence of acids generating the corresponding *trans*-diol (see Section 18.5). However, the reactions of epoxides are not identical to that of diols. For example, epoxides can react with sodium methoxide generating a new methoxy and a hydroxy group. The reaction has been used for the analysis of 1,2-benzanthracene-5,6-epoxide [149].

Trichloromethylsilane (TMCS) reacts with epoxides forming chlorinated trimethylsilyl (TMS) derivative as follows:

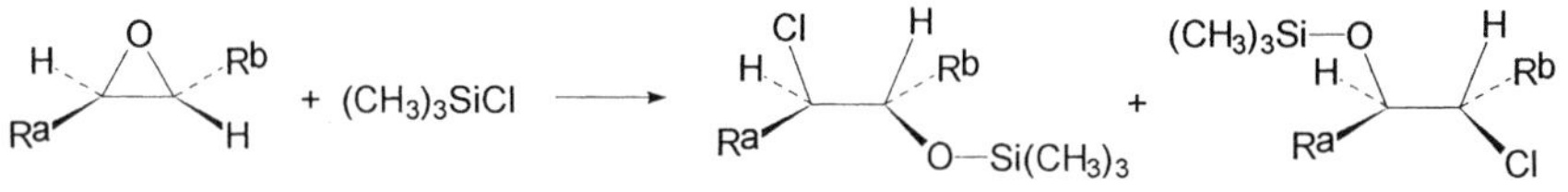

Symmetrical epoxides form unique derivatized compounds. The reaction can be done using other silylation reagents such as HMDS + TMCS. The silylation reaction with TMCS + N,O-bis(trimethylsilyl)acetamide (BSA) lead to the TMS ether of the chlorohydrine [150]. The analysis of the derivatized compounds is usually done by GC.

A reaction that works well for epoxides is the derivatization with diethyldithiocarbamates. The reaction with sodium diethyldithiocarbamate can be written as follows;

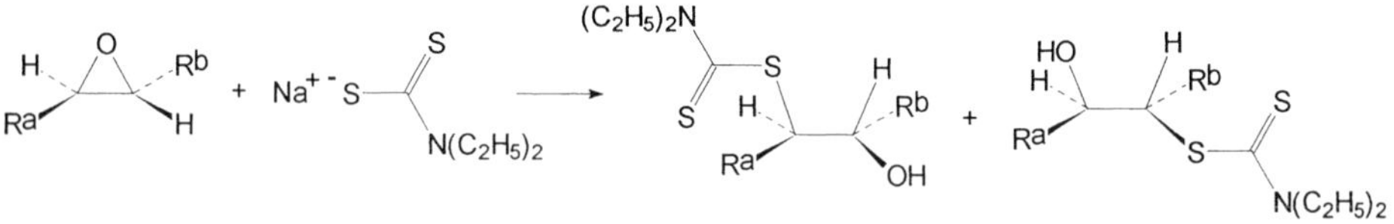

Sodium diethyldithiocarbamate can be used for the derivatization of dianhydrogalactitol [151] by reaction at room temperature for 1 h in aqueous solution in a phosphate buffer at pH = 7. The derivatized compound can be extracted in chloroform, washed with a 33% water solution of NaCl, dried and evaporated, and then dissolved in heptane:chloroform (70:30, v:v) for injection in an HPLC system. The separation can be done on a nitrile column in direct phase separation. Epoxides in polymers have been analyzed using the same reaction but with diethylammonium diethyldithiocarbamate as reagent [152].

Another reaction used for epoxides derivatization is that with isopropylamine, followed by reaction with 2,3,4,6-tetra-O-benoyl-β-D-glucopyranosyl isothiocyanate. The reaction with the amine takes place as follows:

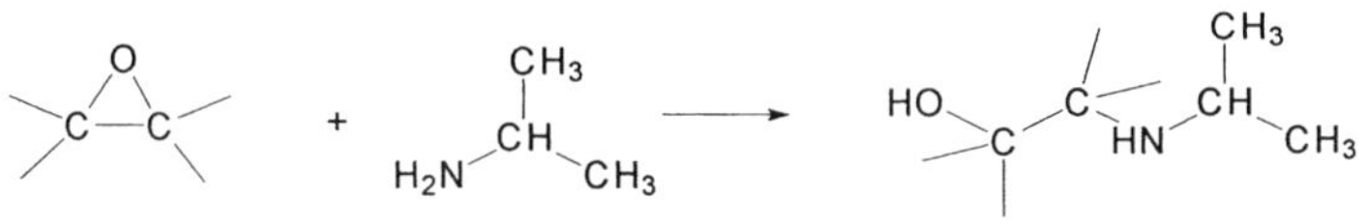

The reaction with the isothiocyanate takes place at the amino group and the analysis is done by HPLC [1].

- Derivatization of ethers

The ether group does not contain active hydrogens and poses no problems in GC analysis. In fact, many derivatizations are performed for the purpose of generating ethers from compounds with active hydrogens. Newly formed ethers have better chromatographic behavior, low polarity, no hydrogen bonds, and higher thermal stability compared to the initial compound. On the other hand, detectability of certain ethers, mainly by HPLC, can be difficult due to the lack of chromophores and low sensitivity. If other groups that allow derivatization are present in the molecule of the ether analyte, usually the ether group is left unaffected. In cases where the only group in the molecule is ether, direct chromatographic analysis is usually performed. However, ethers can be hydrolyzed with strong acids such as HI or HBr. The reaction is an acidolysis and takes place as follows:

$$R^a\text{-O-}R^b + HBr \longrightarrow R^a\text{-OH} + R^bBr$$

Depending on the radical, the alcohol group may go to R^a, R^b, or a mixture can be obtained. The newly formed alcohol can be further derivatized for the desired purpose.

19.2 DERIVATIZATION REACTIONS FOR PHENOLS

Phenols have OH functional groups, and most reactions that were described for alcohols also work for phenols. Some differences result from two characteristics of phenols. The first is the acidity of the hydrogen in the OH group, which has a pKa = 8–11 for phenols as compared to the pKa = 16–18 for alcohols (see Table 18.1.2). The second difference comes from the special chemical properties of the aromatic ring.

The derivatization of phenols is in many cases directed toward the replacement of the active hydrogen in the OH group. The reactions at the aromatic ring are less commonly used for derivatization. Common reactions applied for the replacement of the active hydrogen in the OH group are alkylations with active halides (or reactions with active aryl halides), acylations, and silylations.

The reaction with methyl halides to form methyl ethers takes place much more easily for phenols than for alcohols [15]. A basic compound is typically used to remove the excess of halogenated acid (K_2CO_3, tetrabutylammonium phosphate buffer at pH 11, etc.). For example, phenolic steroids are more easily methylated at the phenolic OH than hydroxy steroids using dimethyl sulfate or using tetramethylammonium hydroxide and thermal decomposition. Other alkyl halides also alkylate more effectively phenols, a typical case being the use of pentafluorobenzyl bromide for the analysis of pesticide residues. Parathion, for example, hydrolyzes with the formation of p-nitrophenol, which can be derivatized with pentafluorobenzyl bromide. The reactions are the following:

For the analysis of the pesticide residues, including phenols and aromatic carboxylic acids, the sample containing less than 50 µg analyte is extracted with acetone and evaporated in the presence of 50 mg K_2CO_3 to 0.5 mL (if water is present, this must be eliminated to dryness and the final volume brought to 0.5 mL with acetone). Pentafluorobenzyl bromide (1 mL sol. with 20 mg/mL reagent) is added and the mixture is heated at 50° C for 15 min. To the resulting material 1.5 mL of isooctane is added, and the volume is brought back to 1.5 mL by evaporation. A preliminary separation is then used for the elimination of the excess reagent. For this purpose the sample is passed through a small silica gel column containing 1 g material pre-wet with hexane. The excess of reagent is eliminated by washing the column with 8 mL 5% benzene in hexane, and the analytes are eluted with 6 mL of 25% benzene in hexane and then with 8 mL 75% benzene in hexane. The analysis is done using GC/ECD on the two eluates separately after they are concentrated by evaporation (appropriate internal standards must be added). Compounds analyzed using this procedure include 1-naphthol, sec-pentyl phenols (that are found in the first eluate), p-nitrophenol, (2,4-dichlorophenoxy)-acetic acid, and p,p'-dichlorodiphenylacetic acid [153], [154], [155].

Among the reactions frequently used for phenols derivatization is that with 2,4-dinitrofluorobenzene (see Section 18.1) [156]. Used in similar derivatizations is 2-chloro-1,3-dinitro-5-(trifluoromethyl)benzene (4-chloro-α, α, α -trifluoro-3,5-dinitrotoluene), which reacts with phenols in the presence of K_2CO_3. The derivatized compounds show good sensitivity in GC/ECD determinations [157].

Diazomethane, which is not successfully used for alcohol derivatization, can be used with good results for the formation of phenol ethers [158]. The analysis of the resulting methyl derivatives is usually done by GC.

Several typical reagents used for phenols derivatization followed by GC analysis are shown in Table 19.2.1

TABLE 19.2.1. *Typical reagents used for phenols derivatization for GC analysis.*

Reagent type	Reagent	Conditions	Phenol type/matrix	Reference
acylation	acetic anhydride	pyridine, heat 45° C, dimethylamino-pyridine	acetaminophen, diphenols, phenolic steroids	[15], [159]
acylation	acetic anhydride	combined with pressurized liquid extraction	phenolic steroids	[160]
acylation	benzoyl chloride	Na$_2$CO$_3$ (combined with silylation of the amino group)	acetaminophen	[161]
alkylation	bromoacetonitrile	Na$_2$CO$_3$ 60 min at 60° C	bisphenol A, 2,2'-bisphenol	[162]
acylation	chloroacetic anhydride	NaOH	phenols	[15]
acylation	dichloroacetyl chloride	NaOH	diethylstilbestrol	[15]
acylation	heptafluorobutyric anhydride	trimethylamine	phenols, phenolic sterols	[15], [163]
acylation	monofluoroacetyl chloride	NaH	thymol	[15]
acylation	pentadecafluorooctanoic anhydride	trimethylamine	diethylstilbestrol, clopidol	[15]
acylation	pentafluoropropionic anhydride	NaH	thymol	[15]
acylation	pentafluorobenzoyl chloride	NaH	thymol	[15]
acylation	trifluoroacetic anhydride	NaOH	phenols	[164]
alkylation	iodoacetonitrile	Na$_2$CO$_3$ 60 min at 60° C	bisphenol A, 2,2'-bisphenol	[165]
alkylation	tetrabutylammonium hydrogen sulfate	300° C in the injection port of the GC	chlorophenols, alkylphenols	[166], [167]
alkylation	tetramethylammonium hydroxide	300° C in the injection port of the GC	alkylphenols	[168]
ester	dimethylaminodimethyl-phosphine + oxygen	acetonitrile	phenolic steroids	[169]
silylation	N,O-bis(trimethylsilyl)-trifluoroacetamide (BSTFA)	overnight at room temp., 30 min. at 80° C, etc.	phenols	[15], [170]
silylation	N,O-bis(trimethylsilyl)-trifluoroacetamide (BSTFA)	presence of CF$_3$COOH promotes silylation	sterically hindered phenols	[15]
silylation	N,O-bis(trimethylsilyl)-acetamide (BSA)	trimethylamine	sterically hindered phenols	[15]
silylation	bromomethyldimethyl-chlorosilane	diethylamine 30 min at 65° C	phenols herbicides metabolites	[171]
silylation	chloromethyldimethyl-chlorosilane (CMSMS) + bis(chloromethyl)-tetra-methyldisilazane (1:3, v:v)	pyridine, 15 min at 25° C	naphthalenediols	[15]
silylation	hexamethyldisilazane + trimethylchlorosilane (2:1, v:v)	DMSO/dioxane (1:1, v:v)	anthocyanins	[172]
silylation	N-methyl-N-trimethylsilyl heptafluorobutyramide (MSHFBA)	trichloromethylsilane present	phenols	[15]
silylation	N-*tert*-butyldimethylsilyl-N-methyltrifluoroacetamide		phenols	[173]
silylation	*tert*-butyldimethylsilyl-chlorosilane	imidazole DMF	phenolic steroids	[174]

Silylation of phenols typically takes place easily and gives unique derivatized compounds. This explains the wide utilization of silylation for GC analysis of phenols. For example, for the determination of phenol levels in cigarette smoke, 10 cigarettes are smoked using FTC (Federal Trade Commission) recommendations [175]. The smoke is collected on a smoke pad and extracted in 10 mL *tert*-butyl methyl ether. An aliquot of

500 µL of the extract is derivatized with 1 mL BSTFA by heating the mixture at 76° C for 30 min. The GC separation can be done on a methyl silicone 5% phenyl silicone column 30 m x 0.25 mm i.d., 0.25-µm film thickness, with a GC temperature gradient from 50° C to 300° C [176], [177]. The detection can be done using MS or SIM-MS. This procedure applied using SIM-MS to a solution of 0.5 µg/mL standards of several phenols leads to the chromatogram shown in Figure 19.2.1.

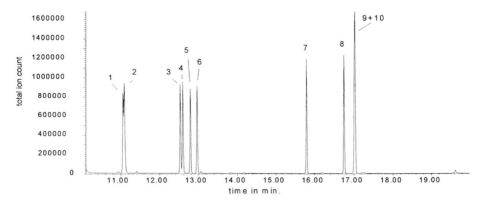

FIGURE 19.2.1. *SIM chromatogram for the standards and deuterated standards of phenols TMS derivatives in a 0.5 µg/mL solution.* The peak identification is: (1) [$^2H_6$]-phenol, (2) phenol, (3) [$^2H_8$]-o-cresol, (4) o-cresol, (5) m-cresol, (6) p-cresol, (7) catechol, (8) resorcinol, (9+10) [$^2H_6$]-hydroquinone + hydroquinone.

Some other reagents are used for phenol derivatization with GC analysis. An example is diethylchlorophosphate ($C_2H_5O)_2P(O)Cl$, which reacts with phenols in the presence of sodium *tert*-butoxide to generate the corresponding diethylphosphate of the phenol. The reaction has been used with good results and offers the potential of using a phosphorus sensitive detector [178].

In the analysis of polyfunctional phenols, reactions with formation of cyclic derivatives can be applied successfully [15] for GC analysis. For example, catecholamines or other diphenols such as hydroxysalicylanilides can be analyzed using butaneboronic acid [179].

- Derivatization of phenols for HPLC analysis

Similarly to the derivatization for GC, many derivatization reactions used for HPLC analysis of alcohols can be applied to phenols. One example is the derivatization with dansyl chloride, which has been used for the derivatization of hydroxybiphenyls, chlorophenols, hydroxychlorobiphenyls, estrogens, and cannabinoids [15] followed by HPLC analysis with UV or fluorescence detection. The reaction with tetrahydro-6,6,9-trimethyl-3-pentyl-6H-dibenzo[b,d]pyran-1-ol (Δ^1-3,4-trans isomer of tetrahydro-cannabinol is the major active constituent in marijuana) takes place as follows [180]:

Other reagents used for phenols derivatization and HPLC analysis with UV detection are shown in Table 19.2.2.

TABLE 19.2.2. *Reagents used for phenols derivatization in HPLC with UV detection*

Reagent		Phenol type, matrix, sensitivity	Detection λ nm, separation	Reference
4-aminoantipyrine in the presence of K₃[Fe(CN)₆] (Emerson reaction)		phenols 5 ppb	470	[181]
anisoyl chloride		phenols 30 ppb	254, silica	[182]
4-cyanophenyldiazonium chloride		phenols 10 ppb	370, C18	[1]
2-fluorenesulfonyl chloride		phenols 50 pg	254, C18	[1]
3-methyl-2-benzothiazolinone hydrazone		phenols in water 2-20 ng	500 postcolumn, C8	[183]
4-nitrobenzenediazonium chloride		phenols 50 pg	365, C18	[1]
2-(N-phthalimido)ethyl 2-(dimethylamino)ethane-sulfonate		chlorophenols 10-20 pmol	225, C18	[1]

Some of the reagents shown in Table 19.2.2 replace active hydrogens in the OH phenolic group. However, other reagents participate in a nucleophilic substitution at the aromatic ring. This is the case for the reaction with 3-methyl-2-benzothiazolinone hydrazone or with diazonium compounds (see Section 18.6).

Fluorescence detection is also common for HPLC analysis of phenols. Several reagents used for phenols derivatization and HPLC analysis with fluorescence detection are shown in Table 19.2.3.

TABLE 19.2.3. *Reagents used for phenols derivatization in HPLC with fluorescence detection.*

Reagent		Phenol type, matrix, sensitivity	Detection/ separation	Reference
2-(9-anthryl)ethyl-chloroformate		phenols 7-10 nmol	ex 256, em 418, RP-18	[1]
1,2-bis(4-methoxyphenyl)-ethylenediamine		catecholamines	ex. 350, em. 480, C18, size excl.	[1]
1,2-(difluoro-1,3,5-triazinyl)benz[f]-isoindolo-[1,,2-b][1,3]-benzothiazolidine		phenolic steroids	ex. 540, em. 560, RP	[2]
2-fluorenesulfonyl chloride		phenols	ex 280, em 325, C18	[184]
pyrenesulfonyl chloride		phenolic steroids 0.2 fmol	laser, ex 325	[185]

Determination of polyfunctional phenols gives various reactions with formation of new cycles that can have properties useful for detection in HPLC. For compounds such as peptides with tyrosine-containing residue, this type of reaction can be exploited by using tyrosinase, which adds one supplementary OH group to the aromatic ring of the amino acid. Reactions with bifunctional amines such as 1,2-bis(4-methoxyphenyl)-ethylenediamine can follow this process for further HPLC analysis [1]. The addition of a second functional group to tyrosine-containing peptides also can be achieved using a formylation reaction [186].

A reagent used for estradiol analysis with CL detection at 550 nm is lissamine rhodamine B sulfonyl chloride:

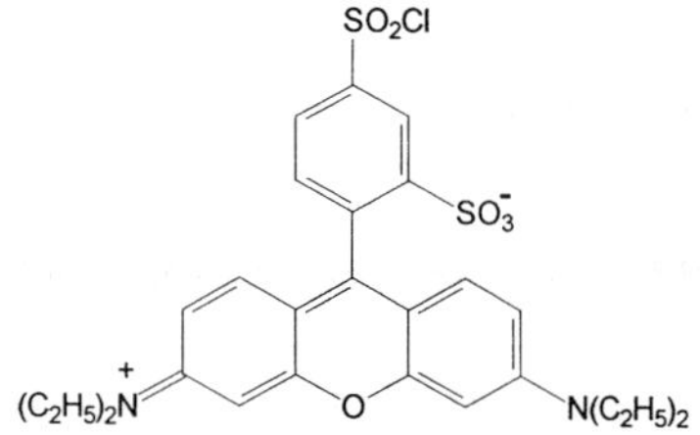

The analysis is performed with postcolumn addition of hydrogen peroxide and bis(2-nitrophenyl)oxalate [1]. Various catecholamines also can be analyzed with CL detection by condensation of the analyte with ethylenediamine and using a postcolumn reaction with bis[4-nitro-2-(3,6,9-trioxadecyloxycarbonyl)phenyl]oxalate [1]. The advantage of this type of detection is the capability to measure as low as 100–200 fmol of analyte.

19.3   DERIVATIZATION REACTIONS FOR THIOLS AND SULFIDES

The thiols (mercaptans) containing SH group(s) on an aliphatic chain are stronger acids than the corresponding alcohols and have the pKa = 10–11. The thiols with the SH group on an aromatic ring are stronger acids than the corresponding phenols and have the pKa = 6–8. The replacement of the active hydrogens in the SH group is therefore the common derivatization procedure for thiols, and several reactions take place more easily for thiols than for the equivalent alcohols or phenols. This is, for example, the case for alkylation reactions. Other derivatizations such as silylation or acylations are also performed with good results on thiols.

Besides derivatization for enhancing some desired analytical property, thiols are in some analyses derivatized for protecting the group from oxidation. Thiols undergo easily a reaction of the type:

$$2 \ R{-}SH \ \xrightarrow{\text{oxidant}} \ R{-}S{-}S{-}R$$

This reaction occurring inadvertently may modify the initial content of a thiol during the analysis. For this reason, derivatization with reagents such as iodoacetic acid or iodomethane can be performed on thiols [1] to protect the SH group. The resulting compounds cannot be derivatized using the same reagents as the thiol, and other groups must be present if further derivatization is needed. One such example is the analysis of a mixture of glutathione, N-(N-L-γ-glutamyl-L-cysteinyl)glycine, and oxidized glutathione. This analysis can be done using first the protection of the SH group in glutathione using the reaction:

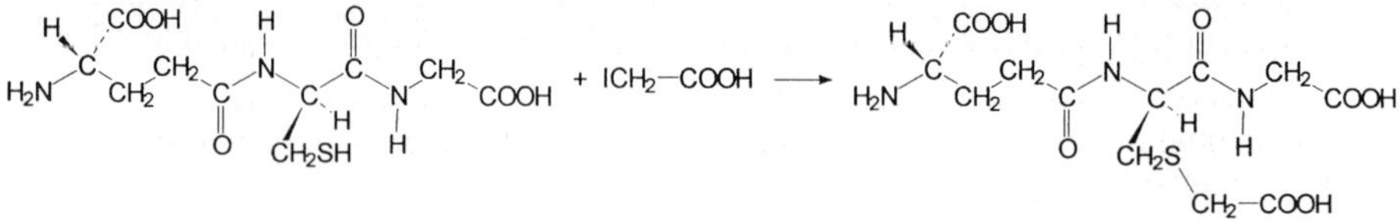

The compounds are further derivatized with 2,4-dinitrofluorobenzene. After the mixture is allowed to stand for 20 h at room temperature, it is followed by HPLC separation on a NH_2 column and detection in UV at 365 nm [187].

Thiol alkylation with active alkyl halides is a common derivatization reaction for direct analytical purposes. For example, pentafluorobenzyl bromide can be used for thiols derivatization by refluxing for two hours the analytes dissolved in acetone in the presence of K_2CO_3 [15]. A similar reaction occurs with 1-chloro-2,4-dinitrobenzene [188]. The resulting substituted thiol from this reaction can be oxidized with $KMnO_4$ in acetic acid. The reaction takes place as follows:

Both the alkylthio-2,4-dinitrobenzene and the sulfone resulting when the oxidation is performed can be analyzed by GC.

Silylation is used for the derivatization of thiols, not always for analytical purposes but with good yields of the reaction. A summary of silylating reagents known to react with thiols is given in Table 19.3.1 [189]. Some of the silylation reagents such as BSTFA are used for the GC analysis of thiols.

TABLE 19.3.1. *Silylation reagents used for thiols derivatization.*

Reagent	Analyte	Conditions	Reference
allyltrimethylsilane	thiols, thiophenols	CF_3SO_3H catalyst	[190]
BSTFA	thiols	DMF solvent	
(ethylthio)trimethylsilane	thiols		[189]
ethyl trimethylsilylacetate	thiols	tetrabutylammonium fluoride catalyst	[189]
hexamethyldisilthiane	thiols		[189]
1-methoxy-2-methyl-1-trimethylsiloxypropene	thiols		[189
trimethylchlorosilane	thiols	hexamethyldisilazane catalyst	[189]

- ***Derivatization of thiols for HPLC analysis***

Many thiols are bioactive compounds and are naturally found in biological systems or are used as drugs. The analysis of these thiols is frequently done by HPLC, and their derivatization is practiced mainly for better detectability (see Section 17.6). Various reagents reported in literature as being used for thiol analysis are given in Table 19.3.2.

TABLE 19.3.2. *Reagents used for thiols HPLC analysis.*

Reagent		Reaction type[1]	Thiol type, matrix, sensitivity	Detection, Separation	Ref.
N-(9-acridinyl)-maleimide		Add	thiols, cysteine	fluorescence ex 355 em 465	[2], [191]
N-[6-(7-amino-4-methylcoumarin-3-acetamido)hexyl)]-3'-(2'-pyridylthio)-propionamide		Dis	6-mercapto-purine 60 pmol/g DNA	fluorescence ex 345 em 450, C8	[1]
4-(aminosulfonyl)-7-fluoro-2,1,3-benzoxadiazole		Alk	thiols in blood, amino acids with thiol groups, etc. 0.5 μmol (UV), 1.9 pmol (fluorescence)	UV 220, fluorescence ex 380 em 510, C18, CE	[1], [192]
N-[p-(2-benzoxazolyl)-phenyl]-maleimide		Add	penicillamine 200 nmol	fluorescence ex 319 em 360, C18	[193]
N-(4-benzoyl-phenyl)-maleimide		Add	captopril 50 ng/mL	UV 254, C18	[194]
1-benzyl-2-chloropyridinium bromide		Alk	captopril in blood (S)-1-(mercapto-2-methyl-1-oxo-propyl)-L-proline 10 ng/mL	UV 314, silica	[1]
1,1-bis(phenyl-sulfonyl)-ethylene		Add	thiol drugs in formulations 100 pmol	UV 254, C18	[195]
7-chloro-2,1,3-benzoxadiazole-4-sulfonic acid (ammonium salt)		Alk	disulfide containing octapeptide reduced with thiothreitol 0.5 ng/mL	fluorescence ex 380 em 470, RP-8	[196]

TABLE 19.3.2 (continued). *Reagents used for thiols HPLC analysis.*

Reagent	Reaction type[1]	Thiol type, matrix, sensitivity	Detection, Separation	Ref.
2-chloro-4,5-bis(p-N,N-dimethyl-aminosulfonyl-phenyl)oxazole (SAOX-Cl)	Alk	thiols, glutathione, cysteine, etc. 1 – 8 fmol	fluorescence ex 330 em 425, C8	[1]
1-chloro-2,4-dinitrobenzene	Alk	thiols in tissue	UV 340, C18	[1]
4-(N-chloroformyl-methyl-N-methyl)-amino-7-N,N-dimethylamino-sulfonyl-2,1,3-benzoxadiazole	Acy	α-mercapto-N,2-naphthyl-acetamide 100 fmol	fluorescence ex 437 em 544, C18	[1]
2-chloro-1-methylquinolinium-tetrafluoroborate	Alk	cysteine 15.4 pmol	UV, C8	[197]
4-chloro-7-nitro-2,1,3-benzoxadiazole	Alk	various, acetylcysteine reduced with dithioerythritol 1 ppm	Vis 425, 470, C18	[198] [199]
CY5.4a-IA	Alk	anticancer drug metabolite, etc. 5 nmol	laser induced fluorescence ex. 670	[1]
dansyl aziridine	Add	thiols 120 pg/mL	fluorescence ex 339 em 540, C18	[200]
2,4'-dibromoaceto-phenone or 2-bromo-1-(4-bromo-phenyl)ethan-1-one	Alk	captopril in blood 5 ng/mL	UV 254, C18	[1]
4-(N,N-dimethyl-amino-sufonyl)-7-fluoro-2,1,3-benzoxadiazole	Alk	thiols in tissues 0.9 pmol	fluorescence ex 390 em 520, ODS	[1]

TABLE 19.3.2 (continued). *Reagents used for thiols HPLC analysis.*

Reagent		Reaction type[1]	Thiol type, matrix, sensitivity	Detection, Separation	Ref.
N-[4-(6-dimethyl-amino-2-benzo-furanyl)phenyl]-maleimide		Add	penicillamine 290 fmol (D), 350 fmol (L)	fluorescence ex 360 em 455, Pirkle type	[1]
N-(4-dimethyl-amino-3,5-dinitro-phenyl)maleimide		Add	captopril 100 ng/mL	UV 254, C18	[1]
7-diethylamino-3-(4'-maleimidyl-phenyl)-4-methylcoumarin		Add	thiols 0.7 nmol/g	fluorescence ex 387 em 465, C18	[1]
N-(7-dimethyl-amino-4-methyl-3-coumarinyl)maleimide		Add	acetylcysteine 53 fmol, thiols in urine 38-53 fmol	fluorescence ex 400 em 480, RP-8	[1], [201]
2,4-dinitrofluoro-benzene		Alk	acetylcysteine 50 ng/mL	UV 360	[1]
5,5'-dithiobis(2-nitrobenzoic acid)		Dis	thiols 0.2-0.5 nmol	UV 280, C18, etc.	[1]
4,4'-dithiodipyridine		Dis	thiols	postcolumn UV 324	[1]
6,6'-dithionicotinic acid		Dis	thiols 0.1 nmol	postcolumn UV 344	[1]
ethacrynic acid		Add	cysteine, glutathione 500 ng/mL	UV 270, C18	[1]
N-ethylmaleimide		Add	6-mercapto-purine 10 ng/mL	UV 280, C18	[202]

TABLE 19.3.2 (continued). *Reagents used for thiols HPLC analysis.*

Reagent	Reaction type[1]	Thiol type, matrix, sensitivity	Detection, Separation	Ref.
Eu^{3+} chelate of 1-[p-(((5-male-imidopentyl)-carbonyl)amino)-benzyl]-ethylene-diaminotetraacetic acid	Add	thiols, cysteine, etc., 13-19 fmol	postcolumn fluorescence ex 344 em 617	[203]
7-fluoro-2,1,3-benzoxadiazole-4-sulfonic acid (ammonium salt)	Alk	thiols in blood 2 μmol	fluorescence ex 385 em 515, C18, CE	[1]
4-fluoro-7-nitro-2,1,3-benzoxadiazole	Alk	various, amino acids 10-150 pmol	fluorescence ex 450, em 520, postcolumn, strong cation exchange	[204]
2-hexanal	Add	glutathione in blood 12 pmol	postcolumn fluorescence ex 340, em 450	[1]
(R)-(-)-4-isothio-cyanatopyrrolidin-1-yl)-7-nitro-2,1,3-benzoxadiazole	Iso	thiols, chiral, 0.5 pm	fluorescence ex 455 em 568, ODS	[1]
2-(4-N-male-imidephenyl)-6-methyl-benzothiazole	Add	thiols in blood, tissues 2-20 fmol	fluorescence ex 320 em 405, ODS	[1]
methyl acrylate	Add	thiols of 1,3-thiazaperhydro-epin-4-one 2.5 ng/mL	LC-MS	[205], [205a]
methyl 4-(6-methoxy-naphthalen-2-yl)-4-oxo-2-butenoate	Add	thiols 0.5 pmol	fluorescence ex 310 em 450, C18, RP8	[206]
4-(6-methyl-naphthalene-2-yl)-4-oxo-2-butenoic acid	Add	glutathione	fluorescence ex 300 em 460, C8	[207]

TABLE 19.3.2 (continued). *Reagents used for thiols HPLC analysis.*

Reagent		Reaction type[1]	Thiol type, matrix, sensitivity	Detection, Separation	Ref.
monobromo-bimane		Alk	thiols in blood, 2'-deoxy-6-thioguanosine, acetylcysteine, glutathione, etc. 500 nmol/g DNA	fluorescence ex 394 em 480, etc., C8, RP18	[1], [208] [209] [210]
monobromotri-methyl-ammonio-bimane		Alk	thiols in blood 1 pmol	fluorescence, cation exch.	[1]
phthalaldehyde + L valine		Cyc	acetylcysteine, chiral 0.4% of major enantiomer	UV 214, CE	[1]
N-(1-pyrenyl) maleimide		Add	captopril 10 ng/mL	fluorescence ex 340 em 389, ODS or postcolumn	[1]
sodium N-chlorodansylamide		Oxi	glutathione 0.1 nmol	postcolumn fluorescence ex 360 em 510, C18	[1]
2,3,4,6-tetra-O-acetyl-β-D-glucopyranosyl isothiocyanate		Iso	thiols chiral	UV 250, ODS	[211]

[1] Acy = acylation reaction, Add = addition to an activated double bond (α,β-unsaturated carbonyls or aziridine), Alk = substitution of the thiol active hydrogen with an alkyl or aryl group, Cyc= reaction with formation of a new cycle, Dis= reaction with a disulfide reagent to form a new disulfide, Iso= addition to an isothiocyanate group, Oxi= oxidation reaction.

In addition to the reagents indicated in Table 19.3.2, some other reagents are applied for thiols labeling for HPLC analysis. For example, thiols such as mercapto-benzothiazole can be detected as low as 200 amol using derivatization with a cyanine type reagent (CY5.4a-IA) with the structure:

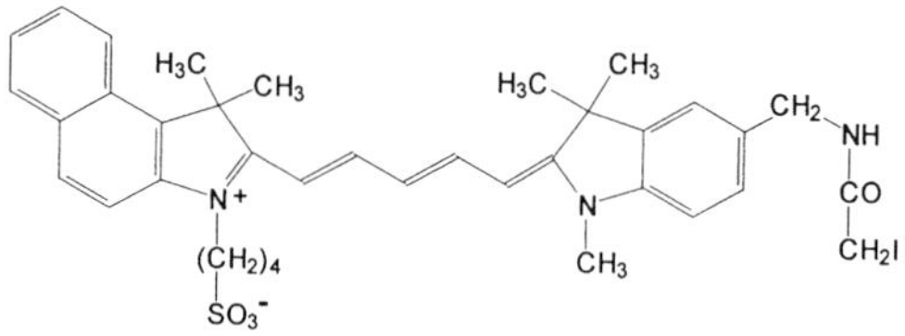

With this reagent, the thiols can be analyzed using HPLC separation with laser induced fluorescence (LIF) detection (ex. 665, em 689 nm) [212]. Also, electrochemical detection in HPLC is utilized in the analysis of a number of thiols with excellent sensitivity. Various reagents are used for derivatization before the electrochemical analysis. The formulas of these reagents are shown below:

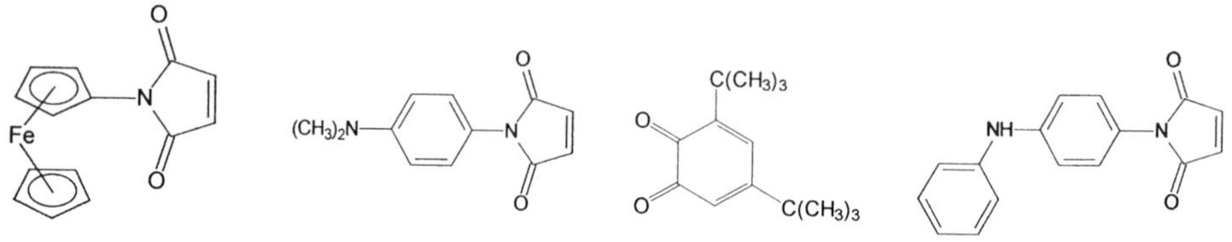

N-(Ferrocenyl)maleimide is used for the analysis of glutathione and L-cysteine, allowing the determination of as low as 0.06 pmol of analyte [213], and N-(4-dimethylamino-phenyl) maleimide is used for the derivatization of (S)-1-(mercapto-2-methyl-1-oxo-propyl)-L-proline (captopril), allowing the amperometric detection of 10 ng/mL analyte [214]. The use of 3,5-di-*tert*-butyl-1,2-benzoquinone allows the determination of various thiols in tissues with the limit of quantitation of 100 nmol, and the use of N-(4-anilinophenyl)maleimide allows the amperometric analysis of (2R,4R)-2-(2-hydroxy-phenyl)-3-(3-mercaptopropionyl)-4-thiazolidinecarboxylic acid (a hypertensive agent) with the limit of quantitation of 2 ng/mL.

- *Derivatization of sulfides and disulfides*

Sulfides are equivalent to ethers, having the oxygen replaced by sulfur. Similarly to ethers, sulfides are not polar, and the compounds with low molecular weight are analyzed by GC without derivatization. Sulfide group may be the only functional group in a compound, but more frequently it is associated with other groups, such as in the amino acid methionine. Besides simple sulfides, a number of cyclic compounds contain a sulfur atom. Examples are thiophenes, as well as compounds containing sulfur and an additional heteroatom such as nitrogen. The analysis of these types of compounds is important, some of them having significant biological activity.

Of practical importance are, for example, phenothiazines that possess a wide range of pharmaceutical properties. The analysis of traces of phenothiazines by HPLC can be done following a derivatization with peroxyacetic acid that generates a colored radical cation or fluorescent sulfoxides. The reaction with the formation of a sulfoxide can be written as follows:

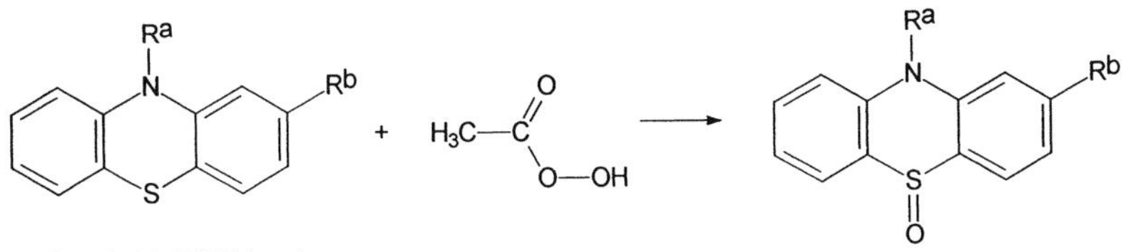

$R^a = (CH_2)_3N(CH_3)_2$, etc.
$R^b = COCl$, Cl, etc.

The oxidation can be done postcolumn, and the procedure allows the analysis of a number of phenothiazines used as pharmaceutical drugs, such as triflupromazine, chlorpromazine, trimeprazine, etc., at levels between 4 nmol to 300 nmol [215].

Disulfides are also rather common among the compounds with important biological activity. Disulfides can be transformed into thiols using an oxidation-reduction reaction. The thiols formed in the first step can be further derivatized, as shown previously. A typical example is the cleavage of a disulfide bond in proteins using dithiothreitol (1,4-dimercapto-2,3-butandiol) in an oxidation-reduction reaction used for the generation of thiol groups:

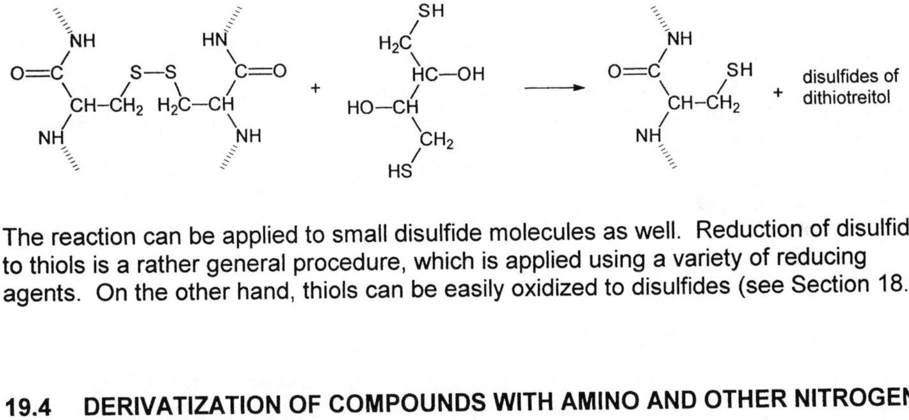

The reaction can be applied to small disulfide molecules as well. Reduction of disulfides to thiols is a rather general procedure, which is applied using a variety of reducing agents. On the other hand, thiols can be easily oxidized to disulfides (see Section 18.6).

19.4 DERIVATIZATION OF COMPOUNDS WITH AMINO AND OTHER NITROGEN-CONTAINING GROUPS

A wide variety of derivatization reactions are used for amine analysis. This is explained by the presence of many amines in nature, mainly in biological materials, and by the diversity of chemical reactions applicable to amines.

Amines are Lewis bases, and because nitrogen is not as electronegative as oxygen, amines have a stronger tendency to react with protons. Amines are much more basic than water, and their aqueous solutions have basic character. Basicity of amines is usually expressed by the equilibrium constant K_b (basicity constant) for the base-acid reaction $RNH_2 + H_2O \rightleftharpoons RNH_3^+ + OH^-$ or by the acidity constant $K_a = 14 - K_b$. The K_b and K_a values for a few amines are given in Table 19.4.1 (see also Table 18.1.3).

TABLE 19.4.1. *Acidity and basicity constants of several amines.*

Amine	Conjugated acid	pK$_b$ (25° C)	pK$_a$ (25° C)
NH$_3$	NH$_4$$^+$	4.76	9.24
CH$_3$NH$_2$	CH$_3$NH$_3$$^+$	3.38	10.62
CH$_3$CH$_2$NH$_2$	CH$_3$CH$_2$NH$_3$$^+$	3.36	10.64
(CH$_3$)$_2$NH	(CH$_3$)$_2$NH$_2$$^+$	3.27	10.73
(CH$_3$CH$_2$)$_2$NH	(CH$_3$CH$_2$)$_2$NH$_2$$^+$	3.06	10.94
(CH$_3$)$_3$N	(CH$_3$)$_3$NH$^+$	4.21	9.79
(CH$_3$CH$_2$)$_3$N	(CH$_3$CH$_2$)$_3$NH$^+$	3.25	10.75
C$_6$H$_5$NH$_2$	C$_6$H$_5$NH$_3$$^+$	9.40	4.6
o-Cl-C$_6$H$_4$NH$_2$	o-Cl-C$_6$H$_4$NH$_3$$^+$	11.35	2.65
m-Cl-C$_6$H$_4$NH$_2$	m-Cl-C$_6$H$_4$NH$_3$$^+$	10.48	3.52
C$_5$H$_5$N (pyridine)	C$_5$H$_5$NH$^+$	8.75	5.25
C$_4$H$_4$N$_2$ (pyrazine)	C$_4$H$_4$N$_2$H$^+$	13.35	0.65
C$_4$H$_4$N$_2$ (pyrimidine)	C$_4$H$_4$N$_2$H$^+$	12.70	1.30
C$_4$H$_4$N$_2$ (pyridazine)	C$_4$H$_4$N$_2$H$^+$	11.70	2.30
C$_3$H$_4$N$_2$ (imidazole)	C$_3$H$_4$N$_2$H$^+$	7.05	6.95
C$_5$H$_4$N$_4$ (purine)	C$_5$H$_4$N$_4$H$^+$	11.7	2.30

The aromatic amines are less basic than the aliphatic ones, but they too can be protonated in diluted acidic solutions. The higher is the σ character of the lone pair electrons of the nitrogen, the higher is the basicity of the amine. The variation of pK$_a$ varies from about 10 for (CH$_3$)$_3$N to about 5 for pyridine (pK$_a$ = 5.25). Some amines such as pyrrole, indole, carbazole, and o-nitroaniline are not basic but acidic (for pyrrole pK$_a$ = -4.4). The acidic character increases when the involvement of the lone pair electrons in the delocalized π system increases. Other heterocycles such as pyrazole (pK$_a$ = 2.5), thiazole (pK$_a$ = 2.4), and oxazole (pK$_a$ = 0.8) are weak bases.

Because the nucleophile character for the atoms in the same row of the periodic table is approximately in order of basicity, usually amines are stronger nucleophiles than alcohols or ethers. The approximate order of nucleophilicity is NH$_2$$^-$ > RO$^-$ > OH$^-$ > R$_2$NH > ArO$^-$ > NH$_3$ > Pyridine > F$^-$ > H$_2$O, and for another series of compounds is R$_3$C$^-$ > R$_2$N$^-$ > RO$^-$ > F$^-$. The parallel between basicity and nucleophilicity is not followed strictly because basicity is thermodynamically controlled and nucleophilicity is kinetically controlled. In general, amines (including tertiary amines) can be easily alkylated (or derivatized with aryl groups) and also acylated. Both these reactions take place as nucleophile substitutions.

A differentiation must be made between primary and secondary amines on one hand and tertiary amines on the other hand. Tertiary amines do not have an active hydrogen to be replaced. This makes a number of reactions typical for primary and secondary amines impossible for tertiary amines. However, some of the amines derivatization reactions still take place with tertiary amines. The alkylation of tertiary amines leads to the formation of quaternary ammonium salts, but this reaction is not commonly used as a derivatization. An unexpected disadvantage appears for GC due to the basic character of tertiary amines (even if they are aromatic). The residual acidic groups in the chromatographic columns or in other parts along the path of the analyte in the GC affect the chromatographic peak shape. This leads frequently to problems in analysis of amines, even if no active hydrogen is bound to the nitrogen atom (pyrrole and indole being acidic do not have this problem). For this reason, other derivatizations, mainly acylations, are preferred for amines.

The N-acyl derivatives of amines are more stable than O-acyl derivatives and easier to form [216]. This makes acylation the derivatization of choice for amines. For example, derivatization using anhydrides of fluorinated short chain acids (pentafluoropropionyl anhydride or heptafluorobutyryl anhydrides) are very common for GC analysis of amines. The formation of amides also can be facilitated using special reagents able to facilitate peptide bond formation (see Table 18.3.3).

Amides having a more neutral character have fewer problems than amines regarding the GC behavior. However, the acylation of primary amines leading to monosubstituted amides still leaves a free hydrogen in the amide group, as shown below for the reaction with an acyl chloride:

$$Ra-NH_2 \ + \ R-COCl \ \longrightarrow \ \underset{H}{\overset{Ra}{\diagdown}}N-\overset{\overset{\textstyle O}{\|}}{C}-R \ + \ HCl$$

In specific analyses it has been found that the hydrogen in the amide group must be further replaced for better analytical results. This is mainly the case of GC analysis of traces, when any residual activity of the analyte may lead to poor results. One such example is that of analysis of heterocyclic amines (see Section 19.4) when, after the acylation of the first hydrogen, the remaining hydrogen in the amide group must be further derivatized. In this case, a different derivatization, which is methylation of the second hydrogen, is preferred. Some primary amines can be acylated to both hydrogens. Used for this purpose are small acyl groups, longer reaction times, higher reaction temperatures, as well as the addition in the reaction medium of basic compounds such as trimethylamine [217], [218], [219] (see Section 18.3). The influence of the concentration of trimethylamine on the yield of acylation of phenylethylamine with trifluoroacetic anhydride has been shown in Table 18.3.2.

Silylation of amines is another common derivatization. However, N-trimethylsilyl derivatives are much less stable than O-trimethylsilyl derivatives (see Section 18.2). This leads to significant problems regarding the hydrolysis of N-trimethylsilyl derivatives that occur when traces of water are present in the reaction medium. For this reason, *tert*-butyldimethylsilyl derivatives of amines, which are less sensitive to hydrolysis, are more successfully used in amine (and amino acid) derivatization [220].

Due to the capability of amines to participate in various condensation reactions, such as reactions with ketones, isocyanates, isothiocyanates (see Section 18.4), or the capability to form aromatic cycles in condensation reactions with γ-diketones (see Section 18.5), these reactions also are used in amine derivatization. However, in many of these reactions, there are differences between primary and secondary amines. Some other differences between primary, secondary, and tertiary amines are discussed in this section.

The amine groups are present in numerous compounds together with other groups such as OH, COOH, etc. Some of the derivatization reactions take place at the amino group independently of the existence of another active group present in the molecule. As an example, derivatization with 9-fluorenylmethyl chloroformate of ampicillin and of cephalosporin C takes place in the same way, although ampicillin has an independent amino group while cephalosporin C is a typical amino acid:

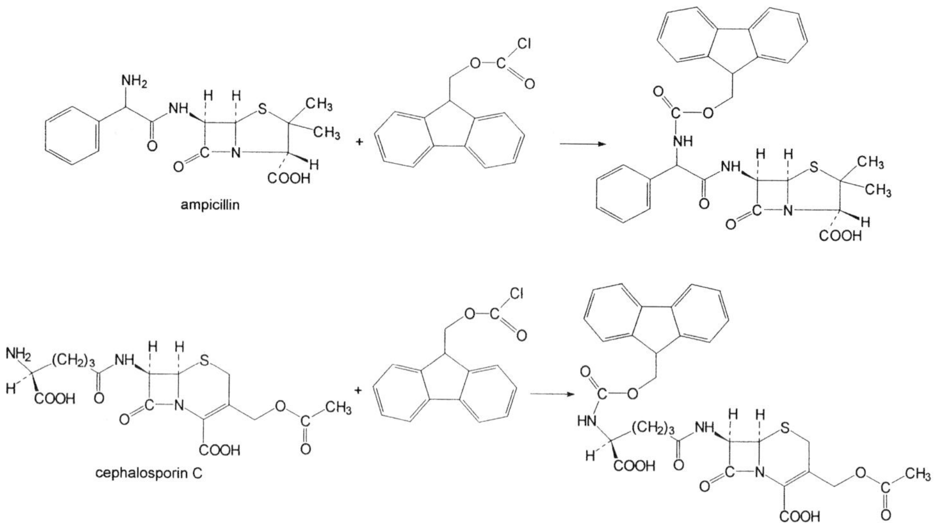

The reaction is used for the HPLC determination of both antibiotics in fermentation media [221]. For the analysis, 1 mL filtered fermentation medium is mixed with 100 μL 200 mM sodium borate buffer of pH 7.7 and 2 mL acetone at 4° C. The mixture is centrifuged at 2600 g in a refrigerated centrifuge for 10 min., and 400 μL of 15 μM 9-fluorenylmethyl chloroformate in dry acetone is added to the supernatant. The solution is washed twice with 3.4 mL portions of pentane, and an aliquot is injected in the HPLC system.

When amino group is present together with other functional groups, the derivatization may take place only at the amine or may involve the other groups. The derivatization of multifunctional compounds involving several groups is discussed separately.

- Derivatization of primary and secondary amines for GC analysis

Primary and secondary amines can be alkylated with alkyl halides more easily than alcohols. For example, methylation using methyl iodide in the presence of NaH is applied for the GC analysis of triazine herbicides containing amino groups [222]. For the same purpose can be used N,N'dimethylformamide dimethyl acetal $(CH_3)_2N-CH(OCH_3)_2$ (Methyl 8®) [223], [224].

Derivatization with activated aryl halides also is applied for amine derivatization. A typical reagent for this type of derivatization is 2,4-dinitrofluorobenzene [15], [225]. Other reagents used for the same purpose are 2-chloro-1,3-dinitro-5-(trifluoromethyl)benzene (DNT-Cl), 1-fluoro-2-nitro-4-(trifluoromethyl)-benzene [15], and sodium 2,4-dinitrobenzensulfonate (DNBS) [226]. With these reagents, the reactions are performed in aqueous solution in the presence of $Na_2B_4O_7$ using heating (55° C for

DNT-Cl and 100° C for DNBS) for about 2 h. For DNBS the reaction can be written as follows:

$$R-NH_2 + Na^+ SO_3^- \cdots \text{(aryl: } O_2N, -NO_2) \longrightarrow \text{(aryl: } O_2N, R-HN, -NO_2)$$

As low as 50 pg derivatized amine-carrying CF_3 groups can be detected using ECD [227].

Permethylation of amines can be achieved using a reductive alkylation with HCHO and $NaBH_4$ or formic acid. The reaction in the presence of $NaBH_4$ takes place by dissolution of the amine in 3M aqueous H_2SO_4 followed by addition of a solution of 40% formaldehyde (3.5 moles/mole NH). To this solution, solid $NaBH_4$ is added (2.1 moles/mole NH), while the temperature is maintained at 0–20° C. After the addition of $NaBH_4$, the mixture is strongly acidified with H_2SO_4. The mixture is washed with ethyl ether and alkalinized with KOH. The derivatized analyte is extracted with ethyl ether and can be analyzed by GC. The procedure is reported for the derivatization of polyamines [228]. The overall reaction can be written as follows:

$$R-NH_2 + HCHO \xrightarrow{NaBH_4} R-N(CH_3)_2$$

Silylation of amines is performed for GC analysis using various reagents. One typical silylation reagent for amines is bis(trimethylsilylacetamide (BSA). The derivatization of a primary amine such as 3,4-dimethoxyphenylethylamine (50 µg) with BSA (40 µL) and pyridine (2:1, v:v) takes place at room temperature in 10 min. and at 60° C in 5 min., leading to a mono-TMS derivative. The addition of trimethylchlorosilane (TMCS) leads to the formation of a di-TMS product when the reaction is done in the presence of pyridine at 60° C for 20 min. [229]. Also, in the presence of TMCS, BSA reacts with histamine, generating a tri-TMS derivative allowing the measurement of as low as 100 ng analyte [230]. Various amines are derivatized using N-methyl-N-(*tert*-butyl-dimethylsilyl) trifluoroacetamide (MTBSTFA) with the formation of their *tert*-butyldimethylsilyl derivative. Primary amines react in solvents such as DMF or CH_3CN, usually at mild heating. Secondary amines react only in CH_3CN. Other reagents such as (*tert*-butyldimethyl) chlorosilane or *tert*-butyldimethylsilyldimethylamine are used as silylating reagents to form *tert*-butyldimethylsilyl derivatives of amines [15].

Most derivatizations for primary and secondary amines with the purpose of GC analysis are acylation reactions done with acid chlorides, acyl imidazole, or anhydrides, in reactions that are shown below:

$$\begin{array}{c} R^a \\ {}^{\textstyle\diagdown}NH + R-COX \\ R^b{}^{\textstyle\diagup} \end{array} \longrightarrow \begin{array}{c} R^a \quad\; O \\ {}^{\textstyle\diagdown}N-\overset{\|}{C}R + HX \\ R^b{}^{\textstyle\diagup} \end{array} \quad \begin{array}{l} X = Cl, OH, CN, imidazolyl, \\ N\text{-hydroxysuccinimidyl, etc.} \end{array}$$

$$\begin{array}{c} R^a \quad RCO \\ {}^{\textstyle\diagdown}NH + \quad\quad O \\ R^b{}^{\textstyle\diagup} \quad RCO \end{array} \longrightarrow \begin{array}{c} R^a \quad\; O \\ {}^{\textstyle\diagdown}N-\overset{\|}{C}R \\ R^b{}^{\textstyle\diagup} \end{array} \quad \begin{array}{l} R = alkyl, perfluoroalkyl, trichloromethyl, \\ etc. \quad\quad R^a, R^b = H, alkyl, aryl. \end{array}$$

A list of various reagents used in acylation for GC analysis of primary and secondary amines is given in Table 19.4.2.

TABLE 19.4.2. *Acylation reagents for GC analysis of primary and secondary amines.*

Reagent	Formula	Analyte	Reference
acetic anhydride	$(CH_3CO)_2O$	amines	[15]
acetyl chloride	CH_3COCl	amines (amphetamines)	[15]
chlorodifluoroacetic anhydride	$(CClF_2CO)_2O$	amines	[15]
3,5-dinitrobenzoyl chloride		diethylamines	[231]
ethyl chloroformate	C_2H_5OCOCl	polyamines	[232]
heptafluorobutyric anhydride (HFBA)	$(C_3F_7CO)_2O$	amines (primary and secondary), amphetamines	[15]
N-methyl-bis(trifluoro-acetamide) (MBTFA)	$(CF_3CO)_2NCH_3$	amines (lofexidine)	[15]
4'-nitroazobenzene-4-carboxylic acid chloride		amines	[15]
4-nitrobenzoyl chloride		arylalkylamines	[233]
pentafluorobenzoyl chloride		phenethyl amine	[234]
pentafluoropropionic anhydride (PFPA)	$(C_2F_5CO)_2O$	anilines, amines, etc.	[235], [236]
propionic anhydride	$(C_2H_5CO)_2O$	piperidine derivatives	[237]
N-succinimidyl-p-nitrophenyl acetate		amines (also amino acids)	[15]
trichloroacetyl chloride	CCl_3COCl	1-α-acetylmethadol metabolites, amphetamines	[15], [238], [239]
trifluoroacetyl Nylon 66		amines	[240]
N-trifluoroacetyl-L-prolyl chloride		amines	[15]
trifluoroacetylimidazole (TFAI)		amines (desipramine)	[15]
trifluoroacetic anhydride	$(CF_3CO)_2O$	amines (desipramine), polyamines, aryl amines, amphetamines	[15], [241]

A typical derivatization using heptafluorobutyric anhydride starts with 1 mL of CH_2Cl_2 solution containing up to 2 μg/mL of each amine. This sample is treated with 10 μL anhydrous pyridine and 2 μL HFBA. The sample is heated at 76° C for 30 min, allowed to cool, and injected into the GC-MS system using negative chemical ionization (NCI) with methane (2 mL/min. flow into the GC-MS source). Higher sensitivity can be obtained using detection in SIM mode. The separation can be done using a methylsilicone column (ZB1, 12 m x 0.25 mm i.d., 0.25 μm film thickness) with the GC temperature programmed between 80° C to 310° C [242].

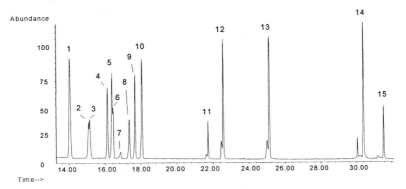

FIGURE 19.4.1 *The chromatogram obtained by GC-NCI-MS from a solution of 0.2 μg/mL standards and deuterated standards of aromatic amines.* Peak identification: (1) aniline + d_5-aniline, (2) d_7-o-toluidine, (3) o-toluidine, (4) m-toluidine, (5) p-toluidine, (6) 2-ethylaniline, (7) 2,5-dimethylaniline, (8) 2,4-dimethylaniline, (9) 3-ethylaniline, (10) 4-ethylaniline, (11) 1-aminonaphthalene, (12) 2-aminonaphthalene + d_7-2-aminonaphthalene, (13) 4-aminobiphenyl + d_9-4-aminobiphenyl, (14) benzidine + d_8-benzidine, (15) tolidine.

The quantitation is done using peak area ratios of the analyte and of the internal standards (deuterated amines) and response factors (see Section 3.8). The reaction of the aromatic amines and HFBA can be written as follows:

$$Ar-NH_2 \ + \ (F_7C_3CO)_2O \ \longrightarrow \ Ar-NH-CO-C_3F_7$$

The NCI-MS spectra of the acylated amines with heptafluorobutyryl groups are characterized by fragments showing the loss of HF (m/z = M - 20). The monosubstituted amides generated in this reaction have good shape for the chromatographic peaks. However, attempts for complete derivatization were done, for example, using a sequence of two derivatizations, one using ethyl chloroformate and the other using PFPA [243]. The sequence of reactions in this case is the following:

$$R-NH_2 \ \xrightarrow{C_2H_5-O-CO-Cl} \ R-NH-CO-OC_2H_5 \ \xrightarrow{(F_5C_2CO)_2O} \ R-N \begin{smallmatrix} CO-C_2F_5 \\ \\ CO-OC_2H_5 \end{smallmatrix}$$

The EI+ mass spectra of these compounds shows ions with m/z = M - 73 corresponding to the loss of $C_2H_5-OC=O$ and with m/z = M - 119 corresponding to the loss of C_2F_5.

Several other reactions and reagents are used in the derivatizations of amines for further GC analysis. For example, aldehydes and ketones react with primary amines forming Schiff bases as follows (see Section 18.4):

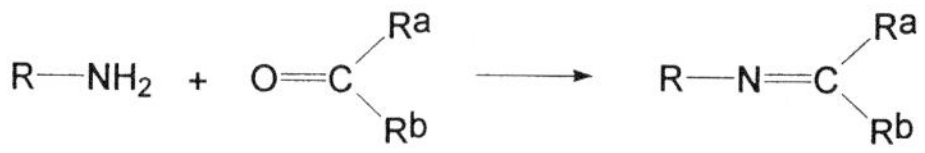

A list of several carbonyl reagents used for the derivatization of primary amines is given in Table 19.4.3.

TABLE 19.4.3. *Aldehydes and ketones used as reagents for GC analysis of primary amines.*

Reagent	Formula	Analyte	Reference
cyclohexanone	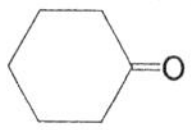	amphetamine	[15]
pentafluorobenzaldehyde	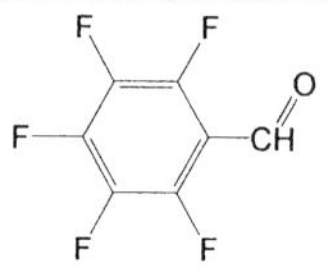	aliphatic primary amines	[15], [244]
propanone	CH_3COCH_3	primary amines, amphetamine,	[15]
2-thiophene aldehyde	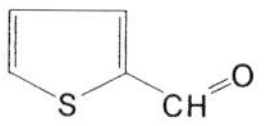	aliphatic primary amines (C_1-C_7)	[245]

Other reactions of amines with compounds containing carbon-hetero multiple bonds are used for derivatization. Among these are the reactions of primary and secondary amines with *tert*-butylisocyanate [246], diethylpyrocarbonate [247], and CS_2 [15]. The reaction with diethylpyrocarbonate takes place at pH 9.5 in 30-40 min. at room temperature and can be written as follows:

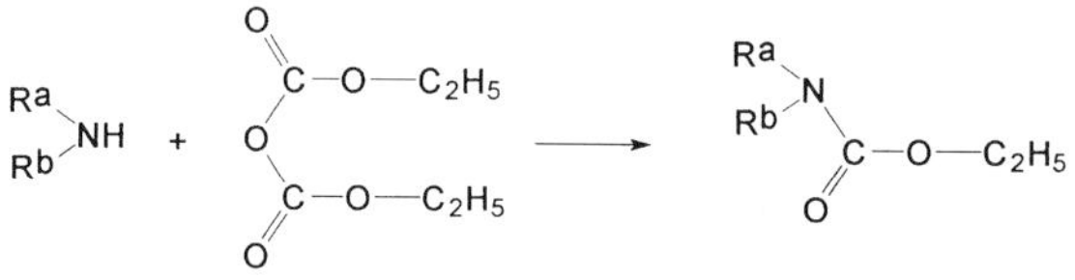

Reactions with formation of cycles are also used in amine derivatization for the purpose of GC analysis. For example, 2,5-hexanedione reacts with primary amines when heated at 60° C for 30 min. in hexane or methanol. The reaction takes place as follows [248]:

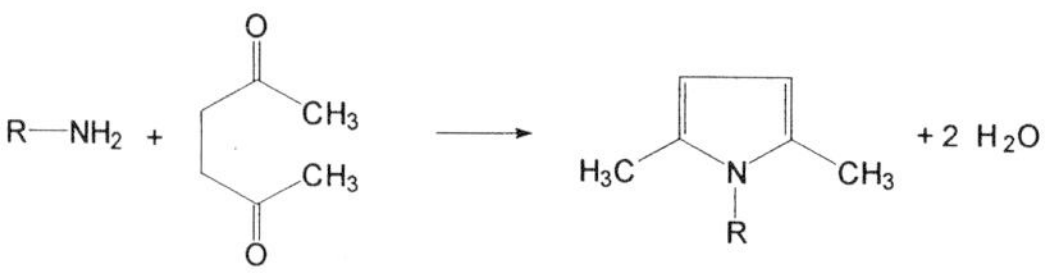

More frequently the formation of new cycles is used in derivatization reactions for the purpose of generating chromophores or fluorescence carrying groups and applied in HPLC analysis.

- *Derivatization of primary and secondary amines for HPLC analysis*

Most reagents used for amine derivatization for HPLC analysis contain chromophore groups or groups that fluoresce (see Section 17.6) allowing better detectability of the analyte. Similarly to alcohol derivatization for HPLC analysis, for a limited number of analytes that are already strongly colored, have a large UV absorption coefficient, or are fluorescent, the derivatization of choice is acetylation or silylation. This is done with the purpose of improving (or modifying) the separation. Some reagents used for amine derivatization for HPLC analysis are given in Table 19.4.4.

TABLE 19.4.4. *Reagents used for primary and secondary amine analysis by HPLC.*

Reagent		Amine type, matrix, sensitivity	React type[1]	Detection λ, Separation	Ref.
acetic anhydride		acebutolol[2]	Anhy	UV 254, C18	[1]
6-aminoquinolyl-N-hydroxysuccinimidyl carbamate		amines in beverages 100-500 ng/mL, amino acids 250 nmol etc.	Acyl	fluorescence ex 250, em 395, ODS	[249], [250]
4-amino-sulfonyl-7-fluoro-2,1,3-benzoxadiazol		amino acids	Alkyl	CE with UV detection	[251]
anthracene-2,3-di-carboxaldehyde + CN⁻ H₂O₂ +bis(2-nitrophenyl)-oxalate		amphetamine 2 fmol	Cycl	postcolumn CL, RP-18	[1]
2-anthroyl chloride		chiral, anti-arrhythmic drug mexiletine 0.5 ng/mL	Acyl	fluorescence ex 270, em 420, Pirkle + silica	[1]
2-(9-anthryl)ethyl chloroformate		polyamines 1-10 pg	Acyl	fluorescence ex 258, em 418, RP-18	[252]
benzoyl chloride		various amines, antibiotics, etc	Acyl	UV 237, etc., silica, C18, various	[1]
3-benzoyl-2-quinoline-carboxaldehyde + CN⁻		amino acids amol range	Cycl	fluorescence ex 457 laser em 552, CE	[2]

TABLE 19.4.4 (continued). *Reagents used for primary and secondary amine analysis by HPLC.*

Reagent		Amine type, matrix, sensitivity	React type[1]	Detection λ, Separation	Ref.
benzyl isothiocyanate	NCS	2-amino-1-(p-nitrophenyl)-1,3-propandiol amino acids	NCS	UV 254, ODS	[1], [253]
2-bromacetophenone		cytosine derivatives in plasma 5 ng/mL	Cycl	fluorescence ex 305, em 370, ODS	[254]
bromoacetaldehyde	CH_2BrCHO	adenine deriv.	Cycl	postcolumn fluorescence ex 254, em 400, RP	[1]
4-(bromomethyl)-6,7-dimethoxycoumarin		fluorouracyls, 6-100 ng/mL	Alkyl	fluorescence ex 340, em 420, ODS	[255]
4-(bromomethyl)-7-methoxycoumarin		fluorouracyls, 100-400 fmol, polycarpine, etc.	Alkyl	fluorescence ex 346, em 395, C18	[1], [256]
N-(*tert*-butoxycarbonyl)-L-leucine + dicyclohexylcarbo-diimide (DCCI)	+ DCCI	alprenolol[2] propranolol, etc., chiral 0.5 ng/mL tyroxines	Anhy	fluorescence ex 228, em 290, C18, UV 230	[1], [257]
N-(*tert*-butoxycarbonyl)-L-leucine-N'-hydroxysuccinimide ester		vigabatrin[2] chiral 0.1% of major enantiomer	Acyl	UV 210 RP-8	[258]
(1S)-(+)-10-camphorsulfonyl chloride		amines, chiral	Acyl	UV 254, silica + RP-18	[1]
3-(4-carboxybenzoyl-2-quinolinecarbox-aldehyde + CN[-]	+ CN[-]	amino sugars 0.5 -2.3 amol	Cycl	fluorescence ex 457 laser em 552, CE	[1]

TABLE 19.4.4 (continued). *Reagents used for primary and secondary amine analysis by HPLC.*

Reagent	Amine type, matrix, sensitivity	React type[1]	Detection λ, Separation	Ref.
5-carboxytetramethyl-rhodamine succinimidyl ester	aminated mono-saccharides, angiotensins	Acyl	UV 280, CE, C18	[1]
cerium trihydroxy peroxide Ce(OH)$_3$OOH	methotrexate[2] 2.78 ng/mL	Oxid	fluorescence ex 367, em 463 on-line oxid.	[1]
chloroacetaldehyde CH$_2$ClCHO	adenosine 200 ng/mL, adenine nucleotides, etc.	Cycl	fluorescence ex 280, em 380, etc., C18	[1]
4-(N-chloroformyl-methyl)-N-methyl)amino-7-N,N-dimethylamino-sulfonyl-2,1,3-benzoxadiazole	amines 50-90 fmol	Acyl	fluorescence ex 445, em 555, C18	[1]
4-(2-chloroformyl-pyrrolidin-1-yl)-7-N,N-dimethylamino-sulfonyl-2,1,3-benzoxadiazole	amines, chiral 500 nmol (laser fluorescence)	Acyl	fluorescence ex 450, em 560, ODS	[1]
4-(2-chloroformyl-pyrrolidin-1-yl)-7-nitro-2,1,3-benzoxadiazole	amines, chiral	Acyl	fluorescence ex 470, em 540, silica	[259]
9-chloro-10-methylacridinium trifluoromethyl-sulfonate (triflate)	cycloserine 150 ng/mL	Alkyl	fluorescence ex 257, em 475, silica	[1]
4-chloro-7-nitro 2,1,3 benzoxadiazole	amines in beverages, 9-130 μmol amino acids 20 μmol	Alkyl	fluorescence ex 488, em 540 CE, etc. ex 450, em 530	[1], [260]

TABLE 19.4.4 (continued). *Reagents used for primary and secondary amine analysis by HPLC.*

Reagent	Amine type, matrix, sensitivity	React type[1]	Detection λ, Separation	Ref.
(S)-(+)-2-(4-chloro-phenyl)-α-methyl-5-benzoxazol-acetic acid chloride (benoxa-profen chloride)	amines chiral	Acyl	fluorescence ex 312, em 365, silica	[1], [261]
4-(2-cyanoisoindolyl)-phenyl isothiocyanate	amines	NCS	fluorescence ex 260, em 410	[2]
dabsyl chloride, or 4-dimethylaminoazo-benzene-4'- sulfonyl chloride, or phenyl-1-azo-(4'-dimethyl-aminobenzene)-4-sulfonyl chloride	taurine, amino sugars, etc. 0.2 nmol, 4 pmol amino acids 400-500 pmol	Acyl	UV 430, ODS, CE, etc.	[1], [262], [263], [264]
dansyl chloride or 5-dimethylamino-naphthalene-1-sulfonyl chloride	various primary and secondary amines, amine drugs, etc. 50 μg/mL taurine, tetrahydro-β-carbolines amino acids	Acyl	UV 254, fluorescence ex 330, etc. em 530, etc. nitrile, C18, C8, silica CE	[1], [265], [266], [267]
dansyl-L-proline + diethylcyano phosphonate	amines chiral 40-50 fmol amino acids methyl esters chiral	Acid	fluorescence ex 345, em 515, ODS	[268]
(1R,2R)-1,3-diacetoxy-1-(4-nitrophenyl)-2-propyl isothiocyanate	amino alcohols, β-blockers chiral	NCS	UV 245, C18	[269]
O,O-dibenzoyl-L-tartaric anhydride	3-amino-quinuclidine chiral, 1% enantiomer amino acids	Anhy	UV 254 C8, CE	[270]
5-dibutylamino-naphthalene-1-sulfonyl chloride (BANS-Cl)	amosulalol[2] 20 ng/mL, amino acids	Acyl	fluorescence ex 356, em 500, silica	[271], [272], [273]

TABLE 19.4.4 (continued). *Reagents used for primary and secondary amine analysis by HPLC.*

Reagent		Amine type, matrix, sensitivity	React type[1]	Detection λ, Separation	Ref.
3-(4,6-difluoro-triazinyl)-amino-7-methoxycoumarin		1-amino-adamantan (amantadine) 2.5 nmol (urine)	Alkyl	fluorescence ex 345, em 410, ODS	[1]
3,4-dihydro-6,7-dimethoxy-4-methyl-3-oxoquinoxaline-2-carbonyl chloride		β-phenyl-ethyl-amine, etc. 300 pmol	Acyl	fluorescence ex 406, em 485, ODS	[1]
4-(6,7-dihydro-5,8-dioxothiazolo[4,5-g]-phthalazin-2-yl) benzoic acid N-hydroxysuccinimide ester K$_3$[Fe(CN)$_6$] and H$_2$O$_2$		amines 0.8 fmol	Acyl	CL	[274]
2-(5',6'-dimethoxybenzo-thiazolyl)benzene-sulfonyl chloride		amines 10-13 fmol	Acyl	fluorescence ex 330, em 450, ODS	[275]
4-(5',6'-dimethoxy-benzothiazol)benzoyl fluoride		amines 1.4-2 fmol	Acyl	fluorescence ex 350, em 450, ODS	[276]
4-(5',6'-dimethoxy-benzothiazol)-phenyl isothiocyanate		amines 1.4-2 fmol	NCS	fluorescence ex 355, em 420, ODS	[2]
3,4-dimethoxyphenyl-glyoxal		cyclic guanosine phosphate 4-14 pmol	Cycl	fluorescence ex 400, em 510, C18	[1]
4-(dimethylamino)-benzaldehyde		amine in sulfonamides 5-10 ng/mL	CO	Vis 450, C18	[1]
3,4-dimethylamino-coumarn-3-carbonyl fluoride		amines	Acyl	fluorescence	[2]
3,4-dimethylamino-coumarn-3-carboxy succinimidyl ester		amines	Acyl	fluorescence	[2]

TABLE 19.4.4 (continued). *Reagents used for primary and secondary amine analysis by HPLC.*

Reagent	Amine type, matrix, sensitivity	React type[1]	Detection λ, Separation	Ref.
4-(N,N-dimethyl-amino)-1-naphthyl isothiocyanate	amines	NCS	fluorescence	[2]
4-(N,N-dimethyl-amino-naphthalene-5-sulfonylamino)-phenyl isothiocyanate	amines	NCS	fluorescence	[2]
4-(N,N-dimethyl-amino-sulfonyl)-7-fluoro-2,1,3-benzoxadiazol	antihypertensive drugs, metoprolol, etc. in blood 0.8 ng/mL, amino acids	Alkyl	CL postcol., ODS, fluorescence ex 450, em 590, C18	[277], [278]
dimethylformamide diethyl acetal	5,6-dihydro-5-azacytidine 25 ng/mL	Alkyl	UV 264, ODS	[279]
(3-S-cis)-2,2-dimethyl-5-isothiocyanato-4-phenyl-1,3-dioxane	labetalol[2], chiral	NCS	UV 254, C18	[280]
(R,R)-N-(3,5-dinitrobenzoyl)-2-aminocyclohexyl isothiocyanate	amines, chiral	NCS	UV254, C18	[281]
2,4-dinitrofluoro-benzene	sulfonylurea drugs, etc. 2-5 ng/mL amino acids, glutathione	Alkyl	UV 350, C8	[1], [282]
3,5-dinitrobenzoyl chloride	hypotaurine, amino acids, etc.	Acyl	UV 254, RP-18, C18	[1], [283], [284]

TABLE 19.4.4 (continued). *Reagents used for primary and secondary amine analysis by HPLC.*

Reagent		Amine type, matrix, sensitivity	React type[1]	Detection λ, Separation	Ref.
3,5-dinitrophenyl isocyanate		fenfluramine[2] 10 ng/mL, phenylethyl-amine	NCO	UV 235, chiral phase	[1]
N-[4-[(2,5-dioxo-1-pyrrolidinyl)-oxycarboxyamino]-butyl]-N-ethylisoluminol + K₃[Fe(CN)₆] + H₂O₂		methamphet-amine 20 pmol	Acyl	postcolumn chemilumin. + H₂O₂, ODS	[285]
disuccinimido carbonate		amines, amino acids	Acyl	UV 240	[2]
ethyl chloroformate		amines in air 1-2 nmol	Acyl	UV 240, C18	[286]
Eu³⁺ chelate of N-(p-isothyocyanatobenzyl) diethylenetriamine-tetraacetic acid		amines	NCS	postcolumn fluorescence ex 344, em 617, C18	[287]
Eu³⁺ chelate of 1-(4-isothyocyanatobenzyl) ethylenediamine-N,N,N',N'-tetraacetic acid		3-phenyl-1-propylamine 100 nmol	NCS	postcolumn fluorescence ex 290, em 615, C18	[288]
ferrocenyl isothiocyanate		4-aminobutyric acid in tissues 0.05 pmol	NCS	electrochem. glassy carbon at + 0.5 V	[1]
(R)-1-(9-fluorenyl)-ethyl chloroformate		atenolol[2], chiral 10 ng/mL, methyl 4-propranolol, 0.5 ng/mL, glyphosate[2], etc.	Acyl	fluorescence ex 227, em 310, C18	[1], [289], [290]
(S)-1-(9-fluorenyl)ethyl chloroformate		amphetamine chiral 5 ng/mL, etc. amino acids less than 1 pmol	Acyl	fluorescence ex 265, em 330 or UV, etc., C18	[1], [291], [292]

TABLE 19.4.4 (continued). *Reagents used for primary and secondary amine analysis by HPLC.*

Reagent		Amine type, matrix, sensitivity	React type[1]	Detection λ, Separation	Ref.
N-(9-fluorenylmeth-oxycarbonyl)glycine + O-benzo-triazol-1-yl-N,N,N',N'-bis(tetra-methylene)uronium hexafluorophosphate		glycosylamines	Acid	UV 266, amino	[1]
9-fluorenylmethyl chloroformate (FMOC)		various amines, cephalosporins, ampicillin, etc. 250 ng/mL amino acids 0.6 pmol	Acyl	fluorescence ex 260, em 311, C18, ODS, fluorescence ex 254, em 315	[1], [293], [294], [295], [296]
N-α-(9-fluorenyl-methyloxycarbonyl)-L-leucine-N-carboxy anhydride		amines chiral amino acids chiral	Anhy	fluorescence ex 263, em 313, silica	[297]
fluorescamine		various amines low ng/mL amino acids, proline, hydroxyproline small peptides	Cycl	fluorescence ex 370-380, em 460-490 various, ODS, CN, etc., CE	[1]
fluorescein isothiocyanate		amines in cheese, amphetamine, etc. a few nmol, proline, hydroxyproline amino acids 0.2 nmol	NCS	fluorescence ex 496, em 518, etc. ODS, etc., CE	[1], [298], [299]
2-[5-fluoro-2,4-dinitrophenyl)amino]-propane amide (Marfey's reagent)		drugs, baclofen[2] chiral 2,6-diamino-pimelic acid 2.5 μmol, peptides amino acids	Alkyl	UV 340, C8+cyclo-dextrin in mobile phase	[1], [300], [301], [302]
1-fluoro-2-nitro-4(trifluoromethyl)-benzene		amines 5-25 pmol	Alkyl	UV 242	[303]

TABLE 19.4.4 (continued). *Reagents used for primary and secondary amine analysis by HPLC.*

Reagent		Amine type, matrix, sensitivity	React type[1]	Detection λ, Separation	Ref.
4-fluoro-7-nitro-2,1,3-benzoxadiazole		drugs, ethambutol, etc. in blood 10 ng/mL (plasma) amino acids 2 nmol	Alkyl	fluorescence ex 490, em 540, ODS, fluorescence ex 470, em 530	[1], [304], [305]
(S)-(+)-2-(4-fluorophenyl)-α-methyl-5-benzoxazol-acetic acid chloride (flunoxaprofen chloride)		*trans*-(+)-2-phenylcyclo-propanamine chiral 2 ng/mL amino acids	Acyl	fluorescence ex 305, em 355, silica, UV 305	[306], [307]
(S)-(-)-1-[(4-fluoro-phenyl)-α-methyl-5-benzoxazo-1-yl]-ethyl isocyanate (flunoxa-profen isocyanate)		propranolol[2] 1-2 ng/mL chiral	NCO	fluorescence ex 305, em 355, C18	[1]
formaldehyde + acetylacetone (Hantzch reaction)		various primary amines, aminolevulinic acid 10 ng/mL	Cycl	fluorescence ex 246, em 458, cation exchange	[308]
3-furoylquinoline -2-carboxaldehyde + CN⁻		ovalbumin 0.01 nmol amino groups in phospholipid	Cycl	fluorescence ex 488 laser, CE	[1], [309]
glyoxal		tryptophan containing peptides 55-4200 fmol	Cycl	fluorescence ex 275, em 465, ODS	[310]
6-isothiocyanato-benzo-[g]-phthalazine-1,4(2H,3H)-dione		amines 0.8-1.2 fmol	NCS	postcolumn chemilumin., ODS	[1]
(R)-(-)-4-(3-isothio-cyanatopyrrolidin-1-yl)-7-(N,N-dimethyl-aminosulfonyl)-2,1,3-benzoxadiazole		propranolol[2] chiral 25-29 fmol, amines, peptides, etc.	NCS	fluorescence ex 460 em 550, etc., ODS	[1]
5-isothiocyanato-1,3-dioxo-2-p-tolyl-2,3-dihydro-1H-benzo[d,e] isoquinoline		amines	NCS	fluorescence ex 300 em 530	[2]

TABLE 19.4.4 (continued). *Reagents used for primary and secondary amine analysis by HPLC.*

Reagent		Amine type, matrix, sensitivity	React type[1]	Detection λ, Separation	Ref.
(R)-(-)-4-(3-isothiocyanato-pyrrolidin-1-yl)-7-nitro-2,1,3-benzoxadiazole		amines, chiral, propranolol	NCS	fluorescence ex 490 em 530, etc., ODS	[311]
Luminarin 1 or 1-[[(2,3,6,7-tetrahydro-11-oxo-1H,5H,11H-[1]-benzopyrano-[6,7,8-ij]quinolizin-9-yl)acetyl]oxy]-2,5-pyrrolidinedione. For chemiluminescence add postcolumn bis-(2,4,6-trichlorophenyl) oxalate + H$_2$O$_2$		pentylamine, amines, 160-300 fmol fluorescence 15-100 fmol chemilumin.	Acyl	fluorescence ex 392 em 470, C18, postcolumn chemilumin. + H$_2$O$_2$	[312], [313]
Luminarin 2 or N-[6-[2,5-dioxo-1-pyrrolidin-yl)-oxy]-6-oxohexyl]-2,3,6,7-tetra hydro-11-oxo-1H,5H, 11H-[1]-benzopyrano [6,7,8-ij]-quinolizine-9-acetamide. For chemiluminescence add postcolumn bis-(2,4,6-trichlorophenyl) oxalate + H$_2$O$_2$		histamine, 100 fmol fluorescence 50 fmol chemilumin	Acyl	fluorescence ex 390 em 490, C18, postcolumn chemilumin. + H$_2$O$_2$	[1]
maleic anhydride		phytotoxins 500 pmol	Anhy	UV 250, C18	[314]
(-)-menthyl chloroformate		atenolol[2], chiral 12.5 ng/mL benzylamine, etc	Acyl	fluorescence ex 230, em 305, etc., ODS	[1]
N-[4-(6-methoxy-2-benzoxazolyl)benzoyl-L-phenylalanine + 2,2'-dipyridyl disulfide and triphenylphosphine		amines, chiral	Acid	fluorescence ex 325, em 403, silica	[1]
N-[4-(6-methoxy-2-benzoxazolyl)benzoyl-L-proline + 2,2'-dipyridyl disulfide and triphenylphosphine		amines, chiral 30 fmol	Acid	fluorescence ex 325, em 403, silica	[1]

TABLE 19.4.4 (continued). *Reagents used for primary and secondary amine analysis by HPLC.*

Reagent		Amine type, matrix, sensitivity	React type[1]	Detection λ, Separation	Ref.
1-methoxycarbonyl-indolizine-3,5-dicarboxaldehyde	H₃COOC— (structure with N, CHO, CHO)	amikacin[2] 5 µg/mL	Cycl	fluorescence ex 414, em 482, UV 280, CE	[315]
7-methoxycoumarin-3-carbonyl fluoride	H₃CO (coumarin structure with C(=O)F)	amines 100 fmol	Acyl	fluorescence ex 350, em 405, etc., ODS	[1]
2-methoxy-2,4-diphenyl-3(2H)-furanone	(furanone structure with OCH₃ and two phenyls)	primary amines 0.5-50 pmol	Cycl	fluorescence ex 390, em 480	[1]
2-methoxy-2,4-diphenyl-3(2H)-furanone + postcolumn ethanolamine	(furanone structure with OCH₃ and two phenyls) + HO—NH₂	secondary amines 0.5-50 pm amino acids, peptides	Cycl	fluorescence ex 390, em 480, postcolumn	[1], [316]
6-methoxy-2-methyl-sulfonylquinolone-4-carbonyl chloride	H₃CO (quinolone structure with C(=O)Cl, SO₂CH₃)	amines 0.5-2 pmol (primary), 100 pmol (secondary)	Acyl	fluorescence ex 342, em 448, etc., ODS	[1]
(S)-2-(6-methoxy-2-naphthyl)propanoyl chloride (naproxen chloride)	H₃CO (naphthyl structure with CH₃, H, Cl, C=O)	tocainide[2] chiral	Acyl	fluorescence ex 313, em 365, silica	[1]
(S)-(-)-1-(6-methoxynaphth-2-yl)ethyl isocyanate or (S)-(-)-naproxen isocyanate	H₃CO (naphthyl structure with NCO, CH₃, H)	propranolol, etc., chiral	NCO	fluorescence ex 276, em 356, C18	[1]
(S)-1-(6-methoxy-2-naphthyl)ethyl isothiocyanate	H₃CO (naphthyl structure with H, CH₃, NCS)	amines, propranolol[2], alprenolol[2], etc., chiral	NCS	UV 230, fluorescence ex 490, em 530, etc., ODS	[317]
(R)-4-(6-methoxy-2-naphthyl)-2-butyl chloroformate	H₃CO (naphthyl structure with (CH₂)₂, O, C(=O)Cl, CH₃, H)	metoprolol[2], etc., chiral 0.9 ng/mL	Acyl	fluorescence ex 270, em 350 or UV 230, silica	[1]

TABLE 19.4.4 (continued). *Reagents used for primary and secondary amine analysis by HPLC.*

Reagent		Amine type, matrix, sensitivity	React type[1]	Detection λ, Separation	Ref.
(R)-2-(6-methoxy-2-naphthyl)-1-propyl chloroformate		alprenolol[2], etc., chiral 0.2 ng/mL	Acyl	fluorescence ex 270, em 365 or UV 230, silica	[318]
(S)-(+)-α-methoxy-phenylacetyl chloride		phenylethyl-amine 0.2% of other enantiomer	Acyl	UV 255, silica	[1]
(R)-(-)-α-methoxy-phenylacetyl chloride		ethambutol {[2]}	Acyl	UV 254, silica	[1]
(R)-α-methoxy-α-(trifluoromethyl)phenyl acetyl chloride		amphetamine, etc. 100 ng	Acyl	UV 254, C18	[319]
2-methylanylino-naphthalene-5-sulfonyl chloride		amines	Acyl	fluorescence ex 350 em 50	[320]
D-α-methylbenzyl isothiocyanate		chloramphenicol derivatives chiral	NCS	UV 254	[1]
(R)-(+)-α-methylbenzyl isocyanate		propranolol 2 ng/mL chiral, in blood, bufuralol[2]	NCO	fluorescence ex 220, em 300, C18	[321]
(S)-(-)-α-methylbenzyl isocyanate		3-amino-quinuclidine, chiral	NCO	UV 254, silica	[322]
N-methylisatoic acid anhydride		peptides 0.1-0.5 pmol	Anhy	fluorescence ex 360, em 420, C18	[1]
2-methyl-3-oxo-4-phenyl-2,3-dihydrofuran-2-yl acetate		peptides less than 1 pmol	Cycl	fluorescence ex 390 em 470, ODS, CE	[324]

TABLE 19.4.4 (continued). *Reagents used for primary and secondary amine analysis by HPLC.*

Reagent		Amine type, matrix, sensitivity	React type[1]	Detection λ, Separation	Ref.
naphthalene-2,3-dicarboxaldehyde	CHO CHO (naphthalene)	glutathione	Cycl	fluorescence ex 458 em 532, CE	[325]
naphthalene-2,3-dicarboxaldehyde + N-acetyl-D-penicillamine	CHO CHO + COCH₃ HN COOH HS CH₃ CH₃	cyclic heptapeptide 1 ng/mL	Cycl	fluorescence ex 436 em 440	[1]
naphthalene-2,3-dicarboxaldehyde + CN⁻	CHO CHO + CN⁻	catecholamines 5 ng/mL, etc. amines, amino acids	Cycl	fluorescence ex 442, em 490, CE	[1], [326]
naphthalene-2,3-dicarboxaldehyde + dithiothreitol	CHO CHO + HS HO OH SH	proteins	Cycl	UV 280, CE	[327]
naphthalene-2,3-dicarboxaldehyde + 2-mercaptoethanol	CHO CHO + HS OH	enkephalins 600 nmol	Cycl	postcolumn fluorescence ex 457.9, C18	[1]
1-naphthalenemethyl isothiocyanate	NCS	labetalol[2],	NCS	UV 254, C18	[328]
2-naphthalene-sulfonyl chloride	SO₂Cl	spectinomycin[2] 40 ng/mL	Acyl	UV 250, 254, silica	[329]
2-naphthoyl chloride	O Cl	amphetamine mexiletine[2] 50 ng/mL chiral	Acyl	UV 254, fluorescence ex 230 em 340, Pirkle	[1]
2-naphthyl chloroformate	O Cl O	amines in food (putrescine) 5-750 ng/g amino acids 25-500 fmol	Acyl	fluorescence ex 274 em 335, RP-18	[1], [330]
(R)-(-)-1-(1-naphthyl)ethyl isocyanate	H₃C NCO H	propafenone[2] 6.25 ng/mL, nadolol[2], etc.	NCO	UV 220, silica fluorescence ex 285, em 340, ODS	[1]
(S)-(+)-1-(1-naphthyl)ethyl isocyanate	H NCO H₃C	tocainide[2] 25 ng/mL in blood, metoprolol[2], amino acids chiral	NCO	fluorescence ex 220, em 345, silica, etc.	[1], [331]

TABLE 19.4.4 (continued). *Reagents used for primary and secondary amine analysis by HPLC.*

Reagent		Amine type, matrix, sensitivity	React type[1]	Detection λ, Separation	Ref.
1-naphthyl isocyanate	NCO structure	nadolol[2] chiral amino acids in blood	NCO	UV 239, chiral phase, fluorescence ex 228, em 320	[1]
1-naphthyl isothio-cyanate + tetraoctyl-ammonium bromide for ion pair separation	NCS structure	pamidronate 1 ng/mL	NCS	fluorescence ex 285, em 390, etc., C18	[332]
(S)-(-)-N-1-(2-naththylsulfonyl)-2-pyrrolidine-carbonyl chloride	structure	quinolone drug, chiral 5 ng/mL	Acyl	UV 280, silica	[333]
4-nitrobenzoyl chloride	structure	amines, amphetamine	Acyl	UV 254, C18	[334]
trans-4-nitrocinnamoyl chloride	structure	perhexiline[2] 30 ng/mL	Acyl	UV 340, phenyl	[335]
p-nitrophenyl-2,5-dihydroxyphenyl-acetate 2,5-bis-tetrahydropyranyl ether	structure	amines 100 fmol	Acyl	electro-chemical +0.2 V	[336]
1-[(4-nitrophenyl)sulfonyl]-L-prolyl chloride	structure	flecainide[2] 50 ng/mL chiral amphetamine	Acyl	UV 280, C18	[1]
4-(2-phenanthrox-azolyl)-phenyl isothiocyanate	structure	amines, amino acids	NCS	fluorescence ex 330 em 400	[2]
phthalaldehyde or benzene-1,2-dicarboxaldehyde + various thiols[3]	structure + R—SH	various drugs with amino groups, amines, pesticides, amino acids	Cycl	various fluorescence including postcolumn	[1], [337] through [353]

TABLE 19.4.4 (continued). *Reagents used for primary and secondary amine analysis by HPLC.*

Reagent		Amine type, matrix, sensitivity	React type[1]	Detection λ, Separation	Ref.
pentafluoropropionic anhydride (PFPA)	$(C_2F_5CO)_2O$	4,4'-methylene-dianiline 8 ng/mL	Anhy	UV 258, C18	[1]
pentafluorobenzoyl chloride		cytosine	Acyl	UV 254, C8	[1]
phenylglyoxal		tryptophan in blood 72 nmol, guanine nucleosides	Cycl	fluorescence ex 385 em 460, ODS	[1]
phenyl isothiocyanate		various amines, baclofen[2], peptides	NCS	UV 254, C18, etc.	[1]
1-phenylsulfonyl-3,3,3-trifluoropropene		polyamines 0.3-0.6 pmol	Add	UV 223, ODS	[1]
4-(2-phthalimidyl)-benzoyl chloride		amines 0.4-1.6 pmol	Acyl	fluorescence ex 299, em 426, C18	[354]
propionic anhydride	$(C_2H_5CO)_2O$	5-amino-saliciclic acid 1 ng	Anhy	fluorescence ex 315, em 430, ODS	[355]
propyl chloroformate		molsidomine[2] metabolite 0.5 ng/mL	Acyl	UV 312, C18	[356]
2-(1-pyrenyl)ethyl chloroformate		biogenic polyamines 0.9-2.9 pg	Acyl	fluorescence ex 275 em 389, RP-18	[357]
5-(4-pyridyl)-2-thyophenecarbox-aldehyde		alkylamines 0.1-0.15 pmol	CO	postcolumn fluorescence ex 340, em 395, ODS	[1]
8-quinolinesulfonyl chloride		amines	Acyl	UV 254, UV 280, C18	[358]

TABLE 19.4.4 (continued). *Reagents used for primary and secondary amine analysis by HPLC.*

Reagent		Amine type, matrix, sensitivity	React type[1]	Detection λ, Separation	Ref.
salicylaldehyde + diphenylboric anhydride		polyamines	CO and Est	UV 234, fluorescence ex 385, em 470, RP-18	[359]
salicyl chloride		amines 1-5 pmol electrochem. 30-200 UV	Acyl	electrochem. +1.2 V., UV 254, RP-18	[360]
sodium nitrite + acid 2-naphthol-3,6-disulfonic		aromatic amines 0.15-0.3 ng	Diazo	Vis 512	[1]
succinimido-1-naphthyl carbamate		amines 3-8 pg	Acyl	fluorescence ex 305 em 378, ODS	[361]
succinimidophenyl carbamate		amines less than 1 ng	Acyl	UV 240, ODS	[361]
N-succinimidyl-3-ferrocenyl propionate		putrescine 3 pmol	Acyl	electrochem. + 0.45 V vs. Ag/AgCl, C8	[362]
N-succinimidyl-1-fluorenyll carbamate		amines	Acyl	fluorescence ex 286 em 343	[2]
(S)-N-succinimidyl-α-methoxy-phenyl acetate		amines, chiral	Acyl	UV 254, silica	[1]

TABLE 19.4.4 (continued). *Reagents used for primary and secondary amine analysis by HPLC.*

Reagent		Amine type, matrix, sensitivity	React type[1]	Detection λ, Separation	Ref.
N-succinimidyl p-nitrophenyl acetate		amines	Acyl	UV 254, silica	[15]
N-succinimidyl-1-naphthyl carbamate		amines	Acyl	fluorescence ex 305 em 378	[2]
N-succinimidyl-propionate		serotonin	Acyl	electrochem. + 0.30 V vs. Ag/AgCl, C18	[1]
N-succinimidyl-tetrathiafulvene-2-carboxylate		β-phenethyl-amine 21 fmol	Acyl	electrochem. glassy carbon + 0.7 V Ag/AgCl	[1]
sulfosuccinimidyl-3-(4-hydroxyphenyl) propionate		amphetamines in urine 2 ng, etc.	Acyl	UV 248, UV 280, silica, C18	[1]
2,3,4,6-tetra-O-acetyl-β-D-glucopyranosyl isothiocyanate	Ac = COCH₃	amino drugs, albuterol[2], etc, chiral 0.25 ng/mL	NCS	UV 220, fluorescence electrochem., C18, C8 etc.	[1], [363]
2,3,4,6-tetra-O-benzoyl-β-D-glucopyranosyl isothiocyanate	Bz = COC₆H₅	amino acids, etc chiral	NCS	UV 231, C18	[364]
p-toluenesulfonyl chloride		glyphosate[2] 10 ng/mL	Acyl	UV 240, ODS	[365]
(R)-N-(p-toluene-sulfonyl)pyrrolidinyl isocyanate		drugs, amphetamine chiral	NCO	UV 254, silica	[1]

TABLE 19.4.4 (continued). *Reagents used for primary and secondary amine analysis by HPLC.*

Reagent		Amine type, matrix, sensitivity	React type[1]	Detection λ, Separation	Ref.
2,3,4-tri-O-acetyl-α-D-arabinoyranosyl isothiocyanate	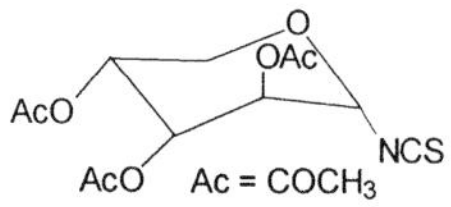 AcO—, OAc, AcO, AcO, NCS, Ac = COCH₃	1-phenyl-2-aminopropane chiral 100 ng	NCS	UV 254	[1]
trichloroethyl chloroformate	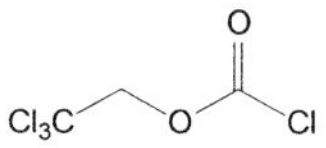 Cl₃C—O—C(=O)—Cl	promethazine[2] 1 ng/mL	Acyl	UV 254	[1]
trifluoroacetic anhydride	(CF₃CO)₂O	nomifensine[2] chiral mobile phase 25 ng	Anhy	UV 254, silica	[1]
1,2,3-triketo-hydrindene hydrate (ninhydrin)	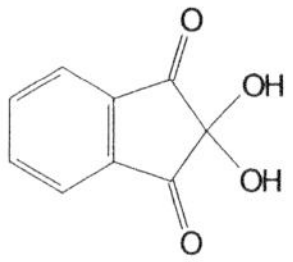 [structure with OH, OH]	primary amines, amino acids, peptides	Oxid	Vis 570 Vis 440 for proline and hydroxy-proline	[367]
2,4,6-trinitrobenzene-sulfonic acid	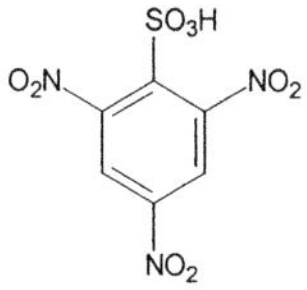 O₂N—, SO₃H, NO₂, NO₂	tobramycin[2], gentamycin[2], kanamycin[2], etc. 1200 ng/mL, 500 ng/mL	Acyl	UV 340, C8	[1]
vanadium pentoxide in phosphoric acid	V₂O₅ in H₃PO₄	reserpine[2], 70 pg/mL, thiamine, etc.	Oxid	fluorescence ex 390, em 470	[368]

[1] Acyl = acylation reaction including reactions with chloroformates or sulfonyl chlorides, Alkyl = alkylation or aryl derivative formation, Cycl = formation of new cycles, Oxid = oxidation, Diazo = diazotation and coupling, NCS = reaction with an isothiocyanate, NCO = reaction with isocyanate, CO = addition to a carbonyl, Acid = acylation with an acid, Anhy = acylation with an anhydride.

[2] For information and chemical structure regarding various drugs see e.g. M. O'Neil et al. (eds.), *The Merck Index*, Merck & Co., Inc., 13th ed., Whitehouse Station, 2001.

[3] Among the thiols used in this reaction are acetylcysteine, *tert*-butanethiol, N,N-dimethyl-2-mercaptoethylamine, ethanethiol, mercaptoacetic acid, 2-mercaptoethanol, 3-mercaptopropionic acid, thioglucose tetraacetate. The reaction takes place as follows (see Section 18.5):

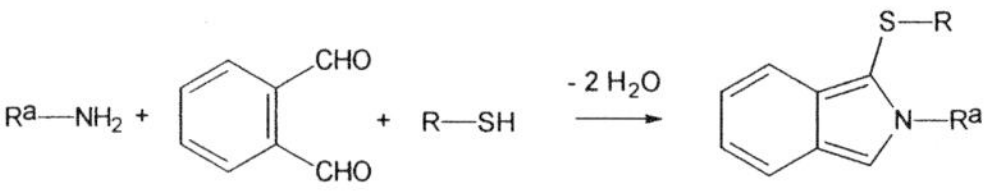

$$R^a{-}NH_2 + \begin{array}{c}CHO\\ \\CHO\end{array} + R{-}SH \xrightarrow{-2\,H_2O} \begin{array}{c}S{-}R\\ \\N{-}R^a\end{array}$$

Sodium sulfite is also used in the same reaction.

The number of reagents used for the amine group derivatization, either in amines or in amino acids is very large. Table 19.4.4 lists only some of these reagents. Not included in Table 19.4.4 are several reagents specially designed for laser induced fluorescence (LIF) detection of amines. These reagents contain fluorophores such as pyronin, thionin, cyanine and groups such as NCS or succinimidyl to react with the amines [2], [369], [370], [371], [372]. The structures of some of these reagents are shown below:

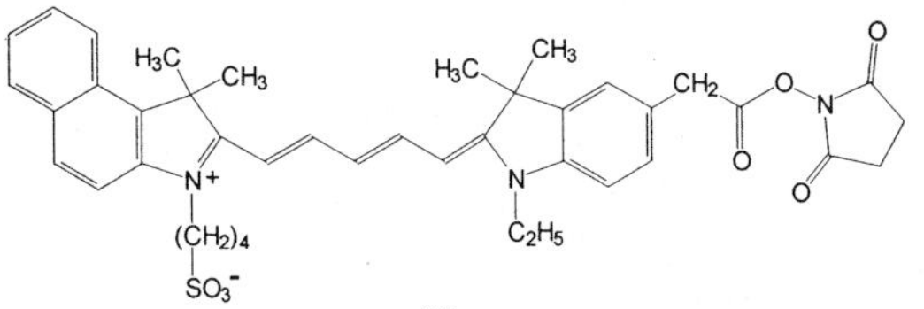

(A)

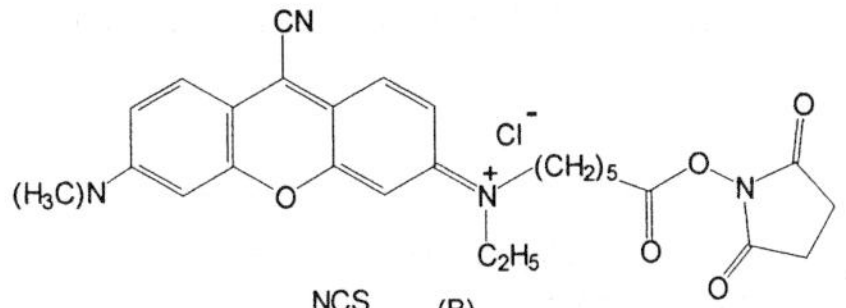

(B)

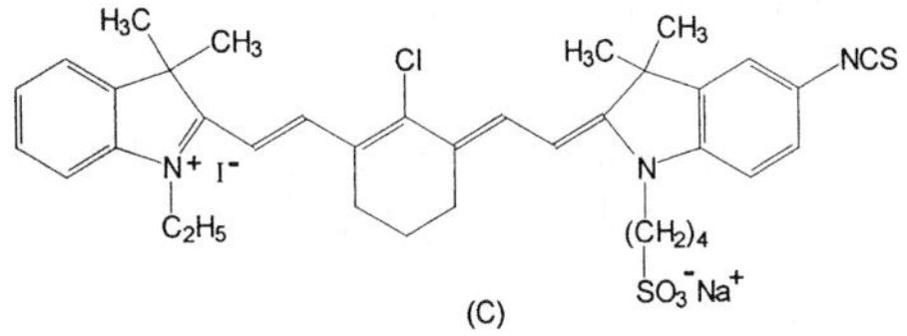

(C)

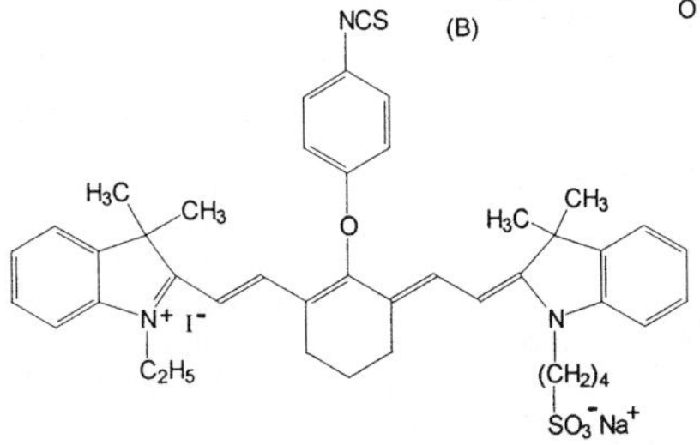

(D)

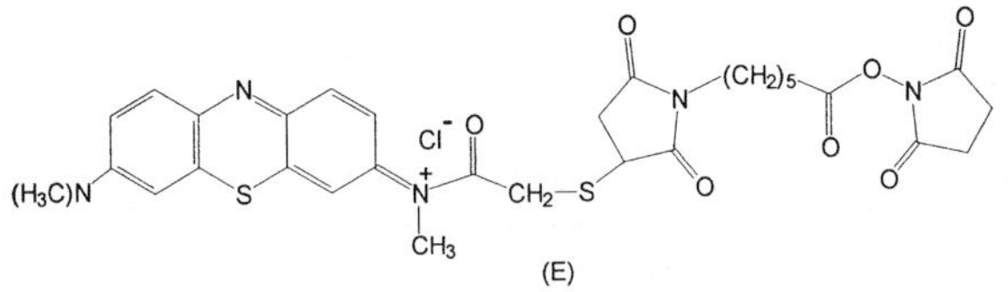

(E)

Less than 1 amol detection limit for amines can be obtained with reagents (B) and (E) using CE [369].

Not included in Table 19.4.4 is electro-chemiluminescence analysis using oxidation of amines with tris(2,2'-bipyridyl)ruthenium (III) complex $[Ru(bpy)_3]^{3+}$. The procedure leads to low nanomolar range of detection. Ruthenium (III) complex is generated in a postcolumn electrochemical oxidation [373], [374]. The reaction sequence in this type of determination of an amine is shown below (see also Section 17.6):

$$[Ru(bpy)_3]^{2+} - e^- \longrightarrow [Ru(bpy)_3]^{3+} \text{ (electrochemical process)}$$
$$[Ru(bpy)_3]^{3+} + R^aR^bNCH_2R^c \longrightarrow [Ru(bpy)_3]^{2+} + R^aR^bN^{+\cdot}CH_2R^c$$
$$R^aR^bN^{+\cdot}CH_2R^c \longrightarrow R^aR^bNC^{\cdot}HR^c + H^+$$
$$R^aR^bNC^{\cdot}HR^c + [Ru(bpy)_3]^{3+} \longrightarrow \{[Ru(bpy)_3]^{2+}\}^* + \text{oxidized amine}$$
$$\{[Ru(bpy)_3]^{2+}\}^* \longrightarrow [Ru(bpy)_3]^{2+} + h\nu$$

Several solid phase reagents are used for amine derivatization for HPLC analysis. A summary of these reagents is given in Table 19.4.5.

Typically the solid phase material is mixed directly with the sample, heated at a temperature between 60° C and 80° C for 10 to 15 min., and extracted with a solvent for the derivatized material such as CH_3CN. An aliquot of this extract is analyzed by HPLC. The solid phase reagents can be reused for a number of times after washing with an appropriate solvent, as long as enough reactive groups are still present.

A number of compounds can be transformed into amines for a more sensitive determination, and these are not captured as amine analysis. One such example is the reduction of nitro-polycyclic aromatic hydrocarbons (nitro-PAHs) to amino-PAHs, followed by peroxyoxalate chemiluminescence analysis with HPLC separation [383].

TABLE 19.4.5. *Solid phase reagents used for primary and secondary amine analysis by HPLC.*

Polymer base	Anchor	Group to be replaced	Reference
polystyrene divinylbenzene (Amberlite)	$N(CH_3)_3^+$ SO_3^-	benzoxazole	[375]
silica	3-[bis(2-hydroxyethyl)amino]propyl-diethoxysilane + 4-hydroxy-3-nitrobenzoyl chloride	3,5-dinitrobenzoate	[376]
silica	3-[bis(2-hydroxyethyl)amino]propyl-diethoxysilane + 4-hydroxy-3-nitrobenzoyl chloride	fluorenylacetyl	[1]
Porapak Q	benzotriazole	acetylsalicylate	[377]
polystyrene divinylbenzene	benzotriazole	6-quinoline-carbamate	[378]
polystyrene divinylbenzene	benzotriazole	3,5-dinitrobenzoate	[379]
polystyrene divinylbenzene	benzotriazole	3,5-dinitrophenyl-carbamate	[1]
polystyrene divinylbenzene	benzotriazole	9-fluorenyl carbonate	[380]
polystyrene divinylbenzene	o-nitrobenzophenone	3,5-dinitrobenzoate	[379]
polystyrene divinylbenzene	o-nitrobenzophenone	9-fluorenacetate	[381]
polystyrene divinylbenzene	o-nitrobenzophenone	fluorenylmethoxy-carbonyl proline	[1]
polystyrene divinylbenzene	o-nitrobenzophenone	fluorenylmethyl carbonate	[382]
polystyrene divinylbenzene	o-nitrobenzophenone	4-nitrobenzoate	[1]

- Derivatization of amino group in amino acids

The need to analyze amino acids is frequent due to their ubiquitous presence in biological samples, drugs, etc. Because amino acids contain at least two reactive groups (NH_2 and COOH), their GC analysis usually requires the derivatization of both those groups for the elimination of active hydrogens. Derivatization of amino acids in general is presented in Section 19.10. However, for HPLC analysis it is common to derivatize only one of the functional groups, and NH_2 is more frequently used than COOH for attaching chromophore or fluorophore groups to the analyte molecule. This explains why among the most common reagents used for amino acid analysis are reagents for amines, which were already included in Table 19.4.4. The reagents most frequently applied to derivatize amino acids are probably ninhydrin [384], dabsyl chloride, dansyl chloride, o-phthalaldehyde (and a thiol), and 9-fluorenylmethyl chloroformate. In addition, phenyl isothiocyanate that forms thiohydantoins and reacts with both NH_2 group and COOH group is frequently used, the application of this reagent being discussed in Section 19.10.

Derivatizations with dabsyl chloride, dansyl chloride, and 9-fluorenylmethyl chloroformate are acylations of the amino group (see also Section 18.3). A typical analysis of amino acids in plasma using dabsyl chloride starts with 250 µL plasma, to which 5 µL of 2 mM norleucine is added as internal standard. To the sample is added 2 mL 10% trichloroacetic acid for the precipitation of proteins. The mixture is centrifuged at 10000 g for 10 min, and the supernatant containing the amino acids is removed. The pH is adjusted to 9.0 with KOH. An aliquot of 40 µL is removed and 40 µL of 100 mM pH 8.3 NaHCO$_3$ is added. To this solution is added 80 µL 4 mM dabsyl chloride in CH$_3$CN and the solution is heated at 70° C for 12 min. with occasional mixing. After cooling at room temperature, 440 µL 50/50 ethanol/phosphate buffer with pH = 7 is added. An aliquot of 50 µL is injected in the HPLC system (C18 column, UV 436 nm detection) [385]. For the analysis of the amino acids from proteins, these must be hydrolyzed, for example, by heating at 110° C for 6 to 24 hours in HCl 6 N for hydrolysis (see Section 20.2).

The derivatization of amino acids with o-phthalaldehyde (and a thiol) is a condensation of the amino group with the dicarbonyl compounds. Several similar reagents are included in Table 19.4.4, and others are given in Table 19.4.6 when their use has been reported only for amino acids. The reaction of o-phthalaldehyde (OPA) in the presence of actylcysteine is shown below:

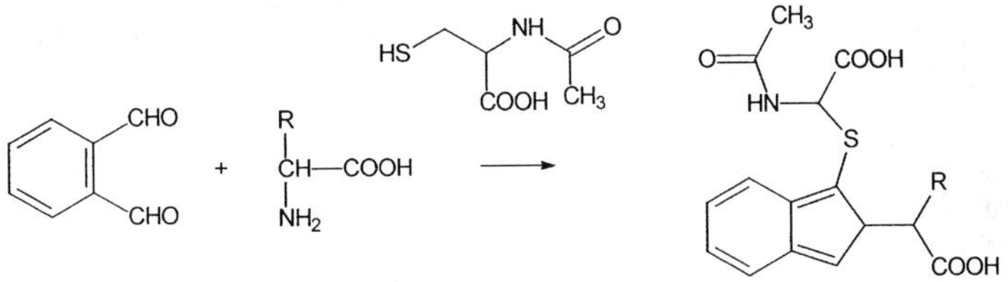

A typical procedure for the analysis starts with 35 µL solution of amino acids to which is added 35 µL 3.3 mg/mL o-phthalaldehyde in methanol/water 50:50, 35 µL 4 mg/mL N-acetyl-L-cysteine in methanol/water 50:50, and 175 µL 400 mM pH 9.4 potassium borate buffer. The solutions are mixed well and allowed to stand at room temperature for 2–3 min. The solution is then neutralized with 140 µL 1 M pH 3.5 sodium phosphate buffer, and an aliquot is injected in HPLC. The separation can be done on a C18 column and the measurement can be done using a fluorescence detector (ex 338, em 415) or an UV detector (338 nm) [386]. Besides actylcysteine, other thiols are used in the reaction of amino acids and OPA. Among these are acetylpenicillamine [387], *tert*-butoxycarbonyl-L-cysteine (BOC-L-cysteine) [1], N, N-dimethyl-2-mercaptoethylamine [1], ethandiol [388], N-isobutyryl-L-cysteine, 2-mercaptoethanol [389], [390], [391], 3-mercapto-propionic acid [392], 1-thio-β-D-galactopyranose [1], 1-thio-β-D-glucose [393], thioglucose tetraacetate [1], etc. Also, sodium sulfite [1] and sodium cyanide are used in the reaction.

Some reagents are used exclusively in amino acid analysis, although only the amino group is used for derivatization. Table 19.4.6 gives several reagents applied in HPLC analysis used for the derivatization of amino group in amino acids, which were not used in other amine derivatizations.

In Table 19.4.6 are included several reactions with isothiocyanates. In specific cases, this reaction can involve the COOH group leading to the formation of thiohydantoins (see Section 18.5). However, thiohydantoin formation usually requires a conversion step performed in acidic conditions. The reactions with isothiocyanates indicated in Table 19.4.6 are not continued to thiohydantoin formation.

Derivatization of some amino acids involves other groups besides NH$_2$ or COOH. For example, cysteine can be derivatized at the SH group, serine at the OH group, arginine at the guanido group. Some of these derivatizations are indicated in the sections for the specific functional group.

TABLE 19.4.6. *Reagents used exclusively for the derivatization of amino group in amino acids for analysis by HPLC.*

Reagent	Matrix, sensitivity	Reaction type[1]	Detection λ Separation	Ref.
(R)-2-(9-anthryl)propyl chloroformate	amino acids in proteins and peptides	Acyl	laser fluorescence ex 351 em 351, CE	[394]
(S)-2-(9-anthryl)propyl chloroformate	amino acids in proteins and peptides	Acyl	laser fluorescence ex 351 em 351, CE	[394]
benzyl chloroformate	amino acids in proteins	Acyl	UV 254, C18	[395]
benzyl isothiocyanate	amino acids in soybean 3.9 pmol	NCS	UV 246, C18	[396]
butyl isothiocyanate (C_4H_9NCS)	amino acids 0.5 nmol	NCS	UV 250, C18	[397]
N-[4-(S)-1-carbamoyl-2-methylpropylamino)-6-chloro-[1,3,5]triazin-2-yl]-L-phenylalanine amide	amino acids chiral	Alkyl-active halide	UV 254, C18	[398]
3-(4-carboxybenzoyl)-2-quinolincarboxaldehyde	amino acids 4.6-13.8 amol (CE)	Cycl	laser fluorescence ex 442 em 560	[399]
(1R,2R)-1,3-diacetoxy-1-(4-nitrophenyl)-2-propyl-isothiocyanate	amino acids	NCS	UV 245	[400]
(1S,2S)-1,3-diacetoxy-1-(4-nitrophenyl)-2-propyl-isothiocyanate	amino acids	NCS	UV 245	[400]

TABLE 19.4.6 (continued). *Reagents used exclusively for the derivatization of amino group in amino acids for analysis by HPLC.*

Reagent		Matrix, sensitivity	Reaction type[1]	Detection λ Separation	Ref.
N,N-diethyl-2,4-dinitro-5-fluoroaniline		amino acids in blood 1 pmol	Alkyl-active halide	UV 360, C18	[401]
diethylethoxymethylene-malonate		amino acids less than 1 pmol		UV 280, C18	[402]
4-(5',6'-dimethoxy-2-phthalimidinyl)phenyl-sulfonyl		proline, hydroxy-proline 10 fmol sec. amino acids 1 μmol	Acyl	fluorescence ex 315 em 385, ODS	[403]
2,4-dinitrophenyl-1-fluoro-5-L-aniline		amino acids chiral	Alkyl-active halide	UV 340	[1]
4-N,N-dimethylamino-1-naphthyl isothiocyanate		amino acids 0.2 pmol (fluorescence)	NCS	UV 254, fluorescence ex 345 em 435, ODS, postcolumn NaOH in CH₃CN	[1]
ethylthioltrifluoroacetate	CF₃COSC₂H₅	amino acids	Acyl		[404]
9-fluorenylethyl chloroformate		amino acids less than 1 pmol	Acyl	fluorescence ex 260 em 310, C18	[1], [405]
N-(9-fluorenylmethoxy-carbonyl)glycyl chloride		amino acids	Acyl	UV 265 cyclodextrin type	[1]
N-α-(9-fluorenylmethoxy-carbonyloxy)-succinimide		amino acids	Acyl	fluorescence ex 254 em 313, ODS	[1]

TABLE 19.4.6 (continued). *Reagents used exclusively for the derivatization of amino group in amino acids for analysis by HPLC.*

Reagent		Matrix, sensitivity	Reaction type[1]	Detection λ Separation	Ref.
2-fluoro-4,5-diphenyloxazole		amino acids 19-64 fmol	Alkyl-active halide	fluorescence ex 320 em 420, C8	[1]
isothiocyanatoacridine		amino acids 100-400 pmol	NCS	UV 280, C18	[1]
1-methoxycarbonyl-indolizine-3,5-dicarbaldehyde		amino acids in hair, proteins 0.2-200 fmol	Cycl	fluorescence ex 414 em 482	[406]
(-)-α-methoxy-α-methyl-1-naphthaleneacetic acid		amino acids methyl esters chiral	Acid	UV 280, silica	[407]
3-β-naphthapyrone-4-acetic acid-N-hydroxysuccinimidyl ester		amino acids, oligo-peptides 6-140 fmol		fluorescence ex 352 em 422, ODS	[408]
2-nitrobenzyloxycarbonyl chloride		amino acids	Acyl	UV 265, RP-8	[398]
4-nitrobenzyloxycarbonyl chloride		amino acids	Acyl	UV 265, RP-8	[398]
4-nitrophenyl isothiocyanate		amino acids 0.5-1 pmol	NCS	UV 254, RP-8	[409]
4-phenylazobenzoyloxy-carbonyl chloride		amino acids 1-10 pmol	Acyl	UV 320, RP-18	[398]

TABLE 19.4.6 (continued). *Reagents used exclusively for the derivatization of amino group in amino acids for analysis by HPLC.*

Reagent		Matrix, sensitivity	Reaction type[1]	Detection λ Separation	Ref.
4-(N-phthalimidinyl)-phenylsulfonyl		amino acids less than 0.2 pmol	Acyl	fluorescence ex 295 em 425, ODS	[410]
pyridoxal + NaBH₄. Reduction with NaBT₄ also applied.		amino acids 10^{-13} mol detection with radio-chemical detection	CO	UV 254, radio-chemical detection	[411]
succinimido-p-bromophenyl carbamate		amino acids 0.15-0.3 ng	Acyl	UV 250	[412]
succinimido-1-maphthylcarbamate		amino acids 75-1200 fmol	Acyl	fluorescence ex 290 em 370, ODS	[413]
sodium 1,2-naphthoquinone-4-sulfonate		amino acids 40-100 pmol	Acyl	UV 305, ODS	[414]
trimethylacetic anhydride	[(CH₃)₃CO]₂O	liothyronine	Anhy	UV 214, C8	[1]
2,4,6-trinitrobenzene-sulfonic acid		amino acids	Acyl	electrochem. glassy carbon -0.85 V Ag/AgCl	[415]

[1] Acyl = acylation reaction including reactions with chloroformates or sulfonyl chlorides, Alkyl = alkylation or aryl derivative formation, Anhy = acylation with an anhydride, Acid = acylation with an acid, CO = carbonyl addition, NCS = reaction with thiocyanate, Cycl = formation of new cycles.

- Derivatization of pyrroles and indoles

Pyrrole and indole derivatives deserve a special discussion because they contain NH groups, but they are acidic compounds. In the same group could be included compounds such as carbazole, 5H-pyrido[4,3-b]indole (γ-carboline), and 9H-pyrido[3,4-b]indole (β-carboline). Derivatives of these heterocycles are common in nature, especially those of indole. Because tryptophan (L-2-amino-3-indolylpropanoic acid) is an amino acid present in many proteins, the analysis of its derivatives is very important. When an amino group is associated with the indole such as in tryptamine, typical reactions for the amines can be used for derivatization. Also, the derivatization of peptides containing a N-terminal tryptophan using glyoxal was described in Section 18.5 [416] and is further presented in Section 19.10.

The derivatization of tryptamines with acetic anhydride in the presence of pyridine leads to the acetylation of the primary amino group while the indole nitrogen is not derivatized. However, with trifluoroacetic anhydride in the presence of trimethylamine in hexane as a solvent, the derivatization takes place quantitatively at the indole nitrogen in less than 30 min. [417] as follows:

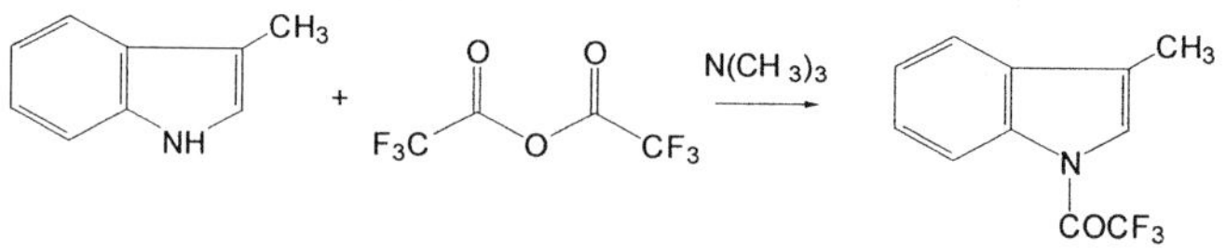

Besides trifluoroacetic anhydride, pentafluoropropionyl anhydride and heptafluorobutyryl anhydride are used for the derivatization of the indole nitrogen and at the same time of other amino or alcohol groups in tryptamine, serotonin or 3-(2-aminoethyl)-1H-indol-5-ol, tryptophol, etc. [15]. Heptafluoro-butyrylimidazole is used for the same purpose [418].

5-Hydroxyindoles react with benzylamine with the formation of oxazoles in the presence of an oxidant such as $K_3[Fe(CN)_6]$ as shown in Section 18.5. The resulting compounds are fluorescent, and the reaction has been used for the analysis of 3-(2-aminoethyl)-5-hydroxyindole (serotonin), 5-hydroxyindole-3-acetic acid, etc. [419]. Also, phenylglyoxal reacts for example with tryptophan as follows:

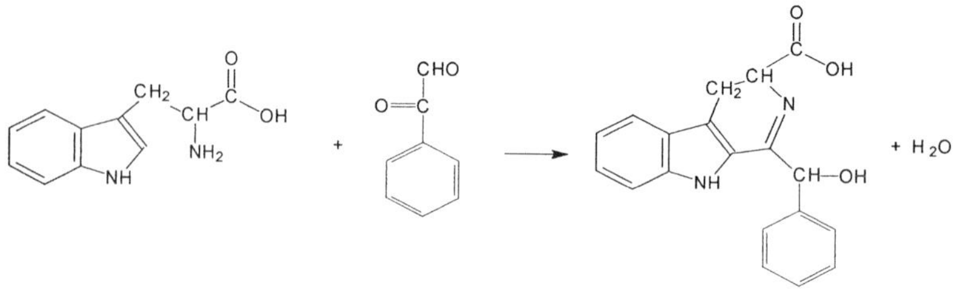

The resulting product shows fluorescence using excitation at 385 nm and emission at 460 nm [420].

The derivatization of indolylamines may lead in some cases to unexpected reactions. For example it was shown (see Section 18.3) that N-acetylserotonin generates by derivatization with pentafluoropropionic anhydride a β-carboline derivative. Melatonin derivatization using the same reagent in benzene is reported [421] to form a different compound, as shown below:

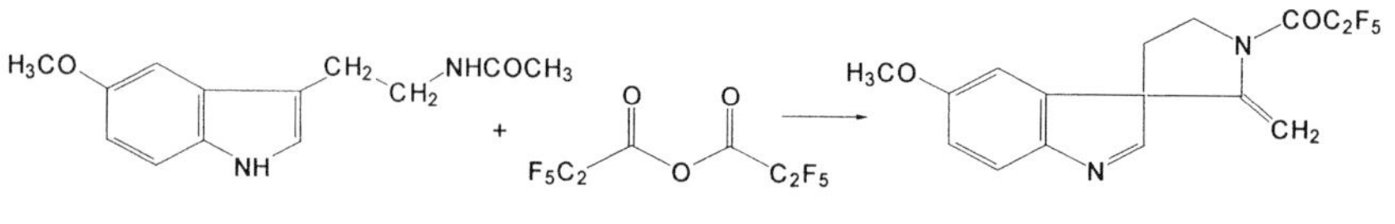

Acylation of tryptamine in the presence of acetic anhydride also leads to a β-carboline derivative as shown below:

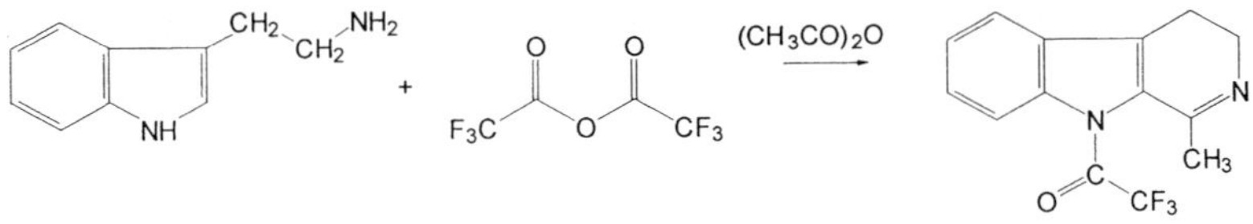

Silylation of indole amines can be performed using various silylation reagents. The indole ring NH group reacts well with silylating reagents due to its acidity. Pyrrole and its derivatives also can be silylated at the NH group using for example bis(trimethylsilyl)-trifluoroacetamide (BSTFA) in DMF [422]. Usually a reagent mixture is preferred in order to enhance the silylation efficiency. The silylated materials can be analyzed by GC or GC/MS. Several silylation reactions applied to this class of compounds are shown in Table 19.4.7.

TABLE 19.4.7. *Trimethylsilylation reagents used for indole amines derivatization.*

Reagent	Amine	Nitrogen derivatized	Reference
N-methyl-N-trimethylsilyltrifluoro-acetamide (MSTFA) + trimethylsilylimidazole (TMSI	indole derivatives	at indole ring	[423]
bis(trimethylsilyl)trifluoro-acetamide (BSTFA) + trimethylchlorosilane (TMCS)	3-(2-aminoethyl-N-methyl)-1H-indole	only at indole ring	[424], [425]
bis(trimethylsilyl)acetamide (BSA) + TMCS + TMSI	melatonin	only at indole ring	[15]
BSTFA	psilocin, psilocybin[1]	all active hydrogens	[15]

[1] For information and chemical structure regarding various drugs see e.g. M. O'Neil et al. (eds.), *The Merck Index*, Merck & Co., Inc., 13th ed., Whitehouse Station, 2001.

As seen in Table 19.4.5 the silylation at the nitrogen from the indole ring is efficient and for specific compounds takes place more easily than for the hydrogen in secondary amino or amido groups. However, in a reaction of serotonin with hexamethyldisilazane (HMDS) followed by a reaction with acetone, only the OH group is silylated and the primary amino group forms a Schiff base in a reaction as follows:

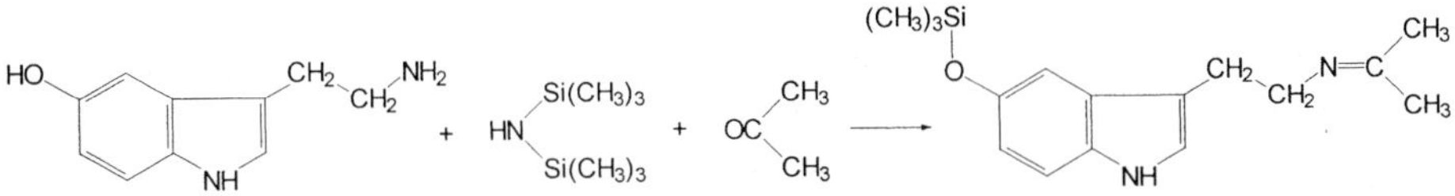

Another reaction that can occur in a sequence of derivatizations of amines is the replacement of the silyl group with a new derivatizing group. This can take place when an acylating reagent is added after silylation, as shown in Section 18.3. For example, serotonin can be derivatized using a mixture of MSTFA and TMSI (in the presence of N-methyltrifluoroacetamide or MTFA), the reaction taking place as follows:

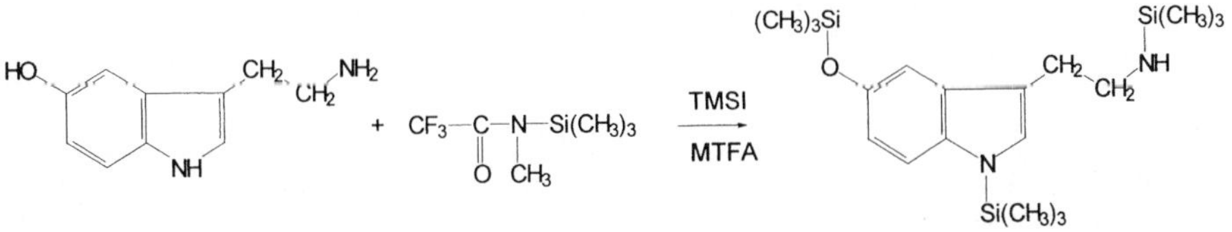

With the addition of N-methylbis(trifluoroacetamide) (MBTFA), the reaction continues as shown below:

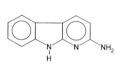

MBTFA

The derivatization of serotonin or tryptamine can be done beginning with an acetylation reaction followed by silylation. For this derivatization, a water solution containing 1–10 µg serotonin is saturated with $NaHCO_3$ followed by the addition of 500 µL acetic anhydride (or 1.5 mL heptafluorobutyric anhydride for the preparation of the fluorinated amide). The mixture is stirred for 5 min. after the elimination of excess CO_2 ceases and then extracted four times with 0.5 mL CH_2Cl_2. The CH_2Cl_2 extract is separated, and after the evaporation of the solvent, 25 µL of MSTFA is added. The mixture is heated for 30 min. at 80° C in a closed vial and then analyzed by GC/MS [15]. The resulting product is acetylated at the aliphatic amine and silylated at the OH group and NH indole group.

- *Derivatization of heterocyclic amines*

Heterocyclic aromatic amines (HAAs) are considered strong mutagens and are present in pyrolyzed food [426], [427], [428], [429]. Some of these substances are generated during the pyrolysis of Maillard polymers [430]. This group includes 2-amino-9H-pyrido[2,3-b]indole (AαC), 2-amino-3-methyl-9H-pyrido[2,3-b]indole (MeAαC), 2-amino-3-methylimidazolo-[4,5-f]-quinoline (IQ), 2-amino-3,4-dimethylimidazolo[4,5-f]quinoline (MeIQ), 2-amino-3,8-dimethylimidazo[4,5-f]quinoxaline (MeIQx), and 2-amino-3,4,8-trimethylimidazo[4,5-f]-quinoxaline (MeIQx), with the structures shown below:

2-amino-9H-pyrido[2,3-b]indole (AαC), from soybean globulin pyrolysate

2-amino-3-methyl-9H-pyrido-[2,3-b]indole, (MeAαC) from soybean globulin pyrolysate

2-amino-3-methylimidazo[4,5-f]-quinoline, (IQ) from broiled sardines

2-amino-3,4-dimethylimidazo[4,5-f]-quinoline (MeIQ) from broiled sardines

2-amino-3,8-dimethylimidazo[4,5-f]-quinoxaline (MeIQx) from fried beef

2-amino-3,4,8-trimethylimidazo[4,5-f]-quinoxaline (4,8-diMeIQx) from fried beef

Another heterocyclic amine, 2-amino-1-methyl-6-phenylimidazolo-[4,5-b]pyridine (PhiP) was detected in sidestream smoke of cigarettes [428] and has the structure:

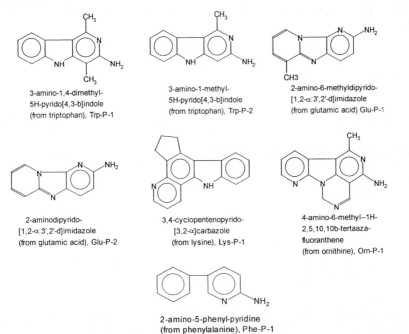

2-amino-1-methyl-6-phenylimidazo[4,5-b]-
pyridine (PhIP) from cigarette sidestream smoke

A second group of heterocyclic amines has been detected in the pyrolysate of various amino acids. Among these are 3-amino-1,4-dimethyl-5H-pyrido[4,3-b]indole (Trp-P-1), 3-amino-1-methyl-5H-pyrido[4,3-b]indole (Trp-P-2), 2-amino-6-methyldipyrido[1,2-α:3'2'-d]imidazole (Glu-P-1) and 2-amino-dipyrido[1,2-α:3'2'-d]imidazole (Glu-P-2),3,4-cyclopentenopyrido[3,2-α]carbazole (Lys-P-1), 4-amino-6-methyl-1H-2,5,10,10b-tetraazafluoranthene (Orn-P-1), and 2-amino-5-phenyl-pyridine (Phe-P-1) with the structures shown below:

3-amino-1,4-dimethyl-
5H-pyrido[4,3-b]indole
(from triptophan), Trp-P-1

3-amino-1-methyl-
5H-pyrido[4,3-b]indole
(from triptophan), Trp-P-2

2-amino-6-methyldipyrido-
[1,2-α:3',2'-d]imidazole
(from glutamic acid) Glu-P-1

2-aminodipyrido-
[1,2-α:3',2'-d]imidazole
(from glutamic acid), Glu-P-2

3,4-cyclopentenopyrido-
[3,2-α]carbazole
(from lysine), Lys-P-1

4-amino-6-methyl-1H-
2,5,10,10b-tertaaza-
fluoranthene
(from ornithine), Orn-P-1

2-amino-5-phenyl-pyridine
(from phenylalanine), Phe-P-1

Because only very low levels of HAAs may be present in a food sample, the analytical procedure used for determination must be very sensitive. The analysis in food of the underivatized compounds is very difficult if not impossible [431], [432], [433]. However, by derivatization of the amino groups of these molecules using heptafluorobutyric anhydride (HFBA) followed by NCI-MS detection, the analysis can be done successfully. Because the acylation is efficient at only one of the active hydrogens of the primary amino group, the chromatographic behavior of the acylated compounds is still unsatisfactory [428]. The second step of derivatization is necessary, and a methyl group can be introduced into the molecule [434]. This can be done using diazomethane

[435] or N,N-dimethylformamide dimethyl acetal (Methyl-8) [434]. The two-step derivatization reactions for PhiP (and similarly for other amines) are the following:

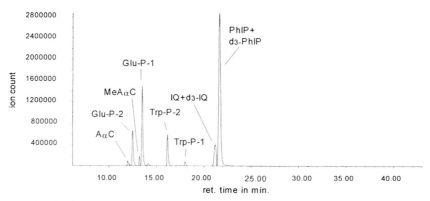

The GC chromatogram for a solution containing 100 ng/mL from several standards of HAAs including d_3-IQ and d_3-PhiP derivatized using HFBA and Methyl-8 is shown in Figure 19.4.2.

FIGURE 19.4.2. *Chromatogram of a 100-ng/mL solution of derivatized HAAs standards.*

The GC separation shown in this figure was done on a 15 m long ZB-1 column (Phenomenex), 0.25 mm id, 0.25 µm film. The oven temperature started at 120° C and was increased to 310° C with a rate of 5° C/min. The detection was done using a MS system in NCI mode. The chemical ionization gas was methane (at 2 mL/min.). As seen in Figure19.4.1, although at equal concentrations, the peak intensity of certain compounds is significantly different from the other (for IQ and PhiP each compound and its deuterated equivalent are coeluting). Good sensitivity can be noticed for PhiP, IQ, Trp-P-2, Glu-P-1, and Glu-P-2. At the same time, AαC, MeAαC and Trp-P-1 showed extremely weak response in NCI-MS.

For quantitation peak areas were measured for the analytes and for deuterated internal standards d_3-PhiP and d_3-IQ, monitoring the ions with m/z = 434 for PhiP, m/z = 437 for d_3-PhiP, m/z = 408 for IQ, and m/z = 411 for d_3-IQ. A typical SIM chromatogram for determining PhiP and IQ in a particulate phase cigarette smoke sample showing IQ (at about 2 ng/cig.) and PhiP (at about 4 ng/cig.) together with their d_3-standards is given in

Figure 19.4.3a. No chromatographic separation was obtained between the natural and deuterated compounds. Using extracted ions for m/z = 408 and m/z = 434, the quantitation of IQ and PhiP can be done using the ratio of the areas for the unknown and the deuterated corresponding standard. The traces of the extracted ions are shown in Figure 19.4.3b.

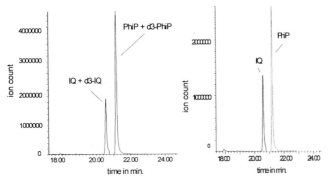

FIGURE 19.4.3a. *Typical SIM chromatogram for a sample of particulate phase cigarette smoke showing IQ and PhiP together with d₃-IQ and d₃-PhiP.*

FIGURE 19.4.3b. *Traces for extracted ions with m/z = 408 and m/z = 434 for a sample of particulate phase cigarette smoke showing IQ and PhiP at levels around 2 and 4 ng/cigarette respectively.*

Other methods for the analysis of heterocyclic amines require a sample preconcentration step using adsorption on a Blue-Rayon fiber followed by derivatization [436]. Besides GC analysis, various HPLC techniques were also developed for HAAs analysis, but usually no derivatization is required for the analysis [437], [438], [439], [440], [441], [442].

- Derivatization of tertiary amines and quaternary ammonium compounds

Tertiary amines are much less frequently derivatized as compared to amines. Their derivatization reactions can be classified in the following types: a) formation of quaternary ammonium compounds, b) replacement of one of the substituents at the amino nitrogen with a different one, c) elimination of the amino group, and d) reaction at a different part of the analyte molecule as a result of the presence of the tertiary amino group.

Reactive alkylating reagents may form quaternary ammonium salts with tertiary amines, as shown for some alkyl fluoromethylsulfonates (see Section 18.1). A similar derivatization may take place with other active alkyl halides. For example, 4-nitrobenzyl bromide reacts with pilocarpine as follows:

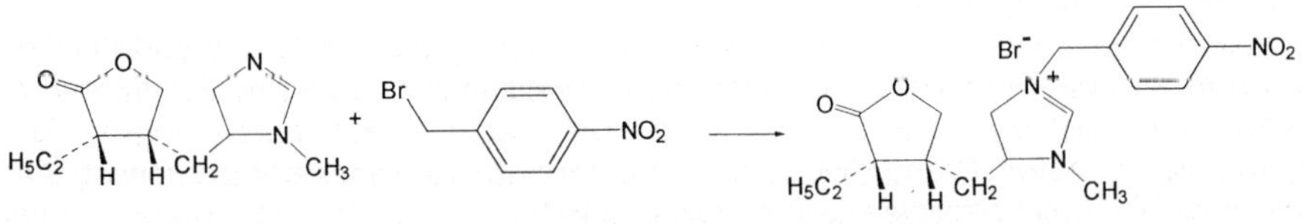

The reaction has been used to quantitate as low as 50 ng/mL pilocarpine using HPLC analysis with detection in UV at 254 nm [1].

The replacement of a substituent group in the tertiary amine can be done with some more reactive compounds such as chloroformates (usually in the presence of $NaHCO_3$), which can be used in a reaction generating carbamates as follows:

The reagents used in this reaction include pentafluorobenzyl chloroformate [443], trichloroethyl chloroformate [444], etc. The procedure has been used for the derivatization of amitriptyline, nortriptyline [445], and pethidine [446], followed by GC-ECD analysis.

The carbamate formed with methyl chloroformate can be further treated with HBr (in CH_3COOH, by heating at $100°$ C for 10 min.), and it decomposes forming a secondary amine as follows:

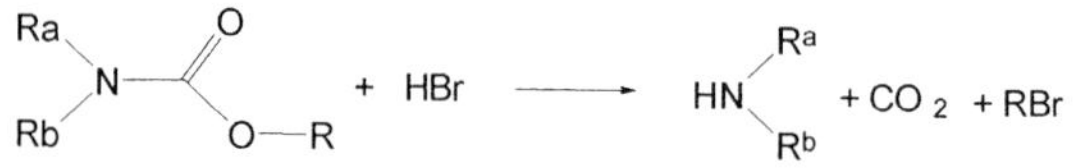

The resulting secondary amine can be derivatized for example with heptafluorobutyric anhydride [447].

A typical reaction with the elimination of the amino group is Hofmann degradation. The elimination reaction takes place for quaternary ammonium hydroxides when they are heated. The analysis of amitriptyline [448] by this procedure using CH_3I in the presence of Ag_2O followed by heating was described in Section 18.6. Other compounds such as haloperidol, droperidol, and penfluoridol are analyzed similarly [449].

Quaternary amines can be derivatized, usually by the elimination of one of the nitrogen substituents. As an example, acetylcholine reacts with sodium benzenethiolate as follows [450]:

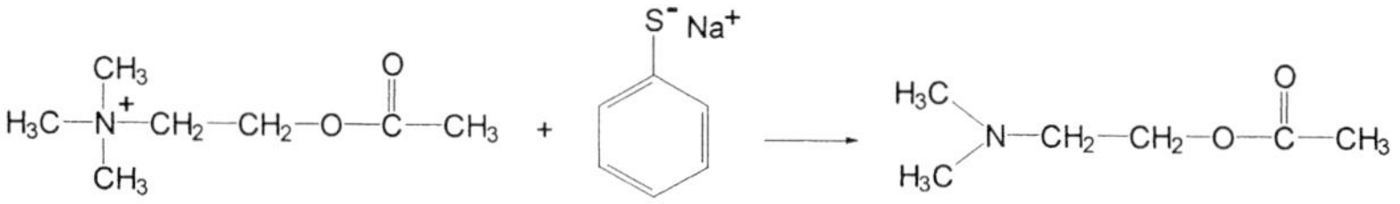

Other reactions use specific features of the amine molecule for derivatization and are not always applicable [15].

- Derivatization of hydrazines

Hydrazines are reactive compounds and their derivatization may be needed for better analysis. The reactivity of hydrazines toward aldehydes has been used for their derivatization for both GC and HPLC analysis. Various hydrazines can be derivatized with 5-nitro-2-hydroxy-benzaldehyde as follows [451]:

For HPLC analysis the reaction with various aldehydes of both hydrazines and hydrazones can be used for enhancing the UV absorption. A few aldehydes used for hydrazine derivatization are given in Table 19.4.8.

TABLE 19.4.8. *Aldehydes used for hydrazines derivatization.*

Reagent		Analyte matrix, sensitivity	Detection λ Separation	Reference
anisaldehyde	H₃CO—⟨ ⟩—CHO	hydralazine[1] 1 ng/mL	UV 365, CN	[452]
benzaldehyde	⟨ ⟩—CHO	hydrazine, phenelzine[1] 0.6 ng	UV 313, silica	[1]
p-dimethylaminobenzaldehyde	(H₃C)₂N—⟨ ⟩—CHO	hydrazine	UV 455	[453]
2-hydroxy-1-naphthaldehyde	CHO, OH (naphthalene)	hydralazine[1] dihydralazine 1 ng/mL	Vis 406, ODS	[1]
2-nitrocinnamaldehyde	CHO, NO₂	hydralazine[1], hydrazine 800 ng/mL	UV 350, C18	[1]

[1] For information and chemical structure regarding various drugs see e.g. M. O'Neil et al. (eds.), *The Merck Index*, Merck & Co., Inc., 13th ed., Whitehouse Station, 2001.

Another reaction used for hydrazine derivatization is based on the capability of forming a pyrazole cycle with β-diketones (see Section 18.5). The condensation with acetylacetone is used for the analysis of hydrazine and methylhydrazine [454]. Also, the condensation with phthalaldehyde, mercaptoethanol and a hydrazine are used to form 3,4-benzopyrrole cycle [455].

Hydrazides give similar reactions with hydrazines, because most reactions with compounds containing the -NH-NH₂ group take place at the NH₂ end. The influence of an alkyl or acyl substituent at the other nitrogen is not significant. A few derivatizations for hydrazides that are considered derivatives of organic acids are discussed in Section 19.7.

- Derivatization of hydroxylamines and oximes

Substituted hydroxylamines can react at the active hydrogen in the group >N-OH using methylation with dimethylsulfate [1], or methyl iodide [456]. The reaction with CH_3I can be written as follows:

The same group can be silylated for example with BSTFA. The reaction takes place at 40–50° in 1 hour [457].

Hydroxylamines containing an active hydrogen bonded to the nitrogen may react similarly to amines using for derivatization an isocyanate. For example, 5-hydroxyaminoindan can be derivatized with methyl isocyanate in a reaction as follows [458]:

Trifluoroacetic anhydride (TFAA) can react at both active hydrogens in a NH-OH group [15].

Oximes, can be derivatized using silylation for example using BSTFA in a reaction of the type:

$$R_2C=N-OH + BSTFA \longrightarrow R_2C=N-O-Si(CH_3)_3$$

This reaction can also be applied after the derivatization of carbonyl compounds with hydroxylamine.

19.5 DERIVATIZATION OF ALDEHYDES AND KETONES

The derivatization of aldehydes and ketones is commonly done for better detectability. The separation by GC or HPLC usually does not raise special problems for carbonyl compounds. This can be attributed to the absence of hydrogen bonding, which aldehydes and ketones cannot form because of the lack of active hydrogens. Some aldehydes or ketones containing an α-H may generate enols, and then the active hydrogens are present. However, even when enols are generated, the equilibrium between the enol and the carbonyl usually favors the carbonyl form.

Aldehydes and ketones are reactive compounds, and this must be taken into account during sample preparation. Condensation reactions may take place between two molecules of a carbonyl compound (see. e.g. [459]). Also, reactions with solvents such as water or alcohols may occur inadvertently. In particular, low molecular weight aldehydes are found in water solutions as hydrates [460] or as hemiacetals and acetals in alcohol solutions. Hemiacetal formation is also characteristic for carbohydrates that contain carbonyl and alcohol groups in the same molecule. The derivatization of aldehydes and ketones must take into account possible equilibrium of the free carbonyl compound with the compounds previously generated from the carbonyl group.

Most derivatizations of carbonyl compounds are additions to the C=O bond. The addition reaction may begin as a nucleophile attack to the carbon or as an electrophile attack of a proton to the heteroatom (see Section 18.4). The reaction is the following:

$$Ra—\overset{+\delta}{\underset{\underset{-\delta}{:O:}}{\overset{|}{C}}}—Rb \quad + \quad R:H \quad \longrightarrow \quad Ra—\overset{R}{\underset{OH}{\overset{|}{\underset{|}{C}}}}—Rb$$

The addition may be followed by a subsequent elimination reaction. In this case, a new double bond is generated as follows:

$$Ra—\overset{\underset{O}{\|}}{C}—Rb \quad + \quad RH_2 \quad \longrightarrow \quad Ra—\overset{RH}{\underset{OH}{\overset{|}{\underset{|}{C}}}}—Rb \quad \longrightarrow \quad Ra—\overset{R}{\overset{\|}{C}}—Rb \quad + \quad H_2O$$

When the enolization takes place, aldehydes or ketones with an α-H generate an OH group as shown below:

$$HC\overset{Ra}{\underset{Rb}{\diagdown}}C{=}O \quad \rightleftharpoons \quad C\overset{Ra}{\underset{Rb}{\diagdown}}C{-}OH$$

The formation of the enol group leads to the possibility of performing on carbonyl compounds some of the derivatizations available for alcohols (e.g. silylation).

The derivatization of carbonyl groups present in complex compounds containing various other functional groups such as OH or COOH may be influenced by these groups. Some of these compounds will be discussed separately (see e.g. derivatization of carbohydrates).

- Derivatization of carbonyl compounds for GC analysis.

The derivatization for GC analysis of carbonyl compounds usually applies reactions of the type described in Section 18.4. The addition to the carbonyl double bond of hydrazines and hydroxylamines is particularly used because these reactions take place with high efficiency. For example, the reaction with phenylhydrazine of various C_1 to C_{11} carbonyl compounds occurs at room temperature overnight or in 30 min. at 50° C [461]. For the analysis, 10 mL of 0.5 M phenylhydrazine hydrochloride in 0.25 M aqueous Na_3PO_4 is first washed with diethylether. Then 10 mL of methanol is added to the aqueous solution, which is further diluted to 100 mL with water. To this solution, 5–50 μL of carbonyl compounds is added. The reaction is performed at room temperature or by heating, and then repeated extractions are performed using 10, 5, and 5 mL of hexane, ethyl ether, and ethyl acetate respectively. The extracts are combined and evaporated to dryness under a stream of nitrogen. The residue is then dissolved in 1 mL of hexane, 5 mL of ethyl ether, or 10 mL of ethyl acetate and dried over anhydrous $MgSO_4$. The extract can be analyzed by GC.

Substituted phenylhydrazines such as 2,4-dinitrophenylhydrazine, although frequently used for HPLC analysis, has been applied for GC analysis of carbonyl compounds [462]. Similarly to oximes, the 2,4-dinitrophenylhydrazones from various carbonyl compounds generate double peaks in GC chromatograms. Only formaldehyde, acrolein, isobutyraldehyde and crotonaldehyde showed single peaks from a group of 22 carbonyl compounds. The double peaks are generated by isomers and are not a result of other interactions [462].

The reaction with hydroxylamines forming oximes with the carbonyl compounds generate *syn* and *anti* isomers as shown below:

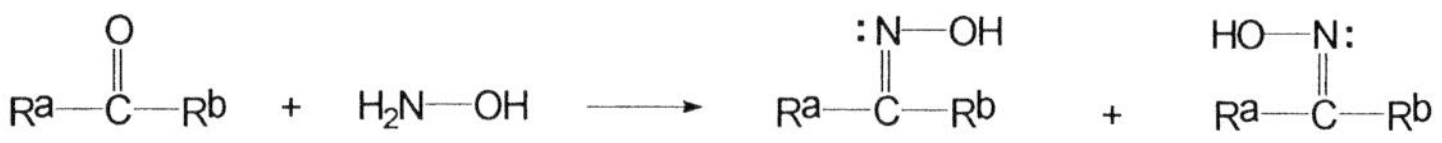

The two diastereoisomers may lead to two chromatographic peaks in a chromatogram. This could be a disadvantage because it complicates the chromatogram, but may also help in the identification of the oximes [463].

Various hydrazines or hydroxylamines used for carbonyl compounds derivatization with GC analysis are given in Table 19.5.1.

TABLE 19.5.1. *Various hydrazines or hydroxylamines used for carbonyl compounds derivatization.*

Reagent		Carbonyl type	Reference
phenylhydrazine	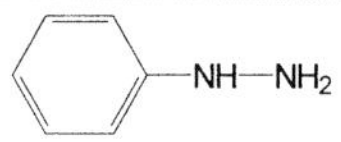	aldehydes, ketones	[461]
2,4-dinitrophenylhydrazine (DNPH)	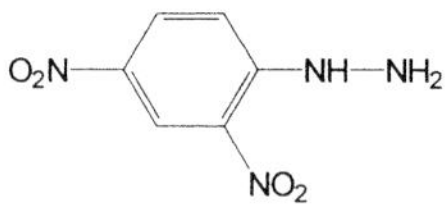	aldehydes, ketones	[15], [462]
N-aminopiperidine	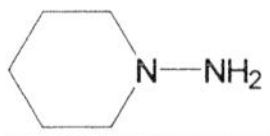	ketones CH_3COOH catalyst	[463]
N-aminohexamethylene-imine or N-amino-homopiperidine	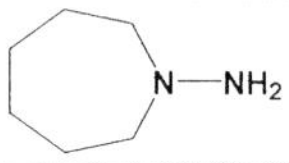	ketones CH_3COOH catalyst	[464]
pentafluorophenyl hydrazine	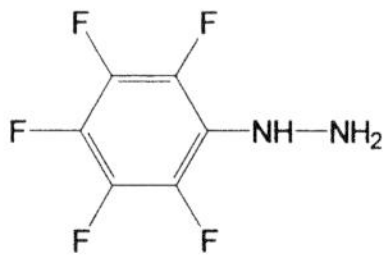	ketones CH_3COOH catalyst	[464], [465]
hydroxylamine	NH_2OH	aldehydes, ketones	[463]
(2,3,4,5,6-pentafluorobenzyl)-hydroxylamine (HCl form)	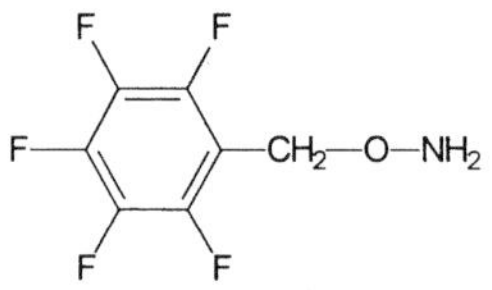	formaldehyde (with simultaneous use of SPME)	[466], [467]
p-nitrobenzyloxime	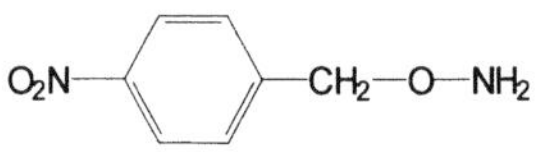	aldehydes, ketones	[15]

Besides the direct analysis of the hydrazones, an oxidation using ozone has been used to transform, for example, dinitrophenylhydrazones in the free carboxylic acid [468].

Among other reactions used for carbonyl compounds derivatization is the acetal or ketal formation. The formation of a hemiacetal is also used in some derivatizations, but because of the formation of an OH group, hemiacetals must be further derivatized. One example is the derivatization of pyridoxal using ethanol followed by derivatization with N-methylbis(trifluoroacetamide) or MBTFA. The reaction takes place as follows [469]:

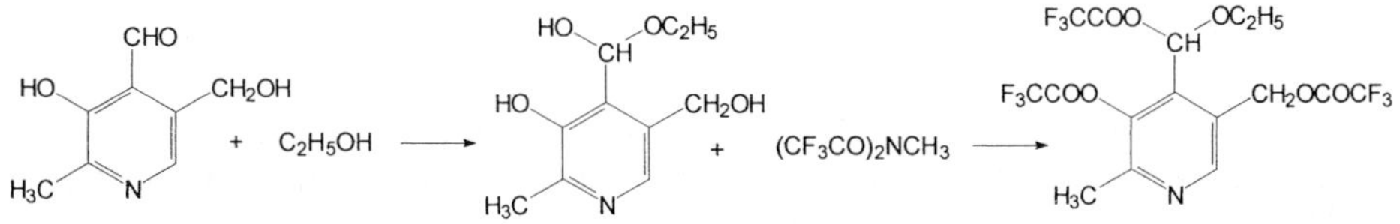

Other similar derivatives of pyridoxal were also used for the GC analysis. Among these are the acetyl derivative of the ethyl hemiacetal, the heptafluorobutyrate of the methyl hemiacetal [470], etc. It is interesting that the treatment of pyridoxal with acetic anhydride in pyridine leads to the reaction:

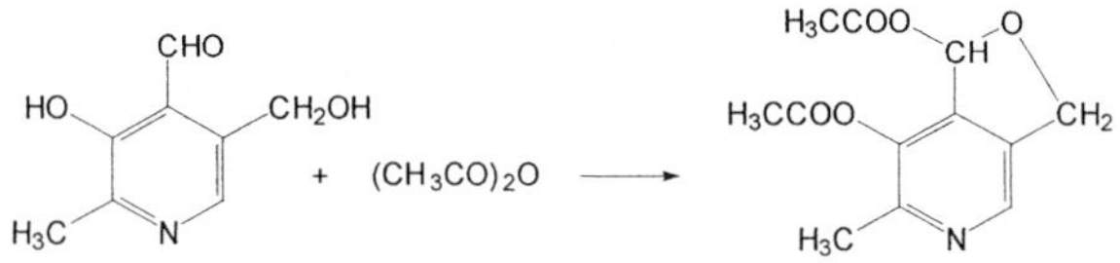

Methanol or isopropanol in the presence of toluenesulfonic acid (TsOH) is used for the derivatization of various aldehydes and ketones with acetal formation. The reaction takes place as follows:

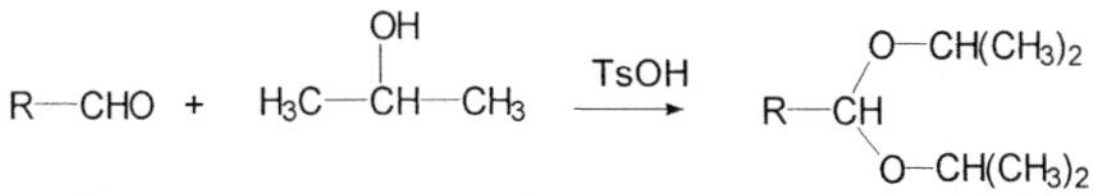

A typical procedure uses 100 mg sample containing the carbonyl compound to which is added 5 mL of alcohol and 10 mg TsOH. The mixture is kept for 1 hr. at -10° C followed by the addition of 5–10 mg of NaOCH$_3$, which stops the reaction. An aliquot is then analyzed by GC. Acetal formation of other carbonyl containing compounds that contain additional groups besides carbonyl also is used for analytical purposes. Typical examples are the derivatizations of carbohydrates, and the subject is presented separately in Section 19. 9.

Silylation is used for the derivatization of particular aldehydes and ketones that are able to form enols. The displacement of the enolization reaction toward the enol formation is probably favored by the silylation that eliminates the enol. In some cases, the silylation at an enol group formed from the carbonyl compound is not intended and the reaction may appear as an artifact (see Section 18.2). The reaction is used for analytical purposes, and also for preparative ones [471]. The list of silylation reagents used for derivatization of carbonyl compounds is quite long, including allyltrimethylsilane, N,O-bis(trimethylsilyl) acetamide, N,N-bis(trimethylsilyl)formamide, bis(trimethylsilyl)sulfate, bromotrimethyl-silane, chlorotrimethylsilane, ethyl trimethylsilylacetate [472], hexamethyl-disilane, (isopropenyloxy)trimethylsilane, 1-methoxy-2-methyl-1-trimethylsiloxypropene, trimethyl-silyl acetate, trimethylsilyl benzenesulfonate, N-trimethylsilylimidazole, trimethylsilyl methanesulfonate, trimethylsilyl trifluoromethanesulfonate (triflate) [15], etc. [189].

Silylation is used in combination with a reaction with urea for the analysis of malonaldehyde. The first step in the reaction is the formation of a pyrimidone and can be written as follows:

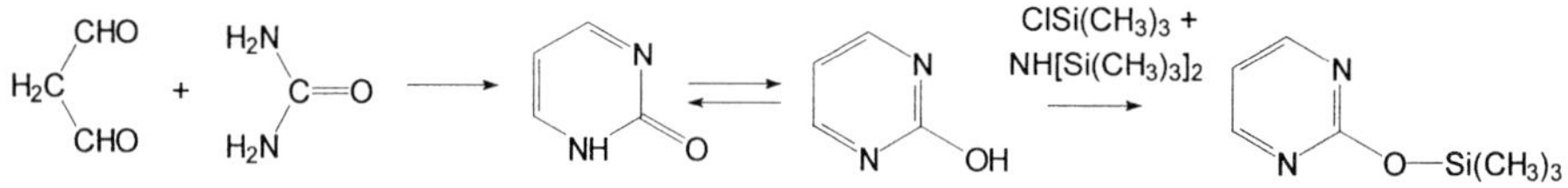

The silylated product can be analyzed by GC/MS for malonaldehyde in tissue homogenates [473].

Amines can also be used for derivatization, usually in the presence of $NaCNBH_3$. The reaction can be written as follows:

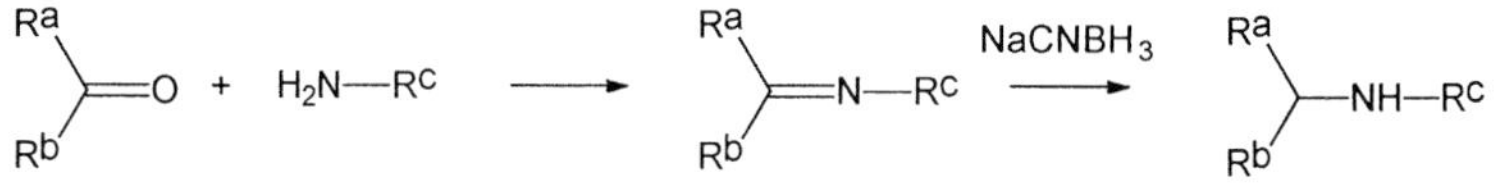

Reagents such as 2-amino-N-[[3,5-(bistrifluoromethyl)-phenyl]methyl]N-methyl-acetamide can be used for derivatization, followed by GC with ECD or EI- MS detection [474].

- Derivatization of carbonyl compounds for HPLC analysis

The derivatizations for HPLC analysis are usually performed to improve UV or fluorescence detection, and the reagents carry groups with chromophores or that fluoresce (fluorophores). The reacting moiety of the reagents consists of functional groups that typically react with aldehydes or ketones such as hydrazines, amines or hydroxylamines (see also Section 18.4). Table 19.5.2 lists some of these reagents. More reagents that react with the carbonyl group, which are applied particularly for the analysis of carbohydrates, are given in Table 19.5.3.

Luminarin 11 and Luminarin 12 [526] can be used for the analysis of acetylacetone.

A number of other derivatizations of carbonyl compounds have good potential for chromatography but are reported as applied without a separation. One such example is the use of N-amino-N'-(1-hexylheptyl)perylene-3,4:9,10-tetracarboxylbisimide for the analysis of propionaldehyde [527].

TABLE 19.5.2. *Reagents used for carbonyl compounds analysis by HPLC.*

Reagent	Analyte, matrix, sensitivity	Reaction type[1]	Detection λ, Separation	Reference
acetylacetone + NH₃	formaldehyde 400 ppb	Cycl	postcolumn, Vis 410	[1]
2-amino-4,5-ethyl-enedioxy-phenol or 6-amino-7-hydroxy-1,4-benzodioxol	aldehydes 5-10 pmol	Cycl	fluorescence ex 330, em 390, ODS	[475]
4-amino-3-hydrazino-5-mercapto-1,2,4-triazole or purpald	aldehydes 15-80 ng	Hydraz	postcolumn, Vis 540	[476]
5-amino-4-thiolo-phthalhydrazide + H₂O₂ + K₃[Fe(CN)₆)]	aromatic aldehydes 0.2-0.4 fmol	Cycl	CL	[477]
4-(aminosulfonyl)-7-hydrazino-2,1,3-benzoxadiazole	heptan-4-one 18.8 μmol, propanal 6.7 μmol	Hydraz	fluorescence ex 450, em 558	[478]
(S)-benzylmethyl-semicarbazide	synthetic peptide	Hydraz	UV 310, ODS, CZE	[1]
(R)-(+)-4-(2-carbazolylpyrrolidin-1-yl)-7-(N,N-dimethyl-amino-sulfonyl)-2,1,3-benzoxadiazole	ketones chiral	Hydraz	fluorescence ex 450, em 540, ODS	[479]
4-(2-carbazolyl-pyrrolidin-1-yl)-7-nitro-2,1,3-benzoxadiazole	aldehydes in food 10 fmol	Hydraz	laser fluorescence ex 488, em 540, ODS	[1]
(carboxymethyl) trimethyl-ammonium chloride hydrazide (Girard T)	ketosteroids	Hydraz	CE, UV 280 or 250	[1]
2-cyanoacetamide	mono-saccharides, aldoses 0.1-1 nmol, etc.	Meth	fluorescence ex 331, em 383, UV 276, postcolumn anion exch.	[1]

TABLE 19.5.2 (continued). *Reagents used for carbonyl compounds analysis by HPLC.*

Reagent		Analyte, matrix, sensitivity	Reaction type[1]	Detection λ, Separation	Reference
1,3-cyclohexandione + NH$_4$$^+$	+ NH$_4^+$	aldehydes 2 pmol	Cycl	fluorescence ex 305-395 em 450, etc., ODS	[1]
dansyl hydrazine	SO$_2$—NHNH$_2$ N(CH$_3$)$_2$	aldehydes in air 10-150 pg, ketones, etc.	Hydraz	fluorescence ex 355, em 525, RP-18	[1], [480], [481]
1,2-diamino-4,5-ethylenedioxy-benzene	NH$_2$ NH$_2$	L-(4-formyl-3-hydroxyphenyl)-glycine, aldehydes 0.01 - 10 pmol	Cycl	fluorescence ex 350, em 400, ODS	[482], [483]
2-diethylamino-benzoic acid hydrazide or deayl hydrazide	NHNH$_2$ N(C$_2$H$_5$)$_2$	abscisic acid[2] (ketone)	Hydraz	fluorescence ex 355, em 445, C18	[1]
7-diethylamino-coumarin-3-carboxylic acid hydrazide or deccyl hydrazide	(H$_5$C$_2$)$_2$N NHNH$_2$	abscisic acid[2] (ketone)	Hydraz	fluorescence ex 440, em 483, C13	[1]
4-(1-(4,5-dihydro-3-phenylhydrazolyl) benzenesulfonyl hydrazide or darpsyl hydrazide	NHNH$_2$ SO$_2$	abscisic acid[2] (ketone)	Hydraz	fluorescence ex 330, em 435, C18	[1]
2,5-dihydroxybenzo-hydrazide	OH NHNH$_2$ OH	aldehydes 100-200 fmol	Hydraz	electrochem . porous graphite, RP-18	[484]
2-diisopropylamino-benzoic acid hydrazide or diayl hydrazide	NHNH$_2$ N(CH(CH$_3$)$_2$)$_2$	abscisic acid[2] (ketone)	Hydraz	fluorescence ex 330, em 435, C18	Lunn
6,7-dimethoxy-1-methyl-2-oxo-1,2-dihydroquinoxalin-3-yl-propionohydrazide	CH$_3$ H$_3$CO NHNH$_2$ H$_3$CO (CH$_2$)$_2$	aldehydes 13,55 fmol	Hydraz	fluorescence ex 362, em 442, ODS	[485]
2-dimethylamino-benzoic acid hydrazide or dmayl hydrazide	NHNH$_2$ N(CH$_3$)$_2$	abscisic acid[2], jasmonic acid (ketone)	Hydraz	fluorescence ex 354, em 450, C18	[1]

TABLE 19.5.2 (continued). *Reagents used for carbonyl compounds analysis by HPLC.*

Reagent		Analyte, matrix, sensitivity	Reaction type[1]	Detection λ, Separation	Reference
2-dimethylamino-6-naphthalenesulfonic hydrazide	(structure) NHNH₂, SO₂, (H₃C)₂N	abscisic acid[2] (ketone)	Hydraz	fluorescence ex 355, em 445, C18	[1]
4-(N,N-dimethyl-aminosulfonyl)-7-hydrazino-2,1,3-benzoxadiazole	(structure) NHNH₂, N–O–N, SO₂N(CH₃)₂	blood acetaldehyde 300 nmol, etc. heptan-4-one 17.1 µmol	Hydraz	fluorescence ex 445, em 560, ODS	[486]
4,5-dimethyl-1,2-phenylenediamine	(structure) H₃C, NH₂, H₃C, NH₂	aldehydes 0.05–2.5 ng/mL	Cycl	fluorescence ex 330, em 400, ODS	[1]
5,5-dimethyl-1,3-cyclohexandione (dimedone) + NH₄⁺	(structure) H₃C, H₃C, + NH₄⁺	aldehydes 30 fmol	Cycl	fluorescence ex 385, em 460, ODS	[487], [488]
3,4-dinitrobenzyl-oxamine	(structure) O₂N, NO₂, ONH₂	4'-mono-phosphoryl lipid A from *Salmonella minnesota* R595,7 ng	Hydrox	UV 254, C18	[489]
2,4-dinitrophenyl-hydrazine	(structure) NHNH₂, NO₂, NO₂	aldehydes in air, 0.2 ppb, aldehydes in blood, ketosterols, ketones 1.5-2.6 ppb	Hydraz	UV 360, C18, etc.	[490] through [500]
2-diphenylacetyl-1,3-indandione-1-hydrazone	(structure) O, O, CH(C₆H₅)₂, N-NH₂	formaldehyde, acetaldehyde 0.25 ppb	Hydraz	UV 360, fluorescence ex 420, em 525, ODS	[501]
diphenylthiobarbituric acid	(structure) H₅C₆–N, CH₂, O, O, C₆H₅	malon-dialdehyde in blood 100 nmol	Meth	Vis 537 PR-18	[1]
2,2'-dithiobis(1-amino-4,5-dimethoxybenzene	(structure) H₃CO, S—S, OCH₃, H₃CO, NH₂ H₂N, OCH₃	aldehydes 8-20 fmol	Cycl	fluorescence ex 335, em 430, ODS	[502]

TABLE 19.5.2 (continued). *Reagents used for carbonyl compounds analysis by HPLC.*

Reagent		Analyte, matrix, sensitivity	Reaction type[1]	Detection λ, Separation	Reference
7-hydrazino-4-nitro-2,13-benzoxadiazole		aldehydes, 5-10 ng	Hydraz	UV 254	[503]
hydrocyanic acid	HCN	pyridoxal-5-phosphate[2] 50 fmol	Add	fluorescence ex 318, em 418, ODS	[504]
luminarin 3 or 2,3,6,7-tetrahydro-11-oxo-1H,5H,11H-[1]benzopyrano[6,7,8-ij]-quinolizine-9-acetic acid hydrazide		aldehydes 158-1950 fmol, ketones 156-1950 fmol	Hydraz	fluorescence ex 399, em 485, ODS	[505]
(R)-(+)-α-methyl-benzylamine		ketones chiral	Amine	UV280, silica	[1]
N-(4-(6-methyl-2-benzothiazolyl)-phenylglycine hydrazide or apmayl hydrazide		abscisic acid[2] (ketone)	Hydraz	fluorescence ex 345, em 413, C18	[1]
3-methyl-2-benzothiazolinone hydrazone		aldehydes 0.3-13.5 pmol, ketones 0.3-15.5 pmol	Hydraz	electrochem glassy carbon +1.05 V Ag/AgCl C18	[506], [507]
1-methyl-1-(2,4-dinitrophenyl)hydrazine		aldehydes ketones in air	Hydraz	UV 368	[508]
N-methyl-4-hydrazino-7-nitrobenzofurazan		aldehydes, ketones	Hydraz	UV 474	[509]
4-(1-methyl-phenanthro[9,10-d]-imidazol-2-yl)benzo-hydrazide		aldehydes 1.4-4.4 fmol (conv.) 0.18-0.80 fmol (laser)	Hydraz	laser fluorescence ex 325, em 460, ODS	[1]

TABLE 19.5.2 (continued). *Reagents used for carbonyl compounds analysis by HPLC.*

Reagent	Analyte, matrix, sensitivity	Reaction type[1]	Detection λ, Separation	Reference
acetylacetone + NH₃ (Hantzch reaction) or Nash reagent (NH₄OOCCH₃ + CH₃COOH + 2,4-pentandione) H_3C CH_3 NH_3	formaldehyde aldehydes	Cycl	fluorescence postcolumn	[510], [511], [512], [513]
2-nitrophenyl-hydrazine $NHNH_2$ NO_2	formaldehyde in urine 100 ng/mL, ketones	Hydraz	UV 254, C18	[514]
4-nitrophenyl-hydrazine $NHNH_2$ O_2N	ketosteroids in blood 80 ng/mL	Hydraz	electrochem. C18	[1], [515]
4-nitrobenzylhydroxyl-amine O_2N ONH_2	xyloglucans 10 pmol, prostaglandins	Hydrox	UV 275, ampero-metric, UV 254	[516]
pararosaniline base H_2N OH NH_2 NH_2	formaldehyde	Amine	Vis	[517], [518]
L-phenylalanine methyl ester NH_2 H $COOCH_3$	gossypol[2], chiral 30 ng/mL	Amine	UV 250, ODS	[519]
phenylhydrazine $NHNH_2$	glyoxylic acid, glycolic acid 500 nmol	Hydraz	UV 324, RP-18	[520]
semicarbazide O H_2N $NHNH_2$	pyridoxal[2] 2 ng/mL	Hydraz	fluorescence ex 365, em 480, C18	[1]
thiobarbituric acid O HN CH_2 O NH O	malon-dialdehyde in blood 100 nmol	Meth	fluorescence ex 532, em 553, C18	[521]

[1] Add = addition, Amine = condensation with amine, Cond = condensation with other groups, Cycl = formation of a new cycle, Hydraz = condensation with substituted hydrazine or hydrazide, Hydrox = condensation with hydroxylamine derivative, Meth = condensation with active methylene.

[2] For information and chemical structure regarding various drugs see e.g. M. O'Neil et al. (eds.), *The Merck Index*, Merck & Co., Inc., 13th ed., Whitehouse Station, 2001.

The application of each reagent indicated in Table 19.5.2 has specific conditions for obtaining the best results, these conditions depending on both the reagent and the analyte. For example, propafenone, an antiarrhythmic drug, and its metabolite 5-hydroxypropafenone in human plasma samples can be analyzed by derivatization with a solution of 2,4-dinitrophenylhydrazine in acetonitrile, in the presence of H_3PO_4. The initial compounds are hydrophilic and their separation is difficult on a reverse phase column. After derivatization, the hydrophobicity of the derivatives is increased and the two compounds can be separated on a C18 column. Also the presence of acetonitrile as a solvent and of H_3PO_4 used to catalyze the condensation reaction contribute to the plasma deproteinization, which takes place at the same time with the derivatization process. A chromatogram of a human plasma sample spiked with propafenone and 5-hydroxypropafenone derivatized with 2,4-dinitrophenylhydrazine is shown in Figure 19.5.1 [522]. The separation is done on a C18 Inertsil column using acetonitrile/ aqueous solution of 0.1% H_3PO_4 in the ratio 75 : 25 (v/v) as a mobile phase, T = 25°C, and detection at λ = 375 nm.

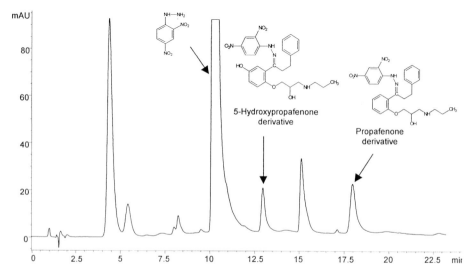

FIGURE 19.5.1. *Chromatogram of a human plasma sample spiked with propafenone and 5-hydroxypropafenone derivatized with 2,4-dinitrophenylhydrazine and separated on a C18 column.*

Some reagents are specific for the analysis of dicarbonylic compounds. Among these are several aromatic diamines that react with vicinal dicarbonylic compounds as shown below:

Besides o-phenylenediamine [522a], among the aromatic diamines reported for diketone derivatization are 1,2-diamino-4,5-dichlorobenzene [523], 2,3-diaminonaphthalene [1], 4,5-dimethyl-1,2-phenilenediamine [524], 4-ethoxy-1,2-phenilenediamine [525], and 4-methoxy-1,2-phenilenediamine [525]. For many resulting compounds the measurement can be done by fluorescence detection (ex. 375 nm, em. 475 nm) with excellent sensitivity.

Hydrazones can be used as reagents for 1,3-diketones in a reaction as shown below:

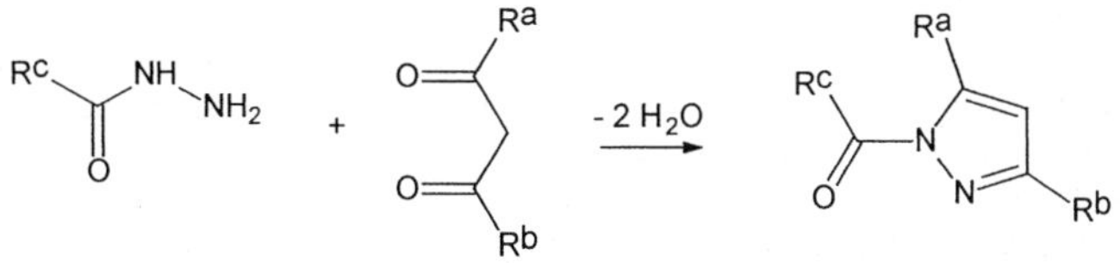

- Derivatization of the carbonyl group in carbohydrates

The free carbonyl form of mono- or disaccharides exists at very low concentrations in aqueous solutions. However, carbohydrates are reactive toward reagents with groups such as hydrazine or hydroxylamine, which give condensation reactions with the carbonyls. More carbonyl form can be generated from the equilibrium with the hemiacetal when the reaction product is highly stable. In addition, the derivatization of all possible groups is not necessary for HPLC analysis, because it is sufficient to attach one chromophore or fluorophore to the molecule of the analyte to achieve the desired improvement in detectability. The derivatization of carbohydrates involving both C=O and OH groups is further presented in Section 19.9, and specific chemical modifications for polymeric carbohydrates are discussed in Sections 20.1 and 20.7. A number of reagents used in the derivatization of carbohydrates with reaction at the carbonyl group are given in Table 19.5.3.

TABLE 19.5.3. *Reagents used for carbohydrates analysis by HPLC with reaction at the carbonyl group.*

Reagent		Analyte, matrix, sensitivity	Reaction type[1]	Detection λ Separation	Ref.
2-aminoacridone followed by reduction with NaBH$_3$CN	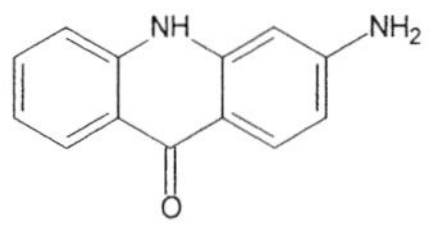	oligo-saccharides 50 pmol	Amine	fluorescence ex 405, em 525, amino	[1]
3-aminobenzamide followed by reduction with NaBH$_3$CN	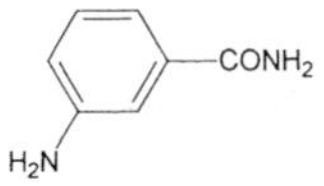	saccharides 4-80 µmol	Amine	CE, UV 285	[528]
4-aminobenzoic acid followed by reduction with NaBH$_3$CN	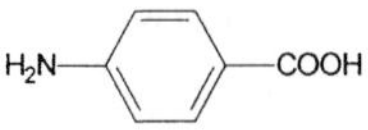	saccharides 4-80 µmol	Amine	CE, UV 285	[1], [529]
4-aminobenzonitrile followed by reduction with NaBH$_3$CN		saccharides	Amine	CE, UV 285	[1]
4-aminobenzoic hydrazide	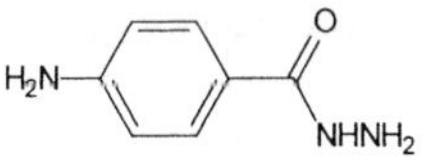	carbohydrates 20 ng	Hydraz	postcolumn UV 410, mediated ion Pb^{2+}	[530]

TABLE 19.5.3 (continued). *Reagents used for carbohydrates analysis by HPLC with reaction at the carbonyl group.*

Reagent		Analyte, matrix, sensitivity	Reaction type[1]	Detection λ Separation	Ref.
N-(4-aminobenzoyl)-L-glutamic acid followed by reduction with NaBH₃CN		saccharides	Amine	CE, UV 291	[531]
4-(2-aminoethyl)-phenol or tyramine followed by reduction with NaBH₃CN		trigalacturonic acid	Amine	UV 274, electrochem. pulsed amperom.	[532]
8-aminonaphthalene-1,3,6-trisulfonic acid followed by reduction with NaBH₃CN		oligo-saccharides 50 nmol	Amine	CE, UV 214, laser fluorescence ex 325, em 520	[1], [533]
aminopyrazine followed by reduction with NaBH₃CN		mono-saccharides	Amine	fluorescence ex 245, em 410, RP	[1]
9-aminopyrene-1,4,6-trisulfonic acid followed by reduction with NaBH₃CN		mono-saccharides	Amine	CE, laser fluorescence ex 488, em 520	[1]
2-aminopyridine followed by reduction with NaBH₃CN		mono-saccharides 10 pmol	Amine	CE, HPLC, fluorescence ex 320, em 400, ODS	[1]
6-aminoquinoline followed by reduction with NaBH₃CN		mono-saccharides 1 μmol	Amine	CE, UV 245	[1]
ammonia followed by reduction with NaBH₃CN and reaction with 3-(4-carboxy-benzoyl)-2-quinolin-carboxaldehyde		saccharides 0.5-2.3 amol (fluorescence), 50 pmol UV	Amine	CE, HPLC, laser fluorescence ex 457, em 552, UV 254, C18	[1]
aniline followed by reduction with NaBH₃CN		oligo-saccharides	Amine	UV 254	[1]

TABLE 19.5.3 (continued). *Reagents used for carbohydrates analysis by HPLC with reaction at the carbonyl group.*

Reagent		Analyte, matrix, sensitivity	Reaction type[1]	Detection λ Separation	Ref.
benzoic hydrazide or benzoyl hydrazine followed by reduction with NaBH₃CN		mono-saccharides 30-40 fmol	Hydraz	CE, UV 220	[534]
4-(biotinamido) phenylacetyl hydrazide		oligo-saccharides 330 fmol	Hydraz	UV 252, ODS	[535]
(R)-(+)-4-(2-carbazolyl-pyrrolidin-1-yl)-7-(N,N-dimethyl-aminosulfonyl)-2,1,3-benzoxadiazole		mono and oligo-saccharides chiral	Hydraz	fluorescence ex 450, em 540, ODS	[536]
2-cyanoacetamide		mono-saccharides, aldoses 0.1-1 nmol, etc.	Meth	fluorescence ex 331, em 383, UV 276, postcolumn anion exch.	[1]
dansyl hydrazine		carbohydrates, etc.	Hydraz	fluorescence ex 355, em 525, RP-18	[537]
2-(diethylamino)ethyl-4-aminobenzoate or procaine (HCl form) followed by reduction with NaBH₃CN		oligo-saccharides	Amine	UV 310, LC-MS, C18	[538], [539]
1,1-diphenylhydrazine		dicarbonyl sugars	Hydraz	CE, UV 220	[540]
ethyl 4-aminobenzoate followed by reduction with NaBH₃CN		oligo-saccharides 50 pmol UV 229	Amine	UV 229, 254 CE	[1]
9-fluorenylmethoxy-carbonyl hydrazide (FMOC-hydrazide)		xylose 0.1 pmol, oligo-saccharides, sugars 20-110 nmol	Hydraz	fluorescence ex 270, em 320, UV 254, ODS	[1], [541], [542]

TABLE 19.5.3 (continued). *Reagents used for carbohydrates analysis by HPLC with reaction at the carbonyl group.*

Reagent		Analyte, matrix, sensitivity	Reaction type[1]	Detection λ Separation	Ref.
hexyl-p-aminobenzoate followed by reduction with NaBH₃CN	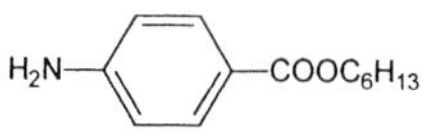 H_2N—⟨⟩—$COOC_6H_{13}$	oligo-saccharides	Amine	UV 304	[1]
hydroxylamine	H_2NOH	mono-saccharides 3 µg	Hydrox	UV 207, C18	[543]
1-(4-methoxy)-phenyl-3-methyl-5-pyrazolone	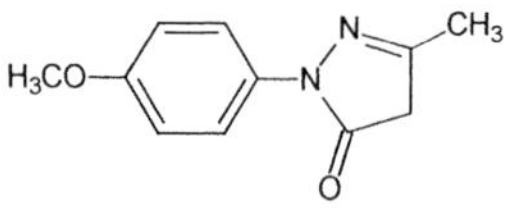 (structure) H_3CO— ...N–N...CH₃	oligo-saccharides	Cond	UV 245, ODS	[1], [544]
3-methyl-1-phenyl-5-pyrazolone	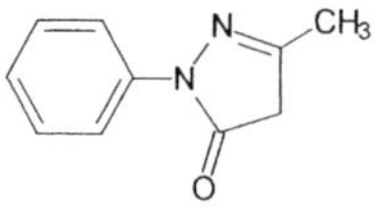 (structure) ...N–N...CH₃	oligo-saccharides	Cond	UV 245, UV 249, ODS	[545]
octylamine followed by reduction with NaBH₃CN	C_8H_{17}—NH_2	oligo-saccharides	Amine	UV 392, ODS	[1]
1-maltoheptaos-yl-1,5-diamino-naphthalene followed by reduction with NaBH₃CN	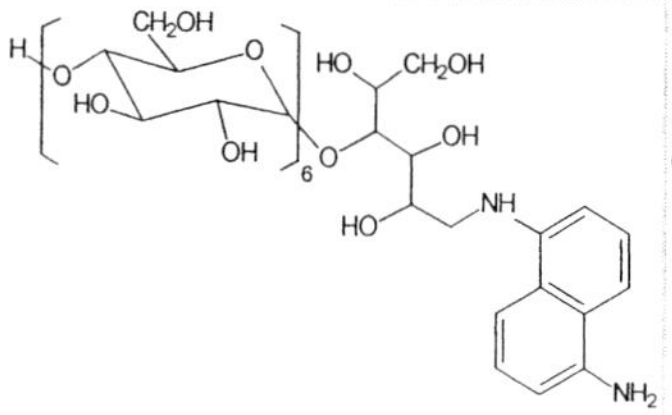 (structure)	heparin (separation of small oligomers)	Amine	CE, laser fluorescence ex 235, em 360	[546]

[1] Add = addition, Amine = condensation with amine, Cond = condensation with other groups, Cycl = formation of a new cycle, Hydraz = condensation with substituted hydrazine or hydrazide, Hydrox = condensation with hydroxylamine derivative, Meth = condensation with active methylene.

Common derivatizations of carbohydrates using only the carbonyl reactivity are those with 3-methyl-1-phenyl-pyrazolone (PMP), or 1-(p-methoxy)phenyl-3-methyl-5-pyrazolone (see Table 19.5.3). This reaction with PMP can be written as follows:

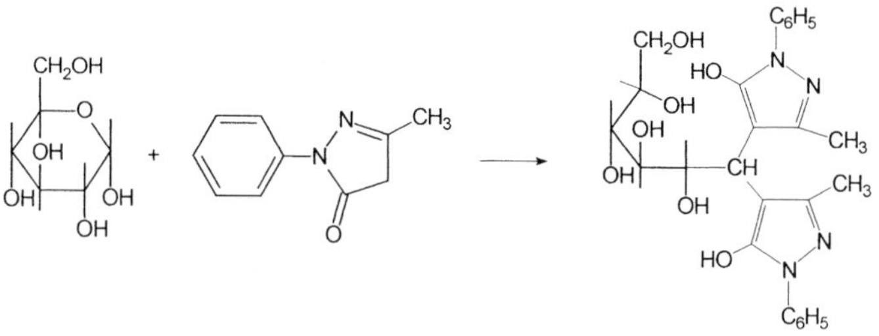

The derivatization with PMP can be used for HPLC analysis with UV detection at 249 nm, and as low as 500 fmol sugar can be analyzed by this procedure.

19.6 DERIVATIZATION OF CARBOXYLIC ACIDS

The hydrogen in COOH group is able to dissociate in water and form enough H_3O^+ ions to confer an acidic character to aqueous solutions of carboxylic compounds. The acidity constant K_a for the equilibrium $R-COOH + H_2O \rightleftharpoons RCOO^- + H_3O^+$ depends on the nature of the radical R, but in general carboxylic acids are weak acids, as shown for several acids in Table 19.6.1. Trichloroacetic acid and trifluoroacetic acid are among the strongest organic acids.

TABLE 19.6.1. *Several acidity constants for carboxylic acids.*

Acid	K_a (25° C)	pK_a (25° C)
$CH_3 COOH$	$1.75 \cdot 10^{-5}$	4.74
$CF_3 COOH$	0.59	0.23
$CCl_3 COOH$	0.23	0.64
$HOCH_2 COOH$	$1.5 \cdot 10^{-4}$	3.83
$C_2H_5 COOH$	$1.33 \cdot 10^{-5}$	4.87
$C_5H_{11} COOH$	$1.32 \cdot 10^{-5}$	4.88
$C_6H_5 COOH$	$6.3 \cdot 10^{-5}$	4.20

In addition to their acidity, the hydrogen in the carboxyl group has a notable tendency to form hydrogen bonds, the acids acting as very polar compounds. On the other hand, the C=O group in the carboxyl does not participate in most of the reactions characteristic for the carbonyl from aldehydes and ketones.

Classification of carboxylic acids can be based on the nature of the radical R in aliphatic and aromatic acids and on the number of COOH groups in mono-, di-, tri-, etc. carboxylic acids. Long chain aliphatic carboxylic acids are also known as fatty acids. Acids containing other groups in their molecules are common in nature and therefore frequently analyzed. Among these are hydroxy acids, amino acids, keto acids, etc. A number of derivatizations for these compounds are presented in this section if the only group involved in the reaction is the carboxyl. Derivatizations involving other groups or more than one group type are presented in other sections.

Carboxylic acids form various classes of derivatives such as esters (with glycerides as an important subclass), amides, nitriles, etc. The analysis of these compounds using derivatization is discussed in Section 19.7. One additional class of compounds related to carboxylic acids are the vinylogous acids, with the general formula R-C(O)-C=C-OH. A few derivatizations for vinylogous acids are discussed in this section.

Free monocarboxylic acids with relatively low molecular weight (below C_{20}) are in general stable and volatile enough for GC analysis. However, problems such as unsatisfactory peak shapes and losses in the chromatographic system are frequent for free acids. Dicarboxylic acids are even more polar and difficult to analyze by GC. For HPLC analysis, the addition of moieties for improving detectability frequently makes use of the reactivity of the carboxyl.

- *Derivatization of carboxylic acids for GC analysis*

The alkylation reaction of carboxylic acids is commonly used for eliminating the active hydrogen in the COOH group (see Section 18.1). The alkylation can be achieved as an esterification reaction with an alcohol such as methanol in the presence of a catalyst or using alkyl halides, diazomethane, etc. Methylation is more common than the substitution with other alkyl groups because the increase in the molecular weight of the acid is minimal, and the reduction in polarity is significant. A typical esterification with methanol starts with 1–10 mg acid sample, which is dissolved in 4 mL 0.5 N HCl in CH_3OH and 0.5 mL benzene. The HCl solution in CH_3OH must be practically anhydrous, and its preparation can be done by bubbling dry HCl gas into dry methanol. The HCl solution in CH_3OH can also be prepared by adding CH_3COCl in anhydrous methanol. However, this procedure generates methyl acetate in addition to HCl [15], [547]. The derivatization of the acids takes place at 80° C in 1 hour. The reaction mixture is then cooled at room temperature and treated with 10 mL water. The mixture is extracted three times with 3 mL ethyl ether. The combined ether extracts are dried over a mixture of anhydrous $Na_2SO_4/NaHCO_3$ (4:1 w/w), concentrated if needed, and analyzed by GC (see also Section 18.1). The reaction can be written as follows:

$$R\text{-COOH} + CH_3OH + H^+ \longrightarrow R\text{-COOCH}_3 + H_3O^+$$

Various other alkylation and some silylation reactions used in acid derivatization for GC analysis are shown in Table 19.6.2.

TABLE 19.6.2. *Derivatization reactions used for GC analysis of acids.*

Reagent	Catalyst/added compounds	Analyte	Reaction type[1]	Ester type	Ref.
aniline + thionyl chloride		C_1-C_8 carboxylic acids	Amide		[548], [549]
benzyl bromide	reaction in SPME fiber	acetic acid	Ester	benzyl	[550]
N,O-bis(trimethylsilyl)-acetamide or BSA + HMDS/TMCS	pyridine	fatty acids	Silyl	TMS	[15]
N,O-bis(trimethylsilyl)-trifluoroacetamide or BSTFA	1% TMCS in pyridine	fatty acids	Silyl	TMS	[551], [552]
p-bromobenzyl bromide	KOH	carboxylic acids (C_1 - C_3)	Alkyl	p-bromo-benzyl	[553]
(bromomethyl)dimethyl-chlorosilane	diethylamine	carboxylic acids herbicides metabolites	Silyl	bromomethyl dimethylsilyl	[554]
(bromomethyl)dimethyl-chlorosilane + NaI	diethylamine	carboxylic acids	Silyl	iodomethyl-dimethylsilyl	[555]
butanol	BF_3	short chain fatty acids	Ester	butyl	[556]
butanol	HCl	citric and nitrilotriacetic acid	Ester	butyl	[557], [558]
butanol	H_2SO_4	2,4,5-trichloro-phenoxyacetic acid	Ester	butyl	[559]
2-butoxyethanol	HCl	2,4-dichlorophenoxyacetic acid	Ester	2-butoxyethyl	[560]
tert-butyldimethylsilyl chloride	imidazole	fatty acids	Silyl	*tert*-butyl-dimethylsilyl	[561]
2-chloroethanol	BF_3	2-methyl-4-chloro-phenoxy acetic acid	Ester	2-chloroethyl	[15]

TABLE 19.6.2 (continued). *Derivatization reactions used for GC analysis of acids.*

Reagent	Catalyst/added compounds	Analyte	Reaction type[1]	Ester type	Ref.
2-chloroethanol	BCl$_3$	2,4-dichlorophenoxy acetic acid (acid herbicides)	Ester	2-chloroethyl	[15]
2-chloroethanol	dicyclohexyl-carbodiimide (DCC)	2,4-dichlorophenoxy acetic acid (acid herbicides)	Ester	2-chloroethyl	[15]
decyl alcohol	HCl	short chain fatty acids	Ester	decyl	[15]
diazomethane		fatty acids, dicarboxylic acids, hippuric acid	Alkyl	methyl	[15]
diazoethane		pesticide related acids such as 2,4-dichloro-phenylacetic acid, indomethacin[2]	Alkyl	ethyl	[562], [563]
2,4'-dibromo-acetophenone or 2-bromo-1-(4-bromo-phenyl)-ethanone	KOH	carboxylic acids	Alkyl	p-bromo-phenacyl	[564]
N, N'-dicyclohexyl-O-benzylisourea		fatty acids	Alkyl	benzyl	[565]
diethyl sulfate (C$_2$H$_5$O)$_2$SO$_2$	SPME derivatization	halogenated acids	Alkyl	methyl	[566]
dimethylformamide dialkyl acetal (alkyl = methyl, ethyl, propyl, butyl, *tert*-butyl)		fatty acids	Alkyl	methyl, ethyl, propyl, butyl, *tert*-butyl	[15]
dimethyl-α-hydroxy-methanephosphonate	dicyclohexyl-carbodiimide (DCC)	carboxylic acids	Ester	α-hydroxy-phosphonate	[567]
dimethyl sulfate (CH$_3$O)$_2$SO$_2$	K$_2$CO$_3$	probenecid[2] or 4-[(dipropyl-amino)-sulfonyl]benzoic acid halogenated acids	Alkyl	methyl	[566], [568], [569]
ethanol	HCl	Krebs cycle acids	Ester	ethyl	[15]
ethanol	H$_2$SO$_4$	haloacetic acids	Ester	ethyl	[570]
hexamethyldisilazane/tri-methylchlorosilane or HMDS/TMCS		fatty acids	Silyl	TMS	[15]
1-iodobutane	trimethylphenyl ammonium hydroxide or (CH$_3$)$_3$NC$_6$H$_5^+$OH$^-$	fatty acids	Alkyl	butyl (also methyl, ethyl, 2-butyl)	[571]
isopropanol	BF$_3$	carboxylic acids	Ester	isopropyl	[572]
methanol	dry HCl	fatty acids	Ester	methyl	[15]
methanol	HCl	C$_6$ - C$_{18}$ acids in tobacco	Ester	methyl	[573]
methanol	dry HCl	formic acid, acetic acid in cigarette smoke	Ester	methyl	[574]
methanol	dry HCl	free fatty acids in fats (collected first on Amberlite IRA 400 OH form)	Ester	methyl	[575]
methanol	Cu(OOCCH$_3$)$_2$ + dry HCl	fatty acids	Ester	methyl	[576]
methanol	H$_2$SO$_4$	haloacetic acids	Ester	methyl	[577]
methanol	BCl$_3$	fatty acids	Ester	methyl	[578]
methanol	BCl$_3$	cyclopropane fatty acids	Ester	methyl	[579]

TABLE 19.6.2 (continued). *Derivatization reactions used for GC analysis of acids.*

Reagent	Catalyst/added compounds	Analyte	Reaction type[1]	Ester type	Ref.
methanol	BF$_3$	fatty acids and lipids, 2,4-dichlorophenoxy-acetic acid; hydroxy acids, pyruvic acid.	Ester	methyl	[580], [581], [582]
methanol	N,N'-carbonyldi-imidazole (CDI) in triethylamine	fatty acids	Ester	methyl	[583]
methanol	dicyclohexylcarbo-diimide (DCC) / pyridine	carboxylic acids	Ester	methyl	[584], [585]
methanol	SOCl$_2$	amino acids, carboxylic acids, hydroxy acids	Ester	methyl	[586], [587]
N-methyl-N-(*tert*-butyldi-methylsilyl)trifluoro-acetamide (MTBSTFA)	imidazole	fatty acids	Silyl	*tert*-butyl-dimethylsilyl	[15]
methyl iodide	K$_2$CO$_3$	fatty acids	Alkyl	methyl	[588]
methyl iodide	AgNO$_3$ + KOH	fatty acids	Alkyl	methyl	[589]
pentafluorobenzyl bromide	K$_2$CO$_3$	carboxylic acids, flurbiprofen[2] or 2-(2-fluoro-4-biphenyl)-propionic acid	Alkyl	pentafluoro-benzyl	[15], [590]
pentafluorobenzyl bromide	(C$_4$H$_9$)$_4$N$^+$HSO$_4^-$	fatty acids,	Alkyl	pentafluoro-benzyl	[591]
pentafluorobenzyl bromide	(C$_5$H$_{11}$)$_4$N$^+$OH$^-$	carboxylic acids	Alkyl	pentafluoro-benzyl	[15]
pentafluorobenzyl bromide	(C$_6$H$_5$)$_4$N$^+$I$^-$ + NaOH	oxyphenonium bromide	Alkyl	pentafluoro-benzyl	[592]
phenyldiazomethane		fatty acids, trifluoroacetic acid	Alkyl	benzyl	[593], [594], [595], [596]
propanol	H$_3$PO$_4$ + BF$_3$	mono and dicarboxylic fatty acids	Ester	propyl	[15]
propanol	BCl$_3$	lactic and succinic acids	Ester	propyl	[15]
propyl iodide + NaHCO$_3$	(C$_4$H$_9$)$_4$N$^+$HSO$_4^-$ + NaHCO$_3$	indomethacin[2]	Alkyl	propyl	[597]
tetramethylammonium hydroxide	heat	carboxylic acids	Alkyl	methyl	[15]
2,2,2-trichloroethanol	trifluoroacetic anhydride + H$_2$SO$_4$	2,4-dichlorophenoxy acetic acid (acid herbicides)	Ester	2,2,2-tri-chloroethyl	[598], [599]
2,2,2-trichloroethanol	heptafluorobutyric anhydride + H$_2$SO$_4$	short chain aliphatic acids	Ester	2,2,2-tri-chloroethyl	[600]
(m-trifluoromethylphenyl)-trimethylammonium hydroxide	heat	fatty acids	Alkyl	methyl	[15]
trimethyloxonium tetrafluoroborate		hemins	Alkyl	methyl	[601]
trimethylphenyl-ammonium hydroxide	heat	carboxylic acids	Alkyl	methyl	[15]
N-trimethylsilylimidazole or TMSI		low molecular weight acids, hydroxy acids	Silyl	TMS	[15]

[1] Ester = esterification, Alkyl = other types of alkylation, Silyl = silylation, Amide = formation of an amide.
[2] For information and chemical structure regarding various drugs see e.g. M. O'Neil et al. (eds.), *The Merck Index*, Merck & Co., Inc., 13th ed., Whitehouse Station, 2001.

As seen in Table 19.6.2, diverse reagents and techniques have been developed to provide appropriate techniques for the analysis of various types of samples. Silylation is one of the most common derivatizations. The results for the analysis of long chain fatty acids using silylation with the formation of TBDMS derivatives are better than those with the formation of TMS derivatives that are less stable above $300°$ C, the temperature required for the GC analysis of compounds with relatively high MW [602].

Besides alkylation and silylation other derivatization can be used for GC analysis, such as formation of amides. An example is the use of 2-amino-N-[3,5-bis(trifluoromethyl)-phenyl]methyl]N-methyl-acetamide, which reacts with acids in the presence of 1-ethyl-3-(3-dimethylaminopropyl)-carbodiimide (EDAC) as shown below:

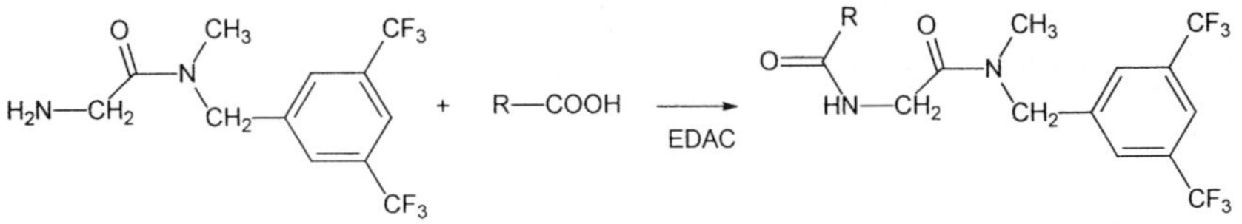

The reagent can be used with ECD or EI- MS detection [603].

Acid derivatization involving strong acids may affect unfavorably other groups that can be present besides COOH. For example, acids containing a cyclopropane ring, vicinal hydroxyls, oxirane oxygen, conjugated dienols or conjugated trienes may undergo secondary reactions in the presence of catalysts such as HCl or BF_3. Some methods using, for example, BCl_3 as a catalyst or CDI as an intermediate reagent (see Section 18.1) are milder and recommended in such cases. Another problem in the analysis of fatty acids is their derivatization in the presence of triglycerides without the triglyceride hydrolysis or transesterification. The determination of the level of free acids in fats is important in numerous practical applications. The transesterification is particularly favored by basic compounds such as KOH or CH_3ONa and may take place simultaneously with the esterification reaction as shown below:

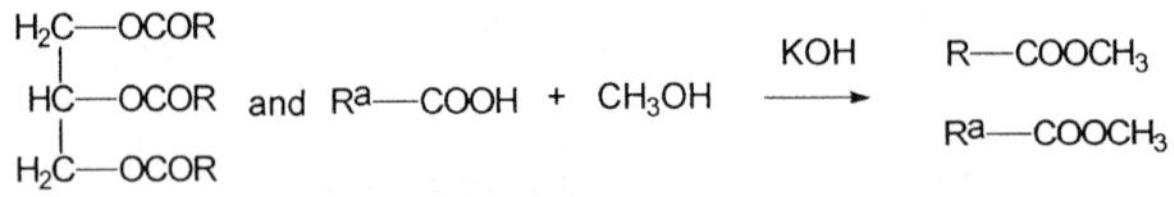

This type of reaction can be avoided by using appropriate conditions where triglycerides are not affected by the acid derivatization. Some procedures require a preliminary separation of the fatty acids from fats before derivatization (see also Section 19.7).

- *Derivatization of vinylogous acids*

Some relatively common compounds such as warfarin or bitter acids in hops are vinylogous acids. Due to the acidity of the hydrogen at the OH group, these compounds are derivatized with reagents typical for acids. For example, diazomethane has been used for the derivatization of acenocoumarol [15], pentafluorobenzyl bromide has been used for the derivatization of warfarin, etc. The reaction for warfarin derivatization can be written as follows:

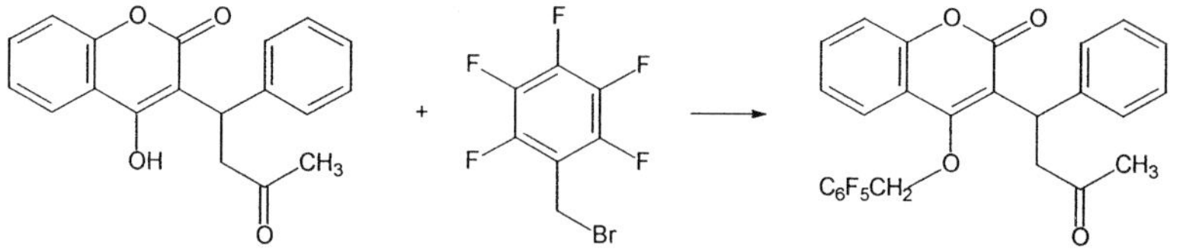

Silylation reaction works similarly to the silylation of acids (or alcohols), substituting the active hydrogen with a TMS group. A common silylation reagent used for derivatization of vinylogous acids is HMDS + TMCS. HMDS in 5,5-dimethoxypropane (DMP) has been used for the derivatization of bitter acids such as humulon or lupulon [604]. The reaction with lupulon (β-lupulic acid) takes place as follows:

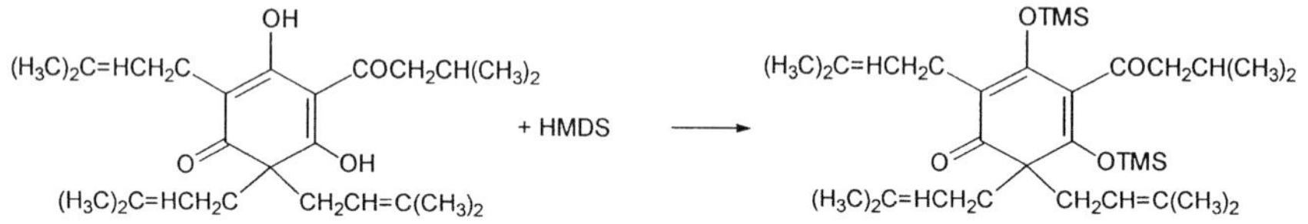

Vinylogous acids may also undergo reactions typical for alcohols, such as in the derivatization of 4-hydroxycoumarine with trifluoroacetamide, which takes place as follows:

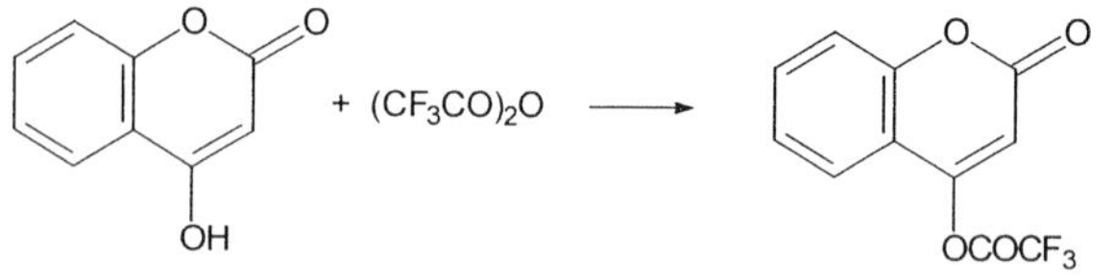

The derivatized product can be analyzed by GC [605].

- Derivatization of carboxylic acids for HPLC analysis

Similarly to other derivatizations for HPLC analysis, most reagents used for carboxylic acid derivatization attach chromophore or fluorophore groups to the acid molecule. Some reagents that have been used for this purpose are given in Table 19.6.3.

Specific conditions are applied for the reactions indicated in Table 19.6.3. In general, reaction equilibria are shifted in the desired direction using compounds such as 1-(3-dimethylaminopropyl)-3-ethylcarbodiimide (see Section 18.3) for the amides formation or dicyclohexylcarbodiimide (see Section 18.1) for the alkylation reactions with activated halides. Hydrazines behave in the reaction of formation of hydrazones similarly with amines, and 1-(3-dimethylaminopropyl)-3-ethylcarbodiimide is typically used to favor the reaction.

TABLE 19.6.3. *Reagents used for carboxylic acids analysis by HPLC.*

Reagent	Acid type, matrix, sensitivity	Reaction type[1]	Detection λ Separation	Ref.
N-(9-acridinyl)-bromoacetate	carboxylic acids 10 fmol	Alkyl	fluorescence ex 375.5, em 482 C18	[1]
L-alanine β-naphthylamide	Imidapril[2] chiral 0.05% of larger isomer acetyl carnitine	Amide	UV 254 silica	[606], [607]
4-aminomethyl-6,7-dimethoxycoumarin	fatty acids in blood 20-50 fmol	Amide	fluorescence ex 348, em 429 C18	[608]
1-aminoanthracene	carnitine[2] 0.25-5 μmol	Amide	fluorescence ex 248, em 418 C18	[609]
4-(2-aminoethylamino)-7-nitro-2,1,3-benzoxadiazole	carnitine[2] 10-100 fmol	Amide	fluorescence ex 485, em 540 ODS	[610]
(S)-(+)-5-(1-amino-ethyl)-2-(4-fluorophenyl)benz-oxazole or S-FLOPA	ibuprofen, chiral etc.	Amide	fluorescence ex 305, em 355 silica	[1]
(-)-2-[4-(1-aminoethyl)phenyl]-6-methoxybenzoxaole	ibuprofen[2] chiral 400 ng/mL (R), 200 ng/mL (S) carboxylic acids	Amide	fluorescence ex 320, em 380 ODS	[1]
1-aminonaphthalene	fatty acids in blood, 4 ng, ketoprofen[2], etc	Amide	UV 280 C18	[1], [611]
7-amino-1,3-naphthalene-disulfonic acid	acids of mono-saccharides herbicides	Amide	fluorescence ex 315, em 400 CE,	[1]

TABLE 19.6.3 (continued). *Reagents used for carboxylic acids analysis by HPLC.*

Reagent		Acid type, matrix, sensitivity	Reaction type[1]	Detection λ Separation	Ref.
9-aminophenanthrene		fatty acids 10-15 pM	Amide	fluorescence ex 303, em 376 C18	[612]
4-aminophenol		bile acids, 2 ng fatty acids, 0.5 ng prostaglandins	Amide	electrochem. + 0.75 V C18	[1]
2-(4-aminophenyl)-6-methylbenzothyazole in the presence of 1-(3-dimethylaminopropyl)-3-ethylcarbodiimide (EDAC)		lipoic and dihydrolipoic acid	Amide	fluorescence ex 343, em 423 C8	[613]
(S)-(+)-4-(3-aminopyrrolidin-1-yl)-7-aminosulfonyl-2,1,3-benzoxadiazole		naproxen[2] chiral 30 fmol	Amide	fluorescence ex 470, em 585 ODS	[614]
(S)-(+)-4-(3-aminopyrrolidin-1-yl)-7-(N,N-dimethylamino-sulfonyl-2,1,3-benzoxadiazole		carboxylic acids chiral 10 fmol	Amide	fluorescence ex 470, em 585 ODS	[614]
(R)-(-)- 4-(3-aminopyrrolidin-1-yl)-7-(N,N-dimethylamino-sulfonyl)-2,1,3-benzoxadiazole		carboxylic acids chiral 10 fmol	Amide	fluorescence ex 470, em 585 ODS	[614]
(R)-(-)- 4-(3-aminopyrrolidin-1-yl)-7-nitro-2,1,3-benzoxadiazole		carboxylic acids chiral 15 fmol	Amide	fluorescence ex 470, em 540 ODS	[614]

TABLE 19.6.3 (continued). *Reagents used for carboxylic acids analysis by HPLC.*

Reagent		Acid type, matrix, sensitivity	Reaction type[1]	Detection λ Separation	Ref.
(S)-(+)- 4-(3-aminopyrrolidin-1-yl)-7-nitro-2,1,3-benzoxadiazole		carboxylic acids chiral 15 fmol	Amide	fluorescence ex 470, em 540 ODS	[614]
4-(aminosulfonyl)-7-(2-aminoethylamino)-2,1,3-benzoxadiazole		fatty acids 23-50 fmol	Amide	fluorescence ex 429, em 573 ODS	[615]
4-(aminosulfonyl)-7-(5-aminopenthylamino)-2,1,3-benzoxadiazole		fatty acids 11-45 fmol	Amide	fluorescence ex 438, em 570 ODS	[615]
4-(aminosulfonyl)-7-(1-piperazinyl)-2,1,3-benzoxadiazole		fatty acids 10-20 fmol	Amide	fluorescence ex 440, em 580 ODS	[615]
p-anisidine		fatty acids prostaglandins 45-75 pg	Amide	UV 254, C18 electrochem. glassy carbon 1.1 V	[1]
2,3-(anthracene-dicarboximido)ethyl trifluoromethane-sulfonate		fatty acids 1.4-3.8 pmol, shellfish toxins 20 ng/g, okadaic acid	Alkyl	fluorescence ex 298, em 456 ODS	[616], [617]
9-anthracenemethanol		ibuprofen[2] 100 fmol fatty acids in blood 50 ng	Ester	fluorescence ex 365, em 415 RP-18	[1]
p-(9-anthroyloxy)-phenacyl bromide		prostaglandins 50 pg (fluorescence) 280 pg (UV)	Alkyl	fluorescence ex 249, em 413 UV 254 silica	[618], [619]

TABLE 19.6.3 (continued). *Reagents used for carboxylic acids analysis by HPLC.*

Reagent		Acid type, matrix, sensitivity	Reaction type[1]	Detection λ Separation	Ref.
9-anthryldiazomethane	CHN₂	quinapril[2] 5 ng/mL, carboxylic acids 50 pg, prostaglandins, etc.	Alkyl	fluorescence ex 360, em 440, UV 256 ODS	[1]
L-1-(1-anthryl)-ethylamine	H₃C, NH₂, H	naproxen[2] chiral 100 fmol	Amide	fluorescence ex 260, em 400 silica	[620]
benzylamine	NH₂	2-arylpropionic acids chiral	Amide	UV 230 chiral phase	[1]
benzylchloromethyl ether	O, Cl	carboxylic acids	Alkyl	UV	[15]
1-benzyl-3-p-tolyltriazene	H₃C—N=N-NH	stearic acid	Alkyl	UV 254 silica	[621]
2-bromo-2'-acetonaphthone	O, Br	valproic acid[2] 3.5 µg/mL, fatty acids 10 ng	Alkyl	UV 280, UV 254 RP-18	[1], [622]
2-bromoacetophenone	O, Br	valproic acid[2] 3.5 µg/mL, fatty acids 10 ng, lactic acid 36 pM	Alkyl	UV 245, UV 320, etc., C18	[1]
5-bromoacetyl acenaphthene	O, Br	ibuprofen[2] 2.5 pM	Alkyl	fluorescence ex 250, em 450 C18	[623]
N-(bromoacetyl)-N'-[5-(dimethylamino)-naphthalene-1-sulfonyl]-piperazine or dansyl-BAP	O, SO₂-N, N, Br, N(CH₃)₂	carboxylic acids 0.8-1 pM	Alkyl	fluorescence ex 246, em 490 RP-18	[624]

TABLE 19.6.3 (continued). *Reagents used for carboxylic acids analysis by HPLC.*

Reagent		Acid type, matrix, sensitivity	Reaction type[1]	Detection λ Separation	Ref.
2-bromoacetyl-6-methoxynaphthalene		bile acids 1-2 pM, enprostil free acid[2], azelaic acid[2]	Alkyl	fluorescence ex 300,em 460 C18, etc.	[1], [625]
1-(bromoacetyl)pyrene		bile acids 10 pM, polyether antibiotics 200 ng/mL	Alkyl	fluorescence ex 370, em 440 etc. C18	[626]
2-bromo-4'-bromoacetophenone in the presence of KOH and 16-crown-6		carboxylic acids	Alkyl	UV 254	[627]
2-bromo-4'-chloroacetophenone		fatty acids	Alkyl	UV 254 C18	[628]
4-bromomethyl-7-acetoxycoumarin		fatty acids 5 pM carboxylic acids 10 fmol	Alkyl	fluorescence ex 365, em 460 etc. RP-18	[629], [630]
9-bromomethylacridine		fatty acids in blood 300 nmol	Alkyl	fluorescence ex 362, em 418, UV 254 C18	[1]
4-(bromomethyl)-6,7-dimethoxycoumarin		hydroxylauric acid, 75 pg carboxylic acids	Alkyl	fluorescence ex 340, em 420, etc. ODS	[631], [632]
3-bromomethyl-6,7-dimethoxy-1-methyl-2(1H)-quinoxalinone		carboxylic acids 0.3-1 fmol, etc	Alkyl	fluorescence ex 370, em 450, C18	[633], [634]

TABLE 19.6.3 (continued). *Reagents used for carboxylic acids analysis by HPLC.*

Reagent		Acid type, matrix, sensitivity	Reaction type[1]	Detection λ Separation	Ref.
5-(bromomethyl)-fluorescein		palmitic acid, 6 nmol (conventional) 0.8 nmol (laser) etc.	Alkyl	fluorescence ex 488, em 520, C18	[635]
3-bromomethyl-7-methoxy-1,4-benzoxazin-2-one		glycine-conjugated bile acids, 20 fmol, carboxylic acids 2-10 fmol	Alkyl	fluorescence ex 355, em 430, etc. C8	[636], [637]
4-bromomethyl)-7-methoxycoumarin		bile acids 0.5 pM, atropine hydrolyzed to tropic acid 108 ng/mL, etc.	Alkyl	fluorescence C18, fluorescence ex 328, em 389 ODS	[638], [639]
4-bromomethyl-6,7-methylendioxycoumarin		loxoprofen[2] in blood, urine 10 ng/mL (plasma), 50 ng/mL (urine)	Alkyl	fluorescence ex 355, em 435 ODS	[1], [640]
3-bromomethyl-6,7-methylenedioxy-1-methyl-2(1H)-quinoxalinone		fatty acids 0.2-0.8 fmol	Alkyl	fluorescence ex 363, em 437 ODS	[1]
3-[4-(bromomethyl)-phenyl]7-(diethyl-amino)-2H-1-benzopyran-2-one		fatty acids 1.5 nmol	Alkyl	fluorescence ex 403, em 474 ODS	[1]
2-bromo-4'-nitroacetophenone		prostaglandins	Alkyl	UV 254 silica	[1]
4'-bromophenacyl trifluoromethane-sulfonate		butyrobetaine 1 pmol, carnitine, carboxylic acids 10 μmol	Alkyl	UV 254 ODS	[1], [641], [642]

TABLE 19.6.3 (continued). *Reagents used for carboxylic acids analysis by HPLC.*

Reagent	Acid type, matrix, sensitivity	Reaction type[1]	Detection λ Separation	Ref.
2-bromo-4'-phenylazo-acetophenone	fatty acids	Alkyl	UV 330	[643]
4-N-cadaverino-7-(N,N-dimethylaminosulfonyl)-2,1,3-benzoxadiazole or 4-(N,N-dimethyl-aminosulfonyl)-7-(5-aminopenthyl-amino)-2,1,3-benzoxadiazole	fatty acids	Amide	fluorescence ex 437, em 561 ODS	[644]
4-(2-carbazolyl-pyrrolidin-1-yl)-7-(N,N-dimethylamino-sulfonyl)-2,1,3-benzoxadiazole	fatty acids	Hydrz	fluorescence ex 450, em 550, ODS	[645]
9-(chloromethyl)-anthracene	carboxylic acids 0.4 ng/mL (fluorescence), etc.	Alkyl	UV 254 fluorescence ex 365, em 412 C18	[646]
1-chloromethylisatin	carboxylic acids 1-10 ng	Alkyl	UV 240 RP-8	[647]
N-(chloromethyl)-phthalimide	carboxylic acids	Alkyl	UV 254 RP-8	[648]
(S)-1-(4-dansylamino-phenyl)ethylamine	carboxylic acids chiral 170 fmol	Amide	fluorescence ex 338, em 535 ODS	[649]
dansyl ethanolamine	fatty acids	Ester	fluorescence ex 360, em 420 ODS	[650]

TABLE 19.6.3 (continued). *Reagents used for carboxylic acids analysis by HPLC.*

Reagent		Acid type, matrix, sensitivity	Reaction type[1]	Detection λ Separation	Ref.
dansyl semipiperazide		fatty acids in blood	Amide	fluorescence ex 350, em 530 ODS	[651]
9,10-diamino-phenanthrene		fatty acids	Cycl	fluorescence ex 255, em 382 silica	[652]
diazomethane	CH₂N₂	prostaglandins	Alkyl	LC/MS ODS	[653]
4-diazomethyl-7-methoxycoumarin		acids	Alkyl	fluorescence ex 325, em 385	[2]
2,4'-dibromoaceto-phenone		valproic acid[2] in blood, 5 μg/mL, fatty acids, prostaglandins	Alkyl	UV 254 etc.	[1]
1-(2,5-dihydroxyphenyl)-2-bromoethanone		bile acids 0.8-0.9 nmol carboxylic acids 1 pM	Alkyl	electro-chemical porous graphite +0.6V	[654], [655]
N,N'-diisopropyl-O-(p-nitrobenzyl)isourea		carboxylic acids in beverages, fatty acids 4 pmol, etc.	Alkyl	UV 265 RP-18 UV 254, silica	[1], [656]
2,4-dimethoxyaniline		prostaglandins 40-70 pg 1.2-2.2 ng (UV)	Amide	electrochem. glassy carbon 1.1V UV 249 C18	[1], [657]
4-(5,6-dimethoxy-2-benzimidazoyl)benzohydrazide		fatty acids 1-3 fmol	Hydrz	fluorescence ex 360, em 460, ODS	[1]

TABLE 19.6.3 (continued). *Reagents used for carboxylic acids analysis by HPLC.*

Reagent		Acid type, matrix, sensitivity	Reaction type[1]	Detection λ Separation	Ref.
5,6-dimethoxy-2-(4-hydrazincarbonyl-phenyl)benzothiazole		fatty acids in blood 1-2 fmol	Hydrz	fluorescence ex 365, em 447, ODS	[658]
6,7-dimethoxy-1-methyl-2(1H)-quinoxalinone-2-propionylcarboxylic acid hydrazide or DMEQ-hydrazide		fatty acids in blood 2-7 fmol, glucuronide conjugates, etc.	Hydrz	fluorescence ex 360, em 435, C8	[1], [659]
1-(5-dimethylamino-1-naphthalenesulfonyl)-(S)-3-aminopyrrolidine		carboxylic acids chiral 0.1 pM	Amide	fluorescence ex 340, em 530 ODS	[1]
(1R)-1-(dimethylamino-naphthalen-1-yl)-ethylamine		naproxen[2] chiral 100 pg	Amide	fluorescence ex 320, em 410 silica	[660]
(1S)-1-(dimethylamino-naphthalen-1-yl)-ethylamine		loxoprofen[2] chiral, 5 ng, N-acetyl amino acids, carboxylic acids, etc.	Amide	fluorescence ex 313, em 420 silica	[1]
4-(N,N-dimethylamino-sulfonyl)-7-(1-piperazinyl)-2,1,3-benzoxadiazole		carboxylic acids 3.9-14 fmol	Amide	fluorescence ex 313, em 420 silica	[644]
3,5-dimethylaniline		nonsteroid anti-inflammatory drugs such as carprofen[2], fenoprofen[2], etc. chiral	Amide	UV 254, UV 280, chiral phase	[661]
diphenylamine		2-(4-chloro-2-methylphenoxy)propanoic acid chiral	Amide	UV 260 chiral phase	[662]
2-ferrocenylethylamine		estrogen glucuronides 0.5 pmol	Amide	electrochem. +0.5 V vs. Ag/AgCl	[1]

TABLE 19.6.3 (continued). *Reagents used for carboxylic acids analysis by HPLC.*

Reagent		Acid type, matrix, sensitivity	Reaction type[1]	Detection λ Separation	Ref.
4-fluoro-7-nitro-2,1,3-benzoxadiazole		domoic acid 0.04 µg/g	Alkyl	fluorescence ex 470, em 530 C18	[663]
2-(5-hydrazincarbonyl-2-furyl)-5,6-dimethoxybenzothiazole		carboxylic acids in blood 50 fmol	Hydrz	fluorescence ex 363, em 452, C18	[664]
hydrazine	N_2H_4	fatty acids 50 ng	Hydrz	UV 229	[1]
2-(5-hydrazinocarbonyl-2-furyl)-5,6-methylenedioxy-benzofuran		prostaglandins in seminal fluid 0.1 pmol	Hydrz	fluorescence ex 362, em 462, ODS	[665]
2-(5-hydrazinocarbonyl-2-oxazolyl)-5,6-dimethoxybenzothiazole		carboxylic acids in blood	Hydrz	fluorescence ex 369, em 451, C18	[666]
2-(5-hydrazinocarbonyl-2-oxazolyl)-5,6-methylenedioxy-benzofuran		prostaglandins 0.1 pmol	Hydrz	fluorescence ex 350, em 450, ODS	[667]
2-(4-hydrazincarbonyl-phenyl)-4,5-diphenylimidazole		fatty acids in blood 7-57 fmol, carnitine[2] 0.24-1.97 µmol	Hydrz	fluorescence ex 335, em 455, ODS	[1], [668]
2-(5-hydrazinocarbonyl-2-thienyl)-5,6-methylenedioxy-benzofuran		prostaglandins 0.1 pmol	Hydrz	fluorescence ex 373, em 483, ODS	[665]
4-(N-hydrazinoformyl-methyl-N-methyl)-amino-7-(N,N-dimethylaminosulfonyl)-2,1,3-benzoxadiazole		carboxylic acids 3-9 fmol	Hydrz	fluorescence ex 440, em 550 ODS	[1]

TABLE 19.6.3 (continued). *Reagents used for carboxylic acids analysis by HPLC.*

Reagent		Acid type, matrix, sensitivity	Reaction type[1]	Detection λ Separation	Ref.
L-leucinamide		indobufen, indoprofen[2], ketoprofen[2], chiral, etc.	Amide	UV 275	[1]
L-leucine-(4-methyl-7-coumarinylamide)		carboxylic acids	Amide	fluorescence ex 330, em 390 C8	[1]
luminarin 4 or N-(4-aminobutyl)-2,3,6,7-tetrahydro-11-oxo-1H,5H,11H-[1]benzo-pyrano[6,7,8-ij]-quinolizine-9-acetamide with bis-(2,4,6-trichlorophenyl) oxalate and H_2O_2		carboxylic acids 50 fmol (chemilumin.) 300 fmol (fluorescence)	Amide	postcolumn chemilumin. 470 nm pass-filter fluorescence ex 390, em 470 ODS	[669]
methanol	CH₃OH	carboxylated poly(ethylene glycols) 0.11-0.28 ng/L	Ester	LC/MS	[1]
methanol for the acid followed by N-a-9-fluorenylmethyloxy-carbonyl-amino acid-N-carboxyanhydrides	CH₃OH	amino acids	Ester	fluorescence ex 263, em 313 silica	[297]
S-(-)-α-methylbenzyl-amine		ibuprofen[2] chiral, ketorolac[2], ketoprofen[2], 10 ng/mL, etc.	Amide	UV 216, UV 245 silica, ODS etc.	[1]
R-(+)-α-methylbenzyl-amine		ibuprofen[2] chiral, ketorolac[2], ketoprofen[2], 10 ng/mL, etc.	Amide	UV 216, UV 245 silica, ODS etc.	[1]
2-[p-(5,6-methylenedioxy-2H-benzotriazol-2-yl]-phenetylamine		ibuprofen[2] 1.5 pg	Amide	fluorescence ex 333, em 372 C18	[670]
(S)-(+)-1-methyl-2-(2,3-naphthalimido)ethyl trifluoromethane-sulfonate		carboxylic acids chiral	Alkyl	fluorescence ex 259, em 394 ODS	[1]

TABLE 19.6.3 (continued). *Reagents used for carboxylic acids analysis by HPLC.*

Reagent		Acid type, matrix, sensitivity	Reaction type[1]	Detection λ Separation	Ref.
(R)-(+)-α-methyl-4-nitrobenzylamine		3,7-dimethyl-6-actenoic acid chiral	Amide	UV 254 silica	[1]
(S)-(-)-α-methyl-4-nitrobenzylamine		2-methyl-hexanoic acid chiral	Amide	UV 254 silica	[1]
4-(1-methylphenanthro-[9,10-d]imidazol-2-yl)-benzohydrazide		carboxylic acids in blood 2.2-12.5 fmol 0.4-2.3 fmol (laser)	Hydrz	fluorescence ex 360, em 460, ex 325, em 460 laser ODS	[671]
monodansyl cadaverine		fatty acids	Amide	fluorescence ex 340, em 518 C18	[1]
1-naphthalene-methylamine		2-arylpropionic acids chiral	Amide	UV 230 chiral phase	[1]
2-(2-naphthoxy)ethyl-2-[1-(4-benzyl)piperazyl]-ethane-sulfonate		caproic acid 0.1 pmol	Alkyl	fluorescence ex 305, em 354 ODS	[672]
2-(2,3-naphthalimino)-ethyl trifluoromethane-sulfonate		carboxylic acids 4 fmol (fluoresc.) 100 fmol (UV)	Alkyl	fluorescence ex 259, em 394 UV 259 C8	[673]

TABLE 19.6.3 (continued). *Reagents used for carboxylic acids analysis by HPLC.*

Reagent		Acid type, matrix, sensitivity	Reaction type[1]	Detection λ Separation	Ref.
1-naphthyldiazo-methane	CHN₂ [structure]	bile acids 20-30 ng	Alkyl	UV 280 silica	[1]
(R)-(+)-1-(1-naphthyl)-ethylamine	H₃C, NH₂ [structure]	ibuprofen[2] loxoprofen[2] chiral 100 ng/mL	Amide	fluorescence ex 290, em 330 C18	[1]
N-(1-naphthyl)-ethylenediamine	HN—(CH₂)₂—NH₂ [structure]	acid chlorides obtained from acid + SOCl₂	Amide		[2]
(S)-(-)-1-(1-naphthyl)-ethylamine	H₃C, NH₂ [structure]	ibuprofen[2] chiral, 100 ng/mL, 2-arylpropionic acids, etc.	Amide	UV 232 fluorescence ex 280, em 320 C18	[1]
Nile Blue + 2-chloro-1-methylpyridinium chloride	[structure] H₂N, O, N⁺(C₂H₅)₂	phenylacetic acid 1 nmol	Amide	postcolumn fluorescence ex 635, em 650	[1]
4-nitrobenzylamine	[structure] NH₂ O₂N	ibuprofen in blood chiral	Amide	UV 235 chiral phase	[1]
1-p-nitrobenzyl-3-p-tolyltriazene	H₃C—[structure]—N=N-NH—[structure]—NO₂	carboxylic acids	Alkyl	UV	[15]
2-nitrophenylhydrazine	[structure] NHNH₂ NO₂	carboxylic acids in beverages 1-4 pmol, fatty acids in blood 200-400 fmol, etc.	Hydrz	Vis 400, UV 230 C8	[1], [674], [675], [676]
(S)-(+)-2-octanol	CH₃CH(OH)(CH₂)₅CH₃	ketorolac[2]	Ester	UV 325	[1]
pentafluorobenzyl bromide	[structure] F F F F F Br	retinoic acid 0.05 ng/mL, fatty acids	Alkyl	UV 369 C18, UV 254	[1], [677]

TABLE 19.6.3 (continued). *Reagents used for carboxylic acids analysis by HPLC.*

Reagent		Acid type, matrix, sensitivity	Reaction type[1]	Detection λ Separation	Ref.
2-(phthalimino)ethyl trifluoromethane-sulfonate	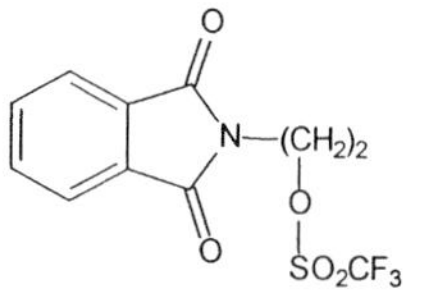 N—(CH₂)₂—O—SO₂CF₃ structure	fatty acids in tissue 200 fmol	Alkyl	UV 219 C8	[678]
N-(1-pyrenyl)bromo-acetamide	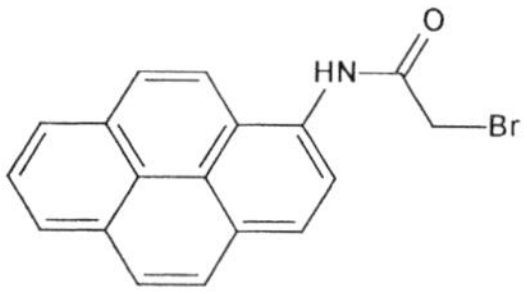 pyrene structure with HN—C(=O)—CH₂—Br	fatty acids	Alkyl	fluorescence ex 344, em 386.5 C18	[1]
1-pyrenyldiazomethane	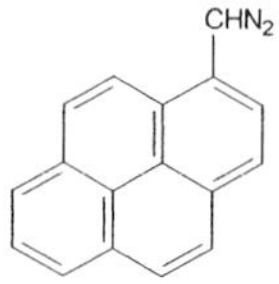 pyrene structure with CHN₂	biotin 100 fmol, S-carboxymethyl-L-cysteine and metabolites, carboxylic acids 20-30 fmol, thiodiglycolic acid	Alkyl	fluorescence ex 340, em 395, UV 240 silica, etc	[1]
5-(4-pyridyl)-2-thiophenemethanol	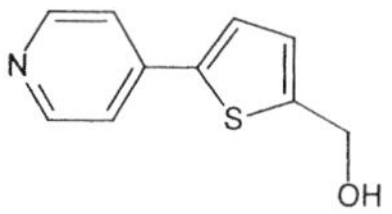 pyridyl-thiophene structure with OH	carboxylic acids 5-5800 fmol	Ester	fluorescence ex 300, em 360 ODS	[1]
sulfanilic acid	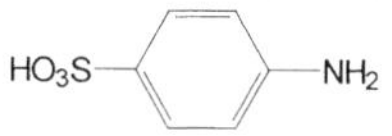 HO₃S—⟨ ⟩—NH₂	gangliosides	Amide	fluorescence ex 315, em 400 UV 247 CE	[679]
triphenylphosphine + p-methoxyaniline	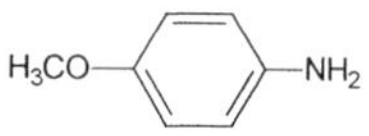 (C₆H₅)₃P + CCl₄ + H₃CO—⟨ ⟩—NH₂	fatty acids	Amide	UV 254	[680]

[1] Alkyl = alkylation with active halide, Ester = esterification, Amide = formation of an amide, Hydrz = reaction with a hydrazine.
[2] For information and chemical structure regarding various drugs see e.g. M. O'Neil et al. (eds.), *The Merck Index*, Merck & Co., Inc., 13th ed., Whitehouse Station, 2001.

Acids at very low levels are analyzed with specific reagents designed for laser-induced fluorescence (LIF). These reagents contain fluorophores similar to those used for amines but with different reactive groups, usually amino, as shown below [2], [681] [682]:

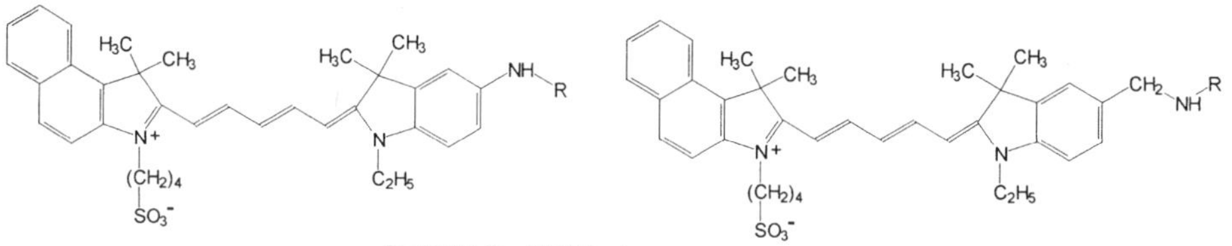

R = COCH₂Br, COCH₂I, etc.

- *Derivatization of unsaturated, or cyclopropane containing carboxylic acids for structural determination*

Several derivatization reactions of carboxylic acids containing double bonds or cyclopropane rings are performed with the purpose of the elucidation of the nature and position of these specific structural features (see also Sections 17.5 and 18.5). The presence of a double bond is generally detected from the mass spectra of the compound. However, its position is sometimes less obvious from the spectra, as shown in some examples in Section 17.5. Further assessment of the presence of the double bond and its position can be made by derivatization with the cleavage of the bond, for example, by strong oxidation. The oxidation products are commonly further derivatized. The reaction with $NaIO_4$ + $KMnO_4$ followed by further derivatization of the resulting acids can be used for the determination of the position of the double bond based on the length of formed fragments. A reaction continued with methylation with tetramethylammonium hydroxide in the GC injection port can be written as follows:

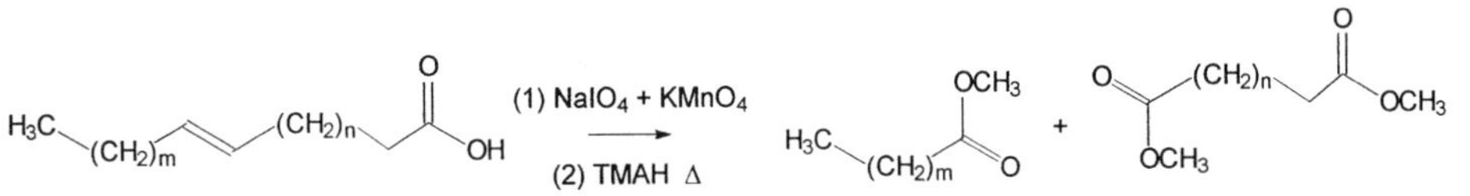

Similar result is obtained in a reaction with $NaIO_4$ + $KMnO_4$ followed by derivatization with CH_3OH + BF_3 [15], in a reaction with CrO_3 followed by derivatization with CH_3OH + H_2SO_4 [683], or using ozone, followed by derivatization with CH_3OH + BF_3.

An interesting combination of reduction and oxidation can be used for the determination of the position of double bonds in polyunsaturated fatty acids. In a first step, one of the double bonds is eliminated by reduction with hydrazine. The resulting mixture of acids is further oxidized with ozone and the ozonide hydrolyzed in reducing conditions. The sequence of reactions can be written as follows for linolic acid [684], [15]:

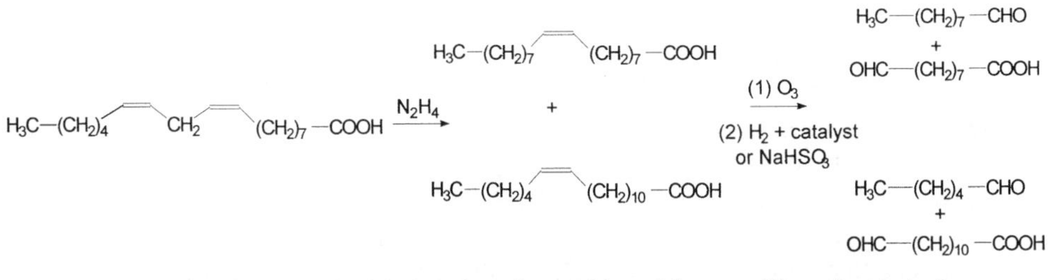

An oxidation reaction of the double bond performed in milder conditions leads to the formation of diols as shown in Section 18.6. The common reagents used for this oxidation are OsO_4, or $KMnO_4$ in alkaline solution, which form initially a cyclic ester. From this ester the diols are generated by hydrolysis with alcoholic solutions of Na_2SO_3 or H_2S. The diols can be further derivatized by methylation, silylation, or formation of acetonides. The sequence of reactions using oxidation with OsO_4 in pyridine/dioxane (1:8, v/v), hydrolysis with a suspension of Na_2SO_3 in water/methanol (8.5 mL 16% Na_2SO_3 in water and 2.5 mL methanol), and silylation with bis(trimethylsilyl) acetamide (BSA) is shown below [15] for a methyl ester of a monounsaturated acid:

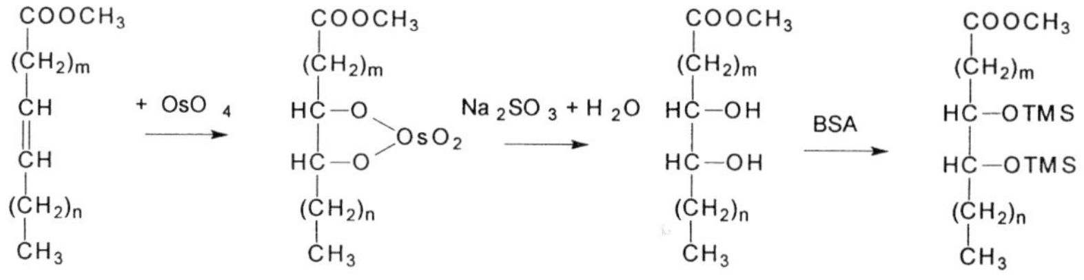

$$
\begin{array}{ccccc}
\begin{array}{c}COOCH_3\\|\\(CH_2)_m\\|\\CH\\\|\!|\\CH\\|\\(CH_2)_n\\|\\CH_3\end{array}
&\xrightarrow{+\ OsO_4}&
\begin{array}{c}COOCH_3\\|\\(CH_2)_m\\|\\HC\!-\!O\\\quad\quad\diagdown OsO_2\\HC\!-\!O\diagup\\|\\(CH_2)_n\\|\\CH_3\end{array}
&\xrightarrow{Na_2SO_3+H_2O}&
\begin{array}{c}COOCH_3\\|\\(CH_2)_m\\|\\HC\!-\!OH\\|\\HC\!-\!OH\\|\\(CH_2)_n\\|\\CH_3\end{array}
\xrightarrow{BSA}
\begin{array}{c}COOCH_3\\|\\(CH_2)_m\\|\\HC\!-\!OTMS\\|\\HC\!-\!OTMS\\|\\(CH_2)_n\\|\\CH_3\end{array}
\end{array}
$$

The GC-MS analysis of derivatized acids is usually diagnostic for the position of the double bond, as shown in Section 17.5.

Another reported reagent for the silylation of oxidized unsaturated acids or acid esters is HMDS/TMCS [685], [686]. Methylation can be done using CH_3I in the presence of methylsulfinyl carbanion solution [687], and the formation of isopropylidine derivative can be done with acetone/$CuSO_4$ [688]. The oxidation of the double bond also can be done with peroxyacids. The reaction generates an epoxide, which can be analyzed as is [689] or can be further isomerized in a ketone, for example, in the presence of NaI + CH_3I [15].

A different derivatization can be done with $Hg(OOCCH_3)_2$. As an example, dimethylene interrupted dienoic esters react with mercuric salts, and the intermediate adduct is further reduced with $NaBH_4$ as shown below:

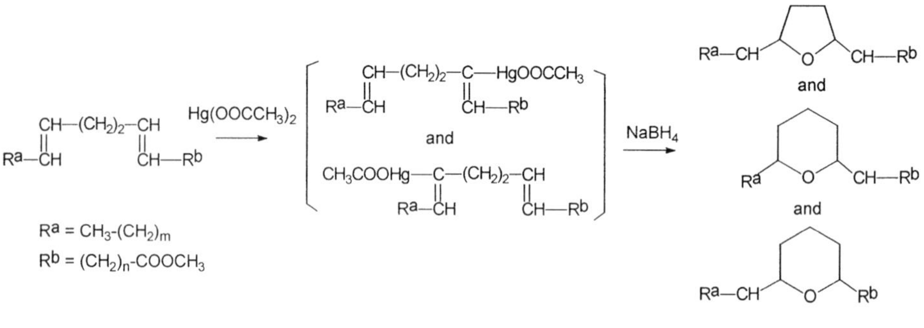

$$R^a = CH_3\text{-}(CH_2)_m$$
$$R^b = (CH_2)_n\text{-}COOCH_3$$

Fatty acids (or fatty acid esters) containing cyclopropane rings are commonly analyzed for the determination of the position of the ring using two types of reactions. In one type the ring is opened using reduction with hydrogen (Pt catalyst) or methoxylated with $CH_3OH + BF_3$ in a reaction as follows:

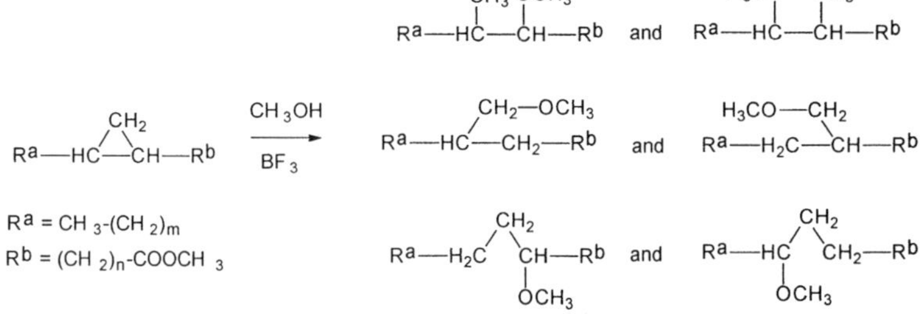

$$R^a = CH_3\text{-}(CH_2)_m$$
$$R^b = (CH_2)_n\text{-}COOCH_3$$

In the other type of derivatization, the carbon next to the cycle is oxidized with CrO_3 or with O_3 [15] as shown below:

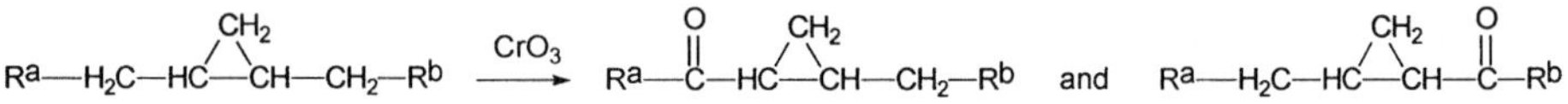

The GC-MS analysis of the derivatized products is usually diagnostic for the position of the cyclopropane ring. However, the analysis by this procedure of mixtures of acids may lead to difficulties in the interpretation of the results.

19.7 DERIVATIZATION OF AMIDES, ESTERS, AND OTHER DERIVATIVES OF CARBOXYLIC ACIDS, AND OF CARBONIC ACID

Organic acids form a number of functional derivatives. Some of these are common natural compounds such as the lipids, and others are used as drugs or pesticides. Table 19.7.1 lists several of these compounds.

TABLE 19.7.1. *Some of the derivatives of carboxylic acids.*

acyl halide	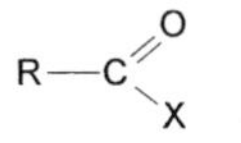 X = F, Cl, Br, I	dithionic acid	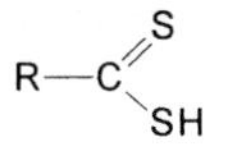	imino ether	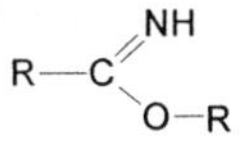
amide	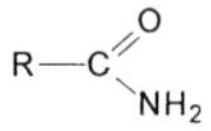	ester	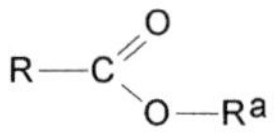	nitrile	R—C≡N
anhydride	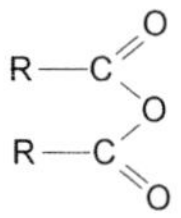	hydrazide	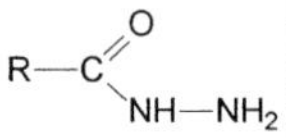	peracid	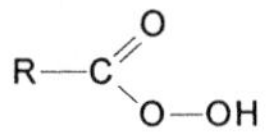
azide	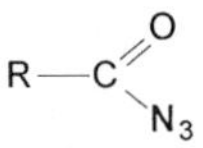	hydroxamic acid	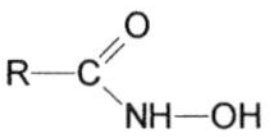	substituted amide	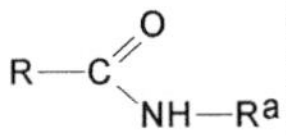

Derivatives of carbonic acid H_2CO_3, which can be seen as HO-COOH or $O{=}C(OH)_2$, are also encountered in practice quite frequently. A list with a few derivatives of carbonic acid is given in Table 19.7.2.

TABLE 19.7.2. *Some of the derivatives of carbonic acid.*

carbamate or urethane	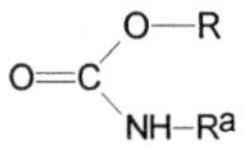	urea	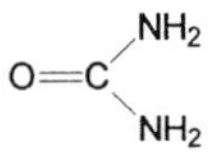	isocyanate	R—NCO
guanidine	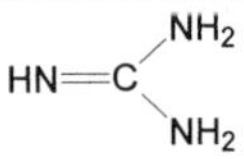	esters of carbonic acid	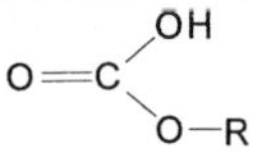	thiocyanate	R—SCN

Among the compounds from the classes given in Tables 19.7.1 and 19.7.2, probably the most common in nature are the amides and the esters. A special class of amides is generated from amino acids, which being bifunctional compounds can form polyamides and proteins. The analysis of polyamides and proteins using chemical modifications is discussed in Chapter 20. The esters common in nature are represented mainly by lipids.

The derivatization of some of the compounds indicated in Tables 19.7.1 and 19.7.2 can be performed using specific procedures, but a hydrolysis can be used first to free, for example, the COOH group. A second derivatization is commonly applied after hydrolysis. Hydrolysis as first derivatization can be used on anhydrides, esters, amides, nitriles, carbamates, etc. Specific derivatizations can be performed for compounds that have active hydrogens such as hydrazide, hydroxamic acids, urea, or guanidine. Also, peracids are frequently analyzed based on their oxidative properties. Some of the specific derivatizations are discussed in this section.

- Acyl halide derivatization

Acyl halides are reactive compounds that are commonly used as reagents for derivatization of other analytes. The derivatization of acyl halides as analytes can be done using a hydrolysis followed by subsequent derivatization of the acid. Some examples of direct derivatization of the acyl halide are also known. Phosgene can be derivatized using reagents such as 1-(2-pyridyl)piperazine [690] or tryptamine [691] followed by HPLC analysis. Solutions of acyl halides have been analyzed using derivatization on a polymer-immobilized 8-amino-2-naphthoxide [692], reaching a limit of detection of 0.84–1.5 ppb when using HPLC analysis with UV detection at 214 nm.

- Amide derivatization

Amides as analyte are frequently analyzed without derivatization. Many derivatizations of amines lead in fact to amides, which have a better behavior in the chromatographic column compared to amines (see Section 19.4). However, amide derivatization still may be needed. Although the polarity of the hydrogen in -C(O)-NH- group is not high, its elimination may still improve the GC separations. Also, addition of a chromophore or fluorophore may be needed for the analysis of traces of an amide.

For the GC analysis, monosubstituted amides can be derivatized with strong alkylating regents such as CH_2N_2, N,N-dimethylformamide dimethyl acetal (Methyl-8®), methyl iodide, benzyl iodide, etc. [15], [693]. The reaction with diazomethane, for example, takes place as follows:

By these procedures can be derivatized the amide group either present in the initial compound such as in barbiturates, diazepines etc., or generated after a previous derivatization by acylation such as in the case of heterocyclic amines (see also Section 19.4). As an example, derivatization of 5,5-diethylbarbituric acid can be done with pentafluorobenzyl bromide in triethylamine as shown below:

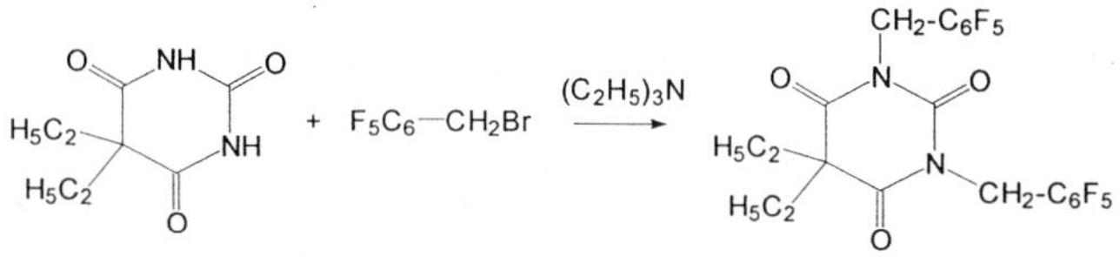

The analysis can be done using GC with ECD detection to measure pg levels of barbiturates [694].

Silylation can be applied to monosubstituted amides. For example, diphenylhydantoin and p-hydroxyphenylphenylhydantoin were derivatized using bis(trimethylsilyl)acetamide (BSA) leading to the complete silylation of the active hydrogens [15]. Various fatty acid amides such as oleamide, myristamide, and palmitamide were analyzed by GC-MS at levels between 1 pmol and 0.7 pmol using silylation with bis-trimethylsilyltrifluoro-acetamide (BSTFA) or with N-methyl-N-trimethylsilyltrifluoroacetamide (MSTFA). The reaction takes place as follows [695]:

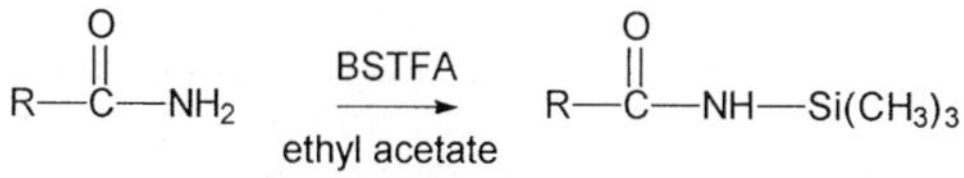

$$R\text{-}CO = C18{:}0,\ C18{:}1^{6},\ C18{:}1^{9},\ C18{:}1^{11},\ C18{:}2^{9,12},\ C18{:}3^{9,12,15}$$
(Ca:b^{c} , a = number of C, b = number of double bonds, c = double bond position)

Some amides can be derivatized by acylation with trifluoroacetic anhydride, as shown below for acetaminophen [218]:

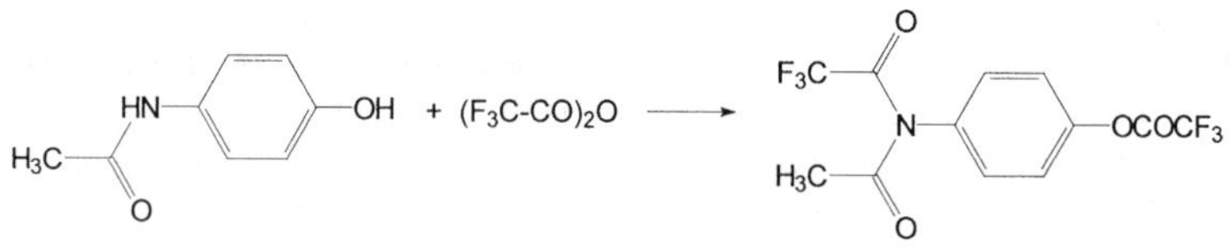

The reactivity of amides to the acylation depends on the nature of the amide and also on reaction conditions (see Section 18.3 and Table 18.3.2). The reactivity of certain amines to the second step acylation at the same nitrogen has been studied, and higher temperature or addition of compounds such as trimethylamine may increase the yield of acylation [217] [218].

For HPLC analysis, amides frequently are derivatized with the intention of improving detectability. The same types of reactions as for GC are used in these derivatizations, including alkylations with reactive alkyl halides, acylation or silylation. Also, hydrolysis followed by other derivatization has been applied [1], [696]. Some reagents used for amide derivatization for HPLC analysis are given in Table 19.7.3.

TABLE 19.7.3. *Reagents used for amide derivatization for HPLC analysis.*

Reagent	Amine type, matrix, sensitivity	Reaction type[1]	Detection λ Separation	Ref.
2-bromo-2'-acetonaphthone	barbiturates[2] 1 ng	Alkyl	UV 249, C18	[697]
3-bromomethyl-6,7-dimethoxy-1-methyl-2(1H)-quinoxalinone	5-fluorouracil, floxuridine[2] in serum 12.5-25 ng/mL	Alkyl	fluorescence ex 370, em 455 C18	[1]
4-bromomethyl)-7-methoxycoumarin	etho-suximide[2] 7 pmol	Alkyl	UV 320 C18	[698]
N-chloromethyl-4-nitrophthalimide	barbiturates[2] 4 ng	Alkyl	UV 254 RP8	[699]
9-fluorenylmethyl chloroformate	hexosamines etc., 4.5-20 pmol (UV), 0.4-3.8 pmol (fluorescence)	Acyl	fluorescence ex 270, em 320 UV 254 C8	[700]
4-nitrobenzoyl chloride	glucosyl-ceramide	Acyl	UV 254	[1]
dansyl chloride	barbiturates[2] in blood	Acyl	fluorescence ex 360, em 520 C18	[701]

[1] Acyl = acylation reaction including reactions with chloroformates or sulfonyl chlorides, Alkyl = alkylation or aryl derivative formation,
[2] For information and chemical structure regarding various drugs see e.g. M. O'Neil et al. (eds.), *The Merck Index*, Merck & Co., Inc., 13th ed., Whitehouse Station, 2001.

- Derivatization of lipids and of other esters

The need for derivatization of this class of compounds usually comes from the fact that many lipids have complex molecules and are not simple esters. Also, some of the lipids have relatively large molecules that can be analyzed by GC only using high temperatures. On the other hand, at high temperatures many lipids are not stable. In HPLC analysis, the lack of chromophores or fluorophores in the lipids may require derivatization. A number of derivatizations of lipids are performed for structure determination and not for quantitation.

Based on their chemical composition, lipids can be classified as: a) simple lipids, which include glycerol esters, cholesterol esters, waxes, hydrocarbons, some vitamins, etc., b) phosphoglycerides, which include phosphatidic acids, phosphate diesters with choline, ethanolamine, serine, etc., and phosphate triesters, c) sphingolipids, which include sphingomielins, cerebrosides, gangliosides, sulfatides, etc., d) glycolipids, e) peptidolipids, and f) peptidoglycolipids.

One important group of compounds that are classified as simple lipids consists of glycerol esters. These substances are very common in plants and animals. Glycerol esters include mono-, di-, or triglycerides with glycerol esterified by fatty acids. There are significant differences in the distribution of acids in fats between land animals, aquatic animals, microorganisms, plants, etc. and even between different species [702]. The acids comprise both saturated and unsaturated acids, usually with C_{14} to C_{20} carbon atoms. Among the most common acids are (Z)-9-octadecenoic acid (oleic acid), hexadecanoic acid (palmitic acid), octadecanoic acid (stearic acid), (Z)-9-hexadecenoic acid (palmitoleic acid), and (Z,Z)-9,12-octadecadienoic acid (linoleic acid), which are shown below:

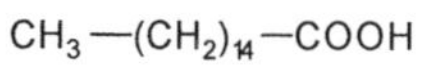

$CH_3—(CH_2)_{14}—COOH$

hexadecanoic acid

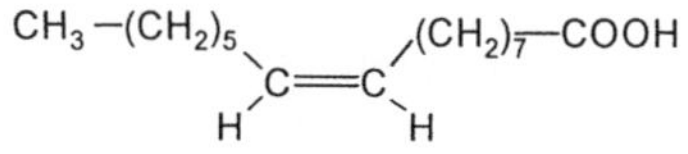

$CH_3—(CH_2)_5$ $(CH_2)_7—COOH$
$C=C$
H H

(Z)-9-hexadecenoic acid

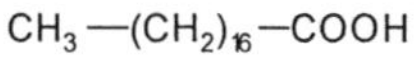

$CH_3—(CH_2)_{16}—COOH$

octadecanoic acid

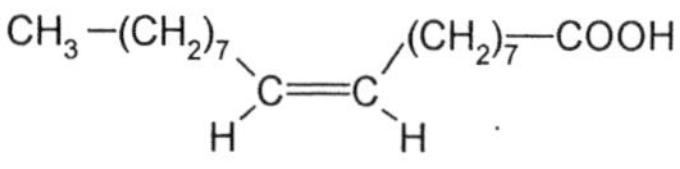

$CH_3—(CH_2)_7$ $(CH_2)_7—COOH$
$C=C$
H H .

(Z)-9-octadecenoic acid

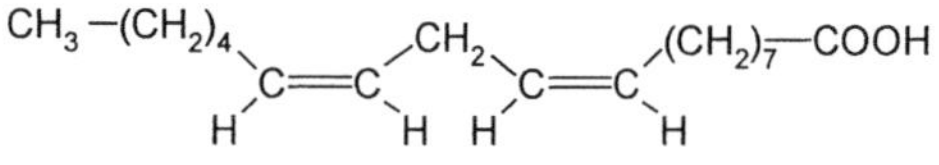

$CH_3—(CH_2)_4$ CH_2 $(CH_2)_7—COOH$
$C=C$ $C=C$
H H H H

(Z,Z)-9,12-octadecadienoic acid

Acids containing a wider range of carbon atoms between C_4 to C_{34} or higher are also found as glycerides. Besides simple acids, more complex ones such as branched and/or hydroxylated acids are common mainly in microorganisms. One group of such acids is, for example, that of mycolic acids:

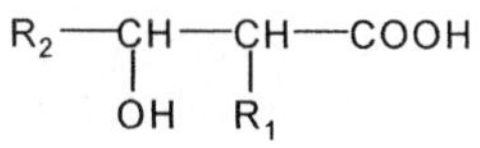

$R_2—CH—CH—COOH$
OH R_1

$R_1 = C_{20}$ to C_{24} $R_2 = C_{30}$ to C_{60}

mycolic acids

In the class of simple lipids are also included cholesterol, phytosterols and their esters with acids such as palmitic, stearic, or oleic. A number of derivatizations for cholesterol and other hydroxysteroids were discussed in Section 19.1. The waxes include a variety of esters of long chain aliphatic acids with long chain aliphatic alcohols, more frequently containing C_{12} to C_{34} carbon atoms.

Phosphoglycerides are fats in which glycerol is esterified with two fatty acids and one phosphoric acid (see also Section 19.8 for the derivatization of phosphate esters). Glycerol esters of phosphoric acid containing a free OH- glycerol group and only one fatty acid, as well as compounds with an ether linkage at one OH- glycerol group named plasmalogens are also present in nature. Phosphoric acid moiety can be further present as an ester. This esterification can be done with ethanolamine (in cephalins), choline (in lecithins), serine, inositol, etc., as shown below in a few examples:

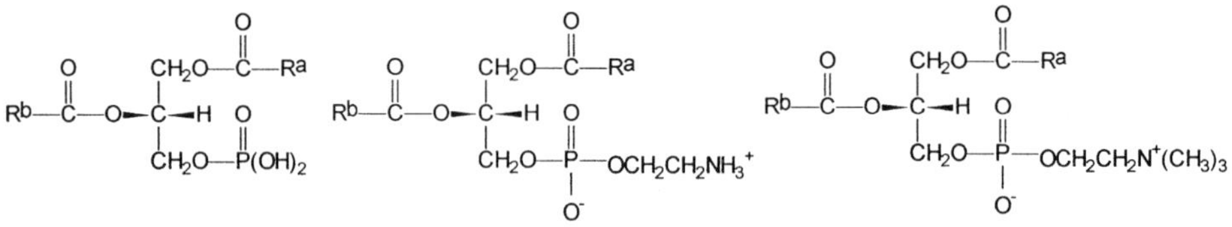

A phosphatidic acid
R^a, R^b = long aliphatic chains

Phosphatidyl ethanolamine

Phosphatidyl choline

Phosphoglycerides with the phosphoric acid moiety further esterified with inositol may generate more complex compounds where inositol is connected with other glycosyl residues.

Sphingolipids form a lipid class that does not contain glycerol. In this class are included sphingomielins, cerebrosides, gangliosides, and sulfatides. The compounds from this class can be considered as derived from sphingosine (shown below), or related substances such as dehydrosphingosine (sphinganine), or 4-hydroxysphinganine.

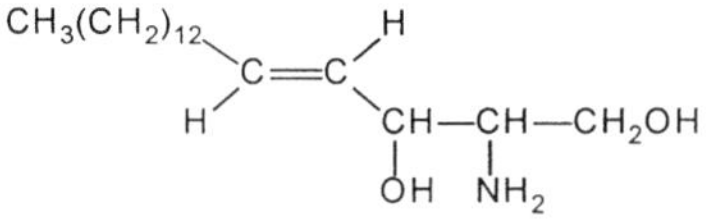

Sphingosine

Ceramides are derivatives of sphingosine where the amino group is acylated with a long chain acyl such as stearyl, palmityl, etc. Sphingomyelins are ceramide phosphatides where the primary alcohol group forms a phosphatide (primary alcohol is esterified with a phosphoric unit that is itself esterified with choline at the second site).

Cerebrosides (glycosylceramides) are formed from ceramides where the primary alcohol group forms an ether bond with a sugar residue. Di-, tri- and tetraglycosylceramides are found in mammalian tissues, mainly in the central nervous system. When the glycosylceramides contain residues of N-acetylneuraminic acid, the substances are considered to form a separate class known as gangliosides.

The length of the sugar chain may vary, and there are some gangliosides with more than one acetylneuraminic acid residue:

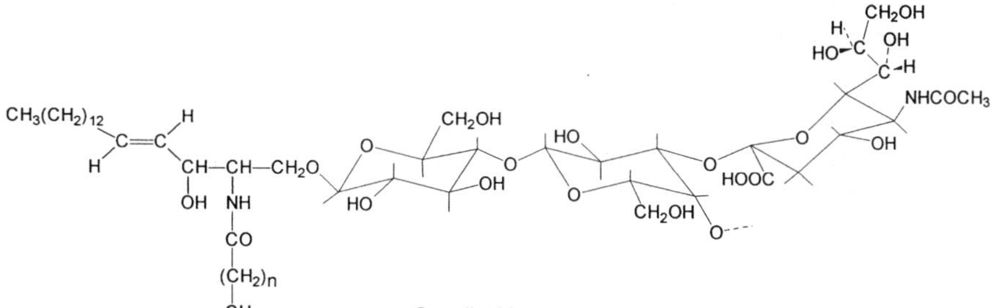

Ganglioside

Glycolipids can be simple glycosyl diacylglycerols, such as mono-D-galactopyranosyl diacylglycerols, where the common acyl groups are stearyl, oleyl, linoleyl, etc. More complex glycolipids are also found in biological samples. Most of them are analogous to other lipids described previously but having a sugar moiety attached through ester or ether bonds. Other glycolipids are carbohydrate esters of fatty acids (such as mycolic acids). Mycosides are part of this group, and they contain a branched aliphatic chain with hydroxyl groups esterified by fatty acids and terminated at one end by a phenol group to which the carbohydrate moiety is linked.

Peptidolipids are another type of lipid present in bioorganic materials. Hydrolysis of bacterial phospholipids frequently generates amino acids such as lysine, ornitine, and arginine. One proposed structure for this type of compound is shown below [703].

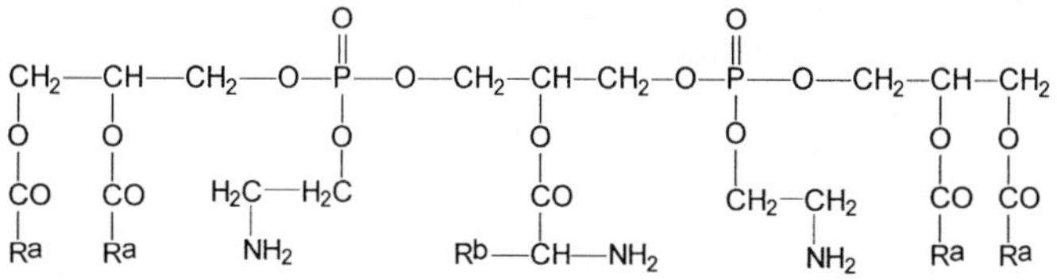

Ra = fatty acid residue   Rb = amino acid residue

As seen from the previous summary of lipid types, these substances cover a variety of molecular structures. The use of derivatization for the analysis of lipids has been done for both quantitation and also for structural determinations. Only a limited number of derivatizations for lipids are given in this book, a more detailed discussion of lipid analysis being available in dedicated literature [704], [705].

Glycerol esters can be derivatized to generate methyl esters for the fatty acids using a combination of hydrolysis/methylation by the following procedure: 200–500 mg lipid is boiled with 5 mL 0.5 N NaOH or KOH in methanol for 3–5 min. To this mixture is added 15 mL of an esterification solution, and the mixture is refluxed for 3 min. The esterification solution is prepared from 2 g NH_4Cl added in 60 mL methanol and 3 mL conc. H_2SO_4 that were refluxed together for 15 min. The esterified acids are transferred into a separation funnel containing 25 mL petroleum ether and 50 mL water. The water is discarded and the organic phase is washed twice with 25 mL water. The resulting organic phase can be concentrated, dried on Na_2SO_4, and analyzed by GC. The reactions can be written as follows [15]:

$$
\begin{array}{c}
CH_2-OCOR \\
| \\
CH-OCOR \\
| \\
CH_2-OCOR
\end{array}
\quad
\xrightarrow[\text{in } CH_3OH]{\text{NaOH}}
\quad
RCOONa
\quad
\xrightarrow[H_2SO_4 \,/\, NH_4Cl]{CH_3OH \,+}
\quad
RCOO-CH_3
$$

Methyl esters of the fatty acids in glycerides can also be obtained using as reagents methanol + $NaOCH_3$, methanol + $KOCH_3$ [15], methanol + KOH, methanol and H_2SO_4, [706], [707], methanol and BF_3, [708], [709], or methanol and HCl [710]. The analysis of the methyl esters of fatty acids from glycerides is commonly done by GC. Derivatization of a triglyceride with 2,2-dimethoxypropane (acetone dimethylacetal) and 10% HCl in methanol followed by the neutralization and drying of the mixture with $NaHCO_3/Na_2CO_3/Na_2SO_4$ (2:1:2 w/w/w) leads to the methyl ester of the acid and isopropylidene glycerol [711].

The analysis of fatty acids from triglycerides also can be done after hydrolysis using formation of *tert*-butyldimethylsilyl derivatives of the fatty acids. A typical procedure starts with 0.8 mg lipid that is hydrolyzed with 200 µL 1M KOH in ethanol using 1 hour heating at 75° C. An alternative hydrolysis uses 0.16 mg lipid and 200 µL 1M TMAH in methanol. The resulting material is dissolved in 500 µL water and washed with diethyl ether to remove nonsaponifiable components. The water solution is acidified with 6N HCl, and the fatty acids are extracted in hexane. The hexane extract is dried, redissolved in 200 µL hexane, and derivatized by adding 75 µL MTBSTFA and 5 µL triethylamine and heating the mixture at 75° C for 30 min. [712].

Another derivatization of the fatty acids from triglycerides for GC analysis has been done with dibutyl carbonate + sodium butoxide [15]. This procedure leads to the formation of butyl esters of the fatty acids, and probably the reaction mechanism is a transesterification because the free acids are not derivatized [713]. Butyl esters of the fatty acids in triglycerides can also be obtained in a reaction with NaOH in butanol, followed by reaction with butanol + BF_3 [714].

The differentiation by analysis between fatty acids and triglycerides is not always simple, because the triglycerides may hydrolyze during the derivatization process and generate free fatty acids, or because of the possibility of transesterification of the acids from triglycerides. Some procedures avoid this problem, such as by the use of tetramethyl-ammonium acetate (TMAAc) and thermal decomposition in the injection port of the GC. This reagent allows pyrolytic methylation for free fatty acids but does not affect esters such as 1-heneicosanyl oleate, colesteryl oleate, or glyceryl tripalmitate [715]. At the same time, TMAH generates methyl esters from both acids and acid esters of fatty acids in the GC injection port. An HPLC analysis of triacyl glycerols containing unsaturated acid residues allows a clear differentiation of the free acids from the triglyceride, and even differentiate positional isomer pairs. For this purpose, the triglyceride is derivatized with bromine, which is added to the analytes and allowed to react at room temperature for 30 min. The separation is further done using HPLC with FID detection [716].

Monoglyceryl esters can be derivatized using HMDS + TMCS in pyridine for the silylation of the free OH groups without the hydrolysis of the ester. The procedure allows GC-MS analysis of 1- and 2-isomers [717], [15]. By the same procedure are analyzed various monoalkyl glycerols (monoglyceryl ethers). Monoglyceryl ethers also can be methylated with CH_2N_2 in the presence of BH_3 [718]. Monoglycerides and monoglycerol ethers presenting two vicinal OH groups can be derivatized with methaneboronic acid [719], similar to the diols. Also, the acetonides are used for GC analysis of glyceryl ethers [720].

The derivatization of waxes is typically done when OH groups are present on the side chain of the alcohol (esters of diols or polyols) or of the acid. The use of N-methyl-N-*tert*-butyldimethylsilyltrifluoroacetamide (MTBSTFA) as a derivatization reagent was proven to give better results than the reagents forming TMS derivatives [721].

Sphingosine and ceramides can be derivatized in more than one step, using methaneboronic acid as the first derivatization step with the formation of cyclic boronates [722], [723]. The reaction for sphingosine can be written as follows:

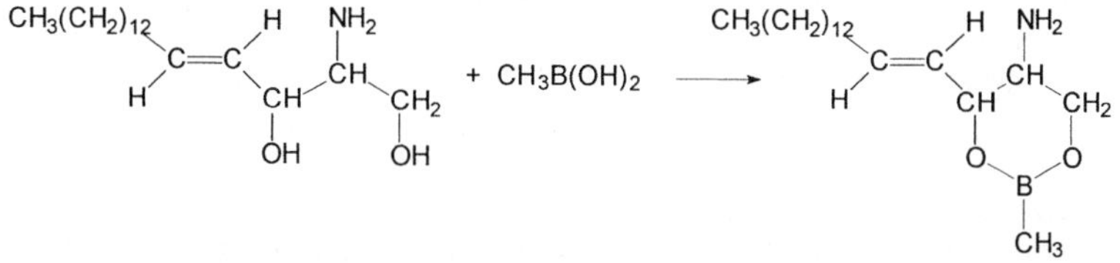
CH3(CH2)12
H
NH2
C=C
H
CH
CH
CH2
OH
OH
+ CH3B(OH)2
CH3(CH2)12
H
NH2
C=C
H
CH
CH
CH2
O
O
B
CH3

The NH_2 group in sphingosine can be further derivatized, for example by acylation. Ceramides can also be derivatized by silylation using HMDS + TMCS. The reaction takes place in pyridine at room temperature in 30 min., followed by evaporation to dryness. The residue can be dissolved in CS_2 and analyzed by GC-MS [15], [724].

Cerebrosides can be analyzed after silylation using direct probe MS [725]. The resulting penta TMS compound obtained from a monoglycosylceramide cannot be analyzed by GC, but a TLC separation followed by direct probe MS can be used for the separation/identification of the analytes. Either HPLC of intact molecules or GC associated with various degradation techniques for reducing the size of the molecule is used for analysis. Acylation of the OH groups with acetic anhydride in pyridine, followed by acylation of the NH group with p-nitrobenzoyl chloride is used as two-step derivatization for HPLC analysis of cerebrosides [726]. The reactions take place as shown below:

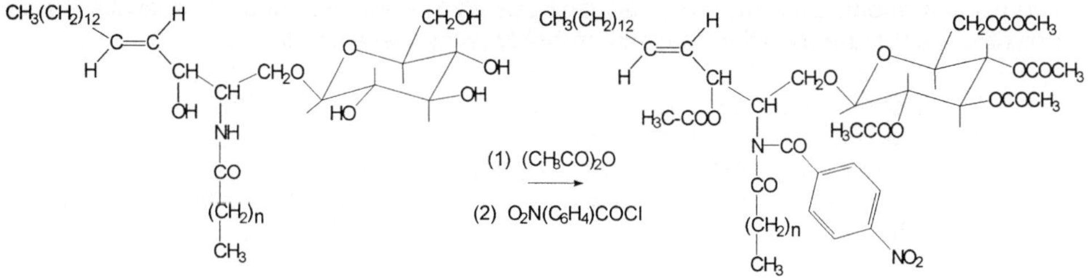
CH3(CH2)12
H
C=C
H
CH
CH2O
CH
OH
NH
CO
(CH2)n
CH3
CH2OH
O
OH
OH
HO
(1) (CH3CO)2O
(2) O2N(C6H4)COCl
CH3(CH2)12
H
C=C
H
CH
CH2O
CH
H3C-COO
N-CO
CO
(CH2)n
CH3
NO2
CH2OCOCH3
O
OCOCH3
OCOCH3
H3CCOO

The analysis of many other compounds classified as lipids is of considerable complexity. Usually the analysis is done following specific methods of hydrolysis and determination of resulting compounds.

- Derivatization of hydrazides

Hydrazides behave similarly to hydrazines, because an alkyl or an acyl radical bound to the NH end of NH-NH$_2$ does not influence significantly the reactivity of the NH$_2$ group. Carbonyl compounds are frequently used as derivatization reagents. For example, isoniazid reacts with an aldehyde as follows:

(isoniazid)

A few aldehydes used for hydrazide derivatization are given in Table 19.7.4.

TABLE 19.7.4. *Aldehydes used for hydrazides derivatization.*

Reagent		Analyte matrix, sensitivity	Detection λ Separation	Reference
cinnamaldehyde		Isoniazid[1] in blood 20 ng/mL	UV 340 C18	[727]
2-nitrobenzaldehyde		3-amino-2-oxazolidinone 5 ng/g	UV 275 C18	[728]
salicilaldehyde		hydrazides, isoniazid[1] 200 ng/mL	UV 320 C8	[729]

[1] For information and chemical structure regarding various drugs see e.g. M. O'Neil et al. (eds.), *The Merck Index*, Merck & Co., Inc., 13th ed., Whitehouse Station, 2001.

Maleic acid hydrazide, a compound that acts as a plant growth regulator, can be analyzed using reductive hydrolysis generating hydrazine, which can be further analyzed by several procedures [730]. A different procedure uses derivatization of maleic acid hydrazide with dimethylsulfate and GC analysis [731]. Silylation and GC analysis is also possible [732]. The silylation reaction probably takes place as follows:

As shown above, maleic acid hydrazide may also exist in enol form as 3,6-dihydroxypyridazine, but the silylated compound generates in GC-MS a unique peak with the mass spectrum given in Figure 19.7.1. This spectrum shows the fragment with m/z = 113, which is more likely to be generated from the hydrazide form, and not from the pyridazine form, which is more likely to generate the ion with m/z = 111.

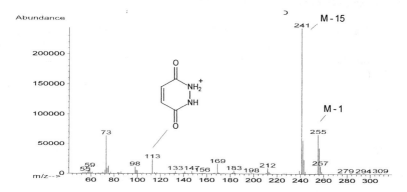

FIGURE 19.7.1. *The EI+ mass spectrum of the silylated maleic hydrazide.*

- Derivatization of hydroxamic acids

Hydroxamic acids have acidic properties (weak acids) such that they can be alkylated or silylated. They also may be acylated similarly to the alcohols of phenols. With strong mineral acids they hydrolyze in carboxylic acid and NH_2OH. Among the derivatizations used for the GC analysis of hydroxamic acids are the alkylation with propyl iodide in the presence of tetramethylammonium hydroxide [733], alkylation with CH_2N_2, silylation with BSA [734], etc. The reaction with BSA can be written as follows:

For the analysis of p-butoxyphenylacethydroxamic acid (bufexamac) using silylation, the compound is first extracted from plasma, and then brought to pH 8 with phosphate buffer using twice 16 mL methylisobutyl ketone. The extract is dried and evaporated, followed by addition of 40 μL BSA to the residue. The mixture is kept at room temperature over night and analyzed by GC.

Other derivatizations include reaction with acetic anhydride in pyridine [735] and reaction with trifluoroacetic anhydride (TFAA). The reaction with TFAA was used to analyze N-hydroxy-2-fluorenylacetamide [736].

- Derivatization of carbamates

Carbamates can be analyzed after hydrolysis or intact. The active hydrogen in the NH group can be acylated using for example acetyl chloride [737], acetic anhydride [738], trifluoroacetic anhydride [739], pentafluoropropionic anhydride [15], heptafluorobutyric anhydride (HFBA) [740], pentadecafluorooctanoic anhydride [741], or chloroacetic anhydride [15]. For example, the reaction of 2,3-dihydro-2,2-dimethyl-7-benzofuranol methylcarbamate (carbofuran) with HFBA can be written as follows:

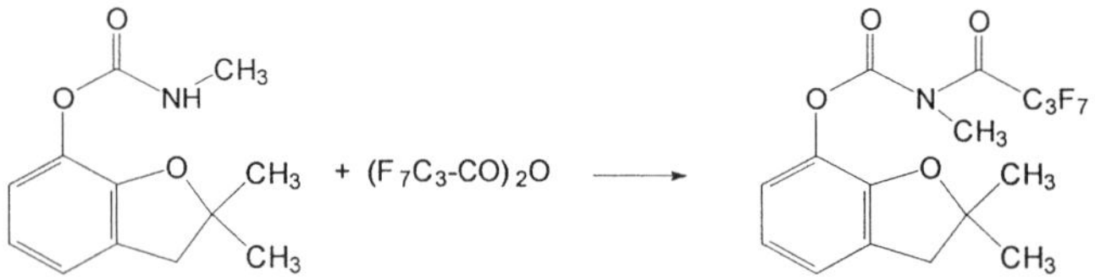

For the analysis, about 100 μg sample are dissolved in 5 mL benzene containing 4 drops of pyridine and 0.1 mL HFBA. The reaction takes place at room temperature over night. The material is washed three times with 5 mL water, dried over Na_2SO_4, and analyzed by GC or GC-MS.

Carbamates can be methylated using CH_3I and NaH or using trimethylphenylammonium hydroxide and formation of methyl derivative in the hot injection port of the GC [742]. Also they can be silylated using HMDS + TMCS [743].

For hydrolysis, carbamates can be treated with basic solutions such as NaOH or with acids such as H_2SO_4. Most carbamates are easier to hydrolyze with basic solutions. The reaction takes place as follows:

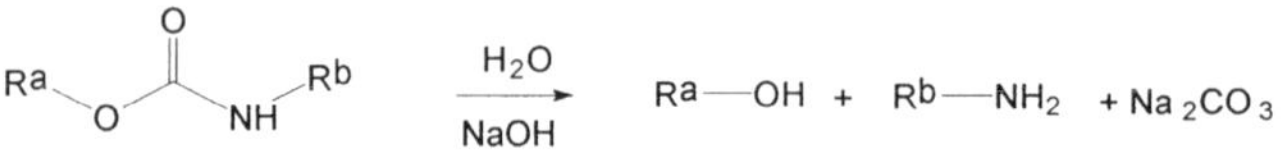

Depending on the carbamate, the alcohol or the amine are further derivatized and analyzed. For example, analysis of carbofuran and its metabolites in crop material starts by adding to the residue obtained from 100 g crop, 100 mL water, 2 mL 0.5 N KOH, and 1 mL 2,4-dinitrofluorobenzene (DNP) in acetone (1.5 mL/25 mL acetone). The mixture is homogenized for 20 min., then 10 mL of 5% aqueous $Na_2B_4O_7$ is added, and the solution is heated on a steam bath for 20 min. The solution is cooled and extracted with 5 mL isooctane. The organic phase can be concentrated or directly analyzed by GC. The reactions that take place can be written as follows:

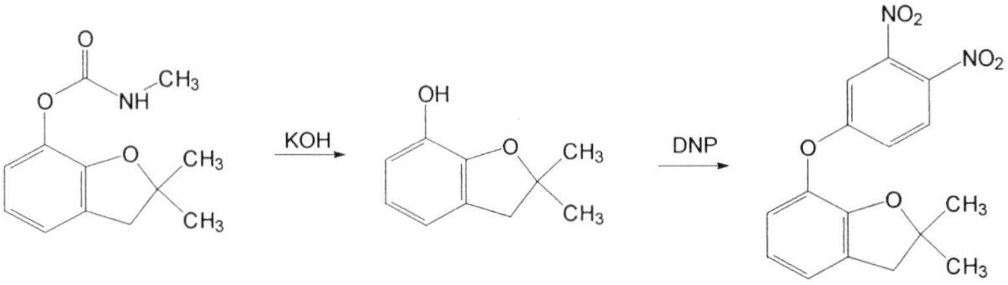

Other derivatizations after hydrolysis can be done with appropriate reagents for the OH or for the NH_2 group [744]. As most carbamates lead by hydrolysis to phenols, the derivatization can be done with 2,5-dichlorobenzenesulfonyl chloride, dimethylchloro-thiophosphate, or pentafluorobenzyl bromide for GC analysis, or with dansyl chloride [745] for HPLC analysis. Procedures using HPLC and postcolumn derivatization with o-phthalaldehyde + mercaptoethanol and fluorescence detection are also reported [746].

- Derivatization of urea and guanidine derivatives

Urea itself can be derivatized using TFAA when only one hydrogen is replaced with a trifluoroacetyl group. For the analysis of urea in plasma, 5 µL material is dried under N_2 at 70° C. To the residue 100 µL TFAA in CH_2Cl_2 (1:3, v/v) is added, and the mixture is sonicated for 1 min. and then allowed to stand at 25° C for 30 min. The solution is analyzed by GC [747].

Various pesticides derived from substituted ureas are analyzed using methylation of the reactive NH group or acylation. For example, phenylurea herbicides such as diuron can be methylated with CH_3I in the presence of potassium *tert*-butoxide and analyzed by GC. The reaction takes place as follows:

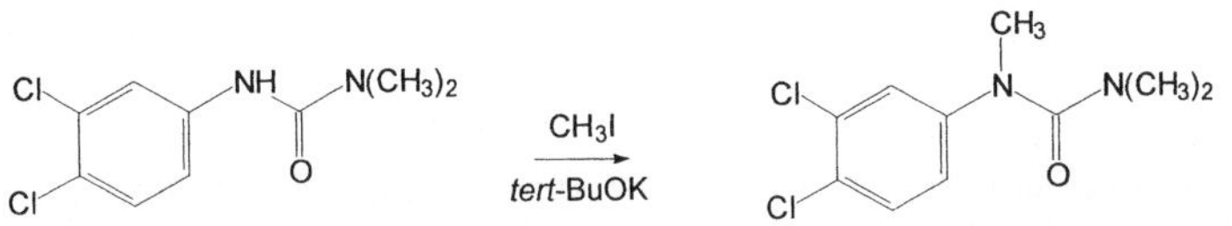

A number of other pesticides can be analyzed using similar procedures [748], [749], for example using CH_3I and NaH in DMSO, etc. Derivatizations with trimethylphenyl-ammonium hydroxide and heat are also used [750], and ethylenethiourea is analyzed after derivatization with 3-trifluoromethylbenzyl chloride. Substituted ureas can also be analyzed after derivatization using TFAA [749].

The hydrolyzed products of substituted ureas can be analyzed after various derivatizations of the amine fragment obtained from ureas. The hydrolysis can be done either in acid or basic medium, but acid hydrolysis is more common. The hydrolysis with HCl followed by derivatization with 2,4-dinitrofluorobenzene (DNFB) takes place as follows, for example, for 3-(4-chlorophenyl)-1-methoxy-1-methylurea (monolinuron) [751]:

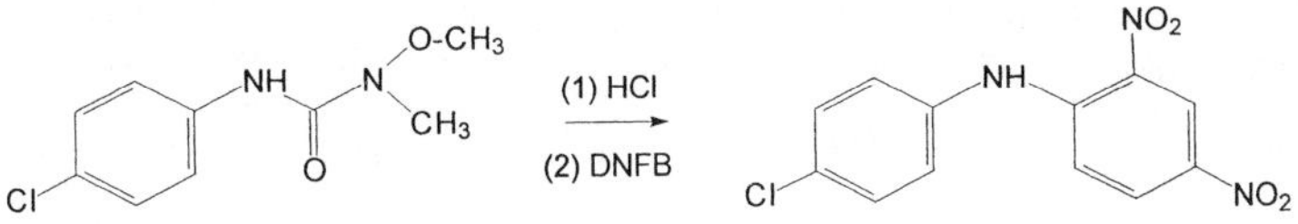

The derivatizations after hydrolysis can be done for HPLC analysis using dansyl chloride or other fluorogenic reagents [15].

Guanidine derivatives can be analyzed using hydrolysis [752], but the use of condensation reactions forming new cycles with hexafluoroacetylacetone (HFAA) has been used successfully for both GC and HPLC determinations. The condensation takes place as follows (see Section 18.5):

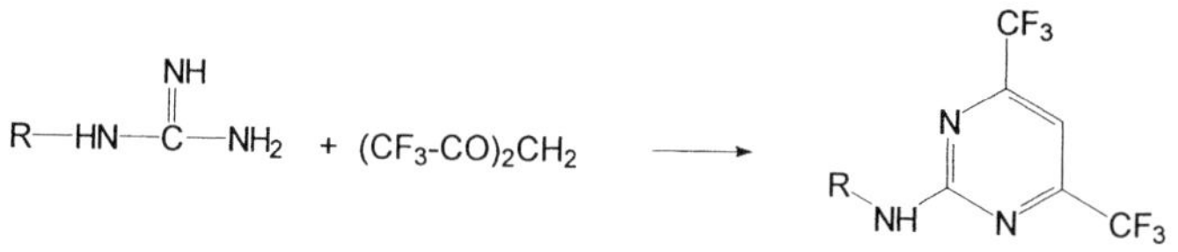

$$R-HN-\overset{\overset{\text{NH}}{\|}}{C}-NH_2 \; + \; (CF_3-CO)_2CH_2 \quad \longrightarrow \quad$$

This type of reaction is used for the analysis of the antihypertensive drug debrisoquin or 3,4-dihydro-2(1H)-isoquinolinecarboximidamide, agmatine, methylguanidine, etc. [753].

Another condensation reaction is used for the analysis of the antidiabetic drug phenformin or 1-phenethylbiguanidine. The biguanidine is condensed with trifluoroacetic anhydride (TFAA) generating a triazine in a reaction as follows:

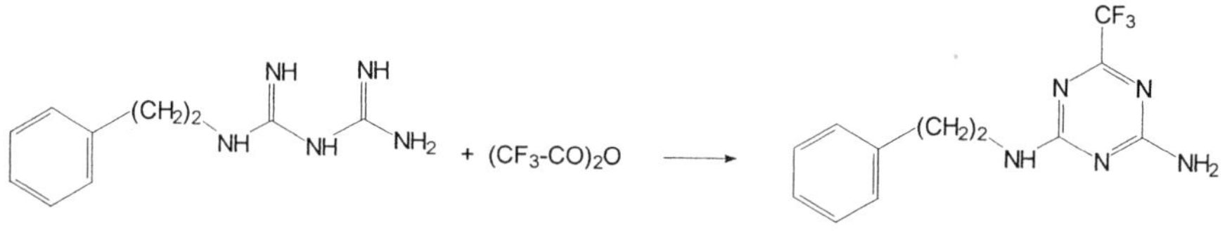

$$+ \; (CF_3-CO)_2O \quad \longrightarrow \quad$$

For the analysis of phenformin in plasma, 0.2–0.5 mL material is treated with 1 mL 10% trichloroacetic acid in 1N HCl for the precipitation of proteins. The solution is neutralized with 10 mL 10 N NaOH and extracted with CH_2Cl_2. The extract is then treated with 50 μL TFAA and heated at 50° C until CH_2Cl_2 evaporates. The resulting material is treated with 1 mL 1N NaOH and extracted with 25 μL pentyl acetate. The organic solution is then analyzed by GC-MS using CI in SIM mode [754]. A similar condensation takes place using p-nitrobenzoyl chloride with 1,1-dimethylbiguanide (metformin) [755]. The derivatization can be conducted in a two-phase system [756], with 1,1-dimethylbiguanide hydrochloride in aqueous phase and p-nitrobenzoyl chloride in the organic phase (dichloromethane). The derivatized metformin is extractable in CH_2Cl_2, while the excess of the unreacted p-nitrobenzoyl chloride is hydrolyzed and extracted into the aqueous phase. The reaction can be written as shown below:

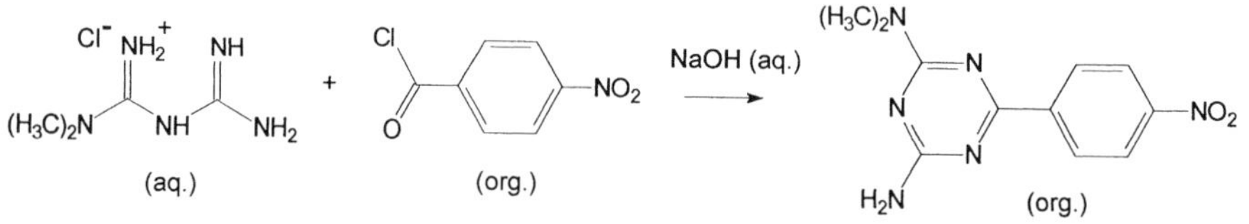

The organic phase can be evaporated and the derivatized metformin redissolved in methanol and analyzed by HPLC using a reverse phase column (ODS) and UV detection at 280 nm. The chromatogram of a derivatized plasma sample spiked with metformin and of a blank are shown in Figure 19.7.2.

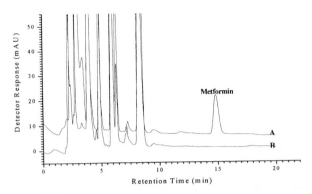

FIGURE 19.7.2. *The chromatogram of a derivatized plasma sample spiked with metformin (A) and of a blank (B) separated on an ODS-2 column, 250 mm x 4.6 mm i.d., 5 μm particle size, using a mobile phase H₂O/CH₃OH (65:35)*

Other compounds containing only one imino group may lead to simple acylations, such as in the derivatization of creatinine, which can be done in two step reaction, one with propylene oxide and the second with trifluoroacetic anhydride, as shown below:

For this analysis, the sample is treated with 0.5 mL 1,2-epoxypropane and heated at 70° C for 30 min. in a sealed tube. The resulting material is evaporated to dryness, and then 100 μL TFAA and 200 μL ethyl acetate are added. The sample is heated at 37° C for 30 min., evaporated, redissolved in 50 μL hexane, and analyzed by GC-MS in SIM mode [757].

Guanidine compounds also can be analyzed with HPLC using derivatization with reagents, generating fluorescent compounds. Among this type of reagent are 9,10-phenanthraquinone and benzoin. The reaction with phenanthraquinone takes place as follows:

The reaction product shows a strong fluorescence with excitation at 380 nm and emission at 510 nm [758].

- *Derivatization of isocyanates*

Isocyanates can be derivatized based on their reactivity toward amines. The reaction with benzylamine, for example, has been used for GC analysis of isocyanates. For the analysis, 1 mL solution 0.5–10 mM isocyanate is treated with excess (several times molar) of benzylamine in acetone and kept for 15 min. at room temperature. The reaction can be written as follows [15]:

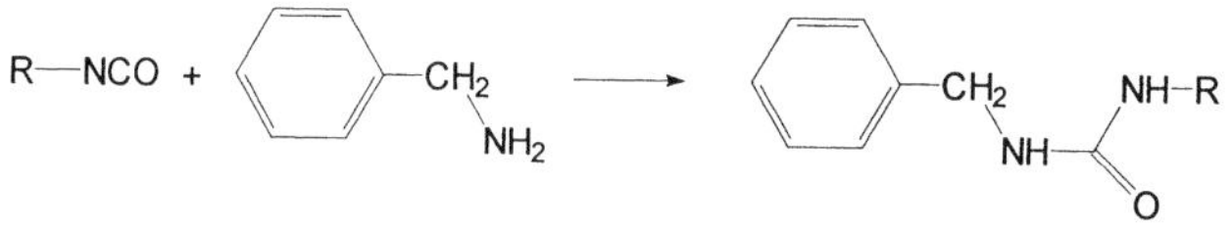

For HPLC analysis, groups carrying chromophores can be attached to the isocyanate. For example, N-(p-nitrobenzyl)-N-propylamine has been used to derivatize toluenediisocyanate and 4,4'-diphenylmethane diisocyanate from air at 5 $\mu g/m^3$ levels [759]. Also, 9-methylamino-methylanthracene has been used for HPLC analysis of isocyanates using UV or fluorescence detection [760].

19.8 DERIVATIZATION OF OTHER TYPES OF ORGANIC GROUPS AND OF ORGANOMETALLIC AND INORGANIC COMPOUNDS

Many organic groups were not captured in the series of derivatizations presented in this chapter. Among these are functional groups containing phosphorus, arsenic, boron or sulfur. The derivatization of dienes was presented only related to unsaturated acids (see Section 19.6). Also, organometallic compounds of mercury, tin, etc. and a number of inorganic compounds, which are derivatized for analysis using chromatography [761], were not yet discussed. Some selected derivatizations of these types of compounds are presented in this section.

- *Derivatization of dienes*

A number of compounds including some of biological interest such as vitamins from A series and D series, vitamin K_2, carotenes, etc. contain the diene group in their molecule. Dienes are able to participate in Diels-Alder cycloadditions with appropriate reagents that can be used for derivatization leading to the desired analytical properties. Especially reactive are the 4-substituted 1,2,4-triazolin-3,5-diones, a number of this type of reagent being indicated in Table 19.8.1.

TABLE 19.8.1. *Reagents used for diene derivatization for analysis by HPLC.*

Reagent		Analyte	Detection λ Separation	Reference
4-(1-anthryl)-1,2,4-triazolin-3,5-dione		7-dihydro-cholesterol 12 pmol (fluorescence) 0.06 pmol (UV)	UV 253, fluorescence ex 330, em 410 C8	[762]
4-[2-(6,7-dimethoxy-4-methyl-3-oxo-3,4-dihydroquinoxalinyl)-ethyl]-1,2,4 triazolin-3,5-dione		vitamin D metabolites	fluorescence ex 370, em 440 RP18	[1]
4-[4-(6-methoxybenz-oxazoyl)phenyl]-1,2,4 triazolin-3,5-dione		vitamin D₃ glucuronide, 7-dehydro-cholesterol 2 fmol, vitamin D	fluorescence ex 320, em 380 C8	[763]
pentafluorobenzyl-1,2,4 triazolin-3,5-dione		sterols 25-200 pg/mL	LC/MS	[764]
4-phenyl-1,2,4-triazolin-3,5-dione		7-dehydro-cholesterol	UV 265	[762]
4-[2-(1-pyrenyl)ethyl]-1,2,4-triazolin-3,5-dione		7-dehydro-cholesterol 0.4 pmol (UV) 0.025 pmol (fluorescence)	UV 245, fluorescence ex 342, em 397 C8	[762]
4-(1-pyrenyl)-1,2,4 triazolin-3,5-dione		7-dehydro-cholesterol 0.12 pmol (UV) 2.4 pmol (fluorescence)	UV 240, fluorescence ex 270, em 370 C8	[762]

- *Derivatization of phosphonate and phosphate esters*

Pentavalent phosphorus forms two main series of organic compounds with active hydrogen. These are the organic phosphonates R-P(=O)(OH)$_2$ and organic phosphates R-O-P(=O)(OH)$_2$ where R can be alkyl or aryl. Phosphonic and phosphoric acids are relatively strong acids (for methyl phosphate, pK$_a$ = 1.54). The derivatization of these compounds is usually done with the purpose of replacing the active hydrogens with alkyl, aryl, or silyl groups. One typical derivatization uses diazomethane for methylation. For example, 2-aminoethylphosphonic acid is derivatized with CH$_2$N$_2$. The reaction is done in two steps, one targeting the derivatization of the amino group using trifluoroacetic anhydride (TFAA), and the second using diazomethane to derivatize the phosphonic acid. The second step of this reaction can be written as follows [765]:

(generated from first step derivatization
of 2-aminoethyl phosphonic acid with TFAA)

For the derivatization of other aminoalkylphosphonic acids, a similar combination, TFAA and diazobutane, has been used [766].

Active alkyl halides react well with organic phosphates and phosphonates that can be alkylated in a reaction as follows:

$$R^a\text{-P}(=O)(OH)_2 + R^b\text{-X} \longrightarrow R^a\text{-P}(=O)(OR^b)_2$$

For HPLC analysis with fluorescence detection, various alkylphosphonic acids have been derivatized using p-(9-anthroyloxy)phenacyl bromide [767]. The procedure allowed the analysis of as low as 20 fmol acid.

Methylation for GC analysis can be done using trimethyl orthoacetate CH$_3$-C(OCH$_3$)$_3$. Various pesticides with amino and phosphoric groups are analyzed using this methylation procedure for the phosphoric groups and acetylation for the amino groups [768].

Silylation is probably the derivatization of choice for phosphonic and phosphoric acid derivatives. The reagents used for these derivatizations include hexamethyldisilazane (HMDS) + chlorotrimethylsilane (TMCS) [769], N,O-bis(trimethylsilyl)trifluoroacetamide (BSTFA), BSTFA + TMCS [770], N-trimethylsilylimidazole (TMSI), BSTFA + N,O-bis(trimethylsilyl)acetamide (BSA) [771], [15], etc. For example, pyridoxine-5-phosphate can be derivatized by treating the sample with HMDS + TMCS in pyridine (2:1:10, v/v/v) at room temperature [15]. The reactions can be written as follows:

The mixture BSTFA/TMCS/pyridine (10:2:5, v/v/v) has been used to derivatize phosphoryl and phosphono ethanolamine. Aminoalkylphosphonic acids are derivatized with BSTFA for the active hydrogens at the phosphonic group and with CS_2 for the amino group that generates an isocyanate [772].

Silylation has been used successfully for the derivatization of various phospholipids and steroid phosphates. The derivatization can be performed on the lipid mixture as is or after specific hydrolysis. Typically, the hydrolysis is performed for the ester group toward the fatty acid. Fatty acids can be analyzed separately (see Section 19.6). The analysis of the remaining groups can be done using silylation. For example, derivatization of phosphatidyl ethanolamine, phosphatidyl choline, phosphatidyl serine, and phosphatidyl inositol can be done using first a hydrolysis with NaOH, followed by the silylation. The reactions can be written for phosphatidyl ethanolamine as follows (R^a and R^b long aliphatic chains):

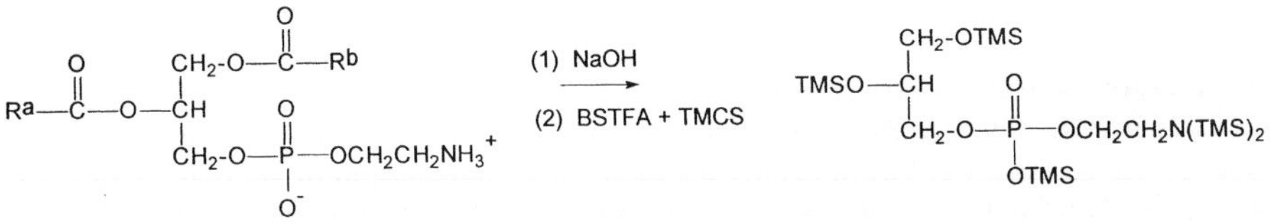

The procedure starts with 10–15 μmol glycerophospholipid that is dissolved in 1 mL $CHCl_3/CH_3OH$ (1:4, v/v). The solution is treated with 0.1 mL 1 N aqueous NaOH and heated at 37° C for 15 min. The mixture is then neutralized with 0.1 mL 1N CH_3COOH, and 2 mL $CHCl_3/CH_3OH$ (9:1, v/v), 1 mL butanol, and 1 mL water are added. The aqueous phase containing the phosphate is separated and washed twice with a few mL ethyl ether. The water solution is concentrated under vacuum at 35° C and applied to a column made from 1–2 g Dowex 50W-X4 in H form. The column is eluted with water for ethanolamine or with 0.01 M HCl for choline phosphatides. The eluate is dried under vacuum, and 0.1 mL pyridine and 0.1 mL BSTFA (with 1% TMCS) are added. The reaction takes place in 1 hour at room temperature, and the material can be analyzed by GC or GC-MS [15], [773].

The determination of other fragments following hydrolysis has been reported [774]. For example, by heating a phospholipid at 250° C for 5 min. with diphenylether, the hydrolysis takes place between the glyceryl and the phosphate residue. The resulting material can be further silylated using BSTFA as shown in the reaction:

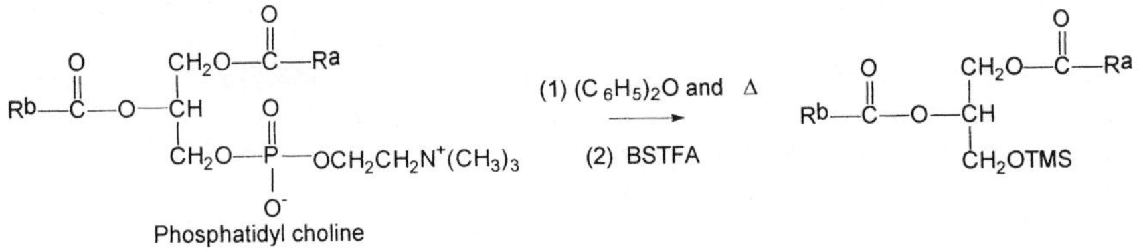

Numerous other analyses using derivatization by silylation were reported for the analysis of phospholipids [15].

Among other derivatizations, phosphoric acid esters with active hydrogens can be derivatized using the formation of a phosphoroamide as follows:

$$R^aO\text{-}P(=O)(OH)_2 + R^b\text{-}NH_2 \longrightarrow R^aO\text{-}P(=O)(OH)\text{-}NH\text{-}R^b$$

The use of dansylethylenediamine allows the derivatization of several 2'-deoxynucleoside-5'-monophosphates at levels as low as 7 pmol, using HPLC with fluorescence detection [775]. Various nucleotides are derivatized using the same procedure [776]. Diazomethane is reported to be applied in a derivatization of phosphatidylcholine after enzymatic hydrolysis of the choline group [777]. Aluminum/morin with indirect fluorescence is used for the analysis of biphosphonates [778], [779].

- Derivatization of sulfonic acids, sulfate esters and related compounds

Sulfonic acids $R\text{-}S(O)_2OH$ (R = alkyl or aryl) are strong acids and very soluble in water. Sulfuric acid forms both mono and diesters, alkylsulfuric acids $RO\text{-}S(O)_2OH$ being approximately as acidic as sulfuric acid itself. The derivatization of these compounds is directed toward replacement of acidic hydrogens. The methylation of sulfonic acids can be easily performed using tetramethylammonium hydroxide and heat in the GC injection port or by direct probe MS [780]. Using this procedure and tetramethylammonium hydrogensulfate or tetrabutylammonium hydrogensulfate as a reagent, various alkylbenzenesulfonates can be derivatized in the injection port of the GC at 300° C. The reaction can be written as shown below [781]:

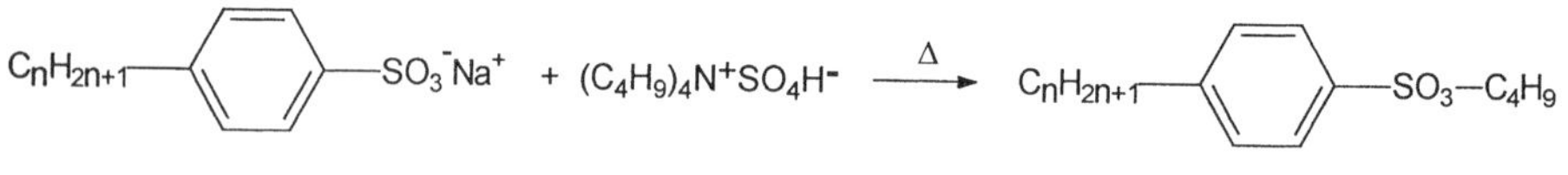

Diazomethane also can be used to form methyl esters. Frequently used is the silylation [782]. The treatment of sulfonic acids with PCl_5 or $SOCl_2$ leads to the formation of sulfonyl chlorides that can be analyzed by GC [783].

Diazomethane has been used for the methylation of sulfate esters such as sugar sulfates [784]. Some other derivatizations were applied after hydrolysis of the sulfate, sometimes done simultaneously with a second derivatization such as acylation of the new OH group, as shown below for (3-methoxy-4-sulfoxyphenyl)ethylene glycol in a reaction with TFAA:

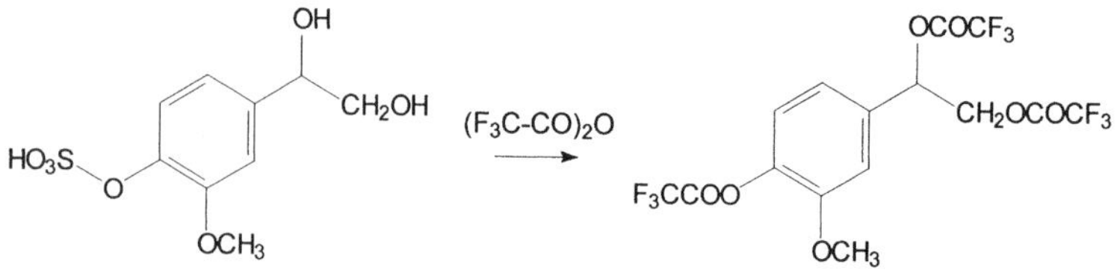

Simultaneous hydrolysis and derivatization are also obtained during the derivatization with BSTFA of sulfate esters of 4-(2aminoethyl)-1-2-benzendiol. The reaction can be written as follows:

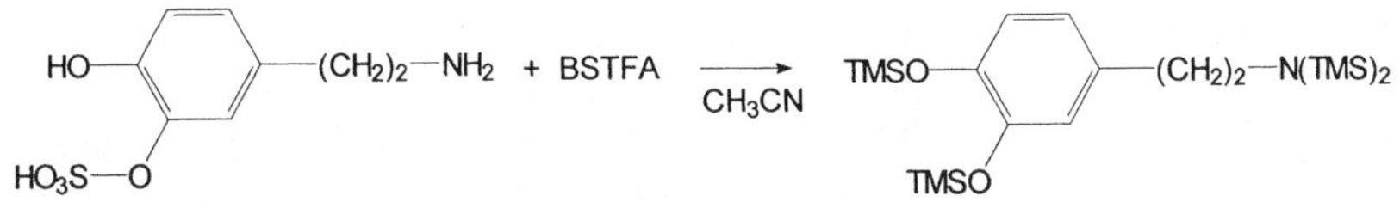

Sulfonamides have been analyzed using derivatization at the NH_2 group, which behaves similarly to amides. One typical derivatization is the methylation using CH_3I in the presence of tetrahexylammonium sodium sulfate or tetrahexylammonium hydrogensulfate. The reaction takes place as follows:

$$R\text{-}SO_2\text{-}NH_2 + CH_3I \longrightarrow R\text{-}SO_2\text{-}N(CH_3)_2$$

With this reagent are derivatized various diuretic drugs such as furosemide or 5-(aminosulfonyl)-4-chloro-2-[(2-furanylmethyl)amino]benzoic acid [785], chlorthalidone or 2-chloro-5-(1-hydroxy-3-oxo-1-isoindolinyl)benzenesulfonamide [786], and neosulfalepsine or 2-chloro-4-(2,5-dioxo-3-phenylpyrrolidinyl)benzene sulfonamide [787]. Another reported alkylation uses as reagent pentafluorobenzyl bromide in the presence of tetrabutylammonium hydrogen sulfate [788]. Other methylation reagents are trimethylbenzylammonium hydroxide with heating in the injection port of the GC [789] and dimethylformamide dimethylacetal (Methyl-8®), which has been used for the methylation of primary sulfonamides [790].

Acylation of sulfonamides can be done using trifluoroacetic anhydride in the presence of trimethylamine. Various antibacterial drugs are analyzed by this derivatization using ECD-GC. The reaction for N-phenyl benzenesulfonamide is written as follows [791]:

$$C_6H_5\text{-}SO_2\text{-}NH\text{-}C_6H_5 + (F_3C\text{-}CO)_2O \longrightarrow C_6H_5\text{-}SO_2\text{-}N(CO\text{-}CF_3)\text{-}C_6H_5$$

Various sulfonylureas used as antidiabetic agents, such as tolbutamide N-butyl-(p-toluene)sulfonylurea and chlorpropamide or N-propyl-(4-chlorobenzene)sulfonylurea, can be methylated using dimethylsulfate. The reaction with chlorpropamide is written as follows [792]:

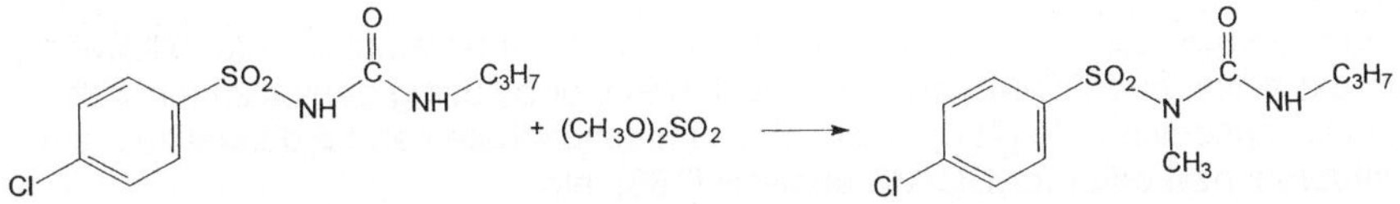

The sulfonamide group from the same drugs can be methylated using diazomethane [15]. Methylated drugs are not extremely stable at higher temperatures and can decompose in the injector port of the GC [793]. Two-step derivatization has been used for the substitution of both active hydrogens in the sulfonylureas, using CH_2N_2 and TFAA in a sequence of reactions as follows:

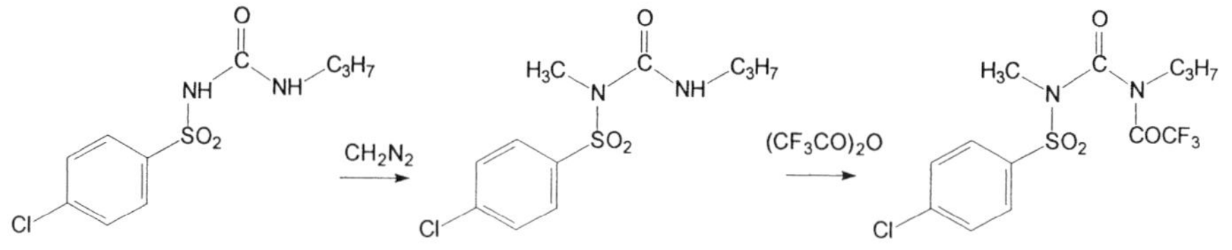

A typical procedure consists of dissolving 100 nmol sulfonylurea in 100 μL methanol and adding 0.5 mL solution of CH₂N₂ in ethyl ether. The solution is kept for 5 min at room temperature and then evaporated under N₂. The residue is dissolved in ethyl acetate/pyridine (9:1 v/v), and 50 μL TFAA is added. The mixture is heated at 65° C for 30 min. and evaporated under N₂, and the residue is dissolved in 100 μL cyclohexane and analyzed by GC-MS or ECD-GC [15], [794].

- Derivatization of organometallic and inorganic compounds

A number of inorganic compounds can be analyzed using derivatization followed by either GC or HPLC analysis. For example, 1,1,1-trifluoroacetylacetone can be used to form compounds with ions such as Be^{2+}, Al^{3+}, Ga^{3+} etc., which are sufficiently stable and volatile for GC analysis. Numerous organometallic compounds such as organotin or organomercuric compounds are also analyzed by GC. For example, a number of organotin compounds in sludges and sediments can be analyzed by extraction using a 0.25% etheral solution of tropolone (2-hydroxy-2,4,6-cyclopentatrien-1-one) followed by derivatization using ethylmagnesium bromide. The reagent converts mono-, di-, and trisubstituted organotin compounds into tetrasubstituted ones. The excess reagent is destroyed with HCl. The resulting compounds are analyzed by GC-MS or GC-FPD [795]. The reactions taking place in this analysis are probably the following:

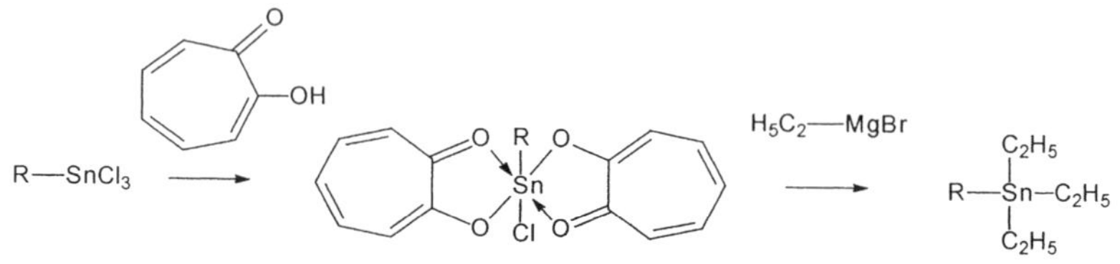

Similarly, methylmercury in fish can be derivatized using tetraethylborate followed by SPME extraction and GC-MS analysis [796], [797], or by using derivatization with sodium tetraphenylborate [798]. Also, zinc dithiophosphate can be derivatized to a pentafluorobenzyl ester for GC-MS analysis [799], etc.

The analysis of inorganic or organometallic compounds also can be done using derivatization and HPLC separation. For example, boric acid can be derivatized with 4,5-dihydroxynaphthalene-2,7-disulfonic acid disodium salt [1]. Bromides and iodides can be oxidized using 2-iodosobenzoic acid to bromine and iodine, respectively, followed by a reaction with 2,6-dimethylphenol [800] and analyzed by HPLC on a C18 column with UV detection. As low as 0.2 ng bromine and 0.5 ng iodine can be measured by this procedure. The sequence of reactions used for this analysis is the following:

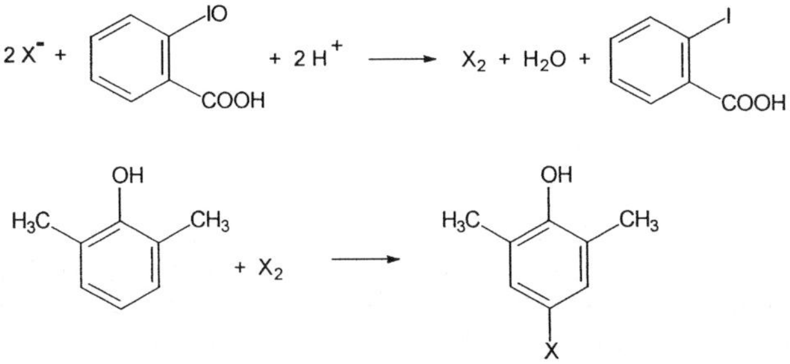

Other compounds can be analyzed using HPLC with postcolumn derivatization. For example, organoselenium compounds can be derivatized with 18-crown-6 ether, followed by MS analysis [801]. Several inorganic ions can be analyzed after complexation with specific organic reagents and HPLC separation with pre- or postcolumn derivatization and UV detection. A large number of reagents are known to form stable chelate compounds with different metallic ions [802], [803], [804]. Some of the analytical techniques using complexation of the metal ions and HPLC analysis are shown in Table 19.8.2

TABLE 19.8.2. *Reagents used for complexation of inorganic ions and analysis by HPLC.*

Reagent		Ion/level	Detection λ Separation	Ref.
2-acetylpyridine-4-phenyl-3-thiosemicarbazide	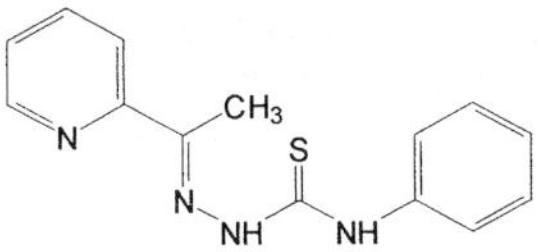	Cu^{2+}, Co^{2+} and Fe^{2+} 120-250 ng	UV 254 C18	[805]
ammonium bis-(2-hydroxy-ethyl)dithiocarbamate	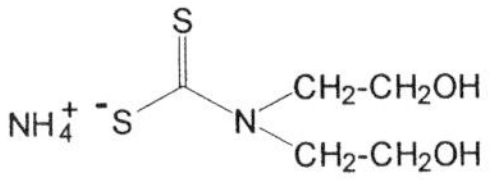	Cu^{2+}, Co^{2+}, Ni^{2+}, and Hg^{2+} 5ng/mL	UV 255 C18	[806]
arsenazo III or 2,2'-(1,8-dihydroxy-3,6-disulfonaphthalene-2,7-bisazo)bisbenzenearsonic acid with 2-hydroxy-isobutyric acid in mobile phase	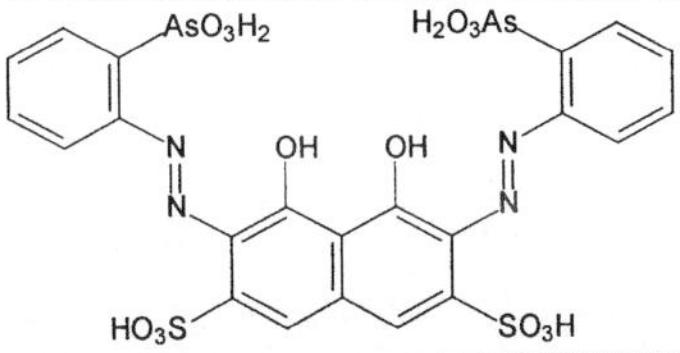	Dy^{3+}	Vis 658 postcolumn	[1]
2-(2-arsenophenylazo)-1,8-dihydroxy-7-(2,6-dibromo-4-fluorophenylazo)naphthalene-3,6-disulfonic acid or dibromofluoro-arsenazo	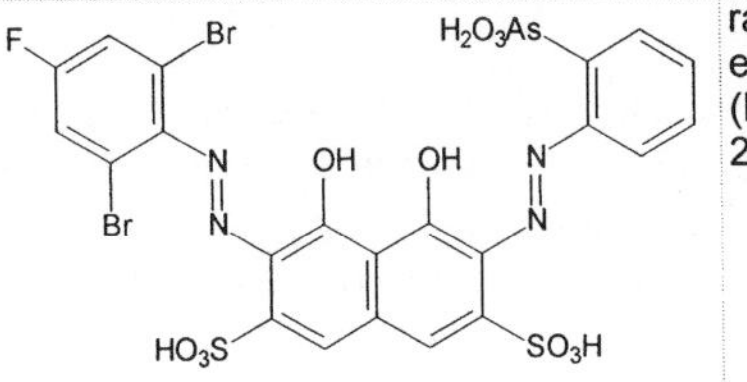	rare-earth elements, Th (IV), Cr^{3+} 2-10 ng/mL	Vis 630 ion pair	[807]

TABLE 19.8.2 (continued). *Reagents used for complexation of inorganic ions and analysis by HPLC.*

Reagent		Ion/level	Detection λ Separation	Ref.
2-[(5-bromo-2-pyridyl)azo]-5-diethylaminophenol		Cu^{2+}, Ni^{2+} Zn^{2+}, Cd^{2+} Co^{2+}, Mn^{2+} Pb^{2+} a few µg/L	Vis 560 postcolumn ion exchange	[808]
4-benzoyl-3-methyl-1-phenyl-2-pyrazolin-5-one		Al^{3+} and Fe^{3+} 13-28 ng/mL	UV 245 ODS	[809]
N-benzoyl-N-phenylhydroxylamine		V (V), Mo (VI) 2.1 ng/mL (V), 3.3 ng/mL (Mo)	UV 360 CN	[810]
bis(salicylaldehyde)tetra-methyl-ethylenediimine		Cu^{2+}, Ni^{2+} Pd^{2+}, Fe^{2+} UO_2^{2+}, Pt^{2+} (as cisplatin) 1 µg/mL	UV 254 ODS	[811]
2-(5-bromo-2-pyridylazo)-5-(diethylamino)phenol		V(V), Co^{2+}, Ni^{2+}, Fe^{2+} 0.72-98.8 pg	Vis 595	[812]
2-(5-bromo-2-pyridylazo)-5-(N-propyl-N-sulfopropyl-amino)-phenol		V(V), Cu^{2+}, Ni^{2+} 26-90 ng/L	Vis 575 RP-18	[813]
chlorophosphonazo III		rare-earth elements	Vis 660 postcolumn	[814]

TABLE 19.8.2 (continued). *Reagents used for complexation of inorganic ions and analysis by HPLC.*

Reagent		Ion/level	Detection λ Separation	Ref.
18-crown-6		alkali metals, alkaline earth	LC-MS post-column	[815]
2,6-diacetylpyridine-bis(benzoylhydrazone)		UO_2^{2+}, 5 ng	UV 265 RP	[816]
2,6-diacetylpyridine-bis(N-methylenepyridinilhydrazone) dichloride		various metals 0.2-5 µM	UV 340 polymer	[1]
4,5-dihydroxybenzene-1,3-disulfonic acid		Mo(VI), Zr(IV) 3.6-9 ppb	UV 315 C18 ion pair	[817]
1-(2,4-dihydroxy-1-phenylazo)-8-hydroxy-3,6-naphthalenedisulfonic acid		Be^{2+}, Al^{3+}, Fe^{3+} 7.2 pg/mL	Vis 500	[818]
ethylenetetraaminoacetic acid		Cu^{2+}, Pb^{2+}, Ni^{2+}, Fe^{2+} 0.5-3 ppm	CE, UV 242	[819]
N,N'-ethylene-bis(salicylaldimine)		UO_2^{2+}, Pt^{2+}, Ni^{2+}, Fe^{2+} Cu^{2+}, Pd^{2+} 2.5 µg/mL	UV 260 ODS	[820]
8-hydroxyquinoline (oxine)		Mg^{2+}, Al^{3+} 18 ng/mL	fluorescence ex 370, em 516	[821]

Table 19.8.2 (continued). *Reagents used for complexation of inorganic ions and analysis by HPLC.*

Reagent		Ion/level	Detection λ Separation	Ref.
luminol or 5-amino-2,3-dihydro-1,4-phthalazinedione + H_2O_2		Cr^{3+}, Cr (VI) 2 pg/L	CL postcolumn	[822]
lumogallion or 4-chloro-3-(2,4-dihydroxy-phenylazo)-2-hydroxybenzene-1-sulfonic acid		Al^{3+} 2.2 ng/mL	fluorescence ex 505, em 574	[823]
5,5'-methylenedisalicylo-hydroxamic acid		Ti(IV) 12-18 ng/mL	UV 360	[824]
1,10-phenanthroline		Co^{2+}, Zn^{2+}, Ni^{2+}, Fe^{2+}, Cu^{2+}, Cd^{2+} CE 0.3 ppm (Fe^{2+})	UV 265 ion exchange CE (Fe^{2+})	[825]
N,N'-o-phenylenebis(3,5-dichlorosalicylaldimine) (EDTA in mobile phase)		Cu^{2+}, Ni^{2+}, 40 nmol (Ni) 100 nmol (Cu)	UV 376 ODS	[826]
N,N'-o-phenylenebis(4-diethylaminosalicylal-dimine) (EDTA in mobile phase)		Cu^{2+}, Ni^{2+}, V(V), 10 nmol (V), 40 nmol (Ni), (Cu)	Vis 448 ODS	[826]
N,N'-o-phenylenebis-(salicylaldimine) (EDTA in mobile phase)		Cu^{2+}, Ni^{2+}, Co^{2+}, V(V), 0.6 ppb (Cu)	UV 380 ODS	[826]
o-phenylenediamine		cisplatin 400 ng/mL Pt	Vis 703 C18	[827]
picolinaldehyde-4-phenyl-3-thiosemicarbazone		Cu^{2+}, Ni^{2+}, Co^{2+}, Fe^{2+}, Bi^{3+}, Cd^{2+}, 0.6-3 ng/mL	UV 390 RP-18	[828]

TABLE 19.8.2 (continued). *Reagents used for complexation of inorganic ions and analysis by HPLC.*

Reagent		Ion/level	Detection λ Separation	Ref.
2-(2-pyridylazo)-1-naphthol-4-sulfonate		Ni^{2+}, Co^{2+}, Fe^{2+}	UV ODS	[1]
2-(2-pyridylazo)-1-naphthol-4-sulfonate		Ni^{2+}, Co^{2+}, Fe^{2+}	UV ODS	[1]
4-(2-pyridylazo)resorcinol (PAR)		Ni^{2+}, Cu^{2+}, Fe^{2+}, Zn^{2+}, Co^{2+}, Pb^{2+}, 0.3-7 ppb, etc.	Vis 520 postcolumn	[1], [829]
sodium diethyldithiocarbamate		Pt^{2+} and also Sb(V), Cd^{2+}, Co^{2+}, Fe^{2+}, Au(III), Fe^{3+}, Hg^{2+}, Ni^{2+}, Ag^+	UV 254 CN	[830], [831]
sodium pyrrolidinedithiocarbamate		CH_3Hg^+, $C_2H_5Hg^+$, $C_6H_5Hg^+$, Hg^{2+}, 30-100 ng/mL	electro-chemical, carbon paste +1.15 V	[832]
tartaric acid + 4-(2-pyridylazo)resorcinol		Cu^{2+}, Co^{2+}, Cd^{2+}, Fe^{2+}, Fe^{3+}, Pb^{2+}, Ni^{2+}, Zn^{2+}, Mn^{2+}	Vis 510 postcolumn	[833]
zinc butyl-2-naphthylmethyl-dithiocarbamate		Cu^{2+}, Co^{2+}, Ni^{2+}, Fe^{2+}, Hg^{2+}, 1-10 ng	UV 221 C18 either pre- or postcolumn	[1]

A number of applications of the reaction of organic compounds with metals can be used having the metal as a reagent and the organic compound as analyte. Flavonoids can be analyzed, for example, using HPLC separation with postcolumn derivatization using the formation of ternary complexes with divalent transition metals and 2,2'-dipyridine, followed by electrospray ionization MS [834], [835].

- Other derivatizations

Many other analytical procedures for sample preparation in chromatography utilize pre- or postcolumn derivatizations. From the large number of derivatizations reported in various original papers, only a limited number were captured as illustrative examples in this book. Some procedures are applied to less common compounds or are related to less frequently used chromatographic techniques and were not included by purpose. Derivatizations with potential use for chromatography but reported as applied without a separation are not covered in this material. Very likely, a number of involuntary omissions are also present.

19.9 DERIVATIZATION OF CARBOHYDRATES

Carbohydrates (sugars or saccharides) form an important group of naturally occurring compounds. Based on the number of sugar units in their molecule, carbohydrates are classified as monosaccharides, oligosaccharides, or polysaccharides. The general formula for monosaccharides is $C_nH_{2n}O_n$ (n=3 for trioses, n=4 for tetroses, n=5 for pentoses, n=6 for hexoses . . .). Monosaccharides contain a carbonyl group (aldehyde in aldoses or ketone in ketoses) and two or more -OH groups. Many derivatizations of simple monosaccharides are reactions at their OH groups. These derivatizations were already discussed in Section 19.1.

The free carbonyl form exists in equilibrium with the hemiacetal but at very low concentrations in aqueous solutions, regardless of whether it comes from an aldose or a ketose. For example, aldohexoses at $25°$ C and pH = 7 in a 0.25 M solution contain only $2–8 \ 10^{-4}$ M free carbonyl form for 1 M cyclic form. For aldopentoses the proportion of free carbonyl is $2–4 \ 10^{-3}$ M for 1 M cyclic form, with the exception of D-ribose that contains $8.5 \ 10^{-2}$ M free carbonyl for 1 M cyclic form. This explains why most sugars are analyzed without opening their cyclic form.

Derivatizations involving the carbonyl group of the carbohydrates still can be used because more carbonyl form can be generated in the equilibrium with the hemiacetal when the carbonyl is consumed in a subsequent reaction. The derivatization of the carbonyl leaves in the sugar molecule a number of free OH groups with active hydrogens. For this reason, for GC analysis the derivatization of the carbonyl is commonly followed by subsequent derivatization to eliminate the active hydrogens. The derivatization of the carbonyl alone is more common in HPLC when the only purpose is to attach a chromophore or fluorophore group. The derivatization of the carbonyl group in carbohydrates was discussed in Section 19.5.

The most common oligosaccharides are disaccharides formed from two monosaccharide units connected by an ether group. Polysaccharides are built by bonding numerous monosaccharide residues through the elimination of water and the formation of a series of ether bonds. There are also various carbohydrate type compounds that contain in their molecule additional groups such as amino, acid, ester, etc. Some of these compounds are common in nature and may play important biological roles [836].

The analysis of polysaccharides and of some oligosaccharides is frequently done using a chain fragmentation, performed with the intention to generate smaller molecules from the polymer. This can be achieved either using chemical degradations [104] or pyrolysis. The fragments are easier to analyze with standard analytical procedures. Chain fragmentation procedures are discussed in Chapter 20.

- Derivatization involving both OH and C=O groups for simple monosaccharides and disaccharides

Pentoses and hexoses are the most common monosaccharides. As shown in Section 19.1, they exist mainly in their hemiacetal form with various possible structures including furanose and pyranose forms each with two anomers. The derivatization at the carbonyl group in carbohydrates implies the modification of the hemiacetal structure of the carbohydrate. This eliminates the problem of two anomers and may lead to simplified chromatographic profiles. For the analysis of complex mixtures of sugars or for quantitation based on peak areas, this simplification can be very useful. Derivatizations for the carbonyls are the reduction, the formation of stable acetals or thioacetals, and the reaction with hydroxylamines with or without the formation of an aldonitrile.

The reduction of the carbonyl to an alcohol is usually followed by a second derivatization. For example, aldoses react with $NaBH_4$, and the resulting polyol can be acetylated with acetic anhydride generating the corresponding alditol acetate. For glucose this reaction can be written as follows:

In a typical procedure, a few mg sugar sample is dissolved in 0.5 mL water with 0.1 mL solution $NaBH_4$ (5 mg/mg sugar) added at low temperature (about $0°$ C). The mixture is stored over night at $4°$ C, and then the excess $NaBH_4$ is destroyed with a few drops of CH_3COOH. The solution is passed over a column of Dowex 50 H form to remove the Na^+ ions. The collected effluent is evaporated to dryness under vacuum. The residue is treated with 3 mL CH_3OH, heated at $50°$ C for 30 s, and evaporated to dryness to eliminate the borate. This step is repeated twice. The residue is then dissolved in 300 μL pyridine and 200 μL acetic anhydride and heated under reflux for 20 min. at $100°$ C. The residue is then evaporated, redissolved in pyridine, and analyzed by GC [15].

Reduction of carbohydrates may lead to some undesired results regarding loss of information about the initial sugar. For example, reduction of D-glucose and D-gulose lead to D- and L-glucitol, respectively. However, glucitol enantiomers are not separated on regular chromatographic columns. Another example is that of D-arabinose and D-lyxose, which both generate D-arabitol. Another problem with the reduction procedure when applied to ketoses is the generation of two isomers from the same compound. For example, reduction of fructose leads to D-glucitol and D-mannitol [15].

Reduction of disaccharides may lead to the reduction of only one of the cycles, as shown below for lactose in a reduction with $NaBD_4$ [837]. The reaction occurs with $NaBD_4$ in D_2O at room temperature in about 2 hours. The reduced disaccharide can be further hydrolyzed, for example, by heating in 1 N H_2SO_4 at $100°$ C for 5 min. After hydrolysis, the monosaccharides can be further reduced and derivatized for GC analysis.

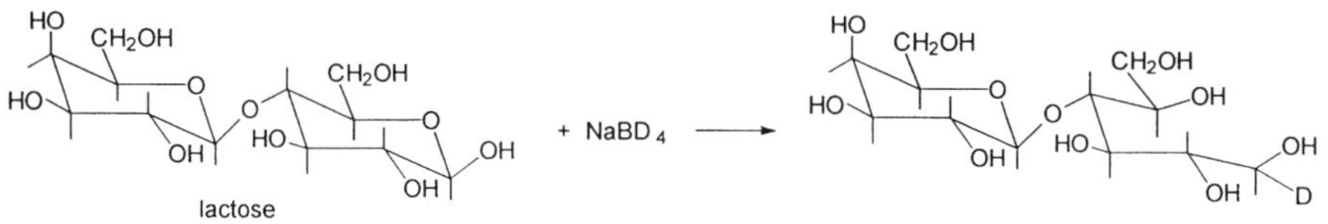

The reaction with ethanthiol leading to a thioacetal can be used instead of the typical reduction with NaBH$_4$ for eliminating the problems of anomers. This reaction can be written as follows:

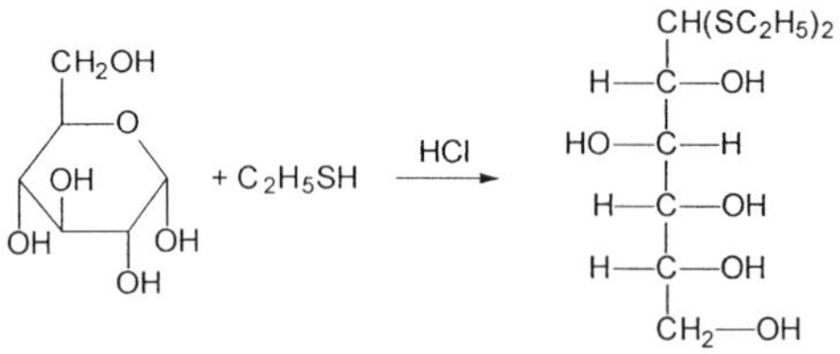

The dithioacetal can be further acetylated using acetic anhydride in pyridine. The preparation of the dithioacetal from 1 g aldose can be done by dissolving the sample in 1 mL conc. HCl at 0° C in a vial that can be closed tightly. To this solution 1 mL ethanthiol is added and allowed to stay for 10 min. at low temperature. The mixture is allowed to warm to room temperature, where the dithioacetal crystallizes. Also, the dithioacetal can be silylated, the reaction taking place only at the OH groups [838], [839].

Hydroxylamines are probably the most common reagents used to derivatize the carbohydrates carbonyl group. The reaction is done either with hydroxylamine or with alkyl-O-hydroxylamines, where the alkyl group can be methyl, benzyl, butyl, etc. After this derivatization, a number of active hydrogens are left in the sugar molecule, either in the OH groups or in =N-OH. Subsequent derivatization by acylation or silylation is common, especially for GC analysis. The oximes are usually present as both *syn* and *anti* isomers, which leads to two peaks in many separations (see Section 18.4). Therefore, this derivatization does not simplify the chromatographic profile for sugars. The use of a Wohl degradation type reaction, changing the two oxime diastereoisomers into one nitrile with simultaneous acetylation (see Section 18.4) can help in the simplification of the chromatographic profile [104]. This reaction has been used in several analytical determinations of sugars [840]. The sequence of reactions used for this derivatization is written as follows:

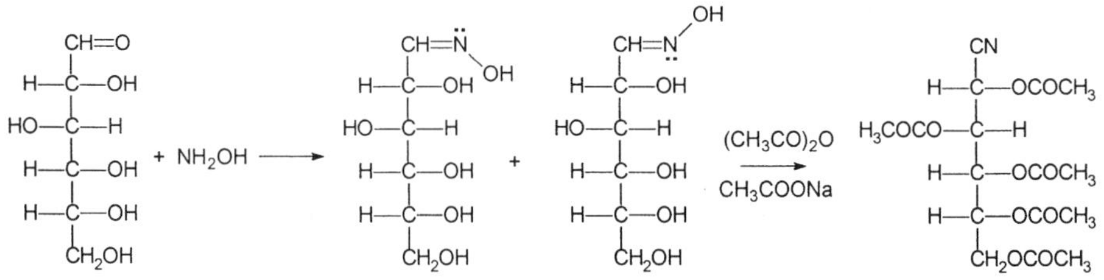

The generation of the nitrile from the aldoxime also can be done, simultaneously with acetylation, in the presence of 1-methylimidazole [841] or 1-dimethylamino-2-propanol [842]. The use of 4-(dimethylamino)pyridine as a catalyst for the acylation with acetic anhydride also leads to nitrile formation [843]. The nitrile also can be generated prior to acetylation by the formation of the oxime using hydroxylamine-O-sulfonic acid [844].

The presence of multiple functional groups makes possible the formation of compounds with various degrees of derivatization. This is in most cases highly undesirable, and especially for GC analysis the derivatizations must eliminate all the active hydrogens. Some derivatization reactions involving both types of groups and used for further GC analysis of mono or disaccharides are given in Table 19.9.1.

TABLE 19.9.1. *Reagents used for the analysis of carbohydrates by GC involving both carbonyl and OH groups.*

Reagent		Carbohydrate type	Reaction type[1]	Derivative type	Ref.
(1) sodium borohydride and (2) acetic anhydride	(1) $NaBH_4$, (2) $(CH_3CO)_2O$	carbohydrates	Red. + Acyl	acylated alditols	[845]
(1) sodium borohydride and (2) acetic anhydride in pyridine	(1) $NaBH_4$, (2) $(CH_3CO)_2O$	carbohydrates	Red. + Acyl	acylated alditols	[846]
(1) sodium borohydride (2) trifluoroacetic anhydride	(1) $NaBH_4$ (2) $(CF_3CO)_2O$	aldoses	Red. + Acyl	trifluoro-acetyl alditols	[15]
(1) hydroxylamine (2) trimethylsilylimidazole	(1) NH_2OH (2) $(CH_3)_3Si$—N⟨⟩N	sugars	Oxime Silyl	oxime, TMS ether	[15], [847]
(1) hydroxylamine (2) hexamethyldisilazane (HMDS)	(1) NH_2OH (2) $[(CH_3)_3Si]_2 NH$	sugars	Oxime Silyl	oxime, TMS ether	[848]
(1) hydroxylamine-O-sulfonic acid and elimination of H_2SO_4 (2) N,O-bis(trimethylsilyl)-trifluoroacetamide (BSTFA)	(1) NH_2OSO_3H (2) CF_3—C(O—$Si(CH_3)_3$)=N—$Si(CH_3)_3$	sugars	Oxime/ CN Silyl	CN TMS ether	[849]
(1) methoxylamine (2) N,O-bis(trimethylsilyl)-trifluoroacetamide (BSTFA)	(1) CH_3ONH_2 (2) CF_3—C(O—$Si(CH_3)_3$)=N—$Si(CH_3)_3$	sugars	Oxime Silyl	oxime, TMS ether	[15], [850], [851]
(1) ethanethiol + HCl (2) acetic anhydride in pyridine	(1) $C_2H_5SH + HCl$ (2) $(CH_3CO)_2O$ + pyridine	sugars	Ketal Acyl	diethyl-dithioacetal peracetate	[15], [852]
(1) ethanethiol + F_3C-COOH (2) TMCS + HMDS	(1) $C_2H_5SH + F_3C$-COOH (2) $(CH_3)_3SiCl + [(CH_3)_3Si]_2 NH$	sugars	Ketal Silyl	diethyl-dithioacetal TMS ether	[853]
(1) hydroxylamine + sodium acetate (2) acetic anhydride	(1) $NH_2OH + NaOOCCH_3$ (2) $(CH_3CO)_2O$	aldoses	Oxime Acyl	aldonitrile acetate	[840]
(1) hydroxylamine hydrochloride+ pyridine (2) acetic anhydride	(1) $NH_2OH{\cdot}HCl$ + pyridine (2) $(CH_3CO)_2O$	aldoses	Oxime Acyl	aldonitrile acetate	[15]

[1] Red = reduction, Acyl = acylation, Silyl = silylation, Ketal = ketal/acetal formation, Oxime = oxime formation

Carbohydrates can be analyzed at levels as low as 50 pmol, using HPLC separations followed by postcolumn derivatization. Among the reagents used in these procedures is 2-cyanoacetamide [854], which reacts with reducing carbohydrates in borate buffer to give an intense fluorescence (ex. 331 nm, em. 383 nm). Borate complexes of carbohydrates are separated using ion exchange chromatography. Similar analyses can be done using reagents such as ethylenediamine [855], 2-ethanolamine [856], benzylamidine [2], p-methoxy-benzylamidine [2], and *meso*-1,2-bis(4-methoxyphenyl)-ethylenediamine [857], [858], which react with reducing sugars leading to fluorescent compounds. Guanidine [859] allows the analysis of both reducing and nonreducing sugars forming a fluorescent derivative. Nonreducing carbohydrates are oxidized with periodate prior to the formation of the fluorescent compound. The fluorescent products of these reactions have not been isolated.

- *Derivatization of sugar related compounds, sugar acids, amino sugars and sugar phosphates.*

Sugars form numerous related compounds such as sugar acids, amino sugars, sugar esters, etc. These may be initially present in a given sample or may be the result of a previous derivatization or sample chemical modification.

Acids related carbohydrates are common in nature, either as monosaccharides or as units for natural polysaccharides such as pectin. Also, they may result in carbohydrate oxidations. There are three types of acids associated with monosaccharides: a) aldonic acids generated by the oxidation of the aldehyde group of a monosaccharide, b) uronic acids generated by the replacement of the primary alcohol group of a monosaccharide by a carboxyl group, and c) saccharic acids generated by simultaneous oxidation of the carbonyl and replacement of the primary alcohol. Aldonic acids have a marked tendency to eliminate water and form lactones. More complex sugar acids are also common in nature. The structural formulas of two hexuronic acids and of KDO are shown below:

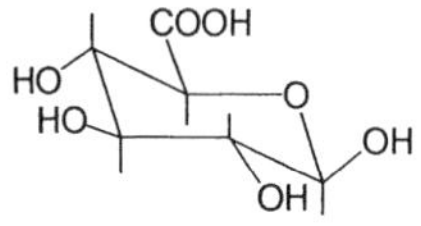

β-D-Glucopyranuronic acid (GlcA) 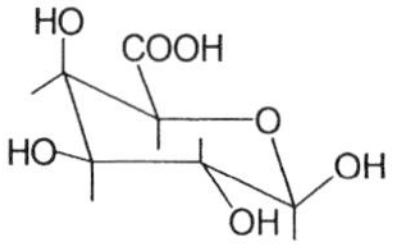   β-D-Galactopyranuronic acid (GalA) 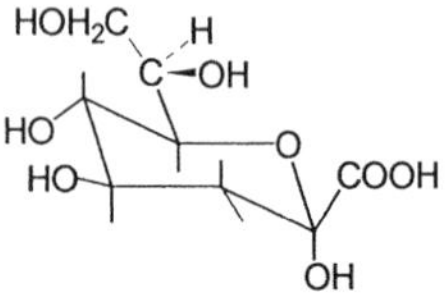   3-Deoxy-β-D-manno-octo-pyranulosonic acid (KDO)

Acid esters, acids derived from deoxysugars, amino sugars, N-acetylated amino sugars, sialic acids (nonulosaminic acids that are N- or O- substituted derivatives of neuraminic acid), etc. are among the compounds with significant biological importance. Some of these compounds are the units of various natural polysaccharides such as chitin or glycosaminoglycans. Three of these monosaccharide-related compounds are shown below.

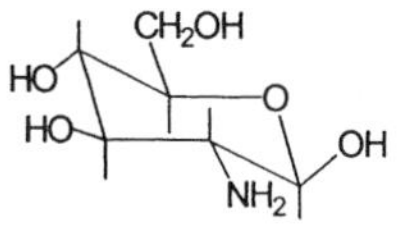

2-amino-2-deoxyβ-D-
glucopyranose

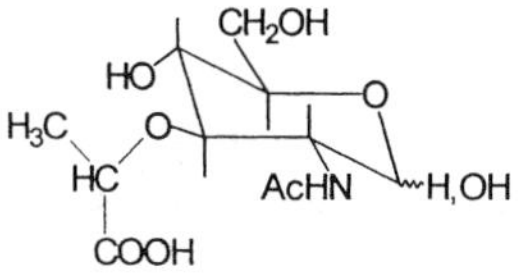

2-acetamido-2-deoxy-
muramic acid

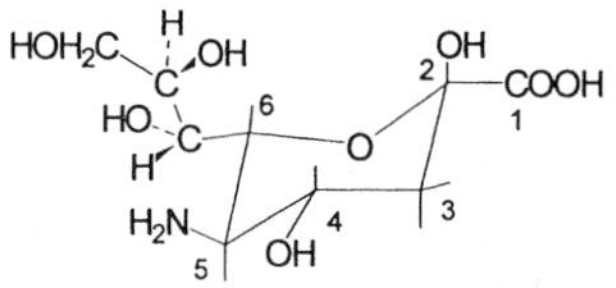

5-amino-3,5-dideoxy-D-
glycero-α-D-galacto-2-
nonulopyranonic acid
(neuraminic acid)

Derivatization procedures have been developed for these compounds, either with the purpose of GC analysis or attaching groups to enhance detectability for HPLC analysis.

In derivatizations for GC analysis, the carbonyl groups may remain underivatized, the only problem in this case being the presence of anomers that generate multiple peaks in the separation. It is desirable for the other groups containing active hydrogens to be derivatized. The OH and COOH groups are easily silylated, which is common for this group of compounds. The derivatization of choice for the amino group is the acylation.

The silylation of sugar acids can be done with HDMS + TMCS in solvents such as DMSO or DMF [860]. Also, BSTFA/TMCS in pyridine has been used for the same purpose. For the analysis, 1 mg sugar is mixed with 100 µL dry pyridine and 100 µL BSTFA (BSTFA + TMCS, 85:15 v/v). The mixture is heated at $65°$ C for 1 hour and then analyzed by GC or GC-MS. Applied on aldonic acids, the reaction may be completed or may generate a mixture of silylated acid and silylated lactone, as shown below [861]:

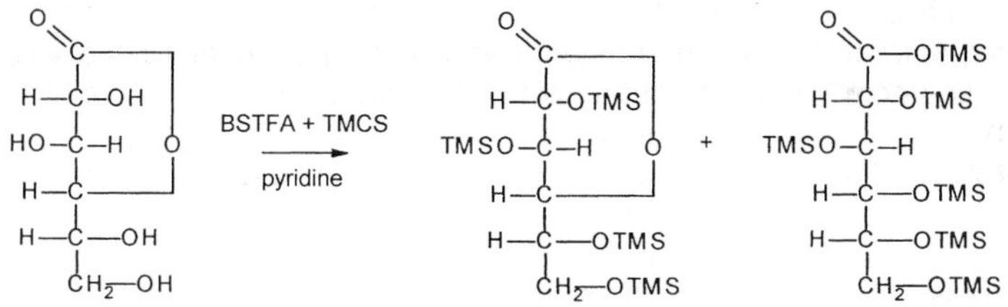

The derivatization of heptono-1,4-lactone with BSTFA + TMCS in CH_3CN does not lead to the opening of the lactone cycle [862]. Also, the derivatization of ascorbic acid with N-trimethylsilylacetamide does not open the lactone [863].

Acetylation has been used for the derivatization of 1,2,3-O-methylated α-D-(glucopyranosid)uronic acid methyl ester. The unique free 4-hydroxyl can be derivatized with acetic anhydride without affecting the methylation [864].

Some derivatizations for sugar acids reported in literature [15] are done with a first step derivatization of the carbonyl group. Among these, the derivatization of various hexuronic acids can be done by reduction with $NaBH_4$ in the presence of $BaCO_3$. After reduction, $NaBH_4$ is destroyed using an ion exchange H form (Rexyn 101). Boric acid is then eliminated by repeated addition and evaporation of CH_3OH. The remaining material is dissolved in HCl and evaporated to dryness and can be derivatized with HMDS + TMCS in pyridine, dissolved in $CHCl_3$, and analyzed by GC. The sequence of reactions in this derivatization can be written as follows:

Instead of silylation, the acids reduced with NaBH$_4$ can be derivatized with butaneboronic acid [865]. Derivatizations using as a first step reaction with NH$_2$OH followed by silylation is also used [866].

Derivatization of sialic acids for HPLC analysis can be done with 1,2-diamino-4,5-methylenedioxybenzene [867] forming a fluorescent quinoxaline derivative. The reaction takes place in the presence of Na$_2$S$_2$O$_4$ and with addition of β-mercaptoethanol as follows:

R = H, COCH$_3$, SO$_4$H
R' = COCH$_3$, COCH$_2$OH

The derivatized compound can be detected using fluorescence with λ_{em} = 448 nm and λ_{ex} = 375 nm, or electrospray ionization mass spectrometry.

The derivatization of amino sugars for GC analysis is done frequently using the acetylation of the amino group followed by silylation of the OH groups, leaving the carbonyl group unaffected [15]. The acetylation is commonly done with acetic anhydride in pyridine at room temperature for 30 min., and the silylation with HMDS + TMCS or with BSTFA. These derivatizations lead to N-acetyl-O-trimethylsilyl derivatives. The derivatization using trifluoroacetic anhydride in toluene with heating at 100° C for 15–20 min. leads to N,O-trifluoroacetyl derivative of hexosamines with all active OH groups acylated [867a].

Different derivatization at the amino group and at the OH also can be achieved using ethyl chloroformate for the derivatization of the amine and hexamethyldisilazane (HMDS) + TMCS in pyridine for the derivatization of the alcohol. The cyclic form of the sugar is not affected in this sequence of reactions that is written as follows:

The reaction takes place using about 1 mmol sugar in 5 mL solution in water to which there are added 0.3 mL saturated $NaHCO_3$ and 25 µL ethyl chloroformate. The mixture is kept for 1 hour at room temperature and then passed over a column made from 2 cm of Dowex 50 W (H form) and 2 cm of De-acidite G [868]. The first 2.5 mL eluate is discarded, and the next 2 mL is collected, evaporated to dryness, and silylated with HMDS/TMCS/pyridine (3:2:10, v/v/v). The silylation takes place at room temperature in 30 min.

Silylation as a unique derivatization step was also studied [869]. The number of groups in hexosamines that are silylated depends on the conditions of the reaction. A mixture of bis(trimethylsilyl)acetamide (BSA) + TMCS + pyridine (100:1:400 v/v/v) analyzed immediately leads to the penta-TMS derivative as shown below:

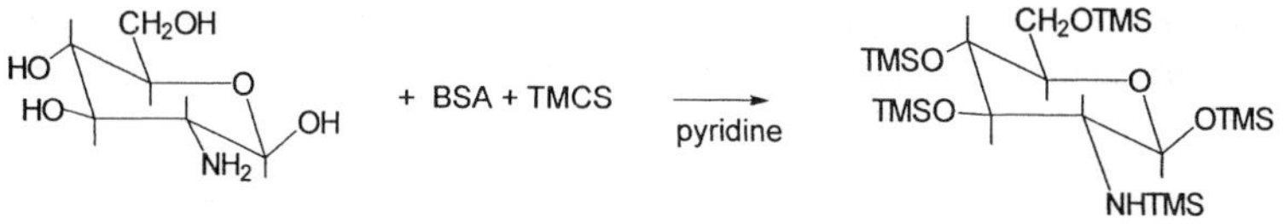

The same result is obtained with a mixture HMDS/BSA/TMCS in pyridine (1:0.5:1:10) [870] when kept for 30 min. at room temperature. A mixture of one part BSA/TMCS (100:1 v/v) in four parts CH_3CN at 60° C for 10 min. or longer leads to all six groups silylated (two at the amino group). Other reagents, such as BSTFA/TMCS in pyridine heated at 60° C for 3 hours, form with hexosamines a mixture of tetra and pentasilyl derivatives, while HMDS/TMCS in pyridine at room temperature for 20 min. forms only tetrasilyl derivative.

Similarly to other sugars, some derivatization procedures involve the carbonyl group. The use of $NaBH_4$ followed by derivatization with trifluoroacetic anhydride leads to a peracylated sugar alcohol [871]. Ethane thiol forms with amino sugars in the presence of HCl a dithioacetal similar to that formed by other sugars [872], [873].

Another possible modification of the amino sugars is the replacement of the amino group with hydroxyl. This can be achieved by treatment with HNO_2 ($NaNO_2$ in acetate buffer at pH = 3.6). However, it was shown [874] that depending on the configuration of the amino sugar, either the expected alcohol or an intramolecular reaction product may be formed as shown below:

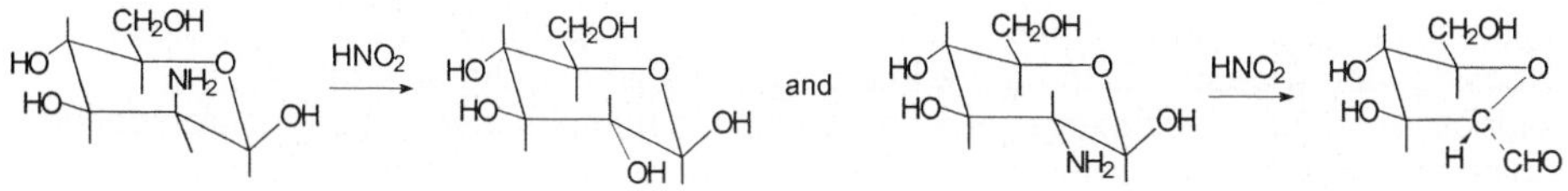

After this reaction, further derivatization can be done using for example $NaBH_4$ and then HMDS/TMCS [875]. The reaction may be used to determine the configuration of the amino group.

Many amino sugars are present in nature in their N-acylated form. These can be hydrolyzed with acids. When the amino sugar is present in a polysaccharide that is hydrolyzed for the analysis of the constituent monosaccharides, the hydrolysis of N-acetylated sugars occurs unintentionally. The amino group can be further derivatized, but silylation to N-TMS derivative may be incomplete [869]. A re-N-acetylation reaction is usually preferred before silylation.

Several derivatizations were reported for sialic acids such as N-acetylneuraminic acid. One derivatization uses BSTFA in CH_3CN. The acetyl group is not affected, while all the active hydrogens are transformed into their TMS ethers [876]. The silylation of N-acetylneuraminic acid can be preceded by a methylation of the COOH group with CH_2N_2. For this purpose, the sample is treated with CH_2N_2 in CH_3OH, evaporated immediately, dissolved in pyridine and silylated with HDMS + TMCS within 2 hours at room temperature. The acetyl group and the methyl ester are not affected by silylation.

Phosphorilated sugars are compounds that play an important role in biological systems, for example in the use of energy from carbohydrates through the glycolysis process. Typical compounds in the glycolysis process are glucose 6-phosphate, fructose 6-phosphate, fructose 1,6-biphosphate. A common derivatization of sugar phosphates with the purpose of GC analysis is the silylation. The reaction for glucose 6-phosphate is written as follows:

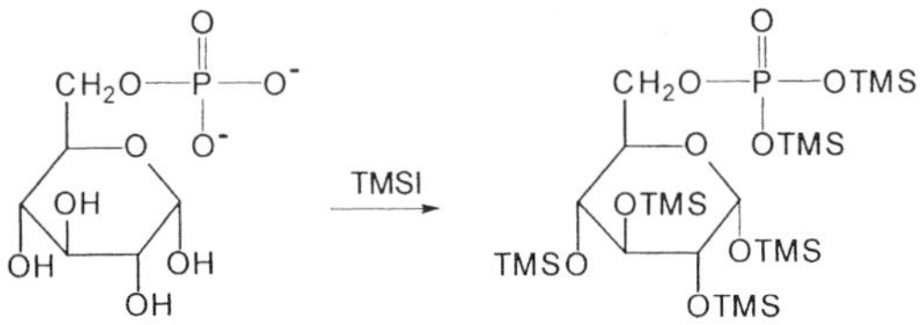

The reaction can be done using various silylating reagents such as trimethylsilyl-imidazole (TMSI) in CH_3CN [877], BSA + TMCS mixture, d18-BSA, BSTFA + TMCS [878], etc.

In addition to silylation, other reactions have been used for derivatization of sugar phosphates. Due to the acid character of the hydrogens in the phosphate group, they can be easily methylated with diazomethane. The remaining groups in the sugar molecule can be further silylated without disturbing the methylated phosphate group [879]. Also, after methylation of the phosphate group, the rest of OH groups from the sugar can be derivatized using tert-butyldimethylchlorosilane [15] or methaneboronic acid [880].

Sugars also can be found connected to a variety of compounds forming glycosides. Glycosides are widespread in nature in plants and animals. As an example, the metabolism of numerous exogenous compounds in humans involves a glucuronidation. Uridine diphosphate glucuronic acid (UDP-glucuronic acid), being part of intermediary metabolism and closely related to glycogen synthesis, is found in all tissues of the body. This compound is an energy-rich intermediate for the transfer of the glucuronic acid moiety to many exogenous compounds such as drugs. This explains the interest in the analysis of numerous glucuronides related to drug metabolism. The analyses of these compounds can be done for the whole molecule, but also it is common to perform a cleavage of the glycosidic bond and analyze the sugar moiety separately. The cleavage can be done similarly to that of polysaccharides, for example by hydrolysis in the presence of a catalyst such as HCl, H_2SO_4, CF_3COOH. For special purposes the hydrolysis can be done with an alkaline catalysts. The cleavage also can be done using methanolysis, acetolysis, or formolysis. Following the cleavage of the glycosidic bond, further derivatization is common. For example, the carboxyl group from a glucuronic acid that is generated by hydrolysis can be derivatized using methylation with CH_2N_2 or silylation. The hydroxyl groups can be methylated with methyl iodide, silylated, or acetylated [15]. For HPLC analysis, the presence of OH or COOH groups in glycosides can be utilized for attaching chromophores or fluorophores for improved detectability [1] without the hydrolysis step of the glycosidic bond.

For compounds with multiple groups some of them already derivatized, the subsequent derivatization may generate problems regarding replacement of the previous group. Some observations on the effect of alkylation on sugar derivatives were given in Section 18.1. Sugars may be found completely or partially O-alkylated, O-acylated, or N-acylated (in case of amino sugars). Also, trimethylsilyl groups may be present after a preliminary silylation. Some derivatives can be more stable than others, depending on the reagent, the sugar and the substitution position. For example, during the alkylation in the presence of Ag_2O with alkyl halides, N-acetyl groups in amino sugars are not affected by alkylation, while O-acetyl groups (or other acyl groups) may migrate and be replaced. The use of NaH as a catalyst and a polar aprotic solvent as a medium leads to a complete replacement of the O-acetyl groups, while N-acetyls may survive. When the BaO + Ba(OH)$_2$ is used as a catalyst, the O-acetyl groups are replaced by alkyl, and N-acetyl groups may also be replaced with alkyl but only when the alkylation process is prolonged. In strongly basic conditions using methylsulfinyl carbanion solution in DMSO, the permethylation of acylated carbohydrates may be expected. As an example, the alkylation of a partly acetylated 2-amino-2-deoxy-β-D-glucose will generate, under correct conditions with NaH as a catalyst and DMF as a solvent, the O-alkylated-N-acetylated compound:

Another possible effect of alkylation using alkyl halides in the presence of NaH is the C-alkylation [881].

Silylation is also known to lead to replacements of preexistent groups on a monosaccharide structure. For example, specific glycosides can be silylated to form persilylated monosaccharides with the elimination of the aglycon (see Section 20.1).

- Derivatization of polymeric carbohydrates

Macromolecular carbohydrates can be derivatized at the free OH or NH_2 groups with alkylating reagents or acylating reagents, which transform these groups, for example, in

OCH$_3$ or NHCOCH$_3$ groups. After this step, the polymer can be subject to a hydrolysis of the bonds between the sugar residues without the lost of the protection groups at OH or NH$_2$. The partially derivatized monosaccharides obtained by hydrolysis are analyzed (possible after a new different derivatization) for the determination of component monosaccharides and of the linkage positions (see Section 20.1).

Other derivatizations are utilized for tagging different groups in polysaccharides such as the NH$_2$ groups in deoxyaminosugar residues or the terminal reducing groups. These derivatization can be followed by HPLC separations of the whole polymer using various detection systems such as fluorescence, LC-MS, etc. [882], [883]. The procedure is commonly utilized in the study of glycoproteins or other materials of biological interest [884].

19.10 DERIVATIZATION OF AMINO ACIDS AND RELATED COMPOUNDS

Amino acids form an important class of bifunctional compounds. There are numerous known amino acids, and certain α-amino acids are the building blocks from which proteins are constructed. These protein amino acids have the general formula R-CH(NH$_2$)-COOH with the exception of proline and hydroxyproline, which have secondary amino groups. With a few exceptions, amino acids have chiral molecules, the amino acids from proteins having L (S) configuration at the chiral α carbon:

L-amino acids (S)   L-proline

The names and the structures of the α-amino acids naturally present in proteins are shown in Table 19.10.1.

Most amino acids are not soluble in nonpolar solvents and are soluble in water. They have amphoteric properties due to the existence in their molecule of a COOH group and a NH$_2$ group, many existing as zwitterions in the form R-CH(NH$_3^+$)-COO$^-$. Amino acids with one COOH and one NH$_2$ have two acidity constants, K$_1$ for the reaction R-CH(NH$_3^+$)-COOH + H$_2$O $\rightleftharpoons$ R-CH(NH$_3^+$)-COO$^-$ + H$_3$O$^+$, and K$_2$ for the reaction R-CH(NH$_3^+$)-COO$^-$ + H$_2$O $\rightleftharpoons$ R-CH(NH$_2$)-COO$^-$ + H$_3$O$^+$. Table 19.10.2 gives the acidity constants for some common amino acids.

TABLE 19.10.1. *Common amino acids present in proteins.*

Name	Abbreviation (three letters)	Abbreviation (one letter)	Radical R connected to the α-carbon [1]	MW	Formula
alanine	La	A	CH3—	89.09	C3H7NO2
arginine	Arg	R		174.20	C6H14N4O2
asparagine	Asn	N	H2N-CO-CH2—	132.12	C4H8N2O3
aspartic acid	Asp	D	HOOC-CH2—	133.10	C4H7NO4
cysteine	Cys	C	HS-CH2—	121.16	C3H7NO2S
cystine	Cys-Cys		—CH2-S-S-CH2—	240.30	C6H12N2O4S2
glutamic acid	Glu	E	HOOC-CH2CH2—	147.13	C5H9NO4
glutamine	Gln	Q	H2N-CO-CH2CH2—	146.15	C5H10N2O3
glycine	Gly	G	H—	75.07	C2H5NO2
histidine	His	H		155.16	C6H9N3O2
hydroxylysine	Hyl		H2N-CH2-CH(OH)-(CH2)2—	162.19	C6H14N2O3
hydroxyproline	Hyp			131.13	C5H9NO3
isoleucine	Ile	I		131.17	C6H13NO2
leucine	Leu	L	(CH3)2CHCH2—	131.17	C6H13NO2
lysine	Lys	K	H2N-(CH2)4—	146.19	C6H14N2O2
methionine	Met	M	CH3SCH2CH2—	149.21	C5H11NO2S
phenylalanine	Phe	F	(C6H5)-CH2—	165.19	C9H11NO2
proline	Pro	P		115.13	C5H9NO2
serine	Ser	S	HO-CH2—	105.09	C3H7NO3
threonine	Thr	T	CH3-CH(OH) —	119.12	C4H9NO3
tryptophan	Trp	W		204.22	C11H12N2O2
tyrosine	Tyr	Y	HO-(C6H4)-CH2—	181.19	C9H11NO3
valine	Val	V	(CH3)2CH—	117.15	C5H11NO2

[1] Whole formulas shown for proline and hydroxyproline; MW calculated considering natural isotope abundance.

TABLE 19.10.2. *Acidity constants (-log K_a) of common amino acids.*

Amino acid	pK_1	pK_2	pK_3	Amino acid	pK_1	pK_2	pK_3
alanine	2.35	9.87		leucine	2.33	9.74	
arginine	1.82	8.99	13.20	lysine	2.16	9.20	10.80
asparagine	2.02	8.80		methionine	2.17	9.27	
aspartic acid	1.99	3.90	10.00	phenylalanine	2.58	9.24	
cysteine	1.86	8.35	10.34	proline	1.95	10.64	
glutamic acid	2.13	4.32	9.95	serine	2.19	9.44	
glutamine	2.17	9.13		threonine	2.09	9.10	
glycine	2.35	9.78		tryptophan	2.43	9.44	
histidine	1.81	6.05	9.15	tyrosine	2.20	9.11	10.07
isoleucine	2.32	9.76		valine	2.29	9.72	

Comparing the acidity constants of amino acids with those of acids and amines, it can be seen that they are rather strong organic acids (comparing pK_1 values, for example, to acetic acid $pK_a = 4.76$) and similar to amines (comparing pK_2 values, for example, to methylamine $pK_a = 10.62$). The pH of the solution of an amino acid in pure water is called the isoelectric point (see also Section 14.1).

Amino acids as bifunctional compounds can form amide bonds (or peptide bonds) between the amine of one amino acid and the carboxyl of another amino acid. The peptide bond contains the planar group of atoms C_aNCC. When a small number of amino acids are linked through peptide bonds, they form simple peptides (dipeptides, tripeptides, etc.), and these can be analyzed using typical GC or HPLC methods. Simple chains containing a larger number of amino acids are named polypeptides. Proteins are formed from polypeptides and may have more complex structures. Polymer chain fragmentations for polypeptides and proteins are further discussed in Section 20.2.

Nonprotein amino acids can be derivatized, in principle, using the same methods as protein amino acids. Some differences can be generated by the molecule of the analyte mainly in the separation or detection.

- *Derivatization of amino acids for GC analysis*

Many derivatizations of amino acids and small peptides for GC analysis are double step reactions, including an alkylation of the COOH group and an acylation of the NH_2. The reaction can be written as follows:

$$R-CH(NH_2)-COOH \ + \ R^aX \ \longrightarrow \ R-CH(NH_2)-COOR^a \ + \ R^bCOX \ \longrightarrow \ R-CH(NHCOR^b)-COOR^a$$

One typical procedure, using methanol + HCl for the methylation and trifluoroacetic anhydride (TFAA) for acylation, starts with 2 mg amino acids, which are dissolved in 2 mL dry methanol. In this solution, dry HCl is bubbled for 30 min. The solution is then dried under vacuum, and 0.2 mL TFAA is added. The mixture is kept for 30 min. at room temperature and evaporated under vacuum for about 4 min. while cooling in an ice bath. The residue is dissolved in CH_2Cl_2 and analyzed by GC or GC-MS. The derivatization of simple amino acids using this type of procedure is straightforward when R does not have additional functionalities. However, there are problems when other groups are present in the amino acid molecule.

The OH group in protein amino acids serine, threonine, hydroxyproline, and tyrosine can be acylated using acyl anhydrides after the methylation of the acid group, and these amino acids do not pose special problems (hydoxylysine can also be acylated at the OH group). With trifluoroacetic anhydride (TFAA), for example, the OH-containing amino acid esters form di-TFA derivatives. However, in a procedure using TFAA as a first derivatization step, the acylation may not occur at the OH groups [885]. This would suggest that methylation must be done first. On the other hand, if methylation performed with CH_2N_2 is the first step, the free OH group in tyrosine, which is more reactive than an aliphatic alcohol, can be methylated (at least in part). This shows that for amino acids with several functional groups, the order of multiple derivatizations and reactivity of the reagent are important for uniform derivatization.

The SH group in cysteine is sensitive to oxidation. The preservation of cysteine can be done adding antioxidants such as BHT during sample preparation. Cystine can be derivatized at both NH_2 and COOH without breaking the disulfide bond, but also can be reduced to cysteine with $SnCl_2$, ethanthiol, or other reducing reagents. The SH groups are usually acylated forming di-acyl derivatives. Methionine usually does not pose special problems, although lower reproducibility in some analyses has been reported [886].

Several amino acids have a second amino group. In lysine and ornithine (α,δ-diaminovaleric acid) the amino group usually can be acylated, leading to di-acyl derivatives. Ornithine (Orn) is not isolated from proteins except after hydrolysis with alkali when it is formed from the decomposition of arginine. Arginine, which contains a guanidine group, can be converted to ornithine for analytical purposes by conversion with arginase [887]. Tryptophan and histidine contain NH groups, but in these amino acids the NH group is part of a heterocycle. Usually, the acylation of these NH groups and that of additional NH_2 group from arginine occurs with difficulty. Arginine may form a tri-acyl derivative, for example by heating the material in a sealed tube at 140° C for 10 min. with TFAA. Histidine can be converted to asparagine by ozonolysis. Leaving the NH groups underivatized may lead to problems in the separation and detection [15], affecting the quantitation results. Strong derivatization conditions such as high HCl concentrations may lead to some tryptophan or arginine decomposition.

Aspartic and glutamic acids have two carboxyl groups. These can be esterified without additional problems. However, asparagine and glutamine, which are the amides of aspartic and glutamic acid, respectively, may create problems due to the hydrolysis of the amide group. This leads to incorrect levels of the acid and its amide in quantitations. Both strong acid and basic media of reaction must be avoided to preserve the amide not hydrolyzed. The amide group is not usually derivatized in the conditions described for amino and carboxyl groups.

Iodoamino acids do not pose special problems in derivatization. Thyroxine, or O-(4-hydroxy-3,5-diiodophenyl)-3,5-diiodotyrosine, and diiodotyrosine can be derivatized using typical derivatization reagents, ensuring that the phenolic OH is also derivatized. Table 19.10.3 shows some of the derivatization techniques using alkylation/acylation.

The derivatizations indicated in Table 19.10.3 leading to various esterified acids and acylated amines may fit specific needs of a given sample or of a specific separation. The volatility of the derivatized amino acids decreases when the molecular mass of the alkyl group forming the ester increases. However, the retention times on a nonpolar column is usually in the order heptafluorobutyrate (HFB) < pentafluoropropionate (PFP) < trifluoroacetate (TFA), PFP and HFB derivatives not differing significantly. HFB derivatives with better ECD sensitivity [932] are preferred when using this type of detection.

TABLE 19.10.3. *Some reagents used for the alkylation/acylation double step derivatization of common amino acids.*

Alkylating reagent	Acylating reagent	Amino acid type, matrix, sensitivity	Comments	Ref.
butanol + HCl	pentafluorobenzoyl chloride in triethylamine	amino acids		[888]
butanol + HCl	N-trifluoroacetyl-L-prolyl chloride	46 amino acids	Hydroxylated amino acids were further silylated at the OH groups	[889]
butanol (second step)	trifluoroacetic anhydride (TFAA) (first step)	tranexamic[1] acid		[890]
butanol + HCl + dibutoxypropane to remove water	trifluoroacetic anhydride (TFAA)	19 amino acids		[891]
butanol + HCl	trifluoroacetic anhydride (TFAA)	amino acids	Some Trp s destroyed. Possible ECD detection. Addition of CH_2Cl_2 to the butanol + HCl improved solubility.	[15], [892], [893], [894]
butanol + HCl	(1) trifluoroacetic anhydride (TFAA) (2) ethoxyformic anhydride	histidine	NH_2 nitrogen gives mono-trifluoroacetyl; NH nitrogen gives ethoxyformyl.	[895]
diazomethane (second step)	trifluoroacetic anhydride (TFAA) (first step)	phenylalanine, tyrosine, nonprotein amino acids	The OH groups are not acylated. The phenolic OH in iodoamino acids is derivatized.	[896]
diazomethane (second step)	isobutyl chloroformate (first step)	amino acids	Arg is not deriv.	[897], [898]
diazomethane	methyl trifluoroacetate	small peptides		[899]
diazopropane	-	N-acetyl-aspartic acid	Butyl esters were prepared from diazobutane.	[900]
dimethylsulfate	trifluoroacetic anhydride (TFAA)	amino acids, hydroxy amino acids	Cys-Cys is reduced to Cys by reflux with C_2H_5SH. Arg derivatization has problems. OH is also acylated	[901]
isopropanol + acetyl chloride	pentafluoropropionic anhydride (PFPA)	amino acids	Separation on chiral GC stationary phases	[220]
isopropanol + HCl	trifluoroacetic anhydride (TFAA)	amino acids	Separation on chiral GC stationary phases	[902]
isopropanol + HCl	pentafluoropropionic anhydride (PFPA)	amino acids	Separation on chiral GC stationary phases	[903], [904]
isobutanol + HCl	heptafluorobutyric anhydride (HFBA)	amino acids	Some procedures add BHT as antioxidant. Ethoxyformic anhydride $(C_2H_5OCO)_2O$ may be used for His second deriv.	[905], [906], [907], [908]
(1) methanol + acetyl chloride (AcCl) (2) isoamyl alcohol + acetyl chloride (AcCl)	heptafluorobutyric anhydride (HFBA)	amino acids	His, Cys-Cys, Cys, Trp cannot be analyzed by this procedure	[909]
methanol + Dowex 50	methyl trifluoroacetate	amino acids	No side reaction from HCl	[910]
methanol + HCl	methyl trifluoroacetate	amino acids, dipeptides	Separates dipeptides from amino acids group	[911], [912], [913]
methanol + HCl	trifluoroacetic anhydride (TFAA)	amino acids	Acylation of Arg at 140° C for 10 min. His gives low yield of deriv. The OH in iodoamino acids is also derivatized.	[914], [915], [916]
methanol + HCl	acetic acid + acetic anhydride	small peptides		[917]

TABLE 19.10.3 (continued). *Some reagents used for the alkylation/acylation double step derivatization of common amino acids.*

Alkylating reagent	Acylating reagent	Amino acid type, matrix, sensitivity	Comments	Ref.
methanol + HCl	methyl trifluoroacetate	small peptides		[918]
methanol + HCl	pentafluoropropionic anhydride (PFPA)	small peptides	Methylation at room temp. for 4 hours does not affect glutamine and asparagine residues; 16 hours at 45° C leads to hydrolysis.	[918]
(1) methanol + HCl (2) butanol + HCl	trifluoroacetic anhydride (TFAA)	20 amino acids		[919]
(1) methanol + HCl (2) isoamyl alcohol + HCl	heptafluorobutyric anhydride (HFBA)	amino acids	Lower volatility allows sample concentration.	[920]
methanol + SOCl$_2$	trifluoroacetic anhydride (TFAA)	amino acids		[15]
methanol + SOCl$_2$	pentafluoropropionic anhydride (PFPA)	small peptides	Useful for dipeptide analysis	[918]
methanol	acetic anhydride	amino acids	One-step derivatization	[921]
(S)-(+)-3-methyl-2-butanol + acetyl chloride (AcCl)	trifluoroacetic anhydride (TFAA)	amino acids	Chiral separation	[922]
propanol + HCl	acetic anhydride in triethylamine	21 amino acids at nmole level	Prepare propyl ester of acid, Cys-Cys reduced to Cys with SnCl$_2$. Arg and His are derivatized.	[923]
propanol + HCl	acetic anhydride in pyridine	16 amino acids in urine	Analysis of Arg and His gives poor results.	[924]
propanol + HCl	heptafluorobutyric anhydride (HFBA)	20 amino acids	Problem with His peak. Addition of acetic anhydride (Ac$_2$O) allows acylation on column.	[925], [926], [927]
propanol + HCl	heptafluorobutyric anhydride (HFBA)	amino acids	Addition of 2,6-di-*tert*-butyl-4-methylphenol (BHT) as antioxidant	[928]
pentanol + HCl	trifluoroacetic anhydride (TFAA)	amino acids	Hydroxyamino acids form N,O-di-TFA deriv. Cys forms N,S-di-TFA deriv.	[15], [929]
pentanol + HBr	acetic anhydride (Ac$_2$O)	amino acids	Poor yields obtained for Arg and His. Trp not detected. Met converted to the sulfoxide.	[930]
tetramethylammonium hydroxide	methyl chloroformate	amino acids		[931]

[1] For information and chemical structure regarding various drugs see e.g. M. O'Neil et al. (eds.), *The Merck Index*, Merck & Co., Inc., 13th ed., Whitehouse Station, 2001.

A low level of amino acids in specific samples may require a preconcentration step. A simple procedure for the analysis of free amino acids in a plant material starts with 2 g ground substance (also dried or kept at a desired moisture content), which is extracted for 30 min. with 50 mL 0.1N HCl containing norleucine as internal standard. The extract is filtered, and 5 mL is passed through a cation exchange column (Poly Prep AG 50W-X8) H form. The column is washed with 15 mL water, and the amino acids are eluted with 15 mL of 1N NH$_4$OH. The eluent is evaporated to dryness and treated with 1 mL isopropanol and 100 mL acetyl chloride. The mixture is then heated for 2 hours at

100° C in a sealed vial and evaporated at 80–90° C. To the residue is added 0.2 mL PFPA, which is then heated in a sealed vial at 70° C for 30 min. The mixture is then dissolved in CH_2Cl_2 and analyzed by GC or GC-MS. The reaction with isopropanol (IP) + acetyl chloride (as a water scavenger) followed by reaction with pentafluoropropionic anhydride (PFPA) is written below:

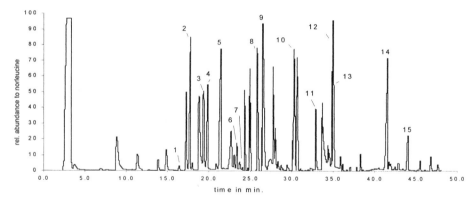

The separation can be done on a chiral column (Chirasil-L-Val) with temperature gradient between 50° C and 200° C [933]. The analysis of a tobacco sample by this method gives the chromatogram shown in Figure 19.10.1. The trace of D-alanine detected in the chromatogram is unusual; only L-amino acids are naturally occurring in tobacco.

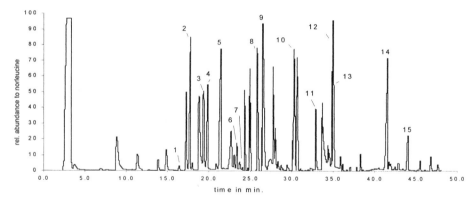

FIGURE 19.10.1. *The chromatogram (TIC) for the analysis of PFP-IP derivatized amino acids from a tobacco sample. Separation done on a Chirasil-L-Val column.* The amino acids seen in the chromatogram are 1) D-alanine, 2) L-alanine, 3) L-threonine, 4) L-valine, 5) glycine, 6) L-serine, 7) L-leucine, 8) L-proline, 9) L-norleucine (internal standard), 10) L-asparagine + aspartic acid, 11) L-glutamine, 12) L-glutamic acid, 13) L-phenylalanine 14) L-ornithine, 15) L-lysine.

The quantitation of amino acids based on chromatographic peak areas can be done using calibrations for each amino acid, but more common is to use a response factor for each amino acid relative to the internal standard (see Section 3.8). As low as 250 fmol amino acid can be analyzed by this procedure.

Typical for the mass spectra of PFP-IP derivatives of the amino acids are the ions with m/z = M - 87, which is the base peak for alanine and proline and rather intense for other amino acids. The ions with m/z = M - 59, as well as the ion with m/z = 119 generated by $C_2F_5^+$ fragment, are also present in the spectra. Some fragmentations obtained from other derivatives of the type fluoroacyl (at NH_2) and alkyl (at COOH) are shown below:

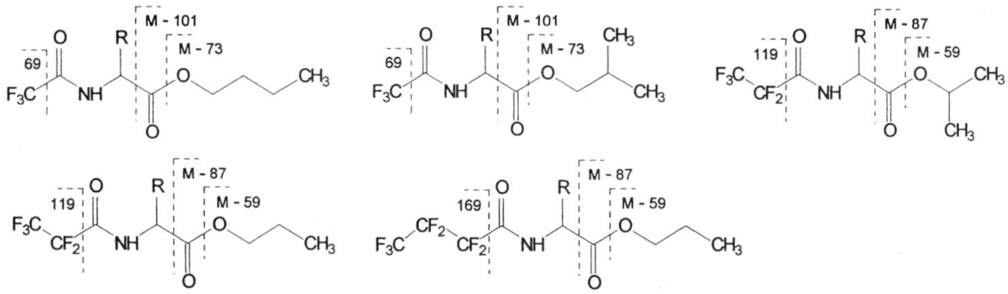

Although the peak intensities vary considerably from compound to compound, it is common to have an intense peak corresponding to the elimination of the -COOR group.

Other derivatizations reported for the GC analysis of amino acids include reaction with isobutylchloroformate at -NH$_2$ and with CH$_2$N$_2$ at -COOH [934], [935], reaction with methanol + SOCl$_2$ at -COOH and with pivaldehyde (2,2-dimethylpropanal) at -NH$_2$, [936], reaction with methanol + HCl at -COOH and with diethylchlorophosphate in triethylamine at -NH$_2$ [15], reaction with trimethylphenylammonium hydroxide at -COOH and with pivaldehyde at -NH$_2$, [937], etc. Dipeptides were also analyzed using the alkylation of the COOH group and a Schiff base formation at the NH$_2$ using for example trimethylphenylammonium hydroxide and acetylacetone [15], etc. A less common derivatization uses a reaction with propanol + HCl at -COOH followed by reaction at -NH$_2$ with CS$_2$ in triethylamine leading to a N-alkylated dithiocarbamic acid, which is further acylated with methyl chloroformate and decomposed with heat to an isothiocyanate (senevol) as shown below [15]:

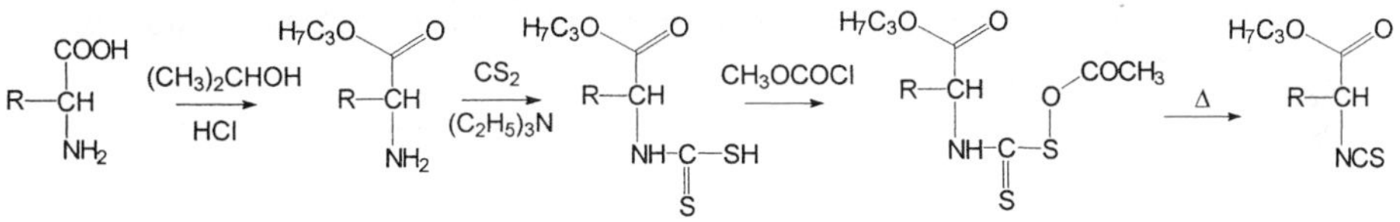

Some other derivatizations use one reagent for both -COOH and -NH$_2$ groups. Among these are N,N-dimethylformamide dimethyl acetal (Methyl-8®), which methylates the COOH group and forms an imine with the NH$_2$ as shown below:

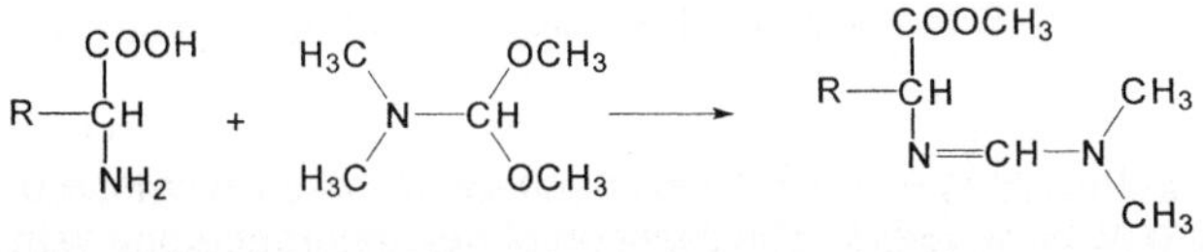

The reagent does not react with alcoholic or phenolic OH, with SH, or with indole NH groups. However, it reacts with the additional -C=NH and -NH$_2$ groups of arginine and with -NH$_2$ group of citruline (2-amino-5-ureidovaleric acid) [938], [939].

Thiohydantoin derivatives are extensively used for HPLC analysis of amino acids and sequencing of proteins using Edman's procedure (see Section 20.2). The analysis by GC is also possible but less common because of the presence of active hydrogens in thiohydantoin derivatives. The reaction for the formation of thiohydantoin derivatives or 5-substituted-2-thioxoimidazolin-4-ones can be written as follows (see Section 18.5):

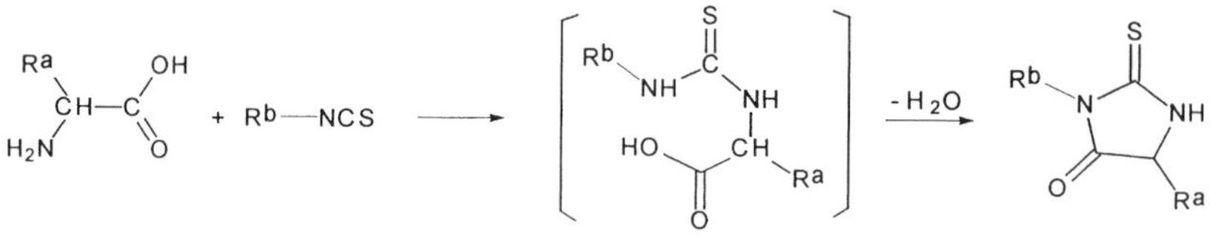

Various isothiocyanates were used in this reaction, isothiocyanic acid, methyl and phenyl isothiocyanates being the most common. The reaction leads to an intermediate thiocarbamyl derivative, which is further transformed into the thiohydantoin usually using an acid treatment. A typical procedure including the hydrolysis of a peptide starts with 0.5–1.0 μmol material, which is dissolved in 250 μL 0.4 M dimethylallylamine in 40 % aqueous pyridine with the pH adjusted at 9.5 with trifluoroacetic acid. To this solution 10 μL methylisothiocyanate (melted at 40° C) is added, and the mixture is heated at 50° C for 30 min. After cooling, the solution is washed twice with 2 mL benzene, which is discarded. The aqueous solution is evaporated under N_2. The residue is treated with 50–100 μL CF_3COOH, heated at 50° C for 10 min., and then evaporated. The residue is extracted with 2 mL CH_2Cl_2 and the extract separated and evaporated. The dry residue is treated with 200 μL 1N HCl and heated at 80° C for 10 min. to generate the methylthiohydantoins. The material is extracted with ethyl acetate and the extract is evaporated to dryness. The residue is dissolved in 1–5 μL ethyl acetate and analyzed by GC or GC-MS. The GC analysis can be used only for methylthiohydantoins of alanine, valine, leucine, isoleucine, and methionine. Other amino acids derivatives cannot be analyzed directly. However, after trimethylsilylation the thiohydantoin derivatives have been used successfully for GC analysis of various amino acids (see further). The same procedure can be used for the preparation of phenylthiohydantoins. Subsequent derivatization for GC analysis of phenylthiohydantoins can be done using silylation or acylation with TFAA. The reaction with TFAA takes place as follows [15]:

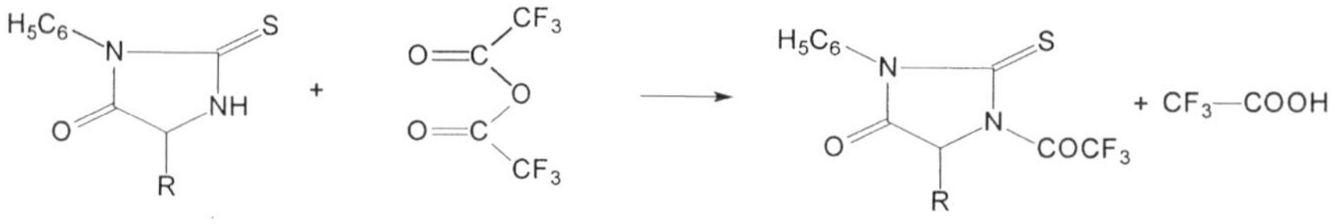

Complete reaction between thiohydantoin derivative and TFAA takes place at room temperature in 30 min. The analysis of serine, threonine, asparagine, glutamine and the basic amino acids gives poor results by this procedure [940].

Another reaction with formation of cycles, which can be used for derivatization followed by GC analysis, is that using active ketones such as hexafluoroacetone or 1,3-dichloro-1,1,3,3-tetrachloro acetone [941]. The reaction of hexafluoroacetone with amino acids takes place in pyridine as follows:

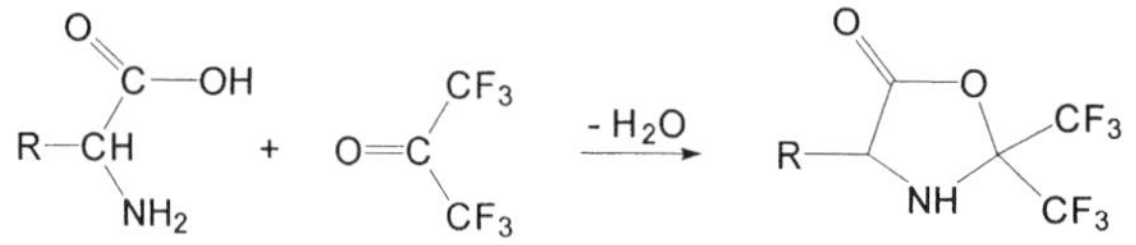

Hydroxylated amino acids do not react with the ketones at the OH group, and further derivatization is needed [941a], [942], [943]. Formation of oxazolin-5-one derivatives also can be used as a derivatization for GC analysis, and the reaction takes place in special conditions by vigorous heating the amino acids with TFAA in a reaction as follows:

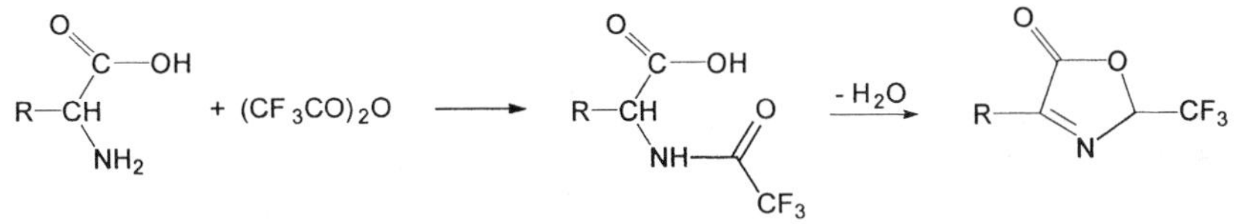

Peptide amides in the same reaction lead to an imidazolin-5(4)-one as shown below:

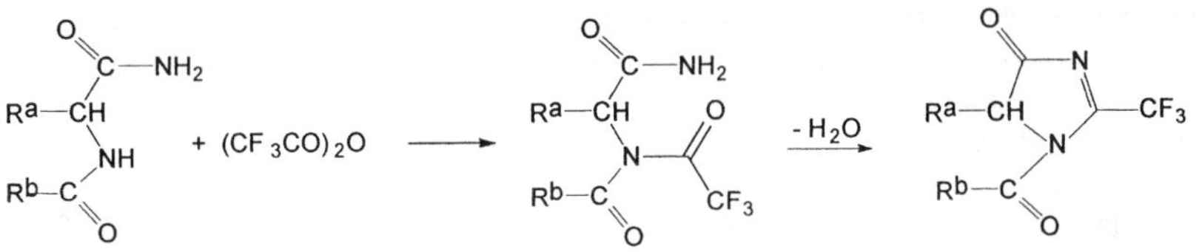

An oxazolinone derivative that can be analyzed by GC or GC-MS is also obtained by refluxing N-acyl derivatives of amino acids or acylated dipeptides with dicyclohexyl-carbodiimide (DCC) [944], [945]. The reactions can be written as follows:

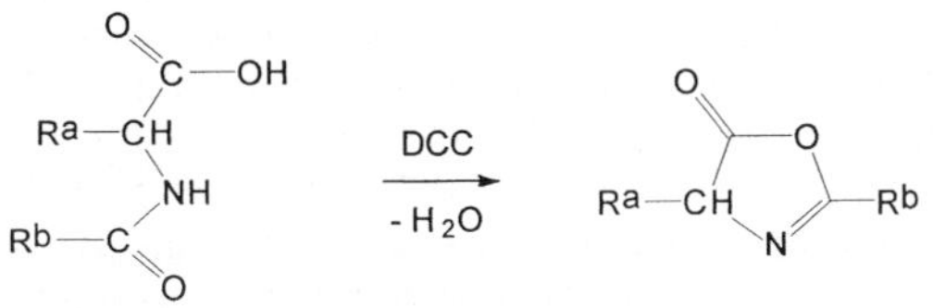

Another type of derivative obtained from amino acids is that of N-alkyloxycarbonyl alkyl esters [946]. This reaction takes place between the amino acid and a chloroformate, such as ethylchloroformate within 1 min. in water/pyridine medium and can be written as follows:

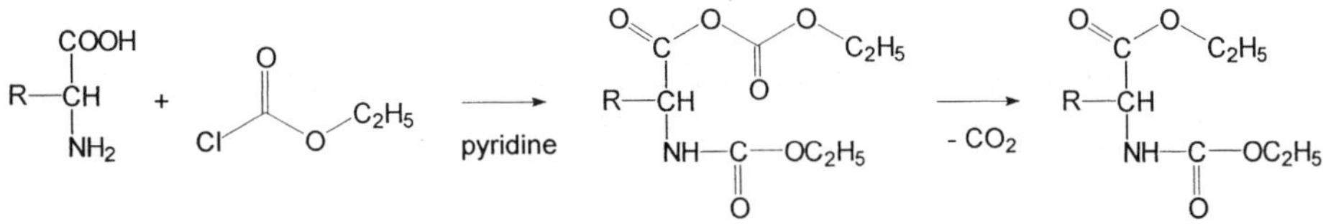

Various chloroformates including 2,2,2-trifluoroethylchloroformate may be used as reagents [947], [948], [949]. A modification of the technique uses the derivatization with ethylchloroformate in the presence of heptafluorobutanol when the reaction takes place as follows:

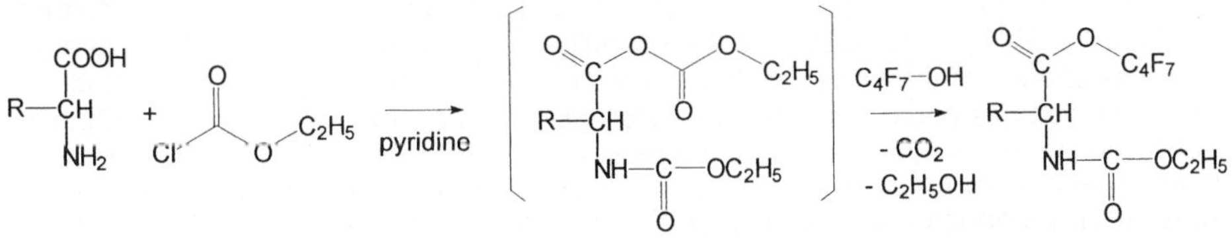

For this reaction, 30 μL of a mixture alcohol:pyridine (2:1 v/v) is added to 100 μL 0.1 N HCl aqueous solution of amino acids containing 4 μmol/mL of each amino acid. This solution is reacted with 7 μL chloroformate for a few min., then treated with 10 mg NaCl and extracted with 100 μL CHCl$_3$. The organic layer is analyzed by GC or GC-MS. The enantiomer separation is possible using a Chirasil-L-Val column [950]. In this procedure, the decomposition of the intermediate compound with the formation of diethyl esters of the amino acid is also possible as a secondary reaction with the resulting ethanol.

A common derivatization of amino acids is the silylation [951]. Trimethylsilyl derivatives were obtained using different silylating reagents and conditions, but as a general rule, TMS derivatives are easily hydrolysable and not very stable [952]. They need to be analyzed immediately after derivatization. On the other hand, very good results were obtained using *tert*-butyl-dimethylsilyl derivatives. Table 19.10.4 lists several reagents and conditions for obtaining TMS derivatives of amino acids.

TABLE 19.10.4. *Some reagents used to form TMS derivatives of amino acids.*

Reagent	Conditions	Analyte/comments	Reference
BSTFA	CH$_3$CN heat at 150° C for 5 min.	Sulfur amino acids, SH group silylated, dipeptides	[953]
BSTFA	CH$_3$CN heat at 125° C for 15 min	Amino acids, Cys and Asn requires 150° C for 30 min., Arg gives irreproducible results.	[15], [931]
BSTFA	CH$_3$CN heat at 150° C for 2.5 hours	Further silylation, Glu-2-TMS, Arg-4-TMS, Lys-4-TMS, His-3-TMS, Tyr-3-TMS, Cys-4-TMS. Some mixtures.	[952], [954]
BSTFA	CH$_3$CN or CH$_2$Cl$_2$ heat at 135° C for 15 min	Various amino acids. Results depend on amino acid.	[955]
BSTFA + TMSDEA + TMCS	pyridine, heat at 100° for 10 min.	Tryptophan, tyrosine and metabolites are derivatized including TMS at the indole ring nitrogen	[956]
BSA + TMCS	pyridine, heat at 60° C for 12-14 hours	Mono-TMS are formed at α-amino except for Gly.	[15]
HMDS	pyridine 60 min at 100° C	Analysis in the presence of sugars	[957]
MSA	dissolve with heating in excess reagent	Various amino acids	[958]
MSTFA	use CF$_3$COOH as solvent at room temp.	Various amino acids including hydroxy amino acids that are silylated at the OH group (not esterified)	[959]
MSTFA + TMCS	CH$_3$CN heat at 120° C for 1 hour	Used on amino acids separated by TLC with silylation of the scraped spot.	[960]
TMSDEA	reflux for 1 hour under N$_2$ in excess reagent	Various amino acids	[15]

Better results for amino acid derivatization are obtained using MTBSTFA. The formation of *tert*-butyldimethylsilyl- (TBDMS) derivatives takes place at the COOH and one of the active hydrogens at the NH$_2$ group [961], [962], [931]. A typical procedure consists of adding to the amino acid sample dissolved in DMF or in CH$_3$CN:pyridine (1:1 v/v) an excess of MTBSTFA (5–6 times the molar ratio to amino acids) and heating the mixture for 30 min. at 80° C. The solution is then directly injected in the GC or GC-MS system. The mass spectra of the amino acids usually include the ions m/z = M - 57, which is frequently the base peak, m/z = M - 15, m/z = M - 85. m/z = M - 159, m/z = M - 302, as well as the ion m/z = 302. Molecular ion peak may be present but in general is very weak. Some possible fragmentations are shown in Figure 19.10.2, together with the spectrum of 2-TBDMS valine (MW = 317):

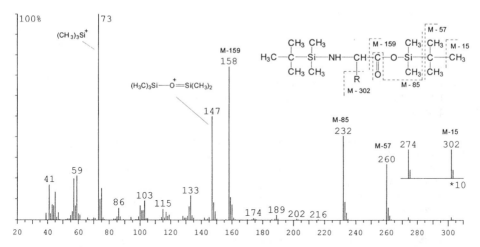

FIGURE 19.10.2. *Typical fragmentation for TBDMS amino acid derivatives, and the MS spectrum of 2-TBDMS-valine (MW = 317).*

Other solvents such as DMSO and shorter heating times can be used for the same derivatization. Threonine and histidine require the heating at 70° C for at least 30 min. for complete derivatization [961]. The use of MTBSTFA with 1% *tert*-butyldimethylsilylchloride reduced the derivatization time but interfered with the analysis of arginine and histidine. The analysis of amino acids by GC or GC-MS using the formation of TBDMS derivatives leads to excellent results provided that the amino acids are soluble in the derivatization medium. MTBSTFA itself and CH_3CN, although good silylation media, are not good solvents for several amino acids, while DMF is a better choice as a solvent.

Silylation also is used in connection with other derivatizations. There are several procedures of this type. For example, a two-step derivatization can be done with MSTFA followed by reaction with MBTFA to form trifluoroacetyl derivative at the NH_2 group and TMS at the COOH and possible OH groups [963]. Alkylation of the COOH group with butanol + HCl and silylation of the NH_2 with BSTFA [964] is possible. Tryptophan and tryptamines can be analyzed by a condensation with formaldehyde followed by silylation of the COOH group and of the NH groups from the β-carboline formed in this reaction (see Section 18.5) [965].

Trimethylsilyl thiohydantoins of amino acids have been used with good results for the GC and GC-MS analysis. The analysis using thiohydantoin as a first step derivatization may be a result of peptide or protein sequential analysis. In this case, further GC-MS is an excellent procedure for the identification of the amino acids resulting in the degradation procedure (see Section 20.2). The silylation reactions for the derivative with isothiocyanic acid takes place as follows (reaction by-products not shown):

BSTFA   pyridine

The formation of two silylated compounds, one with two TMS groups and the other with three TMS groups, explains the presence of doublet peaks in the chromatograms [966]. The silylation of methyl or phenyl thiohydantoins also has been practiced [15]. The reaction takes place as follows:

BSA

+ by-products of BSA

If the amino acid radical R contains active hydrogens, these are also silylated. Various amino acids have been analyzed using GC-MS following this derivatization [967], or using BSTFA as silylating reagent and sulfur photometric detection [968].

Other multiple step derivatizations including silylation are applied in particular cases. For example, hydroxyamino acids can be derivatized in a three-step reaction using dinitrofluorobenzene for reaction at the NH_2 group, CH_2N_2 for reaction at the COOH group, and silylation at the OH group [15].

Some of the nonprotein amino acids may have a tertiary amino group, such as ethylenediaminetetraacetic acid (EDTA). Usually, the derivatization of these acids is done only at the COOH group using typical acylation methods [15] or silylation [969].

Special derivatizations of amino acids are also used, followed by GC-MS analysis. For example, a reduction of amino acids with $NaBH_4$ in tetrahydrofuran in the presence if I_2 leads to the formation of amino alcohols. These can be analyzed using a GC-MS system for high precision isotopic analysis without further derivatization that contaminates the analytes with external carbon [970].

- Derivatization of amino acids for HPLC analysis

HPLC analysis requiring only the attachment of a chromophore or fluorophore group to the analyte molecule can use only one of the functional groups of the amino acids for derivatization. Several reagents used for the derivatization of only the amino group in amino acids were described in Section 19.4. Probably the most common derivatizations of amino acids besides the one with phenyl isothiocyanate are done with o-phthalaldehyde (and a thiol) or with 9-fluorenylmethyl chloroformate, which react only at the amino group.
The derivatization of both groups is also applied in a number of analyses. This type of derivatization can be done in a sequence of two reactions such as CH_3OH + HCl for the methylation of the COOH and reaction of the amine with N-α-9-fluorenylmethyloxy-carbonyl-amino acid-N-carboxyanhydride [297] or reaction with (1S)-(+)-10-camphor-sulfonyl chloride for the amine derivatization followed by derivatization of the COOH with

4-nitrobenzyl bromide [971]. Also, some of the derivatizations described for amines and given in Section 19.4 can be followed by a derivatization of the COOH group, for example, using CH_2N_2.

Several derivatization reactions of amino acids for HPLC analysis involve both NH_2 and COOH groups, but they lead to the formation of new cycles that may have strong light absorption properties or may fluoresce. One such reaction is that using isothiocyanates (or isothiocyanic acid) with the formation of thiohydantoins. The reaction already has been mentioned as a possible derivatization of amino acids for GC analysis (see Section 18.5). More frequently the formation of thiohydantoins is used for HPLC analysis of amino acids. As indicated previously, the reaction with thiocyanates takes place in two steps, the first being the formation of a thiocarbamyl derivative. This derivative can be stable and used in amino acid analysis without further transformation in the corresponding thiohydantoin. Several isothiocyanates that react with the amino group without involving further condensation were given in Table 19.4.6. A number of isothiocyanates, after the first step of the reaction, undergo a second step and form thiohydantoins, usually in acidic conditions. The overall reaction can be written as follows:

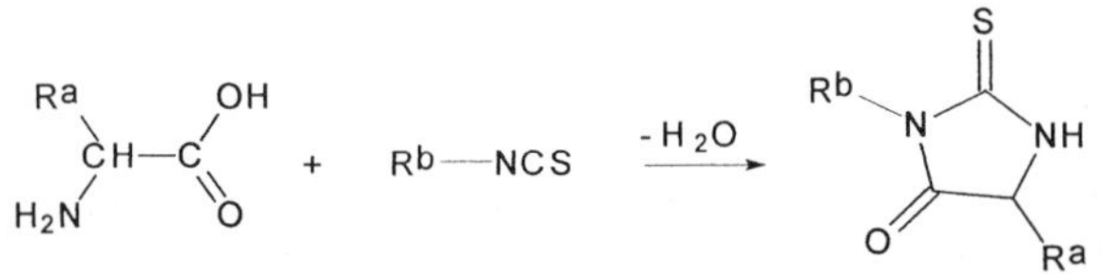

Several procedures can be used to transform the thiocarbamyl derivative in thiohydantoin such as: heating the compound in 1N HCl either in water or in methanol, heating the compound in 25% trifluoroacetic acid (TFA) in water or methanol, thermal conversion such as in a direct insert probe of an MS, 10% TFA in ethyl acetate containing 15 mg/L dithiothreitol by heating at $80°$ C for 10 min. [972], etc. A number of reagents used for thiohydantoin formation are given in Table 19.10.5.

Another derivatization leading to cycles and involving both groups in amino acids is the reaction with borinic acid that takes place as follows [985]:

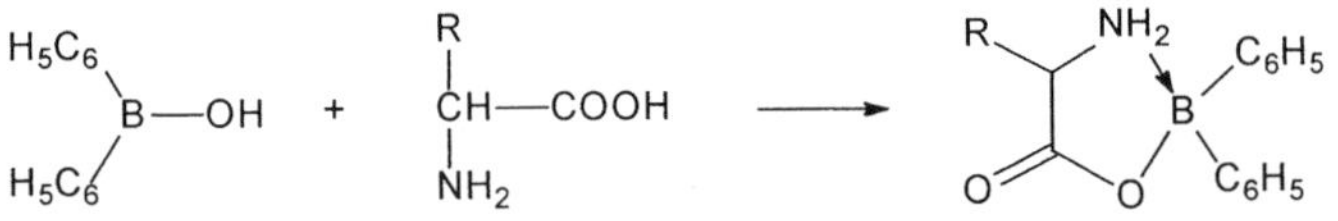

The derivatization allows the analysis of various amino acids and phosphorilated amino acids with separation on a C18 type column with UV detection. Also, 4-nitro-benzaldehyde forms a new oxazolinone cycle in the reaction with acylated glycine allowing determination using HPLC with UV detection [986]. Another cyclization takes place between tryptophan and formaldehyde with the formation of a fluorescent compound, which leads to the HPLC analysis of as low as 300 nmol analyte.

TABLE 19.10.5. *Isothiocyanates used for the formation of thiohydantoins with amino acids for analysis by HPLC.*

Reagent		Matrix, sensitivity	Detection λ Separation	Reference
4-N,N-dimethylamino-azobenzene-4'-isothiocyanate		amino acids less than 1 pmol homocysteic acid, etc.	UV 436 C18	[973]
p-N,N-dimethylamino-phenyl isothiocyanate		amino acids 0.2-0.6 ng	electrochem. glassy carbon 0.85 V Ag/AgCl	[974]
4-(N,N-dimethylamino-sulfonyl)-7-isocyanato-2,1,3-benzoxadiazole		amino acids	fluorescence ex 387, em 524 β-cyclo-dextrin type	[975]
diphenylphosphoro-isothiocyanate to form a 5-substituted-2-thioxoimidazolin-4-one		amino acids	UV 265	[1]
(R)-(-)-4-(3-isocyanato-pyrrolydin-1-yl)-7-(N,N-dimethylaminosulfonyl)-2,1,3-benzoxadiazole		amino acids chiral	fluorescence ex 460, em 540 ODS	[976]
(R)-(-)-4-(3-isocyanato-pyrrolydin-1-yl)-7-nitro-2,1,3-benzoxadiazole		amino acids chiral	fluorescence ex 490, em 530 ODS	[976]
1-naphthyl isothiocyanate		amino acids	UV 254, fluorescence ex 250, em 410 chiraldex	[973]
p-phenylazophenyl isothiocyanate		peptides	UV 254 C18	[977]
phenyl isothiocyanate		amino acids in blood, tissue, urine, ascitic fluid/3 μmol	UV 254 C18, CE, LC/MS	[962], [978], [979], [980], [981], [982], [983], [984]

Other reagents used for amino acid analysis are reported in various publications dealing with the subject in more detail [987].

- *Derivatization of small peptides for MS analysis*

The development of MS and MS/MS techniques for the detection and identification of peptides required some derivatizations that are useful for better identification of specific peptide chains. The procedures usually involve a two-step derivatization with the formation of a methyl ester at the terminal COOH group and a long chain acyl at the terminal NH_2. The esterification of the carboxyl can be done using methanol + HCl, diazomethane, methyl iodide and methylsulfinyl carbanion, triethyloxonium fluoroborate, etc. The acylation can be done with 2,2-dimethylpropanoic acid chloride (pivalyl chloride), HFBA, heptadecanoic acid and octadecanoic acid in the presence of ethyl chlorocarbonate [987a], N-(3-hydroxydecanoyl)succinimide, caproic ethylcarbonic anhydride, N-decanoyloxysuccinimide [15], succinimido dimethylaminobenzoate [988], N-hydroxy-succinimido-β-napthoate, benzoic ethylcarbonic anhydride [989], etc. for the same purpose there were used alkylation of the COOH group and formation of a Schiff base at the NH_2. A number of carbonyl compounds such as acetylacetone, p-dimethylamino benzaldehyde, etc. have been used for this purpose.

Specific amino acid residues can also be modified in small peptides. For example, the arginine residue can be transformed into a pyrimidyl ornithine group by treatment with $[(C_2H_5O)_2CH]_2CH_2$ (tetraethoxypropane) in methanol or can be transformed into a dimethylpyrimidyl ornithine group by treatment with acetylacetone in a reaction as follows [990]:

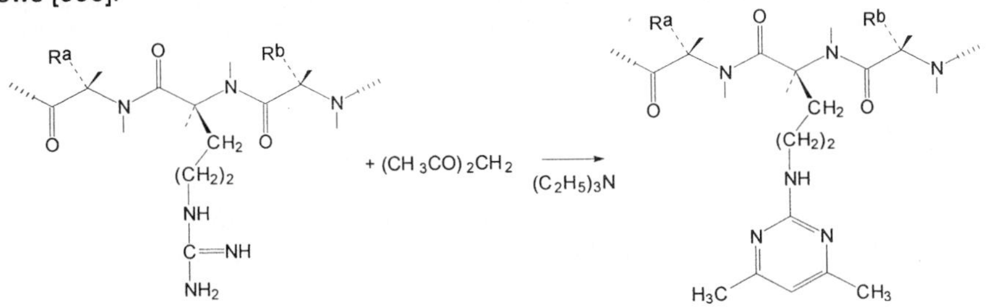

Peptides containing a methionine residue can be desulfurized by treatment with Raney nickel [991], those containing glutamic or aspartic acid can be methylated using CH_3I and methylsulfinyl carbanion, etc. Also a number of reactions can be applied to modify the terminal groups of peptides.

- *Derivatization of proteins for chromatographic analysis*

A significant number of studies are dedicated to analytical procedures for the detection, identification and quantitation of proteins. Protein labeling is a common procedure, and either the terminal NH_2 or COOH groups or other functional groups of certain amino acids in the protein chain such as SH, OH, COOH can be used for derivatization. The derivatization can be done using compounds with reactive groups. As reagents for the amino group, for example, are used active halides, N-hydroxysuccinimide esters, N-hydroxysulfo-succinimide esters (that can be water soluble), imidoesters, isocyanates, acyl azides, azidophenyl, etc. For the SH groups are used N-substituted maleimides, active halides, pyridyl disulfide, etc. For COOH group are used carbodiimides such as 1-ethyl-3-(dimethylaminopropyl)carbodiimide (EDC), etc. To the reactive groups can be attached moieties that carry chromophores, fluorophores, or other groups of interest such as biotin. As an example, N-hydroxysulfosuccinimidyl-biotin reacts with the NH_2 group of a protein as shown below:

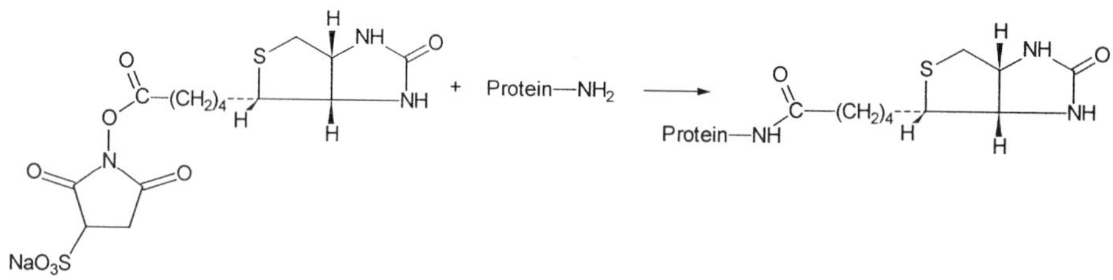

Biotinilated proteins can further bind with avidin or streptavidin. Extensive chemical modifications have little effect on the activity of avidin, and several avidin conjugated compounds are commercially available. Many protein assays may have a significant increase in sensitivity using these coupling procedures, including detection in chromatographic separations. Besides derivatization, the use of noncovalent dyes in protein staining for analysis is also very common [992], [993], [994], [995]. The subject of protein labeling and analysis is very complex and beyond the purpose of this book.

19.11 DERIVATIZATION OF OTHER MULTIFUNCTIONAL COMPOUNDS

A large number of compounds, either natural or synthetic have more than one type of functional group. The combinations of alcohol and carbonyl in carbohydrates and amine and acid in amino acids were previously discussed in Sections 19.9 and 19.10, respectively. Other combinations of groups that are rather common in practice are OH and NH_2 in amino alcohols or amino phenols, C=O and COOH in keto acids, OH and COOH in hydroxyacids, etc. Compounds with significant biological role such as catecholamines, prostaglandins, steroids, and nucleotides have multifunctional molecules. Derivatization of some of these compounds is discussed in this section. A systematic description of the derivatization of multifunctional compounds based on functional groups is rather difficult to make. Specific classes of chemicals such as herbicides, pesticides, specific drugs, and specific pollutants are extensively studied and analyzed using methods described in dedicated reviews and monographs [429], [996], [997], etc.

The presence of more functional groups in the same molecule can be an advantage in HPLC analysis. The derivatization for HPLC typically aims to attach to the molecule a group that facilitates the detection, and the most convenient active group in a multifunctional molecule can be chosen for this purpose. The other functional groups can be left underivatized because usually they do not give problems with the separation. On the other hand, for the GC analysis, any free functional group may affect undesirably the separation and must be derivatized.

The derivatization of multifunctional compounds can be done using specific reagents for each group in multistep derivatizations or using reagents that react with more than one group for a complete derivatization in one single step. For example, amino alcohols or amino thiols can be derivatized with reagents such as anhydrides or silylating reagents that react with both groups. The derivatization of more than one group with the same reagent must consider the differences in reactivity of different groups. It is important to achieve either a complete derivatization of all groups or complete derivatization of a certain group type and no derivatization for other types. Partial derivatization is not desirable. However, this cannot be always excluded. For this reason the derivatization of multifunctional compounds with a unique reagent is not always preferred, although it may seem convenient.

- Derivatization of amino alcohols and amino phenols

Amino alcohols can be derivatized at both amino and alcohol groups using acylating reagents. Acetic anhydride [998] or trifluoroacetic anhydride (TFAA) is used for this purpose, with GC analysis of the reaction products. For example, for GC analysis, the derivatization of 1-[(1-methyl-ethyl)amino]-3-(1-naphthalenyloxy]-2-propanol or propranolol with TFAA can be done in 5 min. at 50–60° C in the presence of trimethylamine. The reaction is the following [999]:

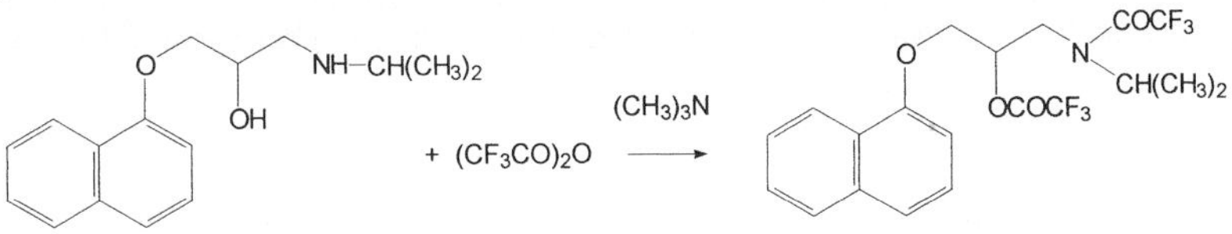

Other acylating reactions may lead only to the reaction at the amino group, which can be more easily derivatized. For example, heptafluorobutyric anhydride (HFBA) [1000] or pentafluorobenzoyl chloride [1001] reacts with ephedrine and related compounds derivatizing only the NH₂ group as shown below:

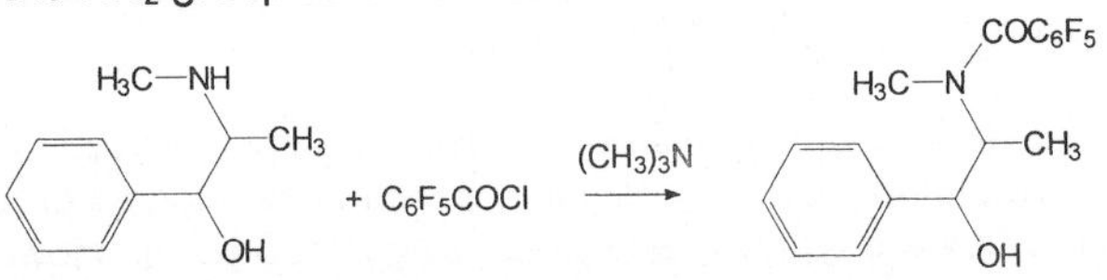

The derivatized products are analyzed by GC, the addition of fluorine in the molecule allowing good ECD detection. Phenolic amines are easier to acylate at both OH and NH₂ groups because of the higher acidity of the phenols compared to alcohols [15]. For example, dopamine can be analyzed by GC-MS after derivatization with pentafluoropropionic anhydride in a reaction as follows [1002], [1003]:

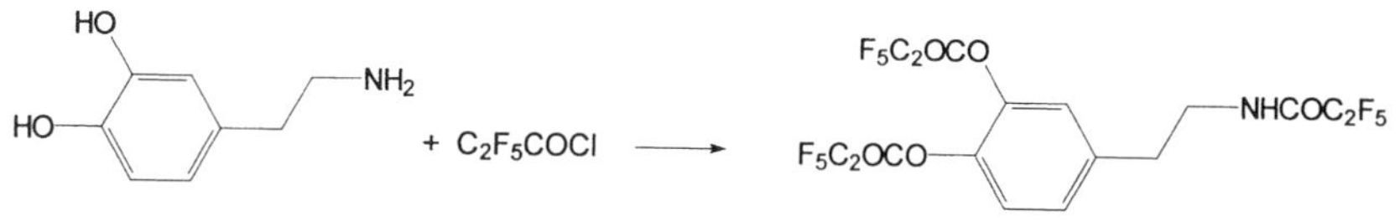

Silylation can be done for all active hydrogens or only for the OH groups, the amines being more difficult to silylate. Silylation of the OH (alcohol or phenol) and acylation of the NH$_2$ in a two-step derivatization also can be applied for this class of compounds [1004]. As an example, the silylation of norepinephrine using trimethylsilylimidazole (TMSI), by heating for 2–3 hours at 60° C and using CH$_3$CN as a solvent, takes place as follows:

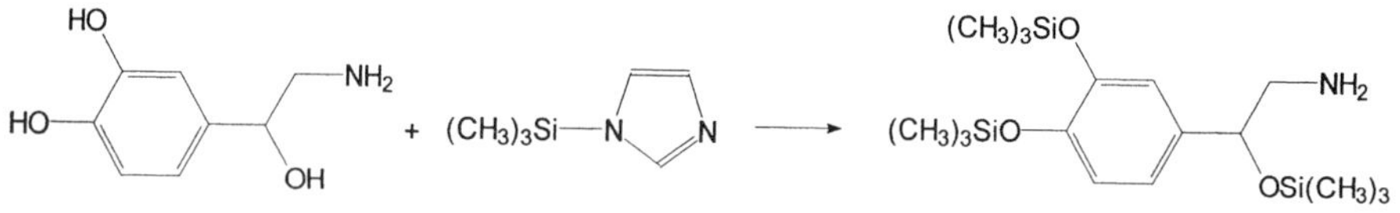

When using a mixture of BSA + TMCS as a silylation reagent, the silylation takes place at both the OH and NH$_2$ groups [1005]. When left underivatized, the amino group can be acylated, for example with TFAA.

A different type of derivatization that is possible for amino alcohols is the reaction with aldehydes or ketones, which leads to the formation of an oxazolidine cycle (see Section 18.5). For example, pseudoephedrine reacts when heated at 60° C for 1 hour with pentafluorobenzaldehyde as follows:

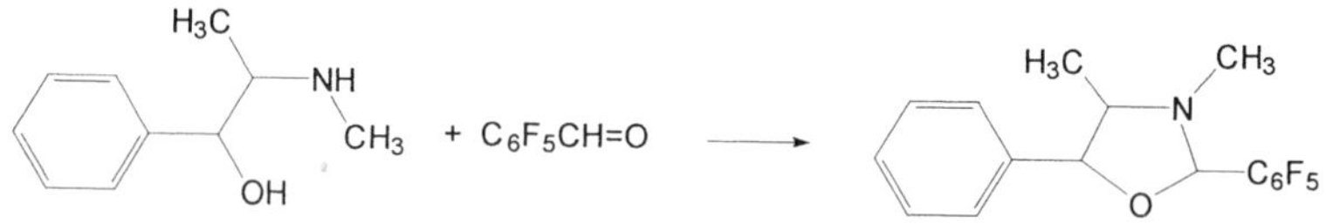

The derivatized product can be analyzed using GC or GC-MS [1006].

Other reactions are also used for amino alcohol analysis. Among these are the reaction with CS$_2$ at the amine group with the formation of an isothiocyanate and derivatization of the OH groups by silylation or alkylation [15], formation of cyclic dimethylsilyl-methylenes, formation of boronates [1007], etc.

HPLC analysis of derivatized amino alcohols is also frequently utilized. A number of β-blocking agents are amino alcohols, including compounds such as propranolol, acebutolol, etc., and they are typically analyzed using HPLC [269]. However, HPLC derivatization does not require the derivatization at all active groups, and only the most convenient group is typically utilized for derivatization. It is common that the amino alcohols are derivatized at the NH$_2$ group for HPLC analysis (see Section 19.4).

- Derivatization of hydroxy and keto acids

Hydroxy acids can be derivatized using two-step reactions such as alkylation of the COOH group and acylation of the OH group. The reagents typically used for the alkylation of the COOH were discussed in Section 19.6, and some of these reagents are used for the alkylation of the COOH group in hydroxy acids. Examples are CH_2N_2, CH_3OH in the presence of BF_3, C_4H_9OH in the presence of BF_3, etc. Among the reagents used for the acylation of the alcohol group in hydroxy acids are acetic anhydride, heptafluorobutyrylimidazole, heptafluorobutyric anhydride, etc. [15]. Silylation can be used as a unique derivatization step for both COOH and OH groups [1008] in nonaromatic hydroxy acids as well as in phenolic acids.

A number of α-hydroxyacids are derivatized using butylboronic acid. Cyclic boronates are formed in the reaction, as shown below [1009]:

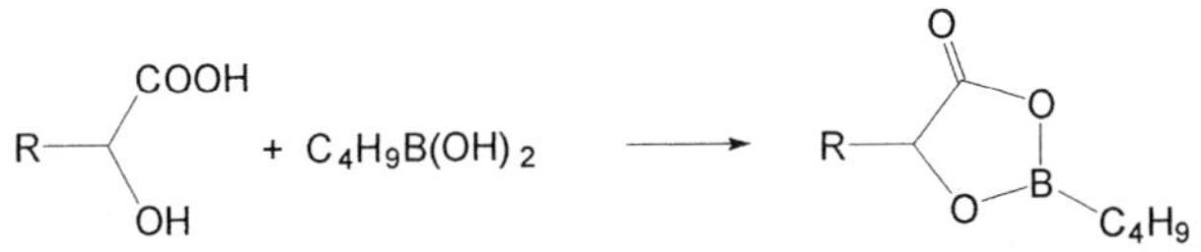

The derivatization of keto acids can be done using alkylation or silylation reactions at the COOH group and derivatizations such as oxime formation for the carbonyl [1010], [1011]. The formation of quinoxalines derivatives, as shown below, can also be used for the derivatization of α-ketoacids (see also Section 18.5):

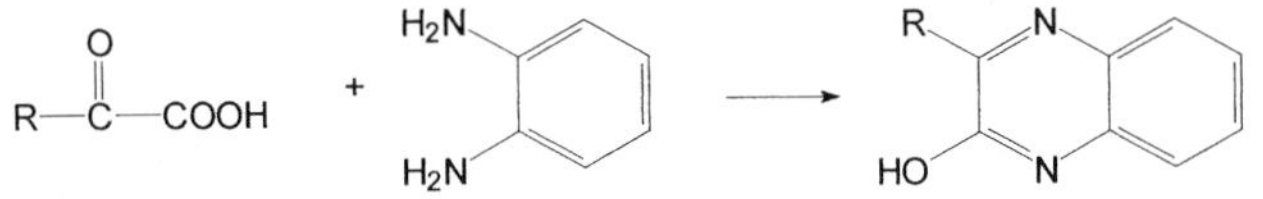

The resulting quinoxalines contain an OH group that can be further derivatized, for example, by silylation when the GC analysis is intended [1012].

A special group of hydroxy or keto acids is that of prostaglandins. The structures of prostaglandin E_1 and $F_{2\alpha}$ are shown below:

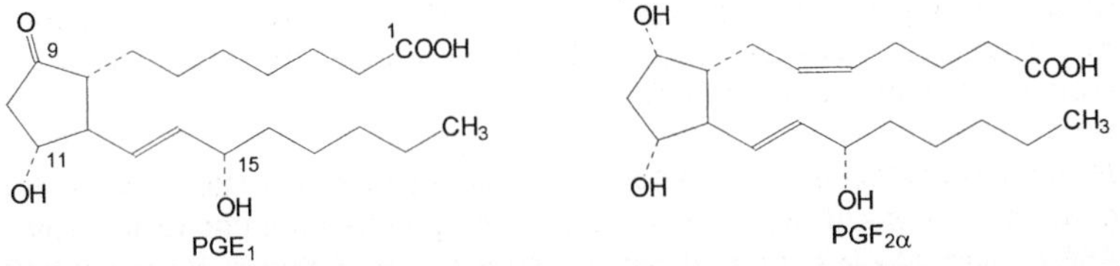

Prostaglandins can be alkylated, acylated or silylated. A permethylation is possible with CH_2N_2 in the presence of BF_3 (etherate) [15], although the acid group is usually easier to methylate, while the alcohol may lead to partial methylation. The OH group can be acylated in a second derivatization step. For example, a two-step derivatization can be done with CH_2N_2 followed by derivatization of the OH groups with HFBA. In this procedure, 0.01–100 µg $PGF_{1\alpha}$ and $PGF_{2\alpha}$ are dissolved in ethyl ether containing 10% methanol, and a stream of CH_2N_2 is passed through the solution for 2 min. The material is then evaporated to dryness and 10 µL HFBA is added. After 5–15 min. the solution is evaporated to dryness, and the residue is redissolved in hexane and analyzed by GC or GC-MS [1013].

The methylation with CH_2N_2 of COOH group can be associated with the silylation of the OH groups. The silylation can be done for example using BSA. The prostaglandins containing a keto group may form enols, and the silylation also takes place at the enol group. To avoid this problem, the derivatization can be done in three steps: methylation of the COOH with CH_2N_2, reaction with a substituted hydroxylamine at the C=O group, and silylation of the OH groups [1014]. The reactions of this type using pentafluoro-benzylhydroxylamine for the derivatization of the carbonyl can be written as follows.

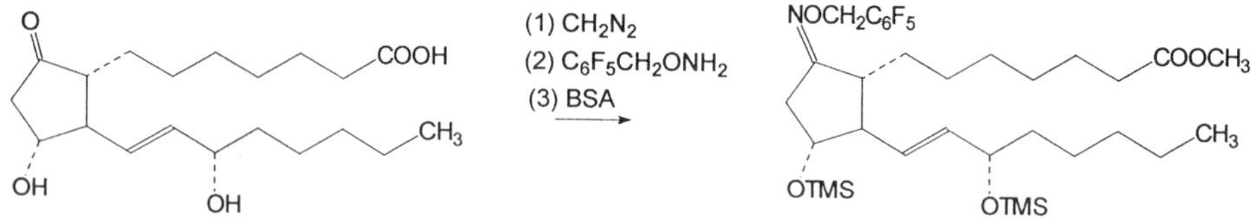

(1) CH_2N_2
(2) $C_6F_5CH_2ONH_2$
(3) BSA

One-step derivatization of prostaglandins for GC analysis can be done using silylation. This can be applied to PGF_1 and PGF_2, using for example BSTFA + TMCS at room temperature within 2 hours [1015], or to PGE_1 and PGE_2 by heating the sample with BSTFA in piperidine at 60° C for 1 hour [1016], [1017]. In the reaction with BSTFA, the carbonyl group in PGE forms an enol that is also silylated as shown below:

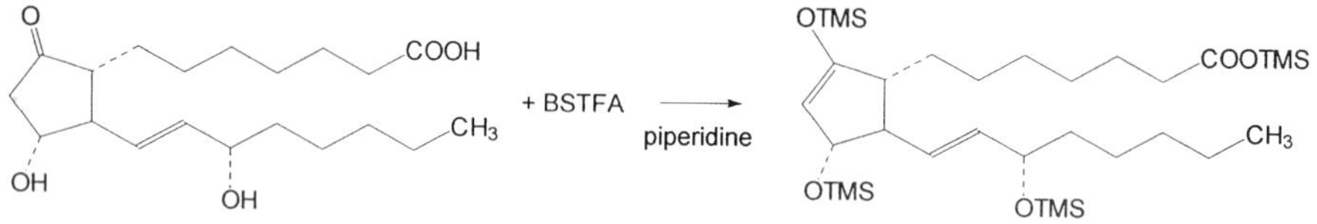

+ BSTFA

piperidine

The silylation of the enol forms can be avoided by preliminary derivatization of the carbonyl, for example, with CH_3ONH_2 [1015]. Other volatile derivatizations of prostaglandins include formation of boronates, degradative hydrogenation, etc. [15].

One other group of compounds containing OH groups possibly associated with C=O or COOH groups is that of steroids (see also Section 19.1). Steroids are compounds common in most animal and plant cells. They derive from two common structures, colestane (*alo* series) and coprostane (*normal* series) containing three cyclohexane (chair conformation) and one cyclopentane fused cycles. Colestane is the *trans*-homologue of coprostane. All sterols contain an alcohol group. The structures of cholesterol (cholest-5-en-3β-ol) that belongs to *alo* series and of cholic acid (3α,7α,12α-trihydroxy-5β-cholan-24-oic acid) that belongs to the *normal* series are given below:

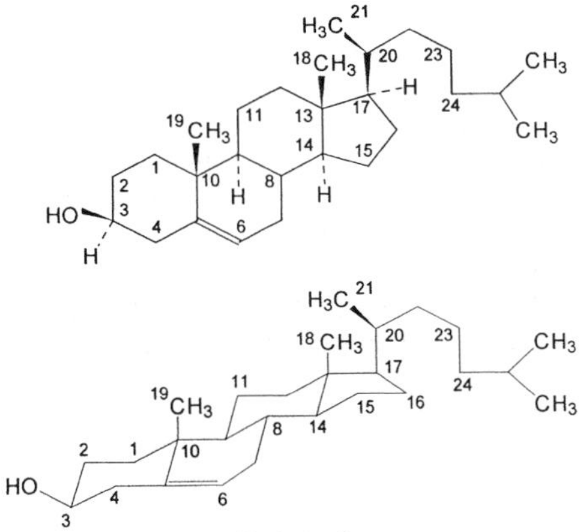

Cholesterol
(alo-series)

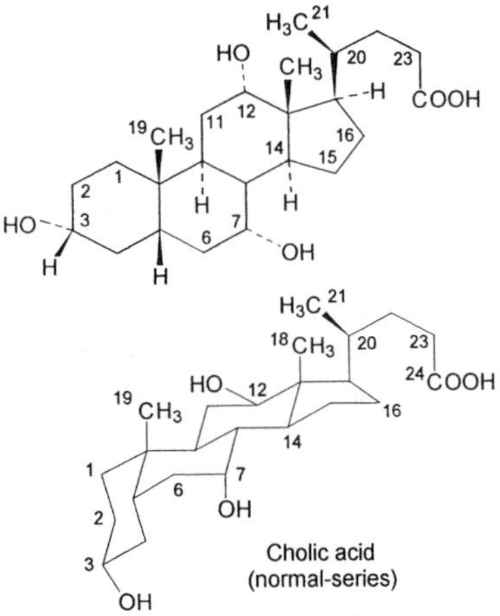

Cholic acid
(normal-series)

Due to the presence of one or more OH groups in steroid structure, most derivatizations are reactions to the alcohol group. For this reason, many steroid derivatizations were already discussed in Section 19.1, and some examples of derivatization for phenolic steroids were given in Section 19.2. Other sterols include ketosteroids such as testosterone, bile acids that contain a carboxyl group in addition to hydroxyl, digitalis steroids, etc. Vitamins from the D series are also related to steroids. In nature, sterols can be found free or esterified with fatty acids. In this section are discussed only derivatizations involving besides the OH group the C=O group in ketosteroids or COOH group in bile acids.

Ketosteroids can be derivatized at the C=O functionality using typical reagents for this group (see Section 19.5) including hydroxylamine [1018], methoxylamine (O-methylhydroxylamine) [15], [1019], and various other O-alkyl-hydroxylamines such as O-sec-butylhydroxylamine, O-isobutylhydroxylamine, O-pentylhydroxylamine, O-benzylhydroxylamine, O-pentafluorobenzylhydroxylamine [1020], etc. In these reactions the OH group is not affected, and for GC analysis it is common to use a second derivatization such as a silylation. For example, testosterone can be analyzed using a derivatization with hydroxylamine hydrochloride followed by derivatization with bis(trimethylsilyl)acetamide (BSA). The reactions can be written as follows:

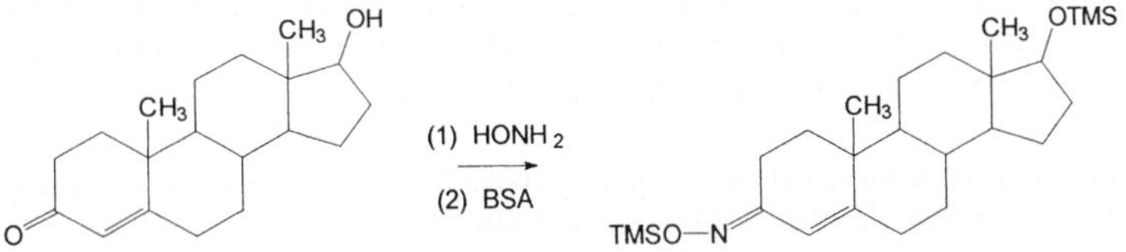

In this reaction, the active hydrogen from the oxime group is also silylated. When the first derivatization is done with O-alkylhydroxylamines, the silylation is applied only to the OH groups from the sterol. The reactions with hydroxylamines usually require long reaction times and mild heating. An example of derivatization starts with 1 mg steroid that is dissolved in 0.5 mL solution of O-benzylhydroxylamine hydrochloride in pyridine (20–40 mg/mL). The solution is heated for 15–20 hours at 60–70° C, then evaporated to dryness under N_2, and processed for the silylation. The derivatization of all OH groups can be done by adding 0.5 mL TMSI or TMSI/BSA (1:1, v/v) to the dried mixture and heating at 150° C for 3 hours. Derivatization of nonhindered or moderately hindered OH groups can be achieved using 0.5 mL BSTFA and heating at 100° C for 1 hour. The analysis can be done by GC or GC-MS. The procedure allows clear separation of ketosteroids from those containing only OH groups because benzoyloxime derivatives have long retention times [P. G. Devaux et al., Anal. Lett., 4 (1971) 151]. Some oxime derivatives may display double peaks due to *syn* and *anti* isomers.

Other derivatizations of ketosteroids have been done with hydrazine derivatives such as N,N-dimethylhydrazine, pentafluorophenylhydrazine [1021], [1022]. 2,4-Dinitro-phenylhydrazine has been used in derivatizations for HPLC analysis (see Section 19.5). The free OH groups can be further derivatized for GC analysis either by silylation or methylation (in phenolic steroids). Keto groups can also be derivatized using ethanedithiol in the presence of p-toluenesulfonic acid (TsOH) in a reaction as follows [1023]:

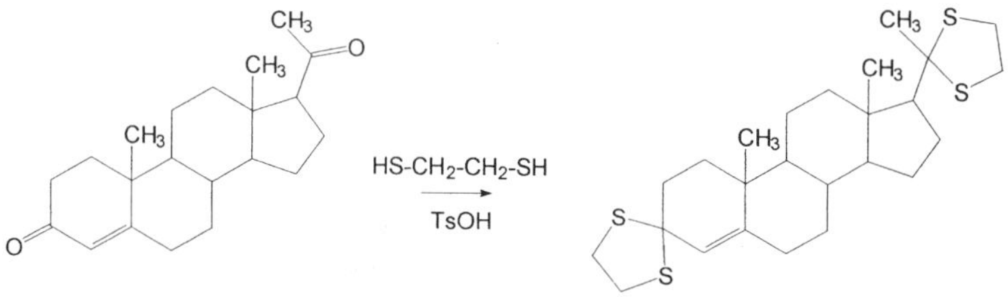

One feature of many ketosteroids is their capability to form enough enol form that some derivatizations such as silylations or acylations displace the equilibrium toward the enol and form at least in part enol derivatives [1024]. This is for example the case of silylation with various silylating reagents such as BSTFA or TMSI performed in the presence of potassium acetate. The silylation of hindered OH groups in sterols requires strong silylation conditions, and addition of KOOC-CH3 has been used to increase silylation yields (see also Section 18.2). A side effect of this procedure is the formation of artifacts from the carbonyl groups. One procedure consists of dissolving 30 μg steroid in a few drops of CS$_2$ followed by the addition of 100 μL BSTFA and 0.5-1 mg KOOC-CH$_3$. The mixture is heated at 60° C for 40 min., evaporated and redissolved in hexamethyl-disilazane (HMDS), and then analyzed by GC-MS [1025]. In these conditions OH groups in positions 3α, 3β or 20α are completely derivatized, while 17α-tertiary OH and 11β-OH are not reactive. Partial enolization is obtained from 17-keto group, while 3-keto group is unreactive. In the presence of 21-hydroxy group, the 20-keto group is also reactive although not reactive otherwise.

The capability to form enols is used for the derivatization of certain keto-enols with heptafluorobutyric anhydride followed by GC analysis with ECD detection. For example, testosterone can be derivatized with heptafluorobutyric anhydride (HFBA) [1026], [1027], [1028], pentafluorobenzoic anhydride + HFBA, etc.

Various other derivatizations are performed on steroids such as oxidation reaction for the elimination of the side chain in position 17 and conversion into a keto group [15], oxidation followed by the formation of lactones, etc.

Bile acids are derivatized usually using two-step derivatizations, one for the OH groups and the other for COOH. For example, CH$_2$N$_2$ can be used for the formation of the methyl ester and HMDS + TMCS for the formation of the silylation of the OH groups. Two-step derivatization using diazomethane and TFAA has been used for ECD-GC analysis of various bile acids [1029]. Other derivatizations for the COOH group also have been applied such as reaction with methanol + HCl or with hexafluoroisopropanol. The OH groups can be derivatized using silylation or acylation [15].

Several derivatizations for HPLC analysis of sterols were already indicated in Sections 19.1, 19.2 and 19.6. Because the attachment of a chromophore or fluorophore usually requires only one functional group, specific derivatizations were presented in connection with the group chosen for derivatization (alcohol, phenol or acid). For example, bile acids are commonly derivatized as acids, and any of the following reagents can be used for derivatization in HPLC: 4-aminophenol [1], 2-bromoacetyl-6-methoxynaphthalene [1030], 1-(bromoacetyl)pyrene [1031], 3-bromomethyl-7-methoxy-1,4-benzoxazin-2-one [1032], [1033], 4-bromomethyl)-7-methoxycoumarin [1034], [1035], 1-(2,5-dihydroxyphenyl)-2-bromoethanone [1036], [1037], 1-naphthyldiazomethane [1], 4-aminophenol, [1].

- *Derivatization of nucleotides and related compounds*

Nucleotides are the monomeric units of nucleic acids. A nucleotide is formed from a carbohydrate residue connected to a heterocyclic base by a β-D-glycosidic bond and to a phosphate group at C-5' (compounds containing the phosphate group at C-3' are also known). The molecules derived from nucleotides by removing the phosphate group are the nucleosides. The carbohydrate in a nucleotide resulting from a deoxyribonucleic acid (DNA) is 2-deoxy-D-ribose, and the one resulting from ribonucleic acid (RNA) is ribose. The heterocyclic bases present in nucleotides are certain purines (adenine and guanine) and pyrimidines (cytosine, thymine and uracil).

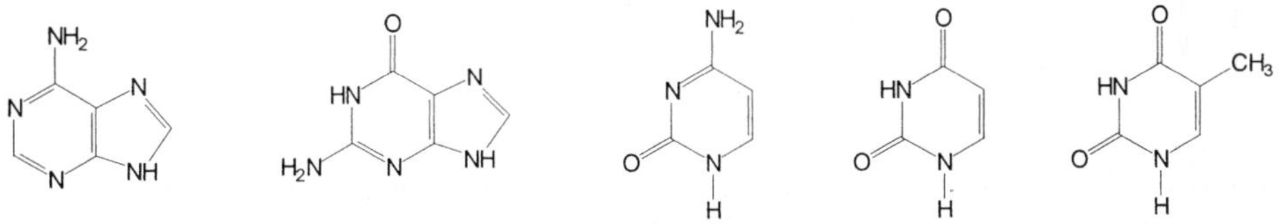

Adenine (A)   Guanine (G)   Cytosine (C)   Uracil (U)   Thymine (T)

Trace amounts of other bases are occasionally present in DNAs and RNAs. For example, the DNA may contain 5-hydroxymethyl cytosine and several N-methyl purines. The transfer RNA (tRNA) may contain several unusual bases such as certain methyl purines or hydroxymethyl pyrimidines. The structures of adenosine-5'-phosphoric acid (5'-adenylic acid) and uridine-5'-phosporic acid (5'-uridylic acid) are shown below:

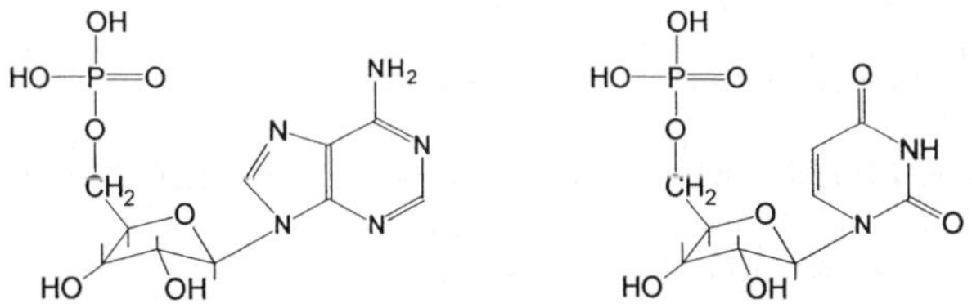

Different hydrolysis conditions of the nucleic acids may lead to nucleotides, nucleosides and phosphate, or even to the five-carbon monosaccharide, the phosphate, and the heterocyclic base.

The derivatization of the heterocyclic bases from nucleic acids is typically done using either methylation or silylation. Methylation in the injection port of the GC using tetramethylammonium hydroxide (TMAH) or trimethylphenylammonium hydroxide is applied to both purines and pyrimidines [1038]. Silylation is also used as a derivatization procedure. As an example, the reaction with guanine occurs as follows:

The reaction takes place by heating at $150°$ C for 30 min. 500 ng base, 50 μL BSTFA, 25 μL CH_3CN and 25 μL $C_2H_4Cl_2$ [1039], [1040], [1041]. Other silylation reagents are applied for derivatization, such as BSA, BSTFA + TMCS, and TMSDEA + TMCS [15]. For these silylations it is typical for the C=O group, when present in the base, to be silylated in its enol form.

Guanine and other bases also can be derivatized after a deamination step. This can be done using a reaction with *tert*-butyl nitrite in HCl as follows:

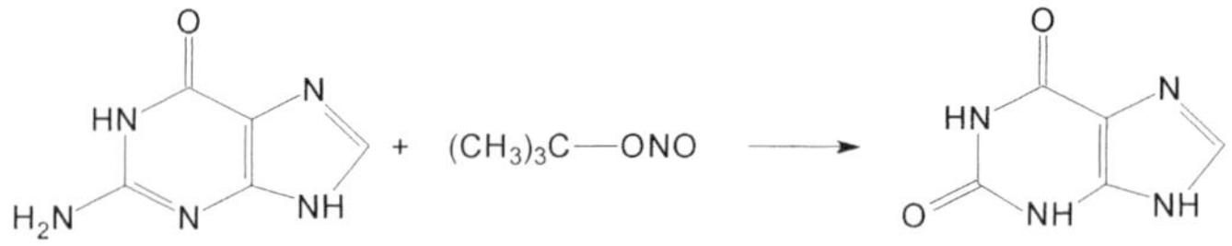

The resulting 1,3,7-trihydropurine-2,6-dione can be further derivatized using for example pentafluorobenzylbromide, which reacts forming di or three substituted amides. The reaction with the purine resulting from the deamination of 7-(2'-hydroxyethl)-guanine takes place as shown below:

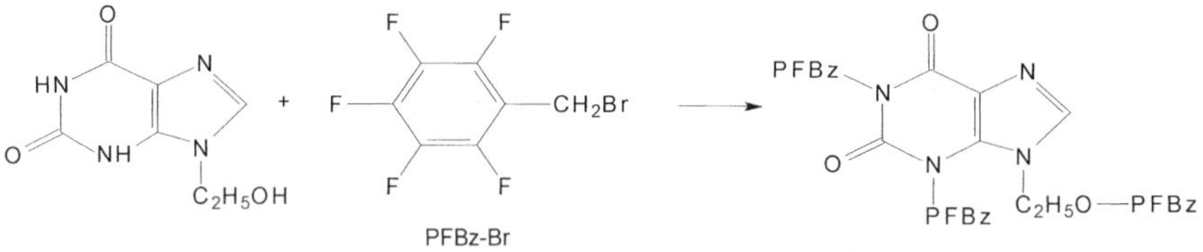

The resulting fluorinated compounds can be analyzed using GC and NCI-MS [1042].

For the derivatization of nucleosides and deoxynucleosides, methylation and silylation are the most common procedures. Permethylation can be done, for example, using CH_3I and Ag_2O, or CH_3I and $CH_3SOCH_2^-$ [1043]. The reaction for adenosine is written below:

Silylation can be done using typical reagents such as HMDS + TMCS, BSTFA, BSTFA + TMCS, etc. Similarly to the silylation of the free heterocyclic bases, the carbonyl groups may react in their enol form [1044]. Two-step derivatizations are also applied, using for example silylation with BSTFA + TMCS and acylation of the amino groups with trifluoroacetylimidazole (TFAI) [1045]. The formation of an acetonide at the two vicinal OH groups in ribose followed by silylation of the other groups that contain active hydrogens can be used for GC or GC-MS analysis of nucleosides or deoxynucleosides.

Nucleotides and deoxynucleotides are also derivatized using methylation or silylation. The methylation can be done, for example, using TMAH and GC analysis with the injector port of the GC set at 300° C. The phosphate group is also methylated in these conditions at the two active hydrogens [1046]. Silylation can be done using HMDS + TMCS, BSA + TMCS, or BSTFA + TMCS. The reaction with deoxyadenosine-5'-phosphoric acid (or phosphate) takes place at all active hydrogens in the molecule. In a typical procedure, 20–50 μg nucleotide is dissolved in 30 μL pyridine, 30 μL BSTFA, and 1 μL TMCS. Depending on the nature of the nucleotide, for the reaction the mixture can be heated up to 150° C, but also lower temperatures may give satisfactory results [1047]. The phosphate group also can be attached to a nucleoside in position 3'. The silylation of 3'-uridylic acid with trimethylsilylimidazole takes place again at all active hydrogens in the molecule. A significant number of other reagents are available for the derivatization of nucleotide fragments in nucleic acids.

REFERENCES 19

1. G. Lunn, L. C. Hellwig, *Handbook of Derivatization Reactions for HPLC*, J. Wiley, New York, 1998.
2. T. Toyo'oka (ed.), *Modern Derivatization Methods for Separation Sciences*, J. Wiley, Chichester, 1999.
3. K. Blau, J. Halket (eds.), *Handbook of Derivatives for Chromatography*, J. Wiley, Chichester, 1993.
4. I. Ciucanu, F. Kerek, Carbohydr. Res., 13 (1984) 209.
5. R. N. Shah et al., J. Carbohydr. Chem., 6 (1987) 645.
6. S. Hakomori, J. Biochem. (Tokyo), 55 (1964) 205.
7. M. Neeman et al., Tetrahedron, 6 (1959) 36.
8. E. Muler, W. Rundel, Angew. Chem., 70 (1958) 105.
9. R. Ryhage, B. Samuelson, Biochem. Biophys. Res. Commun., 19 (1965) 279.
10. A. E. Pierce, *Silylation of Organic Compounds*, Pierce Chem. Co., Rockford, 1982.
11. M. Dönike, J. Chromatogr., 42 (1969) 103.
12. R. Piekos et al., J. Chromatogr., 116 (1976) 315.
13. M. Dönike, L. Jaenike, Angew. Chem. Ind. Ed., 8 (1969) 974.

14. R. Piekos et al., J. Chromatogr., 117 (1976) 431.
15. D. R. Knapp, *Handbook of Analytical Derivatization Reactions*, J. Wiley, New York, 1979.
16. Q. Lang, C. M. Wai, Anal. Chem., 71 (1999) 2929.
17. W. J. Chamberlain et al., J. Chromatogr., 513 (1990), 55.
18. J. A. McCloskey et al., Anal. Chem., 40 (1968) 233.
19. S. C. Moldoveanu, unpublished results.
20. S.-L. Li et al., J. Chromatogr., A, 859 (1999) 183.
21. W. R. Supina et al., Am. Oil Chem. Soc., 44 (1967) 74.
22. W. J. Richter, D. H. Hunnemann, Helv. Chim. Acta, 57 (1974) 1131.
23. P. M. Burkinshaw, E. D. Morgan, J. Chromatogr., 132 (1977) 548.
24. S. C. Moldoveanu, unpublished results.
25. R. B. Watts, R. G. O. Kekwick, J. Chromatogr., 88 (1974) 15.
26. D. Borrey et al., J. Chromatogr., A, 910 (2001) 105.
27. J. S. Fritz, G. H. Schenk, Anal. Chem., 31 (1959) 1808.
28. D. P. Schwartz, Anal. Biochem., 71 (1967) 24.
29. A. Zlatkis, B. C. Petit, Chromatographia, 2 (1969) 484.
30. M. Dönike, J. Chromatogr., 78 (1973) 273.
31. J. M. E. Quirke, G. J. Van Berkel, J. Mass Spectrom., 36 (2001) 179.
32. R. Vilceanu, P. Schultz, J. Chromatogr., 82 (1973) 279.
33. Z.-P. Lin, W. A. Aue, J. Chromatogr., A, 855 (1999) 538.
34. C. J. W. Brooks et al., Anal. Chem., 45 (1973) 896.
35. W. Pereira et al., Anal. Lett., 3 (1970) 23.
36. R. S. Burden et al., J. Chromatogr., 391 (1987) 273.
37. J. M. Cross et al., J. Chromatogr. Sci., 8 (1970) 679.
38. P. Michelsen, G. Odham, J. Chromatogr., 331 (1985) 295.
39. Y. Yamazaki, H. Maeda, Agric. Biol. Chem., 50 (1986) 79.
40. P. E. Sonnet et al., J. Chromatogr., 436 (1988) 205.
41. S. Hammarstrom, M. Hamberg, Anal. Biochem., 52 (1973) 169.
42. M. A. J. Bayliss et al., J. Chromatogr., 445 (1988) 393.
43. M. Holcapek et al., Anal. Chem., 71 (1999) 2288.
44. O. R. Idowu et al., J. Liq. Chromatogr., Relat. Technol., 20 (1997) 1553.
45. S.-H. Chen et al., J. Liq. Chromatogr., Relat. Technol., 20 (1997) 1967.
46. E. E. Knaus et al., J. Chromatogr. Sci., 14 (1976) 525.
47. S. Tirendi et al., Farmaco., 49 (1994) 427.
48. Y. Nakagawa et al., Microbiol. Immunol., 41 (1997) 27.
49. O. R. Idowu et al., J. Liq. Chromatogr., Relat. Technol., 20 (1997) 1553.
50. L. Virag et al., J. Chromatogr., B, 681 (1996) 263.
51. D. Gosh et al., Chromatographia, 37 (1993) 543.
52. R. G. Harvey, H. Cho, Anal. Biochem., 80 (1977) 540.
53. R. E. Doolittle, R. R. Health, J. Org. Chem., 49 (1984) 5041.
54. M. Aycard et al., J. Liq. Chromatogr., 15 (1992) 2175.
55. S. R. Almquist et al., J. Chromatogr., A, 697 (1994) 139.
56. T. Wolski et al., J. Chromatogr., 362 (1986) 217.
57. M. Petchey, M. J. C. Crabbe, J. Chromatogr., 307 (1984) 180.
58. S. Nojiri et al., J. Chromatogr., A, 893 (2000) 195.
59. J. Doehl, T. Greibrokk, J. Chromatogr., 529 (1990) 21.
60. H. E. Indyk, D. C. Woolard, Analyst, 119 (1994) 397.
61. J. Dohl, T. Greybrokk, J. Chromatogr., 282 (1983) 435.
62. K. Imai et al., Biomed. Chromatogr., 8 (1994) 107.

63. J. C. Motte et al., J. Chromatogr., A, 728 (1996) 333.
64. J. Gato et al., J. Chromatogr., 276 (1982) 289.
65. J.-H. Kim et al., J. Chromatogr., A, 693 (1995) 241.
66. Y. Nashida, J. Carbohydr. Chem., 13 (1994) 1003.
67. J. You et al., J. Chromatogr., A, 909 (2001) 171.
68. J. M. You et al., Chromatographia, 49 (1999) 657.
69. P. Assaf et al., J. Chromatogr., A, 869 (2000) 243.
70. A. I. Haj-Yehia, L. Z. Benet, J. Chromatogr., A, 724 (1996) 107.
71. M. Novotny et al., J. Chromatogr., 292 (1984) 159.
72. H. Spahn et al., J. Chromatogr., 310 (1984) 167.
73. M. Jiménez et al., J. Chromatogr., A, 870 (2000) 473.
74. M. Yamaguchi, Anal. Chim. Acta, 193 (1987) 209.
75. J. Goto et al., Chem. Pharm. Bull., 29 (1981) 899.
76. H. Fujino, S. Goya, Anal. Sci., 5 (1989) 105.
77. H. Fujino, S. Goya, Chem. Pharm. Bull., 37 (1989) 1939.
78. G. Huang et al., Anal. Chem., 71 (1999) 4245.
79. J. Gato et al., Anal. Sci., 7 (1991) 723.
80. J. Gato et al., Anal. Sci., 7 (1991) 645.
81. J. Gato et al., Anal. Sci., 6 (1990) 261.
82. K. E. Karlsson et al, Anal. Chem., 57 (1985) 229.
83. S. Uchiyama et al., Anal. Chem., 71 (1999) 5367.
84. T. Toyo'oka et al, J. Chromatogr., A, 675 (1994) 79.
85. J. Ishida et al., J. Chromatogr., B, 692 (1997) 31.
86. C. X. Gao, I. S. Krull, J. Chromatogr., 515 (1990) 337.
87. P. M. J. Kwakman et al., J. Pharm. Biomed. Anal., 9 (1991) 753.
88. R. Wintersteiger, M. J. Sepuldeva, Anal. Chim. Acta, 273 (1993) 383.
89. M. Nakajima et al., J. Chromatogr. 641 (1993) 176.
90. L. R. Phillips et al., Proc. Am. Soc. Mass Spectrom., 43 (1995) 163.
91. O. R. Idowu et al., J. Liq. Chromatogr., Relat. Technol., 20 (1997) 1553.
92. H. Zegota, J. Chromatogr., A, 663 (1999) 227.
93. Y. Daali et al., J. Chromatogr., A, 903 (2000) 237.
94. E. O. A. Haahti, H. M. Fales, J. Lipid Res., 8 (1967) 131.
95. C. J. W. Brooks, I. Maclean, J. Chromatogr. Sci., 9 (1971) 18.
96. P. Bournot, J. Chromatogr., 57 (1971) 55.
97. F. Eisenberg, Carbohydr. Res., 19 (1971) 135.
98. P. J. Wood et al., Carbohydr. Res., 42 (1975) 1.
99. K. Gamoh et al., Anal. Chim. Acta, 228 (1990) 101.
100. K. Gamoh et al., J. Chromatogr., 515 (1990) 227.
101. S. T Wu et al., J. Chromatogr., B., 692 (1997) 149.
102. K. Gamoh et al., J. Chromatogr., 469 (1989) 424.
103. J. Pis, J. Harmatha, J. Chromatogr., 596 (1992) 271.
104. C. J. Biermann, G. D. McGinnis, *Analysis of Carbohydrates by GLC and MS*, CRC Press, Boca Raton, 1989.
105. S. C. Moldoveanu, unpublished results.
106. D. C. DeJongh et al., J. Am. Chem. Soc., 91 (1969) 1728.
107. T. Okuda et al., Carbohydr. Res., 68 (1979) 1.
108. R., N. Shah et al., J. Carbohydr. Chem., 6 (1987) 645.
109. G. J. Gerwig, Carbohydr. Res., 62 (1983) 349.
110. D. C. DeJong, K. Biemann, J. Am. Chem. Soc., 85 (1963) 2289.
111. C. Bosso et al., Org. Mass Spectrom., 12 (1977) 493.

112. J. Wiecko, W. R. Sherman, J. Am. Chem. Soc., 98 (1976) 7631.
113. G. Eklund et al., J. Chromatogr., 142 (1977) 575.
114. J. E. Sullivan, L. R. Schewe, J. Chromatogr. Sci., 15 (1977) 196.
115. S. C. Moldoveanu, unpublished results.
116. V. N. Reinhold et al., Carbohydr. Res., 37 (1974) 203.
117. R. Greenhalgh, J. P. Wood, J. Chromatogr., 82 (1973) 410.
118. D. C. DeJongh, K. Biemann, J. Am. Chem. Soc., 86 (1964) 67.
119. S. Morgenlie, Carbohydr. Res., 41 (1975) 285.
120. M. Kiso, A. Hasegawa, Carbohydr. Res., 52 (1976) 87.
121. M. L. Wolfrom et al., J. Org. Chem., 33 (1968) 1067.
122. Z. I. Nikolov, P. J. Reilly, J. Chromatogr., 254 (1983) 157.
123. C. Bosso et al., Org. Mass Spectrom., 12 (1977) 493.
124. V. N. Reinhold et al., Carbohydr. Res., 37 (1974) 203.
125. G. J. Gerwig, Carbohydr. Res., 62 (1983) 349.
126. F. Nachtmann, K. W. Budna, J. Chromatogr., 136 (1977) 279.
127. H. Gleispach, J. Chromatogr., 91 (1974) 407.
128. E. Heftmann (ed.), *Modern Methods of Steroid Analysis*, Academic Press, New York, 1973.
129. E. D. Morgan. C. F. Poole, J. Chromatogr., 116 (1976) 333.
130. P. Voros, D. J. Harvey, Anal. Chem., 45 (1973) 7.
131. W. J. A. VandenHeuval, J. Chromatogr., 85 (1967) 85.
132. G. Phillipou, J. Chromatogr., 129 (1976) 384.
133. I. A. Blair, G. Phillipou, J. Chromatogr. Sci., 15 (1977) 478.
134. G. Philipou et al., Steroids, 26 (1975) 516.
135. C. J. W. Brooks, B. S. Middleditch, Anal. Lett., 5 (1972) 611.
136. J. R. Chapman, E. Bailey, J. Chromatogr., 89 (1974) 215.
137. B. S. Thomas, J. Chromatogr., 56 (1971) 37.
138. E. Symes, B. S. Thomas, J. Chromatogr., 116 (1976) 163.
139. E. D. Morgan, C. F. Poole, J. Chromatogr., 104 (1975) 351.
140. R. W. Kelly, Steroids, 13 (1969) 507.
141. W. J. A. VandenHeuval et al., Biochim. Biophys. Acta, 48 (1961) 596.
142. I. R. Sarda et al., Steroids, 12 (1968) 607.
143. C. F. Poole, E. D. Morgan, J. Chromatogr., 90 (1974) 380.
144. N. Ikekawa et al., J. Chromatogr. Sci., 10 (1972) 233.
145. H. Miyazaki et al., Anal. Chem., 45 (1973) 1164.
146. P. A. Leclercq, D. M. Desiderio, Anal. Lett., 4 (1971) 305.
147. E. Bayley, Steroids, 10 (1967) 527.
148. C. J. W. Brooks, D. J. Harvey, J. Chromatogr., 54 (1971) 193.
149. S. K. Yang et al., J. Chromatogr., 461 (1989) 377.
150. J. D. Harvey et al., Anal. Lett, 5 (1972) 745.
151. D. Munger et al., J. Chromatogr., 143 (1977) 375.
152. F. Van Damme, A. C. Oomens, J. Chromatogr., A, 696 (1995) 41.
153. L. G. Johnson, J. Assoc. Off. Anal. Chem., 56 (1973) 1503.
154. F. K. Kawahara, Anal. Chem., 40 (1968) 1009.
155. F. K. Kawahara, Anal. Chem., 40 (1968) 2073.
156. I. C. Cohen et al., J. Chromatogr., 44 (1969) 251.
157. J. N. Seiber et al., J. Chromatogr., 73 (1972) 89.
158. S. F. Howard, G. Yip, J. Assoc. Off. Anal. Chem., 54 (1971) 970.
159. H. Wotiz, S. Chattoraj, Anal. Chem., 36 (1964) 1466.
160. J. Porschmann et al., J. Chromatogr., A, 909 (2001) 95.

161. H. V. Street, J. Chromatogr., 109 (1975) 29.
162. H.-S. Shin et al., J. Chromatogr., A, 912 (2001) 119.
163. S. J. Clark, H. H. Wotiz, Steroids, 2 (1963) 535.
164. A. T. Shulgin, Anal. Chem., 36 (1964) 920.
165. H.-S. Shin et al., J. Chromatogr., A, 912 (2001) 119.
166. W.-H. Ding et al., J. Chromatogr., A, 896 (2000) 111.
167. W.-H. Ding et al., J. Chromatogr., A, 862 (1999) 113.
168. S. C. Moldoveanu, unpublished results.
169. W. Vogt et al., J. Chromatogr. Sci., 12 (1974) 658.
170. J. Porschmann et al., J. Chromatogr., A. 909 (2001) 95.
171. C. A. Bache et al., Anal. Chem., 40 (1968) 1241.
172. E. Bombardelli et al., J. Chromatogr., 139 (1977) 111.
173. H. G. J. Mol et al., J. Chromatogr., A, 879 (2000) 97.
174. H. Hosada et al., Chem. Pharm. Bull., 23 (1975) 2118.
175. H. C. Pillsbury et al., J. Assoc. Off. Anal. Chem., 52 (1969) 458.
176. J. B. Forehand, S. C. Moldoveanu, unpublished results.
177. M. F. Borgerding et al., Food Chem. Toxicol., 36 (1997) 169.
178. M. P. Heenan, N. K. McCallum, J. Chromatogr. Sci., 12 (1974) 89.
179. W. J. A. VandenHeuval, Anal. Lett., 6 (1973) 51.
180. R. M. Cassidy et al., J. Chromatogr. Sci., 12 (1974) 85.
181. M. K. Fayad et al., Chromatographia, 28 (1989) 465.
182. J. P. Porcaro, P. Shubiak, Anal. Chem., 44 (1972) 1865.
183. O. Fiehn, M. Jekel, J. Chromatogr., A, 769 (1997) 189.
184. R. M. Carlson et al., J. Chromatogr. Sci., 22 (1984) 272.
185. K. H. deSilva et al., Biomed. Chromatogr., 10 (1996) 318.
186. J. Ishida et al., J. Chromatogr., 356 (1986) 171.
187. T. Yoshida, J. Chromatogr., B, 678 (1996) 157.
188. L. Gasco, R. Barrera, Anal. Chim. Acta, 61 (1972) 253.
189. G. van Look et al., *Silylating Agents*, Fluka Chemie AG, Buchs, 1995.
190. G. A. Olah et al., J. Org. Chem., 46 (1981) 5212.
191. J. Schafer et al., Anal. Biochem., 209 (1993) 53.
192. T. Toyo'oka, K. Imai, Anal. Chem., 56 (1984) 2451.
193. J. O. Miners et al., J. Chromatogr., 275 (1983) 89.
194. K. Hayashi et al., J. Chromatogr., 338 (1985) 161.
195. V. Cavrini et al., Chromatographia, 42 (1996) 515.
196. G. R. Rhodes et al., J. Chromatogr., 488 (1989) 456.
197. E. Bald et al., J. Chromatogr., A, 913 (2001) 319.
198. D. Valdez, J. C. Reier, J. Liq. Chromatogr., 10 (1987) 2133.
199. H. Frank et al., J. Chromatogr., 309 (1984) 261.
200. C. D. Orford et al., J. Chromatogr., 481 (1989) 245.
201. B. Kagedal, M. Kallberg, J. Chromatogr., 229 (1982) 409.
202. K. Tsutsumi et al., J. Chromatogr., 231 (1982) 393.
203. Y. Okabayashi, T. Kitagawa, Anal. Chem., 66 (1994) 1448.
204. Y. Watanabe, K. Imai, Anal. Chem., 55 (1983) 1786.
205. M. Jemal, D. J. Hawthorne, J. Chromatogr., B, 693 (1997) 109.
205a. M. Jemal et al., Anal. Chem., 73 (2001) 5450.
206. R. Gotti et al., J. Chromatogr., 507 (1990) 451.
207. R. Gotti et al., Chromatographia, 39 (1994) 23.
208. S. Velury, S. B. Howell, J. Chromatogr., 424 (1988) 141.
209. C.-S. Yang et al., J. Chromatogr., B, 674 (1995) 23.

210. A. R. Ivanov et al., J. Chromatogr., A., 913 (2001) 315.
211. S. Ito et al., J. Chromatogr., 626 (1992) 187.
212. A. J. G. Mank et al., Anal. Chem., 65 (1993) 2197.
213. K. Shimada et al., J. Chromatogr., 419 (1987) 17.
214. K. Shimada et al., J. Chromatogr., 227 (1982) 445.
215. G. Diehl, U. Karst, J. Chromatogr., A, 890 (2000) 281.
216. W. J. A. VandenHeuval et al., Anal. Chem., 36 (1964) 1550.
217. H. Ehrsson, H. Brötell, Acta Pharm. Suec., 8 (1971) 591.
218. H. Ehrsson, B. Mellström, Acta Pharm. Suec., 9 (1972) 107.
219. P. A. Metz et al., J. Chromatogr., 479 (1989) 107.
220. T. H. Burch et al., 44th TCRC, Winston-Salem, 1990.
221. A. J. Shah, M. W. Adlard, J. Chromatogr., 424 (1988) 325.
222. J. F. Lawrence, J. Agr. Food. Chem., 22 (1974) 936.
223. Y. Zhang et al., Anal. Biochem., 212 (1993) 481.
224. M. Scoggins, J. Chromatogr. Sci., 13 (1975) 146.
225. E. W. Day et al., Anal. Chem., 38 (1966) 1053.
226. S. Baba et al., J. Chromatogr., 88 (1974) 373.
227. D. G. Crosby, J. B. Bowers, J. Agric. Food Chem., 16 (1968) 839.
228. A. G. Giumanini et al., Anal. Chem., 48 (1976) 484.
229. W. J. A. VandenHeuval, J. Chromatogr., 36 (1968) 354.
230. N. Mahy, E. Gelpi, J. Chromatogr., 130 (1977) 237.
231. D. H. Neiderhiser et al., J. Chromatogr., 117 (1976) 187.
232. M. H. Choi et al., J. Chromatogr., A, 897 (2000) 295.
233. C. R. Clark et al., Anal. Chem., 49 (1977) 912.
234. A. C. Moffat et al., J. Chromatogr., 66 (1972) 225.
235. D. E. Bradway, T. Shafik, J. Chromatogr. Sci., 15 (1977) 322.
236. J. A. Corbin, L. B. Rogers, Anal. Chem., 42 (1970) 974.
237. T. A. Bryce, J. L. Burrows, J. Chromatogr., 136 (1977) 401.
238. D. H. Law, G. H. Henderson, J. Chromatogr., 129 (1976) 329.
239. E. Anggard, A. Hankey, Acta Chem. Scand., 23 (1969) 3110.
240. H. Schuttenberg, R. C. Schultz, Angew. Chem. Int. Ed. Eng., 15 (1976) 777.
241. K. Matsubayashi et al., J. Chromatogr., 143 (1977) 571.
242. J. B. Forehand et al., J. Chromatogr., A, 898 (2000) 111.
243. M. H. Choi et al., J. Chromatogr., A, 897 (2000) 295.
244. Y. Hoshika, Anal. Chem., 49 (1977) 541.
245. Y. Hoshika, J. Chromatogr., 136 (1977) 253.
246. I. Nitsche et al., J. Chromatogr., 94 (1974) 65.
247. T. Gejvall, J. Chromatogr., 90 (1974) 157.
248. T. Wahle, Acta Pharm. Suec., 5 (1968) 353.
249. O. Busto et al., J. Chromatogr., A, 737 (1996) 205.
250. G.-D. Li et al., J. Chromatogr., A, 724 (1996) 147.
251. I.-J. Kim et al., J. Chromatogr., A, 877 (2000) 217.
252. A. J. Faulkner et al., Anal. Chem., 63 (1991) 292.
253. K.-L. Woo, Y. K. Ahan, J. Chromatogr., A, 740 (1996) 41.
254. E. J. Eisenberg, K. C. Kundy, J. Chromatogr., B, 679 (1996) 119.
255. S. Yoshida et al., J. Chromatogr., 430 (1988) 156.
256. M. Iwamoto et al., J. Chromatogr., 310 (1984) 151.
257. E. P. Lynkmayr et al., J. Chromatogr., 198 (1980) 471.
258. T.-M. Chen, J. J. Contario, J. Chromatogr., 314 (1984) 495.
259. T. Toyo'oka et al., J. Chromatogr., A, 675 (1994) 79.

260. A. Casirano, J. Chromatogr., 318 (1985) 132.
261. D. A. Durden et al., Adv. Mass Spectrom. Biochem. Med., 2 (1976) 597.
262. I. Krause et al., J. Chromatogr., A, 715 (1995) 67.
263. A. Casini et al., J. Chromatogr., 249 (1982) 187.
264. R. Romero et al., J. Chromatogr., A, 871 (2000) 75.
265. A. M. Rizzi et al., J. Chromatogr., A, 710 (1995) 287.
266. H. Tsuchiya et al., J. Chromatogr., 339 (1985) 59.
267. A. M. Rizzi et al., J. Chromatogr., 582 (1992) 35.
268. K. Gamoh, H. Sawamoto, Anal. Chim. Acta, 243 (1991) 251.
269. M. Péter et al., J. Chromatogr., A, 910 (2001) 247.
270. W. Schutzner et al., Anal. Chem., 67 (1995) 3866.
271. H. Kamimura et al., J. Chromatogr., 225 (1981) 115.
272. N. Seiler et al., J. Chromatogr., 84 (1973) 95.
273. N. Seiler et al., J. Chromatogr., 97 (1973) 286.
274. H. Yoshida et al., J. Chromatogr., A, (907 (2001) 39.
275. S. Hara et al., Analyst, 122 (1997) 475.
276. S. Hara et al., Analyst, 122 (1997) 475.
277. S. Uzu et al., Analyst, 116 (1991) 1353.
278. T. Toyo'oka et al., Analyst, 114 (1989) 1233.
279. P. N. Huguenin et al., J. Liq. Chromatogr., 7 (1984) 1433.
280. D. M. Dasai, J. Gal, J. Chromatogr., 579 (1992) 165.
281. O. P. Kleidernigg et al., J. Chromatogr., A, 729 (1996) 33.
282. M. Lämmerhofer et al., J. Chromatogr., B, 689 91997) 123.
283. C.-C. Chen, C.-E. Lin, J. Chromatogr. Sci., 33 (1995) 229.
284. J. Kirschbaum et al., J. Chromatogr., A, 881 (2000) 517.
285. K. Nakashima et al., J. Chromatogr., 530 (1990) 154.
286. H. Tinnerberg et al., Analyst, 121 (1996) 1101.
287. Y. Okabayashi, T. Kitagawa, Anal. Chem., 66 (1994) 1448.
288. T. Iwata et al., Anal. Chem., 69 (1997) 1861.
289. M.T. Rosseel et al., J. Chromatogr., 568 (1991) 239.
290. A. Royer et al., Anal. Chem., 72 (2000) 3826.
291. E. Okuma, H. Abe, J. Chromatogr., B, 660 (1994) 243.
292. D. Shangguan et al., Anal Chem., 73 (2001) 2054.
293. A. J. Shah, M. W. Adlard, J. Chromatogr., 424 (1988) 325.
294. B. Staffeldt et al., J. Chromatogr., 571 (1991) 133.
295. A. H. Moye, A. J. Boning, Anal. Lett., 12 (1979) 25.
296. S. Einarsson et al., J. Chromatogr., 282 (1983) 609.
297. M. Pugniere et al., J. Chromatogr., A, 767 (1997) 69.
298. P. Britz-Mckibbin et al., Anal. Chem., 71 (1999) 1633.
299. L. D. Hutt et al., Anal. Chem., 71 (1999) 4000.
300. K. Shimada et al., J. Liq. Chromatogr., 16 (1993) 3311.
301. D. R. Goodlett et al., J. Chromatogr., A, 707 (1995) 233.
302. A. Péter et al., J. Chromatogr., A, 871 (2000) 105.
303. B. P. Spragg, A.D Hutchings, J. Chromatogr., 258 (1983) 289.
304. M. Breda et al., J. Chromatogr., A, 729 (1996) 301.
305. S. Hu, P. C. H. Li, J. Chromatogr., A, 876 (2000) 183.
306. H. S. Spahn, J. Chromatogr., 427 (1988) 131.
307. P. Languth et al., J. Chromatogr., 528 (1990) 55.
308. A. Okayama, J. Chromatogr., 426 (1988) 365.
309. L. Zhang et al., J. Chromatogr., A, 894 (2000) 129.

310. M. Kai et al., J. Chromatogr., A, 653 (1993) 235.
311. T. Toyo'oka, Y.-M. Liu, Analyst, 120 (1995) 385.
312. M. Tod et al., Anal. Chim. Acta, 223 (1989) 309.
313. H. Kouwatli et al., Anal. Chim. Acta, 266 (1992) 243.
314. D. J. Siler, D. G. Gilchrist, J. Chromatogr., 238 (1982) 167.
315. S. Oguri, Y. Miki, J. Chromatogr., B, 686 (1996) 205.
316. M. Weigele et al., Biochem. Biophys Res. Commun., 54 (1973) 899.
317. R. Büschges et al., J. Chromatogr., A, 725 (1996) 323.
318. R. Büschges et al., J. Chromatogr., A 725 (1996) 323.
319. K. L. Miller et al., J. Chromatogr., 307 (1984) 335.
320. N. N. Osborne et al, J. Chromatogr., 123 (1976) 212.
321. S. Laganière et al., J. Chromatogr., 488 (1989) 407.
322. I. Demian, D. F. Gripshover, J. Chromatogr., 466 (1989) 415.
324. P. Chen, M. V. Novotny, Anal. Chem., 69 (1997) 2806.
325. O. Orwar et al., Anal. Chem., 67 (1995) 4261.
326. L. A. Dawson et al., J. Chromatogr., B, 694 (1997) 455.
327. E. L. Gump, C. A. Monnig, J. Chromatogr., A, 715 (1995) 167.
328. D. M. Desai, J. Gal, J. Chromatogr., 579 (1992) 165.
329. K. Tsuji, K. M. Jenkins, J. Chromatogr., 369 (1986) 105.
330. H. Brückner, M. Lüpke, J. Chromatogr., 697 (1995) 295.
331. D. S. Dunlop, A. Neidle, Anal. Biochem., 165 (1987) 38.
332. R. W. Sparidans et al., J. Chromatogr., B, 696 (1997) 137.
333. M. Matsuoka et al., J. Chromatogr., B, 676 (1996) 117.
334. R. C. Clark et al., Anal. Chem., 49 (1977) 912.
335. N. Grigurinocich, J. Chromatogr., B, 696 (1997) 75.
336. M. J. Rose et al., Anal. Chem., 71 (1999) 2221.
337. Z. Abolfathi et al., J. Chromatogr., 579 (1992) 366.
338. V. L. Lanchote et al., J. Chromatogr., B, 685 (1996) 281.
339. L. Milerioux et al., J. Chromatogr., A, 729 (1996) 309.
340. T. Takagi et al., J. Chromatogr., 272 (1983) 279.
341. A. Sabala et al., J. Chromatogr., A, 778 (1997) 103.
342. A. Taga, S. Honda, J. Chromatogr., A, 742 (1996) 243.
343. M. Catalá-Icardo et al., J. Liq. Chromatogr., 18 (1995) 2827.
344. P. Furst et al., J. Liq. Chromatogr., 12 (1989) 2733.
345. B. H. Jones, J. P. Gilligan, J. Chromatogr., 266 (1983) 471.
346. M. D. Oates, J. W. Jorgenson, Anal. Chem., 62 (1981) 2056.
347. I. Molnár-Perl, J. Chromatogr., A, 913 (2001) 283.
348. D. Fekes et al., J. Chromatogr., B, 669 (1995) 177.
349. H. Brückner et al., J. Chromatogr., A, 697 (1995) 229.
350. Y. V. Tcherkas, A. D., Denisenko, J. Chromatogr., A., 913 (2001) 309.
351. E. Kaale et al., J. Chromatogr., A, 895 (2000) 67.
352. A. Vasanits et al., J. Chromatogr., A, 870 (2000) 271.
353. R. Harráez-Hernández, P. Campíns-Falcó, J. Chromatogr., A, 893 (2000) 69.
354. M. Zheng et al., Analyst, 118 (1993) 269.
355. M. De Vos et al., J. Chromatogr., 564 (1991) 296.
356. C. Dutot et al., J. Chromatogr., 528 (1990) 435.
357. M. A. Cichy et al., J. Chromatogr., 613 (1993) 15.
358. M. I. Saleh, F. W. Pok, J. Chromatogr., A, 763 (1997) 173.
359. E. Winkler et al., J. Chromatogr., 436 (1988) 447.
360. R. Wintersteiger et al., Anal. Chim. Acta, 306 (1995) 273.

361. N. Nimura et al., Anal. Chem., 48 (1986) 2372.
362. K. Shimada et al., J. Chromatogr., 487 (1989) 247.
363. A. Péter et al., J. Chromatogr., A, 871 (2000) 105.
364. M. Lobell, M. P. Schneider, J. Chromatogr., 633 (1993) 287.
365. S. Kawai et al., J. Chromatogr., 540 (1991) 411.
366. S. L. Wellons, M. A. Carrey, J. Chromatogr., 154 (1978) 219.
367. S. Blackburn, *Amino Acid Determination,* M. Dekker, New York, 1968.
368. R. Sams, Anal. Lett., B11 (1978) 697.
369. H. Higashijima et al., Anal. Chem., 64 (1992) 711.
370. A. J. G. Mank, E. S. Yeng, J. Chromatogr., 708 (1995) 309.
371. T. Fuchigami et al., Anal. Chim. Acta, 282 (1993) 209.
372. R. J. Williams et al., Anal. Chem., 65 (1993) 601.
373. S. J. Woltman et al., Anal. Chem., 72 (2000) 4928.
374. S. J. Woltman et al., Anal. Chem., 71 (1999) 1504.
375. O. R. Idowu, G. O. Adewuyi, J. Liq. Chromatogr., 16 (1993) 3773.
376. F.-X. Zhou et al., J. Liq. Chromatogr., 14 (1991) 1325.
377. C.-X. Gao et al., J. Chromatogr. Sci., 26 (1988) 449.
378. J. H. Yu et al., J. Chromatogr., B, 658 (1994) 249.
379. A. J. Bourque, I. S. Krull, J. Chromatogr., 537 (1991) 123.
380. T.-Y. Chou et al., J. Chromatogr., 454 (1988) 169.
381. F.-X. Zhou et al., J. Chromatogr., 619 (1993) 93.
382. C.-X. Gao et al., J. Chromatogr. Sci., 28 (1990) 102.
383. K. W. Sigvardson, J. W. Birks, J. Chromatogr., 316 (1984) 507.
384. R. L. Cunico, T. Schlabach, J. Chromatogr., 266 (1983) 461.
385. D. Drnevich, T. C. Vary, J. Chromatogr., 613 (1993) 137.
386. H. Brückner et al., J. Chromatogr., 476 (1989) 73.
387. R. H. Buck, H. Krummen, J. Chromatogr., 387 (1987) 255.
388. J. P. Liu et al., J. Chromatogr., 468 (1989) 55.
389. B. H. Klein, J. H. Dudenhausen, J. Liq. Chromatogr., 18 (1995) 4007.
390. A. M. Uhe et al., J. Chromatogr., 564 (1991) 81.
391. A. Bazzanella et al., Chromatographia, 45 (1997) 59.
392. H. G. Worthen, H. Liu, J. Liq. Chromatogr., 15 (1992) 3323.
393. A. Jegorov et al., J. Chromatogr., 434 (1988) 417.
394. T. Gunnar et al., Anal. Chem., 73 (2001) 2625.
395. T. A. Egorova, J. Chromatogr., B, 665 (1995) 53.
396. K.-L. Woo, Y.-K. Ahan, J. Chromatogr., A, 740 (1996) 41.
397. K.-L. Woo, S.-H. Lee, J. Chromatogr., A, 667 (1994) 105.
398. H. Brückner, M. Wachsmann, J. Chromatogr., A, 728 (1996) 447.
399. J. Liu et al., Anal. Chem., 63 (1991) 408.
400. M. Péter et al., J. Chromatogr., A, 871 (2000) 115.
401. I. Fermo et al., J. Chromatogr., 433 (1988) 53.
402. M. Alaiz et al., J. Chromatogr., 591 (1992) 181.
403. H. Inoue et al., Analyst, 120 (1995) 1141.
404. E. E. Schallenberg, M. Calvin, J. Am. Chem. Soc., 77 (1955) 2779.
405. E. Okuma, H. Abe, J. Chromatogr., B, 660 (1994) 243.
406. S. Oguri et al., J. Chromatogr., A, 724 (1996) 169.
407. J. Goto et al., J. Chromatogr., 152 (1978) 413.
408. X. Liu et al., Chromatographia, 53 (2001) 326.
409. S. A. Cohen, J. Chromatogr., 512 (1990) 283.
410. Y. Tsuruta et al., J. Chromatogr., 502 (1990) 178.

411. H.-W. Lange et al., Z. Anal. Chem., 261 (1972) 337.
412. N. Nimura et al., Anal. Chem., 48 (1986) 2372.
413. K. Iwaki et al., J. Chromatogr., 407 (1987) 273.
414. J. Saurina, S. Hernández-Cassou, J. Chromatogr., A, 740 (1996) 21.
415. W. A. Jacobs, P. T. Kissinger, J. Liq. Chromatogr., 5 (1982) 881.
416. M. Kai et al., J. Chromatogr., A, 653 (1993) 235.
417. H. Ehrson, Acta Pharm. Suec., 9 (1972) 419.
418. J. J. Franken, M. M. F. Trijbels, J. Chromatogr., 91 (1974) 425.
419. J. Ishida et al., Clin. Chem., 39 (1993) 165.
420. E. Kojima et al., J. Chromatogr., 612 (1993) 187.
421. K. Blau et al., Biomed. Mass Spectrom., 4 (1977) 232.
422. S. C. Moldoveanu, unpublished results.
423. M. Donike, Chromatographia, 9 (1970) 440.
424. N. Narasimhachari et al., J. Chromatogr. Sci., 9 (1971) 502.
425. N. Narasimhachari, K. Leiner, J. Chromatogr. Sci., 15 (1977) 181.
426. G. R. Waller, M. S. Feather (eds.), *The Maillard Reaction in Food and Nutrition*, ACS Symp. Ser. 215, Washington, 1983.
427. D. Bacu, *The Anti-Humans*, T.L.C., Englewood, 1971.
428. H. Kataoka et al., Bull. Environ. Contam. Toxicol., 60 (1998) 60.
429. R. H. Adamson et al. (eds.), *Heterocyclic Amines in Cooked Foods: Possible Human Carcinogens,* Princeton Scientific Publishing, Princeton, 1995.
430. S. C. Moldoveanu, *Analytical Pyrolysis of Natural Organic Polymers*, Elsevier, Amsterdam, 1998.
431. M. Yamashita et al., Jap. J. Cancer Res., 77 (1986) 419.
432. S. Manabe et al., J. Chromatogr., 529 (1990) 125.
433. S. Manabe et al., Carcinogenesis, 12 (1991) 1945.
434. T. A. Sasaki et al., Anal. Lett. 34 (2001) 1749.
435. R. Reistad et al., Food Chem. Toxicol., 35 (1997) 955.
436. H. Kataoka, J. Chromatogr., A, 774 (1997) 121.
437. S. Grivas, T. Nyahammar, Mutat. Res., 142 (1985) 5.
438. X.-M. Zhang et al., Mutat. Res., 201 (1988) 181.
439. S. Manabe et al., Cancer Res., 47 (1987) 6150.
440. G. A. Gross, Carcinogenesis, 11 (1990) 1597.
441. G. A. Gross, A. Gruter, J. Chromatogr., 592 (1992) 271.
442. R. Schwartzenbach, D. Gubler, J. Chromatogr., 624 (1992) 491.
443. P. Hartvig et al., Anal. Chem., 48 (1976) 390.
444. P. Hartvig, B. Nasland, J. Chromatogr., 133 (1977) 367.
445. P. Hartvig, J. Vessman, J. Chromatogr. Sci., 12 (1974) 722.
446. P. Hartvig et al., J. Chromatogr., 121 (1976) 235.
447. P. Hartvig, J. Vessman, Acta Pharm. Suec., 11 (1974) 115.
448. H. B. Hucker, J. K. Miller, J. Chromatogr., 32 (1968) 408.
449. J. Rosenfeld et al., J. Chromatogr., 129 (1976) 387.
450. D. J. Jenden et al., Anal. Chem., 44 (1972) 1879.
451. G. Neurath, W. Lüttich, J. Chromatogr., 34 (1968) 257.
452. T. M. Ludden, J. Pharm. Sci., 72 (1983) 693.
453. M. Ihnat et al., J. Assoc. Off. Anal. Chem., 56 (1973) 1164.
454. L. A. Dee, Anal. Chem., 43 (1971) 1416.
455. I. N. Acworth et al., J. Liq. Chromatogr., 17 (1994) 685.
456. F. Beliardo et al., J. Chromatogr., 553 (1991) 41.
457. B. Lindeke et al., Acta Pharm. Suec., 10 (1973) 493.

458. L. A. Sternson et al., J. Chromatogr., 153 (1978) 481.
459. A. Streitwiesser Jr., C. H. Heathcock, *Introduction to Organic Chemistry*, Macmillan Pub. Co., New York, 1981.
460. H. C. Sutton, T. M. Downes, J. Chem. Soc. Chem. Commun., 1 (1972) 1.
461. J. Korolczuk et al., J. Chromatogr., 88 (1974) 177.
462. Y. Hoshika, Y. Takata, J. Chromatogr., 120 (1976) 379.
463. J. W. Vogt, Anal. Chem., 43 (1971) 1618.
464. W. J. A. VandenHeuval et al., J. Chromatogr., 18 (1965) 391.
465. E. E. Stashenko et al., J. Chromatogr., A, 886 (2000) 175.
466. H. A. Martos, J. Pawliszyn, Anal. Chem., 70 (1998) 2311.
467. S. Strassing et al., J. Chromatogr., A, 891 (2000) 267.
468. P. Ronkainen et al., J. Chromatogr., 28 (1967) 270.
469. E. M. Patzer, D. M. Kilker, J. Chromatogr., 135 (1977) 489.
470. A. K. Williams, J. Agr. Food Chem., 22 (1974) 107.
471. R. J. Fessenden, J. S. Fessenden, J. Org. Chem., 32 (1967) 3535.
472. E. Nakamura et al., J. Am. Chem. Soc., 98 (1976) 2346.
473. Y. Sasaki, T. Hashizume, Anal. Biochem., 16 (1966) 1.
474. R. G. Lu, R. W. Giese, Anal. Chem., 72 (2000) 1798.
475. H. Nohta et al., Anal. Chim. Acta, 287 (1994) 223.
476. N. W. Jacobsen, R. G. Dikinson, Anal. Chem., 46 (1974) 298.
477. H. Yoshida et al., J. Chromatogr., A, 898 (2000) 1.
478. S. Uzu et al., Analyst, 117 (1990) 1477.
479. T. Toyo'oka, A. Kuze, Biomed. Chromatogr., 11 (1997) 132.
480. W. Schmeid et al., Fresenius Z. Anal. Chem., 335 (1989) 464.
481. T. Seki et al., J. Chromatogr., 424 (1988) 410.
482. W.-F. Chao et al., J. Chromatogr., 430 (1988) 361.
483. W.-F. Chao et al., Anal. Chim. Acta, 215 (1988) 259.
484. E. Bousquet et al., J. Liq. Chromatogr., 18 (1995) 1933.
485. T. Iwata et al., Analyst, 118 (1993) 517.
486. K. Nakashima et al., J. Chromatogr., B, 661 (1994) 205.
487. K. Mopper et al., J. Chromatogr., 256 (1983) 243.
488. G. Zurek, U. Karst, J. Chromatogr., A, 864 (1999) 191.
489. S. R. Hagen et al., J. Chromatogr., A, 767 (1997) 53.
490. C. Dye, M. Oehme, J. High Res. Chromatogr., 15 (1992) 5.
491. R. A. Henry et al., J. Chromatogr. Sci., 9 (1971) 513.
492. F. A. Fitzpatrick et al., Anal. Chem., 44 (1972) 2211.
493. I. Vanhees et al., J. Chromatogr., A, 915 (2001) 75.
494. J. Lehotay, K. Hromulakova, J. Liq. Chromatogr., 17 (1994) 597.
495. J. J. Michels, J. Chromatogr., A., 914 (2001) 123.
496. E. Grojsen et al., Anal. Chem., 71 (1999) 1851.
497. S. Velikonja et al., J. Chromatogr., A, 704 (1995) 449.
498. M. Vogel et al., J. Chromatogr., A, 886 (2000) 303.
499. D. L. Manning et al., J. Assoc. Off. Anal. Chem., 66 (1983) 8.
500. P. R. Houlgate et al., Analyst, 114 (1989) 355.
501. M. Possanzini, V. Di Palo, Chromatographia, 46 (1997) 235.
502. S. Hara et al., Anal. Chim. Acta, 291 (1994) 189.
503. G. Guebitz et al., J. Liq. Chromatogr., 7 (1984) 839.
504. M. Naoi et al., J. Chromatogr., 434 (1988) 209.
505. F. Traoré et al., Anal. Chim. Acta, 269 (1992) 211.
506. G. Chiavari et al., J. Chromatogr., 387 (1987) 459.

507. S. Velikonja et al., J. Chromatogr., A, 704 (1995) 449.
508. A. Büldt, U. Karst, Anal. Chem., 69, (1997) 3617.
509. A. Büldt, U. Karst, Anal. Chem., 71 (1999) 1893.
510. J. J. Michels, J. Chromatogr., A., 914 (2001) 123.
511. S. B. Jones et al., Anal. Chem., 71 (1999) 4030.
512. H. Engelhardt, R. Klinker, Chromatographia, 20 (1985) 559.
513. G. Zurek, U. Karst, J. Chromatogr., A, 864 (1999) 191.
514. J. E. van Doorn et al., J. Chromatogr., 489 (1989) 419.
515. S. Zaraga et al., Steroid Biochem., 4 (1973) 417.
516. M. Pauly, Carbohydr. Res., 282 (1996) 1.
517. R. R. Miksch et al., Anal. Chem., 53 (1981) 2118.
518. M. Pedrero-Muñoz et al., Analyst, 114 (1989) 1469.
519. S. A. Matlin et al., J. Liq. Chromatogr., 13 (1990) 2261.
520. M. Pertarulo et al., J. Chromatogr., 432 (1988) 37.
521. D. Londero, P. Lo Greco, J. Chromatogr., A, 729 (1996) 207.
522. V. David et al., Anal. Chim. Acta, submitted for publication.
522a. R. W. Keating, P. R. Haddad, J. Chromatogr., 245 (1982) 249.
523. G. Filek et al., J. Chromatogr., A, 732 (1996) 291.
524. A. Lopez-Anaya, M. Mayersohn, Clin. Chem., 33 (1987) 1874.
525. N. Bilic, J. Chromatogr., 543 (1991) 367.
526. F. Traoré et al, Chromatographia, 36 (1993) 96.
527. G. J. Mohr et al., Anal. Chem., 72 (2000) 1084.
528. K. Kakehi et al., J. Chromatogr., A, 863 (1999) 205.
529. A. Meyer et al., Anal. Chem., 73 (2001) 2377.
530. R. A. Famia, R. Weinberger, J. Chromatogr., 402 (1987) 127.
531. J. Plocek, M. V. Novotny, J. Chromatogr., A, 757 (1997) 215.
532. M. D. Spiro et al., Carbohydr. Res., 290 (1996) 147.
533. C. Chiesa, C. Horvath, J. Chromatogr., 645 (1993) 337.
534. Q. Liu et al., J. Liq. Chromatogr., Relat. Technol., 20 (1997) 1123.
535. Y. Shinohara et al., Anal. Chem., 68 (1996) 2573.
536. T. Toyo'oka, A. Kuze, Biomed. Chromatogr., 11 (1997) 132.
537. M. Takeda et al., J. Chromatogr., 244 (1982) 347.
538. K. Yoshino et al., Anal. Chem., 67 (1995) 4028.
539. W. Mo et al., Anal. Chem., 71 (1999) 4100.
540. I. Miksik et al., J. Chromatogr., A, 772 (1977) 297.
541. Z. Zhang et al., J. Chromatogr., A, 728 (1996) 343.
542. D. Shangguan et al., J. Chromatogr., A, 910 (2001) 367.
543. D. Velasco et al., J. Chromatogr., 519 (1990) 228.
544. A. Taga et al., J. Chromatogr., A, 911 (2001) 259.
545. K. Kakehi et al., Anal. Biochem., 199 (1991) 256.
546. J. Sudor, M. V. Novotny, Anal. Chem., 69 (1997) 3199.
547. E. G. Perkins (ed.), Analysis of Lipids and Lipoproteins, Am. Oil Chem. Soc.,
     Champaign, 1975.
548. E. O. Umeh, J. Chromatogr., 51 (1970) 147.
549. E. O. Umeh, J. Chromatogr., 51 (1970) 139.
550. G. Wittmann et al., J. Chromatogr., A, 874 (2000) 225.
551. M. R. Guerin et al., Anal. Chem, 46 (1974) 761.
552. J. Porschmann et al., J. Chromatogr., A, 909 (2001) 95.
553. R. Watson, P. Crescuolo, J. Chromatogr., 52 (1970) 63.
554. C. A. Bache et al., Anal. Chem., 40 (1968) 1241.

555. J. B. Brooks et al., Anal. Chem., 47 (1975) 1960.
556. C. Choudhary, C. W. Moss, J. Chromatogr., 128 (1976) 261.
557. C. Choudhary, C. W. Moss, J. Chromatogr., 128 (1976) 261.
558. W. A. Aue et al., J. Chromatogr., 72 (1972) 259.
559. C. E. McCone, R. J. Hance, J. Chromatogr., 69 (1972) 204.
560. W. R. Meagher, J. Agr. Food Chem., 14 (1966) 374.
561. G. Phillipou et al., Lipids, 10 (1975) 714.
562. C. W. Stanley, J. Agric. Food Chem., 14 (1966) 321.
563. L. Helleberg, J. Chromatogr., 117 (1976) 167.
564. E. O. Omeh, J. Chromatogr., 56 (1971) 29.
565. H.-P. Klemm et al., J. Chromatogr., 75 (1973) 19.
566. N. M. Sarrión et al, Anal. Chem., 72 (2000) 4865.
567. P. Schulz, R. Vîlceanu, J. Chromatogr. 111 (1975) 105.
568. K. Sahib et al., J. Pharm. Sci., 60 (1971) 745.
569. M. I. Catalina et al., J. Chromatogr., A, 877 (2000) 153.
570. M. N. Sarrión et al., J. Chromatogr., A, 859 (1999) 159.
571. R. H. Greely, J. Chromatogr., 88 (1974) 229.
572. P. A. Bondi, M. Cagnasso, J. Chromatogr., 109 (1975) 389.
573. M. X. Wang et al., 48th TCRC, Greensboro, 1994.
574. E. T. Oakley et al., Anal. Chem., 37 (1965) 380.
575. I. Hornstein et al., Anal. Chem., 32 (1960) 540.
576. M. Hoshi et al., J. Lipid Res., 14 (1973) 599.
577. M. N. Sarrión et al., J. Chromatogr., A, 859 (1999) 159.
578. W. E. Klopfenstein, J. Lipid. Res., 12 (1971) 773.
579. B. L. Brian et al., J. Chromatogr., 66 (1972) 138.
580. L. D. Metcalfe, A. A. Schmitz, Anal. Chem., 33 (1961) 363.
581. L. D. Metcalfe, A. A. Schmitz, Anal. Chem., 38 (1966) 514.
582. M. A. Harmon, H. W. Doelle, J. Chromatogr., 42 (1969) 157.
583. H. Ko, M. E. Royer, J. Chromatogr., 88 (1974) 253.
584. E. Felder et al., Chromatogr., 82 (1973) 291.
585. E. Felder et al., Chromatogr., 82 (1973) 390.
586. M. Brenner, W. Huber, Helv. Chim. Acta, 36 (1953) 1109.
587. M. Gee, Anal. Chem., 37 (1965) 926.
588. A. Grünert, K. H. Bassler, Z. Anal. Chem., 267 (1973) 342.
589. C. W. Gehrke, D. F. Goerlitz, Anal. Chem., 35 (1963) 76.
590. F. K. Kawahara, Anal. Chem., 40 (1968) 2073.
591. O. Gyllenhaal et al., J. Chromatogr., 129 (1976) 295.
592. J. E. Greving et al., J. Chromatogr., 142 1977) 611.
593. H.-P. Klemm et al., J. Chromatogr., 75 (1973) 481.
594. H.-P. Klemm et al., J. Chromatogr., 75 (1973) 19.
595. D. L. Corina, J. Chromatogr., 87 (1973) 254.
596. E. K Doms, J. Chromatogr., 140 (1977) 29.
597. A. Arbin, J. Chromatogr., 144 (1977) 85.
598. S. Mierzwa, S. Witek, J. Chromatogr., 136 (1977) 105.
599. R. V. Smith, S. L. Tsai, J. Chromatogr., 29 (1971) 29.
600. C. C. Alley et al., Anal. Chem., 48 (1976) 387.
601. R. T. Dean et al., Anal. Biochem., 75 (1976) 1.
602. R. Aichholz, E. Lorbeer, J. Chromatogr., A, 855 (1999) 601.
603. R. G. Lu, R. W. Giese, Anal. Chem., 72 (2000) 1798.
604. M. Verzele et al., Anal. Chem., 45 (1973) 1594.

605. F. Deckert, J. Chromatogr., 69 (1972) 201.
606. H. Nashi et al., J. Chromatogr., A, 672 (1994) 125.
607. M. Kagawa et al., J. Chromatogr., A, (1999) 127.
608. K. Sasamoto et al., Anal. Sci., 12 (1996) 189.
609. A. Longo et al., J. Chromatogr., B, 686 (1996) 129.
610. K. Matsumoto et al., J. Chromatogr., A, 678 (1994) 214.
611. M. Ikeda et al., J. Chromatogr., 272 (1983) 251
612. M. Ikeda et al., J. Chromatogr., 305 (1984) 261.
613. A. I. Haj-Yehia et al., J. Chromatogr., A, 870 (2000) 381.
614. T. Toyo'oka et al., Analyst, 117 (1992) 727.
615. T. Toyo'oka et al., Analyst, 116 (1991) 609.
616. K. Akasaka et al., Analyst, 118 (1993) 765.
617. K. Akasaka et al., J. Chromatogr., A, 729 (1996) 381.
618. T. A. Stein et al., J. Chromatogr., 385 (1987) 377.
619. W. D. Watkins, M. B. Peterson, Anal. Biochem., 125 (1982) 30.
620. J. Goto et al., J. Liq. Chromatogr., 9 (1986) 683.
621. I. R. Politzer et al., Anal. Lett., 6 (1973) 539.
622. R. Alric et al., J. Chromatogr., 224 (1981) 289.
623. L. A. Gifford et al., J. Chromatogr., A, 715 (1995) 201.
624. P. J. M. Kwakman et al., Analyst, 116 (1991) 1385.
625. C. H. Kiang et al., J. Chromatogr., 567 (1991) 195.
626. H. Asukabe et al., J. Chromatogr., A, 657 (1993) 349.
627. H. D. Durst et al., Anal. Chem., 47 (1975) 1797.
628. H. C. Jordi, J. Liq. Chromatogr., 1 (1978) 215.
629. R. A. Kelly et al., J. Chromatogr., 416 (1987) 247.
630. H. Tsuchiya et al., J. Chromatogr., 234 (1982) 121.
631. Y. Amet et al., J. Chromatogr., B, 681 (1996) 233.
632. R. Farinotti et al., J. Chromatogr., 269 (1983) 81.
633. M. Yamaguchi et al., J. Chromatogr., 346 (1985) 227.
634. W.-F. Chao et al., J. Chromatogr., 526 (1990) 77.
635. P. S. Mukherjee et al., Pharm. Res., 12 (1995) 930.
636. K. Shimada et al., J. Liq. Chromatogr., 16 (1993) 3965.
637. A. Nakanishi et al., J. Chromatogr., 591 (1992) 159.
638. M. T. Tsamis et al, J. Chromatogr., 277 (1983) 61.
639. S. Li, S. K. Wahba Khalil, J. Liq. Chromatogr., 13 (1990) 1339.
640. H. Naganuma, Y. Kawahara, J. Chromatogr., 530 (1990) 387.
641. S. Krahenbuhl et al., J. Chromatogr., 573 (1992) 3.
642. L. Vernez et al., J. Chromatogr., A, 895 (2000) 309.
643. E. Vioque et al., J. Chromatogr., 331 (1985) 187.
644. T. Toyo'oka et al., J. Chromatogr., 588 (1991) 61.
645. T. Toyo'oka, A. Kuze, Biomed. Chromatogr., 9 (1995) 162.
646. H. Kaneda et al., J. Agric. Food Chem., 38 (1990) 1363.
647. G. Gübitz, J. Chromatogr., 187 (1980) 208.
648. W. Linder, W. Santi, J. Chromatogr., 176 (1979) 55.
649. K. Iwaki et al., J. Chromatogr., A, 662 (1994) 87.
650. P. J. Ryan, T. W. Honeyman, J. Chromatogr., 312 (1984) 461.
651. I. Yanagisawa et al., J. Chromatogr., 345 (1985) 229.
652. J. B. F. Loyd, J. Chromatogr., 189 (1980) 357.
653. J. Abián et al., Biomed. Environ. Mass. Spectrom., 16 (1988) 215.
654. E. Bousquet et al., J. Liq. Chromatogr., Relat. Technol., 20 (1997) 757.

655. R. K. Munns et al., J. Chromatogr., 442 (1988) 209.
656. R. Badoud, G. Pratz, J. Chromatogr., 360 (1986) 119.
657. J. Knospe et al., J. Chromatogr., 442 (1988) 444.
658. M. Yamaguchi et al., J. Liq. Chromatogr., 18 (1995) 2991.
659. T. Uwata et al., J. Chromatogr., 654 (1994) 171.
660. J. Goto et al., J. Chromatogr., 239 (1982) 559.
661. W. H. Pirkle, P. G. Murray, J. Liq. Chromatogr., 13 (1990) 2123.
662. B. Blessington et al., J. Chromatogr., 396 (1987) 177.
663. K. J. James et al., J. Chromatogr., A, 871 (2000) 1.
664. M. Saito et al., Anal. Sci., 11 (1995) 103.
665. M. Saito et al., Anal. Chim. Acta, 300 (1995) 243.
666. M. Saito et al., J. Chromatogr., 674 (1995) 167.
667. M. Saito et al., Anal. Sci., 10 (1994) 679.
668. K. Nakashima et al., J. Chromatogr., 619 (1993) 1.
669. M. Tod et al., J. Chromatogr., 542 (1991) 295.
670. S. Narita, T. Kitagawa, Anal. Sci., 5 (1989) 31.
671. T. Iwata et al., Analyst, 119 (1994) 1747.
672. H. Wu et al., J. Chromatogr., A, 769 (1997) 201.
673. Y. Yasaka et al., J. Chromatogr., 508 (1990) 133.
674. H. Miwa, M. Yamamoto, J. Chromatogr., A, 721 (1996) 261.
675. H. Miwa, M. Yamamoto, J. Chromatogr., 421 (1987) 33.
676. H. Miwa, J. Chromatogr., A, 881 (2000) 365.
677. A. G. Netting, A. M. Duffield, J. Chromatogr., 257 (1983) 174.
678. Y. Yasaka et al., Anal. Sci., 6, (1990) 49.
679. Y. Mechref et al., J. Chromatogr., A, 695 (1995) 83.
680. N. E. Hoffman, J. C. Liao, Anal. Chem., (1976) 1104.
681. A. J. B. Mank et al., Anal. Chim. Acta, 315 (1995) 209.
682. D. L. Gallaher, M. E. Johnson, Anal. Chem., 72 (2000) 2080.
683. B. Hallgren, S. Larson, J. Lipid Res., 3 (1962) 31.
684. O. S. Privett, E. C. Nickell, Lipids, 1 (1966) 98.
685. V. Dommes et al., J. Chromatogr. Sci., 14 (1976) 360.
686. P. Capella, C. M. Zorzut, Anal. Chem., 40 (1968) 1458.
687. W. G. Niehaus, Jr., R. Ryhage, Anal. Chem., 40 (1968) 1840.
688. B. M. Johnson, J. W. Taylor, Anal. Chem., 44 (1972) 1438.
689. L. M. McDonough, D. A. George, J. Chromatogr. Sci., 8 (1970) 158.
690. R. J. Rango et al., J. Liq. Chromatogr., 16 (1993) 3291.
691. W. S. Wu, V. S. Gaind, Analyst, 118 (1993) 1285.
692. S. T. Colgan et al., J. Chromatogr. Sci., 26 (1988) 501.
693. J. A. F. de Silva et al., Anal. Chem., 48 (1976) 10.
694. T. Walle, J. Chromatogr., 114 (1975) 345.
695. A. J. Gee et al., J. Chromatogr., A, 849 (1999) 541.
696. H. Shintani, S. Ube, J. Chromatogr., 344 (1985) 145.
697. A. Hulshoff et al., J. Chromatogr., 186 (1979) 535.
698. S.-H. Chen et al., J. Liq. Chromatogr., Relat. Technol., 20 (1997) 1579.
699. W. Linder, W. Santi, J. Chromatogr., 176 (1979) 55.
700. Z. Zhang et al., J. Chromatogr., A, 730 (1996) 107.
701. W. Dünges et al., J. Chromatogr. Sci., 12 (1974) 655.
702. J. Asselineau, *The Bacterial Lipids*, Hermann, Paris, 1962.
703. D. B. Shina, W. L. Gaby, J. Biol. Chem, 239 (1964) 3668.
704. H. J. Deuel, *The Lipids*, Interscience Pub., New York, 1951.

705. C. Fenselau (ed.), *Mass Spectrometry for the Characterization of Microorganisms*, ACS Symposium Ser. 541, ACS, Washington DC, 1994.
706. K. V. Peisker, J. Am. Oil Chem. Soc., 41 (1964) 87.
707. G. W. McGinnis, L. R. Dugan, J. Am. Oil Chem. Soc., 42 (1965) 305.
708. E. A. Davidowicz, T. E. Thompson, J. Lipid Res., 12 (1971) 636.
709. W. R. Morrison, L. M. Smith, J. Lipid Res., 5 (1964) 600.
710. W. Stoffel et al., Anal. Chem., 31 (1959) 307.
711. P. G. Simmonds, A. Zlatkis, Anal. Chem., 37 (1965) 302.
712. K.-L. Woo, J.-I. Kim, J. Chromatogr., A, 862 (1999) 199.
713. J. Sampugna et al., J. Dairy Sci., 49 (1966) 1462.
714. J. L. Iverson, A. J. Sheppard, J. Assoc. Off. Anal. Chem., 60 (1977) 284.
715. H.-L. Hardell, N.-O. Nilvebrant, J. Anal. Appl. Pyrol., 52 (1999) 1.
716. W. E. Neff et al., J. Chromatogr., A, 912 (2001) 187.
717. K. Satouchi et al., Biomed. Mass Spectrom., 5 (1978) 87.
718. B. Halgren, S. Larson, J. Lipid Res., 3 (1962) 31.
719. S. J. Gaskell et al., Anal. Lett., 9 (1976) 325.
720. D. G. Hanahan et al., Biochemistry, 2 (1963) 630.
721. R. Aichholz, E. Lorbeer, J. Chromatogr., A, 855 (1999) 601.
722. S. J. Gaskell, C. J. W. Brooks, J. Chromatogr., 122 (1976) 415.
723. S. J. Gaskell et al., J. Chromatogr., 126 (1976) 591.
724. B. Samuelsson, K. Samuelsson, J. Lipid Res., 10 (1969) 41.
725. M. Oshima et al., Chem. Phys. Lipids, 19 (1977) 289.
726. A. Suzuki et al., J. Biochem., 82 (1977) 1185.
727. N. Sadeg et al., J. Chromatogr., B, 675 (1996) 113.
728. E. Horne et al., Analyst, 121 (1996) 1463.
729. A. Walubo et al., J. Chromatogr., 567 (1991) 261.
730. S. S. Yang, J. Chromatogr., 595 (1992) 346.
731. J. M. Renaud et al., J. Chromatogr., 604 (1992) 243.
732. A. F. Haeberer, O. T. Chortyk, J. Assoc. Off. Anal. Chem., 62 (1979) 171.
733. I. M. Kapetanovic et al., Anal. Chem., 49 (1977) 1843.
734. R. Roncucci et al., J. Chromatogr., 57 (1971) 410.
735. P. R. Vagelos et al., Anal. Biochem., 2 (1961) 50.
736. G. Lhoest et al., Biomed. Mass Spectrom., 3 (1976) 21.
737. A. J. Epstein et al., Anal. Chem., 39 (1967) 721.
738. L. J. Sullivan et al., J. Agr. Food Chem., 15 (1967) 927.
739. S. Khalifa, R. O. Mumma, J. Agr. Food Chem., 20 (1972) 632.
740. J. F. Lawrence, J. Chromatogr., 123 (1976) 287.
741. J. J. Ryan, J. F. Lawrence, J. Chromatogr., 135 (1977) 117.
742. R. G. Wien, F. S. Tanaka, J. Chromatogr., 130 (1977) 55.
743. L. Fishbein, W. L. Zelinski, J. Chromatogr., 20 (1965) 9.
744. Y. S. Wu et al., Anal. Chem., 72 (2000) 1441.
745. R. W. Frei, J. F. Lawrence, J. Chromatogr., 61 (1971) 174.
746. D. Chaput, J. Assoc. Off. Anal. Chem., 71 (1988) 542.
747. J. M. Mee, J. Chromatogr., 94 (1974) 302.
748. D. Hartley, H. Kidd (eds.), *The Agrochemical Handbook*, Royal Society of Chemistry, Nottingham, 1987.
749. D. G. Saunders, L. E. Vanatta, Anal. Chem., 46 (1974) 1319.
750. F. S. Tanaka, R. G. Wien, J. Chromatogr., 87 (1973) 85.
751. I. C. Cohen, B. B. Wheals, J. Chromatogr., 43 (1969) 233.
752. J. H. Hengstmann et al., Anal. Chem., 46 (1974) 34.

753. T. Kawabata et al., J. Chromatogr., 140 (1977) 47.
754. S. B. Matin et al., Biomed. Mass Spectrom., 1 (1974) 320.
755. M. S. F. Ross, J. Chromatogr., 133 (1977) 408.
756. F. Take et al., Microchem. J., submitted for publication.
757. I. Björkhem et al., Clin. Chem., 23 (1977) 2114.
758. S. Yamada, H. A. Itano, Biochim. Biophys. Acta, 130 (1966) 538.
759. K. L. Dunlap et al., Anal. Chem., 48 (1976) 497.
760. R. J. Rando et al., J. Liq. Chromatogr., 16 (1993) 3977.
761. A. M. Caricchia et al., Anal. Chim. Acta, 286 (1994) 329.
762. K. Shimada et al., Analyst, 116 (1991) 1393.
763. K. Shimada et al., J. Liq. Chromatogr., 18 (1995) 2885.
764. K. Wang et al., Anal. Biochem., 243 (1996) 28.
765. E. D. Korn et al., J. Biol. Chem., 248 (1973) 2257.
766. M. L. Rueppel et al., Biomed. Mass Spectrom., 3, (1976) 28.
767. M. C. Roach et al., Anal. Chem., 59 (1987) 1056.
768. C. D. Stalikas, G. A. Pilidis, J. Chromatogr., A, 872 (2000) 215.
769. L. T. Sennello, C. J. Argoudelis, Anal. Chem., 41 (1969) 171.
770. K. A Karlsson, Biochem. Biophys. Res. Commun., 39 (1970) 847.
771. J.-J. Shieh, D. Desiderio, Anal. Lett., 10 (1977) 831.
772. D. J. Harvey, M. G. Horning, J. Chromatogr., 79 (1973) 65.
773. J. H. Duncan et al., Biochemistry, 10 (1971) 927.
774. M. G. Horning et al., J. Chromatogr. Sci., 7 (1969) 267.
775. S. Soniki et al., J. Liq. Chromatogr., 16 (1993) 2731.
776. S. Soniki et al., J. Liq. Chromatogr., 17 (1994) 1057.
777. Y. Nakagawa, K. Waku, J. Chromatogr., 381 (1986) 225.
778. M. J. Lovdahl, D. J. Pietrzyk, J. Chromatogr., A, 850 (1999) 143.
779. R. W. Sparidans et al., J. Chromatogr., A, 868 (2000) 141.
780. A. Heywood et al., Anal. Chem., 42 (1970) 1272.
781. W.-H. Ding, C.-T. Chen, J. Chromatogr., A 857 (1999) 359.
782. J. Eagles, M. E. Knowles, Anal. Chem., 43 (1971) 1697.
783. J. S. Parsons, J. Chromatogr. Sci., 11 (1973) 659.
784. A. I. Kalinovski et al., Carbohydr. Res., 58 (1977) 473.
785. B. Lindtröm, M. Molander, J. Chromatogr., 101 (1974) 219.
786. M. Ehrvick, K. Gustavii, Anal. Chem., 46 (1974) 39.
787. P. H. Degen, A. Schweitzer, J. Chromatogr., 142 (1977) 549.
788. O. Gyllenhaal, H. Ehrsson, J. Chromatogr., 107 (1975) 327.
789. A. Frigerio, C. Pantarotto, J. Chromatogr., 130 (1977) 361.
790. W. J. A. VandenHeuval, V. F. Gruber, J. Chromatogr., 112 (1975) 513.
791. O. Gyllenhaal, H. Ehrsson, J. Chromatogr., 107 (1975) 327.
792. K. Sabih, Biomed. Mass Spectrom., 1 (1974) 163.
793. D. L. Simmons et al., J. Chromatogr., 71 (1972) 421.
794. W. E. Brazelton et al., Anal. Chem., 48 (1976) 1386.
795. M. D. Muler, Anal. Chem., 59 (1987) 617.
796. Y. Cai, J. M. Bayona, J. Chromatogr., A, 696 (1995) 113.
797. Y. Cai et al., J. Chromatogr., A, 876 (2000) 147.
798. M. Abuin et al., J. Chromatogr., A, 889 (2000) 185.
799. M. Becchi et al., J. Chromatogr., A, 905 (2001) 207.
800. K. K. Verma et al., Anal. Chem., 64 (1992) 1484.
801. W. Z. Shou et al. , Anal. Chem., 72 (2000) 3266.

802. G. Popa, S. C. Moldoveanu, *Reactivii Organici In Chimia Analitica*, Ed. Academiei, Bucuresti, 1976.
803. G. Popa, S. C. Moldoveanu, *Analiza Chimica Cantitativa cu Reactivi Organici*, Ed. Tehnica, Bucuresti, 1969.
804. G. G. Talanova et al., Anal. Chem., 71 (1999) 3106.
805. M. Y. Khuhawar et al., Chromatographia, 41 (1995) 236.
806. J. N. King, J. S. Fritz, Anal. Chem., 59 (1987) 703.
807. X. Zhang et al., Anal. Chim. Acta, 237 (1990) 311.
808. H. Lu et al., J. Chromatogr., A, 857 (1999) 343.
809. Y. Akama, A. Tong, J. Chromatogr., 633 (1993) 129.
810. G. Bagur et al., J. Chromatogr., A, 730 (1996) 241.
811. M. Y. Khuhawar et al., J. Chromatogr., B, 693 (1997) 175.
812. J. Miura, N. Itoh, J. Liq. Chromatogr., Relat. Technol., 20 (1997) 2367.
813. Y. Shijo et al., Analyst, 121 (1996) 325.
814. Y. Inoue et al., Anal. Chem., 68 (1996) 1517.
815. W. Z. Shou, R. F. Browner, Anal. Chem., 71 (1999) 3365.
816. A. Casoli, Anal. Chem., 57 (1985) 561.
817. S. J. J. Tsai, H. T. Yan, Analyst, (1993) 521.
818. H. Hoshino et al., Anal. Chem., 68 (1996) 1960.
819. I. Haumann, K. Bachmann, J. Chromatogr., A, 717 (1995) 385.
820. M. Y. Khuhawar et al., J. Chromatogr., A, 740 (1996) 296.
821. T. Takeuchi et al., J. Chromatogr., A, 910 (2001) 373.
822. M. Derbyshire et al., Anal. Chem., 71 (1999) 4203.
823. J. Wu et al., J. Chromatogr., B., 663 (1995) 247.
824. G. Bagur et al., J. Chromatogr. Sci., 35 (1997) 131.
825. J. W. O'Laughlin, Anal. Chem., 54 (1982) 178.
826. M. Kanbayashi et al., J. Chromatogr., 386 (1987) 191.
827. H. Hasson, A. Warshawsky, J. Chromatogr., 530 (1990) 219.
828. N. Uehara et al., Analyst, 117 (1992) 977.
829. S. Motelier, H. Pitsch, J. Chromatogr., A, 739 (1996) 119.
830. P. A. Reece, J. Chromatogr., 306 (1984) 417.
831. A. K. Malik et al., J. Chromatogr., A, 587 (1999) 365.
832. M. Pilar da Silva et al., J. Chromatogr., A, 761 (1997) 139.
833. S. Zappoli et al., J. Chromatogr., A, 721 (1996) 269.
834. M. Satterfield, J. S. Brodbelt, Anal. Chem., 72 (2000) 5898.
835. J. Shen, J. S. Brodbelt, Rapid Commun. Mass Spectrom., 13 (1999) 1381.
836. J. F. Kennedy, C. A. White, *Bioactive Carbohydrates In Chemistry, Biochemistry and Biology*, E. Horwood, Chichester, 1983.
837. A. Kamei et al., Chem. Pharm. Bull., 24 (1976) 1108.
838. S. Honda et al., J. Chromatogr., 169 (1979) 287.
839. S. Honda et al., J. Chromatogr., 176 (1979) 367.
840. C. D. Pfaffenberger et al., J. Chromatogr., 126 (1976) 535.
841. G. D. McGinnis, Carbohydr. Res., 108 (1982) 284.
842. T. P. Mawhinney et al., Anal. Biochem., 101 (1980) 112.
843. G. O. Guerrant, C. W. Moss, Anal. Chem., 56 (1984) 633.
844. F. M. Rubino, J. Chromatogr., 473 (1989) 125.
845. D. H. Shaw, G. W. Moss, J. Chromatogr., 41 (1967) 350.
846. W. F. Lenhardt, R. J. Winzler, J. Chromatogr., 34 (1968) 471.
847. D. Anderle et al., Anal. Chem., 49 (1977) 137.
848. Zs. F. Katona et al., J. Chromatogr., A, 847 (1999) 91.

849. F. M. Rubino, J. Chromatogr., 473 (1989) 125.
850. R. A. Laine, C. C. Sweeley, Carbohydr. Res., 27 (1973) 199.
851. O. Peletier, S. Cadieux, J. Chromatogr., 231 (1982) 225.
852. D. C. DeJongh, K. Biemann, J. Am. Chem. Soc., 86 (1964) 3149.
853. S. Honda et al., J. Chromatogr., 176 (1979) 367.
854. S. Honda et al., Anal. Chem., 52 (1980) 1079.
855. K. Moper et al., Anal. Chem., 52 (1980) 2018.
856. M. J. DelNozal et al., J. Chromatogr., 607 (1992) 191.
857. Y. Umegae et al., Anal. Chim. Acta, 217 (1989) 263.
858. Y. Umegae et al., Chem. Pharm. Bull., 38 (1990) 963.
859. S. Yamaguchi et al., Analyst, 118 (1993) 769.
860. J. F. Kennedy et al., Carbohydr. Res., 49 (1976) 243.
861. J. Szafranek et al., J. Chromatogr., 88 (1974) 149.
862. M. B. Perry et al., J. Chromatogr., 44 (1969) 614.
863. M. Vecchi, K. Kaiser, J. Chromatogr., 26 (1967) 22.
864. D. Anderle, P. Kovac, J. Chromatogr., 91 (1974) 463.
865. F. Eisenberg, Anal. Biochem., 60 (1974) 181.
866. G. Petersson, Carbohydr. Res., 33 (1974) 47.
867. N. Morimoto et al., Anal. Chem., 73 (2001) 5422.
867a. J. M. L. Mee, J. Chromatogr., 94 (1974) 298.
868. M. D. G. Oates, J. Schrager, J. Chromatogr., 28 (1967) 232.
869. R. E. Hurst, Carbohydr. Res., 30 (1973) 143.
870. J. Kärkkäinen et al., J. Chromatogr., 20 (1965) 457.
871. Z. Tamura et al., Chem. Pharm. Bull., 16 (1968) 1864.
872. D. C. DeJongh, S. Hanessian, J. Am. Chem. Soc., 87 (1965) 1408.
873. D. C. DeJongh, S. Hanessian, J. Am. Chem. Soc., 87 (1965) 3744.
874. S. Hase, Y. Matsushima, J. Biochem., 66 (1969) 57.
875. I. Mononen, Carbohydrate Res., 88 (1981) 39.
876. D. A. Craven., C. W. Gehrke, J. Chromatogr., 37 (1968) 414.
877. D. J. Harvey, M. G. Horning, J. Chromatogr., 76 (1973) 51.
878. D. J. Harvey et al., Anal. Lett., 3 (1970) 489.
879. W. W. Wells et al., Biochim. Biophys. Acta, 82 (1964) 408.
880. J. Wiecko, W. R. Sherman, Org. Mass Spectrom., 10 (1975) 1007.
881. K. J. A. Jackson, J. K. A. Jones, Can. J. Chem., 43 (1965) 450.
882. Y. Sato et al., Anal. Chem., 72 (2000) 1207.
883. J. C. Bigge et al., Anal. Biochem., 230 (1995) 229.
884. J. Charlwood et al., Anal. Chem., 72 (2000), 1453.
885. M.-J. Zagalak et al., J. Chromatogr., 142 (1977) 523.
886. A. J. Cliffe et al., J. Chromatogr., 78 (1973) 333.
887. R. F. McGregor, Clin. Chim. Acta, 48 (1973) 65.
888. H. Iwase, A. Murai, Anal. Biochem., 78 (1977) 340.
889. H. Iwase, A. Murai, Chem. Pharm. Bull., 25 (1977) 285.
890. H. Miyazaki et al., Chem. Pharm. Bull., 23 (1975) 1806.
891. C. Zomzely et al., Anal. Chem., 34 (1962) 1414.
892. D. Roach, C. W. Gehrke, J. Chromatogr., 44 (1969) 269.
893. K. R. Leimer et al., J. Chromatogr., 141 (1977) 121.
894. R. W. Zumwalt et al., J. Chromatogr., 57 (1971) 193.
895. I. M. Moodie, J. Chromatogr., 99 (1974) 495.
896. M.-J. Zagalak et al., J. Chromatogr., 142 (1977) 523.
897. M. Makita et al., J. Chromatogr., 120 (1976) 129.

898. M. Makita et al., J. Chromatogr., 124 (1976) 92.
899. E. Bayer, W. A. Koenig, J. Chromatogr., 7 (1969) 95.
900. F. Marcucci, E. Mussini, J. Chromatogr., 25 (1966) 11.
901. P. A. Cruickshank, J. C. Sheehan, Anal. Chem., 36 (1964) 1191.
902. W. Parr, P. Howard, Chromatographia, 4 (1971) 162.
903. W. Parr et al., J. Chromatogr, 50 (1970) 510.
904. W. A. Koenig et al., J. Chromatogr. Sci., 8 (1970) 183.
905. S. L. MacKenzie, D. Tenaschuk, J. Chromatogr., 97 (1974) 19.
906. S. L. MacKenzie, D. Tenaschuk, J. Chromatogr., 111 (1975) 413.
907. S. L. MacKenzie, D. Tenaschuk, J. Chromatogr., 104 (1975) 176.
908. R. J. Pearce, J. Chromatogr., 136 (1977) 113.
909. P. Felker, R. S. Bandurski, Anal. Biochem., 67 (1975) 245.
910. H. A. Saroff, A. Karmen, Anal. Biochem., 1 (1960) 344.
911. F. Weygand et al., Z. Naturforsch., 18B (1963) 1924.
912. F. Weygand et al., Z. Naturforsch., 20B (1965) 1169.
913. E. Bayer et al., Z. Naturforsch., 22B, (1967) 924.
914. A. Islam, A. Darbre, J. Chromatogr., 43 (1969) 11.
915. B. Teuwissen, A. Darbre , J. Chromatogr., 49 (1970) 298.
916. A. J. Cliffe et al., J. Chromatogr., 78 (1973) 333.
917. M. Senn et al., J. Am. Chem. Soc., 88 (1966) 5593.
918. H. Lindley, P.C. Davis, J. Chromatogr., 100 (1974) 117.
919. G. E. Pollack, Anal. Chem., 39 (1967) 1194.
920. J. P. Vanetta, G. Vincendon, J. Chromatogr., 76 (1973) 91.
921. J. M. L. Mee et al., Biomed. Mass Spectrom., 4 (1977) 178.
922. K.-R. Kim et al., J. Chromatogr., A, 874 (2000) 91.
923. R. F. Adams, J. Chromatogr., 95 (1974) 189.
924. J. R. Coulter, C. S. Hann, J. Chromatogr., 36 (1968) 42.
925. C. W. Moss et al., J. Chromatogr., 60 (1971) 134.
926. C. W. Moss, M. S. Lambert, J. Chromatogr., 57 (1968) 42.
927. M. A. Kirkman, J. Chromatogr., 97 (1974) 175.
928. J. F. March, Anal. Biochem., 69 (1975) 420.
929. A. Darbre, K. Blau, J. Chromatogr., 17 (1965) 31.
930. D. E. Johnson et al., Anal. Chem., 33 (1961) 669.
931. C. Rodier et al., J. Chromatogr., A, 915 (2001) 199.
932. G. E. Pollack, Anal. Chem., 39 (1967) 1194.
933. G. Y. Tang, S. C. Moldoveanu, unpublished results.
934. M. Makita et al., J. Chromatogr., 120 (1976) 129.
935. M. Makita et al., J. Chromatogr., 124 (1976) 92.
936. E. Jellum et al., Anal. Biochem., 31 (1969) 227.
937. K. M. Williams, B. Halpern, Anal. Lett., 6 (1973) 839.
938. J. P. Thenot, E. C. Horning, Anal. lett., 5 (1972) 519.
939. I. Horman, F. J. Hesford, Biomed. Mass Spectrom., 1 (1974) 115.
940. G. Roda, A. Zamorani, J. Chromatogr., 46 (1970) 315.
941. R. D. Zumwalt et al. (eds.), *Amino Acid Analysis by Gas Chromatography*, CRC Press, Boca Raton, 1987.
941a. P. Hušek, J. Chromatogr., 91 (1974) 475.
942. P. Hušek, J. Chromatogr., 91 (1974) 483.
943. F. Weygand et al., Chem. Ber., 99 (1966) 1461.
944. O. Grahl-Nielsen, E. Solheim, J. Chromatogr., 69 (1972) 366.
945. O. Grahl-Nielsen, J. Chromatogr., 93 (1974) 229.

946. P. Hušek, J. Chromatogr., 552 (1991) 289.
947. J. Wang et al., J. Chromatogr., A, 663 (1994) 71.
948. P. Hušek, J. Chromatogr., B, 717 (1998) 57.
949. I. Abe et al., J. Chromatogr., A, 722 (1996) 221
950. S. Casal et al., J. Chromatogr., A, 866 (2000) 221.
951. K. Blau, *Biomedical Applications of Gas Chromatography*, vol. 2, Plenum, New York, 1968.
952. C. W. Gehrke, K. Leimer, J. Chromatogr., 57 (1971) 219.
953. F. Shahrokhi, C. W. Gehrke, J. Chromatogr., 36 (1968) 31.
954. C. W. Gehrke et al., J. Chromatogr., 45 (1969) 24.
955. C. W. Gehrke, K. Leimer, J. Chromatogr., 53 (1970) 201.
956. P. W. Albro, L. Fishbein, J. Chromatogr., 55 (1971) 297.
957. Zs. F. Katona et al., J. Chromatogr., A, 847 (1999) 91.
958. L. Birkhofer, M. Donike, J. Chromatogr., 26 (1967) 270.
959. M. Dönike, J. Chromatogr., 85 (1973) 1.
960. M. Dönike et al., J. Chromatogr., 85 (1973) 9.
961. T. P. Mawhiney et al., J. Chromatogr., 358 (1986) 231.
962. J.. H. Lauterbach, S. C. Moldoveanu, 42 TCRC, Lexington, 1988.
963. M. Dönike, J. Chromatogr., 115 (1975) 591.
964. J. P. Hardy, S. L. Kerrin, Anal. Chem., 44 (1972) 1497.
965. B. S. Middledich, J. Chromatogr., 126 (1976) 581.
966. M. Rangarajan et al., J. Chromatogr., 87 (19730 499.
967. R. E. Harman et al., Anal. Biochem., 25 (1968) 452.
968. M. R. Guerin, W. D. Shults, J. Chromatogr. Sci., 7 (1969) 701.
969. R. J. Stolzberg, D. N. Hume, Anal. Chem., 49 (1977) 374.
970. B. I. Zaideh et al., Anal Chem., 73 (2001) 799.
971. H. Furukawa et al., Chem Pharm. Bull., 23 (1975) 1625.
972. A. S. Brown, J. C. Bennett, Anal. Biochem., 150 (1985) 457.
973. A. M. Rizzi et al., J. Chromatogr., A, 710 (1995) 287.
974. T. J. Mahachi et al., J. Chromatogr., 289 (1984) 279.
975. H. Matsunaga et al., Biomed. Chromatogr., 10 (1996) 95.
976. T. Toyo'oka, Y.-M. Liu, J. Chromatogr., A, 689 (1995) 23.
977. S. Datta et al., Biochem. Biophys. Res. Commun., 72 (1976) 1296.
978. V. Fireabracci et al., J. Chromatogr., 570 (1991) 285.
979. L. E. Lavi et al., J. Chromatogr., 377 (1986) 155.
980. G. Buzzigoli et al., J. Chromatogr., 507 (1990) 85.
981. J. F. Davey, R. S. Ersser, J. Chromatogr., 528 (1990) 9.
982. B. A. Bidlingmeyer et al., J. Chromatogr., 336 (1984) 93.
983. C. M. Noyes, J. Chromatogr., 266 (1983) 451.
984. H. Kanazawa et al., Anal. Chem., 72 (2000) 5961.
985. C. J. Strang et al., Anal. Biochem., 178 (1989) 276.
986. C. Goddard, R. L. Felsted, Biochem. J., 253 (1988) 839.
987. P. Husek, P. Simek, LC-GC, 19 (2001) 986.
987a. E. Bricas et al., Biochemistry, 4 (1965) 2254.
988. H. Haralambidou, R. A. Day, Org. Mass. Spectrom., 10 (1975) 683.
989. P. Kamerling et al., Org. Mass Spectrom., 1 (1968) 345.
990. H. R. Morris et al., Biochem. Biophys. Res. Commun., 51 (1973) 247.
991. D. W. Thomas et al., Biochem. Biophys. Res. Commun., 32 (1968) 519.
992. G. Patonay, M. D. Antoine, Anal. Chem., 63 (1991) 321A.
993. R. Williams et al., Anal. Chem., 65 (1993) 601.

994. K. Sauda et al., Anal. Chem., 58 (1986) 2649.
995. A Guttman et al., J. Chromatogr., A, 894 (2000) 329.
996. J. L. Tadeo et al., J. Chromatogr., A 882 (2000) 175.
997. T. Vo-Dinh (ed.), *Chemical Analysis of Polyciclyc Aromatic Compounds*, J. Wiley, New York, 1989.
998. C. J. W. Brooks, E. C. Horning, Anal. Chem., 36 (1964) 1540.
999. D. A. Saelens et al., J. Chromatogr., 123 (1976) 185.
1000. L. M. Cummins, M. J. Fourier, Anal. Lett., 2 (1969) 403.
1001. G. R. Wilkinson, Anal. Lett., 3 (1970) 289.
1002. H. Ko et al., Anal. Lett., 7 (1974) 243.
1003. H. Miyazaki et al., J. Chromatogr., 99 (1974) 575.
1004. R. J. Francis et al., Biomed. Mass Spectrom., 3 (1976) 281.
1005. M. Horning et al., Biochim. Biophys. Acta, 148 (1967) 597.
1006. A. C. Moffat et al., J. Chromatogr., 66 (1972) 255.
1007. G. M. Anthony et al., J. Chromatogr. Sci, 7 (1969) 623.
1008. J. P. Shyluk et al., J. Chromatogr., 26 (1976) 268.
1009. G. M. Anthony et al., J. Chromatogr. Sci., 7 (1969) 623.
1010. M. G. Horning et al., Anal. Lett., 1 (1968) 713.
1011. G. Lancaster et al., Clin. Chim. Acta, 48 (1973) 279.
1012. A. Frigerio et al., J. Chromatogr., 81 (1973) 139.
1013. M. J. Levitt, Anal. Chem., 45 (1973) 618.
1014. K. Green et al., Anal. Biochem., 54 (1973) 434.
1015. F. Vane, M. G. Horning, Anal. Lett., 2 (1969) 357.
1016. J. Rosello et al., J. Chromatogr., 122 (1976) 469.
1017. J. Rosello et al., J. Chromatogr., 130 (1977) 65.
1018. D. S. Millington et al., Biomed. Mass Spectrom., 2 (1975) 219.
1019. H. M. Fales, T. Luukkainen, Anal. Chem., 37 (1965) 955.
1020. K. T. Koshy et al., J. Chromatogr. Sci., 13 (1975) 97.
1021. R. A. Mead et al., J. Chromatogr. Sci., 7 (1969) 554.
1022. J. Attal et al., Anal. Biochem., 20 (1967) 394.
1023. A. Migrod, H. R. Lindner, Steroids, 8 (1966) 119.
1024. E. M. Chambaz et al., Anal. Lett., 1 (1968) 749.
1025. S. Zaraga et al., J. Steroid Biochem., 4 (1973) 417.
1026. J. Chamberlain, J. Chromatogr., 28 (1967) 404.
1027. J. R. G. Challis et al., J. Chromatogr., 50 (1970) 228.
1028. W. P. Collins et al, J. Chromatogr., 37 (1968) 33.
1029. T. Kanno et al., J. Chromatogr. Sci., 9 (1971) 53.
1030. C. H. Kiang et al., J. Chromatogr., 567 (1991) 195.
1031. H. Asukabe et al., J. Chromatogr., A, 657 (1993) 349.
1032. K. Shimada et al., J. Liq. Chromatogr., 16 (1993) 3965.
1033. A. Nakanishi et al., J. Chromatogr., 591 (1992) 159.
1034. M. T. Tsamis et al, J. Chromatogr., 277 (1983) 61.
1035. S. Li, S. K. Wahba Khalil, J. Liq. Chromatogr., 13 (1990) 1339.
1036. E. Bousquet et al., J. Liq. Chromatogr., Relat. Technol., 20 (1997) 757.
1037. R. K. Munns et al., J. Chromatogr., 442 (1988) 209.
1038. J. M. Rosenfeld et al., Anal. Chem., 49 (1977) 725.
1039. D. B. Lakings et al, J. Chromatogr., 116 (1976) 69.
1040. C. W. Gehrke, D. B. Lakings, J. Chromatogr., 61 (1971) 45.
1041. D. B. Lakings, C. W. Gehrke, J. Chromatogr., 62 (1971) 347.
1042. M. Saha et al., J. Chromatogr., A, 712 (1995) 345.

1043. A. P. DeLeenheer, C. F. Gelijkens, Anal. Chem., 48 (1976) 2203.
1044. C. W. Gehrke, A. B. Patel, J. Chromatogr., 130 (1977) 103.
1045. U. I. Kramer et al., Anal. Biochem., 82 (1977) 217.
1046. G. R. Petit et al., Biomed. Mass Spectrom., 5 (1978) 153.
1047. A. M. Lowson et al., J. Am. Chem. Soc., 93 (1971) 1014.

CHAPTER 20

Chemical Degradation of Polymers and Pyrolysis

20.1 CHEMICAL DEGRADATION OF POLYMERIC CARBOHYDRATES

The analysis of intact polymers is a rather difficult task. Polymers are not volatile, some of them have low solubility in most solvents, and some decompose easily during heating. Therefore the direct application of powerful analytical tools such as GC-MS cannot be done directly on most polymers. The use of LC and LC-MS in polymer analysis is becoming more common [1], [2], but problems are still encountered regarding structure elucidation. For the analysis of polymeric materials including carbohydrates, a number of chemical modifications can be done to the initial polymer, such as hydrolysis, methanolysis, pyrolysis, etc., that lead to smaller molecules much easier to analyze compared to the initial material. Structure elucidation by controlled chemical degradation of carbohydrates is, however, a very complex process, and an in-depth discussion is beyond the purpose of this book (see e.g. [3], [4]. Only a general overview of the subject is included herein.

Polysaccharides are formed from a number of monosaccharide residues connected by ether bonds. Typical monosaccharide residues are pentoses or hexoses with different types of ether links [5]. Polysaccharide structure can be described similarly to protein structure. A primary structure refers to the sequence of connected monosaccharides. This structure can be more complicated than that of proteins, because the bonding of the sugar units can be done at different points and can involve branching. A secondary and tertiary structure refers to the overall shape of the macromolecule, and a quaternary structure refers to the aggregation of polysaccharide chains [5].

A practical classification of natural polysaccharides is based on their source, such as plants, algae, microbes, fungi, or animals. The problem with this classification is that some polysaccharides are very common in nature and others are not. A group such as plant polysaccharides is much larger than other groups such as that of fungal polysaccharides. Therefore, for a large group, a more detailed classification is needed. One such classification, considering both the source and how common in nature are the individual polysaccharides, recognizes the following groups: a) cellulose and its derivatives, b) amylose, amylopectin, starch and their derivatives, c) pectins, d) gums and mucilages including plant exudates such as gum arabic, gum ghatti, gum karaya, gum tragacanth, and seed gums such as guar gum, locust bean gum, tamarind kernel powder, etc., e) hemicelluloses and other plant polysaccharides such as larch arabinogalactan, f) algal polysaccharides such as agar, alginic acid, carrageenan, fucoidan, furcellaran, and laminaran, g) microbial polysaccharides and biosynthetic gums such as xanthan and dextran, h) lipopolysaccharides from the cell surface of bacteria, i) fungal polysaccharides, j) glycogen, k) chitin, and l) proteoglycans.

Some natural polysaccharides are homopolysaccharides, and they consist of unique monomeric units interconnected by identical links. Table 20.1.1. lists a number of natural homopolysaccharides.

TABLE 20.1.1. *Natural homopolysaccharides.*

Type of bond	Common name
L-arabinans	
(1 → 3)-α-L-, (1 → 5)-α-L- branched	plant pectins
D-fructans	
(2 → 1)-β-D- linear	inulin
(2 → 6)-β-D- linear	levans
(2 → 6)-β-D-, (2 → 1)-β-D- branched	levans
L-fucans	
(1 → 2)-α-L-, (1 → 4)-α-L- branched	fucoidan
D-galactans	
(1 → 3)-β-D-, (1 → 4)-α-D- linear	carrageenan
(1 → 3)-β-D-, (1 → 6)-β-D-branched	
(1 → 4)-β-D- linear	plant pectins
(1 → 5)-β-D- linear	galactocarolose
D-galacturonans	
(1 → 4)-α-D- linear	pectic acid
D-glucans	
(1 → 2)-β-D- linear	
(1 → 3)-α-D-, (1 → 4)-α-D- linear	nigeran, isolichenan
(1 → 3)-β-D- linear	laminaran
(1 → 3)-β-D-, (1 → 6)-β-D- branched	scleroglucan
(1 → 3)-β-D-, (1 → 4)-β-D- linear	lichenan
(1 → 4)-α-D- linear	amylose
(1 → 4)-α-D-, (1 → 6)-α-D- linear	pullulan
(1 → 4)-α-D-, (1 → 6)-α-D- branched	glycogen, amylopectin
(1 → 4)-β-D- linear	cellulose
(1 → 6)-β-D-, (1 → 3)-α-D- branched	dextran
(1 → 6)-β-D- linear	pustulan
2-amino-2-deoxy-D-glucans	
(1 → 4)-β-D- linear	chitin
D-mannans	
(1 → 2)-α-D-, (1 → 6)-α-D- branched	
(1 → 4)-β-D- linear	
D-xylans	
(1 → 3)-β-D- linear	rhodymenan
(1 → 3)-β-D-, (1 → 4)-β-D- linear	
(1 → 4)-β-D- linear	

Many natural polysaccharides are heteropolysaccharides formed from two or more monosaccharide residues. As an example, guar gum has as the main component (about 85%) a well-characterized, relatively simple galactomannan (guaran). This contains a main backbone of (1→4)-linked β-D-mannopyranosyl units with α-D-galactopyranosyl attached by (6→1) links as shown below:

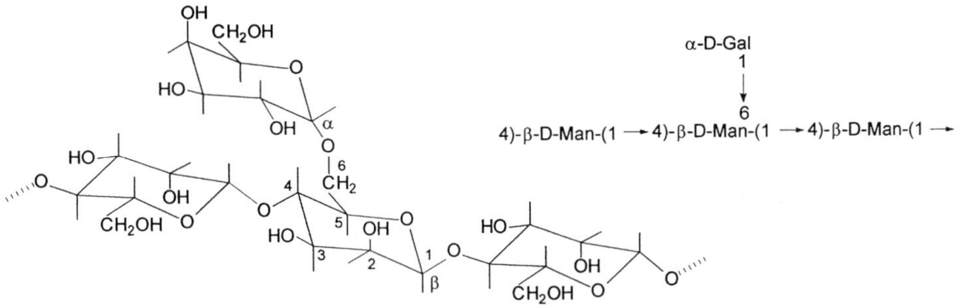

$$\text{4)-}\beta\text{-D-Man-(1} \longrightarrow \text{4)-}\beta\text{-D-Man-(1} \longrightarrow \text{4)-}\beta\text{-D-Man-(1} \longrightarrow$$

A significant number of biopolymers found in plants and animals contain chains of carbohydrates associated with lipids or with proteins. Lipopolysaccharides are commonly found on the cell surface of bacteria and are responsible for type specific immunological reactions. Their structure shows three distinct parts: O-antigen, core, and lipid A. The O-antigen portion is the outer part of the capsule. It is usually a complex polysaccharide, although a number of common repeating oligosaccharides formed from D-mannopyranosyl, D-galactopyranosyluronic acid, or D-glucopyranosyluronic acid residues can be present. The core polysaccharide is usually simpler and consists of a linear chain with few short branches, containing D-glucose, D-galactose, and 2-acetamido-2-deoxy-D-glucose. The core is connected to the lipid A portion by a heptose and an octose such as 3-deoxy-β-D-manno-octopyranulosonic acid (KDO). The lipid A consists of repeating units of an amido-disaccharide connected through phosphodiester bonds. This disaccharide has O- and N-substitution with long chain fatty acids.

The biopolymers containing both a protein chain and carbohydrates in the same molecule are classified as either glycoproteins or proteoglycans. The differentiation of the two classes is based on the number of carbohydrate units per unit length of protein backbone, with the protein predominant in glycoproteins and carbohydrate predominating in proteoglycans. A third class, carbohydrate-protein complexes, is also known, but in these compounds the protein and the carbohydrates are not covalently linked, and certain separation procedures can be applied without destroying the molecular entities. Also, some proteins such as albumins contain low amounts of carbohydrates without being classified as glycoproteins.

Glycoproteins contain a protein chain with covalently attached carbohydrate segments usually made from hetero-oligosaccharides. These segments are usually branched and can contain neutral monosaccharides, basic monosaccharides (such as 2-amino-2-deoxy-D-glucose), and a unique nine-carbon sugar, 5-amino-3,5-dideoxy-D-glycero-D-galacto-2-nonulopyranuronic acid (neuraminic acid). Because the amino sugars in glycoproteins are frequently acetylated, glycoproteins are commonly slightly acidic due to the neuraminic acid.

Proteoglycans are essential parts of the connective tissue in mammals and are also present to some extent in fish and bacteria. Proteoglycans are formed from a protein chain, a linkage (carbohydrate) region, and a considerably large carbohydrate component consisting of a glycosaminoglycan region, connected as shown below:

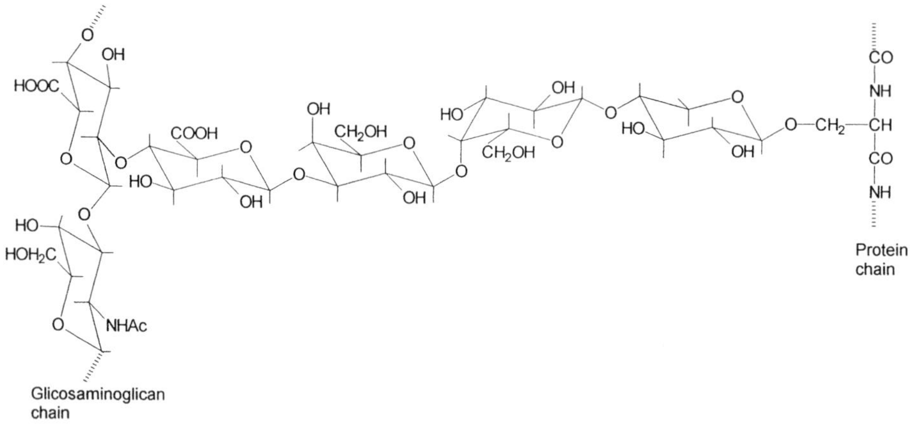

Glicosaminoglican
chain

Eight main glycosaminoglycans have been identified in proteoglycans including galactosaminoglucuronans, glucosaminoglucuronans, glucosaminogalactans, etc. [5]. The term mucopolysaccharide was used in the past to describe polysaccharide materials of animal origin containing 2-amino-2-deoxyhexoses.

Both N-glycosidic and O-glycosidic linkages may be present in carbohydrates associated with lipids or with proteins (glycoconjugates). Typical O-glycosidic linkages take place, for example, between D-xylose or D-galactose residues and the OH group in L-serine or L-threonine, D-galactose and the OH group in L-hydroxy-lysine, L-arabinose and L-hydroxyproline. N-Glycosylic linkages are formed, for example, between N-acetyl-glucosamine residues and the amido group in asparagine as shown below:

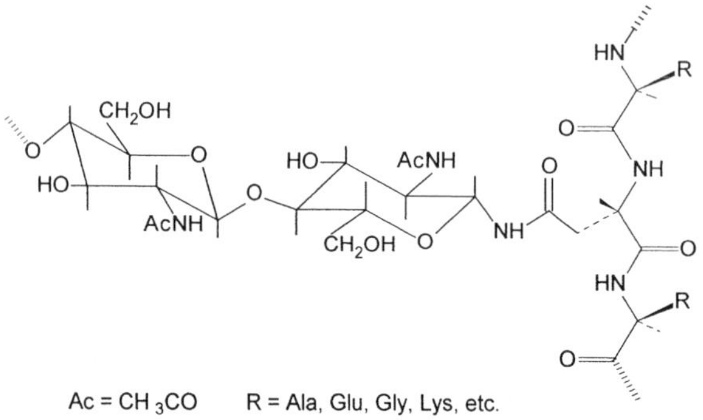

Ac = CH₃CO   R = Ala, Glu, Gly, Lys, etc.

In some peptidoglycans containing 2-acetamido-2-deoxy-D-glucose and 2-acetamido-3-O-(1-carboxyethyl)-2-deoxy-D-glucose (2-acetamido-2-deoxymuramic acid), the linkage is done through the carboxyl group of the 2-acetamido-2-deoxymuramic acid and the free terminal NH₂ group of the protein chain, as shown below:

Ac = CH₃CO   R = Ala, Glu, Gly, Lys, etc.

In the analysis of a polysaccharide, a number of steps are followed. The first important problem is the isolation of the material in a pure form. Microheterogeneity can be present in natural materials, and purification can be a very challenging problem. Changes in the structure of the initial polymer may be inflicted when acids or bases are used during the purification process. Some procedures used in sample preparation already discussed in Part 2 of the book are applicable for this step in sample preparation.

The progress made in the MS analysis of large molecules, using for example MALDI-TOF MS procedures, allows the analysis of some polysaccharides without chemical degradations (see e.g. [6], [7] [8]). However, chemical cleavage of large molecules is still a common practice for structural identifications. Typically, once the polysaccharide is obtained in a pure form, the following analyses are done: a) identification of the constituent monosaccharides, b) determination of the D and L configuration of component monosaccharides, c) determination of the degree of polymerization, d) determination of the position of glycosidic linkages, and e) determination of the sequence of monosaccharide residues. More detailed information can be of interest, such as the determination of the ring structure of component monosaccharides or the determination of the anomeric configuration of the glycosidic linkage [9]. Chromatography is a key method for the analysis of reaction products following chemical degradations, and from this point of view polysaccharide cleavage can be considered a sample preparation procedure for chromatographic analysis.

The identification of constituent monosaccharides in polysaccharides and glycoconjugates can be done by the cleavage of the glycosidic linkages followed by further analysis of the monosaccharides. The cleavage can be done by hydrolysis or other solvolysis procedures. Depending on the goal of the analysis, the cleavage of the carbohydrate moiety may be complete, leading to the formation of monosaccharides, or can be limited, leading to oligomers, or in the case of glycoconjugates to the isolation of the intact carbohydrate apart from the protein or lipid.

The determination of L and D configuration of component monosaccharides is usually done using a preparative chromatographic separation followed by the measurement of optical specific rotation $[\alpha]^T_\lambda$, or using enzymes with specificity for the D or L isomer. The degree of polymerization is commonly measured using physical techniques such as ultracentrifugation or ultrafiltration.

The determination of the position of glycosidic linkage can be done for example using the permethylation of the carbohydrate, followed by hydrolysis. The hydrolyzed products are further derivatized by acylation or silylation and analyzed by GC or GC-MS and identified. The position of the acyl or silyl groups indicate the position of glycosidic linkage, while the methyl groups remain attached to the previously free OH groups in the polymer.

The determination of the sequence of monosaccharide residues in the polysaccharide can be done using additional reactions such as periodate oxidation, followed by reduction with $NaBH_4$ and further derivatization and analysis. Various step hydrolysis and derivatization can be applied for obtaining more detailed structural information. Acetolysis can also be applied for structural determinations [10]. Different derivatizations and hydrolysis procedures also provide information regarding the ring structure of the individual monosaccharide residues.

The anomeric configuration of glycosidic linkages is usually done using enzymatic hydrolysis that is specific for a precise anomeric configuration, oxidation with CrO_3, or NMR studies.

- *Hydrolysis and other solvolysis procedures of polymeric carbohydrates*

Hydrolysis is the most common cleavage procedure for the determination of constituent monosaccharides in polysaccharides and in specific cases also provides further structural information. Typically it is carried out in aqueous solution in the presence of a catalyst. Acids such as HCl, H_2SO_4, and CF_3COOH or for special purposes alkali can be used as catalysts. Hydrolysis with strong acids, as well as with alkali, must be done with precaution, because monosaccharides suffer isomerizations and degradations with acids and bases. Enzymatic hydrolysis can be very useful for the analysis of more labile carbohydrates that suffer further modifications in strong acidic or alkaline conditions. Other solvents, besides water, can be used for the cleavage of the ether bonds. Among these are methanol, acetic anhydride, and formic acid.

Optimum hydrolysis of polymeric carbohydrates performed in the presence of a catalyst depends mainly on the nature of the polysaccharide. The problem during hydrolysis is either incomplete cleavage of the glycosidic linkage or the decomposition of the monosaccharide in strong conditions (heat, very low or high pH) or when the hydrolysis time is too long. Minimization of decomposition of liberated monosaccharide is important for structural determinations. Specific studies are done for determining optimum conditions for hydrolysis, for example, by determining the amount of monosaccharide liberated as a function of time or as a function of reaction temperature. During solvolysis, the concentration of the carbohydrate should be kept below 0.5% to avoid acid catalyzed condensations of the fragments (reversion), which may lead to a misinterpretation of results. As a rule, the stability of a specific residue is higher in the polymer form than as a monosaccharide. Aldoses are more stable toward acids than toward alkali, therefore acid hydrolysis is preferable when it is efficient enough. In strong acid conditions, typical decomposition products are several furans and acids such as levulinic and formic. For homopolysaccharides the elimination of the end unit can be the predominant reaction, although complex mixtures of oligosaccharides also can be generated. The compounds with α configuration of the glycosidic linkage are more susceptible to hydrolysis than compounds with β-glycosidic linkage.

Hydrolysis with H_2SO_4 as a catalyst can be performed in a 1M aqueous H_2SO_4 solution (50 mg polymer in 5 mL solution) with heating for 4 hours at 100° C. Longer heating time may lead to monosaccharide decomposition. Figure 20.1.1 shows the variation in the recoveries of several monosaccharides from a neutral polysaccharide when the heating at 100° C is prolonged for 5 and for 18 hours in 1M aqueous H_2SO_4 [11].

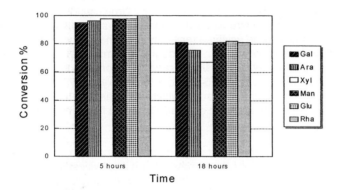

FIGURE 20.1.1. *Variation in the recoveries of several monosaccharides from a neutral polysaccharide with heating at 100° C for 5 and for 18 hours in 1M aqueous H_2SO_4.*

The removal of H_2SO_4 from the reaction medium can be done either using ion exchange resins or with $BaCO_3$ followed by filtration. $BaCO_3$ may lead to some losses due to the absorption of carbohydrates.

The hydrolysis also can be catalyzed by HCl. The conditions depend to a great extent on the type of sugar, and, for example, the hydrolysis of glycosaminoglycans requires 4M HCl at 100° C for 9 hours. It was found that the presence of oxygen increases the decomposition of the resulting monosaccharide [12], [13].

Trifluoroacetic acid (TFA) at conc. 2M and temperatures at 100° C for 15 hours are the type of conditions used for hydrolysis of plant cell-wall polysaccharides [14]. TFA can be removed from the hydrolysate by evaporation under reduced pressure or in a stream of dry N_2.

Hydrolysis with strong alkali (0.1–0.5 N NaOH) also can be applied. Alkalies usually cleave polysaccharides starting from the reducing end, leading to complex reactions. The amide groups in the protein chains are more stable in alkali-catalyzed hydrolysis compared to acid hydrolysis. For this reason, glycosidic linkages in glycoconjugates are usually hydrolyzed using alkali catalysis that leaves the protein unmodified. A solution of 0.1 M NaOH at 50° C can hydrolyze O-glycosidic bonds in 4 hours, while the N-glycosidic linkages may require 1 M NaOH at 100° C. The hydrolysis in the presence of a reducing agent such as $NaBH_4$ is recommended to prevent other degradations [15]. Alditols are more stable than monosaccharides in acids or bases. The procedure allows the identification of the linking monosaccharide residue. This identification can be facilitated by the reduction with NaBD$_4$ when the analysis is done by GC-MS [16]. Strong hydrolysis conditions in the presence of alkali are likely to lead to complex degradation products [4].

Polysaccharides made from neutral monosaccharide units are in general easier to hydrolyze, with decompositions of the liberated monosaccharides usually less than 10%. Liberated aldopentoses and deoxysugars are more susceptible to decompositions compared to aldohexoses. The use of 0.25 M H$_2$SO$_4$ at 70° C may be required for the hydrolysis of a pentose-containing polysaccharide. Uronic acids are relatively stable, although they still may be decarboxylated. Amino sugars are also relatively stable, but in many samples they may be acylated (acetylated) at the amino group. The amides are more sensitive to acid hydrolysis. Peptidoglycans (mureins) that are common in cell walls of bacteria generate by hydrolysis N-acetylated monosaccharides such as 2-acetamido-2-deoxy-muramic acid or 5-amino-3,5-dideoxy-D-glycero-α-D-galacto-2-nonulopyranuronic acid (neuraminic acid), which are sensitive to hydrolysis. The decomposition of N-acetylneuraminic acid in 0.01 N HCl for 30 min. at 100° C can be as high as 20% [17].

The hydrolysis conditions depend not only on the chemical composition of the polysaccharide, but also on the type of material and its physical appearance (powder, block, etc.). Woody materials, for example, usually require two-step hydrolysis. In the first step, the sample is swelled with 72% H$_2$SO$_4$ (w/w) for 1 hour at 30° C, or 2 hours at 20° C. In the second step, the H$_2$SO$_4$ is diluted to 1M and refluxed for 2 to 5 hours, depending on the material [18]. The remaining insoluble material is Klason lignin [19]. Other combinations of two-step hydrolysis are also applied to woody materials [17]. Gums require milder conditions such as 75 mM H$_2$SO$_4$ at 96° C for 36 hours [20], [21].

Many glycoconjugates are hydrolyzed using HCl at concentrations of 1M or 2 M, with heating at 100° C for various periods of time depending on the material [22], [23], [24]. Trifluoroacetic acid as well as NaOH are also used for hydrolysis. Preliminary derivatization to generate first a permethylated or peracetylated polymeric carbohydrate followed by hydrolysis can be applied for the determination of the structure of the polymer.

Among the most common solvolysis procedures besides hydrolysis is methanolysis. This can be carried out with dry HCl in anhydrous methanol. For example, a glycoside can be subjected to methanolysis [25] by heating the sample (5–10 mg, lyophilized) with 0.5 mL of 0.5 M HCl in anhydrous methanol for 20 hours at 50° C. The resulting products can be further derivatized by adding 150 μL pyridine and 100 μL acetic anhydride and allowing the mixture to stand for 30 min., and then evaporating to dryness under vacuum for the acetylation of possible amino groups in the sugars. The OH groups of the monosaccharides can be further silylated. For this purpose, the dry material is treated with 0.5 mL BSTFA/TMSI/pyridine (10:1:0.4 v/v/v) and allowed to stand a few min. The derivatized material can be analyzed by GC or GC-MS [17].

During the methanolysis, one of the liberated fragments contains the methoxy group and the other a hydroxyl. Also, several concurrent reactions may take place following protonation at different positions in the sugar ring. For these reasons methanolysis may lead to a complex mixture of methylated monosaccharides [26]. Methanolysis of simple polysaccharides such as starch lead mainly to methyl-α-D-glucopyranoside.

Acetolysis is another solvolysis procedure applied to polysaccharides. Following acetolysis, a complete acetylation of the free hydroxyl groups of the polysaccharide and of the resulting monosaccharide takes place. The rate constants for the solvolysis of glycosidic bonds may be different for different polysaccharides. Typically the reaction is done in acetic anhydride/acetic acid/sulfuric acid (10:10:1 v/v/v). Acetolysis has selectivity toward (1→6)-linkages that are less stable than (1→2)-, (1→3)- and (1→4)-linkages. Also, compounds with α configuration of the glycosidic linkage are more susceptible to acetolysis than compounds with β-glycosidic linkage.

Formolysis in formic acid + H_2SO_4 and mercaptolysis are also utilized in polysaccharide studies [4], [27].

The hydrolysis process may be stopped before the reaction is completed, such that fragments of oligosaccharides are generated. The analysis of the oligosaccharide fragments may provide important information on the initial polymer. The hydrolysis of homopolysaccharides will lead to a number of oligosaccharides of the same monomer with different degrees of polymerization (DP), because the rates of hydrolysis are the same for all glycosidic linkages in the initial polymer. However, for hetero-polysaccharides the hydrolysis rates of different types of linkages may be different. For example, furanosides are hydrolyzed 10 to 1000 times faster than pyranosides. Also, during acid hydrolysis, (1→4)-linkages can be less stable than (1→6)-linkages, while in acetic anhydride with 5% H_2SO_4, (1→6)-linkages are less stable.

Enzymes offer an excellent procedure for controlled hydrolysis of polysaccharides. The use of enzymes provides information in two ways, one being from the analysis of hydrolysis fragments, and the other from finding enzymes to which the polysaccharide is resistant or partially resistant, leading to an elimination process.

The enzymes used for polysaccharide hydrolysis are classified as *endo-* and *exo-* polysaccharide hydrolases [5]. The *endo*-polysaccharide hydrolases such as α-amylase are specific for linkage and monosaccharide residue and cause random fragmentation of homopolysaccharides. The *exo*-polysaccharide hydrolases are specific for a mono- or disaccharide unit and stereochemistry at C1, but do not differentiate between the residues attached glycosidically at C1. The cleavage typically occurs sequentially starting from one end of the polymer (usually the nonreducing end). For example β-amylase removes maltose units, cleaving amylose and generating almost quantitatively maltose. Enzymes that act specifically on glycoproteins (glycopeptidases) are also known. The purity of the enzymes is, however, a key factor in the success of enzymatic hydrolysis [28].

- Reactions for the determination of the glycosidic linkage position

Hydrolysis and acetolysis provide information regarding monosaccharide composition of polysaccharides and to a certain extent may indicate the position of the glycosidic linkage based on relative resilience to hydrolysis. Hydrolysis in basic medium may provide, in specific cases, more direct information for the position of the glycosidic linkage. For example, if the reaction takes place at a glucose end of a polysaccharide, the degradation products indicate the substitution position, as shown below for a 3-substituted residue and for a 4-substituted residue. The 3-substituted residue leads to a mixture of 3-deoxy-D-arabino and -D-ribo hexonic acids:

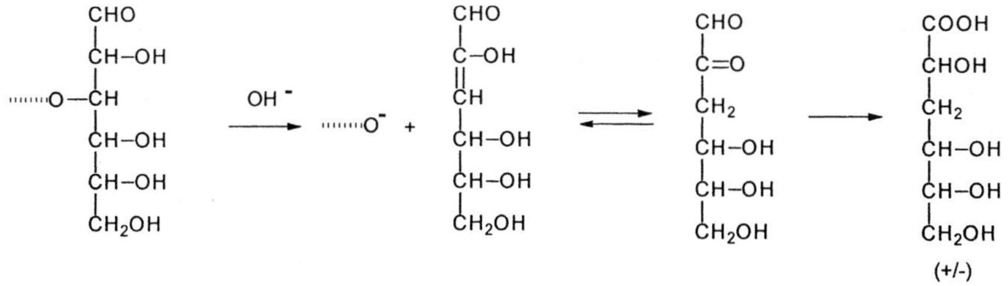

```
   CHO                     CHO                  CHO              COOH
   |                       |                    |                |
   CH-OH                   C-OH                 C=O              CHOH
   |            OH⁻        ||                   |                |
,,,,,O-CH      ────→       CH      +  ,,,,,O⁻   CH₂      ───→     CH₂
   |                       |                    |                |
   CH-OH            ,,,,,O⁻  CH-OH              CH-OH            CH-OH
   |                       |                    |                |
   CH-OH                   CH-OH                CH-OH            CH-OH
   |                       |                    |                |
   CH₂OH                   CH₂OH                CH₂OH            CH₂OH
                                                                 (+/-)
```

The 4-substituted residue leads to a mixture of 3-deoxy-2-C-(hydroxymethyl)-D-threo and D-erithro pentonic acids:

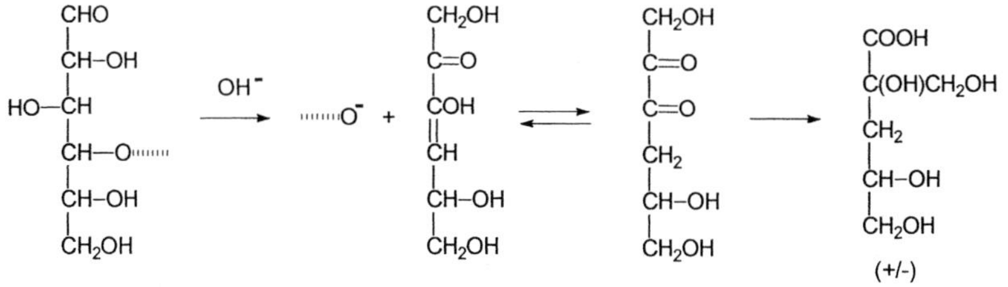

```
   CHO                     CH₂OH                CH₂OH
   |                       |                    |              COOH
   CH-OH                   C=O                  C=O            |
   |            OH⁻        |                    |              C(OH)CH₂OH
HO-CH         ────→        COH      +  ,,,,,O⁻   C=O           |
   |                       ||                   |              CH₂
   CH-O,,,,,        ,,,,,O⁻  CH                 CH₂      ───→   |
   |                       |                    |              CH-OH
   CH-OH                   CH-OH                CH-OH          |
   |                       |                    |              CH₂OH
   CH₂OH                   CH₂OH                CH₂OH          (+/-)
```

Another procedure for the determination of the position of the glycosidic linkage is based on the permethylation of the polysaccharide followed by hydrolysis. For example, a procedure may start with the polymer that is first dissolved or dispersed in an appropriate solvent such as DMSO, a solution of SO_2 in DMSO, LiCl-N,N-dimethyl-acetamide, etc. The solubilization of cellulosic materials is usually difficult, other polymers being easier to dissolve. The permethylation can be done with CH_3I in the presence of methylsulfinylmethanide anion or by adding powder of NaOH or KOH to the DMSO solution followed by CH_3I addition. Equal molar amounts of methylsulfinyl-methanide anion and CH_3I are typically used, in up to double excess of the molar requirement for methylation. The reactions occurring in this step can be written as follows:

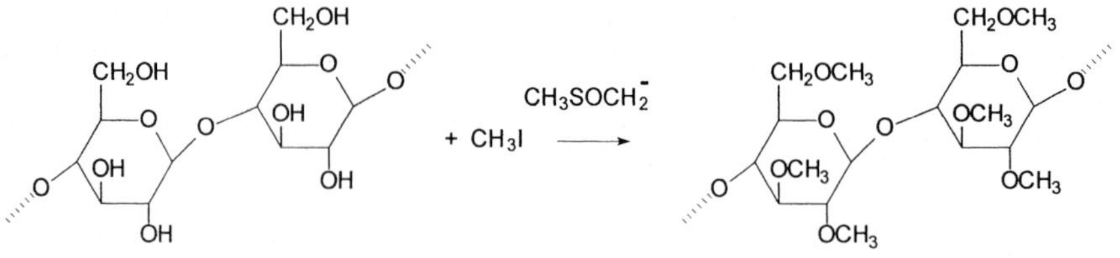

$CH_3SOCH_2^-$ + CH_3I ⟶

Partial methylation is still possible and usually unwanted because it leads to inconclusive results. Polysaccharides containing uronic acids or hexosamine residues are more difficult to methylate. Uronic acids may generate ketal derivatives, and hexosamines may lead to incomplete methylation or if acetylated may form N-methyl-N-acetamido derivatives.

The hydrolysis of the permethylated compound (without the cleavage of the methyl groups) can be done using for example H_2SO_4 or CF_3COOH. However, the solubility of permethylated polysaccharides in water may be very poor, and other solvolysis can be used in these cases such as formolysis. The OH groups of the partially methylated monosaccharide liberated from hydrolysis can be further derivatized, for example by silylation or acylation. Reduction with $NaBH_4$ and the derivatization of the alditols is also practiced [9]. The use of $NaBD_4$ may provide additional information regarding the position of the bond. The sequence of reactions using $NaBD_4$ as a reducing agent can be written as follows:

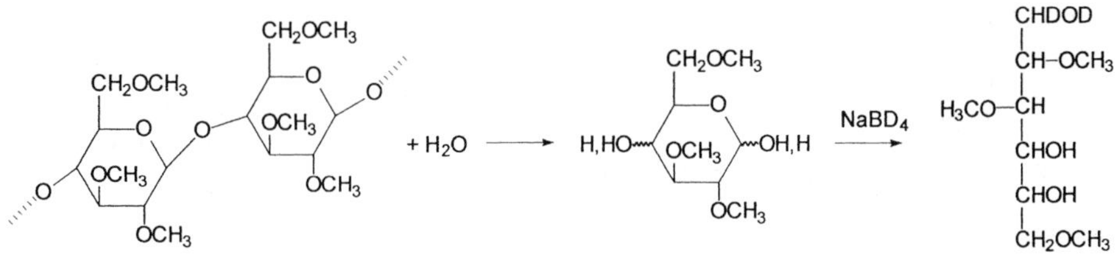

The resulting compounds may form rather complex mixtures because the anomeric structure is usually lost in these reactions. The analysis of the acetylated or silylated compounds resulting after further derivatization is typically done using chromatographic techniques [17].

- *Reactions for the determination of the sequence of monosaccharide residues*

The determination of the sequence of monosaccharide residues is not a simple task. Selective hydrolysis, mainly using enzymes, is of significant help. In addition, the use of oxidation procedures is a classic technique in structure determination of carbohydrates. The oxidation is commonly done using a periodate such as $NaIO_4$. The conditions of the oxidation may vary depending on the polymeric carbohydrate and the purpose of analysis. Typical oxidations are done with 0.01 to 0.1 M solution of $NaIO_4$, at temperatures chosen between 20 and 30° C and reaction times up to 70 hours [4]. The periodate reacts with vicinal hydroxyl groups and cleaves the linkage between them with the consumption of one mole of periodate per diol. Vicinal triol groups generate formic acid, and a primary OH group adjacent to a secondary OH group generates formaldehyde. Information on the carbohydrate structure can be obtained from the measurement of the amount of reagent consumed and the amount of formic acid and formaldehyde. The result of the oxidation on different types of monosaccharide units is shown schematically below [5]. Structure (A) contains a nonreducing terminal unit, structures (B)–(E) contain reducing terminal units with different points of substitution, and structures (F)–(I) contain nonterminal units.

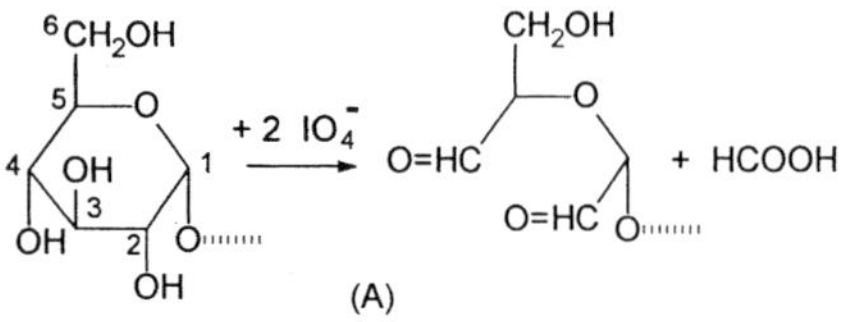

(A)

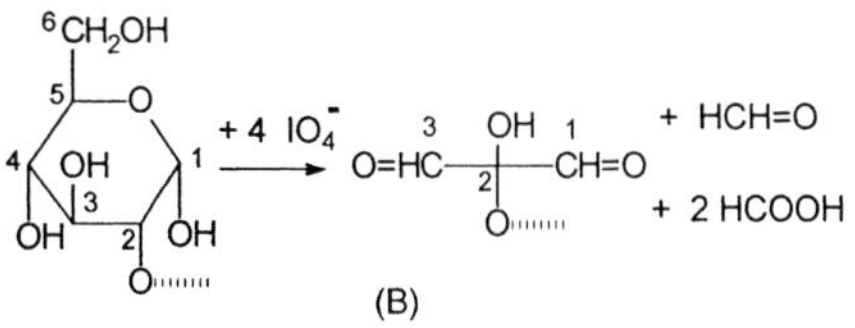

(B)

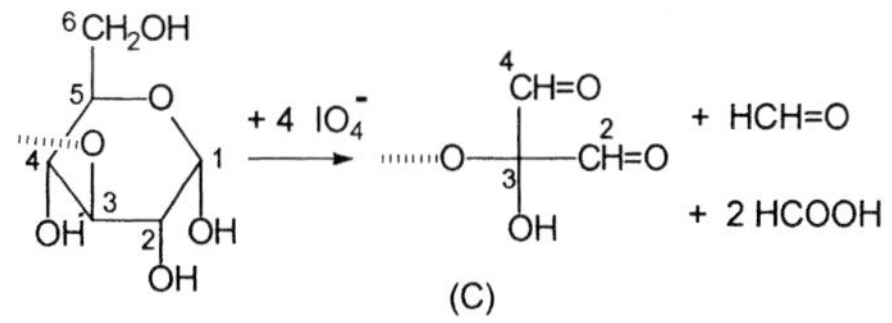

(C)

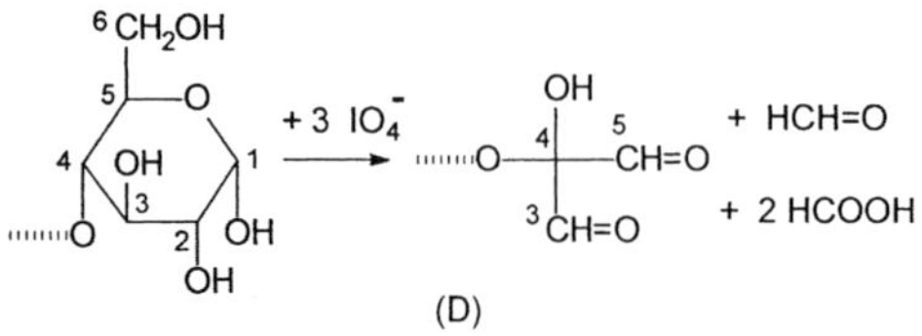

(D)

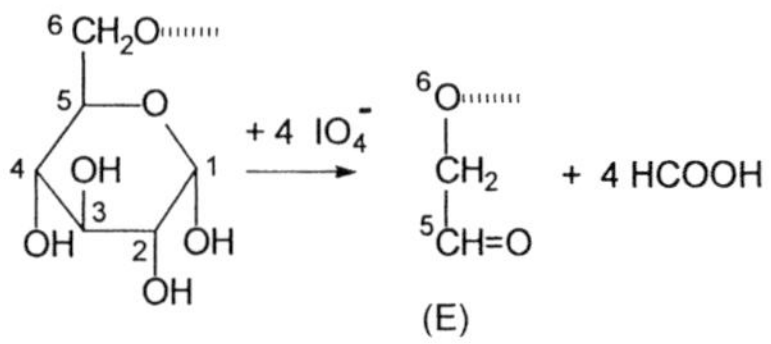

(E)

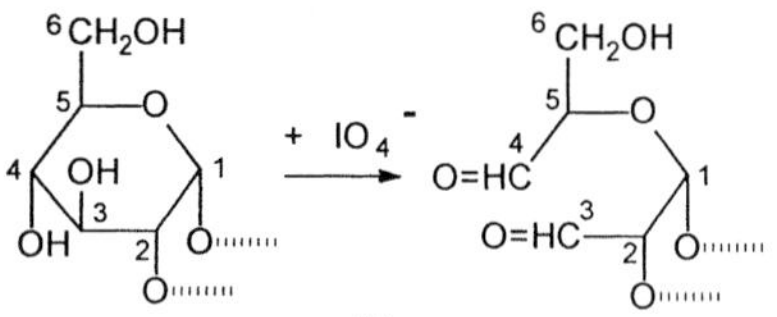

(F)

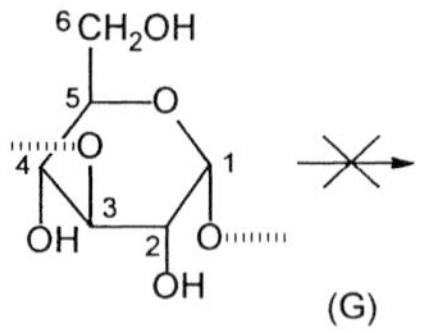

(G)

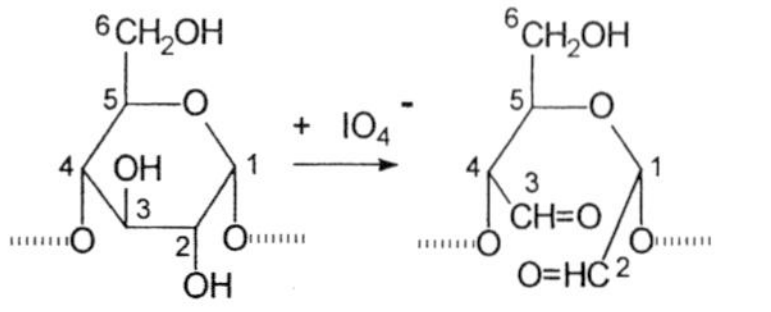

(H)

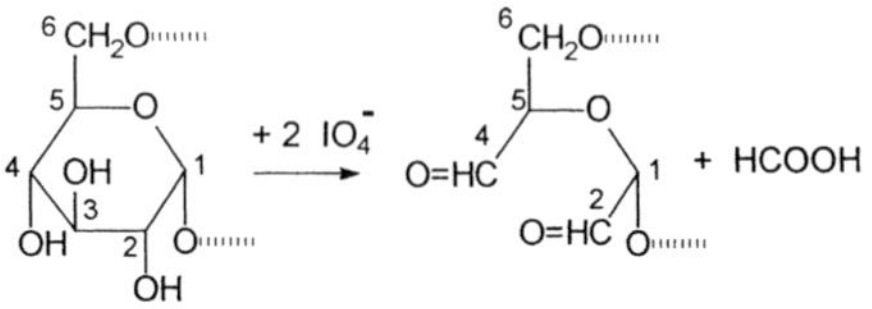

(I)

More structural information is obtained using the hydrolysis of the oxidized compounds. In order to protect the dialdehydes formed in the reaction, a reductive hydrolysis is applied, either in the presence of NaBH$_4$ or using ethanthiol. Mild conditions of reductive hydrolysis can be achieved for example with TFA and ethanthiol [29]. Also, the oxidized polysaccharide can react with phenylhydrazine in diluted acetic acid to form phenylhydrazones. The hydrolysis products of the modified polysaccharide can be further derivatized using for example silylation and then analyzed by GC-MS. Some information such as anomeric configuration is lost in these experiments, but very good data can be obtained regarding the primary structure of polysaccharides.

Anomeric information can be obtained by oxidation with CrO$_3$ in glacial acetic acid, followed by methylation and hydrolysis. A polysaccharide with α-glycosidic linkages is stable to CrO_3 oxidation and after methylation and hydrolysis generates the same products as the polysaccharide untreated with CrO_3. Polysaccharides containing β-glycosidic linkages are more easily oxidized and modified with CrO_3.

The reduction can be applied for the terminal reducing ends of the polysaccharide. The reaction can be done using, for example, a 0.1% aqueous solution of $NaBH_4$ or $NaBD_4$. The reduction can be associated with further hydrolysis, followed by other derivatizations and chromatographic analysis.

20.2 CHEMICAL DEGRADATION OF PROTEINS FOR ANALYTICAL PURPOSES

Proteins are basically formed from one or more chains of amino acids that are linked through an amide bond (or peptide bond). The particular order of amino acids in a chain (forming a polypeptide) is known as the primary structure of the protein (this also includes the location of disulfide bridges). The molecular weights of proteins start around 5 kDa (kilodalton) and go higher than 10^3 kDa. The chains of amino acids in proteins, being very long, can coil and fold. This spatial arrangement of amino acids is described by the secondary and tertiary structures of proteins. Some proteins have a periodic structure such as helical structure (α-helix) where the amino acid chain forms a three-dimensional coil, and the amino acids that are four units apart can have hydrogen bonds between their N-H and C=O groups. The secondary structure describes this steric relation of amino acid residues that are close to one another in the linear sequence and characterize structures such as α-helix or β-plated sheet. Tertiary structure refers to the steric relation of amino acid residues that are far apart to one another in the linear sequence (such as folding of parts of the protein). The protein structure can be quite complex. As an example, acetylcholinesterase (AChE) is a protein that exists in the synaptic cleft associated with collagen and proteoglycans and plays an important role in the hydrolysis of acetylcholine. Its molecular weight is estimated around 260 kDa, and it has an $\alpha_2\beta_2$ structure. The amino acid sequence (see Table 19.10.1 for amino acids single-letter abbreviation), the disulfide bridges between the cysteine residues, and secondary structure of the monomer of AChE are shown below:

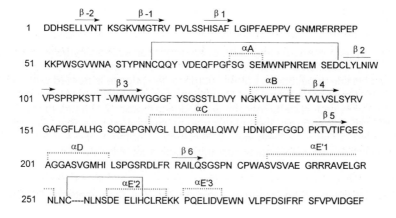

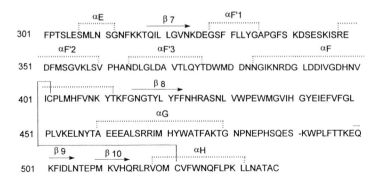

```
                αE                β 7              αF'1
        301  FPTSLESMLN SGNFKKTQIL LGVNKDEGSF FLLYGAPGFS KDSESKISRE
              αF'2              αF'3                            αF
        351  DFMSGVKLSV PHANDLGLDA VTLQYTDWMD DNNGIKNRDG LDDIVGDHNV
                                  β 8
        401  ICPLMHFVNK YTKFGNGTYL YFFNHRASNL VWPEWMGVIH GYEIEFVFGL
                                  αG
        451  PLVKELNYTA EEEALSRRIM HYWATFAKTG NPNEPHSQES -KWPLFTTKEQ
              β 9        β 10              αH
        501  KFIDLNTEPM KVHQRLRVOM CVFWNQFLPK LLNATAC
```

The molecular structure of AChE is known from amino acid sequencing and x-ray study of the enzyme obtained from *Torpedo californica* electric organ [30]. The enzyme has been obtained crystallized as a homodimer anchored to the plasma membrane through covalently attached phosphatidylinositol [30]. The monomer contains 537 amino acids and consists of a 12-stranded central mixed β sheet surrounded by 14 α helices.

Protein structure determination starts with the determination of the primary structure, which is typically determined using chemical or enzymatic procedures. Secondary, tertiary, and quaternary structure are usually determined from physico-chemical determinations such as x-ray analysis.

The primary protein structure determination alone covers a large body of scientific literature (see e.g. [31], [32], [33], [34], [35], [36], [37], [38], etc.), information in data banks (e.g. Brookhaven Protein Data Bank [39], SWISS-PROT, Protein Identification Resource [40]), and computer programs developed to assist in protein analysis (SEQUEST-SNP [41], etc.). The subject of protein structure determination is beyond the purpose of this book, and only a brief discussion is given here.

A classical procedure for the determination of a protein structure starts with the amino acid analysis of a protein hydrolysate. Hydrolysis can be done with 6M HCl using heating at 110° C for 24 hours [36]. The procedure has a number of alternatives and problems related to deamination of glutamine and asparagine or to oxidation of the cysteine. Amino acid analysis can be done further using procedures as described in Sections 19.10 and 19.4.

Specific hydrolysis procedures also have been developed such as hydrolysis associated with preliminary reduction of disulfide bonds. This can be done for the analysis of cysteine and cystine in peptides. For this purpose, the peptide is treated with 2–3 times molar excess of mercaptoethanol in 8 M solution of urea (under N_2). After the reduction, the SH groups are protected by reaction with iodoacetic acid (2–3 times molar excess in 1 N NaOH) [42]. After 15 min. the reaction can be considered complete, and the peptide can be purified on a Sephadex G-75 column. For the analysis of the cysteine residue, about 0.5–1 μmol peptide is hydrolyzed with 1 mL 8 M HCl in propanol. The mixture is heated at 110° C for 20 min. and then evaporated to dryness under N_2. The residue is treated with 50 mg anhydrous Na_2CO_3 and 1 mL acetone/triethylamine/acetic anhydride (5:2:1, v/v/v). The mixture is sonicated for 15 s, then heated at 60° C for 3 min., and evaporated to dryness under N_2 at $60°$ C, and the residue is redissolved in 0.5 mL ethyl acetate, which is used for GC or GC-MS analysis [43].

Peptide mapping is another common technique geared toward protein analysis [44], [45]. In peptide mapping a purified protein is cleaved selectively by enzymes. Among the enzymes used for this purpose, trypsin is common (tryptic digest) [46]. It cleaves the amide bond at the carboxyl side of arginine, lysine, and aminoethyl cysteine as shown below:

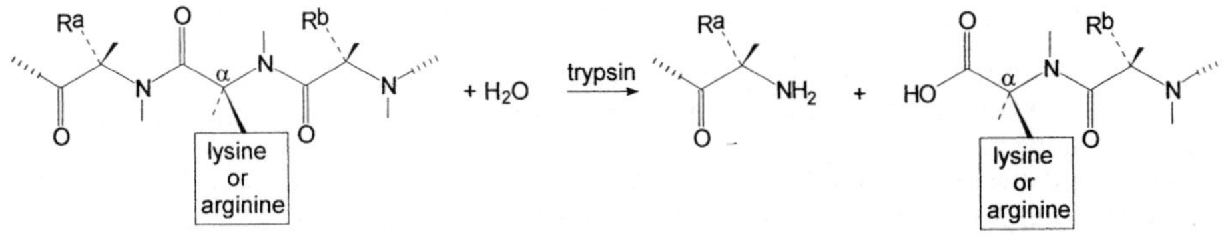

Due to the relatively high frequency of arginine and lysine in proteins, it is common that a tryptic digest contains mainly peptides with 7–8 amino acids. Other enzymes used for protein cleavage are chimotrypsin, endoproteinase Lys-C, V8 proteinase from *Staphylococcus aureus*, pepsin, immobilized pepsin [47], subtilsin, clostripain, etc. Also, cyanogen bromide can be used for the cleavage of proteins at the carboxyl side of a methionine residue as follows:

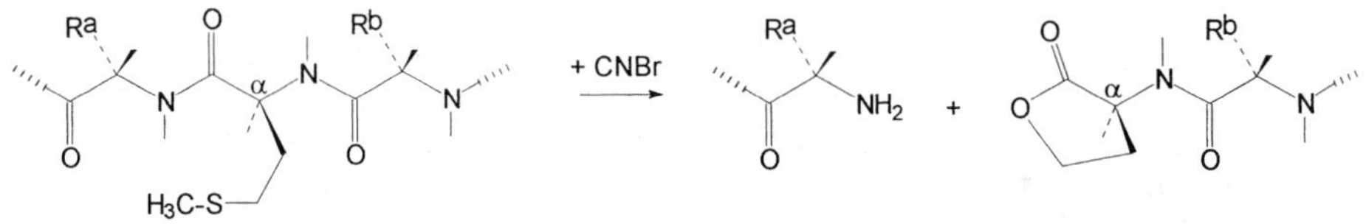

A number of other cleavage possibilities are available [48].

Many large proteins contain internal disulfide linkages that tend to restrict the access of the enzyme to parts of the molecule. This problem is eliminated by reducing the disulfides with dithiothreitol and then attaching a carboxymethyl to the SH group using iodoacetic acid (see Section 19.3) for protection. However, incomplete digestion, nonspecific cleavage, hydrolysis of glutamine and asparagine residues, etc. may affect the protein cleavage.

Peptides generated by this procedure are further separated by HPLC or CE [37] and identified using for example MS/MS techniques [40], [38] or other MS type techniques [49]. Dedicated computer programs are available for peptide MS/MS spectra analysis such as SEQUEST program [50]. A large number of proteins have been analyzed by this procedure.

The earlier procedures for peptide analysis were successful only in the determination of the terminal amino acid, using a derivatization for example with fluorodinitrobenzene or 4-dimethylaminoazobenzene-4'-sulfonyl chloride (dabsyl chloride) and hydrolysis. The scheme of this procedure can be written as follows:

R—X + H$_2$N—(R^a)—CONH—(R^b)—CONH—(R^c)⋯⋯ ⟶ R—HN—(R^a)—CONH—(R^b)—CONH—(R^c)⋯⋯

R—HN—(R^a)—CONH—(R^b)—CONH—(R^c)⋯⋯ $\xrightarrow{\text{hydrolysis}}$ R—HN—(R^a)—COOH + $\begin{array}{l} \text{H}_2\text{N—(R}^b\text{)—COOH} \\ \text{H}_2\text{N—(R}^c\text{)—COOH} \end{array}$

etc.

A classical procedure for protein analysis is protein sequencing using Edman degradation technique [51]. This procedure can be applied to a purified protein or to peptides obtained by preparative HPLC from tryptic digests. The initial Edman procedure uses phenyl isothiocyanate as a reagent [51] for the terminal amino group in the protein. After derivatization a hydrolysis can be conducted such that it takes place only for the derivatized amino acid residue. This is further separated from the rest of the protein and analyzed. Similarly can be used p-bromophenyl isothiocyanate. The procedure starts with 1.1 mmol reagent that is added to about 1 mmol peptide in 5 mL water:pyridine (1:1 v/v) at pH = 9 (adjusted with 1N NaOH) [52]. The reaction takes place with the formation of a thiocarbamoyl peptide derivative. The excess of reagent reacts with the NaOH added for the pH adjustment and forms N,N'-bis(bromophenyl)-thiourea, which is removed by extraction with benzene. The solution of derivatized peptide is evaporated and redissolved in water. The hydrolysis is done in 24 hours using 1–1.5 mL 1N HCl. The bromophenylthiohydantoin of the terminal amino acid precipitates and can be separated from the peptide and further analyzed by a chromatographic procedure such as HPLC, GC, or GC-MS for identification. The reactions can be written as follows:

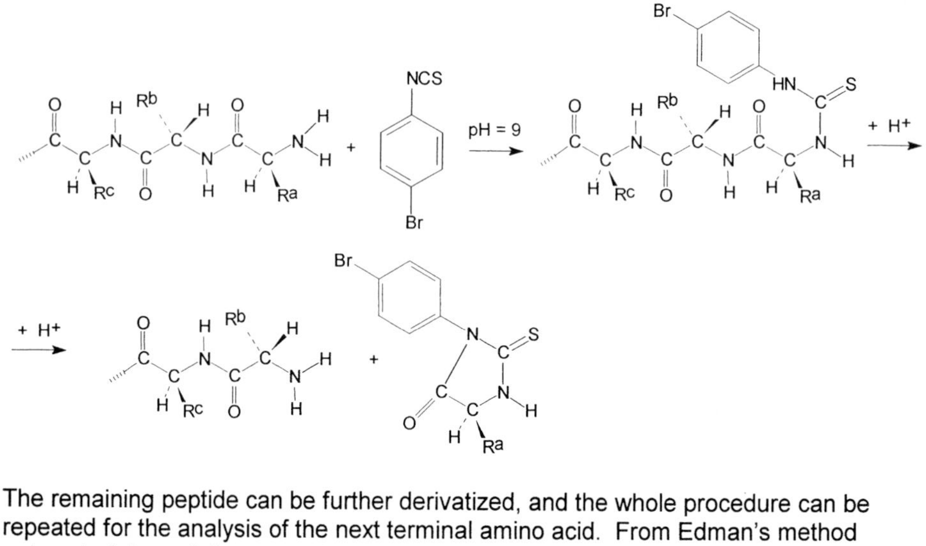

The remaining peptide can be further derivatized, and the whole procedure can be repeated for the analysis of the next terminal amino acid. From Edman's method automated degradation procedures have been developed, and dedicated instrumentation for the process is available [53].

Methyl isothiocyanate can be used similarly allowing a GC-MS detection of the hydrolyzed amino acid. The reaction of 0.5–1 µmol peptide dissolved in 250 µL 0.4 M dimethylallylamine (in 40% aqueous pyridine with the pH adjusted to 9.5 using trifluoroacetic acid) takes place by addition of 10 μL methylisothiocyanate (melted) followed by heating at 50° C for 30 min. The mixture is then extracted twice with 2 mL benzene and dried under N_2. To the solid material, 50–100 μL CF_3COOH is added, and the mixture is heated at 50° C for 10 min. The mixture is evaporated to dryness under N_2, and the residue is extracted with 2 mL CH_2Cl_2. The extract is dried and 200 μL 1 N HCl is added followed by heating at 80° C for 10 min to hydrolyze the thiocarbamoyl peptide derivative. The resulting material is extracted with 3 mL ethyl acetate, and the extract contains the thiohydantoin of the hydrolyzed amino acid residue. The extract is evaporated to dryness, redissolved in 1–5 μL ethyl acetate, and analyzed by GC-MS. The remaining material after the initial extraction with ethyl acetate is dried and redissolved in 0.4 M dimethylallylamine solution for further analysis [54].

Some other reagents leading to thiohydantoin derivatives were experimented for peptide analysis, such as pentafluorophenyl isothiocyanate, p-phenylazophenyl isothiocyanate [55], 4-N,N-dimethylaminoazobenzene-4'-isothyocyanate [56], and isocyanates that form fluorescent compounds, such as 4-(7-N,N-dimethylamino-sulfonyl-2,1,3-benzoxadiazolyl) isothiocyanate [57] [58], or 7-methylthyo-4-(2,1,3-benzoxadiazolyl) isothiocyanate [59].

The analysis of the N-terminal amino acid in peptides can be done using other reactions such as that with pivaloyl chloride (2,2-dimethylpropanoic acid chloride) or benzoyl chloride. The adjacent amido group to the pivaloyl or benzoyl terminal amido group is more prone to acid catalyzed cleavage than the rest of the peptide, probably due to the formation of an oxazolinone intermediate [55]. The hydrolysis can be done using HCl in methanol. The hydrolyzed terminal amino acid can be further methylated at the COOH group, for example with CH_2N_2, and analyzed by GC or GC-MS [60]. The scheme of this procedure can be written as follows:

R—X + H$_2$N—(R^a)—CONH—(R^b)—CONH—(R^c)'''''' ⟶ R—HN—(R^a)—CONH—(R^b)—CONH—(R^c)''''''

hydrolysis (1)

R—HN—(R^a)—CONH—(R^b)—CONH—(R^c)'''''' ⟶ R—HN—(R^a)—COOH + H$_2$N—(R^b)—CONH—(R^c)''''''

analysis of R—HN—(R^a)—COOH

R—X + H$_2$N—(R^b)—CONH—(R^c)'''''' ⟶ R—HN—(R^b)—CONH—(R^c)''''''

hydrolysis (2)

R—HN—(R^b)—CONH—(R^c)'''''' ⟶ R—HN—(R^b)—COOH + H$_2$N—(R^c)''''''

analysis of R—HN—(R^b)—COOH , etc.

A similar procedure can be applied sequencing the peptide from the carboxylic end group, but it is more complicated and less frequently used. The reaction sequence in this case is the following:

More details about amino acid sequencing using thiohydantoins can be found in dedicated reviews [61], [62], [63].

Labeling with colored, fluorogenic, or luminescence producing groups of the amino acid residues in proteins is a common practice in protein analysis, associated for example with enzymatic cleavage and chromatographic separations. A significant amount of information on this subject can be found in the dedicated literature (see e.g. [64], [64a]).

The study of tertiary and quaternary structure of proteins is usually done using physical methods. However, hydrogen/deuterium exchange at peptide amide linkage is also used for the same purpose. The dynamics of the hydrogen/deuterium exchange depends on the protein structure and can be measured using MS, typically performed after an enzymatic cleavage and HPLC separation of the fragments [65].

20.3 CHEMICAL DEGRADATION OF OTHER MACROMOLECULAR COMPOUNDS

Controlled chemical degradation, usually associated with further derivatization, is applied to many macromolecular compounds. These include molecules with a well-defined structure such as nucleic acids, but also less well-defined macromolecular compounds such as lignins, tannins [66], humin, humic acids, and fulvic acids. After a controlled degradation step of this type of compound, the resulting material is separated and analyzed. Chromatographic techniques are usually used for this purpose, and in this sense, specific polymer degradations can be considered chemical procedures for sample preparation. Only some selected examples of this type of procedure are given in this section.

One example is the use of controlled chemical degradation for structure determination of lignin. Lignin makes up 15% to 36% of the dry weight of wood. It appears in nature together with cellulose, from which it is difficult to separate due to the fact that besides physical mixing some covalent bonds exist between lignin and cellulose. Lignin is obtained from wood either using the hydrolysis or solubilization of cellulose with the lignin left as an amorphous material insoluble in water or by dissolving lignin with specific solvents or reagents followed by reprecipitation. Some of these procedures cause a significant chemical modification of lignin.

Several idealized structures for lignin were proposed, for example as a polymer of coniferyl alcohol (4-hydroxy-3-methoxycinnamyl alcohol) [67]. However, lignin is not a uniform polymer, and depending on the lignin source and the method of preparation, lignin has different structures. One typical procedure for lignin characterization is the oxidative hydrolysis of lignin with a solution of 2N NaOH in the presence of nitrobenzene. This procedure leads to a mixture of aromatic carbonyl compounds and carboxylic acids such as vanillin, vanillic acid, syringaldehyde, p-hydroxybenzaldehyde, and guaiacol. The process is exemplified for coniferyl alcohol:

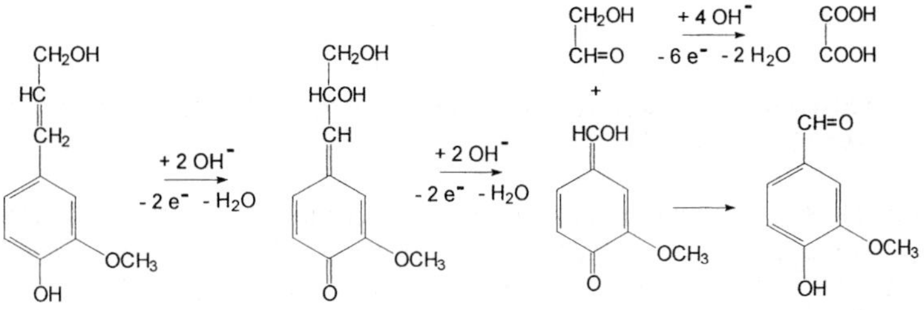

Three main phenolic compounds are present in lignin structure: phenol (H), guaiacol (G), and syringol (S). Based on the occurrence of these basic units, lignins can be classified as G lignins or guaiacyl lignins (also known as type N), G/S lignins or guaiacyl-syringyl lignins (also known as type L), and H/G/S lignins.

The analysis of lignin structure using acidolysis performed, for example, by refluxing the material for 4 hours with 0.2 M HCl in a mixture dioxane/water leads to a large number of compounds that allow the identification of the main repeating structures of lignin [68]. Typical hydrolysis products imply two of these structures, arylglycerol-β-aryl ether and phenylcoumaran, which are shown below:

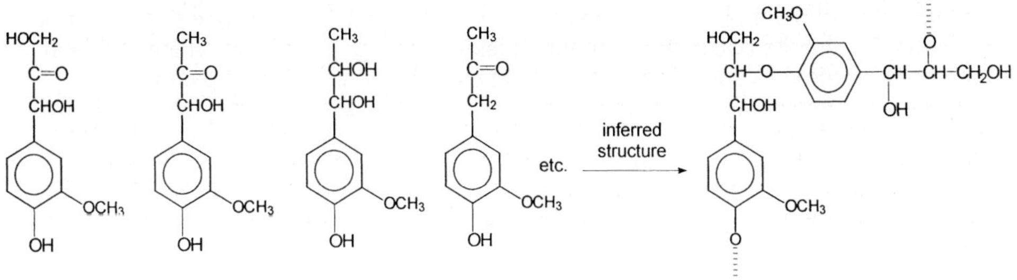

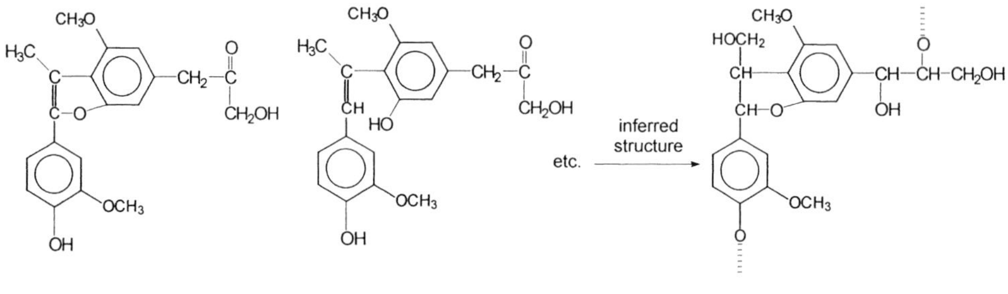

The structure of nucleic acids is another example of the use of chemical degradations and derivatization for structure determination. Nucleic acids (DNA and RNA) are biopolymers present in every living cell, with monomeric units consisting of a carbohydrate (2-deoxy-D-ribose in DNA and D-ribose in RNA) linked via a β-D-glycosidic bond to a heterocyclic base and interconnected by phosphodiester bonds at positions C-3' and C-5'. The monomeric units of nucleic acids can be considered the nucleotides (see also Section 19.11).

Three types of RNA are recognized, and they are messenger RNA (mRNA), transfer RNA (tRNA), and ribosomal RNA (rRNA), which is the most abundant in cells. Values between 10^6 and 10^9 Dalton have been reported for the molecular weight of DNA, and the molecular weight is about 10^6 for rRNA, 10^5 for mRNA, and 10^4 for tRNA. The simplified structures of DNA and RNA are the following:

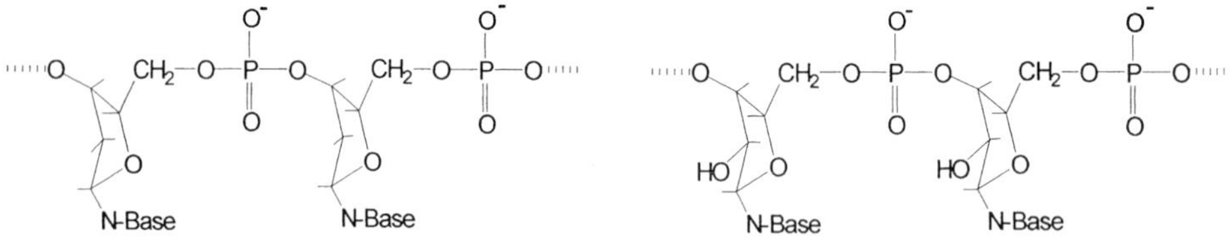

DNA chain   RNA chain

The bases always present in DNA are adenine (A), guanine (G), cytosine (C), and thymine (T), while in RNA the bases are adenine, guanine, cytosine, and uracil (U).

Similarly to proteins, both DNA and RNA have a secondary and a tertiary structure. The secondary structure of DNA shows two chains running in opposite directions, coiled in a left-handed (double) helix about the same axis. All the bases are inside the helix, and the sugar phosphate backbone is on the outside (see e.g. [69]). The chains are held together by hydrogen bonds between the bases with adenine always paired with thymine and guanine always paired with cytosine. The base pairing in DNA is shown below:

Pairing of adenine and thymine

Pairing of guanine and cytosine

A tertiary structure is also known for DNA, some DNA molecules assuming a circular or more complicated global shape (highly looped) in their intact state. RNA molecules are commonly single stranded, but by the formation of loops RNA can also contain portions with double helical structure. The bases are typically paired in RNA only in about 50% of the molecule.

The base sequence along the chain of DNA contains encoded genetic information, and an enormous research effort is being done regarding DNA sequencing (see e.g. [70], [71], [72], [73], etc.). The problem is beyond the subject of this book, and only a very simplistic approach is presented here.

In the first step the DNA molecule is cleaved into specific fragments using restriction enzymes [74]. A number of techniques are used for the sequencing of DNA fragments [75], [76], [77], [78]. A common procedure uses sequencing with specific chemical cleavage [79]. In this procedure, the first step is usually the labeling of the terminal 5'-OH group with a phosphate that contains radioactive ^{32}P. This step can be done using an enzymatic procedure with polynucleotide kinase. The labeled DNA is then cleaved preferentially at a specific base, each in separate experiments. The cleavage is conducted in such conditions that only one cleavage is made per chain, and fragments of different lengths are generated. For the cleavage at cytosine (C) and thymine (T), the DNA is treated with hydrazine, which reacts at the C=O group, followed by treatment with piperidine, which will cause cleavage at the 5'-side of C and T residues. If the experiment is performed in the presence of 2N NaCl, the reaction with thymine does not take place and only fragments connected to C can be obtained. Guanine (G) and adenine (A) are methylated with dimethylsulfate at positions N-7 and N-3 respectively. The glycosidic bond of the methylated purine is then cleaved by heating at neutral pH, leaving the sugar without the base. The backbone is further cleaved by heating with alkali, generating fragments connected to G or A. However, guanine is methylated much more rapidly than adenine, and from the amount of the material generated, the two bases can be differentiated. The purines also can be cleaved with formic acid or diethylpyrocarbonate.

The product of each experiment is then separated using gel electrophoresis, and the gel is placed in contact with a photographic plate, generating an autoradiogram. All fragments from each experiment are separated in descending order of their mass, but only labeled fragments are seen in the autoradiogram. Typical experiments are done to find segments connected with C, with C or T, with G, and with G or A. From these experiments the structure of the DNA fragment can be obtained. A simple example of the procedure is given for a DNA fragment with the structure GCAATCACGTC (with the order $G^{(1)}C^{(2)}A^{(3)}A^{(4)}T^{(5)}C^{(6)}A^{(7)}C^{(8)}G^{(9)}T^{(10)}C^{(11)}$). This fragment is labeled to ^{32}P-GCAATCACGTC. The first base (G) connected to ^{32}P phosphate is not determined by the procedure. The evaluation of the autoradiogram indicates the following fragments:

Fragments cleaved at A: ^{32}P-GCAATC, ^{32}P-GCA, ^{32}P-GC
Fragments cleaved at G: ^{32}P-GCAATCAC
Fragments cleaved at C: ^{32}P-GCAATCACGT, ^{32}P-GCAATCA, ^{32}P-GCAAT, ^{32}P-G
Fragments cleaved at T: ^{32}P-GCAATCACGT, ^{32}P-GCAA.

From these fragments and their distribution in the autoradiogram, an array indicating the position of each base can be generated, as shown in Table 20.3.1.

TABLE 20.3.1. *Base distribution in a DNA fragment ^{32}P-GCAATCACGTC.*

A	G	C	T	Position
		X		2
X				3
X				4
			X	5
		X		6
X				7
		X		8
	X			9
			X	10
		X		11

The procedure has been significantly extended and automated. Sequencing instrumentation including computer programs for data interpretation is available. Also, the labeling can be done with colored, fluorescent [80], [81], chemiluminescent dyes or biotin [82]. Other procedures are also applied in DNA sequencing [83], such as the use of specific enzymes for cleavage, controlled interruption of enzymatic replication, or use of PCR amplification with modified nucleotides as delimiters. The determination of the fragments obtained in DNA sequencing is done using various non-chromatographic and chromatographic techniques. A large amount of information on this subject exists in the dedicated literature (see e.g. [84], [85] [86]).

20.4 OVERVIEW OF ANALYTICAL PYROLYSIS

Pyrolysis is defined as the chemical transformation of a sample when heated at a temperature significantly higher than ambient. The heating is usually done between 500° C and 800° C [87]. Following pyrolysis, many molecules undergo decompositions and generate smaller, volatile molecules. These molecules are usually easier to analyze compared to the initial molecule. Therefore, pyrolysis can be seen as a sample preparation procedure.

Analytical pyrolysis has been developed as a collection of techniques involving pyrolysis performed with the purpose of obtaining analytical information on a given sample. The pyrolytic process, being a chemical reaction, does not provide analytical data unless it is associated with some kind of measurement. The measurement is commonly part of a typical analytical method, and chromatography is very frequently utilized for this purpose. The combination pyrolysis-GC-MS is also common, and analytical pyrolysis in itself has been developed as a versatile stand-alone analytical technique. Various books (e.g. [87], [88], [89]) and even a dedicated journal (J. Anal. Appl. Pyrol.) present the subject in detail.

There are many applications of analytical pyrolysis, most of them being geared toward polymer analysis or composite material analysis. Pyrolysis products of polymers, composite organic materials, etc. can easily be analyzed using GC or LC procedures. From the "fingerprint" of the pyrolysis products, valuable information can be obtained about the initial sample. Also, analytical pyrolysis can be used as a sample preparation technique that is able to simulate at micro scale the burning or industrial pyrolytic processes [90], [91].

Commonly, analytical pyrolysis is performed as *flash pyrolysis*. Heating of the sample cannot be done instantaneously, but in flash pyrolysis a fast rate of temperature increase is used, of the order of 10,000° K/s. After attaining a specific temperature, this is maintained essentially constant (*isothermal pyrolysis*) for a specified period of time. The isothermal condition targeted for performing flash pyrolysis is referred to as *equilibrium temperature* (T_{eq}). The T_{eq} is also named *final pyrolysis temperature*. The time period of temperature increase is called *temperature rise time* (TRT), and the total time for pyrolysis is called *total heating time* (THT). The choice of T_{eq}, TRT and THT depends on the material to be pyrolyzed and on the scope of the analysis.

Generally, as the pyrolysis temperature T_{eq} increases, smaller and less characteristic fragments begin to dominate the pyrogram. Dehydration processes, even for nonpolymeric substances, and formation of small molecules more stable at higher temperatures are common above 600° C. The temperature value at which the pyrolysis is performed is usually chosen to assure the desired decompositions but not so high as to form only very small molecules. Larger molecules usually retain more information about the initial compound and therefore may be desirable in the pyrolysis products.

Besides flash pyrolysis, several special types of analytical pyrolysis are sometimes used. One example is *stepwise pyrolysi,s* in which the sample temperature is raised stepwise and the pyrolysis products are analyzed between each step. *Temperature-*

programmed pyrolysis, in which the sample is heated at a controlled rate within a temperature range, is another special type.

A set of conditions such as good reproducibility, formation of stable reaction products, etc. is very important for pyrolysis to make it adequate for providing correct analytical information. The experimental conditions used for performing pyrolytic reactions play an important role for the end result of the process. For this reason, the pyrolytic process in analytical pyrolysis must be strictly controlled regarding the temperature, pyrolysis time, atmosphere, etc. Pyrolysis is commonly carried out in an inert atmosphere such as in a stream of helium. *Oxidative pyrolysis* (a pyrolysis that occurs in the presence of an oxidative atmosphere) and *reductive pyrolysis* (a pyrolysis that occurs in the presence of a reducing atmosphere) are sometimes utilized.

Several chromatographic techniques are used for the analysis of pyrolysates. One of the most common is gas chromatography (Py-GC). In this technique the volatile pyrolysates are conducted into a gas chromatograph for separation and detection. Less volatile or too polar compounds for a chosen chromatographic column are not seen with GC analysis. Also, compounds that are too volatile such as CH_4, HCHO, etc., may be a problem to detect if the chromatographic conditions are not appropriate. Pyrolysis-gas chromatography-mass spectrometry (Py-GC-MS) is also frequently utilized and has the advantage of compound identification in the pyrolysate. Pyrolysis followed by LC is less common, although it has been applied for specific analyses [92], [93]. Among the advantages of analytical pyrolysis (mainly as Py-GC-MS) are the capabilities to analyze very small amounts of sample (between 0.1 to 1 mg), to analyze compounds that are not volatile and not soluble, and to provide both qualitative and quantitative information about the sample.

- Thermodynamic and kinetic factors in pyrolysis

The pyrolytic reactions usually take place at temperatures higher than $250–300°$ C, commonly between $500°$ C and $800°$ C. Although taking place at higher temperatures, pyrolytic reactions are not different in principle from any other chemical reactions. For a nonisolated system in isothermal conditions, spontaneous processes take place with a negative variation of free enthalpy ΔG where

$$\Delta G = \Delta H - T \Delta S \qquad (20.4.1)$$

and ΔH is the variation in enthalpy of the system and T is the absolute temperature. For some compounds, the decomposition reactions at standard temperature of $25°$ C have negative values for the standard free enthalpy ΔG^0. These reactions should, therefore, occur spontaneously. However, their reaction rates are in most cases slow enough to assure the chemical stability of numerous organic compounds. This points out the importance of the kinetic factors over the thermodynamic ones in pyrolytic reactions. Each term in rel. (20.4.1) depends on temperature. At lower temperatures, enthalpy will play a more important role. At higher temperatures, the entropy term $T \Delta S$ will be more important. For most chemical reactions, equilibrium is attained at a certain temperature when $\Delta G = 0$, due to the opposite contribution of products and reactants to the total value of ΔG. From the expression for ΔG the following equation occurs at equilibrium:

$$\Delta H = T\, \Delta S \qquad\qquad (20.4.2)$$

or for the standard values at constant pressure of 1 atm.:

$$\Delta H^{0} = T\, \Delta S^{0} \qquad\qquad (20.4.3)$$

For a pyrolytic process, the temperature T satisfying relation (20.4.3) is defined as the *ceiling temperature* T_C:

$$T_C = \Delta H^{0} / \Delta S^{0} \qquad\qquad (20.4.4)$$

The ceiling temperature T_C can be considered the upper temperature at which a pyrolytic process will reach equilibrium. It may be seen, therefore, as a recommended temperature for pyrolysis. However, in practice, the application of rel. (20.4.4) is not straightforward. The theory was developed for ideal systems (in gas phase), and although in principle this theory should hold true for any system, its application to condensed phases or polymeric materials may be accompanied by effects difficult to account for (phase change, melting, cage effect [89], etc.). The reaction rate could be low at calculated T_C values. For this reason, temperatures 50° C or 100° C higher than T_C must be used as practical values of the temperature used in pyrolysis. Too high temperatures are avoided because of the formation of noncharacteristic small molecules.

The free enthalpy of a chemical process is related to the equilibrium constant K of that process. This constant allows (in principle) the calculation of the concentrations of the products given the concentration of the reactants. An equilibrium constant K is related to the variation of the standard free enthalpy ΔG^{0} of the system by the relation:

$$\Delta G^{0} = -R\,T \ln K \qquad\qquad (20.4.5)$$

Both ΔG^{0} and K are temperature dependent (R is the gas constant and in IS units $R = 8.31451$ J deg^{-1} mol^{-1} = 1.987 cal deg^{-1} mol^{-1}). Because the value for ΔG^{0} is not always known at pyrolysis temperature, the practical utility of this relation for pyrolytic processes is limited. However, the dependence of the equilibrium constant on temperature indicates that in analytical pyrolysis a good reproducibility can be obtained only if the temperature is accurately controlled.

The use of thermochemical data in pyrolysis is commonly limited to some predictions regarding the possible mechanism of a pyrolytic process. For example, the thermodynamic factors may allow prediction of the structure of the most likely radical to be formed in a pyrolytic process. This calculation can be based on the strength of a bond that breaks. The enthalpy (energy) for the dissociation of a bond A-B is not always available. However, for a homolytic dissociation with formation of free radicals, the enthalpy can be derived [94] from tabulated heats of formations ΔH_f^{0}, using the relation:

$$\Delta H_f^{0}\,(A\text{-}B) = \Delta H_f^{0}\,(A\cdot) + \Delta H_f^{0}\,(B\cdot) - \Delta H_f^{0}\,(AB) \qquad\qquad (20.4.6)$$

The ΔH_f^{0} (AB) values are easily available. Tables with heats of formation for radicals are less common but still available in literature (e.g. [95]), sometimes together with homolytic bond dissociations calculated based on rel. (20.4.6). Several average bond energies (evaluated at 25° C) for different bond types are given in Table 20.4.1, and some dissociation energies for specific R-R bonds (at 25° C) are given in Table 20.4.2.

TABLE 20.4.1. *Average bond energies at 25° C [96].*

Bond	E kcal/mol	E kJ/mol
O-H	110-111	460-464
C-H	96-99	400-415
N-H	93	390
S-H	82	340
C-O	85-91	355-380
C-C	83-85	345-355
C-Cl	79	330
C-N	69-75	290-315
C-Br	66	275
C-S	61	255
C-I	52	220
$C \equiv C$	199-200	835
C=C	146-151	610-630
$C \equiv N$	204	854
C=O	173-181	724-757
C=N	143	598

TABLE 20.4.2. *Several dissociation energies for R-R bonds at 25° C in kcal/mol [96], [89].*

R-/-R	-H	-CH$_3$	-C$_2$H$_5$	-C$_3$H$_7$	(CH$_3$)$_2$ -CH-	(CH$_3$)$_3$ -C-	CH$_2$=CH -CH$_2$-	C$_6$H$_5$ -CH$_2$-	-CH$_2$ =CH$_2$	-C$_6$H$_5$
CH$_3$-	104.0	88.4	84.5	84.9	83.8	80.5	73.6	71.9	93.7	94.0
C$_2$H$_5$-	98.0	84.5	81.6	81.4	80.2	76.9	70.3	88.8	90.0	90.6
C$_3$H$_7$-	97.6	84.9	81.4	81.4	80.0	76.8	70.3	69.0	98.7	90.8
C$_4$H$_9$-	98.0	84.7	81.4	81.2						
(CH$_3$)$_2$CH-	94.5	83.8	80.2	80.0	77.7	73.4	67.7	67.8	88.5	87.7
(CH$_3$)$_3$C-	91.1	80.5	76.9	76.8	73.4	67.6			81.1	82.2
CH$_2$=CH-CH$_2$-	86.8	73.6	70.3	70.3	67.7		60.0	58.0	78.4	
C$_6$H$_5$CH$_2$-	85.1	71.9	88.8	69.0	67.8			58.0	56.0	
CH$_2$=CH$_2$-	104.0	93.7	90.0	98.7	88.5	81.1	78.4		101.7	100.8
C$_6$H$_5$-	104.0	94.0	90.6	90.8	87.7	82.2		72.7	100.8	100.5
CH$_2$=CH-CH(CH$_3$)-	85.7	72.3	67.8							
C$_6$H$_5$CH-(CH$_3$)-	81.6	68.7	66.5							
C$_6$H$_5$-C(CH$_3$)$_2$-	76.5	65.7								
(CH$_3$)C=CH$_2$-	100.0									
cyclohexyl-	95.5									
HCO-	87.0									
CF$_3$-		90.0								
CCl$_3$-	96.0									

An example of the use of thermodynamic data for predicting the reaction path is related to the thermal decomposition of aliphatic hydrocarbons. For the alkane pyrolysis, two main reaction paths are possible, one with dehydrogenation and the other with fragmentation. The free enthalpies for these reactions are as follows:

$$C_{m+n}H_{2(m+n)+2} \rightarrow C_{m+n}H_{2(m+n)} + H_2 \qquad\qquad \Delta G = 30.20 - 0.0338\ T\ \text{(kcal)}$$

$$C_{m+n}H_{2(m+n)+2} \rightarrow C_mH_{2m+2} + C_nH_{2n} \qquad\qquad \Delta G = 18.94 - 0.0338\ T\ \text{(kcal)}$$

The ΔG expressions for this case are practically independent of the number of carbons in the molecule. At low temperatures, both reactions have positive ΔG values, and therefore do not take place. However, thermodynamically the fragmentation is favored because for this reaction $\Delta G = 0$ at about $287°$ C, while for dehydrogenation $\Delta G = 0$ only at $620°$ C. Therefore, during pyrolysis at $450°$ C–$550°$ C, it is unlikely that hydrogen will be formed from hydrocarbons.

Kinetic factors play a more important role than thermodynamic ones in the pyrolytic process. Both theoretical and practical aspects require information on reaction kinetics. The implications of kinetic factors for analytical pyrolysis are mainly related to the proper choice of certain experimental conditions such as pyrolysis temperature or time. Also, because the rate of reactions during pyrolysis is significantly affected by temperature, it is much easier to control reproducibility in isothermal conditions than in a temperature gradient. This explains why the analytical pyrolysis is commonly conducted in isothermal conditions.

For pyrolytic reactions, the variation of the molar concentration of a substance during the pyrolysis is not always the most appropriate parameter to be monitored. A more convenient parameter for monitoring pyrolytic reactions is the sample weight. For a reaction of the first order, the reaction rate is described by the expression:

$$- \frac{dW}{dt} = kW$$

(20.4.7)

where W is the weight (mass) of the sample at any time during the reaction. This type of equation can approximate (with good results) the kinetics for many pyrolytic processes. Equation (20.4.7) can be easily integrated to give the following:

$$\ln W \Big|_{t=0}^{t} = -kt \quad \text{or} \quad \ln \frac{W}{W_o} = -kt \quad \text{or} \quad \frac{W}{W_o} = \exp(-kt)$$

(20.4.8)

where W_o is the initial sample weight (at $t = 0$).

In some cases, the pyrolysis leaves a residue of undecomposed sample of the weight W_f (final weight). In this case relation (20.4.8) will be

$$- \frac{dW}{dt} = k(W - W_f)$$

(20.4.9)

This type of relation can be applied to pyrolytic processes even if the reaction is not of the first order. In this case the reaction rate is described by the relation:

$$- \frac{dW}{dt} = k(W - W_f)^n$$

(20.4.10)

Other parameters can replace weight in rel. (20.4.7), (20.4.9), or (20.4.10), such as residual mass fraction (W/W_o) where W_o is the initial sample weight, the volatilized mass fraction ($1 - W/W_o$), or for uniform polymers the number of decomposed polymer molecules.

In principle, it is assumed that the chemical reactions may take place only when the molecules collide. Following this collision, an intermediate state called an activated complex is formed. The reaction rate will depend on the difference between the energy of the reactants and the energy of the activated complex. This energy $E^{\#}$ is called activation energy. The reaction rate will also depend on the frequency of collisions. Based on these assumptions it was shown (e.g. [97]) that k has the following expression (Arrhenius reaction rate equation):

$$k = A \exp\left(-\frac{E^{\#}}{RT}\right)$$

(20.4.11)

where A is a parameter related to the collision number and is called *frequency factor*, and R is the *gas constant*. Relation (20.4.11) indicates the explicit dependence of the rate constant k on temperature (expressed in $^\circ$K).

Because pyrolysis reactions occur only under the influence of heat, it can be expected that these are unimolecular reactions. However, chemical reactions are supposed to take place only when the molecules collide. It is necessary, therefore, to understand how the molecular collision concept is applicable to a unimolecular reaction when it takes place in gas phase and possibly at low pressure. For unimolecular reactions the reacting molecule still acquires energy by collision with another molecule. The energized molecule does not, however, react immediately, and it is still able to lose its energy by collision with another molecule. A redistribution of the energy takes place afterwards. In this process the acquired energy is passed to one or more bonds that are weak enough to break. The process can continue with the chemical decomposition. Based on these assumptions the kinetics of many unimolecular reactions can be described with relatively good approximation by expression of the type (20.4.7). The activation energy $E^{\#}$ in the Arrhenius equation will be, in this case, the difference between the energy of the molecule and the energy of the activated complex.

Arrhenius rel. (20.4.11) can be used to describe the kinetics of a single chemical process or an overall reaction kinetics. Measurements of the variation of W in time in isothermal conditions allows the calculation of the constant k and order n for which a best fit of the experimental data can be obtained. For different reaction orders, Arrhenius formula for k given by relation (20.4.11) is usually still applicable. Once the values for k and n are known for a given reaction, the reaction kinetics can be described in a wider range of conditions.

The need for a precise temperature control during pyrolysis is illustrated in the example shown in Figure 20.4.1. This figure gives the values for the residual mass fraction W/W_0 for two hypothetical pyrolytic processes that occur with a THT of 1.0 s at a fixed temperature having two different reaction kinetics, both of the first order. The first process has $E^{\#} = 100.7$ kJ/mol and $A = 9.6 \ 10^5$ sec^{-1}, and the second process has $E^{\#} = 65$ kJ/mol and $A = 5.5 \ 10^3$ sec^{-1} (the kinetics parameters were selected from data for cellulose pyrolysis).

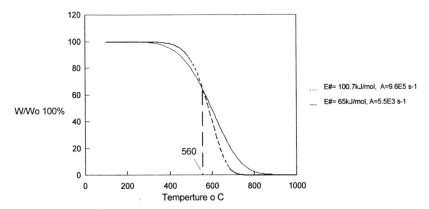

FIGURE 20.4.1. *The ratio W/W_O for two hypothetical pyrolytic processes of the same material controlled by different kinetics.*

At temperatures up to 560° C, the second process will dominate the pyrolysis, while at temperatures higher than 560° C, the pyrolysis will be dominated by the first process. If the pyrolysis products for the first process are different from those for the second process, it can be seen that a small variation in the temperature profile may significantly modify the analytical results, producing more of the products from one process or the other. Flash pyrolysis with very short TRT is used to avoid variations in the pyrolysis kinetics.

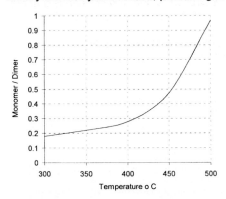

FIGURE 20.4.2. *Ratio of monomer/dimer isoprene in the pyrolysis of natural rubber.*

The dependence of the composition of the pyrolysis products on temperature can be exemplified by the study of flash pyrolysis of natural rubber at different temperatures [98]. Natural rubber forms as a main decomposition product isoprene and isoprene dimer. Figure 20.4.2 shows the plot of monomer/dimer ratio for flash pyrolysis of rubber at discrete temperatures between 300° C and 500° C. The figure shows the increase in monomer formation at higher temperatures.

Ideally, the decomposition of the sample should occur at a constant temperature. However, before attaining T_{eq}, the temperature increases. It can be assumed that this increase is linear with the rate q and starting at To. The temperature during TRT is therefore given at the moment t by the formula:

$$T = To + q\,t \qquad (20.4.12)$$

The variation of W/W_O for three compounds having the same frequency factor $A = 10^6$ sec^{-1} and different activation energies $E^{\#}$ (100 kJ/mol, 80 kJ/mol, and 60 kJ/mol) during pyrolysis is shown in Figure 20.4.3. In these examples, To = 200° C, q = 10° C/ms, TRT = 40 ms, and THT = 500 ms.

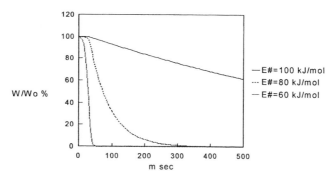

FIGURE 20.4.3. *Variation of W/W_O with time for several activation energies $E^{\#}$ in a hypothetical reaction.* Temperature ramp 10^0 C/msec, starting temperature 200^0 C, TRT 40 msec, THT 500 msec.

It can be seen in Figure 20.4.3 that for compounds with lower activation energies $E^{\#}$, most decomposition takes place during the TRT in nonisothermal conditions. The values for TRT and q and the reproducibility of the heating can become very important factors. The information about the activation energies is also important for the understanding of the kinetics of other reactions such as free radical propagation process (see Table 20.5.1).

When the pyrolysis does not occur in gas phase, different other problems appear. Although equations of the type (20.4.7) with k expressed by rel. (20.4.11) or (20.4.12) can be used in certain cases, these may lead to incorrect results in many cases. Various empirical models were developed for describing the reaction kinetics during the pyrolysis of solid samples. Most of these models attempt to establish equations that will globally describe the kinetics of the process and fit the pyrolysis data. Relation (20.4.11) can be used for the understanding of the common choices for the pyrolysis parameters.

The kinetics equation of the type (20.4.7) is commonly applied for describing the overall reaction kinetics during pyrolysis. However, this equation provides only an approximation when the process is not composed of a single reaction. Also, the theory described so far is commonly applied for small molecules and usually in gas phase. The pyrolysis of solid samples is in fact a complicated process, and the use of rel. (20.4.7) may lead to erroneous results. The simpler relations valid for the kinetics in homogeneous systems do not fit well the experimental data for solid samples. Factors related to heterogeneous reactions must be taken into account in this case. For the reactions in solid state, several empirical equations were proposed to better simulate the dependence on temperature of the reaction rate. Considering F the mass fraction of the unreacted substance at the time t, the empirical kinetics equation for heterogeneous systems can be expressed in the general form:

$$\frac{dF}{dt} = - k \ f(F)$$

(20.4.13)

where k is a constant and f(F) is a function that can be chosen of the form:

$$f(F) = \ F^{m} (1 - F)^{n} \ [- \ln (1 - F)]^{p}$$

(20.4.14)

The terms in equation (20.4.14) attempt to describe the reaction rate controlled by the movement of the phase boundaries, diffusion, nucleation in solid state, etc., and different values (including zero values) for m, n and p are proposed in literature (see e.g. [99]). A series of even more elaborate models have been developed for studying the kinetics of polymer pyrolysis [87], [89], [100], [101]. Extensively studied is the thermal decomposition of uniform repetitive polymers. This has been done mainly in relation to the thermal stability of synthetic polymers [102], [103].

20.5 PYROLYSIS AS A CHEMICAL MODIFICATION TECHNIQUE

As a chemical modification technique, pyrolysis is typically used for the formation of fragments from a larger molecule. The technique is more frequently used in polymer analysis, but pyrolysis of small molecules is also studied. The pyrolysis of a molecular species may consist of one or more reactions occurring simultaneously or sequentially at high temperature. The path of a pyrolytic process depends on the experimental conditions. After a first decomposition reaction step, it is common to have subsequent steps of pyrolysis. For example, side chain reactions from a polymer chain are followed by the polymer chain scission generating small molecules from the polymer. Therefore, pyrolysis of both small and large molecules occurs in the pyrolysis of a polymer. The result is a complex sequence of chemical reactions with a variety of compounds generated. When composite materials are pyrolyzed, more than one molecular species is subject to thermal degradation.

Various reaction types may occur during pyrolysis. The pyrolytic elimination is the reaction that probably dominates most pyrolytic processes. The reaction usually takes place as a β elimination with two groups lost from adjacent atoms. For the gas phase pyrolysis an E_i mechanism is common [96]. However, for polymers where the pyrolysis takes place in condensed phase, E_2 and E_1 mechanisms are not excluded. There are also several other mechanisms that have been found to operate in pyrolytic eliminations. A typical E_i elimination with a six-membered cyclic transition state takes place as follows:

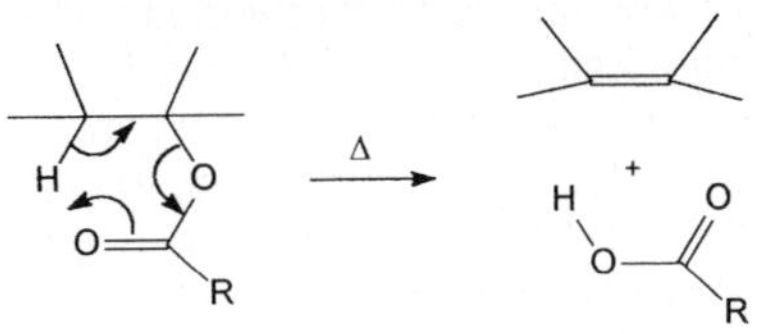

This type of reaction occurs, for example, during the pyrolysis of various esters such as polymeric vinyl esters or cellulose sulfate as shown below:

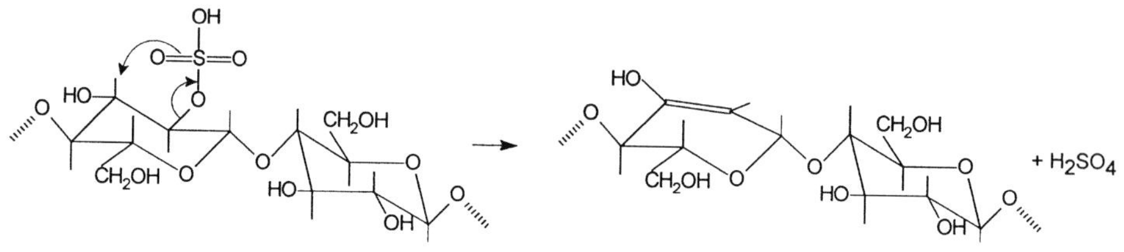

Elimination reactions are common for many other pyrolytic decompositions. For example, the formation of diketopiperazines during peptide and protein pyrolysis probably has the following mechanism:

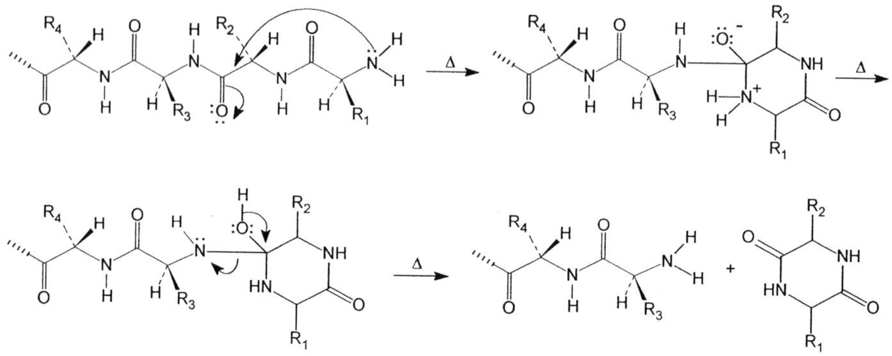

During pyrolytic reactions of E_i type, if a double bond is present, the formation of a conjugate system is preferred if sterically possible. Otherwise, the orientation in the pyrolytic elimination is statistical and is determined by the number of β hydrogens. The newly formed double bond goes mainly toward the least highly substituted carbon (Hofmann's rule). In the bridged systems, the double bond is formed away from the bridgehead. Also, for the E_i mechanism, a cis β hydrogen is required. Therefore, in cyclic systems, if there is a cis hydrogen on only one side, the double bond will go that way. However, when there is a six-membered transition state, this does not necessarily mean that the leaving groups must be cis to each other, since such transition states do not need to be completely coplanar. If the leaving group is axial, then the hydrogen must be equatorial and cis to the leaving group, since the transition state cannot be realized when the groups are both axial. But if the leaving group is equatorial, it can form a transition state with a β hydrogen that is either axial (cis) or equatorial (trans).

In some cases, an E_1 mechanism appears to be followed and the more stable olefin is formed. Instead of Hofmann's rule, Zaitsev's rule is followed (the double bond goes mainly toward the most highly substituted carbon). Also, in some reactions the direction of elimination is determined by the need to minimize steric interactions, sometimes even when the steric hindrance appears only during the transition state. Besides β eliminations, 1,3 or 1,n eliminations may also take place during pyrolysis with the formation of cycles. An example of this type of reaction occurs during the pyrolysis of certain peptides (and proteins). A glutamic acid unit, for example, can eliminate water by the following reaction:

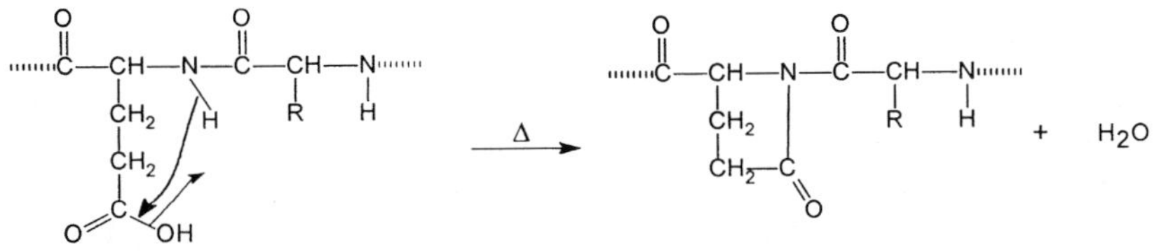

The same type of reaction takes place during the pyrolysis of nylon 6,6:

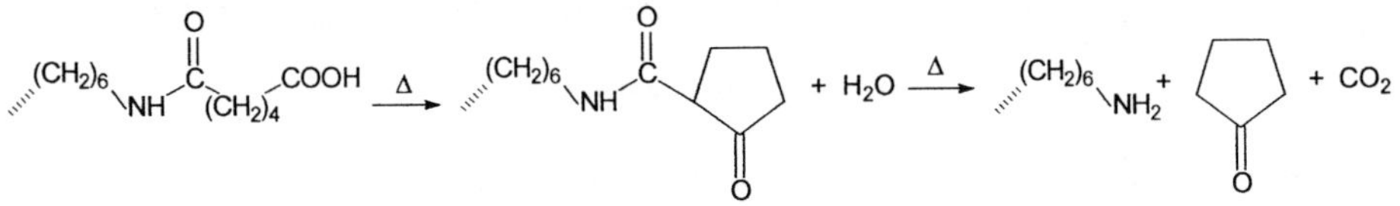

Not only small molecules are eliminated in this type of reaction. Polymer scission may occur with this mechanism, such as during cellulose pyrolysis in a transglycosidation reaction:

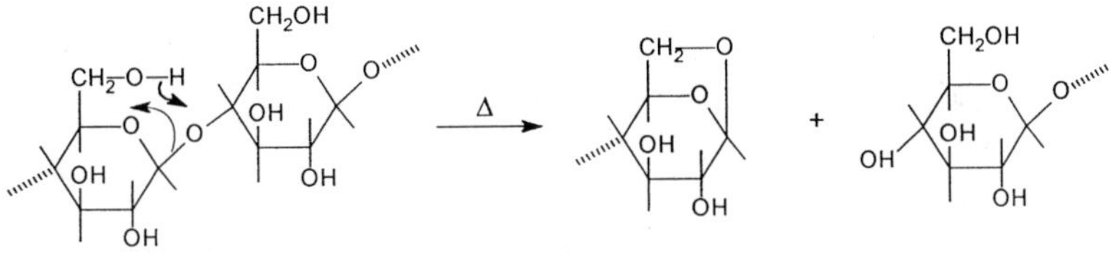

In certain eliminations, one carbocation can be a leaving group. In this situation, the reaction is called a fragmentation. The reaction commonly takes place in substances of the form Y-C-C-X, where X could be halogen, OH_2^+, OTs, NR_3^+, etc. (Ts is p-toluenesulfonate or tosylate.)

Another common type of mechanism found to operate in pyrolytic eliminations involves free radicals. Initiation occurs by pyrolytic cleavage. A schematic example of this type of reaction is shown below:

Initiation $R_2CH-CH_2X \xrightarrow{\Delta} R_2CH-CH_2^{\bullet} + X^{\bullet}$

Propagation

$R_2CH-CH_2X + X^{\bullet} \longrightarrow R_2C^{\bullet}-CH_2X + HX$

$R_2C^{\bullet}-CH_2X \longrightarrow R_2C=CH_2 + X^{\bullet}$

Termination $2\ R_2C^{\bullet}-CH_2X \longrightarrow R_2C=CH_2 + R_2CX-CH_2X$

Free radical eliminations are frequent during pyrolytic reactions, and they are common for linear chain polymers. At higher temperatures (600° C–900° C) this type of reaction is also common for small molecules and explains the formation of unsaturated or aromatic hydrocarbons from aliphatic ones.

The free radical formation as well as the propagation process with the formation of smaller molecules from the free radical chains usually takes place with the dissociation of the weaker bonds, which are expected to dissociate first. Besides the weaker bonds, other bonds can also be dissociated, most commonly when there are small differences between the bond dissociation energies. The strength of the bond being broken is commonly unknown, but it can be derived from tabulated heats of formation (see Tables 20.4.1 and 20.4.2). For the understanding of the propagation reaction, a more accurate picture of the process can be obtained based on the activation energies of the process:

$$R\text{-}H + X^{\bullet} \longrightarrow XH + R^{\bullet}$$

The higher an activation energy, the lower is the reaction rate, and therefore the reaction is less likely to occur. Several calculated activation energies for the reaction of hydrogen transfer with hydrocarbon radicals are given in Table 20.5.1:

TABLE 20.5.1. *Several activation energies for the reaction R-H + X* $\longrightarrow$ *XH + R* *in kcal/mol [89].*

$X^{\bullet}$ vs. R-H	C_2H_5-H	$isoC_3H_7$-H	$(CH_3)_3C$-H	$C_6H_5CH_2$-H	C_6H_5C-$(CH_3)_2$-H	CH_2=CH-CH_2-H
$C_2H_5^{\bullet}$	11.5	10.6	9.8	8.9	6.1	8.7
$isoC_3H_7^{\bullet}$	14.1	11.5	10.6	9.2	7.0	
$(CH_3)_3C^{\bullet}$	16.0	14.2	11.5	10.0	7.9	
$C_6H_5CH_2^{\bullet}$	19.3	18.6	16.0	11.5	9.4	
$C_6H_5C\text{-}(CH_3)_2^{\bullet}$	27.6	25.0	22.4	17.9	11.5	

Another type of reaction in pyrolysis is the rearrangement. A rearrangement is a reaction in which a group moves (migrates) from one atom to another in the same molecule. A variety of rearrangements can take place during pyrolysis such as migration of a group, electrocyclic rearrangements, and sigmatropic rearrangements.

The hydrogen elimination is a typical oxidation type reaction that is not uncommon in pyrolysis. Some other oxidations or reductions may take place during pyrolysis as a subsequent reaction to the initial process. Certain free-radical substitutions that involve the transfer of a hydrogen atom can also be considered oxidation/reduction reactions. It should be noted that oxidation due to the presence of oxygen (intended or accidental) may also take place during pyrolysis (below ignition temperature). As an example, substituted ethyl celluloses degrade oxidatively. The reaction probably starts with the initiation step at free aldehyde groups and has a free radical mechanism. This explains the formation of formic acid, acetaldehyde, ethanol, ethyl formate, ethane, CO_2, CO, etc. from this material.

Either as a first step of pyrolysis or as a result of the interaction of molecules from previous pyrolysis steps, substitutions and additions are also common reactions during the pyrolytic process. The nucleophilic substitution takes place with the attack of a reagent that brings an electron pair to the substrate. This pair is used to form a new bond. The leaving group retains its electron pair. Decarboxylation mechanism of aromatic acids is probably an electrophilic substitution.

Free radical substitutions are very common in pyrolytic reactions. An example of this type is the formation of biphenyl from benzene at 700° C. This reaction can be viewed as an oxidation because of the hydrogen elimination.

Some pyrolytic reactions can be seen as a reverse (retrograde) addition. Diels-Alder reaction for example is known to be reversible, and retro Diels-Alder reactions are rather common. The retro-ene reaction (retro hydro-allyl addition), for example, takes place by the following mechanism:

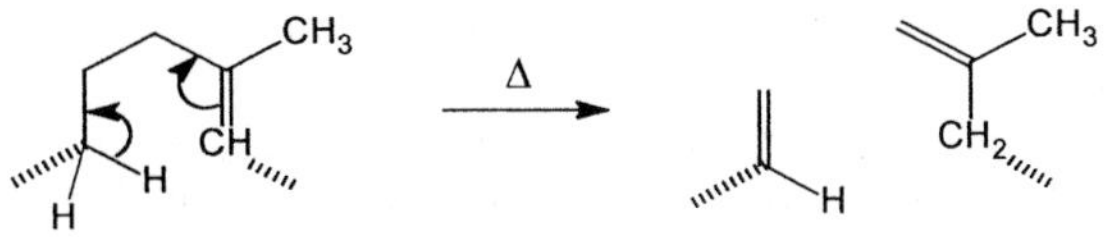

An example of such reaction probably occurs during lignin degradation as shown below for a model of lignin [104]:

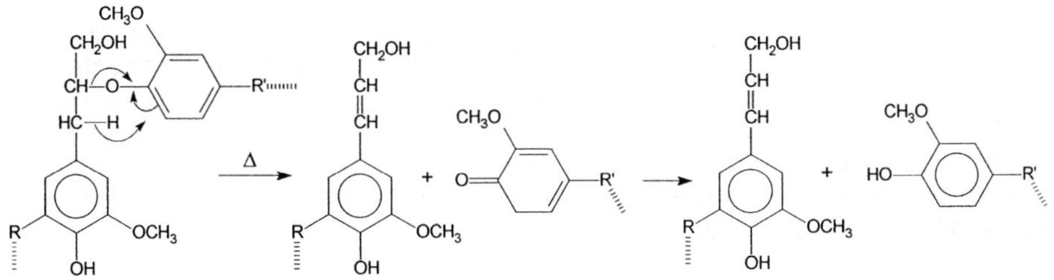

Retro-aldol condensations are also known to take place during pyrolysis. The mechanism of these reactions can be written as follows:

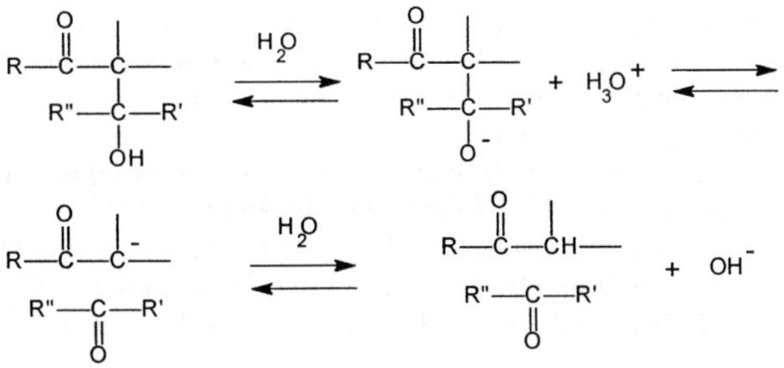

An example of a retrograde aldol reaction (retroaldolization) is the pyrolytic decomposition of cellulose with formation of hydroxyacetaldehyde:

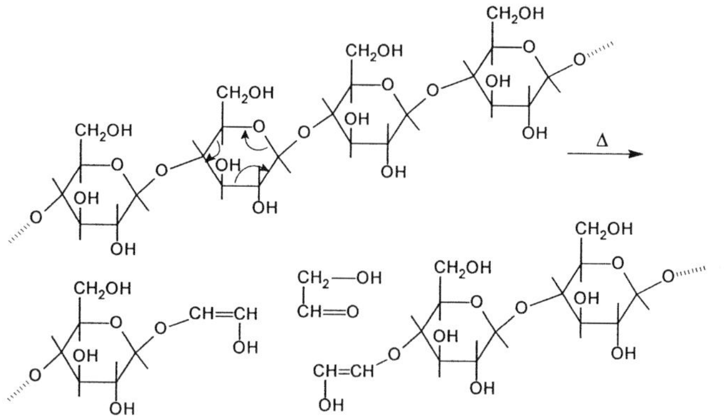

Other mechanisms for pyrolysis of cellulose are also possible. More paths for the same process is a common occurrence in pyrolysis, and more than one mechanism is frequently needed to explain the variety of reaction products.

- *Polymeric chain scission during pyrolysis*

Any polymer degradation during pyrolysis consists of chemical reactions of the types described previously. However, for a better understanding of the expected pyrolysis products of a polymer, a specific classification can be made allowing the correlation of the nature of the reaction products with the structure of the polymer. It is possible to categorize polymer degradation reactions as follows: a) polymeric chain scission, b) side group reactions, c) combined reactions.

The polymeric chain scission is an elimination reaction that takes place by breaking the bonds that form the polymeric chain. When the reaction takes place as a successive removal of the monomer units from the polymeric chain, it is called a *depolymerization*. It also may occur as a *random cleavage* of the polymer chain, and this happens mainly when the bonding energies are similar along the chain. If no intramolecular rearrangement takes place, the result of random cleavage is the formation of oligomers. If the chain scission is followed by secondary reactions, this leads to a variety of compounds such as cyclic oligomers.

Many chain scissions reactions have a free radical mechanism [103], [105]. As an example, the formation of isoprene from natural rubber falls in this class:

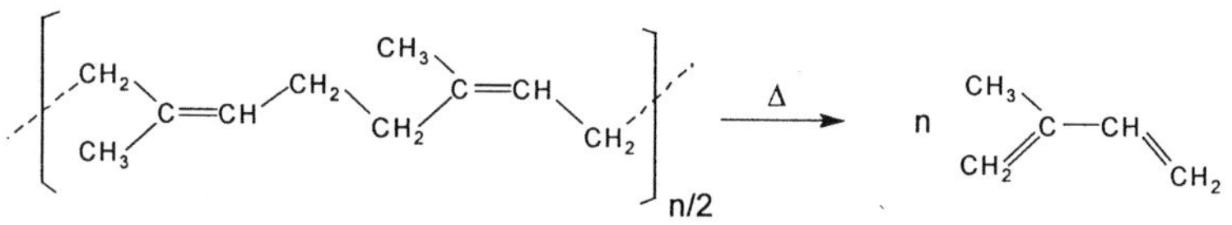

Only up to 58% of natural rubber can be practically depolymerized to isoprene during pyrolysis. This is also the case for many other synthetic and natural polymers. The yield of monomer for pyrolysis (different temperatures) for various polymers is shown in Table 20.5.2.

TABLE 20.5.2. *Yield of monomer for pyrolysis (different temperatures) of various polymers.*

Polymer	Monomer yield weight %	Polymer	Monomer yield weight %
methyl methacrylate	95	vinylcyclohexane	0.1
methyl acrylate	2	tetrafluoroethylene	95
α-methylstyrene	95	trifluorostyrene	75
m-methylstyrene	45	trifluorochloroethylene	28
styrene	42	trifluoroethylene	1
isoprene (rubber)	58	vinylidene fluoride	1
isobutylene	20	p-xylylene	0
propylene	2	benzyl	0
ethylene (linear)	0.1	vinyl chloride	1
ethylene (branched)	0.025		

A random chain scission may also take place along the polymeric chain. The result is the formation of molecules of lower molecular weight. However, in order to be volatile enough to be analyzed by typical analytical techniques associated with analytical pyrolysis, these fragments have to be relatively small. The formation of monomers as a final step in the random chain scission is not uncommon, and sometimes it is difficult to decide if a depolymerization or a random chain scission was the first step in pyrolysis.

For the initiation step, the free radicals formed may consist of one free radical chain plus one monomeric free radical, one free radical chain plus one low molecular weight free radical different from the monomer, or may consist of two free radical chains. The random chain scission could take place truly randomly or at the weaker link (see Table 20.4.2). It has been noticed that the sp^3 bonds to a carbon next to a double bond or an aromatic ring (not the bond to the sp^2 carbon) are in general weaker than other C-C or C-H bonds. However, other reactions are not excluded in the free radical formation. Some possibilities are α-chain scission, β-chain scission, and methyl scission.

The α-chain scission refers to the breaking of the bond to an sp^2 carbon and takes place as follows:

For polyisoprene this bond is estimated to have a dissociation energy of about 83–94 kcal mol^{-1}.

The β-chain scission refers to the breaking of the bond to an sp^3 carbon that is connected to the allyl carbon (not the sp^2 bond). For polyisoprene this sp^3 bond is estimated to have a dissociation energy of about 61.5–63 kcal mol^{-1}. Having a lower energy than α-scission, this is the most common chain scission for polyisoprene. The reaction takes place as follows:

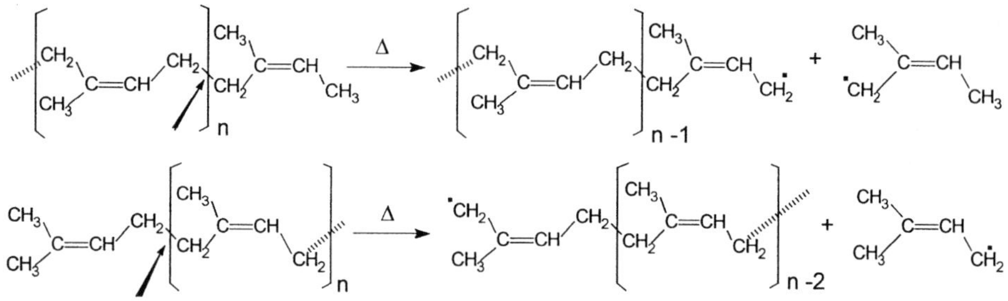

The methyl scission takes place as follows:

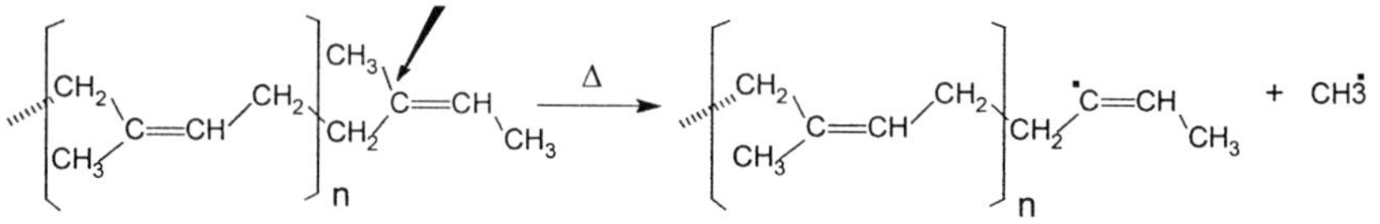

The scission may also take place in the middle of the chain. However, in some polymers scission at chain ends is preferred.

Propagation is the second step in the free radical chain reaction. The free radicals generated by β-chain scission can eliminate the monomer by scission of another β-link and shorten the macromolecular radical chain by the reaction:

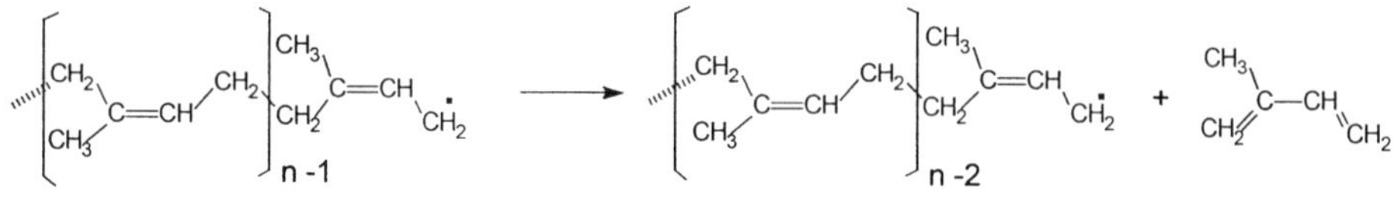

The free radicals formed by α-scission can also eliminate the monomer by the reaction:

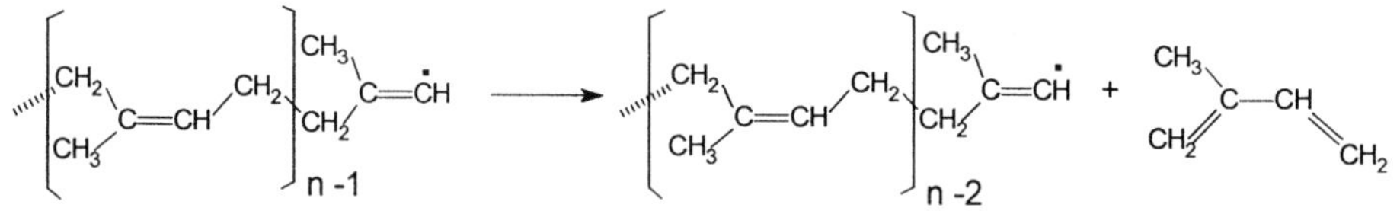

As a rule, the stability of the free radical chains is higher than that of a small free radical. The result is a simple propagation reaction with the formation of monomers by the following scheme:

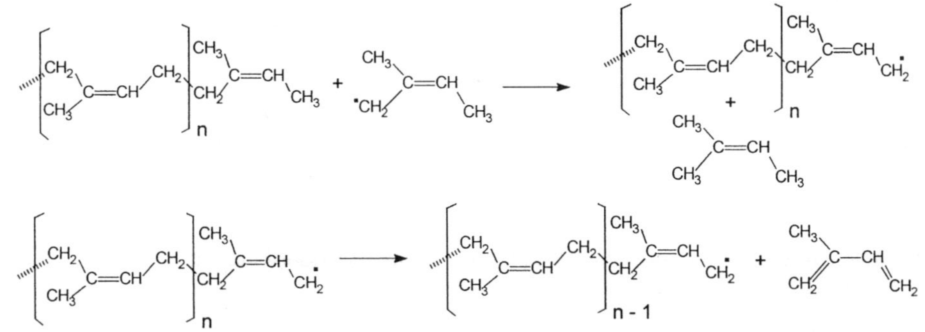

The disproportionation reaction of the free radical chain can generate the monomer as a successive process. However, in addition to the "regular" propagation step, different reactions may occur in a so-called transfer step. In this step, the free radical chain reacts with another molecule and generates a different radical chain and a new polymeric molecule. There are two possible types of transfer reactions. The transfer step can be an intermolecular chain transfer or an intramolecular chain transfer. In an intermolecular chain transfer the unpaired electron moves from one molecule to another. The intramolecular free radical chain transfer takes place as an intramolecular rearrangement, and an example of this kind of transfer is shown below:

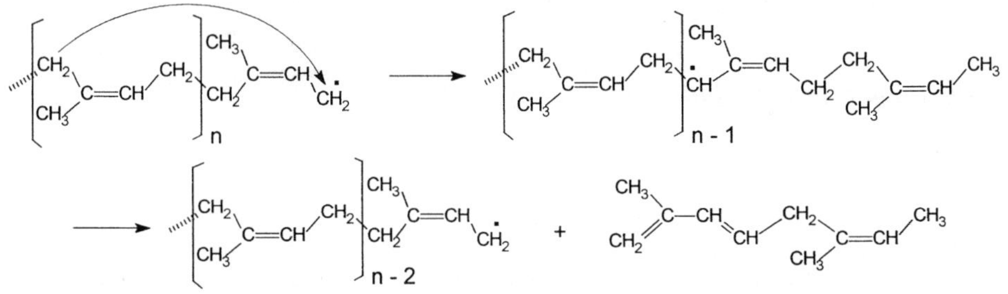

This type of reaction explains (in part) the formation of the isoprene dimer during rubber pyrolysis. The same compound may also be formed from other reactions such as isoprene dimerization.

The compounds with the formula C_5H_7-$(C_5H_8)_k$-C_5H_9 will have the molecular weight $(68)_k$ where $k = 0,1,2 \ldots$. The formation of 1-methyl-4-isopropenylcyclohexene (DL-limonene) during the pyrolysis of polyisoprene also comes from an intramolecular transfer step:

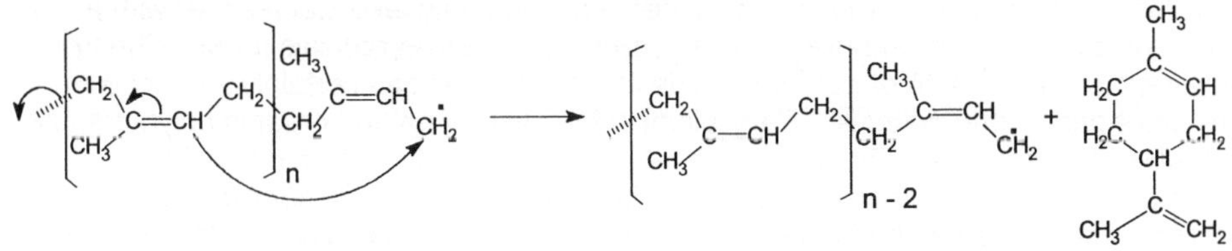

Other cyclic dimers can be formed by the following scheme:

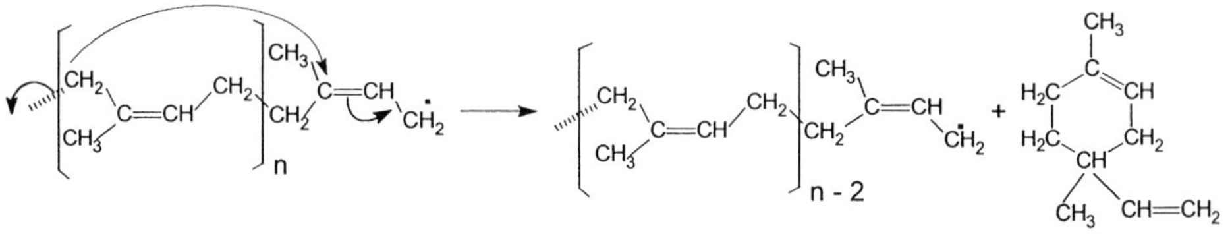

The radical reactions can be terminated by the usual disproportionation:

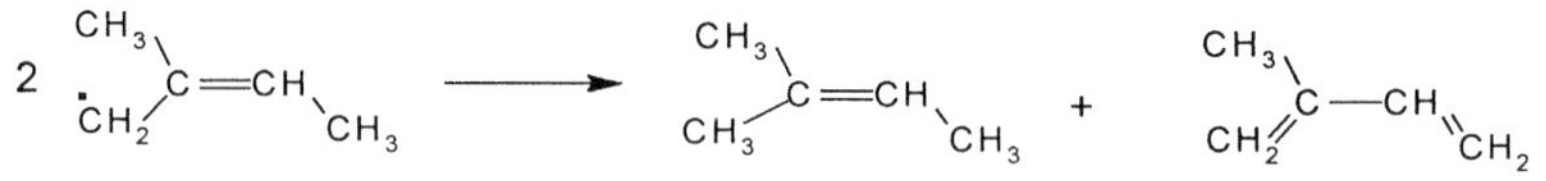

The recombination of the two free radicals is also possible:

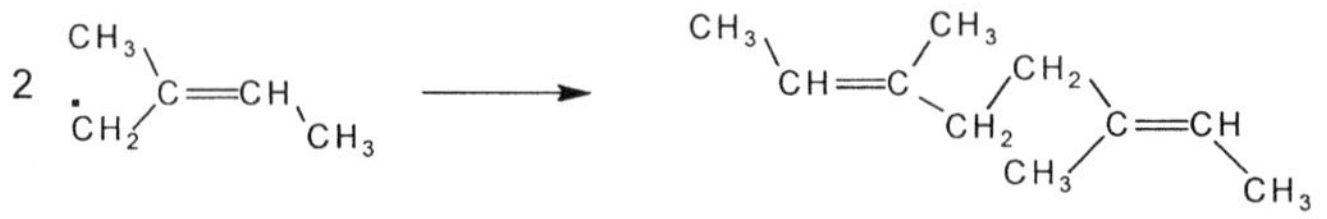

The same types of reactions may take place for the free radical chains. Either a disproportionation or a recombination may take place. In the discussion of the example chosen above, not all the possibilities were considered.

- Side group reactions during polymer pyrolysis

Side group reactions are common during polymer pyrolysis and may take place before chain scission. The presence of water and carbon dioxide as main pyrolysis products in numerous pyrolytic processes can be explained by this type of reaction. The reaction can have either an elimination mechanism, or it can have a substitution mechanism. Side eliminations are common for many linear polymers. However, because these reactions generate smaller molecules but do not affect the chain of the polymeric materials, the heating continues to affect the chain. Therefore the initial reactions are continued with chain scission reactions, and the end result appears as combined types of reactions. Several side chain reactions are discussed in Sections 20.6 and 20.7.

- Combined reactions during pyrolysis

Eliminations and other reactions do not necessarily take place only on the polymeric chain or only on the side groups. Combined reactions may take place either with a cyclic transition state or with free radical formation. The free radicals formed during polymeric chain scission or during the side chain reactions can certainly interact with any other part of the molecule. Particularly in the case of natural organic polymers, the products of pyrolysis and the reactions that occur can be of extreme diversity. A common result in the pyrolysis of polymers is, for example, the carbonization. The carbonization is the result of a sequence of reactions of different types. Various other small molecules are generated by multiple step reactions. As an example, pyrolysis of pectin generates 2-furancarboxaldehyde and 4-(hydroxymethyl)-1,4-butyrolactone. The proportion of 4-(hydroxymethyl)-butyrolactone compared to that of furancarboxaldehyde in the pyrolysis products of pectin correlates with the methylation degree of pectin [106]. The reaction mechanism for formation of furancarboxaldehyde is probably the following:

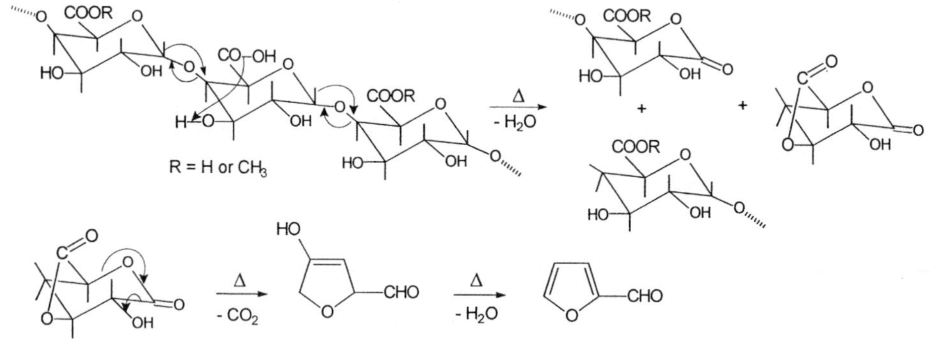

and that of 4-(hydroxymethyl)-butyrolactone is probably the following:

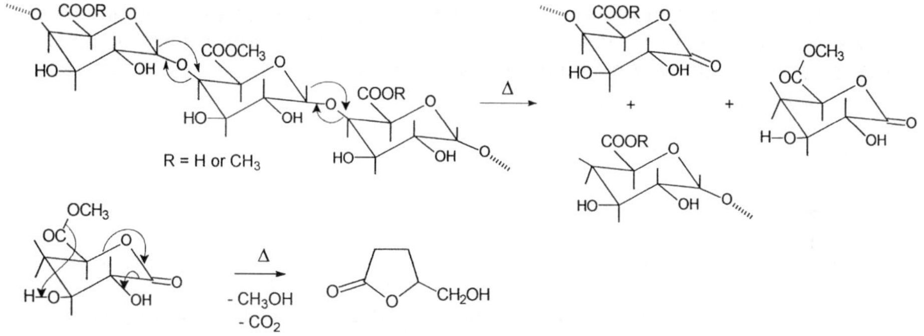

The two reaction paths shown for pectin illustrate the complexity of decompositions that can take place during the pyrolysis of one single polymer.

Pyrolytic reactions can appear much more complicated compared to other reactions. However, this is mainly due to subsequent reactions taking place after the initial elimination step. A common cause of this problem is related to the fact that the reactions do not actually take place in ideal gas phase. Some pyrolytic processes may take place in true condensed phase. Multiple reaction paths and the interaction of the resulting molecules are, therefore, inevitable. Also, additional issues may affect the practical results of a pyrolysis. Some are related to the fact that the true pyrolysis can be associated with reactions caused by the presence (intentional or not) of noninert gases such as oxygen or hydrogen that may be present during the heating. Also, the pyrolyzed materials may be in contact with noninert surfaces that can have catalytic effects. In order to diminish these effects in the pyrolysis done for analytical purposes, an inert gas frequently is present during the pyrolytic reaction.

- Pyrolysis in the presence of additional reactants or with catalysts

Pyrolytic reactions, mainly for analytical purposes, are commonly done in a helium atmosphere. Sometimes these reactions are done, intentionally or not, in the presence of additional reactants or in the presence of catalysts. The most common additional reactants are probably oxygen, hydrogen, water, and quaternary N-alkyl (or aryl) ammonium hydroxides. Oxygen from the air and water are sometimes unintentionally present during pyrolysis. The presence of an additional reactant can modify the result of the pyrolytic reaction.

Oxygen can be a participant in the pyrolysis in two different ways: it can be present as atmospheric oxygen, or it may be already reacted with part of the sample in an autoxidation process by exposure of the sample to air and light over a period of time. In the presence of excess air and at temperatures above the flaming point of the utilized material, burning and not pyrolysis will take place. Although burning is commonly associated with secondary pyrolytic processes, this is not the subject of interest here. The results of the pyrolytic process in the presence of air (below the flaming temperature) can be seen more like a vacuum pyrolysis catalyzed by oxygen [107]. In this type of situation, the pyrolysis products are not significantly different from those obtained without oxygen, but the rate of the reaction is different. However, oxidations may also take place. Free oxygen has an unusual molecule. In its ground state each of the two highest occupied molecular orbitals, which are degenerated, contain unpaired electrons (a triplet state of the molecule). This means that ordinary oxygen has the properties of a diradical. Although not extremely reactive, this diradical will react rapidly with many radicals in a chain oxidation as follows:

$$R^{\bullet} + O_2 \longrightarrow ROO^{\bullet}$$

$$ROO^{\bullet} + RH \longrightarrow ROOH + R^{\bullet}$$

In an excited electronic state (singlet oxygen) oxygen is much more reactive. The oxygen in a singlet state can be generated by a photochemical reaction and may react with a wide variety of materials by a so-called autoxidation process. The singlet oxygen may react with the double bond forming a dioxetane intermediate:

$$\text{C}=\text{C} \quad + \quad O_2 \longrightarrow$$

Polymers exposed to air and light may contain oxidized groups such as peroxides (~OOR). The O-O bond is weak (30–50 kcal/mol) and, upon heating, dissociates to form free RO radicals and radical polymeric chains. These radicals may influence the composition of the pyrolysis products. Autoxidation may take place in food, paint, rubber, etc. It is important, therefore, to consider this possibility when evaluating the composition of the pyrolysis products of a material that was exposed to air and light although the pyrolysis is performed in an inert gas.

Pyrolysis in the presence of hydrogen has been done intentionally [108]. In principle hydrogen can react with numerous chemical compounds. However, molecular hydrogen as such is not very reactive. In most chemical reactions, only the hydrogen generated directly in the reaction medium is active (i.e. from Zn and HCl). Pyrolysis in molecular hydrogen proceeds in most cases in a manner similar to the pyrolysis in an inert gas (helium or nitrogen). In order to make use of the hydrogen reactivity, a catalyst must be used. Common catalysts are metals such as platinum or nickel. In analytical pyrolysis, hydrogen and a catalyst can be used with the purpose of diminishing the number of species resulting in pyrolysis. When the pyrolytic process is followed by a chromatographic separation, the chromatogram of the pyrolysate (the pyrogram) can appear to be too complicated. If this pyrogram consists, for example, of groups of compounds with the same carbon chain but containing single and multiple bonds, this can be simplified by hydrogenation. For each group of compounds, only the saturated one will appear after a catalytic hydrogenation. The procedure can be useful only in some particular cases, and it is not commonly used.

The presence of water as a reaction product from the pyrolytic processes or as adsorbed water on the material to be pyrolyzed is not unusual. However, in analytical pyrolysis, water is not commonly added to the sample. During some pyrolytic processes with industrial applications such as wood pyrolysis, water is sometimes added intentionally. The main effect of water during pyrolysis is hydrolysis. This takes place as the temperature elevates. For polymers like cellulose or starch, the chain scission by hydrolysis (instead of transglycosidation) is the main effect of water addition. This can be seen in the modification of the yields of different final pyrolysis products. Therefore, the reproducibility in analytical pyrolysis may be influenced by the variability of water content of the initial sample [109].

Pyrolysis in the presence of quaternary N alkyl (or alkyl, aryl) ammonium hydroxides is also applied in practice. This reaction was previously discussed in Section 20.1 [110]. The procedure was initially applied by the addition of the reagent in the hot injection port of a gas chromatograph and later with different pyrolytic techniques by directly adding tetramethyl ammonium hydroxide (TMAH) together with the material to be pyrolyzed [111], [112]. The derivatization reagents are applied on the sample either as an aqueous solution (e.g. for tetramethyl ammonium hydroxide) or as methanolic solutions. Not only tetramethyl ammonium hydroxide is used as an "in situ" derivatization reagent. Other quaternary N alkyl (or alkyl, aryl) ammonium hydroxides are successfully used as derivatization reagents. Such reagents are tetrabutyl ammonium hydroxide [113], phenyltrimethyl ammonium hydroxide (or trimethyl anilinium hydroxide), and (m-trifluoromethylphenyl)trimethyl ammonium hydroxide (or trimethyl-trifluoro-m-tolyl ammonium hydroxide) [114]. Also, other pyrolytic derivatizations with the formation of ethyl, propyl, hexyl, etc. derivatives are known.

Although pyrolysis in the presence of TMAH generates methylated compounds, the reaction does not take place identically for all analytes. The pyrolysis/methylation takes place as expected for example in the analysis of lignin and pitch deposits of Kraft pulp mills [115]. The phenolic compounds generated for these materials are easily methylated being sufficiently acidic, and lignin is more resistant to basic hydrolysis (see Section 20.3). For cellulose this pyrolysis in the presence of TMAH is not equivalent with the methylation of the compounds generated from the simple pyrolysis of cellulose. Some compounds are not methylated although they contain active hydrogens, or only part of the material is methylated and part remains not methylated. The OH groups in cellulose are less acidic than phenolic OH groups and are not readily methylated. Also, the amount of different compounds generated during pyrolysis is modified, and the chemical nature of the pyrolysate is significantly different. These types of differences are also seen for the pyrolysis of monosaccharides [116]. The chemical reactions involved in this type of pyrolysis (thermochemolysis) are a combination of thermally assisted base hydrolysis (peeling reaction), pyrolytic bond cleavage, and methylation of acidic functional groups. Some typical reactions taking place during pyrolysis of cellulose in the presence of TMAH are shown in the following scheme [116]:

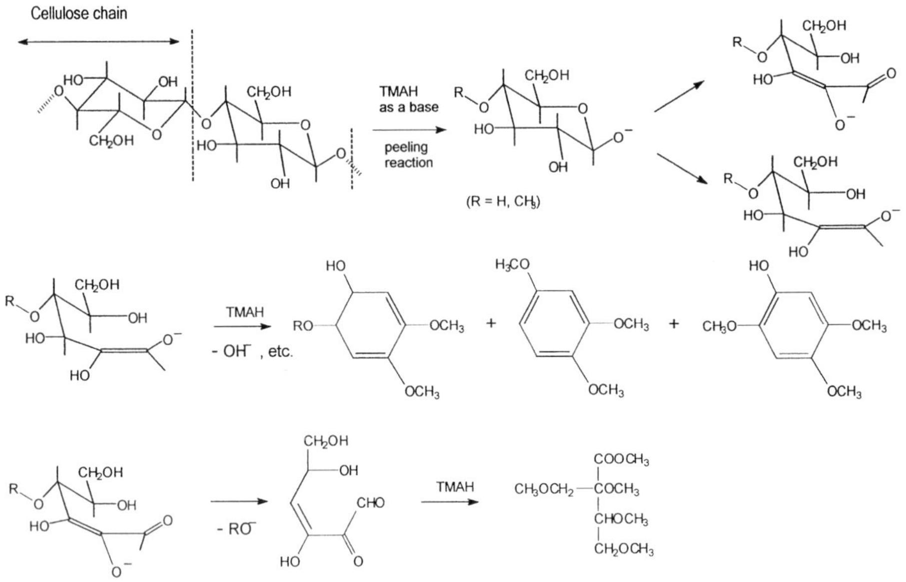

The strong basic character of substituted ammonium hydroxide type reagents puts some limitations to their use. For this reason, the use of tetramethylammonium fluoride (TMAF), phenyltrimethylammonium fluoride, or phenyltrimethylammonium acetate as pyrolytic methylation reagents without the additional effect of the basic character of TMAH is very promising [117].

A comparison between the pyrolysis of cellulose and that of cellulose in the presence of TMAH can be made by comparing the chromatograms in Figure 20.5.1.and 20.5.2.

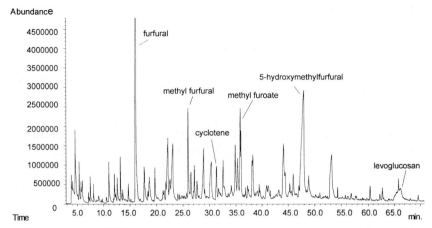

FIGURE 20.5.1. *The SPME chromatogram of a cellulose pyrolysate obtained off-line at 600° C with a filament flash pyrolyzer.*

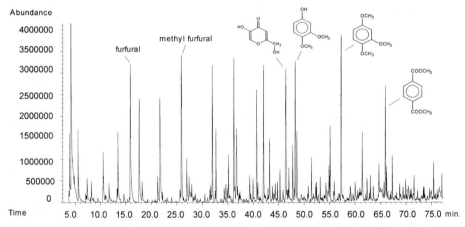

FIGURE 20.5.2. *The SPME chromatogram of a cellulose + TMAH pyrolysate obtained off-line at 600° C with a filament flash pyrolyzer.*

These chromatograms were obtained with a SPME technique for analysis of pyrolysates [117]. The pyrolysis was performed at 600° C for both cellulose and cellulose in the presence of TMAH. The GC separation was done using a dimethyl-polysiloxane + 5% diphenylpolysiloxane column (SPB-5), 60 m long, 0.25 mm i.d., and 0.25 µm film. The GC temperature gradient started at 50° C and ended at 300° C with a rate of 10° C/min. The SPME fiber was Carboxen/Polydimethylsiloxane (CAR/PDMS) with 75-µm film thickness. By comparing the two chromatographic profiles for the pyrolysates without and in the presence of TMAH, the discrepancy between the chemical composition of a cellulose pyrolysate and that of a pyrolysate obtained by thermochemolysis in the presence of TMAH is obvious. Only a very limited number of compounds are common or are the methylated analog of a compound from cellulose pyrolysate.

- *Stereochemistry preservation during pyrolysis*

One important feature that should be noticed for pyrolytic reactions is that the
preexistent isomerism may remain unaffected during pyrolysis (if the particular bonds
remain in the pyrolysate). The preservation of certain stereochemical features in the
fragments is important for relating the fragments to the initial polymer structure. As an
example, during the pyrolysis of polysaccharides, common pyrolysis products are the
anhydrosugars of the specific monosaccharide units that form the polysaccharide. The
anhydrosugar maintains the stereoisomerism of the monosaccharide unit. For example,
the pyrogram of a (1→4)-linked glucose-containing polysaccharide, such as cellulose,
shows as a main pyrolysis product 1,6-anhydroglucopyranose:

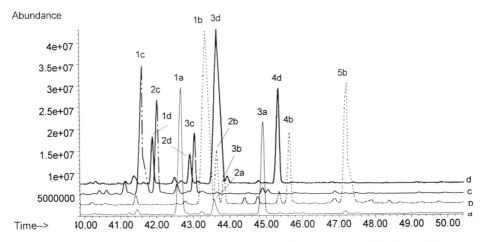

At the same time, the pyrolysis of a (1→4)-linked galactose-containing polysaccharide
gives as a main product 1,6-anhydrogalactopyranose:

Four GC traces for the same time interval for silylated pyrolysates of allose, galactose,
glucose, and mannose are shown in Figure 20.5.3. As can be seen from Figure 20.5.3,
the traces are different. Assignments for the peaks are given in Table 20.5.3.

FIGURE 20.5.3. *Four traces of silylated pyrolysate of (a) mannose, (b) allose, (c)
galactose, and (d) glucose.* Peak identification is given in Table 20.5.3.

TABLE 20.5.3. *Peak assignments for anhydrosugars found in several monosaccharide silylated pyrolysates shown in Figure 20.5.3.*

Saccharide	Peak	Assignment
mannose	1a	1,6-anhydromannopyranose
	2a	1,6-anhydromannofuranose
	3a	1,4-anhydromannopyranose
allose	1b	1,6-anhydroallopyranose
	2b	1,6-anhydroglucopyranose (impurity in allose)
	3b	an anhydrofuranose ?
	4b	1,6-anhydroallofuranose
	5b	1,4-anhydroallopyranose
galactose	1c	1,6-anhydrogalactopyranose
	2c	1,6-anhydrogalactofuranose
	3c	1,4-anhydrogalactopyranose
glucose	1d	1,6-anhydroglucofuranose
	2d	1,4-anhydroglucopyranose type ?
	3d	1,6-anhydroglucopyranose (levoglucosan)
	4d	1,4-anhydroglucopyranose

The preservation of some stereochemical features during pyrolysis is in fact expected. Certain bond breaking during pyrolysis may occur before or at the same time as the change in the stereochemical structure. However, depending on the activation energy of the reaction, either the decomposition or the stereochemical modification may occur first. An example where the stereochemical modification occurs during pyrolysis is that of glycyrrhizic acid. The pyrolysis of this compound generates a variety of small molecules including about 1–2 % of glycyrrhetinic acid. Glycyrrhetinic acid, the aglycon of glycyrrhizic acid, is generated by breaking the glycosidic bond as shown below:

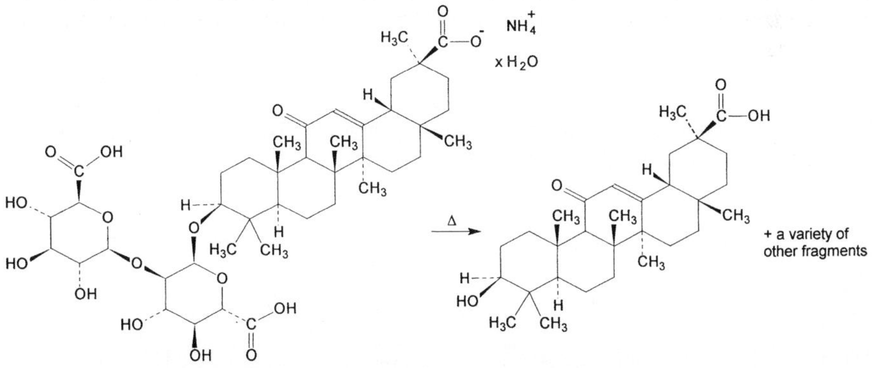

glycyrrhizic acid monoammonium salt   glycyrrhetinic acid

The analysis of the pyrolysate performed in order to identify heavier fragments must be done using a derivatization procedure such as silylation. The silylation of glycyrrhetinic acid leads very likely to one silylated compound with only one other possible derivative, as follows:

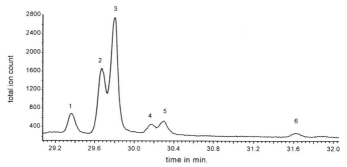

MW = 614   unlikely artifact

However, the chromatographic trace of the silylated pyrolysate of glycyrrhizic acid contains six peaks with a mass spectrum corresponding to silylated glycyrrhetinic acid as shown in Figure 20.5.4.

FIGURE 20.5.4. *Portion of chromatogram of the silylated pyrolysate of glycyrrhizic acid showing six peaks with similar spectra, corresponding to silylated glycyrrhetinic acid.*

The separation of the silylated pyrolysate was performed on a 30 m long, 0.25 mm i.d., and 0.25-μm film SPB-5 column. The GC temperature gradient started at 50° C and ended at 300° C using a 10° C/min temperature gradient. Examples of the spectra for peaks 2 and 3 from Figure 20.5.4 are given in Figure 20.5.5.

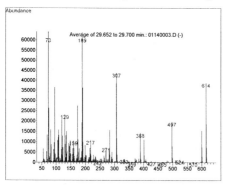
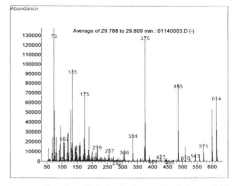

FIGURE 20.5.5. *Example of two spectra corresponding to silylated glycyrrhetinic acid for peaks 2 and 3 in Figure 20.5.4.*

It is very likely that the presence of six compounds in the chromatogram having the same spectrum corresponding to silylated glycyrrhetinic acid is due to the formation of different stereoisomers of this compound. In conclusion, pyrolysis preserves the steric configuration in certain compounds, but in some other cases there are possible changes.

20.6 INSTRUMENTATION USED IN ANALYTICAL PYROLYSIS

The pyrolytic process is done in a pyrolysis unit (pyrolyzer), which commonly interfaces with an analytical instrument such as a GC. The pyrolysis unit consists of a controller and the pyrolyzer itself. The controller provides the appropriate electrical energy needed for heating. A simplified scheme of a pyrolyzer (based on the design of a flash heated filament system made by CDS Inc.) is shown in Figure 20.6.1.

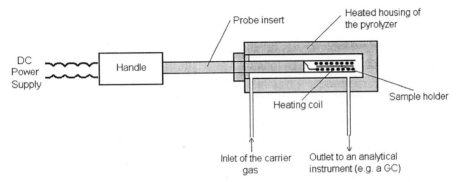

FIGURE 20.6.1. *The simplified scheme of a pyrolyzer (based on the design of a heated filament system made by CDS Inc.).*

The main heating element where pyrolysis occurs is represented in Figure 20.6.1 as a coil that can be heated at high temperatures. This heating element can differ depending on the pyrolyzer principle and instrument type. The pyrolyzer body (sometimes called interface) is a housing for the main heating element connected to an analytical instrument. The interface is usually heated at a temperature necessary to avoid condensation of the pyrolysis products. An inert gas flow is commonly passed through the pyrolyzer to sweep the pyrolysis products into the analytical instrument. Inside the interface, a probe insert can be introduced, which contains the sample as is or in a sample holder.

Besides assuring a reproducible heating, various other requirements are imposed to an analytical pyrolyzer. One of these is the capability to heat the sample in a very short time or TRT (temperature rise time). Various reasons require short TRT values, including good reproducibility, possibility to use a GC system without the need of a cryo-focusing step, etc. One way to produce a rapid heating of the sample is to diminish the sample size [118]. This implies that the amount of heat required by the sample to reach a certain temperature is small and that the heat can be transferred rapidly. Typical sample sizes in analytical pyrolysis vary from a few μg to a few mg. A small sample size is, however, related to other effects, some advantageous and some not. Secondary reactions during pyrolysis are diminished for a small sample, but the contact with metal surfaces may increase (relative to the amount of sample), which is not desirable because of possible catalytic effects. One determining factor that does not allow a significant decrease in the sample size is the limited sensitivity of the analytical procedure following pyrolysis. The small sample size needed in analytical pyrolysis is an advantage in many analyses. However, when working with heterogeneous samples, the small size must be of concern because the sample must be representative for a large amount of material.

Among the types of pyrolyzers that are more frequently used in practice are a) resistively heated filament pyrolyzers, b) Curie point pyrolyzers, c) furnace pyrolyzers, and d) laser pyrolyzers. A short discussion of the main pyrolyzer types follows.

Resistively heated filament pyrolyzers are probably the most common [119], [88]. The principle of this type of pyrolyzer is that an electric current passing through a resistive conductor generates heat in accordance with Joule's law:

$$Q = I^2 R t = (V^2 t) / R \qquad\qquad\qquad (20.6.1)$$

where Q is the amount of heat (in J), I is the current intensity (in A), R is the electrical resistance of the conductor (in ohms), t is the time in sec, and V is the voltage (in V). A simple flash pyrolysis unit that operates at a fixed voltage could easily be constructed. However, such a unit operating within common values for the current intensity and voltage will have a TRT that is too long to be appropriate for flash pyrolysis. Systems with boosted current or boosted voltage are used to achieve a more rapid heating [120]. These systems apply a superimposed constant current with an initial boost pulse to assure a rapid temperature increase at the beginning of the heating period and then enough to maintain a constant temperature for the rest of total heating time (THT). Several procedures for a precise temperature control of the filament are available [121], [122].

Curie point pyrolyzers are based on the fact that ferromagnetic material can be rapidly heated by interaction with a high frequency (radio frequency, RF) electromagnetic field. The sample to be pyrolyzed is placed in close contact with the ferromagnetic materials, which can be shaped into different forms such as a wire, ribbon, or cylinder to properly hold the sample. Heating of the ferromagnetic material and subsequently of the sample can occur in a short time interval (TRT), commonly between 10 and 100 ms. Eddy currents in the ferromagnetic material surface (skin) and hysteresis losses due to changes in the magnetic polarity cause the temperature to increase rapidly when the material is placed in the high frequency electromagnetic field. Increase in temperature is limited to the Curie point temperature for these ferromagnetic conductors [123]. This is temperature specific for each material where the transition from ferromagnetic to paramagnetic properties occurs and the heating in the electromagnetic field is ended. In this way, besides a rapid heating, a well-defined end temperature is attained. This end temperature (Curie point) depends on the composition of the ferromagnetic metal or alloy. The rate of temperature rise depends on the conductor mass and specific heat, as well as on the power consumption of the ferromagnetic conductor.

Furnace pyrolyzers are devices used in both flash pyrolysis and slow gradient pyrolysis. For flash pyrolysis, the common principle of use is to keep the furnace at the desired temperature and to introduce the sample suddenly into the furnace. Heating of the furnace is commonly done using electrical heating, which can be controlled using thermocouples and feedback systems for maintaining the correct temperature. For analytical purposes it is preferable to have small furnaces with low dead volumes. Several designs are used for furnace pyrolyzers, a successful one being a vertical furnace that allows the sample to be dropped from a cool zone into a heated zone [124]. Laser pyrolyzers are also used in analytical pyrolysis. In these instruments a laser beam is focused onto a small spot of a sample to deliver the necessary radiative energy for pyrolysis [87].

The pyrolysate generated in a pyrolyzer must be further transferred to an analytical instrument for analysis. There are two main types of transfer of the pyrolysate to the analytical instrument: on-line and off-line. The choice between these types depends on the pyrolyzing instrument and also on the procedure used for analysis of the pyrolysate [125].

Gas chromatography (GC) and gas chromatography-mass spectrometry (GC-MS) are the most common techniques used for the on-line or off-line analysis of pyrolysates. The clear advantages of these techniques, such as sensitivity and capability to identify unknown compounds, explain their use. The limitations of GC to process nonvolatile samples and the fact that larger molecules in a pyrolysate commonly retain more structural information on a polymer would make HPLC more appropriate for pyrolysate analysis. However, not many results on HPLC analysis of pyrolysates are reported.

- *Pyrolysis gas chromatography (Py-GC)*

Pyrolysis gas chromatography with on-line transfer of pyrolysate, which can be seen as a stand-alone analytical technique, has numerous applications. Dedicated instruments are available, where the pyrolysis products are swept by a flow of gas from the pyrolyzer into the gas chromatograph. The pyrolysate is usually transferred into the injection port of the GC. In other systems the injection port is bypassed, and the pyrolysate is carried through a transfer line and further into the analytical column. Also, some systems have the capability to automatically isolate the GC when the insertion probe is removed and air can penetrate into the GC. The flow of an inert gas in which the pyrolysis is performed is commonly used as the carrier gas for the chromatography. Some pyrolyzers have the capability to perform pyrolysis in a different gas from the carrier gas. A complete transfer of pyrolysate into the GC is desirable. However, higher boiling point compounds are sometimes difficult to transfer, and the compounds associated with the char are not analyzed [87].

When a pyrolyzer is used at the front end of a chromatograph, it is also common to have a mass spectrometer used as a detector. The complexity of pyrolysates makes the presence of the mass spectrometer especially useful for peak identification. No special problems related to the GC-MS analysis are really added in this case. Pyrolysis-gas chromatography-mass spectrometry (Py-GC-MS) is an excellent tool for polymer analysis. Automation is also available with PY-GC instrumentation, allowing the analysis of multiple samples without significant operator intervention.

- *Off-line pyrolysis techniques*

Off-line pyrolysis can be applied before chromatography as a sample preparation procedure. Although not utilized as frequently as on-line techniques and involving more manual operations, off-line procedures have several advantages. One of these advantages is the possibility to perform additional sample preparation on pyrolysates. For example, specific derivatizations of the pyrolysate can be applied, which may extend the range of compounds analyzed. Silylation of pyrolysate has been proven very useful mainly in the analysis of pyrolysates from natural polymers that generate numerous polar fragment molecules [87]. For performing the silylation, the pyrolysate is first collected in a setup as shown in Figure 20.6.2. In this setup, the sample is pyrolyzed in the interface and purged with helium gas such that the effluent is passed through a piece of fused silica capillary. A loop of the capillary is immersed in ice water. After pyrolysis, the capillary is flushed with helium. The capillary is then removed from the ice water bath and the ends inserted into two sealed GC autosampler vials. The first vial contains the reagent, in this case 50 μL of N,O-bis(trimethylsilyl)trifluoroacetamide (BSTFA). After the capillary is warmed to room temperature, the material trapped in the tube is dissolved in the silylation reagent by repeatedly forcing the liquid through the silica capillary from one vial to the other using pressure. The vial that finally collects the material is further heated for 15 min. at 76° C for completing the silylation.

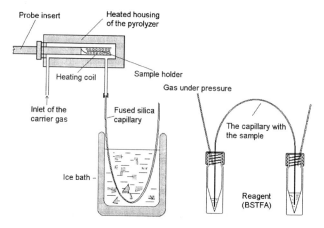

FIGURE 20.6.2. *A simple setup for collecting the sample and adding a derivatization agent to the pyrolysate.*

The results obtained by this procedure for a cellulose sample pyrolyzed at 590° C, silylated, and analyzed using GC-MS are shown in Figure 20.6.3. The separation for both chromatograms was done using a dimethylpolysiloxane + 5% diphenylpolysiloxane column (SPB-5), 60 m long, 0.25 mm i.d., and 0.25 μm film. The GC temperature gradient started at 50° C and ended at 300° C with a rate of 2° C/min.

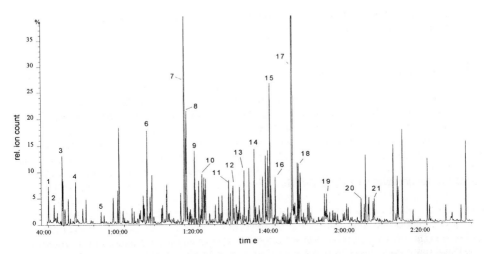

FIGURE 20.6.3. *The total ion chromatogram (TIC) of a cellulose pyrolysate obtained off-line at 590° C with a filament flash pyrolyzer [87] followed by silylation and separated on a SPB-5 column.* #1 1,2-dihydroxyethane 2 TMS, #2 2-hydroxypropionic (lactic) acid 2 TMS, #3 hydroxyacetic (glycolic) acid 2 TMS, #4 furanmethanol TMS, #5 hydroxyfuranone TMS ?, #6 1,3-dihydroxypropanone 2 TMS, #7 1,4-dioxane-2,5-diol 2 TMS, #8 1,3-dioxolane-4,5-diol 2 TMS, #9 1,3-dihydroxybenzene 2 TMS, #10 2-methyl-1,4-dioxane-2,5-diol 2 TMS, #11 1,4-dihydroxybenzene 2 TMS, #12 3-hydroxy-2-(hydroxymethyl)-2-cyclopenten-1-one 2 TMS, #13 2-hydroxy-5-(hydroxymethyl)-4(H)-pyran-4-one 2 TMS, #14 1,2,3-trihydroxybenzene 3 TMS, #15 anhydrosugar 3 TMS, #16 anhydrosugar 3 TMS, #17 levoglucosan (1,6-anhydro-β-D-glucopyranose) 3 TMS, #18 monosaccharide 5 TMS, #19 inositol 6 TMS , #20 anhydrosugar TMS ?, #21 1,6-anhydro-β-D-glucofuranose TMS ?.

Solid phase microextraction (SPME) followed by GC also can be used successfully as an off-line procedure for the analysis of pyrolysates. A simple setup for this procedure is shown in Figure 20.6.4.

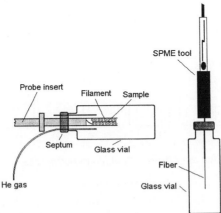

FIGURE 20.6.4. *A simple setup for using SPME in the analysis of a pyrolysate.*

Typical GC analysis can be performed after the desorption of the SPME fiber in the injection port of a GC instrument [117].

Besides derivatization or selective analysis using SPME, off-line pyrolysis before a chromatographic analysis can be used for various other purposes. One such purpose is the use in the pyrolyzer of larger samples, which would not be appropriate to send directly into a GC instrument. Larger sample must be used in specific applications, for example in the analysis of trace components in pyrolysates. Another application of off-line systems is found when a dedicated GC or GC-MS instrument cannot be afforded only for pyrolysate analysis.

- *Data interpretation in Py-GC-MS*

The interpretation of the results in Py-GC-MS depends on the purpose of the analysis. Some results do not need interpretation beyond common GC or GC/MS data analysis. Other results require specific interpretation in order to link the results of the analysis with the initial material subject to pyrolysis. This is true mainly in polymer analysis and in taxonomic applications [87], [126], [89] when the nature or the composition of a given material is in question. The results, usually consisting of a chromatogram of the pyrolysate (pyrogram), can be interpreted using different procedures. One common way is based on matching the peaks in the chromatogram of the sample with that of a standard ("fingerprint" comparison). The matching can be limited to the retention times or, in the case of utilization of a GC-MS, the matching of the nature of the peaks. Other procedures are based on the identification of key compounds or on matching cumulative spectra. A cumulative spectrum is the sum of all spectra of all the peaks in a pyrogram [127] of a specific polymer. Particular peak information is lost in this technique, but characteristic cumulative spectra can be generated for different polymers. An example of a cumulative spectrum for the pyrolysate of cellulose is shown in Figures 20.6.5.

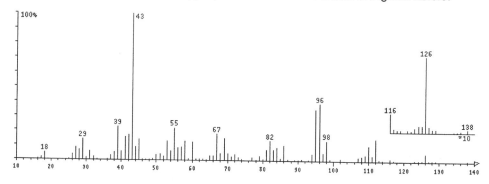

FIGURE 20.6.5. *Cumulative spectrum for cellulose pyrolysate.*

20.7 APPLICATION OF ANALYTICAL PYROLYSIS

Analytical pyrolysis has a variety of applications, mainly related to polymer analysis and the study of the burning process. Use of pyrolysis-GC-MS is common for fingerprinting of samples for taxonomic purposes or for other types of characterization, for modeling thermal decompositions, for the investigation of pyrolysis products of industrial and municipal waste, for the analysis of composite materials such as paints, inks, and subfossil plant materials, for the detection of various types of biomass, for cell and tissue characterization, for the analysis of forensic, archeological or art samples, etc. Not only polymeric materials are amenable for studies using analytical pyrolysis, but also composite samples containing both small and polymeric molecules. Pyrolysis-GC-MS studies are also important for understanding the formation of compounds that may represent health hazards during various cooking practices such as broiling [128], [129]. Also the formation of cigarette smoke is closely related to pyrolysis [130], [131], [132].

- *Pyrolysis of natural organic polymers*

Many organic polymers are common in nature, and they can be grouped into repetitive polymers generated from a unique monomer molecule, nonrepetitive polymers, and more complex composite materials. The repetitive polymers include common materials such as rubber, cellulose, amylose, amylopectin, chitin, etc. Nonrepetitive polymers are also common and include lignins, Maillard browning polymers, peptides, proteins, nucleic acids, humic substances, etc. Natural polymeric composite materials include wood, other plant materials, whole microorganisms, and various biological materials [87]. This classification is not always free from ambiguity, and many materials fit between two classes. For example, pectins are polysaccharides based on $(1\rightarrow4)$-α-linked D-galacturonic acid units, but their structure may include units of methyl esters of galacturonic acid, amidated galacturonic acid [133], as well as units of β-D-xylose, α-D-galactose, α-L-arabinose, α-L-fucose, β-D-galactose, etc. [87]. A large number of natural polymers fit between repetitive and nonrepetitive groups. Many other materials fit between nonrepetitive polymers and composite materials. Examples of this type are conjugated proteins and various geopolymers such as coal or kerogens.

While the pyrograms of uniform polymers are quite reproducible from laboratory to laboratory when the same analytical conditions are used, the pyrograms of nonrepetitive polymers and composite materials may be quite different due to large variability in the source materials. As an example, the variability in pectins can be seen in the pyrolysis profile of pectins from different sources. Figures 20.7.1 to 20.7.3 show the total ion chromatograms (TIC) generated in a GC-MS analysis of the pyrolysates of poly-galacturonic acid and two pectins, one from apple and one from citrus fruits, both with a degree of methylation of about 7%. The pyrolysates were generated at 590° C. The separation of the pyrolysates was on a Carbowax 20 M chromatographic column, 60 m long, 0.32 mm i.d., 0.25 μ film thickness. The GC separation was performed with a temperature gradient of 2° C/min. between 35° C and 240° C. As expected, a significant number of compounds are the same in pyrolysates of the three materials, and some identifications are given in Table 20.7.1. However, the intensities of most peaks in Figures 20.7.1 to 20.7.3 are not identical, and detailed analysis of each chromatogram using the electronic data system of the mass spectrometer indicates various differences.

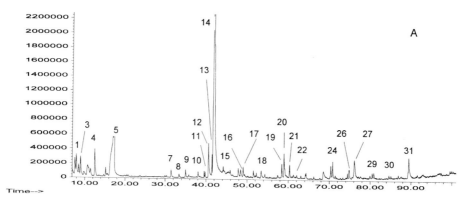

FIGURE 20.7.1. *TIC trace for pyrolysates obtained at 590° C from polygalacturonic acid.* Peak identifications are shown in Table 20.7.1.

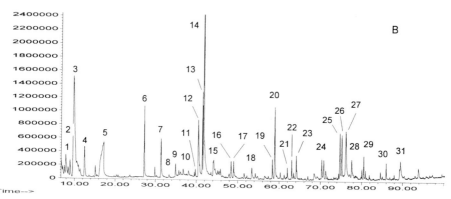

FIGURE 20.7.2. *TIC trace for pyrolysates obtained at 590° C from apple pectin.* Peak identifications are shown in Table 20.7.1.

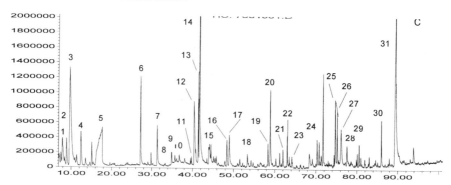

FIGURE 20.7.3. *TIC trace for pyrolysates obtained at 590° C from citrus pectin.* Peak identifications are shown in Table 20.7.1.

TABLE 20.7.1. *Some peak identifications from pectin pyrolysate at 590° C.*

Peak No.	Compound	MW	Characteristic ions (relative intensity in parentheses)
1	propanone	58	43(100), 58(33)
2	2-methylfuran	82	82(100), 53(80), 81(75)
3	3-methyl-2-butanone	86	43(100), 86(21),41(15)
4	methylbenzene	92	91(100), 92(73), 65(11)
5	water	18	18(100), 17(21)
6	2-oxopropanoic acid methyl ester ?	102	43(100), 102(14), 59(8)
7	1-hydroxy-2-propanone	74	43(100), 74(25), 31(20)
8	2,3-dihydro-1,4-dioxin	86	86(100), 29(42), 57(48), 30(10)
9	2-cyclopenten-1-one	82	82(100), 39(40), 54(32), 53(30)
10	3-methyl-2-cyclopenten-1-one	96	96(100), 67(88), 81(45), 53(28)
11	5-methyl-2(3H)-furanone	98	55(100), 98(88), 43(86)
12	acetic acid	60	60(100), 45(85), 43(80)
13	3-oxopropanoic acid methyl ester	102	43(100), 102(32), 59(4)
14	furancarboxaldehyde	96	96(100), 95(91), 39(40)
15	1-(2-furanyl)-ethanone	110	95(100), 110(44), 39(13)
16	5-methyl-2-furfural	110	110(100), 109(95), 53(40)
17	4-cyclopenten-1,3-dione	96	96(100), 42(83), 68(82), 54(50)
18	furanmethanol	98	98(100), 97(60), 81(60), 69(40)
19	2(5H)-furanone	84	55(100), 84(76), 27(65)
20	2-hydroxycyclopent-2-en-1-one	98	98(100), 55(63), 42(55), 69(31)
21	2-hydroxy-3-methyl-2-cyclopenten-1-one	112	112(100), 97(15), 83(20), 69(22)
22	unknown	128 ?	113(100), 128(93), 58(90), 87(55)
23	unknown	114	114(100), 58(90), 85(5)
24	phenol	94	94(100), 66(34), 65(26)
25	unknown	154 ?	123(100), 154(60), 95(24), 39(15)
26	2-butendioic acid diethyl ester	172	128(100), 99(61), 54(29)
27	5-ethyldihydro-2-(3H)-furanone	114	85(100), 42(18), 57(12), 114(11)
28	butandioic acid dimethyl ester	146	115(100), 55(62), 59(45)
29	3-methyl-2,4-(3H, 5H)-furandione	114	114(100), 56(92), 28(78), 85(12)
30	butandioic acid monomethyl ester	132	101(100), 55(42), 45(25)
31	mix: 2-furancarboxylic acid, benzoic acid and 2-hydroxypyridine in traces B and C		112, 95 and 122, 105

A different set of problems comes from the similarities between the pyrograms of compounds that are related but not the same. In this case, pyrolysis as a procedure for processing a sample with the purpose of identification may not be appropriate. For example, glucose and starch, although different, generate very similar pyrograms. A pyrogram of glucose is shown in Figure 20.7.4, and a pyrogram of starch is shown in Figure 20.7.5.

Abundance

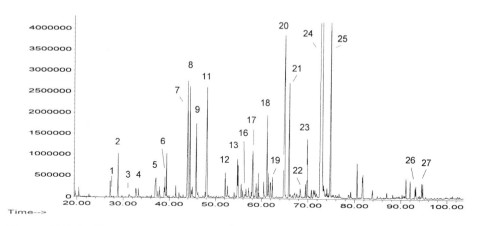

FIGURE.20.7.4. *Glucose pyrolysate at 590° C derivatized (silylated) and separated on a DB5 column.* Peak identification is indicated in Table 20.7.2.

Abundance

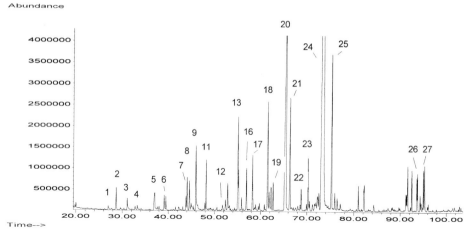

FIGURE 20.7.5. *Starch pyrolysate at 590° C derivatized (silylated) and separated on a DB5 column.* Peak identification is indicated in Table 20.7.2.

The pyrolysis for both samples was performed at 590° C off-line, followed by trimethylsilylation (with BSTFA), and the separation was done on a DB5 column (60 m long, 0.32 mm i.d. and 0.25 μ film thickness). Some peak identifications are given in Table 20.7.2.

Obvious similarities can be seen in Figures 20.7.4 and 20.7.5, indicating that the influence of α-glucosidic (1→4) links in starch does not have a significant contribution to the result of pyrolysis.

TABLE 20.7.2. *Peak identification for glucose and starch pyrolysates silylated and separated on a DB5 column as shown in Figure 20.7.4. and 20.7.5, respectively.*

Peak #	Compound	Formula	MW
1	lactic acid-di-TMS	C9H22O3Si2	234
2	hydroxyacetic acid-di-TMS	C8H20O3Si2	220
3	methoxy furan	C5H6O2	98
4	2-(TMS-oxy)-2-cyclopenten-1-one	C8H14O2Si	170
5	reagent		ion 99
6	dihydroxyacetone di-TMS	C9H22O3Si2	234
7	glycerol tris-(TMS)	C12H32O3Si3	308
8	triethylene glycol-2-TMS	C12H30O4Si2	294
9	2-methyl-3-(TMS-oxy)-pyranone	C9H14O3Si	198
10	dihydroxypropanoic acid-tri-TMS	C12H30O4Si3	322
11	cyclopentanecarboxylic acid-TMS ?	C9H18O2Si	186
12	dihydroxybenzene-di-TMS	C12H22O2Si2	254
13	dihydroxymethylfuran-di-TMS	C11H22O3Si2	258
14	hydroxy-(hydroxymethyl)furan-di-TMS	C11H22O3Si2	258
15	pyranone	C5H4O2	96
16	hydroxycyclopentadiene carboxylic acid-di-TMS	C12H22O3Si2	270
17	(hydroxymethyl)-furoic acid-di-TMS	C12H22O4Si2	286
18	hydroxy-(hydroxymethyl)-cyclopentenone-di-TMS	C12H26O2Si2	258
19	2,3,5-trihydroxybenzene tri-TMS	C15H30O3Si3	342
20	hydroxy-cyclohexanoic acid-di-TMS	C13H28O3Si2	288
21	internal standard		
22	1,4:3,6-dianhydro-α-D-glucopyranose-tri-TMS	C15H32O4Si3	360
23	deoxyhexonic acid-γ-lactone-tri-TMS	C15H34O5Si3	378
24	levoglucosan-tri-TMS	C15H34O5Si3	378
25	anhydrosugar x-TMS		
26	monosaccharide 5-TMS	C21H52O6Si5	540
27	monosaccharide 5-TMS	C21H52O6Si5	540

As seen in the previous examples, peak identification and chromatographic profiles are the main tools for the interpretation of a pyrogram. However, for specific details in a given analytical problem, some other characteristics can be exploited, such as peak ratio for specific compounds, amount of char left after pyrolysis, etc. [87].

- Pyrolysis of synthetic polymers

Use of analytical pyrolysis for the study of synthetic polymers is a field of considerable interest. A large number of applications have been found for pyrolysis of synthetic polymers. The pyrograms of uniform synthetic polymers are rather typical and very useful for the identification of a given polymer. However, the practical use of a large number of copolymers makes the pyrograms of these compounds rather complex and more difficult to interpret without preliminary information on the polymer. On the other hand, the evaluation of differences between copolymers from the same class frequently is done based on details in pyrograms [89]. Only a few specific characteristics for various groups of synthetic polymers are discussed here.

Polyolefins represent a large class of synthetic polymers. A variety of pyrolysis studies have been performed on polyethylenes, polypropylenes, polybutadiene, and their copolymers [134], [135]. Also, numerous studies have been done on synthetic rubber

[136], [137]. Pyrolysis studies allow different levels of microstructure characterization. A typical pyrogram for a high-density polyethylene is shown in Figure 20.7.6.

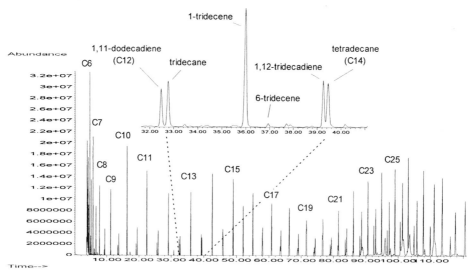

FIGURE 20.7.6. *High density polyethylene; pyrogram of obtained at 600° C and separated on a Carbowax column.* Peak identification is indicated in the figure. Peaks labeled correspond to an alkene.

The pyrogram shows the formation of a series of hydrocarbons, the most intense peaks corresponding to 1-alkenes. Based on various peak intensities or diene content in the pyrograms, the percent of low-density and high-density polyethylene has been determined [138]. Studies of polymer microstructure also have been done for polypropylene [139], [140], polybutadiene [141], etc.

Vinyl polymers have various degradation mechanisms, depending on the strength of specific carbon-carbon bonds and the stability of resulting radicals (see Section 20.5). A variety of compounds may be formed upon pyrolysis. As shown in Table 20.5.2, some polymers from this group suffer depolymerization during heating, and the main pyrolysis products are the monomers. Examples are the polymers of methyl methacrylate, α-methylstyrene, trifluorostyrene, and tetrafluoroethylene. Some polymers have a lower yield of monomer such as polystyrene, and others generate very low levels or even no monomer. The results of depolymerization are related to the reaction mechanism and the stability of the generated fragments versus that of the polymeric chain.

Some vinyl polymers are characterized by elimination of a specific side chain group. As an example, polyvinyl chloride generates very little vinyl chloride but forms HCl by a side chain elimination reaction. It also forms various other compounds, some as indicated in Table 20.7.3. A pyrogram of polyvinyl chloride is shown in Figure 20.7.7. The pyrolysis was at 600° C, and the separation was on a Carbowax column, 60 m long, 0.32 mm i.d., and 0.25 µ film thickness. The GC oven had a temperature gradient of 35° C to 240° C.

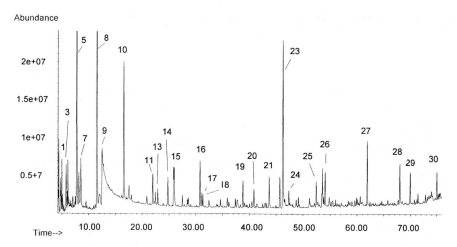

FIGURE 20.7.7. *Pyrogram of polyvinyl chloride obtained at 600° C and separated on a Carbowax column.* Peak identification is indicated in the Table 20.7.3.

TABLE 20.7.3. *Peak identification for polyvinyl chloride pyrolysates separated on a Carbowax column as shown in Figure 20.7.7.*

Peak #	Compound	Formula	Molecular ion
1	vinyl chloride	C2H3Cl	62
2	1,3-pentadiene	C5H8	68
3	1,3-cyclopentadiene	C5H6	66
4	1-methylcyclopentene	C6H10	82
5	2-ethyl-1-hexene	C8H16	112
6	3-methyl-2-heptene	C8H16	112
7	3-ethyl-4-methyl-1-pentene	C8H16	112
8	benzene	C6H6	78
9	hydrochloric acid	HCl	36
10	toluene	C7H8	92
11	ethylbenzene	C8H10	106
12	1,4-dimethylbenzene	C8H10	106
13	1,2-dimethylbenzene	C8H10	106
14	3-chloromethylheptane	C8H17Cl	148
15	1,3-dimethylbenzene	C8H10	106
16	styrene	C8H8	104
17	1-ethyl-2-methylbenzene	C9H12	120
18	1-propenylbenzene	C9H10	118
19	2,3-dihydro-1H-indene	C9H10	118
20	1-propenylbenzene	C9H10	118
21	acetic acid	C2H4O2	60
22	2,3-dihydro-5-methyl-1H-indene	C10H12	132
23	1H-indene	C9H8	116
24	2,3-dihydro-1-methyl-1H-indene	C10H12	132
25	1-methyl-1H-indene	C10H10	130
26	1,2-dihydronaphthalene	C01H10	130
27	naphthalene	C10H8	128
28	2-methylnaphthalene	C11H10	142
29	1-methylnaphthalene	C11H10	142
30	1,1'-biphenyl	C12H10	154

The presence of acetic acid in the pyrogram can be explained by its presence as an impurity in the polyvinyl chloride. As indicated previously, the use of GC may not make possible the detection of larger fragment molecules. These are not detected because they do not elute from the chromatographic column. When using higher temperatures for the GC oven or less polar chromatographic columns, larger molecules compared to those shown in Table 20.7.3 (such as anthracene) were detected in the pyrolysate of vinyl chloride [89].

Other vinyl polymers may generate specific compounds, but these are a result of both polymer chain scission and reaction with the side chain groups. For example, polyacrylamide generates piperidinedione and pyrrolidindione derivatives in a reaction of the type:

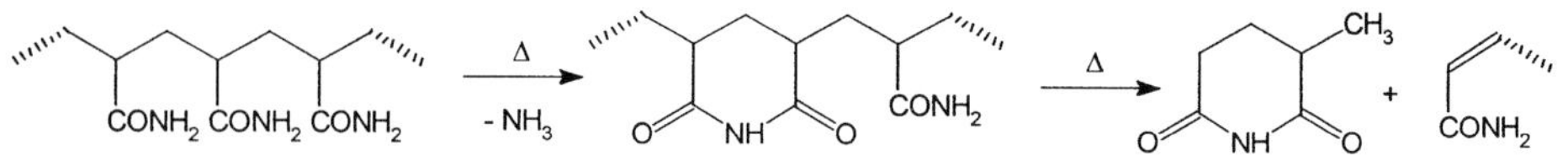

A wide variety of condensation polymers are also known. These include phenol-formaldehyde resins, polyesters, polyamides, polyethers, polycarbonates, and polyurethanes.

Phenol formaldehyde resins form as major pyrolysis products phenol and methyl substituted phenols such as 2-and 4-methylphenol, 2,6- and 2,4-dimethylphenol, as well as 2,4,6-trimethylphenol.

Polyesters generate a variety of compounds depending on their molecular structure. A typical mechanism of pyrolytic decomposition of a terephthalate is shown below:

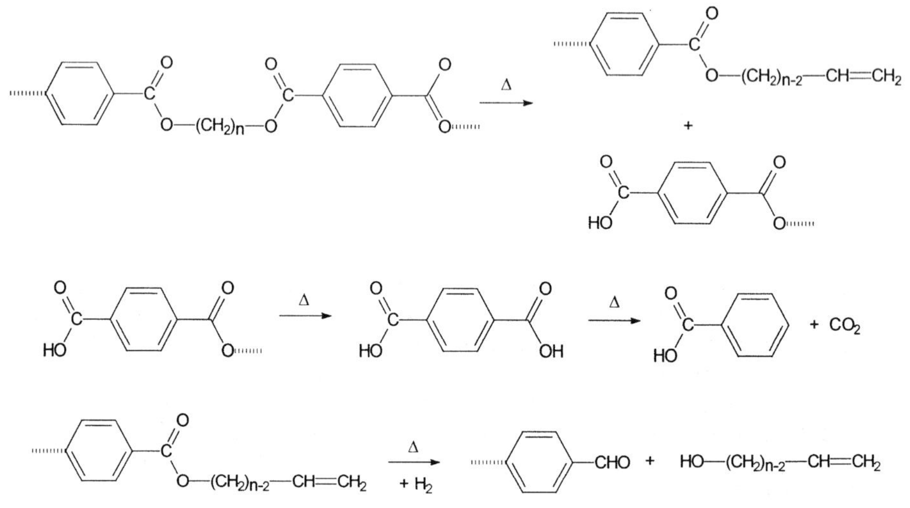

As a result, terephthalic acid, benzoic acid, as well as compounds derived from the diol used in the polyester are formed (n=2 is a common case). Larger fragments can also be detected in pyrolysates [142], [143]. The same mechanism is valid for polyesters of hydroxyacids. Fully aromatic polyesters [144] have a similar decomposition path generating phenol, benzene, phenylbenzoate, phenyl-(p-hydroxybenzoate), etc. Aromatic rings are relatively more stable to heating than the aliphatic chains, and their integrity is commonly kept in the pyrolysis process. At the same time, there are various points where an aliphatic chain can break, and this leads to more complex pyrograms for the polymers with aliphatic chains than those for the aromatic polymers.

Polyamides also generate a variety of fragments by pyrolysis, which depend on the nature of the polyamide. Some pyrolysis mechanisms for this group were discussed in Section 20.5. A number of results are reported in literature [89], [145]. A schematic description of the possible reactions is shown below:

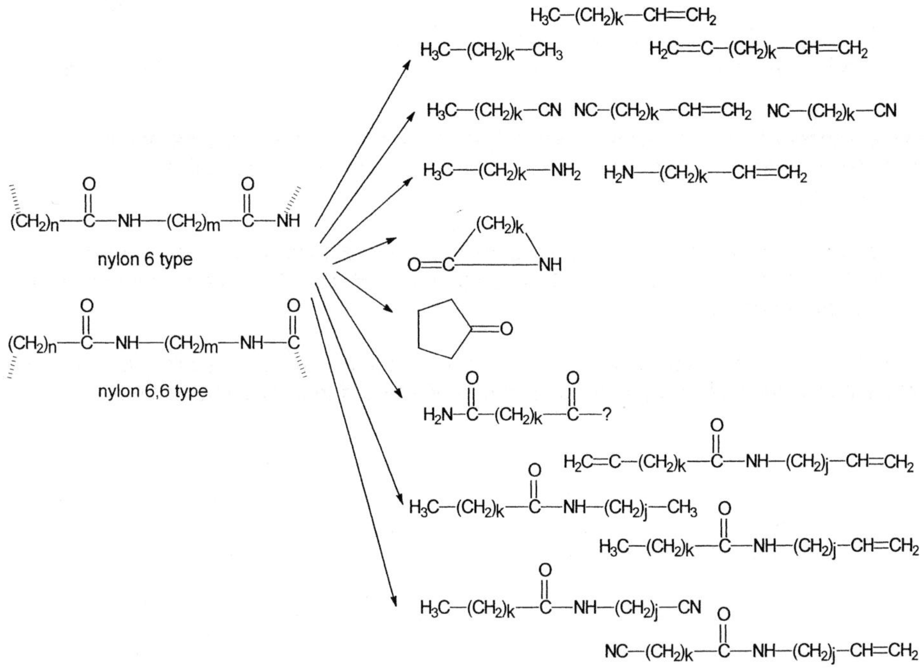

$$H_3C-(CH_2)_k-CH=CH_2$$
$$H_3C-(CH_2)_k-CH_3 \qquad H_2C=C-(CH_2)_k-CH=CH_2$$

$$H_3C-(CH_2)_k-CN \quad NC-(CH_2)_k-CH=CH_2 \quad NC-(CH_2)_k-CN$$

$$H_3C-(CH_2)_k-NH_2 \quad H_2N-(CH_2)_k-CH=CH_2$$

nylon 6 type: $(CH_2)_n-\overset{O}{\overset{||}{C}}-NH-(CH_2)_m-\overset{O}{\overset{||}{C}}-NH$

$$O=C\overset{(CH_2)_k}{\diagdown}-NH$$

nylon 6,6 type: $(CH_2)_n-\overset{O}{\overset{||}{C}}-NH-(CH_2)_m-NH-\overset{O}{\overset{||}{C}}$

(cyclopentanone) $=O$

$$H_2N-\overset{O}{\overset{||}{C}}-(CH_2)_k-\overset{O}{\overset{||}{C}}-?$$

$$H_2C=C-(CH_2)_k-\overset{O}{\overset{||}{C}}-NH-(CH_2)_j-CH=CH_2$$

$$H_3C-(CH_2)_k-\overset{O}{\overset{||}{C}}-NH-(CH_2)_j-CH_3$$

$$H_3C-(CH_2)_k-\overset{O}{\overset{||}{C}}-NH-(CH_2)_j-CH=CH_2$$

$$H_3C-(CH_2)_k-\overset{O}{\overset{||}{C}}-NH-(CH_2)_j-CN$$

$$NC-(CH_2)_k-\overset{O}{\overset{||}{C}}-NH-(CH_2)_j-CH=CH_2$$

The major pyrolysis products for nylon 6,6 are cyclopentanone associated with some dinitriles and mononitriles containing one amide group. The number of carbon atoms in the fragments can differentiate for example nylon 6,10 from nylon 6,6. Nylon 6 generates ε-caprolactame and some mononitriles. Wholly aromatic polyamides such as poly(1,3-phenyleneisophthalamide) (Nomex) and poly(1,4-phenyleneterephthalamide) (Kevlar) generate benzonitrile, benzene, toluene, benzodinitrile, etc. [146]. The molecular fragments of these polymers are fewer and easier to interpret, as in the case of aromatic esters.

Polyethers have lower thermal stability compared to the analogous polyolefins, although C-O and C-C bonds have about the same energy. Most pyrolysis products show cleavage of C-O bonds and only of a few C-C bonds, as shown below:

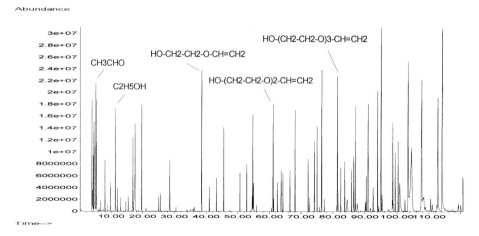

A typical pyrogram for polyethylene glycol, performed at 600° C and separated on a Carbowax column, is shown in Figure 20.7.8.

FIGURE 20.7.8. *Pyrogram of polyethylene glycol obtained at 600° C and separated on a Carbowax column.*

Pyrolysis of polymeric aromatic carbonates follows the same rules as for other aromatic polymers showing simpler pyrograms. The typical structure of these polymers is shown below:

The stability of the aromatic ring leads to the formation of phenolic compounds and some other related molecules such as 2,3-dihydrobenzofuran. The pyrogram of an aromatic polycarbonate resin is shown in Figure 20.7.9.

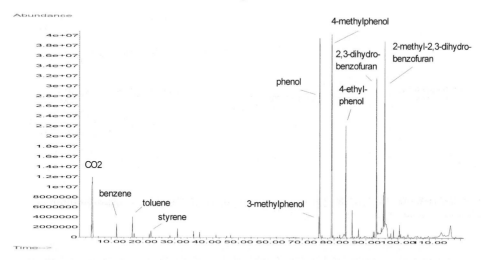

FIGURE 20.7.9. *Pyrogram of a polycarbonate obtained at 600° C and separated on a Carbowax column.*

Polyurethanes are another common class of polymers generated from a diisocyanate and a diol. Various diols are used, but a common type is a low molecular weight polymer made from ethyleneglycol or propylene glycol and adipic acid. The structure of this type of polyurethane is shown below:

Depending on the proportion of diisocyanate and diol and also on the length of diol oligomer molecules, various pyrolysis products may be formed. A number of nonaromatic molecules, such as short alkenes, acetaldehyde, ethers and keto ethers such as 1-methylethoxy-2-propanone, usually dominate the pyrograms.

Another group of polymers that are analyzed using pyrolysis are the silicones. Most silicones generate by pyrolysis cyclic alkyl siloxanes. An example of a pyrogram obtained from a silicone grease at 600° C and separated on a Carbowax column is shown in Figure 20.7.10.

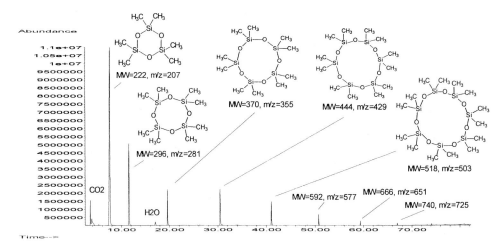

FIGURE 20.7.10. *Pyrogram of a silicone grease obtained at 600° C and separated on a Carbowax column.*

More details regarding the study of synthetic polymers or copolymers using pyrolysis can be found in the dedicated literature [88], [89]. Further discussion on the subject is beyond the purpose of this book.

- *Pyrolysis of small molecules*

Although the analysis of small molecules using pyrolysis as a technique to introduce the sample into a GC is not very common, a significant number of studies regarding the pyrolysis of small molecules is available [147], [148], [149], [150], [151], [117], [152], [153], [154]. Pyrolysis of small molecule analytes is usually performed in order to study the behavior of the analyte during pyrolysis (possibly associated with burning). These studies are done either for practical purposes or for understanding various decomposition mechanisms.

A particular application of pyrolysis of small molecules is the pyrolytic methylation using tetramethylammonium hydroxide (TMAH) or related compounds. However, this type of pyrolysis can be performed in the injection port of the GC, and does not require a pyrolyzer. Also, the pyrolysis is done on the reagent and not on the analyte. Pyrolytic methylation was discussed in Section 18.1.

- *Pyrolysis of composite materials*

Composite materials containing a mixture of polymers and possible small molecules can be analyzed without the separation of components using pyrolysis. Detailed presentation of the subject is beyond the purpose of this book, and a number of dedicated papers and books [87], [88], [155] present the subject.

When composite materials are pyrolyzed, more than one molecular species is subject to thermal degradation. However, for composite materials each component can be considered as starting the pyrolytic process independently, which reduces somewhat the complexity of the problem. Pyrolysis products of different components may interact to a certain extent, but some components remain unaffected by the presence of other molecular species, because the temperature of the pyrolysate decreases abruptly after the pyrolytic process ends. The possible interaction of some components in the pyrolysis products depends mainly on the nature of those components [156], [157], [158]. Two model cases can be considered. One is the pyrolysis of cellulose in the presence of tetramethylammonium hydroxide (TMAH), and the other is pyrolysis of cellulose in the presence of N,O-bis(trimethylsilyl)-trifluoroacetamide (BSTFA). The pyrolysis products of cellulose in the presence of TMAH are significantly different from that of cellulose, as shown in Section 20.5. At the same time, pyrolysis in the presence of BSTFA with BSTFA added on the cellulose before pyrolysis reacts only when the residence of the pyrolysis products in the transfer lines is long enough. The effect of various additives to polymer pyrolysis has been studied in a number of particular cases [87], [89], [160] and has importance in many practical applications [161].

REFERENCES 20

1. C. Fenselau (ed.), *Mass Spectrometry for the Characterization of Microorganisms*, ACS Symp. Ser. 541, Washington, 1994.
2. L. Huang, R. M. Riggin, Anal. Chem., 72 (2000) 3539.
3. W. Pigman, D. Horton (eds.), *The Carbohydrates. Chemistry and Biochemistry*, Academic Press, New York, 1972.
4. R. L. Whistler, M. L. Wolfrom (eds.), *Methods in Carbohydrate Chemistry*, Academic Press, New York, 1964.
5. J. F. Kennedy, C. A. White, *Bioactive Carbohydrates in Chemistry, Biochemistry and Biology*, E. Horwood., Chichester, 1983.
6. K. A.Thomsson et al., Anal. Chem., 72 (2000) 4543.
7. H. Karlsson et al., Anal. Biochem., 182 (1989) 438.
8. M. T. Cancilla et al., Anal. Chem., 72 (2000) 2901.
9. M. F. Chaplin, J. F. Kennedy (eds.), *Carbohydrate Analysis, A Practical Approach*, IRL Press, Oxford, 1986.
10. B. Lindberg et al., Adv. Carbohydr. Chem. Biochem., 31 (1975) 185.
11. R. R. Selverdan et al., Anal. Biochem., 96 (1979) 282.
12. L. A. Torello et al., J. Chromatogr., 202 (1980) 195.
13. L. E. Griggs et al., Anal. Biochem., 43 (1971) 369.
14. P. Albersheim et al., Carbohydr. Res., 5 (1967) 340.
15. S. I. Ogata, K. O. Loyd, Anal. Biochem., 119 (1982) 351.
16. T. P. Mawhinney, J. Chromatogr., 351 (1986) 91.
17. C. J. Biermann, G. D. McGinnis, *Analysis of Carbohydrates by GLC and MS*, CRC Press, Boca Raton, 1989.
18. J. F. Saeman et al., TAPPI, 37 (1954) 336.
19. I. A. Pearl, *The Chemistry of Lignin*, M. Dekker, New York, 1967.
20. R. Varma et al., J. Chromatogr., 77 (1972) 926.
21. J. F. Lawrence, J. R. Lyengar, J. Chromatogr., 350 (1985) 237.
22. W. Niedermeier, M. Tomana, Anal. Biochem., 57 (1974) 363.

23. R. Kannan et al., J. Chromatogr., 92 (1974) 95.
24. W. F. Alpenfels et al., J. Liq. Chromatogr., 5 (1982) 1711.
25. E. Bombardelly et al., J. Chromatogr., 279 (1983) 593.
26. R. E. Chambers, J. R. Clamp, Biochem. J., 125 (1971) 1009.
27. J. Conchie et al., Carbohydr. Res., 103 (1982) 129.
28. B. V. McCleary, N. K. Matheson, Adv. Carbohydr. Chem. Biochem., 44 (1986) 147.
29. S. Honda et al., Anal. Chem., 50 (1978) 55.
30. J. L. Sussman et al., Science, 253 (1991) 872.
31. B. Wittmann-Liebold (ed.), *Methods in Protein Sequence Analysis*, Springer, Berlin 1989.
32. T. S. Work, R. H. Burdon (eds.), *Sequencing of Proteins and Peptides. Laboratory Techniques in Biochemistry and Molecular Biology*, Elsevier, Amsterdam, 1981.
33. M. W. Dong, A. D. Tran, J. Chromatogr., 499 (1990) 125.
34. P. A. Hartman et al., J. Chromatogr., 360 (1986) 385.
35. L. Leadbeater, F. B. Ward, J. Chromatogr., 397 (1987) 435.
36. L. D. Ward et al, J. Chromatogr., 519 (1990) 199.
37. S. H. Kang et al., Anal. Chem., 72 (2000) 3014.
38. C. L. Gatlin et al., Anal. Chem., 72 (2000) 757.
39. F. C. Bernstein et al., J. Mol. Biol., 112 (1977) 535.
40. M. Mann, M. Wilm, Anal. Chem., 66 (1994) 4390.
41. J. K. Eng et al., J. Am. Soc. Mass Spectrom., 5 (1994) 976.
42. A. M. Crestfield et al., J. Biol. Chem., 238 (1963) 622.
43. R. F. Adams, J. Chromatogr., 95 (1974) 189.
44. S. Borman, Anal. Chem., 59 (1987) 969.
45. W. A. Schroeder, W. S. Hancock (eds.), *CRC Handbook of the Separation of Amino Acids, Peptides and Proteins*, CRC Press, Boca Raton, 1984.
46. I. M. Lazar et al., Anal. Chem., 73 (2001) 1733.
47. G. Marie et al., Anal. Chem., 72 (2000) 5423.
48. A. Kishiyama et al., Anal. Chem., 72 (2000) 5431.
49. R. Guevremont et al., Anal. Chem., 72 (2000) 4577.
50. J. K. Eng et al., J. Am. Soc. Mass Spectrom., 5 (1994) 976.
51. P. Edman, G. Begg, Eur. J. Biochem., 1 (1967) 80.
52. M. Schneider, H. Tschesche, Z. Physiol. Chem., 357 (1976) 1339.
53. R. S. Simpson et al., Eur. J. Biochem., 176 (1988) 187.
54. M. Waterfield, E. Haber, Biochemistry, 9 (1970) 832.
55. D. R. Knapp, *Handbook of Analytical Derivatization Reactions*, J. Wiley, New York, 1979.
56. E. H. Creaser, Biochem. J., 157 (1976) 77.
57. K. Imai et al., Biomed. Chromatogr., 7 (1993) 56.
58. H. Matsunaga et al., Anal. Chem., 67 (1995) 4276.
59. A. Toriba et al., Anal. chem., 72 (2000) 740.
60. J. C. Cavadore et al., Anal. Biochem., 60 (1974) 608.
61. J. Rosmus, Z. Deyl, Chromatogr. Rev., 13 (1970) 163.
62. J. Rosmus, Z. Deyl, J. Chromatogr., 70 (1972) 221.
63. Z. Deyl, J. Chromatogr., 127 (1976) 91.
64. G. T. Harmanson, *Bioconjugate Techniques*, Academic Press, San Diego, 1996.
64a. V. Ghetie et al., Bioconjugate Chem., 1 (1990) 24.
65. J. R. Engen, D. L. Smith, Anal. Chem., 73 (2001) 257A.
66. P. J. Hernes, J. I. Hedges, Anal. Chem., 72 (2000), 5115.

67. K. V. Sarkanen, C. H. Ludwig (eds.), *Lignins, Occurrence, Formation, Structure and Reactions*, J. Wiley, New York, 1971.
68. J. Marton (ed.), *Lignins Structure and Reactions*, ACS Ser. 59, Washington, 1966.
69. L. Stryer, *Biochemistry*, Freeman, New York, 1988.
70. International Human Genome Sequencing Consortium, Nature, 409 (2001) 860.
71. D. D. Shoemaker et al., Nature, 409 (2001) 922.
72. The International SNP Map Working Group, Nature, 409 (2001) 928.
73. R. R. Bentley et al., Nature, 409 (2001) 942.
74. T. K. Christopoulos, Anal. Chem., 71 (1999) 425R.
75. F. Sanger et al., Proc. Natl. Acad. Sci. USA, 74 (1977) 5463.
76. F. Sanger et al., Nature, 246 (1977) 687.
77. A. P. Feinberg, B. Vogelstein, Anal. Biochem., 132 (1983) 6.
78. A. P. Feinberg, B. Vogelstein, Anal. Biochem., 137 (1983) 266.
79. A. M. Maxam, W. Gilbert, Proc. Natl. Acad. Sci. USA, 74 (1977) 560.
80. L. M. Smith et al., Nature, 321 (1986) 674.
81. H. He, L. B. McCown, Anal. Chem., 72 (2000) 5865.
82. N. R. Isola et al., Anal. Chem., 71 (1999) 2266.
83. V. E. Velculescu et al., Science, 270 (1995) 484.
84. L. J. Krika et al. (eds.), *Nonisotopic Probing, Blotting and Sequencing*, Academic Press, New York, 1995.
85. D. Savage et al., *Avidin-Biotin Chemistry: A Handbook*, Pierce Chemical Co., Rockford , 1992.
86. D. H. Phillips et al., *Postlabelling Methods for Detection of DNA Adducts*, IARC Sci. Pub. No. 124, IARC, Lyon, 1993.
87. S. C. Moldoveanu, *Analytical Pyrolysis of Natural Organic Polymers*, Elsevier, Amsterdam, 1998.
88. T. P. Wampler (ed.), *Applied Pyrolysis Handbook*, M. Dekker, New York, 1995.
89. S. A. Liebman, E. J. Levy (eds.), *Pyrolysis and GC in Polymer Analysis*, M. Dekker, New York, 1985.
90. A. M. Cunliffe, P. T. Williams, J. Anal. Appl. Pyrol., 44 (1998) 131.
91. W. S. Schlotzhauer, O. T. Chortyk, J. Anal. Appl. Pyrol., 12 (1987) 193.
92. E. R. E. van der Hage, J. J. Boon, J. Chromatogr., A, 736 (1996) 61.
93. P. W. Arisz et al., Anal. Chem., 62 (1990) 1519.
94. S. W. Benson, J. Chem. Ed., 42 (1965) 502.
95. J. A. Settula et al., J. Am. Chem. Soc., 112 (1990) 1347.
96. J. March, *Advanced Organic Chemistry*, J. Wiley, New York, 1992.
97. R. S. Berry et al., *Physical Chemistry*, Oxford Univ. Press, Oxford, 2000.
98. S. A. Groves et al., J. Anal. Appl. Pyrol., 19 (1991) 301.
99. J. Sestak, G. Berggren, Thermochimica Acta, 3 (1971) 1.
100. J. R. MacCallum, J. Anal. Appl. Pyrol., 11 (1987) 65.
101. R. Font, A. N. Garcia, J. Anal. Appl. Pyrol., 35 (1995) 249.
102. R. T. Conley (ed.), *Thermal Stability of Polymers*, M. Dekker, New York, 1970.
103. H. H. G. Jellinek (ed.), *Aspects of Degradation and Stabilization of Polymers*, Elsevier, Amsterdam, 1980.
104. R. J. Evans et al., J. Anal. Appl. Pyrol., 9 (1986) 207.
105. P. W. Arisz, J. J. Boon, J. Anal. Appl. Pyrol., 25 (1993) 371.
106. R. E. Aries et al., Anal. Chem., 60 (1988) 1498.
107. A. Basch, M. Lewin, J. Polymer Sci., 11 (1973) 3095.
108. T. S. Tsuge et al., J. Anal. Appl. Pyrol., 1 (1980) 221.

109. G. Verhegyi et al., J. Anal. Appl. Pyrol., 26 (1993) 159.
110. W. C. Kossa et al., J. Chromatog. Sci., 17 (1979) 177.
111. H. Hardell, J. Anal. Appl. Pyrol., 27 (1993) 73.
112. H. Ohtani et al., J. Anal. Appl. Pyrol., 25 (1993) 1.
113. J. Pecci, T. J. Giovanniello, J. Chromatog., 109 (1975) 163.
114. K. O. Gerhardt, C. W. Gehrke, J. Chromatog., 143 (1977) 335.
115. J. C. del Río et al., J. Chromatogr., A, 874 (2000) 235.
116. D. Fabbri, R. Helleur, J. Anal. Appl. Pyrol., 49 (1999) 277.
117. S. C. Moldoveanu, J. Microcolumn Sep.,13 (2001) 102.
118. A. van der Kaaden et al., J. Anal. Appl. Pyrol., 9 (1986) 267.
119. C. E. R. Jones, A. F. Moyels, Nature, 189 (1961) 222.
120. R. S. Lehrle, J. C. Robb, J. Gas Chromatogr., 5 (1967) 89.
121. I. Tyden-Ericsson, Chromatographia, 6 (1973) 353.
122. I. Ericsson, J. Anal. Appl. Pyrol., 2 (1980) 187.
123. C. Buchler, W. Simon, J. Chromatogr. Sci., 8 (1970) 323.
124. S. Tsuge, T. Takeuki, Anal. Chem., 49 (1977) 348.
125. W. J. Irwin, *Analytical Pyrolysis*, M. Dekker, New York, 1982.
126. C. S. Gutteridge, in *Methods in Microbiology*, vol. 19, Academic Press, New York, 1987.
127. CDS Analytical Inc., Polymer Library of Global Spectra.
128. G. R. Waller, M. S. Feather (eds.), *The Maillard Reaction in Food and Nutrition*, ACS Symp. Ser. 215, Washington, 1983.
129. T. Sugimura et al., Proc. Japan Acad., 53B (1977) 58.
130. R. R. Baker, J. Anal. Appl. Pyrol., 11 (1987) 555.
131. R. R. Baker, J. Anal. Appl. Pyrol., 4 (1983) 297.
132. W. W. Weeks et al., J. Agr. Food Chem., 41 (1993) 1321.
133. J. C. E. Reitsma, W. Pilnik, Carbohydr. Polym., 10 (1989) 315.
134. L. Michajlov et al., Polymer, 12 (1975) 170.
135. M. Seeger et al., Z. Anal. Chem., 276 (1971) 267.
136. M. Galin, J. Makromol. Sci. -Chem., A-7 (1973) 873.
137. J. L. Savoca et al., J. Anal. Appl. Pyrol., 9 (1985) 19.
138. D. Deur-Siftar, J. Gas Chromatog., 5 (1967) 72.
139. G. Audisio, G. Bojo, Makromol. Chem., 176 (1975) 991.
140. H. Seno et al., Makromol. Chem., 161 (1972) 185.
141. G. S. Perry, J. Gas Chromatogr., 5 (1967) 77.
142. I. Luderwald, Makromol. Chem., 178 (1977) 2603.
143. H. R. Krickeldorf, I. Luderwald, Makromol. Chem., 179 (1978) 421.
144. K. Sueoka et al., J. Polym. Sci. Part A, 29 (1991) 1903.
145. H. Ohtani et al., J. Anal. Appl. Pyrol., 4 (1982) 117.
146. D. A. Chatfield et al., J. Polym. Sci., Polym. Chem. Ed., 17 (1979) 1367.
147. G. Chiavari, G. Galletti, J. Anal. Appl. Pyrol., 24 (1992) 123.
148. S. Tsuge, H. Matsubara, J. Anal. Appl. Pyrol., 8 (1985) 49.
149. R. J. Heleur et al., Anal. Chim. Acta, 192 (1987) 367.
150. N Mitsuo et al., Chem. Pharm. Bull., 37, (1989) 1624.
151. S. L. Morgan, C. A. Jacques, Anal. Chem., 54 (1982) 741.
152. T. Sugimura et al., Proc. Japan Acad., 53B (1977) 58.
153. T. Yamamoto et al., Proc. Japan Acad., 54B (1978) 248.
154. G. Rouzet et al., J. Anal. Appl. Pyrol., 57 (2001) 153.
155. K. J. Voorhees, *Analytical Pyrolysis, Techniques and Applications*, Butterworths, London, 1984.

156. E. Jakab et al., J. Anal. Appl. Pyrol., 58-59 (2001) 49.
157. Z. Czégény, M. Blazsó, J. Anal. Appl. Pyrol., 58-59 (2001) 95.
158. C. Vasile et al., J. Anal. Appl. Pyrol., 57 (2001) 287.
159. S. C. Moldoveanu, unpublished results.
160. J. C. J. Bart, J. Anal. Appl. Pyrol., 58-59 (2001) 3.
161. I. Glassman, *Combustion*, Academic Press, Orlando, 1987.

SYMBOLS, UNITS, AND CONSTANTS

- *Symbols*

a	1) activity, 2) various constants, 3) accuracy
a_c	centrifugal acceleration
A	1) peak area, 2) light absorbance
A	frequency factor
$\mathcal{A}$	coefficient of liquid thermal expansion
$\mathcal{A}_m$	the surface area occupied on an adsorbent by a solvent molecule
α	separation factor
α	polarizability
α̲	degree of dissociation
bp	boiling point temperature
β	phase ratio
c	molar concentration
C	1) value of molar concentration, 2) concentration
C	number of components in a system
d_f	film thickness of stationary phase
d_p	particle diameter
D	diffusion coefficient
$\mathcal{D}$	dialysance
D_S, D_M	diffusion coefficients in stationary or mobile phase
δ	1) solubility parameter, 2) solute loss
e	electron charge
E	1) internal energy, 2) electrochemical potential, 3) extraction fraction
$E_{1/2}$	half wave electrochemical potential
$E^{\#}$	activation energy
E_m	free energy of adsorption for a group of atoms
ε	1) electric permittivity (dielectric constant), 2) molar absorption coefficient of light, 3) porosity
ε"	microwave loss factor
$ε^o$	elutropic strength
$ε_o$	electric permittivity of vacuum
$\mathcal{E}$	electric field
η	viscosity
f	frequency
F	1) fluorescence intensity, 2) Faraday constant
F̄	force
F	number of degrees of freedom
ΔF^o	standard free energy
Φ	1) quantum fluorescence efficiency, 2) centrifugal force
Φ_{CL}	chemiluminescence efficiency
g_n	acceleration of gravity
G	free enthalpy (Gibbs)
ΔG^o	standard free enthalpy
γ'	activity coefficient relative to mole fraction
γ	1) activity coefficient relative to concentration, 2) activity coefficient relative to fugacity
γ	surface tension
h	Planck constant
H	1) enthalpy, 2) height equivalent to a theoretical plate, 3) informational entropy
H	partial molar enthalpy
i	intensity of current
I	1) chromatographic retention index, 2) light intensity, 3) ionization potential, 4) electric current intensity

- *Symbols (continued)*

I_{nf}	information
J	flux
k	1) capacity factor, 2) reaction rate constant, 3) other constants
k	Henry's law constant relative to mole fraction
k	Henry's law constant relative to concentrations
k_B	Boltzmann constant
K	1) partition coefficient, 2) distribution constant, 3) equilibrium constant, 4) adsorption constant
K_a	acidity constant
K_b	basicity constant
K_{sp}	solubility product
K_s	salting parameter
K	equilibrium constant
K	thermodynamic distribution constant
L	1) length, 2) chromatographic column length
LTPRI	linear temperature programmed retention index
λ	wavelength
m	1) mass, 2) average (mean), 3) number of particles, 4) mass transfer coefficient
M	molecular mass
MW	molecular weight
μ	1) chemical potential, 2) electrophoretic mobility
μ	dipole moment
μ	mean of a *population*
n	1) number of moles, 2) number of theoretical plates, 3) refractive index, 4) number of measurements, 5) number of electrons in an electrochemical reactions
n_c	peak capacity
N	1) number of moles, 2) effective plate number
N	Avogadro's constant
N	normal distribution
ν	frequency
ω	angular velocity
p	1) pressure, 2) fraction of particles
p	number of visible peaks in a chromatogram
p_c	critical pressure
P	1) probability, 2) polarity (McReynolds)
P	number of phases
P'	polarity parameter for solvents
Pc	parachor
ψ	electric potential
Π	osmotic pressure, osmotic pressure of a resin
q	quantity
Q	1) heat, 2) ion exchange resin capacity, 3) electrical charge
r	radius
R	1) fraction of moles in the mobile phase, 2) correlation coefficient, 3) recovery, 4) electric resistance
R	gas constant
R_f	relative migration rate
Rm	molar refraction
R_s	resolution
ρ	density
$ρ_c$	critical density
s	1) standard deviation of a *sample* of measurements, 2) ionic strength
σ	1) *population* standard deviation, 2) surface charge density

- *Symbols (continued)*

S	1) entropy, 2) sensitivity
S	partial molar entropy
S°	1) measure of adsorption energy, 2) synergy coefficient
S	molar solubility
SI	similarity index
S/N	signal to noise
t	time
t_g	retention time of an unretained component
t_R	(absolute) retention time
t'_R	time spent by the solute in the stationary phase
T	temperature
T_b	boiling point temperature
T_f	fusion (melting) temperature
tan (δ)	dissipation factor
T_c	critical temperature
τ	dielectric relaxation time
u	linear velocity (linear flow rate)
U	volumetric flow rate
U_s	ionic mobility
V	1) rate of mass transfer in electrochemical reactions, 2) partial molar volume
v_{eff}	effective volume of a plate
$\bar{v}$	speed
V	1) volume, 2) molar volume
V	molecular volume of a macromolecule
$\overline{V}$	molecular diffusion volume
V_c	critical volume
V_R	retention volume
V_M	volume of mobile phase
W	1) width, 2) chromatographic peak width, 3) weight, 4) work
x	mole fraction
x	solvent characterization parameter
z	electrical charge

- Units

Å	ångström 1 Å = 10^{-10} m
C	coulomb
cP	centipoise = mPa s
D	debye = 3.33564 10^{-30} C m
eq	equivalent gram
eV	electron volt = 1.60218 10^{-19} J
h	hour
Herzberg	time for 100 mL to pass through 10 cm^2 filter paper with 10 cm water head pressure
Hz	hertz
J	joule = 0.239 cal, 1 cal = 4.1868 J (international), 1 cal = 4.184 J (thermochemical)
m	meter
meq	milliequivalent
min.	minute
N	newton kg m s^{-2}
Pa	pascal = N m^{-2} = kg m^{-1} s^{-2} = 9.869 10^{-6} atm = 7.5 10^{-3} Torr = 1.45 10^{-4} psi =
s	second
V	volt
W	watt = kg m^2 s^{-3} = J s = 0.239 cal s

- Constants

c_{light}, speed of light in vacuum	299792458 m s^{-1}
e, electron charge	1.60217733 10^{-19} C
F, Faraday's constant	9.6485309 10^4 C mol^{-1}
g_n, acceleration of gravity	9.80665 m s^{-2}
h, Plank constant	6.6260755 10^{-34} J s
k_B, Boltzmann constant	1.380658 10^{-23} J K^{-1}
N, Avogadro's constant	6.0221367 10^{23}
R, gas constant	8.31451 J deg^{-1} mol^{-1} = 1.987 cal deg^{-1} mol^{-1}.

INDEX

JOURNAL OF CHROMATOGRAPHY LIBRARY

A Series of Books Devoted to Chromatographic and Electrophoretic Techniques and their Applications

Although complementary to the Journal of Chromatography, each volume in the library series is an important and independent contribution in the field of chromatography and electrophoresis. The library contains no material reprinted from the journal itself.

Other volumes in this series